STUDENT SOLUTIONS MANUAL

Calculus for Engineers
Fourth Edition

Donald Trim
University of Manitoba

PEARSON

Prentice
Hall

Toronto

ISBN-13 978-0-13-241329-9
ISBN-10 0-13-241329-9

Acquisitions Editor: Michelle Sartor
Developmental Editor: Kimberley Hermans
Production Editor: Marisa D'Andrea
Production Coordinator: Avinash Chandra

Printed in the United States of America

PREFACE

This manual is designed to help you in your calculus studies. It contains solutions to all even-numbered exercises (except those where the text answer constitutes a complete solution, and those with two asterisks). The following suggestions will help you gain the greatest benefit from the manual.

After your instructor has covered a section in the text, study it for yourself. This is important. Do not attempt exercises without first reading lecture notes and/or covering material in the text. Then, and only then, should you turn to the exercises.

When you feel that you have a clear understanding of the ideas, start the exercises. It is essential to do exercises. You will never master calculus by listening and reading—you must do exercises. Only when you solve problems do you think about what you have read or heard, and then you truly begin to understand.

The first few exercises in each section are routine, drill-type problems and should therefore be straightforward. These have no asterisk. If you do encounter difficulties at this stage, reread the text material in order to learn which fundamentals you have misunderstood. If difficulties persist, seek help from a tutorial assistant or your instructor.

Each set of exercises is graded in difficulty, becoming more and more difficult with increasing number. Exercises with one asterisk are more difficult than those without an asterisk for various reasons. They may involve more difficult calculations; they may require you to think about the material in the section in a novel way; they may require you to relate the material in the section to material from previous sections; or they may involve a combination of these. Exercises with two asterisks are challenging; they should be attempted only when material in earlier exercises has been mastered. Exercises with a calculator icon require the use of a calculator to solve. Do not attempt to solve these questions by hand. Resist the temptation to consult this manual too soon. You learn more by solving a problem incorrectly and benefiting from these mistakes than from reading my solution.

The most difficult part of any solution is the first line, the initial idea that gets you going. Before writing anything, plan your attack; decide what steps must be taken in order to get from given to required. Often this means working backwards from required to given. It is surprising to students when I tell them that although I have presented a logical progression of ideas in the solution to a problem, this was probably not the way I solved it in the first place. It is more likely that I worked the problem, at least mentally, in reverse, and then presented it as shown. My point is that you must analyze before writing. If you read the first line of my solution, you lose the opportunity to think for yourself. And after all, is not "thinking" what learning is all about?

Students always ask how many exercises they are expected to do. The only answer can be: as many as possible. Each exercise has its purpose. Early exercises reinforce fundamentals, and later exercises deepen your understanding of calculus as a whole and delve into the important area of applications. Work as many problems as your schedule permits.

I wish you every success in your calculus studies.

D.W. Trim

CONTENTS

This manual contains detailed solutions to all even-numbered (but not double-asterisk) exercises in the text. We would appreciate being made aware of any errors.

CHAPTER 1

EXERCISES 1.2

2. The solution is $x = -5/14$.

4. Since $12x^2 + 11x - 5 = (3x - 1)(4x + 5)$, solutions are $x = 1/3$ and $x = -5/4$.

6. Quadratic formula 1.5 gives $x = \dfrac{-10 \pm \sqrt{100 - 4(-4)(9)}}{-8} = \dfrac{-10 \pm \sqrt{244}}{-8} = \dfrac{5 \pm \sqrt{61}}{4}$.

8. Since $4x^2 - 36x + 81 = (2x - 9)^2$, the solution is $x = 9/2$ with multiplicity 2.

10. Since the discriminant $(-8)^2 - 4(4)(9) = -80$ is negative, the equation has no real solutions.

12. Possible rational solutions are ± 1, $\pm 1/2$, $\pm 1/4$, $\pm 1/8$. We find that $x = -1/2$ is a solution. We factor $2x + 1$ from the cubic,

$$8x^3 + 12x^2 + 6x + 1 = (2x + 1)(4x^2 + 4x + 1) = (2x + 1)(2x + 1)^2 = (2x + 1)^3.$$

The only solution is $x = -1/2$ with multiplicity 3.

14. Possible rational solutions are $x = \pm 1$, ± 3, ± 9, but it should also be clear that no positive value of x can satisfy the equation. We find that $x = -1$ is a solution. We factor $x + 1$ from the cubic,

$$x^3 + 4x^2 + 12x + 9 = (x + 1)(x^2 + 3x + 9) = 0.$$

Since the discriminant of the quadratic is negative, the other two solutions are complex.

16. Possible rational solutions are ± 1, ± 2, ± 3, ± 4, ± 6, ± 9, ± 12, ± 18, ± 36. We find that $x = -3$ is a solution. We factor $x + 3$ from the quartic,

$$x^4 + 7x^3 + 9x^2 - 21x - 36 = (x + 3)(x^3 + 4x^2 - 3x - 12).$$

Possible rational zeros of the cubic are ± 1, ± 2, ± 3, ± 4, ± 6, ± 12. We find that $x = -4$ is a zero. We factor $x + 4$ from the cubic,

$$x^4 + 7x^3 + 9x^2 - 21x - 36 = (x + 3)(x + 4)(x^2 - 3).$$

The solutions are $x = -3, -4, \pm\sqrt{3}$.

18. Possible rational solutions are ± 1, ± 3, ± 5, ± 15, $\pm 1/2$, $\pm 3/2$, $\pm 5/2$, $\pm 15/2$. We find that $x = -5$ is a solution. We factor $x + 5$ from the quartic,

$$2x^4 + 9x^3 - 6x^2 - 8x - 15 = (x + 5)(2x^3 - x^2 - x - 3).$$

Possible rational zeros of the cubic are ± 1, ± 3, $\pm 1/2$, $\pm 3/2$. We find that $x = 3/2$ is a zero. We factor $2x - 3$ from the cubic,

$$2x^4 + 9x^3 - 6x^2 - 8x - 15 = (x + 5)(2x - 3)(x^2 + x + 1).$$

Since the quadratic has a negative discriminant, the only real solutions are $x = -5$ and $x = 3/2$.

20. No real numbers can satisfy this equation.

22. Possible rational solutions are ± 1, ± 2, ± 3, ± 4, ± 6, ± 8, ± 9, ± 12, ± 16, ± 18, ± 24, ± 32, ± 36, ± 48, ± 64, ± 72, ± 96, ± 144, ± 192, ± 192, ± 288, ± 576. We find that $x = -4$ is a solution. We factor $x + 4$ from the quartic,

$$x^4 - 4x^3 - 44x^2 + 96x + 576 = (x + 4)(x^3 - 8x^2 - 12x + 144).$$

Possible rational zeros of the cubic are ± 1, ± 2, ± 3, ± 4, ± 6, ± 8, ± 9, ± 12, ± 16, ± 18, ± 24, ± 36, ± 48, ± 72, ± 144. We find that $x = -4$ is a zero. We factor $x + 4$ from the cubic,

$$x^4 - 4x^3 - 44x^2 + 96x + 576 = (x + 4)(x + 4)(x^2 - 12x + 36) = (x + 4)^2(x - 6)^2.$$

Thus, $x = -4$ and $x = 6$ are solutions, each of multiplicity 2.

24. Possible rational solutions are $\pm 1, \pm 2, \pm 4, \pm 5, \pm 10, \pm 20, \pm 1/2, \pm 5/2, \pm 1/3, \pm 2/3, \pm 4/3, \pm 5/3,$ $\pm 10/3, \pm 20/3, \pm 1/6, \pm 5/6$. We find that $x = 5/6$ is a zero. We factor $6x - 5$ from the cubic,

$$6x^3 + x^2 + 19x - 20 = (6x - 5)(x^2 + x + 4).$$

Since the quadratic has a negative discriminant, $x = 5/6$ is the only real solution.

26. Possible rational solutions are $\pm 1, \pm 2, \pm 3, \pm 4, \pm 5, \pm 6, \pm 8, \pm 10, \pm 12, \pm 15, \pm 20, \pm 24, \pm 30, \pm 40, \pm 60,$ ± 120. We find that $x = 1$ is a solution. We factor $x - 1$ from the polynomial,

$$x^5 - 15x^4 + 85x^3 - 225x^2 + 274x - 120 = (x - 1)(x^4 - 14x^3 + 71x^2 - 154x + 120).$$

We use the same set of rational possibilities for the quartic. We find that $x = 2$ is a zero. When we factor $x - 2$ from the quartic,

$$x^5 - 15x^4 + 85x^3 - 225x^2 + 274x - 120 = (x - 1)(x - 2)(x^3 - 12x^2 + 47x - 60).$$

Possible rational zeros of the cubic are $\pm 1, \pm 2, \pm 3, \pm 4, \pm 5, \pm 6, \pm 10, \pm 12, \pm 15, \pm 20, \pm 30, \pm 60$. We find that $x = 3$ is a zero. We factor $x - 3$ from the cubic,

$$\begin{aligned} x^5 - 15x^4 + 85x^3 - 225x^2 + 274x - 120 &= (x - 1)(x - 2)(x - 3)(x^2 - 9x + 20) \\ &= (x - 1)(x - 2)(x - 3)(x - 4)(x - 5). \end{aligned}$$

Solutions are therefore $x = 1, 2, 3, 4, 5$.

28. Possible rational solutions are $\pm 1, \pm 2, \pm 4, \pm 1/5, \pm 2/5, \pm 4/5, \pm 1/25, \pm 2/25, \pm 4/25$. We find that $x = 2/5$ is a solution. We factor $5x - 2$ from the quartic,

$$25x^4 - 120x^3 + 109x^2 - 36x + 4 = (5x - 2)(5x^3 - 22x^2 + 13x - 2).$$

Possible rational zeros of the cubic are $\pm 1, \pm 2, \pm 1/5, \pm 2/5$. We find that $x = 2/5$ is a zero. We factor $5x - 2$ from the cubic,

$$25x^4 - 120x^3 + 109x^2 - 36x + 4 = (5x - 2)(5x - 2)(x^2 - 4x + 1).$$

Thus, $x = 2/5$ is a real solution with multiplicity 2 and the quadratic formula gives the remaining two solutions

$$x = \frac{4 \pm \sqrt{16 - 4}}{2} = 2 \pm \sqrt{3}.$$

30. Possible rational solutions are $\pm 1, \pm 2, \pm 3, \pm 4, \pm 6, \pm 8, \pm 9, \pm 12, \pm 16, \pm 18, \pm 24, \pm 27, \pm 36, \pm 48, \pm 54,$ $\pm 72, \pm 81, \pm 108, \pm 144, \pm 162, \pm 216, \pm 324, \pm 432, \pm 648, \pm 1296$. We find that $x = 3$ is a solution. When we factor $x - 3$ from the polynomial,

$$x^6 + 16x^4 - 81x^2 - 1296 = (x - 3)(x^5 + 3x^4 + 25x^3 + 75x^2 + 144x + 432).$$

When we note that no positive value can satisfy the fifth degree polynomial, possible rational zeros are $-1, -2, -3, -4, -6, -8, -9, -12, -16, -18, -24, -27, -36, -48, -54, -72, -108, -144, -216,$ -432. We find that $x = -3$ is a zero. When we factor $x + 3$ from the polynomial,

$$x^6 + 16x^4 - 81x^2 - 1296 = (x - 3)(x + 3)(x^4 + 25x^2 + 144).$$

Since there can be no real zeros of the quartic, the real solutions are $x = \pm 3$.

32. $2[x - (3 + \sqrt{65})/4][x - (3 - \sqrt{65})/4]$ **34.** $24(x + 5/3)(x - 1/4)(x - 1/2)$

36. $16(x - 1/2)^2(x + 1/2)^2$

38. According to the rational root theorem, possible rational zeros must be divisors of a_0 divided by divisors of 1. This means that possible rational zeros are divisors of a_0.

EXERCISES 1.3

2. With formula 1.10, the distance is $\sqrt{6^2 + (-3)^2} = 3\sqrt{5}$.

4. With formula 1.10, the distance is $\sqrt{(-7)^2 + (-3)^2} = \sqrt{58}$.

6. With formula 1.11, the midpoint is $\left(\dfrac{-2+4}{2}, \dfrac{1-2}{2}\right) = \left(1, -\dfrac{1}{2}\right)$.

8. With formula 1.11, the midpoint is $\left(\dfrac{3-4}{2}, \dfrac{2-1}{2}\right) = \left(-\dfrac{1}{2}, \dfrac{1}{2}\right)$.

10. The line is horizontal.
Its equation is $y = -6$.

12. $y + \dfrac{3}{2} = -\dfrac{1}{2}(x-1) \implies x + 2y + 2 = 0$.

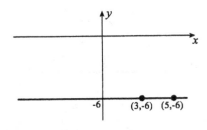

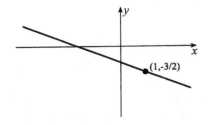

14. The equation of the x-axis is $y = 0$.

16. With $m = (4+2)/(0+1) = 6$, equation 1.13 gives $y - 4 = 6(x - 0) \implies y = 6x + 4$.

18. The midpoint of the line segment is
$\left(\dfrac{3-7}{2}, \dfrac{4+8}{2}\right) = (-2, 6)$. With
$m = (6-0)/(-2-0) = -3$, equation 1.13
gives $y - 0 = -3(x - 0) \implies y = -3x$.

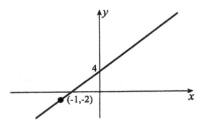

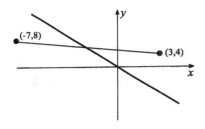

20. Since slopes of both lines are $-1/3$, they are parallel.

22. Since slopes $-2/3$ and $3/2$ are negative reciprocals, the lines are perpendicular.

24. Since slopes are 1 and $-2/3$, the lines are neither parallel nor perpendicular.

26. Since slopes are -1 and 3, the lines are neither parallel nor perpendicular.

28. The point of intersection is $(1, 2)$.

30. When we substitute $y = 2x + 6$ into the second equation, $x = (2x + 6) + 4 \implies x = -10$. The point of intersection is $(-10, -14)$.

32. If we multiply the first equation by 5 and add the result to the second equation, we obtain $73x = 37 \implies x = 37/73$. The point of intersection is $(37/73, 153/146)$.

34. Formula 1.16 gives $\dfrac{|1 - 6 - 3|}{\sqrt{1+4}} = \dfrac{8}{\sqrt{5}}$.

36. Formula 1.16 gives $\dfrac{|3 + 1|}{\sqrt{1+1}} = \dfrac{4}{\sqrt{2}} = 2\sqrt{2}$.

38. Formula 1.16 gives $\dfrac{|60 + 4 + 3|}{\sqrt{225 + 4}} = \dfrac{67}{\sqrt{229}}$.

40. Since the slope of $x - y = 4$ is 1, and the point of intersection of $2x + 3y = 3$ and $x - y = 4$ is $(3, -1)$, the required equation is $y + 1 = -1(x - 3) \implies x + y = 2$.

42. Since the slope of the line through $(-3, 4)$ and $(1, -2)$ is $(4+2)/(-3-1) = -3/2$, the equation of the required line is $y + 2 = (2/3)(x + 3) \implies 2x = 3y$.

44. When we equate slopes of line segments AC and AB, $-\dfrac{b}{a} = \dfrac{5}{3-a} \implies b = \dfrac{5a}{a-3}$.

Combine this with the fact that the area of $\triangle OAB = (1/2)ab = 30$, and we obtain $a = 6$ and $b = 10$. The equation of the line is $y = -(5/3)(x - 6) \implies 5x + 3y = 30$.

46. If the equation is $y = c$, then $A = (c, c)$ and $B = (4 - c, c)$. Since triangle ABD has area 9, it follows that $9 = \dfrac{1}{2}(2 - c)(4 - 2c)$, solutions of which are 5, -1. The required equation is therefore $y = -1$.

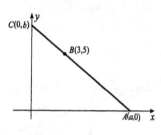

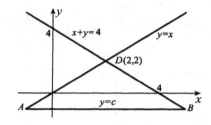

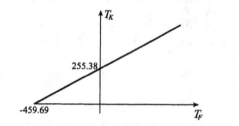

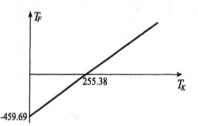

48. (a) The conversion equation is $T_K = 5(T_F - 32)/9 + 273.16$.
(b) The conversion equation is $T_F = 9(T_K - 273.16)/5 + 32$.
(c) They are one and the same line if we plot T_F along the horizontal axis and T_K along the vertical axis (left figure below). If we plot T_F along the vertical axis and T_K along the horizontal axis in part (b), we obtain the right figure below.

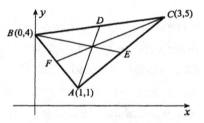

(d) If we set $T_F = T_K$ in the equation from part (a), we obtain $T_K = 5(T_K - 32)/9 + 273.16 \implies 4T_K = 2298.44 \implies T_K = 574.61$.

50. Coordinates of E, the midpoint of side AC, are $(2, 3)$. Since the slope of BE is $-1/2$, the equation of median BE is $y - 4 = -(1/2)x \implies x + 2y = 8$. Similarly, equations for medians AD and CF are $7x - y = 6$ and $y = x + 2$, respectively. Line segments AD and CF intersect in the point $(4/3, 10/3)$, and this point also satisfies $x + 2y = 8$. Thus, the three medians intersect at the point $(4/3, 10/3)$.

52. The midpoint of the line segment is $(1, -1)$, and its slope is $-3/2$. The equation of the perpendicular bisector is $y + 1 = (2/3)(x - 1) \implies 2x = 3y + 5$.

54. Suppose two nonvertical, parallel lines have slopes m_1 and m_2. Their equations can be expressed in the form $y = m_1 x + b_1$ and $y = m_2 x + b_2$. To find their point of intersection we set $m_1 x + b_1 = m_2 x + b_2$. Since the lines do not intersect, there must be no solution of this equation. This happens only if $m_1 = m_2$; that is, the lines have the same slope. Conversely, if two nonvertical lines have the same slope, their equations must be of the form $y = mx + b_1$ and $y = mx + b_2$, where $b_1 \neq b_2$. When we attempt to find their point of intersection by setting $mx + b_1 = mx + b_2$, we find that $b_1 = b_2$, a contradiction. In other words, the lines do not intersect.

56. No. Definition 1.1 defines parallelism only for different lines.

EXERCISES 1.4A

2. Factored in the form $y = -(x-3)(x-1)$, the x-intercepts of the parabola are 1 and 3. Its maximum is at $x = 2$.

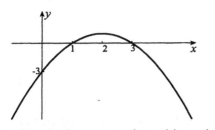

4. This is the parabola $x = 4y^2/3$ shifted $1/3$ unit to the left.

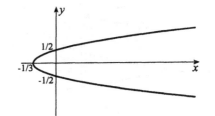

6. Factored in the form $y = -(x+1)(x-4)/2$, x-intercepts of the parabola are -1 and 4. Its maximum is at $x = 3/2$.

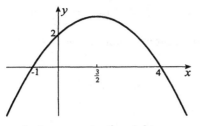

8. Factored in the form $x = -(2y-5)(y+1)$, y-intercepts of the parabola are -1 and $5/2$. Its maximum in the x-direction occurs for $y = 3/4$.

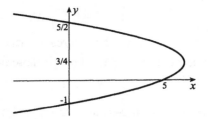

10. This parabola opens to the right.

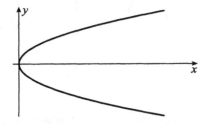

12. This parabola opens to the left and touches the y-axis at $y = -4$.

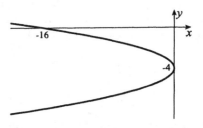

14. The equation must be of the form $y = ax^2 + 1$. Since $(2,3)$ is on the parabola, $3 = 4a + 1 \implies a = 1/2$.

16. Since the parabola crosses the x-axis at $x = -1$ and $x = 3$, it must be of the form $y = a(x+1)(x-3)$. Since $(0, -1)$ is on the parabola, $-1 = a(1)(-3) \implies a = 1/3$.

18. We set $x + 1 = 1 - x^2 \implies 0 = x^2 + x$ $= x(x+1) \implies x = 0, -1$. Points of intersection are $(-1, 0)$ and $(0, 1)$.

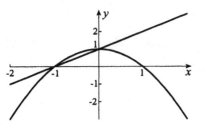

20. We set $2x - x^2 - 6 = 5 + x/5 \implies$ $5x^2 - 9x + 55 = 0$. With equation 1.5, $x = \left(9 \pm \sqrt{81 - 4(5)(55)}\right)/10$. Since $81 - 4(5)(55) < 0$, there are no points of intersection

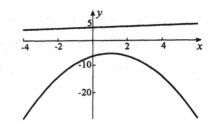

22. We set $-y^2 + 1 = y^2 + 2y - 3$
$\implies 0 = 2y^2 + 2y - 4 = 2(y+2)(y-1)$
$\implies y = 1, -2.$ Points of intersection are
$(0,1)$ and $(-3,-2)$.

24. The range of the shell is
$R = (v^2/9.81)\sin 2\theta.$ Since $0 \le \theta \le \pi/2$,
range is a maximum when $\sin 2\theta = 1$
$\implies 2\theta = \pi/2 \implies \theta = \pi/4$ radians.

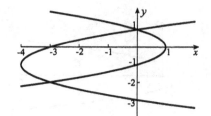

26. We set $5x = (x-2)^4 + 4 \implies (x-2)^4 - 5x + 4 = 0 \implies x^4 - 8x^3 + 24x^2 - 37x + 20 = 0.$ Possible rational solutions are $\pm 1, \pm 2, \pm 4, \pm 5, \pm 10, \pm 20.$ One solution is $x = 1$, so that

$$x^4 - 8x^3 + 24x^2 - 37x + 20 = (x-1)(x^3 - 7x^2 + 17x - 20) = 0.$$

We find that $x = 4$ is a zero of the cubic, so that

$$x^4 - 8x^3 + 24x^2 - 37x + 20 = (x-1)(x-4)(x^2 - 3x + 5).$$

Since $x^2 - 3x + 5 = 0$ has no real solutions, the only points of intersection are $(1,1)$ and $(4,4)$.

28. If the parabola is to pass through these points, then

$$2 = a + b + c, \qquad 10 = 9a - 3b + c, \qquad 4 = 9a + 3b + c.$$

The solution of these equations is $a = 1/2$, $b = -1$, $c = 5/2$.

30. If $P(a,b)$ is the point at which the rope meets the parabola, the equation of the line PQ is $y - 4 = [(b-4)/(a-3)](x-3).$ The x-coordinates of points of intersection of this line with the parabola are given by the equation

$$x^2 - 1 = \left(\frac{b-4}{a-3}\right)(x-3) + 4$$
$$\implies$$
$$x^2 - \left(\frac{b-4}{a-3}\right)x + 3\left(\frac{b-4}{a-3}\right) - 5 = 0.$$

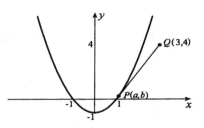

Since the line is to meet the parabola in only one point, the discriminant of this quadratic must be zero

$$\left(\frac{b-4}{a-3}\right)^2 - 4\left[3\left(\frac{b-4}{a-3}\right) - 5\right] = 0 \implies (b-4)^2 - 12(b-4)(a-3) + 20(a-3)^2 = 0.$$

Since $P(a,b)$ is on the parabola, $b = a^2 - 1$, and when we substitute this into the above equation,

$$0 = (a^2 - 5)^2 - 12(a^2 - 5)(a - 3) + 20(a - 3)^2 = a^4 - 12a^3 + 46a^2 - 60a + 25.$$

Possible rational solutions of this equation are $\pm 1, \pm 5, \pm 25.$ We find that $a = 1$ is a solution. When it is factored from the quartic,

$$0 = a^4 - 12a^3 + 46a^2 - 60a + 25 = (a-1)(a^3 - 11a^2 + 35a - 25).$$

Once again $a = 1$ is a zero of the cubic, so that

$$0 = a^4 - 12a^3 + 46a^2 - 60a + 25 = (a-1)(a-1)(a^2 - 10a + 25) = (a-1)^2(a-5)^2.$$

Clearly, $a = 5$ is inadmissible, and the required point is $(1,0)$.

EXERCISES 1.4B

2. The centre of the circle is $(-5, 2)$ and its radius is $\sqrt{6}$.

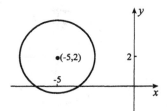

4. When we complete the square on the y-terms, $x^2 + (y-2)^2 = 3$. The centre of the circle is $(0, 2)$ and its radius is $\sqrt{3}$.

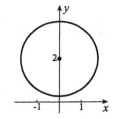

6. When we complete the square on the x-terms, $(x + 3/2)^2 + y^2 = 59/4$. The centre of the circle is $(-3/2, 0)$ and its radius is $\sqrt{59}/2$.

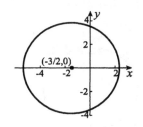

8. When we complete squares on x- and y-terms, $(x + 2)^2 + (y - 1)^2 = 10$. The centre of the circle is $(-2, 1)$ and its radius is $\sqrt{10}$.

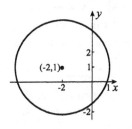

10. When we complete squares on x- and y-terms, $(x + 3)^2 + (y + 3/2)^2 = -35/4$. No point can satisfy this equation.

12. Since the centre is $(1, 0)$ and its radius is 1, the equation of the circle is $(x - 1)^2 + y^2 = 1$.

14. The figure indicates that the centre of the circle is $(3/2, -3/2)$. Its radius is then $\sqrt{(3/2)^2 + (-3/2)^2} = 3/\sqrt{2}$. The equation of the circle is therefore $(x - 3/2)^2 + (y + 3/2)^2 = 9/2$.

16. If we take the equation for the circle in form 1.22, and substitute each of the points $(1, 3)$, $(5, 1)$ and $(2, -2)$,

$$
\begin{aligned}
1 + 9 + f + 3g + e &= 0, & f + 3g + e &= -10, \\
25 + 1 + 5f + g + e &= 0, & \implies \qquad 5f + g + e &= -26, \\
4 + 4 + 2f - 2g + e &= 0, & 2f - 2g + e &= -8.
\end{aligned}
$$

The solution of these equations is $f = -14/3$, $g = -4/3$, and $e = -4/3$. The equation of the circle is $x^2 + y^2 - 14x/3 - 4y/3 - 4/3 = 0$ or $3x^2 + 3y^2 - 14x - 4y - 4 = 0$.

18. If we take the equation of the circle in the form $(x - h)^2 + (y - k)^2 = r^2$, and substitute the two points

$$(3 - h)^2 + (4 - k)^2 = r^2, \qquad (1 - h)^2 + (-10 - k)^2 = r^2.$$

When these equations are subtracted, the result is

$$0 = (3 - h)^2 - (1 - h)^2 + (4 - k)^2 - (-10 - k)^2 \implies h + 7k = -19.$$

(a) When the centre of the circle is on the line $2x + 3y + 16 = 0$, we must also have $2h + 3k + 16 = 0$. When these equations are solved, $h = -5$ and $k = -2$. The radius of the circle is $r = \sqrt{(3 + 5)^2 + (4 + 2)^2} = 10$. (b) When the centre is on the line $x + 7y + 19 = 0$, we must have $h + 7k + 19 = 0$. But this is the same equation obtained from the two points. In other words, there is an infinity of circles with centres on the line $x + 7y + 19 = 0$ passing through the points $(3, 4)$ and $(1, -10)$. Any equation of the form $(x + 7k + 19)^2 + (y - k)^2 = r^2$, where $r^2 = (3 + 7k + 19)^2 + (4 - k)^2 = 50k^2 + 300k + 500$.

20. We set $x^2 + (1 - 2x)^2 - 4(1 - 2x) + 1 = 0 \implies 5x^2 + 4x - 2 = 0$. Solutions are $x = (-4 \pm \sqrt{16 + 40})/10 = (-2 \pm \sqrt{14})/5$. Intersection points are $\left((-2 \pm \sqrt{14})/5, (9 \mp 2\sqrt{14})/5\right)$.

22. We set $(x+3)^2 + 16(x+1) = 25 \implies 0 = x^2 + 22x = x(x+22)$. Thus, $x = 0$ and $x = -22$. From $x = 0$ we obtain the points $(0, \pm 4)$, while $x = -22$ yields no points.

24. If we choose a coordinate system with origin at the centre of the circle, the equation of the circle is $x^2 + y^2 = r^2$. Let $P(a, b)$ and $Q(c, d)$ be any two points on the circle. The midpoint of line segment PQ has coordinates $((a+c)/2, (b+d)/2)$. Since the slope of the perpendicular bisector of PQ is $-(c-a)/(d-b)$, the equation of the perpendicular bisector is

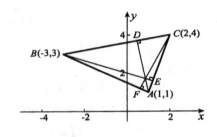

$$y - \frac{b+d}{2} = -\left(\frac{c-a}{d-b}\right)\left(x - \frac{a+c}{2}\right).$$

This line passes through the origin if and only if

$$-\frac{b+d}{2} = -\left(\frac{c-a}{d-b}\right)\left(-\frac{a+c}{2}\right) \iff -(b+d)(d-b) = (c-a)(a+c) \iff a^2 + b^2 = c^2 + d^2.$$

Since P and Q are on the circle, it follows that $a^2 + b^2 = r^2$ and $c^2 + d^2 = r^2$, and therefore $a^2 + b^2 = c^2 + d^2$.

26. If L is the loudness of the speaker at $(0, 20)$, then the amount of sound received at point (x, y) from this speaker is

$$A_1 = \frac{kL}{x^2 + (y-20)^2},$$

where k is a constant of proportionality. The amount of sound received from the speaker at the origin is

$$A_2 = \frac{k(0.7L)}{x^2 + y^2}.$$

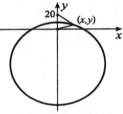

For $A_1 = A_2$, $\dfrac{kL}{x^2 + (y-20)^2} = \dfrac{7kL}{10(x^2 + y^2)} \implies 10(x^2 + y^2) = 7[x^2 + (y-20)^2]$. This gives

$$3x^2 + 3y^2 + 280y = 2800 \implies x^2 + \left(y + \frac{140}{3}\right)^2 = \frac{2800}{3} + \frac{140^2}{9} = \frac{28\,000}{9}.$$

We have a circle centred at $(0, -140/3)$ and radius $20\sqrt{70}/3$.

28. Since the slope of BC is $1/5$, the equation of altitude AD is $y - 1 = -5(x-1) \implies 5x + y = 6$. Since the slope of AC is 3, the equation of altitude BE is $y - 3 = -(1/3)(x+3) \implies x + 3y = 6$. The point of intersection of these altitudes is $(6/7, 12/7)$. The equation of altitude CF is $y - 4 = 2(x-2)$, and the point $(6/7, 12/7)$ satisfies this equation. Hence, the altitudes intersect at $(6/7, 12/7)$.

EXERCISES 1.4C

2. The ellipse is centred at the origin. Its x- and y-intercepts are $x = \pm 4/\sqrt{7}$ and $y = \pm 4/\sqrt{3}$.

4. The ellipse is centred at the origin. Its x- and y-intercepts are $x = \pm\sqrt{7}$ and $y = \pm\sqrt{7/2}$.

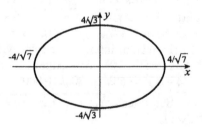

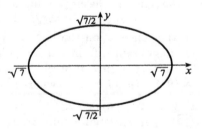

6. When we complete squares on x- and y-terms,
$$\frac{(x+1)^2}{4} + (y-2)^2 = 1.$$
The centre of the ellipse is $(-1, 2)$. It intersects the line $y = 2$ when $x = -3$ and $x = 1$, and intersects the line $x = -1$ when $y = 1$ and $y = 3$.

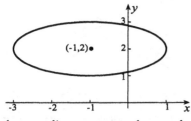

8. When we complete squares on x- and y-terms,
$$\frac{(x+2)^2}{4} + \frac{(y+4)^2}{2} = 1.$$
The centre of the ellipse is $(-2, -4)$. It intersects the line $y = -4$ when $x = -4$ and $x = 0$, and intersects the line $x = -2$ when $y = -4 \pm \sqrt{2}$.

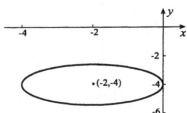

10. With the coordinate system shown, the equation of the ellipse must be of the form $x^2/a^2 + y^2/b^2 = 1$. Since $(0, 4)$ and $(2, 7/2)$ are points on the ellipse,
$$\frac{16}{b^2} = 1, \qquad \frac{4}{a^2} + \frac{49/4}{b^2} = 1.$$
These imply that $a^2 = 256/15$ and $b^2 = 16$. The width of the arch is therefore $2a = 32/\sqrt{15}$.

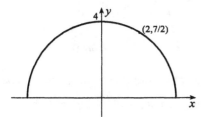

12. If we substitute $y = x + 3$ into the equation of the ellipse, $16x^2 + 9(x+3)^2 = 144$, from which $0 = 25x^2 + 54x - 63 = (x+3)(25x - 21)$. Points of intersection of the curves are $(-3, 0)$ and

$(21/25, 96/25)$.

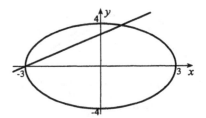

14. If we put $y = \sqrt{3}(5 - x)/2$ into the equation of the ellipse, $9x^2 - 18x + 4\left[\dfrac{\sqrt{3}}{2}(5-x)\right]^2 = 27$, from which $0 = 12x^2 - 48x + 48 = 12(x-2)^2$. The line and ellipse therefore touch at the point $(2, 3\sqrt{3}/2)$.

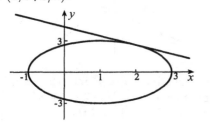

16. If we substitute $y = x^2 - 4$ into the equation of the ellipse, $x^2 + 4(x^2 - 4)^2 = 4$, from which
$$0 = 4x^4 - 31x^2 + 60 = (x^2 - 4)(4x^2 - 15).$$
Solutions are $x = \pm 2$, and $\pm\sqrt{15}/2$. Points of intersection are $(\pm\sqrt{15}/2, -1/4)$ and $(\pm 2, 0)$.

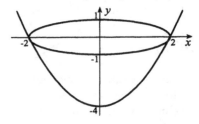

EXERCISES 1.4D

2. Asymptotes for the hyperbola are $y = \pm x$, intersecting at the origin. The hyperbola intersects the x-axis at $x = \pm 1$.

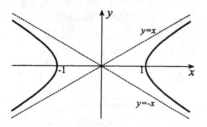

4. Asymptotes for the hyperbola are $y = \pm 2x/5$, intersecting at the origin. The hyperbola intersects the y-axis at $y = \pm 2$.

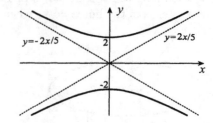

6. Asymptotes for the hyperbola are $y = \pm\sqrt{3}x/2$, intersecting at the origin. The hyperbola intersects the x-axis at $x = \pm 5/\sqrt{3}$.

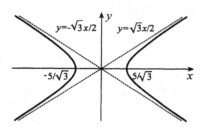

8. When we complete squares on x- and y-terms, the equation becomes $\dfrac{(x-1)^2}{4} - \dfrac{(y+2)^2}{9/4} = 1$. Asymptotes are $y = -2 \pm 3(x-1)/4$ intersecting in the point $(1, -2)$. The hyperbola cuts the line $y = -2$ when $x = -1$ and $x = 3$.

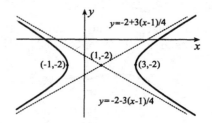

10. When we complete squares on x- and y-terms, the equation becomes $\dfrac{(x+1)^2}{16} - (y-2)^2 = 1$. Asymptotes are $y = 2 \pm (x+1)/4$ intersecting in the point $(-1, 2)$. The hyperbola cuts the line $y = 2$ when $x = -5$ and $x = 3$.

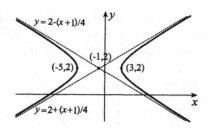

12. If we substitute $x = 2y$ into the equation of the hyperbola, $(2y)^2 - 2y^2 = 1 \implies y = \pm 1/\sqrt{2}$. Points of intersection are therefore

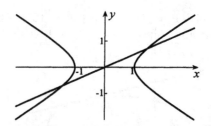

14. The figure suggests that the line and hyperbola do not intersect. This can be verified algebraically. If we substitute $x = 3y$ into the equation of the hyperbola, $36 = 9y^2 - 4(3y)^2 = -27y^2$, an impossibility.

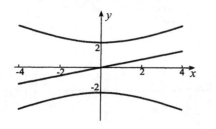

16. If we substitute $x = y^2$ into the equation of the hyperbola, $0 = (y^2)^2 - 2(y^2) - y^2 = y^2(y^2 - 3)$. Thus, $y = 0, \pm\sqrt{3}$, and these give the points of intersection $(0, 0)$ and $(3, \pm\sqrt{3})$.

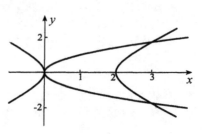

18. If we substitute $(y-1)^2 = 27x/5$ into the equation

of the hyperbola, $36 = 9(x-1)^2 - 4\left(\dfrac{27x}{5}\right)$.

This equation reduces to

$$0 = 5x^2 - 22x - 15 = (x-5)(5x+3).$$

Thus, $x = 5$ or $x = -3/5$. Since x cannot be negative (the equation of the parabola demands this), the solution $x = 5$ leads to the two points $(5, 1 \pm 3\sqrt{3})$.

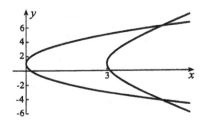

EXERCISES 1.5

2. Since $x - 2$ cannot be 0, the function is defined for all $x \neq 2$.

4. Since $x^2 - 4$ must be positive, x must be greater than 2 or less than -2. Hence, $|x| > 2$.

6. The first two lines of the diagram indicate when the expressions $x^2 - 4$ and $9 - x^2$ are positive and negative. The third line combines these to give the sign of $(x^2 - 4)/(9 - x^2)$. It indicates that x must be in one of the intervals $-3 < x \leq -2$ or $2 \leq x < 3$; that is, $2 \leq |x| < 3$.

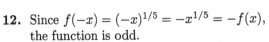

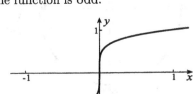

8. Since $x^3 - x^2 = x^2(x-1) \geq 0$ for $x = 0$ and $x \geq 1$, the largest domain consists of the point $x = 0$ and the interval $x \geq 1$.

10. Since $f(-x) = (-x)^5 - (-x) = -x^5 + x$
$= -(x^5 - x) = -f(x)$, the function is odd.

12. Since $f(-x) = (-x)^{1/5} = -x^{1/5} = -f(x)$, the function is odd.

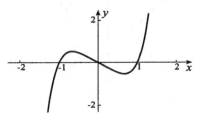

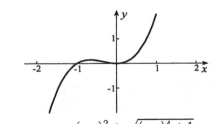

14. Since $f(-x) = (-x)^3 + (-x)^2 = -x^3 + x^2$, and this is neither $f(x)$ nor $-f(x)$, the function is neither even nor odd.

16. Since the function is not defined for $x < 0$, it cannot be even or odd.

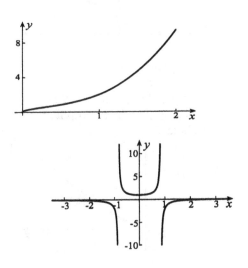

18. Since $f(-x) = \dfrac{(-x)^2 + \sqrt{(-x)^4 + 1}}{(-x)^6 + 3(-x)^2}$

$= \dfrac{x^2 + \sqrt{x^4 + 1}}{x^6 + 3x^2} = f(x)$, the function is even.

20. Even and odd parts of this function are

$$f_e(x) = \frac{1}{2}\left(\frac{x-2}{x+5} + \frac{-x-2}{-x+5}\right) = \frac{x^2 + 10}{x^2 - 25}, \qquad f_o(x) = \frac{1}{2}\left(\frac{x-2}{x+5} - \frac{-x-2}{-x+5}\right) = \frac{-7x}{x^2 - 25}.$$

22. Since this function is odd, its odd part is itself and its even part is zero.

24. Since this function is only defined for $x \geq 1$, it does not have even and odd parts.

26. (a) Let $f(x)$ and $g(x)$ be any two even functions and set $h(x) = f(x)g(x)$. Then,

$$h(-x) = f(-x)g(-x) = [f(x)][g(x)] = f(x)g(x) = h(x).$$

Thus, $h(x)$ is an even function. The proof for two odd functions is similar.
(b) When $f(x)$ is even and $g(x)$ is odd, and $h(x) = f(x)g(x)$,

$$h(-x) = f(-x)g(-x) = f(x)[-g(x)] = -f(x)g(x) = -h(x).$$

Thus, $h(x)$ is an odd function.

28. This is a semicircle.

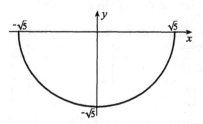

30. This is the upper half of the hyperbola $x^2 - y^2 = 5$.

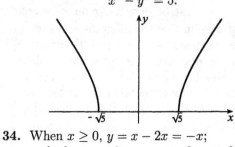

32. This is the lower half of the ellipse $4x^2 + y^2 = 5$.

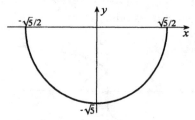

34. When $x \geq 0$, $y = x - 2x = -x$; and when $x < 0$, $y = -x - 2x = -3x$.

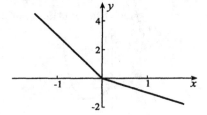

36. We rewrite the equation of the curve in the form $x = y^3$.

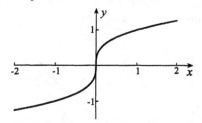

38. This function is only defined for $x \geq 0$.

40. We multiply ordinates of the curves $y = x$ and $y = \sqrt{x+1}$, the top half of a parabola.

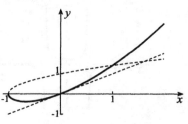

42. We multiply ordinates of the curves $y = -x$ and $y = \sqrt{4 - 9x^2}$, the top half of an ellipse.

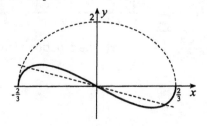

44. We multiply ordinates of the curves $y = x^2$ and $y = \sqrt{x^2 - 4}$, the top half of a hyperbola.

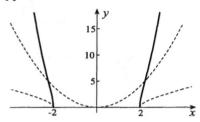

46. We draw the parabola $y = x^2 - x - 12$ $= (x - 4)(x + 3)$ and turn that part the x-axis, upside down.

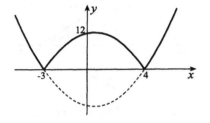

48. This is the top half of the ellipse $16(x - 1/4)^2 + 4y^2 = 1$.

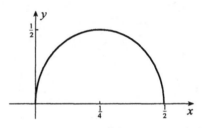

50. If we set $y = \sqrt{2x - x^2 - 4}$, square the equation, and complete square on the x-terms, the result is $(x - 1)^2 + y^2 = -3$. No real values of x and y can satisfy this equation.

52. The square root $\sqrt{9 - 4x^2}$ is defined only for $|x| \leq 3/2$, whereas the square root $\sqrt{4x^2 - 9}$ is defined for $|x| \geq 3/2$. The only points common to these intervals are $x = \pm 3/2$. The graph therefore consists of the two points $(\pm 3/2, 0)$

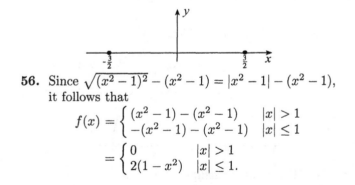

54. Since $\sqrt{(x^2 - 4)} = |x^2 - 4|$, we draw the parabola $y = x^2 - 4$, and then turn that part of the parabola between ± 2 upside down.

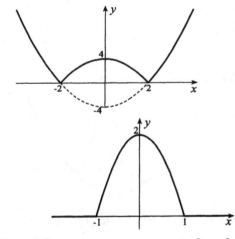

56. Since $\sqrt{(x^2 - 1)^2} - (x^2 - 1) = |x^2 - 1| - (x^2 - 1)$, it follows that

$$f(x) = \begin{cases} (x^2 - 1) - (x^2 - 1) & |x| > 1 \\ -(x^2 - 1) - (x^2 - 1) & |x| \leq 1 \end{cases}$$

$$= \begin{cases} 0 & |x| > 1 \\ 2(1 - x^2) & |x| \leq 1. \end{cases}$$

58. $y = x^2/9$ is a stretch of $y = x^2$ by a factor of 3 in the x-direction. It is also a compression by a factor of 9 in the y-direction.

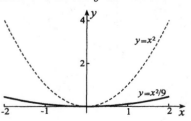

60. $x^2 + 16y^2 = 1$ is a compression of $x^2 + y^2 = 1$ by a factor of 4 in the y-direction.

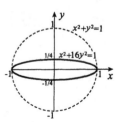

62. $|x| + |y/2| = 1$ is a stretch of $|x| + |y| = 1$ by a factor of 2 in the y-direction.

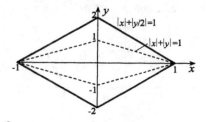

64. $y = -x^3 - 3x^2$ is $y = x^3 - 3x^2$ reflected in the y-axis.

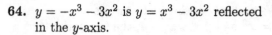

66. $x = y^2 + 2y$ is $x = y^2 - 2y$ reflected in the x-axis.

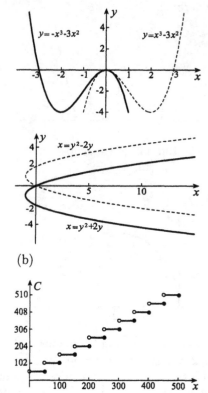

68. (a)

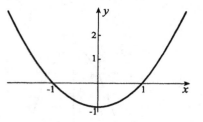

(b)

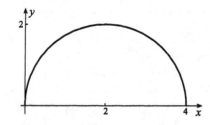

(c) Let $C(x)$ be the cost for mailing an item of x grams ($0 \le x \le 500$). When $x/50$ is an integer, $C(x) = 51(x/50)$. When $x/50$ is not an integer,

$$C(x) = 51\left(1 + \text{integer part of } \frac{x}{50}\right) = 51\left(1 + \lfloor x/50 \rfloor\right) = 51\lfloor 1 + x/50 \rfloor.$$

70. This defines a function.

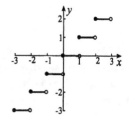

72. This defines a function.

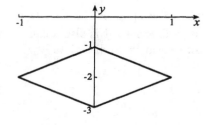

74. This does not define a function.

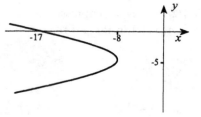

76. This does not define a function.

78. This does not define a function.

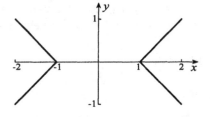

80. This does not define a function.

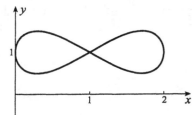

82. (a) Since the equation of the parabola is $s = 4t^2$ and that of the line is $s = -2(t-3)$, the algebraic definition of the signal is $s(t) = \begin{cases} 0, & t < 0 \\ 4t^2, & 0 \le t \le 1 \\ 2(3-t), & 1 < t \le 3 \\ 0, & t > 3 \end{cases}$.

(b) Graphs of $s(t+1/2)$ and $s(t-3)$ are shown below.

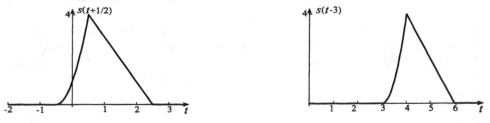

(c) The algebraic representations can be obtained by replacing t by $t+1/2$ and $t-3$ in $s(t)$,

$$s(t+1/2) = \begin{cases} 0, & t+1/2 < 0 \\ 4(t+1/2)^2, & 0 \le t+1/2 \le 1 \\ 2(3-t-1/2), & 1 < t+1/2 \le 3 \\ 0, & t+1/2 > 3 \end{cases} = \begin{cases} 0, & t < -1/2 \\ 4(t+1/2)^2, & -1/2 \le t \le 1/2 \\ 5-2t, & 1/2 < t \le 5/2 \\ 0, & t > 5/2 \end{cases};$$

$$s(t-3) = \begin{cases} 0, & t-3 < 0 \\ 4(t-3)^2, & 0 \le t-3 \le 1 \\ 2(3-t+3), & 1 < t-3 \le 3 \\ 0, & t-3 > 3 \end{cases} = \begin{cases} 0, & t < 3 \\ 4(t-3)^2, & 3 \le t \le 4 \\ 2(6-t), & 4 < t \le 6 \\ 0, & t > 6 \end{cases}.$$

84. The area of the rectangle shown is $A = 4xy$. When we solve the equation of the ellipse for the positive value of y, the result is $y = (b/a)\sqrt{a^2 - x^2}$. The area of the rectangle can therefore be expressed in the form

$$A = f(x) = \frac{4bx}{a}\sqrt{a^2 - x^2}, \quad 0 \le x \le a.$$

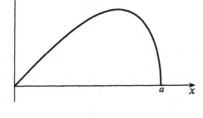

86. The amounts of A and B used to produce an amount x of C are $2x/3$ and $x/3$, respectively. It follows that

$$R = k\left(20 - \frac{2x}{3}\right)\left(40 - \frac{x}{3}\right)$$

$$= \frac{2k}{9}(30 - x)(120 - x), \quad 0 \le x \le 30.$$

The rate is a maximum at $x = 0$.

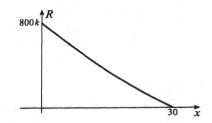

88. The depth d of the lowest corner of the square below the surface of the water is given by $d = t/2$. For $0 \le t \le 4\sqrt{2}$ (when the lower half of the square is being submerged), the area submerged at time t is $A = d^2 = (t/2)^2 = t^2/4$ (left figure below). For $4\sqrt{2} < t \le 8\sqrt{2}$ (when the top half is being submerged),

$$A = 16 - \left(4\sqrt{2} - \frac{t}{2}\right)^2 = -16 + 4\sqrt{2}t - \frac{t^2}{4}.$$

A graph of the function $A(t)$ is shown in the right figure below.

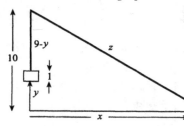

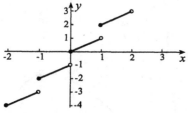

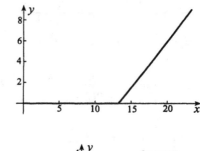

90. The rope becomes taut when $x = \sqrt{256 - 81} = 5\sqrt{7}$ and the box reaches the pulley when $x = \sqrt{625 - 81} = 4\sqrt{34}$. Between these values of x, the left figure below indicates that $z^2 = x^2 + 81$ and $z + (9 - y) = 25$. When we eliminate z between these equations, the result is $x^2 + 81 = (16 + y)^2 \implies y = -16 + \sqrt{x^2 + 81}$, the graph of which is shown in the right figure below.

92.

94.

96. If we solve the equation for r in terms of h,

the result is $r = \dfrac{ah}{\sqrt{h^2 - b^2}}$. When $h = 2b$,

$r = 2a/\sqrt{3}$. By writing the function in the form

$$r = \frac{ah}{\sqrt{h^2 - b^2}} = \frac{a}{\sqrt{1 - \frac{b^2}{h^2}}},$$

we see that as h increases, r decreases, and for large h, r is very close to a.

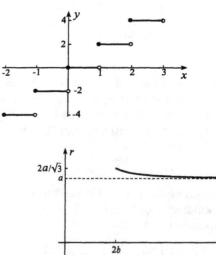

EXERCISES 1.6

2. Since the function is increasing for $x \ge -1$, it has an inverse. When we solve $y = \sqrt{x + 1}$, for x, we obtain $x = y^2 - 1$, and therefore the inverse function is $f^{-1}(x) = x^2 - 1$, $x \ge 0$.

4. Since horizontal lines intersect the graph of $f(x)$ in only one point, the function has an inverse. When we solve $y = (x + 5)/(2x + 4)$ for x, we obtain $x = (5 - 4y)/(2y - 1)$. The inverse function is $f^{-1}(x) = (5 - 4x)/(2x - 1)$.

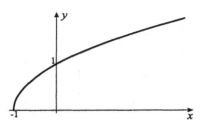

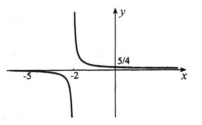

6. Since $f(x)$ is increasing for all x, the function has an inverse. Because $x = [(y-2)/3]^{1/3}$, the inverse function is $f^{-1}(x) = [(x-2)/3]^{1/3}$.

8. Because $f(x)$ is increasing for all x, the function has an inverse. Graphically, the inverse function is $f^{-1}(x) = x$ for $x < 0$ and $x/3$ for $x \geq 0$.

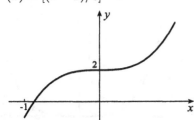

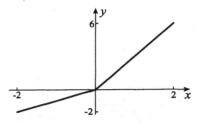

10. Since horizontal lines $y = c$, $0 \leq c < 1$, intersect the graph in two points, there is no inverse function.

12. Because $f(x)$ is increasing for $x \geq 1$, it has an inverse function. If we write $x^2 - 2x + (4-y) = 0$, then $x = \left[2 \pm \sqrt{4 - 4(4-y)}\right]/2 = 1 \pm \sqrt{y-3}$. Since $x > 1$, we must choose the positive sign, and the inverse function is $f^{-1}(x) = 1 + \sqrt{x-3}$, $x \geq 3$.

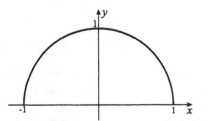

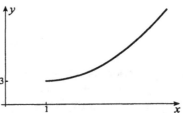

14. Since almost all horizontal lines that intersect the graph do so twice, there is no inverse function.

16. Since horizontal lines $y = c > 0$ intersect the graph of the function twice, the function does not have an inverse. On the intervals $x < 0$ and $x > 0$, it does have inverses. For $x > 0$, the inverse function is $f^{-1}(x) = 1/x^{1/4}$, and for $x < 0$, the inverse is $f^{-1}(x) = -1/x^{1/4}$.

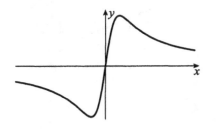

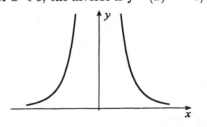

18. Because horizontal lines $y = c > 2$ intersect the graph of
the function twice, the function does not have an inverse.
On the intervals $x \leq 0$ and $x \geq 0$, it does have inverses.
To find them, we set $y = x^4 + 4x^2 + 2 = (x^2 + 2)^2 - 2$,
and solve for $x^2 + 2 = \pm\sqrt{y+2}$, but only the positive
result is acceptable. Hence $x^2 + 2 = \sqrt{y+2}$, and therefore
$x = \pm\sqrt{\sqrt{y+2}-2}$. For $x \geq 0$, the inverse function
is $f^{-1}(x) = \sqrt{\sqrt{x+2}-2}$, and for $x \leq 0$, the inverse
is $f^{-1}(x) = -\sqrt{\sqrt{x+2}-2}$.

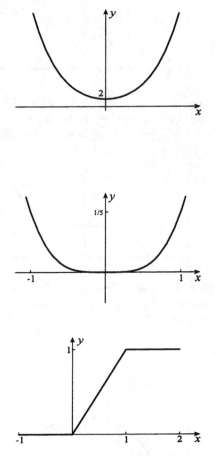

20. Because horizontal lines $y = c > 0$ intersect the graph of the
function twice, the function does not have an inverse. On
the intervals $x \leq 0$ and $x \geq 0$, it does have inverses. To
find them, we set $y = x^4/(x^2 + 4)$, from which
$x^4 - yx^2 - 4y = 0$. This quadratic in x^2 has solutions
$x^2 = (y \pm \sqrt{y^2 + 16y})/2$, but only the positive root
is acceptable. Thus, $x = \pm\sqrt{(y + \sqrt{y^2 + 16y})/2}$. The
inverse function for $x \geq 0$ is
$f^{-1}(x) = \sqrt{(x + \sqrt{x^2 + 16x})/2}$, and for $x \leq 0$ the
inverse is $f^{-1}(x) = -\sqrt{(x + \sqrt{x^2 + 16x})/2}$.

22. The function $f(x) = \begin{cases} 0, & x < 0 \\ x, & 0 \leq x \leq 1 \quad \text{(shown to the right)} \\ 1, & x > 1 \end{cases}$
is one-to-one on $0 \leq x \leq 1$, but on no other interval.

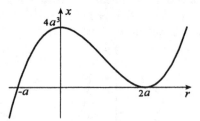

24. The demand function factors as $f(r) = (r + a)(r - 2a)^2$, and therefore its graph must be somewhat
as shown in the left figure below. It is decreasing from $r = 0$ to $r = 2a$ for the following reasons.
First $f(0) = 4a^3$. Secondly, when written in the form $f(r) = 4a^3 - r^2(3a - r)$, we see that because
$r^2(3a - r) > 0$ for $0 < r < 2a$, we must have $f(r) < 4a^3$ on this interval. The function therefore has
an inverse function for $0 < r < 2a$. The domain of the inverse function is the range of $f(r)$, namely,
$0 < x < 4a^3$. We can sketch its graph (right figure below) by reflecting the graph of $x = f(r)$ in the line
$x = r$.

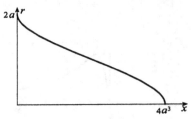

EXERCISES 1.7

In questions 1–10 we multiply the degree measure by $\pi/180$ to find the radian measure of the angle. The
answers are:

2. $\pi/3$ **4.** $-\pi/2$ **6.** $17\pi/4$ **8.** $-32\pi/45$ **10.** $-213\pi/180$

In questions 11–20 we multiply the radian measure by $180/\pi$ to find the degree measure of the angle. The answers are:

12. $-225°$ **14.** $1440°$ **16.** $180/\pi°$ **18.** $450/\pi°$ **20.** $1980/\pi°$

22. Since the height from the top of the transit to the top of the building is $30\tan 1.30$, the height of the building is $2 + 30\tan(1.30) = 110.1$ m.

24. From the figure below, the height of the flagpole is $20\tan 0.8 = 20.6$ m.

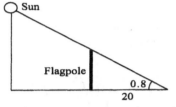

26. This is a right angle-angled triangle. Since $\sin A = 4/5$, it follows that $A = 0.927$ radians. From $\sin B = 3/5$, we obtain $B = 0.644$ radians.

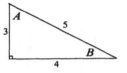

28. The remaining angle is $\pi - \pi/6 - \pi/5 = 19\pi/30$ radians. The sine law gives the remaining two sides,

$$\frac{a}{\sin(\pi/5)} = \frac{4}{\sin(19\pi/30)} \implies a = \frac{4\sin(\pi/5)}{\sin(19\pi/30)} = 2.57.$$

$$\frac{b}{\sin(\pi/6)} = \frac{4}{\sin(19\pi/30)} \implies b = \frac{4\sin(\pi/6)}{\sin(19\pi/30)} = 2.19.$$

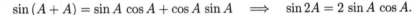

30. To verify 1.45a, we set $B = A$ in 1.43a,

$$\sin(A + A) = \sin A \cos A + \cos A \sin A \implies \sin 2A = 2\sin A \cos A.$$

To verify 1.46a, set $B = A$ in 1.43c. Then use the fact that $\sin^2 A + \cos^2 A = 1$ to derive 1.46b,c. To verify 1.47, set $B = A$ in 1.44.

32. If we set $X = A+B$ and $Y = A-B$, and solve for A and B, results are $A = (X+Y)/2$ and $B = (X-Y)/2$. If we substitute these into 1.48b,

$$\sin\left(\frac{X+Y}{2}\right)\cos\left(\frac{X-Y}{2}\right) = \frac{1}{2}(\sin X + \sin Y) \implies \sin X + \sin Y = 2\sin\left(\frac{X+Y}{2}\right)\cos\left(\frac{X-Y}{2}\right).$$

Proofs of 1.49b,c,d are similar.

34. The period is π.

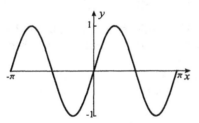

36. The sine curve is shifted $\pi/4$ units to the left.

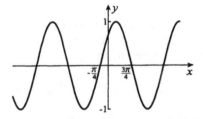

38. The period is π and the curve is shifted $\pi/8$ units to the left.

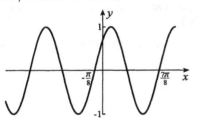

40. The amplitude is 4 and the period is 6π.

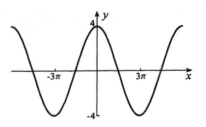

42. The amplitude is 5, the period is $2\pi/3$, and the curve is shifted $\pi/6$ units to the right.

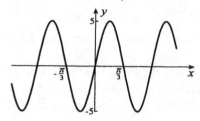

44. The period is $\pi/3$.

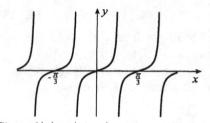

46. The cotangent curve is shifted $\pi/4$ units to the left.

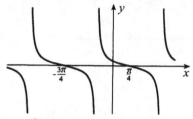

48. Since $f(x) = |\sin x|$, we invert that part of the sine curve below the x-axis.

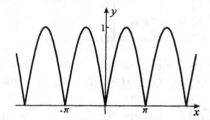

50. We invert the secant curve, double ordinates, and shift the curve vertically 5 units.

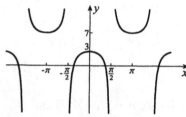

52. We reflect that part of the tangent curve to the right of the y-axis in the y-axis.

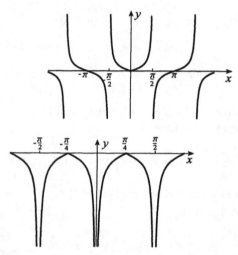

54. The period is 4π and ordinates are multiplied by 3.

56. One angle satisfying the equation is $x = \pi/3$ radians. All solutions can be expressed in the form $x = \pi/3 + 2n\pi, 2\pi/3 + 2n\pi$, where n is an integer. Following the lead of Example 1.38, these angles can be combined in the form

$$\frac{\pi}{2} \pm \frac{\pi}{6} + 2n\pi = \frac{(4n+1)\pi}{2} \pm \frac{\pi}{6}, \quad n \text{ an integer.}$$

58. One angle satisfying the equation is $x = 2\pi/3$ radians. All solutions can be expressed in the form $x = \pm 2\pi/3 + 2n\pi$, where n is an integer.

60. If we divide by $\cos x$, then $\tan x = 1$. One solution of this equation is $x = \pi/4$ radians. All solutions can be expressed in the form $x = \pi/4 + n\pi$, where n is an integer.

62. One solution of this equation for $2x$ is $2x = 3\pi/4$. All solutions can be expressed in the form

$$2x = \pm\frac{3\pi}{4} + 2n\pi \quad \Longrightarrow \quad x = \pm\frac{3\pi}{8} + n\pi, \quad n \text{ an integer.}$$

64. One solution of the equation $\sin 3x = 1/2$ for $3x$ is $3x = \pi/6$. All solutions can be expressed in the form

$$3x = \frac{\pi}{2} \pm \frac{\pi}{3} + 2n\pi \quad \Longrightarrow \quad x = \frac{\pi}{6} \pm \frac{\pi}{9} + \frac{2n\pi}{3}, \quad n \text{ an integer.}$$

66. If $\sin 2x = \sin x$, then $2 \sin x \cos x = \sin x \Longrightarrow (2 \cos x - 1) \sin x = 0$, and therefore either $\sin x = 0$ or $\cos x = 1/2$. Solutions of the former are $x = n\pi$, where n is an integer, and solutions of the latter are $x = \pm \pi/3 + 2n\pi$, where n is an integer.

68. This equation implies that $\cot x = \pm 1/\sqrt{3}$. One solution of $\cot x = 1/\sqrt{3}$ is $\pi/3$ radians and one solution of $\cot x = -1/\sqrt{3}$ is $-\pi/3$ radians. All solutions can be expressed in the form $x = \pm \pi/3 + n\pi$, where n is an integer.

70. (a) Using identity 1.49c,

$$x(t) = \cos(440\pi t) + \cos(360\pi t)$$
$$= 2 \cos(400\pi t) \cos(40\pi t).$$

(b)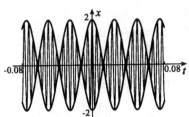

72. If we set $f(x) = 3 \sin 3x + 3 \cos 3x = A \sin(3x + \phi)$, and expand the right side,

$$3 \sin 3x + 3 \cos 3x = A(\sin 3x \cos \phi + \cos 3x \sin \phi).$$

This will be true for all x if we set $A \cos \phi = 3$ and $A \sin \phi = 3$. When these are squared and added, $3^2 + 3^2 = A^2 \cos^2 \phi + A^2 \sin^2 \phi = A^2$. If we choose $A = 3\sqrt{2}$, then $\sin \phi = 1/\sqrt{2}$ and $\cos \phi = 1/\sqrt{2}$. These are satisfied by $\phi = \pi/4$. The amplitude is $3\sqrt{2}$, the period is $2\pi/3$, and the phase shift is $-\pi/12$.

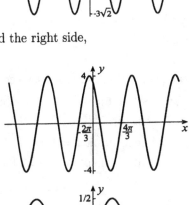

74. If we set $f(x) = -2 \sin x + 2\sqrt{3} \cos x = A \sin(x + \phi)$, and expand the right side,

$$-2 \sin x + 2\sqrt{3} \cos x = A(\sin x \cos \phi + \cos x \sin \phi).$$

This will be true for all x if we set $A \cos \phi = -2$ and $A \sin \phi = 2\sqrt{3}$. When these are squared and added, $(-2)^2 + (2\sqrt{3})^2 = A^2 \cos^2 \phi + A^2 \sin^2 \phi = A^2$. If we choose $A = 4$, then $\sin \phi = \sqrt{3}/2$ and $\cos \phi = -1/2$. These are satisfied by $\phi = 2\pi/3$. The amplitude is 4, the period is 2π, and the phase shift is $-2\pi/3$.

76. This is simply done using equation 1.45, $f(x) = (1/2) \sin 2x$. The amplitude is $1/2$, the period is π, and the phase shift is 0.

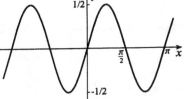

78. We expand $\cos 3x$,

$$\cos 3x = \cos(2x + x) = \cos 2x \cos x - \sin 2x \sin x$$
$$= (2 \cos^2 x - 1) \cos x - (2 \sin x \cos x) \sin x = 2 \cos^3 x - \cos x - 2 \cos x (1 - \cos^2 x)$$
$$= 4 \cos^3 x - 3 \cos x.$$

80. We expand $\tan 3x$,

$$\tan 3x = \tan(2x + x) = \frac{\tan 2x + \tan x}{1 - \tan 2x \tan x} = \frac{\dfrac{2\tan x}{1 - \tan^2 x} + \tan x}{1 - \dfrac{2\tan^2 x}{1 - \tan^2 x}}$$

$$= \frac{2\tan x + \tan x - \tan^3 x}{1 - \tan^2 x - 2\tan^2 x} = \frac{3\tan x - \tan^3 x}{1 - 3\tan^2 x}.$$

82. We expand the right side, $\tan\left(x + \dfrac{\pi}{4}\right) = \dfrac{\tan x + \tan(\pi/4)}{1 - \tan x \tan(\pi/4)} = \dfrac{\tan x + 1}{1 - \tan x}.$

84. No. The two equations define the quadrant for ϕ, but the equation $\tan\phi = A/B$ does not.

86. If we use identity 1.49c, $0 = \cos x + \cos 3x = 2\cos 2x \cos x$. Hence, $\cos 2x = 0$ or $\cos x = 0$. Solutions of the first are defined by $2x = (2n+1)\pi/2 \implies x = (2n+1)\pi/4$, where n is an integer. Solutions of $\cos x = 0$ are $(2n+1)\pi/2$. The only solutions between 0 and 2 are $\pi/4$ and $\pi/2$.

88. If we square the equation,

$$\sin^2 x + 2\sin x \cos x + \cos^2 x = 3\sin^2 x \cos^2 x \implies 3\sin^2 x \cos^2 x - 2\sin x \cos x - 1 = 0$$

$$\implies (3\sin x \cos x + 1)(\sin x \cos x - 1) = 0 \implies \sin 2x = -2/3 \quad\text{or}\quad \sin 2x = 2.$$

The second of these is impossible. From the first,

$$2x = \begin{cases} -0.7297 + 2n\pi \\ -2.412 + 2n\pi \end{cases} \implies x = \begin{cases} -0.365 + n\pi \\ -1.21 + n\pi \end{cases},$$

where n is an integer. Only $x = -0.365 + (2n+1)\pi, -1.21 + 2n\pi$ satisfy the original equation. None of the solutions are between 0 and 2.

EXERCISES 1.8

2. $\operatorname{Sin}^{-1}(1/4) = 0.253$ **4.** $\operatorname{Csc}^{-1}(-2/\sqrt{3}) = -2\pi/3$

6. $\operatorname{Cos}^{-1}(3/2)$ does not exist since the domain of $\operatorname{Cos}^{-1} x$ is $-1 \le x \le 1$.

8. $\operatorname{Tan}^{-1}(-1) = -\pi/4$

10. $\tan\left(\operatorname{Sin}^{-1}3\right)$ does not exist since the domain of $\operatorname{Sin}^{-1} x$ is $-1 \le x \le 1$.

12. $\operatorname{Tan}^{-1}[\sin(1/6)] = \operatorname{Tan}^{-1}[0.165\,896] = 0.164$ **14.** $\operatorname{Sin}^{-1}[\sin(3\pi/4)] = \operatorname{Sin}^{-1}[1/\sqrt{2}] = \pi/4$

16. $\operatorname{Sin}^{-1}\left[\cos\left(\operatorname{Sec}^{-1}(-\sqrt{2})\right)\right] = \operatorname{Sin}^{-1}\left[\cos(-3\pi/4)\right] = \operatorname{Sin}^{-1}\left[-1/\sqrt{2}\right] = -\pi/4$

18. From the solution $x = \operatorname{Tan}^{-1}(-1.2) = -0.876$, we obtain $x = n\pi - 0.876$, where n is an integer.

20. One solution of $\cot 4x = -2.2$ for $4x$ is $4x = \operatorname{Cot}^{-1}(-2.2) = 2.715$. All solutions are given by

$$4x = 2.715 + n\pi \implies x = .679 + \frac{n\pi}{4}, \quad n \text{ an integer.}$$

22. One solution of $\tan 3x = -3.2/3$ for $3x$ is $3x = \operatorname{Tan}^{-1}(-3.2/3) = -0.8176$. All solutions are given by

$$3x = -0.8176 + n\pi \implies x = -0.273 + \frac{n\pi}{3}, \quad n \text{ an integer.}$$

24. Since $1 = 4\sin^2 x + 2(1 - \sin^2 x)$, we require $2\sin^2 x = -1$, an impossibility.

26. This quadratic equation in $\sin x$ has solutions $\sin x = \left(3 \pm \sqrt{9+20}\right)/2 = (3 \pm \sqrt{29})/2$. Because neither of these numbers is between -1 and 1, there are no solutions to the equation.

28. We add ordinates of $y = \sqrt{\text{Tan}^{-1} x}$ and $y = \sqrt{\text{Sec}^{-1} x}$.

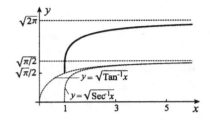

30. Since the domain of $\text{Sin}^{-1} x$ is $|x| \leq 1$, and that of $\text{Csc}^{-1} x$ is $|x| \geq 1$, the function is defined only for $x = \pm 1$. Its graph is therefore two points.

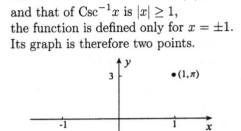

32. Using equation 1.59, $\tan \phi = -1/2 \implies \phi = 2.68$ radians.

34. Using equation 1.59, $\tan \phi = 3 \implies \phi = 1.25$ radians.

36. Since the line is horizontal, $\phi = 0$.

38. Since slopes of both lines are $-1/3$, the lines are parallel.

40. Since slopes of the lines are $-2/3$ and $3/2$, the lines are perpendicular.

42. Since slopes of the lines are 1 and $-2/3$, formula 1.60 gives

$$\theta = \text{Tan}^{-1} \left| \frac{1 + 2/3}{1 + (-2/3)} \right| = 1.37 \text{ radians.}$$

44. Since slopes of the lines are -1 and 3, formula 1.60 gives

$$\theta = \text{Tan}^{-1} \left| \frac{-1 - 3}{1 - 3} \right| = 1.11 \text{ radians.}$$

46. Since $0 = \tan^2 x (\sin x + 1) - 3(\sin x + 1) = (\tan^2 x - 3)(\sin x + 1)$, it follows that $\tan x = \pm\sqrt{3}$ or $\sin x = -1$. The solutions of these equations are $x = \pm\pi/3 + n\pi$ and $x = -\pi/2 + 2n\pi$, where n is an integer, but only $x = n\pi \pm \pi/3$ satisfy the original equation.

48. This equation implies that $\sin x = \pm 3\pi/4 + 2n\pi$, where n is an integer. But for no n are these values between between ± 1. Hence, there are no solutions.

50. This equation implies that $\tan x = \pm 3\pi/4 + 2n\pi$. Therefore, $x = \text{Tan}^{-1}(2n\pi \pm 3\pi/4) + m\pi$ where m and n are integers.

52. (a) Since $\text{Sin}^{-1} x$ is defined only for $-1 \leq x \leq 1$, and on this interval, the sine function is the inverse of $\text{Sin}^{-1} x$, it follows that $f(x) = x$. Its graph is shown to the left below.
(b) On the interval $-\pi/2 \leq x \leq \pi/2$, $\text{Sin}^{-1} x$ is the inverse of $\sin x$ and therefore $f(x) = x$ on this interval. For $\pi/2 \leq x \leq 3\pi/2$, $f(x) = \pi - x$. These define $f(x)$ on the interval $-\pi/2 \leq x \leq 3\pi/2$ of length 2π. Since $\sin x$ is 2π-periodic, so also is $f(x)$, and the graph is shown to the right below.

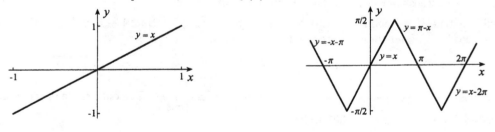

54. If we expand $R\sin(2x+\phi)$ according to 1.43a and equate it to $f(x)$, we obtain

$$R\sin 2x\,\cos\phi + R\cos 2x\,\sin\phi = 4\sin 2x + \cos 2x.$$

This equation is satisfied for all x if R and ϕ satisfy $R\cos\phi = 4$ and $R\sin\phi = 1$. When these are squared and added, the result is $R^2 = 17$. Consequently, $R = \sqrt{17}$, and

$$\cos\phi = \frac{4}{\sqrt{17}}, \qquad \sin\phi = \frac{1}{\sqrt{17}}.$$

The only angle in the range $0 < \phi < \pi$ satisfying these is $\phi = 0.245$ radians. Thus, $f(x)$ can be expressed in the form $\sqrt{17}\sin(2x + 0.245)$.

56. If we expand $R\sin(2x+\phi)$ according to 1.43a and equate it to $f(x)$, we obtain

$$R\sin 2x\,\cos\phi + R\cos 2x\,\sin\phi = -2\sin 2x + 4\cos 2x.$$

This equation is satisfied for all x if R and ϕ satisfy $R\cos\phi = -2$ and $R\sin\phi = 4$. When these are squared and added, the result is $R^2 = 20$. Consequently, $R = 2\sqrt{5}$, and

$$\cos\phi = \frac{-1}{\sqrt{5}}, \qquad \sin\phi = \frac{2}{\sqrt{5}}.$$

The only angle in the range $0 < \phi < \pi$ satisfying these is $\phi = 2.03$ radians. Thus, $f(x)$ can be expressed in the form $2\sqrt{5}\sin(2x + 2.03)$.

58. We set $x(t) = A\sin(\omega t + \phi) = A(\sin\omega t\,\cos\phi + \cos\omega t\,\sin\phi) = f(t) + g(t)$

$$= 4[\cos\omega t\,\cos(2\pi/3) - \sin\omega t\,\sin(2\pi/3)] + 3[\sin\omega t\,\cos(\pi/3) + \cos\omega t\,\sin(\pi/3)]$$

$$= \left(-2\sqrt{3} + \frac{3}{2}\right)\sin\omega t + \left(-2 + \frac{3\sqrt{3}}{2}\right)\cos\omega t.$$

This will be true if we choose A and ϕ to satisfy $\quad A\cos\phi = \dfrac{3}{2} - 2\sqrt{3}\quad$ and $\quad A\sin\phi = \dfrac{3\sqrt{3}}{2} - 2$. When these are squared and added, the result is

$$A^2 = \left(\frac{3}{2} - 2\sqrt{3}\right)^2 + \left(\frac{3\sqrt{3}}{2} - 2\right)^2 = 25 - 12\sqrt{3} \implies A = \sqrt{25 - 12\sqrt{3}}.$$

Hence, $\quad\cos\phi = \dfrac{3/2 - 2\sqrt{3}}{\sqrt{25 - 12\sqrt{3}}}$ and $\sin\phi = \dfrac{3\sqrt{3}/2 - 2}{\sqrt{25 - 12\sqrt{3}}}$. The only angle in the interval $-\pi < \phi < \pi$ satisfying these is $\phi = 2.846$ radians.

60. We set $x(t) = A\cos(\omega t + \phi) = A(\cos\omega t\,\cos\phi - \sin\omega t\,\sin\phi) = f(t) + g(t)$

$$= 2(\sin\omega t\,\cos 4 + \cos\omega t\,\sin 4) + 3(\sin\omega t\,\cos 1 + \cos\omega t\,\sin 1)$$

$$= (2\cos 4 + 3\cos 1)\sin\omega t + (2\sin 4 + 3\sin 1)\cos\omega t.$$

This will be true if we choose A and ϕ to satisfy $\quad A\cos\phi = 2\sin 4 + 3\sin 1\quad$ and $\quad -A\sin\phi = 2\cos 4 + 3\cos 1$. When these are squared and added, the result is

$$A^2 = (2\sin 4 + 3\sin 1)^2 + (2\cos 4 + 3\cos 1)^2 \implies A = \sqrt{13 + 12\cos 3}.$$

Hence, $\quad\cos\phi = \dfrac{2\sin 4 + 3\sin 1}{\sqrt{13 + 12\cos 3}}$ and $\sin\phi = -\dfrac{2\cos 4 + 3\cos 1}{\sqrt{13 + 12\cos 3}}$. The only angle in the interval $-\pi < \phi < \pi$ satisfying these is $\phi = -0.301$ radians.

62. We set $x(t) = A\sin(\omega t + \phi) = A(\sin\omega t\,\cos\phi + \cos\omega t\,\sin\phi) = f(t) + g(t) + h(t)$

$$= 5\sin\omega t + 4[\cos\omega t\,\cos(\pi/3) - \sin\omega t\,\sin(\pi/3)] + 2[\sin\omega t\,\cos(\pi/4) + \cos\omega t\,\sin(\pi/4)]$$

$$= (5 - 2\sqrt{3} + \sqrt{2})\sin\omega t + (2 + \sqrt{2})\cos\omega t.$$

This will be true if we choose A and ϕ to satisfy $\quad A\cos\phi = 5 - 2\sqrt{3} + \sqrt{2}\quad$ and $\quad A\sin\phi = 2 + \sqrt{2}$. When these are squared and added, the result is

$$A^2 = (5 - 2\sqrt{3} + \sqrt{2})^2 + (2 + \sqrt{2})^2 \implies A = \sqrt{45 + 14\sqrt{2} - 20\sqrt{3} - 4\sqrt{6}}.$$

Hence, $\cos\phi = \dfrac{5 - 2\sqrt{3} + \sqrt{2}}{\sqrt{45 + 14\sqrt{2} - 20\sqrt{3} - 4\sqrt{6}}}$ and $\sin\phi = \dfrac{2 + \sqrt{2}}{\sqrt{45 + 14\sqrt{2} - 20\sqrt{3} - 4\sqrt{6}}}$. The only angle in the interval $-\pi < \phi < \pi$ satisfying these is $\phi = 0.858$ radians.

64. If we let x be the distance from slider C to line OA, then

$x = R\sin\phi$ and $x = (L + R\cos\phi)\tan\theta$.

Equating these gives

$$R\sin\phi = (L + R\cos\phi)\tan\theta \; Longrightarrow \; \tan\theta = \frac{R\sin\phi}{L + R\cos\phi}.$$

Thus, $\theta = \mathrm{Tan}^{-1}\left(\dfrac{R\sin\phi}{L + R\cos\phi}\right).$

66. The equation

$$\theta = \theta_0 \cos\omega t + \frac{v_0}{\omega L}\sin\omega t = R\sin(\omega t + \phi) = R\sin\omega t\cos\phi + R\cos\omega t\sin\phi$$

is satisfied if R and ϕ satisfy

$$R\cos\phi = \frac{v_0}{\omega L} \quad \text{and} \quad R\sin\phi = \theta_0.$$

When these are squared and added, $R^2 = \theta_0^2 + v_0^2/(\omega^2 L^2)$. If we choose $R = \sqrt{\theta_0^2 + v_0^2/(\omega^2 L^2)}$, then

$$\cos\phi = \frac{v_0}{\omega L\sqrt{\theta_0^2 + v_0^2/(\omega^2 L^2)}} = \frac{v_0}{\sqrt{v_0^2 + \omega^2 L^2\theta_0^2}}, \qquad \sin\phi = \frac{\theta_0}{\sqrt{\theta_0^2 + v_0^2/(\omega^2 L^2)}} = \frac{\omega L\theta_0}{\sqrt{v_0^2 + \omega^2 L^2\theta_0^2}}.$$

Because $v_0 > 0$, it follows that $\cos\phi > 0$, and we may take $-\pi/2 < \phi < \pi/2$. Since $\sin\phi$ has the same sign as θ_0, angle ϕ is in $0 < \phi < \pi/2$ when $\theta_0 > 0$, and is in $-\pi/2 < \phi < 0$ when $\theta_0 < 0$. Now

$$\tan\phi = \frac{\sin\phi}{\cos\phi} = \frac{\omega L\theta_0}{\sqrt{v_0^2 + \omega^2 L^2\theta_0^2}}\frac{\sqrt{v_0^2 + \omega^2 L^2\theta_0^2}}{v_0} = \frac{\omega L\theta_0}{v_0}.$$

We can write $\phi = \mathrm{Tan}^{-1}(\omega L\theta_0/v_0)$ since principal values are between $-\pi/2$ and 0 when $\theta_0 < 0$, and between 0 and $\pi/2$ when $\theta_0 > 0$.

68. The equation

$$y = y_0 \cos\omega t + \frac{v_0}{\omega}\sin\omega t = R\sin(\omega t + \phi) = R\sin\omega t\cos\phi + R\cos\omega t\sin\phi$$

is satisfied if R and ϕ satisfy

$$R\cos\phi = \frac{v_0}{\omega}, \qquad R\sin\phi = y_0.$$

When these are squared and added, $R^2 = y_0^2 + v_0^2/\omega^2$. If we choose $R = \sqrt{y_0^2 + v_0^2/\omega^2}$, then

$$\cos\phi = \frac{v_0}{\omega\sqrt{y_0^2 + v_0^2/\omega^2}} = \frac{v_0}{\sqrt{v_0^2 + \omega^2 y_0^2}}, \qquad \sin\phi = \frac{y_0}{\sqrt{y_0^2 + v_0^2/\omega^2}} = \frac{\omega y_0}{\sqrt{v_0^2 + \omega^2 y_0^2}}.$$

Because $v_0 > 0$, it follows that $\cos\phi > 0$, and we may take $-\pi/2 < \phi < \pi/2$. Since $\sin\phi$ has the same sign as y_0, angle ϕ is in $0 < \phi < \pi/2$ when $y_0 > 0$, and is in $-\pi/2 < \phi < 0$ when $y_0 < 0$. Now

$$\tan\phi = \frac{\sin\phi}{\cos\phi} = \frac{\omega y_0}{\sqrt{v_0^2 + \omega^2 y_0^2}}\frac{\sqrt{v_0^2 + \omega^2 y_0^2}}{v_0} = \frac{\omega y_0}{v_0}.$$

We can write $\phi = \mathrm{Tan}^{-1}(\omega y_0/v_0)$ since principal values are between $-\pi/2$ and 0 when $y_0 < 0$, and between 0 and $\pi/2$ when $y_0 > 0$.

70. If we expand $A\cos(\omega t - \phi)$ and equate it to $f(x)$, we obtain

$$A\cos\omega t\,\cos\phi + A\sin\omega t\,\sin\phi = \frac{E_0}{\sqrt{R^2 + \left(\omega L - \frac{1}{\omega C}\right)^2}}\left[R\cos\omega t + \left(\omega L - \frac{1}{\omega C}\right)\sin\omega t\right].$$

This equation is satisfied for all t if A and ϕ satisfy

$$A\cos\phi = \frac{E_0 R}{\sqrt{R^2 + \left(\omega L - \frac{1}{\omega C}\right)^2}}, \qquad A\sin\phi = \frac{E_0\left(\omega L - \frac{1}{\omega C}\right)}{\sqrt{R^2 + \left(\omega L - \frac{1}{\omega C}\right)^2}}.$$

When these are squared and added, the result is $A^2 = E_0^2$. Consequently, $A = E_0$, and

$$\cos\phi = \frac{R}{\sqrt{R^2 + \left(\omega L - \frac{1}{\omega C}\right)^2}}, \qquad \sin\phi = \frac{\omega L - \frac{1}{\omega C}}{\sqrt{R^2 + \left(\omega L - \frac{1}{\omega C}\right)^2}}.$$

Because $R > 0$, it follows that $\cos\phi > 0$, and this is consistent with the demand that $-\pi/2 < \phi < \pi/2$. We could use the equation in $\sin\phi$ to define ϕ, or we can also write that $\tan\phi = [\omega L - 1/(\omega C)]/R$ so that $\phi = \mathrm{Tan}^{-1}\{[\omega L - 1/(\omega C)]/R\}$.

72. If $5\cos\omega t = A(\cos\omega t\,\cos 1 - \sin\omega t\,\sin 1) + 5(\sin\omega t\,\cos\phi + \cos\omega t\,\sin\phi)$, then

$$5 = A\cos 1 + 5\sin\phi \qquad \text{and} \qquad 0 = -A\sin 1 + 5\cos\phi.$$

These imply that $(5\sin\phi)^2 + (5\cos\phi)^2 = (5 - A\cos 1)^2 + (A\sin 1)^2$, from which $25 = 25 - 10A\cos 1 + A^2 \implies A = 0$ or $A = 10\cos 1$. Since A must be positive, we choose $A = 10\cos 1$, in which case

$$\sin\phi = 1 - 2\cos^2 1 = -\cos 2 \qquad \text{and} \qquad \cos\phi = 2\sin 1\cos 1 = \sin 2.$$

From the second of these, we may write $\cos\phi = \cos(\pi/2 - 2)$. We conclude that $\phi = \pm(\pi/2 - 2) + 2n\pi$, where n is an integer. But, $\sin(\pi/2 - 2 + 2n\pi) = \sin(\pi/2 - 2) = \cos 2$, which is not true. Hence, we must take $\phi = -(\pi/2 - 2) + 2n\pi = 2 + (4n - 1)\pi/2$.

74. When $0 \leq x \leq 1$, we set $y = \mathrm{Sin}^{-1}x$, in which case $0 \leq y \leq \pi/2$. It follows that $x = \sin y$, and because $\sin^2 y + \cos^2 y = 1$, we have

$$\cos y = \pm\sqrt{1 - \sin^2 y} = \pm\sqrt{1 - x^2}.$$

Since y is an angle in the first quadrant, its cosine must be nonnegative, and therefore $\cos y = \sqrt{1 - x^2}$. When we apply the inverse cosine function to both sides of this equation, we obtain $\mathrm{Cos}^{-1}(\cos y) = y = \mathrm{Cos}^{-1}\sqrt{1 - x^2}$. When $-1 \leq x < 0$, we continue to set $y = \mathrm{Sin}^{-1}x$, and once again obtain $\cos y = \sqrt{1 - x^2}$, because $-\pi/2 \leq y < 0$. Application of the inverse cosine function gives $\mathrm{Cos}^{-1}(\cos y) = \mathrm{Cos}^{-1}\sqrt{1 - x^2}$, but the left side is not equal to y because y is not in the principal value range of the inverse cosine function. This is easily adjusted by noting that with $-\pi/2 \leq y < 0$, we have $\cos y = \cos(-y)$. Hence,

$$\mathrm{Cos}^{-1}(\cos y) = \mathrm{Cos}^{-1}[\cos(-y)] = -y = \mathrm{Cos}^{-1}\sqrt{1 - x^2};$$

that is, $y = -\mathrm{Cos}^{-1}\sqrt{1 - x^2}$.

76. When $x > 0$, we set $y = \mathrm{Cot}^{-1}x$, in which case $0 < y < \pi/2$. It follows that $x = \cot y$, and

$$\frac{1}{x} = \frac{1}{\cot y} = \tan y.$$

If we apply the inverse tangent function to both sides of this equation, the result is

$$\mathrm{Tan}^{-1}\left(\frac{1}{x}\right) = \mathrm{Tan}^{-1}(\tan y) = y,$$

because y is in the principal value range of the inverse tangent function. Hence, when $x > 0$, we can say that $\text{Cot}^{-1}x = \text{Tan}^{-1}(1/x)$.

When $x < 0$, we again set $y = \text{Cot}^{-1}x$, and obtain

$$\text{Tan}^{-1}\left(\frac{1}{x}\right) = \text{Tan}^{-1}(\tan y).$$

But in this case the right side is not equal to y, because $\pi/2 < y < \pi$. To remedy this, we note that $\tan y = \tan(y - \pi)$, and when $\pi/2 < y < \pi$, $y - \pi$ is in the principal value range for the inverse tangent function. It follows that

$$\text{Tan}^{-1}\left(\frac{1}{x}\right) = \text{Tan}^{-1}(\tan y) = \text{Tan}^{-1}[\tan(y - \pi)] = y - \pi = \text{Cot}^{-1}x - \pi.$$

Thus, $\text{Cot}^{-1}x = \pi + \text{Tan}^{-1}(1/x)$.

78. When $x \geq 1$, we set $y = \text{Csc}^{-1}x$, in which case $0 < y \leq \pi/2$. It follows that $x = \csc y$, and because $1 + \cot^2 y = \csc^2 y$, we have $\cot y = \pm\sqrt{\csc^2 y - 1} = \pm\sqrt{x^2 - 1}$. Since y is an angle in the first quadrant, its cotangent is positive, and therefore $\cot y = \sqrt{x^2 - 1}$. When we apply the inverse cotangent function to both sides of this equation, we obtain $\text{Cot}^{-1}(\cot y) = y = \text{Cot}^{-1}\sqrt{x^2 - 1}$; that is, $\text{Csc}^{-1}x = \text{Cot}^{-1}\sqrt{x^2 - 1}$.

When $x \leq -1$, we again set $y = \text{Csc}^{-1}x$, and obtain $\cot y = \sqrt{x^2 - 1}$, because $-\pi < y \leq -\pi/2$. Application of the inverse cotangent function gives $\text{Cot}^{-1}(\cot y) = \text{Cot}^{-1}\sqrt{x^2 - 1}$, but the left side is not equal to y because y is not in the principal value range of the inverse cotangent function. This is easily adjusted by noting that $\cot y = \cot(\pi + y)$, and when $-\pi < y \leq -\pi/2$, $\pi + y$ is in the principal value range of the inverse cotangent function. Hence $\text{Cot}^{-1}(\cot y) = \text{Cot}^{-1}[\cot(\pi + y)] = \pi + y = \text{Cot}^{-1}\sqrt{x^2 - 1}$; that is, $y = -\pi + \text{Cot}^{-1}\sqrt{x^2 - 1}$.

EXERCISES 1.9

2. If $10^{3x} = 5$, then $3x = \log_{10} 5 \implies x = (1/3)\log_{10} 5$.

4. If $\ln(x^2 + 2x + 10) = 1$, then $x^2 + 2x + 10 = e^1 \implies x = \left[-2 \pm \sqrt{4 - 4(10 - e)}\right]/2$. These are not real.

6. If $10^{1-x^2} = 100 = 10^2$, then $1 - x^2 = 2$. This equation does not have real solutions.

8. We write $1 = \log_{10}[(3 - x)x]$, and take exponentials, $10 = (3 - x)x \implies 0 = x^2 - 3x + 10$. This equation has no real solutions.

10. We write $\log_{10}[x^2(x - 1)] = 2 \implies x^2(x - 1) = 10^2 = 100 \implies 0 = x^3 - x^2 - 100 = (x - 5)(x^2 + 4x + 20) \implies x = 5$.

12. We take exponentials to obtain $x(x + 2) = a^2 \implies x^2 + 2x - a^2 = 0$. Solutions of this quadratic equation are $x = (-2 \pm \sqrt{4 + 4a^2})/2 = -1 \pm \sqrt{1 + a^2}$.

14.

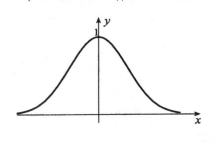

16.

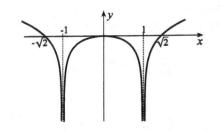

18.

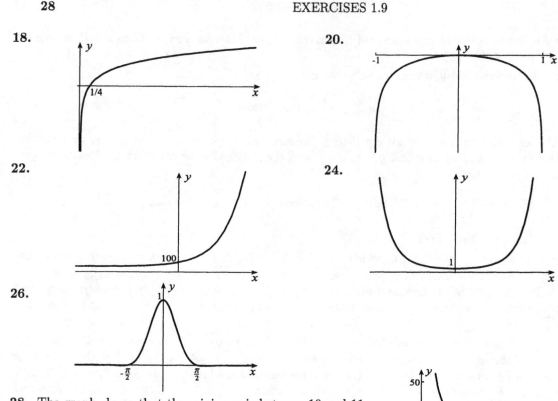

20.

22.

24.

26.

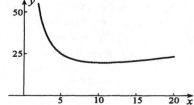

28. The graph shows that the minimum is between 10 and 11. Values $y(10) = 19.562$ and $y(11) = 19.557$ indicate that the minimum occurs for $x = 11$.

30. The graph of $f(x)$ is symmetric about the y-axis, whereas the graph of $g(x)$ exists only for $x > 0$. They are identical to the right of the y-axis.

32. After the first year its value is $20\,000(3/4)$. After two years, the value is $20\,000(3/4)^2$, and after t years, it is $20\,000(3/4)^t$.

34. If y is the logarithm of x to base a, $y = \log_a x$, then $x = a^y$. It follows that $x = (1/a)^{-y}$, and this implies that $-y$ is the logarithm of x to base $1/a$.

36. No. Both x_1 and x_2 must be positive.

38. (a) After one interest period the accumulated value is $P[1 + i/(100n)]$. After two interest periods, it is $P[1 + i/(100n)]^2$. Continuing, the accumulated value after t years, or nt interest periods is $P[1 + i/(100n)]^{nt}$.

(b) When $A = 2P$, $i = 8$ and $n = 2$,

$$2P = P\left(1 + \frac{8}{200}\right)^{2t} = P\left(\frac{26}{25}\right)^{2t} \quad \Longrightarrow \quad 2t \log_{10}(26/25) = \log_{10} 2 \quad \Longrightarrow \quad t = \frac{\log_{10} 2}{2 \log_{10}(26/25)} = 8.84.$$

Thus, money doubles in 9 years.

(c) If we write that $\quad A = P\left(1 + \dfrac{i}{100n}\right)^{nt} = P\left[\left(1 + \dfrac{i}{100n}\right)^{100n/i}\right]^{it/100}$, $\quad$ and note that

$\left(1 + \dfrac{i}{100n}\right)^{100n/i} \quad$ gets closer and closer to e as n gets larger and larger, we conclude that as n gets larger and larger, A approaches $P\,e^{it/100}$.

(d) The accumulated value is $P = 1000\,e^{6(10)/100} = 1822.12$. For the accumulated value at 6% compounded only once each year, we obtain $1000(1.06)^{10} = 1790.85$.

40. (a) If the current is i_0/e at time τ, then

$$\frac{i_0}{e} = i_0 e^{-R\tau/L}.$$

Division by i_0 and logarithms give

$$-1 = -\frac{R\tau}{L} \quad\Longrightarrow\quad \tau = \frac{L}{R}.$$

(b) The current at time $t + \tau$ is $i_0 e^{-R(t+\tau)/L} = i_0 e^{-Rt/L} e^{-R\tau/L} = i e^{-\tau/\tau} = i/e$.

42. If we write $2^x + 2^{2x} = 2^{3x}$, and divide by 2^x, then

$$1 + 2^x = 2^{2x} \quad\Longrightarrow\quad (2^x)^2 - 2^x - 1 = 0 \quad\Longrightarrow\quad 2^x = \frac{1 \pm \sqrt{1+4}}{2} = \frac{1 \pm \sqrt{5}}{2}.$$

Since 2^x must be positive, we choose only $(1 + \sqrt{5})/2$, and take logarithms,

$$x \log_{10} 2 = \log_{10}\left(\frac{1+\sqrt{5}}{2}\right) \quad\Longrightarrow\quad x = \frac{\log_{10}\left(\frac{1+\sqrt{5}}{2}\right)}{\log_{10} 2} = 0.694.$$

44. If we set $y = \log_x 2$, then $2 = x^y$. But, then $y = \log_{2x} 8$ also, and this implies that $8 = (2x)^y = 2^y x^y = 2^y(2) = 2^{y+1}$. Consequently, $y + 1 = 3$ or $y = 2$. Thus, $2 = x^2$, and since x must be positive, $x = \sqrt{2}$.

46. Repair costs for the second year are $50(1.2)$, for the third year, $50(1.2)^2$, and so on. If $R(t)$ represents repair costs for t years, then $R(t) = 50 + 50(1.2) + \cdots + 50(1.2)^{t-1}$. If we multiply this by 1.2, then $1.2R(t) = 50(1.2) + 50(1.2)^2 + \cdots + 50(1.2)^t$, and

$$1.2R(t) - R(t) = 50(1.2)^t - 50 \quad\Longrightarrow\quad R(t) = \frac{50(1.2)^t - 50}{1.2 - 1} = 250[(1.2)^t - 1].$$

Thus, the average yearly cost asociated with owning the car for t years is

$$C(t) = \frac{1}{t}\left\{ 20\,000\left[1 - \left(\frac{3}{4}\right)^t\right] + 250\left[\left(\frac{6}{5}\right)^t - 1\right]\right\}.$$

48. If we multiply by e^{2x},

$$(e^{2x})^2 - 2y(e^{2x}) - 1 = 0 \quad\Longrightarrow\quad e^{2x} = \frac{2y \pm \sqrt{4y^2 + 4}}{2} = y \pm \sqrt{y^2 + 1}.$$

Since e^{2x} must be positive, $e^{2x} = y + \sqrt{y^2 + 1} \Longrightarrow x = (1/2)\ln\left(y + \sqrt{y^2 + 1}\right)$.

50. If we cross multiply,

$$y(e^x + e^{-x}) = e^x - e^{-x} \quad\Longrightarrow\quad e^x(1 - y) = e^{-x}(1 + y) \quad\Longrightarrow\quad e^{2x} = \frac{1+y}{1-y} \quad\Longrightarrow\quad x = \frac{1}{2}\ln\left(\frac{1+y}{1-y}\right).$$

EXERCISES 1.10

2. $\sinh(\pi/2) = \dfrac{e^{\pi/2} - e^{-\pi/2}}{2} = 2.30$

4. $\mathrm{Sin}^{-1}(\mathrm{sech}\,10) = \mathrm{Sin}^{-1}\left(\dfrac{2}{e^{10} + e^{-10}}\right) = 9.08 \times 10^{-5}$

6. $\coth(\sinh 5) = \dfrac{e^{\sinh 5} + e^{-\sinh 5}}{e^{\sinh 5} - e^{-\sinh 5}} = 1.00$ **8.** $\mathrm{sech}[\sec(\pi/3)] = \mathrm{sech}\,2 = \dfrac{2}{e^2 + e^{-2}} = 0.266$

10. $\sinh\left[\mathrm{Cot}^{-1}(-3\pi/10)\right] = \sinh 2.3266 = \dfrac{e^{2.3266} - e^{-2.3266}}{2} = 5.07$

12. If we write the equations in the form

$$A(\cos kL - \cosh kL) = -B(\sin kL - \sinh kL), \quad A(\cos kL + \cosh kL) = -B(\sin kL + \sinh kL),$$

and divide one by the other,

$$\frac{\cos kL - \cosh kL}{\cos kL + \cosh kL} = \frac{\sin kL - \sinh kL}{\sin kL + \sinh kL}.$$

Hence,

$$(\cos kL - \cosh kL)(\sin kL + \sinh kL) = (\cos kL + \cosh kL)(\sin kL - \sinh kL)$$

or, $2\cos kL \sinh kL = 2\sin kL \cosh kL$. Division by $2\cos kL \cosh kL$ gives $\tanh kL = \tan kL$.

EXERCISES 1.11

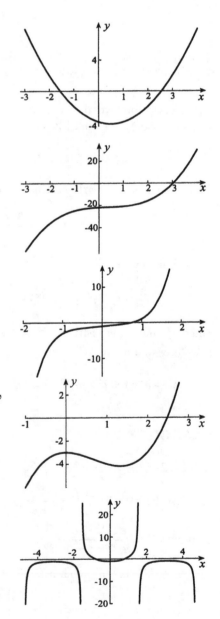

2. The plot shows two roots. My electronic device gives $-1.561\,552\,8$ and $2.561\,552\,8$ for roots of $f(x) = x^2 - x - 4 = 0$. To verify that $2.561\,553$ is accurate to six decimals, we calculate $f(2.561\,552\,5) = -1.3 \times 10^{-6}$ and $f(2.561\,553\,5) = 2.8 \times 10^{-6}$. A similar calculation verifies the accuracy of $-1.561\,553$.

4. The plot shows one root. My electronic device gives $3.044\,723\,1$ for the root of $f(x) = x^3 - x^2 + x - 22 = 0$. To verify that $3.044\,723$ is accurate to six decimals, we calculate $f(3.044\,722\,5) = -1.5 \times 10^{-5}$ and $f(3.044\,723\,5) = 8.0 \times 10^{-6}$.

6. The plot shows one root. My electronic device gives $0.754\,877\,7$ for the root of $f(x) = x^5 + x - 1 = 0$. To verify that $0.754\,878$ is accurate to six decimals, we calculate $f(0.754\,877\,5) = -4.4 \times 10^{-7}$ and $f(0.754\,878\,5) = 2.2 \times 10^{-6}$.

8. We rewrite the equation in the form $f(x) = x^3 - 2x^2 - 3 = 0$, the graph of which is shown to the right. My electronic calculator gives the root $2.485\,584\,0$. To verify that $2.485\,584$ is accurate to six decimals, we calculate $f(2.485\,583\,5) = -4.3 \times 10^{-6}$ and $f(2.485\,584\,5) = 4.3 \times 10^{-6}$.

10. The plot shows two roots. My electronic device gives $\pm 0.795\,323\,9$ for the roots of $f(x) = \sec x - 2/(1 + x^4) = 0$. To verify that $0.795\,324$ is accurate to six decimals, we calculate $f(0.795\,323\,5) = -1.4 \times 10^{-6}$ and $f(0.795\,324\,5) = 2.1 \times 10^{-6}$.

12. The plot shows six roots. My electronic device
gives $-2.931\,137\,1$, $-2.467\,517\,5$, $-1.555\,365\,0$,
$-0.787\,652\,8$, $0.056\,257\,6$ and $0.642\,850\,7$ for
the roots of $f(x) = (x+1)^2 - 5\sin 4x = 0$. To
verify that $-2.931\,137$ is accurate to six decimals,
we calculate $f(-2.931\,136\,5) = -1.1 \times 10^{-5}$ and
$f(-2.931\,137\,5) = 6.7 \times 10^{-6}$. Verification
that $-2.467\,518$, $-1.555\,365$, $-0.787\,653$,
$0.056\,258$, and $0.642\,851$ are accurate to
six decimals is similar.

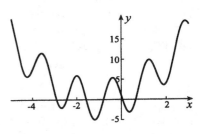

14. The plot shows one root. My electronic device
gives $4.188\,760\,1$ for the root of $f(x) = x\ln x - 6 = 0$.
To verify that $4.188\,760$ is accurate to six
decimals, we calculate $f(4.188\,759\,5) = -1.5 \times 10^{-6}$
and $f(4.188\,760\,5) = 9.3 \times 10^{-7}$.

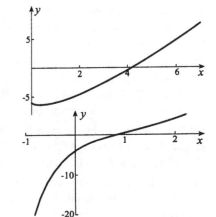

16. The plot shows one root. My electronic device
gives $0.852\,605\,5$ for the root of $f(x) = x^2 - 4e^{-2x} = 0$.
To verify that $0.852\,606$ is accurate to six decimals,
we calculate $f(0.852\,605\,5) = -6.4 \times 10^{-9}$ and
$f(0.852\,606\,5) = 3.2 \times 10^{-6}$.

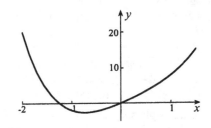

18. The plot shows two roots, one of which is $x = 0$.
My electronic device gives $-1.248\,30$ for the other
root of $f(x) = x^4 - x^3 + 2x^2 + 6x = 0$. To verify
that $-1.248\,3$ has error no greater than 10^{-4},
we calculate $f(-1.248\,2) = -1.1 \times 10^{-3}$ and
$f(-1.248\,4) = 1.2 \times 10^{-3}$.

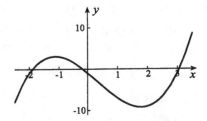

20. We rewrite the equation in the form
$f(x) = x^3 - x^2 - 6x - 1 = 0$. The plot shows
three roots. My electronic device gives
$-1.892\,0$, $-0.172\,5$, and $3.064\,4$ for the
roots. To verify that -1.892 has error no greater
than 10^{-3}, we calculate $f(-1.893) = -8.9 \times 10^{-3}$
and $f(-1.891) = 8.1 \times 10^{-3}$. Verification
that -0.172 and 3.064 also have the required
accuracy is similar.

22. The plot shows two roots. My electronic device gives
$\pm 1.098\,59$ for the roots of $f(x) = \cos^2 x - x^2 + 1 = 0$.
To verify that $1.098\,6$ has error no greater than
10^{-4}, we calculate $f(1.098\,5) = 2.6 \times 10^{-4}$
and $f(1.098\,7) = -3.4 \times 10^{-4}$.

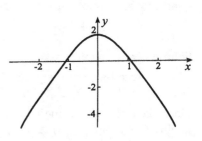

24. The plot shows one root. My electronic device gives
0.321 21 for the root of $f(x) = e^{3x} + e^x - 4 = 0$.
To verify that 0.321 2 has error no greater than
10^{-4}, we calculate $f(0.321\,1) = -1.0 \times 10^{-3}$ and
$f(0.321\,3) = 8.2 \times 10^{-4}$.

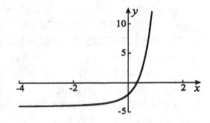

26. To find x-coordinates of the points of intersection, we set $(x+1)^2 = x^3 - 4x$. My electronic device gives
$x = -1.891\,954$, -0.172480, and $3.064\,435$ as solutions of $f(x) = x^3 - 4x - (x+1)^2 = 0$. The values
$f(-1.891\,954\,5) = -5.1 \times 10^{-7}$ and $f(-1.891\,953\,5) = 8.0 \times 10^{-6}$ confirm six-decimal accuracy of the
first. Both equations $y = x^3 - 4x$ and $y = (x+1)^2$ give the same four decimals $y = 0.7956$. A point of
intersection is therefore $(-1.8920, 0.7956)$. Similar procedures lead to the other points of intersection,
$(-0.1725, 0.6848)$ and $(3.0644, 16.1596)$.

28. To find x-coordinates of the points of intersection, we set $x/(x+1) = x^2 + 2$. My electronic device gives
$x = -1.353\,210$ as the only solution of $f(x) = x^3 + x^2 + x + 2 = 0$. The values $f(-1.353\,210\,5) = -2.0 \times$
10^{-6} and $f(-1.353\,209\,5) = 1.8 \times 10^{-6}$ confirm six-decimal accuracy. Both equations $y = x/(x+1)$ and
$y = x^2 + 2$ give the same four decimals $y = 3.8312$. The point of intersection is therefore $(-1.3532, 3.8312)$.

30. (a) My electronic device gives $t = 3.833$ as the solution of $f(t) = 1181(1 - e^{-t/10}) - 98.1t = 0$. Since
$y(3.825) = 0.14$ and $y(3.835) = -0.03$, it follows that to 2 decimals $t = 3.83$ s.
(b) When we set $0 = y = 20t - 4.905t^2$, the positive solution is 4.08 s.

32. To simplify calculations, we set $z = c/\lambda$. Then, z must satisfy the equation $f(z) = (5 - z)e^z - 5 = 0$.
My electronic device gives $z = 4.965\,114\,232$. With this approximation for z, we obtain $\lambda = c/z =$
$0.000\,028\,974$. For a seven decimal answer, we use $g(\lambda) = (5\lambda - c)e^{c/\lambda} - 5\lambda$ to calculate $g(0.000\,028\,95) =$
-1.7×10^{-5} and $g(0.000\,029\,05) = 5.1 \times 10^{-5}$. Thus, to 7 decimals, $\lambda = 0.000\,029\,0$.

34. Suppose we let T_1 and T_2 be the temperatures at the points $(r, 0)$ and $(-r, 0)$. Then,
$F(r) = f(r) - g(-r) = T_1 - T_2$ and
$F(-r) = f(-r) - g(r) = T_2 - T_1$.
If $T_1 = T_2$, then temperatures are the same at
the points $(r, 0)$ and $(-r, 0)$. Otherwise, values
of $F(x)$ have opposite signs at $x = r$ and $x = -r$.

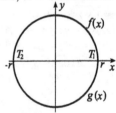

This implies that there is a value of x between $-r$ and r at which $F(x) = 0$. At this value, $f(x) = g(-x)$,
and these give equal temperatures at points opposite each other on the equator.

REVIEW EXERCISES

2. Possible rational solutions are ± 1, ± 3, ± 9, ± 27, $\pm 1/2$, $\pm 3/2$, $\pm 9/2$, $\pm 27/2$. We find that $x = 3$ is a
solution. We factor $x - 3$ from the cubic,

$$2x^3 - 9x^2 + 27 = (x - 3)(2x^2 - 3x - 9) = (x - 3)(2x + 3)(x - 3) = (x - 3)^2(2x + 3).$$

Solutions are $x = 3$ with multiplicity 2 and $x = -3/2$.

4. The list of possible rational solutions here is formidable. Perhaps we can be a little ingenious. The fact
that $36x^4$ and 225 are perfect squares suggests investigating whether the poynomial is the square of a
quadratic expression. A little experimentation reveals that

$$36x^4 + 12x^3 - 179x^2 - 30x + 225 = (6x^2 + x - 15)^2 = (3x + 5)^2(2x - 3)^2.$$

Solutions are therefore $x = -5/3$ and $x = 3/2$ each of multiplicity 2.

6. The distance between the points is $\sqrt{(2+3)^2 + (1+4)^2} = 5\sqrt{2}$. The midpoint of the line segment is
$((2-3)/2, (1-4)/2) = (-1/2, -3/2)$.

8. Since the slope of the line joining $(-2, 1)$ and the origin is $-1/2$, and the midpoint of the line segment
joining $(1, 3)$ and $(-1, 5)$ is $(0, 4)$, the equation of the required line is $y - 4 = 2(x - 0)$, or, $y = 2x + 4$.

10. If we substitute $y = x^2$ into the second equation, $5x = 6 - x^4$, or $x^4 + 5x - 6 = 0$. Possible rational solutions are ± 1, ± 2, ± 3, ± 6. We find that $x = 1$ is a solution and factor $x - 1$ from the quartic

$$x^4 + 5x - 6 = (x - 1)(x^3 + x^2 + x + 6) = 0.$$

We now see that $x = -2$ is a zero of the cubic so that

$$x^4 + 5x - 6 = (x - 1)(x + 2)(x^2 - x + 3) = 0.$$

The curves therefore intersect at the points $(-2, 4)$ and $(1, 1)$. The equation of the line joining these points is $y - 1 = -(x - 1) \implies x + y = 2$.

12. For $x^2 - 5$ to be nonnegative, $|x| \geq \sqrt{5}$.

14. Since $x^3 + 2x^2 + x = x(x + 1)^2$, we cannot set $x = 0$ or $x = -1$.

16. $x \geq 0$

18. Since $2x^2 + 4x - 5 = 0$ when $x = \left(-4 \pm \sqrt{16 + 40}\right)/4 = -1 \pm \sqrt{14}/2$, and is negative between these values, we must restrict x to the intervals $x < -1 - \sqrt{14}/2$ and $x > -1 + \sqrt{14}/2$. These can be combined into $|x + 1| > \sqrt{14}/2$.

20. We require
$$0 \leq x - \frac{1}{x} = \frac{x^2 - 1}{x} = \frac{(x - 1)(x + 1)}{x}.$$
The sign diagram indicates that this occurs for $-1 \leq x < 0$, and $x \geq 1$.

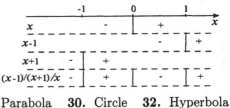

22. Parabola   **24.** Ellipse   **26.** None of these   **28.** Parabola   **30.** Circle   **32.** Hyperbola

34. With formula 1.16, the distance is $\left| \dfrac{(-2) + 3(-5) - 4}{\sqrt{1^2 + 3^2}} \right| = \dfrac{21}{\sqrt{10}}$.

36. This is the circle $x^2 + y^2 = 4$ with centre $(0, 0)$ and radius 2.

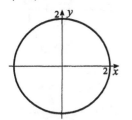

38. This equation describes two straight lines $y = \pm x$.

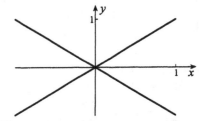

40. This is an ellipse with x-intercepts $\pm\sqrt{6}$ and y-intercepts $\pm\sqrt{2}$.

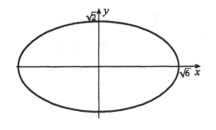

42. When we complete squares on x- and y-terms, $(y - 2)^2 - (x - 1)^2 = 2$. Asymptotes for this hyperbola are $y = 2 \pm (x - 1)$ intersecting at $(1, 2)$.

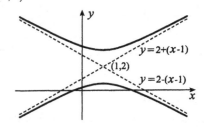

44. When we expand the cosine function,
$$y = -\sin 2x.$$

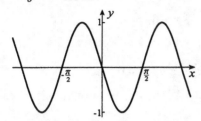

46. When we expand the sine function,
$$y = 2\cos 3x.$$

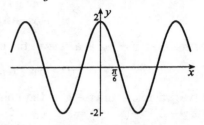

48. If we square the equation $y^2 = -x^2 + 4x + 4$, and then complete the square on the x-terms, $(x-2)^2 + y^2 = 8$. This is a circle with centre $(2,0)$ and radius $2\sqrt{2}$. The original equation describes the top half of the circle.

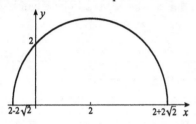

50. When we square the equation, $y^2 = |x-1| - 1$, or, $y^2 + 1 = |x-1|$. Thus, $x - 1 = \pm(y^2 + 1)$, or, $x = 1 \pm (y^2 + 1) = -y^2, y^2 + 2$. This is two parabolas, the top halves of which are shown below.

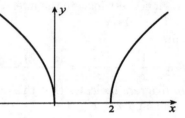

52. We must have $x > -4/3$. The x-intercept is -1.

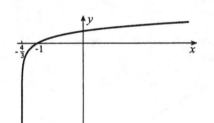

54. We graph this even function by reflecting that part of $y = \sin x$ to the right of the y-axis in the y-axis.

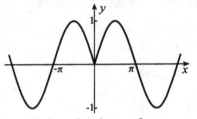

56. We take square roots of ordinates of $y = \sin 2x$.

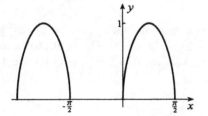

58. The graph has the shape of $\tanh x$ with asymptotes $y = \pm 4$.

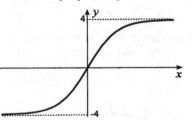

60. With slopes $1/4$ and $-2/3$, equation 1.60 gives $\theta = \text{Tan}^{-1}\left|\dfrac{1/4 + 2/3}{1 - 1/6}\right| = 0.833$ radians.

62. Since $(x+1)(2-x)$ is nonnegative only for $-1 \leq x \leq 2$, the function $\sqrt{(x+1)(2-x)}$ is only defined for these values of x.

64. Since $\sqrt{-x}$ is defined only for $x \leq 0$, a function is $1 + \sqrt{-x}$.

66. The graph shows that $f(x)$ has an inverse on the intervals $x \le -2$, $-2 \le x \le 0$, $0 \le x \le 2$, and $x \ge 2$. If we set $y = x^4 - 8x^2$, then $(x^2)^2 - 8(x^2) - y = 0$. The quadratic formula gives

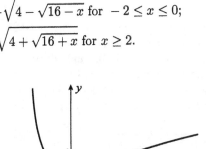

$$x^2 = \frac{8 \pm \sqrt{64 + 4y}}{2} = 4 \pm \sqrt{16 + y},$$

from which $x = \pm\sqrt{4 \pm \sqrt{16 + y}}$. Since $x = -\sqrt{4 + \sqrt{16 + y}}$, $x = -\sqrt{4 - \sqrt{16 - y}}$, $x = \sqrt{4 - \sqrt{16 - y}}$, and $x = \sqrt{4 + \sqrt{16 + y}}$ describe parts of the graph on each of the above intervals, respectively, it follows that inverse functions on these intervals are:

$$f^{-1}(x) = -\sqrt{4 + \sqrt{16 + x}} \text{ for } x \le -2; \quad f^{-1}(x) = -\sqrt{4 - \sqrt{16 - x}} \text{ for } -2 \le x \le 0;$$
$$f^{-1}(x) = \sqrt{4 - \sqrt{16 - x}} \text{ for } 0 \le x \le 2; \quad f^{-1}(x) = \sqrt{4 + \sqrt{16 + x}} \text{ for } x \ge 2.$$

68. The graph shows that $f(x)$ has an inverse for $-1 < x \le 0$ and for $x \ge 0$. If we set $y = \sqrt{x^2/(x+1)}$, then $x^2 = (x+1)y^2$, from which $x^2 - y^2x - y^2 = 0$. The quadratic formula gives $x = (y^2 \pm \sqrt{y^4 + 4y^2})/2$. Since $x = (y^2 - \sqrt{y^4 + 4y^2})/2$ describes the graph for $-1 < x \le 0$, and $x = (y^2 + \sqrt{y^4 + 4y^2})/2$ describes the graph for $x \ge 0$, inverse functions are $f^{-1}(x) = (x^2 - \sqrt{x^4 + 4x^2})/2$ for $-1 < x \le 0$, and $f^{-1}(x) = (x^2 + \sqrt{x^4 + 4x^2})/2$ for $x \ge 0$.

70. Since $2 \sin 2x$ has period π, and $3 \cos 3x$ has period $2\pi/3$, the function $f(x)$ has period 2π.

72. One solution for $2x$ of $\sin 2x = 1/4$ is $2x = \mathrm{Sin}^{-1}(1/4) = 0.253$. All solutions are given by

$$2x = \frac{\pi}{2} \pm \left(\frac{\pi}{2} - 0.253\right) + 2n\pi \quad \Longrightarrow \quad x = \frac{\pi}{4} \pm 0.659 + n\pi, \quad n \text{ an integer}.$$

74. Since $5 - 2\pi$ is in the principal value range for the inverse tangent function, we may take tangents of both sides of the equation,

$$3x + 2 = \tan(5 - 2\pi) \quad \Longrightarrow \quad x = -\frac{2}{3} + \frac{1}{3}\tan(5 - 2\pi) = -1.79.$$

76. If we write $\ln[\sin x(1 + \sin x)] = \ln(3/2)$, and exponentiate both sides to base e, we obtain $\sin x(1 + \sin x) = 3/2$, or, $2\sin^2 x + 2\sin x - 3 = 0$. Thus, $\sin x = \frac{-2 \pm \sqrt{4 + 24}}{4} = \frac{-1 \pm \sqrt{7}}{2}$. Since $\sin x$ must be nonnegative in the original equation, we take $\sin x = (\sqrt{7} - 1)/2$. Solutions of this equation are

$$x = \{\mathrm{Sin}^{-1}[(\sqrt{7} - 1)/2] + 2n\pi, \quad \pi - \mathrm{Sin}^{-1}[(\sqrt{7} - 1)/2] + 2n\pi\} = \{0.966 + 2n\pi, \quad 2.175 + 2n\pi\}.$$

These can be expressed in the form $x = \frac{\pi}{2} \pm \left(\frac{\pi}{2} - 0.966\right) + 2n\pi = \left(\frac{4n + 1}{2}\right)\pi \pm 0.604$.

78. If $3\sin(e^{x+2}) = 2$, then $\sin(e^{x+2}) = 2/3$. This implies that

$$e^{x+2} = \{\mathrm{Sin}^{-1}(2/3) + 2n\pi, \quad \pi - \mathrm{Sin}^{-1}(2/3) + 2n\pi\} = \{0.73 + 2n\pi, \quad \pi - 0.73 + 2n\pi\}.$$

These values can be represented more compactly as

$$e^{x+2} = \frac{\pi}{2} \pm 0.84 + 2n\pi = \left(\frac{4n + 1}{2}\right)\pi \pm 0.84.$$

Because e^{x+2} must be positive, n must be a nonnegative integer. Finally then,

$$x = \ln\left[\left(\frac{4n + 1}{2}\right)\pi \pm 0.84\right] - 2, \text{ where } n \ge 0.$$

80. If $4 = \sinh x = (e^x - e^{-x})/2$, then multiplication by e^x gives $e^{2x} - 8e^x - 1 = 0$. Thus,

$$e^x = \frac{8 \pm \sqrt{64 + 4}}{2} = 4 \pm \sqrt{17}.$$

Since e^x must be positive, $e^x = 4 + \sqrt{17}$, and $x = \ln(4 + \sqrt{17}) = 2.09$.

82. To solve the equation $x(t) = 0$, we divide by $\cos 4t$,

$$\tan 4t = -\frac{1}{10} \quad \Longrightarrow \quad 4t = \mathrm{Tan}^{-1}(-0.1) + n\pi = -0.09967 + n\pi \quad \Longrightarrow \quad t = \frac{1}{4}(-0.09967 + n\pi),$$

where n is an integer. The smallest positive solution is 0.760 (when $n = 1$).

84. My electronic device gives 1.5260 for the only root of $f(x) = x^3 - 2x^2 + 4x - 5 = 0$. To verify that 1.526 is accurate to three decimals, we calculate $f(1.5255) = -2.2 \times 10^{-3}$ and $f(1.5265) = 2.6 \times 10^{-3}$.

86. My electronic device gives -11.61869, -0.87380, and 0.49249 for roots of $f(x) = x^3 + 12x^2 + 4x - 5 = 0$. To verify that 0.4925 has error less than 10^{-4}, we calculate $f(0.4924) = -1.5 \times 10^{-3}$ and $f(0.4926) = 1.8 \times 10^{-3}$. A similar calculation confirms -11.6187 and -0.8738 as the other roots.

88. Let us take a coordinate system as shown, in which case $a^2 = b^2 + d^2$. Slopes of AB and OC are $d/(b - a)$ and $d/(b + a)$. The product of these is

$$\left(\frac{d}{b - a}\right)\left(\frac{d}{b + a}\right) = \frac{d^2}{b^2 - a^2} = -1.$$

Hence the diagonals are perpendicular.

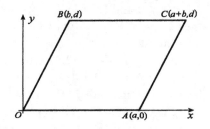

CHAPTER 2

EXERCISES 2.1

2. $\lim\limits_{x \to -2} \dfrac{x^3 + 8}{x + 5} = \dfrac{0}{3} = 0$

4. $\lim\limits_{x \to 0} \dfrac{x^2 + 3x}{3x^2 - 2x} = \lim\limits_{x \to 0} \dfrac{x(x + 3)}{x(3x - 2)} = \lim\limits_{x \to 0} \dfrac{x + 3}{3x - 2} = \dfrac{3}{-2} = -\dfrac{3}{2}$

6. $\lim\limits_{x \to 2^-} \dfrac{2x - 4}{3x + 2} = \dfrac{0}{8} = 0$ $\qquad\qquad$ **8.** $\lim\limits_{x \to 2^+} \dfrac{x^2 + 2x + 4}{x - 3} = \dfrac{4 + 4 + 4}{-1} = -12$

10. $\lim\limits_{x \to 3^+} \dfrac{x^2 - 9}{x - 3} = \lim\limits_{x \to 3^+} \dfrac{(x - 3)(x + 3)}{x - 3} = \lim\limits_{x \to 3^+} (x + 3) = 6$

12. $\lim\limits_{x \to 3} \dfrac{x^2 - 2x - 3}{3 - x} = \lim\limits_{x \to 3} \dfrac{(x - 3)(x + 1)}{3 - x} = \lim\limits_{x \to 3} [-(x + 1)] = -4$

14. $\lim\limits_{x \to 2} \dfrac{x^3 - 6x^2 + 12x - 8}{x^2 - 4x + 4} = \lim\limits_{x \to 2} \dfrac{(x - 2)^3}{(x - 2)^2} = \lim\limits_{x \to 2} (x - 2) = 0$

16. $\lim\limits_{x \to 2} \dfrac{x^3 - 6x^2 + 11x - 6}{x^2 - 3x + 2} = \lim\limits_{x \to 2} \dfrac{(x - 1)(x - 2)(x - 3)}{(x - 1)(x - 2)} = \lim\limits_{x \to 2} (x - 3) = -1$

18. $\lim\limits_{x \to 3^-} \dfrac{x^3 - 6x^2 + 11x - 6}{x^2 - 3x + 2} = \dfrac{27 - 54 + 33 - 6}{9 - 9 + 2} = 0$ $\quad$ **20.** $\lim\limits_{x \to -1} \dfrac{12x + 5}{x^2 - 2x + 1} = \dfrac{-7}{4} = -\dfrac{7}{4}$

22. $\lim\limits_{x \to 5} \dfrac{\sqrt{1 - x^2}}{3x + 2}$ $\quad$ does not exist $\qquad\qquad$ **24.** $\lim\limits_{x \to \pi/4} \dfrac{\sin x}{\tan x} = \dfrac{1/\sqrt{2}}{1} = \dfrac{1}{\sqrt{2}}$

26. $\lim\limits_{x \to 0^+} \dfrac{\sin 6x}{\sin 3x} = \lim\limits_{x \to 0^+} \dfrac{2 \sin 3x \cos 3x}{\sin 3x} = \lim\limits_{x \to 0^+} (2 \cos 3x) = 2$

28. $\lim\limits_{x \to 2} \dfrac{x - 2}{\sqrt{x} - \sqrt{2}} = \lim\limits_{x \to 2} \dfrac{(\sqrt{x} + \sqrt{2})(\sqrt{x} - \sqrt{2})}{\sqrt{x} - \sqrt{2}} = \lim\limits_{x \to 2} (\sqrt{x} + \sqrt{2}) = 2\sqrt{2}$

30. $\lim\limits_{x \to 5^+} \dfrac{|x^2 - 25|}{x^2 - 25} = \lim\limits_{x \to 5^+} \dfrac{x^2 - 25}{x^2 - 25} = \lim\limits_{x \to 5^+} (1) = 1$

32. Since $\lim\limits_{x \to 5^+} \dfrac{|x^2 - 25|}{x^2 - 25} = 1$ (Exercise 30) and $\lim\limits_{x \to 5^-} \dfrac{|x^2 - 25|}{x^2 - 25} = \lim\limits_{x \to 5^-} \dfrac{-(x^2 - 25)}{x^2 - 25} = \lim\limits_{x \to 5^-} (-1) = -1$, it follows that the given limit does not exist.

34. $\lim\limits_{x \to 0} \dfrac{1 - \sqrt{x^2 + 1}}{2x^2} = \lim\limits_{x \to 0} \left[\dfrac{1 - \sqrt{x^2 + 1}}{2x^2} \dfrac{1 + \sqrt{x^2 + 1}}{1 + \sqrt{x^2 + 1}} \right] = \lim\limits_{x \to 0} \dfrac{1 - (x^2 + 1)}{2x^2 \left(1 + \sqrt{x^2 + 1}\right)}$

$\qquad = \lim\limits_{x \to 0} \dfrac{-1}{2 \left(1 + \sqrt{x^2 + 1}\right)} = \dfrac{-1}{2(2)} = -\dfrac{1}{4}$

36. $\lim\limits_{x \to 0} \dfrac{\sqrt{1 + x} - \sqrt{1 - x}}{x} = \lim\limits_{x \to 0} \left[\dfrac{\sqrt{1 + x} - \sqrt{1 - x}}{x} \dfrac{\sqrt{1 + x} + \sqrt{1 - x}}{\sqrt{1 + x} + \sqrt{1 - x}} \right]$

$\qquad = \lim\limits_{x \to 0} \dfrac{(1 + x) - (1 - x)}{x \left(\sqrt{1 + x} + \sqrt{1 - x}\right)} = \lim\limits_{x \to 0} \dfrac{2}{\sqrt{1 + x} + \sqrt{1 - x}} = \dfrac{2}{1 + 1} = 1$

38. $\lim\limits_{x \to 0} \dfrac{x}{\sqrt{x + 4} - 2} = \lim\limits_{x \to 0} \left[\dfrac{x}{\sqrt{x + 4} - 2} \dfrac{\sqrt{x + 4} + 2}{\sqrt{x + 4} + 2} \right] = \lim\limits_{x \to 0} \dfrac{x \left(\sqrt{x + 4} + 2\right)}{(x + 4) - 4} = \lim\limits_{x \to 0} \left(\sqrt{x + 4} + 2\right) = 4$

40. $\lim\limits_{x \to 0} \dfrac{\sqrt{x + 1} - \sqrt{2x + 1}}{\sqrt{3x + 4} - \sqrt{2x + 4}} = \lim\limits_{x \to 0} \left(\dfrac{\sqrt{x + 1} - \sqrt{2x + 1}}{\sqrt{3x + 4} - \sqrt{2x + 4}} \dfrac{\sqrt{x + 1} + \sqrt{2x + 1}}{\sqrt{x + 1} + \sqrt{2x + 1}} \dfrac{\sqrt{3x + 4} + \sqrt{2x + 4}}{\sqrt{3x + 4} + \sqrt{2x + 4}} \right)$

$\qquad = \lim\limits_{x \to 0} \dfrac{(x + 1 - 2x - 1)(\sqrt{3x + 4} + \sqrt{2x + 4})}{(3x + 4 - 2x - 4)(\sqrt{x + 1} + \sqrt{2x + 1})}$

$\qquad = \lim\limits_{x \to 0} \left(-\dfrac{\sqrt{3x + 4} + \sqrt{2x + 4}}{\sqrt{x + 1} + \sqrt{2x + 1}} \right) = -2$

42. $\displaystyle\lim_{x \to a} \frac{x^2 - a^2}{x - a} = \lim_{x \to a} \frac{(x + a)(x - a)}{x - a} = \lim_{x \to a} (x + a) = 2a$

44. $\displaystyle\lim_{x \to -a} \frac{x + a}{x^2 + ax - x - a} = \lim_{x \to -a} \frac{x + a}{(x + a)(x - 1)} = \lim_{x \to -a} \frac{1}{x - 1} = -\frac{1}{a + 1}$

46. $\displaystyle\lim_{x \to a} \frac{\sqrt{x} - \sqrt{a}}{x - a} = \lim_{x \to a} \frac{\sqrt{x} - \sqrt{a}}{(\sqrt{x} + \sqrt{a})(\sqrt{x} - \sqrt{a})} = \lim_{x \to a} \frac{1}{\sqrt{x} + \sqrt{a}} = \frac{1}{2\sqrt{a}}$

48. $\displaystyle\lim_{x \to 0} \frac{\sqrt{a + x} - \sqrt{a - x}}{x} = \lim_{x \to 0} \left[\frac{\sqrt{a + x} - \sqrt{a - x}}{x} \frac{\sqrt{a + x} + \sqrt{a - x}}{\sqrt{a + x} + \sqrt{a - x}} \right]$

$\displaystyle = \lim_{x \to 0} \frac{(a + x) - (a - x)}{x\left(\sqrt{a + x} + \sqrt{a - x}\right)} = \lim_{x \to 0} \frac{2}{\sqrt{a + x} + \sqrt{a - x}} = \frac{2}{\sqrt{a} + \sqrt{a}} = \frac{1}{\sqrt{a}}$

50. Although the plot does not show it, there should be a hole in the graph at $x = 0$. The plot suggests that the limit is 1.

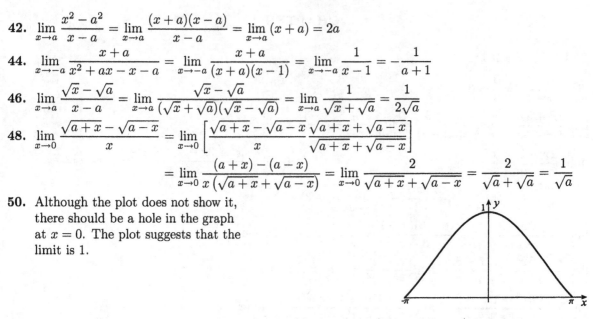

52. Since $-1 \le \cos(3/x) \le 1$ for all x, it follows that $-x^4 \le x^4 \cos(3/x) \le x^4$. Because $\lim_{x \to 0} (-x^4) = \lim_{x \to 0} x^4 = 0$, the squeeze theorem implies that $\lim_{x \to 0} x^4 \cos(3/x) = 0$ also.

54. This statement is false. For example, $x^2 < 2x^2$ for all $x \ne 0$, but $\lim_{x \to 0} x^2 = \lim_{x \to 0} 2x^2 = 0$.

56. $\displaystyle\lim_{h \to 0} \frac{\sqrt{x + h} - \sqrt{x}}{h} = \lim_{h \to 0} \left[\frac{\sqrt{x + h} - \sqrt{x}}{h} \frac{\sqrt{x + h} + \sqrt{x}}{\sqrt{x + h} + \sqrt{x}} \right] = \lim_{h \to 0} \frac{(x + h) - x}{h\left(\sqrt{x + h} + \sqrt{x}\right)}$

$\displaystyle = \lim_{h \to 0} \frac{1}{\sqrt{x + h} + \sqrt{x}} = \frac{1}{2\sqrt{x}}$

58. (a) Our calculator gave

x	0.1	0.01	0.001	0.0001	0.00001	0.000001	0.0000001
$(\sin x - x)/x^3$	-0.16658	-0.16667	-0.16667	-0.1667	-0.17	0.0	0.0

It would appear that the limit is 0.

(b) Plots of the function on the suggested intervals are shown below.

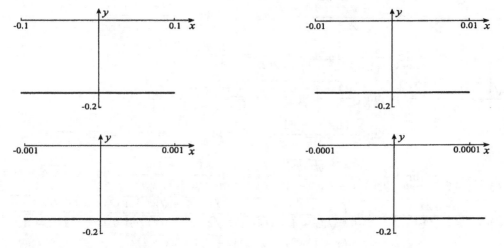

They suggest that the limit is approximately -0.17.

60. If we set $z = -x$, then $\displaystyle\lim_{x \to -a} f(x) = \lim_{z \to a} f(-z) = \lim_{z \to a} f(z) = L$.

62. This limit cannot be determined.

64. If we set $z = -x$, then $\displaystyle\lim_{x \to -a^-} f(x) = \lim_{z \to a^+} f(-z) = \lim_{z \to a^+} [-f(z)] = -\lim_{z \to a^+} f(z) = -L.$

66. The limit is 0 if $F = 0$; it does not exist if $F \neq 0$.

EXERCISES 2.2

2. $\displaystyle\lim_{x \to 2^-} \frac{1}{x - 2} = -\infty$ **4.** $\displaystyle\lim_{x \to 2^+} \frac{1}{(x - 2)^2} = \infty$

6. $\displaystyle\lim_{x \to 2} \frac{1}{(x - 2)^2} = \infty$

8. $\displaystyle\lim_{x \to 1/2} \frac{6x^2 + 7x - 5}{2x - 1} = \lim_{x \to 1/2} \frac{(3x + 5)(2x - 1)}{2x - 1} = \lim_{x \to 1/2} (3x + 5) = \frac{13}{2}$

10. $\displaystyle\lim_{x \to 2} \frac{x - 2}{x^2 - 4x + 4} = \lim_{x \to 2} \frac{x - 2}{(x - 2)^2} = \lim_{x \to 2} \frac{1}{x - 2}$

Since $\displaystyle\lim_{x \to 2^+} \frac{1}{x - 2} = \infty$ and $\displaystyle\lim_{x \to 2^-} \frac{1}{x - 2} = -\infty$, the given limit does not exist.

12. $\displaystyle\lim_{x \to \pi/4} \sec (x - \pi/4) = 1$ **14.** $\displaystyle\lim_{x \to 0^+} \cot x = \infty$

16. $\displaystyle\lim_{x \to \pi/2^-} \tan x = \infty$

18. $\displaystyle\lim_{x \to 0} \frac{\sqrt{1 + x} - 1}{x^2} = \lim_{x \to 0} \left[\frac{\sqrt{1 + x} - 1}{x^2} \frac{\sqrt{1 + x} + 1}{\sqrt{1 + x} + 1} \right] = \lim_{x \to 0} \frac{(1 + x) - 1}{x^2 (\sqrt{1 + x} + 1)} = \lim_{x \to 0} \frac{1}{x (\sqrt{1 + x} + 1)}$

Since $\displaystyle\lim_{x \to 0^+} \frac{1}{x (\sqrt{x + 1} + 1)} = \infty$ and $\displaystyle\lim_{x \to 0^-} \frac{1}{x (\sqrt{x + 1} + 1)} = -\infty$, the given limit does not exist.

20. Since $\displaystyle\lim_{x \to 4^+} \frac{|4 - x|}{x^2 - 8x + 16} = \lim_{x \to 4^+} \frac{x - 4}{(x - 4)^2} = \lim_{x \to 4^+} \frac{1}{x - 4} = \infty$

and $\displaystyle\lim_{x \to 4^-} \frac{|4 - x|}{x^2 - 8x + 16} = \lim_{x \to 4^-} \frac{4 - x}{(x - 4)^2} = \lim_{x \to 4^-} \frac{-1}{x - 4} = \infty$, the given limit does not exist.

22. $\displaystyle\lim_{x \to 1} \frac{1}{\ln |x - 1|} = 0$ **24.** $\displaystyle\lim_{x \to 0} e^{1/|x|} = \infty$

26. $\displaystyle\lim_{x \to a} \frac{|x - a|}{x^2 - 2ax + a^2} = \lim_{x \to a} \frac{|x - a|}{(x - a)^2} = \lim_{x \to a} \frac{1}{|x - a|} = \infty$

28. Since $\displaystyle\lim_{x \to -a^-} e^{1/(|x| - a)} = \infty$ and $\displaystyle\lim_{x \to -a^+} e^{1/(|x| - a)} = 0$, the limit does not exist.

30.

x	1.0	0.1	0.05	0.01	0.005
x^{10}	1	10^{-10}	9.77×10^{-14}	10^{-20}	9.77×10^{-24}
$e^{1/x}$	2.72	2.20×10^4	4.85×10^8	2.69×10^{43}	7.23×10^{86}
$x^{10} e^{1/x}$	2.72	2.20×10^{-6}	4.74×10^{-5}	2.69×10^{23}	7.06×10^{63}

Thus, $\displaystyle\lim_{x \to 0^+} x^{10} e^{1/x} = \infty$.

EXERCISES 2.3

2. $\displaystyle\lim_{x \to \infty} \frac{1 - x}{3 + 2x} = \lim_{x \to \infty} \frac{\frac{1}{x} - 1}{\frac{3}{x} + 2} = -\frac{1}{2}$ **4.** $\displaystyle\lim_{x \to \infty} \frac{1 - 4x^3}{3 + 2x - x^2} = \lim_{x \to \infty} \frac{\frac{1}{x^2} - 4x}{\frac{3}{x^2} + \frac{2}{x} - 1} = \infty$

6. $\displaystyle\lim_{x \to -\infty} \frac{x^3 - 2x^2}{3x^3 + 4x^2} = \lim_{x \to -\infty} \frac{1 - \frac{2}{x}}{3 + \frac{4}{x}} = \frac{1}{3}$

8. $\displaystyle\lim_{x \to -\infty} \frac{x^3 - 2x^2 + x + 1}{x^2 - x + 1} = \lim_{x \to -\infty} \frac{x - 2 + \frac{1}{x} + \frac{1}{x^2}}{1 - \frac{1}{x} + \frac{1}{x^2}} = -\infty$

10. $\displaystyle\lim_{x\to\infty}\frac{3x-1}{\sqrt{5+4x^2}}=\lim_{x\to\infty}\frac{3-\frac{1}{x}}{\sqrt{\frac{5}{x^2}+4}}=\frac{3}{2}$

12. $\displaystyle\lim_{x\to-\infty}\frac{\sqrt{1-2x}}{x+2}=\lim_{x\to-\infty}\frac{\frac{\sqrt{1-2x}}{\sqrt{-x}}}{\frac{x+2}{\sqrt{-x}}}=\lim_{x\to-\infty}\frac{\sqrt{-\frac{1}{x}+2}}{-\sqrt{-x}+\frac{2}{\sqrt{-x}}}=0$

14. $\displaystyle\lim_{x\to\infty}\frac{\sqrt{3+x}}{\sqrt{x}}=\lim_{x\to\infty}\sqrt{\frac{3}{x}+1}=1$ **16.** $\displaystyle\lim_{x\to\infty}\left(x+\frac{1}{x}\right)=\infty$

18. $\displaystyle\lim_{x\to-\infty}\frac{x^2}{\sqrt{3-x}}=\lim_{x\to-\infty}\frac{\frac{x^2}{\sqrt{-x}}}{\frac{\sqrt{3-x}}{\sqrt{-x}}}=\lim_{x\to-\infty}\frac{-x\sqrt{-x}}{\sqrt{1-\frac{3}{x}}}=\infty$

20. $\displaystyle\lim_{x\to\infty}\frac{3x}{\sqrt[3]{2+4x^3}}=\lim_{x\to\infty}\frac{3}{\sqrt[3]{\frac{2}{x^3}+4}}=\frac{3}{\sqrt[3]{4}}$ **22.** $\displaystyle\lim_{x\to-\infty}\frac{1}{2x}\cos x=0$

24. $\displaystyle\lim_{x\to\infty}\frac{\sin^2 x}{x}=0$ **26.** $\displaystyle\lim_{x\to\infty}\frac{1}{x}\tan x$ does not exist

28. $\displaystyle\lim_{x\to\infty}\left(\sqrt{x^2+4}-x\right)=\lim_{x\to\infty}\left[\left(\sqrt{x^2+4}-x\right)\frac{\sqrt{x^2+4}+x}{\sqrt{x^2+4}+x}\right]=\lim_{x\to\infty}\left[\frac{(x^2+4)-x^2}{\sqrt{x^2+4}+x}\right]=0$

30. $\displaystyle\lim_{x\to-\infty}\left(\sqrt{2x^2+1}-x\right)=\infty$ **32.** $\displaystyle\lim_{x\to\infty}\frac{\sqrt{4x^2+7}}{2x+3}=\lim_{x\to\infty}\frac{\sqrt{4+\frac{7}{x^2}}}{2+\frac{3}{x}}=1$

34. $\displaystyle\lim_{x\to-\infty}\frac{\sqrt{4x^2+7}}{2x+3}=\lim_{x\to-\infty}\frac{\frac{\sqrt{4x^2+7}}{x}}{\frac{2x+3}{x}}=\lim_{x\to-\infty}\frac{-\sqrt{4+\frac{7}{x^2}}}{2+\frac{3}{x}}=-1$

36. $\displaystyle\lim_{x\to\infty}\left(\sqrt[3]{1+x}-\sqrt[3]{x}\right)=\lim_{x\to\infty}\left\{\left[(1+x)^{1/3}-x^{1/3}\right]\frac{(1+x)^{2/3}+(1+x)^{1/3}x^{1/3}+x^{2/3}}{(1+x)^{2/3}+(1+x)^{1/3}x^{1/3}+x^{2/3}}\right\}$

$\displaystyle=\lim_{x\to\infty}\left[\frac{(1+x)-x}{(1+x)^{2/3}+(1+x)^{1/3}x^{1/3}+x^{2/3}}\right]=0$

38. $\displaystyle\lim_{x\to-\infty}\left(\sqrt{x^2+x}-x\right)=\infty$ **40.** $\displaystyle\lim_{x\to\infty}\frac{x}{\sqrt{ax^2+3x+2}}=\lim_{x\to\infty}\frac{1}{\sqrt{a+\frac{3}{x}+\frac{2}{x^2}}}=\frac{1}{\sqrt{a}}$

42. $\displaystyle\lim_{x\to-\infty}\frac{\sqrt{ax^2+7}}{x-3a}=\lim_{x\to-\infty}\frac{\frac{\sqrt{ax^2+7}}{x}}{\frac{x-3a}{x}}=\lim_{x\to-\infty}\frac{-\sqrt{a+\frac{7}{x^2}}}{1-\frac{3a}{x}}=-\sqrt{a}$

44. A vertical asymptote is $x=5/2$. Since $\displaystyle\lim_{x\to\pm\infty}\frac{x+3}{2x-5}=\frac{1}{2}$, the horizontal asymptote is $y=1/2$. With $f(x)$ expressed in the form $f(x)=\dfrac{1}{2}+\dfrac{11/2}{2x-5}$, we can say that the for large negative x, $f(x)<1/2$, and for large positive x, $f(x)>1/2$. Hence, the graph approaches the horizontal asymptote from below as $x\to-\infty$ and from above as $x\to\infty$.

46. A vertical asymptote is $x=-3/2$. Since $\displaystyle\lim_{x\to\infty}\frac{\sqrt{5x^2+7}}{2x+3}=\lim_{x\to\infty}\frac{\sqrt{5+7/x^2}}{2+3/x}=\frac{\sqrt{5}}{2}$, $y=\sqrt{5}/2$ is a horizontal asymptote as $x\to\infty$. Since $\displaystyle\lim_{x\to-\infty}\frac{\sqrt{5x^2+7}}{2x+3}=\lim_{x\to-\infty}\frac{-\sqrt{5+7/x^2}}{2+3/x}=-\frac{\sqrt{5}}{2}$, $y=-\sqrt{5}/2$ is a horizontal asymptote as $x\to-\infty$. To determine whether the graph approaches $y=\sqrt{5}/2$ from above or below as $x\to\infty$, we write

$$f(x)=\frac{\sqrt{5x^2+7}}{2x+3}=\sqrt{\frac{5x^2+7}{(2x+3)^2}}=\sqrt{\frac{5x^2+7}{4x^2+12x+9}}=\sqrt{\frac{5}{4}-\frac{15x+17/4}{4x^2+12x+9}}.$$

This shows that $f(x) < \sqrt{5}/2$ for large x, and the graph therefore approaches the asymptote from below. Similarly, for large negative values of x, we express $f(x)$ in the form $f(x) = -\sqrt{\dfrac{5}{4} - \dfrac{15x + 17/4}{4x^2 + 12x + 9}}$, and this shows that the graph of $f(x)$ approaches $y = -\sqrt{5}/2$ from below as $x \to -\infty$.

48. Since $x^2 - 3x + 1 = 0$ for $x = (3 \pm \sqrt{9 - 4})/2 = (3 \pm \sqrt{5})/2$, vertical asymptotes occur at these values of x. With $f(x)$ expressed in the form $f(x) = \dfrac{3x^3 + 2x - 1}{1 - 3x + x^2} = 3x + 9 + \dfrac{26x - 10}{x^2 - 3x + 1}$, we see that $y = 3x + 9$ is an oblique asymptote that is approached from above as $x \to \infty$, and from below as $x \to -\infty$.

EXERCISES 2.4

2. For $x \neq -4$, $f(x) = \dfrac{(4 - x)(4 + x)}{x + 4} = 4 - x$. The graph of the function is therefore the straight line $y = 4 - x$ with the point at $x = -4$ deleted. The computer does not show the hole at the removable discontinuity $x = -4$.

4. The function has no discontinuities.

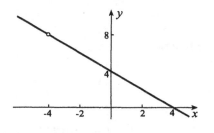

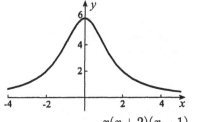

6. The function has no discontinuities.

8. For $x \neq 0, 1$, $f(x) = \dfrac{x(x + 2)(x - 1)}{x(x - 1)} = x + 2$. The graph of the function is therefore the straight line $y = x + 2$ with the points at $x = 0, 1$ deleted. The computer does not show holes at the removable discontinuities $x = 0, 1$.

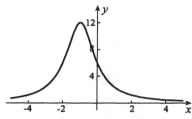

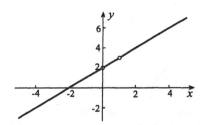

10. The tangent function has infinite discontinuities at $x = (2n + 1)\pi/2$, where n is an integer.

12. The function is discontinuous at $x = 0$. The discontinuity is not removable, jump, or infinite.

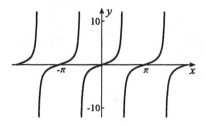

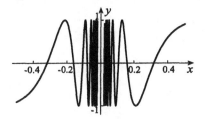

14. The function has infinite
discontinuities at $x = \pm 3$.

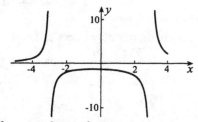

16. The function has a jump discontinuity
at $x = 3$. The computer does not show
the empty circles at $x = 3$.

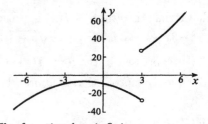

18. The function has infinite
discontinuities at $x = -1, 2$.

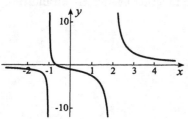

20. The function has infinite
discontinuities at $x = -1, 7$.

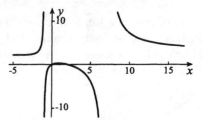

22. The function has an infinite
discontinuity at $x = 1$.

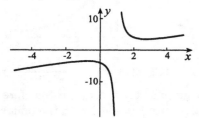

24. The function has an infinite
discontinuities at $x = 0, -5$.

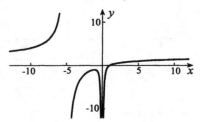

26. The function has an infinite
discontinuity at $x = -5$.

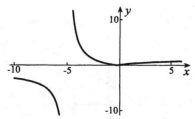

28. The function is continuous for $x > -5$.

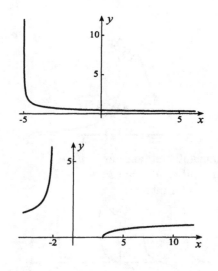

30. The function is continuous for
$x < -2$ and $x \geq 3$.

32. Right- and left-limits as $x \to 3/4$ and $x \to \pm\infty$ lead to the vertical and horizontal asymptotes in the left drawing below. With x- and y-intercepts, we finish the graph as shown to the right.

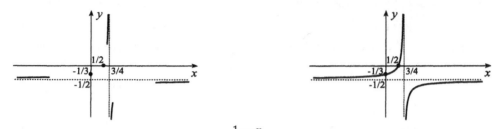

34. First we factor the denominator $f(x) = \dfrac{1-x}{(2x-1)(x+3)}$. Right- and left-limits as $x \to -3$ and $x \to 1/2$, and limits as $x \to \pm\infty$ lead to the vertical and horizontal asymptotes in the left drawing below. With x- and y-intercepts, we finish the graph as shown to the right.

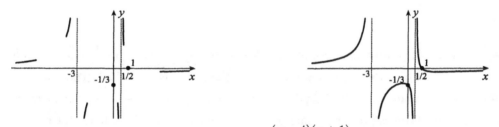

36. First we factor numerator and denominator $f(x) = \dfrac{(x-4)(x+1)}{(3x+1)(x-5)}$. Right- and left-limits as $x \to -1/3$ and $x \to 5$ lead to the vertical asymptotes in the left drawing below. To take limits as $x \to \pm\infty$, we use long division to write $f(x)$ in the form $f(x) = \dfrac{1}{3} + \dfrac{5x/3 - 7/3}{3x^2 - 14x - 5}$. Limits as $x \to \pm\infty$ lead to the horizontal asymptote in the same drawing. With x- and y-intercepts, we finish the graph as shown to the right.

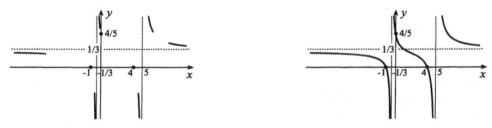

38. Right- and left-limits as $x \to 1$ lead to the vertical asymptote in the left drawing below. The graph has an oblique asymptote that we can identify with long division, $f(x) = -3x - 1 + \dfrac{6}{1-x}$. The line $y = -3x - 1$ is the oblique asymptote. With no x-intercepts, and y-intercept equal to 5, we finish the graph as shown to the right.

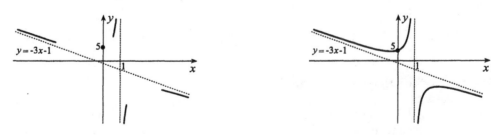

40. First we factor numerator and denominator $f(x) = \dfrac{(x-2)(x^2+x+4)}{(x-1)^3}$. Right- and left-limits as $x \to 1$ lead to the vertical asymptote in the left drawing below. To take limits as $x \to \pm\infty$, we use long division to write $f(x)$ in the form $f(x) = 1 + \dfrac{2x^2 - x - 7}{x^3 - 3x^2 + 3x - 1}$. Limits as $x \to \pm\infty$ lead to the horizontal asymptote in the same drawing. With x- and y-intercepts, we finish the graph as shown to the right.

42. First we factor the numerator and denominator $f(x) = \dfrac{(x-4)(x^2+4x+16)}{(x-2)(x-3)}$. Right- and left-limits as $x \to 2$ and $x \to 3$ lead to the vertical asymptotes in the left drawing below. The graph as an oblique asymptote that we can identify with long division, $f(x) = x + 5 + \dfrac{19x - 94}{x^2 - 5x + 6}$. The line $y = x+5$ is the oblique asymptote. With x-intercept 4, and y-intercept equal to $-32/3$, we finish the graph as shown to the right.

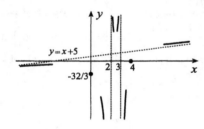

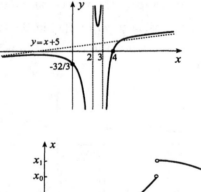

44. No. If it were to have a discontinuity at $t = t_0$ as in the figure to the right, the particle would disappear at position x_0 and reappear instantaneously at position x_1.

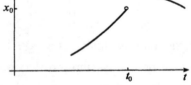

46. If we set $h = x - a$ in the left side of 2.4a, then $f(a) = \lim\limits_{x \to a} f(x) = \lim\limits_{h+a \to a} f(a+h) = \lim\limits_{h \to 0} f(a+h)$.

48. (a) The function is discontinuous at $x = n/10$, where n is an integer.
(b) Let a positive number x be denoted by $z.abc\cdots$ where z is the integer part, and a, b, and c are the first three decimals. Then

$$f(z.abc\cdots) = (1/10)\lfloor za.bc\cdots\rfloor = \tfrac{1}{10}(za) = z.a.$$

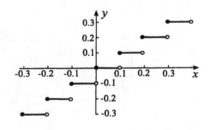

50. (a) The function is discontinuous at $x = n + 1/2$, where n is an integer.

(b) Let a positive number x be denoted by $z.a \cdots$ where z is the integer part, and a is the first decimal.
Suppose that a is equal to 0, 1, 2, 3, or 4.
The integer part of $x + 1/2$ is z and the
first decimal in the number $x + 1/2$ is
5, 6, 7, 8, or 9, and therefore
$f(x) = \lfloor x + 1/2 \rfloor = z$. On the other hand,
suppose that a is equal to 5, 6, 7, 8, or 9.
Then the integer part of $x + 1/2$ is $z + 1$,
and its first decimal is 0, 1, 2, 3, or 4.
Hence, $f(x) = \lfloor x + 1/2 \rfloor = z + 1$.

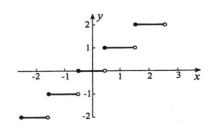

EXERCISES 2.5

2. $f(x) = (1 - x^2)[1 - h(x)] + x^2 h(x)$
$\quad\quad = 1 - x^2 + (2x^2 - 1)h(x)$

4. $f(x) = (x - 2)[h(x + 1) - h(x - 3)]$
$\quad\quad\quad + (x - 4)[h(x - 3) - h(x - 5)]$
$\quad\quad = (x - 2)h(x + 1) - 2\,h(x - 3)$
$\quad\quad\quad + (4 - x)h(x - 5)$

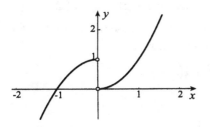

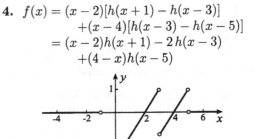

6. $f(x) = \sin x[h(x) - h(x - \pi)]$
$\quad\quad\quad + 2\sin(x - \pi)[h(x - \pi) - h(x - 2\pi)]$
$\quad\quad\quad + 3\sin(x - 2\pi)[h(x - 2\pi) - h(x - 3\pi)]$
$\quad\quad\quad + 4\sin(x - 3\pi)[h(x - 3\pi) - h(x - 4\pi)]$
$\quad\quad = \sin x\,[h(x) - h(x - \pi)]$
$\quad\quad\quad - 2\sin x\,[h(x - \pi) - h(x - 2\pi)]$
$\quad\quad\quad + 3\sin x\,[h(x - 2\pi) - h(x - 3\pi)]$
$\quad\quad\quad - 4\sin x\,[h(x - 3\pi) - h(x - 4\pi)]$
$\quad\quad = \sin x\,[h(x) - 3h(x - \pi) + 5h(x - 2\pi)$
$\quad\quad\quad - 7h(x - 3\pi) + 4h(x - 4\pi)]$

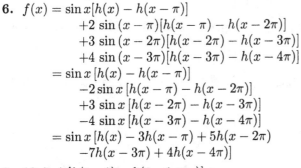

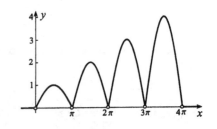

8. $10\sin 4t[h(t - 1) - h(t - 1 - \pi)]$

10. $[100 + 2(t - 10)][h(t - 10) - h(t - 60)] = (80 + 2t)[h(t - 10) - h(t - 60)]$

12. $-(2mg/L)[h(x) - h(x - L/2)]$

14. $F_1\delta(x - x_1) - F_2\delta(x - x_2)$

16. $h(x - a) - h(x - b) + h(x - c)$

18. Yes, except at $x = a$ where $h(x - a)h(x - b)$ is undefined whereas $h(x - b) = 0$.

20.   **22.**

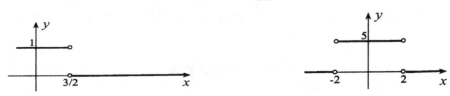

24.

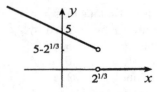

EXERCISES 2.6

2. Suppose $\epsilon > 0$ is given. We must show that we can choose x sufficiently close to 2 so that $|(2x-3)-1| < \epsilon$. To do this, we rewrite the inequality with the x's in the combination $x - 2$,

$$|(2x - 3) - 1| = |2(x - 2)| = 2|x - 2|.$$

We must now choose x so that $2|x - 2| < \epsilon$. But this will be true if $|x - 2| < \epsilon/2$. In other words, if we choose x to satisfy $0 < |x - 2| < \epsilon/2$, then

$$|(2x - 3) - 1| = 2|x - 2| < 2\left(\frac{\epsilon}{2}\right) = \epsilon.$$

We have shown that we can make $2x - 3$ within ϵ of 1 by choosing x within $\epsilon/2$ of 2.

4. Suppose $\epsilon > 0$ is given. We must show that we can choose x sufficiently close to 1 so that $|(x^2+4)-5| < \epsilon$. To do this, we rewrite the inequality with the x's in the combination $x - 1$,

$$|(x^2 + 4) - 5| = |(x - 1)^2 + 2(x - 1)|.$$

We must now choose x so that

$$|(x - 1)^2 + 2(x - 1)| < \epsilon.$$

Since $|(x - 1)^2 + 2(x - 1)| \leq |x - 1|^2 + 2|x - 1|$, the above inequality is satisfied if x is chosen so that

$$|x - 1|^2 + 2|x - 1| < \epsilon.$$

Suppose we set $z = |x - 1|$, and consider the parabola $Q(z) = z^2 + 2z - \epsilon$ in the figure. It crosses the z-axis when

$$z = \frac{-2 \pm \sqrt{4 + 4\epsilon}}{2} = -1 \pm \sqrt{1 + \epsilon}.$$

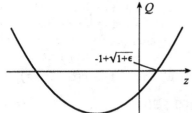

The graph shows that $Q(z) < 0$ whenever $0 < z < -1 + \sqrt{1 + \epsilon}$. In other words, if $0 < |x - 1| < \sqrt{1 + \epsilon} - 1$, then $|x - 1|^2 + 2|x - 1| < \epsilon$, and therefore $|(x - 1)^2 + 2(x - 1)| < \epsilon$.

6. Suppose $\epsilon > 0$ is given. We must show that we can choose x sufficiently close to 3 so that $|(x^2-7x)+12| < \epsilon$. To do this, we rewrite the inequality with the x's in the combination $x - 3$,

$$|(x^2 - 7x) + 12| = |x^2 - 7x + 12| = |(x - 3)^2 - (x - 3)|.$$

We must now choose x so that

$$|(x - 3)^2 - (x - 3)| < \epsilon.$$

Since $|(x - 3)^2 - (x - 3)| \leq |x - 3|^2 + |x - 3|$, the above inequality is satisfied if x is chosen so that

$$|x - 3|^2 + |x - 3| < \epsilon.$$

Suppose we set $z = |x - 3|$, and consider the parabola $Q(z) = z^2 + z - \epsilon$ in the figure. It crosses the z-axis when

$$z = \frac{-1 \pm \sqrt{1 + 4\epsilon}}{2}.$$

The graph shows that $Q(z) < 0$ whenever
$0 < z < \left(-1 + \sqrt{1 + 4\epsilon}\right)/2$. In other words,
if $0 < |x - 3| < \left(\sqrt{1 + 4\epsilon} - 1\right)/2$, then
$|x - 3|^2 + |x - 3| < \epsilon$, and therefore
$|(x - 3)^2 - (x - 3)| < \epsilon$.

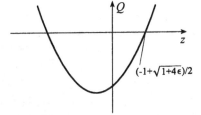

8. Suppose $\epsilon > 0$ is given. We must show that we can choose x sufficiently close to 1 so that $|(x^2 + 3x + 5) - 9| < \epsilon$. To do this, we rewrite the inequality with the x's in the combination $x - 1$,

$$|(x^2 + 3x + 5) - 9| = |x^2 + 3x - 4| = |(x - 1)^2 + 5(x - 1)|.$$

We must now choose x so that

$$|(x - 1)^2 + 5(x - 1)| < \epsilon.$$

Since $|(x - 1)^2 + 5(x - 1)| \le |x - 1|^2 + 5|x - 1|$, the above inequality is satisfied if x is chosen so that

$$|x - 1|^2 + 5|x - 1| < \epsilon.$$

Suppose we set $z = |x - 1|$ and consider the parabola $Q(z) = z^2 + 5z - \epsilon$ in the figure. It crosses the z-axis when

$$z = \frac{-5 \pm \sqrt{25 + 4\epsilon}}{2}.$$

The graph shows that $Q(z) < 0$ whenever
$0 < z < \left(\sqrt{25 + 4\epsilon} - 5\right)/2$. In other words,
if $0 < |x - 1| < \left(\sqrt{25 + 4\epsilon} - 5\right)/2$, then
$|x - 1|^2 + 5|x - 1| < \epsilon$, and therefore
$|(x - 1)^2 + 5(x - 1)| < \epsilon$.

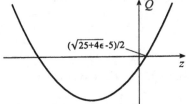

10. $\lim_{x \to a^+} f(x) = L$ if given any $\epsilon > 0$, there exists a $\delta > 0$ such that $|f(x) - L| < \epsilon$ whenever $a < x < a + \delta$.

12. $\lim_{x \to \infty} f(x) = L$ if given any $\epsilon > 0$, there exists an $X > 0$ such that $|f(x) - L| < \epsilon$ whenever $x > X$.

14. $\lim_{x \to a} f(x) = \infty$ if given any $M > 0$, there exists a $\delta > 0$ such that $f(x) > M$ whenever $0 < |x - a| < \delta$.

16. $\lim_{x \to \infty} f(x) = \infty$ if given any $M > 0$, there exists an $X > 0$ such that $f(x) > M$ whenever $x > X$.

18. $\lim_{x \to -\infty} f(x) = \infty$ if given any $M > 0$, there exists an $X < 0$ such that $f(x) > M$ whenever $x < X$.

20. Suppose to the contrary that $f(x)$ has two limits L_1 and L_2 as x approaches a where $L_2 > L_1$ and $L_2 - L_1 = \epsilon$. Since $\lim_{x \to a} f(x) = L_1$, there exists a δ_1 such that $|f(x) - L_1| < \epsilon/3$ when $0 < |x - a| < \delta_1$; that is, when x is in the interval $0 < |x - a| < \delta_1$, the function is within $\epsilon/3$ of L_1. On the other hand, since $\lim_{x \to a} f(x) = L_2$, there exists a δ_2 such that $|f(x) - L_2| < \epsilon/3$ when $0 < |x - a| < \delta_2$; that is, the function is within $\epsilon/3$ of L_2 in the interval $0 < |x - a| < \delta_2$. But this is impossible because L_1 and L_2 are ϵ apart. Consequently, $f(x)$ can have at most one limit as x approaches a.

22. We use the definition in Exercise 15: $\lim_{x \to -2} [-1/(x + 2)^2] = -\infty$ if given any number $M < 0$, there exists a $\delta > 0$ such that $-1/(x + 2)^2 < M$ whenever $0 < |x + 2| < \delta$. The inequality $-1/(x + 2)^2 < M \iff (x + 2)^2 < -1/M \iff |x + 2| < 1/\sqrt{-M}$. Consequently, if we choose $\delta = 1/\sqrt{-M}$, then whenever $0 < |x + 2| < \delta = 1/\sqrt{-M}$, we must have $-1/(x + 2)^2 < M$.

24. We use the definition in Exercise 17: $\lim_{x \to \infty} (5 - x^2) = -\infty$ if given any number $M < 0$, there exists an $X > 0$ such that $5 - x^2 < M$ whenever $x > X$. The inequality $5 - x^2 < M \iff x^2 > 5 - M$, which for positive x implies that $x > \sqrt{5 - M}$. Consequently, if we choose $X = \sqrt{5 - M}$, then whenever $x > X$, we must have $5 - x^2 < M$.

26. We use the definition in Exercise 13: $\lim_{x \to -\infty} (x + 2)/(x - 1) = 1$ if given any number $\epsilon > 0$, there exists an $X < 0$ such that $|(x + 2)/(x - 1) - 1| < \epsilon$ whenever $x < X$. The inequality $|(x + 2)/(x - 1) - 1| < \epsilon \iff 3/|x - 1| < \epsilon \iff |x - 1| > 3/\epsilon$. This inequality is satisfied if $x < 1 - 3/\epsilon$. In other words, if we choose $X = 1 - 3/\epsilon$, then whenever $x < X$, $|(x + 2)/(x - 1) - 1| < \epsilon$.

28. We use the definition in Exercise 19: $\lim_{x\to-\infty}(3+x-x^2)=-\infty$ if given any number $M<0$, there exists an $X<0$ such that $3+x-x^2<M$ whenever $x<X$. The inequality $3+x-x^2<M \iff M > -(x-1/2)^2+13/4 \iff (x-1/2)^2 > 13/4-M \iff |x-1/2| > \sqrt{13/4-M}$. This is satisfied for negative x if $x-1/2 < -\sqrt{13/4-M}$, or, $x < 1/2 - \sqrt{13/4-M}$. Thus, if we choose $X = 1/2 - \sqrt{13/4-M}$, then whenever $x<X$, we must have $3+x-x^2<M$.

30. No. A graph of the function $g(x) = x\sin(1/x)$ is shown to the right. It has limit 0 as $x \to 0$. If values at $x = 1/(n\pi)$ are redefined as 1, then all values of $f(x)$ no longer approach 0 as x approaches 0. Every interval around $x = 0$ contains an infinity of points at which the value of $f(x)$ is equal to 1.

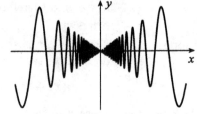

32. Since $\lim_{x\to a} f(x) = F$, there exists a $\delta_1 > 0$ such that

$$|f(x) - F| < \epsilon/2 \qquad \text{whenever } 0 < |x-a| < \delta_1.$$

Since $\lim_{x\to a} g(x) = G$, there exists a $\delta_2 > 0$ such that

$$|g(x) - G| < \epsilon/2 \qquad \text{whenever } 0 < |x-a| < \delta_2.$$

It follows that whenever $0 < |x-a| < \delta$, where δ is the smaller of δ_1 and δ_2,

$$\big|[f(x) - g(x)] - (F-G)\big| = \big|[f(x)-F] - [g(x)-G]\big| \le |f(x)-F| + |g(x)-G| < \frac{\epsilon}{2} + \frac{\epsilon}{2} = \epsilon.$$

REVIEW EXERCISES

2. $\displaystyle\lim_{x\to-1}\frac{x^2-1}{x+1} = \lim_{x\to-1}\frac{(x+1)(x-1)}{x+1} = \lim_{x\to-1}(x-1) = -2$

4. $\displaystyle\lim_{x\to\infty}\frac{x+5}{x-3} = \lim_{x\to\infty}\frac{1+\frac{5}{x}}{1-\frac{3}{x}} = 1$

6. $\displaystyle\lim_{x\to-\infty}\frac{5-x^3}{3+4x^3} = \lim_{x\to-\infty}\frac{\frac{5}{x^3}-1}{\frac{3}{x^3}+4} = -\frac{1}{4}$

8. $\displaystyle\lim_{x\to\infty}\frac{4-3x+x^2}{3+5x^3} = \lim_{x\to\infty}\frac{\frac{4}{x^2}-\frac{3}{x}+1}{\frac{3}{x^2}+5x} = 0$

10. $\displaystyle\lim_{x\to 2^-}\frac{x^2-4x+4}{x-2} = \lim_{x\to 2^-}\frac{(x-2)^2}{x-2} = \lim_{x\to 2^-}(x-2) = 0$

12. $\displaystyle\lim_{x\to 1}\frac{x^2+5x}{(x-1)^2} = \infty$

14. $\displaystyle\lim_{x\to 1}\frac{\sqrt{x}-1}{x-1} = \lim_{x\to 1}\frac{\sqrt{x}-1}{(\sqrt{x}-1)(\sqrt{x}+1)} = \lim_{x\to 1}\frac{1}{\sqrt{x}+1} = \frac{1}{2}$

16. $\displaystyle\lim_{x\to\infty}\frac{\cos 5x}{x} = 0$

18. $\displaystyle\lim_{x\to-\infty}\frac{\sqrt{3x^2+4}}{2x+5} = \lim_{x\to-\infty}\frac{\frac{\sqrt{3x^2+4}}{x}}{\frac{2x+5}{x}} = \lim_{x\to-\infty}\frac{-\sqrt{3+\frac{4}{x^2}}}{2+\frac{5}{x}} = -\frac{\sqrt{3}}{2}$

20. $\displaystyle\lim_{x\to\infty}\left(\sqrt{2x+1}-\sqrt{3x-1}\right) = \lim_{x\to\infty}\left[\left(\sqrt{2x+1}-\sqrt{3x-1}\right)\frac{\sqrt{2x+1}+\sqrt{3x-1}}{\sqrt{2x+1}+\sqrt{3x-1}}\right]$

$$= \lim_{x\to\infty}\frac{(2x+1)-(3x-1)}{\sqrt{2x+1}+\sqrt{3x-1}} = \lim_{x\to\infty}\frac{2-x}{\sqrt{2x+1}+\sqrt{3x-1}}$$

$$= \lim_{x\to\infty}\frac{\frac{2-x}{\sqrt{x}}}{\frac{\sqrt{2x+1}+\sqrt{3x-1}}{\sqrt{x}}} = \lim_{x\to\infty}\frac{\frac{2}{\sqrt{x}}-\sqrt{x}}{\sqrt{2+\frac{1}{x}}+\sqrt{3-\frac{1}{x}}} = -\infty$$

22. The limits $\lim\limits_{x \to 2^+} f(x) = \infty$, $\lim\limits_{x \to 2^-} f(x) = -\infty$, $\lim\limits_{x \to \infty} f(x) = 1^+$, and $\lim\limits_{x \to -\infty} f(x) = 1^-$ lead to the graph to the right. The discontinuity at $x = 2$ is infinite.

24. For $x \neq 6$, $f(x) = x + 6$. Consequently, the graph is a straight line with the point at $x = 6$ removed. The discontinuity at $x = 6$ is removable.

26. The limits $\lim\limits_{x \to 1^+} f(x) = \infty$, $\lim\limits_{x \to 1^-} f(x) = \infty$, $\lim\limits_{x \to \infty} f(x) = 1^+$, and $\lim\limits_{x \to -\infty} f(x) = 1^-$ lead to the graph to the right. The discontinuity at $x = 1$ is infinite.

28. The limits $\lim\limits_{x \to 1^+} f(x) = \infty$, $\lim\limits_{x \to 1^-} f(x) = \infty$, $\lim\limits_{x \to \infty} f(x) = 1^+$, and $\lim\limits_{x \to -\infty} f(x) = -1^+$ lead to the graph to the right. The discontinuity at $x = 1$ is infinite.

30. With $f(x) = \dfrac{2x^2}{(x-4)(x+1)}$, we calculate
$\lim\limits_{x \to 4^+} f(x) = \infty$, $\lim\limits_{x \to 4^-} f(x) = -\infty$, $\lim\limits_{x \to -1^+} f(x) = -\infty$,
and $\lim\limits_{x \to -1^-} f(x) = \infty$. Furthermore, with
$f(x) = 2 + \dfrac{6x + 8}{x^2 - 3x - 4}$, we find $\lim\limits_{x \to \infty} f(x) = 2^+$
and $\lim\limits_{x \to -\infty} f(x) = 2^-$. The graph is shown to the right. The discontinuities at $x = -1$, 4 are infinite.

32. For $x \neq 1$, $f(x) = \dfrac{(x-1)^3}{(x-1)^2} = x - 1$
The graph of the function is therefore the straight line $y = x - 1$ with the point at $x = 1$ deleted. The function has a removable discontinuity at $x = 1$.

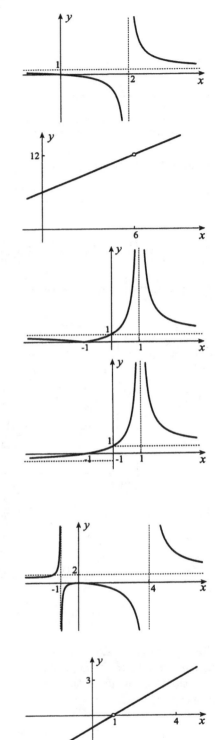

34. The function can be expressed in the form

$$f(x) = (3 + x^3)[1 - h(x+1)] + (x^2 + 2)[h(x+1) - h(x-2)] + 4h(x-2)$$
$$= 3 + x^3 - (x^3 - x^2 + 1)h(x+1) + (2 - x^2)h(x-2).$$

CHAPTER 3

EXERCISES 3.1

2. $f'(x) = \lim\limits_{h \to 0} \dfrac{f(x+h) - f(x)}{h} = \lim\limits_{h \to 0} \dfrac{[3(x+h)^2 + 5] - (3x^2 + 5)}{h}$

$\quad = \lim\limits_{h \to 0} \dfrac{3x^2 + 6xh + 3h^2 + 5 - 3x^2 - 5}{h} = \lim\limits_{h \to 0} (6x + 3h) = 6x$

4. $f'(x) = \lim\limits_{h \to 0} \dfrac{f(x+h) - f(x)}{h} = \lim\limits_{h \to 0} \dfrac{[(x+h)^3 + 2(x+h)^2] - (x^3 + 2x^2)}{h}$

$\quad = \lim\limits_{h \to 0} \dfrac{x^3 + 3x^2 h + 3xh^2 + h^3 + 2x^2 + 4xh + 2h^2 - x^3 - 2x^2}{h} = \lim\limits_{h \to 0} \dfrac{h(3x^2 + 3xh + h^2 + 4x + 2h)}{h}$

$\quad = 3x^2 + 4x$

6. $f'(x) = \lim\limits_{h \to 0} \dfrac{f(x+h) - f(x)}{h} = \lim\limits_{h \to 0} \dfrac{1}{h} \left(\dfrac{x+h+4}{x+h-5} - \dfrac{x+4}{x-5} \right)$

$\quad = \lim\limits_{h \to 0} \dfrac{1}{h} \left[\dfrac{(x^2 + xh + 4x - 5x - 5h - 20) - (x^2 + xh - 5x + 4x + 4h - 20)}{(x+h-5)(x-5)} \right]$

$\quad = \lim\limits_{h \to 0} \dfrac{-9}{(x+h-5)(x-5)} = \dfrac{-9}{(x-5)^2}$

8. If we write $f(x) = x^3 + 2x^2$, then this is the same function as in Exercise 4. Consequently, $f'(x) = 3x^2 + 4x$.

10. $f'(x) = \lim\limits_{h \to 0} \dfrac{f(x+h) - f(x)}{h} = \lim\limits_{h \to 0} \dfrac{1}{h} \left\{ \dfrac{(x+h)^2 - (x+h) + 1}{(x+h)^2 + (x+h) + 1} - \dfrac{x^2 - x + 1}{x^2 + x + 1} \right\}$

$\quad = \lim\limits_{h \to 0} \dfrac{1}{h} \left\{ \dfrac{(x^2 + x + 1)[(x+h)^2 - (x+h) + 1] - (x^2 - x + 1)[(x+h)^2 + (x+h) + 1]}{(x^2 + x + 1)[(x+h)^2 + (x+h) + 1]} \right\}$

$\quad = \lim\limits_{h \to 0} \left\{ \dfrac{h(2x^2 + 2xh - 2)}{h(x^2 + x + 1)[(x+h)^2 + (x+h) + 1]} \right\} = \dfrac{2x^2 - 2}{(x^2 + x + 1)^2}$

12. Since $A = f(r) = \pi r^2$, $\quad \dfrac{dA}{dr} = \lim\limits_{h \to 0} \dfrac{f(r+h) - f(r)}{h} = \lim\limits_{h \to 0} \dfrac{\pi(r+h)^2 - \pi r^2}{h} = \lim\limits_{h \to 0} \dfrac{\pi h(2r+h)}{h} = 2\pi r.$

14. Since $V = f(r) = (4/3)\pi r^3$,

$$\dfrac{dV}{dr} = \lim\limits_{h \to 0} \dfrac{f(r+h) - f(r)}{h} = \lim\limits_{h \to 0} \dfrac{(4/3)\pi(r+h)^3 - (4/3)\pi r^3}{h}$$

$$= \dfrac{4\pi}{3} \lim\limits_{h \to 0} \dfrac{r^3 + 3r^2 h + 3rh^2 + h^3 - r^3}{h} = \dfrac{4\pi}{3} \lim\limits_{h \to 0} (3r^2 + 3rh + h^2) = 4\pi r^2.$$

16. Since $f'(x)$ is the slope of the tangent line to the straight line, and the tangent line is the line itself, $f'(x) = m$.

18. Since $f'(1)$ is the slope of the tangent line to the parabola at $x = 1$, it follows that $f'(1) = 0$.

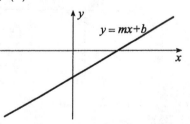

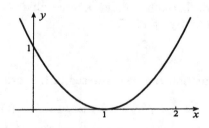

20. Since $\dfrac{dy}{dx} = \lim\limits_{h \to 0} \dfrac{[(x+h)^2 + 3] - [x^2 + 3]}{h} = \lim\limits_{h \to 0} \dfrac{h(2x+h)}{h} = 2x$, the slope of the tangent line to the parabola at $(1, 4)$ is $2(1) = 2$. The equation of the tangent line is $y - 4 = 2(x - 1) \implies y = 2x + 2$.

22. Since $\dfrac{dy}{dx} = \lim\limits_{h\to 0} \dfrac{\frac{1}{(x+h)^2} - \frac{1}{x^2}}{h} = \lim\limits_{h\to 0} \dfrac{1}{h}\left[\dfrac{x^2 - (x+h)^2}{x^2(x+h)^2}\right] = \lim\limits_{h\to 0} \dfrac{-(2x+h)}{x^2(x+h)^2} = -\dfrac{2}{x^3}$, the slope of the

tangent line at $(2, 1/4)$ is $-2/(2^3) = -1/4$. The equation of the tangent line is $y - 1/4 = -(1/4)(x-2) \implies x + 4y = 3$.

24. Since $\dfrac{dy}{dx} = \lim\limits_{h\to 0} \dfrac{(x+h)^2 - x^2}{h} = \lim\limits_{h\to 0} \dfrac{h(2x+h)}{h} = 2x$, the slope of the tangent line to the parabola at $(1,1)$ is 2. The inclination is defined by $\tan\phi = 2$, and is therefore equal to 1.107 radians.

26. According to Exercise 22, the slope of the tangent line at $(2, 1/4)$ is $-1/4$. Since the inclination satisfies $\tan\phi = -1/4$, it follows that $\phi = 2.897$ radians.

28. $f'(x) = \lim\limits_{h\to 0} \dfrac{f(x+h) - f(x)}{h} = \lim\limits_{h\to 0} \dfrac{(x+h)^8 - x^8}{h}$

$= \lim\limits_{h\to 0} \dfrac{(x^8 + 8x^7h + 28x^6h^2 + \cdots + h^8) - x^8}{h}$ (using the binomial theorem)

$= \lim\limits_{h\to 0} (8x^7 + 28x^6h + \cdots + h^7) = 8x^7$

30. $f'(x) = \lim\limits_{h\to 0} \dfrac{f(x+h) - f(x)}{h} = \lim\limits_{h\to 0} \dfrac{1}{h}\left\{\dfrac{1}{(x+h-2)^4} - \dfrac{1}{(x-2)^4}\right\}$

$= \lim\limits_{h\to 0}\left\{\dfrac{(x-2)^4 - [(x-2)^4 + 4(x-2)^3h + 6(x-2)^2h^2 + 4(x-2)h^3 + h^4]}{h(x-2)^4(x+h-2)^4}\right\}$

$= \lim\limits_{h\to 0}\left\{\dfrac{-4(x-2)^3 - 6(x-2)^2h - 4(x-2)h^2 - h^3}{(x-2)^4(x+h-2)^4}\right\} = \dfrac{-4(x-2)^3}{(x-2)^8} = \dfrac{-4}{(x-2)^5}$

32. $f'(x) = \lim\limits_{h\to 0} \dfrac{f(x+h) - f(x)}{h} = \lim\limits_{h\to 0} \dfrac{(x+h)\sqrt{(x+h)+1} - x\sqrt{x+1}}{h}$

$= \lim\limits_{h\to 0}\left[\dfrac{(x+h)\sqrt{x+h+1} - x\sqrt{x+1}}{h} \dfrac{(x+h)\sqrt{x+h+1} + x\sqrt{x+1}}{(x+h)\sqrt{x+h+1} + x\sqrt{x+1}}\right]$

$= \lim\limits_{h\to 0} \dfrac{(x+h)^2(x+h+1) - x^2(x+1)}{h[(x+h)\sqrt{x+h+1} + x\sqrt{x+1}]}$

$= \lim\limits_{h\to 0} \dfrac{x^2 + 2x(x+h+1) + h(x+h+1)}{(x+h)\sqrt{x+h+1} + x\sqrt{x+1}} = \dfrac{3x^2 + 2x}{2x\sqrt{x+1}} = \dfrac{3x+2}{2\sqrt{x+1}}$

34. Since $V = (4/3)\pi r^3$ and $A = 4\pi r^2$, we have $V = f(A) = \dfrac{4\pi}{3}\left(\dfrac{A}{4\pi}\right)^{3/2} = \dfrac{1}{6\sqrt{\pi}} A^{3/2}$. The derivative of this function is

$\dfrac{dV}{dA} = \lim\limits_{h\to 0} \dfrac{f(A+h) - f(A)}{h} = \lim\limits_{h\to 0} \dfrac{(A+h)^{3/2}/(6\sqrt{\pi}) - A^{3/2}/(6\sqrt{\pi})}{h}$

$= \dfrac{1}{6\sqrt{\pi}} \lim\limits_{h\to 0}\left[\dfrac{(A+h)^{3/2} - A^{3/2}}{h} \dfrac{(A+h)^{3/2} + A^{3/2}}{(A+h)^{3/2} + A^{3/2}}\right]$

$= \dfrac{1}{6\sqrt{\pi}} \lim\limits_{h\to 0} \dfrac{A^3 + 3A^2h + 3Ah^2 + h^3 - A^3}{h[(A+h)^{3/2} + A^{3/2}]} = \dfrac{1}{6\sqrt{\pi}} \lim\limits_{h\to 0} \dfrac{3A^2 + 3Ah + h^2}{(A+h)^{3/2} + A^{3/2}} = \dfrac{1}{6\sqrt{\pi}} \dfrac{3A^2}{2A^{3/2}} = \dfrac{\sqrt{A}}{4\sqrt{\pi}}$.

36. If $x > 0$, then $f(x) = x$, and $f'(x) = \lim\limits_{h\to 0} \dfrac{f(x+h) - f(x)}{h} = \lim\limits_{h\to 0} \dfrac{(x+h) - x}{h} = \lim\limits_{h\to 0} 1 = 1$.

When $x < 0$, $f(x) = -x$, and $f'(x) = \lim\limits_{h\to 0} \dfrac{f(x+h) - f(x)}{h} = \lim\limits_{h\to 0} \dfrac{-(x+h) - (-x)}{h} = \lim\limits_{h\to 0} (-1) = -1$.

Finally, $f'(0) = \lim\limits_{h\to 0} \dfrac{f(0+h) - f(0)}{h} = \lim\limits_{h\to 0} \dfrac{|0+h| - 0}{h} = \lim\limits_{h\to 0} \dfrac{|h|}{h}$.

Since this limit does not exist, $f'(0)$ does not exist. These three results are combined in the one formula $f'(x) = |x|/x$.

38. The graph of the function suggests that $f'(R)$ does not exist since there appears to be no tangent line when $r = R$. To verify this we calculate

$$f'(R) = \lim_{h \to 0} \frac{f(R+h) - f(R)}{h}.$$

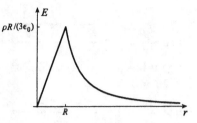

When $h < 0$, we find that $\quad \lim_{h \to 0^-} \frac{1}{h} \left[\frac{\rho(R+h)}{3\epsilon_0} - \frac{\rho R}{3\epsilon_0} \right] = \lim_{h \to 0^-} \frac{\rho}{3\epsilon_0} = \frac{\rho}{3\epsilon_0}.$ When $h > 0$,

$$\lim_{h \to 0^+} \frac{1}{h} \left[\frac{\rho R^3}{3\epsilon_0 (R+h)^2} - \frac{\rho R}{3\epsilon_0} \right] = \frac{\rho R}{3\epsilon_0} \lim_{h \to 0^+} \frac{1}{h} \left[\frac{R^2 - (R+h)^2}{(R+h)^2} \right] = \frac{\rho R}{3\epsilon_0} \lim_{h \to 0^+} \frac{-2R - h}{(R+h)^2} = \frac{-2\rho}{3\epsilon_0}.$$

Since these limits are not the same, $f'(R)$ does not exist.

EXERCISES 3.2

2. $f'(x) = 9x^2 + 4$ **4.** $f'(x) = 20x^4 - 30x^2 + 3$

6. $f'(x) = 2(-3x^{-4}) = -6/x^4$

8. $f'(x) = -(1/2)(-2x^{-3}) + 3(-4x^{-5}) = 1/x^3 - 12/x^5$

10. $f'(x) = 20x^3 + \dfrac{1}{4}(-5x^{-6}) = 20x^3 - \dfrac{5}{4x^6}$ **12.** $f'(x) = \dfrac{1}{2}x^{-1/2} = \dfrac{1}{2\sqrt{x}}$

14. $f'(x) = -(3/2)x^{-5/2} + (3/2)x^{1/2} = -\dfrac{3}{2x^{5/2}} + \dfrac{3\sqrt{x}}{2}$

16. $f'(x) = \pi(\pi x^{\pi-1}) = \pi^2 x^{\pi-1}$

18. Since $f(x) = 4x - x^{-3}$, we obtain $f'(x) = 4 + 3x^{-4} = 4 + 3/x^4$.

20. Since $f(x) = 8x^3 + 60x^2 + 150x + 125$, we obtain $f'(x) = 24x^2 + 120x + 150 = 6(2x + 5)^2$.

22. Since $f'(x) = (1/2)x^{-1/2}$, the slope of the tangent line at $(4, 7)$ is $(1/2)4^{-1/2} = 1/4$. Equations for the tangent and normal lines are $y - 7 = (1/4)(x - 4)$ and $y - 7 = -4(x - 4)$, or, $4y = x + 24$ and $4x + y = 23$. The tangent and normal lines do not appear to intersect at right angles because there are different scales along the axes.

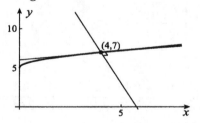

24. Since $y = x^2 - 1$, $dy/dx = 2x$. The slope of the the tangent line is therefore 6, and equations for the tangent and normal lines are $y - 8 = 6(x - 3)$ and $y - 8 = -(1/6)(x - 3)$, or, $y = 6x - 10$ and $x + 6y = 51$. The tangent and normal lines do not appear to intersect at right angles because the axes have different scales.

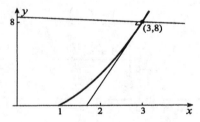

26. Since the slope of the tangent line at the point (x_0, y_0) is $2ax_0$, the equation of the tangent line is $y - y_0 = 2ax_0(x - x_0)$. The x-intercept can be found by setting $y = 0$ and solving for x,

$$-y_0 = 2ax_0(x - x_0) \quad \Longrightarrow \quad x = x_0 - \frac{y_0}{2ax_0}.$$

Since $y_0 = ax_0^2$, it follows that the x-intercept is $x = x_0 - \dfrac{ax_0^2}{2ax_0} = x_0 - \dfrac{x_0}{2} = \dfrac{x_0}{2}$, which is halfway between the origin and the point $x = x_0$ on the x-axis.

28. (a) The graph is to the right.

(b) Since $\dfrac{d}{dt}(1000\pi t^2 + 100\pi t) = 2000\pi t + 100\pi$,
the frequencies at $t = 0$ and $t = 0.1$ are
$100\pi/(2\pi) = 50$ Hz and $300\pi/(2\pi) = 150$ Hz.

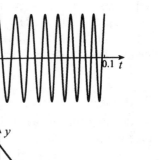

30. If $P(a, b)$ is any point on the curve, the slope
of the tangent line at P is $f'(a) = 3a^2 + 2a - 22$.
The equation of the tangent line at P is
$y - b = (3a^2 + 2a - 22)(x - a)$, and this line will pass
through the origin if $-b = (3a^2 + 2a - 22)(-a)$. Since
(a, b) is on the curve, its coordinates must satisfy
the equation of the curve, $b = a^3 + a^2 - 22a + 20$.
When we equate these two expressions for b,

$$a^3 + a^2 - 22a + 20 = 3a^3 + 2a^2 - 22a$$

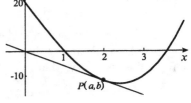

which simplifies to $0 = 2a^3 + a^2 - 20 = (a - 2)(2a^2 + 5a + 10)$. The only solution is $a = 2$, giving the
point $(2, -12)$.

32. The slope of the tangent line to $\sqrt{x} + \sqrt{y} = \sqrt{a}$, or, $y = f(x) = (\sqrt{a} - \sqrt{x})^2 = a - 2\sqrt{a}\sqrt{x} + x$ at any

point (c, d) is $f'(c) = -\dfrac{\sqrt{a}}{\sqrt{c}} + 1$.

The equation of the tangent line at this point is

$$y - d = \left(1 - \frac{\sqrt{a}}{\sqrt{c}}\right)(x - c),$$

and its x- and y-intercepts are

$$c - \frac{d}{1 - \sqrt{a}/\sqrt{c}} \quad \text{and} \quad d - c(1 - \sqrt{a}/\sqrt{c}).$$

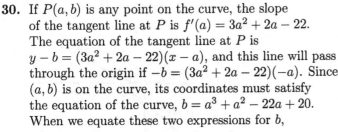

Since (c, d) is on the curve, it follows that $d = (\sqrt{a} - \sqrt{c})^2$, and the sum of the intercepts is

$$c - \frac{\sqrt{c}\,d}{\sqrt{c} - \sqrt{a}} + d - \sqrt{c}(\sqrt{c} - \sqrt{a}) = c - \frac{\sqrt{c}(\sqrt{a} - \sqrt{c})^2}{\sqrt{c} - \sqrt{a}} + (\sqrt{a} - \sqrt{c})^2 - \sqrt{c}(\sqrt{c} - \sqrt{a})$$

$$= c + \sqrt{c}(\sqrt{a} - \sqrt{c}) + (\sqrt{a} - \sqrt{c})^2 - \sqrt{c}(\sqrt{c} - \sqrt{a})$$

$$= c + 2\sqrt{c}(\sqrt{a} - \sqrt{c}) + a - 2\sqrt{c}\sqrt{a} + c = a.$$

34. The required position occurs at the point R
where the tangent line from $T(-120, 30)$ to
the parabola representing the hill intersects
the x-axis. Let the point of tangency be
$P(a, b)$. If $y = cx^2 + d$ is the equation of
the parabola, then using the points $(100, 0)$
and $(0, 20)$, we obtain $0 = 100^2 c + d$ and

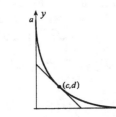

$20 = c(0)^2 + d$. These give $c = -1/500$ and $d = 20$, so that the equation of the parabola is $y = f(x) = 20 - x^2/500$, $-100 \le x \le 100$. Since $f'(x) = -x/250$, the slope of the tangent line at $P(a, b)$ is $f'(a) = -a/250$. The slope of this line is also the slope of PT, namely, $(b - 30)/(a + 120)$, and therefore

$$\frac{b - 30}{a + 120} = -\frac{a}{250} \quad \Longrightarrow \quad b = 30 - \frac{a(a + 120)}{250}.$$

Since $P(a, b)$ is on the parabola, it also follows that $b = 20 - a^2/500$. When we equate these expressions
for b,

$$30 - \frac{a(a + 120)}{250} = 20 - \frac{a^2}{500} \quad \Longrightarrow \quad a^2 + 240a - 5000 = 0.$$

Of the two solutions $-120 \pm 10\sqrt{194}$, only $a = 10\sqrt{194} - 120$ is positive. The y-coordinate of P is $b = 20 - (10\sqrt{194} - 120)^2/500 = (24\sqrt{194} - 238)/5$. The equation of the tangent line is therefore $y - b = (-a/250)(x - a)$, and its x-intercept is given by $-b = (-a/250)(x - a) \implies x = 250b/a + a$. When we substitute the calculated values for a and b, the result is

$$x = \frac{250(24\sqrt{194} - 238)/5}{10\sqrt{194} - 120} + 10\sqrt{194} - 120 = 15(4 + \sqrt{194}) \text{ m.}$$

36. If we denote coordinates of the two points by $P(a, b)$ and $Q(c, d)$, then the fact that P and Q are on the curves requires

$$b = a^2, \quad d = -c^2 + 2c - 2.$$

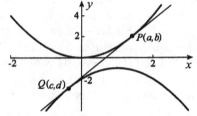

Since $d(x^2)/dx = 2x$, and $d(-x^2 + 2x - 2)/dx = -2x + 2$, and P and Q share a common tangent line, it follows that

$$2a = -2c + 2, \quad \frac{d - b}{c - a} = 2a.$$

To find P and Q, we solve these four equations in a, b, c, and d. If we substitute from the first and second equations into the fourth,

$$a^2 - (-c^2 + 2c - 2) = 2a(a - c) \quad \implies \quad c^2 - 2c + 2 - a^2 + 2ac = 0.$$

From the third equation, $a = 1 - c$, and therefore $c^2 - 2c + 2 - (1 - c)^2 + 2c(1 - c) = 0$. This reduces to $2c^2 - 2c - 1 = 0$ with solutions $c = (1 \pm \sqrt{3})/2$. These give the pairs of points $P((1+\sqrt{3})/2, (2+\sqrt{3})/2)$ and $Q((1 - \sqrt{3})/2, (-4 - \sqrt{3})/2)$, and $P((1 - \sqrt{3})/2, (2 - \sqrt{3})/2)$ and $Q((1 + \sqrt{3})/2, (-4 + \sqrt{3})/2)$.

38. If $x > 0$, then $\frac{d}{dx}|x|^n = \frac{d}{dx}x^n = nx^{n-1} = n|x|^{n-1}$. On the other hand, if $x < 0$, then

$$\frac{d}{dx}|x|^n = \frac{d}{dx}(-x)^n = (-1)^n\frac{d}{dx}x^n = (-1)^n nx^{n-1} = -n(-x)^{n-1} = -n|x|^{n-1}.$$

Graphs of $y = f(x) = |x|^n$ indicate that $f'(0) = 0$ when $n > 1$. This can also be verified algebraically,

$$f'(0) = \lim_{h \to 0} \frac{|0 + h|^n - |0|^n}{h} = \lim_{h \to 0} \frac{|h|^n}{h} = 0 \quad \text{when } n > 1.$$

These three situations are all encompassed by the formula $\frac{d}{dx}|x|^n = n|x|^{n-1}\text{sgn}(x)$, where $\text{sgn}(x)$ is the signum function of Exercise 47 in Section 2.4.

EXERCISES 3.3

2. Since $f(x)$ is not defined for $x < 0$, it cannot have a left-hand derivative at $x = 0$. Its right-hand derivative at $x = 0$ is defined by

$$f'_+(0) = \lim_{h \to 0^+} \frac{f(0 + h) - f(0)}{h} = \lim_{h \to 0^+} \frac{h^{3/2} - 0}{h} = \lim_{h \to 0^+} \sqrt{h} = 0.$$

4. The graph of $f(x) = \text{sgn}\, x$ makes it clear that the function does not have a left-hand derivative at $x = 0$ or a right-hand derivative, and cannot therefore have a derivative. We can also verify this algebraically.

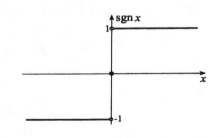

$$f'_-(0) = \lim_{h \to 0^-} \frac{f(0 + h) - f(0)}{h} = \lim_{h \to 0^-} \frac{-1 - 0}{h} = \infty,$$

and

$$f'_+(0) = \lim_{h \to 0^+} \frac{f(0 + h) - f(0)}{h} = \lim_{h \to 0^+} \frac{1 - 0}{h} = \infty.$$

6. The graph of $\lfloor x \rfloor$ in Exercise 68 of Section 1.5 shows that the right derivative is equal to 0, but there is no left derivative, and therefore no derivative.

8. If $h(a)$ is defined as 1, then $h'_+(a) = 0$, but $h'_-(a)$ and $h'(a)$ are still undefined.

10. True   **12.** True

14. By equation 3.2, $f'(0) = \lim\limits_{h \to 0} \dfrac{f(0+h) - f(0)}{h}$

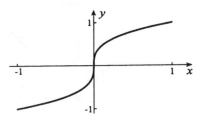

$$= \lim\limits_{h \to 0} \dfrac{h^{1/3} - 0}{h} = \lim\limits_{h \to 0} \dfrac{1}{h^{2/3}} = \infty,$$

The graph of the function confirms this.

16. Since $f(x)$ is not defined for $x < 0$, it cannot have a derivative at $x = 0$. We can show that it does not have a right-hand derivative at $x = 0$ by calculating

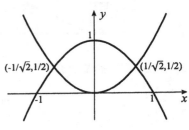

$$f'_+(0) = \lim\limits_{h \to 0^+} \dfrac{f(0+h) - f(0)}{h}$$

$$= \lim\limits_{h \to 0^+} \dfrac{h^{1/4} - 0}{h} = \lim\limits_{h \to 0^+} \dfrac{1}{h^{3/4}} = \infty.$$

The graph of the function confirms this.

18. To find point(s) of intersection, we set $x^2 = 1 - x^2$, solutions of which are $x = \pm 1/\sqrt{2}$. These give the points of intersection $(\pm 1/\sqrt{2}, 1/2)$. Slopes of the curves $y = f(x) = x^2$ and $y = g(x) = 1 - x^2$ at $(1/\sqrt{2}, 1/2)$ are

$$f'(1/\sqrt{2}) = \{2x\}_{|x=1/\sqrt{2}} = \sqrt{2}$$

and

$$g'(1/\sqrt{2}) = \{-2x\}_{|x=1/\sqrt{2}} = -\sqrt{2}.$$

Equation 1.60 gives for the angle θ between the curves at this point

$$\theta = \mathrm{Tan}^{-1} \left| \dfrac{\sqrt{2} + \sqrt{2}}{1 + (\sqrt{2})(-\sqrt{2})} \right| = 1.231 \text{ radians.}$$

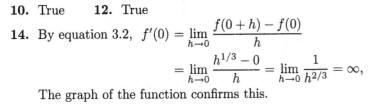

The same angle is obtained at the other point of intersection.

20. To find point(s) of intersection of the curves we set $3 - x^2 = (7 + x^2)/4$, solutions of which are $x = \pm 1$. Points of intersection are therefore $(\pm 1, 2)$. Slopes of the curves $y = f(x) = 3 - x^2$ and $y = g(x) = (x^2 + 7)/4$ at $(1, 2)$ are

$$f'(1) = \{-2x\}_{|x=1} = -2 \quad \text{and} \quad g'(1) = \{x/2\}_{|x=1} = 1/2.$$

Since these slopes are negative reciprocals, the curves intersect orthogonally at the point $(1, 2)$. The curves are also orthogonal at $(-1, 2)$.

22. Slopes of the curves $y = f(x) = x^3$ and $y = g(x) = x^2 + x - 1$ at the point $(1, 1)$ are

$$f'(1) = \{3x^2\}_{|x=1} = 3 \quad \text{and} \quad g'(1) = \{2x + 1\}_{|x=1} = 3.$$

The curves are therefore tangent at the point.

24. Since each straight line in the first family has slope 2/3 and each line in the second family has slope $-1/k$, the families are orthogonal trajectories if $(2/3)(-1/k) = -1 \Longrightarrow k = 2/3$.

26. According to equation 3.2, $f'(0) = \lim\limits_{h \to 0} \dfrac{f(0+h) - f(0)}{h} = \lim\limits_{h \to 0} \dfrac{h|h| - 0}{h} = \lim\limits_{h \to 0} |h| = 0.$

28. Sometimes. The function $f(x) = x$ is differentiable, but $|x|$ does not have a derivative at $x = 0$ (left figure below). On the other hand, $f(x) = x^3$ is differentiable, and $|x^3|$ does have a derivative at $x = 0$ (right figure below).

30. (a) For $X = L/2$,

$$G(x; L/2) = \frac{1}{L\tau}[x(L - L/2)h(L/2 - x) + (L/2)(L - x)h(x - L/2)]$$

$$= \frac{1}{\tau}\begin{cases} x/2 & 0 \le x < L/2 \\ (L - x)/2, & L/2 < x \le L \end{cases}.$$

Its graph in the left figure below is symmetric about $x = L/2$ as would be expected.

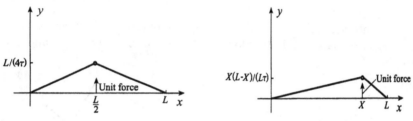

(b) When $L/2 < X < L$,

$$G(x; X) = \frac{1}{L\tau}\begin{cases} x(L - X), & 0 \le x < X \\ X(L - x), & X < x \le L \end{cases}.$$

This is shown in the right figure above.

(c) Since the functions x, $h(x - X)$, $L - x$, and $h(X - x)$ are continuous for $0 \le x \le L$, except at $x = X$ for the Heaviside functions, it follows that $G(x; X)$ is continuous for all $x \ne X$. Since

$$\lim_{x \to X+} G(x; X) = \frac{X(L - X)}{L\tau} = \lim_{x \to X-} G(x; X),$$

the discontinuity at $x = X$ is removable. The graphs in parts (a) and (b) illustrate this.

(d) The jump in the discontinuity of dG/dx at $x = X$ is

$$\lim_{x \to X+} \frac{dG}{dx} - \lim_{x \to X-} \frac{dG}{dx} = \frac{1}{L\tau} \lim_{x \to X+} [(L - X)\, h(X - x) - X\, h(x - X)]$$

$$- \frac{1}{L\tau} \lim_{x \to X-} [(L - X)\, h(X - x) - X\, h(x - X)]$$

$$= \frac{1}{L\tau}[-X] - \frac{1}{L\tau}[L - X] = -\frac{1}{\tau}.$$

This is the change in the slope of the graph of $G(x; X)$ at X.

32. According to equation 3.2, $f'(0) = \lim_{h \to 0} \dfrac{f(0 + h) - f(0)}{h} = \lim_{h \to 0} \dfrac{h \sin(1/h) - 0}{h} = \lim_{h \to 0} \sin\left(\dfrac{1}{h}\right)$. Since this limit does not exist, there is no derivative at $x = 0$.

EXERCISES 3.4

2. $f'(x) = (2 - x^2)(2x + 4) + (-2x)(x^2 + 4x + 2) = 8 - 12x^2 - 4x^3$

4. $f'(x) = \dfrac{(4x^2 - 5)(2x) - x^2(8x)}{(4x^2 - 5)^2} = \dfrac{-10x}{(4x^2 - 5)^2}$ **6.** $f'(x) = \dfrac{(4x^2 + 1)(3x^2) - x^3(8x)}{(4x^2 + 1)^2} = \dfrac{x^2(4x^2 + 3)}{(4x^2 + 1)^2}$

8. $f'(x) = \dfrac{(3x + 2)\left(\dfrac{1}{2\sqrt{x}}\right) - \sqrt{x}(3)}{(3x + 2)^2} = \dfrac{2 - 3x}{2\sqrt{x}(3x + 2)^2}$

10. $f'(x) = \dfrac{(2x^2 - 1)(1) - (x + 5)(4x)}{(2x^2 - 1)^2} = -\dfrac{2x^2 + 20x + 1}{(2x^2 - 1)^2}$

12. $f'(x) = \dfrac{(x^2 - 5x + 1)(2x + 2) - (x^2 + 2x + 3)(2x - 5)}{(x^2 - 5x + 1)^2} = \dfrac{-7x^2 - 4x + 17}{(x^2 - 5x + 1)^2}$

14. If we write the function in the form $f(x) = \dfrac{(x + 1)^3 + 9}{(x + 1)^3} = 1 + \dfrac{9}{(x + 1)^3}$, its derivative is

$f'(x) = \dfrac{(x + 1)^3(0) - 9\dfrac{d}{dx}(x + 1)^3}{(x + 1)^6}$. Since $\dfrac{d}{dx}(x + 1)^3 = \dfrac{d}{dx}(x^3 + 3x^2 + 3x + 1) = 3x^2 + 6x + 3$

$= 3(x + 1)^2$, we obtain $f'(x) = \dfrac{-9(3)(x + 1)^2}{(x + 1)^6} = \dfrac{-27}{(x + 1)^4}$.

16. $f'(x) = \dfrac{(\sqrt{x} - 4)\left(\dfrac{1}{2\sqrt{x}} + 2\right) - (\sqrt{x} + 2x)\left(\dfrac{1}{2\sqrt{x}}\right)}{(\sqrt{x} - 4)^2}$

$= \dfrac{(\sqrt{x} - 4)(1 + 4\sqrt{x}) - (\sqrt{x} + 2x)}{2\sqrt{x}(\sqrt{x} - 4)^2} = \dfrac{2x - 16\sqrt{x} - 4}{2\sqrt{x}(\sqrt{x} - 4)^2} = \dfrac{x - 8\sqrt{x} - 2}{\sqrt{x}(\sqrt{x} - 4)^2}$

18. (a) The plot is shown to the right.

(b) $f'(x) = \dfrac{(x^2 + x - 2)(2x) - x^2(2x + 1)}{(x^2 + x - 2)^2} = \dfrac{x^2 - 4x}{(x^2 + x - 2)^2} = \dfrac{x(x - 4)}{(x + 2)^2(x - 1)^2}$

This expression shows that:

$f'(x) > 0$ for $x < -2$,
$f'(x) > 0$ for $-2 < x < 0$,
$f'(x) < 0$ for $0 < x < 1$,
$f'(x) < 0$ for $1 < x < 4$,
$f'(x) > 0$ for $x > 4$.

We can see the first three of these on the graph, but not the last two. The expression for $f'(x)$ also indicates that $f'(x) = 0$ at $x = 0$ and $x = 4$. This is clear on the graph when $x = 0$, but not when $x = 4$.

20. $\dfrac{d}{dx}[f(x)g(x)h(x)] = f(x)\dfrac{d}{dx}[g(x)h(x)] + f'(x)[g(x)h(x)]$

$= f(x)[g(x)h'(x) + g'(x)h(x)] + f'(x)[g(x)h(x)]$

$= f(x)g(x)h'(x) + f(x)g'(x)h(x) + f'(x)g(x)h(x)$

22. To find the point(s) of intersection, we set $2x + 2 = x^2/(x - 1)$ from which $x = \pm\sqrt{2}$. These give the points of intersection $(\pm\sqrt{2}, 2 \pm 2\sqrt{2})$. Slopes of the curves $y = f(x) = 2x + 2$ and $y = g(x) = x^2/(x - 1)$ at the point $(\sqrt{2}, 2 + 2\sqrt{2})$ are

$$f'(\sqrt{2}) = 2 \quad \text{and} \quad g'(\sqrt{2}) = \left[\dfrac{(x - 1)(2x) - x^2(1)}{(x - 1)^2}\right]_{|x=\sqrt{2}} = \dfrac{2}{1 - \sqrt{2}}.$$

Angle θ between the curves at this point is given by equation 1.60,

$$\theta = \text{Tan}^{-1}\left|\frac{2 - \left(\dfrac{2}{1 - \sqrt{2}}\right)}{1 + 2\left(\dfrac{2}{1 - \sqrt{2}}\right)}\right| = 0.668 \text{ radians.}$$

Slopes of the curves at the other point of intersection $(-\sqrt{2}, 2 - 2\sqrt{2})$ are

$$f'(-\sqrt{2}) = 2 \quad \text{and} \quad g'(-\sqrt{2}) = \left[\frac{(x-1)(2x) - x^2(1)}{(x-1)^2}\right]_{|x=-\sqrt{2}} = \frac{2}{1 + \sqrt{2}}.$$

The angle between the curves at this point is

$$\theta = \text{Tan}^{-1}\left|\frac{2 - \left(\dfrac{2}{1 + \sqrt{2}}\right)}{1 + 2\left(\dfrac{2}{1 + \sqrt{2}}\right)}\right| = 0.415 \text{ radians.}$$

24. The slope of the tangent line to
$y = f(x) = (5 - x)/(6 + x)$ at $P(a, b)$ is
$$f'(a) = \left[\frac{(6+x)(-1) - (5-x)(1)}{(6+x)^2}\right]_{|x=a} = \frac{-11}{(6+a)^2}.$$
The equation of the tangent line at P is
$$y - b = [-11/(6+a)^2](x - a),$$
and this line passes through $(0, 0)$ if
$$-b = [-11/(6+a)^2(-a).$$

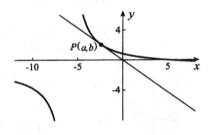

Since $b = (5-a)/(6+a)$, it follows that $\quad -\dfrac{5-a}{6+a} = \dfrac{11a}{(6+a)^2} \implies a^2 - 10a - 30 = 0.$ The two solutions

of this equation are $a = 5 \pm \sqrt{55}$. The points at which the tangent line passes through $(0, 0)$ are therefore $(5 \pm \sqrt{55}, \mp\sqrt{55}/(11 \pm \sqrt{55}))$.

26. (a) $f'(x) = \begin{cases} 0, & x < -1 \\ 2x + 1, & x > -1 \end{cases}$

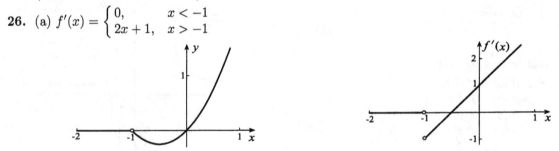

(b) No. The left-hand derivative would be 0 and the right-hand derivative would be -1.

EXERCISES 3.5

2. Since $f'(x) = 3x^2 - 6x + 2$, $f''(x) = 6x - 6$, and $f'''(x) = 6$.

4. Since $f'(x) = 4x^3 - 6x - 1/x^2$, $f''(x) = 12x^2 - 6 + 2/x^3$, and $f'''(x) = 24x - 6/x^4$. Hence, $f'''(1) = 24(1) - 6/(1)^4 = 18$.

6. Since $f'(t) = 3t^2 + 3/t^4$ and $f''(t) = 6t - 12/t^5$, we obtain $f'''(t) = 6 + 60/t^6$.

8. Since $f'(u) = \dfrac{(u+1)\left(\dfrac{1}{2\sqrt{u}}\right) - \sqrt{u}(1)}{(u+1)^2} = \dfrac{1-u}{2\sqrt{u}(u+1)^2}$, it follows that

$$f''(u) = \frac{1}{2}\left\{\frac{\sqrt{u}(u+1)^2(-1) - (1-u)\left[\dfrac{1}{2\sqrt{u}}(u+1)^2 + \sqrt{u}\dfrac{d}{du}(u^2+2u+1)\right]}{u(u+1)^4}\right\}$$

$$= \frac{\dfrac{-2u(u+1)^2 - (1-u)[(u+1)^2 + 2u(2u+2)]}{2\sqrt{u}}}{2u(u+1)^4} = \frac{3u^2 - 6u - 1}{4u^{3/2}(u+1)^3}.$$

10. Since $f'(x) = \dfrac{(\sqrt{x}+1)(1) - x\left(\dfrac{1}{2\sqrt{x}}\right)}{(\sqrt{x}+1)^2} = \dfrac{\sqrt{x}+2}{2(\sqrt{x}+1)^2}$, it follows that

$$f''(x) = \frac{1}{2}\left\{\frac{(\sqrt{x}+1)^2\left(\dfrac{1}{2\sqrt{x}}\right) - (\sqrt{x}+2)\dfrac{d}{dx}(x+2\sqrt{x}+1)}{\left(\sqrt{x}+1\right)^4}\right\}$$

$$= \frac{\dfrac{(\sqrt{x}+1)^2 - 2\sqrt{x}(\sqrt{x}+2)\left(1+\frac{1}{\sqrt{x}}\right)}{2\sqrt{x}}}{2(\sqrt{x}+1)^4} = -\frac{\sqrt{x}+3}{4\sqrt{x}\left(\sqrt{x}+1\right)^3}.$$

12. (a) $\dfrac{d}{dr}\left(r^2\dfrac{dT}{dr}\right) = \dfrac{d}{dr}\left[r^2\left(\dfrac{kr}{3} - \dfrac{c}{r^2}\right)\right] = \dfrac{d}{dr}\left[\dfrac{kr^3}{3} - c\right] = kr^2$

(b) Since $f(a) = T_a$ and $f(b) = T_b$, it follows that $T_a = \dfrac{ka^2}{6} + \dfrac{c}{a} + d$, $\quad T_b = \dfrac{kb^2}{6} + \dfrac{c}{b} + d$. The solution of these equations is

$$c = \frac{ab}{b-a}\left[(T_a - T_b) + \frac{k}{6}(b^2 - a^2)\right], \qquad d = \frac{bT_b - aT_a}{b-a} - \frac{k}{6}(a^2 + ab + b^2).$$

14. (a) Using the result in Example 3.19,

$$\frac{d^3}{dx^3}(uv) = \frac{d}{dx}\left(v\frac{d^2u}{dx^2} + 2\frac{du}{dx}\frac{dv}{dx} + u\frac{d^2v}{dx^2}\right)$$

$$= \frac{dv}{dx}\frac{d^2u}{dx^2} + v\frac{d^3u}{dx^3} + 2\frac{d^2u}{dx^2}\frac{dv}{dx} + 2\frac{du}{dx}\frac{d^2v}{dx^2} + \frac{du}{dx}\frac{d^2v}{dx^2} + u\frac{d^3v}{dx^3}$$

$$= v\frac{d^3u}{dx^3} + 3\frac{dv}{dx}\frac{d^2u}{dx^2} + 3\frac{d^2v}{dx^2}\frac{du}{dx} + u\frac{d^3v}{dx^3}.$$

(b) $\dfrac{d^4}{dx^4}(uv) = v\dfrac{d^4u}{dx^4} + 4\dfrac{dv}{dx}\dfrac{d^3u}{dx^3} + 6\dfrac{d^2v}{dx^2}\dfrac{d^2u}{dx^2} + 4\dfrac{d^3v}{dx^3}\dfrac{du}{dx} + v\dfrac{d^4u}{dx^4}$

16. (a) When $X = L/2$,

$$G(x; L/2) = \frac{1}{6EI}(x - L/2)^3 h(x - L/2) - \frac{x^3}{6EI} + \frac{Lx^2}{4EI}$$

$$= \frac{1}{12EI}\begin{cases} -2x^3 + 3Lx^2, & 0 \le x < L/2 \\ 2(x - L/2)^3 - 2x^3 + 3Lx^2, & L/2 < x \le L \end{cases}$$

$$= \frac{1}{12EI}\begin{cases} -2x^3 + 3Lx^2, & 0 \le x < L/2 \\ 3xL^2/2 - L^3/4, & L/2 < x \le L \end{cases}.$$

A plot is shown to the right. It has a
removable discontinuity at $x = L/2$. It is
straight for $L/2 < x \le L$, as would be expected
since no forces act on the board for $x > L/2$.

(b) For $x > X$,

$$G(x; X) = \frac{1}{6EI}(x - X)^3 - \frac{x^3}{6EI} + \frac{Xx^2}{2EI}$$

$$= \frac{1}{6EI}(x^3 - 3x^2X + 3xX^2 - X^3) - \frac{x^3}{6EI} + \frac{Xx^2}{2EI} = \frac{X^2}{2EI}x - \frac{X^3}{6EI},$$

and this is a straight line.

(c) Since the functions $(x - X)^3$, $h(x - X)$, x^3, and x^2 are continuous for $0 \le x \le L$, except at $x = X$
for the Heaviside function, it follows that $G(x; X)$ is continuous for all $x \ne X$. Since

$$\lim_{x \to X^+} G(x; X) = -\frac{X^3}{6EI} + \frac{X^3}{2EI} = \frac{X^3}{3EI} = \lim_{x \to X^-} G(x; X),$$

the discontinuity at $x = X$ is removable. Limits of $dG/dx = [(x - X)^2h(x - X) - x^2 + 2Xx]/(2EI)$ and
$d^2G/dx^2 = [(x - X)h(x - X) - x + X]/(EI)$, show that they are also continuous except for a removable
discontinuity at $x = X$.

(d) Since $d^3G/dx^3 = [h(x - X) - 1]/(EI)$, the jump in d^3G/dx^3 at $x = X$ is

$$\lim_{x \to X^+} \frac{d^3G}{dx^3} - \lim_{x \to X^-} \frac{d^3G}{dx^3} = 0 - \left(-\frac{1}{EI}\right) = \frac{1}{EI}.$$

EXERCISES 3.6

2. The velocity of the particle is $v(t) = \dfrac{2t^3}{3} - \dfrac{32t^2}{5} + \dfrac{50t}{3} - \dfrac{251}{30}$.

(a) Since $x(1) = -2$, the particle is to the left of the origin at $t = 1$. Since $x(4) = 6$, the particle is to
the right of the origin at $t = 4$.

(b) Since $v(1/2) = -31/20$, the particle is moving to the left at $t = 1/2$. Since $v(3) = 61/30$, the particle
is moving to the right at $t = 3$.

(c) The graph of $v(t)$ to the right shows that
$v(t)$ changes sign three times. Hence, the
particle changes direction three times.

(d) Since $v(7/2) = 3/20$ and $v(9/2) = -133/60$,
the velocity is greater at $t = 7/2$.

(e) Since $|v(7/2)| = 3/20$ and $|v(9/2)| = 133/60$,
the speed is greater at $t = 9/2$.

4. The velocity of the particle is $v(t) = \dfrac{14t^2}{15} - \dfrac{202t}{45} + \dfrac{132}{45}$.

(a) Since $x(1) = 3$, the particle is to the right of the origin at $t = 1$. Since $x(4) = -34/15$, the particle
is to the left of the origin at $t = 4$.

(b) Since $v(1/2) = 83/90$, the particle is moving to the right at $t = 1/2$. Since $v(3) = -32/15$, the
particle is moving to the left at $t = 3$.

(c) The graph of $v(t)$ to the right shows that
$v(t)$ changes sign twice. Hence, the particle
changes direction twice.

(d) Since $v(7/2) = -121/90$ and $v(9/2) = 49/30$,
the velocity is greater at $t = 9/2$.

(e) Since $|v(7/2)| = 121/90$ and $|v(9/2)| = 49/30$,
the speed is greater at $t = 9/2$.

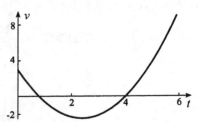

6. The jerk is $x'''(t) = 28/15$.

8. Graphs of $x(t)$, $v(t) = 2t - 7$, $|v(t)| = |2t - 7|$, and $a(t) = 2$ are shown below.

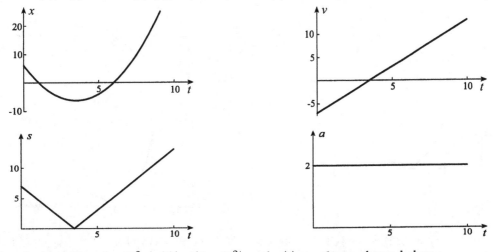

10. Graphs of $x(t)$, $v(t) = 4 - 3t^2$, $|v(t)| = |4 - 3t^2|$, and $a(t) = -6t$ are shown below.

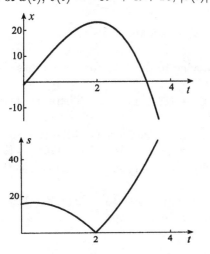
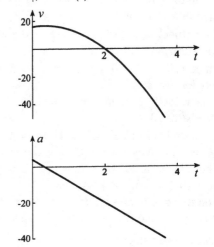

12. Plots of $x(t)$, $v(t) = -6t^2 + 4t + 16$, $|v(t)| = |-6t^2 + 4t + 16|$, and $a(t) = -12t + 4$ are shown below.

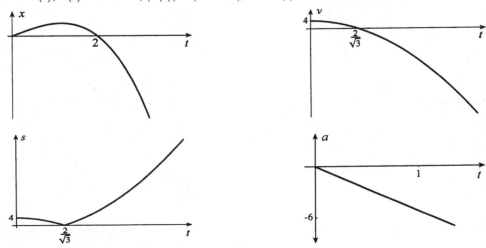

14. Plots of $x(t)$, $v(t) = 1 - 4/t^2$, $|v(t)| = |1 - 4/t^2|$, and $a(t) = 8/t^3$ are shown below.

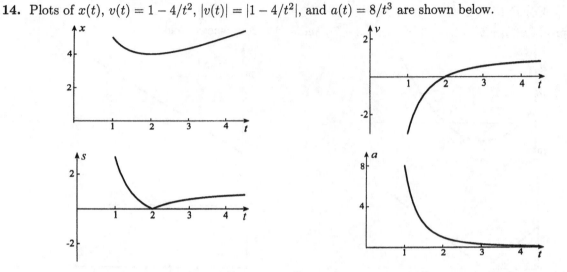

16. Plots of $x(t)$, $v(t) = (5/2)t^{3/2} - 3\sqrt{t} + (1/2)/\sqrt{t} = (5t^2 - 6t + 1)/(2\sqrt{t})$, $|v(t)| = |(5t^2 - 6t + 1)/(2\sqrt{t})|$, and $a(t) = (15/4)\sqrt{t} - 3/(2\sqrt{t}) - 1/(4t^{3/2}) = (15t^2 - 6t - 1)/(4t^{3/2})$ are shown below.

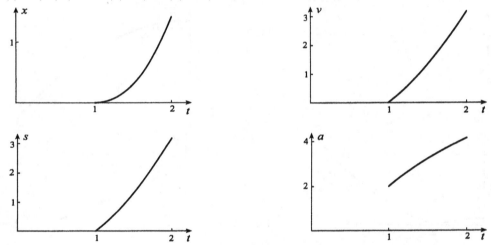

18. We caluclate that $v(t) = 3t^2 - 18t + 15 = 3(t - 1)(t - 5)$ and $a(t) = 6t - 18 = 6(t - 3)$.

(a) $x(3) = -11$ m, $v(3) = -12$ m/s, $|v(3)| = 12$ m/s, $a(3) = 0$ m/s^2

(b) The object is at rest when $v(t) = 0$ and this occurs at $t = 1$ s and $t = 5$ s.

(c) Acceleration vanishes at $t = 3$ s.

(d) Since $v(t) > 0$ for $0 \leq t < 1$ and $t > 5$, the object is moving to the right for these times. It moves to the left for $1 < t < 5$.

(e) The velocity is 1 m/s if $1 = 3t^2 - 18t + 15 \implies 3t^2 - 18t + 14 = 0$. There are two solutions $t = (9 \pm \sqrt{39})/3$.

(f) The speed is 1 m/s if $|3t^2 - 18t + 15| = 1$. This implies that $3t^2 - 18t + 15 = 1$ or $3t^2 - 18t + 15 = -1$. The first gives the times in part (e). For the second possibility, we solve $3t^2 - 18t + 16 = 0$ for $t = (9 \pm \sqrt{33})/3$.

(g) The velocity is 20 m/s if $20 = 3t^2 - 18t + 15 \implies 3t^2 - 18t - 5 = 0$. Of the two solutions $t = (9 \pm 4\sqrt{6})/3$, only $t = (9 + 4\sqrt{6})/3$ is positive.

(h) The speed is 20 m/s if $|3t^2 - 18t + 15| = 20$. This implies that $3t^2 - 18t + 15 = 20$ or $3t^2 - 18t + 15 = -20$. The first gives the time in part (g). The second quadratic has no solutions.

20. (a) Its average velocity is

$$\frac{x_2 - x_1}{t_2 - t_1} = \frac{(at_2^2/2 + bt_2 + c) - (at_1^2/2 + bt_1 + c)}{t_2 - t_1} = \frac{(a/2)(t_2^2 - t_1^2) + b(t_2 - t_1)}{t_2 - t_1} = \frac{a}{2}(t_1 + t_2) + b.$$

(b) The instantaneous velocity is equal to the average velocity when

$$at + b = \frac{a}{2}(t_1 + t_2) + b \quad \Longrightarrow \quad t = \frac{t_1 + t_2}{2}.$$

The position of the object at this time is $x\big((t_1 + t_2)/2\big) = \frac{a}{2}\left(\frac{t_1 + t_2}{2}\right)^2 + b\left(\frac{t_1 + t_2}{2}\right) + c.$

If we subtract this from $(x_1 + x_2)/2$, we obtain

$$\frac{x_1 + x_2}{2} - x\big((t_1 + t_2)/2\big) = \frac{1}{2}\left(\frac{1}{2}at_1^2 + bt_1 + c + \frac{1}{2}at_2^2 + bt_2 + c\right) - \frac{a}{2}\left(\frac{t_1 + t_2}{2}\right)^2 - b\left(\frac{t_1 + t_2}{2}\right) - c$$

$$= \frac{a}{8}(t_1 - t_2)^2.$$

Since this quantity is positive, the object is closer to x_1.

22. At point A, we must have

$$y_1 = a\theta_1^2, \qquad y_1 = m\theta_1 + b, \qquad m = 2a\theta_1.$$

From these,

$$a = \frac{y_1}{\theta_1^2}, \qquad m = 2\left(\frac{y_1}{\theta_1^2}\right)\theta_1 = \frac{2y_1}{\theta_1}, \qquad b = y_1 - 2\theta_1\left(\frac{y_1}{\theta_1^2}\right)\theta_1 = -y_1.$$

At point B, we must have

$$y_2 = A(\theta_2 - \theta_3)^2 + y_3, \qquad y_2 = m\theta_2 + b, \qquad m = 2A(\theta_2 - \theta_3).$$

From these,

$$A = \frac{y_2 - y_3}{(\theta_2 - \theta_3)^2}, \qquad m = 2\left[\frac{y_2 - y_3}{(\theta_2 - \theta_3)^2}\right](\theta_2 - \theta_3) = \frac{2(y_2 - y_3)}{\theta_2 - \theta_3},$$

and

$$b = y_3 + \frac{(y_2 - y_3)}{(\theta_2 - \theta_3)^2}(\theta_2 - \theta_3)^2 - \frac{2(y_2 - y_3)\theta_2}{\theta_2 - \theta_3} = \frac{y_2(\theta_2 - \theta_3) - 2(y_2 - y_3)\theta_2}{\theta_2 - \theta_3} = \frac{2y_3\theta_2 - y_2\theta_2 - y_2\theta_3}{\theta_2 - \theta_3}.$$

If we equate the values of b,

$$-y_1 = \frac{2y_3\theta_2 - y_2\theta_2 - y_2\theta_3}{\theta_2 - \theta_3} \quad \Longrightarrow \quad \theta_2 = \frac{(y_1 + y_2)\theta_3}{y_1 - y_2 + 2y_3}.$$

When we equate values of m, $\dfrac{2y_1}{\theta_1} = \dfrac{2(y_2 - y_3)}{\theta_2 - \theta_3},$ from which

$$\theta_1 = \frac{y_1(\theta_2 - \theta_3)}{y_2 - y_3} = \frac{y_1}{y_2 - y_3}\left[\frac{(y_1 + y_2)\theta_3}{y_1 - y_2 + 2y_3} - \theta_3\right] = \frac{y_1}{y_2 - y_3}\left[\frac{(y_1 + y_2)\theta_3 - \theta_3(y_1 - y_2 + 2y_3)}{y_1 - y_2 + 2y_3}\right]$$

$$= \frac{y_1}{y_2 - y_3}\left[\frac{2\theta_3(y_2 - y_3)}{y_1 - y_2 + 2y_3}\right] = \frac{2y_1\theta_3}{y_1 - y_2 + 2y_3}.$$

These now give

$$a = \frac{y_1}{\theta_1^2} = \frac{y_1(y_1 - y_2 + 2y_3)^2}{4y_1^2\theta_3^2} = \frac{(y_1 - y_2 + 2y_3)^2}{4y_1\theta_3^2},$$

$$m = \frac{2y_1}{\theta_1} = \frac{2y_1(y_1 - y_2 + 2y_3)}{2y_1\theta_3} = \frac{y_1 - y_2 + 2y_3}{\theta_3},$$

$$b = -y_1,$$

$$A = \frac{y_2 - y_3}{(\theta_2 - \theta_3)^2} = \frac{y_2 - y_3}{\left[\dfrac{(y_1 + y_2)\theta_3}{y_1 - y_2 + 2y_3} - \theta_3\right]^2} = \frac{(y_2 - y_3)(y_1 - y_2 + 2y_3)^2}{[(y_1 + y_2)\theta_3 - \theta_3(y_1 - y_2 + 2y_3)]^2}$$

$$= \frac{(y_2 - y_3)(y_1 - y_2 + 2y_3)^2}{4\theta_3^2(y_2 - y_3)^2} = \frac{(y_1 - y_2 + 2y_3)^2}{4\theta_3^2(y_2 - y_3)}.$$

EXERCISES 3.7

2. $\dfrac{dy}{dx} = \dfrac{dy}{du}\dfrac{du}{dx} = \left[\dfrac{(u+1)(1) - u(1)}{(u+1)^2}\right]\left(\dfrac{1}{2\sqrt{x}}\right) = \dfrac{1}{2\sqrt{x}(u+1)^2}$

4. $\dfrac{dy}{dx} = \dfrac{dy}{ds}\dfrac{ds}{dx} = \left[\dfrac{(s^2-2)(1) - s(2s)}{(s^2-2)^2}\right](2x-2) = \dfrac{2(1-x)(s^2+2)}{(s^2-2)^2}$

6. $\dfrac{dy}{dx} = \dfrac{dy}{dt}\dfrac{dt}{dx} = \left[\dfrac{(t-4)(1) - (t+3)(1)}{(t-4)^2}\right]\left[\dfrac{(x+1)(1) - (x-2)(1)}{(x+1)^2}\right] = \dfrac{-21}{(t-4)^2(x+1)^2}$

8. $\dfrac{dy}{dx} = \dfrac{dy}{du}\dfrac{du}{dx} = \left(2u + \dfrac{5}{2}u^{3/2}\right)\left[\dfrac{(x-x^2)(1) - (x+1)(1-2x)}{(x-x^2)^2}\right] = \dfrac{u\left(4 + 5\sqrt{u}\right)(x^2 + 2x - 1)}{2(x-x^2)^2}$

10. $f'(x) = (1)\sqrt{x+1} + x\left(\dfrac{1}{2\sqrt{x+1}}\right) = \dfrac{3x+2}{2\sqrt{x+1}}$

12. $f'(x) = \dfrac{\sqrt{2x+1}(1) - x\left(\dfrac{2}{2\sqrt{2x+1}}\right)}{2x+1} = \dfrac{x+1}{(2x+1)^{3/2}}$

14. $f'(x) = \dfrac{(3x+5)(2)(2x-1)(2) - (2x-1)^2(3)}{(3x+5)^2} = \dfrac{(2x-1)(6x+23)}{(3x+5)^2}$

16. $f'(x) = 3x^2(2-5x^2)^{1/3} + x^3(1/3)(2-5x^2)^{-2/3}(-10x)$

$= 3x^2(2-5x^2)^{1/3} - \dfrac{10x^4}{3(2-5x^2)^{2/3}} = \dfrac{x^2(18 - 55x^2)}{3(2-5x^2)^{2/3}}$

18. $f'(x) = (x+1)^2(3)(3x+1)^2(3) + 2(x+1)(3x+1)^3 = (3x+1)^2(x+1)[9(x+1) + 2(3x+1)]$

$= (3x+1)^2(x+1)(15x+11)$

20. $f'(x) = \dfrac{x^2(1/2)(2-3x)^{-1/2}(-3) - \sqrt{2-3x}(2x)}{x^4} = \dfrac{\dfrac{-3x^2}{2\sqrt{2-3x}} - 2x\sqrt{2-3x}}{x^4}$

$= \dfrac{-3x - 4(2-3x)}{2x^3\sqrt{2-3x}} = \dfrac{9x-8}{2x^3\sqrt{2-3x}}$

22. $f'(x) = \dfrac{1}{4}\left(\dfrac{2-x}{2+x}\right)^{-3/4}\left[\dfrac{(2+x)(-1) - (2-x)(1)}{(2+x)^2}\right] = \dfrac{1}{4}\left(\dfrac{2+x}{2-x}\right)^{3/4}\left[\dfrac{-4}{(2+x)^2}\right] = \dfrac{-1}{(2-x)^{3/4}(2+x)^{5/4}}$

24. $f'(x) = 4(x+5)^3\sqrt{1+x^3} + (x+5)^4(1/2)(1+x^3)^{-1/2}(3x^2)$

$= (x+5)^3\left[4\sqrt{1+x^3} + \dfrac{3x^2(x+5)}{2\sqrt{1+x^3}}\right] = (x+5)^3\left[\dfrac{8(1+x^3) + 3x^2(x+5)}{2\sqrt{1+x^3}}\right]$

$= \dfrac{(x+5)^3(11x^3 + 15x^2 + 8)}{2\sqrt{1+x^3}}$

26. With the product rule of Exercise 20 in Section 3.4,

$f'(x) = (1)(x+5)^4\sqrt{1+x^3} + x(4)(x+5)^3\sqrt{1+x^3} + x(x+5)^4(1/2)(1+x^3)^{-1/2}(3x^2)$

$= (x+5)^3\left[(x+5)\sqrt{1+x^3} + 4x\sqrt{1+x^3} + \dfrac{3x^3(x+5)}{2\sqrt{1+x^3}}\right]$

$= (x+5)^3\left[\dfrac{2(x+5)(1+x^3) + 8x(1+x^3) + 3x^3(x+5)}{2\sqrt{1+x^3}}\right] = \dfrac{(x+5)^3(13x^4 + 25x^3 + 10x + 10)}{2\sqrt{1+x^3}}$

28. $f'(x) = (1)\sqrt{1 + x\sqrt{1+x}} + x(1/2)\left(1 + x\sqrt{1+x}\right)^{-1/2}[\sqrt{1+x} + x(1/2)(1+x)^{-1/2}]$

$= \sqrt{1 + x\sqrt{1+x}} + \dfrac{x}{2\sqrt{1 + x\sqrt{1+x}}}\left[\dfrac{2(1+x) + x}{2\sqrt{1+x}}\right]$

$= \dfrac{4\sqrt{1+x}\left(1 + x\sqrt{1+x}\right) + 3x^2 + 2x}{4\sqrt{1+x}\sqrt{1 + x\sqrt{1+x}}} = \dfrac{7x^2 + 6x + 4\sqrt{1+x}}{4\sqrt{1+x}\sqrt{1 + x\sqrt{1+x}}}$

30. $\dfrac{dy}{dx} = \dfrac{dy}{ds}\dfrac{ds}{dx} = \left[\dfrac{1}{3}(2s - s^2)^{-2/3}(2 - 2s)\right]\left[\dfrac{-2x}{(x^2+5)^2}\right] = \dfrac{4x(s-1)}{3(2s-s^2)^{2/3}(x^2+5)^2}$

32. $\dfrac{dy}{dx} = \dfrac{dy}{du}\dfrac{du}{dx} = \left[\dfrac{(u+5)(1) - u(1)}{(u+5)^2}\right]\left[\dfrac{x(1/2)(x-1)^{-1/2} - \sqrt{x-1}(1)}{x^2}\right]$

$= \dfrac{5}{(u+5)^2}\left[\dfrac{x - 2(x-1)}{2x^2\sqrt{x-1}}\right] = \dfrac{5(2-x)}{2x^2\sqrt{x-1}(u+5)^2}$

34. $\dfrac{dy}{dx} = \dfrac{dy}{dt}\dfrac{dt}{dx} = \left[1 + \dfrac{1}{2\sqrt{t+\sqrt{t}}}\left(1 + \dfrac{1}{2\sqrt{t}}\right)\right]\left[\dfrac{(x^2-1)(2x) - (x^2+1)(2x)}{(x^2-1)^2}\right]$

$= \left[1 + \dfrac{2\sqrt{t}+1}{4\sqrt{t}\sqrt{t+\sqrt{t}}}\right]\left[\dfrac{-4x}{(x^2-1)^2}\right]$

36. $\dfrac{dy}{dx} = \dfrac{dy}{dk}\dfrac{dk}{dx} = \left[\dfrac{(1+k+k^2)(1/2)k^{-1/2} - \sqrt{k}(1+2k)}{(1+k+k^2)^2}\right]\left[(x^2+5)^5 + 5x(x^2+5)^4(2x)\right]$

$= \left[\dfrac{1 + k + k^2 - 2k(1+2k)}{2\sqrt{k}(1+k+k^2)^2}\right]\left[(x^2+5)^4(x^2+5+10x^2)\right]$

$= \dfrac{(1 - k - 3k^2)(x^2+5)^4(11x^2+5)}{2\sqrt{k}(1+k+k^2)^2}$

38. Since F is a function of r, and r is a function of t, the chain rule gives $\dfrac{dF}{dt} = \dfrac{dF}{dr}\dfrac{dr}{dt} = \dfrac{-2Qq}{4\pi\epsilon_0 r^3}\dfrac{dr}{dt}$.

Since $dr/dt = 2$ m/s, we obtain when $r = 2$, $\dfrac{dF}{dt} = -\dfrac{2(3\times 10^{-6})(5\times 10^{-6})}{4\pi(8.85\times 10^{-12})(2)^3}(2) = -0.067$ N/s.

40. If we use the chain rule on $y = f(x)$, $x = g(y)$, then $\dfrac{dy}{dy} = 1 = \dfrac{dy}{dx}\dfrac{dx}{dy} \implies \dfrac{dy}{dx} = \dfrac{1}{\dfrac{dx}{dy}}$.

42. The proof relies on the fact that $\Delta u \neq 0$. If u is a constant function, then $\Delta u \equiv 0$. In addition, even when u is not a constant function, we might have $\Delta u = 0$ for arbitrarily small Δx. A complete proof of the chain rule must account for both these possibilities.

44. Since $\dfrac{dy}{dx} = \dfrac{dy}{du}\dfrac{du}{dx} = \left[3(u+1)^2 + \dfrac{1}{u^2}\right]\left[1 + \dfrac{1}{2\sqrt{x+1}}\right]$, it follows that

$\dfrac{d^2y}{dx^2} = \dfrac{d}{dx}\left[3(u+1)^2 + \dfrac{1}{u^2}\right]\left[1 + \dfrac{1}{2\sqrt{x+1}}\right] + \left[3(u+1)^2 + \dfrac{1}{u^2}\right]\dfrac{d}{dx}\left[1 + \dfrac{1}{2\sqrt{x+1}}\right]$

$= \left[6(u+1) - \dfrac{2}{u^3}\right]\dfrac{du}{dx}\left[1 + \dfrac{1}{2\sqrt{x+1}}\right] + \left[3(u+1)^2 + \dfrac{1}{u^2}\right]\left[\dfrac{-1}{4(x+1)^{3/2}}\right]$

$= \left[\dfrac{6u^3(u+1) - 2}{u^3}\right]\left[1 + \dfrac{1}{2\sqrt{x+1}}\right]^2 - \dfrac{3u^2(u+1)^2 + 1}{4u^2(x+1)^{3/2}}$.

46. Since $\dfrac{dy}{dx} = \dfrac{dy}{ds}\dfrac{ds}{dx} = \left[\dfrac{(s+6)(1) - s(1)}{(s+6)^2}\right]\left[\dfrac{(1+\sqrt{x})(1/2)x^{-1/2} - \sqrt{x}(1/2)x^{-1/2}}{(1+\sqrt{x})^2}\right]$

$= \left[\dfrac{6}{(s+6)^2}\right]\left[\dfrac{1}{2\sqrt{x}(1+\sqrt{x})^2}\right],$

it follows that

$\dfrac{d^2y}{dx^2} = \dfrac{d}{dx}\left[\dfrac{3}{(s+6)^2}\right]\left[\dfrac{1}{\sqrt{x}(1+\sqrt{x})^2}\right] + \left[\dfrac{3}{(s+6)^2}\right]\dfrac{d}{dx}\left[\dfrac{1}{\sqrt{x}(1+\sqrt{x})^2}\right]$

$= \left[\dfrac{-6}{(s+6)^3}\dfrac{ds}{dx}\right]\left[\dfrac{1}{\sqrt{x}(1+\sqrt{x})^2}\right] + \left[\dfrac{3}{(s+6)^2}\right]\left[\dfrac{-1}{2x^{3/2}(1+\sqrt{x})^2} - \dfrac{2}{\sqrt{x}(1+\sqrt{x})^3 2\sqrt{x}}\right]$

$= \left[\dfrac{-6}{(s+6)^3\sqrt{x}(1+\sqrt{x})^2}\right]\left[\dfrac{1}{2\sqrt{x}(1+\sqrt{x})^2}\right] + \left[\dfrac{3}{(s+6)^2}\right]\left[\dfrac{-1 - \sqrt{x} - 2\sqrt{x}}{2x^{3/2}(1+\sqrt{x})^3}\right]$

$= \dfrac{-6\sqrt{x} - 3(s+6)(1+\sqrt{x})(1+3\sqrt{x})}{2x^{3/2}(s+6)^3(1+\sqrt{x})^4} = \dfrac{-3[2\sqrt{x} + (s+6)(1+4\sqrt{x}+3x)]}{2x^{3/2}(1+\sqrt{x})^4(s+6)^3}$.

48. If we set $u = 3 - 4x$, then the chain rule gives

$$\frac{d}{dx}[f(3-4x)]^2 = \frac{d}{du}[f(u)]^2\frac{du}{dx} = 2f(u)f'(u)(-4) = -8f(3-4x)f'(3-4x).$$

When $f(u) = u^3 - 2u$, then $f'(u) = 3u^2 - 2$, and

$$\frac{d}{dx}[f(3-4x)]^2 = -8[(3-4x)^3 - 2(3-4x)][3(3-4x)^2 - 2]$$
$$= 8(4x-3)(16x^2 - 24x + 7)(48x^2 - 72x + 25).$$

50. If we set $u = x + 1/x$, then

$$\frac{d}{dx}f\left(x + \frac{1}{x}\right) = f'(u)\frac{du}{dx} = f'(u)\left(1 - \frac{1}{x^2}\right) = \left(1 - \frac{1}{x^2}\right)f'\left(x + \frac{1}{x}\right).$$

When $f(u) = u^3 - 2u$, then $f'(u) = 3u^2 - 2$, and

$$\frac{d}{dx}f\left(x + \frac{1}{x}\right) = \left(1 - \frac{1}{x^2}\right)\left[3\left(x + \frac{1}{x}\right)^2 - 2\right] = \frac{(x^2-1)(3x^4 + 4x^2 + 3)}{x^4}.$$

52. If we set $u = 1 - 3x$, then

$$\frac{d}{dx}\sqrt{3 - 4[f(1-3x)]^2} = \frac{d}{du}\sqrt{3 - 4[f(u)]^2}\frac{du}{dx} = \frac{1}{2\sqrt{3-4[f(u)]^2}}[-8f(u)f'(u)](-3)$$

$$= \frac{12f(u)f'(u)}{\sqrt{3-4[f(u)]^2}} = \frac{12f(1-3x)f'(1-3x)}{\sqrt{3-4[f(1-3x)]^2}}.$$

When $f(u) = u^3 - 2u$, then $f'(u) = 3u^2 - 2$, and

$$\frac{d}{dx}\sqrt{3-4[f(1-3x)]^2} = \frac{12[(1-3x)^3 - 2(1-3x)][3(1-3x)^2 - 2]}{\sqrt{3 - 4[(1-3x)^3 - 2(1-3x)]^2}}$$

$$= \frac{12(1-3x)(9x^2 - 6x - 1)(27x^2 - 18x + 1)}{\sqrt{3 - 4(1-3x)^2(9x^2 - 6x - 1)^2}}.$$

54. If we set $u = x - f(x)$, then

$$\frac{d}{dx}\left[f\bigl(x - f(x)\bigr)\right] = f'(u)\frac{du}{dx} = f'(u)[1 - f'(x)] = f'\bigl(x - f(x)\bigr)[1 - f'(x)].$$

When $f(u) = u^3 - 2u$, then $f'(u) = 3u^2 - 2$, and

$$\frac{d}{dx}\left[f\bigl(x - f(x)\bigr)\right] = \{3[x - f(x)]^2 - 2\}[1 - (3x^2 - 2)] = \{3[x - (x^3 - 2x)]^2 - 2\}(3 - 3x^2)$$

$$= \{3[3x - x^3]^2 - 2\}(3 - 3x^2) = [27x^2 - 18x^4 + 3x^6 - 2](3 - 3x^2)$$

$$= 3(1 - x^2)(3x^6 - 18x^4 + 27x^2 - 2).$$

56. The slope of the tangent line at any point on any of the parabolas in the first family is

$$\frac{dy}{dx} = 2ax = 2\left(\frac{y}{x^2}\right)x = \frac{2y}{x}.$$

When we solve for y in the family of ellipses, $y = \pm(1/\sqrt{2})\sqrt{c^2 - x^2}$. The slope of the tangent line at any point on any ellipse is

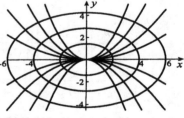

$$\frac{dy}{dx} = \pm\frac{1}{\sqrt{2}}\frac{-2x}{2\sqrt{c^2 - x^2}} = \mp\frac{x}{\sqrt{2}\sqrt{c^2 - x^2}} = \mp\frac{x}{\sqrt{2}(\pm\sqrt{2}y)} = -\frac{x}{2y}.$$

Since these slopes are negative reciprocals, the families are orthogonal.

58. When we solve for y in the family of ellipses, $y = \pm\frac{1}{\sqrt{3}}\sqrt{C^2 - 2x^2}$. The slope of the tangent line at any point on any ellipse is

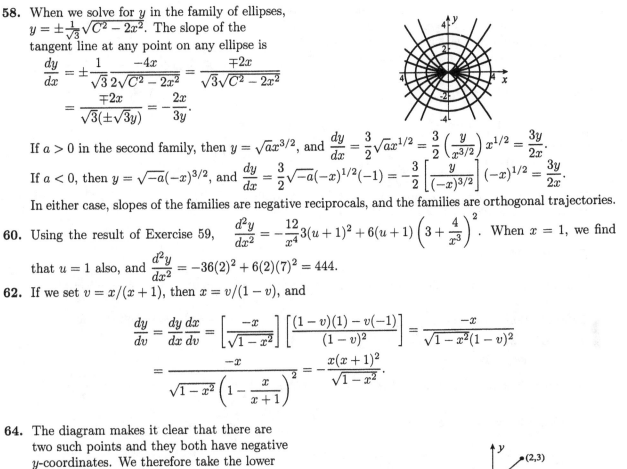

$$\frac{dy}{dx} = \pm\frac{1}{\sqrt{3}}\frac{-4x}{2\sqrt{C^2 - 2x^2}} = \frac{\mp 2x}{\sqrt{3}\sqrt{C^2 - 2x^2}}$$
$$= \frac{\mp 2x}{\sqrt{3}(\pm\sqrt{3}y)} = -\frac{2x}{3y}.$$

If $a > 0$ in the second family, then $y = \sqrt{a}x^{3/2}$, and $\dfrac{dy}{dx} = \dfrac{3}{2}\sqrt{a}x^{1/2} = \dfrac{3}{2}\left(\dfrac{y}{x^{3/2}}\right)x^{1/2} = \dfrac{3y}{2x}$.

If $a < 0$, then $y = \sqrt{-a}(-x)^{3/2}$, and $\dfrac{dy}{dx} = \dfrac{3}{2}\sqrt{-a}(-x)^{1/2}(-1) = -\dfrac{3}{2}\left[\dfrac{y}{(-x)^{3/2}}\right](-x)^{1/2} = \dfrac{3y}{2x}$.

In either case, slopes of the families are negative reciprocals, and the families are orthogonal trajectories.

60. Using the result of Exercise 59, $\dfrac{d^2y}{dx^2} = -\dfrac{12}{x^4}3(u+1)^2 + 6(u+1)\left(3 + \dfrac{4}{x^3}\right)^2$. When $x = 1$, we find

that $u = 1$ also, and $\dfrac{d^2y}{dx^2} = -36(2)^2 + 6(2)(7)^2 = 444$.

62. If we set $v = x/(x+1)$, then $x = v/(1-v)$, and

$$\frac{dy}{dv} = \frac{dy}{dx}\frac{dx}{dv} = \left[\frac{-x}{\sqrt{1-x^2}}\right]\left[\frac{(1-v)(1) - v(-1)}{(1-v)^2}\right] = \frac{-x}{\sqrt{1-x^2}(1-v)^2}$$

$$= \frac{-x}{\sqrt{1-x^2}\left(1 - \dfrac{x}{x+1}\right)^2} = -\frac{x(x+1)^2}{\sqrt{1-x^2}}.$$

64. The diagram makes it clear that there are two such points and they both have negative y-coordinates. We therefore take the lower half of the hyperbola in the form $y = f(x) = -\sqrt{x^2 - 16}/4$. The slope of the tangent line to the hyperbola at $P(a, b)$ is

$$f'(a) = \frac{-x}{4\sqrt{x^2 - 16}}\Big|_{x=a} = \frac{-a}{4\sqrt{a^2 - 16}}.$$

Since the tangent line must pass through the point $(2, 3)$ the slope of the tangent line is also given by $(b - 3)/(a - 2)$. Consequently,

$$\frac{b-3}{a-2} = \frac{-a}{4\sqrt{a^2 - 16}}.$$

We combine this with $b = -\dfrac{1}{4}\sqrt{a^2 - 16}$, since $P(a, b)$ is on the hyperbola. Substitution from the second into the first gives

$$-\frac{1}{4}\sqrt{a^2 - 16} - 3 = \frac{-a(a-2)}{4\sqrt{a^2 - 16}}, \qquad \text{or,} \qquad (a^2 - 16) + 12\sqrt{a^2 - 16} = a(a-2).$$

This equation simplifies to $6\sqrt{a^2 - 16} = 8 - a$, and squaring leads to the quadratic $35a^2 + 16a - 640 = 0$. The two solutions are $a = (-8 \pm 24\sqrt{39})/35$. The y-coordinates of these points are $(-12 \pm \sqrt{39})/35$.

66. The conchoid is shown to the right. There are two points
P and Q at which the tangent line is horizontal. To find
P the x-coordinate of which is between -2 and -1,
we first solve the equation $x^2y^2 = (x+1)^2(4-x^2)$ for
$y = [(x+1)/x]\sqrt{4-x^2}$, and set the
derivative equal to zero,

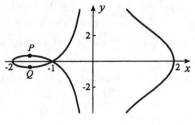

$$0 = \frac{d}{dx}\left[\left(1+\frac{1}{x}\right)\sqrt{4-x^2}\right] = \frac{-1}{x^2}\sqrt{4-x^2} + \left(1+\frac{1}{x}\right)\frac{-x}{\sqrt{4-x^2}}$$

$$= \frac{-(4-x^2) - x^2(x+1)}{x^2\sqrt{4-x^2}} = \frac{-(x^3+4)}{x^2\sqrt{4-x^2}}.$$

The only solution of this equation is $x = -4^{1/3}$. The coordinates of P are therefore $(-4^{1/3}, (4^{1/3}-1)^{3/2})$.
Point Q has coordinates $(-4^{1/3}, -(4^{1/3}-1)^{3/2})$.

EXERCISES 3.8

2. If we differentiate with respect to x, we find $4x^3 + 2y\dfrac{dy}{dx} + 3y^2\dfrac{dy}{dx} = 0 \implies \dfrac{dy}{dx} = \dfrac{-4x^3}{3y^2+2y}$.

4. Differentiation with respect to x gives $6x^2 - 3y^4 - 12xy^3\dfrac{dy}{dx} + 5y + 5x\dfrac{dy}{dx} = 0$, and therefore
$\dfrac{dy}{dx} = \dfrac{6x^2 - 3y^4 + 5y}{12xy^3 - 5x}$.

6. When we differentiate with respect to x, we obtain $2(x+y)\left(1+\dfrac{dy}{dx}\right) = 2 \implies \dfrac{dy}{dx} = \dfrac{1-x-y}{x+y}$.

8. Differentiation with respect to x gives $\dfrac{1}{2\sqrt{x+y}}\left(1+\dfrac{dy}{dx}\right) + 2y\dfrac{dy}{dx} = 24x + \dfrac{dy}{dx}$, from which

$$\left(\frac{1}{2\sqrt{x+y}} + 2y - 1\right)\frac{dy}{dx} = 24x - \frac{1}{2\sqrt{x+y}}. \text{ Thus,}$$

$$\frac{dy}{dx} = \left(\frac{48x\sqrt{x+y}-1}{2\sqrt{x+y}}\right)\left(\frac{2\sqrt{x+y}}{1+2(2y-1)\sqrt{x+y}}\right) = \frac{48x\sqrt{x+y}-1}{1+2(2y-1)\sqrt{x+y}}.$$

10. If we write $x^2 - xy - y^2 = 4x(x+y)$, or, $3x^2 + 5xy + y^2 = 0$, and differentiate with respect to x, we
obtain $6x + 5y + 5x\dfrac{dy}{dx} + 2y\dfrac{dy}{dx} = 0$. Therefore, $\dfrac{dy}{dx} = -\dfrac{6x+5y}{5x+2y}$.

12. When we differentiate with respect to x, we find $2x + 3y^2\dfrac{dy}{dx} + \dfrac{dy}{dx} = 0$, and therefore
$\dfrac{dy}{dx} = -\dfrac{2x}{3y^2+1}$. Differentiation of this equation gives

$$\frac{d^2y}{dx^2} = -\frac{(3y^2+1)(2) - 2x\left(6y\dfrac{dy}{dx}\right)}{(3y^2+1)^2} = -\frac{2(3y^2+1) + 12xy\left(\dfrac{2x}{3y^2+1}\right)}{(3y^2+1)^2} = -\frac{2(3y^2+1)^2 + 24x^2y}{(3y^2+1)^3}.$$

14. If we differentiate with respect to x, we obtain $2y\dfrac{dy}{dx} + 2\dfrac{dy}{dx} = 5$, and this equation can be solved for
$\dfrac{dy}{dx} = \dfrac{5}{2(y+1)}$. A second differentiation gives

$$\frac{d^2y}{dx^2} = \frac{-5}{2(y+1)^2}\frac{dy}{dx} = \frac{-5}{2(y+1)^2}\left[\frac{5}{2(y+1)}\right] = \frac{-25}{4(y+1)^3}.$$

16. If we differentiate with respect to x, we obtain $3x^2y + x^3\dfrac{dy}{dx} + y^3 + 3xy^2\dfrac{dy}{dx} = 0$. Therefore,

$\dfrac{dy}{dx} = -\dfrac{3x^2y + y^3}{x^3 + 3xy^2}$. When $x = 1$, we have $y + y^3 = 2$, and the only solution of this equation is $y = 1$. Thus, $\dfrac{dy}{dx}\Big|_{x=1} = -\dfrac{3+1}{1+3} = -1$.

18. Differentiation with respect to x leads to $2xy^3 + 3x^2y^2\dfrac{dy}{dx} + 2 + 4\dfrac{dy}{dx} = 0$, from which $\dfrac{dy}{dx} = -\dfrac{2 + 2xy^3}{3x^2y^2 + 4}$.

A second differentiation gives

$$\frac{d^2y}{dx^2} = -\frac{(3x^2y^2 + 4)\left(2y^3 + 6xy^2\dfrac{dy}{dx}\right) - (2 + 2xy^3)\left(6xy^2 + 6x^2y\dfrac{dy}{dx}\right)}{(3x^2y^2 + 4)^2}$$

$$= -\frac{(3x^2y^2 + 4)\left[2y^3 - 6xy^2\left(\dfrac{2 + 2xy^3}{3x^2y^2 + 4}\right)\right] - (2 + 2xy^3)\left[6xy^2 - 6x^2y\left(\dfrac{2 + 2xy^3}{3x^2y^2 + 4}\right)\right]}{(3x^2y^2 + 4)^2}$$

$$= \frac{-2y^3(3x^2y^2 + 4)^2 + 12xy^2(2 + 2xy^3)(3x^2y^2 + 4) - 6x^2y(2 + 2xy^3)^2}{(3x^2y^2 + 4)^3}$$

20. Differentiation with respect to x gives $1 = \dfrac{dy}{dx}\sqrt{1-y^2} + y\dfrac{-y}{\sqrt{1-y^2}}\dfrac{dy}{dx} = \dfrac{1 - y^2 - y^2}{\sqrt{1-y^2}}\dfrac{dy}{dx} = \dfrac{1 - 2y^2}{\sqrt{1-y^2}}\dfrac{dy}{dx}$.

Thus, $\dfrac{dy}{dx} = \dfrac{\sqrt{1-y^2}}{1 - 2y^2}$. Another differentiation gives

$$\frac{d^2y}{dx^2} = \frac{(1 - 2y^2)(1/2)(1 - y^2)^{-1/2}\left(-2y\dfrac{dy}{dx}\right) - \sqrt{1-y^2}\left(-4y\dfrac{dy}{dx}\right)}{(1 - 2y^2)^2}$$

$$= \frac{-y(1 - 2y^2) + 4y(1 - y^2)}{\sqrt{1-y^2}(1 - 2y^2)^2}\frac{dy}{dx} = \frac{3y - 2y^3}{\sqrt{1-y^2}(1 - 2y^2)^2}\frac{\sqrt{1-y^2}}{1 - 2y^2} = \frac{3y - 2y^3}{(1 - 2y^2)^3}.$$

22. When we differentiate with respect to x, we obtain $2xy^3 + 3x^2y^2\dfrac{dy}{dx} + y + x\dfrac{dy}{dx} = 0$, and therefore

$\dfrac{dy}{dx} = -\dfrac{2xy^3 + y}{3x^2y^2 + x}$. When $x = 1$, we find that $y^3 + y = 2$, the only solution of which is $y = 1$. Thus, $\dfrac{dy}{dx}\Big|_{x=1} = -\dfrac{2+1}{3+1} = -\dfrac{3}{4}$.

24. If we differentiate with respect to x, we find $2x + 2y + 2x\dfrac{dy}{dx} + 6y\dfrac{dy}{dx} = 0$. Therefore,

$\dfrac{dy}{dx} = -\dfrac{2x + 2y}{2x + 6y} = -\dfrac{x + y}{x + 3y}$. When $y = 1$, we find that $x^2 + 2x + 1 = 0$, and the only solution of this equation is $x = -1$. Thus, $dy/dx|_{x=-1} = -(-1+1)/(-1+3) = 0$. A second differentiation gives

$$\frac{d^2y}{dx^2} = -\frac{(x + 3y)\left(1 + \dfrac{dy}{dx}\right) - (x + y)\left(1 + 3\dfrac{dy}{dx}\right)}{(x + 3y)^2},$$

and when we substitute $x = -1$, $y = 1$, and $dy/dx = 0$, $\dfrac{d^2y}{dx^2}\Big|_{x=-1} = -\dfrac{(-1+3)(1) - (-1+1)(1)}{(-1+3)^2} = -\dfrac{1}{2}$.

26. If we differentiate with respect to x, then $2x + \dfrac{2}{3}y^{-1/3}\dfrac{dy}{dx} = 0 \implies \dfrac{dy}{dx} = -3xy^{1/3}$. Since

$$\frac{d^2y}{dx^2} = -3y^{1/3} - xy^{-2/3}\frac{dy}{dx} = -3y^{1/3} - xy^{-2/3}(-3xy^{1/3}) = -3y^{1/3} + 3x^2y^{-1/3} = 3y^{-1/3}(-y^{2/3} + x^2),$$

the second derivative vanishes if $x^2 = y^{2/3}$. When we substitute this into $x^2 + y^{2/3} = 2$, we obtain $x^2 + x^2 = 2$, and therefore $x = \pm 1$. The corresponding y-values are $y = \pm 1$, and the required points are $(1, \pm 1)$ and $(-1, \pm 1)$.

28. (a) Since $\dfrac{dy}{dx} = \dfrac{(x+2)(2x+1) - (x^2+x)(1)}{(x+2)^2} = \dfrac{x^2 + 4x + 2}{(x+2)^2}$, it follows that

$$\frac{Ey}{Ex} = \frac{x}{\dfrac{x(x+1)}{x+2}}\,\frac{x^2+4x+2}{(x+2)^2} = \frac{x^2+4x+2}{(x+1)(x+2)}.$$

(b) If we differentiate with respect to x, $1 = \dfrac{(3-y)\left(400\dfrac{dy}{dx}\right) - (400y+200)\left(-\dfrac{dy}{dx}\right)}{(3-y)^2} = \dfrac{1400\dfrac{dy}{dx}}{(3-y)^2}$, from

which $\dfrac{dy}{dx} = \dfrac{(3-y)^2}{1400}$. Thus, $\dfrac{Ey}{Ex} = \left[\dfrac{400y+200}{y(3-y)}\right]\dfrac{(3-y)^2}{1400} = \dfrac{(2y+1)(3-y)}{7y}$.

30. If we differentiate $2x^2 + 3y^2 = 14$ with respect to x, $4x + 6y\dfrac{dy}{dx} = 0$.
The slope of the tangent line at P is therefore
$$\frac{dy}{dx}\Big|_{(a,b)} = -\frac{2a}{3b}.$$
Since the slope of AP is $(b-5)/(a-2)$, and this
line is perpendicular to the tangent line at P,
$$-\frac{2a}{3b} = -\frac{a-2}{b-5}.$$

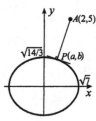

When this equation is solved for b, the result is $b = 10a/(6-a)$. Because P is on the ellipse, we may write that

$$14 = 2a^2 + 3b^2 = 2a^2 + 3\left(\frac{10a}{6-a}\right)^2 \implies (14 - 2a^2)(6-a)^2 = 300a^2.$$

Our diagram makes it clear that there is one and only one point in the first quadrant which satisfies the requirements, and $a = 1$ is a solution of the above equation. Thus, the required point is $(1, 2)$.

32. If we differentiate the equation of the circle with respect to x,

$$2(x-h) + 2(y-k)\frac{dy}{dx} = 0 \implies \frac{dy}{dx} = -\frac{x-h}{y-k}.$$

Thus, $\dfrac{d^2y}{dx^2} = -\dfrac{(y-k)(1) - (x-h)dy/dx}{(y-k)^2} = \dfrac{-1}{(y-k)^2}\left[(y-k) + (x-h)\left(\dfrac{x-h}{y-k}\right)\right]$

$\qquad = \dfrac{-1}{(y-k)^3}\left[(y-k)^2 + (x-h)^2\right] = \dfrac{-r^2}{(y-k)^3}.$

We now calculate that $\left|\dfrac{d^2y/dx^2}{[1 + (dy/dx)^2]^{3/2}}\right| = \left|\dfrac{-r^2/(y-k)^3}{\left[1 + \left(-\dfrac{x-h}{y-k}\right)^2\right]^{3/2}}\right| = \left|\dfrac{r^2/(y-k)^3}{\left[\dfrac{(y-k)^2 + (x-h)^2}{(y-k)^2}\right]^{3/2}}\right|$

$$= \left| \frac{r^2/(y-k)^3}{\left[\frac{r^2}{(y-k)^2} \right]^{3/2}} \right| = \frac{r^2}{|y-k|^3} \frac{|y-k|^3}{r^3} = \frac{1}{r}.$$

34. The product rule gives $R'(x) = r(x) + xr'(x)$. To obtain $r'(x)$ we differentiate $x = 4a^3 - 3ar^2 + r^3$ with respect to x,

$$1 = -6ar\frac{dr}{dx} + 3r^2\frac{dr}{dx}.$$

Thus, $\dfrac{dr}{dx} = \dfrac{1}{3r^2 - 6ar}$, and $R'(x) = r + \dfrac{x}{3r^2 - 6ar} = \dfrac{3r^3 - 6ar^2 + x}{3r^2 - 6ar}$.

36. (a) When we differentiate with respect to x, we obtain $\sqrt{1+2y} + \dfrac{x}{\sqrt{1+2y}}\dfrac{dy}{dx} = 2x - \dfrac{dy}{dx}$, and therefore

$\dfrac{dy}{dx} = \dfrac{2x - \sqrt{1+2y}}{\dfrac{x}{\sqrt{1+2y}} + 1}$. Since $y = 0$ when $x = 0$, we obtain $f'(0) = -1/1 = -1$.

(b) When the equation is squared, $x^2(1+2y) = x^4 - 2x^2y + y^2 \implies x^2 + 4x^2y = x^4 + y^2$. If we differentiate this equation with respect to x,

$$2x + 8xy + 4x^2\frac{dy}{dx} = 4x^3 + 2y\frac{dy}{dx} \quad \implies \quad \frac{dy}{dx} = \frac{2x + 8xy - 4x^3}{2y - 4x^2} = \frac{x + 4xy - 2x^3}{y - 2x^2}.$$

In this case we cannot simply set $x = 0$ and $y = 0$ to obtain $f'(0)$. To see why $x^2 + 4x^2y = x^4 + y^2$ cannot be used to find $f'(0)$, we write the equation in the form $y^2 - 4x^2y + x^4 - x^2 = 0$. This is a quadratic equation in y with solutions

$$y = \frac{4x^2 \pm \sqrt{16x^4 - 4(x^4 - x^2)}}{2} = 2x^2 \pm \sqrt{3x^4 + x^2}.$$

Thus, the equation $x^2 + 4x^2y = x^4 + y^2$ does not define y as a function of x; there are two values of y satisfying the equation for each value of x. We can also see this graphically. A plot of $x\sqrt{1+2y} = x^2 - y$ is shown in the left figure below. It defines y as a function of x near $x = 0$. A plot of $x^2 + 4x^2y = x^4 + y^2$ in the right figure shows that it does not define y as a function of x near $x = 0$.

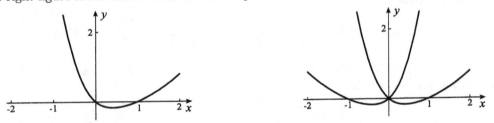

38. The slope of the tangent line at any point on any of the parabolas in the first family is $dy/dx = 2ax = 2x(y/x^2) = 2y/x$. To find the slope of the tangent line at any point on any of the ellipses, we differentiate with respect to x, getting $2x + 4y(dy/dx) = 0 \implies dy/dx = -x/(2y)$. Since this is the negative reciprocal of the slope of the parabola, the families are orthogonal trajectories.

40. Differentiation of $2x^2 + 3y^2 = C^2$ with respect to x gives $4x + 6y(dy/dx) = 0 \implies dy/dx = -2x/(3y)$.
Differentiation of $y^2 = ax^3$ gives $2y\dfrac{dy}{dx} = 3ax^2 \implies \dfrac{dy}{dx} = \dfrac{3ax^2}{2y} = \dfrac{3x^2(y^2/x^3)}{2y} = \dfrac{3y}{2x}$. Since these slopes are negative reciprocals, the families are orthogonal trajectories.

42. (a) If we differentiate with respect to x, we obtain $\dfrac{1}{2\sqrt{x}} + \dfrac{1}{2\sqrt{y}}\dfrac{dy}{dx} = 0$, from which $\dfrac{dy}{dx} = -\dfrac{\sqrt{y}}{\sqrt{x}}$.

(b) When $x \geq 0$ and $y \geq 0$, the equation of the curve reduces to $\sqrt{x} + \sqrt{y} = 1$. According to the formula in part (a),

$$\lim_{x \to 1^-} \frac{dy}{dx} = -\frac{0}{1} = 0^-, \quad \text{and} \quad \lim_{x \to 0^+} \frac{dy}{dx} = -\infty.$$

This enables us to sketch the first quadrant part of the curve. The remaining parts are obtained by symmetry.

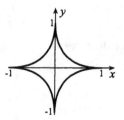

44. The chain rule gives

$$\frac{dy}{dx} = \frac{dy}{du}\frac{du}{dx} = \left[\frac{\sqrt{u^2-1}(1) - u(1/2)(u^2-1)^{-1/2}(2u)}{u^2-1}\right]\frac{du}{dx} = \frac{u^2-1-u^2}{(u^2-1)^{3/2}}\frac{du}{dx} = \frac{-1}{(u^2-1)^{3/2}}\frac{du}{dx}.$$

To obtain du/dx we differentiate $x^2u^2 + \sqrt{u^2-1} = 4$ with respect to x,

$$2xu^2 + 2x^2u\frac{du}{dx} + \frac{u}{\sqrt{u^2-1}}\frac{du}{dx} = 0 \quad \Longrightarrow \quad \frac{du}{dx} = \frac{-2xu^2}{2x^2u + \frac{u}{\sqrt{u^2-1}}} = \frac{-2xu\sqrt{u^2-1}}{1 + 2x^2\sqrt{u^2-1}}.$$

Thus, $\dfrac{dy}{dx} = \left[\dfrac{-1}{(u^2-1)^{3/2}}\right]\left[\dfrac{-2xu\sqrt{u^2-1}}{1 + 2x^2\sqrt{u^2-1}}\right] = \dfrac{2xu}{(u^2-1)\left(1 + 2x^2\sqrt{u^2-1}\right)}.$

46. If we differentiate the equation with respect to x,

$$\frac{2xy^3 - 3x^2y^2\dfrac{dy}{dx}}{y^6} - 1 = 0 \quad \Longrightarrow \quad \frac{dy}{dx} = \frac{2xy^3 - y^6}{3x^2y^2} = \frac{2xy - y^4}{3x^2}.$$

Consequently, $3x^2\dfrac{dy}{dx} + y^4 = 3x^2\left(\dfrac{2xy - y^4}{3x^2}\right) + y^4 = 2xy.$

48. If $n = a/b$, where a and b are integers, then when $y = x^n = x^{a/b}$, we may set $y^b = x^a$. Differentiation with respect to x gives

$$by^{b-1}\frac{dy}{dx} = ax^{a-1} \quad \Longrightarrow \quad \frac{dy}{dx} = \frac{a}{b}\frac{x^{a-1}}{y^{b-1}} = \frac{a}{b}\frac{x^{a-1}}{(x^{a/b})^{b-1}} = \frac{a}{b}x^{a-1-a+a/b} = nx^{n-1}.$$

50. If we express the first family in the form $y^2(a - x) = x^3$, we may solve for $a = x + x^3/y^2$. Differentiation of this equation with respect to x gives

$$0 = 1 + \frac{3y^2x^2 - 2x^3y\dfrac{dy}{dx}}{y^4} \quad \Longrightarrow \quad \frac{dy}{dx} = \frac{3x^2y + y^3}{2x^3}.$$

When we solve the second family for $b = (x^2 + y^2)^2/(2x^2 + y^2)$, and differentiate

$$0 = \frac{(2x^2 + y^2)(2)(x^2 + y^2)\left(2x + 2y\dfrac{dy}{dx}\right) - (x^2 + y^2)^2\left(4x + 2y\dfrac{dy}{dx}\right)}{(2x^2 + y^2)^2}.$$

This implies that $4(2x^2 + y^2)\left(x + y\dfrac{dy}{dx}\right) - 2(x^2 + y^2)\left(2x + y\dfrac{dy}{dx}\right) = 0$, and we can solve for

$$\frac{dy}{dx} = \frac{4x(2x^2 + y^2) - 4x(x^2 + y^2)}{2y(x^2 + y^2) - 4y(2x^2 + y^2)} = \frac{-2x^3}{3x^2y + y^3}.$$

Since these expressions for dy/dx are negative reciprocals, the families are orthogonal trajectories.

52. If we differentiate the equation with respect to x, $3x^2 + 3y^2(dy/dx) = 3ay + 3ax(dy/dx)$. When we solve this for dy/dx and equate the result to -1,

$$-1 = \frac{dy}{dx} = \frac{ay - x^2}{y^2 - ax} \implies a(y - x) = x^2 - y^2 = (x - y)(x + y).$$

Thus, $y = x$ or $x + y = -a$. When $y = x$, the equation of the folium implies that $x^3 + x^3 = 3ax^2 \implies x = 3a/2$. A contradiction is obtained when we put $y = -a - x$ into the equation of the folium. Thus, the only point at which the slope of the tangent line is equal to -1 is $(3a/2, 3a/2)$.

EXERCISES 3.9

2. $\dfrac{dy}{dx} = -\sin x - 4\cos 5x\,(5) = -\sin x - 20\cos 5x$ **4.** $\dfrac{dy}{dx} = -3\tan^{-4} 3x\,\sec^2 3x\,(3) = -\dfrac{9\sec^2 3x}{\tan^4 3x}$

6. $\dfrac{dy}{dx} = -\csc(4 - 2x)\cot(4 - 2x)\,(-2) = 2\csc(4 - 2x)\cot(4 - 2x)$

8. $\dfrac{dy}{dx} = \cot x^2 - x\csc^2 x^2\,(2x) = \cot x^2 - 2x^2\csc^2 x^2$

10. $\dfrac{dy}{dx} = \dfrac{(x + 1)(\sin x + x\cos x) - x\sin x\,(1)}{(x + 1)^2} = \dfrac{\sin x + x(1 + x)\cos x}{(x + 1)^2}$

12. Since $y = \dfrac{1}{2}\sin 4x$, it follows that $\dfrac{dy}{dx} = \dfrac{1}{2}\cos 4x\,(4) = 2\cos 4x$.

14. $\dfrac{dy}{dx} = \dfrac{1}{4}(1 + \tan^3 x)^{-3/4}(3\tan^2 x\,\sec^2 x) = \dfrac{3\tan^2 x\,\sec^2 x}{4(1 + \tan^3 x)^{3/4}}$

16. Differentiation with respect to x gives $\cos y - x\sin y\dfrac{dy}{dx} + y\sin x - \cos x\dfrac{dy}{dx} = 0$, and therefore

$$\frac{dy}{dx} = \frac{\cos y + y\sin x}{\cos x + x\sin y}.$$

18. If we differentiate with respect to x, we find $\sec^2(x + y)\left(1 + \dfrac{dy}{dx}\right) = \dfrac{dy}{dx}$, and therefore

$$\frac{dy}{dx} = \frac{\sec^2(x + y)}{1 - \sec^2(x + y)}.$$

20. Differentiation with respect to x gives $3x^2 y + x^3\dfrac{dy}{dx} + 2\tan y\,\sec^2 y\dfrac{dy}{dx} = 3$, and therefore

$$\frac{dy}{dx} = \frac{3(1 - x^2 y)}{x^3 + 2\tan y\,\sec^2 y}.$$

22. $\dfrac{dy}{dx} = -\sin(\tan x)\sec^2 x = -\sec^2 x\,\sin(\tan x)$

24. Since $y = (\sin^2 x^2 + \cos^2 x^2)(\sin^2 x^2 - \cos^2 x^2) = -(\cos^2 x^2 - \sin^2 x^2) = -\cos 2x^2$, we find that $dy/dx = \sin 2x^2\,(4x) = 4x\sin 2x^2$.

26. $\dfrac{dy}{dx} = \dfrac{dy}{dv}\dfrac{dv}{dx} = \left(\dfrac{-\sec v\,\tan v}{2\sqrt{3 - \sec v}}\right)\left(\sec^2\sqrt{x}\,\dfrac{1}{2\sqrt{x}}\right) = -\dfrac{\sec v\,\tan v\,\sec^2\sqrt{x}}{4\sqrt{x}\sqrt{3 - \sec v}}$.

28. $\dfrac{dy}{dx} = \dfrac{dy}{du}\dfrac{du}{dx} = \left[\dfrac{1}{3}(1 + \sec^3 u)^{-2/3}3\sec^2 u\,\sec u\,\tan u\right]\left[\dfrac{1}{2}(1 + \cos x^2)^{-1/2}(-\sin x^2\,(2x))\right]$

$$= -\frac{x\sin x^2\,\sec^3 u\,\tan u}{\sqrt{1 + \cos x^2}(1 + \sec^3 u)^{2/3}}$$

30. $\dfrac{dy}{dx} = \dfrac{x^2\sin x[3\tan^2(3x^2 - 4)\sec^2(3x^2 - 4)\,(6x)] - [1 + \tan^3(3x^2 - 4)][2x\sin x + x^2\cos x]}{x^4\sin^2 x}$

$$= \frac{18x^2\tan^2(3x^2 - 4)\sec^2(3x^2 - 4) - (2 + x\cot x)[1 + \tan^3(3x^2 - 4)]}{x^3\sin x}$$

32. If we differentiate with respect to x, we obtain $\sec^2 y \dfrac{dy}{dx} = 1 + x\dfrac{dy}{dx} + y \implies \dfrac{dy}{dx} = \dfrac{1+y}{\sec^2 y - x}$. A second differentiation now gives

$$\frac{d^2y}{dx^2} = \frac{(\sec^2 y - x)(dy/dx) - (1+y)[2\sec^2 y \tan y \,(dy/dx) - 1]}{(\sec^2 y - x)^2}$$

$$= \frac{\left[\sec^2 y - x - 2(1+y)\sec^2 y \tan y\right]\left(\dfrac{1+y}{\sec^2 y - x}\right) + (1+y)}{(\sec^2 y - x)^2}$$

$$= \frac{(1+y)[(\sec^2 y - x) - 2(1+y)\sec^2 y \tan y + (\sec^2 y - x)]}{(\sec^2 y - x)^3}$$

$$= \frac{2(1+y)[\sec^2 y - x - (1+y)\sec^2 y \tan y]}{(\sec^2 y - x)^3}.$$

34. $\displaystyle\lim_{x\to 0} \frac{1 - \cos x}{x} = \lim_{x\to 0} \frac{1 - [1 - 2\sin^2(x/2)]}{x} = \lim_{x\to 0}\left[\frac{\sin(x/2)}{x/2}\sin(x/2)\right] = (1)(0) = 0$

36. $\displaystyle\lim_{x\to\infty} \frac{\sin(2/x)}{\sin(1/x)} = \lim_{x\to\infty} \frac{2\sin(1/x)\cos(1/x)}{\sin(1/x)} = \lim_{x\to\infty}[2\cos(1/x)] = 2$

38. $\displaystyle\lim_{x\to\pi/2} \frac{\cos x}{(x - \pi/2)^2} = \lim_{x\to\pi/2}\frac{\sin(\pi/2 - x)}{(x - \pi/2)^2} = \lim_{x\to\pi/2}\left[\frac{\sin(\pi/2 - x)}{\pi/2 - x}\frac{1}{\pi/2 - x}\right]$

Since $\displaystyle\lim_{x\to\pi/2}\frac{\sin(\pi/2 - x)}{\pi/2 - x} = 1$ and $\displaystyle\lim_{x\to\pi/2}\frac{1}{\pi/2 - x}$ does not exist, the original limit does not exist.

40. For $x \neq 0$, $g'(x) = \sin\left(\dfrac{1}{x}\right) + x\cos\left(\dfrac{1}{x}\right)\left(\dfrac{-1}{x^2}\right) = \sin\left(\dfrac{1}{x}\right) - \dfrac{1}{x}\cos\left(\dfrac{1}{x}\right)$. The limit of this function as x approaches 0 does not exist.

42. Since $\dfrac{dy}{dt} = A\cos\left(\sqrt{\dfrac{k}{m}}t\right)\sqrt{\dfrac{k}{m}} - B\sin\left(\sqrt{\dfrac{k}{m}}t\right)\sqrt{\dfrac{k}{m}} = \sqrt{\dfrac{k}{m}}\left[A\cos\left(\sqrt{\dfrac{k}{m}}t\right) - B\sin\left(\sqrt{\dfrac{k}{m}}t\right)\right]$,
a second differentiation gives

$$\frac{d^2y}{dt^2} = \sqrt{\frac{k}{m}}\left[-A\sin\left(\sqrt{\frac{k}{m}}t\right)\sqrt{\frac{k}{m}} - B\cos\left(\sqrt{\frac{k}{m}}t\right)\sqrt{\frac{k}{m}}\right] = -\frac{k}{m}\left[A\sin\left(\sqrt{\frac{k}{m}}t\right) + B\cos\left(\sqrt{\frac{k}{m}}t\right)\right].$$

Hence,

$$m\frac{d^2y}{dt^2} + ky = -k\left[A\sin\left(\sqrt{\frac{k}{m}}t\right) + B\cos\left(\sqrt{\frac{k}{m}}t\right)\right] + k\left[A\sin\left(\sqrt{\frac{k}{m}}t\right) + B\cos\left(\sqrt{\frac{k}{m}}t\right)\right] = 0.$$

44. To get an idea of how many values of x satisfy $f'(x) = 0$ and approximations to them, we plot a graph of $f(x)$. The plot on the interval $0.1 \leq x \leq 20$ in the left figure below suggests regular behaviour of the function for large values of x, but wild oscillations as $x \to 0$. This is consistent with the fact that for large x, the term $1/x$ becomes less and less significant, and $f(x)$ can be approximated by $\cos x$. On the other hand, when x is close to 0, $f(x)$ should behave much like $\sin(1/x)$ in Figure 2.8b. This is illustrated in the right figure below where we have plotted the function on the interval $0.01 \leq x \leq 0.1$.

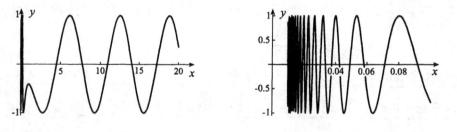

With these ideas in mind, we now solve $0 = f'(x) = -\left(1 - \dfrac{1}{x^2}\right)\sin\left(x + \dfrac{1}{x}\right)$. There are two possibilities:

$$1 - \frac{1}{x^2} = 0 \quad \text{or} \quad \sin\left(x + \frac{1}{x}\right) = 0.$$

The first gives $x = \pm 1$, and it is clear that the tangent line at $x = 1$ in the left figure is indeed horizontal. The second equation above implies that $x + \dfrac{1}{x} = n\pi$, where n is an integer. Multiplication by x leads to the quadratic equation

$$x^2 - n\pi x + 1 = 0 \quad \Longrightarrow \quad x = \frac{n\pi \pm \sqrt{n^2\pi^2 - 4}}{2}.$$

Clearly, we must choose $n > 0$ (else $n^2\pi^2 - 4 < 0$ when $n = 0$, and $x < 0$ when $n < 0$). When $n = 1$, the solution is $x = (\pi + \sqrt{\pi^2 - 4})/2 \approx 2.8$. This is the first point to the right of $x = 1$ at which the tangent line is horizontal in the left figure. As n increases, the 4 becomes less and less significant and $x \approx (n\pi \pm n\pi)/2$. When we choose the positive sign, we obtain $x \approx n\pi$, and we can indeed see that points on the graph in the left figure where the tangent line is horizontal do indeed seem to be multiples of π. When we choose the negative sign, we obtain points to the left of $x = 1$ at which the tangent line is horizontal (right figure). For large n, they can be approximated by $1/(n\pi)$, but this is not an easy fact to show.

EXERCISES 3.10

2. $\dfrac{dy}{dx} = \dfrac{-1}{1 + (x^2 + 2)^2}(2x) = \dfrac{-2x}{1 + (x^2 + 2)^2}$ **4.** $\dfrac{dy}{dx} = \dfrac{1}{1 + (2 - x^2)^2}(-2x) = \dfrac{-2x}{1 + (2 - x^2)^2}$

6. $\dfrac{dy}{dx} = \mathrm{Csc}^{-1}(x^2 + 5) - \dfrac{x}{(x^2 + 5)\sqrt{(x^2 + 5)^2 - 1}}(2x) = \mathrm{Csc}^{-1}(x^2 + 5) - \dfrac{2x^2}{(x^2 + 5)\sqrt{(x^2 + 5)^2 - 1}}$

8. $\dfrac{dy}{dx} = \dfrac{1}{1 + (x + 2)}\dfrac{1}{2\sqrt{x + 2}} = \dfrac{1}{2\sqrt{x + 2}(x + 3)}$ **10.** $\dfrac{dy}{dx} = \dfrac{-1}{1 + x^2 - 1}\dfrac{x}{\sqrt{x^2 - 1}} = \dfrac{-1}{x\sqrt{x^2 - 1}}$

12. $\dfrac{dy}{dx} = 2x\,\mathrm{Sec}^{-1}x + \dfrac{x^2}{x\sqrt{x^2 - 1}} = 2x\,\mathrm{Sec}^{-1}x + \dfrac{x}{\sqrt{x^2 - 1}}$

14. $\dfrac{dy}{dx} = \dfrac{-1}{1 + \left(\dfrac{1 + x}{1 - x}\right)^2}\left[\dfrac{(1 - x)(1) - (1 + x)(-1)}{(1 - x)^2}\right] = \dfrac{-(1 - x)^2}{(1 - x)^2 + (1 + x)^2}\left[\dfrac{2}{(1 - x)^2}\right] = \dfrac{-2}{2 + 2x^2} = \dfrac{-1}{1 + x^2}$

16. $\dfrac{dy}{dx} = \dfrac{1}{\sqrt{1 - \left(\dfrac{1 - x}{1 + x}\right)^2}}\left[\dfrac{(1 + x)(-1) - (1 - x)(1)}{(1 + x)^2}\right] = \dfrac{|x + 1|}{\sqrt{(1 + x)^2 - (1 - x)^2}}\left[\dfrac{-2}{(x + 1)^2}\right]$ Since x must

be greater than 0 in order that $-1 \le \dfrac{1 - x}{1 + x} \le 1$, it follows that $\dfrac{dy}{dx} = \dfrac{(x + 1)(-2)}{2\sqrt{x}(x + 1)^2} = \dfrac{-1}{\sqrt{x}(x + 1)}$.

18. $\dfrac{dy}{dx} = \dfrac{dy}{dt}\dfrac{dt}{dx} = \left(\mathrm{Cos}^{-1}t - \dfrac{t}{\sqrt{1 - t^2}}\right)\left(\dfrac{-x}{\sqrt{1 - x^2}}\right) = \dfrac{x(t - \sqrt{1 - t^2}\,\mathrm{Cos}^{-1}t)}{\sqrt{1 - t^2}\sqrt{1 - x^2}}$

20. If we differentiate with respect to x, we obtain $\dfrac{1}{\sqrt{1 - x^2y^2}}\left(x\dfrac{dy}{dx} + y\right) = 5 + 2\dfrac{dy}{dx}$, and therefore

$$\frac{dy}{dx} = \frac{5 - \dfrac{y}{\sqrt{1 - x^2y^2}}}{\dfrac{x}{\sqrt{1 - x^2y^2}} - 2} = \frac{5\sqrt{1 - x^2y^2} - y}{x - 2\sqrt{1 - x^2y^2}}.$$

22. $\dfrac{dy}{dx} = \dfrac{-1/3}{\dfrac{x}{3}\sqrt{\dfrac{x^2}{9} - 1}}\left(\dfrac{1}{3}\right) - \dfrac{x^2(1/2)(x^2 - 9)^{-1/2}(2x) - \sqrt{x^2 - 9}(2x)}{x^4}$

$= \dfrac{-1}{x\sqrt{x^2 - 9}} - \dfrac{x^2 - 2(x^2 - 9)}{x^3\sqrt{x^2 - 9}} = \dfrac{-x^2 - (-x^2 + 18)}{x^3\sqrt{x^2 - 9}} = \dfrac{-18}{x^3\sqrt{x^2 - 9}}$

24. $\dfrac{dy}{dx} = -\dfrac{1}{x^2}\text{Csc}^{-1}(3x) + \dfrac{1}{x}\dfrac{(-1)(3)}{3x\sqrt{9x^2 - 1}} + \dfrac{\sqrt{9x^2 - 1}}{x^2} - \dfrac{9x}{x\sqrt{9x^2 - 1}}$

$= -\dfrac{1}{x^2}\text{Csc}^{-1}(3x) + \dfrac{-1 + (9x^2 - 1) - 9x^2}{x^2\sqrt{9x^2 - 1}} = -\dfrac{1}{x^2}\text{Csc}^{-1}(3x) - \dfrac{2}{x^2\sqrt{9x^2 - 1}}$

26. $\dfrac{dy}{dx} = (\text{Cos}^{-1}x)^2 - \dfrac{2x\,\text{Cos}^{-1}x}{\sqrt{1 - x^2}} - 2 + \dfrac{2x}{\sqrt{1 - x^2}} = (\text{Cos}^{-1}x)^2 - 2 + \dfrac{2x}{\sqrt{1 - x^2}}(1 - \text{Cos}^{-1}x)$

28. $\dfrac{dy}{dx} = \sqrt{4x - x^2} + \dfrac{x - 2}{2\sqrt{4x - x^2}}(4 - 2x) + \dfrac{4}{\sqrt{1 - \left(\dfrac{x - 2}{2}\right)^2}}\left(\dfrac{1}{2}\right)$

$= \dfrac{(4x - x^2) + (x - 2)(2 - x)}{\sqrt{4x - x^2}} + \dfrac{4}{\sqrt{4x - x^2}} = \dfrac{4x - x^2 - x^2 + 4x - 4 + 4}{\sqrt{4x - x^2}} = 2\sqrt{4x - x^2}$

30. $\dfrac{dy}{dx} = \dfrac{1}{1 + \dfrac{2x^2}{1 + x^4}}\left[\dfrac{\sqrt{2}}{\sqrt{1 + x^4}} - \dfrac{\sqrt{2}x}{2(1 + x^4)^{3/2}}(4x^3)\right]$

$= \dfrac{1 + x^4}{1 + x^4 + 2x^2}\left[\dfrac{\sqrt{2}(1 + x^4) - 2\sqrt{2}x^4}{(1 + x^4)^{3/2}}\right] = \dfrac{\sqrt{2}(1 - x^4)}{(1 + x^2)^2\sqrt{1 + x^4}} = \dfrac{\sqrt{2}(1 - x^2)}{(1 + x^2)\sqrt{1 + x^4}}$

32. (a) If we set $y = \text{Sec}^{-1}x$, then $x = \sec y$, and implicit differentiation gives

$$1 = \sec y\,\tan y\,\dfrac{dy}{dx} \quad\Longrightarrow\quad \dfrac{dy}{dx} = \dfrac{1}{\sec y\,\tan y}.$$

Now, $\tan y = \pm\sqrt{\sec^2 y - 1} = \pm\sqrt{x^2 - 1}$. If $x > 0$, then $0 < y < \pi/2$, and $\tan y = \sqrt{x^2 - 1}$. If $x < 0$, and principal values are chosen as stated, then $\pi/2 < y < \pi$, and $\tan y = -\sqrt{x^2 - 1}$. Thus,

$$\dfrac{dy}{dx} = \begin{cases} \dfrac{1}{x\sqrt{x^2 - 1}}, & x > 0 \\[2ex] \dfrac{-1}{x\sqrt{x^2 - 1}}, & x < 0 \end{cases} = \dfrac{1}{|x|\sqrt{x^2 - 1}}.$$

(b) A similar analysis gives $\dfrac{d}{dx}\text{Csc}^{-1}x = \dfrac{-1}{|x|\sqrt{x^2 - 1}}$.

34. To verify 3.37c, we set $y = \text{Tan}^{-1}x$, in which case $x = \tan y$. Differentiation gives $1 = \sec^2 y\,(dy/dx) \Longrightarrow$ $dy/dx = 1/\sec^2 y = 1/(1 + \tan^2 y) = 1/(1 + x^2)$. Verification of 3.37d is similar. To verify 3.37e, we set $y = \text{Sec}^{-1}x$, in which case $x = \sec y$. Differentiation gives $1 = \sec y\,\tan y\,(dy/dx) \Longrightarrow dy/dx = $ $1/(\sec y\,\tan y)$. Now $\tan y = \pm\sqrt{\sec^2 y - 1} = \pm\sqrt{x^2 - 1}$. When $x > 0$, we obtain $0 < y < \pi/2$, so that $\tan y > 0$. On the other hand, when $x < 0$, principal values give $-\pi < y < -\pi/2$, but $\tan y$ is again positive. Thus, in either case, $\tan y = \sqrt{x^2 - 1}$, and $dy/dx = 1/(x\sqrt{x^2 - 1})$. Verification of 3.37f is similar.

EXERCISES 3.11

2. $\dfrac{dy}{dx} = \dfrac{1}{3x^2 + 1}(6x) = \dfrac{6x}{3x^2 + 1}$ **4.** $\dfrac{dy}{dx} = e^{1 - 2x}(-2) = -2e^{1 - 2x}$

6. $\dfrac{dy}{dx} = \ln x + x\left(\dfrac{1}{x}\right) = 1 + \ln x$ **8.** $\dfrac{dy}{dx} = \dfrac{1}{3 - 4x}(-4)\log_{10}e = \dfrac{-4\log_{10}e}{3 - 4x}$

10. $\dfrac{dy}{dx} = \dfrac{1}{3\cos x}(-3\sin x) = -\tan x$

12. $\dfrac{dy}{dx} = 2x + 3x^2 e^{4x} + x^3 e^{4x}(4) = 2x + (3x^2 + 4x^3)e^{4x}$

14. $\dfrac{dy}{dx} = \cos\left(e^{2x}\right)\left(2e^{2x}\right) = 2e^{2x}\cos\left(e^{2x}\right)$

16. $\dfrac{dy}{dx} = -2e^{-2x}\sin 3x + e^{-2x}\cos 3x\,(3) = e^{-2x}(3\cos 3x - 2\sin 3x)$

18. $\dfrac{dy}{dx} = \dfrac{(e^x + e^{-x})(e^x + e^{-x}) - (e^x - e^{-x})(e^x - e^{-x})}{(e^x + e^{-x})^2} = \dfrac{4}{(e^x + e^{-x})^2}$

20. $\dfrac{dy}{dx} = \sin\left(\ln x\right) - \cos\left(\ln x\right) + x\left[\dfrac{1}{x}\cos\left(\ln x\right) + \dfrac{1}{x}\sin\left(\ln x\right)\right] = 2\sin\left(\ln x\right)$

22. $\dfrac{dy}{dx} = \dfrac{dy}{dv}\dfrac{dv}{dx} = \dfrac{1}{\cos v}(-\sin v)\,2\sin x\cos x = -\tan v\sin 2x$

24. Differentiation with respect to x gives $e^y + xe^y\dfrac{dy}{dx} + 2x\ln y + \dfrac{x^2}{y}\dfrac{dy}{dx} + \sin x\dfrac{dy}{dx} + y\cos x = 0$, and there-

fore $\dfrac{dy}{dx} = -\dfrac{y\cos x + 2x\ln y + e^y}{\sin x + x^2/y + xe^y}$.

26. $\dfrac{dy}{dx} = \sqrt{x^2+1} + \dfrac{x^2}{\sqrt{x^2+1}} - \dfrac{1}{x+\sqrt{x^2+1}}\left(1 + \dfrac{x}{\sqrt{x^2+1}}\right)$

$= \dfrac{x^2+1+x^2}{\sqrt{x^2+1}} - \dfrac{1}{x+\sqrt{x^2+1}}\left(\dfrac{\sqrt{x^2+1}+x}{\sqrt{x^2+1}}\right) = \dfrac{2x^2}{\sqrt{x^2+1}}$

28. $\dfrac{dy}{dx} = 1 - \dfrac{1}{4(1+5e^{4x})}(5e^{4x})(4) = \dfrac{1+5e^{4x}-5e^{4x}}{1+5e^{4x}} = \dfrac{1}{1+5e^{4x}}$

30. If we differentiate with respect to x, we obtain $e^{1/x}\left(-\dfrac{1}{x^2}\right) + e^{1/y}\left(-\dfrac{1}{y^2}\right)\dfrac{dy}{dx} = -\dfrac{1}{x^2} - \dfrac{1}{y^2}\dfrac{dy}{dx}$, and

therefore $\dfrac{dy}{dx} = \dfrac{e^{1/x}(1/x^2) - 1/x^2}{1/y^2 + e^{1/y}(-1/y^2)} = \dfrac{e^{1/x}-1}{x^2}\dfrac{y^2}{1-e^{1/y}} = \dfrac{y^2(e^{1/x}-1)}{x^2(1-e^{1/y})}$.

32. (a) Since $dT/dr = k + c/r$, it follows that $\dfrac{d}{dr}\left(r\dfrac{dT}{dr}\right) = \dfrac{d}{dr}(kr+c) = k$.

(b) If temperatures on inner and outer edges of the insulation are T_a and T_b, then

$$T_a = ka + c\ln a + d, \qquad T_b = kb + c\ln b + d.$$

These can be solved for $c = \dfrac{T_b - T_a + k(a-b)}{\ln(b/a)}$ and $d = \dfrac{T_a\ln b - T_b\ln a + k(b\ln a - a\ln b)}{\ln(b/a)}$.

34. (a) If we substitute $x(t)$ into the left and right sides of the differential equation, we obtain

$$\dfrac{dx}{dt} = \dfrac{(3-2e^{-15t})(-90e^{-15t})(-15) - 90(1-e^{-15t})(-2e^{-15t})(-15)}{(3-2e^{-15t})^2}$$

$$= \dfrac{1350e^{-15t}(3-2e^{-15t}-2+2e^{-15t})}{(3-2e^{-15t})^2} = \dfrac{1350e^{-15t}}{(3-2e^{-15t})^2}$$

$$(30-x)(45-x) = \left[30 - \dfrac{90(1-e^{-15t})}{3-2e^{-15t}}\right]\left[45 - \dfrac{90(1-e^{-15t})}{3-2e^{-15t}}\right]$$

$$= \dfrac{[90-60e^{-15t}-90(1-e^{-15t})][135-90e^{-15t}-90(1-e^{-15t})]}{(3-2e^{-15t})^2}$$

$$= \dfrac{(30e^{-15t})(45)}{(3-2e^{-15t})^2} = \dfrac{1350e^{-15t}}{(3-2e^{-15t})^2}$$

(b) $\lim\limits_{t\to\infty} x(t) = 30$ This is reasonable
since 10 g of A will eventually
combine with 20 g of B to produce
30 g of C.

36. (a) The energy balance equation is

$$1257 + 20t = \frac{1257T}{10} + 419000\frac{dT}{dt} \quad\Longrightarrow\quad \frac{dT}{dt} + \frac{3T}{10\,000} = \frac{3}{1000} + \frac{t}{20\,950}.$$

(b) If we substitute $T(t)$ into the left side of the differential equation,

$$\frac{dT}{dt} + \frac{3T}{10\,000} = \frac{200}{1257} + \frac{2\,000\,000}{3771}\left(-\frac{3}{10\,000}\right)e^{-3t/10\,000}$$

$$+\frac{3}{10\,000}\left(\frac{200t}{1257} - \frac{1\,962\,290}{3771} + \frac{2\,000\,000}{3771}e^{-3t/10\,000}\right)$$

$$= \frac{3}{1000} + \frac{t}{20950}.$$

The value of the function at $t = 0$ is $T(0) = 10$.

38. We calculate:

$$\frac{dx}{dt} = -e^{-t}(\sin 2t - \cos 2t) + e^{-t}(2\cos 2t + 2\sin 2t) = e^{-t}(\sin 2t + 3\cos 2t)$$

and

$$\frac{d^2x}{dt^2} = -e^{-t}(\sin 2t + 3\cos 2t) + e^{-t}(2\cos 2t - 6\sin 2t) = e^{-t}(-7\sin 2t - \cos 2t).$$

Consequently,

$$\frac{d^2x}{dt^2} + 2\frac{dx}{dt} + 5x = e^{-t}(-7\sin 2t - \cos 2t) + 2\,e^{-t}(\sin 2t + 3\cos 2t) + 5\,e^{-t}(\sin 2t - \cos 2t) = 0.$$

In addition, $x(0) = -1$ and $x'(0) = 3$.

40. (a) Since solution is added at 4 L/s and removed at 2 L/s, the amount of solution in the tank is $10\,000+2t$.
In this case,

$$\frac{dS}{dt} = \frac{1}{5} - \frac{2S}{10\,000 + 2t},$$

subject to the condition that $S(0) = 100$.

(b) $\dfrac{dS}{dt} + \dfrac{S}{5000 + t} = \dfrac{1}{10} + \dfrac{2\times 10^6}{(5000 + t)^2}$

$$+\frac{1}{5000 + t}\left(500 + \frac{t}{10} - \frac{2\times 10^6}{5000 + t}\right) = \frac{1}{5}$$

The graph is asymptotic to the line $S = 500 + t/10$.

42. (a) Since alcohol enters the vat at the rate of 4/25 L/s and leaves at the rate of $2(A/2000)$, it follows
that $\dfrac{dA}{dt} = \dfrac{4}{25} - \dfrac{A}{1000}$. We must also have $A(0) = 80$.

(b) $\dfrac{dA}{dt} + \dfrac{A}{1000} = \dfrac{80}{1000}e^{-t/1000} + \dfrac{1}{1000}(160 - 80e^{-t/1000}) = \dfrac{4}{25}$

A plot of the function is shown to the right.
It is asymptotic to the line $A = 160$.

(c) The beer is 5% alcohol when

$$\frac{5}{100}(2000) = 160 - 80e^{-t/1000}.$$

This implies that $e^{-t/1000} = 3/4$,
from which $t = 1000\ln{(4/3)}$.

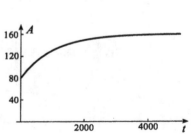

44. The function is linear in h (left figure below). To draw the right graph, we express the function in the form $\Phi = \dfrac{\mu_0 h i}{2\pi}\left[\ln\left(\dfrac{R}{r}\right) + \ln\left(\dfrac{r+w}{R+w}\right)\right]$. It has value 0 at $w = 0$; it is increasing; and $\lim\limits_{w\to\infty} \Phi = \dfrac{\mu_0 h i}{2\pi}\ln\left(\dfrac{R}{r}\right)$.

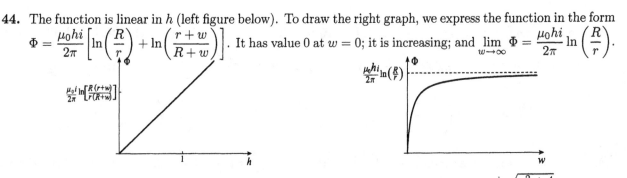

46. (a) If we multiply the equation by e^y, then $(e^y)^2 - x\,e^y - 1 = 0 \implies e^y = \dfrac{x \pm \sqrt{x^2+4}}{2}$. Since e^y must be positive, we choose $e^y = \dfrac{x + \sqrt{x^2+4}}{2} \implies y = \ln\left(\dfrac{x+\sqrt{x^2+4}}{2}\right)$. Consequently,

$$\frac{dy}{dx} = \frac{2}{x+\sqrt{x^2+4}}\left(\frac{1}{2}\right)\left(1 + \frac{x}{\sqrt{x^2+4}}\right) = \frac{1}{x+\sqrt{x^2+4}}\left(\frac{\sqrt{x^2+4}+x}{\sqrt{x^2+4}}\right) = \frac{1}{\sqrt{x^2+4}}.$$

(b) If we differentiate implicity with respect to x, then $1 = e^y\dfrac{dy}{dx} + e^{-y}\dfrac{dy}{dx}$, from which $\dfrac{dy}{dx} = \dfrac{1}{e^y + e^{-y}}$. To show that this is the same derivative as in part (a), we note that

$$e^y + e^{-y} = \frac{x+\sqrt{x^2+4}}{2} + \frac{2}{x+\sqrt{x^2+4}} = \frac{x+\sqrt{x^2+4}}{2} + \frac{2}{x+\sqrt{x^2+4}}\left(\frac{x-\sqrt{x^2+4}}{x-\sqrt{x^2+4}}\right)$$

$$= \frac{x+\sqrt{x^2+4}}{2} + \frac{2(x-\sqrt{x^2+4})}{x^2-(x^2+4)} = \sqrt{x^2+4}.$$

This shows that $1/(e^y + e^{-y}) = 1/\sqrt{x^2+4}$.

48. Implicit differentiation gives $\dfrac{1}{x^2+y^2}\left(2x + 2y\dfrac{dy}{dx}\right) = \dfrac{2}{1+(y^2/x^2)}\left(\dfrac{1}{x}\dfrac{dy}{dx} - \dfrac{y}{x^2}\right)$, from which

$$\frac{1}{x^2+y^2}\left(x + y\frac{dy}{dx}\right) = \frac{1}{x^2+y^2}\left(x\frac{dy}{dx} - y\right) \implies \frac{dy}{dx} = \frac{x+y}{x-y}.$$

A second differentiation now gives

$$\frac{d^2y}{dx^2} = \frac{(x-y)\left(1+\dfrac{dy}{dx}\right) - (x+y)\left(1-\dfrac{dy}{dx}\right)}{(x-y)^2} = \frac{(x-y)\left(1+\dfrac{x+y}{x-y}\right) - (x+y)\left(1-\dfrac{x+y}{x-y}\right)}{(x-y)^2}$$

$$= \frac{2(x^2+y^2)}{(x-y)^3}.$$

EXERCISES 3.12

2. Natural logarithms of $y = x^{4\cos x}$ give $\ln y = 4\cos x \ln x$, and differentiation with respect to x yields

$$\frac{1}{y}\frac{dy}{dx} = 4\left(-\sin x \ln x + \frac{1}{x}\cos x\right) \implies \frac{dy}{dx} = 4x^{4\cos x}\left(\frac{1}{x}\cos x - \sin x \ln x\right).$$

4. Natural logarithms of $y = (\sin x)^x$ give $\ln y = x\ln(\sin x)$ and differentiation with respect to x yields

$$\frac{1}{y}\frac{dy}{dx} = \ln(\sin x) + x\frac{\cos x}{\sin x} \implies \frac{dy}{dx} = (\sin x)^x[\ln(\sin x) + x\cot x].$$

6. Natural logarithms of $y = \left(1 + \dfrac{1}{x}\right)^{x^2}$ give $\ln y = x^2 \ln\left(1 + \dfrac{1}{x}\right) = x^2[\ln(x+1) - \ln x]$, and differentiation with respect to x leads to $\dfrac{1}{y}\dfrac{dy}{dx} = 2x[\ln(x+1) - \ln x] + x^2\left(\dfrac{1}{x+1} - \dfrac{1}{x}\right)$. Thus,

$$\frac{dy}{dx} = xy\left[2\ln\left(1 + \frac{1}{x}\right) + x\left(\frac{x - (x+1)}{x(x+1)}\right)\right] = x\left(1 + \frac{1}{x}\right)^{x^2}\left[2\ln\left(1 + \frac{1}{x}\right) - \frac{1}{x+1}\right].$$

8. If we take natural logarithms of $y = \left(\dfrac{2}{x}\right)^{3/x}$, then $\ln y = \dfrac{3}{x}\ln\left(\dfrac{2}{x}\right) = \dfrac{3}{x}(\ln 2 - \ln x)$, and differentiation with respect to x gives

$$\frac{1}{y}\frac{dy}{dx} = -\frac{3}{x^2}(\ln 2 - \ln x) + \frac{3}{x}\left(-\frac{1}{x}\right) \implies \frac{dy}{dx} = -\frac{3y}{x^2}\left[\ln\left(\frac{2}{x}\right) + 1\right] = -\frac{3}{x^2}\left(\frac{2}{x}\right)^{3/x}\left[1 + \ln\left(\frac{2}{x}\right)\right].$$

10. When we take natural logarithms of $y = (\ln x)^{\ln x}$, we find $\ln y = \ln x \ln(\ln x)$, and differentiation with respect to x yields

$$\frac{1}{y}\frac{dy}{dx} = \frac{1}{x}\ln(\ln x) + \ln x\left(\frac{1}{x\ln x}\right) \implies \frac{dy}{dx} = \frac{y}{x}[\ln(\ln x) + 1] = \frac{1}{x}(\ln x)^{\ln x}[1 + \ln(\ln x)].$$

12. If we take natural logarithms of $y = \dfrac{\sqrt{x}(1 + 2x^2)}{\sqrt{1 + x^2}}$, then $\ln y = \dfrac{1}{2}\ln x + \ln(1 + 2x^2) - \dfrac{1}{2}\ln(1 + x^2)$. Differentiation with respect to x gives $\dfrac{1}{y}\dfrac{dy}{dx} = \dfrac{1}{2x} + \dfrac{4x}{1 + 2x^2} - \dfrac{x}{1 + x^2}$, and therefore

$$\frac{dy}{dx} = y\left[\frac{(1 + 2x^2)(1 + x^2) + 8x^2(1 + x^2) - 2x^2(1 + 2x^2)}{2x(1 + 2x^2)(1 + x^2)}\right] = \frac{6x^4 + 9x^2 + 1}{2\sqrt{x}(1 + x^2)^{3/2}}.$$

14. If we take natural logarithms of $|y| = |x^2 + 3x|^3(x^2 + 5)^4$, then $\ln|y| = 3\ln|x^2 + 3x| + 4\ln(x^2 + 5)$. Differentiation with respect to x using formula 3.46 gives $\dfrac{1}{y}\dfrac{dy}{dx} = 3\left(\dfrac{2x + 3}{x^2 + 3x}\right) + 4\left(\dfrac{2x}{x^2 + 5}\right)$, and therefore

$$\frac{dy}{dx} = y\left[\frac{3(2x + 3)(x^2 + 5) + 8x(x^2 + 3x)}{(x^2 + 3x)(x^2 + 5)}\right] = (x^2 + 3x)^3(x^2 + 5)^4\left[\frac{14x^3 + 33x^2 + 30x + 45}{(x^2 + 3x)(x^2 + 5)}\right]$$
$$= (x^2 + 3x)^2(x^2 + 5)^3(14x^3 + 33x^2 + 30x + 45).$$

16. When we take natural logarithms of $y = x^{3/2}e^{-2x}$, we have $\ln y = (3/2)\ln x - 2x$, and differentiation with respect to x results in

$$\frac{1}{y}\frac{dy}{dx} = \frac{3}{2x} - 2 \implies \frac{dy}{dx} = x^{3/2}e^{-2x}\left(\frac{3 - 4x}{2x}\right) = \frac{1}{2}\sqrt{x}(3 - 4x)e^{-2x}.$$

18. Natural logarithms of $|y| = \dfrac{e^x}{|\ln(x - 1)|}$ give $\ln|y| = x - \ln|\ln(x - 1)|$. Differentiation with respect to x now gives $\dfrac{1}{y}\dfrac{dy}{dx} = 1 - \dfrac{1}{(x - 1)\ln(x - 1)}$. Thus, $\dfrac{dy}{dx} = \dfrac{e^x}{\ln(x - 1)}\left[1 - \dfrac{1}{(x - 1)\ln(x - 1)}\right]$.

20. If we take natural logarithms of $|y| = \dfrac{\sqrt{x}|1 - x^2|}{\sqrt{1 + x^2}}$, then $\ln|y| = \dfrac{1}{2}\ln x + \ln|1 - x^2| - \dfrac{1}{2}\ln(1 + x^2)$. Differentiation with respect to x using formula 3.46 gives $\dfrac{1}{y}\dfrac{dy}{dx} = \dfrac{1}{2x} - \dfrac{2x}{1 - x^2} - \dfrac{x}{1 + x^2}$, and therefore

$$\frac{dy}{dx} = \frac{\sqrt{x}(1 - x^2)}{\sqrt{1 + x^2}}\left[\frac{(1 - x^2)(1 + x^2) - 4x^2(1 + x^2) - 2x^2(1 - x^2)}{2x(1 - x^2)(1 + x^2)}\right] = \frac{1 - 6x^2 - 3x^4}{2\sqrt{x}(1 + x^2)^{3/2}}.$$

22. Natural logarithms of $|y| = |x|^3 |x^2 - 4x| \sqrt{1 + x^3}$ give $\ln |y| = 3 \ln |x| + \ln |x^2 - 4x| + \dfrac{1}{2} \ln (1 + x^3)$. Differentiation with respect to x using formula 3.46 results in $\dfrac{1}{y} \dfrac{dy}{dx} = \dfrac{3}{x} + \dfrac{2x - 4}{x^2 - 4x} + \dfrac{3x^2}{2(1 + x^3)}$, and therefore

$$\frac{dy}{dx} = x^3 (x^2 - 4x) \sqrt{1 + x^3} \left[\frac{6(x^2 - 4x)(1 + x^3) + 2x(2x - 4)(1 + x^3) + 3x^3(x^2 - 4x)}{2x(x^2 - 4x)(1 + x^3)} \right]$$

$$= \frac{x^3 (13x^4 - 44x^3 + 10x - 32)}{2\sqrt{1 + x^3}}.$$

24. Natural logarithms of $|y| = \dfrac{|\sin 2x| |\sec 5x|}{|1 - 2 \cot x|^3}$ give $\ln |y| = \ln |\sin 2x| + \ln |\sec 5x| - 3 \ln |1 - 2 \cot x|$. Differentiation with respect to x using 3.46 yields $\dfrac{1}{y} \dfrac{dy}{dx} = \dfrac{2 \cos 2x}{\sin 2x} + \dfrac{5 \sec 5x \tan 5x}{\sec 5x} - \dfrac{3(2 \csc^2 x)}{1 - 2 \cot x}$, from which $\dfrac{dy}{dx} = \dfrac{\sin 2x \sec 5x}{(1 - 2 \cot x)^3} \left[2 \cot 2x + 5 \tan 5x - \dfrac{6 \csc^2 x}{1 - 2 \cot x} \right]$.

26. (a) If we take natural logarithms, $\ln x = a \ln r - b(r + c)$. Differentiation with respect to r gives $\dfrac{1}{x} \dfrac{dx}{dr} = \dfrac{a}{r} - b \implies \dfrac{dx}{dr} = \dfrac{x}{r}(a - br)$. Since $r > a/b$, it follows that the derivative $dx/dr < 0$. But if the slope of the tangent line is always negative, x must decrease as r increases. Therefore x increases as r decreases.

(b) If we differentiate $\ln x = a \ln r - b(r + c)$ with respect to x, $\dfrac{1}{x} = \dfrac{a}{r} \dfrac{dr}{dx} - b \dfrac{dr}{dx} = \left(\dfrac{a - br}{r} \right) \dfrac{dr}{dx}$. Thus,

$$\frac{dr}{dx} = \frac{r}{x(a - br)} \implies \frac{Er}{Ex} = \frac{x}{r} \frac{r}{x(a - br)} = \frac{1}{a - br}.$$

EXERCISES 3.13

2. $\dfrac{dy}{dx} = \sinh (x/2) + x \cosh (x/2) (1/2) = \sinh (x/2) + (x/2) \cosh (x/2)$

4. $\dfrac{dy}{dx} = \operatorname{sech}^2 (\ln x) \left(\dfrac{1}{x} \right)$

6. Differentiation with respect to x gives $\dfrac{dy}{dx} - \operatorname{csch}^2 x = \dfrac{1}{2\sqrt{1 + y}} \dfrac{dy}{dx}$, from which

$$\frac{dy}{dx} = \frac{\operatorname{csch}^2 x}{1 - \dfrac{1}{2\sqrt{1 + y}}} = \frac{2\sqrt{1 + y} \operatorname{csch}^2 x}{2\sqrt{1 + y} - 1}.$$

8. $\dfrac{dy}{dx} = \dfrac{dy}{dt} \dfrac{dt}{dx} = [\sec^2 (\cosh t) \sinh t][-\sin (\tanh x) \operatorname{sech}^2 x] = -\sinh t \operatorname{sech}^2 x \sec^2 (\cosh t) \sin (\tanh x)$

10. Since $y = (1/2) \ln (\tanh 2x)$,

$$\frac{dy}{dx} = \frac{1}{2 \tanh 2x} \operatorname{sech}^2 2x (2) = \frac{1}{\cosh^2 2x} \frac{\cosh 2x}{\sinh 2x} = \frac{1}{\sinh 2x \cosh 2x} = \frac{1}{(\sinh 4x)/2} = 2 \operatorname{csch} 4x.$$

12. (a) $f'(x) \quad = -Ak \sin kx + Bk \cos kx + Ck \sinh kx + Dk \cosh kx$

$f''(x) \quad = -Ak^2 \cos kx - Bk^2 \sin kx + Ck^2 \cosh kx + Dk^2 \sinh kx$

$f'''(x) = Ak^3 \sin kx - Bk^3 \cos kx + Ck^3 \sinh kx + Dk^3 \cosh kx$

$f''''(x) = Ak^4 \cos kx + Bk^4 \sin kx + Ck^4 \cosh kx + Dk^4 \sinh kx$

Thus, $\dfrac{d^4 y}{dx^4} - k^4 y = 0$.

(b) These conditions imply that

$$0 = f(0) = A + C,$$
$$0 = f'(0) = Bk + Dk = k(B + D),$$
$$0 = f(L) = A \cos kL + B \sin kL + C \cosh kL + D \sinh kL,$$
$$0 = f''(L) = -Ak^2 \cos kL - Bk^2 \sin kL + Ck^2 \cosh kL + Dk^2 \sinh kL.$$

Thus, $C = -A$, $D = -B$, and

$$0 = A \cos kL + B \sin kL - A \cosh kL - B \sinh kL = A(\cos kL - \cosh kL) + B(\sin kL - \sinh kL);$$
$$0 = -A \cos kL - B \sin kL - A \cosh kL - B \sinh kL = A(\cos kL + \cosh kL) + B(\sin kL + \sinh kL).$$

(c) If we write the conditions in part (b) in the form

$$A(\cos kL - \cosh kL) = -B(\sin kL - \sinh kL), \quad A(\cos kL + \cosh kL) = -B(\sin kL + \sinh kL),$$

and divide one by the other,

$$\frac{\cos kL - \cosh kL}{\cos kL + \cosh kL} = \frac{\sin kL - \sinh kL}{\sin kL + \sinh kL}.$$

Hence,

$$(\cos kL - \cosh kL)(\sin kL + \sinh kL) = (\cos kL + \cosh kL)(\sin kL - \sinh kL)$$

or, $2 \cos kL \sinh kL = 2 \sin kL \cosh kL$. Division by $2 \cos kL \cosh kL$ gives $\tanh kL = \tan kL$.

EXERCISES 3.14

2. Since $f(x)$ and $f'(x)$ are continuous on $1 \le x \le 3$, we may apply the mean value theorem on this interval. Equation 3.49 states that $3 - 4c = \dfrac{-5 - 5}{3 - 1} = -5$. The only solution is $c = 2$.

4. Since $f'(x)$ is not defined at $x = 0$, we cannot apply the mean value theorem on the interval $-1 \le x \le 1$.

6. Since $f(x)$ and $f'(x)$ are continuous on $-3 \le x \le 2$, we may apply the mean value theorem on this interval. Equation 3.49 states that $3c^2 + 4c - 1 = \dfrac{12 + 8}{2 + 3} = 4$. Of the two solutions $c = (-2 \pm \sqrt{19})/3$, only $(-2 + \sqrt{19})/3$ is within the interval $-3 \le x \le 2$.

8. Since $f(x)$ and $f'(x)$ are continuous on $2 \le x \le 4$, we may apply the mean value theorem on this interval. Equation 3.49 states that

$$\left[\frac{(x - 1)(1) - (x + 2)(1)}{(x - 1)^2} \right]_{|x=c} = \frac{2 - 4}{4 - 2}; \text{ that is,} \quad \frac{-3}{(c - 1)^2} = -1.$$

Of the two solutions $c = 1 \pm \sqrt{3}$ of this equation, only $c = 1 + \sqrt{3}$ lies in the given interval.

10. Since $f(x)$ and $f'(x)$ are continuous on $-2 \le x \le 3$, we may apply the mean value theorem on this interval. Equation 3.49 states that

$$\left[\frac{(x + 3)(2x) - x^2(1)}{(x + 3)^2} \right]_{|x=c} = \frac{3/2 - 4}{3 + 2}; \text{ that is,} \quad \frac{c^2 + 6c}{(c + 3)^2} = -\frac{1}{2}.$$

Of the two solutions $c = -3 \pm \sqrt{6}$ of this equation, only $c = -3 + \sqrt{6}$ lies in the given interval.

12. Since $\ln(2x + 1)$ and its derivative are continuous on $0 \le x \le 2$, we may apply the mean value theorem. Equation 3.49 states that $\dfrac{2}{2c + 1} = \dfrac{\ln 5 - \ln 1}{2} = \dfrac{1}{2} \ln 5$. The solution of this equation is $c = (4 - \ln 5)/(2 \ln 5)$.

14. Since $\sec x$ is not defined at $x = \pi/2$, the mean value theorem cannot be applied on the interval $0 \le x \le \pi$.

16. Since $g'(0) = 0$, Cauchy's generalized mean value theorem cannot be applied on the interval $-1 \le x \le 1$.

18. Since $f(x)$, $f'(x)$, $g(x)$, and $g'(x)$ are continuous on $-3 \le x \le -2$, and $g'(x) = -1/(x-1)^2$ is never zero, we may apply Cauchy's generalized mean value theorem. Equation 3.48 states that

$$\frac{2-3/2}{2/3-3/4} = \frac{\dfrac{1}{(c+1)^2}}{\dfrac{-1}{(c-1)^2}} \qquad \Longrightarrow \qquad 6 = \left(\frac{c-1}{c+1}\right)^2.$$

Of the two solutions $c = (1 \pm \sqrt{6})/(1 \mp \sqrt{6})$, only $c = (1+\sqrt{6})/(1-\sqrt{6})$ is in the given interval.

20. Equation 3.49 for $f(x) = dx^2 + ex + g$ on $a \le x \le b$ states that

$$2dc + e = \frac{(db^2 + eb + g) - (da^2 + ea + g)}{b-a} = \frac{d(b^2 - a^2) + e(b-a)}{b-a} = d(b+a) + e,$$

and therefore $c = (a+b)/2$.

22. If we define a function $F(x) = f(x) - g(x)$, then $F(x)$ is continuous and has a derivative at each point in $a \le x \le b$. We may therefore apply the mean value theorem to $F(x)$ and write $F'(c) = \dfrac{F(b) - F(a)}{b-a}$. Since $F'(x) = f'(x) - g'(x)$, this equation states that there exists a point c between a and b such that $f'(c) - g'(c) = \dfrac{[f(b) - g(b)] - [f(a) - g(a)]}{b-a} = 0$; that is, $f'(c) = g'(c)$.

REVIEW EXERCISES

2. $\dfrac{dy}{dx} = 6x + 2 - \dfrac{1}{x^2}$

4. $\dfrac{dy}{dx} = \dfrac{1}{3x^{2/3}} - \dfrac{10}{9}x^{2/3}$

6. $\dfrac{dy}{dx} = 2(x^2+2)(2x)(x^3-3)^3 + (x^2+2)^2(3)(x^3-3)^2(3x^2) = x(x^2+2)(x^3-3)^2(13x^3 + 18x - 12)$

8. $\dfrac{dy}{dx} = \dfrac{(x+5)(3) - (3x-2)(1)}{(x+5)^2} = \dfrac{17}{(x+5)^2}$

10. $\dfrac{dy}{dx} = \dfrac{(x^2+5x-2)(4) - 4x(2x+5)}{(x^2+5x-2)^2} = \dfrac{-4(x^2+2)}{(x^2+5x-2)^2}$

12. We first write the equation in the form $x^2 + y^2 = x^2 y$, and then differentiate with respect to x, $2x + 2y\dfrac{dy}{dx} = 2xy + x^2\dfrac{dy}{dx}$. Thus, $\dfrac{dy}{dx} = \dfrac{2xy - 2x}{2y - x^2}$.

14. Differentiation with respect to x gives $2xy + x^2\dfrac{dy}{dx} + \dfrac{dy}{dx}\sqrt{1+x} + \dfrac{y}{2\sqrt{1+x}} = 0$. Thus,

$$\frac{dy}{dx} = -\frac{2xy + \dfrac{y}{2\sqrt{1+x}}}{x^2 + \sqrt{1+x}} = -\frac{4xy\sqrt{1+x} + y}{2\sqrt{1+x}\,(x^2 + \sqrt{1+x})}.$$

16. $\dfrac{dy}{dx} = 2\sec(1-4x)\sec(1-4x)\tan(1-4x)(-4) = -8\sec^2(1-4x)\tan(1-4x)$

18. $\dfrac{dy}{dx} = \sec(\tan 2x)\tan(\tan 2x)\sec^2 2x(2) = 2\sec^2 2x \sec(\tan 2x)\tan(\tan 2x)$

20. Since $y = \left(\dfrac{1}{2}\sin 2x\right)^2 = \dfrac{1}{4}\sin^2 2x = \dfrac{1}{4}\left(\dfrac{1-\cos 4x}{2}\right) = \dfrac{1}{8}(1 - \cos 4x)$, it follows that $\dfrac{dy}{dx} = \dfrac{1}{8}\sin 4x(4) = \dfrac{1}{2}\sin 4x$.

22. $\dfrac{dy}{dx} = \dfrac{dy}{dt}\dfrac{dt}{dx} = (1 - 2\sin 2t)(1 + 2\sin 2x)$

24. $\dfrac{dy}{dx} = \dfrac{dy}{dv}\dfrac{dv}{dx} = [\cos^2 v + v(2\cos v)(-\sin v)]\left(\dfrac{-x}{\sqrt{1-x^2}}\right) = \dfrac{x(2v\sin v\cos v - \cos^2 v)}{\sqrt{1-x^2}} = \dfrac{x(v\sin 2v - \cos^2 v)}{v}$

26. If we differentiate the equation with respect to x, the result is $1 = e^{2y}\left(2\dfrac{dy}{dx}\right) \implies \dfrac{dy}{dx} = \dfrac{1}{2}e^{-2y}$.

28. $\dfrac{dy}{dx} = 2x\ln\left(x^2 - 1\right) + (x^2 + 1)\dfrac{2x}{x^2 - 1}$

30. If we differentiate with respect to x, then $-5\sin(x - y)\left(1 - \dfrac{dy}{dx}\right) = 0 \implies \dfrac{dy}{dx} = 1$.

32. $\dfrac{dy}{dx} = 3\cosh(x^2)(2x) = 6x\cosh(x^2)$

34. $\dfrac{dy}{dx} = \dfrac{1}{1 + \left(\dfrac{1}{x} + x\right)^2}\left(-\dfrac{1}{x^2} + 1\right) = \dfrac{1}{1 + \dfrac{x^4 + 2x^2 + 1}{x^2}}\left(\dfrac{x^2 - 1}{x^2}\right) = \dfrac{x^2 - 1}{x^4 + 3x^2 + 1}$

36. If we differentiate with respect to x, we obtain $\cosh y\,\dfrac{dy}{dx} = \cos x \implies \dfrac{dy}{dx} = \dfrac{\cos x}{\cosh y} = \cos x\,\text{sech}\,y$.

38. $\dfrac{dy}{dx} = \text{Csc}^{-1}\left(\dfrac{1}{x^2}\right) - \dfrac{x}{\dfrac{1}{x^2}\sqrt{\dfrac{1}{x^4} - 1}}\left(\dfrac{-2}{x^3}\right) = \text{Csc}^{-1}\left(\dfrac{1}{x^2}\right) + \dfrac{2x^2}{\sqrt{1 - x^4}}$

40. If $\ln\left[\text{Tan}^{-1}(x + y)\right] = 1/10$, then $\text{Tan}^{-1}(x+y) = e^{1/10}$, from which $x+y = \tan e^{1/10}$ or $y = -x+\tan e^{1/10}$. Consequently, $dy/dx = -1$.

42. If we square the equation, $x^2 = \dfrac{4 + y}{4 - y}$. This equation can be solved for $y = \dfrac{4(x^2 - 1)}{x^2 + 1}$. Differentiation now gives $\dfrac{dy}{dx} = 4\left[\dfrac{(x^2 + 1)(2x) - (x^2 - 1)(2x)}{(x^2 + 1)^2}\right] = \dfrac{16x}{(x^2 + 1)^2}$.

44. If we take natural logarithms of $|y| = \dfrac{|x|^2\sqrt{1 - x}}{|x + 5|}$, then $\ln|y| = 2\ln|x| + \dfrac{1}{2}\ln(1 - x) - \ln|x + 5|$. Differentiation with respect to x using formula 3.46 gives $\dfrac{1}{y}\dfrac{dy}{dx} = \dfrac{2}{x} - \dfrac{1}{2(1 - x)} - \dfrac{1}{x + 5}$. Therefore,

$$\dfrac{dy}{dx} = \dfrac{x^2\sqrt{1 - x}}{x + 5}\left[\dfrac{4(1 - x)(x + 5) - x(x + 5) - 2x(1 - x)}{2x(1 - x)(x + 5)}\right] = \dfrac{x(20 - 23x - 3x^2)}{2\sqrt{1 - x}(x + 5)^2}.$$

46. If we write $x^2 - xy = xy + y^2$, or, $y^2 + 2xy - x^2 = 0$, and differentiate with respect to x, we obtain $2y\dfrac{dy}{dx} + 2y + 2x\dfrac{dy}{dx} - 2x = 0$. Thus, $\dfrac{dy}{dx} = \dfrac{x - y}{x + y}$.

48. If we use the result of Exercise 40 in Section 3.7,

$$\dfrac{dy}{dx} = \dfrac{1}{\dfrac{dx}{dy}} = \dfrac{1}{\dfrac{(y^3 + 4y + 6)(2y - 2) - (y^2 - 2y)(3y^2 + 4)}{(y^3 + 4y + 6)^2}} = \dfrac{(y^3 + 4y + 6)^2}{-y^4 + 4y^3 + 4y^2 + 12y - 12}.$$

50. Since $y = \dfrac{(\sqrt{x} - \sqrt{2})(\sqrt{x} + \sqrt{2})}{\sqrt{x} - \sqrt{2}} = \sqrt{x} + \sqrt{2}$, it follows that $\dfrac{dy}{dx} = \dfrac{1}{2\sqrt{x}}$, provided $x \neq 2$.

52. If we take natural logarithms, then $\ln y = x\ln\cos x$, and differentiation with respect to x gives

$$\dfrac{1}{y}\dfrac{dy}{dx} = \ln(\cos x) + x\left(\dfrac{-\sin x}{\cos x}\right) \implies \dfrac{dy}{dx} = (\cos x)^x\left[\ln(\cos x) - x\tan x\right].$$

54. $\dfrac{dy}{dx} = \dfrac{1}{\log_{10} x}\log_{10} e\dfrac{d}{dx}\log_{10} x = \dfrac{1}{\log_{10} x}\log_{10} e\left(\dfrac{1}{x}\right)\log_{10} e = \dfrac{(\log_{10} e)^2}{x\log_{10} x}$

56. Differentiation with respect to x gives $1 = e^y\dfrac{dy}{dx} - e^{-y}\dfrac{dy}{dx} \implies \dfrac{dy}{dx} = \dfrac{1}{e^y - e^{-y}}$.

58. If we differentiate with respect to x, we obtain $2xy + x^2\dfrac{dy}{dx} + \dfrac{1}{x+y}\left(1 + \dfrac{dy}{dx}\right) = 1$. Thus,

$$\frac{dy}{dx} = \frac{1 - 2xy - \dfrac{1}{x+y}}{x^2 + \dfrac{1}{x+y}} = \frac{(x+y)(1-2xy)-1}{x^2(x+y)+1}.$$

60. Since $dy/dx = -1/(x+5)^2$, the slope of the tangent line at $(0, 1/5)$ is $-1/25$. Equations for the tangent and normal lines are $y - 1/5 = -(1/25)(x-0)$ and $y - 1/5 = 25(x-0)$, or $x + 25y = 5$ and $5y - 125x = 1$.

62. Since $\dfrac{dy}{dx} = \dfrac{(2x-5)(2x+3) - (x^2+3x)(2)}{(2x-5)^2} = \dfrac{2x^2 - 10x - 15}{(2x-5)^2}$, the slope of the tangent line at $(1, -4/3)$ is $-23/9$. Equations for the tangent and normal lines are $y + 4/3 = -(23/9)(x-1)$ and $y + 4/3 = (9/23)(x-1)$, or $23x + 9y = 11$ and $27x - 69y = 119$.

64. If we first expand the equation, we obtain $x^2 - 5xy + y^2 = 0$. Differentiation with respect to x now gives $2x - 5y - 5x\dfrac{dy}{dx} + 2y\dfrac{dy}{dx} = 0 \implies \dfrac{dy}{dx} = \dfrac{2x - 5y}{5x - 2y}$. A second differentiation yields

$$\frac{d^2y}{dx^2} = \frac{(5x - 2y)\left(2 - 5\dfrac{dy}{dx}\right) - (2x - 5y)\left(5 - 2\dfrac{dy}{dx}\right)}{(5x - 2y)^2}$$

$$= \frac{(5x - 2y)\left[2 - 5\left(\dfrac{2x-5y}{5x-2y}\right)\right] - (2x - 5y)\left[5 - 2\left(\dfrac{2x-5y}{5x-2y}\right)\right]}{(5x - 2y)^2}$$

$$= \frac{2(5x-2y)^2 - 5(2x-5y)(5x-2y) - 5(2x-5y)(5x-2y) + 2(2x-5y)^2}{(5x-2y)^3} = \frac{-42(x^2 - 5xy + y^2)}{(5x-2y)^3}.$$

66. The graph to the right was obtained by joining the points shown. The function is not periodic.

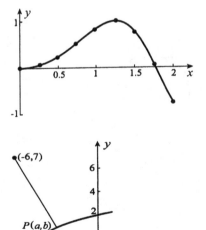

68. Let $P(a, b)$ be the required point. If we differentiate the equation of the parabola with respect to x, the result is $1 = 2y(dy/dx)$. The slope of the tangent line at P is therefore $1/(2b)$, and that of the normal line is $-2b$. Since the slope of the normal line at P is also $(b-7)/(a+6)$, it follows that $(b-7)/(a+6) = -2b$. Because $P(a, b)$ is on the parabola, a and b

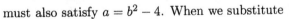

must also satisfy $a = b^2 - 4$. When we substitute this into the above equation, we obtain $(b-7)/(b^2 - 4 + 6) = -2b$. Thus, $0 = 2b^3 + 5b - 7 = (b-1)(2b^2 + 2b + 7)$. The required point is therefore $(-3, 1)$. The length of the line segment joining $(-6, 7)$ and $(-3, 1)$ is the shortest distance from $(-6, 7)$ to the parabola.

70. Since $x = L\sin(\pi/3) = \sqrt{3}L/2$, the area of the triangle is

$$A = \frac{xL}{2} = \frac{\sqrt{3}L^2}{4}.$$

Consequently,

$$\frac{dA}{dL} = \frac{\sqrt{3}L}{2}.$$

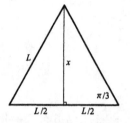

72. Differentiation with respect to x gives $2(x^2 + y^2)\left(2x + 2y\dfrac{dy}{dx}\right) = 2x - 2y\dfrac{dy}{dx}$, and therefore

$\dfrac{dy}{dx} = \dfrac{x - 2x(x^2 + y^2)}{y + 2y(x^2 + y^2)}$. The tangent line is horizontal when $0 = x - 2x(x^2 + y^2) = x(1 - 2x^2 - 2y^2)$. Since $x = 0$ is not a solution to our problem, we must set $x^2 + y^2 = 1/2$. We substitute this into the equation for the lemniscate, obtaining thereby $1/4 = x^2 - y^2$. Addition of this and $1/2 = x^2 + y^2$ gives $2x^2 = 3/4$, or, $x = \pm\sqrt{3/8} = \pm\sqrt{6}/4$. These values yield the points $(\sqrt{6}/4, \pm\sqrt{2}/4)$ and $(-\sqrt{6}/4, \pm\sqrt{2}/4)$.

74. For this $f(x)$ and $g(x)$, equation 3.48 gives $\dfrac{6c - 2}{3c^2 + 2} = \dfrac{5 - 9}{3 + 3} = -\dfrac{2}{3}$. Of the two solutions $c = (-9\pm\sqrt{93})/6$ of this equation, only $c = (\sqrt{93} - 9)/6$ is in the interval $-1 \le x \le 1$.

CHAPTER 4

EXERCISES 4.1

2. Newton's iterative procedure defines

$$x_{n+1} = x_n - \frac{x_n^2 - x_n - 4}{2x_n - 1}.$$

For the root between 2 and 3 we use an
initial approximation $x_1 = 2.5$. The
resulting iterations give

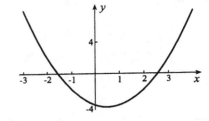

$$x_2 = 2.562\,5, \qquad x_3 = 2.561\,553\,0,$$
$$x_4 = 2.561\,552\,8, \quad x_5 = 2.561\,552\,8.$$

With $f(x) = x^2 - x - 4$, we calculate that $f(2.561\,552\,5) = -1.3 \times 10^{-6}$ and $f(2.561\,553\,5) = 2.8 \times 10^{-6}$.
Hence, to six decimals, a root is $x = 2.561\,553$. A similar procedure for the root between -2 and -1
gives $x = -1.561\,553$.

4. Newton's iterative procedure defines

$$x_{n+1} = x_n - \frac{x_n^3 - x_n^2 + x_n - 22}{3x_n^2 - 2x_n + 1}.$$

The only root of the equation is slightly
larger than 3. To find it we use $x_1 = 3.1$.
Iteration gives

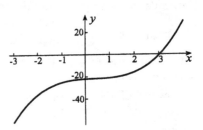

$$x_2 = 3.045\,789\,3, \quad x_3 = 3.044\,7236,$$
$$x_4 = 3.044\,723\,1, \quad x_5 = 3.044\,723\,1.$$

With $f(x) = x^3 - x^2 + x - 22$, we calculate that $f(3.044\,722\,5) = -1.5 \times 10^{-5}$ and $f(3.044\,723\,5) = 8.0 \times 10^{-6}$. Thus, to six decimals, the root is $x = 3.044\,723$.

6. Newton's iterative procedure defines

$$x_{n+1} = x_n - \frac{x_n^5 + x_n - 1}{5x_n^4 + 1}.$$

The only root of the equation is between
0 and 1. To find it we use $x_1 = 0.5$.
Iteration gives

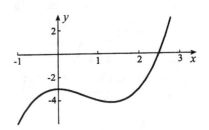

$$x_2 = 0.857\,142\,9, \quad x_3 = 0.770\,682\,2,$$
$$x_4 = 0.755\,283\,0, \quad x_5 = 0.754\,877\,9,$$
$$x_6 = 0.754\,877\,7, \quad x_7 = 0.754\,877\,7.$$

With $f(x) = x^5 + x - 1$, we calculate that $f(0.754\,877\,5) = -4.4 \times 10^{-7}$ and $f(0.754\,878\,5) = 2.2 \times 10^{-6}$.
Thus, to six decimals, the root is $x = 0.754\,878$.

8. The equation can be rearranged into the form
$x^3 - 2x^2 - 3 = 0$. Newton's iterative procedure defines

$$x_{n+1} = x_n - \frac{x_n^3 - 2x_n^2 - 3}{3x_n^2 - 4x_n}.$$

The only root of the equation is between
2 and 3. To find it we use $x_1 = 2.5$.
Iteration gives

$$x_2 = 2.485\,714\,3, \quad x_3 = 2.485\,584\,0,$$
$$x_4 = 2.485\,584\,0.$$

With $f(x) = x^3 - 2x^2 - 3$, we calculate that $f(2.485\,583\,5) = -4.3 \times 10^{-6}$ and $f(2.485\,584\,5) = 4.3 \times 10^{-6}$.
Thus, to six decimals, the root is $x = 2.485\,584$.

10. If we rearrange the equation into
the form $f(x) = (1 + x^4) \sec x - 2 = 0$, Newton's
iterative procedure defines

$$x_{n+1} = x_n - \frac{(1 + x_n^4) \sec x_n - 2}{4x_n^3 \sec x_n + (1 + x_n^4) \sec x_n \tan x_n}.$$

For the root near 1 we use $x_1 = 1$. Iteration
gives

$$x_2 = 0.870\,777\,3, \quad x_3 = 0.807\,267\,5,$$
$$x_4 = 0.795\,650\,2, \quad x_5 = 0.795\,324\,2,$$
$$x_6 = 0.795\,323\,9, \quad x_7 = 0.795\,323\,9.$$

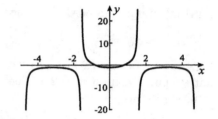

We now calculate that $f(0.795\,323\,5) = -2.0 \times 10^{-6}$ and $f(0.795\,324\,5) = 2.9 \times 10^{-6}$. Thus, to six decimals, the root is $x = 0.795\,324$. Because of the symmetry of the graph, the other root is $-0.795\,324$.

12. The graph suggests that there
are 6 roots of the equation.
Newton's iterative procedure defines

$$x_{n+1} = x_n - \frac{(x_n + 1)^2 - 5 \sin 4x_n}{2(x_n + 1) - 20 \cos 4x_n}.$$

For the smallest positive root we use $x_1 = 0$.
Iteration gives

$$x_2 = 0.055\,555\,6, \quad x_3 = 0.056\,257\,3,$$
$$x_4 = 0.056\,257\,6, \quad x_5 = 0.056\,257\,6.$$

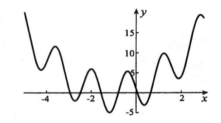

With $f(x) = (x + 1)^2 - 5 \sin 4x$, we calculate that $f(0.056\,257\,5) = 1.9 \times 10^{-6}$ and $f(0.056\,258\,5) = -1.6 \times 10^{-5}$. Thus, to six decimals, the root is $x = 0.056\,258$. Similar procedures lead to the other 5 roots $-2.931\,137$, $-2.467\,518$, $-1.555\,365$, $-0.787\,653$, $0.642\,851$.

14. Newton's iterative procedure defines

$$x_{n+1} = x_n - \frac{x_n \ln x_n - 6}{\ln x_n + 1}.$$

The only root of the equation is between 4 and 5.
To find it we use $x_1 = 4.3$. Iteration gives

$$x_2 = 4.189\,350\,5, \quad x_3 = 4.188\,760\,1,$$
$$x_4 = 4.188\,760\,1.$$

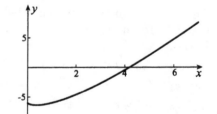

With $f(x) = x \ln x - 6$, we calculate that $f(4.188\,759\,5) = -1.5 \times 10^{-6}$ and $f(4.188\,760\,5) = 9.3 \times 10^{-7}$. Thus, to six decimals, the root is $x = 4.188\,760$.

16. Newton's iterative procedure defines

$$x_{n+1} = x_n - \frac{x_n^2 - 4e^{-2x_n}}{2x_n + 8e^{-2x_n}}.$$

With $x_1 = 0.8$, we obtain

$$x_2 = 0.852\,123\,5, \quad x_3 = 0.852\,605\,5,$$
$$x_4 = 0.852\,605\,5.$$

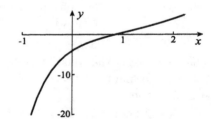

With $f(x) = x^2 - 4e^{-2x}$, we calculate that $f(0.852\,605\,5) = -6.4 \times 10^{-9}$ and $f(0.852\,606\,5) = 3.2 \times 10^{-6}$. Hence, to six decimals, the root is $x = 0.852\,606$.

18. Obviously $x = 0$ is a solution of the equation. The remaining solutions must satisfy $f(x) = x^3 - x^2 + 2x + 6 = 0$. The graph indicates only one root between -2 and -1. To find it we use, $x_1 = -1.2$, and

$$x_{n+1} = x_n - \frac{x_n^3 - x_n^2 + 2x_n + 6}{3x_n^2 - 2x_n + 2}.$$

Iteration gives

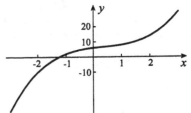

$$x_2 = -1.249\,541, \quad x_3 = -1.248\,299,$$
$$x_4 = -1.248\,298.$$

Since $f(-1.248\,4) = -9.4 \times 10^{-4}$ and $f(-1.248\,2) = 8.9 \times 10^{-4}$, we can say that the root is $x = -1.2483$.

20. We rearrange the equation into the form $f(x) = x^3 - x^2 - 6x - 1 = 0$. The graph suggests three solutions. To find the positive one, we use $x_1 = 3$ and

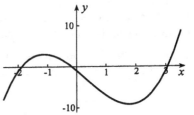

$$x_{n+1} = x_n - \frac{x_n^3 - x_n^2 - 6x_n - 1}{3x_n^2 - 2x_n - 6}.$$

Iteration gives

$$x_2 = 3.066\,67, \quad x_3 = 3.064\,44, \quad x_4 = 3.064\,43.$$

Since $f(3.063) = -2.3 \times 10^{-2}$ and $f(3.065) = 9.1 \times 10^{-3}$, the root is $x = 3.064$. A similar procedure leads to the other two roots -1.892 and -0.172.

22. The graph indicates two roots. To find the positive one, we use $x_1 = 1$ and

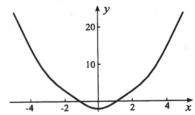

$$x_{n+1} = x_n - \frac{x_n^2 - 1 - \cos^2 x_n}{2x_n + 2\cos x_n \sin x_n}.$$

Iteration gives

$$x_2 = 1.100\,343, \quad x_3 = 1.098\,587, \quad x_4 = 1.098\,587.$$

With $f(x) = x^2 - 1 - \cos^2 x$, we calculate that $f(1.098\,5) = -2.6 \times 10^{-4}$ and $f(1.098\,7) = 3.4 \times 10^{-4}$. Thus, the root is $x = 1.098\,6$. Symmetry gives the other root as $-1.098\,6$.

24. The graph indicates a solution of this equation between 0 and 1. To find it we use $x_1 = 0.3$ and

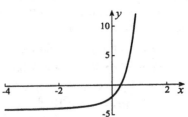

$$x_{n+1} = x_n - \frac{e^{3x_n} + e^{x_n} - 4}{3e^{3x_n} + e^{x_n}}.$$

Iteration gives

$$x_2 = 0.321\,83, \quad x_3 = 0.321\,21, \quad x_4 = 0.321\,21.$$

With $f(x) = e^{3x} + e^x - 4$, we calculate that $f(0.3211) = -1.0 \times 10^{-3}$ and $f(0.3213) = 8.2 \times 10^{-4}$. Thus, the root is $x = 0.3212$.

26. To find x-coordinates of the points of intersection, we set $(x + 1)^2 = x^3 - 4x$, and this reduces to $f(x) = x^3 - x^2 - 6x - 1 = 0$. With $x_1 = 3$, and

$$x_{n+1} = x_n - \frac{x_n^3 - x_n^2 - 6x_n - 1}{3x_n^2 - 2x_n - 6},$$

iteration gives

$$x_2 = 3.066\,7, \quad x_3 = 3.064\,435, \quad x_4 = 3.064\,435.$$

Since $f(3.064\,434\,5) = -5.4 \times 10^{-7}$ and $f(3.064\,435\,5) = 1.6 \times 10^{-5}$, the root is $x = 3.064\,435$. In either of the original equations, $x = 3.064\,435$ yields the same four decimals, $y = 16.519\,6$. A point of intersection is therefore $(3.064\,4, 16.519\,6)$. Similar procedures lead to the other points of intersection $(-1.892\,0, 0.795\,6)$ and $(-0.172\,5, 0.684\,8)$.

28. When we equate the two expressions for y, and rearrange the equation, x must satisfy $f(x) = x^3 + x^2 + x + 2 = 0$. The graph indicates that there is only one solution, and to find it we use $x_1 = -1.3$ and

$$x_{n+1} = x_n - \frac{x_n^3 + x_n^2 + x_n + 2}{3x_n^2 + 2x_n + 1}.$$

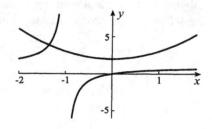

Iteration gives

$$x_2 = -1.368, \qquad x_3 = -1.353\,38,$$
$$x_4 = -1.353\,210, \quad x_5 = -1.353\,210.$$

Since $f(-1.353\,210\,5) = -2.0 \times 10^{-6}$ and $f(-1.353\,209\,5) = 1.8 \times 10^{-6}$, we can say that $x = -1.353\,210$. In either of the original equations, $x = -1.353\,210$ yields the same four decimals, $y = 3.831\,2$. The point of intersection is therefore $(-1.353\,2, 3.831\,2)$.

30. If x_1 is the initial approximation to the solution of $f(x) = x^{7/5} = 0$, then the second approximation as defined by Newton's iterative procedure is

$$x_2 = x_1 - \frac{x_1^{7/5}}{(7/5)x_1^{2/5}} = x_1 - \frac{5x_1}{7} = \frac{2}{5}x_1.$$

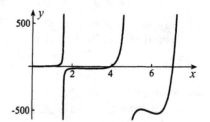

What this implies is that every approximation is 2/5 times the previous one. Approximations therefore must approach zero, the root of the equation. This is illustrated in the graph to the right.

32. If we write the equation in the form $f(x) = e^x(-1 + \tan x) + e^{-x}(1 + \tan x) = 0$, Newton's iterative procedure defines

$$x_{n+1} = x_n - \frac{e^{x_n}(-1 + \tan x_n) + e^{-x_n}(1 + \tan x_n)}{e^{x_n}(-1 + \tan x_n + \sec^2 x_n) + e^{-x_n}(-1 - \tan x_n + \sec^2 x_n)}.$$

For the root just larger than 7, we use an initial approximation $x_1 = 7$. The resulting iterations give

$$x_2 = 7.068\,8, \qquad x_3 = 7.068\,789,$$
$$x_4 = 7.068\,583, \quad x_5 = 7.068\,583.$$

Since $f(7.068\,582\,5) = -5.8 \times 10^{-4}$ and $f(7.068\,583\,5) = 1.8 \times 10^{-3}$, we can say

that $x = 7.068\,583$. When this is divided by 20π, the result to four decimals is 0.1125. Similarly, the smallest frequency is 0.0625.

34. To simplify calculations, we set $z = c/\lambda$. Then, z must satisfy the equation $f(z) = (5 - z)e^z - 5 = 0$. Since $f(4) = e^4 - 5$ and $f(5) = -5$, the solution for z is slightly less than 5. To find it more accurately, we use

$$z_1 = 4.9, \qquad z_{n+1} = z_n - \frac{(5 - z_n)e^{z_n} - 5}{-e^{z_n} + (5 - z_n)e^{z_n}} = z_n - \frac{(5 - z_n)e^{z_n} - 5}{(4 - z_n)e^{z_n}}.$$

Iteration gives

$$z_2 = 4.969\,741\,205, \quad z_3 = 4.965\,135\,924, \quad z_4 = 4.965\,114\,232, \qquad z_5 = 4.965\,114\,232.$$

With this approximation for z, we obtain $\lambda = c/z_5 = 0.000\,028\,974$. For a seven decimal answer, we use $g(\lambda) = (5\lambda - c)e^{c/\lambda} - 5\lambda$ to calculate $g(0.000\,028\,95) = -1.7 \times 10^{-5}$ and $g(0.000\,029\,05) = 5.1 \times 10^{-5}$. Thus, to 7 decimals, $\lambda = 0.000\,029\,0$.

36. Let $P(x)$ be a cubic polynomial with roots a, b, and c. Then, $P(x) = k(x-a)(x-b)(x-c)$, where k is a constant. Suppose we use Newton's iterative procedure with $x_1 = (a+b)/2$. We require

$$P(x_1) = k\left(\frac{a+b}{2} - a\right)\left(\frac{a+b}{2} - b\right)\left(\frac{a+b}{2} - c\right) = \frac{k}{8}(b-a)(a-b)(a+b-2c) = -\frac{k}{8}(b-a)^2(a+b-2c),$$

and

$$P'(x_1) = k\left[\left(\frac{a+b}{2} - a\right)\left(\frac{a+b}{2} - b\right) + \left(\frac{a+b}{2} - a\right)\left(\frac{a+b}{2} - c\right) + \left(\frac{a+b}{2} - b\right)\left(\frac{a+b}{2} - c\right)\right]$$

$$= \frac{k}{4}\left[(b-a)(a-b) + (b-a)(a+b-2c) + (a-b)(a+b-2c)\right]$$

$$= -\frac{k}{4}(b-a)^2.$$

Therefore,

$$x_2 = \frac{a+b}{2} - \frac{-(k/8)(b-a)^2(a+b-2c)}{-(k/4)(b-a)^2} = \frac{a+b}{2} - \frac{1}{2}(a+b-2c) = c.$$

EXERCISES 4.2

2. Since $f'(x) = -5$, the function is decreasing for all x.

4. Since $f'(x) = -4x + 5$, it follows that $f'(x) \leq 0$ for $x \geq 5/4$ and $f'(x) \geq 0$ for $x \leq 5/4$. The function is therefore decreasing for $x \geq 5/4$ and increasing for $x \leq 5/4$.

6. Since $f'(x) = 2 - 8x$, it follows that $f'(x) \leq 0$ for $x \geq 1/4$ and $f'(x) \geq 0$ for $x \leq 1/4$. The function is therefore decreasing for $x \geq 1/4$ and increasing for $x \leq 1/4$.

8. Since $f'(x) = 3x^2 + 12x + 12 = 3(x+2)^2 \geq 0$ for all x, the function is always increasing.

10. Since $f'(x) = -18 - 18x - 6x^2 = -6(3 + 3x + x^2) < 0$ for all x, the function is always decreasing.

12. Since $f'(x) = 12x^3 - 12x^2 + 48x - 48 = 12(x-1)(x^2+4)$, it follows that $f'(x) \leq 0$ for $x \leq 1$ and $f'(x) \geq 0$ for $x \geq 1$. The function is therefore decreasing for $x \leq 1$ and increasing for $x \geq 1$.

14. Since $f'(x) = 5x^4 - 5 = 5(x-1)(x+1)(x^2+1)$, we find that $f'(x) \leq 0$ when $-1 \leq x \leq 1$, and $f'(x) \geq 0$ when $x \leq -1$ and $x \geq 1$. The function is therefore decreasing for $-1 \leq x \leq 1$, and it is increasing on the intervals $x \leq -1$ and $x \geq 1$.

16. The sign diagram below evaluates the sign of $f'(x) = 2x - 2/x^3 = 2(x-1)(x+1)(x^2+1)/x^3$. The function is decreasing on the intervals $x \leq -1$ and $0 < x \leq 1$, and increasing on the intervals $-1 \leq x < 0$ and $x \geq 1$. The graph corroborates this.

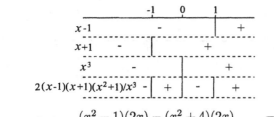
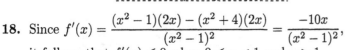
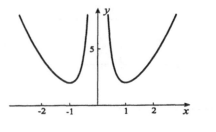

18. Since $f'(x) = \dfrac{(x^2-1)(2x) - (x^2+4)(2x)}{(x^2-1)^2} = \dfrac{-10x}{(x^2-1)^2}$,

it follows that $f'(x) \leq 0$ when $0 \leq x < 1$ and $x > 1$, and $f'(x) \geq 0$ when $x < -1$ and $-1 < x \leq 0$. The function is decreasing on the intervals $0 \leq x < 1$ and $x > 1$, and it is increasing on $x < -1$ and $-1 < x \leq 0$. The graph corroborates this.

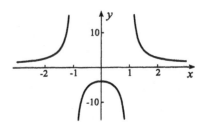

20. The graph of $f(x)$ indicates that $f(x)$
is decreasing on the intervals $x \leq -1$
and $0 \leq x \leq 1$, and it is increasing on
$-1 \leq x \leq 0$ and $x \geq 1$.

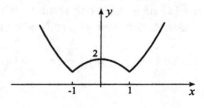

22. Since $f'(x) = 2xe^{-x} - x^2 e^{-x} = x(2-x)e^{-x}$, it follows that $f'(x) \leq 0$ for $x \leq 0$ and $x \geq 2$, and $f'(x) \geq 0$ for $0 \leq x \leq 2$. The function is decreasing on the intervals $x \leq 0$ and $x \geq 2$, and it is increasing for $0 \leq x \leq 2$.

24. Since $f'(x) = \ln x + 1$, we find that $f'(x) \leq 0$ when $0 < x \leq 1/e$, and $f'(x) \geq 0$ when $x \geq 1/e$. The function is therefore decreasing for $0 < x \leq 1/e$ and increasing for $x \geq 1/e$.

26. Since $f(x) = \dfrac{(x-1)(x+1)(x+2)}{(2+x)(1-x)} = -x - 1$, except for $x = -2$ and $x = 1$, where $f(x)$ is undefined, the function is decreasing for $x < -2$, $-2 < x < 1$, and $x > 1$.

28.

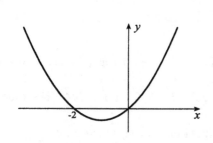 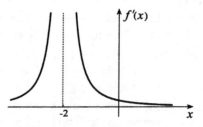

30.

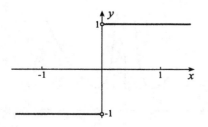 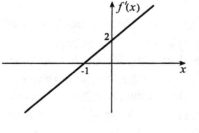

32.

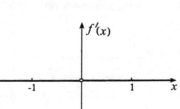

34.

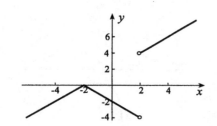 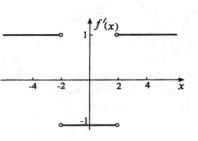

36. We first consider points where $0 = f'(x)$
$= 4x^3 + 4x - 6$. The plot of $f(x)$ indicates
that $f'(x)$ has a zero between 0 and 1. To find
it more accurately, we use Newton's iterative
procedure with $x_1 = 0.7$ and

$$x_{n+1} = x_n - \frac{4x_n^3 + 4x_n - 6}{12x_n^2 + 4}.$$

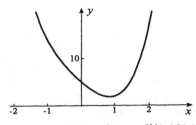

Iteration gives $x_2 = 0.885\,02$, $x_3 = 0.861\,67$, $x_4 = 0.861\,22$, $x_5 = 0.861\,22$. Since $f'(0.861\,15) = -9.6 \times 10^{-4}$ and $f'(0.861\,25) = 3.3 \times 10^{-4}$, it follows that $x = 0.861\,2$ is the solution of $f'(x) = 0$ to four decimals. The graph makes it clear that $f(x)$ is decreasing for $x \le 0.861\,2$ and increasing for $x \ge 0.861\,2$.

38. We first consider points where $0 = f'(x) =$
$2x \sin x + x^2 \cos x = x(2 \sin x + x \cos x)$.
This equation implies that $x = 0$ or
$2 \sin x + x \cos x = 0$. The graph of $f(x)$
indicates that the second of these has a
solution between $x = 2$ and $x = 3$. To find
it we use Newton's iterative procedure with

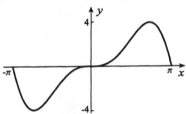

$$x_1 = 2, \qquad x_{n+1} = x_n - \frac{2 \sin x_n + x_n \cos x_n}{3 \cos x_n - x_n \sin x_n}.$$

Iteration gives $x_2 = 2.321\,58$, $x_3 = 2.289\,13$, $x_4 = 2.288\,93$, and $x_5 = 2.288\,93$. Since $f'(2.288\,85) = 6.7 \times 10^{-4}$ and $f'(2.288\,95) = -1.7 \times 10^{-4}$, we can say that a solution of $f'(x) = 0$ to four decimals is $x = 2.288\,9$. The graph makes it clear that so also is $x = -2.288\,9$. We conclude that $f(x)$ is decreasing on the intervals $-\pi \le x \le -2.288\,9$ and $2.288\,9 \le x \le \pi$, and it is increasing on $-2.288\,9 \le x \le 2.288\,9$.

40. For $f(x) = x^{23} + 3x^{15} + 4x + 1$, we find that $f(-1) = -7$ and $f(0) = 1$. By the zero intermediate value theorem, there exists at least one solution of $f(x) = 0$ between $x = -1$ and $x = 0$. Since $f'(x) = 23x^{22} + 45x^{14} + 4 > 0$, the function is increasing for all x. Hence, there can be only one solution of the equation.

42. Since $f(x) = x^n + ax - 1$ is continuous, and $f(0) = -1$ and $\lim_{x \to \infty} f(x) = \infty$, it follows that the graph of $f(x)$ must cross the x-axis at least once for positive x. Since $f'(x) = nx^{n-1} + a > 0$ for $x > 0$, the function is increasing for $0 \le x < \infty$. Hence, the graph of $f(x)$ can cross the x-axis only once for $x > 0$.

44. If $f(x) = x - \sin x$, then $f(0) = 0$. Since $f'(x) = 1 - \cos x \ge 0$, it follows that the function $f(x)$ is increasing for all $x > 0$. This means that $f(x) > 0$ for all $x > 0$, and therefore $x > \sin x$.

46. If $f(x) = \sin x - x + x^3/6$, then $f(0) = 0$. Since $f'(x) = \cos x - 1 + x^2/2$, Exercise 45 implies that $f'(x) \ge 0$ for $x > 0$. Consequently, $f(x)$ is increasing for $x > 0$, and $\sin x > x - x^3/6$.

48. If we define a function $f(x) = \dfrac{1}{\sqrt{1 + 3x}} - 1 + \dfrac{3x}{2}$, then $f(0) = 0$, and

$$f'(x) = \frac{-3}{2(1 + 3x)^{3/2}} + \frac{3}{2} = \frac{3}{2}\left[1 - \frac{1}{(1 + 3x)^{3/2}}\right].$$

This derivative is clearly positive for $x > 0$, and therefore the function $f(x)$ is increasing for $x > 0$. Thus, $f(x) > 0$ for $x > 0$, and the required result follows.

50. Yes. Since $f(x) > 0$, $f'(x) \ge 0$, $g(x) > 0$, and $g'(x) \ge 0$ on I, it follows that $[f(x)g(x)]' = f'(x)g(x) + f(x)g'(x) \ge 0$ on I also; that is, $f(x)g(x)$ is increasing on I.

52. Consider the function $g(x) = x - f(x)$. Since $g'(x) = 1 - f'(x)$, and $f'(x) < 1$ for all x, it follows that $g(x)$ is a decreasing function for all x. Consequently, it can have value 0 at most once; that is, there can exist at most one point x_0 at which $g(x_0) = 0$, and therefore at which $f(x_0) = x_0$.

EXERCISES 4.3

2. Since $f'(x) = 6x^2 + 30x + 24 = 6(x+1)(x+4)$, critical points are $x = -1$ and $x = -4$. Because $f'(x)$ changes from positive to negative as x increases through -4, $x = -4$ yields a relative maximum. Since $f'(x)$ changes from negative to positive as x increases through -1, $x = -1$ gives a relative minimum.

4. Since $f'(x) = 5(x-1)^4$, the only critical point is $x = 1$. Because $f'(x) \geq 0$ for all x, the function is always increasing, and $x = 1$ does not give a relative maximum or minimum.

6. Since $f'(x) = \dfrac{(x-1)(2x) - (x^2+1)(1)}{(x-1)^2} = \dfrac{x^2 - 2x - 1}{(x-1)^2}$, critical points are $x = (2 \pm \sqrt{4+4})/2 = 1 \pm \sqrt{2}$. The derivative does not exist at $x = 1$, but this is not a critical point because the function is not defined at $x = 1$. Since $f'(x)$ changes from positive to negative as x increases through $1 - \sqrt{2}$, this critical point gives a relative maximum. Because $f'(x)$ changes from negative to positive as x increases through $1 + \sqrt{2}$, there is a relative minimum at this point.

8. Since $f'(x) = 3x^2 - 12x + 12 = 3(x-2)^2$, the only critical point is $x = 2$. Since $f'(x) \geq 0$ for all x, it follows that $f(x)$ is increasing for all x, and there cannot be a relative extremum at $x = 2$.

10. Since $f'(x) = (1/3)x^{-2/3}$, and this derivative does not exist at $x = 0$, this is a critical point. Because $f'(x)$ is positive for all $x \neq 0$, it follows that the function is always increasing, and $x = 0$ cannot therefore yield a relative extremum.

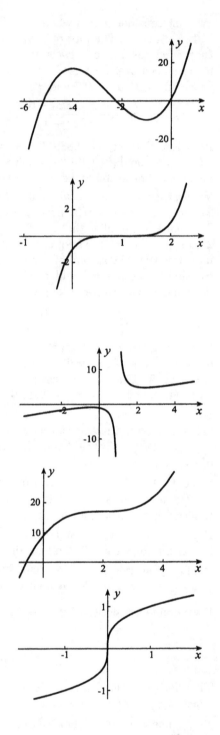

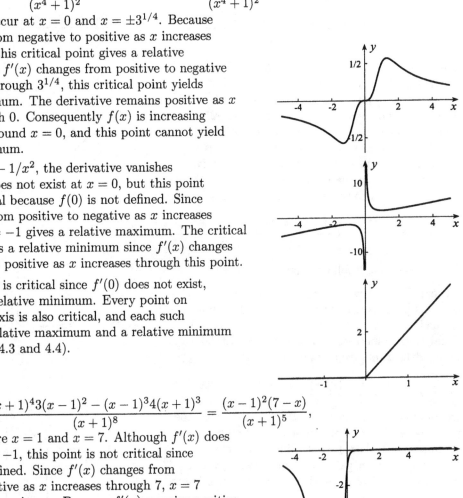

12. Since $f'(x) = \dfrac{(x^4+1)(3x^2) - x^3(4x^3)}{(x^4+1)^2} = \dfrac{x^2(3^{1/4}-x)(3^{1/4}+x)(\sqrt{3}+x^2)}{(x^4+1)^2}$,

critical points occur at $x = 0$ and $x = \pm 3^{1/4}$. Because $f'(x)$ changes from negative to positive as x increases through $-3^{1/4}$, this critical point gives a relative minimum. Since $f'(x)$ changes from positive to negative as x increases through $3^{1/4}$, this critical point yields a relative maximum. The derivative remains positive as x increases through 0. Consequently $f(x)$ is increasing in an interval around $x = 0$, and this point cannot yield a relative extremum.

14. Since $f'(x) = 1 - 1/x^2$, the derivative vanishes at $x = \pm 1$. It does not exist at $x = 0$, but this point cannot be critical because $f(0)$ is not defined. Since $f'(x)$ changes from positive to negative as x increases through -1, $x = -1$ gives a relative maximum. The critical point $x = 1$ gives a relative minimum since $f'(x)$ changes from negative to positive as x increases through this point.

16. The point $x = 0$ is critical since $f'(0)$ does not exist, and it yields a relative minimum. Every point on the negative x-axis is also critical, and each such point yields a relative maximum and a relative minimum (see Definitions 4.3 and 4.4).

18. Since $f'(x) = \dfrac{(x+1)^4 3(x-1)^2 - (x-1)^3 4(x+1)^3}{(x+1)^8} = \dfrac{(x-1)^2(7-x)}{(x+1)^5}$,

critical points are $x = 1$ and $x = 7$. Although $f'(x)$ does not exist at $x = -1$, this point is not critical since $f(-1)$ is not defined. Since $f'(x)$ changes from positive to negative as x increases through 7, $x = 7$ gives a relative maximum. Because $f'(x)$ remains positive as x increases through 1, this point does not give a relative extremum.

20. Since $f'(x) = 1 + 2\cos x$, the derivative vanishes for all values of x satisfying $\cos x = -1/2$. Solutions of this equation are $x = 2\pi/3 + 2n\pi$ and $x = 4\pi/3 + 2n\pi$, where n is an integer. Because $f'(x)$ changes from positive to negative as x increases through the values $2(3n+1)\pi/3$, these critical points give relative maxima. On the other hand, $f'(x)$ changes from negative to positive as x increases through the remaining critical points $x = 2(3n+2)\pi/3$, which therefore give relative minima.

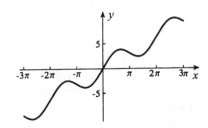

22. For critical points we first solve

$$0 = f'(x) = 2x + \frac{(x-2)^2(50x) - 25x^2(2)(x-2)}{(x-2)^4} = 2x + \frac{50x(x-2-x)}{(x-2)^3} = 2x\left[\frac{(x-2)^3 - 50}{(x-2)^3}\right].$$

Thus, either $x = 0$, or $(x-2)^3 = 50$, and the latter equation requires $x = 2 + 50^{1/3}$. The derivative does not exist at $x = 2$, but neither does $f(2)$, and therefore $x = 2$ is not critical. Since $f'(x)$ changes from negative to positive as x increases through 0, $x = 0$ gives a relative minimum. Because $f'(x)$ changes from negative to positive as x increases through $2 + 50^{1/3}$, this critical point also yields a relative minimum.

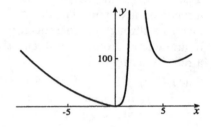

24. Since $f'(x) = \dfrac{(1 + x^3)(1 + 2x + 3x^2) - (1 + x + x^2 + x^3)(3x^2)}{(1 + x^3)^2} =$

$\dfrac{(1 - x)(1 + x)^3}{(1 + x^3)^2}$, it follows that $x = 1$ is critical. The derivative does not exist at $x = -1$, but this point is not critical since $f(-1)$ is not defined. Because $f'(x)$ changes from positive to negative as x increases through 1, this critical point gives a relative maximum.

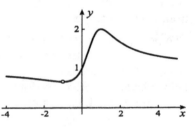

26. Since $f'(x) = \dfrac{(x^2 - 4)(2x) - x^2(2x)}{(x^2 - 4)^2} = \dfrac{-8x}{(x^2 - 4)^2}$, the only critical point is $x = 0$. The derivative does not exist at $x = \pm 2$, but these are not critical points because $f(\pm 2)$ do not exist. Since $f'(x)$ changes from positive to negative as x increases through $x = 0$, this critical point yields a relative maximum.

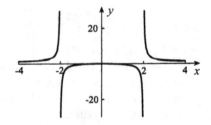

28. With the function written in the form $f(x) = \dfrac{2x^2 - 17x + 8}{x^2 - 5x + 4}$, we set

$0 = f'(x) = \dfrac{(x^2 - 5x + 4)(4x - 17) - (2x^2 - 17x + 8)(2x - 5)}{(x^2 - 5x + 4)^2} =$

$\dfrac{7(x^2 - 4)}{(x - 1)^2(x - 4)^2}$. Critical points are $x = \pm 2$. Since $f'(x)$ changes from positive to negative as x increases through -2, this critical point gives a relative maximum. The critical point $x = 2$ gives a relative minimum as $f'(x)$ changes from negative to positive through this point.

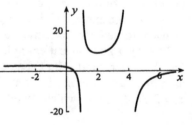

30. Since $f'(x) = \cos x - \sin x$, critical points are $x = \pi/4 + n\pi$, where n is an integer. Since $f'(x)$ changes from positive to negative as x increases through the critical points $\pi/4 + 2n\pi$, these points yield relative maxima. The remaining critical points $5\pi/4 + 2n\pi$ give relative minima as $f'(x)$ changes from negative to positive through these points.

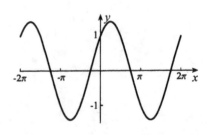

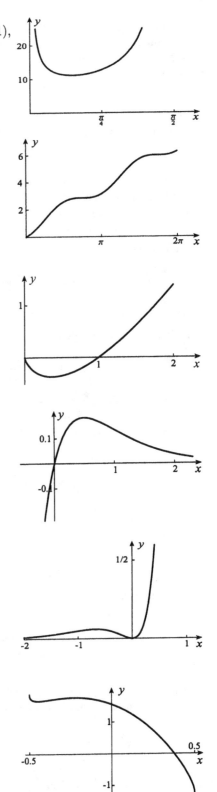

32. Since $f'(x) = -\csc x \cot x + 8 \sec x \tan x = \dfrac{\cos x}{\sin^2 x}(8 \tan^3 x - 1)$, critical points satisfy $\tan x = 1/2$. The only solution of this equation between 0 and $\pi/2$ is 0.464. Because $f'(x)$ changes from negative to positive as x increases through this critical point, it yields a relative minimum.

34. Since $f'(x) = 1 + 2 \sin x \cos x = 1 + \sin 2x$, critical points occur when $\sin 2x = -1$. The only solutions of this equation in the interval $0 < x < 2\pi$ are $x = 3\pi/4$ and $x = 7\pi/4$. Because $f'(x)$ is always nonnegative, the function is increasing, and the critical points do not give relative extrema.

36. The derivative $f'(x) = \ln x + 1$ vanishes when $x = 1/e$. This critical point yields a relative minimum because $f'(x)$ changes from negative to positive as x increases through $1/e$.

38. Since $f'(x) = e^{-2x} - 2xe^{-2x} = (1 - 2x)e^{-2x}$, the only critical point is $x = 1/2$. Since $f'(x)$ changes from positive to negative as x increases through $1/2$, the critical point gives a relative maximum.

40. Since $f'(x) = 2xe^{3x} + 3x^2 e^{3x} = x(3x + 2)e^{3x}$, we have two critical points, $x = 0$ and $x = -2/3$. Because $f'(x)$ changes from negative to positive as x increases through 0, this critical point gives a relative minimum. The critical point $x = -2/3$ gives a relative maximum since $f'(x)$ changes from positive to negative as x increases through this point.

42. For critical points we solve
$$0 = f'(x) = \frac{-2}{\sqrt{1 - 4x^2}} - 10x = \frac{-2(1 + 5x\sqrt{1 - 4x^2})}{\sqrt{1 - 4x^2}}.$$
When we set $1 + 5x\sqrt{1 - 4x^2} = 0$, we obtain
$$25x^2(1 - 4x^2) = 1$$
$$100x^4 - 25x^2 + 1 = 0$$
$$(20x^2 - 1)(5x^2 - 1) = 0$$

We can only accept the negative solutions of this equation, namely, $x = -1/\sqrt{5}$ and $x = -\sqrt{5}/10$. Since $f'(x)$ changes from negative to positive as x increases through $-1/\sqrt{5}$, this critical point gives a relative minimum. Since $f'(x)$ changes from positive to negative as x increases through $-\sqrt{5}/10$, this critical point gives a relative maximum.

44. Every point is critical. Every integer gives a
relative maximum. Every other value of x
gives both a relative maximum and a relative
minimum.

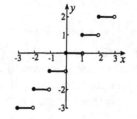

46. Implicit differentiation gives

$$2x + 3y^2 \frac{dy}{dx} + \frac{dy}{dx} = 0 \qquad \Longrightarrow \qquad \frac{dy}{dx} = \frac{-2x}{3y^2 + 1}.$$

The only critical point at which the derivative vanishes is $x = 0$. Since the denominator $3y^2 + 1$ is always
positive, the derivative changes from positive to negative as x increases through 0. Hence, $x = 0$ gives
a relative maximum.

48. Implicit differentiation gives

$$4y^3 \frac{dy}{dx} + y^3 + 3xy^2 \frac{dy}{dx} = 0 \qquad \Longrightarrow \qquad \frac{dy}{dx} = \frac{-y^3}{3xy^2 + 4y^3}.$$

For this to vanish we set $y = 0$, but this is not permitted by the original equation $y^4 + xy^3 = 1$.

50. Implicit differentiation gives

$$2xy^4 + 4x^2y^3 \frac{dy}{dx} + 3y^2 \frac{dy}{dx} = 0 \qquad \Longrightarrow \qquad \frac{dy}{dx} = \frac{-2xy^4}{4x^2y^3 + 3y^2}.$$

For this to vanish we set either $x = 0$ or $y = 0$. The original equation $x^2y^4 + y^3 = 1$ does not permit
$y = 0$. Since the denominator $4x^2y^3 + 3y^2$ is always positive, the derivative changes from positive to
negative as x increases through 0. Hence, $x = 0$ gives a relative maximum.

52. Implicit differentiation gives

$$4x^3y + x^4 \frac{dy}{dx} + 5y^4 \frac{dy}{dx} = 4 \qquad \Longrightarrow \qquad \frac{dy}{dx} = \frac{4 - 4x^3y}{x^4 + 5y^4}.$$

For this to vanish we set $4 - 4x^3y = 0$, from which $y = 1/x^3$. When this is substituted into the original
equation

$$\frac{x^4}{x^3} + \frac{1}{x^{15}} = 4x \qquad \Longrightarrow \qquad 3x^{16} = 1 \qquad \Longrightarrow \qquad x = \pm \frac{1}{3^{1/16}}.$$

54. (a) Implicit differentiation gives

$$\frac{dy}{dx} = 2x\sqrt{1 - y^2} + x^2 \left(\frac{-y}{\sqrt{1 - y^2}} \right) \frac{dy}{dx} \qquad \Longrightarrow \qquad \frac{dy}{dx} = \frac{2x\sqrt{1 - y^2}}{1 + \dfrac{x^2 y}{\sqrt{1 - y^2}}} = \frac{2x(1 - y^2)}{x^2 y + \sqrt{1 - y^2}}.$$

Since y cannot be equal to ± 1, the only critical point at which the derivative vanishes is $x = 0$.
(b) If we square the equation, we obtain

$$y^2 = x^4(1 - y^2) \qquad \Longrightarrow \qquad y^2(1 + x^4) = x^4 \qquad \Longrightarrow \qquad y = \pm \sqrt{\frac{x^4}{1 + x^4}}.$$

Since y must be positive, the explicit definition of the function is $y = \dfrac{x^2}{\sqrt{1 + x^4}}$. For critical points we
solve

$$0 = \frac{dy}{dx} = \frac{\sqrt{1 + x^4}(2x) - x^2(1/2)(1 + x^4)^{-1/2}(4x^3)}{1 + x^4} = \frac{2x(1 + x^4) - 2x^5}{(1 + x^4)^{3/2}} = \frac{2x}{(1 + x^4)^{3/2}}.$$

The only solution is $x = 0$.

56. (a) Since $f'_+(0) = 0$, it follows that $x = 0$ is critical. The point $x = 1$ is not critical since $f'_-(1) = 2$.
(b) No. End points of intervals cannot be relative extrema (see Definitions 4.3 and 4.4).

58. False   According to Definitions 4.3 and 4.4 they can be neither relative maxima nor relative minima.

60. True   All values of x give relative maxima and minima for the function $f(x) = 1$.

62. False   The discontinuous function to
the right has two relative maxima but no
relative minima.

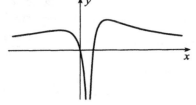

64. False   The function in Figure 2.9(b) has an infinite number of such critical points in the interval $0 < x < 1$.

66. (a) Critical points of $P(V) = \dfrac{RTe^{-a/(RTV)}}{V - b}$ are defined by

$$0 = P'(V) = RT \left[\frac{(V - b)e^{-a/(RTV)}[a/(RTV^2)] - e^{-a/(RTV)}}{(V - b)^2} \right].$$

This implies that

$$0 = \frac{a(V - b)}{RTV^2} - 1 \qquad \Longrightarrow \qquad (RT)V^2 - aV + ab = 0,$$

a quadratic equation with solutions $V = \dfrac{a \pm \sqrt{a^2 - 4abRT}}{2RT}$. These exist provided $a^2 - 4abRT > 0 \Longrightarrow$
$T < a/(4bR)$.
(b) When $T = T_c = a/(4bR)$, there is one critical point, $V = \dfrac{a}{2RT_c} = \dfrac{a}{2R[a/(4bR)]} = 2b$,   at which

$$P = \frac{RT_c e^{-\frac{a}{RT_c(2b)}}}{2b - b} = \frac{R\left(\frac{a}{4bR}\right)e^{-2}}{b} = \frac{a}{4b^2e^2}.$$

68. (a) A plot is shown to the right.
(b) Since $f'(x) = e^{-x}\cos x - e^{-x}\sin x$
$= e^{-x}(\cos x - \sin x)$, critical points are
$x = \pi/4 + n\pi$, where $n \geq 0$ is an integer.
The graph indicates (as would the first
derivative test) that relative maxima
occur at $x = \pi/4 + 2n\pi$ and relative
minima occur at $x = \pi/4 + (2n + 1)\pi$.

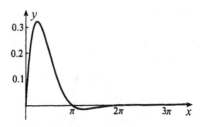

70. Critical points are defined by $0 = f'(x) = 4x^3 - 20x - 4$. The graph of $f(x)$ indicates that there are
three critical points. To find the positive one, we use
Newton's iterative procedure with

$$x_1 = 2.2, \qquad x_{n+1} = x_n - \frac{x_n^3 - 5x_n - 1}{3x_n^2 - 5}.$$

Iteration gives

$$x_2 = 2.342\,02, \qquad x_3 = 2.330\,15,$$
$$x_4 = 2.330\,06, \qquad x_5 = 2.330\,06.$$

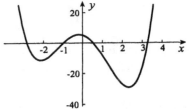

Since $f'(2.330\,05) = -3.9 \times 10^{-4}$ and $f'(2.330\,15) = 4.1 \times 10^{-3}$, it follows that $2.330\,1$ is a critical
point. Since $f'(x)$ changes from negative to positive as x increases through this value, there is a relative
minimum at this critical point. Similar procedures give a relative maximum at $-0.201\,6$ and a relative
minimum at $-2.128\,4$.

72. For critical points we solve

$$0 = f'(x) = \frac{(x^2 - 5x + 4)^2(2x) - (x^2 - 4)(2)(x^2 - 5x + 4)(2x - 5)}{(x^2 - 5x + 4)^4} = \frac{-2(x^3 - 12x + 20)}{(x - 1)^3(x - 4)^3}.$$

The graph of $x^3 - 12x + 20$ to the right indicates that the only critical point is slightly less than -4. To find it we use Newton's iterative procedure with

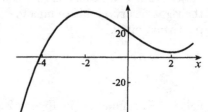

$$x_1 = -4, \quad x_{n+1} = x_n - \frac{x_n^3 - 12x_n + 20}{3x_n^2 - 12}.$$

Iteration gives

$$x_2 = -4.111, \qquad x_3 = -4.107\,25,$$
$$x_4 = -4.107\,24, \quad x_5 = -4.107\,25.$$

Since $f'(-4.107\,25) = 7.4 \times 10^{-9}$ and $f'(-4.107\,15) = -1.0 \times 10^{-7}$, the critical point is $-4.107\,2$. It yields a relative maximum.

74. If we differentiate the equation of the cardioid with respect to x, we obtain

$2(x^2 + y^2 + x)\left(2x + 2y\dfrac{dy}{dx} + 1\right) = 2x + 2y\dfrac{dy}{dx}$. When we solve this for dy/dx and set it equal to 0, we

obtain $0 = \dfrac{dy}{dx} = \dfrac{(2x + 1)(x^2 + y^2 + x) - x}{y - 2y(x^2 + y^2 + x)}$. We now set the numerator equal to 0, and this implies that $x^2 + y^2 + x = x/(2x + 1)$. Substitution of this into the equation of the cardioid gives

$$\frac{x^2}{(2x + 1)^2} = x^2 + y^2 \qquad \Longrightarrow \qquad \frac{x^2}{(2x + 1)^2} = \frac{x}{2x + 1} - x.$$

Since $x = 0$ does not lead to a maximum for y, we set $x/(2x + 1)^2 = 1/(2x + 1) - 1$, and this equation simplifies to $4x^2 + 3x = 0$, with solution $x = -3/4$. The y-coordinate of the point on the cardioid in the second quadrant corresponding to this value of x is $3\sqrt{3}/4$.

76. (a) By equation 3.2, $f'(0) = \lim\limits_{h \to 0} \dfrac{f(h) - f(0)}{h} = \lim\limits_{h \to 0} \dfrac{h \sin(1/h)}{h} = \lim\limits_{h \to 0} \sin(1/h)$, and this limit does not exist (see Example 2.9). Since $f'(0)$ does not exist, but $f(0)$ does, $x = 0$ is a critical point.
(b) The plot of $f'(x) = \sin(1/x) - (1/x)\cos(1/x)$ in the left figure below clearly shows that the derivative does not change sign as x increases through 0; the sign oscillates more and more rapidly the closer x is to zero.
(c) No   In every interval around $x = 0$, the function takes on both positive and negative values. (right graph below).

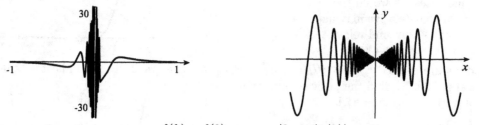

78. (a) By definition 3.2, $f'(0) = \lim\limits_{h \to 0} \dfrac{f(h) - f(0)}{h} = \lim\limits_{h \to 0} \dfrac{-|h \sin(1/h)|}{h}$, and this limit does not exist. Since the derivative does not exist at $x = 0$, but the function has a value there, $x = 0$ is a critical point.

(b) The plot of $f'(x) = -\dfrac{|x \sin(1/x)|}{x \sin(1/x)}[\sin(1/x) - (1/x)\cos(1/x)]$
clearly shows that the derivative does not
change from positive to negative nor from
negative to positive as x increases through 0;
its sign oscillates more and more rapidly the
closer x is to zero. In addition, there is an
increasing number of values of x at which the
derivative does not exist, namely $x = 1/(n\pi)$,
where n is an integer.

(c) Since function values are always negative for $x \neq 0$, and $f(0) = 0$, it follows that a relative maximum
occurs at $x = 0$.

EXERCISES 4.4

2. Since $f''(x) = 36x^2 + 24x = 12x(3x + 2)$, it follows that $f''(x) \leq 0$ for $-2/3 \leq x \leq 0$, and $f''(x) \geq 0$
for $x \leq -2/3$ and $x \geq 0$. Consequently, the graph of the function is concave downward on the interval
$-2/3 \leq x \leq 0$, and concave upward on the intervals $x \leq -2/3$ and $x \geq 0$. The points $(-2/3, 470/27)$
and $(0, 2)$ that separate these intervals are points of inflection.

4. Since $f'(x) = \dfrac{(x^2 - 1)(2x) - (x^2 + 4)(2x)}{(x^2 - 1)^2} = \dfrac{-10x}{(x^2 - 1)^2}$, the second derivative vanishes when

$$0 = f''(x) = -10\left[\frac{(x^2 - 1)^2(1) - x(2)(x^2 - 1)(2x)}{(x^2 - 1)^4}\right] = \frac{10(3x^2 + 1)}{(x^2 - 1)^3}.$$

Since $f''(x) > 0$ for $x < -1$ and for $x > 1$, and $f''(x) < 0$ for $-1 < x < 1$, the graph is concave upward
on the intervals $x < -1$ and $x > 1$, and concave downward on $-1 < x < 1$. Since $f(1)$ and $f(-1)$ are
not defined, there are no points of inflection.

6. Since $f''(x) = 2 + \sin x$, it follows that $f''(x) > 0$ for all x. Hence the graph is always concave upward.

8. Since $f''(x) = 2 + 4\sin x$, the second derivative vanishes when $\sin x = -1/2$. The angles in the interval
$|x| < 2\pi$ that satisfy this equation are $x = -5\pi/6$, $-\pi/6$, $7\pi/6$, and $11\pi/6$. The graph of the function
indicates that the graph is concave upward on the
intervals $-2\pi < x \leq -5\pi/6$, $-\pi/6 \leq x \leq 7\pi/6$,
and $11\pi/6 \leq x < 2\pi$; it is concave downward on
$-5\pi/6 \leq x \leq -\pi/6$ and $7\pi/6 \leq x \leq 11\pi/6$. The points
which separate these intervals are points of inflection,
namely, $(-5\pi/6, 2 + 25\pi^2/36)$, $(-\pi/6, 2 + \pi^2/36)$,
$(7\pi/6, 2 + 49\pi^2/36)$, and $(11\pi/6, 2 + 121\pi^2/36)$.

10. Since $f'(x) = 2x\ln x + x$, we obtain $f''(x) = 2\ln x + 3$. It follows that $f''(x) = 0$ when $x = e^{-3/2}$. Since
$f''(x) \geq 0$ for $x \geq e^{-3/2}$, and $f''(x) \leq 0$ for $0 < x \leq e^{-3/2}$, it follows that the graph is concave downward
on the interval $0 < x \leq e^{-3/2}$, and it is concave upward for $x \geq e^{-3/2}$. The point $(e^{-3/2}, -3/(2e^3))$
which separates these intervals is a point of inflection.

12. Since $f'(x) = e^{-2x} - 2x e^{-2x}$, we find that $f''(x) = -4e^{-2x} + 4x e^{-2x} = 4(x-1)e^{-2x}$. Because $f''(x) \leq 0$
for $x \leq 1$, and $f''(x) \geq 0$ for $x \geq 1$, the graph is concave downward for $x \leq 1$, and concave upward for
$x \geq 1$. The point of inflection is $(1, e^{-2})$.

14. Since $f'(x) = 2x + e^{-x}$, we obtain $f''(x) = 2 - e^{-x}$. It follows that $f''(x) = 0$ when $x = -\ln 2$. Because
$f''(x) \leq 0$ when $x \leq -\ln 2$, and $f''(x) \geq 0$ when $x \geq -\ln 2$, the graph is concave downward on the
interval $x \leq -\ln 2$ and concave upward on $x \geq -\ln 2$. The point $(-\ln 2, (\ln 2)^2 - 2)$ separating these
intervals is a point of inflection.

16. Since $f'(x) = 1 - 1/x^2$, critical points occur at $x = \pm 1$. With $f''(x) = 2/x^3$, we calculate that
$f''(-1) = -2$ and $f''(1) = 2$. Hence, $x = -1$ yields a relative maximum and $x = 1$ a relative minimum.

18. Since $f'(x) = (5/4)x^{1/4} - 1/4x^{-3/4} = \dfrac{5x-1}{4x^{3/4}}$, the only critical point at which $f'(x) = 0$ is $x = 1/5$. Because $f''(x) = (5/16)x^{-3/4} + (3/16)x^{-7/4} = \dfrac{5x+3}{16x^{7/4}}$, it follows that $f''(1/5) > 0$, and therefore $x = 1/5$ yields a relative minimum.

20. Since $f'(x) = 2x \ln x + x = x(2 \ln x + 1)$, the only critical point at which $f'(x) = 0$ is $x = 1/\sqrt{e}$. Because $f''(x) = 2 \ln x + 3$, it follows that $f''(1/\sqrt{e}) = 2(-1/2) + 3 = 2 > 0$, and the critical point yields a relative minimum.

22. Since $f'(x) = 2xe^{-2x} - 2x^2e^{-x} = 2x(1-x)e^{-2x}$, critical points at which $f'(x) = 0$ are $x = 0$ and $x = 1$. Because $f''(x) = (2 - 4x)e^{-2x} - 2(2x - 2x^2)e^{-2x} = (2 - 8x + 4x^2)e^{-2x}$, it follows that $f''(0) = 2$ and $f''(1) = -2e^{-2}$. Hence, $x = 0$ gives a relative minimum and $x = 1$ gives a relative maximum.

24. For points of inflection we solve $0 = \dfrac{d^2y}{dx^2} = \dfrac{d}{dx}(\sin x + x \cos x) = 2 \cos x - x \sin x$. If (x_0, y_0) is a point of inflection, then x_0 satisfies $0 = 2 \cos x_0 - x_0 \sin x_0$, and y_0 is given by $y_0 = x_0 \sin x_0$. It follows that $0 = 2 \cos x_0 - y_0 \implies y_0 = 2 \cos x_0$. In other words, (x_0, y_0) is on the curve $y = 2 \cos x$.

26.

28.

30.

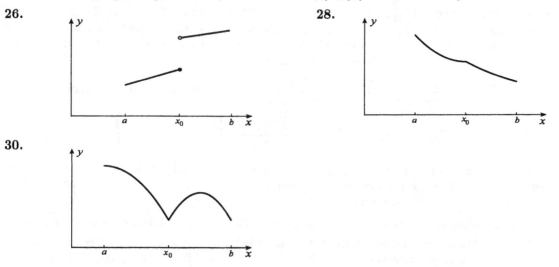

32. According to Example 3.19 in Section 3.5, $[f(x)g(x)]'' = f''(x)g(x) + 2f'(x)g'(x) + f(x)g''(x)$. Since the graphs of $f(x)$ and $g(x)$ are concave upward on I, we can state that $f''(x) \geq 0$ and $g''(x) \geq 0$ on I. This does not, however, guarantee that $[f(x)g(x)]'' \geq 0$ on I.

34. For a horizontal point of inflection we set first and second derivatives equal to 0,

$$0 = P'(V) = \frac{-RT}{(V-b)^2}e^{-a/(RT^{3/2}V)} + \frac{RT}{V-b}e^{-a/(RT^{3/2}V)}\left(\frac{a}{RT^{3/2}V^2}\right)$$

$$= RTe^{-a/(RT^{3/2}V)}\left[\frac{-1}{(V-b)^2} + \frac{a}{RT^{3/2}V^2(V-b)}\right]$$

$$= \frac{RTe^{-a/(RT^{3/2}V)}}{RT^{3/2}V^2(V-b)^2}[a(V-b) - RT^{3/2}V^2],$$

$$0 = P''(V) = RTe^{-a/(RT^{3/2}V)}\left(\frac{a}{RT^{3/2}V^2}\right)\left[\frac{-1}{(V-b)^2} + \frac{a}{RT^{3/2}V^2(V-b)}\right]$$

$$+ RTe^{-a/(RT^{3/2}V)}\left[\frac{2}{(V-b)^3} - \frac{2a}{RT^{3/2}V^3(V-b)} - \frac{a}{RT^{3/2}V^2(V-b)^2}\right].$$

Because $P'(V) = 0$, the term in the first set of brackets of $P''(V)$ vanishes, and we therefore set

$$0 = \frac{RTe^{-a/(RT^{3/2}V)}}{RT^{3/2}V^3(V-b)^3}[2RT^{3/2}V^3 - 2a(V-b)^2 - aV(V-b)].$$

From $0 = P'(V)$, we obtain $\quad V - b = \dfrac{RT^{3/2}V^2}{a}, \quad$ which we substitute into $0 = P''(V)$,

$$0 = 2RT^{3/2}V^3 - 2a\left(\frac{R^2T^3V^4}{a^2}\right) - aV\left(\frac{RT^{3/2}V^2}{a}\right) = \frac{RT^{3/2}V^3}{a}(2a - 2RT^{3/2}V - a)$$

$$= \frac{RT^{3/2}V^3}{a}(a - 2RT^{3/2}V).$$

Thus, $a = 2RT^{3/2}V$. It now follows that

$$b = V - \frac{RT^{3/2}V^2}{2RT^{3/2}V} = \frac{V}{2}.$$

We now have a and b in terms of T and V. To replace V with P, we use Dieterici's equation to write

$$P = \frac{RT}{V - b}e^{-a/(RT^{3/2}V)}.$$

When we substitute the expressions for a and b into this equation we obtain

$$P = \frac{RT}{V - V/2}e^{-2RT^{3/2}V/(RT^{3/2}V)} = \frac{2RT}{e^2V} \quad\Longrightarrow\quad V = \frac{2RT}{e^2P}.$$

This then gives

$$a = 2RT^{3/2}\left(\frac{2RT}{e^2P}\right) = \frac{4R^2T^{5/2}}{e^2P}, \qquad b = \frac{RT}{e^2P};$$

that is, expressions for a and b in terms of critical temperature and pressure are

$$a = \frac{4R^2T_c^{5/2}}{e^2P_c}, \qquad b = \frac{RT_c}{e^2P_c}.$$

36. $f'(x) = \dfrac{(x^2 + k^2)(-1) - (k - x)(2x)}{(x^2 + k^2)^2} = \dfrac{x^2 - 2kx - k^2}{(x^2 + k^2)^2}$ For points of inflection we solve

$$0 = f''(x) = \frac{(x^2 + k^2)^2(2x - 2k) - (x^2 - 2kx - k^2)(2)(x^2 + k^2)(2x)}{(x^2 + k^2)^4},$$

and this simplifies to $0 = \dfrac{-2(x + k)(x^2 - 4kx + k^2)}{(x^2 + k^2)^3}$. Thus, $x = -k$ and $x = (4k \pm \sqrt{16k^2 - 4k^2})/2 = (2 \pm \sqrt{3})k$. Since $f''(x)$ changes sign as x passes through each of these values, each gives a point of inflection, and these points are

$$P\left(-k, \frac{1}{k}\right), \quad Q\left((2 + \sqrt{3})k, \frac{1 - \sqrt{3}}{4k}\right), \quad R\left((2 - \sqrt{3})k, \frac{1 + \sqrt{3}}{4k}\right).$$

These points are collinear if the slopes of line segments PR and PQ are equal. This is indeed the case since

$$\text{Slope of } PR = \frac{\dfrac{1 + \sqrt{3}}{4k} - \dfrac{1}{k}}{(2 - \sqrt{3})k + k} = \frac{1 + \sqrt{3} - 4}{4k^2(2 - \sqrt{3} + 1)} = -\frac{1}{4k^2},$$

$$\text{Slope of } PQ = \frac{\dfrac{1 - \sqrt{3}}{4k} - \dfrac{1}{k}}{(2 + \sqrt{3})k + k} = \frac{1 - \sqrt{3} - 4}{4k^2(2 + \sqrt{3} + 1)} = -\frac{1}{4k^2}.$$

EXERCISES 4.5

2. Since $f'(x) = 3x^2 - 12x + 12 = 3(x-2)^2$,
the only critical point is $x = 2$. Because $f'(x)$
remains positive as x passes through 2, the
critical point gives a horizontal point of
inflection at $(2, 17)$. Since $f''(x) = 6(x-2)$,
there are no other points of inflection.
The graph is shown to the right.

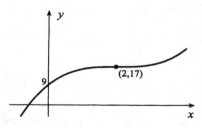

4. Since $f'(x) = 6x^2 - 30x + 6 = 6(x^2 - 5x + 1)$, the critical points are $x = (5 \pm \sqrt{25 - 4})/2 = (5 \pm \sqrt{21})/2$. Since $f'(x)$ changes from positive to negative as x increases through $(5 - \sqrt{21})/2$, there is a relative maximum at $\big((5 - \sqrt{21})/2, (-87 + 21\sqrt{21})/2\big)$. Similarly, there is a relative minimum at $\big((5 + \sqrt{21})/2, (-87 - 21\sqrt{21})/2\big)$. Since $0 = f''(x) = 12x - 30$ at $x = 5/2$, and $f''(x)$ changes sign as x passes through $5/2$, there is a point of inflection at $(5/2, -87/2)$. This information is shown in the left figure below. We complete the graph of this cubic polynomial as shown in the right figure.

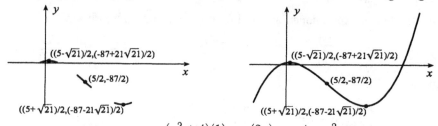

6. For critical points we solve $0 = f'(x) = \dfrac{(x^2 + 4)(1) - x(2x)}{(x^2 + 4)^2} = \dfrac{4 - x^2}{(x^2 + 4)^2}$. Solutions are $x = \pm 2$. We now calculate $f''(x) = \dfrac{(x^2 + 4)^2(-2x) - (4 - x^2)(2)(x^2 + 4)(2x)}{(x^2 + 4)^4} = \dfrac{2x(x^2 - 12)}{(x^2 + 4)^3}$. Since $f''(-2) = 1/16$ and $f''(2) = -1/16$, there is a relative minimum at $x = -2$ equal to $f(-2) = -1/4$, and a relative maximum at $x = 2$ of $f(2) = 1/4$.

Because $f''(x) = 0$ at $x = 0, \pm 2\sqrt{3}$, and $f''(x)$ changes sign as x passes through each of these, points of inflection occur at $(0, 0)$, $(2\sqrt{3}, \sqrt{3}/8)$, and $(-2\sqrt{3}, -\sqrt{3}/8)$.

This information is shown in the left diagram below, together with the limits $\lim_{x \to -\infty} f(x) = 0^-$ and $\lim_{x \to \infty} f(x) = 0^+$. The final graph is shown to the right. We could have shortened the analysis by considering only the right half of the graph and using the fact that the function is odd.

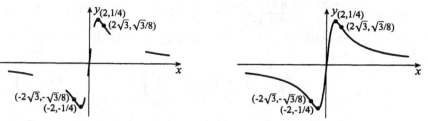

8. For critical points we solve $0 = f'(x) = (2/3)x^{-1/3}(8 - x) - x^{2/3} = \dfrac{16 - 5x}{3x^{1/3}}$. Clearly, $x = 16/5$ is critical, but so also is $x = 0$ because $f'(0)$ does not exist and $f(0) = 0$. We now calculate

$$f''(x) = \frac{3x^{1/3}(-5) - (16 - 5x)x^{-2/3}}{9x^{2/3}} = \frac{-2(5x + 8)}{9x^{4/3}}.$$

Since $f''(16/5) < 0$, there is a relative maximum of $f(16/5) = 10.42$. Because $f'(x)$ changes from negative to positive as x increases through 0, this critical point gives a relative mimumum of $f(0) = 0$.

Since $f''(-8/5) = 0$, and $f''(x)$ changes sign as x passes through $-8/5$, there is a point of inflection at $(-1.6, 13.13)$.

We have shown this information in the left figure below along with the additional facts that $\lim_{x \to 0^-} f'(x) = -\infty$ and $\lim_{x \to 0^+} f'(x) = \infty$. The final graph is shown to the right.

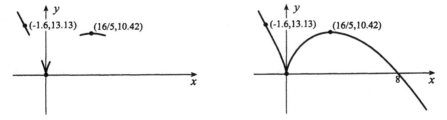

10. For critical points we solve

$$0 = f'(x) = 3x^{1/2} - 9 + 6x^{-1/2} = \frac{3(x - 3\sqrt{x} + 2)}{\sqrt{x}} = \frac{3(\sqrt{x} - 1)(\sqrt{x} - 2)}{\sqrt{x}}.$$

Solutions are $x = 1$ and $x = 4$. The point $x = 0$ is also critical since $f'(0)$ is not defined but $f(0) = 0$. To classify these critical points we calculate

$$f''(x) = \frac{3}{2\sqrt{x}} - \frac{3}{x^{3/2}} = \frac{3(x - 2)}{2x^{3/2}}.$$

Since $f''(1) = -3/2$ and $f''(4) = 3/8$, $f(1) = 5$ is a relative maximum and $f(4) = 4$ is a relative minimum. Because $f(x)$ is not defined for $x < 0$, the critical point $x = 0$ does not give a relative extrema. For graphical purposes we note that $\lim_{x \to 0+} f'(x) = \infty$.

Since $f''(2) = 0$, and $f''(x)$ changes sign as x passes through 2, there is a point of inflection at $(2, 16\sqrt{2} - 18)$.

This information is shown in the left figure below. The final graph is shown to the right.

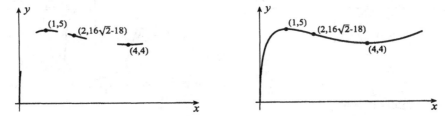

12. The function is discontinuous at $x = 0$. Left- and right-hand limits of the function as $x \to 0$ are $\lim_{x \to 0-} f(x) = -\infty$ and $\lim_{x \to 0+} f(x) = \infty$. For critical points we express the function in the form $f(x) = x + 1 + \dfrac{1}{x}$, and solve $0 = f'(x) = 1 - \dfrac{1}{x^2}$. Clearly, $x = \pm 1$. The second derivative is $f''(x) = 2/x^3$. Since $f''(1) = 2$, there is a relative minimum at $x = 1$ of $f(1) = 3$. With $f''(-1) = -2$, there is a relative maximum of $f(-2) = -1$. Since $f''(x)$ is never zero, there can be no points of inflection. Finally, we note that $y = x + 1$ is an oblique asymptote. This information is shown in the left figure below. The final graph is shown to the right.

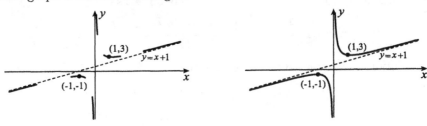

14. The function is discontinuous at $x = \pm 2$. We take left-and right-hand limits of $f(x) = \dfrac{x^3}{(x+2)(x-2)}$ as $x \to \pm 2$

$$\lim_{x \to -2^-} f(x) = -\infty, \quad \lim_{x \to -2^+} f(x) = \infty, \quad \lim_{x \to 2^-} f(x) = -\infty, \quad \lim_{x \to 2^+} f(x) = \infty.$$

For critical points we solve

$$0 = f'(x) = \frac{(x^2 - 4)(3x^2) - x^3(2x)}{(x^2 - 4)^2} = \frac{x^2(x^2 - 12)}{(x^2 - 4)^2}.$$

Solutions are $x = 0, \pm 2\sqrt{3}$. The second derivative is

$$f''(x) = \frac{(x^2 - 4)^2(4x^3 - 24x) - (x^4 - 12x^2)(2)(x^2 - 4)(2x)}{(x^2 - 4)^4} = \frac{8x(x^2 + 12)}{(x^2 - 4)^3}.$$

Since $f''(2\sqrt{3}) > 0$, we have a relative minimum $f(2\sqrt{3}) = 3\sqrt{3}$. Similarly, $f''(-2\sqrt{3}) < 0$ indicates that $f(-2\sqrt{3}) = -3\sqrt{3}$ is a relative maximum. Since $f''(x) = 0$ only at $x = 0$, and $f''(x)$ changes sign as x passes through 0, there is a horizontal point of inflection at $(0,0)$.

By writing $f(x)$ in the form $f(x) = x + \dfrac{4x}{x^2 - 4}$, we see that the graph is asymptotic to the line $y = x$. This information is shown in the left figure below. The final graph is shown to the right.

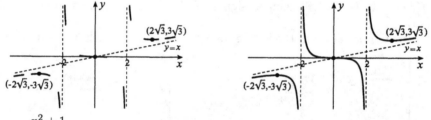

16. With $f(x) = \dfrac{x^2 + 1}{(x-1)(x+1)}$, we see that the function is discontinuous at $x = \pm 1$. Left-and right-hand limits at these values of x are

$$\lim_{x \to -1^-} f(x) = \infty, \quad \lim_{x \to -1^+} f(x) = -\infty, \quad \lim_{x \to 1^-} f(x) = -\infty, \quad \lim_{x \to -1^+} f(x) = \infty.$$

For critical points, we solve

$$0 = f'(x) = \frac{(x^2 - 1)(2x) - (x^2 + 1)(2x)}{(x^2 - 1)^2} = \frac{-4x}{(x^2 - 1)^2}.$$

The only solution is $x = 0$. Since $f'(x)$ changes from positive to negative as x increases through 0, there is a relative maximum of $f(0) = -1$. For points of inflection, we solve

$$0 = f''(x) = \frac{(x^2 - 1)^2(-4) - (-4x)(2)(x^2 - 1)(2x)}{(x^2 - 1)^4} = \frac{12x^2 + 4}{(x^2 - 1)^3}.$$

There are no solutions indicating no points of inflection.

By writing $f(x)$ in the form $f(x) = 1 + \dfrac{1}{x^2 - 1}$, we see that $y = 1$ is a horizontal asymptote, and

$$\lim_{x \to -\infty} f(x) = 1^+, \quad \lim_{x \to \infty} f(x) = 1^+.$$

This information is shown in the left figure below. The final graph is shown to the right.

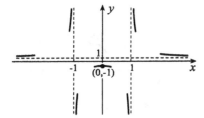

18. For critical points we solve
$f'(x) = (2/3)(x-1)^{-1/3} = 0$. There
is no solution, but $x = 1$ is critical
since $f(1) = 0$ and $f'(1)$ does not exist.
Because $f(x) > 0$ for all $x \neq 1$, it follows
that $f(1)$ is a relative minimum. Since
$f''(x) = -(2/9)(x-1)^{-4/3}$ never vanishes,
there are no points of inflection. The graph
is shown to the right.

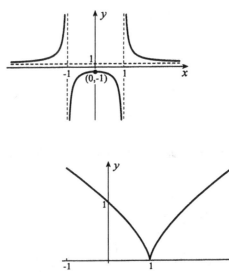

20. For critical points we solve $0 = f'(x) = 1 + 2\cos x$. Solutions are $x = \pm 2\pi/3 + 2n\pi$, where n is an integer. To classify them we calculate $f''(x) = -2\sin x$. Since $f''(2\pi/3 + 2n\pi) < 0$, the critical points $x = 2\pi/3 + 2n\pi$ give relative maxima $f(2\pi/3 + 2n\pi) = \sqrt{3} + 2\pi/3 + 2n\pi$. Similarly, there are relative minima at $(-2\pi/3 + 2n\pi, -\sqrt{3} - 2\pi/3 + 2n\pi)$.

For points of inflection, we solve $0 = f''(x) = -2\sin x$. Solutions are $x = n\pi$, where n is an integer. Since $f''(x)$ changes sign at each of these, they all give points of inflection. This information is shown in the left figure below. The final graph is shown to the right. It oscillates about the line $y = x$.

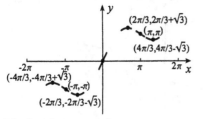

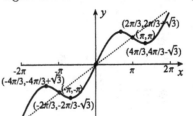

22. For critical points we solve

$$0 = f'(x) = \text{Sin}^{-1}x + \frac{x}{\sqrt{1-x^2}} - \frac{x}{\sqrt{1-x^2}}$$
$$= \text{Sin}^{-1}x.$$

The only solution is $x = 0$. For points of inflection
we solve $0 = f''(x) = \dfrac{1}{\sqrt{1-x^2}}$.
There are no solutions. Because $f''(0) = 1$, $f(x)$ has
a relative minimum of $f(0) = 1$.

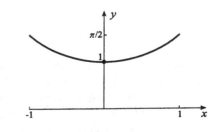

24. For critical points we solve

$$0 = f'(x) = 4x^3 + 30x^2 + 12x - 64 = 2(x+2)(2x^2 + 11x - 16).$$

Solutions are $x = -2$ and $x = (-11 \pm \sqrt{121 + 128})/4 = (-11 \pm \sqrt{249})/4$. Since $f'(x)$ changes from positive to negative as x increases through -2, there is a relative maximum at this critical point of $f(-2) = 93$. Since $f'(x)$ changes from negative to positive as x increases through the remaining critical points, they give relative minima of $f\big((-11-\sqrt{249})/4\big) = f(-6.69) = -289.4$ and $f\big((-11+\sqrt{249})/4\big) = f(1.19) = -43.8$. This information is shown in the left figure below. The final graph is shown to the right.

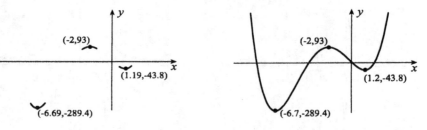

26. The function is discontinuous at $x = 5$, and therefore we calculate

$$\lim_{x \to 5^-} f(x) = -\infty \qquad \text{and} \qquad \lim_{x \to 5^+} f(x) = \infty.$$

For critical points we solve

$$0 = f'(x) = \frac{(x-5)(2x) - (x^2 - 8)(1)}{(x-5)^2} = \frac{x^2 - 10x + 8}{(x-5)^2}.$$

Solutions are $x = \dfrac{10 \pm \sqrt{100 - 32}}{2} = 5 \pm \sqrt{17}$. Since $f'(x)$ changes from positive to negative as x increases through $5 - \sqrt{17}$, there is a relative maximum at this critical point of $f(5 - \sqrt{17}) = 10 - 2\sqrt{17}$. Since $f'(x)$ changes from negative to positive as x increases through $5 + \sqrt{17}$, there is a relative minimum at this critical point of $f(5 + \sqrt{17}) = 10 + 2\sqrt{17}$. Since $f(x) = x + 5 + 17/(x-5)$, the graph has oblique asymptote $y = x + 5$. This information, along with x- and y-intercepts, is shown in the left figure below. The final graph is shown to the right.

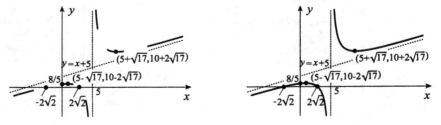

28. For critical points we solve

$$0 = f'(x) = \frac{5}{4}x^{1/4} - \frac{1}{4}x^{-3/4} = \frac{5x - 1}{4x^{3/4}}.$$

The solution is $x = 1/5$. Since $f'(x)$ changes from negative to positive as x increases through the critical point, there is a relative minimum of $f(1/5) = -4/5^{5/4}$. There is also critical point at $x = 0$, but because the function is not defined for $x < 0$, it cannot yield a relative maximum. We note that $f(0) = 0$ and $\lim_{x \to 0^+} f'(x) = -\infty$. This information, along with the x-intercept $x = 1$, is shown in the left figure below. The final graph is shown to the right.

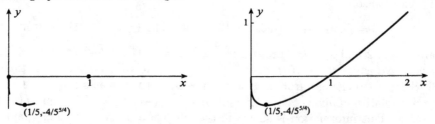

30. The function is discontinuous at $x = 1$ and $x = 4$, and we therefore calculate

$$\lim_{x \to 1^-} f(x) = -\infty, \quad \lim_{x \to 1^+} f(x) = \infty, \quad \lim_{x \to 4^-} f(x) = \infty, \quad \lim_{x \to 4^+} f(x) = -\infty.$$

To find critical points we solve

$$0 = f'(x) = \frac{(x^2 - 5x + 4)(4x - 17) - (2x^2 - 17x + 8)(2x - 5)}{(x^2 - 5x + 4)^2} = \frac{7(x - 2)(x + 2)}{(x^2 - 5x + 4)^2}.$$

Solutions are $x = -2, 2$. Since $f'(x)$ changes from positive to negative as x increases through -2, there is a relative maximum of $f(-2) = 25/9$. Since $f'(x)$ changes from negative to positive as x increases through 2, there is a relative minimum of $f(2) = 9$.

The graph has horizontal asymptote $y = 2$, and to determine how the asymptote is approached, we express $f(x)$ in the form

$$f(x) = \frac{2x^2 - 17x + 8}{x^2 - 5x + 4} = 2 - \frac{7x}{x^2 - 5x + 4}.$$

This shows that $y = 2$ is approached from below as $x \to \infty$ and from above as $x \to -\infty$. This information is shown in the left figure below. The final graph is shown to the right.

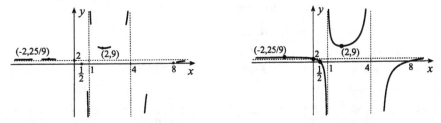

32. For critical points we solve $0 = f'(x) = \cos x - \sin x$. Solutions are $x = \pi/4 + 2n\pi$, where n is an integer. Since $f'(x)$ changes from positive to negative as x increases through $\pi/4 + 2n\pi$, there are a relative maxima of $f(\pi/4 + 2n\pi) = \sqrt{2}$. Since $f'(x)$ changes from negative to positive as x increases through $\pi/4 + (2n + 1)\pi$, there are a relative minima of $f(\pi/4 + (2n + 1)\pi) = -\sqrt{2}$. This information is shown in the left figure below. The final graph is shown to the right.

The function could also be graphed by expressing it as a general sine function $A \sin(x + \phi)$ (see Example 1.45 in Section 1.8),

$$\sin x + \cos x = A \sin(x + \phi) = A(\sin x \cos \phi + \cos x \sin \phi).$$

To satisfy this equation we set $A \cos \phi = 1$ and $A \sin \phi = 1$. These are satisfied if we choose $A = \sqrt{2}$ and $\phi = \pi/4$. Thus, $f(x)$ can be expressed in the form $f(x) = \sqrt{2} \sin(x + \pi/4)$.

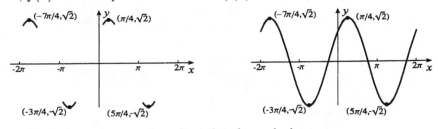

34. The function is discontinuous at $x = 0$, and we therefore calculate

$$\lim_{x \to 0^-} f(x) = -\infty, \qquad \lim_{x \to 0^+} f(x) = \infty.$$

To identify critical points we solve

$$0 = f'(x) = -\frac{8}{x^2}\sqrt{x^2 + 100} + \left(1 + \frac{8}{x}\right)\frac{x}{\sqrt{x^2 + 100}} = \frac{x^3 - 800}{x^2\sqrt{x^2 + 100}}.$$

The only critical point is $800^{1/3}$. Since $f'(x)$ changes from negative to positive as x increases through this critical point, there is a relative minimum of $f(800^{1/3}) = \left(\frac{8 + 800^{1/3}}{800^{1/3}}\right)\sqrt{100 + 800^{2/3}} \approx 25.4$. This information is shown in the left figure below. The final graph is shown to the right.

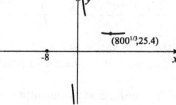

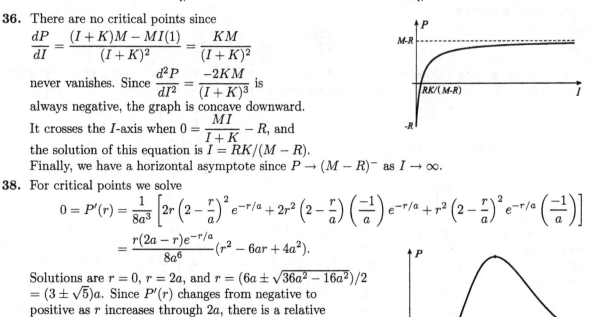

36. There are no critical points since

$$\frac{dP}{dI} = \frac{(I+K)M - MI(1)}{(I+K)^2} = \frac{KM}{(I+K)^2}$$

never vanishes. Since $\dfrac{d^2P}{dI^2} = \dfrac{-2KM}{(I+K)^3}$ is
always negative, the graph is concave downward.
It crosses the I-axis when $0 = \dfrac{MI}{I+K} - R$, and
the solution of this equation is $I = RK/(M-R)$.
Finally, we have a horizontal asymptote since $P \to (M-R)^-$ as $I \to \infty$.

38. For critical points we solve

$$0 = P'(r) = \frac{1}{8a^3}\left[2r\left(2 - \frac{r}{a}\right)^2 e^{-r/a} + 2r^2\left(2 - \frac{r}{a}\right)\left(\frac{-1}{a}\right)e^{-r/a} + r^2\left(2 - \frac{r}{a}\right)^2 e^{-r/a}\left(\frac{-1}{a}\right)\right]$$

$$= \frac{r(2a-r)e^{-r/a}}{8a^6}(r^2 - 6ar + 4a^2).$$

Solutions are $r = 0$, $r = 2a$, and $r = (6a \pm \sqrt{36a^2 - 16a^2})/2$
$= (3 \pm \sqrt{5})a$. Since $P'(r)$ changes from negative to
positive as r increases through $2a$, there is a relative
minimum of $f(2a) = 0$ at $r = 2a$. Since $P'(r)$ changes
from positive to negative as r increases through both
$(3 \pm \sqrt{5})a$, they give relative maxima of $P\big((3 \pm \sqrt{5})a\big)$,
and these simplify to $(2/a)(9 \pm 4\sqrt{5})e^{-3\mp\sqrt{5}}$. When
combined with the fact that $\lim_{r\to\infty} P(r) = 0^+$, the graph to the right is obtained.

40. For critical points we solve $0 = P'(E) = \dfrac{-e^{(E-E_f)/(kT)}}{kT[e^{(E-E_f)/(kT)} + 1]^2}$. There are no solutions to this equation.

For points of inflection we solve

$$0 = P''(E) = \frac{-1}{kT}\left\{\frac{e^{(E-E_f)/(kT)}}{[e^{(E-E_f)/(kT)} + 1]^2}\left(\frac{1}{kT}\right) - \frac{2e^{(E-E_f)/(kT)}}{[e^{(E-E_f)/(kT)} + 1]^3}e^{(E-E_f)/(kT)}\left(\frac{1}{kT}\right)\right\}$$

$$= \frac{e^{(E-E_f)/(kT)}[e^{(E-E_f)/(kT)} - 1]}{k^2T^2[e^{(E-E_f)/(kT)} + 1]^3}.$$

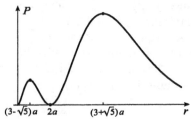

The only solution is $E = E_f$. Since $P''(E)$
changes sign as E passes through E_f, there
is a point of inflection at $(E_f, 1/2)$. The graph
is asymptotic to the E-axis.

42. (a) The plots are shown to the right.
   (b) For critical points we solve

$$0 = 2v\, e^{-Mv^2/(2RT)} + v^2 e^{-Mv^2/(2RT)}\left(\frac{-2Mv}{2RT}\right)$$

$$= v\left(2 - \frac{Mv^2}{RT}\right)e^{-Mv^2/(2RT)}.$$

The positive solution is $v = \sqrt{2RT/M}$.
For points of inflection we solve

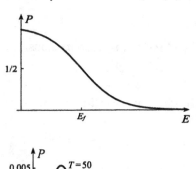

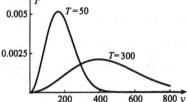

$$0 = \left(2 - \frac{Mv^2}{RT}\right)e^{-Mv^2/(2RT)} + v\left(\frac{-2Mv}{RT}\right)e^{-Mv^2/(2RT)} + v\left(2 - \frac{Mv^2}{RT}\right)e^{-Mv^2/(2RT)}\left(\frac{-Mv}{RT}\right)$$

$$= e^{-Mv^2/(2RT)}\left[2 - \frac{Mv^2}{RT} - \frac{2Mv^2}{RT} - \frac{Mv^2}{RT}\left(2 - \frac{Mv^2}{RT}\right)\right]$$

$$= e^{-Mv^2/(2RT)}\left(\frac{M^2v^4}{R^2T^2} - \frac{5Mv^2}{RT} + 2\right).$$

Solutions of this quadratic in $Mv^2/(RT)$ are $\dfrac{Mv^2}{RT} = \dfrac{5 \pm \sqrt{25 - 8}}{2} \implies v = \sqrt{\dfrac{5 \pm \sqrt{17}}{2}}\sqrt{\dfrac{RT}{M}}$. Since the second derivative changes sign as v passes through these values, they yield points of inflection.

44. The function can be expressed in the form

$$f(x) = \frac{(x+1)(x^2+1)}{(x+1)(x^2-x+1)} = \begin{cases} \dfrac{x^2+1}{x^2-x+1}, & x \neq -1, \\ \text{undefined}, & x = -1. \end{cases}$$

We therefore draw a graph of the function $g(x) = (x^2+1)/(x^2-x+1)$ and then delete the point at $x = -1$. We have a horizontal asymptote for $y = g(x)$ since

$$\lim_{x \to -\infty} \frac{x^2+1}{x^2-x+1} = 1^- \quad \text{and} \quad \lim_{x \to \infty} \frac{x^2+1}{x^2-x+1} = 1^+.$$

For critical points of $g(x)$ we solve

$$0 = g'(x) = \frac{(x^2-x+1)(2x) - (x^2+1)(2x-1)}{(x^2-x+1)^2} = \frac{1-x^2}{(x^2-x+1)^2}.$$

There are two solutions $x = \pm 1$. Since $g'(x)$ changes from negative to positive as x increases through -1, $g(x)$ has a relative minimum $g(-1) = 2/3$. There is a relative maximum $g(1) = 2$ because $g'(x)$ changes from positive to negative as x increases through 1. The graph of $g(x)$ is shown in the left diagram below. That of $f(x)$ is to the right.

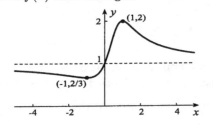
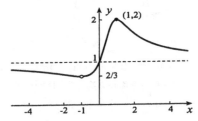

EXERCISES 4.6

2. The plot suggests a horizontal point of inflection. To confirm this we calculate $f'(x) = 3x^2 - 12x + 12 = 3(x-2)^2$. The only critical point is $x = 2$. Because $f'(x)$ remains positive as x passes through 2, the critical point does indeed give a horizontal point of inflection at $(2, 17)$. Since $f''(x) = 6(x-2)$, there are no other points of inflection.

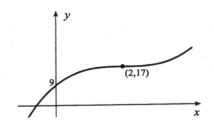

4. The plot in the left figure below indicates one relative maximum, one relative minimum, and one point of inflection. To confirm this, we first find critical points. Since $f'(x) = 6x^2 - 30x + 6 = 6(x^2 - 5x + 1)$, the critical points are $x = (5 \pm \sqrt{25 - 4})/2 = (5 \pm \sqrt{21})/2$. Since $f'(x)$ changes from positive to negative as x increases through $(5 - \sqrt{21})/2$, there is a relative maximum at $((5 - \sqrt{21})/2, (-87 + 21\sqrt{21})/2)$. Similarly, there is a relative minimum at $((5 + \sqrt{21})/2, (-87 - 21\sqrt{21})/2)$. Since $0 = f''(x) = 12x - 30$ at $x = 5/2$, and $f''(x)$ changes sign as x passes through $5/2$, there is a point of inflection at $(5/2, -87/2)$. These are shown in the right figure.

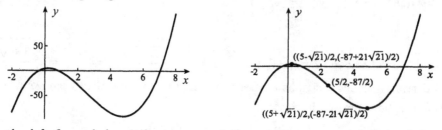

6. The plot in the left figure below indicates one relative maximum, one relative minimum, and three points of inflection. To confirm this, we first find critical points. For critical points we solve the equation $0 = f'(x) = \dfrac{(x^2 + 4)(1) - x(2x)}{(x^2 + 4)^2} = \dfrac{4 - x^2}{(x^2 + 4)^2}$. Solutions are $x = \pm 2$. We now calculate the second derivative $f''(x) = \dfrac{(x^2 + 4)^2(-2x) - (4 - x^2)(2)(x^2 + 4)(2x)}{(x^2 + 4)^4} = \dfrac{2x(x^2 - 12)}{(x^2 + 4)^3}$. Since $f''(-2) = 1/16$ and $f''(2) = -1/16$, there is a relative minimum at $x = -2$ equal to $f(-2) = -1/4$, and a relative maximum at $x = 2$ of $f(2) = 1/4$.

Because $f''(x) = 0$ at $x = 0, \pm 2\sqrt{3}$, and $f''(x)$ changes sign as x passes through each of these, points of inflection occur at $(0, 0)$, $(2\sqrt{3}, \sqrt{3}/8)$, and $(-2\sqrt{3}, -\sqrt{3}/8)$.

The final graph is shown to the right. We could have shortened the analysis by considering only the right half of the graph and using the fact that the function is odd.

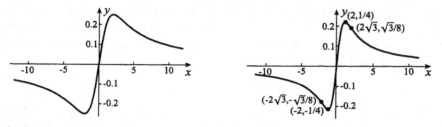

8. The plot in the left figure below indicates one relative minimum one relative maximum, perhaps a point of inflection for negative x, and perhaps a point of inflection at $x = 0$. To decide, we first find critical points. Since $0 = f'(x) = (2/3)x^{-1/3}(8 - x) - x^{2/3} = \dfrac{16 - 5x}{3x^{1/3}}$, $x = 16/5$ is critical, but so also is $x = 0$ because $f'(0)$ does not exist and $f(0) = 0$. We now calculate

$$f''(x) = \frac{3x^{1/3}(-5) - (16 - 5x)x^{-2/3}}{9x^{2/3}} = \frac{-2(5x + 8)}{9x^{4/3}}.$$

Since $f''(16/5) < 0$, there is a relative maximum of $f(16/5) = 10.42$. Because $f'(x)$ changes from negative to positive as x increases through 0, this critical point gives a relative mimumum of $f(0) = 0$.

Since $f''(-8/5) = 0$, and $f''(x)$ changes sign as x passes through $-8/5$, there is a point of inflection at $(-1.6, 13.13)$. The point $(0, 0)$ is not a point of inflection.

The final graph is shown to the right.

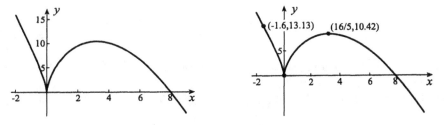

10. The plot in the left figure below indicates one relative minimum one relative maximum, and a point of inflection between them. For critical points we solve

$$0 = f'(x) = 3x^{1/2} - 9 + 6x^{-1/2} = \frac{3(x - 3\sqrt{x} + 2)}{\sqrt{x}} = \frac{3(\sqrt{x} - 1)(\sqrt{x} - 2)}{\sqrt{x}}.$$

Solutions are $x = 1$ and $x = 4$. The point $x = 0$ is also critical since $f'(0)$ is not defined but $f(0) = 0$. To classify these critical points we calculate $f''(x) = \frac{3}{2\sqrt{x}} - \frac{3}{x^{3/2}} = \frac{3(x - 2)}{2x^{3/2}}$. Since $f''(1) = -3/2$ and $f''(4) = 3/8$, $f(1) = 5$ is a relative maximum and $f(4) = 4$ is a relative minimum. Because $f(x)$ is not defined for $x < 0$, the critical point $x = 0$ does not give a relative extrema. We note that $\lim_{x \to 0^+} f'(x) = \infty$.

Since $f''(2) = 0$, and $f''(x)$ changes sign as x passes through 2, there is a point of inflection at $(2, 16\sqrt{2} - 18)$. The final graph is shown to the right.

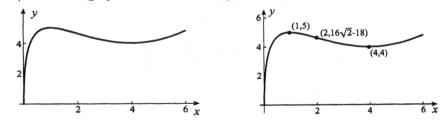

12. The plot in the left figure below indicates one relative minimum, one relative maximum, no points of inflection, and an oblique asymptote. By writing $f(x)$ in the form $f(x) = x + 1 + \frac{1}{x}$, we see that the graph is asymptotic to the line $y = x + 1$. Exact locations of the relative extrema can be determined by solving $0 = f'(x) = 1 - \frac{1}{x^2}$. Clearly, $x = \pm 1$. The second derivative is $f''(x) = 2/x^3$. Since $f''(1) = 2$, there is a relative minimum at $x = 1$ of $f(1) = 3$. With $f''(-1) = -2$, there is a relative maximum of $f(-1) = -1$. Since $f''(x)$ is never zero, there can be no points of inflection. The final graph is shown to the right.

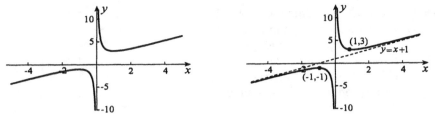

14. The plot in the left figure below indicates one relative minimum, one relative maximum, a horizontal point of inflection at $x = 0$, and an oblique asymptote. By writing $f(x)$ in the form $f(x) = x + \frac{4x}{x^2 - 4}$, we see that the graph is asymptotic to the line $y = x$. There are discontinuities at $x = \pm 2$, which yield vertical asymptotes. To identify the relative extrema, we solve

$$0 = f'(x) = \frac{(x^2 - 4)(3x^2) - x^3(2x)}{(x^2 - 4)^2} = \frac{x^2(x^2 - 12)}{(x^2 - 4)^2}.$$

Solutions are $x = 0, \pm 2\sqrt{3}$. The second derivative is

$$f''(x) = \frac{(x^2-4)^2(4x^3-24x) - (x^4-12x^2)(2)(x^2-4)(2x)}{(x^2-4)^4} = \frac{8x(x^2+12)}{(x^2-4)^3}.$$

The fact that $f''(2\sqrt{3}) > 0$ confirms a relative minimum $f(2\sqrt{3}) = 3\sqrt{3}$. Similarly, $f''(-2\sqrt{3}) < 0$ confirms a relative maximum $f(-2\sqrt{3}) = -3\sqrt{3}$. Since $f''(x) = 0$ only at $x = 0$, and $f''(x)$ changes sign as x passes through 0, there is a horizontal point of inflection at $(0,0)$. The final graph is shown to the right.

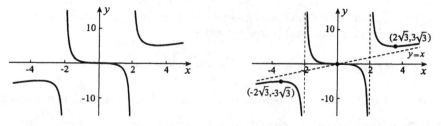

16. The plot in the left figure below suggests one relative maximum, no relative minima, no points of inflection, and a horizontal asymptote. The limits $\lim\limits_{x \to \pm\infty} \dfrac{x^2+1}{x^2-1} = 1^+$ confirm $y = 1$ as the horizontal asymptote. There are discontinuities at $x = \pm 1$, leading to vertical asymptotes. For critical points, we solve

$$0 = f'(x) = \frac{(x^2-1)(2x) - (x^2+1)(2x)}{(x^2-1)^2} = \frac{-4x}{(x^2-1)^2}.$$

The only solution is $x = 0$. Since $f'(x)$ changes from positive to negative as x increases through 0, there is a relative maximum of $f(0) = -1$. For points of inflection, we solve

$$0 = f''(x) = \frac{(x^2-1)^2(-4) - (-4x)(2)(x^2-1)(2x)}{(x^2-1)^4} = \frac{12x^2+4}{(x^2-1)^3}.$$

There are no solutions confirming no points of inflection. The final graph is shown to the right.

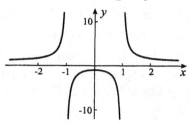

18. There appears to be a relative minimum at $x = 1$ where the derivative does not exist, and no other significant features to the graph. Because $f(x) > 0$ for all $x \neq 1$, it follows that $f(1)$ is indeed a relative minimum. Since $f'(x) = (2/3)(x-1)^{-1/3}$, we see that $f'(x)$ is not defined at the minimum.

20. The plot to the right indicates an infinite number of relative extrema and points of inflection. For critical points we solve $0 = f'(x) = 1 + 2\cos x$. Solutions are $x = \pm 2\pi/3 + 2n\pi$, where n is an integer. Relative maxima occur at $(2\pi/3 + 2n\pi, \sqrt{3} + 2\pi/3 + 2n\pi)$, and relative minima at $(-2\pi/3 + 2n\pi, -\sqrt{3} - 2\pi/3 + 2n\pi)$.

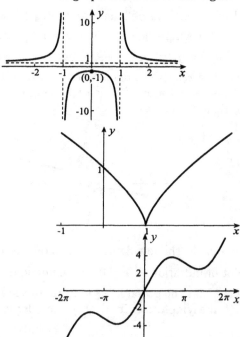

For points of inflection, we solve $0 = f''(x) = -2\sin x$. Solutions are $x = n\pi$, where n is an integer. Since $f''(x)$ changes sign at each of these, they all give points of inflection.

22. For critical points we solve

$$0 = f'(x) = 4x^3 + 30x^2 + 12x - 64$$
$$= 2(x+2)(2x^2 + 11x - 16).$$

Solutions of this equation are $x = -2$ and $x = (-11 \pm \sqrt{121 + 128})/4 = (-11 \pm \sqrt{249})/4$. Hence, we have a relative maximum $f(-2) = 93$ and relative minima $f\big((-11 - \sqrt{249})/4\big) = f(-6.69) = -289.4$ and $f\big((-11 + \sqrt{249})/4\big) = f(1.19) = -43.8$.

Solutions of this equation are $x = -1/2, 1$. We have a

24. Since $f(x) = x + 5 + 17/(x - 5)$, critical points are given by $0 = f'(x) = 1 - 17/(x - 5)^2$. Solutions are $x = 5 \pm \sqrt{17}$. These gives relative extrema equal to $10 \pm 2\sqrt{17}$.

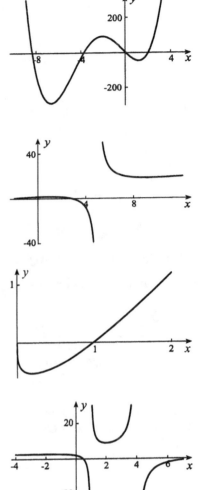

26. For critical points we solve

$$0 = f'(x) = \frac{5}{4}x^{1/4} - \frac{1}{4}x^{-3/4} = \frac{5x - 1}{4x^{3/4}}.$$

Thus, $x = 1/5$ at which $f(x)$ has a relative minimum of $f(1/5) = -4/5^{5/4}$. There is also critical point at $x = 0$, but because the function is not defined for $x < 0$, it cannot yield a relative maximum.

28. To find critical points we solve

$$0 = f'(x)$$
$$= \frac{(x^2 - 5x + 4)(4x - 17) - (2x^2 - 17x + 8)(2x - 5)}{(x^2 - 5x + 4)^2}$$
$$= \frac{7(x - 2)(x + 2)}{(x^2 - 5x + 4)^2}.$$

Solutions are $x = -2, 2$. There is a relative maximum of $f(-2) = 25/9$ and a relative minimum of $f(2) = 9$.

30. This function is most easily analyzed by expressing it as a general sine function $A\sin(x + \phi)$ (see Example 1.45 in Section 1.8),

$$\sin x + \cos x = A\sin(x + \phi)$$
$$= A(\sin x \cos \phi + \cos x \sin \phi).$$

To satisfy this equation we set $A\cos\phi = 1$ and $A\sin\phi = 1$. These are satisfied if we choose $A = \sqrt{2}$ and $\phi = \pi/4$. Thus, $f(x)$ can be expressed in the form $f(x) = \sqrt{2}\sin(x + \pi/4)$, and the relative maxima and minima are $\pm\sqrt{2}$.

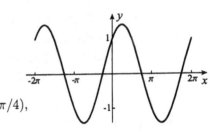

32. To identify critical points we solve

$$0 = f'(x) = -\frac{8}{x^2}\sqrt{x^2 + 100} + \left(1 + \frac{8}{x}\right)\frac{x}{\sqrt{x^2 + 100}}$$
$$= \frac{x^3 - 800}{x^2\sqrt{x^2 + 100}}.$$

The only critical point is $800^{1/3}$ at which there is a relative minimum of

$$f(800^{1/3}) = \left(\frac{8 + 800^{1/3}}{800^{1/3}}\right)\sqrt{100 + 800^{2/3}} \approx 25.4.$$

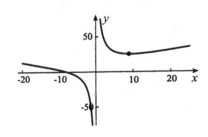

34. (a) A plot is shown in the left figure below. Derivatives of $C(x)$ are

$$C'(x) = \frac{(x+300)(3x^2+200x)-(x^3+100x^2)}{300(x+300)^2} = \frac{x^3+500x^2+30\,000x}{150(x+300)^2},$$

$$C''(x) = \frac{(x+300)^2(3x^2+1000x+30\,000)-(x^3+500x^2+30\,000x)(2x+600)}{150(x+300)^4}.$$

This simplifies to $C''(x) = \dfrac{x^3+900x^2+270\,000x+9\times10^6}{150(x+300)^3} > 0$. Hence the graph is concave upward.

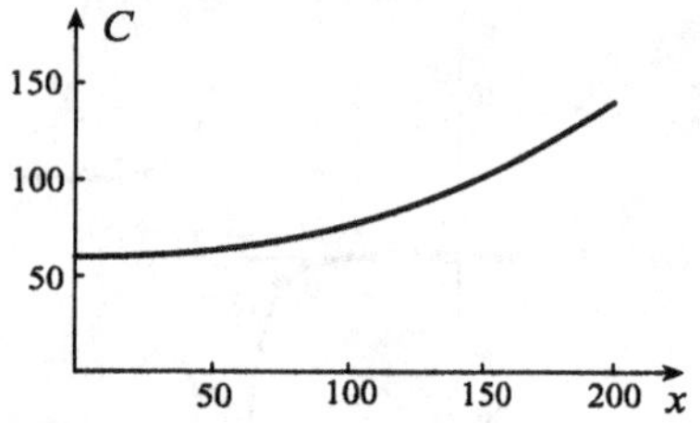

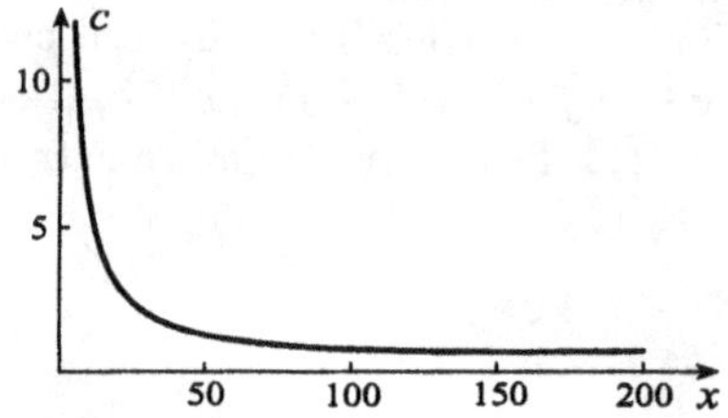

(b) A plot is shown in the right figure above.

(c)(i) Critical points of $c(x) = \dfrac{x}{300}\left(\dfrac{x+100}{x+300}\right) + \dfrac{60}{x}$ are given by

$$0 = \frac{(x+300)(2x+100)-(x^2+100x)(1)}{300(x+300)^2} - \frac{60}{x^2} = \frac{x^2(x^2+600x+30\,000)-18\,000(x+300)^2}{300x^2(x+300)^2}$$

$$= \frac{x^2(x^2+600x+90\,000)-60\,000x^2-18\,000(x+300)^2}{300x^2(x+300)^2} = \frac{(x+300)^2(x^2-18\,000)-60\,000x^2}{300x^2(x+300)^2}.$$

When we equate the numerator to zero, we obtain the required equation.

(ii) The tangent line to $C(x)$ passes through the origin when the slope of the tangent line is equal to $C(x)/x$,

$$\frac{x}{300}\left(\frac{x+100}{x+300}\right) + \frac{60}{x} = \frac{x^3+500x^2+30\,000x}{150(x+300)^2}.$$

If we multiply by $300x(x+300)^2$,

$$0 = 2x(x^3+500x^2+30\,000x) - x^2(x+100)(x+300) - 18\,000(x+300)^2$$
$$= 2x^2(x^2+500x+30\,000) - x^2(x+300)(x+300) + 200x^2(x+300) - 18\,000(x+300)^2$$
$$= 2x^2(x^2+600x+60\,000) - x^2(x+300)^2 - 18\,000(x+300)^2$$
$$= 2x^2(x^2+600x+90\,000) - 60\,000x^2 - x^2(x+300)^2 - 18\,000(x+300)^2$$
$$= (x+300)^2(x^2-18\,000) - 60\,000x^2.$$

EXERCISES 4.7

2. For critical points we solve $0 = f'(x) = \dfrac{(x+1)(1)-(x-4)(1)}{(x+1)^2} = \dfrac{5}{(x-1)^2}$. Since there are no critical points, we evaluate $f(0) = -4$ and $f(10) = 6/11$. These are the absolute minimum and maximum.

4. For critical points we solve $0 = f'(x) = 1 - 2\cos x$. Solutions on the given interval are $\pi/3$, $5\pi/3$, $7\pi/3$, and $11\pi/3$. Since

$$f(0) = 0, \quad f\left(\frac{\pi}{3}\right) = \frac{\pi}{3} - \sqrt{3}, \quad f\left(\frac{5\pi}{3}\right) = \frac{5\pi}{3} + \sqrt{3},$$

$$f\left(\frac{7\pi}{3}\right) = \frac{7\pi}{3} - \sqrt{3}, \quad f\left(\frac{11\pi}{3}\right) = \frac{11\pi}{3} + \sqrt{3}, \quad f(4\pi) = 4\pi,$$

the absolute minimum is $\pi/3 - \sqrt{3}$ and the absolute maximum is $11\pi/3 + \sqrt{3}$.

6. For critical points we solve $0 = f'(x) = \dfrac{-12(2x + 2)}{(x^2 + 2x + 2)^2}$, and obtain $x = -1$. Since

$$\lim_{x \to -\infty} f(x) = 0, \qquad f(-1) = 12, \qquad \lim_{x \to 0^-} f(x) = 6,$$

the absolute maximum is 12 but the function does not have an absolute minimum.

8. For critical points we solve $0 = f'(x) = \dfrac{(x^2 + 3)(1) - x(2x)}{(x^2 + 3)^2} = \dfrac{3 - x^2}{(x^2 + 3)^2}$.

Only the critical point $x = \sqrt{3}$ is positive.
Since

$$\lim_{x \to 0^+} f(x) = 0, \quad f(\sqrt{3}) = \frac{\sqrt{3}}{6}, \quad \lim_{x \to \infty} f(x) = 0,$$

the absolute maximum is $\sqrt{3}/6$ but the
function does not have an absolute minimum.

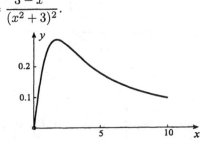

10. We separate the proof into three cases:
CASE 1 – $f(x) = k$, a constant. Then $f'(x) = 0$ for all x, and c can be chosen as any point in the interval.
CASE 2 – $f(x) > f(a)$ for some x in $a < x < b$. Because $f(x)$ is continuous on $a \le x \le b$, it must have an absolute maximum on this interval, and this maximum must occur at a critical point c interior to the interval. Since $f(x)$ is differentiable at every point in $a < x < b$, it follows that the derivative must vanish at the critical point, $f'(c) = 0$.
CASE 3 – $f(x) < f(a)$ for all x in $a < x < b$. In this case the absolute minimum of $f(x)$ must occur at a critical point c between a and b at which $f'(c) = 0$.

12. If x and y represent the length and width of each plot, then the amount of fencing required is $F = 9x + 16y$. Since each plot must have area 9000 m^2, it follows that $xy = 9000$. Thus, $y = 9000/x$, and

$$F = F(x) = 9x + \frac{144\,000}{x}, \quad x > 0.$$

To find critical points of $F(x)$, we solve $0 = F'(x) = 9 - 144\,000/x^2$. The only positive solution is $x = 40\sqrt{10}$. We now evaluate

$$\lim_{x \to 0^+} F(x) = \infty, \qquad F(40\sqrt{10}) = 720\sqrt{10}, \qquad \lim_{x \to \infty} F(x) = \infty.$$

The minimum amount of fencing is therefore $720\sqrt{10}$ m. The graph of $F(x)$ in the right figure also indicates that $F(x)$ is minimized at its critical point.

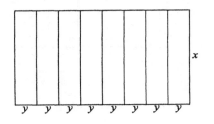

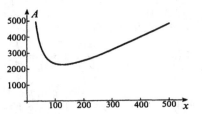

14. The surface area of the box is $A = x^2 + 4xh$. Since the volume of the box must be 6000 L or 6 m³, x and h must satisfy $6 = x^2 h$. Hence, $h = 6/x^2$, and

$$A = A(x) = x^2 + 4x\left(\frac{6}{x^2}\right) = x^2 + \frac{24}{x}, \quad x > 0.$$

To find critical points of $A(x)$, we solve $0 = A'(x) = 2x - 24/x^2$. The positive solution is $12^{1/3}$. We now evaluate

$$\lim_{x \to 0^+} A(x) = \infty, \qquad A(12^{1/3}/2) < \infty, \qquad \lim_{x \to \infty} A(x) = \infty.$$

Minimum area therefore occurs when the base of the box is $12^{1/3} \times 12^{1/3}$ m and the height is $6/12^{2/3} = 12^{1/3}/2$ m. The graph of $A(x)$ in the right figure also shows that $A(x)$ is minimized at its critical point.

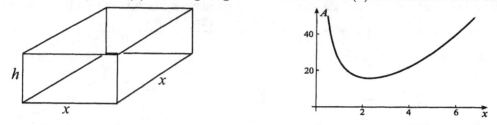

16. The area of the playing field (not including end zones) is $A = 2rl$. Since the perimeter must be 400 m, $400 = 2l + 40 + 2\pi r$. Thus, $l = 180 - \pi r$, and $A = A(r) = 2r(180 - \pi r)$. We must choose $r > 0$, and for l to be positive, we must take $r < 180/\pi$. For critical point of $A(r)$, we solve $0 = A'(r) = 360 - 4\pi r$. The solution is $r = 90/\pi$. We now evaluate

$$\lim_{r \to 0^+} A(r) = 0, \qquad A(90/\pi) > 0, \qquad \lim_{r \to 180/\pi^-} A(r) = 0.$$

Thus, $A(r)$ is maximized when the width of the field is $180/\pi$ m and its length is 90 m. The graph of $A(x)$ in the right figure also indicates that $A(x)$ is maximized at its critical point.

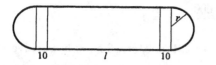

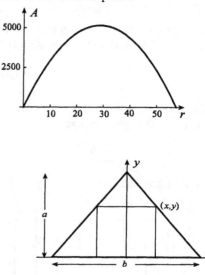

18. The area of the rectangle in the diagram is $A = 2xy$. Since the equation of the line containing the point (x, y) is $y = a - 2ax/b$, we can express A in terms of x as follows

$$A(x) = 2x\left(a - \frac{2ax}{b}\right) = 2a\left(x - \frac{2x^2}{b}\right), \quad 0 \le x \le b/2.$$

For critical points of $A(x)$ we solve

$$0 = A'(x) = 2a\left(1 - \frac{4x}{b}\right) \implies x = \frac{b}{4}.$$

Since $A(0) = 0$, $A(b/4) = 2a\left(\dfrac{b}{4} - \dfrac{b}{8}\right) = \dfrac{ab}{4}$, and $A(b/2) = 0$, maximum area is $ab/4$.

20. When P is at height y, the sum of the lengths of AP, BP, and CP is

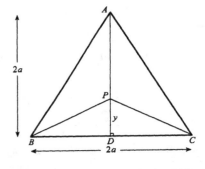

$$L = L(y) = (2a - y) + 2\sqrt{a^2 + y^2}, \quad 0 \le y \le 2a.$$

For critical points of $L(y)$ we solve

$$0 = \frac{dL}{dy} = -1 + \frac{2y}{\sqrt{a^2 + y^2}}.$$

This equation can be expressed in the form $2y = \sqrt{a^2 + y^2}$, and squaring gives $4y^2 = a^2 + y^2$, the positive solutions of which is $y = a/\sqrt{3}$. Since

$$L(0) = 2a + 2a = 4a, \quad L\left(\frac{a}{\sqrt{3}}\right) = 2a - \frac{a}{\sqrt{3}} + 2\sqrt{a^2 + \frac{a^2}{3}} = a(2 + \sqrt{3}), \quad L(2a) = 2\sqrt{a^2 + 4a^2} = 2\sqrt{5}a,$$

minimum length is attained when $y = a/\sqrt{3}$.

22. If length and width of the page are denoted by y and x, then the area of the page is $A = xy$. Since the area of the printed portion of the page must be 150 cm^2, we must have $150 = (x - 5)(y - 7.5)$. This equation can be solved for $x = (10y + 225)/(2y - 15)$, and therefore

$$A = A(y) = \frac{10y^2 + 225y}{2y - 15}, \quad y > \frac{15}{2}.$$

To find critical points, we solve

$$0 = A'(y) = \frac{(2y - 15)(20y + 225) - (10y^2 + 225y)(2)}{(2y - 15)^2}.$$

When we equate the numerator to 0, we obtain $0 = 20y^2 - 300y - 3375 = 5(2y - 45)(2y + 15) \implies y = 45/2$. We now evaluate

$$\lim_{y \to 0^+} A(y) = \infty, \quad A(45/2) < \infty, \quad \lim_{y \to \infty} A(y) = \infty.$$

Hence, area is smallest when $y = 45/2$ cm and $x = 15$ cm. The graph of $A(y)$ in the right figure also indicates that $A(y)$ is minimized at its critical point.

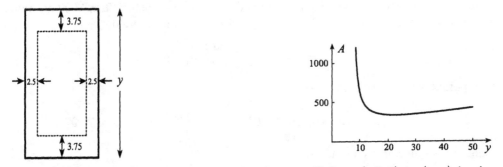

24. If (x, y) is any point on the parabola, then the distance D from $(-2, 5)$ to (x, y) is given by $D^2 = (x + 2)^2 + (y - 5)^2 = (x + 2)^2 + (x^2 - 5)^2$. To minimize D we minimize D^2 on the interval $x \le 0$. For critical points, we solve

$$0 = \frac{d}{dx} D^2 = 2(x + 2) + 2(x^2 - 5)(2x) = 2(2x^3 - 9x + 2) = 2(x - 2)(2x^2 + 4x - 1).$$

The only negative solution is $x = -1 - \sqrt{6}/2$. We now evaluate

$$\lim_{x \to -\infty} D^2(x) = \infty, \quad D^2(-1 - \sqrt{6}/2) = 0.05, \quad D^2(0) = 29.$$

The point closest to $(-2, 5)$ is $(-1 - \sqrt{6}/2, 5/2 + \sqrt{6})$. The graph of D^2 to the right also shows that D^2

is minimized at its critical point.

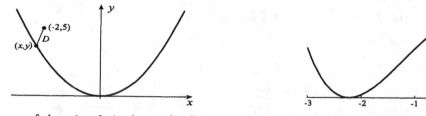

26. The area of the triangle is $A = xy/2$. Since slopes of the line segments joining $(0, y)$ to $(x, 0)$ and $(2, 5)$ to $(x, 0)$ must be the same, we have $\dfrac{y - 0}{0 - x} = \dfrac{0 - 5}{x - 2} \Longrightarrow y = \dfrac{5x}{x - 2}$. Thus,

$$A = A(x) = \frac{5x^2}{2(x - 2)}, \quad x > 2.$$

For critical points, we solve

$$0 = A'(x) = \frac{5}{2}\left[\frac{(x - 2)(2x) - x^2(1)}{(x - 2)^2}\right] = \frac{5x(x - 4)}{2(x - 2)^2}.$$

We now evaluate

$$\lim_{x \to 2^+} A(x) = \infty, \qquad A(4) < \infty, \qquad \lim_{x \to \infty} A(x) = \infty.$$

Area is minimized when $x = 4$ and $y = 10$. The graph of $A(x)$ in the right figure also indicates that $A(x)$ is minimized at its critical point.

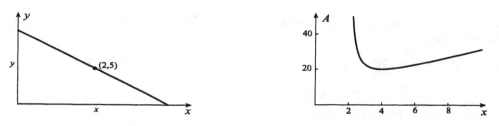

28. The manufacturer's yearly costs for x employees is $C = 20\,000x + 365y$. When 1000 automobiles are to be produced, we must have $1000 = x^{2/5}y^{3/5} \Longrightarrow y = 10^5/x^{2/3}$, and therefore

$$C(x) = 20\,000x + \frac{365 \times 10^5}{x^{2/3}}, \quad x > 0.$$

For critical points of $C(x)$, we set

$$0 = C'(x) = 20\,000 + 365 \times 10^5 \left(\frac{-2}{3}\right) x^{-5/3} \quad \Longrightarrow \quad x = 70.97.$$

Since x must be an integer, we evalulate

$$\lim_{x \to 0^+} C(x) = \infty, \quad C(70) = 3.5490 \times 10^6, \quad C(71) = 3.5487 \times 10^6, \quad \lim_{x \to \infty} C(x) = \infty.$$

It follows that $C(x)$ is minimized for $x = 71$.

30. The printer will choose the number of set types that will minimize production costs. If x set types are used, then the cost for the set types themselves is $2x$ dollars. With this number of set types, the press prints $1000x$ cards per hour. In order to produce $200\,000$ cards, the press must therefore run for $200\,000/(1000x)$ hours at a cost of $[10][200\,000/(1000x)]$ dollars. The total cost of producing the cards when x set types are used is therefore given by

$$C(x) = 2x + \frac{2000}{x}, \quad 1 \leq x \leq 40.$$

To find critical points of $C(x)$, we set

$$0 = C'(x) = 2 - \frac{2000}{x^2}.$$

The only solution of this equation in the interval $1 \leq x \leq 40$ is $x = 10\sqrt{10} = 31.6$. Since x must be an integer, we evaluate

$$C(1) = 2002, \quad C(31) = 126.52, \quad C(32) = 126.50, \quad C(40) = 130.$$

The printer should therefore use 32 set types.

32. (a) If the courier heads to point R, his travel time is

$$T(x) = \frac{\sqrt{x^2 + 36}}{14} + \frac{3 - x}{50}, \quad 0 \leq x \leq 3.$$

To find critical points, we solve

$$0 = T'(x) = \frac{x}{14\sqrt{x^2 + 36}} - \frac{1}{50}.$$

This equation can be expressed in the form $25x = 7\sqrt{x^2 + 36}$, and squaring gives $625x^2 = 49(x^2 + 36)$. The positive solution of this equation is $x = 7/4$.
We now evaluate

$$T(0) = 0.49, \quad T(7/4) = 0.47, \quad T(3) = 0.48.$$

Thus, travel time is minimized when the courier heads to the point on the road $7/4$ km from P. The graph of $T(x)$ function in the right figure also indicates that $T(x)$ is minimized at its critical point.

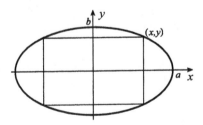

(b) In this case travel time is given by the formula

$$T(x) = \frac{\sqrt{x^2 + 36}}{14} + \frac{1 - x}{50}, \quad 0 \leq x \leq 1.$$

The derivative of this function is identical to that in part (a), but in this case the point $x = 7/4$ must be rejected. Since $T(0) = 6/14 + 1/50 = 0.4486$ and $T(1) = \sqrt{37}/14 = 0.4345$, the courier should head directly to Q.

34. The area of the rectangle is $A = 4xy$. When we solve the equation of the ellipse for the positive value of y, the result is $y = (b/a)\sqrt{a^2 - x^2}$. The area of the rectangle can therefore be expressed in the form

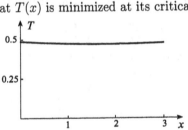

$$A = A(x) = \frac{4bx}{a}\sqrt{a^2 - x^2}, \quad 0 \leq x \leq a.$$

For critical points of $A(x)$ we solve

$$0 = A'(x) = \frac{4b}{a}\left(\sqrt{a^2 - x^2} - \frac{x^2}{\sqrt{a^2 - x^2}}\right).$$

This equation can be expressed in the form $\sqrt{a^2 - x^2} = \frac{x^2}{\sqrt{a^2 - x^2}}$, from which $a^2 - x^2 = x^2$. The positive solution is $x = a/\sqrt{2}$. Since

$$A(0) = 0, \qquad A\left(\frac{a}{\sqrt{2}}\right) > 0, \qquad A(a) = 0,$$

area is maximized when the length of the rectangle in the x-direction is $\sqrt{2}a$ and that in the y-direction is $\sqrt{2}b$.

36. The area of the rectangle is $A = 2xy$. Since $y = e^{-x^2}$, area can be expressed as $A = A(x) = 2xe^{-x^2}$, $x \geq 0$. To find critical points, we solve

$$0 = A'(x) = 2e^{-x^2} - 4x^2 e^{-x^2} = 2(1 - 2x^2)e^{-x^2}.$$

The positive solution of this equation is $x = 1/\sqrt{2}$. We now evaluate

$$A(0) = 0, \qquad T(1/\sqrt{2}) = \sqrt{2/e}, \qquad \lim_{x \to \infty} A(x) = 0.$$

Maximum area is therefore $\sqrt{2/e}$. The graph of $A(x)$ also indicates that it is maximized at its critical point.

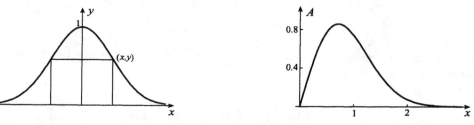

38. We have shown cross-sections of the cone and an inscribed cylinder. The volume of the cylinder is $V = \pi x^2 y$. Since the equation of the line containing the point (x, y) is $y = h - hx/r$, it follows that

$$V = \pi x^2 \left(h - \frac{hx}{r}\right) = \pi h \left(x^2 - \frac{x^3}{r}\right), \quad 0 \leq x \leq r.$$

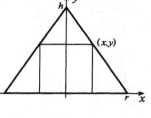

For critical points of $V(x)$ we solve

$$0 = \frac{dV}{dx} = \pi h \left(2x - \frac{3x^2}{r}\right).$$

The positive solution is $x = 2r/3$. Since

$$V(0) = 0, \qquad V\left(\frac{2r}{3}\right) = \pi h \left(\frac{4r^2}{9} - \frac{8r^3}{27r}\right) = \frac{4\pi h r^2}{27}, \qquad V(r) = 0,$$

maximum volume for the cylinder is $4\pi h r^2/27$.

40. For critical points we solve

$$0 = \frac{dR}{d\theta} = \frac{v^2}{g}\left(-\sin^2\theta + \cos^2\theta - \sin\theta\sqrt{\sin^2\theta + \frac{2gh}{v^2}} + \frac{\sin\theta\cos^2\theta}{\sqrt{\sin^2\theta + 2gh/v^2}}\right).$$

From this equation,

$$(\sin^2\theta - \cos^2\theta)\sqrt{\sin^2\theta + \frac{2gh}{v^2}} = \sin\theta\cos^2\theta - \sin\theta\left(\sin^2\theta + \frac{2gh}{v^2}\right)$$

$$= \sin\theta(\cos^2\theta - \sin^2\theta) - \frac{2gh}{v^2}\sin\theta.$$

Hence, $\quad -\cos 2\theta\sqrt{\sin^2\theta + \frac{2gh}{v^2}} = \sin\theta\cos 2\theta - \frac{2gh}{v^2}\sin\theta.$ Squaring this gives

$$\cos^2 2\theta\left(\sin^2\theta + \frac{2gh}{v^2}\right) = \sin^2\theta\cos^2 2\theta - \frac{4gh}{v^2}\sin^2\theta\cos 2\theta + \frac{4g^2h^2}{v^4}\sin^2\theta,$$

from which

$$\cos^2 2\theta = \frac{2gh}{v^2}\sin^2\theta - 2\sin^2\theta\cos 2\theta = \left(\frac{2gh}{v^2} - 2\cos 2\theta\right)\left(\frac{1-\cos 2\theta}{2}\right)$$

$$= \frac{gh}{v^2} - \cos 2\theta - \frac{gh}{v^2}\cos 2\theta + \cos^2 2\theta.$$

Thus, $\cos 2\theta = \dfrac{gh/v^2}{1+gh/v^2} = \dfrac{gh}{v^2+gh}$. The only solution of this equation in the interval $0 < \theta < \pi/2$ is $\theta = \dfrac{1}{2}\mathrm{Cos}^{-1}\left(\dfrac{gh}{v^2+gh}\right)$. It is geometrically clear that there is an angle between $\theta = 0$ and $\theta = \pi/2$ that maximizes R, and since only one critical point has been obtained, it must maximize R. For $v = 13.7$ m/s and $h = 2.25$ m, $\theta = (1/2)\mathrm{Cos}^{-1}\{9.81(2.25)/[13.7^2 + 9.81(2.25)]\} = 0.733$ radians.

42. Length L of the beam is the sum of the lengths L_1 and L_2. Now,

$$L_1 = \|FD\|\sec\theta = \sec\theta\left(3 - \frac{1}{3}\sin\theta\right), \quad \text{and} \quad L_2 = \|GE\|\csc\theta = \csc\theta\left(6 - \frac{1}{3}\cos\theta\right).$$

Hence, $\qquad L = L_1 + L_2 = \dfrac{1}{3}\left[\sec\theta(9 - \sin\theta) + \csc\theta(18 - \cos\theta)\right],$

and this function is defined for $0 < \theta < \pi/2$. The longest beam that can be transported around the corner is represented by the minimum value of $L(\theta)$. To find critical points, we solve

$$0 = \frac{dL}{d\theta} = \frac{1}{3}\left[\sec\theta(-\cos\theta) + \sec\theta\tan\theta(9 - \sin\theta) + \csc\theta(\sin\theta) - \csc\theta\cot\theta(18 - \cos\theta)\right]$$

$$= \frac{1}{3}\left[-1 + \frac{9\sin\theta}{\cos^2\theta} - \frac{\sin^2\theta}{\cos^2\theta} + 1 - \frac{18\cos\theta}{\sin^2\theta} + \frac{\cos^2\theta}{\sin^2\theta}\right] = \frac{1}{3}\left[\frac{9\sin\theta - \sin^2\theta}{\cos^2\theta} - \frac{18\cos\theta - \cos^2\theta}{\sin^2\theta}\right].$$

Critical points therefore satisfy

$$0 = f(\theta) = 9\sin^3\theta - \sin^4\theta - 18\cos^3\theta + \cos^4\theta$$

$$= 9\sin^3\theta - 18\cos^3\theta + (\cos^2\theta + \sin^2\theta)(\cos^2\theta - \sin^2\theta)$$

$$= 9\sin^3\theta - 18\cos^3\theta + 2\cos^2\theta - 1.$$

To solve this equation we use Newton's iterative procedure with

$$\theta_1 = 0.9, \quad \theta_{n+1} = \theta_n - \frac{9\sin^3\theta_n - 18\cos^3\theta_n + 2\cos^2\theta_n - 1}{27\sin^2\theta_n\cos\theta_n + 54\cos^2\theta_n\sin\theta_n - 4\cos\theta_n\sin\theta_n}.$$

Iteration gives $\theta_2 = 0.9091$ and $\theta_3 = 0.9091$. We now evaluate

$$\lim_{\theta\to 0^+} L(\theta) = \infty, \qquad L(0.90) = 11.8, \qquad \lim_{\theta\to\pi/2^-} L(\theta) = \infty.$$

Thus, the length of the longest beam is 11.8 m. The graph of $L(\theta)$ in the right figure also shows that it is minimized at its critical point.

44. For critical points we solve $\quad 0 = f'(x) = \dfrac{(x^2+c)(1) - x(2x)}{(x^2+c)^2} = \dfrac{c - x^2}{(x^2+c)^2}.\quad$ The only solution in $0 \le x \le c$ is $x = \sqrt{c}$. We now calculate

$$f(0) = 0, \qquad f(\sqrt{c}) = \frac{\sqrt{c}}{c+c} = \frac{1}{2\sqrt{c}}, \qquad f(c) = \frac{c}{c^2+c} = \frac{1}{c+1}.$$

Since $c > 0$, we can say that $1/(c+1) \le 1/(2\sqrt{c})$ if and only if

$$c + 1 \ge 2\sqrt{c} \quad \Longleftrightarrow \quad c - 2\sqrt{c} + 1 \ge 0 \quad \Longleftrightarrow \quad (\sqrt{c} - 1)^2 \ge 0,$$

and this is always valid. Hence, the absolute maximum and minimum values of $f(x)$ on $0 \le x \le c$ are $1/(2\sqrt{c})$ and 0.

46. The illumination L at a point which is a distance x from the source with intensity I_1 is

$$L = L(x) = \frac{kI_1}{x^2} + \frac{kI_2}{(d-x)^2}, \quad 0 < x < d,$$

where k is a constant. For critical points of this function we solve

$$0 = \frac{dL}{dx} = -\frac{2kI_1}{x^3} + \frac{2kI_2}{(d-x)^3} \implies I_2 x^3 = I_1(d-x)^3 \implies \left(\frac{d-x}{x}\right)^3 = \frac{I_2}{I_1}.$$

The solution of this equation is $x = dI_1^{1/3}/(I_1^{1/3} + I_2^{1/3})$. Since

$$\lim_{x \to 0^+} L(x) = \infty \qquad \text{and} \qquad \lim_{x \to d^-} L(x) = \infty,$$

illumination is minimized at the critical point $dI_1^{1/3}/(I_1^{1/3} + I_2^{1/3})$ units from the I_1 source.

48. (a) The cost in pennies for the material to construct a tank with base x m and height y m is

$$C = 125(x^2 + 4xy) + 475x^2.$$

Since the tank must hold 100 m^3, x and y must satisfy $100 = x^2 y$. Thus, $y = 100/x^2$, and

$$C = C(x) = 125x^2 + 500x\left(\frac{100}{x^2}\right) + 475x^2$$

$$= 600x^2 + \frac{50\,000}{x}, \quad x > 0.$$

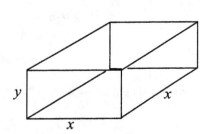

For critical points of $C(x)$ we solve

$$0 = 1200x - \frac{50\,000}{x^2}.$$

The solution is $x = 5/3^{1/3}$. Since $\displaystyle\lim_{x \to 0^+} C(x) = \infty \quad$ and $\quad \displaystyle\lim_{x \to \infty} C(x) = \infty, \quad$ it follows that cost is minimized when $x = 5/3^{1/3}$ m and $y = 4 \cdot 3^{2/3}$ m.

(b) When welding is taken into account the cost is

$$C(x) = 125(x^2 + 4xy) + 475x^2 + 750(8x + 4y) = 600x^2 + \frac{50\,000}{x} + 750\left(8x + \frac{400}{x^2}\right)$$

$$= 600x^2 + \frac{50\,000}{x} + 6000x + \frac{300\,000}{x^2}, \quad x > 0.$$

For critical points of this function we solve

$$0 = 1200x - \frac{50\,000}{x^2} + 6000 - \frac{600\,000}{x^3},$$

and this equation simplifies to $3x^4 + 15x^3 - 125x - 1500 = 0$. The positive solution of this equation, $x = 4.19$, can be found using Newton's iterative procedure (see Example 4.1 in Section 4.1). Since limits of $C(x)$ are once again "infinity" as $x \to 0^+$ and $x \to \infty$, cost is minimized for $x = 4.19$ m and $y = 5.70$ m.

50. Total cost for the pipeline is $\quad C(x) = c_1\sqrt{x^2 + y_1^2} + c_2\sqrt{(x_2 - x)^2 + y_2^2}$, $0 \le x \le x_2$. For critical points of this function we solve $\quad 0 = C'(x) = \dfrac{c_1 x}{\sqrt{x^2 + y_1^2}} + \dfrac{c_2(x_2 - x)(-1)}{\sqrt{(x_2 - x)^2 + y_2^2}}$. Since $\sin\theta_1 = x/\sqrt{x^2 + y_1^2}$ and $\sin\theta_2 = (x_2 - x)/\sqrt{(x_2 - x)^2 + y_2^2}$, it follows that at the critical point $0 = c_1\sin\theta_1 - c_2\sin\theta_2$. We can show that the critical point yields a minimum by calculating

$$C''(x) = \frac{c_1}{\sqrt{x^2 + y_1^2}} - \frac{c_1 x^2}{(x^2 + y_1^2)^{3/2}} + \frac{c_2}{\sqrt{(x_2 - x)^2 + y_2^2}} - \frac{c_2(x_2 - x)^2}{[(x_2 - x)^2 + y_2^2]^{3/2}}$$

$$= \frac{c_1 y_1^2}{(x^2 + y_1^2)^{3/2}} + \frac{c_2 y_2^2}{[(x_2 - x)^2 + y_2^2]^{3/2}}.$$

Since $C''(x)$ is positive for $0 \le x \le x_2$, the graph of $C(x)$ is concave upward. Consequently, the critical point must minimize $C(x)$.

52. The volume of the box is $V = lwh$. Dimensions on the cardboard make it clear that $2l + 2w = 2$, and therefore $w = 1-l$. The fact that the outer flaps must meet in the centre requires $l = 2(1/2)(1-h) = 1-h$. Hence, $w = 1 - (1 - h) = h$ and

$$V = V(h) = (1 - h)hh = h^2 - h^3, \quad 0 \le h \le 1.$$

For critical points we solve $0 = V'(h) = 2h - 3h^2 = h(2 - 3h)$. Since $V(0) = 0$, $V(2/3) > 0$, and $V(1) = 0$, maximum volume occurs when $h = w = 2/3$ m and $l = 1/3$ m. Since the sum of the spaces between inner flaps on top and bottom is $2w + 2l - 2 = 2/3$, the inner flaps are $1/3$ m apart.

54. The volume of the drinking cup is $V = \pi r^2 h/3$.
Since $R^2 = r^2 + h^2$, we may write

$$V = V(h) = \frac{1}{3}\pi(R^2 - h^2)h.$$

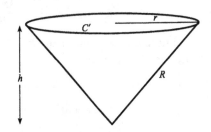

To obtain the domain of this function, we relate θ and h by noting that the length of the arc joining A and B is the same as that of C':

$$R(2\pi - \theta) = 2\pi r = 2\pi\sqrt{R^2 - h^2}.$$

This equation implies that $h = 0$ when $\theta = 0$, and $h = R$ when $\theta = 2\pi$. The appropriate domain for $V(h)$ is therefore $0 \le h \le R$. For critical points of the function we solve

$$0 = V'(h) = \frac{1}{3}\pi(R^2 - 3h^2).$$

The positive solution is $h = R/\sqrt{3}$. Since $V(0) = 0 = V(R)$, it follows that V is maximized for $h = R/\sqrt{3}$. For this h, $R(2\pi - \theta) = 2\pi\sqrt{R^2 - R^2/3}$, and the solution of this equation is $\theta = 2\pi(1 - \sqrt{6}/3)$.

56. The thrust to speed ratio is

$$g(v) = \frac{F}{v} = \frac{1}{2}\rho A v \left(0.000182 + \frac{4w^2}{6.5\pi\rho^2 A^2 v^4}\right) = \frac{1}{2}\rho A\left(0.000182v + \frac{4w^2}{6.5\pi\rho^2 A^2 v^3}\right).$$

For critical points of this function we solve

$$0 = g'(v) = \frac{1}{2}\rho A\left(0.000182 - \frac{12w^2}{6.5\pi\rho^2 A^2 v^4}\right) \quad \Longrightarrow \quad v = \left[\frac{12w^2}{0.000182(6.5)\pi\rho^2 A^2}\right]^{1/4}.$$

Since $g(v)$ becomes infinite as $v \to 0$ and $v \to \infty$, it follows that this value of v must minimize $g(v)$.
(a) At sea level, when $A = 1600$, $w = 150,000$, and $\rho = 0.0238$, the speed is 323 mph.
(b) At 30,000 feet, when $A = 1600$, $w = 150,000$, and $\rho = (0.375)(0.0238)$, the speed is 527 mph.

58. From the figure to the right, the sum
of the distance from P to the vertices is
$$L(y) = (2a - y) + 2\sqrt{b^2 + y^2}, \quad 0 \le y \le 2a.$$
For critical points of this function, we solve
$$0 = L'(y) = -1 + \frac{2y}{\sqrt{b^2 + y^2}}.$$
The solution is $y = b/\sqrt{3}$. To show that this value
minimizes $L(y)$, we evaluate

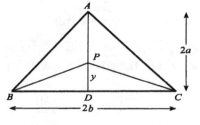

$$L(0) = 2a + 2b, \qquad L(b/\sqrt{3}) = 2a - \frac{b}{\sqrt{3}} + 2\sqrt{b^2 + \frac{b^2}{3}} = 2a + \sqrt{3}b, \qquad L(2a) = 2\sqrt{b^2 + 4a^2}.$$

Certainly the second of these is less than the first. Since the second and third are both positive, the
second is less than the third if, and only if,
$$(2a + \sqrt{3}b)^2 < 4(b^2 + 4a^2) \qquad \Longleftrightarrow \qquad 4a^2 + 4\sqrt{3}ab + 3b^2 < 4b^2 + 16a^2.$$
But this is equivalent to
$$0 < 12a^2 - 4\sqrt{3}ab + b^2 = (2\sqrt{3}a - b)^2,$$
which is obviously true. Thus, $y = b/\sqrt{3}$ does indeed minimize L.

60. The distance D from (x_1, y_1) to any point $P(x, y)$ on the line $Ax + By + C = 0$ is given by
$$D^2 = (x - x_1)^2 + (y - y_1)^2 = (x - x_1)^2 + \left(-\frac{C}{B} - \frac{Ax}{B} - y_1\right)^2 \qquad \text{(provided } B \ne 0\text{)}$$
$$= (x - x_1)^2 + \frac{(C + Ax + By_1)^2}{B^2}, \quad -\infty < x < \infty.$$

To minimize D we minimize D^2. For critical
points of D^2 we solve
$$0 = 2(x - x_1) + 2\frac{(C + Ax + By_1)}{B^2}A.$$
The solution of this equation is
$$x = \frac{B^2 x_1 - AC - ABy_1}{A^2 + B^2}.$$
Since D^2 becomes infinite as $x \to \pm\infty$,
it follows that D^2 is minimized for this value of x,
and the minimum distance is

$$\sqrt{\left(\frac{B^2 x_1 - AC - ABy_1}{A^2 + B^2} - x_1\right)^2 + \frac{1}{B^2}\left(C + \frac{AB^2 x_1 - A^2 C - A^2 By_1}{A^2 + B^2} + By_1\right)^2}$$
$$= \sqrt{A^2\left(\frac{Ax_1 + By_1 + C}{A^2 + B^2}\right)^2 + B^2\left(\frac{Ax_1 + By_1 + C}{A^2 + B^2}\right)^2} = \frac{|Ax_1 + By_1 + C|}{\sqrt{A^2 + B^2}}.$$

If $B = 0$, then the line $Ax + C = 0$ is vertical and the minimum distance is $|C/A + x_1|$. But this is
predicted by the above formula when $B = 0$. Consequently, the formula is correct for any line whatsoever.

62. The distance D from $(4, 13\sqrt{5}/6)$ to any point $P(x, y)$ on the ellipse is given by
$$D^2 = (x - 4)^2 + \left(y - \frac{13\sqrt{5}}{6}\right)^2, \text{ where } x \text{ and } y \text{ must satisfy}$$
$4x^2 + 9y^2 = 36$. For critical points of D^2, we solve
$$0 = 2(x - 4) + 2\left(y - \frac{13\sqrt{5}}{6}\right)\frac{dy}{dx}.$$

We obtain dy/dx by differentiating the equation of the ellipse, $8x + 18y \, dy/dx = 0$. We solve this for dy/dx and substitute into the equation defining critical points, $0 = x - 4 + \left(y - \dfrac{13\sqrt{5}}{6} \right) \left(-\dfrac{4x}{9y} \right)$. This equation can be solved for $y = \dfrac{26\sqrt{5}x}{3(36 - 5x)}$. When this is substituted into the equation of the ellipse,

$$36 = 4x^2 + 9 \left[\frac{26\sqrt{5}x}{3(36 - 5x)} \right]^2,$$

and this equation simplifies to $(36 - 5x)^2(36 - 4x^2) - 3380x^2 = 0$. The only solution between 0 and 3 is $x = 2$. The corresponding y-coordinate of the point is $y = 2\sqrt{5}/3$. Since $D^2(0) = 24.09$, $D^2(2) = 15.25$, and $D^2(3) = 24.47$, the point $(2, 2\sqrt{5}/3)$ is indeed closest to $(4, 13\sqrt{5}/6)$.

64. The area of triangle PQR is $F = \|RS\|\|PQ\|/2$. If we use formula 1.16 for length $\|RS\|$, then

$$F(x) = \frac{\|PQ\|}{2\sqrt{A^2 + B^2}} |Ax + B(ax^2 + bx + c) + C|,$$

where $\|PQ\|$ is constant. For critical points of this function we solve

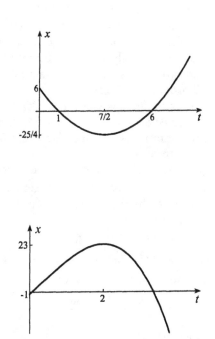

$$0 = F'(x) = \frac{\|PQ\|}{2\sqrt{A^2 + B^2}}(A + 2Bax + bB)\frac{|Ax + B(ax^2 + bx + c) + C|}{Ax + B(ax^2 + bx + c) + C}$$

(see formula 3.13). The only solution of this equation is $x = -(A + bB)/(2aB)$. Now, there must be a value of x between P and Q which maximizes F since coordinates of these points lead to degenerate triangles with zero area. Hence, this critical point maximizes the area.

EXERCISES 4.8

2. The velocity and acceleration are
$$v(t) = 2t - 7 \text{ m/s,}$$
$$a(t) = 2 \text{ m/s}^2.$$
At time $t = 0$, the object is at $x = 6$ m and moving to the left with speed 7 m/s. It continues to move to the left until $t = 7/2$ s when it stops at $x = -25/4$ m. It then moves to the right with ever increasing speed.

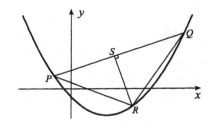

4. The velocity and acceleration are
$$v(t) = -6t^2 + 4t + 16 = -2(3t + 4)(t - 2) \text{ m/s,}$$
$$a(t) = -12t + 4 = -4(3t - 1) \text{ m/s}^2.$$
The object begins at time $t = 0$ at position $x = -1$ m with speed 16 m/s to the right. Its acceleration at this time is 4 m/s^2 so that it is picking up speed. At time $t = 1/3$ s, acceleration is zero and thereafter acceleration is negative. This means that its velocity decreases for $t \geq 1/3$ s. At time $t = 2$ s (and $x = 23$ m), the object's velocity is zero. Thereafter it moves to the left with increasing speed.

6. The velocity and acceleration are
$$v(t) = -12\sin 4t \text{ m/s},$$
$$a(t) = -48\cos 4t \text{ m/s}^2.$$
The object moves back and forth along the x-axis between $x = \pm 3$ m. Its velocity is zero in the turns, and its acceleration is equal to zero each time it passes through $x = 0$. The period of each oscillation is $\pi/2$ s.

8. The velocity and acceleration are
$$v(t) = 1 - 4/t^2 \text{ m/s},$$
$$a(t) = 8/t^3 \text{ m/s}^2.$$
At time $t = 1$, the object is at position $x = 5$ m and moving to the left with speed 3 m/s. Its acceleration is always to the right resulting in an instantaneous stop at $t = 2$ s at $x = 4$ m. The object then moves to the right thereafter with increasing speed. For large t, the velocity of the object approaches 1 m/s, and its acceleration approaches 0.

10. The velocity and acceleration are
$$v(t) = \frac{5}{2}t^{3/2} - 3\sqrt{t} + \frac{1}{2\sqrt{t}} = \frac{(5t-1)(t-1)}{2\sqrt{t}} \text{ m/s},$$
$$a(t) = \frac{15}{4}\sqrt{t} - \frac{3}{2\sqrt{t}} - \frac{1}{4t^{3/2}} = \frac{15t^2 - 6t - 1}{4t^{3/2}} \text{ m/s}^2.$$
The object starts at the origin with zero velocity, but with positive acceleration. It therefore moves to the right, continuing to do so forever with ever increasing speed.

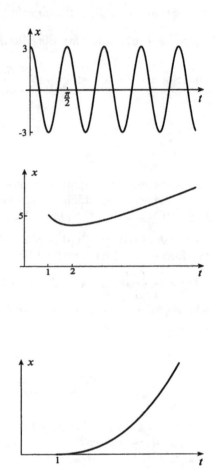

12. A plot of the displacement function is shown to the right. We need its velocity and acceleration:
$$v(t) = -15 + 18t - 3t^2 = -3(t-1)(t-5) \text{ m/s},$$
$$a(t) = 18 - 6t = -6(t-3) \text{ m/s}^2.$$
(a) Since $v(1) = 0$, the object is stopped at $t = 1$, it's speed is neither increasing nor decreasing.
(b) Critical points of $v(t)$ occur when acceleration vanishes, namely, at $t = 3$. Since $v(0) = -15$, $v(3) = 12$, and $v(6) = -15$, maximum and minimum velocities are 12 m/s and -15 m/s.
(c) Maximum speed is 15 m/s and minimum speed is 0.
(d) Since acceleration is linear in t, maximum and minimum acceleration occur at $t = 0$ and $t = 6$. They are ± 18 m/s^2.
(e) Since $a'(t) = -6 < 0$, the acceleration is never increasing.

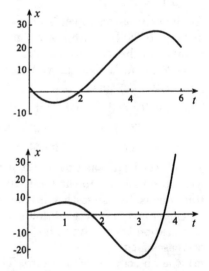

14. A plot of the displacement function is shown to the right. We need the velocity and acceleration functions:
$$v(t) = 12t^3 - 48t^2 + 36t = 12t(t-1)(t-3) \text{ m/s},$$
$$a(t) = 36t^2 - 96t + 36 = 12(3t^2 - 8t + 3) \text{ m/s}^2.$$
(a) Acceleration is zero for
$$t = \frac{8 \pm \sqrt{64 - 36}}{6} = \frac{4 \pm \sqrt{7}}{3}.$$
Since $a(t) \geq 0$ for $0 \leq t \leq (4 - \sqrt{7})/3$ and $(4 + \sqrt{7})/3 \leq t \leq 4$, velocity is increasing on these intervals,

and it is decreasing for $(4 - \sqrt{7})/3 \leq t \leq (4 + \sqrt{7})/3$.

(b) Speed is increasing on the intervals $0 \leq t \leq (4 - \sqrt{7})/3$, $1 \leq t \leq (4 + \sqrt{7})/3$, and $3 \leq t \leq 4$. It is decreasing for $(4 - \sqrt{7})/3 \leq t \leq 1$ and $(4 + \sqrt{7})/3 \leq t \leq 3$.

(c) Critical points of velocity occur where acceleration is zero, namely $t = (4 \pm \sqrt{7})/3$. Since

$$v(0) = 0, \quad v\big((4 - \sqrt{7})/3\big) = \frac{8(-10 + 7\sqrt{7})}{9}, \quad v\big((4 + \sqrt{7})/3\big) = \frac{-8(10 + 7\sqrt{7})}{9}, \quad v(4) = 144,$$

velocity is a maximum at $t = 4$ and a minimum at $t = (4 + \sqrt{7})/3$.

(d) Speed is a maximum at $t = 4$ and a minimum at $t = 0, 1, 3$ where it is zero.

(e) The graph makes it clear that maximum distance from the origin is $x(4) = 34$.

(f) To answer this we should maximize $x(t) - 5$. Critical points of this function are $t = 1, 3$. Since

$$x(0) - 5 = -3, \quad x(1) - 5 = 2, \quad x(3) - 5 = -30, \quad x(4) - 5 = 29,$$

maximum distance from $x = 5$ is 30 at $t = 3$.

16. A horizontal point of inflection.

18. The speed graph to the right indicates maximum speed at $t = 0$ of $|v(0)| = 144$ m/s.

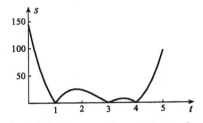

20. This is not always true. Consider the position function shown to the left below together with its absolute value to the right. Position has relative maximum at t_1, whereas its absolute value has relative maximum at t_2.

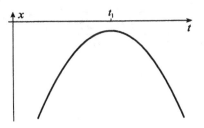 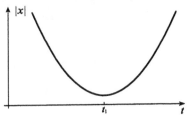

22. This is not always true. Consider the velocity function $v(t) = \sin t$, $t \geq 0$. Graphs of the velocity and speed functions are shown below. Velocity has relative minima at $t = 3\pi/2 + 2n\pi$, where n is a nonnegative integer, whereas speed has relative maxima at these times.

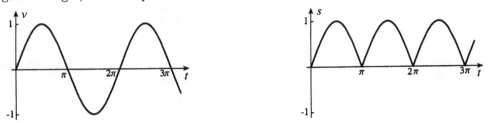

24. (a) Since ϕ lies in the interval $-\pi/2 < \phi < \pi/2$, we can solve $e = L \sin \phi - r \sin \theta$ for

$$\sin \phi = \frac{e + r \sin \theta}{L} \quad \Longrightarrow \quad \cos \phi = \sqrt{1 - \left(\frac{e + r \sin \theta}{L}\right)^2}.$$

Hence,

$$x = r \cos \theta + L\sqrt{1 - \left(\frac{e + r \sin \theta}{L}\right)^2} = r \cos \theta + \sqrt{L^2 - (e + r \sin \theta)^2}.$$

(b) When $L = 9$, $r = 2$, and $e = 1$,

$$x = 2\cos\theta + \sqrt{81 - (1 + 2\sin\theta)^2},$$

a plot of which is shown in the left figure below.

(c) Since the length of the stroke is the difference between maximum and minimum values of x, the graph suggests a stroke length of about 4 cm. The formula in Example 4.28 gives the stroke length

$$s = \sqrt{(9+2)^2 - 1^2} - \sqrt{(9-2)^2 - 1^2} = 2\sqrt{30} - 4\sqrt{3} = 4.03.$$

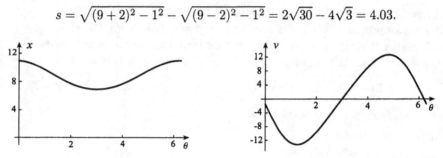

(d) Differentiation of x with respect to t gives

$$v = \frac{dx}{dt} = \frac{dx}{d\theta}\frac{d\theta}{dt} = \omega\left[-r\sin\theta + \frac{-2(e + r\sin\theta)r\cos\theta}{2\sqrt{L^2 - (e + r\sin\theta)^2}}\right] = -\omega r\left[\sin\theta + \frac{\cos\theta(e + r\sin\theta)}{\sqrt{L^2 - (e + r\sin\theta)^2}}\right].$$

If we substitute for $\sin\phi$ and $\cos\phi$ in the velocity formula of Example 4.31,

$$v = \frac{-\omega r(\sin\theta\cos\phi + \cos\theta\sin\phi)}{\cos\phi} = -\omega r\left[\sin\theta + \frac{\cos\theta\left(\dfrac{e + r\sin\theta}{L}\right)}{\sqrt{1 - \left(\dfrac{e + r\sin\theta}{L}\right)^2}}\right]$$

$$= -\omega r\left[\sin\theta + \frac{\cos\theta(e + r\sin\theta)}{\sqrt{L^2 - (e + r\sin\theta)^2}}\right],$$

the above result.

(e) With $\omega = 2\pi$, $L = 9$, $r = 2$, and $e = 1$,

$$v = -4\pi\left[\sin\theta + \frac{\cos\theta(1 + 2\sin\theta)}{\sqrt{81 - (1 + 2\sin\theta)^2}}\right],$$

a plot of which is shown in the right figure above. The velocity would appear to be zero when the displacement graph is at its highest and lowest points. These are the inner and outer dead positions in Example 4.28. Maximum and minimum velocities are approximately ± 13 cm/s.

EXERCISES 4.9

2. For the right-angled triangle, we may write $z^2 = s^2 + 16$. Differentiation with respect to time gives $\quad 2z\dfrac{dz}{dt} = 2s\dfrac{ds}{dt}$. When $s = 6$, we find that $z = 2\sqrt{13}$, and

$$(2\sqrt{13})(-2) = (6)\frac{ds}{dt}.$$

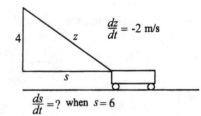

Thus, $ds/dt = -2\sqrt{13}/3$. The cart is therefore moving to the left at $2\sqrt{13}/3$ m/s.

4. When the depth of liquid is h, the volume of liquid in the funnel is $V = \frac{1}{3}\pi r^2 h$. Similar triangles give $\dfrac{r}{h} = \dfrac{15/2}{30}$. Thus, $r = h/4$, and

$$V = \frac{1}{3}\pi \left(\frac{h}{4}\right)^2 h = \frac{\pi}{48}h^3.$$

Differentiation with respect to time gives

$$\frac{dV}{dt} = \frac{\pi}{16}h^2\frac{dh}{dt}.$$

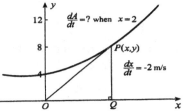

Since the net rate of change of the volume of liquid in the funnel is 65 cm^3/s, we can say that when $h = 20$,

$$65 = \frac{\pi}{16}(20)^2\frac{dh}{dt} \implies \frac{dh}{dt} = \frac{65 \cdot 16}{400\pi} = \frac{13}{5\pi}.$$

The liquid level is therefore rising at $13/(5\pi)$ cm/s.

6. The area of the triangle is

$$A = \frac{1}{2}xy = \frac{1}{2}x(x^2 + x + 4).$$

If we differentiate with respect to time,

$$\frac{dA}{dt} = \frac{1}{2}\left(3x^2\frac{dx}{dt} + 2x\frac{dx}{dt} + 4\frac{dx}{dt}\right)$$
$$= \frac{1}{2}(3x^2 + 2x + 4)\frac{dx}{dt}.$$

When $x = 2$, we obtain

$$\frac{dA}{dt} = \frac{1}{2}[3(2)^2 + 2(2) + 4](-2) = -20.$$

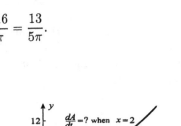

The area is therefore decreasing at 20 m^2/s.

8. If P and V are pressure and volume of the gas, Boyle's law states that $P = k/V$, where k is a constant. Differentiation with respect to time gives

$$\frac{dP}{dt} = -\frac{k}{V^2}\frac{dV}{dt}.$$

Since $V = 1/100$ m^3 when $P = 50$ N/m^2, it follows that $50 = k(100)$, which implies that $k = 1/2$. At this instant,

$$\frac{dP}{dt} = -\frac{1/2}{(1/100)^2}\left(\frac{1}{2000}\right) = -2.5.$$

Pressure is decreasing at 2.5 N/m^2/s.

10. The amount of line between reel and fish is $z^2 = x^2 + y^2$. Differentiation of this equation with respect to time gives $2z\dfrac{dz}{dt} = 2x\dfrac{dx}{dt} + 2y\dfrac{dy}{dt}$. It takes 25 s for the boat to travel the 50 m from the instant the fish struck, and during this time the fish dives 75 m. The amount of line line between fish and reel at this instant is $\sqrt{150^2 + 75^2} = 75\sqrt{5}$. Consequently,

$$75\sqrt{5}\frac{dz}{dt} = (150)(2) + 75(3) \implies \frac{dz}{dt} = \frac{525}{75\sqrt{5}} = \frac{7}{\sqrt{5}}.$$

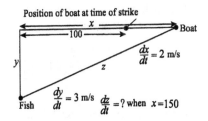

The line is therefore being played out at $7/\sqrt{5}$ m/s.

12. (a) When sand completely covers the bottom of the cylinder, the volume of sand is

$$V = C + \pi \left(\frac{1}{2}\right)^2 h = C + \frac{\pi h}{4},$$

where C is the volume of sand in the cone. Since C is constant, differentiation of this equation with respect to time gives $\dfrac{dV}{dt} = \dfrac{\pi}{4}\dfrac{dh}{dt}$.

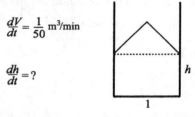

$\dfrac{dV}{dt} = \dfrac{1}{50}$ m³/min

$\dfrac{dh}{dt} = ?$

h

Since $dV/dt = 1/50$, it follows that $dh/dt = 2/(25\pi)$, and the top of the pile is rising at $2/(25\pi)$ m/min.

(b) Since the height of the cone is constant, sand also travels along the side of the cylinder at $2/(25\pi)$ m/min.

14. (a) When Car 2 is on that part of the racetrack between D and B, its x- and y-coordinates must satisfy the equation $(x - 50)^2 + y^2 = 2500$, the equation of a circle of radius 50 centred at the point $(50, 0)$. If we differentiate this equation with respect to time,

$$2(x - 50)\frac{dx}{dt} + 2y\frac{dy}{dt} = 0 \quad \Longrightarrow \quad \frac{dy}{dt} = -\frac{x - 50}{y}\frac{dx}{dt}.$$

Since the rate of change of the x-coordinate of Car 2 is the same as that of Car 1, namely 10 m/s, it follows that $dx/dt = 10$, and

$$\frac{dy}{dt} = \frac{10(50 - x)}{y} \text{ m/s}.$$

(b) When the car is at E, its x-coordinate is 75 and its y-coordinate is $-\sqrt{2500 - (75 - 50)^2} = -25\sqrt{3}$. At this point,

$$\frac{dy}{dt} = \frac{10(50 - 75)}{-25\sqrt{3}} = \frac{10}{\sqrt{3}} \text{ m/s}.$$

(c) As Car 2 approaches B, its x-coordinate approaches 100 and its y-coordinate approaches 0. This means that the rate of change of its y-coordinate is $\displaystyle\lim_{x \to 100} \frac{10(50 - x)}{y} = \infty$. Car 1 therefore suffers the most damage.

16. The cosine law applied to triangle ORQ gives $L^2 = z^2 + 100^2 - 200z \cos \theta$. Differentiation of this equation with respect to time gives

$$2L\frac{dL}{dt} = 2z\frac{dz}{dt} - 200 \cos \theta \frac{dz}{dt} + 200z \sin \theta \frac{d\theta}{dt}.$$

When $z = 10$ and $\theta = \pi/4$, we obtain

$$L = \sqrt{100 + 100^2 - 200(10)(1/\sqrt{2})} = 93.197567.$$

Substitution of these values into the equation involving dL/dt gives

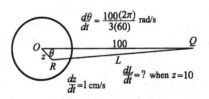

$\dfrac{d\theta}{dt} = \dfrac{100(2\pi)}{3(60)}$ rad/s

100 Q

O z θ L

R

$\dfrac{dz}{dt} = 1$ cm/s

$\dfrac{dL}{dt} = ?$ when $z = 10$

$$2(93.197567)\frac{dL}{dt} = 2(10)(1) - 200\left(\frac{1}{\sqrt{2}}\right)(1) + 200(10)\left(\frac{1}{\sqrt{2}}\right)\left(\frac{10\pi}{9}\right),$$

and this equation can be solved for $dL/dt = 25.83$. The distance is therefore increasing at 25.83 cm/s.

18. The cosine law on triangle OPQ gives

$$L^2 = 61 + 25 - 2(5)\sqrt{61}\cos\theta = 86 - 10\sqrt{61}\cos\theta.$$

Differentiation of this equation with respect to time gives

$$2L\frac{dL}{dt} = 10\sqrt{61}\sin\theta\frac{d\theta}{dt}.$$

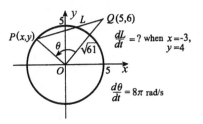

When the particle is at $P(-3,4)$, the length of PQ is $L = \sqrt{(5+3)^2 + (6-4)^2} = 2\sqrt{17}$, and the equation $68 = 86 - 10\sqrt{61}\cos\theta$ gives $\cos\theta = 9/(5\sqrt{61})$. It follows that $\sin\theta = \sqrt{1 - 81/(25\cdot 61)} = 38/(5\sqrt{61})$. Substitution of these into the equation for dL/dt gives

$$2(2\sqrt{17})\frac{dL}{dt} = 10\sqrt{61}\left(\frac{38}{5\sqrt{61}}\right)(8\pi).$$

When this is solved for dL/dt, the result is 115.8 cm/s.

20. (a) The slope of the tangent line to the hyperbola can be obtained by differentiating $x^2 - y^2 = 1$ with respect to x, $2x - 2y(dy/dx) = 0 \implies dy/dx = x/y$. The slope of the normal line at P is therefore $-y/x$. Since the slope of this line is also $(y-0)/(x-x^*)$, it follows that $-\dfrac{y}{x} = \dfrac{y}{x - x^*}$. When wew cross multiply and solve for x^*, we obtain the required result that $x^* = 2x$.

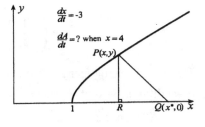

(b) The area of triangle PQR is

$$A = \frac{1}{2}\|PR\|\|RQ\| = \frac{1}{2}y(x^* - x) = \frac{1}{2}\sqrt{x^2 - 1}(2x - x) = \frac{1}{2}x\sqrt{x^2 - 1}.$$

Differentiation with respect to time gives

$$\frac{dA}{dt} = \frac{1}{2}\left[\sqrt{x^2 - 1}\frac{dx}{dt} + x\left(\frac{x}{\sqrt{x^2-1}}\frac{dx}{dt}\right)\right] = \frac{2x^2 - 1}{2\sqrt{x^2-1}}\frac{dx}{dt}.$$

When $x = 4$, we obtain $\dfrac{dA}{dt} = \dfrac{2(4)^2 - 1}{2\sqrt{16-1}}(-3) = -\dfrac{31\sqrt{15}}{10}$. The area is therefore decreasing at $31\sqrt{15}/10$ units of area per unit of time.

22. Similar triangles give $\dfrac{x}{25} = \dfrac{10}{d} \implies x = \dfrac{250}{d} = \dfrac{250}{4.905t^2}$.

Differentiation of this gives $\dfrac{dx}{dt} = -\dfrac{500}{4.905t^3}$.

At $t = 1$ s, we obtain $dx/dt = -500/4.905 = -102$. The shadow is therefore moving at 102 m/s.

24. When we solve the equation of the curve for y in terms of x, we obtain $y = -x \pm 4\sqrt{x}$. Since we are concerned with that part of the curve which contains the point $(4,4)$, we must choose $y = 4\sqrt{x} - x$. Distance D from any point (x,y) on the curve to $(1,2)$ is given by $D^2 = (x-1)^2 + (y-2)^2 = (x-1)^2 + (4\sqrt{x} - x - 2)^2$. Differentiation with respect to time gives

$$2D\frac{dD}{dt} = 2(x-1)\frac{dx}{dt} + 2(4\sqrt{x} - x - 2)\left(\frac{2}{\sqrt{x}} - 1\right)\frac{dx}{dt}.$$

When $(x,y) = (4,4)$, we find $D = \sqrt{(4-1)^2 + (4-2)^2} = \sqrt{13}$, and

$$2\sqrt{13}\frac{dD}{dt} = 2(3)(2) + 2(8 - 4 - 2)(1 - 1)(2) \implies \frac{dD}{dt} = \frac{6}{\sqrt{13}}.$$

The distance is changing at $6/\sqrt{13}$ m/s.

26. (a) When the height of each cone is h, the volume of sand in the container is $V = 2\pi r^2 h/3$. Similar triangles require $r/h = 2/3$, and therefore

$$V = \frac{2}{3}\pi\left(\frac{2h}{3}\right)^2 h = \frac{8}{27}\pi h^3.$$

Differentiation with respect to time gives

$$\frac{dV}{dt} = \frac{8}{9}\pi h^2 \frac{dh}{dt}.$$

When $h = 3/2$, this result yields

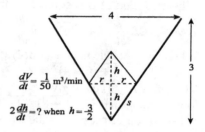

$$\frac{1}{50} = \frac{8}{9}\pi\left(\frac{3}{2}\right)^2 \frac{dh}{dt}.$$

Thus, $dh/dt = 1/(100\pi)$, and the top of the pile is rising at $1/(50\pi)$ metres per minute.

(b) Since $s^2 = h^2 + r^2 = h^2 + (2h/3)^2 = 13h^2/9$, differentiation gives $2s(ds/dt) = (26/9)h(dh/dt)$. When $h = 3/2$, $s = \sqrt{13(3/2)^2/9} = \sqrt{13}/2$, and

$$2\left(\frac{\sqrt{13}}{2}\right)\frac{ds}{dt} = \frac{26}{9}\left(\frac{3}{2}\right)\left(\frac{1}{100\pi}\right) \quad\Longrightarrow\quad \frac{ds}{dt} = \frac{\sqrt{13}}{300\pi}.$$

The sand is rising along the side of the container at $\sqrt{13}/(300\pi)$ metres per minute.

28. If we differentiate $z^2 = y^2 + w^2 = y^2 + (1 + x^2)$, with respect to time t,

$$2z\frac{dz}{dt} = 2y\frac{dy}{dt} + 2x\frac{dx}{dt}.$$

At the instant in question, $x = 2 + 100/60 = 11/3$ km, $y = 200/60 = 10/3$ km, and $z^2 = 100/9 + 1 + 121/9 = 230/9$ km. Hence,

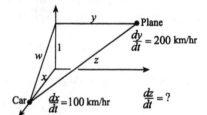

$$\frac{\sqrt{230}}{3}\frac{dz}{dt} = \frac{10}{3}(200) + \frac{11}{3}(100),$$

from which $dz/dt = 3100/\sqrt{230}$ km/hr.

30. As a point on the chain travels around the front sprocket, the distance S that it travels is $S = R\theta$, where θ is the angle through which it turns. Similarly, a point on the rear sprocket travels through a distance $s = r\phi$ where ϕ is the angle through which it turns. Since the chain is inextensible, it follows that $S = s \Longrightarrow R\theta = r\phi$. Since the stone rotates through the same angle ϕ as the point on the chain on the rear sprocket, the distance the stone travels as it rotates through angle ϕ is $\overline{S} = \overline{R}\phi$. Consequently, $\overline{S} = \overline{R}R\theta/r$, and differentiation of this equation with respect to time gives

$$\frac{d\overline{S}}{dt} = \frac{\overline{R}R}{r}\frac{d\theta}{dt} = \frac{\overline{R}R}{r}(2\pi).$$

The stone therefore travels at $2\pi\overline{R}R/r$ m/s when it leaves the tire.

32. If we differentiate $A^2 = s(s-a)(s-b)(s-c)$ with respect to time,

$$2A\frac{dA}{dt} = (s-a)(s-b)(s-c)\frac{ds}{dt} + s(s-b)(s-c)\left(\frac{ds}{dt} - \frac{da}{dt}\right)$$

$$+ s(s-a)(s-c)\left(\frac{ds}{dt} - \frac{db}{dt}\right) + s(s-a)(s-b)\left(\frac{ds}{dt} - \frac{dc}{dt}\right).$$

Furthermore, when we differentiate $s = (a + b + c)/2$,

$$\frac{ds}{dt} = \frac{1}{2}\left(\frac{da}{dt} + \frac{db}{dt} + \frac{dc}{dt}\right).$$

Since $da/dt = db/dt = dc/dt = 1$, it follows that $ds/dt = 3/2$. When $a = 3$, $b = 4$, and $c = 5$, we obtain $s = 6$, and $A = \sqrt{6(6-3)(6-4)(6-5)} = 6$. For these values,

$$2(6)\frac{dA}{dt} = (3)(2)(1)\left(\frac{3}{2}\right) + 6(2)(1)\left(\frac{3}{2} - 1\right) + 6(3)(1)\left(\frac{3}{2} - 1\right) + 6(3)(2)\left(\frac{3}{2} - 1\right).$$

This implies that $dA/dt = 7/2$ square centimetres per minute.

EXERCISES 4.10

2. If we substitute for i into the left side of the equation,

$$R\frac{di}{dt} + \frac{i}{C} = R\left[Ae^{-t/(RC)}\left(\frac{-1}{RC}\right) + \frac{\omega V_0}{Z}\cos(\omega t - \phi)\right] + \frac{A}{C}e^{-t/(RC)} + \frac{V_0}{CZ}\sin(\omega t - \phi)$$

$$= \frac{\omega RV_0}{Z}(\cos\omega t \cos\phi + \sin\omega t \sin\phi) + \frac{V_0}{CZ}(\sin\omega t \cos\phi - \cos\omega t \sin\phi).$$

Because $\tan\phi = -1/(\omega CR)$, it follows that

$$\sin\phi = \frac{-1}{\sqrt{1 + \omega^2 C^2 R^2}} = \frac{-1}{\omega C\sqrt{R^2 + 1/(\omega^2 C^2)}} = \frac{-1}{\omega C Z},$$

$$\cos\phi = \frac{\omega CR}{\sqrt{1 + \omega^2 C^2 R^2}} = \frac{\omega CR}{\omega C Z} = \frac{R}{Z},$$

provided we choose ϕ in the fourth quadrant. With these,

$$R\frac{di}{dt} + \frac{i}{C} = \frac{\omega RV_0}{Z}\left[\frac{R}{Z}\cos\omega t - \frac{1}{\omega CZ}\sin\omega t\right] + \frac{V_0}{CZ}\left[\frac{R}{Z}\sin\omega t + \frac{1}{\omega CZ}\cos\omega t\right]$$

$$= \frac{V_0}{Z^2}\left(\omega R^2 + \frac{1}{\omega C^2}\right)\cos\omega t = \frac{V_0\omega}{Z^2}\left(R^2 + \frac{1}{\omega^2 C^2}\right)\cos\omega t = V_0\omega\cos\omega t = \frac{dV}{dt}.$$

4. If we substitute for i into the left side of the equation,

$$L\frac{di}{dt} + Ri = L\left[Ae^{-Rt/L}(-R/L) + \frac{\omega V_0}{Z}\cos(\omega t - \phi)\right] + RAe^{-Rt/L} + \frac{RV_0}{Z}\sin(\omega t - \phi)$$

$$= \frac{\omega LV_0}{Z}(\cos\omega t \cos\phi + \sin\omega t \sin\phi) + \frac{RV_0}{Z}(\sin\omega t \cos\phi - \cos\omega t \sin\phi).$$

Because $\tan\phi = \omega L/R$, it follows that

$$\sin\phi = \frac{\omega L}{\sqrt{R^2 + \omega^2 L^2}} = \frac{\omega L}{Z}, \qquad \cos\phi = \frac{R}{\sqrt{R^2 + \omega^2 L^2}} = \frac{R}{Z},$$

provided we choose ϕ in the first quadrant. With these,

$$L\frac{di}{dt} + Ri = \frac{\omega LV_0}{Z}\left(\frac{R}{Z}\cos\omega t + \frac{\omega L}{Z}\sin\omega t\right) + \frac{RV_0}{Z}\left(\frac{R}{Z}\sin\omega t - \frac{\omega L}{Z}\cos\omega t\right)$$

$$= \frac{V_0}{Z^2}(\omega^2 L^2 + R^2)\sin\omega t = V_0\sin\omega t.$$

6. Since $\dfrac{di}{dt} = -\dfrac{A}{\sqrt{LC}}\sin\left(\dfrac{t}{\sqrt{LC}}\right) + \dfrac{B}{\sqrt{LC}}\cos\left(\dfrac{t}{\sqrt{LC}}\right)$, it follows that

$$L\frac{d^2 i}{dt^2} + \frac{i}{C} = L\left[-\frac{A}{LC}\cos\left(\frac{t}{\sqrt{LC}}\right) - \frac{B}{LC}\sin\left(\frac{t}{\sqrt{LC}}\right)\right] + \frac{1}{C}\left[A\cos\left(\frac{t}{\sqrt{LC}}\right) + B\sin\left(\frac{t}{\sqrt{LC}}\right)\right] = 0.$$

8. (a) When we substitute the function into the left side of the equation,

$$L\frac{di}{dt} + Ri = L\left\{\frac{\omega LA}{R^2 + \omega^2 L^2}\left[e^{-Rt/L}\left(\frac{-R}{L}\right) + \omega\sin\omega t\right] + \frac{RA\omega}{R^2 + \omega^2 L^2}\cos\omega t\right\}$$

$$+ R\left[\frac{\omega LA}{R^2 + \omega^2 L^2}(e^{-Rt/L} - \cos\omega t) + \frac{RA}{R^2 + \omega^2 L^2}\sin\omega t\right]$$

$$= \left(\frac{\omega^2 L^2 A + R^2 A}{R^2 + \omega^2 L^2}\right)\sin\omega t = A\sin\omega t = V.$$

(b) We express $i(t)$ in the form

$$i(t) = \frac{\omega L A}{R^2 + \omega^2 L^2} e^{-Rt/L} + \frac{A}{R^2 + \omega^2 L^2} (R \sin \omega t - \omega L \cos \omega t),$$

and set

$$R \sin \omega t - \omega L \cos \omega t = B \sin (\omega t - \phi) = B(\sin \omega t \cos \phi - \cos \omega t \sin \phi).$$

This equation is satisfied if B and ϕ are chosen to satisfy

$$R = B \cos \phi, \qquad \omega L = B \sin \phi.$$

These imply that $B^2 = R^2 + \omega^2 L^2$ and $\tan \phi = \omega L / R$, and therefore

$$i(t) = \frac{\omega L A}{R^2 + \omega^2 L^2} e^{-Rt/L} + \frac{A}{R^2 + \omega^2 L^2} \sqrt{R^2 + \omega^2 L^2} \sin (\omega t - \phi)$$

$$= \frac{\omega L A}{R^2 + \omega^2 L^2} e^{-Rt/L} + \frac{A}{\sqrt{R^2 + \omega^2 L^2}} \sin (\omega t - \phi)$$

10. (a) If we set

$$R \cos \omega t + \left(\omega L - \frac{1}{\omega C} \right) \sin \omega t = A \cos (\omega t - \phi) = A(\cos \omega t \cos \phi + \sin \omega t \sin \phi),$$

then, $R = A \cos \phi$ and $\omega L - \dfrac{1}{\omega C} = A \sin \phi$. These imply that $A = \sqrt{R^2 + \left(\omega L - \dfrac{1}{\omega C} \right)^2}$, and therefore

$$i(t) = \frac{V_0}{R^2 + \left(\omega L - \frac{1}{\omega C} \right)^2} \sqrt{R^2 + \left(\omega L - \frac{1}{\omega C} \right)^2} \cos (\omega t - \phi)$$

$$= \frac{V_0}{\sqrt{R^2 + \left(\omega L - \frac{1}{\omega C} \right)^2}} \cos (\omega t + \phi).$$

To maximize the amplitude, we minimize the denominator. This occurs when $\omega L - \dfrac{1}{\omega C} = 0$, and this implies that $\omega = 1/\sqrt{LC}$.

(b) Critical points of $i(t)$ are given by

$$0 = \frac{V_0}{R^2 + \left(\omega L - \frac{1}{\omega C} \right)^2} \left[-R\omega \sin \omega t + \omega \left(\omega L - \frac{1}{\omega C} \right) \cos \omega t \right].$$

If $\bar{t}$ denotes a solution of this equation, then $\tan \omega \bar{t} = \dfrac{\omega L - \frac{1}{\omega C}}{R}$. This implies that

$$\sin \omega \bar{t} = \pm \frac{\omega L - 1/(\omega C)}{\sqrt{R^2 + \left(\omega L - \frac{1}{\omega C} \right)^2}}, \qquad \cos \omega \bar{t} = \pm \frac{R}{\sqrt{R^2 + \left(\omega L - \frac{1}{\omega C} \right)^2}}.$$

Consequently,

$$i(\bar{t}) = \frac{V_0}{R^2 + \left(\omega L - \frac{1}{\omega C} \right)^2} \left[\frac{\pm R^2}{\sqrt{R^2 + \left(\omega L - \frac{1}{\omega C} \right)^2}} \pm \frac{(\omega L - 1/(\omega C))^2}{\sqrt{R^2 + \left(\omega L - \frac{1}{\omega C} \right)^2}} \right] = \frac{\pm V_0}{\sqrt{R^2 + \left(\omega L - \frac{1}{\omega C} \right)^2}}.$$

The value of ω that makes this a maximum is $1/\sqrt{LC}$.

EXERCISES 4.11

Although many of the limits in these exercises can be done without L'Hôpital's rule, we shall demonstrate use of this rule whenever it is applicable.

2. $\displaystyle\lim_{x\to 3}\frac{x^2-9}{x-3}=\lim_{x\to 3}\frac{2x}{1}=6$
$\qquad$ **4.** $\displaystyle\lim_{x\to\infty}\frac{2x^2+3x}{5x^3+4}=\lim_{x\to\infty}\frac{4x+3}{15x^2}=\lim_{x\to\infty}\frac{4}{30x}=0$

6. L'Hôpital's rule is not appliable. $\quad\displaystyle\lim_{x\to 1}\frac{1}{(x-1)^2}=\infty$

8. L'Hôpital's rule is not applicable. $\quad\displaystyle\lim_{x\to -\infty}\frac{\sin x}{x}=0$

10. $\displaystyle\lim_{x\to\pi/2}\frac{\cos x}{(x-\pi/2)^2}=\lim_{x\to\pi/2}\frac{-\sin x}{2(x-\pi/2)}=\pm\infty$ depending on whether x approaches $\pi/2$ from left or right

12. $\displaystyle\lim_{x\to\infty}\frac{\sin(1/x)}{1/x^2}=\lim_{x\to\infty}\frac{-(1/x^2)\cos(1/x)}{-2/x^3}=\lim_{x\to\infty}\frac{x}{2}\cos\left(\frac{1}{x}\right)=\infty$

14. $\displaystyle\lim_{x\to 0}\frac{\sqrt{5+x}-\sqrt{5-x}}{x}=\lim_{x\to 0}\frac{\dfrac{1}{2\sqrt{5+x}}+\dfrac{1}{2\sqrt{5-x}}}{1}=\frac{1}{\sqrt 5}$

16. $\displaystyle\lim_{x\to a}\frac{x^n-a^n}{x-a}=\lim_{x\to a}\frac{nx^{n-1}}{1}=na^{n-1}$
$\qquad$ **18.** $\displaystyle\lim_{x\to 0}\frac{\tan x}{x}=\lim_{x\to 0}\frac{\sec^2 x}{1}=1$

20. $\displaystyle\lim_{x\to 1}\frac{(1-\sqrt{2-x})^{3/2}}{x-1}=\lim_{x\to 1}\frac{\dfrac{3}{2}(1-\sqrt{2-x})^{1/2}\left(\dfrac{1}{2\sqrt{2-x}}\right)}{1}=0$

22. $\displaystyle\lim_{x\to 0}\frac{(1-\cos x)^2}{3x^4}=\lim_{x\to 0}\frac{2(1-\cos x)\sin x}{12x^3}=\lim_{x\to 0}\frac{2\sin x-\sin 2x}{12x^3}=\lim_{x\to 0}\frac{2\cos x-2\cos 2x}{36x^2}$
$\displaystyle\qquad=\lim_{x\to 0}\frac{-2\sin x+4\sin 2x}{72x}=\lim_{x\to 0}\frac{-2\cos x+8\cos 2x}{72}=\frac{1}{12}$

24. It is easier to factor the numerator in this limit than to apply L'Hôpital's rule,

$$\lim_{x\to 2}\frac{(x-2)^{10}}{(\sqrt{x}-\sqrt{2})^{10}}=\lim_{x\to 2}\frac{(\sqrt{x}+\sqrt{2})^{10}(\sqrt{x}-\sqrt{2})^{10}}{(\sqrt{x}-\sqrt{2})^{10}}=\lim_{x\to 2}(\sqrt{x}+\sqrt{2})^{10}=2^{15}.$$

26. L'Hôpital's rule is not applicable. $\quad\displaystyle\lim_{x\to\infty}xe^x=\infty$

28. $\displaystyle\lim_{x\to -\infty}x\sin\left(\frac{4}{x}\right)=\lim_{x\to -\infty}\frac{\sin(4/x)}{1/x}=\lim_{x\to -\infty}\frac{(-4/x^2)\cos(4/x)}{-1/x^2}=\lim_{x\to -\infty}4\cos\left(\frac{4}{x}\right)=4$

30. $\displaystyle\lim_{x\to 0}\csc x\,(1-\cos x)=\lim_{x\to 0}\frac{1-\cos x}{\sin x}=\lim_{x\to 0}\frac{\sin x}{\cos x}=0$

32. If we set $L=\displaystyle\lim_{x\to 0^+}x^{\sin x}$, and take natural logarithms,

$$\ln L=\ln\left[\lim_{x\to 0^+}x^{\sin x}\right]=\lim_{x\to 0^+}\left[\ln(x^{\sin x})\right]=\lim_{x\to 0^+}\left[\sin x\ln x\right]=\lim_{x\to 0^+}\frac{\ln x}{\csc x}$$
$$=\lim_{x\to 0^+}\frac{1/x}{-\csc x\cot x}=\lim_{x\to 0^+}\frac{-\sin x\tan x}{x}=\lim_{x\to 0^+}\frac{-(\sin x\sec^2 x+\cos x\tan x)}{1}=0.$$

Thus, $L=\displaystyle\lim_{x\to 0^+}x^{\sin x}=e^0=1$.

34. If we set $L=\displaystyle\lim_{x\to 0}(1+x)^{\cot x}$, and take natural logarithms,

$$\ln L=\ln\left[\lim_{x\to 0}(1+x)^{\cot x}\right]=\lim_{x\to 0}\left[\cot x\ln(1+x)\right]=\lim_{x\to 0}\frac{\ln(1+x)}{\tan x}=\lim_{x\to 0}\frac{\dfrac{1}{1+x}}{\sec^2 x}=1.$$

Thus, $L=\displaystyle\lim_{x\to 0}(1+x)^{\cot x}=e$.

36. If we set $L = \lim\limits_{x\to 0^+} |\ln x|^{\sin x}$, and take natural logarithms,

$$\ln L = \ln\left[\lim_{x\to 0^+} |\ln x|^{\sin x}\right] = \lim_{x\to 0^+} \ln\left[|\ln x|^{\sin x}\right] = \lim_{x\to 0^+} (\sin x \ln|\ln x|) = \lim_{x\to 0^+} \frac{\ln|\ln x|}{\csc x}$$

$$= \lim_{x\to 0^+} \frac{\dfrac{1}{|\ln x|}\dfrac{1}{x}\dfrac{|\ln x|}{\ln x}}{-\csc x \cot x} = \lim_{x\to 0^+} \frac{-\sin x \tan x}{x \ln x}.$$

To verify that we can use L'Hôpital's rule again, we note that

$$\lim_{x\to 0^+} x\ln x = \lim_{x\to 0^+} \frac{\ln x}{1/x} = \lim_{x\to 0^+} \frac{1/x}{-1/x^2} = \lim_{x\to 0^+}(-x) = 0.$$

Thus, $\ln L = \lim\limits_{x\to 0^+} \dfrac{-(\sin x \sec^2 x + \cos x \tan x)}{x/x + \ln x} = 0,$ and $L = \lim\limits_{x\to 0^+} |\ln x|^{\sin x} = e^0 = 1.$

38. This limit does not exist since $\lim_{x\to 0^+}(\tan x - \csc x) = -\infty$ and $\lim_{x\to 0^-}(\tan x - \csc x) = \infty$.

40. $\lim\limits_{x\to 1}\left(\dfrac{x}{\ln x} - \dfrac{1}{x\ln x}\right) = \lim\limits_{x\to 1}\dfrac{x^2-1}{x\ln x} = \lim\limits_{x\to 1}\dfrac{2x}{\ln x + 1} = 2$

42. $\lim\limits_{x\to 0}\left(\dfrac{1}{x^2} - \dfrac{1}{\sin^2 x}\right) = \lim\limits_{x\to 0}\dfrac{\sin^2 x - x^2}{x^2\sin^2 x} = \lim\limits_{x\to 0}\dfrac{(1-\cos 2x)/2 - x^2}{x^2(1-\cos 2x)/2}$

$$= \lim_{x\to 0}\frac{1-\cos 2x - 2x^2}{x^2 - x^2\cos 2x} = \lim_{x\to 0}\frac{2\sin 2x - 4x}{2x + 2x^2\sin 2x - 2x\cos 2x}$$

$$= \lim_{x\to 0}\frac{4\cos 2x - 4}{2 + 8x\sin 2x + 4x^2\cos 2x - 2\cos 2x}$$

$$= \lim_{x\to 0}\frac{-8\sin 2x}{12\sin 2x + 24x\cos 2x - 8x^2\sin 2x}$$

$$= \lim_{x\to 0}\frac{-16\cos 2x}{48\cos 2x - 64x\sin 2x - 16x^2\cos 2x} = -\frac{1}{3}$$

44. For critical points, we solve $0 = f'(x) = 2x\,e^{3x} + 3x^2\,e^{3x} = x(2+3x)e^{3x}$. Solutions are $x = 0$ and $x = -2/3$. Since $f'(x)$ changes from positive to negative as x increases through $-2/3$, this critical point gives a relative maximum of $f(-2/3) = 4/(9e^2)$. There is a relative minimum of $f(0) = 0$ at $x = 0$ because $f'(x)$ changes from negative to positive as x increases through this value. For points of inflection, we solve $0 = f''(x) = (2+6x)e^{3x} + 3x(2+3x)e^{3x} = (9x^2 + 12x + 2)e^{3x}$. Since $f''(x)$ changes sign as x passes through the solutions $x = (-2 \pm \sqrt{2})/3$, there are points of inflection at $((-2-\sqrt{2})/3, 0.043)$ and $((-2+\sqrt{2})/3, 0.021)$. Since

$$\lim_{x\to -\infty} x^2 e^{3x} = \lim_{x\to -\infty}\frac{x^2}{e^{-3x}} = \lim_{x\to -\infty}\frac{2x}{-3e^{-3x}} = \lim_{x\to -\infty}\frac{2}{9e^{-3x}} = 0^+,$$

the x-axis is a horizontal asymptote. This information, shown in the left figure below, leads to the final graph in the right figure.

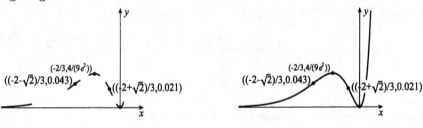

46. The following limits show that $y = 1$ is a horizontal asymptote, and the y-axis is a vertical asymptote,

$$\lim_{x \to -\infty} e^{1/x} = 1^-, \quad \lim_{x \to \infty} e^{1/x} = 1^+, \quad \lim_{x \to 0^+} e^{1/x} = \infty.$$

Since $f'(x) = -(1/x^2)e^{1/x}$, the function has no critical points, but we notice that

$$\lim_{x \to 0^-} f'(x) = \lim_{x \to 0^-} \frac{-1/x^2}{e^{-1/x}} = \lim_{x \to 0^-} \frac{2/x^3}{(1/x^2)e^{-1/x}} = \lim_{x \to 0^-} \frac{2/x}{e^{-1/x}} = \lim_{x \to 0^-} \frac{-2/x^2}{(1/x^2)e^{-1/x}} = \lim_{x \to 0^-} (-2e^{1/x}) = 0^-.$$

We can locate points of inflection by solving $0 = f''(x) = \frac{2}{x^3}e^{1/x} + \frac{1}{x^4}e^{1/x} = \frac{1}{x^4}(2x + 1)e^{1/x}$. Since the only solution is $x = -1/2$, and $f''(x)$ changes sign as x passes through this value, a point of inflection is $(-1/2, 1/e^2)$. This information, shown in the left figure below, leads to the final graph in the right figure.

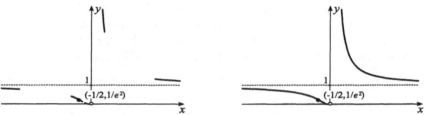

48. For critical points, we solve $0 = f'(x) = 2x \ln x + x^2/x = x(2\ln x + 1)$. The only solution is $x = 1/\sqrt{e}$. Since $f''(x) = 2\ln x + 2x/x + 1 = 2\ln x + 3$, it follows that $f''(1/\sqrt{e}) = 2$. The critical point therefore gives a relative minimum of $f(1/\sqrt{e}) = -1/(2e)$. Since $f''(e^{-3/2}) = 0$, and $f''(x)$ changes sign as x passes through $e^{-3/2}$, there is a point of inflection at $(e^{-3/2}, -3/(2e^3))$. We use L'Hôpital's rule to show that the graph approaches the origin,

$$\lim_{x \to 0^+} x^2 \ln x = \lim_{x \to 0^+} \frac{\ln x}{1/x^2} = \lim_{x \to 0^+} \frac{1/x}{-2/x^3} = \lim_{x \to 0^+} \left(-\frac{x^2}{2} \right) = 0^-.$$

The slope of the graph also approaches zero as $x \to 0^+$,

$$\lim_{x \to 0^+} f'(x) = \lim_{x \to 0^+} [x(2\ln x + 1)] = \lim_{x \to 0^+} \frac{2\ln x + 1}{1/x} = \lim_{x \to 0^+} \frac{2/x}{-1/x^2} = \lim_{x \to 0^+} (-2x) = 0^-.$$

This information, shown in the left figure below, leads to the final graph in the right figure.

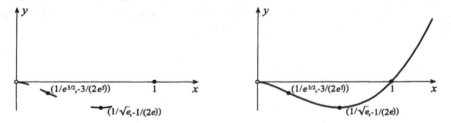

50. The limits

$$\lim_{x \to 0^+} \frac{x^2}{\ln x} = 0^-, \quad \lim_{x \to 1^-} \frac{x^2}{\ln x} = -\infty, \quad \lim_{x \to 1^+} \frac{x^2}{\ln x} = \infty,$$

show that the curve approaches the origin as $x \to 0^+$, and that the line $x = 1$ is a vertical asymptote. For critical points, we solve

$$0 = f'(x) = \frac{2x}{\ln x} - \frac{x^2}{(\ln x)^2}\left(\frac{1}{x} \right) = \frac{x}{(\ln x)^2}(2\ln x - 1).$$

Thus, $x = \sqrt{e}$, and this gives a relative minimum of $f(\sqrt{e}) = 2e$ ($f'(x)$ changing from negative to positive as x increases through $\sqrt{e}$). The following limit shows that the slope of the graph approaches zero as

$x \to 0^+$,

$$\lim_{x \to 0^+} f'(x) = \lim_{x \to 0^+} \left[\frac{2x}{\ln x} - \frac{x}{(\ln x)^2} \right] = 0^-.$$

This information, shown in the left figure below, leads to the final graph in the right figure.

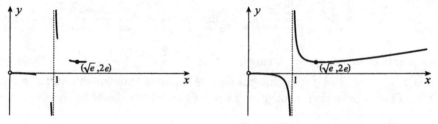

52. For critical points we solve $0 = f'(x) = 10x^9 e^{-x} - x^{10} e^{-x} = x^9(10-x)e^{-x}$. The solutions are $x = 0$ and $x = 10$. Since $f'(x)$ changes from positive to negative as x increases through 10, this critical point gives a relative maximum of $f(10) = 10^{10}/e^{10}$. We have a relative minimum at $f(0) = 0$ since $f'(x)$ changes from negative to positive as x increases through 0. For points of inflection, we solve

$$0 = f''(x) = (90x^8 - 10x^9)e^{-x} - (10x^9 - x^{10})e^{-x} = x^8(x^2 - 20x^9 + 90)e^{-x}.$$

Solutions are $x = \dfrac{20 \pm \sqrt{400 - 360}}{2} = 10 \pm \sqrt{10}$. Since $f''(x)$ changes sign as x passes through these values, there are points of inflection at $(10-\sqrt{10}, 239\,624)$ and $(10+\sqrt{10}, 299\,920)$. Repeated applications of L'Hôpital's rule show that $\lim_{x \to \infty} x^{10} e^{-x} = 0^+$. This information, shown in the left figure below, leads to the final graph in the right figure.

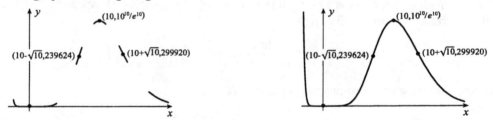

54. For critical points, we solve $0 = f'(x) = -2\csc x \cot x + \csc^2 x = \csc^2 x(1 - 2\cos x)$. The only critical point is $x = \pi/3$. Since $f'(x)$ changes from negative to positive as x increases through $\pi/3$, there is a relative minimum of $f(\pi/3) = \sqrt{3}$ at this critical point. The limit $\lim_{x \to 0^+} (2\csc x - \cot x) = \lim_{x \to 0^+} \frac{2 - \cos x}{\sin x} = \infty$ indicates that the y-axis is a vertical asymptote. This information, shown in the left figure below, leads to the final graph in the right figure.

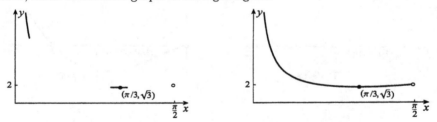

56. If we set $L = \lim_{x \to \infty} (x - \ln x)$, and take exponentials on both sides of the equation,

$e^L = e^{\lim_{x \to \infty} (x - \ln x)}$. If we interchange the limit and exponentiation operations,

$$e^L = \lim_{x \to \infty} e^{x - \ln x} = \lim_{x \to \infty} \frac{e^x}{x} = \lim_{x \to \infty} \frac{e^x}{1} = \infty.$$

It follows therefore that $L = \lim_{x \to \infty} (x - \ln x) = \infty$ also.

58. (a) $\lim\limits_{\lambda\to 0^+} \psi(\lambda) = \lim\limits_{\lambda\to 0^+} \dfrac{-5k\lambda^{-6}}{e^{c/\lambda}(-c/\lambda^2)} = \lim\limits_{\lambda\to 0^+} \dfrac{5k\lambda^{-4}}{ce^{c/\lambda}} = \lim\limits_{\lambda\to 0^+} \dfrac{5k(-4)\lambda^{-5}}{ce^{c/\lambda}(-c/\lambda^2)}$

$\qquad = \lim\limits_{\lambda\to 0^+} \dfrac{20k\lambda^{-3}}{c^2 e^{c/\lambda}} = \cdots = \lim\limits_{\lambda\to 0^+} \dfrac{120k}{c^5 e^{c/\lambda}} = 0$

$\quad \lim\limits_{\lambda\to\infty} \psi(\lambda) = \lim\limits_{\lambda\to\infty} \dfrac{-5k\lambda^{-6}}{e^{c/\lambda}(-c/\lambda^2)} = \lim\limits_{\lambda\to\infty} \dfrac{5k}{c\lambda^4 e^{c/\lambda}} = 0$

(b) For critical points of $\psi(\lambda)$ we solve $\quad 0 = \psi'(\lambda) = \dfrac{-k}{[\lambda^5(e^{c/\lambda}-1)]^2}\left[5\lambda^4(e^{c/\lambda}-1) + \lambda^5 e^{c/\lambda}(-c/\lambda^2)\right]$,

and therefore $0 = \lambda^3[5\lambda(e^{c/\lambda}-1) - ce^{c/\lambda}]$.
Since $\lambda \neq 0$, we must set $(5\lambda - c)e^{c/\lambda} = 5\lambda$.
This equation was solved using Newton's
iterative procedure in Exercise 34 of
Section 4.1. The critical point is
$\lambda = 0.000\,029\,0$.
(c) The limits in (a) together with the fact
that $\psi(\lambda)$ is always positive for $\lambda > 0$
implies that the critical point must give a
relative maximum. The graph of the function
is shown to the right.

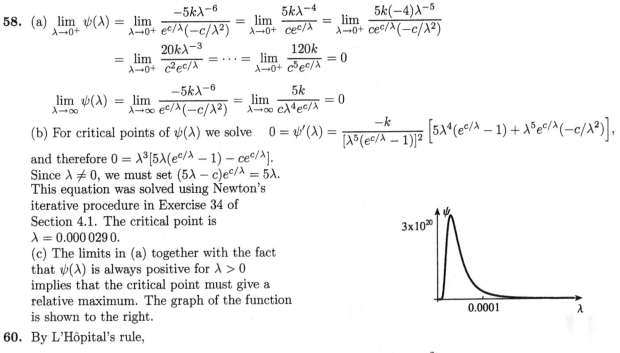

60. By L'Hôpital's rule,

$$5 = \lim\limits_{x\to 0} \dfrac{ae^{ax} - b + (1+2cx)\sin(x+cx^2)}{6x^2 + 10x}.$$

Since the limit of the numerator is $a - b$ and that of the denominator is 0, the only way this limit can be 5 is for $a = b$. In this case,

$$5 = \lim\limits_{x\to 0} \dfrac{ae^{ax} - a + (1+2cx)\sin(x+cx^2)}{6x^2 + 10x}$$

$$= \lim\limits_{x\to 0} \dfrac{a^2 e^{ax} + 2c\sin(x+cx^2) + (1+2cx)^2\cos(x+cx^2)}{12x + 10} = \dfrac{a^2 + 1}{10}.$$

Thus, $a = \pm 7 = b$, and c is arbitrary.

EXERCISES 4.12

2. $dy = f'(x)\,dx = \left[\dfrac{(x-1)(1) - (x+1)(1)}{(x-1)^2}\right]dx = \dfrac{-2}{(x-1)^2}dx$

4. $dy = f'(x)\,dx = [2x\cos(x^2+2) + \sin x]\,dx$

6. $dy = f'(x)\,dx = [3x^2\sqrt{3-4x^2} + (1/2)x^3(3-4x^2)^{-1/2}(-8x)]\,dx$

$\qquad = \dfrac{3x^2(3-4x^2) - 4x^4}{\sqrt{3-4x^2}}dx = \dfrac{x^2(9-16x^2)}{\sqrt{3-4x^2}}dx$

8. Since $f(x) = \dfrac{(x-1)^3 + 6}{(x-1)^2} = x - 1 + \dfrac{6}{(x-1)^2}$, $\quad dy = f'(x)dx = \left[1 - \dfrac{12}{(x-1)^3}\right]dx.$

10. $dy = f'(x)\,dx = \left[\dfrac{(x^3+5x)(3x^2-4x) - (x^3-2x^2)(3x^2+5)}{(x^3+5x)^2}\right]dx$

$\qquad = \dfrac{2x^4 + 10x^3 - 10x^2}{(x^3+5x)^2}dx = \dfrac{2(x^2+5x-5)}{(x^2+5)^2}dx$

12. The approximate percentage change in F is

$$100\dfrac{dF}{F} = \dfrac{100}{F}\left(-\dfrac{2GmM}{r^3}dr\right) = -\dfrac{200GmM}{r^3}dr\left(\dfrac{r^2}{GmM}\right) = -2\left(100\dfrac{dr}{r}\right) = -2(2) = -4.$$

14. The approximate percentage change in H is

$$100\frac{dH}{H} = \frac{100}{H}\left(\frac{2v^2\sin\theta\cos\theta}{19.62}d\theta\right) = \frac{200v^2\sin\theta\cos\theta}{19.62}d\theta\left(\frac{19.62}{v^2\sin^2\theta}\right)$$

$$= 200\cot\theta\, d\theta = 2\theta\cot\theta\left(100\frac{d\theta}{\theta}\right) = 4\theta\cot\theta,$$

since the change in θ is 2%. When $\theta = \pi/3$, this becomes

$$100\frac{dH}{H} = 4\left(\frac{\pi}{3}\right)\left(\frac{1}{\sqrt{3}}\right) = \frac{4\sqrt{3}\pi}{9}.$$

16. The differential of F is $dF = -\dfrac{2GmM}{r^3}dr$. If we set $dF = -0.01m$ for a decrease from $9.81m$ to $9.80m$, then

$$-0.01m = -\frac{2GmM}{r^3}dr.$$

But when $r = 6.37 \times 10^6$, we know that $F = 9.81m$ so that

$$9.81m = \frac{GmM}{(6.37 \times 10^6)^2}.$$

Consequently,

$$-0.01m = \frac{-2}{(6.37 \times 10^6)^3}(9.81m)(6.37 \times 10^6)^2\, dr,$$

and this equation implies that $dr = 3.25 \times 10^3$. Thus, at a height of 3.25 km, the gravitational force of attraction decreases to $9.80m$ N.

18. (a) The approximate percentage error in V due to an $a\%$ error in r is

$$100\frac{dV}{V} = \frac{100}{V}(2\pi rh\, dr) = \frac{200\pi rh}{\pi r^2 h}dr = 2\left(100\frac{dr}{r}\right) = 2a.$$

(b) The approximate percentage error in V due to a $b\%$ error in h is

$$100\frac{dV}{V} = \frac{100}{V}(\pi r^2\, dh) = \frac{100\pi r^2}{\pi r^2 h}dh = 100\frac{dh}{h} = b.$$

(c) The maximum approximate percentage error in V due to errors $a\%$ in r and $b\%$ in h is

$$100\frac{\text{Maximum change in V}}{V} = \frac{100}{\pi r^2 h}(2\pi rh\, dr + \pi r^2\, dh) = 2\left(100\frac{dr}{r}\right) + \left(100\frac{dh}{h}\right) = 2a + b.$$

20. (a) The approximate percentage error in z due to an $a\%$ error in x is

$$100\frac{dz}{z} = \frac{100}{x^n y^m}(nx^{n-1}y^m\, dx) = n\left(100\frac{dx}{x}\right) = na.$$

(b) The approximate percentage error in z due to a $b\%$ error in y is

$$100\frac{dz}{z} = \frac{100}{x^n y^m}(my^{m-1}x^n\, dx) = m\left(100\frac{dy}{y}\right) = mb.$$

(c) The maximum approximate percentage error in z due to errors $a\%$ in x and $b\%$ in y is

$$100\frac{\text{Maximum change in z}}{z} = \frac{100}{x^n y^m}(nx^{n-1}y^m\, dx + my^{m-1}x^n\, dy) = n\left(100\frac{dx}{x}\right) + m\left(100\frac{dy}{y}\right) = na + mb.$$

22. The approximate percentage change in n is

$$100\frac{dn}{n} = \frac{100}{n}\frac{(1/2)\cos\left[(\psi_m + \gamma)/2\right]}{\sin(\gamma/2)}d\psi_m$$

$$= \frac{50\cos\left[(\psi_m + \gamma)/2\right]}{\sin(\gamma/2)}d\psi_m\left\{\frac{\sin(\gamma/2)}{\sin\left[(\psi_m + \gamma)/2\right]}\right\} = \frac{\psi_m}{2}\cot\left[(\psi_m + \gamma)/2\right]\left(100\frac{d\psi_m}{\psi_m}\right).$$

Since the error in ψ_m is 1% when it is $\pi/6$ and $\gamma = \pi/3$,

$$100\frac{dn}{n} = \frac{\pi/6}{2}\cot\left[(\pi/6 + \pi/3)/2\right](1) = \frac{\pi}{12}.$$

REVIEW EXERCISES

2. (a) For critical points we solve $0 = f'(x) = 12x^2 + 2x - 2 = 2(3x - 1)(2x + 1)$. Solutions are $x = 1/3$ and $x = -1/2$. Since $f''(x) = 24x + 2$, it follows that $f''(1/3) = 10$ and $f''(-1/2) = -10$. Consequently, $x = 1/3$ gives a relative minimum of $f(1/3) = 16/27$, and $x = -1/2$ gives a relative maximum of $f(-1/2) = 7/4$. Since $f''(-1/12) = 0$ and $f''(x)$ changes sign as x passes through $-1/12$, there is a point of inflection at $(-1/12, 253/216)$. This information, shown in the left figure below, leads to the final graph in the right figure.

(b) With the function written as $f(x) = \dfrac{(x^2 - 2x + 1) + 3}{(x - 1)^2} = 1 + \dfrac{3}{(x - 1)^2}$, critical points are given by $0 = f'(x) = -6/(x - 1)^3$. There are no solutions and hence no relative extrema. Since the second derivative $f''(x) = 18/(x - 1)^4$ never vanishes, there are no points of inflection. Limits as $x \to \pm\infty$ and right- and left-limits at $x = 1$ give the information in the left figure below. The final graph is to the right.

4. If x and y are any two positive numbers with sum $x + y = c$, their product is $P = P(x) = xy = x(c - x)$, $0 < x < c$. For critical points of $P(x)$, we solve $0 = P'(x) = c - 2x$. The solution is $x = c/2$. Since

$$\lim_{x \to 0^+} P(x) = 0, \qquad P(c/2) = \frac{c^2}{4}, \qquad \lim_{x \to c^-} P(x) = 0,$$

it follows that $P(x)$ has no absolute minimum on the interval $0 < x < c$. The product can be made arbitrarily close to 0 by choosing x or y sufficiently close to 0.

6. If x and y are any two positive numbers with product $xy = c$, their sum is $S = S(x) = x + y = x + (c/x)$, $x > 0$. For critical points of $S(x)$ we solve $0 = S'(x) = 1 - c/x^2$. The positive solution is $x = \sqrt{c}$. Since

$$\lim_{x \to 0^+} S(x) = \infty, \qquad S(\sqrt{c}) = 2\sqrt{c}, \qquad \lim_{x \to \infty} S(x) = \infty,$$

it follows that $S(x)$ has no absolute maximum on the interval $x > 0$. The sum can be made arbitrarily large by choosing x or y sufficiently close to 0.

8. $\displaystyle\lim_{x\to 0}\frac{3x^2+2x^3}{3x^3-2x^2}=\lim_{x\to 0}\frac{6x+6x^2}{9x^2-4x}=\lim_{x\to 0}\frac{6+12x}{18x-4}=-\frac{3}{2}$

10. $\displaystyle\lim_{x\to 4}\frac{x^2-16}{x-4}=\lim_{x\to 4}\frac{2x}{1}=8$ **12.** $\displaystyle\lim_{x\to -\infty}\frac{\sin x^2}{2x}=0$

14. $\displaystyle\lim_{x\to\infty}x^2 e^{-3x}=\lim_{x\to\infty}\frac{x^2}{e^{3x}}=\lim_{x\to\infty}\frac{2x}{3e^{3x}}=\lim_{x\to\infty}\frac{2}{9e^{3x}}=0$

16. $\displaystyle\lim_{x\to 0^+}x^4\ln x=\lim_{x\to 0^+}\frac{\ln x}{1/x^4}=\lim_{x\to 0^+}\frac{1/x}{-4/x^5}=\lim_{x\to 0^+}\left(\frac{-x^4}{4}\right)=0$

18. If we set $L=\displaystyle\lim_{x\to\infty}\left(\frac{x+1}{x-1}\right)^x$ and take natural logarithms,

$$\ln L=\lim_{x\to\infty}x\ln\left(\frac{x+1}{x-1}\right)=\lim_{x\to\infty}\frac{\ln\left(\dfrac{x+1}{x-1}\right)}{1/x}=\lim_{x\to\infty}\frac{\left(\dfrac{x-1}{x+1}\right)\left[\dfrac{(x-1)(1)-(x+1)(1)}{(x-1)^2}\right]}{-1/x^2}$$

$$=\lim_{x\to\infty}\frac{2x^2}{x^2-1}=\lim_{x\to\infty}\frac{4x}{2x}=2.$$

Thus, $L=e^2$.

20. (a) For critical points we solve

$0=f'(x)=4x^3+6x-2=2(2x^3+3x-1).$

The graph of $f'(x)$ to the right indicates
that there is a critical point between $x=0$
and $x=1$. To find it with Newton's iterative
procdure, we use $x_1=1/3$ and

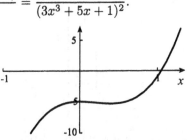

$$x_{n+1}=x_n-\frac{2x_n^3+3x_n-1}{6x_n^2+3}.$$

Iteration gives $x_2=0.313\,131\,3$, $x_3=0.312\,908\,4$, $x_4=0.312\,908\,4$. Since $f'(0.312\,907\,5)=-6.5\times 10^{-6}$ and $f'(0.312\,908\,5)=6.4\times 10^{-7}$, the critical point is $0.312\,908$.

(b) For critical points we solve

$$0=f'(x)=\frac{(3x^3+5x+1)(3x^2)-(x^3+1)(9x^2+5)}{(3x^3+5x+1)^2}=\frac{10x^3-6x^2-5}{(3x^3+5x+1)^2}.$$

Thus, critical points are defined by the
equation $10x^3-6x^2-5=0$. The graph of the
function $10x^3-6x^2-5$ to the right indicates
only one critical point just larger than 1.
To find it we use $x_1=1$ and

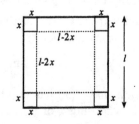

$$x_{n+1}=x_n-\frac{10x_n^3-6x_n^2-5}{30x_n^2-12x_n}.$$

Iteration gives $x_2=1.055\,555\,6$, $x_3=1.051\,904\,7$, $x_4=1.051\,888\,1$, $x_5=1.051\,888\,1$. Since $f'(1.051\,887\,5)=-1.3\times 10^{-7}$ and $f'(1.051\,888\,5)=8.5\times 10^{-8}$, the critical point is $1.051\,888$.

22. The velocity at t_0 abruptly changes from a positive quantity to a negative quantity. This could be caused by a collision with a large object.

24. If squares of side length x are cut from the corners, the resulting box has volume

$V=x(l-2x)^2=4x^3-4lx^2+l^2x,\ 0\le x\le l/2.$

Critical points of V are given by

$0=V'(x)=12x^2-8lx+l^2=(2x-l)(6x-l).$

Thus, $x=l/2$ or $x=l/6$. Since

$$V(0)=0,\quad V(l/6)=\frac{2l^3}{27},\quad V(l/2)=0,$$

maximum volume is $2l^3/27$.

26. (a) We take the following limits at the discontinuities $x = \pm 1$:

$$\lim_{x \to -1^-} f(x) = -\infty, \quad \lim_{x \to -1^+} f(x) = \infty, \quad \lim_{x \to 1^-} f(x) = -\infty, \quad \lim_{x \to 1^+} f(x) = \infty.$$

Critical points are given by

$$0 = f'(x) = \frac{(x^2 - 1)(3x^2) - x^3(2x)}{(x^2 - 1)^2} = \frac{x^2(x^2 - 3)}{(x^2 - 1)^2}.$$

Solutions are $x = 0, \pm\sqrt{3}$. Since $f'(x)$ changes from a positive quantity to a negative quantity as x increases through $-\sqrt{3}$, there is a relative maximum at $x = -\sqrt{3}$ of $-3\sqrt{3}/2$. Since $f'(x)$ changes from a negative quantity to a positive quantity as x increases through $\sqrt{3}$, there is a relative minimum at $x = \sqrt{3}$ of $3\sqrt{3}/2$. Because $f'(x)$ does not change sign at $x = 0$, it gives a horizontal point of inflection $(0, 0)$. To verify that no other points of inflection occur, we calculate

$$f''(x) = \frac{(x^2 - 1)^2(4x^3 - 6x) - (x^4 - 3x^2)(2)(x^2 - 1)(2x)}{(x^2 - 1)^4} = \frac{x(2x^2 + 6)}{(x^2 - 1)^3}.$$

Since $f''(x)$ is 0 only at $x = 0$, there are no other points of inflection. By writing $f(x)$ in the form $f(x) = x + x/(x^2 - 1)$, we see that $y = x$ is an oblique asymptote for the graph. This information, shown in the left figure below, leads to the final graph in the right figure.

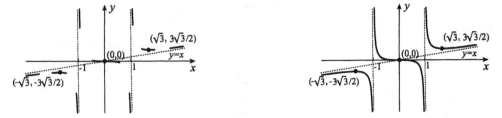

(b) For critical points we consider $0 = f'(x) = 2x + 2\sin x \cos x = 2x + \sin 2x$. The only solution of this equation is $x = 0$. For points of inflection we solve $0 = f''(x) = 2 + 2\cos 2x$. Solutions of this equation are $x = (2n + 1)\pi/2$, where n is an integer. Since $f''(x)$ does not change sign as x passes through these points, the graph has no points of inflection. In addition, since $f''(0) = 4$, there is a relative minimum at $x = 0$ of $f(0) = 0$. This information, little as it is, shown in the left figure below, leads to the final graph in the right figure.

28. If the price is raised x dollars per ticket, then the expected numbers of $10 + x$, $9 + x$, and $8 + x$ dollar tickets the team expects to sell are respectively

$$10\,000(1 - x/10), \quad 20\,000(1 - x/10), \quad 30\,000(1 - x/10).$$

Total revenue at the new prices is therefore

$$R(x) = 10\,000(1 - x/10)(10 + x) + 20\,000(1 - x/10)(9 + x) + 30\,000(1 - x/10)(8 + x)$$
$$= 10\,000(1 - x/10)(6x + 52) = 2000(10 - x)(3x + 26).$$

We must take $x \geq 0$ and x cannot be greater than 10, else no tickets will be sold. For critical points of $R(x)$ we solve $0 = R'(x) = 2000(-6x + 4)$. Thus, $x = 2/3$. Since $R''(x) = -12\,000$, the graph of the function $R(x)$ is always concave downward. This means that $x = 2/3$ must yield an absolute maximum. The price increase should be 67 cents.

30. When $\|PQ\|$ is the shortest distance from P to the parabola $y = x^2$, line PQ is perpendicular to the tangent line to $y = x^2$ at $Q(X, Y)$. It follows therefore that

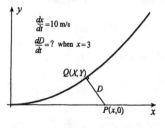

$$2X = -\frac{1}{\dfrac{Y - 0}{X - x}} = \frac{x - X}{Y}.$$

We combine this with $Y = X^2$ to obtain

$$2X(X^2) = x - X,$$

or $2X^3 + X - x = 0$. This equation defines the x-coordinate X of Q in terms of the x-coordinate x of P. The distance D from P to $y = x^2$ is then given by

$$D^2 = (x - X)^2 + Y^2 = (x - X)^2 + X^4 = (2X^3)^2 + X^4 = 4X^6 + X^4.$$

Differentiation of this equation with respect to time gives

$$2D\frac{dD}{dt} = 24X^5\frac{dX}{dt} + 4X^3\frac{dX}{dt}.$$

But differentiation of $2X^3 + X - x = 0$ gives

$$6X^2\frac{dX}{dt} + \frac{dX}{dt} - \frac{dx}{dt} = 0.$$

When $x = 3$, X is defined by $2X^3 + X - 3 = 0$, and the only solution of this equation is $X = 1$. At this instant then

$$6(1)^2\frac{dX}{dt} + \frac{dX}{dt} - 10 = 0,$$

from which $dX/dt = 10/7$. Since $D = \sqrt{(3 - 1)^2 + 1^2} = \sqrt{5}$ at this instant,

$$2\sqrt{5}\frac{dD}{dt} = 24(1)^5\left(\frac{10}{7}\right) + 4(1)^3\left(\frac{10}{7}\right),$$

and therefore $dD/dt = 4\sqrt{5}$. The distance is therefore increasing at $4\sqrt{5}$ m/s.

32. The farmer's losses when x hectares of corn and y hectares of potatoes are planted are $L = pax^2 + qby^2$. Since $x + y = 100$,

$$L = L(x) = pax^2 + qb(100 - x)^2, \quad 0 \le x \le 100.$$

For critical points of $L(x)$ we solve $0 = L'(x) = 2pax - 2qb(100 - x)$. The solution is $x = x_c = 100qb/(pa + qb)$. Now

$$L(0) = 10\,000qb, \qquad L(x_c) = \frac{10\,000abpq}{pa + qb}, \qquad L(100) = 10\,000pa.$$

If we write

$$\frac{1}{L(0)} = \frac{10^{-4}}{qb}, \qquad \frac{1}{L(x_c)} = 10^{-4}\left(\frac{1}{qb} + \frac{1}{pa}\right), \qquad \frac{1}{L(100)} = \frac{10^{-4}}{pa},$$

it is clear that $1/L(x_c)$ is greater than $1/L(0)$ and $1/L(100)$. Consequently, $L(x_c)$ is less than $L(0)$ and $L(100)$, and $L(x)$ is minimized for $x = 100qb/(pa + qb)$ hectares and $y = 100pa/(pa + qb)$ hectares.

Notice that if a increases while p, q, and b remain constant, our results suggest that more and more potatoes should be planted. Conversely large b implies planting more corn. On the other hand, if a, b, and q remain constant but p increases, the farmer should plant more potatoes. The reason is that with a large area in corn, he will suffer a substantial loss of money.

CHAPTER 5

EXERCISES 5.1

2. $\displaystyle\int (x^4 + 3x^2 + 5x)\,dx = \frac{x^5}{5} + x^3 + \frac{5x^2}{2} + C$ **4.** $\displaystyle\int \sin x\,dx = -\cos x + C$

6. $\displaystyle\int \sqrt{x}\,dx = \frac{2}{3}x^{3/2} + C$ **8.** $\displaystyle\int \left(\frac{1}{x^2} - \frac{2}{x^4}\right) dx = -\frac{1}{x} + \frac{2}{3x^3} + C$

10. $\displaystyle\int \left(\frac{1}{x^2} + \frac{1}{2\sqrt{x}}\right) dx = -\frac{1}{x} + \sqrt{x} + C$ **12.** $\displaystyle\int \left(-\frac{1}{2x^2} + 3x^3\right) dx = \frac{1}{2x} + \frac{3x^4}{4} + C$

14. $\displaystyle\int (2\sqrt{x} + 3x^{3/2} - 5x^{5/2})\,dx = \frac{4}{3}x^{3/2} + \frac{6}{5}x^{5/2} - \frac{10}{7}x^{7/2} + C$

16. $\displaystyle\int \sqrt{x}(x+1)\,dx = \int (x^{3/2} + \sqrt{x})\,dx = \frac{2}{5}x^{5/2} + \frac{2}{3}x^{3/2} + C$

18. $\displaystyle\int x^2(1+x^2)^2\,dx = \int (x^2 + 2x^4 + x^6)\,dx = \frac{x^3}{3} + \frac{2x^5}{5} + \frac{x^7}{7} + C$

20. $\displaystyle\int \frac{(x-1)^2}{\sqrt{x}}\,dx = \int \left(x^{3/2} - 2\sqrt{x} + \frac{1}{\sqrt{x}}\right) dx = \frac{2}{5}x^{5/2} - \frac{4}{3}x^{3/2} + 2\sqrt{x} + C$

22. If we take the indefinite integral of $f'(x) = 2x^3 + 4x$ with respect to x, we obtain $f(x) = x^4/2 + 2x^2 + C$. Since $f(0) = 5$, it follows that $5 = C$, and therefore $y = x^4/2 + 2x^2 + 5$.

24. If we take the indefinite integral of $f'(x) = 2 - 4x + 8x^7$ with respect to x, we obtain $f(x) = 2x - 2x^2 + x^8 + C$. Since $f(1) = 1$, it follows that $1 = 2 - 2 + 1 + C$. Thus, $C = 0$, and $y = 2x - 2x^2 + x^8$.

26. Integration of $f''(x) = -5x$ with respect to x gives $f'(x) = -5x^2/2 + C$. Because $f(2) = 3$ is a relative maximum, it follows that $f'(2) = 0$. Thus, $0 = -5(2) + C$, from which $C = 10$, and $f'(x) = -5x^2/2 + 10$. Another integration now gives $f(x) = -5x^3/6 + 10x + D$. Since $f(2) = 3$, we find that $3 = -20/3 + 20 + D$. Thus, $D = -31/3$, and $f(x) = -5x^3/6 + 10x - 31/3$.

In Exercises 28–67 we use the following tabular setup to summarize calculations. The first column is an initial proposal. The second column is the derivative of this proposal. The last column is the final answer.

	Initial proposal	Derivative of proposal	Final answer
28.	$(x+2)^{3/2}$	$\frac{3}{2}(x+2)^{1/2}$	$\frac{2}{3}(x+2)^{3/2} + C$
30.	$(2-x)^{3/2}$	$\frac{3}{2}(2-x)^{1/2}(-1)$	$-\frac{2}{3}(2-x)^{3/2} + C$
32.	$(2x-3)^{5/2}$	$\frac{5}{2}(2x-3)^{3/2}(2)$	$\frac{1}{5}(2x-3)^{5/2} + C$
34.	$(1-2x)^8$	$8(1-2x)^7(-2)$	$-\frac{1}{16}(1-2x)^8 + C$
36.	$\dfrac{1}{(1+3x)^5}$	$\dfrac{-5}{(1+3x)^6}(3)$	$\dfrac{-1}{15(1+3x)^5} + C$
38.	$(2+3x^3)^8$	$8(2+3x^3)^7(9x^2)$	$\frac{1}{72}(2+3x^3)^8 + C$
40.	$\sin 2x$	$2\cos 2x$	$\frac{1}{2}\sin 2x + C$
42.	$\sin^2 2x$	$2\sin 2x \cos 2x(2)$	$\frac{3}{4}\sin^2 2x + C$
44.	$\cot 4x$	$-4\csc^2 4x$	$-\frac{1}{4}\cot 4x + C$
46.	e^{-x^2}	$e^{-x^2}(-2x)$	$-\frac{1}{2}e^{-x^2} + C$

48. $\quad e^{4x-3}$ $\qquad\qquad e^{4x-3}(4)$ $\qquad\qquad \dfrac{1}{4}e^{4x-3}+C$

50. $\quad \ln|7-5x|$ $\qquad\qquad \dfrac{1}{7-5x}(-5)$ $\qquad\qquad -\dfrac{2}{5}\ln|7-5x|+C$

52. $\quad \ln|1-4x^3|$ $\qquad\qquad \dfrac{1}{1-4x^3}(-12x^2)$ $\qquad\qquad -\dfrac{1}{4}\ln|1-4x^3|+C$

54. $\quad 3^{2x}$ $\qquad\qquad 3^{2x}(2)\ln 3$ $\qquad\qquad \dfrac{1}{2\ln 3}3^{2x}=\dfrac{1}{2}(\log_3 e)3^{2x}+C$

56. $\quad (1+\cos x)^5$ $\qquad\qquad 5(1+\cos x)^4(-\sin x)$ $\qquad\qquad -\dfrac{1}{5}(1+\cos x)^5+C$

58. $\quad (1+e^{2x})^4$ $\qquad\qquad 4(1+e^{2x})^3(2e^{2x})$ $\qquad\qquad \dfrac{1}{8}(1+e^{2x})^4+C$

60. $\quad \text{Sin}^{-1}2x$ $\qquad\qquad \dfrac{2}{\sqrt{1-4x^2}}$ $\qquad\qquad \dfrac{1}{2}\text{Sin}^{-1}2x+C$

62. $\quad \text{Sec}^{-1}\sqrt{3}x$ $\qquad\qquad \dfrac{\sqrt{3}}{\sqrt{3}x\sqrt{3x^2-1}}$ $\qquad\qquad \text{Sec}^{-1}\sqrt{3}x+C$

64. $\quad \sinh 4x$ $\qquad\qquad 4\cosh 4x$ $\qquad\qquad \dfrac{1}{4}\sinh 4x+C$

66. $\quad \text{sech}2x$ $\qquad\qquad -2\,\text{sech}2x\tanh 2x$ $\qquad\qquad -\dfrac{1}{2}\text{sech}2x+C$

68. Integration with respect to x gives $y=\dfrac{x^4}{4}+\dfrac{1}{x}+C$.

70. Since $\dfrac{d}{dx}\dfrac{1}{(3x+5)^{1/2}}=\dfrac{-1/2}{(3x+5)^{3/2}}(3)$, it follows that $y=\dfrac{-2}{3(3x+5)^{1/2}}+C$.

72. Since $\dfrac{d}{dx}\dfrac{1}{2+3x^4}=\dfrac{-1}{(2+3x^4)^2}(12x^3)$, it follows that $y=\dfrac{-1}{12(2+3x^4)}+C$.

74. Indefinite integrals on the intervals $x<0$ and $x>0$ give $y=f(x)=\begin{cases} -\dfrac{1}{x}+C, & x<0 \\[2mm] -\dfrac{1}{x}+D, & x>0. \end{cases}$

The conditions $f(-1)=-2$ and $f(1)=1$ require $C=-3$ and $D=2$.

EXERCISES 5.2

2. Integration of $dv/dt=a(t)=6-2t$ gives $v(t)=6t-t^2+C$. Since $v(0)=5$, it follows that $C=5$, and $v(t)=dx/dt=6t-t^2+5$. Integration now gives $x(t)=3t^2-t^3/3+5t+D$. The condition $x(0)=0$ requires $D=0$, and therefore $x(t)=3t^2-t^3/3+5t$.

4. Integration of $dv/dt=a(t)=120t-12t^2$ gives $v(t)=60t^2-4t^3+C$. Since $v(0)=0$, it follows that $C=0$, and $v(t)=dx/dt=60t^2-4t^3$. Integration now gives $x(t)=20t^3-t^4+D$. The condition $x(0)=4$ requires $D=4$, and therefore $x(t)=20t^3-t^4+4$.

6. Integration of $dv/dt=a(t)=t^2+5t+4$ gives $v(t)=t^3/3+5t^2/2+4t+C$. Since $v(0)=-2$, it follows that $C=-2$, and $v(t)=dx/dt=t^3/3+5t^2/2+4t-2$. Integration now gives $x(t)=t^4/12+5t^3/6+2t^2-2t+D$. The condition $x(0)=-3$ requires $D=-3$, and therefore $x(t)=t^4/12+5t^3/6+2t^2-2t-3$.

8. Integration of $dv/dt=3\sin t$ gives $v(t)=-3\cos t+C$. Since $v(0)=1$, it follows that $1=-3+C$. Thus, $C=4$ and $v(t)=dx/dt=4-3\cos t$. Integration now gives $x(t)=4t-3\sin t+D$. The condition $x(0)=4$ requires $D=4$, and therefore $x(t)=4t+4-3\sin t$.

10. (a) Integration of $dv/dt=6t-2$ gives $v(t)=3t^2-2t+C$. Since $v(0)=-3$, we find that $C=-3$ and $v(t)=3t^2-2t-3$. Integration now gives $x(t)=t^3-t^2-3t+D$. The condition $x(0)=1$ requires $D=1$, and therefore $x(t)=t^3-t^2-3t+1$.

(b) The velocity is zero when $3t^2-2t-3=0$, a quadratic equation with solutions $t=(2\pm\sqrt{4+36})/6=(1\pm\sqrt{10})/3$. Only the solution $t=(1+\sqrt{10})/3$ is positive.

12. (a) Integration of $dv/dt = 3 - t/5$ gives $v(t) = 3t - t^2/10 + C$. Since $v(0) = 0$, it follows that $C = 0$, and $v(t) = 3t - t^2/10$. Integration now gives $x(t) = 3t^2/2 - t^3/30 + D$. If we choose a positive x-axis in the direction of motion of the car with $x = 0$ at $t = 0$, then $x(0) = 0$. This condition requires $D = 0$, and therefore $x(t) = 3t^2/2 - t^3/30$. The position of the car after 10 s is $x(10) = 3(100)/2 - (1000)/30 = 350/3$ m.

(b) For $t > 10$, the acceleration is $a(t) = -2$. Integration of this yields $v(t) = -2t + E$. Because $v(10) = 3(10) - 100/10 = 20$, it follows that $20 = -2(10) + E$. Hence, $E = 40$, and $v(t) = 40 - 2t$. Integration now gives $x(t) = 40t - t^2 + F$. Because $x(10) = 350/3$, it follows that $350/3 = 40(10) - 100 + F$. Thus, $F = -550/3$, and $x(t) = 40t - t^2 - 550/3$. The car comes to a stop when $0 = v(t) = 40 - 2t$, and this implies that $t = 20$. The position of the car at this time is $x(20) = 40(20) - (20)^2 - 550/3 = 650/3$ m.

14. We choose y as positive upward with $y = 0$ and $t = 0$ at the point and instant of projection. The acceleration of the stone is $a = -9.81$. Integration gives $v(t) = -9.81t + C$. Since $v(0) = 10$, it follows that $C = 10$, and $v(t) = -9.81t + 10$. Integration now gives $y(t) = -4.905t^2 + 10t + D$. The condition $y(0) = 0$ requires $D = 0$, and therefore $y(t) = -4.905t^2 + 10t$. At the peak height of the stone, $0 = v(t) = -9.81t + 10$, and this occurs when $t = 10/9.81$. The height of the stone at this time is $y(10/9.81) = -4.905(10/9.81)^2 + 10(10/9.81) = 5.1$ m.

16. We choose y as positive upward with $y = 0$ and $t = 0$ at the point and instant the ball is thrown. The acceleration of the ball is $a = -9.81$. Integration gives $v(t) = -9.81t + C$. If v_0 is the initial speed of the ball, then $C = v_0$, and $v(t) = v_0 - 9.81t$. Integration gives $y(t) = v_0 t - 4.905t^2 + D$. The condition $y(0) = 0$ requires $D = 0$, and therefore $y(t) = v_0 t - 4.905t^2$. For the ball just to reach your friend, we must have $v = 0$ when $y = 20$:

$$0 = v_0 - 9.81t, \qquad 20 = v_0 t - 4.905t^2.$$

The first implies that $t = v_0/9.81$, and this can be substituted into the second,

$$20 = v_0 \left(\frac{v_0}{9.81} \right) - 4.905 \left(\frac{v_0}{9.81} \right)^2 .$$

The positive solution of this equation is $v_0 = 19.8$ m/s.

18. The velocity and acceleration are $v(t) = 3t^2 - 12t + 9 = 3(t-1)(t-3)$ and $a(t) = 6t - 12 = 6(t-2)$. The graph of $x(t)$ has critical points at $t = 1$ and $t = 3$, the first giving a relative maximum of $x(1) = -16$, and the second a relative minimum of $x(3) = -20$. The graph has a point of inflection at $(2, -18)$. The graph of $v(t)$ has zeros at $t = 1$ and $t = 3$, the critical points of $x(t)$. The graph of $a(t)$ has a zero at $t = 2$, the point of inflection for $x(t)$.

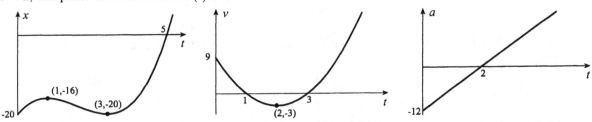

20. We choose $x = 0$ and $t = 0$ at the point and instant the brakes are applied. If we assume that the acceleration of the car is -9.81 m/s^2, and determine the initial speed v_0 which produces a skid mark of 9 m, then v_0 is the maximum possible speed. In other words, we are testifying for the defence. Integration of $a = -9.81$ gives $v(t) = -9.81t + C$. The condition $v(0) = v_0$ requires $C = v_0$, and therefore $v(t) = -9.81t + v_0$. Another integration yields $x(t) = -4.905t^2 + v_0 t + D$. Since $x(0) = 0$, it follows that $D = 0$. Because $x = 9$ when $v = 0$,

$$0 = -9.81t + v_0, \qquad 9 = -4.905t^2 + v_0 t.$$

The first requires $t = v_0/9.81$, and when this is substituted into the second,

$$9 = -4.905 \left(\frac{v_0}{9.81} \right)^2 + v_0 \left(\frac{v_0}{9.81} \right).$$

The positive solution of this equation is $v_0 = 13.3$ m/s or 47.8 km/hr.

22. We choose y as positive downward with $y = 0$ and $t = 0$ at the point and instant the stone is dropped. The acceleration of the stone is $a = 9.81$. Integration gives $v(t) = 9.81t + C$. Since $v(0) = 0$, it follows that $C = 0$, and $v(t) = 9.81t$. Integration gives $y(t) = 4.905t^2 + D$. The condition $y(0) = 0$ requires $D = 0$, and therefore $y(t) = 4.905t^2$. If d is the distance from the top of the well to the surface of the water and T is the time it takes the stone to fall this distance, then $d = 4.905T^2 \implies T = \sqrt{d/4.905}$. Since the time taken for the sound to travel the distance d is $d/340$, it follows that

$$\sqrt{\frac{d}{4.905}} + \frac{d}{340} = 3.1 \implies \sqrt{\frac{d}{4.905}} = 3.1 - \frac{d}{340}.$$

Squaring gives

$$\frac{d}{4.905} = (3.1)^2 - \frac{3.1d}{170} + \frac{d^2}{340^2} \implies \frac{d^2}{340^2} - \left(\frac{3.1}{170} + \frac{1}{4.905} \right) d + (3.1)^2 = 0.$$

Solutions of this quadratic are $d = \dfrac{(3.1/170 + 1/4.905) \pm \sqrt{(3.1/170 + 1/4.905)^2 - 4(3.1)^2/340^2}}{2/340^2}$. Only the solution $d = 43.3$ satisfies the original equation. Hence the depth of the well is 43.3 m.

24. We take y as positive downward from the top of the building with $t = 0$ at the instant the bearing is dropped. Let h be the distance from the top of the building to the top of the window and H be the distance from the bottom of the window to the sidewalk. Integration of $dv/dt = 9.81$ gives $v(t) = 9.81t + C$. Since $v(0) = 0$, we find that $C = 0$ and $v(t) = 9.81t$. A second integration gives $y(t) = 4.905t^2 + D$. With $y(0) = 0$, we obtain $D = 0$, and $y(t) = 4.905t^2$. If T is the time taken to reach the top of the window on the way down, then

$$h = 4.905T^2, \qquad h + 1 = 4.905(T + 1/8)^2, \qquad h + 1 + H = 4.905(T + 9/8)^2.$$

When the first of these is subtracted from the second, and the resulting equation is solved for T, we obtain $T = 0.753$. Substitution of this into the third gives $h + 1 + H = 17.3$. Thus, the building is 17.3 metres high.

26. Let us choose a different coordinate system to that of Example 5.6 by taking y as positive upward with $y = 0$ and $t = 0$ at the point and instant of projection of the first stone. The acceleration of stone 1 is

$$a_1 = -9.81 = \frac{dv_1}{dt},$$

from which $v_1(t) = -9.81t + C$. Since $v_1(0) = 25$, it follows that $25 = C$, and $v_1(t) = -9.81t + 25$, $t \geq 0$. Thus,

$$\frac{dy_1}{dt} = -9.81t + 25,$$

from which we have $y_1(t) = -4.905t^2 + 25t + D$. Since $y_1(0) = 0$, we find that $0 = D$, and $y_1(t) = -4.905t^2 + 25t$, $t \geq 0$.

The acceleration of stone 2 is also -9.81; hence we have

$$a_2 = -9.81 = \frac{dv_2}{dt},$$

from which $v_2(t) = -9.81t + E$. Because $v_2(1) = 20$, we must have $20 = -9.81(1) + E$, or $E = 29.81$. Thus, $v_2(t) = -9.81t + 29.81$, $t \geq 1$. Consequently,

$$\frac{dy_2}{dt} = -9.81t + 29.81,$$

from which $y_2(t) = -4.905t^2 + 29.81t + F$. Since $y_2(1) = 0$, it follows that $0 = -4.905(1)^2 + 29.81(1) + F$, or $F = -24.905$. Thus, $y_2(t) = -4.905t^2 + 29.81t - 24.905$, $t \geq 1$. The stones will pass each other if y_1 and y_2 are ever equal for the same time t; that is, if

$$-4.905t^2 + 25t = -4.905t^2 + 29.81t - 24.905.$$

The solution of this equation is

$$t = \frac{24.905}{4.81} = 5.2.$$

Because the first stone does not strike the base of the cliff for 7.7 s (Example 5.6), it follows that the stones do indeed pass each other 5.2 s after the first stone is projected.

28. Flow rate when distances are a fraction of the safe distance is

$$r(v) = \frac{v}{l + k\left(vT - \dfrac{v^2}{2a}\right)}.$$

The critical point of this function is defined by

$$0 = \frac{\left[l + k\left(vT - \dfrac{v^2}{2a}\right)\right](1) - v\left[k\left(T - \dfrac{v}{a}\right)\right]}{\left[l + k\left(vT - \dfrac{v^2}{2a}\right)\right]^2} \qquad \Longrightarrow \qquad 0 = \frac{1}{2a}(2al + kv^2).$$

Thus, $v = \sqrt{-2al/k}$.

30. During the acceleration stage, $a = dv/dt = 3$ so that $v = 3t + C$. If we choose time $t = 0$ as the vehicle leaves a speed bump with velocity 2.5 m/3, then $C = 2.5$ and $v = 3t + 2.5$. If x measures displacement, then $x = 3t^2/2 + 2.5t + D$. If we take $x = 0$ at the speed bump, then $D = 0$ and $x = 3t^2/2 + 2.5t$. Since speed is to be 10 m/s after the acceleration stage, which we suppose takes T seconds, $10 = 3T + 2.5 \Longrightarrow T = 2.5$ s. The position of the vehicle at this time is $x = 3(2.5)^2/2 + 2.5(2.5) = 15.625$ m. During the deceleration stage, $a = dv/dt = -7$ so that $v = -7t + E$. Since speed is 10 m/s when $t = 2.5$ s, $10 = -7(2.5) + E \Longrightarrow E = 27.5$, and $v = -7t + 27.5$. Displacement during this stage is $x = -7t^2/2 + 27.5t + F$. Since $x = 15.625$ when $t = 2.5$, $15.625 = -7(2.5)^2/2 + 27.5(2.5) + F \Longrightarrow F = -31.25$. Since speed at the end of this stage is to be 2.5 m/s at the second bump, we can find when this occurs by solving $2.5 = -7t + 27.5 \Longrightarrow t = 25/7$. The displacement of the vehicle at this time is $x = -7(25/7)^2/2 + 27.5(25/7) - 31.25 = 22.3$. This is the distance in metres between speed bumps.

EXERCISES 5.3

In Exercises 1–7, it is not necessary to use a substitution; these integrations can be done by adjusting constants.

2. $\displaystyle\int \sqrt{1 - 2x}\, dx = -\frac{1}{3}(1 - 2x)^{3/2} + C$

4. $\displaystyle\int \frac{5}{(5 - 42x)^{1/4}}\, dx = \frac{-(5)(4)}{3(42)}(5 - 42x)^{3/4} + C = -\frac{10}{63}(5 - 42x)^{3/4} + C$

6. $\displaystyle\int \frac{x}{(x^2 + 4)^2}\, dx = \frac{-1}{2(x^2 + 4)} + C$

8. If we set $u = x - 2$, then $du = dx$, and

$$\int \frac{x^2}{(x - 2)^4}\, dx = \int \frac{(u + 2)^2}{u^4}\, du = \int \left(\frac{1}{u^2} + \frac{4}{u^3} + \frac{4}{u^4}\right) du$$

$$= -\frac{1}{u} - \frac{2}{u^2} - \frac{4}{3u^3} + C = \frac{-1}{x - 2} - \frac{2}{(x - 2)^2} - \frac{4}{3(x - 2)^3} + C.$$

10. If we set $u = 2x + 3$, then $du = 2\,dx$, and

$$\int \frac{x}{\sqrt{2x+3}}\,dx = \int \frac{(u-3)/2}{\sqrt{u}}\left(\frac{du}{2}\right) = \frac{1}{4}\int\left(\sqrt{u} - \frac{3}{\sqrt{u}}\right)du$$

$$= \frac{1}{4}\left(\frac{2}{3}u^{3/2} - 6\sqrt{u}\right) + C = \frac{1}{6}(2x+3)^{3/2} - \frac{3}{2}\sqrt{2x+3} + C.$$

12. If we set $u = s^2 + 5$, then $du = 2s\,ds$, and

$$\int s^3\sqrt{s^2+5}\,ds = \int s^2\sqrt{s^2+5}\,s\,ds = \int (u-5)\sqrt{u}\left(\frac{du}{2}\right) = \frac{1}{2}\int (u^{3/2} - 5\sqrt{u})\,du$$

$$= \frac{1}{2}\left(\frac{2}{5}u^{5/2} - \frac{10}{3}u^{3/2}\right) + C = \frac{1}{5}(s^2+5)^{5/2} - \frac{5}{3}(s^2+5)^{3/2} + C.$$

14. If we set $u = 1 - \cos x$, then $du = \sin x\,dx$, and

$$\int \sqrt{1 - \cos x}\,\sin x\,dx = \int \sqrt{u}\,du = \frac{2}{3}u^{3/2} + C = \frac{2}{3}(1 - \cos x)^{3/2} + C.$$

16. If we set $u = y - 4$, then $du = dy$, and

$$\int y^2\sqrt{y-4}\,dy = \int (u+4)^2\sqrt{u}\,du = \int (u^{5/2} + 8u^{3/2} + 16\sqrt{u})\,du = \frac{2}{7}u^{7/2} + \frac{16}{5}u^{5/2} + \frac{32}{3}u^{3/2} + C$$

$$= \frac{2}{7}(y-4)^{7/2} + \frac{16}{5}(y-4)^{5/2} + \frac{32}{3}(y-4)^{3/2} + C.$$

18. If we set $u = 3x^3 - 5$, then $du = 9x^2\,dx$, and

$$\int x^8(3x^3 - 5)^6\,dx = \int (x^3)^2(3x^3 - 5)^6 x^2\,dx = \int \left(\frac{u+5}{3}\right)^2 u^6\left(\frac{du}{9}\right) = \frac{1}{81}\int (u^8 + 10u^7 + 25u^6)\,du$$

$$= \frac{1}{81}\left(\frac{u^9}{9} + \frac{5u^8}{4} + \frac{25u^7}{7}\right) + C = \frac{1}{729}(3x^3 - 5)^9 + \frac{5}{324}(3x^3 - 5)^8 + \frac{25}{567}(3x^3 - 5)^7 + C.$$

20. $\displaystyle\int \frac{x+1}{(x^2 + 2x + 2)^{1/3}}\,dx = \frac{3}{4}(x^2 + 2x + 2)^{2/3} + C$

22. If we set $u = 3 - 4\sin x$, then $du = -4\cos x\,dx$, and

$$\int \frac{\cos^3 x}{(3 - 4\sin x)^4}\,dx = \int \frac{(1 - \sin^2 x)\cos x}{(3 - 4\sin x)^4}\,dx = \int \frac{1 - [(3-u)/4]^2}{u^4}\left(-\frac{du}{4}\right) = \frac{1}{64}\int \frac{-7 - 6u + u^2}{u^4}\,du$$

$$= \frac{1}{64}\int\left(\frac{1}{u^2} - \frac{6}{u^3} - \frac{7}{u^4}\right)du = \frac{1}{64}\left(-\frac{1}{u} + \frac{3}{u^2} + \frac{7}{3u^3}\right) + C$$

$$= \frac{-1}{64(3 - 4\sin x)} + \frac{3}{64(3 - 4\sin x)^2} + \frac{7}{192(3 - 4\sin x)^3} + C.$$

24. If we set $u = 1 + \sqrt{x}$, then $du = \dfrac{1}{2\sqrt{x}}\,dx$, and

$$\int \sqrt{1 + \sqrt{x}}\,dx = \int \sqrt{u}\,2(u-1)\,du = 2\int (u^{3/2} - \sqrt{u})\,du$$

$$= 2\left(\frac{2}{5}u^{5/2} - \frac{2}{3}u^{3/2}\right) + C = \frac{4}{5}(1 + \sqrt{x})^{5/2} - \frac{4}{3}(1 + \sqrt{x})^{3/2} + C.$$

26. $\displaystyle\int \tan x \sec^2 x\,dx = \frac{1}{2}\tan^2 x + C$

28. If we set $u = \ln x$, then $du = \dfrac{1}{x}\,dx$, and $\displaystyle\int \frac{\ln x}{x}\,dx = \int u\,du = \frac{u^2}{2} + C = \frac{1}{2}(\ln x)^2 + C.$

30. If we set $u = \ln(x^2 + 1)$, then $du = \dfrac{2x}{x^2 + 1}dx$, and

$$\int \frac{x}{(x^2+1)[\ln(x^2+1)]^2}dx = \int \frac{1}{u^2}\left(\frac{du}{2}\right) = -\frac{1}{2u} + C = \frac{-1}{2\ln(x^2+1)} + C.$$

32. If $x \geq 0$, then $\displaystyle\int\sqrt{\frac{x^2}{1+x}}\,dx = \int\frac{x}{\sqrt{1+x}}\,dx$. If we set $u = 1 + x$, then $du = dx$, and

$$\int\sqrt{\frac{x^2}{1+x}}\,dx = \int\frac{u-1}{\sqrt{u}}\,du = \int\left(\sqrt{u} - \frac{1}{\sqrt{u}}\right)du = \frac{2}{3}u^{3/2} - 2\sqrt{u} + C = \frac{2}{3}(1+x)^{3/2} - 2\sqrt{1+x} + C.$$

If $-1 < x < 0$, then $\displaystyle\int\sqrt{\frac{x^2}{1+x}}\,dx = \int\frac{-x}{\sqrt{1+x}}\,dx = -\frac{2}{3}(1+x)^{3/2} + 2\sqrt{1+x} + C.$

34. If we set $u = 1/x$, then $du = -\dfrac{1}{x^2}dx$. For $x - x^2 = x(1-x)$ to be nonnegative, x must be in the interval $0 \leq x \leq 1$. It follows that u is positive, and

$$\int\frac{\sqrt{x-x^2}}{x^4}\,dx = \int\frac{\sqrt{\dfrac{1}{u} - \dfrac{1}{u^2}}}{\dfrac{1}{u^2}}(-du) = -\int u\sqrt{u-1}\,du.$$

We now set $v = u - 1$, in which case $dv = du$, and

$$\int\frac{\sqrt{x-x^2}}{x^4}\,dx = -\int(v+1)\sqrt{v}\,dv = -\int(v^{3/2} + \sqrt{v})\,dv = -\left(\frac{2}{5}v^{5/2} + \frac{2}{3}v^{3/2}\right) + C$$

$$= -\frac{2}{5}(u-1)^{5/2} - \frac{2}{3}(u-1)^{3/2} + C = -\frac{2}{5}\left(\frac{1}{x} - 1\right)^{5/2} - \frac{2}{3}\left(\frac{1}{x} - 1\right)^{3/2} + C.$$

36. Since $u^2 = \dfrac{1-x}{1+x} \implies u^2(1+x) = 1 - x \implies x = \dfrac{1-u^2}{1+u^2}$, we obtain

$$dx = \frac{(1+u^2)(-2u) - (1-u^2)(2u)}{(1+u^2)^2}\,du = \frac{-4u}{(1+u^2)^2}\,du, \text{ and}$$

$$\int\frac{1}{3(1-x^2) - (5+4x)\sqrt{1-x^2}}\,dx = \int\frac{1}{3(1-x)(1+x) - (5+4x)\sqrt{(1-x)(1+x)}}\,dx$$

$$= \int\frac{-4u/(1+u^2)^2}{3\left(1 - \dfrac{1-u^2}{1+u^2}\right)\left(1 + \dfrac{1-u^2}{1+u^2}\right) - \left(5 + \dfrac{4-4u^2}{1+u^2}\right)\sqrt{\left(1 - \dfrac{1-u^2}{1+u^2}\right)\left(1 + \dfrac{1-u^2}{1+u^2}\right)}}\,du$$

$$= \int\frac{-4u}{\dfrac{12u^2}{(1+u^2)^2} - \dfrac{9+u^2}{1+u^2}\sqrt{\dfrac{4u^2}{(1+u^2)^2}}}\frac{1}{(1+u^2)^2}\,du$$

$$= \int\frac{2}{(u-3)^2}\,du = -\frac{2}{u-3} + C = \frac{-2}{\sqrt{\dfrac{1-x}{1+x}} - 3} + C.$$

38. If $(x+1)u = \sqrt{4+3x-x^2}$, then $(x+1)^2u^2 = 4 + 3x - x^2 = (4-x)(1+x)$, or, $(x+1)u^2 = 4 - x$. When this equation is solved for x, the result is $x = \dfrac{4-u^2}{1+u^2}$, and therefore

$$dx = \frac{(1+u^2)(-2u) - (4-u^2)(2u)}{(1+u^2)^2}\,du = \frac{-10u}{(1+u^2)^2}\,du.$$

Since $\sqrt{4 + 3x - x^2} = (x+1)u = u\left(\dfrac{4-u^2}{1+u^2} + 1\right) = \dfrac{5u}{1+u^2},$

$$\int \frac{1}{\sqrt{4+3x-x^2}}\,dx = \int \frac{1+u^2}{5u}\frac{-10u}{(1+u^2)^2}\,du = -2\int \frac{1}{1+u^2}\,du,$$

and the integrand is a rational function of u.

EXERCISES 5.4

2. Deflections must satisfy equation 5.8 with $F(x) = -9.81m/L$ subject to conditions $y(0) = y'(0) = 0$ and $y(L) = y'(L) = 0$. Integration of the differential equation gives $y(x) = [-9.81mx^4/(24L) + Ax^3 + Bx^2 + Cx + D]/(EI)$. The boundary conditions require

$$0 = EI\,y(0) = D, \quad 0 = EI\,y'(0) = C,$$

$$0 = EI\,y(L) = \frac{-9.81mL^3}{24} + AL^3 + BL^2 + CL + D, \quad 0 = EI\,y'(L) = \frac{-9.81mL^2}{6} + 3AL^2 + 2BL + C.$$

Solutions are $A = 9.81m/12$, $B = -9.81mL/24$, $C = 0$, and $D = 0$, so that

$$y(x) = \frac{1}{EI}\left(-\frac{9.81mx^4}{24L} + \frac{9.81mx^3}{12} - \frac{9.81mLx^2}{24}\right) = -\frac{9.81m}{24EIL}(x^4 - 2Lx^3 + L^2x^2).$$

4. Deflections must satisfy equation 5.8 with $F(x) = -9.81m/L$ subject to conditions $y(0) = y'(0) = 0$ and $y(L) = y''(L) = 0$. Integration of the differential equation gives $y(x) = [-9.81mx^4/(24L) + Ax^3 + Bx^2 + Cx + D]/(EI)$. The boundary conditions require

$$0 = EI\,y(0) = D, \quad 0 = EI\,y'(0) = C,$$

$$0 = EI\,y(L) = \frac{-9.81mL^3}{24} + AL^3 + BL^2 + CL + D, \quad 0 = EI\,y''(L) = \frac{-9.81mL}{2} + 6AL + 2B.$$

Solutions are $A = 5(9.81)m/48$, $B = -9.81mL/16$, $C = 0$, and $D = 0$, so that

$$y(x) = \frac{1}{EI}\left(-\frac{9.81mx^4}{24L} + \frac{5(9.81)mx^3}{48} - \frac{9.81mLx^2}{16}\right) = -\frac{9.81m}{48EIL}(2x^4 - 5Lx^3 + 3L^2x^2).$$

6. Deflections must satisfy the differential equation $\dfrac{d^4y}{dx^4} = -\dfrac{F}{EI}\delta(x - L/2)$ subject to the boundary conditions $y(0) = y'(0) = 0 = y''(L) = y'''(L)$. Four integrations of the differential equation give

$$y(x) = \frac{1}{EI}\left[-\frac{F}{6}(x - L/2)^3 h(x - L/2) + Ax^3 + Bx^2 + Cx + D\right].$$

The boundary conditions require

$$0 = EIy(0) = D, \ 0 = EIy'(0) = C, \ 0 = EIy''(L) = -F\left(\frac{L}{2}\right) + 6AL + 2B, \ 0 = EIy'''(L) = -F + 6A.$$

These give $A = \dfrac{F}{6}$, $B = \dfrac{-FL}{4}$, and hence

$$y(x) = \frac{1}{EI}\left[-\frac{F}{6}(x - L/2)^3 h(x - L/2) + \frac{Fx^3}{6} - \frac{FLx^2}{4}\right] = \frac{-F}{12EI}\left[2F(x - L/2)^3 h(x - L/2) - 2x^3 + 3Lx^2\right].$$

For $x > L/2$,

$$y = \frac{-F}{12EI}\left[2F(x - L/2)^3 - 2x^3 + 3Lx^2\right] = \frac{FL^2}{48EI}(L - 6x),$$

the equation of a straight line.

8. Deflections must satisfy the differential equation $\dfrac{d^4y}{dx^4} = \dfrac{1}{EI}\left[-F\delta(x-L/2) - \dfrac{mg}{L}\right]$ subject to the boundary conditions $y(0) = y'(0) = 0 = y(L) = y'(L)$. Four integrations of the differential equation give

$$y(x) = \frac{1}{EI}\left[-\frac{F}{6}(x-L/2)^3 h(x-L/2) - \frac{mgx^4}{24L} + Ax^3 + Bx^2 + Cx + D\right].$$

The boundary conditions require

$$0 = EIy(0) = D, \qquad 0 = EIy'(0) = C,$$

$$0 = EIy(L) = -\frac{F}{6}\left(\frac{L}{2}\right)^3 - \frac{mgL^3}{24} + AL^3 + BL^2 + CL + D,$$

$$0 = EIy'(L) = -\frac{F}{2}\left(\frac{L}{2}\right)^2 - \frac{mgL^2}{6} + 3AL^2 + 2BL + C.$$

These give $A = (F + mg)/12$, $B = -L(3F + 2mg)/48$, and therefore

$$y(x) = \frac{1}{EI}\left[-\frac{F}{6}(x-L/2)^3 h(x-L/2) - \frac{mgx^4}{24L} + \frac{(F+mg)x^3}{12} - \frac{L(3F+2mg)x^2}{48}\right].$$

10. Deflections must satisfy the differential equation $\dfrac{d^4y}{dx^4} = \dfrac{1}{EI}\left[-F\delta(x-L/2) - \dfrac{mg}{L}\right]$ subject to the boundary conditions $y(0) = y''(0) = 0 = y(L) = y''(L)$. Four integrations of the differential equation give

$$y(x) = \frac{1}{EI}\left[-\frac{F}{6}(x-L/2)^3 h(x-L/2) - \frac{mgx^4}{24L} + Ax^3 + Bx^2 + Cx + D\right].$$

The boundary conditions require

$$0 = EIy(0) = D, \qquad 0 = EIy''(0) = 2B,$$

$$0 = EIy(L) = -\frac{F}{6}\left(\frac{L}{2}\right)^3 - \frac{mgL^3}{24} + AL^3 + BL^2 + CL + D,$$

$$0 = EIy''(L) = -F\left(\frac{L}{2}\right) - \frac{mgL}{2} + 6AL + 2B.$$

These give $A = (F + mg)/12$, and $C = -L^2(3F + 2mg)/48$, and hence

$$y(x) = \frac{1}{EI}\left[-\frac{F}{6}(x-L/2)^3 h(x-L/2) - \frac{mgx^4}{24L} + \frac{(F+mg)x^3}{12} - \frac{L^2(3F+2mg)x}{48}\right].$$

12. Deflections must satisfy the differential equation $\dfrac{d^4y}{dx^4} = \dfrac{1}{EI}\{-98.1 - 98.1[h(x-5) - h(x-10)]\}$
$= \dfrac{-98.1}{EI}[1 + h(x-5)]$ since $h(x-10) = 0$ if $0 < x < 10$, subject to the boundary conditions $y(0) = y'(0) = 0 = y''(10) = y'''(10)$. Integration of the differential equation four times using equation 5.11 gives

$$y(x) = \frac{-4.0875}{EI}[x^4 + (x-5)^4 h(x-5) + Ax^3 + Bx^2 + Cx + D].$$

The boundary conditions require

$$0 = EIy(0) = -4.0875D, \qquad 0 = EIy'(0) = -4.0875C,$$

$$0 = EIy''(10) = -4.0875[12(10)^2 + 12(5)^2 + 6A(10) + 2B],$$

$$0 = EIy'''(10) = -4.0875[24(10) + 24(5) + 6A].$$

These give $A = -60$ and $B = 1050$, and therefore

$$y(x) = -\frac{4.0875}{EI}[x^4 + (x-5)^4 h(x-5) - 60x^3 + 1050x^2].$$

Deflection at $x = 10$ is greater in this case.

14. When the concentrated force is placed at x_0 just to the left of the end of the beam, the differential equation for displacements is $\dfrac{d^4y}{dx^4} = -\dfrac{F}{EI}\delta(x-x_0)$, subject to the boundary conditions $y(0) = y'(0) = 0 = y''(L) = y'''(L)$. Four integrations of the differential equation give

$$y(x) = \frac{1}{EI}\left[-\frac{F}{6}(x-x_0)^3 h(x-x_0) + Ax^3 + Bx^2 + Cx + D\right].$$

The boundary conditions require

$$0 = EIy(0) = D, \ 0 = EIy'(0) = C, \ 0 = EIy''(L) = -F(L-x_0) + 6AL + 2B, \ 0 = EIy'''(L) = -F + 6A.$$

These give $A = \dfrac{F}{6}$, $B = \dfrac{-Fx_0}{2}$, and hence $y(x) = \dfrac{1}{EI}\left[-\dfrac{F}{6}(x-x_0)^3 h(x-x_0) + \dfrac{Fx^3}{6} - \dfrac{Fx_0 x^2}{2}\right]$. If we now take the limit as $x_0 \to L^-$, we obtain

$$y(x) = \frac{1}{EI}\left[-\frac{F}{6}(x-L)^3 h(x-L) + \frac{Fx^3}{6} - \frac{FLx^2}{2}\right] = \frac{Fx^2(x-3L)}{6EI}.$$

EXERCISES 5.5

2. If $N(t)$ is the number of bacteria at time t, then the fact that they increase at a rate proportional to N can be expressed as $dN/dt = kN$, where k is a constant. This is a separable differential equation $\dfrac{dN}{N} = k\,dt$. Solutions are defined implicitly by

$$\ln|N| = kt + C \quad\Longrightarrow\quad |N| = e^{kt+C} \quad\Longrightarrow\quad N = De^{kt}, \quad (D = e^C).$$

If N_0 is the number of bacteria at time $t = 0$, then $N_0 = D$, and $N(t) = N_0 e^{kt}$. Since $N(3) = 2N_0$, it follows that $2N_0 = N_0 e^{3k}$. Hence, $k = (\ln 2)/3$. If T is the time when the number of bacteria triples, then $3N_0 = N_0 e^{kT}$. This equation can be solved for $T = k^{-1}\ln 3 = \dfrac{3}{\ln 2}\ln 3 = 4.75$ hours.

4. The amount of radioactive material in a sample at any time t is $A = A_0 e^{kt}$ where A_0 is the amount at time $t = 0$. If $A(3) = 0.9A_0$, then $0.9A_0 = A_0 e^{3k}$, from which $k = \ln(0.9)/3$. The half life T of the material is the time at which $A = A_0/2$; that is, $A_0/2 = A_0 e^{kT}$. This can be solved for $T = -k^{-1}\ln 2 = -\dfrac{3}{\ln(0.9)}\ln 2 = 19.74$ seconds.

6. If $A(t)$ is the amount of drug in the body as a function of time t, then $\dfrac{dA}{dt} = kA$, $k < 0$ a constant. This differential equation is separable, $\dfrac{dA}{A} = k\,dt$. Solutions are defined implicitly by $\ln A = kt + C$. Exponentiation yields $A = e^{kt+C} = De^{kt}$, where $D = e^C$. If we choose time $t = 0$ when the original amount is A_0, then $A_0 = D$, and therefore $A = A_0 e^{kt}$. Since $A(1) = 0.95A_0$, we have $0.95A_0 = A_0 e^k$. This equation implies that $k = \ln(0.95)$. The amount of drug in the body will be $A_0/2$ when $A_0/2 = A_0 e^{kt}$, and the solution of this equation for t is $\quad t = -\dfrac{1}{k}\ln 2 = -\dfrac{\ln 2}{\ln 0.95} = 13.51$ hours.

8. (a) Since sugar particles dissolve independently, the time taken for the sugar to dissolve is the time taken for each spherical particle to dissolve. Since dissolving occurs at a rate proportional to the surface area of the particle,

$$\frac{dV}{dt} = k(4\pi r^2), \quad k < 0 \text{ a constant.}$$

Since $V = (4/3)\pi r^3$, it also follows that $4k\pi r^2 = \dfrac{d}{dt}\left(\dfrac{4\pi r^3}{3}\right) = 4\pi r^2 \dfrac{dr}{dt}$. Thus, $\dfrac{dr}{dt} = k$, and integration gives $r = kt + C$. The initial condition $r(0) = r_0$ implies that $C = r_0$, and therefore $r = kt + r_0$. The sugar is completely dissolved when $0 = r = kt + r_0 \implies t = -r_0/k$.

(b) In this case, $dV/dt = kV$. The solution of this differential equation is $V = Ce^{kt}$. The initial condition $V(0) = (4/3)\pi r_0^3 = V_0$ implies that $C = V_0$, and $V = V_0 e^{kt}$. The sugar is completely dissolved when $V = 0$, but this does not occur in finite time.

10. According to the discussion on carbon dating, the amount of C^{14} in the fossil at time t after the creature's death is $A = A_0 e^{kt}$ where A_0 is the amount present at death and $k = -\ln 2/5550$. When $t = 100\,000$, the percentage of C^{14} present is

$$\frac{100A}{A_0} = \frac{100A_0}{A_0}e^{100\,000k} = 100e^{100\,000(-\ln 2)/5550} = 3.8 \times 10^{-4}\%.$$

12. The rate of change dC/dt of the amount of glucose in the blood is equal to the rate at which it is added less the rate at which it is used up,

$$\frac{dC}{dt} = R - kC, \quad k > 0 \text{ a constant.}$$

This is separable, $\dfrac{dC}{R - kC} = dt$, and solutions are therefore defined implicitly by $-\dfrac{1}{k}\ln|R - kC| = t + D$. Thus, $|R - kC| = e^{-k(t+D)}$, from which $R - kC = Ee^{-kt}$, where $E = \pm e^{-kD}$ is a constant. We can now solve for

$$C(t) = \frac{1}{k}(R - Ee^{-kt}).$$

Since $C(0) = C_0$, it follows that $C_0 = (R - E)/k$, and this implies that $E = R - kC_0$. Thus,

$$C(t) = \frac{1}{k}\left[R - (R - kC_0)e^{-kt}\right] = \frac{R}{k}(1 - e^{-kt}) + C_0 e^{-kt}.$$

Graphs in the cases $C_0 < R/k$ and $C_0 > R/k$ are shown below.

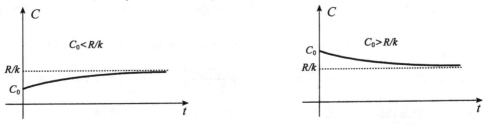

14. Separation of variables leads to $\dfrac{1}{T - 20}\,dT = k\,dt$, and therefore $\ln|T - 20| = kt + C$. The absolute values may be dropped since $T \geq 20$. Exponentiation then gives $T = 20 + De^{kt}$. Since $T(0) = 90$, we find that $D = 70$, and therefore $T = 20 + 70e^{kt}$. Because $T(40) = 60$, it follows that $60 - 20 = 70e^{40k}$, and $k = (1/40)\ln(4/7)$. Hence, $T = 20 + 70e^{(1/40)\ln(4/7)t} = 20 + 70e^{-0.01399t}$.

16. When the boy is x km from school, his velocity is

$$\frac{dx}{dt} = kx^2, \qquad k = \text{a constant.}$$

We separate variables, $\dfrac{dx}{x^2} = k\,dt$, in which case solutions are defined implicitly by

$$-\frac{1}{x} = kt + C \implies x = \frac{-1}{kt + C}.$$

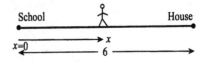

If we choose time $t = 0$ when $x = 6$, then $6 = -1/C$. Thus, $x = -1/(kt - 1/6) = 6/(1 - 6kt)$. Since $x(1) = 3$, it follows that $3 = 6/(1 - 6k) \implies k = -1/6$, and $x(t) = 6/(1 + t)$ km. The boy reaches school when $x = 0$, but this does not happen in finite time.

18. The volume of water in the tank is $V = \frac{1}{3}\pi r^2 D$.

Because $r/D = R/H$, it follows that

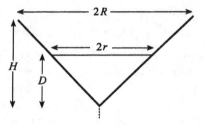

$$V = \frac{1}{3}\pi \left(\frac{RD}{H}\right)^2 D = \frac{\pi R^2}{3H^2}D^3.$$

Thus, $\dfrac{dV}{dt} = \dfrac{\pi R^2 D^2}{H^2}\dfrac{dD}{dt}$.

But the rate at which water exits through the hole is Av. In other words,

$$\frac{\pi R^2 D^2}{H^2}\frac{dD}{dt} = -Av = -Ac\sqrt{2gD}.$$

We separate variables, $D^{3/2}\,dD = -\dfrac{\sqrt{2g}\,AcH^2}{\pi R^2}\,dt$, in which case solutions are defined implicitly by $\dfrac{2}{5}D^{5/2} = -\dfrac{\sqrt{2g}\,AcH^2}{\pi R^2}t + C$. If we choose time $t = 0$ when the tank is full ($D = H$), then $C = (2/5)H^{5/2}$, and

$$\frac{2}{5}D^{5/2} = -\frac{\sqrt{2g}\,AcH^2}{\pi R^2}t + \frac{2}{5}H^{5/2}.$$

The tank empties when $D = 0$, and this occurs when $t = \dfrac{2}{5}H^{5/2}\dfrac{\pi R^2}{\sqrt{2g}\,AcH^2} = \dfrac{\pi R^2}{5cA}\sqrt{\dfrac{2H}{g}}$.

20. When the depth of water in the trough is y, the volume of water is $V = 4(xy) = 4xy$. Similar triangles require $y/x = (1/2)/(1/4) \implies y = 2x$. Thus, $V = 4y(y/2) = 2y^2$. Differentiation of this equation with respect to time t gives $dV/dt = 4y(dy/dt)$. Because water exits through a hole of area 10^{-4} m^2 with speed $\sqrt{gy/2}$, it follows that $dV/dt = -10^{-4}\sqrt{gy/2}$.

Hence, $4y\dfrac{dy}{dt} = -\dfrac{\sqrt{gy/2}}{10^4} \implies \sqrt{y}\,dy = -\dfrac{\sqrt{g}}{10^4(4\sqrt{2})}\,dt$. Solutions of this separated equation are defined implicitly by $\dfrac{2}{3}y^{3/2} = -\dfrac{\sqrt{g}\,t}{10^4(4\sqrt{2})} + C$. If we choose $t = 0$ when the trough is full, then $y(0) = 1/2$, and this implies that $(2/3)(1/2)^{3/2} = C$. Thus, $\dfrac{2}{3}y^{3/2} = \dfrac{-\sqrt{g}\,t}{10^4(4\sqrt{2})} + \dfrac{1}{3\sqrt{2}}$. The tank empties when $y = 0$,

and the time at which this occurs is $t = \dfrac{4 \times 10^4}{3\sqrt{g}} = 4257$ seconds, or 70.95 minutes.

22. If we set (see Exercise 21) $\dfrac{d^2r}{dt^2} = \dfrac{dv}{dt} = \dfrac{dv}{dr}\dfrac{dr}{dt} = v\dfrac{dv}{dr}$, then $mv\dfrac{dv}{dr} = -\dfrac{GmM}{r^2}$, or $v\,dv = -\dfrac{GM}{r^2}\,dr$. Solutions of this separated differential equation are defined implicitly by $\dfrac{v^2}{2} = \dfrac{GM}{r} + C$. Since $v = 0$ when $r = R + h$, where R is the radius of the earth, it follows that $0 = GM/(R + h) + C$, and therefore $\dfrac{v^2}{2} = \dfrac{GM}{r} - \dfrac{GM}{R + h}$. The velocity of m when it strikes the earth ($r = R$) is defined by

$$\frac{v^2}{2} = \frac{GM}{R} - \frac{GM}{R + h} = \frac{GMh}{R(R + h)} \implies v = -\sqrt{\frac{2GMh}{R(R + h)}}.$$

Maximum attainable speed occurs when h becomes infinite; that is,

$$|v_{\max}| = \lim_{h \to \infty} \sqrt{\frac{2GMh}{R(R + h)}} = \sqrt{\frac{2GM}{R}}.$$

24. If V and A represent volume and area of the mothball at any time, then the assumption that evaporation is proportional to surface area is represented by the equation

$$\frac{dV}{dt} = kA,$$

where k is a constant. We have four variables in the problem: t, r, A, and V. By substituting $V = 4\pi r^3/3$ and $A = 4\pi r^2$, we eliminate V and A:

$$\frac{d}{dt}\left(\frac{4}{3}\pi r^3\right) = 4\pi r^2 k \quad \Longrightarrow \quad 4\pi r^2 \frac{dr}{dt} = 4\pi r^2 k \quad \Longrightarrow \quad \frac{dr}{dt} = k.$$

Solutions of this differential equation are $r = kt + C$. Using the conditions $r(0) = R$ and $r(1) = R/2$, we find $C = R$ and $k = -R/2$. Consequently,

$$r(t) = R - \frac{Rt}{2} = R\left(1 - \frac{t}{2}\right).$$

The mothball completely disappears when $r = 0$, and this occurs when $t = 2$ years.

26. The modified Torricelli law in equation 5.22 implies that water exits through the hole with horizontal speed $v_x = c\sqrt{2g(H-h)}$, where $0 < c < 1$ is a constant, and $g = 9.81$ is the acceleration due to gravity. Suppose we follow a droplet of water on its journey to the ground if it exits at time $t = 0$. Because the only force acting on it is gravity, its vertical acceleration is $a_y = -g$, from which $v_y = -gt + C$. Since the initial velocity of the droplet is horizontal, $v_y(0) = 0$, and this implies that $C = 0$. Integration of $dy/dt = -gt$ gives $y = -gt^2/2 + D$. Since $y(0) = h$, it follows that $h = D$, and $y = -gt^2/2 + h$. The horizontal acceleration of the droplet is zero so that its horizontal velocity must always be $v_x = c\sqrt{2g(H-h)}$. Since this is dx/dt, we integrate to get $x = ct\sqrt{2g(H-h)} + E$. Because $x(0) = 0$, we obtain $E = 0$, and $x = ct\sqrt{2g(H-h)}$. The droplet hits the ground when

$$0 = y = -\frac{gt^2}{2} + h \quad \Longrightarrow \quad t = \sqrt{\frac{2h}{g}}.$$

The x-coordinate of the point at which the droplet hits the ground is therefore

$$x = c\sqrt{\frac{2h}{g}}\sqrt{2g(H-h)} = 2c\sqrt{h(H-h)}, \quad 0 \le h \le H.$$

We must find the value of h that maximizes this function. For critical points we solve

$$0 = \frac{dx}{dh} = \frac{c}{\sqrt{h(H-h)}}(H - 2h) \quad \Longrightarrow \quad h = \frac{H}{2}.$$

Since $x(0) = x(H) = 0$, it follows that x is maximized when $h = H/2$.

28. If we multiply the differential equation by r^2, it can be written in the form

$$0 = r^2 \frac{d^2T}{dr^2} + 2r\frac{dT}{dr} = \frac{d}{dr}\left(r^2 \frac{dT}{dr}\right).$$

Integration gives $r^2 \dfrac{dT}{dr} = C$, where C is a constant, from which $dT/dr = C/r^2$. A second integration now gives $T = -C/r + D$. Since $T(1) = 10$ and $T(2) = 20$, it follows that $10 = -C + D$ and $20 = -C/2 + D$. These imply that $C = 20$ and $D = 30$, so that $T(r) = 30 - 20/r$.

30. (a) We can separate the differential equation $\frac{1}{\rho^{2-\delta}}d\rho = -\frac{1}{k\delta}dh$. Solutions are defined implicitly by

$$\frac{1}{(\delta-1)\rho^{1-\delta}} = -\frac{h}{k\delta} + C.$$

Because $\rho = \rho_0$ when $h = 0$, $\frac{1}{(\delta-1)\rho_0^{1-\delta}} = C$, and therefore

$$\frac{1}{(\delta-1)\rho^{1-\delta}} = -\frac{h}{k\delta} + \frac{1}{(\delta-1)\rho_0^{1-\delta}}.$$

This can be written $\rho^{\delta-1} = -\frac{h}{k}\left(\frac{\delta-1}{\delta}\right) + \rho_0^{\delta-1}$.

(b) If $P = k\rho^\delta$, then $\rho^{\delta-1} = \left(\rho^\delta\right)^{(\delta-1)/\delta} = \left(\frac{P}{k}\right)^{(\delta-1)/\delta}$. Because $P = P_0$ when $\rho = \rho_0$, it follows that $\rho_0^{\delta-1} = \left(\frac{P_0}{k}\right)^{(\delta-1)/\delta}$. When these are substituted into the result of part (a),

$$\left(\frac{P}{k}\right)^{1-1/\delta} = -\frac{h}{k}\left(\frac{\delta-1}{\delta}\right) + \left(\frac{P_0}{k}\right)^{1-1/\delta},$$

or,

$$P^{1-1/\delta} = P_0^{1-1/\delta} - \frac{h}{k}\left(1-\frac{1}{\delta}\right)k^{1-1/\delta} = P_0^{1-1/\delta} - h\left(1-\frac{1}{\delta}\right)k^{-1/\delta}.$$

But $\rho_0^{\delta-1} = (P_0/k)^{(\delta-1)/\delta}$ implies that $k^{-1/\delta} = \rho_0 P_0^{-1/\delta}$, and therefore

$$P^{1-1/\delta} = P_0^{1-1/\delta} - h\left(1-\frac{1}{\delta}\right)\rho_0 P_0^{-1/\delta}.$$

(c) If we define the effective height of the atmosphere when $P = 0$, then this occurs for

$$0 = P_0^{1-1/\delta} - h\left(1-\frac{1}{\delta}\right)\rho_0 P_0^{-1/\delta} \quad\Longrightarrow\quad h = \frac{\delta P_0}{(\delta-1)\rho_0}.$$

32. (a) If $x(t)$ represents the amount of C in the mixture at time t, then

$$\frac{dx}{dt} = k\left(20 - \frac{2x}{3}\right)\left(10 - \frac{x}{3}\right) = \frac{2k}{9}(30-x)^2.$$

We can separate this equation, $\frac{dx}{(30-x)^2} = \frac{2k\,dt}{9} \implies \frac{1}{30-x} = \frac{2kt}{9} + C$. For $x(0) = 0$, we must have $1/30 = C$, and therefore $\frac{1}{30-x} = \frac{2kt}{9} + \frac{1}{30} \implies x(t) = \frac{600kt}{20kt+3}$.

REVIEW EXERCISES

2. $\int\left(\frac{1}{x^5} + 2x - \frac{1}{x^3}\right)dx = -\frac{1}{4x^4} + x^2 + \frac{1}{2x^2} + C$ **4.** $\int\left(\frac{1}{x^2} - 2\sqrt{x}\right)dx = -\frac{1}{x} - \frac{4}{3}x^{3/2} + C$

6. $\int x(1+3x^2)^4\,dx = \frac{1}{30}(1+3x^2)^5 + C$

8. $\int\left(\frac{x^2+5}{\sqrt{x}}\right)dx = \int\left(x^{3/2} + \frac{5}{\sqrt{x}}\right)dx = \frac{2}{5}x^{5/2} + 10\sqrt{x} + C$

10. $\int\left(\frac{\sqrt{x}}{x^2} - \frac{15}{\sqrt{x}}\right)dx = \int\left(\frac{1}{x^{3/2}} - \frac{15}{\sqrt{x}}\right)dx = -\frac{2}{\sqrt{x}} - 30\sqrt{x} + C$

12. $\int x\sqrt{1-x^2}\,dx = -\dfrac{1}{3}(1-x^2)^{3/2} + C$

14. $\int x^2(1-2x^2)^2\,dx = \int (x^2 - 4x^4 + 4x^6)\,dx = \dfrac{x^3}{3} - \dfrac{4x^5}{5} + \dfrac{4x^7}{7} + C$

16. If we set $u = 2 - x$, then $du = -dx$, and

$$\int \frac{x}{\sqrt{2-x}}\,dx = \int \frac{2-u}{\sqrt{u}}(-du) = \int \left(\sqrt{u} - \frac{2}{\sqrt{u}}\right)du = \frac{2}{3}u^{3/2} - 4\sqrt{u} + C = \frac{2}{3}(2-x)^{3/2} - 4\sqrt{2-x} + C.$$

18. $\int (2 + \sqrt{x})^2\,dx = \int (4 + 4\sqrt{x} + x)\,dx = 4x + \dfrac{8}{3}x^{3/2} + \dfrac{x^2}{2} + C$

20. $\int \sin^4 x \cos x\,dx = \dfrac{1}{5}\sin^5 x + C$

22. $\int xe^{-4x^2}\,dx = -\dfrac{1}{8}e^{-4x^2} + C$

24. If we set $u = \ln x$, then $du = \dfrac{1}{x}dx$, and $\int \dfrac{1}{5x\ln x}\,dx = \dfrac{1}{5}\int \dfrac{1}{u}\,du = \dfrac{1}{5}\ln|u| + C = \dfrac{1}{5}\ln|\ln x| + C.$

26. If we set $u = \sqrt{7}x$, then $du = \sqrt{7}\,dx$, and

$$\int \frac{3}{1+7x^2}\,dx = \int \frac{3}{1+u^2}\left(\frac{du}{\sqrt{7}}\right) = \frac{3}{\sqrt{7}}\text{Tan}^{-1}u + C = \frac{3}{\sqrt{7}}\text{Tan}^{-1}\sqrt{7}x + C.$$

28. $\int \text{sech}^2 5x\,dx = \dfrac{1}{5}\tanh 5x + C$

30. If $T(t)$ represents the temperature of the water as a function of time t, then according to Newton's law of cooling,

$$\frac{dT}{dt} = k(T + 20) \quad \Longrightarrow \quad \frac{dT}{T+20} = k\,dt, \quad \text{where } k < 0 \text{ is a constant.}$$

This is a separated differential equation with solutions defined implicitly by

$$\ln|T+20| = kt + C \quad \Longrightarrow \quad T + 20 = e^{kt+C} \quad \Longrightarrow \quad T = -20 + De^{kt}$$

where $D = e^C$. If we choose time $t = 0$ when $T = 70$, then $70 = -20 + D$. Thus, $D = 90$, and $T(t) = -20 + 90e^{kt}$. Because $T(10) = 50$, it follows that $50 = -20 + 90e^{4k}$, from which $k = (1/4)\ln(7/9)$. This solution would only be valid until the time at which the water reaches temperature zero and freezes.

32. If we integrate $f''(x) = 12x^2$, the result is $f'(x) = 4x^3 + C$. A second integration gives $f(x) = x^4 + Cx + D$. Since $f(1) = 4$ and $f(-1) = -3$, it follows that

$$4 = (1)^4 + C(1) + D \quad \text{and} \quad -3 = (-1)^4 + C(-1) + D.$$

Solutions of these equations are $C = 7/2$ and $D = -1/2$. The equation of the required curve is therefore $y = x^4 + 7x/2 - 1/2$.

34. When the boy is x km from school, his velocity is

$$\frac{dx}{dt} = k\sqrt{x}, \qquad k = \text{a constant.}$$

If we separate $\dfrac{dx}{\sqrt{x}} = k\,dt$, then

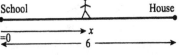

solutions are defined implicitly by

$$2\sqrt{x} = kt + C.$$

If we choose time $t = 0$ when $x = 6$, then $2\sqrt{6} = C$. Thus, $2\sqrt{x} = kt + 2\sqrt{6}$. Since $x(1) = 3$, it follows that $2\sqrt{3} = k + 2\sqrt{6}$, from which $k = 2(\sqrt{3} - \sqrt{6})$, and $2\sqrt{x} = 2(\sqrt{3} - \sqrt{6})t + 2\sqrt{6}$. Thus, $x = [(\sqrt{3} - \sqrt{6})t + \sqrt{6}]^2$ km. The boy reaches school when $x = 0$, and this occurs when $t = \sqrt{6}/(\sqrt{6} - \sqrt{3})$ hours.

36. We choose y as positive downward with $y = 0$ and $t = 0$ at the instant the stone is released. The acceleration of the stone is $dv/dt = 9.81$. Inegration gives $v(t) = 9.81t + C$. If we denote by v_0 the initial speed of the stone, then $v(0) = v_0$, and this implies that $C = v_0$. Thus, $dy/dt = 9.81t + v_0$. Integration now yields $y(t) = 4.905t^2 + v_0 t + D$. The condition $y(0) = 0$ requires $D = 0$. Because $y(2.2) = 50$, it follows that $50 = 4.905(2.2)^2 + v_0(2.2)$, and this implies that $v_0 = 11.9$ m/s.

38. If we set $u = \sqrt{1+x}$, then $du = \dfrac{1}{2\sqrt{1+x}}dx$, and

$$\int \frac{x}{\sqrt{1+x}+1}dx = \int \frac{u^2-1}{u+1}2u\,du = 2\int (u^2 - u)\,du = 2\left(\frac{u^3}{3} - \frac{u^2}{2}\right) + C$$

$$= \frac{2}{3}(1+x)^{3/2} - (1+x) + C = \frac{2}{3}(1+x)^{3/2} - x + D.$$

40. If we set $u = 3 - 2x^3$, then $du = -6x^2\,dx$, and

$$\int x^8(3-2x^3)^6\,dx = \int (x^3)^2(3-2x^3)^6 x^2\,dx = \int \left(\frac{3-u}{2}\right)^2 u^6 \left(-\frac{du}{6}\right) = \frac{1}{24}\int (-9u^6 + 6u^7 - u^8)\,du$$

$$= \frac{1}{24}\left(-\frac{9u^7}{7} + \frac{3u^8}{4} - \frac{u^9}{9}\right) + C = \frac{-3}{56}(3-2x^3)^7 + \frac{1}{32}(3-2x^3)^8 - \frac{1}{216}(3-2x^3)^9 + C.$$

42. If we set $u = \sin x$, then $du = \cos x\,dx$, and

$$\int \sin^3 x \cos^3 x\,dx = \int \sin^3 x(1 - \sin^2 x)\cos x\,dx = \int u^3(1-u^2)\,du$$

$$= \int (u^3 - u^5)\,du = \frac{u^4}{4} - \frac{u^6}{6} + C = \frac{1}{4}\sin^4 x - \frac{1}{6}\sin^6 x + C.$$

44. If we set $u = \cos x$, then $du = -\sin x\,dx$, and

$$\int \tan x\,dx = \int \frac{\sin x}{\cos x}dx = \int \frac{1}{u}(-du) = -\ln|u| + C = -\ln|\cos x| + C.$$

46. Since the slope of $y = f(x)$ at any point is dy/dx, it follows that $\dfrac{dy}{dx} = \dfrac{1}{2}\left(\dfrac{y}{x}\right)^2$. If we write $\dfrac{dy}{y^2} = \dfrac{dx}{2x^2}$, then solutions are defined implicitly by $-\dfrac{1}{y} = -\dfrac{1}{2x} + C$. Since the curve is to pass through $(1,1)$, it follows that $-1/1 = -1/2 + C$. Thus $C = -1/2$, and $-\dfrac{1}{y} = -\dfrac{1}{2x} - \dfrac{1}{2} \Longrightarrow y = 2x/(1+x)$.

48. If let $V(t)$ be the volume of water in the cone (figure to the right), then

$$\frac{dV}{dt} = k(\pi r^2).$$

But $V = (1/3)\pi r^2 h$, and from similar triangles, $r/h = 6/15$. Hence,

$$V = \frac{1}{3}\pi r^2 \left(\frac{5r}{2}\right) = \frac{5\pi r^3}{6}.$$

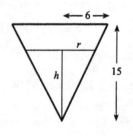

If we substitute this into the differential equation, we obtain

$$\frac{5\pi}{6}(3r^2)\frac{dr}{dt} = k\pi r^2 \qquad \Longrightarrow \qquad \frac{dr}{dt} = \frac{2k}{5}.$$

Integration gives $r = 2kt/5 + C$. If we take $t = 0$ when the cone is full, $6 = C$. Thus, $r = 2kt/5 + 6$. Since the water level drops 1 cm in 6 days, and $h = 5r/2$, it follows that

$$14 = \frac{5}{2}\left[\frac{2k(6)}{5} + 6\right] \qquad \Longrightarrow \qquad k = -\frac{1}{6}.$$

The radius of the surface of the water is therefore $r(t) = -t/15 + 6$. Half the water has evaporated when

$$\frac{1}{3}\pi r^2 h = \frac{1}{2}\left[\frac{1}{3}\pi(6)^2(15)\right].$$

Since $h = 5r/2$, this equation becomes $r^2(5r/2) = 270 \implies r = 108^{1/3}$. We can now find out how long this takes,

$$108^{1/3} = -\frac{t}{15} + 6 \qquad \implies \qquad t = 15(6 - 108^{1/3}) = 18.6 \text{ days.}$$

50. Deflections of the beam must satisfy the differential equation

$$\frac{d^4 y}{dx^4} = -\frac{9.81M}{5EI}[h(x-5) - h(x-10)] = k\,h(x-5), \quad \text{where } k = -9.81M/(5EI),$$

(since $h(x - 10) = 0$ for $0 < x < 10$), subject to the boundary conditions $y(0) = y'(0) = 0 = y''(10) = y'''(10)$. Integration of the differential equation four times gives
$y(x) = k\left[\dfrac{(x-5)^4}{24}h(x-5) + Ax^3 + Bx^2 + Cx + D\right]$. The boundary conditions require

$$0 = y(0) = k(D), \qquad 0 = y'(0) = k(C),$$

$$0 = y''(10) = k\left[\frac{5^2}{2} + 6A(10) + 2B\right], \qquad 0 = y'''(10) = k(5 + 6A).$$

These gives $A = -5/6$ and $B = 75/4$, and therefore the function describing deflections of the beam is

$$y(x) = -\frac{9.81M}{5EI}\left[\frac{(x-5)^4}{24}h(x-5) - \frac{5x^3}{6} + \frac{75x^2}{4}\right].$$

For $x < 5$, $y(x) = -\dfrac{9.81M}{5EI}\left[-\dfrac{5x^3}{6} + \dfrac{75x^2}{4}\right]$. Since this equation is not linear, the beam is not straight on this interval, nor should it be.

CHAPTER 6

EXERCISES 6.1

In trying to find the pattern by which terms are formed it is often beneficial to write the value of the index of summation above each term. We follow this suggestion when appropriate.

2. $\overset{k=1}{\dfrac{1}{2}} + \overset{k=2}{\dfrac{2}{4}} + \overset{k=3}{\dfrac{3}{8}} + \overset{k=4}{\dfrac{4}{16}} + \overset{k=5}{\dfrac{5}{32}} + \cdots + \overset{k=?}{\dfrac{10}{1024}} = \displaystyle\sum_{k=1}^{10} \dfrac{k}{2^k}$

4. Since each term is the square root of a positive integer, we write

$$1 + \sqrt{2} + \sqrt{3} + 2 + \sqrt{5} + \sqrt{6} + \sqrt{7} + \sqrt{8} + 3 + \cdots + 121 = \sum_{k=1}^{14641} \sqrt{k}.$$

6. Disregarding the sign changes, each term is an integer. To have even terms positive and odd terms negative, we use the fact that $(-1)^k$ is 1 when k is an even integer and -1 when k is an odd integer. In other words, $-2 + 3 - 4 + 5 - 6 + 7 - 8 + \cdots - 1020 = \displaystyle\sum_{k=1}^{1019} (-1)^k (k+1).$

8. $\overset{k=1}{\dfrac{\tan 1}{2}} + \overset{k=2}{\dfrac{\tan 2}{1+2^2}} + \overset{k=3}{\dfrac{\tan 3}{1+3^2}} + \overset{k=4}{\dfrac{\tan 4}{1+4^2}} + \cdots + \overset{k=?}{\dfrac{\tan 225}{1+225^2}} = \displaystyle\sum_{k=1}^{225} \dfrac{\tan k}{1+k^2}$

10. $0.9 + 0.99 + 0.999 + \cdots + 0.999\,999\,999 = \dfrac{9}{10} + \dfrac{99}{100} + \dfrac{999}{1000} + \cdots + \dfrac{999\,999\,999}{1\,000\,000\,000} = \displaystyle\sum_{k=1}^{9} \dfrac{10^k - 1}{10^k}$

12. If we set $m = k - 2$, then values of m corresponding to $k = 2$ and $k = 101$ are $m = 0$ and $m = 99$. Thus,

$$\sum_{k=2}^{101} \frac{3k - k^2}{\sqrt{k+5}} = \sum_{m=0}^{99} \frac{3(m+2) - (m+2)^2}{\sqrt{(m+2)+5}} = \sum_{m=0}^{99} \frac{2 - m - m^2}{\sqrt{m+7}}.$$

14. If we set $m = i + 2$, then values of m corresponding to $i = 0$ and $i = 37$ are $m = 2$ and $m = 39$. Thus,

$$\sum_{i=0}^{37} \frac{3^{3i}}{i!} = \sum_{m=2}^{39} \frac{3^{3(m-2)}}{(m-2)!} = \sum_{m=2}^{39} \frac{3^{3m}}{729(m-2)!}.$$

16. $\displaystyle\sum_{n=1}^{12} (3n+2) = 3 \sum_{n=1}^{12} n + 2 \sum_{n=1}^{12} 1 = 3 \left[\frac{(12)(13)}{2} \right] + 2(12) = 258$

18. $\displaystyle\sum_{m=1}^{n} (4m-2)^2 = \sum_{m=1}^{n} (16m^2 - 16m + 4) = 16 \sum_{m=1}^{n} m^2 - 16 \sum_{m=1}^{n} m + 4 \sum_{m=1}^{n} 1$

$$= 16 \left[\frac{n(n+1)(2n+1)}{6} \right] - 16 \left[\frac{n(n+1)}{2} \right] + 4n = \frac{4n(4n^2 - 1)}{3}$$

20. $\displaystyle\sum_{n=1}^{25} (n+5)(n-4) = \sum_{n=1}^{25} (n^2 + n - 20) = \sum_{n=1}^{25} n^2 + \sum_{n=1}^{25} n - 20 \sum_{n=1}^{25} 1$

$$= \frac{(25)(26)(51)}{6} + \frac{25(26)}{2} - 20(25) = 5350$$

22. $\displaystyle\sum_{n=10}^{24} (n^2 - 5) = \sum_{n=1}^{24} n^2 - \sum_{n=1}^{9} n^2 - 5(15) = \frac{24(25)(49)}{6} - \frac{9(10)(19)}{6} - 75 = 4540$

24. $\displaystyle\sum_{k=5}^{n}(k+3)(k+4) = \sum_{k=1}^{n}(k^2+7k+12) - \sum_{k=1}^{4}(k^2+7k+12)$

$$= \sum_{k=1}^{n}k^2 + 7\sum_{k=1}^{n}k + 12n - \sum_{k=1}^{4}k^2 - 7\sum_{k=1}^{4}k - 12(4)$$

$$= \frac{n(n+1)(2n+1)}{6} + 7\left[\frac{n(n+1)}{2}\right] + 12n - 30 - 70 - 48 = \frac{n^3+12n^2+47n-444}{3}$$

26. $\displaystyle\sum_{i=m}^{n}[f(i)+g(i)] = [f(m)+g(m)] + [f(m+1)+g(m+1)] + \cdots + [f(n)+g(n)]$

$$= [f(m)+f(m+1)+\cdots+f(n)] + [g(m)+g(m+1)+\cdots+g(n)] = \sum_{i=m}^{n}f(i) + \sum_{i=m}^{n}g(i)$$

$$\sum_{i=m}^{n}cf(i) = [c(f(m)] + [cf(m+1)] + \cdots + [cf(n)] = c[f(m)+f(m+1)+\cdots+f(n)] = c\sum_{i=m}^{n}f(i)$$

28. $\displaystyle\sum_{i=1}^{n}[f(i)-f(i-1)] = [f(1)-f(0)] + [f(2)-f(1)] + [f(3)-f(2)] + \cdots$

$$+[f(n-1)-f(n-2)] + [f(n)-f(n-1)]$$

$$= f(n) - f(0)$$

30. No   For example if $n=2$, then the left side is $\displaystyle\sum_{i=1}^{2}[f(i)g(i)] = f(1)g(1) + f(2)g(2)$, whereas the right

side is $\left[\displaystyle\sum_{i=1}^{2}f(i)\right]\left[\displaystyle\sum_{i=1}^{2}g(i)\right] = [f(1)+f(2)][g(1)+g(2)].$

32. $\dfrac{1}{8} + \dfrac{1}{16} + \cdots + \dfrac{1}{1\,048\,576} = \dfrac{(1/8)[1-(1/2)^{18}]}{1-1/2} = \dfrac{1-2^{-18}}{4}$

34. $40(0.99) + 40(0.99)^2 + \cdots + 40(0.99)^{15} = \dfrac{40(0.99)[1-(0.99)^{15}]}{1-0.99} = 3960[1-(0.99)^{15}] = 554.2$

36. $\left|\displaystyle\sum_{i=1}^{n}f(i)\right| = |f(1)+f(2)+\cdots+f(n)| \le |f(1)| + |f(2)| + \cdots + |f(n)| = \sum_{i=1}^{n}|f(i)|$

EXERCISES 6.3

2. Since $f(x) = 3x$ is continuous for $0 \le x \le 2$, the definite integral exists, and we may choose any partition and star-points in its evaluation. For n equal subdivisions of length $2/n$, we use $x_i = 2i/n$, $i = 0, \ldots, n$. If we choose the right end of each subinterval as star-point, then $x_i^* = x_i = 2i/n$. Equation 6.10 now gives

$$\int_0^2 3x\,dx = \lim_{\|\Delta x_i\|\to 0}\sum_{i=1}^{n}f(x_i^*)\Delta x_i = \lim_{\|\Delta x_i\|\to 0}\sum_{i=1}^{n}3x_i^*\Delta x_i.$$

Since all subintervals have equal length $\Delta x_i = 2/n$, the norm of the partition is $\|\Delta x_i\| = 2/n$, and taking the limit as $\|\Delta x_i\| \to 0$ is tantamount to letting $n \to \infty$. Thus,

$$\int_0^2 3x\,dx = \lim_{n\to\infty}\sum_{i=1}^{n}3\left(\frac{2i}{n}\right)\left(\frac{2}{n}\right) = \lim_{n\to\infty}\frac{12}{n^2}\sum_{i=1}^{n}i = \lim_{n\to\infty}\frac{12}{n^2}\left[\frac{n(n+1)}{2}\right] = 6.$$

4. Since $f(x) = x^3$ is continuous for $0 \leq x \leq 2$, the definite integral exists, and we may choose any partition and star-points in its evaluation. For n equal subdivisions of length $2/n$, we use $x_i = 2i/n$, $i = 0, \ldots, n$. If we choose the right end of each subinterval as star-point, then $x_i^* = x_i = 2i/n$. Equation 6.10 gives

$$\int_0^2 x^3 \, dx = \lim_{\|\Delta x_i\| \to 0} \sum_{i=1}^n f(x_i^*) \Delta x_i = \lim_{\|\Delta x_i\| \to 0} \sum_{i=1}^n (x_i^*)^3 \Delta x_i.$$

Since all subintervals have equal length $\Delta x_i = 2/n$, the norm of the partition is $\|\Delta x_i\| = 2/n$, and taking the limit as $\|\Delta x_i\| \to 0$ is tantamount to letting $n \to \infty$. Thus,

$$\int_0^2 x^3 \, dx = \lim_{n \to \infty} \sum_{i=1}^n \left(\frac{2i}{n}\right)^3 \left(\frac{2}{n}\right) = \lim_{n \to \infty} \frac{16}{n^4} \sum_{i=1}^n i^3 = \lim_{n \to \infty} \frac{16}{n^4} \left[\frac{n^2(n+1)^2}{4}\right] = 4.$$

6. Since $f(x) = 1 - x$ is continuous for $-1 \leq x \leq 0$, the definite integral exists, and we may choose any partition and star-points in its evaluation. For n equal subdivisions of length $1/n$, we use the points $x_i = -1 + i/n$, $i = 0, \ldots, n$. If we choose the right end of each subinterval as star-point, then $x_i^* = x_i = -1 + i/n$. Equation 6.10 now gives

$$\int_{-1}^0 (-x + 1) \, dx = \lim_{\|\Delta x_i\| \to 0} \sum_{i=1}^n (-x_i^* + 1) \Delta x_i = \lim_{\|\Delta x_i\| \to 0} \sum_{i=1}^n \left(1 - \frac{i}{n} + 1\right) \left(\frac{1}{n}\right).$$

Since all subintervals have equal length $\Delta x_i = 1/n$, the norm of the partition is $\|\Delta x_i\| = 1/n$, and taking the limit as $\|\Delta x_i\| \to 0$ is tantamount to letting $n \to \infty$. Thus,

$$\int_{-1}^0 (-x + 1) \, dx = \lim_{n \to \infty} \frac{1}{n^2} \sum_{i=1}^n (2n - i) = \lim_{n \to \infty} \frac{1}{n^2} \left[2n^2 - \frac{n(n+1)}{2}\right] = \lim_{n \to \infty} \frac{3n - 1}{2n} = \frac{3}{2}.$$

8. Since $f(x) = x^3$ is continuous for $-1 \leq x \leq 1$, the definite integral exists, and we may choose any partition and star-points in its evaluation. For n equal subdivisions of length $2/n$, we use the points $x_i = -1 + 2i/n$, $i = 0, \ldots, n$. If we choose the right end of each subinterval as star-point, then $x_i^* = x_i = -1 + 2i/n$. Equation 6.10 now gives

$$\int_{-1}^1 x^3 \, dx = \lim_{\|\Delta x_i\| \to 0} \sum_{i=1}^n f(x_i^*) \Delta x_i = \lim_{\|\Delta x_i\| \to 0} \sum_{i=1}^n (x_i^*)^3 \Delta x_i.$$

Since all subintervals have equal length $\Delta x_i = 2/n$, the norm of the partition is $\|\Delta x_i\| = 2/n$, and taking the limit as $\|\Delta x_i\| \to 0$ is tantamount to letting $n \to \infty$. Thus,

$$\int_{-1}^1 x^3 \, dx = \lim_{n \to \infty} \sum_{i=1}^n \left(-1 + \frac{2i}{n}\right)^3 \left(\frac{2}{n}\right) = \lim_{n \to \infty} \frac{2}{n^4} \sum_{i=1}^n (-n^3 + 6n^2 i - 12ni^2 + 8i^3)$$

$$= \lim_{n \to \infty} \frac{2}{n^4} \left[-n^4 + \frac{6n^3(n+1)}{2} - \frac{12n^2(n+1)(2n+1)}{6} + \frac{8n^2(n+1)^2}{4}\right]$$

$$= 2(-1 + 3 - 4 + 2) = 0.$$

10. (a) For n equal subdivisions of the interval $0 \leq x \leq 1$, we use the points $x_i = i/n$ where $i = 0, \cdots, n$. With star points chosen as $x_i^* = x_i = i/n$, equation 6.10 gives

$$\int_0^1 2^x \, dx = \lim_{\|\Delta x_i\| \to 0} \sum_{i=1}^n 2^{x_i^*} \Delta x_i = \lim_{n \to \infty} \sum_{i=1}^n 2^{i/n} \left(\frac{1}{n}\right) = \lim_{n \to \infty} \frac{1}{n} \sum_{i=1}^n 2^{i/n}.$$

(b) The formula in question 31(b) of Exercises 6.1 allows us to evaluate the sum in closed form,

$$\int_0^1 2^x \, dx = \lim_{n \to \infty} \left\{\frac{1}{n} \frac{2^{1/n} \left[1 - (2^{1/n})^n\right]}{1 - 2^{1/n}}\right\} = \lim_{n \to \infty} \frac{2^{1/n}}{n(2^{1/n} - 1)}.$$

(c) The limit of the numerator is 1. The denominator is of the indeterminate form $0 \cdot \infty$, and we therefore use L'Hôpital's rule to evaluate

$$\lim_{n \to \infty} [n(2^{1/n} - 1)] = \lim_{n \to \infty} \frac{2^{1/n} - 1}{1/n} = \lim_{n \to \infty} \frac{2^{1/n}(-1/n^2)\ln 2}{-1/n^2} = \lim_{n \to \infty} (2^{1/n} \ln 2) = \ln 2.$$

Hence, $\displaystyle\int_0^1 2^x dx = \frac{1}{\ln 2} = \log_2 e.$

EXERCISES 6.4

2. $\displaystyle\int_1^3 (x^2 - 2x + 3)\, dx = \left\{ \frac{x^3}{3} - x^2 + 3x \right\}_1^3 = (9 - 9 + 9) - \left(\frac{1}{3} - 1 + 3 \right) = \frac{20}{3}$

4. $\displaystyle\int_{-3}^{-1} \frac{1}{x^2}\, dx = \left\{ -\frac{1}{x} \right\}_{-3}^{-1} = (1) - \left(\frac{1}{3} \right) = \frac{2}{3}$

6. $\displaystyle\int_0^{\pi/2} \sin x\, dx = \left\{ -\cos x \right\}_0^{\pi/2} = (0) - (-1) = 1$

8. $\displaystyle\int_{-1}^{-2} \left(\frac{1}{x^2} - 2x \right) dx = \left\{ -\frac{1}{x} - x^2 \right\}_{-1}^{-2} = \left(\frac{1}{2} - 4 \right) - (1 - 1) = -\frac{7}{2}$

10. $\displaystyle\int_0^1 x(x^2 + 1)\, dx = \int_0^1 (x^3 + x)\, dx = \left\{ \frac{x^4}{4} + \frac{x^2}{2} \right\}_0^1 = \left(\frac{1}{4} + \frac{1}{2} \right) - 0 = \frac{3}{4}$

12. $\displaystyle\int_0^{2\pi} \cos 2x\, dx = \left\{ \frac{1}{2} \sin 2x \right\}_0^{2\pi} = (0) - (0) = 0$

14. $\displaystyle\int_0^1 (x^{2.2} - x^\pi)\, dx = \left\{ \frac{x^{3.2}}{3.2} - \frac{x^{\pi+1}}{\pi + 1} \right\}_0^1 = \left(\frac{1}{3.2} - \frac{1}{\pi + 1} \right) - 0 = \frac{5}{16} - \frac{1}{\pi + 1}$

16. $\displaystyle\int_3^4 \frac{(x^2 - 1)^2}{x^2}\, dx = \int_3^4 \left(x^2 - 2 + \frac{1}{x^2} \right) dx = \left\{ \frac{x^3}{3} - 2x - \frac{1}{x} \right\}_3^4 = \left(\frac{64}{3} - 8 - \frac{1}{4} \right) - \left(9 - 6 - \frac{1}{3} \right) = \frac{125}{12}$

18. $\displaystyle\int_{-2}^3 (x - 1)^3\, dx = \left\{ \frac{1}{4}(x - 1)^4 \right\}_{-2}^3 = \frac{1}{4}(16) - \frac{1}{4}(81) = -\frac{65}{4}$

20. $\displaystyle\int_0^{\pi/4} 3\cos x\, dx = \left\{ 3\sin x \right\}_0^{\pi/4} = \frac{3}{\sqrt{2}} - 0 = \frac{3}{\sqrt{2}}$

22. $\displaystyle\int_{\pi/2}^\pi \sin x \cos x\, dx = \left\{ \frac{1}{2} \sin^2 x \right\}_{\pi/2}^\pi = \frac{1}{2}(0) - \frac{1}{2}(1) = -\frac{1}{2}$

24. $\displaystyle\int_{-\pi/4}^{\pi/4} \sec x \tan x\, dx = \left\{ \sec x \right\}_{-\pi/4}^{\pi/4} = (\sqrt{2}) - (\sqrt{2}) = 0$

26. $\displaystyle\int_{-1}^2 e^x\, dx = \left\{ e^x \right\}_{-1}^2 = e^2 - e^{-1}$

28. $\displaystyle\int_{-3}^{-2} \frac{1}{x}\, dx = \left\{ \ln |x| \right\}_{-3}^{-2} = (\ln 2) - (\ln 3) = \ln (2/3)$

30. $\displaystyle\int_0^1 3^{4x}\, dx = \left\{ \frac{1}{4} 3^{4x} \log_3 e \right\}_0^1 = \frac{1}{4} \log_3 e(3^4 - 3^0) = 20 \log_3 e$

32. Since $|x + 1|$ is positive between $x = 0$ and $x = 4$,

$$\int_0^4 x|x + 1|\, dx = \int_0^4 x(x + 1)\, dx = \int_0^4 (x^2 + x)\, dx = \left\{ \frac{x^3}{3} + \frac{x^2}{2} \right\}_0^4 = \left(\frac{64}{3} + 8 \right) - 0 = \frac{88}{3}.$$

34. Because $x + 1$ changes sign at $x = -1$, we divide the integral into two parts,

$$\int_{-2}^{1} x|x+1|\, dx = \int_{-2}^{-1} x(-x-1)\, dx + \int_{-1}^{1} x(x+1)\, dx = \left\{ -\frac{x^3}{3} - \frac{x^2}{2} \right\}_{-2}^{-1} + \left\{ \frac{x^3}{3} + \frac{x^2}{2} \right\}_{-1}^{1}$$

$$= \left(\frac{1}{3} - \frac{1}{2} \right) - \left(\frac{8}{3} - 2 \right) + \left(\frac{1}{3} + \frac{1}{2} \right) - \left(-\frac{1}{3} + \frac{1}{2} \right) = -\frac{1}{6}.$$

36. $\displaystyle \int_{-}^{1} \frac{1}{1+x^2}\, dx = \left\{ \text{Tan}^{-1} x \right\}_{-1}^{1} = \text{Tan}^{-1} 1 - \text{Tan}^{-1}(-1) = \frac{\pi}{2}$

38. $\displaystyle \int_{0}^{1} \cosh 2x\, dx = \left\{ \frac{1}{2} \sinh 2x \right\}_{0}^{1} = \frac{1}{2} \sinh 2$

40. $\displaystyle \int_{0}^{1/2} \frac{1}{1+4x^2}\, dx = \left\{ \frac{1}{2} \text{Tan}^{-1} 2x \right\}_{0}^{1/2} = \frac{1}{2} \text{Tan}^{-1}(1) = \frac{\pi}{8}$

42. $\displaystyle \int_{0}^{5} v(t)\, dt = \int_{0}^{5} (3t^2 - 6t - 105)\, dt = \left\{ t^3 - 3t^2 - 105t \right\}_{0}^{5}$

$$= 125 - 75 - 525 = -475$$

This is the displacement of the particle at
$t = 5$ relative to its position at $t = 0$. Since
$v(t) = 3(t-7)(t+5)$ is negative for $0 \le t \le 5$,
the integral of $|v(t)|$ is 475. This is the distance
travelled by the particle between $t = 0$ and $t = 5$.

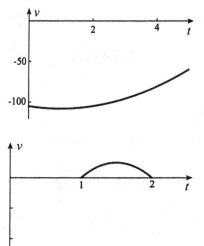

44. $\displaystyle \int_{1}^{2} v(t)\, dt = \int_{1}^{2} (-t^2 + 3t - 2)\, dt$

$$= \left\{ -\frac{t^3}{3} + \frac{3t^2}{2} - 2t \right\}_{1}^{2}$$

$$= \left(-\frac{8}{3} + 6 - 4 \right) - \left(-\frac{1}{3} + \frac{3}{2} - 2 \right) = \frac{1}{6}$$

This is the displacement of the particle at $t = 2$
relative to its position at $t = 1$. Since $v(t) = -(t-1)(t-2)$ is positive between $t = 1$ and $t = 2$, the
integral of $|v(t)|$ is also 1/6. It is the distance travelled by the particle between $t = 1$ and $t = 2$.

46. $\displaystyle \int_{0}^{3} v(t)\, dt = \int_{0}^{3} (t^3 - 3t^2 + 2t)\, dt$

$$= \left\{ \frac{t^4}{4} - t^3 + t^2 \right\}_{0}^{3}$$

$$= \frac{81}{4} - 27 + 9 = \frac{9}{4}$$

This is the displacement of the particle
at $t = 3$ relative to its position at $t = 0$.
Since $v(t) = t(t-1)(t-2)$ changes sign at
$t = 1$ and $t = 2$,

$$\int_{0}^{3} |v(t)|\, dt = \int_{0}^{1} (t^3 - 3t^2 + 2t)\, dt + \int_{1}^{2} (-t^3 + 3t^2 - 2t)\, dt + \int_{2}^{3} (t^3 - 3t^2 + 2t)\, dt$$

$$= \left\{ \frac{t^4}{4} - t^3 + t^2 \right\}_{0}^{1} + \left\{ -\frac{t^4}{4} + t^3 - t^2 \right\}_{1}^{2} + \left\{ \frac{t^4}{4} - t^3 + t^2 \right\}_{2}^{3}$$

$$= \left(\frac{1}{4} - 1 + 1 \right) + (-4 + 8 - 4) - \left(-\frac{1}{4} + 1 - 1 \right) + \left(\frac{81}{4} - 27 + 9 \right) - (4 - 8 + 4) = \frac{11}{4}.$$

This is the distance travelled by the particle between $t = 0$ and $t = 3$.

48. The integrand is clearly nonnegative for $0 \leq x \leq \pi/2$. For these values of x, the largest value of $\sin x$ is 1 and the smallest value of $1 + x$ is 1. It follows that $\sin x/(1+x)$ cannot be larger than 1. By inequality 6.15 then

$$0\left(\frac{\pi}{2}\right) \leq \int_0^{\pi/2} \frac{\sin x}{1+x}\, dx \leq 1\left(\frac{\pi}{2}\right) \quad \Longrightarrow \quad 0 \leq \int_0^{\pi/2} \frac{\sin x}{1+x}\, dx \leq \frac{\pi}{2}.$$

50. The integrand is nonnegative for $\pi/4 \leq x \leq \pi/2$. For these values of x, the largest value of $\sin 2x$ is 1 and the smallest value of $10 + x^2$ is $10 + \pi^2/16$. It follows that $\sin 2x/(10 + x^2)$ cannot be larger than $1/(10 + \pi^2/16)$. By inequality 6.15 then

$$0\left(\frac{\pi}{4}\right) \leq \int_{\pi/4}^{\pi/2} \frac{\sin 2x}{10+x^2}\, dx \leq \frac{1}{10 + \pi^2/16}\left(\frac{\pi}{4}\right) \quad \Longrightarrow \quad 0 \leq \int_{\pi/4}^{\pi/2} \frac{\sin 2x}{10+x^2}\, dx \leq \frac{4\pi}{160 + \pi^2}.$$

52. Because $\sqrt{5} \leq \sqrt{4 + x^3} \leq \sqrt{31}$ for $1 \leq x \leq 3$, it follows that $\quad 2\sqrt{5} \leq \int_1^3 \sqrt{4 + x^3}\, dx \leq 2\sqrt{31}$.

EXERCISES 6.5

2. By equation 6.19, $\dfrac{d}{dx} \displaystyle\int_1^x \dfrac{1}{\sqrt{t^2+1}}\, dt = \dfrac{1}{\sqrt{x^2+1}}$.

4. By reversing limits, $\dfrac{d}{dx} \displaystyle\int_x^{-1} t^3 \cos t\, dt = -\dfrac{d}{dx} \displaystyle\int_{-1}^x t^3 \cos t\, dt = -x^3 \cos x$.

6. We set $u = 2x$ and use the chain rule,

$$\frac{d}{dx} \int_1^{2x} \sqrt{t+1}\, dt = \left[\frac{d}{du} \int_1^u \sqrt{t+1}\, dt\right] \frac{du}{dx} = \sqrt{u+1}\,(2) = 2\sqrt{2x+1}.$$

8. We set $u = 5x + 4$ and use the chain rule,

$$\frac{d}{dx} \int_{-2}^{5x+4} \sqrt{t^3+1}\, dt = \left[\frac{d}{du} \int_{-2}^u \sqrt{t^3+1}\, dt\right] \frac{du}{dx} = \sqrt{u^3+1}\,(5) = 5\sqrt{(5x+4)^3+1}.$$

10. When a is any number between $4x$ and $4x + 4$, we may write

$$\int_{4x}^{4x+4} \left(t^3 - \frac{1}{\sqrt{t}}\right) dt = \int_{4x}^{a} \left(t^3 - \frac{1}{\sqrt{t}}\right) dt + \int_a^{4x+4} \left(t^3 - \frac{1}{\sqrt{t}}\right) dt$$

$$= -\int_a^{4x} \left(t^3 - \frac{1}{\sqrt{t}}\right) dt + \int_a^{4x+4} \left(t^3 - \frac{1}{\sqrt{t}}\right) dt.$$

In these integrals we set $u = 4x$ and $v = 4x + 4$ respectively, and use chain rules,

$$\frac{d}{dx} \int_{4x}^{4x+4} \left(t^3 - \frac{1}{\sqrt{t}}\right) dt = -\left[\frac{d}{du} \int_a^u \left(t^3 - \frac{1}{\sqrt{t}}\right) dt\right] \frac{du}{dx} + \left[\frac{d}{dv} \int_a^v \left(t^3 - \frac{1}{\sqrt{t}}\right) dt\right] \frac{dv}{dx}$$

$$= -\left[u^3 - \frac{1}{\sqrt{u}}\right](4) + \left[v^3 - \frac{1}{\sqrt{v}}\right](4)$$

$$= -4\left[(4x)^3 - \frac{1}{\sqrt{4x}}\right] + 4\left[(4x+4)^3 - \frac{1}{\sqrt{4x+4}}\right]$$

$$= -256x^3 + \frac{2}{\sqrt{x}} + 256(x+1)^3 - \frac{2}{\sqrt{x+1}}$$

$$= 256(3x^2 + 3x + 1) + \frac{2}{\sqrt{x}} - \frac{2}{\sqrt{x+1}}.$$

12. When a is a number between $-x^2$ and $-2x^2$, we may write

$$\int_{-x^2}^{-2x^2} \sec(1-t)\,dt = \int_{-x^2}^{a} \sec(1-t)\,dt + \int_{a}^{-2x^2} \sec(1-t)\,dt$$

$$= -\int_{a}^{-x^2} \sec(1-t)\,dt + \int_{a}^{-2x^2} \sec(1-t)\,dt.$$

In these integrals we set $u = -x^2$ and $v = -2x^2$ respectively, and use chain rules,

$$\frac{d}{dx}\int_{-x^2}^{-2x^2} \sec(1-t)\,dt = -\left[\frac{d}{du}\int_{a}^{u} \sec(1-t)\,dt\right]\frac{du}{dx} + \left[\frac{d}{dv}\int_{a}^{v} \sec(1-t)\,dt\right]\frac{dv}{dx}$$

$$= -\sec(1-u)(-2x) + \sec(1-v)(-4x)$$

$$= 2x\sec(1+x^2) - 4x\sec(1+2x^2).$$

14. When a is a number between $\cos x$ and $\sin x$, we may write

$$\int_{\cos x}^{\sin x} \frac{1}{\sqrt{t+1}}\,dt = \int_{\cos x}^{a} \frac{1}{\sqrt{t+1}}\,dt + \int_{a}^{\sin x} \frac{1}{\sqrt{t+1}}\,dt = -\int_{a}^{\cos x} \frac{1}{\sqrt{t+1}}\,dt + \int_{a}^{\sin x} \frac{1}{\sqrt{t+1}}\,dt.$$

In these integrals we set $u = \cos x$ and $v = \sin x$ respectively, and use chain rules,

$$\frac{d}{dx}\int_{\cos x}^{\sin x} \frac{1}{\sqrt{t+1}}\,dt = -\left[\frac{d}{du}\int_{a}^{u} \frac{1}{\sqrt{t+1}}\,dt\right]\frac{du}{dx} + \left[\frac{d}{dv}\int_{a}^{v} \frac{1}{\sqrt{t+1}}\,dt\right]\frac{dv}{dx}$$

$$= -\frac{1}{\sqrt{u+1}}(-\sin x) + \frac{1}{\sqrt{v+1}}(\cos x) = \frac{\sin x}{\sqrt{\cos x+1}} + \frac{\cos x}{\sqrt{\sin x+1}}.$$

16. If a is a number between $\sqrt{x}$ and $2\sqrt{x}$, we may write

$$\int_{\sqrt{x}}^{2\sqrt{x}} \sqrt{t}\,dt = \int_{\sqrt{x}}^{a} \sqrt{t}\,dt + \int_{a}^{2\sqrt{x}} \sqrt{t}\,dt = -\int_{a}^{\sqrt{x}} \sqrt{t}\,dt + \int_{a}^{2\sqrt{x}} \sqrt{t}\,dt.$$

In these integrals we set $u = \sqrt{x}$ and $v = 2\sqrt{x}$ respectively, and use chain rules,

$$\frac{d}{dx}\int_{\sqrt{x}}^{2\sqrt{x}} \sqrt{t}\,dt = -\left[\frac{d}{du}\int_{a}^{u} \sqrt{t}\,dt\right]\frac{du}{dx} + \left[\frac{d}{dv}\int_{a}^{v} \sqrt{t}\,dt\right]\frac{dv}{dx}$$

$$= -\sqrt{u}\left(\frac{1}{2\sqrt{x}}\right) + \sqrt{v}\left(\frac{1}{\sqrt{x}}\right) = -\frac{x^{1/4}}{2\sqrt{x}} + \frac{\sqrt{2}x^{1/4}}{\sqrt{x}} = \frac{2\sqrt{2}-1}{2x^{1/4}}.$$

18. $\dfrac{d}{dx}\displaystyle\int_{x}^{2} \ln(t^2+1)\,dt = -\dfrac{d}{dx}\int_{2}^{x} \ln(t^2+1)\,dt = -\ln(x^2+1)$

20. If a is a number between $-2x$ and $3x$, then

$$\int_{-2x}^{3x} e^{-4t^2}\,dt = \int_{-2x}^{a} e^{-4t^2}\,dt + \int_{a}^{3x} e^{-4t^2}\,dt = -\int_{a}^{-2x} e^{-4t^2}\,dt + \int_{a}^{3x} e^{-4t^2}\,dt.$$

In these integrals we set $u = -2x$ and $v = 3x$ respectively, and use chain rules,

$$\frac{d}{dx}\int_{-2x}^{3x} e^{-4t^2}\,dt = -\left[\frac{d}{du}\int_{a}^{u} e^{-4t^2}\,dt\right]\frac{du}{dx} + \left[\frac{d}{dv}\int_{a}^{v} e^{-4t^2}\,dt\right]\frac{dv}{dx}$$

$$= -e^{-4u^2}(-2) + e^{-4v^2}(3) = 2e^{-16x^2} + 3e^{-36x^2}.$$

22. $\dfrac{d}{dx}\displaystyle\int_{-2}^{5x+4} \sqrt{t^3+1}\,dt = \sqrt{(5x+4)^3+1}\,(5)$

24. $\dfrac{d}{dx}\displaystyle\int_{-x^2}^{-2x^2} \sec(1-t)\,dt = [\sec(1+2x^2)](-4x) - [\sec(1+x^2)](-2x)$

$$= 2x \sec(1 + x^2) - 4x \sec(1 + 2x^2)$$

26. $\dfrac{d}{dx} \displaystyle\int_{\sqrt{x}}^{2\sqrt{x}} \sqrt{t}\, dt = \sqrt{2\sqrt{x}}\left(\dfrac{1}{\sqrt{x}}\right) - \sqrt{\sqrt{x}}\left(\dfrac{1}{2\sqrt{x}}\right) = \dfrac{2\sqrt{2} - 1}{2x^{1/4}}$

28. $\dfrac{d}{dx} \displaystyle\int_{-2x}^{3x} e^{-4t^2}\, dt = e^{-4(3x)^2}(3) - e^{-4(-2x)^2}(-2) = 3\,e^{-36x^2} + 2\,e^{-16x^2}$

EXERCISES 6.6

2. Since the integrand is an odd function, its average value over the interval $-1 \le x \le 1$ is 0.

4. The average value is $\dfrac{1}{2-1} \displaystyle\int_1^2 x^4\, dx = \left\{\dfrac{x^5}{5}\right\}_1^2 = \dfrac{1}{5}(32 - 1) = \dfrac{31}{5}$.

6. The average value is $\dfrac{1}{2} \displaystyle\int_{-1}^1 \sqrt{x+1}\, dx = \dfrac{1}{2}\left\{\dfrac{2}{3}(x+1)^{3/2}\right\}_{-1}^1 = \dfrac{2\sqrt{2}}{3}$.

8. The average value is $\dfrac{1}{2} \displaystyle\int_0^2 (x^4 - 1)\, dx = \dfrac{1}{2}\left\{\dfrac{x^5}{5} - x\right\}_0^2 = \dfrac{1}{2}\left(\dfrac{32}{5} - 2\right) = \dfrac{11}{5}$.

10. The average value is $\dfrac{1}{\pi/2} \displaystyle\int_0^{\pi/2} \cos x\, dx = \dfrac{2}{\pi}\{\sin x\}_0^{\pi/2} = \dfrac{2}{\pi}$.

12. The average value is $\dfrac{1}{2} \displaystyle\int_0^2 |x|\, dx = \dfrac{1}{2}\displaystyle\int_0^2 x\, dx = \dfrac{1}{2}\left\{\dfrac{x^2}{2}\right\}_0^2 = 1$.

14. The average value is

$$\dfrac{1}{6}\int_{-3}^3 |x^2 - 4|\, dx = \dfrac{1}{6}\left[\int_{-3}^{-2}(x^2 - 4)\, dx + \int_{-2}^2 (4 - x^2)\, dx + \int_2^3 (x^2 - 4)\, dx\right]$$

$$= \dfrac{1}{6}\left[\left\{\dfrac{x^3}{3} - 4x\right\}_{-3}^{-2} + \left\{4x - \dfrac{x^3}{3}\right\}_{-2}^2 + \left\{\dfrac{x^3}{3} - 4x\right\}_2^3\right]$$

$$= \dfrac{1}{6}\left[\left(-\dfrac{8}{3} + 8\right) - (-9 + 12) + \left(8 - \dfrac{8}{3}\right) - \left(-8 + \dfrac{8}{3}\right) + (9 - 12) - \left(\dfrac{8}{3} - 8\right)\right]$$

$$= \dfrac{23}{9}.$$

16. The average value is $\dfrac{1}{4}\displaystyle\int_{-1}^3 \operatorname{sgn} x\, dx = \dfrac{1}{4}\left[\int_{-1}^0 (-1)\, dx + \int_0^3 (1)\, dx\right] = \dfrac{1}{4}\left[\{-x\}_{-1}^0 + \{x\}_0^3\right] = \dfrac{1}{2}$.

18. The average value is $\dfrac{1}{2}\displaystyle\int_0^2 h(x - 4)\, dx = \dfrac{1}{2}\displaystyle\int_0^2 (0)\, dx = 0$.

20. The average value is

$$\dfrac{1}{3.5}\int_0^{3.5} \lfloor x\rfloor\, dx = \dfrac{1}{3.5}\left[\int_0^1 0\, dx + \int_1^2 1\, dx + \int_2^3 2\, dx + \int_3^{3.5} 3\, dx\right] = \dfrac{1}{3.5}\left[0 + 1 + 2 + \dfrac{3}{2}\right] = \dfrac{9}{7}.$$

22. The average value is $\dfrac{1}{R}\displaystyle\int_0^R c(R^2 - r^2)\, dr = \dfrac{c}{R}\left\{R^2 r - \dfrac{r^3}{3}\right\}_0^R = \dfrac{c}{R}\left(R^3 - \dfrac{R^3}{3}\right) = \dfrac{2cR^2}{3}$.

24. According to 6.31, $4(c^3 - 8c) = \displaystyle\int_{-2}^2 (x^3 - 8x)\, dx = \left\{\dfrac{x^4}{4} - 4x^2\right\}_{-2}^2 = (4 - 16) - (4 - 16) = 0$.

The only value of c between -2 and 2 that satisfies this equation is $c = 0$.

26. According to 6.31, $\pi \cos c = \displaystyle\int_0^\pi \cos x\, dx = \{\sin x\}_0^\pi = 0$. The only value of c between 0 and π which satisfies this equation is $c = \pi/2$.

28. From 6.31, $1(c^2)(c+1) = \int_0^1 x^2(x+1)\,dx = \left\{\dfrac{x^4}{4} + \dfrac{x^3}{3}\right\}_0^1 = \dfrac{1}{4} + \dfrac{1}{3} = \dfrac{7}{12}.$

Thus, c must satisfy $g(c) = 12c^3 + 12c^2 - 7 = 0$. There is only one solution of this equation between 0 and 1, which we can find by Newton's iterative procedure,

$$c_1 = 0.5, \qquad c_{n+1} = c_n - \frac{12c_n^3 + 12c_n^2 - 7}{36c_n^2 + 24c_n}.$$

Iteration gives $c_2 = 0.619$, $c_3 = 0.603\,50$, $c_4 = 0.603\,20$, $c_5 = 0.603\,20$. Since $g(0.603\,15) = -0.0015$ and $g(0.603\,25) = 0.0013$, we can say that to 4 decimals, $c = 0.6032$.

30. From 6.31, $1\left(\dfrac{1}{c^2} + \dfrac{1}{c^3}\right) = \int_1^2 \left(\dfrac{1}{x^2} + \dfrac{1}{x^3}\right) dx = \left\{-\dfrac{1}{x} - \dfrac{1}{2x^2}\right\}_1^2 = \left(-\dfrac{1}{2} - \dfrac{1}{8}\right) - \left(-1 - \dfrac{1}{2}\right) = \dfrac{7}{8}.$

Thus, c must satisfy $g(c) = 7c^3 - 8c - 8 = 0$. There is only one solution of this equation between 1 and 2, which we can find by Newton's iterative procedure,

$$c_1 = 1.5, \qquad c_{n+1} = c_n - \frac{7c_n^3 - 8c_n - 8}{21c_n^2 - 8}.$$

Iteration gives $c_2 = 1.408$, $c_3 = 1.399\,8$, $c_4 = 1.399\,8$. Since $g(1.3995) = -0.0086$ and $g(1.4005) = 0.025$, we can say that to 3 decimals, $c = 1.400$.

32. The moving average is

$$\overline{f}(x) = \frac{1}{2\pi} \int_{x-2\pi}^{x} \sin t\,dt = \frac{1}{2\pi}\left\{-\cos t\right\}_{x-2\pi}^{x} = \frac{1}{2\pi}\left[\cos(x - 2\pi) - \cos x\right] = \frac{1}{2\pi}(\cos x - \cos x) = 0.$$

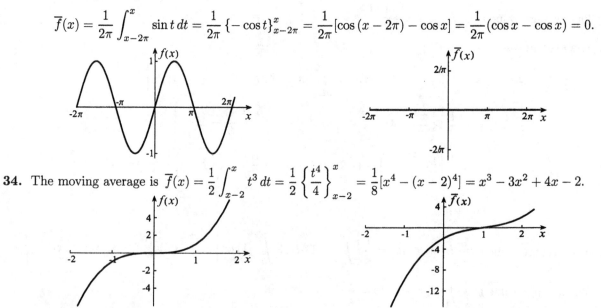

34. The moving average is $\overline{f}(x) = \dfrac{1}{2} \int_{x-2}^{x} t^3\,dt = \dfrac{1}{2}\left\{\dfrac{t^4}{4}\right\}_{x-2}^{x} = \dfrac{1}{8}[x^4 - (x-2)^4] = x^3 - 3x^2 + 4x - 2.$

36. The moving average is $\dfrac{1}{1} \int_{x-1}^{x} f(t)\,dt.$ When $x \le -1$, $\overline{f}(x) = 0$.

When $-1 < x < 0$, $\overline{f}(x) = \int_{-1}^{x} (1+t)\,dt = \left\{t + \dfrac{t^2}{2}\right\}_{-1}^{x} = \dfrac{x^2}{2} + x + \dfrac{1}{2}.$

When $0 \le x < 1$, $\overline{f}(x) = \int_{x-1}^{0} (1+t)\,dt + \int_0^x (1-t)\,dt = \left\{t + \dfrac{t^2}{2}\right\}_{x-1}^{0} + \left\{t - \dfrac{t^2}{2}\right\}_0^x = -x^2 + x + \dfrac{1}{2}.$

When $1 \le x < 2$, $\overline{f}(x) = \int_{x-1}^{1} (1-t)\,dt = \left\{t - \dfrac{t^2}{2}\right\}_{x-1}^{1} = \dfrac{x^2}{2} - 2x + 2.$

When $x \ge 2$, $\overline{f}(x) = 0.$

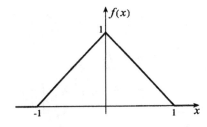

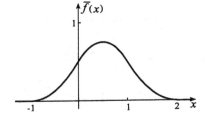

38. The moving average is $\quad \overline{f}(x) = \dfrac{1}{b-a} \displaystyle\int_{x-b+a}^{x} [h(x-a) - h(x-b)]\, dx.$ $\quad$ When $x < a$, $\overline{f}(x) = 0$.

When $a \le x < b$, $\quad \overline{f}(x) = \dfrac{1}{b-a} \displaystyle\int_{a}^{x} 1\, dt = \dfrac{x-a}{b-a}.$

When $b \le x < 2b - a$, $\quad \overline{f}(x) = \dfrac{1}{b-a} \displaystyle\int_{x-b+a}^{b} 1\, dt = \dfrac{2b-x-a}{b-a}.$ $\quad$ When $x \ge 2b - a$, $\quad \overline{f}(x) = 0$.

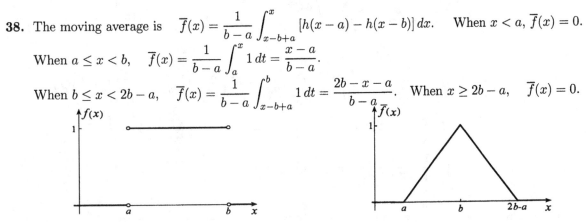

EXERCISES 6.7

2. If we set $u = 1 - z$, then $du = -dz$, and

$$\int_0^1 z\sqrt{1-z}\, dz = \int_1^0 (1-u)\sqrt{u}(-du) = \int_1^0 (u^{3/2} - \sqrt{u})\, du = \left\{ \frac{2}{5}u^{5/2} - \frac{2}{3}u^{3/2} \right\}_1^0 = -\left(\frac{2}{5} - \frac{2}{3} \right) = \frac{4}{15}.$$

4. $\displaystyle\int_{\pi/4}^{\pi/3} \cos^5 x \sin x\, dx = \left\{ -\frac{1}{6}\cos^6 x \right\}_{\pi/4}^{\pi/3} = -\frac{1}{6}\left(\frac{1}{2} \right)^6 + \frac{1}{6}\left(\frac{1}{\sqrt{2}} \right)^6 = \frac{7}{384}$

6. This definite integral does not exist because the integrand is not defined for $-2\sqrt{3} \le x \le 2\sqrt{3}$.

8. If we set $u = 1 + x$, then $du = dx$, and

$$\int_{1/2}^1 \sqrt{\frac{x^2}{1+x}}\, dx = \int_{1/2}^1 \frac{x}{\sqrt{1+x}}\, dx = \int_{3/2}^2 \frac{u-1}{\sqrt{u}}\, du = \int_{3/2}^2 \left(\sqrt{u} - \frac{1}{\sqrt{u}} \right)\, du$$

$$= \left\{ \frac{2}{3}u^{3/2} - 2\sqrt{u} \right\}_{3/2}^2 = \left(\frac{4\sqrt{2}}{3} - 2\sqrt{2} \right) - \left(\sqrt{\frac{3}{2}} - 2\sqrt{\frac{3}{2}} \right) = \sqrt{\frac{3}{2}} - \frac{2\sqrt{2}}{3}.$$

10. If we set $u = x^2 + 2x + 2$, then $du = (2x+2)\, dx$, and

$$\int_{-2}^1 \frac{x+1}{(x^2+2x+2)^{1/3}}\, dx = \int_2^5 \frac{1}{u^{1/3}}\left(\frac{du}{2} \right) = \frac{1}{2}\left\{ \frac{3}{2}u^{2/3} \right\}_2^5 = \frac{3}{4}\left(5^{2/3} - 2^{2/3} \right).$$

12. If we set $u = 2 + 3\sin x$, then $du = 3\cos x\, dx$, and

$$\int_0^{\pi/6} \sqrt{2 + 3\sin x}\, \cos x\, dx = \int_2^{7/2} \sqrt{u}\left(\frac{du}{3} \right) = \frac{1}{3}\left\{ \frac{2}{3}u^{3/2} \right\}_2^{7/2} = \frac{2}{9}\left(\frac{7\sqrt{7}}{2\sqrt{2}} - 2\sqrt{2} \right) = \frac{\sqrt{2}}{18}(7\sqrt{7} - 8).$$

14. $\displaystyle\int_1^4 \frac{(x+1)(x-1)}{\sqrt{x}}\, dx = \int_1^4 \left(x^{3/2} - \frac{1}{\sqrt{x}} \right)\, dx = \left\{ \frac{2}{5}x^{5/2} - 2\sqrt{x} \right\}_1^4 = \left(\frac{64}{5} - 4 \right) - \left(\frac{2}{5} - 2 \right) = \frac{52}{5}$

16. If we set $u = 1 + x$, then $du = dx$, and

$$\int_{-1/2}^{1} \sqrt{\frac{x^2}{1+x}}\,dx = \int_{-1/2}^{0} \frac{-x}{\sqrt{1+x}}\,dx + \int_{0}^{1} \frac{x}{\sqrt{1+x}}\,dx = \int_{1/2}^{1} \frac{-(u-1)}{\sqrt{u}}\,du + \int_{1}^{2} \frac{u-1}{\sqrt{u}}\,du$$

$$= \int_{1/2}^{1} \left(\frac{1}{\sqrt{u}} - \sqrt{u} \right) du + \int_{1}^{2} \left(\sqrt{u} - \frac{1}{\sqrt{u}} \right) du = \left\{ 2\sqrt{u} - \frac{2}{3}u^{3/2} \right\}_{1/2}^{1} + \left\{ \frac{2}{3}u^{3/2} - 2\sqrt{u} \right\}_{1}^{2}$$

$$= \left(2 - \frac{2}{3} \right) - \left(\sqrt{2} - \frac{1}{3\sqrt{2}} \right) + \left(\frac{4\sqrt{2}}{3} - 2\sqrt{2} \right) - \left(\frac{2}{3} - 2 \right) = \frac{1}{6}(16 - 9\sqrt{2}).$$

18. If we set $u = x + 2$, then $du = dx$, and

$$\int_{-1}^{1} \left| \frac{x}{(x+2)^3} \right| dx = \int_{-1}^{0} \frac{-x}{(x+2)^3}\,dx + \int_{0}^{1} \frac{x}{(x+2)^3}\,dx = \int_{1}^{2} \left(\frac{2-u}{u^3} \right) du + \int_{2}^{3} \left(\frac{u-2}{u^3} \right) du$$

$$= \int_{1}^{2} \left(\frac{2}{u^3} - \frac{1}{u^2} \right) du + \int_{2}^{3} \left(\frac{1}{u^2} - \frac{2}{u^3} \right) du = \left\{ -\frac{1}{u^2} + \frac{1}{u} \right\}_{1}^{2} + \left\{ -\frac{1}{u} + \frac{1}{u^2} \right\}_{2}^{3}$$

$$= \left(-\frac{1}{4} + \frac{1}{2} \right) - (-1 + 1) + \left(-\frac{1}{3} + \frac{1}{9} \right) - \left(-\frac{1}{2} + \frac{1}{4} \right) = \frac{5}{18}.$$

20. If we set $u = \ln x$, then $du = \frac{1}{x}\,dx$, and $\displaystyle \int_{1}^{2} \frac{(\ln x)^2}{x}\,dx = \int_{0}^{\ln 2} u^2\,du = \left\{ \frac{u^3}{3} \right\}_{0}^{\ln 2} = \frac{1}{3}(\ln 2)^3.$

22. If we set $u = 4 + 3\tan x$, then $du = 3\sec^2 x\,dx$, and

$$\int_{-\pi/4}^{\pi/4} \frac{\sec^2 x}{\sqrt{4 + 3\tan x}}\,dx = \int_{1}^{7} \frac{1}{\sqrt{u}} \left(\frac{du}{3} \right) = \frac{1}{3} \left\{ 2\sqrt{u} \right\}_{1}^{7} = \frac{2(\sqrt{7} - 1)}{3}.$$

24. We can rewrite the integral as the sum of three integrals:

$$\int_{a}^{a+p} f(x)\,dx = \int_{a}^{0} f(x)\,dx + \int_{0}^{p} f(x)\,dx + \int_{p}^{a+p} f(x)\,dx.$$

In the last integral we change variables according to $u = x - p$. Then $du = dx$, and

$$\int_{a}^{a+p} f(x)\,dx = -\int_{0}^{a} f(x)\,dx + \int_{0}^{p} f(x)\,dx + \int_{0}^{a} f(u+p)\,du$$

$$= -\int_{0}^{a} f(x)\,dx + \int_{0}^{p} f(x)\,dx + \int_{0}^{a} f(u)\,du = \int_{0}^{p} f(x)\,dx.$$

26. If we set $u = 1/x$, then $du = -dx/x^2$, and

$$\int_{-6}^{-1} \frac{\sqrt{x^2 - 6x}}{x^4}\,dx = \int_{-1/6}^{-1} \frac{\sqrt{\frac{1}{u^2} - \frac{6}{u}}}{1/u^4} \left(-\frac{du}{u^2} \right) = -\int_{-1/6}^{-1} u^2 \sqrt{\frac{1 - 6u}{u^2}}\,du = \int_{-1/6}^{-1} u\sqrt{1 - 6u}\,du.$$

We now set $v = 1 - 6u$, in which case $dv = -6\,du$, and

$$\int_{-1/6}^{-1} \frac{\sqrt{x^2 - 6x}}{x^4}\,dx = \int_{2}^{7} \left(\frac{1 - v}{6} \right) \sqrt{v} \left(\frac{dv}{-6} \right) = \frac{1}{36} \int_{2}^{7} (v^{3/2} - \sqrt{v})\,dv$$

$$= \frac{1}{36} \left\{ \frac{2v^{5/2}}{5} - \frac{2v^{3/2}}{3} \right\}_{2}^{7} = \frac{1}{36} \left(\frac{98\sqrt{7}}{5} - \frac{14\sqrt{7}}{3} - \frac{8\sqrt{2}}{5} + \frac{4\sqrt{2}}{3} \right) = \frac{56\sqrt{7} - \sqrt{2}}{135}.$$

REVIEW EXERCISES

2. $\displaystyle \int_{-1}^{1} (x^2 - x^4)\,dx = \left\{ \frac{x^3}{3} - \frac{x^5}{5} \right\}_{-1}^{1} = \left(\frac{1}{3} - \frac{1}{5} \right) - \left(-\frac{1}{3} + \frac{1}{5} \right) = \frac{4}{15}$

4. $\displaystyle\int_0^2 (x^2 - 2x)\,dx = \left\{\frac{x^3}{3} - x^2\right\}_0^2 = \frac{8}{3} - 4 = -\frac{4}{3}$

6. $\displaystyle\int_{-3}^{-2} \frac{1}{x^2}\,dx = \left\{-\frac{1}{x}\right\}_{-3}^{-2} = \frac{1}{2} - \frac{1}{3} = \frac{1}{6}$

8. $\displaystyle\int_0^\pi \cos x\,dx = \left\{\sin x\right\}_0^\pi = 0$

10. $\displaystyle\int_1^2 x^2(x^2 + 3)\,dx = \left\{\frac{x^5}{5} + x^3\right\}_1^2 = \left(\frac{32}{5} + 8\right) - \left(\frac{1}{5} + 1\right) = \frac{66}{5}$

12. $\displaystyle\int_1^5 x\sqrt{x^2 - 1}\,dx = \left\{\frac{1}{3}(x^2 - 1)^{3/2}\right\}_1^5 = \frac{1}{3}(24)^{3/2} = 16\sqrt{6}$

14. If we set $u = x + 1$, then $du = dx$, and

$$\int_{-1}^0 x\sqrt{x+1}\,dx = \int_0^1 (u-1)\sqrt{u}\,du = \int_0^1 (u^{3/2} - \sqrt{u})\,du = \left\{\frac{2}{5}u^{5/2} - \frac{2}{3}u^{3/2}\right\}_0^1 = \frac{2}{5} - \frac{2}{3} = -\frac{4}{15}.$$

16. If we set $u = 2 - x$, then $du = -dx$, and

$$\int_{-4}^{-2} x^2\sqrt{2-x}\,dx = \int_6^4 (2-u)^2\sqrt{u}(-du) = \int_4^6 (4\sqrt{u} - 4u^{3/2} + u^{5/2})\,du = \left\{\frac{8}{3}u^{3/2} - \frac{8}{5}u^{5/2} + \frac{2}{7}u^{7/2}\right\}_4^6$$

$$= \left(16\sqrt{6} - \frac{8(36)\sqrt{6}}{5} + \frac{12(36)\sqrt{6}}{7}\right) - \left(\frac{64}{3} - \frac{256}{5} + \frac{256}{7}\right) = \frac{2112\sqrt{6} - 704}{105}.$$

18. $\displaystyle\int_2^3 x(1 + 2x^2)^4\,dx = \left\{\frac{1}{20}(1 + 2x^2)^5\right\}_2^3 = \frac{1}{20}(19^5 - 9^5) = 120\,852.5$

20. $\displaystyle\int_{-4}^4 |x + 2|\,dx = \int_{-4}^{-2} -(x + 2)\,dx + \int_{-2}^4 (x + 2)\,dx = -\left\{\frac{x^2}{2} + 2x\right\}_{-4}^{-2} + \left\{\frac{x^2}{2} + 2x\right\}_{-2}^4$

$$= -(2 - 4) + (8 - 8) + (8 + 8) - (2 - 4) = 20$$

22. (a) Since $x^2 + 3x$ is nonnegative for $0 \le x \le 4$, it is impossible for the limit summation of equation 6.10 to give a negative number.

(b) Since $1/x$ is negative for $-3 \le x \le -2$, it is impossible for the limit summation of 6.10 to give a positive number.

24. The averge value is $\displaystyle\frac{1}{1}\int_{-2}^{-1} \left(\frac{1}{x^2} - x\right)\,dx = \left\{-\frac{1}{x} - \frac{x^2}{2}\right\}_{-2}^{-1} = \left(1 - \frac{1}{2}\right) - \left(\frac{1}{2} - 2\right) = 2.$

26. If we set $u = \sin x$ and $du = \cos x\,dx$, the average value is

$$\frac{1}{\pi/2}\int_0^{\pi/2} \cos^3 x \sin^2 x\,dx = \frac{2}{\pi}\int_0^{\pi/2} (1 - \sin^2 x)\sin^2 x \cos x\,dx = \frac{2}{\pi}\int_0^1 (1 - u^2)u^2\,du$$

$$= \frac{2}{\pi}\left\{\frac{u^3}{3} - \frac{u^5}{5}\right\}_0^1 = \frac{2}{\pi}\left(\frac{1}{3} - \frac{1}{5}\right) = \frac{4}{15\pi}.$$

28. When we reverse the limits, $\displaystyle\frac{d}{dx}\int_x^{-3} t^2(t + 1)^3\,dt = -\frac{d}{dx}\int_{-3}^x t^2(t + 1)^3\,dt = -x^2(x + 1)^3.$

30. When we reverse the limits, and set $u = 2x$,

$$\frac{d}{dx}\int_{2x}^4 t\cos t\,dt = -\frac{d}{dx}\int_4^{2x} t\cos t\,dt = -\left[\frac{d}{du}\int_4^u t\cos t\,dt\right]\frac{du}{dx} = -u\cos u(2) = -4x\cos 2x.$$

32. When a is a number between $-x^2$ and x^2,

$$\int_{-x^2}^{x^2} \sin^2 t \, dt = \int_{-x^2}^{a} \sin^2 t \, dt + \int_{a}^{x^2} \sin^2 t \, dt = -\int_{a}^{-x^2} \sin^2 t \, dt + \int_{a}^{x^2} \sin^2 t \, dt.$$

In these integrals we set $u = -x^2$ and $v = x^2$ respectively, and use chain rules,

$$\frac{d}{dx} \int_{-x^2}^{x^2} \sin^2 t \, dt = -\left[\frac{d}{du} \int_{a}^{u} \sin^2 t \, dt \right] \frac{du}{dx} + \left[\frac{d}{dv} \int_{a}^{v} \sin^2 t \, dt \right] \frac{dv}{dx}$$

$$= -\sin^2 u(-2x) + \sin^2 v(2x) = 2x[\sin^2(-x^2) + \sin^2(x^2)] = 4x \sin^2 x^2.$$

34. $\displaystyle\int_1^3 v(t) \, dt = \int_1^3 (t^3 - 6t^2 + 11t - 6) \, dt$

$$= \left\{ \frac{t^4}{4} - 2t^3 + \frac{11t^2}{2} - 6t \right\}_1^3$$

$$= \left(\frac{81}{4} - 54 + \frac{99}{2} - 18 \right)$$

$$- \left(\frac{1}{4} - 2 + \frac{11}{2} - 6 \right) = 0$$

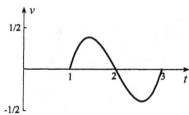

This means that the particle is at the same position at times $t = 1$ and $t = 3$. Since $v(t) = (t-1)(t-2)(t-3)$ changes sign at $t = 2$,

$$\int_1^3 |v(t)| \, dt = \int_1^2 (t^3 - 6t^2 + 11t - 6) \, dt + \int_2^3 (-t^3 + 6t^2 - 11t + 6) \, dt$$

$$= \left\{ \frac{t^4}{4} - 2t^3 + \frac{11t^2}{2} - 6t \right\}_1^2 + \left\{ -\frac{t^4}{4} + 2t^3 - \frac{11t^2}{2} + 6t \right\}_2^3$$

$$= (4 - 16 + 22 - 12) - \left(\frac{1}{4} - 2 + \frac{11}{2} - 6 \right)$$

$$+ \left(-\frac{81}{4} + 54 - \frac{99}{2} + 18 \right) - (-4 + 16 - 22 + 12)$$

$$= \frac{1}{2}.$$

This is the distance travelled by the particle between $t = 1$ and $t = 3$.

36. If we set $u = 1 + \sin x$, then $du = \cos x \, dx$, and

$$\int_0^{\pi/6} \frac{\cos^3 x}{\sqrt{1 + \sin x}} dx = \int_0^{\pi/6} \frac{1 - \sin^2 x}{\sqrt{1 + \sin x}} \cos x \, dx = \int_1^{3/2} \frac{1 - (u-1)^2}{\sqrt{u}} du = \int_1^{3/2} (2\sqrt{u} - u^{3/2}) \, du$$

$$= \left\{ \frac{4}{3} u^{3/2} - \frac{2}{5} u^{5/2} \right\}_1^{3/2} = \left(2\sqrt{\frac{3}{2}} - \frac{9}{10} \sqrt{\frac{3}{2}} \right) - \left(\frac{4}{3} - \frac{2}{5} \right) = \frac{33\sqrt{6} - 56}{60}.$$

38. If we set $u = x + 3$, then $du = dx$, and

$$\int_{-1}^{2} \left| \frac{x}{\sqrt{3 + x}} \right| du = \int_2^5 \left| \frac{u - 3}{\sqrt{u}} \right| du = \int_2^3 \frac{3 - u}{\sqrt{u}} du + \int_3^5 \frac{u - 3}{\sqrt{u}} du$$

$$= \left\{ 6\sqrt{u} - \frac{2}{3} u^{3/2} \right\}_2^3 + \left\{ \frac{2}{3} u^{3/2} - 6\sqrt{u} \right\}_3^5$$

$$= (6\sqrt{3} - 2\sqrt{3}) - \left(6\sqrt{2} - \frac{4\sqrt{2}}{3} \right) + \left(\frac{10\sqrt{5}}{3} - 6\sqrt{5} \right) - (2\sqrt{3} - 6\sqrt{3})$$

$$= \frac{1}{3}(24\sqrt{3} - 14\sqrt{2} - 8\sqrt{5}).$$

40. If we set $u = x^3 + 2x^2 + x$, then $du = (3x^2 + 4x + 1)\,dx$, and

$$\int_1^5 \frac{6x^2 + 8x + 2}{\sqrt{x^3 + 2x^2 + x}}\,dx = \int_4^{180} \frac{1}{\sqrt{u}}\,2du = \left\{4\sqrt{u}\right\}_4^{180} = 4(\sqrt{180} - 2) = 24\sqrt{5} - 8.$$

42. $\displaystyle\int_0^1 \frac{1}{\sqrt{2+x} + \sqrt{x}}\,dx = \int_0^1 \frac{1}{\sqrt{2+x} + \sqrt{x}} \frac{\sqrt{2+x} - \sqrt{x}}{\sqrt{2+x} - \sqrt{x}}\,dx = \int_0^1 \frac{\sqrt{2+x} - \sqrt{x}}{2+x-x}\,dx$

$$= \frac{1}{2}\left\{\frac{2}{3}(2+x)^{3/2} - \frac{2}{3}x^{3/2}\right\}_0^1 = \frac{1}{2}\left(2\sqrt{3} - \frac{2}{3}\right) - \frac{1}{2}\left(\frac{4\sqrt{2}}{3}\right) = \frac{1}{3}(3\sqrt{3} - 1 - 2\sqrt{2})$$

CHAPTER 7

EXERCISES 7.1

2. $A = 2 \int_0^2 [(4x+8) - (x^3+8)]\, dx$

$\qquad = 2 \left\{ 2x^2 - \dfrac{x^4}{4} \right\}_0^2 = 8$

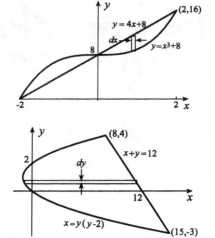

4. $A = \int_{-3}^4 [(12-y) - y(y-2)]\, dy$

$\qquad = \int_{-3}^4 (12 + y - y^2)\, dy$

$\qquad = \left\{ 12y + \dfrac{y^2}{2} - \dfrac{y^3}{3} \right\}_{-3}^4 = \dfrac{343}{6}$

6. $A = \int_1^2 (e^{3x} + x)\, dx$

$\qquad = \left\{ \dfrac{e^{3x}}{3} + \dfrac{x^2}{2} \right\}_1^2$

$\qquad = \left(\dfrac{e^6}{3} + 2 \right) - \left(\dfrac{e^3}{3} + \dfrac{1}{2} \right)$

$\qquad = \dfrac{1}{3}(e^6 - e^3) + \dfrac{3}{2}$

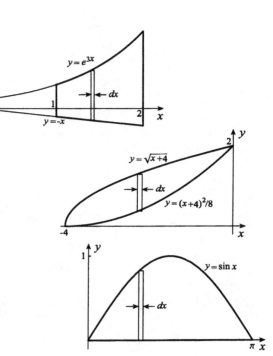

8. $A = \int_{-4}^0 \left[\sqrt{x+4} - \dfrac{(x+4)^2}{8} \right] dx$

$\qquad = \left\{ \dfrac{2}{3}(x+4)^{3/2} - \dfrac{1}{24}(x+4)^3 \right\}_{-4}^0$

$\qquad = \dfrac{8}{3}$

10. $A = \int_0^\pi \sin x\, dx = \left\{ -\cos x \right\}_0^\pi = 2$

12. $A = \int_{-5/3}^{-1/3} [(2x+6)-(1-x)]\, dx + \int_{-1/3}^0 [(5-x)-(1-x)]\, dx$

$\qquad + \int_0^{4/3} [(5-x)-(2x+1)]\, dx$

$\qquad = \int_{-5/3}^{-1/3} (3x+5)\, dx + \int_{-1/3}^0 4\, dx + \int_0^{4/3} (4-3x)\, dx$

$\qquad = \left\{ \dfrac{3x^2}{2} + 5x \right\}_{-5/3}^{-1/3} + \left\{ 4x \right\}_{-1/3}^0 + \left\{ 4x - \dfrac{3x^2}{2} \right\}_0^{4/3} = \dfrac{20}{3}$

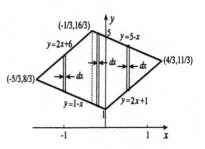

14. $A = \int_{e^{2/3}}^{2} \left(\sqrt{y} - \dfrac{e}{y} \right) dy$

$= \left\{ \dfrac{2}{3} y^{3/2} - e \ln |y| \right\}_{e^{2/3}}^{2}$

$= \left(\dfrac{4\sqrt{2}}{3} - e \ln 2 \right) - \left(\dfrac{2e}{3} - \dfrac{2e}{3} \right)$

$= \dfrac{4\sqrt{2}}{3} - e \ln 2$

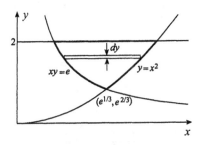

16. $A = \int_{0}^{2} [4 - (y-1)^2 - (y+1)]\, dy$

$= \int_{0}^{2} [3 - y - (y-1)^2]\, dy$

$= \left\{ 3y - \dfrac{y^2}{2} - \dfrac{(y-1)^3}{3} \right\}_{0}^{2}$

$= \left(6 - 2 - \dfrac{1}{3} \right) - \left(\dfrac{1}{3} \right) = \dfrac{10}{3}$

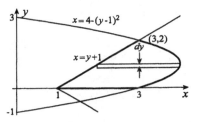

18. (a) The area is $A = \int_{1}^{3} (x^3 + 1)\, dx = \left\{ \dfrac{x^4}{4} + x \right\}_{1}^{3} = 22$.

(b) $A_2 = f(2)(1) + f(3)(1) = 9 + 28 = 37$ The error is $37 - 22 = 15$. It is the areas marked with e in the left figure.

(c) $A_4 = f(3/2)(1/2) + f(2)(1/2) + f(5/2)(1/2) + f(3)(1/2) = \dfrac{1}{2} \left(\dfrac{35}{8} + 9 + \dfrac{133}{8} + 28 \right) = 29$

The error is $29 - 22 = 7$. The extra precision is the deletion of the two rectangles marked with an E in the middle figure.

(d) $A_8 = \dfrac{1}{4}[f(5/4) + f(3/2) + f(7/4) + f(2) + f(9/4) + f(5/2) + f(11/4) + f(3)] = \dfrac{203}{8}$

The error is $203/8 - 22 = 27/8$. The extra precision is the deletion of the four rectangles marked with an E in the right figure.

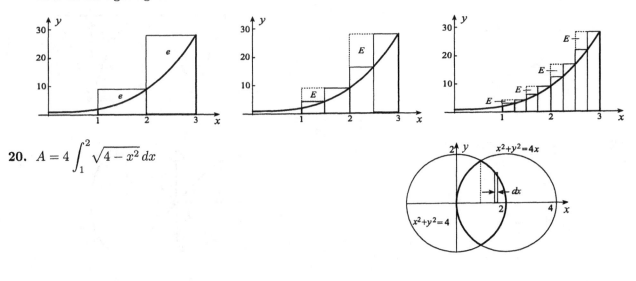

20. $A = 4 \int_{1}^{2} \sqrt{4 - x^2}\, dx$

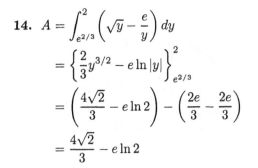

22. $A = 2 \displaystyle\int_0^{\sqrt{(\sqrt{65}-1)/2}} (\sqrt{16 - y^2} - y^2)\, dy$

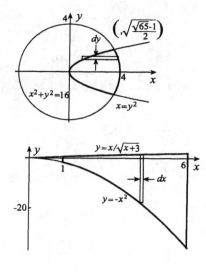

24. $A = \displaystyle\int_1^6 \left(\dfrac{x}{\sqrt{x+3}} + x^2 \right) dx$

If we set $u = x + 3$ in the first term, then $du = dx$, and

$$A = \int_4^9 \frac{u-3}{\sqrt{u}}\, du + \left\{ \frac{x^3}{3} \right\}_1^6$$

$$= \left\{ \frac{2}{3}u^{3/2} - 6\sqrt{u} \right\}_4^9 + \frac{215}{3} = \frac{235}{3}$$

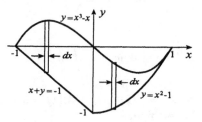

26. $A = \displaystyle\int_{-1}^0 [(x^3 - x) - (-x - 1)]\, dx$

$$+ \int_0^1 [(x^3 - x) - (x^2 - 1)]\, dx$$

$$= \left\{ \frac{x^4}{4} + x \right\}_{-1}^0 + \left\{ \frac{x^4}{4} - \frac{x^2}{2} - \frac{x^3}{3} + x \right\}_0^1$$

$$= \frac{7}{6}$$

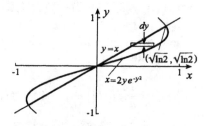

28. $A = 2 \displaystyle\int_0^{\sqrt{\ln 2}} (2y e^{-y^2} - y)\, dy$

$$= 2 \left\{ -e^{-y^2} - \frac{y^2}{2} \right\}_0^{\sqrt{\ln 2}}$$

$$= 2 \left(-e^{-\ln 2} - \frac{\ln 2}{2} \right) - 2(-1)$$

$$= 1 - \ln 2$$

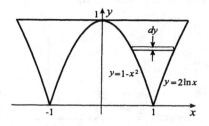

30. $A = 2 \displaystyle\int_0^1 (e^{y/2} - \sqrt{1 - y})\, dy$

$$= 2 \left\{ 2e^{y/2} + \frac{2}{3}(1 - y)^{3/2} \right\}_0^1$$

$$= 2(2\sqrt{e}) - 2 \left(2 + \frac{2}{3} \right)$$

$$= 4\sqrt{e} - \frac{16}{3}$$

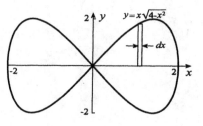

32. $A = 4 \displaystyle\int_0^2 x\sqrt{4 - x^2}\, dx$

$$= 4 \left\{ -\frac{1}{3}(4 - x^2)^{3/2} \right\}_0^2$$

$$= \frac{4}{3}(4)^{3/2} = \frac{32}{3}$$

34. $A = 2 \displaystyle\int_2^5 x\sqrt{x^2 - 4}\,dx$

$\quad = 2\left\{\dfrac{1}{3}(x^2 - 4)^{3/2}\right\}_2^5$

$\quad = \dfrac{2}{3}(21^{3/2}) = 14\sqrt{21}$

36. The equation of the tangent line at any point $P(a, b)$ on the first quadrant part of the parabola is $y - b = -2a(x - a)$ (left figure below). The x- and y-intercepts of this line are $x_1 = a + b/(2a)$, and $y_1 = b + 2a^2$. The area of the triangle is $A = \dfrac{1}{2}x_1 y_1 = \dfrac{1}{2}\left(a + \dfrac{b}{2a}\right)(b + 2a^2)$. Since $b = 2 - a^2$, we can express A in the form

$$A = \frac{1}{2}\left(a + \frac{2 - a^2}{2a}\right)(2 - a^2 + 2a^2) = \frac{1}{4a}(2 + a^2)^2, \quad 0 < a \le \sqrt{2}.$$

The plot of this function in the right figure indicates that its minimum occurs at the critical point. To find it we solve

$$0 = \frac{dA}{da} = \frac{1}{4}\left[\frac{a(2)(2 + a^2)(2a) - (2 + a^2)^2}{a^2}\right] = \frac{(2 + a^2)(3a^2 - 2)}{4a^2}.$$

The only positive solution is $a = \sqrt{2/3}$. Hence the required point is $(\sqrt{2/3}, 4/3)$.

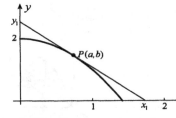

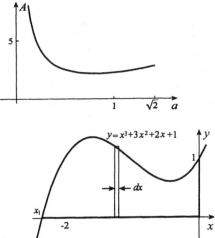

38. Newton's iterative procedure gives the x-intercept of the cubic as $x_1 = -2.324\,718$. Hence,

$A = \displaystyle\int_{x_1}^0 (x^3 + 3x^2 + 2x + 1)\,dx$

$\quad = \left\{\dfrac{x^4}{4} + x^3 + x^2 + x\right\}_{x_1}^0$

$\quad = 2.182.$

40. Since $x^4 - 5x^2 + 5$ is a quadratic in x^2, x-intercepts can be found exactly; they are $\pm\sqrt{(5 + \sqrt5)/2}$ and $\pm\sqrt{(5 - \sqrt5)/2}$. If we set $x_1 = \sqrt{(5 - \sqrt5)/2}$, and $x_2 = \sqrt{(5 + \sqrt5)/2}$, then

$A = 2\displaystyle\int_0^{x_1}(x^4 - 5x^2 + 5)\,dx$

$\quad + 2\displaystyle\int_{x_1}^{x_2}(-x^4 + 5x^2 - 5)\,dx$

$\quad = 2\left\{\dfrac{x^5}{5} - \dfrac{5x^3}{3} + 5x\right\}_0^{x_1} + 2\left\{-\dfrac{x^5}{5} + \dfrac{5x^3}{3} - 5x\right\}_{x_1}^{x_2} = 8.436.$

42. The x-coordinates of points of intersection of the curves are defined by the equation $\cos x = (x+2)/4$. Newton's method gives $x_1 = -4.146\,081$, $x_2 = -1.427\,069$, and $x_3 = 0.796\,591$ as solutions of this equation. Thus,

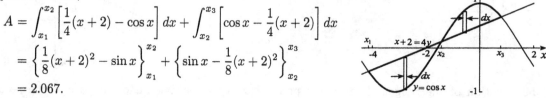

$$A = \int_{x_1}^{x_2} \left[\frac{1}{4}(x+2) - \cos x\right] dx + \int_{x_2}^{x_3} \left[\cos x - \frac{1}{4}(x+2)\right] dx$$

$$= \left\{\frac{1}{8}(x+2)^2 - \sin x\right\}_{x_1}^{x_2} + \left\{\sin x - \frac{1}{8}(x+2)^2\right\}_{x_2}^{x_3}$$

$$= 2.067.$$

44. Points of intersection of the curves have y-coordinates satisfying $y = (4-y^2)^3 - 3(4-y^2)^2 + 4(4-y^2) - 2$, and this equation simplifies to $y^6 - 9y^4 + 28y^2 + y - 30 = 0$. The negative solution is $y = -2$ and the positive solution can be obtained by Newton's method as $y = 1.466\,078$. The equation of the parabola gives the corresponding x-coordinate as $x_1 = 1.850\,616$. The area is

$$A = \int_0^{x_1} (x^3 - 3x^2 + 4x - 2 + \sqrt{4-x})\, dx$$

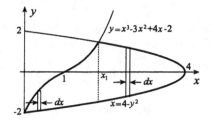

$$+ 2\int_{x_1}^4 \sqrt{4-x}\, dx$$

$$= \left\{\frac{x^4}{4} - x^3 + 2x^2 - 2x - \frac{2}{3}(4-x)^{3/2}\right\}_0^{x_1}$$

$$+ 2\left\{-\frac{2}{3}(4-x)^{3/2}\right\}_{x_1}^4 = 7.177.$$

46. For points of intersection of the curves, we solve

$$mx = \frac{x}{3x^2 + 1}$$

This equation simplifies to $3mx^3 + mx = x$, one solution of which is $x = 0$. Other solutions must satisfy $3mx^2 + m - 1 = 0$, from which

$$x = \pm\sqrt{\frac{1-m}{3m}}.$$

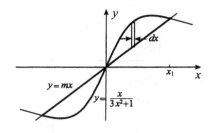

Consequently, an area is defined when $0 < m < 1$. If we set $x_1 = \sqrt{(1-m)/(3m)}$, the required area is

$$A = 2\int_0^{x_1} \left(\frac{x}{3x^2+1} - mx\right) dx = 2\left\{\frac{1}{6}\ln(3x^2+1) - \frac{mx^2}{2}\right\}_0^{x_1} = \frac{1}{3}\ln(3x_1^2+1) - mx_1^2$$

$$= \frac{1}{3}\ln\left[3\left(\frac{1-m}{3m}\right)+1\right] - m\left(\frac{1-m}{3m}\right) = \frac{1}{3}\ln\left[\frac{1-m+m}{m}\right] + \frac{m-1}{3} = -\frac{1}{3}\ln m + \frac{m-1}{3}.$$

48. If x_{i-1} and x_i are the ends of the ith rectangle, then $\Delta x_i = 2/2^n = 1/2^{n-1}$, and $x_i = 1 + i/2^{n-1}$. It follows that

$$A_{2^n} = \sum_{i=1}^{2^n} f(x_i)\Delta x_i = \sum_{i=1}^{2^n}\left[16 - \left(1 + \frac{i}{2^{n-1}}\right)^2\right]\left(\frac{1}{2^{n-1}}\right) = \frac{1}{2^{n-1}}\sum_{i=1}^{2^n}\left[15 - \frac{i}{2^{n-2}} - \frac{i^2}{2^{2n-2}}\right]$$

$$= \frac{1}{2^{n-1}}\left[15(2^n) - \frac{1}{2^{n-2}}\frac{2^n(2^n+1)}{2} - \frac{1}{2^{2n-2}}\frac{2^n(2^n+1)(2^{n+1}+1)}{6}\right] \quad \text{(see equations 6.3, 6.4)}$$

$$= 30 - \left(4 + \frac{1}{2^{n-2}}\right) - \frac{1}{6}\left(1 + \frac{1}{2^n}\right)\left(16 + \frac{1}{2^{n-3}}\right) = \frac{70}{3} - \frac{1}{2^{n-2}} - \frac{1}{6}\left(\frac{3}{2^{n-3}} + \frac{1}{2^{2n-3}}\right).$$

Since $A = 70/3$, we may write $A_{2^n} = A - \left[\frac{1}{2^{n-2}} + \frac{1}{6}\left(\frac{3}{2^{n-3}} + \frac{1}{2^{2n-3}}\right)\right]$, and clearly, $\lim_{n\to\infty} A_{2^n} = A$.

50. If the coordinates of P are (c, ac^3), then the tangent line at P has equation
$$y - ac^3 = 3ac^2(x - c),$$
and the x-coordinate of Q is defined by
$$ax^3 = ac^3 + 3ac^2(x - c).$$
The solution of this equation is $x = -2c$.
The equation of the tangent line at Q is

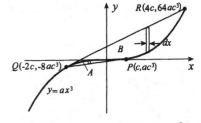

$y + 8ac^3 = 12ac^2(x + 2c)$, and the intersection of this line with $y = ax^3$ is the point $R(4c, 64ac^3)$. Now,
$$A = \int_{-2c}^{c} [ax^3 - ac^3 - 3ac^2(x - c)]\, dx = \left\{ \frac{ax^4}{4} - \frac{3ac^2x^2}{2} + 2ac^3x \right\}_{-2c}^{c} = \frac{27}{4}ac^4,$$

and
$$B = \int_{-2c}^{4c} [-8ac^3 + 12ac^2(x + 2c) - ax^3]\, dx = \left\{ 16ac^3x + 6ac^2x^2 - \frac{ax^4}{4} \right\}_{-2c}^{4c} = 108ac^4.$$

Thus, $B = 16A$.

EXERCISES 7.2

2. $V = 2\int_0^{\sqrt{5}} \pi x^2\, dy = 2\pi \int_0^{\sqrt{5}} (5 - y^2)^2\, dy$

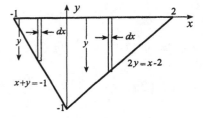

$\qquad = 2\pi \int_0^{\sqrt{5}} (25 - 10y^2 + y^4)\, dy$

$\qquad = 2\pi \left\{ 25y - \frac{10y^3}{3} + \frac{y^5}{5} \right\}_0^{\sqrt{5}} = \frac{80\sqrt{5}\pi}{3}$

4. $V = 2\int_0^2 [\pi(20)^2 - \pi(16 + y^2)^2]\, dy$

$\qquad = 2\pi \int_0^2 (144 - 32y^2 - y^4)\, dy$

$\qquad = 2\pi \left\{ 144y - \frac{32y^3}{3} - \frac{y^5}{5} \right\}_0^2 = \frac{5888\pi}{15}$

6. $V = \int_{-1}^0 \pi(-y)^2\, dx + \int_0^2 \pi(-y)^2\, dx$

$\qquad = \pi \int_{-1}^0 (1 + x)^2\, dx + \pi \int_0^2 \frac{1}{4}(2 - x)^2\, dx$

$\qquad = \pi \left\{ \frac{1}{3}(1 + x)^3 \right\}_{-1}^0 + \pi \left\{ -\frac{1}{12}(2 - x)^3 \right\}_0^2$

$\qquad = \pi$

8. $V = \int_{-1}^3 [\pi(5)^2 - \pi(-2y + y^2 + 2)^2]\, dy$

$\qquad = \pi \int_{-1}^3 (21 + 8y - 8y^2 + 4y^3 - y^4)\, dy$

$\qquad = \pi \left\{ 21y + 4y^2 - \frac{8y^3}{3} + y^4 - \frac{y^5}{5} \right\}_{-1}^3$

$\qquad = \frac{1088\pi}{15}$

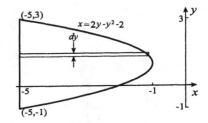

10. $V = \displaystyle\int_0^2 \left[\pi(2 - x^2 + 2x)^2 - \pi(2 - 2x + x^2)^2 \right] dx$

$\qquad = 8\pi \displaystyle\int_0^2 (2x - x^2)\, dx$

$\qquad = 8\pi \left\{ x^2 - \dfrac{x^3}{3} \right\}_0^2 = \dfrac{32\pi}{3}$

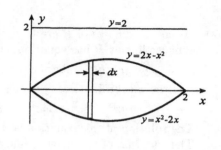

12. $V = \displaystyle\int_0^1 \pi x^2\, dy = \pi \displaystyle\int_0^1 (e^y - 1)^2\, dy$

$\qquad = \pi \displaystyle\int_0^1 (e^{2y} - 2e^y + 1)\, dy$

$\qquad = \pi \left\{ \dfrac{e^{2y}}{2} - 2e^y + y \right\}_0^1$

$\qquad = \pi \left(\dfrac{e^2}{2} - 2e + 1 - \dfrac{1}{2} + 2 \right)$

$\qquad = \dfrac{\pi}{2}(e^2 - 4e + 5)$

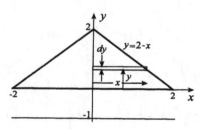

14. $V = \displaystyle\int_{-2}^0 2\pi(-y)x\, dy$

$\qquad = -2\pi \displaystyle\int_{-2}^0 y(4 - y^2)\, dy$

$\qquad = -2\pi \left\{ 2y^2 - \dfrac{y^4}{4} \right\}_{-2}^0 = 8\pi$

16. $V = \displaystyle\int_0^2 2\pi y \left(4 - y - \dfrac{y^2}{4} - 1 \right) dy$

$\qquad = \dfrac{\pi}{2} \displaystyle\int_0^2 (-y^3 - 4y^2 + 12y)\, dy$

$\qquad = \dfrac{\pi}{2} \left\{ -\dfrac{y^4}{4} - \dfrac{4y^3}{3} + 6y^2 \right\}_0^2 = \dfrac{14\pi}{3}$

18. $V = 2 \displaystyle\int_0^2 2\pi(y + 1)x\, dy$

$\qquad = 4\pi \displaystyle\int_0^2 (y + 1)(2 - y)\, dy$

$\qquad = 4\pi \displaystyle\int_0^2 (2 + y - y^2)\, dy$

$\qquad = 4\pi \left\{ 2y + \dfrac{y^2}{2} - \dfrac{y^3}{3} \right\}_0^2 = \dfrac{40\pi}{3}$

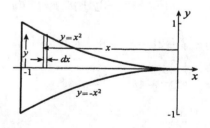

20. $V = 2 \displaystyle\int_{-1}^0 2\pi(x + 1)y\, dx$

$\qquad = 4\pi \displaystyle\int_{-1}^0 (x + 1)x^2\, dx$

$\qquad = 4\pi \displaystyle\int_{-1}^0 (x^3 + x^2)\, dx$

$\qquad = 4\pi \left\{ \dfrac{x^4}{4} + \dfrac{x^3}{3} \right\}_{-1}^0 = \dfrac{\pi}{3}$

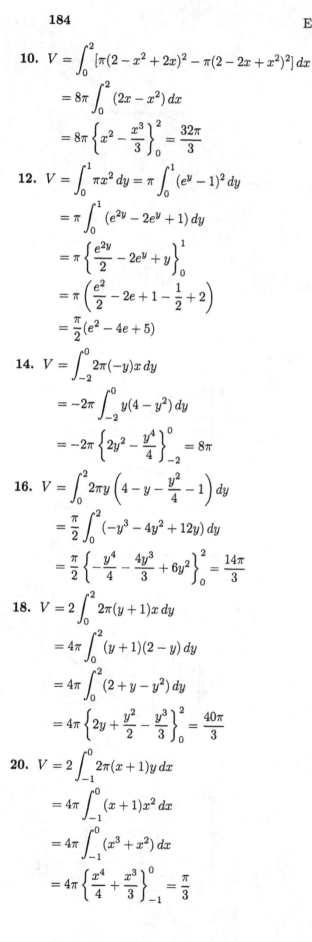

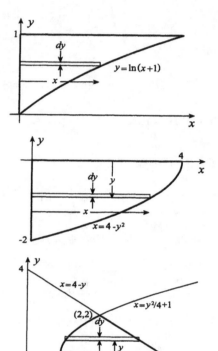

22. $V = \int_3^5 2\pi x \left(x - \frac{9}{x} \right) dx + \int_5^9 2\pi x \left(10 - x - \frac{9}{x} \right) dx$

$= 2\pi \left\{ \frac{x^3}{3} - 9x \right\}_3^5 + 2\pi \left\{ 5x^2 - \frac{x^3}{3} - 9x \right\}_5^9$

$= \frac{344\pi}{3}$

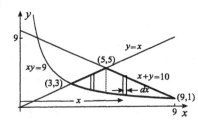

24. $V = 2 \int_1^9 2\pi y \left(\sqrt{10 - y} - \frac{3}{\sqrt{y}} \right) dy$

If we set $u = 10 - y$ and $du = -dy$ in the first term,

$V = 4\pi \int_9^1 (10 - u)\sqrt{u}(-du) + 4\pi \left\{ -2y^{3/2} \right\}_1^9$

$= 4\pi \left\{ \frac{2}{5}u^{5/2} - \frac{20}{3}u^{3/2} \right\}_9^1 - 8\pi(27 - 1)$

$= \frac{1472\pi}{15}.$

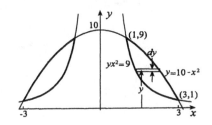

26. $V = \int_0^5 2\pi x \left[\frac{12x}{5} - (x - 1)^2 + 4 \right] dx$

$= \frac{2\pi}{5} \int_0^5 (-5x^3 + 22x^2 + 15x) \, dx$

$= \frac{2\pi}{5} \left\{ -\frac{5x^4}{4} + \frac{22x^3}{3} + \frac{15x^2}{2} \right\}_0^5$

$= \frac{775\pi}{6}$

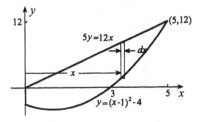

28. $V = 2 \int_0^1 \pi(y + 1)^2 \, dx + 2 \int_1^2 \pi(y + 1)^2 \, dx$

$= 2\pi \int_0^1 (1 - x^2 + 1)^2 \, dx + 2\pi \int_1^2 (x^2 - 1 + 1)^2 \, dx$

$= 2\pi \int_0^1 (4 - 4x^2 + x^4) \, dx + 2\pi \int_1^2 x^4 \, dx$

$= 2\pi \left\{ 4x - \frac{4x^3}{3} + \frac{x^5}{5} \right\}_0^1 + 2\pi \left\{ \frac{x^5}{5} \right\}_1^2$

$= \frac{272\pi}{15}$

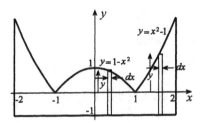

30. $V = \int_{-1}^0 2\pi(-y)(20y + 24 - \sqrt{4 + 12y^2}) \, dy$

$= -2\pi \int_{-1}^0 \left(20y^2 + 24y - y\sqrt{4 + 12y^2} \right) dy$

$= -2\pi \left\{ \frac{20y^3}{3} + 12y^2 - \frac{1}{36}(4 + 12y^2)^{3/2} \right\}_{-1}^0$

$= \frac{68\pi}{9}$

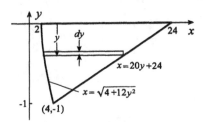

32. To eliminate duplications we reject that part
of the area between $y = -2$ and $y = -1$. We rotate
that part of the area to the right of the y-axis and
double the result. The rectangle $0 \le x \le 2^{1/4}$,
$-1 \le y \le 0$ gives a cylinder, and the other two
parts of the area require integrations,

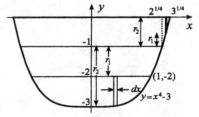

$$V = 2[\pi(1)^2(2^{1/4})] + 2\int_{2^{1/4}}^{3^{1/4}} (\pi r_2^2 - \pi r_1^2)\, dx + 2\int_0^1 (\pi r_2^2 - \pi r_1^2)\, dx$$

$$= 2^{5/4}\pi + 2\pi \int_{2^{1/4}}^{3^{1/4}} [(1)^2 - (x^4 - 3 + 1)^2]\, dx + 2\pi \int_0^1 [(-1 - x^4 + 3)^2 - 1^2]\, dx$$

$$= 2^{5/4}\pi + 2\pi \int_{2^{1/4}}^{3^{1/4}} (-3 + 4x^4 - x^8)\, dx + 2\pi \int_0^1 (3 - 4x^4 + x^8)\, dx$$

$$= 2^{5/4}\pi + 2\pi \left\{-3x + \frac{4x^5}{5} - \frac{x^9}{9}\right\}_{2^{1/4}}^{3^{1/4}} + 2\pi \left\{3x - \frac{4x^5}{5} + \frac{x^9}{9}\right\}_0^1 = \frac{16\pi}{45}(13 + 2^{17/4} - 3^{9/4}).$$

34. Newton's method can be used to solve the equation $e^{-2x} + x^2 - 4 = 0$ for x-coordinates of the points of
intersection of the two curves. The results are $x_1 = -0.639\,263$ and $x_2 = 1.995\,373$. The volume of the
solid of revolution is

$$V = \int_{x_1}^{x_2} [\pi(4 - x^2 + 1)^2 - \pi(e^{-2x} + 1)^2]\, dx$$

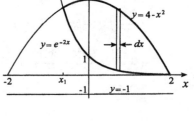

$$= \pi \int_{x_1}^{x_2} (24 - 10x^2 + x^4 - e^{-4x} - 2e^{-2x})\, dx$$

$$= \pi \left\{24x - \frac{10x^3}{3} + \frac{x^5}{5} + \frac{e^{-4x}}{4} + e^{-2x}\right\}_{x_1}^{x_2}$$

$$= 111.303.$$

36. The y-coordinate of the point of intersection of
the curves can be obtained by solving the equation
$y = (4 - y^2)^3 + 1$. Newton's iterative procedure leads
to the solution $y_1 = 1.757\,401\,58$. The volume of
the solid of revolution is

$$V = \int_0^{y_1} 2\pi y[(4 - y^2) - (y - 1)^{1/3}]\, dy$$

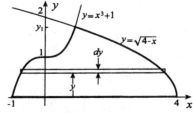

$$= 2\pi \int_0^{y_1} [4y - y^3 - y(y - 1)^{1/3}]\, dy.$$

We set $u = y - 1$ and $du = dy$ in the last term,

$$V = 2\pi \left\{2y^2 - \frac{y^4}{4}\right\}_0^{y_1} - 2\pi \int_{-1}^{y_1 - 1} (u + 1)u^{1/3}\, du = 2\pi \left\{2y^2 - \frac{y^4}{4}\right\}_0^{y_1} - 2\pi \left\{\frac{3}{7}u^{7/3} + \frac{3}{4}u^{4/3}\right\}_{-1}^{y_1 - 1}$$

$$= 2\pi \left\{2y^2 - \frac{y^4}{4} - \frac{3}{7}(y - 1)^{7/3} - \frac{3}{4}(y - 1)^{4/3}\right\}_0^{y_1} = 21.186.$$

38. $V = 2\int_0^4 2\pi yx\, dy = 4\pi \int_0^4 y\left[\frac{y(4 - y)}{64}\right] dy$

$$= \frac{\pi}{16} \left\{\frac{4y^3}{3} - \frac{y^4}{4}\right\}_0^4 = \frac{4\pi}{3}$$

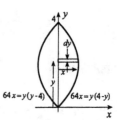

40. (a) For the points $(0,0)$, $(4,2)$, and $(6,0)$ to be on the cubic, a, b, c, and d must satisfy

$$0 = d, \quad 2 = 64a + 16b + 4c + d, \quad 0 = 216a + 36b + 6c + d.$$

In addition, because $y'(4) = 0$, we must have $0 = 48a + 8b + c$. These four equations imply that $a = -1/16$, $b = 3/8$, $c = d = 0$, and therefore $y = -x^3/16 + 3x^2/8 = x^2(6-x)/16$.
(b) The amount of fill required is

$$V = \int_0^6 2\pi(x+5)y\,dx$$

$$= 2\pi \int_0^6 (x+5)\frac{x^2(6-x)}{16}\,dx$$

$$= \frac{\pi}{8} \int_0^6 (30x^2 + x^3 - x^4)\,dx$$

$$= \frac{\pi}{8}\left\{10x^3 + \frac{x^4}{4} - \frac{x^5}{5}\right\}_0^6 = \frac{1161\pi}{10}\ \text{m}^3.$$

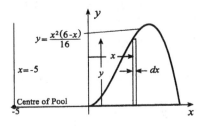

EXERCISES 7.3

2. Small lengths along the curve are approximated by

$$\sqrt{(dx)^2 + (dy)^2} = \sqrt{1 + \left(\frac{dy}{dx}\right)^2}\,dx$$

$$= \sqrt{1 + [2x(x^2+1)^{1/2}]^2}\,dx = \sqrt{1 + 4x^2(x^2+1)}\,dx$$

$$= (2x^2 + 1)\,dx.$$

Total length of the curve is

$$L = \int_{-2}^{-1} (2x^2 + 1)\,dx = \left\{\frac{2x^3}{3} + x\right\}_{-2}^{-1} = \frac{17}{3}.$$

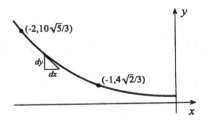

4. We approximate small lengths along the curve by

$$\sqrt{(dx)^2 + (dy)^2} = \sqrt{1 + \left(\frac{dy}{dx}\right)^2}\,dx$$

$$= \sqrt{1 + \left(\frac{3}{2}\sqrt{x-1}\right)^2}\,dx = \frac{1}{2}\sqrt{9x - 5}\,dx.$$

Total length of the curve is

$$L = \int_2^{10} \frac{1}{2}\sqrt{9x-5}\,dx = \frac{1}{2}\left\{\frac{2}{27}(9x-5)^{3/2}\right\}_2^{10} = \frac{85^{3/2} - 13^{3/2}}{27}.$$

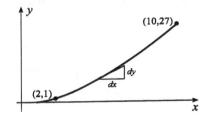

6. Small lengths along the curve are approximated by

$$\sqrt{(dx)^2 + (dy)^2} = \sqrt{1 + \left(\frac{dy}{dx}\right)^2}\,dx$$

$$= \sqrt{1 + \left(\frac{e^x - e^{-x}}{2}\right)^2}\,dx = \sqrt{\frac{e^{2x} + 2 + e^{-2x}}{4}}\,dx$$

$$= \frac{1}{2}(e^x + e^{-x})\,dx.$$

Total length of the curve is $L = \int_0^1 \frac{1}{2}(e^x + e^{-x})\,dx = \frac{1}{2}\left\{e^x - e^{-x}\right\}_0^1 = \frac{1}{2}(e - e^{-1}).$

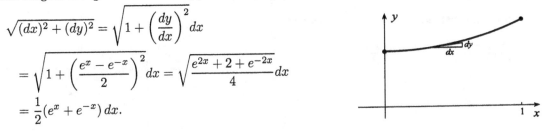

8. We approximate small lengths along the curve by

$$\sqrt{(dx)^2 + (dy)^2} = \sqrt{1 + \left(\frac{dy}{dx}\right)^2}\, dx$$

$$= \sqrt{1 + \left(\frac{x^2}{12} - \frac{3}{x^2}\right)^2}\, dx = \sqrt{1 + \frac{x^4}{144} - \frac{1}{2} + \frac{9}{x^4}}\, dx$$

$$= \sqrt{\left(\frac{x^2}{12} + \frac{3}{x^2}\right)^2}\, dx = \left(\frac{x^2}{12} + \frac{3}{x^2}\right) dx.$$

Total length of the curve is $\quad L = \int_3^4 \left(\frac{x^2}{12} + \frac{3}{x^2}\right) dx = \left\{\frac{x^3}{36} - \frac{3}{x}\right\}_3^4 = \frac{23}{18}.$

10. We approximate small lengths along the curve by

$$\sqrt{(dx)^2 + (dy)^2} = \sqrt{1 + \left(\frac{dy}{dx}\right)^2}\, dx$$

$$= \sqrt{1 + \left(x^4 - \frac{1}{4x^4}\right)^2}\, dx = \sqrt{1 + x^8 - \frac{1}{2} + \frac{1}{16x^8}}\, dx$$

$$= \sqrt{\left(x^4 + \frac{1}{4x^4}\right)^2}\, dx = \left(x^4 + \frac{1}{4x^4}\right) dx.$$

Total length of the curve is $\quad L = \int_1^2 \left(x^4 + \frac{1}{4x^4}\right) dx = \left\{\frac{x^5}{5} - \frac{1}{12x^3}\right\}_1^2 = \frac{3011}{480}.$

12. We approximate small lengths along the curve by

$$\sqrt{(dx)^2 + (dy)^2} = \sqrt{1 + \left(\frac{dy}{dx}\right)^2}\, dx$$

$$= \sqrt{1 + (6x - 4)^2}\, dx = \sqrt{36x^2 - 48x + 17}\, dx.$$

Total length of the curve is given by $L = \int_1^2 \sqrt{36x^2 - 48x + 17}\, dx.$

14. We approximate small lengths along the curve by

$$\sqrt{(dx)^2 + (dy)^2} = \sqrt{1 + \left(\frac{dx}{dy}\right)^2}\, dy$$

$$= \sqrt{1 + \left(\frac{y}{\sqrt{1 + y^2}}\right)^2}\, dy = \sqrt{\frac{1 + 2y^2}{1 + y^2}}\, dy.$$

Total length of the curve is given by $L = \int_{-\sqrt{3}}^{2\sqrt{2}} \sqrt{\frac{1 + 2y^2}{1 + y^2}}\, dy.$

16. We approximate small lengths along the curve by

$$\sqrt{(dx)^2 + (dy)^2} = \sqrt{1 + \left(\frac{dy}{dx}\right)^2}\, dx$$

$$= \sqrt{1 + \left(\frac{-\sin x}{\cos x}\right)^2}\, dx = \sec x\, dx.$$

Total length of the curve is given by $L = \int_0^{\pi/4} \sec x\, dx.$

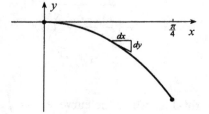

18. Differentiation of $8y^2 = x^2 - x^4$ with respect to x

gives $16y\dfrac{dy}{dx} = 2x - 4x^3$, and from this equation

$\dfrac{dy}{dx} = \dfrac{x - 2x^3}{8y}$. Small lengths along that portion of

the curve in the first quadrant are approximated by

$$\sqrt{(dx)^2 + (dy)^2} = \sqrt{1 + \left(\frac{dy}{dx}\right)^2}\, dx = \sqrt{1 + \left(\frac{x - 2x^3}{8y}\right)^2}\, dx = \frac{\sqrt{64y^2 + x^2 - 4x^4 + 4x^6}}{8y}\, dx$$

$$= \frac{\sqrt{8x^2 - 8x^4 + x^2 - 4x^4 + 4x^6}}{8y}\, dx = \frac{3x - 2x^3}{2\sqrt{2}x\sqrt{1 - x^2}}\, dx = \frac{3 - 2x^2}{2\sqrt{2}\sqrt{1 - x^2}}\, dx.$$

Total length of the curve is four times that in the first quadrant,

$$L = 4\int_0^1 \frac{3 - 2x^2}{2\sqrt{2}\sqrt{1 - x^2}}\, dx = \sqrt{2}\int_0^1 \frac{3 - 2x^2}{\sqrt{1 - x^2}}\, dx.$$

20. We approximate small lengths along that part of the ellipse in the first quadrant by

$$\sqrt{(dx)^2 + (dy)^2} = \sqrt{1 + \left(\frac{dy}{dx}\right)^2}\, dx$$

$$= \sqrt{1 + \left(\frac{-3x}{2\sqrt{4 - x^2}}\right)^2}\, dx = \sqrt{\frac{16 + 5x^2}{4(4 - x^2)}}\, dx.$$

Total length of the curve is four times that in the first quadrant

$$L = 4\int_0^2 \sqrt{\frac{16 + 5x^2}{4(4 - x^2)}}\, dx = 2\int_0^2 \sqrt{\frac{16 + 5x^2}{4 - x^2}}\, dx.$$

22. Differentiation of $x^{2/3} + y^{2/3} = 1$ with respect to x gives $\dfrac{2}{3}x^{-1/3} + \dfrac{2}{3}y^{-1/3}\dfrac{dy}{dx} = 0$, and therefore

$\dfrac{dy}{dx} = -\left(\dfrac{y}{x}\right)^{1/3}$. We approximate small lengths along the curve in the first quadrant by

$$\sqrt{(dx)^2 + (dy)^2} = \sqrt{1 + \left(\frac{dy}{dx}\right)^2}\, dx$$

$$= \sqrt{1 + \left(\frac{y}{x}\right)^{2/3}}\, dx = \frac{\sqrt{x^{2/3} + y^{2/3}}}{x^{1/3}}\, dx = x^{-1/3}\, dx.$$

Total length of the curve is therefore $L = 4\displaystyle\int_0^1 x^{-1/3}\, dx = 4\left\{\dfrac{3}{2}x^{2/3}\right\}_0^1 = 6.$

24. Small lengths along the curve are approximated by

$$\sqrt{(dx)^2 + (dy)^2} = \sqrt{1 + \left(\frac{dy}{dx}\right)^2}\, dx = \sqrt{1 + \left[\frac{(2n+1)x^{2n}}{4(2n-1)} - \frac{(2n-1)}{(2n+1)x^{2n}}\right]^2}\, dx$$

$$= \sqrt{1 + \frac{(2n+1)^2 x^{4n}}{16(2n-1)^2} - \frac{1}{2} + \frac{(2n-1)^2}{(2n+1)^2 x^{4n}}}\, dx$$

$$= \sqrt{\left[\frac{(2n+1)x^{2n}}{4(2n-1)} + \frac{2n-1}{(2n+1)x^{2n}}\right]^2}\, dx = \left[\frac{(2n+1)x^{2n}}{4(2n-1)} + \frac{2n-1}{(2n+1)x^{2n}}\right] dx.$$

The length of the curve is therefore

$$L = \int_a^b \left[\frac{(2n+1)x^{2n}}{4(2n-1)} + \frac{2n-1}{(2n+1)x^{2n}}\right] dx = \left\{\frac{x^{2n+1}}{4(2n-1)} - \frac{1}{(2n+1)x^{2n-1}}\right\}_a^b$$

$$= \frac{b^{2n+1} - a^{2n+1}}{4(2n-1)} + \frac{1}{2n+1}\left(\frac{1}{a^{2n-1}} - \frac{1}{b^{2n-1}}\right).$$

EXERCISES 7.4

2. Let the spring be stretched in the positive x-direction and $x = 0$ correspond to the free end of the spring in the unstretched position (figure below). The restoring force of the spring is $F_s = -kx$. Since $F_s = -10$ N when $x = 0.03$ m, it follows that $-10 = -0.03k$, and therefore $k = 1000/3$ N/m. The force required to counteract the spring force when the spring is stretched an amount x is $F(x) = 1000x/3$. The work done by this force in stretching the spring a further distance dx is $\dfrac{1000x}{3} dx$ J. The total work in stretching the spring from 7 cm to 9 cm is

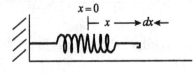

$$W = \int_{0.07}^{0.09} \frac{1000}{3} x\, dx = \frac{1000}{3} \left\{ \frac{x^2}{2} \right\}_{0.07}^{0.09} = \frac{8}{15} \text{ J.}$$

4. When the lower end of the cable is y m above the ground, the force of gravity on that part of the cable hanging from the building is $-9.81(2)(100 - y)$ N. The force required to counter gravity is therefore $19.62(100 - y)$ N. The work done by this force in raising the cable a further distance dy is $19.62(100 - y)\, dy$ J. The total work to raise the cable the 10 m is

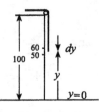

$$W = \int_{50}^{60} 19.62(100 - y)\, dy = 19.62 \left\{ 100y - \frac{y^2}{2} \right\}_{50}^{60} = 8829 \text{ J.}$$

6. When the end of the chain has been lifted a distance y, the force necessary to overcome gravity and hold the chain in this position is $9.81(10)y$ N. The work done by this force in lifting the end of the chain an additional amount dy is $98.1y\, dy$ J. The total work to lift the end of the chain 1 m is

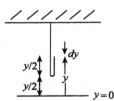

therefore $W = \displaystyle\int_{0}^{1} 98.1y\, dy = 98.1 \left\{ \frac{y^2}{2} \right\}_{0}^{1} = 49.05$ J.

8. (a) When the lower end has been lifted a distance y, the force necessary to overcome gravity is $9.81(3)(y/2)$ N. The work done by this force in lifting the lower end of the chain an additional amount dy is $(29.43y/2)\, dy$ J. The total work done in lifting the end the 5 m is

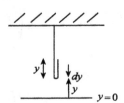

$$W = \int_{0}^{5} \frac{29.43y}{2}\, dy = \frac{29.43}{2} \left\{ \frac{y^2}{2} \right\}_{0}^{5} = 183.9 \text{ J.}$$

(b) In this case the force necessary to counter gravity at the position shown is $9.81(3)y$ N. To move the bend in the cable up a distance dy, the end of the cable must be lifted a distance $2dy$, requiring $29.43y(2dy)$ J of work. Hence, the total work done is

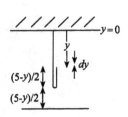

$$W = \int_{0}^{2.5} 58.86y\, dy = 58.86 \left\{ \frac{y^2}{2} \right\}_{0}^{2.5} = 183.9 \text{ J.}$$

(c) When the chain is at the position shown, the force to overcome gravity is $F(y) = -9.81(3)(5 - y)/2$ N. The work to raise the chain through a displacement dy which is in the negative y-direction is $[-29.43(5 - y)/2]dy$. The total work to raise the chain is therefore

$$W = \int_{5}^{0} -\frac{29.43}{2}(5 - y)\, dy = -\frac{29.43}{2} \left\{ -\frac{1}{2}(5 - y)^2 \right\}_{5}^{0} = 183.9 \text{ J.}$$

10. The force of gravity on the disc of water shown is

$$F_g = -9.81(1000)\pi x^2\, dy = -9810\pi \left(\frac{4y}{3}\right) dy \text{ N.}$$

The work that an equal and opposite force would do in raising this disc to a level 2 m above the top of the tank is

$$(14 - y)9810\pi \left(\frac{4y}{3}\right) dy \text{ J.}$$

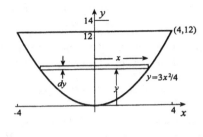

The total work to empty the tank is therefore

$$W = \int_0^{12} (14 - y)9810\pi \left(\frac{4y}{3}\right) dy = \frac{4(9810)\pi}{3}\left\{7y^2 - \frac{y^3}{3}\right\}_0^{12} = 1.78 \times 10^7 \text{ J.}$$

12. The force of gravity on a slab of water dy m thick is

$$-9.81(1000)2x(5)dy = -98\,100y\, dy \text{ N.}$$

To lift this slab to a height 2 m above the trough requires $(3 - y)98\,100y\, dy$ J of work. Hence the work required to empty the trough is

$$W = \int_0^1 98\,100y(3 - y)\, dy$$

$$= 98\,100\left\{\frac{3y^2}{2} - \frac{y^3}{3}\right\}_0^1 = 114\,450 \text{ J.}$$

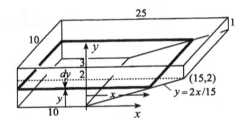

14. The force of gravity on slabs of water dy m thick for $0 \le y \le 2$ is

$$-9.81(1000)(10)(x + 10)\, dy$$

$$= -98\,100\left(\frac{15y}{2} + 10\right) dy \text{ N.}$$

For slabs above $y = 2$, the force is

$$-9.81(1000)(10)(25)\, dy$$

$$= -2\,452\,500\, dy \text{ N.}$$

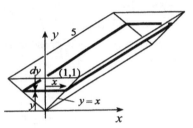

The work to empty the pool over its edge is therefore

$$W = \int_0^2 98\,100\left(\frac{15y}{2} + 10\right)(3 - y)\, dy + \int_2^3 2\,452\,500(3 - y)\, dy$$

$$= 245\,250\int_0^2 (12 + 5y - 3y^2)\, dy + 2\,452\,500\int_2^3 (3 - y)\, dy$$

$$= 245\,250\left\{12y + \frac{5y^2}{2} - y^3\right\}_0^2 + 2\,452\,500\left\{3y - \frac{y^2}{2}\right\}_2^3 = 7.60 \times 10^6 \text{ J.}$$

16. When q_3 is at position x, the total force on it due to q_1 and q_2 is

$$F(x) = \frac{-q_1 q_3}{4\pi\epsilon_0(5 - x)^2} + \frac{q_2 q_3}{4\pi\epsilon_0(x + 2)^2}.$$

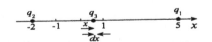

The work done by this force as q_3 moves from $x = 1$ to $x = -1$ is

$$W = \int_1^{-1}\left[\frac{-q_1 q_3}{4\pi\epsilon_0(5 - x)^2} + \frac{q_2 q_3}{4\pi\epsilon_0(x + 2)^2}\right] dx = \frac{-q_1 q_3}{4\pi\epsilon_0}\left\{\frac{1}{5 - x}\right\}_1^{-1} + \frac{q_2 q_3}{4\pi\epsilon_0}\left\{\frac{-1}{x + 2}\right\}_1^{-1}$$

$$= \frac{q_1 q_3}{48\pi\epsilon_0} - \frac{q_2 q_3}{6\pi\epsilon_0} = \frac{q_3}{48\pi\epsilon_0}(q_1 - 8q_2).$$

18. When P is moved x m to the right, the
resultant force of the two springs on P is

$$-k(1+x) + k(1-x) = -2kx \text{ N}.$$

The work done by an equal and opposite force
in moving P a distance b m to the right is

$$W = \int_0^b 2kx \, dx = \left\{kx^2\right\}_0^b = kb^2 \text{ J}.$$

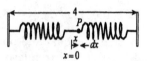

20. When the bucket has been raised y m $(y < 50)$,
the force of gravity on what remains in the bucket
and the hanging cable is

$$-9.81[(100 - y/5) + 5(100 - y)] = -9.81(600 - 26y/5) \text{ N}.$$

When $y \geq 50$, the force of gravity on bucket,
cable and pigeon is

$$-9.81[(100 - y/5) + 5(100 - y) + 2 - (y - 50)]$$
$$= -9.81(652 - 31y/5) \text{ N}.$$

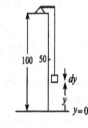

The work to overcome these forces is

$$W = \int_0^{50} 9.81 \left(600 - \frac{26y}{5}\right) dy + \int_{50}^{100} 9.81 \left(652 - \frac{31y}{5}\right) dy$$

$$= 9.81 \left\{600y - \frac{13y^2}{5}\right\}_0^{50} + 9.81 \left\{652y - \frac{31y^2}{10}\right\}_{50}^{100} = 3.22 \times 10^5 \text{ J}.$$

22. (a) Assuming that the earth is a sphere

$$F = \frac{6.67 \times 10^{-11}(m)(4/3)\pi(6.37 \times 10^6)^3(5.52 \times 10^3)}{(6.37 \times 10^6)^2} = 9.82m \text{ N}.$$

(b) The work required is $W = \displaystyle\int_{6.37 \times 10^6}^{6.38 \times 10^6} \frac{G(10)M}{r^2} dr = 10GM \left\{-\frac{1}{r}\right\}_{6.37 \times 10^6}^{6.38 \times 10^6}$

$$= 10(6.67 \times 10^{-11})\frac{4}{3}\pi(6.37 \times 10^6)^3(5.52 \times 10^3)\left(\frac{-1}{6.38 \times 10^6} + \frac{1}{6.37 \times 10^6}\right)$$

$$= 9.8087 \times 10^5 \text{ J}.$$

(c) With a constant $F = 9.82m$, $\quad W = \displaystyle\int_{6.37 \times 10^6}^{6.38 \times 10^6} 9.82(10) \, dy = 9.82 \times 10^5 \text{ J}.$

24. At position x, the force exerted by the piston is

$$F(x) = PA = \frac{C}{V}A = \frac{CA}{A(10 - x)} = \frac{C}{10 - x}.$$

The work done is therefore

$$W = \int_0^5 \frac{C}{10 - x} dx = C\left\{-\ln|10 - x|\right\}_0^5$$
$$= C(-\ln 5 + \ln 10) = C \ln 2.$$

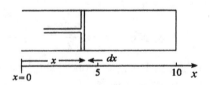

26. Using points A and D we obtain $k_2 = 12\,000$ and $k_1 = 20\,000$. Since the work is the area bounded by
the curves

$$W = \int_{1/5}^{3/5} \left(\frac{k_1}{V} - \frac{k_2}{V}\right) dV = (k_1 - k_2)\left\{\ln V\right\}_{1/5}^{3/5} = 8.8 \times 10^3 \text{ J}.$$

28. Using points B and C we obtain $k_2 = 4.64$ and $k_1 = 6.89$. Since the work is the area bounded by the curves

$$W = \int_{2\times 10^{-4}}^{8\times 10^{-4}} \left(\frac{k_1}{V^{1/4}} - \frac{k_2}{V^{1.4}} \right) dV = (k_1 - k_2) \left\{ \frac{-1}{0.4 V^{0.4}} \right\}_{2\times 10^{-4}}^{8\times 10^{-4}} = 72 \text{ J.}$$

30. Using points A and D we obtain $k_2 = 2.62$ and $k_1 = 32.2$. Since the work is the area bounded by the curves

$$W = \int_{150\,000}^{1\,040\,000} \left[\left(\frac{k_1}{P} \right)^{5/7} - \left(\frac{k_2}{P} \right)^{5/7} \right] dP = (k_1^{5/7} - k_2^{5/7}) \left\{ \frac{7}{2} P^{2/7} \right\}_{150\,000}^{1\,040\,000} = 7.8 \times 10^2 \text{ J.}$$

32. After time t, the mass of liquid remaining in the drop is $M - mt$, so that it completely disappears after time $t = M/m$. If we choose y as positive downward, then the work done by gravity in a small distance dy is $g(M - mt)\,dy$. Total work done is therefore

$$W = \int_0^{t=M/m} g(M - mt)\,dy = \int_0^{M/m} g(M - mt)\frac{dy}{dt}\,dt.$$

To find the velocity $v = dy/dt$ of the drop, we use Newton's second law in the form $F = d[(M - mt)v]/dt$. It requires

$$(M - mt)g = \frac{d}{dt}[(M - mt)v].$$

Integration gives

$$-\frac{g}{2m}(M - mt)^2 = (M - mt)v + C.$$

Since the initial velocity at time $t = 0$ is $v = 0$, we find that $C = -M^2 g/(2m)$, and therefore

$$-\frac{g}{2m}(M - mt)^2 = (M - mt)v - \frac{M^2 g}{2m} \quad \Longrightarrow \quad v = \frac{dy}{dt} = -\frac{g}{2m}(M - mt) + \frac{M^2 g}{2m(M - mt)}.$$

The work done by gravity can now be calculated

$$W = \int_0^{M/m} g(M - mt) \left[-\frac{g}{2m}(M - mt) + \frac{M^2 g}{2m(M - mt)} \right] dt$$

$$= \frac{g^2}{2m} \int_0^{M/m} [-(M - mt)^2 + M^2]\,dt = \frac{g^2}{2m} \left\{ \frac{(M - mt)^3}{3m} + M^2 t \right\}_0^{M/m} = \frac{g^2 M^3}{3m^2} \text{ J.}$$

EXERCISES 7.5

2. (a) If x represents the stretch in the spring, then the sum of the potential energy in the spring and the kinetic energy of the mass must always be constant, $C = kx^2/2 + mv^2/2$. Initially, $C = kx_0^2/2 + mv_0^2/2$. Consequently, x and v are related thereafter by the equation $kx_0^2/2 + mv_0^2/2 = kx^2/2 + mv^2/2 \Longrightarrow kx^2 + mv^2 = kx_0^2 + mv_0^2$.

(b) <u>Maximum stretch</u> occurs when $v = 0$ in which case $kx^2 = kx_0^2 + mv_0^2$. This equation can be solved for $x = \sqrt{x_0^2 + mv_0^2/k}$.

(c) <u>Maximum speed</u> occurs when $x = 0$ in which case $mv^2 = kx_0^2 + mv_0^2$. This equation can be solved for $v = \sqrt{v_0^2 + kx_0^2/m}$.

4. At position x, the force of repulsion on q is $qQ/(4\pi\epsilon_0 x^2)$. The gain in potential energy as q moves from $x = r$ to $x = r/2$ is the work done by an equal and opposite force in causing the motion,

$$W = \int_r^{r/2} \frac{-qQ}{4\pi\epsilon_0 x^2}\,dx = \left\{ \frac{qQ}{4\pi\epsilon_0 x} \right\}_r^{r/2} = \frac{qQ}{4\pi\epsilon_0 r}.$$

6. (a) The ultimate compression x occurs when the energy stored in the spring is equal to the original gravitational potential energy of the mass relative to this position,

$$\frac{1}{2}kx^2 = 9.81mx \implies x = \frac{19.62m}{k}.$$

(b) During the oscillations of the mass, the sum of spring potential energy, gravitational potential energy, and kinetic energy will be constant. If we equate initial values of these energies (taking $x = 0$ at the uncompressed position of the spring, and x positive upward) and values at maximum compression, denoted by x,

$$\frac{1}{2}mv_0^2 = \frac{1}{2}kx^2 + 9.81mx \implies kx^2 + 19.62mx - mv_0^2 = 0.$$

Solutions of this quadratic equation are $x = \dfrac{-19.62m \pm \sqrt{(19.62m)^2 + 4kmv_0^2}}{2k}$. Maximum compression of the spring is therefore $\left(9.81m + \sqrt{(9.81m)^2 + kmv_0^2}\right)/k$.

8. (a) The magnitude of the force of attraction on the mass at distance x from the centre of the earth is GmM/x^2. The work done by a force equal and opposite to this in raising the mass from the earth's surface to height 10^5 m is $W = \displaystyle\int_{6.37\times10^6}^{6.47\times10^6} \frac{GmM}{x^2}\,dx = GmM\left\{-\frac{1}{x}\right\}_{6.37\times10^6}^{6.47\times10^6} = 9.67\times10^6$ J.

(b) If the mass is dropped from this height, this gravitational potential energy is converted into kinetic energy. If it strikes the earth with speed v, then

$$\frac{1}{2}mv^2 = 9.67\times10^6 \implies v = \sqrt{\frac{2(9.67\times10^6)}{\cdot\,10}} = 1.4\times10^3 \text{ m/s}.$$

10. When the right end of the spring is moved a distance x, then $x_1 + x_2 = x$, where $x_2/x_1 = k_1/k_2$. These equations imply that $x_1 = k_2x/(k_1 + k_2)$, and $x_2 = k_1x/(k_1 + k_2)$. Since the force necessary to maintain a combined stretch x is $F(x) = k_1x_1 + k_2x_2 = \dfrac{2k_1k_2x}{k_1 + k_2}$, the work to produce total stretch L is

$$W = \int_0^L \frac{2k_1k_2x}{k_1 + k_2}\,dx = \frac{2k_1k_2}{k_1 + k_2}\left\{\frac{x^2}{2}\right\}_0^L = \frac{k_1k_2L^2}{k_1 + k_2}.$$

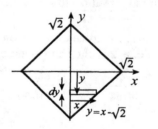

EXERCISES 7.6

2. Since the force on the representative rectangle is $9.81(1000)(180 - y)2x\,dy$ N, the total force on the dam is

$$F = \int_0^{180} 19\,620(180 - y)(45\times10^6)^{1/4}y^{1/4}\,dy$$

$$= 196\,200(4500)^{1/4}\left\{144y^{5/4} - \frac{4}{9}y^{9/4}\right\}_0^{180}$$

$$= 6.78\times10^{10} \text{ N.}$$

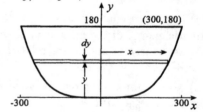

4. The force on the rectangle is $9810(1000)(-y)x\,dy$ N. The total force on the plate is

$$F = 2\int_{-\sqrt{2}}^0 9810(-y)(y + \sqrt{2})\,dy$$

$$= -19\,620\left\{\frac{y^3}{3} + \frac{\sqrt{2}y^2}{2}\right\}_{-\sqrt{2}}^0 = 9.25\times10^3 \text{ N.}$$

6. The force on the deep end of the pool (left figure below) is

$$F = \int_0^3 9810(3-y)(10)\,dy = 98\,100\left\{3y - \frac{y^2}{2}\right\}_0^3 = 4.41 \times 10^5 \text{ N.}$$

The force on the shallow end (right figure) is

$$F = \int_0^1 9810(1-y)(10)\,dy = 98\,100\left\{y - \frac{y^2}{2}\right\}_0^1 = 4.91 \times 10^4 \text{ N.}$$

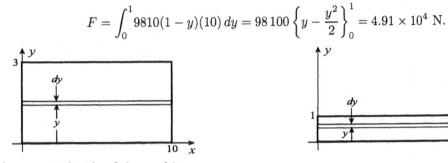

The force on each side of the pool is

$$F = \int_0^2 9810(3-y)(x+10)\,dy + \int_2^3 9810(3-y)(25)\,dy$$

$$= 9810\int_0^2 (3-y)\left(\frac{15y}{2} + 10\right)\,dy + 25(9810)\int_2^3 (3-y)\,dy$$

$$= \frac{5(9810)}{2}\int_0^2 (12 + 5y - 3y^2)\,dy + 25(9810)\int_2^3 (3-y)\,dy$$

$$= \frac{5(9810)}{2}\left\{12y + \frac{5y^2}{2} - y^3\right\}_0^2 + 25(9810)\left\{3y - \frac{y^2}{2}\right\}_2^3$$

$$= 7.60 \times 10^5 \text{ N.}$$

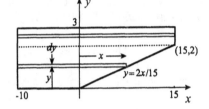

8. Since the force on the tiny rectangle shown is $9.81\rho(-y)(h\,dy)$ N, the force on the long rectangle is

$$F = \int_{y_1}^{y_2} -9.81\rho h y\,dy = -9.81\rho h\left\{\frac{y^2}{2}\right\}_{y_1}^{y_2}$$

$$= \frac{9.81\rho h}{2}(y_1^2 - y_2^2) \text{ N.}$$

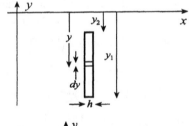

10. The force on the rectangle is

$$9.81\rho(b-y)(x_1 - x_2)\,dy$$

$$= 9.81\rho(b-y)\left[\frac{cy}{b} - \left(\frac{c-a}{b}\right)y\right]\,dy$$

$$= \frac{9.81\rho a}{b}(b-y)y\,dy \text{ N.}$$

The total force on the plate is

$$F = \int_0^b \frac{9.81\rho a}{b}(by - y^2)\,dy = \frac{9.81\rho a}{b}\left\{\frac{by^2}{2} - \frac{y^3}{3}\right\}_0^b = \frac{9.81\rho ab^2}{6} \text{ N.}$$

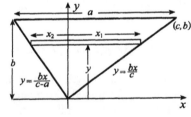

12. The force on the horizontal part of the bottom is the weight of the water directly above it:

$$9810(10)(3)(10) = 2.943 \times 10^6 \text{ N}.$$

For the slanted part of the bottom we notice
that differential dy gives rise to a rectangle
of width $\sqrt{229}dy/2$ across the bottom. The
force on this part is therefore

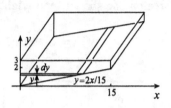

$$F = \int_0^2 9810(3-y)(10)\left(\frac{\sqrt{229}}{2}dy\right)$$

$$= 5\sqrt{229}(9810)\left\{3y - \frac{y^2}{2}\right\}_0^2 = 2.969 \times 10^6 \text{ N}.$$

14. If lengths of AB and BC are L m, the force on face $ABCD$ is

$$F = \int_0^L 9810(L-y)L\,dy$$

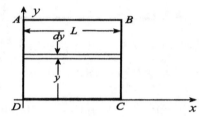

$$= 9810L\left\{Ly - \frac{y^2}{2}\right\}_0^L = 4905L^3 \text{ N}.$$

When we set this equal to 20 000 and solve,
the result is $L = (20\,000/4905)^{1/3} = 1.60$ m.

16. If the top of the cylinder is at depth d, then the magnitude of the force on this face is $9.81\rho d(\pi r^2)$. Since the force on the bottom is $9.81\rho(d+h)\pi r^2$, the resultant vertical force on the cylinder is $9.81\rho(d+h)\pi r^2 - 9.82\rho d(\pi r^2) = 9.81\rho(\pi r^2 h)$. This is the weight of the fluid displaced by the cylinder.

18. (a) Suppose V and V' are the volumes of the object above and below the surface respectively. The buoyant force on the object is therefore $9.81\rho_w V'$. Because the object is floating at this position, this force is equal to that of gravity on the object;

$$9.81\rho_w V' = 9.81\rho_o(V + V').$$

Thus, $\dfrac{V'}{V + V'} = \dfrac{\rho_o}{\rho_w}$.   The percentage of
the volume of the object above water is

$$100\frac{V}{V + V'} = 100\left(1 - \frac{V'}{V + V'}\right) = 100\left(1 - \frac{\rho_o}{\rho_w}\right) = 100\frac{\rho_w - \rho_o}{\rho_w}.$$

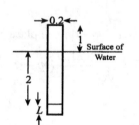

(b) For an iceberg, this percentage is $100\dfrac{1000 - 915}{1000} = 8.5$.

20. Suppose L is the length of the concrete
attachment. Archimedes' principle states
that the weight of the buoy must be equal
to the weight of water displaced by the
buoy,

$$\pi(0.1)^2(3)(500)g + \pi(0.1)^2(L)(3000)g =$$
$$\pi(0.1)^2(2 + L)(1000)g.$$

The solution of this equation is $L = 1/4$ m.

22. (a) The pressure at A due to the mercury in the tube is equal to the weight of a column of mercury of unit cross-sectional area and height h. With density of mercury equal to 13.6×10^3 kg/m^3, this weight is $(9.81)(13.6 \times 10^3)h = 1.33 \times 10^5 h$ N/m^2.

(b) When $h = .761$, atmospheric pressure is $(1.33 \times 10^5)(.761) = 1.01 \times 10^5$ N/m^2.

24. The force on the end of the tank when the radius is r is

$$F = \int_{-r}^{r} 9.81\rho(r-y)2\sqrt{r^2-y^2}\,dy$$

$$= 19.62\rho r \int_{-r}^{r} \sqrt{r^2-y^2}\,dy + 19.62\rho \left\{\frac{1}{3}(r^2-y^2)^{3/2}\right\}_{-r}^{r}.$$

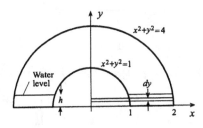

Since the integral represents half the area of the end of the tank,

$$F = 19.62\rho r \left(\frac{1}{2}\pi r^2\right) = 9.81\pi\rho r^3.$$

The radius of the end of the tank must satisfy $40\,000 - 2\pi r(1000) = 9.81\pi\rho r^3$. We use Newton's iterative procedure with initial approximation $r = 1$ to solve this equation,

$$r_1 = 1, \qquad r_{n+1} = r_n - \frac{9.81(1019)\pi r_n^3 - 2000\pi r_n - 40\,000}{29.43(1019)\pi r_n^2 - 2000\pi}.$$

The result is $r = 1.02$ m.

26. (a) If the shell is placed carefully on the water, no water will penetrate the hemispherical cavity. If the depth of the flat edge is h (figure to the right), then the volume of water displaced by the shell is

$$V = \int_0^h \pi x^2\,dy = \pi \int_0^h (4-y^2)\,dy$$

$$= \pi \left\{4y - \frac{y^3}{3}\right\}_0^h = \pi \left(4h - \frac{h^3}{3}\right)$$

$$= \frac{\pi h(12 - h^2)}{3}.$$

According to Archimedes' principle, the weight of water displaced must be equal to the weight of the shell; that is,

$$9.81\left(\frac{2}{3}\pi\right)(2^3 - 1^3)(2) = 9.81(1000)\frac{\pi h(12 - h^2)}{3}.$$

This equation reduces to $250h^3 - 3000h + 7 = 0$. With Newton's iterative procedure

$$h_1 = 0.0025, \qquad h_{n+1} = h_n - \frac{250h_n^3 - 3000h_n + 7}{750h_n^2 - 3000},$$

we obtain $h = 0.00233$. Hence, the shell sinks 2.33 mm.

(b) With a hole in the top of the shell, air will escape from the cavity and the water level inside the cavity will be the same as outside; only the water displaced by the shell itself is taken into account. The volume of water displaced is

$$V = \int_0^h \pi[(4 - y^2) - (1 - y^2)]\,dy = \pi \int_0^h 3\,dy = 3\pi h.$$

In this case, Archimedes' principle requires

$$9.81\left(\frac{2}{3}\pi\right)(2^3 - 1^3)(2) = 9.81(1000)3\pi h,$$

and therefore $h = 0.00311$. The shell now sinks 3.11 mm.

EXERCISES 7.7

2. $M = \int_0^a \rho \left(h - \frac{hx^2}{a^2} \right) dx$

$\qquad = \rho h \left\{ x - \frac{x^3}{3a^2} \right\}_0^a = \frac{2\rho a h}{3}$

Since $M\bar{x} = \int_0^a x\rho \left(h - \frac{hx^2}{a^2} \right) dx$

$\qquad\qquad = \rho h \left\{ \frac{x^2}{2} - \frac{x^4}{4a^2} \right\}_0^a = \frac{\rho h a^2}{4}$,

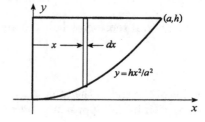

it follows that $\bar{x} = \frac{\rho h a^2}{4} \frac{3}{2\rho a h} = \frac{3a}{8}$. Since

$$M\bar{y} = \int_0^a \rho \left(\frac{h + hx^2/a^2}{2} \right) \left(h - \frac{hx^2}{a^2} \right) dx = \frac{\rho h^2}{2} \int_0^a \left(1 - \frac{x^4}{a^4} \right) dx = \frac{\rho h^2}{2} \left\{ x - \frac{x^5}{5a^4} \right\}_0^a = \frac{2\rho a h^2}{5},$$

we find $\bar{y} = \frac{2\rho a h^2}{5} \frac{3}{2\rho a h} = \frac{3h}{5}$.

4. The mass of the plate is $M = \rho \pi r^2/4$. Since

$\qquad M\bar{x} = \int_0^r x\rho y \, dx = \rho \int_0^r x\sqrt{r^2 - x^2} \, dx$

$\qquad\qquad = \rho \left\{ -\frac{1}{3}(r^2 - x^2)^{3/2} \right\}_0^r = \frac{\rho r^3}{3}$,

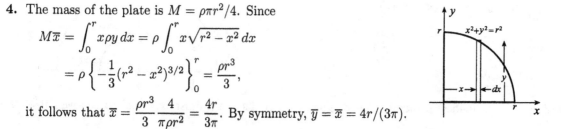

it follows that $\bar{x} = \frac{\rho r^3}{3} \frac{4}{\pi \rho r^2} = \frac{4r}{3\pi}$. By symmetry, $\bar{y} = \bar{x} = 4r/(3\pi)$.

6. $A = \int_{-1}^0 (y_2 - y_1) \, dx = \int_{-1}^0 [-(x+1)^2 - (x^2 - 1)] \, dx = \left\{ -\frac{1}{3}(x+1)^3 - \frac{x^3}{3} + x \right\}_{-1}^0 = \frac{1}{3}$

Since $A\bar{x} = \int_{-1}^0 x(y_2 - y_1) \, dx = \int_{-1}^0 x[-(x+1)^2 - (x^2 - 1)] \, dx$

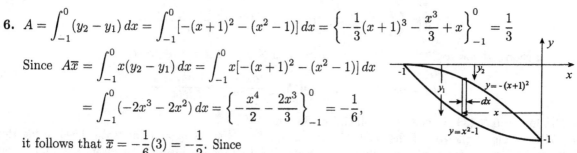

$\qquad = \int_{-1}^0 (-2x^3 - 2x^2) \, dx = \left\{ -\frac{x^4}{2} - \frac{2x^3}{3} \right\}_{-1}^0 = -\frac{1}{6}$,

it follows that $\bar{x} = -\frac{1}{6}(3) = -\frac{1}{2}$. Since

$\qquad A\bar{y} = \int_{-1}^0 \left(\frac{y_1 + y_2}{2} \right) (y_2 - y_1) \, dx = \frac{1}{2} \int_{-1}^0 (y_2^2 - y_1^2) \, dx$

$\qquad = \frac{1}{2} \int_{-1}^0 [(x+1)^4 - (x^4 - 2x^2 + 1)] \, dx = \frac{1}{2} \left\{ \frac{1}{5}(x+1)^5 - \frac{x^5}{5} + \frac{2x^3}{3} - x \right\}_{-1}^0 = -\frac{1}{6}$,

we find $\bar{y} = -3/6 = -1/2$.

8. By symmetry, $\bar{x} = \bar{y} = 0$.

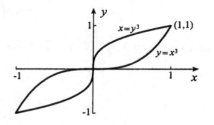

10. $A = \displaystyle\int_{-3}^{4} (x_2 - x_1)\, dy = \int_{-3}^{4} [(12 - y) - (y^2 - 2y)]\, dy$

$\qquad = \left\{ 12y + \dfrac{y^2}{2} - \dfrac{y^3}{3} \right\}_{-3}^{4} = \dfrac{343}{6}$

Since $A\bar{x} = \displaystyle\int_{-3}^{4} \left(\dfrac{x_2 + x_1}{2} \right)(x_2 - x_1)\, dy = \dfrac{1}{2} \int_{-3}^{4} (x_2^2 - x_1^2)\, dy$

$\qquad = \dfrac{1}{2} \displaystyle\int_{-3}^{4} [(12 - y)^2 - (y^4 - 4y^3 + 4y^2)]\, dy = \dfrac{1}{2} \left\{ -\dfrac{1}{3}(12 - y)^3 - \dfrac{y^5}{5} + y^4 - \dfrac{4y^3}{3} \right\}_{-3}^{4} = \dfrac{3773}{10},$

it follows that $\bar{x} = (3773/10)(6/343) = 33/5$. Since

$$A\bar{y} = \int_{-3}^{4} y(x_2 - x_1)\, dy = \int_{-3}^{4} y(12 + y - y^2)\, dy = \left\{ 6y^2 + \dfrac{y^3}{3} - \dfrac{y^4}{4} \right\}_{-3}^{4} = \dfrac{343}{12},$$

we find $\bar{y} = (343/12)(6/343) = 1/2$.

12. $A = \displaystyle\int_{0}^{1} (x_2 - x_1)\, dy = \int_{0}^{1} [(y + 3) - (4y - 4y^2)]\, dy$

$\qquad = \left\{ 3y - \dfrac{3y^2}{2} + \dfrac{4y^3}{3} \right\}_{0}^{1} = \dfrac{17}{6}$

Since $A\bar{x} = \displaystyle\int_{0}^{1} \left(\dfrac{x_2 + x_1}{2} \right)(x_2 - x_1)\, dy = \dfrac{1}{2} \int_{0}^{1} (x_2^2 - x_1^2)\, dy$

$\qquad = \dfrac{1}{2} \displaystyle\int_{0}^{1} [(y + 3)^2 - (16y^2 - 32y^3 + 16y^4)]\, dy = \dfrac{1}{2} \left\{ \dfrac{1}{3}(y + 3)^3 - \dfrac{16y^3}{3} + 8y^4 - \dfrac{16y^5}{5} \right\}_{0}^{1} = \dfrac{59}{10},$

we find that $\bar{x} = (59/10)(6/17) = (177/85)$. Because

$$A\bar{y} = \int_{0}^{1} y(x_2 - x_1)\, dy = \int_{0}^{1} y(3 - 3y + 4y^2)\, dy = \left\{ \dfrac{3y^2}{2} - y^3 + y^4 \right\}_{0}^{1} = \dfrac{3}{2},$$

it follows that $\bar{y} = (3/2)(6/17) = (9/17)$.

14. Since the centre of mass is on the y-axis, the first moment is $5M$ where M is the mass of the plate. Because the plate is symmetric about the y-axis, we find the mass of the right half and double the result:

$\text{Moment} = 5(2) \displaystyle\int_{0}^{1} \rho[(2 - x) - \sqrt{x}]\, dx$

$\qquad = 10\rho \left\{ 2x - \dfrac{x^2}{2} - \dfrac{2}{3}x^{3/2} \right\}_{0}^{1} = \dfrac{25\rho}{3}.$

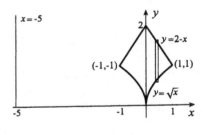

16. The mass of the seesaw itself is $2\rho L$, and this mass may be considered to act at its centre of mass, the midpoint of the seesaw. For balance to occur, the total first moment about the fulcrum must vanish:

$$0 = \sum_{i=1}^{6} m_i(x_i - \bar{x}) + 2\rho L(L - \bar{x}).$$

If we set $M = \displaystyle\sum_{i=1}^{6} m_i$, then $\bar{x}(M + 2\rho L) = \displaystyle\sum_{i=1}^{6} m_i x_i + 2\rho L^2$, and therefore

$$\bar{x} = \frac{1}{M + 2\rho L} \left(\sum_{i=1}^{6} m_i x_i + 2\rho L^2 \right).$$

18. $A = 2 \int_0^1 (y_2 - y_1)\, dx + 2 \int_1^4 (y_2 - y_1)\, dx$

$\qquad = 2 \int_0^1 (2 + x^2)\, dx + 2 \int_1^4 [2 - (x - 2)]\, dx$

$\qquad = 2 \left\{ 2x + \dfrac{x^3}{3} \right\}_0^1 + 2 \left\{ 4x - \dfrac{x^2}{2} \right\}_1^4 = \dfrac{41}{3}.$

By symmetry, $\bar{x} = 0$, and because

$A\bar{y} = 2 \int_0^1 \left(\dfrac{y_2 + y_1}{2} \right) (y_2 - y_1)\, dx + 2 \int_1^4 \left(\dfrac{y_2 + y_1}{2} \right) (y_2 - y_1)\, dx$

$\qquad = \int_0^1 (y_2^2 - y_1^2)\, dx + \int_1^4 (y_2^2 - y_1^2)\, dx = \int_0^1 (4 - x^4)\, dx + \int_1^4 [4 - (x - 2)^2]\, dx$

$\qquad = \left\{ 4x - \dfrac{x^5}{5} \right\}_0^1 + \left\{ 4x - \dfrac{1}{3}(x - 2)^3 \right\}_1^4 = \dfrac{64}{5},$

it follows that $\bar{y} = \dfrac{64}{5} \dfrac{3}{41} = \dfrac{192}{205}.$

20. If ρ is the mass per unit area,

$\qquad M = 2 \int_0^2 \rho \left(1 - \dfrac{y^2}{4} \right) dy + 10\rho + 2 \int_0^4 \rho \left(8 - \dfrac{y^2}{8} - 6 \right) dy$

$\qquad = 2\rho \left\{ y - \dfrac{y^3}{12} \right\}_0^2 + 10\rho + 2\rho \left\{ 2y - \dfrac{y^3}{24} \right\}_0^4 = \dfrac{70\rho}{3}.$

Clearly $\bar{y} = 0$, and because

$M\bar{x} = 2 \int_0^2 \dfrac{1}{2} \left(1 + \dfrac{y^2}{4} \right) \rho \left(1 - \dfrac{y^2}{4} \right) dy + 10\rho \left(\dfrac{7}{2} \right) + 2 \int_0^4 \dfrac{1}{2} \left(8 - \dfrac{y^2}{8} + 6 \right) \rho \left(8 - \dfrac{y^2}{8} - 6 \right) dy$

$\qquad = \rho \int_0^2 \left(1 - \dfrac{y^4}{16} \right) dy + 35\rho + \rho \int_0^4 \left(28 - 2y^2 + \dfrac{y^4}{64} \right) dy = \rho \left\{ y - \dfrac{y^5}{80} \right\}_0^2 + 35\rho + \rho \left\{ 28y - \dfrac{2y^3}{3} + \dfrac{y^5}{320} \right\}_0^4$

$\qquad = \dfrac{1637\rho}{15},$

it follows that $\bar{x} = (1637\rho/15)[3/(70\rho)] = 1637/350.$

22. We find the centre of mass of the plate. Since $M = 2 \int_0^1 \rho(4 - 2x^2 - 2x)\, dx = 2\rho \left\{ 4x - \dfrac{2x^3}{3} - x^2 \right\}_0^1 =$

$\dfrac{14\rho}{3}$, and

$M\bar{y} = 2 \int_0^1 \dfrac{\rho}{2}(4 - 2x^2 + 2x)(4 - 2x^2 - 2x)\, dx = 4\rho \int_0^1 (4 - 5x^2 + x^4)\, dx$

$\qquad = 4\rho \left\{ 4x - \dfrac{5x^3}{3} + \dfrac{x^5}{5} \right\}_0^1 = \dfrac{152\rho}{15},$

it follows that $\bar{y} = \dfrac{152\rho}{15} \cdot \dfrac{3}{14\rho} = \dfrac{76}{35}.$ Symmetry of the plate

indicates that $\bar{x} = 0$. If we concentrate the mass at its centre
of mass, and use formula 1.16 for the distance from the
centre of mass to the line $y = x$, we obtain for the
required moment

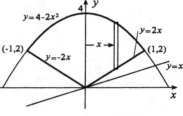

$\qquad\qquad \left(\dfrac{14\rho}{3} \right) \dfrac{|76/35 - 0|}{\sqrt{2}} = \dfrac{76\sqrt{2}\rho}{15}.$

24. We find the centre of mass of the plate. Since $M = \int_0^1 \rho(2 - y - y^3)\, dy = \rho\left\{2y - \dfrac{y^2}{2} - \dfrac{y^4}{4}\right\}_0^1 = \dfrac{5\rho}{4}$, and

$$M\bar{x} = \int_0^1 \frac{\rho}{2}(2 - y + y^3)(2 - y - y^3)\, dy = \frac{\rho}{2}\int_0^1 (4 - 4y + y^2 - y^6)\, dy$$

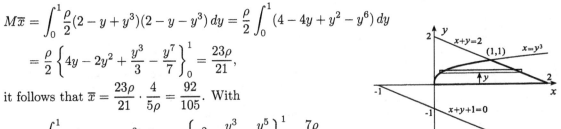

$$= \frac{\rho}{2}\left\{4y - 2y^2 + \frac{y^3}{3} - \frac{y^7}{7}\right\}_0^1 = \frac{23\rho}{21},$$

it follows that $\bar{x} = \dfrac{23\rho}{21} \cdot \dfrac{4}{5\rho} = \dfrac{92}{105}$. With

$$M\bar{y} = \int_0^1 \rho y(2 - y - y^3)\, dy = \rho\left\{y^2 - \frac{y^3}{3} - \frac{y^5}{5}\right\}_0^1 = \frac{7\rho}{15},$$

we find $\bar{y} = \dfrac{7\rho}{15} \cdot \dfrac{4}{5\rho} = \dfrac{28}{75}$. If we concentrate the mass at its centre of mass, and use formula 1.16 for the distance from the centre of mass to the line $x + y + 1 = 0$, we obtain for the required moment

$$\left(\frac{5\rho}{4}\right)\frac{|92/105 + 28/75 + 1|}{\sqrt{2}} = \frac{1181\sqrt{2}\rho}{840}.$$

26. The first moment of A about the y-axis is $\quad A\bar{x} = \displaystyle\sum_{i=1}^{n} A_i \bar{x}_i,$

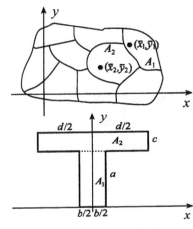

and therefore $\quad \bar{x} = \dfrac{1}{A}\displaystyle\sum_{i=1}^{n} A_i \bar{x}_i.$

Similarly, $\bar{y} = \dfrac{1}{A}\displaystyle\sum_{i=1}^{n} A_i \bar{y}_i.$

28. Symmetry gives $\bar{x} = 0$. If we divide the plate into two parts as shown, then

$$\bar{y} = \frac{1}{A}(A_1\bar{y}_1 + A_2\bar{y}_2) = \frac{1}{ab + cd}\left[\frac{a}{2}(ab) + \left(a + \frac{c}{2}\right)(cd)\right]$$

$$= \frac{a^2 b + cd(2a + c)}{2(ab + cd)}.$$

30. If we divide the plate into three parts as shown, then

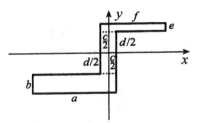

$$\bar{x} = \frac{1}{ab + cd + ef}\left[\left(\frac{c/2 - a + c/2}{2}\right)(ab)\right.$$

$$\left. + \left(\frac{f - c/2 - c/2}{2}\right)(ef)\right]$$

$$= \frac{ab(c - a) + ef(f - c)}{2(ab + cd + ef)};$$

and

$$\bar{y} = \frac{1}{ab + cd + ef}\left[\left(-\frac{d}{2} - \frac{b}{2}\right)(ab) + \left(\frac{d}{2} + \frac{e}{2}\right)(ef)\right] = \frac{ef(d + e) - ab(b + d)}{2(ab + cd + ef)}.$$

32. To six decimal places, the x-intercept of the curve is $a = -2.324\,718$. The area of the region is

$$A = \int_a^0 (x^3 + 3x^2 + 2x + 1)\, dx$$

$$= \left\{ \frac{x^4}{4} + x^3 + x^2 + x \right\}_a^0 = 2.182\,258.$$

Since $A\bar{x} = \int_a^0 x(x^3 + 3x^2 + 2x + 1)\, dx$

$$= \left\{ \frac{x^5}{5} + \frac{3x^4}{4} + \frac{2x^3}{3} + \frac{x^2}{2} \right\}_a^0 = -2.652\,017,$$

we get $\bar{x} = -2.652\,017/2.182\,258 = -1.215$. Since

$$A\bar{y} = \int_a^0 \frac{1}{2}(x^3 + 3x^2 + 2x + 1)^2\, dx = \frac{1}{2}\int_a^0 (x^6 + 6x^5 + 13x^4 + 14x^3 + 10x^2 + 4x + 1)\, dx$$

$$= \frac{1}{2}\left\{ \frac{x^7}{7} + x^6 + \frac{13x^5}{5} + \frac{7x^4}{2} + \frac{10x^3}{3} + 2x^2 + x \right\}_a^0 = 1.140\,899,$$

it follows that $\bar{y} = 1.140\,899/2.182\,258 = 0.523$.

34. The y-coordinates of the points of intersection are $a = -0.354740$ and $b = 2.310040$. The area of the region is

$$A = \int_a^b (\sqrt{2y + 1} - y^3 + y^2 + 2y)\, dy$$

$$= \left\{ \frac{1}{3}(2y + 1)^{3/2} - \frac{y^4}{4} + \frac{y^3}{3} + y^2 \right\}_a^b = 6.608\,233.$$

Since

$$A\bar{x} = \int_a^b \left(\frac{\sqrt{2y + 1} + y^3 - y^2 - 2y}{2} \right) (\sqrt{2y + 1} - y^3 + y^2 + 2y)\, dy$$

$$= \frac{1}{2}\int_a^b (1 + 2y - y^6 + 2y^5 + 3y^4 - 4y^3 - 4y^2)\, dy = \frac{1}{2}\left\{ y + y^2 - \frac{y^7}{7} + \frac{y^6}{3} + \frac{3y^5}{5} - y^4 - \frac{4y^3}{3} \right\}_a^b$$

$$= 1.448\,074,$$

it follows that $\bar{x} = 1.448\,074/6.608\,233 = 0.219$. If we set $u = 2y + 1$ and $du = 2\, dy$ in the first term of the following integral, then

$$A\bar{y} = \int_a^b y(\sqrt{2y + 1} - y^3 + y^2 + 2y)\, dy = \int_{2a+1}^{2b+1} \left(\frac{u - 1}{2} \right) \sqrt{u} \left(\frac{du}{2} \right) + \left\{ -\frac{y^5}{5} + \frac{y^4}{4} + \frac{2y^3}{3} \right\}_a^b$$

$$= \frac{1}{4}\left\{ \frac{2u^{5/2}}{5} - \frac{2u^{3/2}}{3} \right\}_{2a+1}^{2b+1} + \left\{ -\frac{y^5}{5} + \frac{y^4}{4} + \frac{2y^3}{3} \right\}_a^b = 7.494\,397.$$

Hence, $\bar{y} = 7.494\,397/6.608\,233 = 1.134$.

36. If we choose the y-axis along the axis of rotation, the volume generated is

$$V = \int_a^b 2\pi x[f(x) - g(x)]\, dx$$

$$= 2\pi \int_a^b x[f(x) - g(x)]\, dx$$

$$= 2\pi(A\bar{x}) = (2\pi\bar{x})A.$$

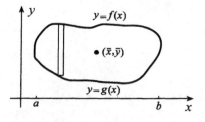

38. We find the centroid of the area. Since

$$A = \int_0^1 (1 - x^2)\, dx = \left\{ x - \frac{x^3}{3} \right\}_0^1 = \frac{2}{3},$$

and

$$A\overline{x} = \int_0^1 x(1 - x^2)\, dx = \left\{ \frac{x^2}{2} - \frac{x^4}{4} \right\}_0^1 = \frac{1}{4},$$

we find $\overline{x} = (1/4)(3/2) = 3/8$. Since

$$A\overline{y} = \int_0^1 \frac{1}{2}(1 - x^2)^2\, dx = \frac{1}{2} \int_0^1 (1 - 2x^2 + x^4)\, dx = \frac{1}{2} \left\{ x - \frac{2x^3}{3} + \frac{x^5}{5} \right\}_0^1 = \frac{4}{15},$$

we get $\overline{y} = (4/15)(3/2) = 2/5$. Using the result of Exercise 36 and distance formula 1.16, the volume is

$$V = 2\pi \left(\frac{|3/8 - 2/5 + 1|}{\sqrt{2}} \right) \left(\frac{2}{3} \right) = \frac{13\sqrt{2}\pi}{20}.$$

40. For the plate shown,

$$F = \int_a^b 9.81\rho(-y)[f(y) - g(y)]\, dy$$

$$= -9.81\rho \int_a^b y[f(y) - g(y)]\, dy$$

$$= -9.81\rho(A\overline{y}) = 9.81\rho A(-\overline{y}).$$

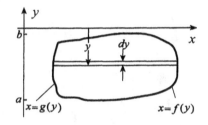

EXERCISES 7.8

2. $I = \displaystyle\int_{-4}^0 (-y)^2 \rho(x_2 - x_1)\, dy$

$$= \rho \int_{-4}^0 y^2 \left[y - \left(\frac{y-4}{2} \right) \right] dy = \frac{\rho}{2} \int_{-4}^0 (y^3 + 4y^2)\, dy$$

$$= \frac{\rho}{2} \left\{ \frac{y^4}{4} + \frac{4y^3}{3} \right\}_{-4}^0 = \frac{32\rho}{3}$$

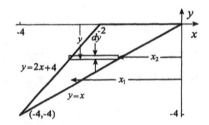

4. $I = \displaystyle\int_{-1}^2 (x + 2)^2 \rho(y_2 - y_1)\, dx$

$$= \rho \int_{-1}^2 (x + 2)^2 (2x - x^2 - x^2 + 4)\, dx$$

$$= \rho \int_{-1}^2 (16 + 24x + 4x^2 - 6x^3 - 2x^4)\, dx$$

$$= \rho \left\{ 16x + 12x^2 + \frac{4x^3}{3} - \frac{3x^4}{2} - \frac{2x^5}{5} \right\}_{-1}^2 = \frac{603\rho}{10}$$

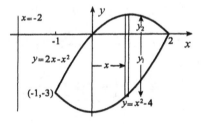

6. $I = 2 \displaystyle\int_0^1 x^2 \rho(y_2 - y_1)\, dx$

$$= 2\rho \int_0^1 x^2 (2 - x^{1/3} - x^{1/3})\, dx$$

$$= 4\rho \int_0^1 (x^2 - x^{7/3})\, dx$$

$$= 4\rho \left\{ \frac{x^3}{3} - \frac{3}{10}x^{10/3} \right\}_0^1 = \frac{2\rho}{15}$$

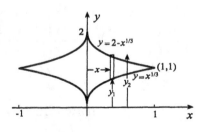

8. If we divide the long horizontal rectangle with width dy into tiny rectangles with length dx, then the moment of inertia of the long rectangle about the line $x = -1$ is

$$\int_{x_1}^{x_2} (x+1)^2 \rho dy\, dx = \rho\, dy \left\{ \frac{1}{3}(x+1)^3 \right\}_{x_1}^{x_2} = \frac{\rho}{3}[(x_2+1)^3 - (x_1+1)^3]\, dy.$$

The moment of inertia of the plate is

$$I = \frac{2\rho}{3} \int_0^1 [(1-y^2+1)^3 - (y^2-1+1)^3]\, dy$$

$$= \frac{2\rho}{3} \int_0^1 (8 - 12y^2 + 6y^4 - 2y^6)\, dy$$

$$= \frac{2\rho}{3} \left\{ 8y - 4y^3 + \frac{6y^5}{5} - \frac{2y^7}{7} \right\}_0^1 = \frac{344\rho}{105}$$

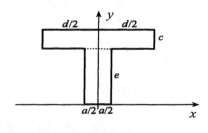

10. $I = \int_{-2}^{1} (y-3)^2 \rho(x_2-x_1)\, dy = \rho \int_{-2}^{1} (y-3)^2[(2-y)-y^2]\, dy$

$$= \rho \int_{-2}^{1} (18 - 21y - y^2 + 5y^3 - y^4)\, dy$$

$$= \rho \left\{ 18y - \frac{21y^2}{2} - \frac{y^3}{3} + \frac{5y^4}{4} - \frac{y^5}{5} \right\}_{-2}^{1} = \frac{1143\rho}{20}$$

12. The product of the mass and the square of the distance from the x-axis to the centre of mass of the rectangle is $\rho(y_2 - y_1)h \left(\dfrac{y_1 + y_2}{2} \right)^2$. This is not the same as 7.42. For example, if $y_1 = 0$, so that the rectangle has its base on the x-axis, then 7.42 gives $\rho h y_2^3/3$ whereas the above expression gives $\rho h y_2^3/4$.

14. If we divide the right half of the section into two parts as shown, and use formula 7.42 with $\rho = 1$, we obtain

$$I_x = \frac{2}{3} \left(\frac{d}{2} \right) [(c+e)^3 - e^3] + \frac{2}{3} \left(\frac{a}{2} \right) e^3$$

$$= \frac{1}{3}[ae^3 + cd(c^2 + 3ce + 3e^2),$$

and

$$I_y = 2 \left(\frac{c}{3} \right) \left(\frac{d}{2} \right)^3 + 2 \left(\frac{e}{3} \right) \left(\frac{a}{2} \right)^3 = \frac{1}{12}(cd^3 + ea^3),$$

16. If we divide the right half of the section into subareas as shown, and set $\rho = 1$ in formula 7.42,

$$I_x = \frac{2}{3} \left(\frac{c}{2} \right) \left[\left(d + \frac{a}{2} \right)^3 - \left(\frac{a}{2} \right)^3 \right] + \frac{4}{3} \left(\frac{d}{2} \right) \left(\frac{a}{2} \right)^3$$

$$+ \frac{2}{3} \left(\frac{b}{2} \right) \left[\left(d + \frac{a}{2} \right)^3 - \left(\frac{a}{2} \right)^3 \right]$$

$$= \frac{1}{24}[2a^3 d + (b+c)(6a^2 d + 12ad^2 + 8d^3)],$$

and

$$I_y = \frac{2}{3}(d) \left(\frac{c}{2} \right)^3 + \frac{4}{3} \left(\frac{a}{2} \right) \left(\frac{d}{2} \right)^3 + \frac{2}{3}(d) \left(\frac{b}{2} \right)^3 = \frac{d}{12}(ad^2 + b^3 + c^3).$$

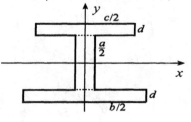

18. The mass of the plate is

$$M = \int_{-1}^{0} \rho(y_2 - y_1)\, dx + \int_{0}^{1} \rho(y_2 - y_1)\, dx$$

$$= \rho \int_{-1}^{0} \left(\frac{x+3}{2} + x^3 \right) dx + \rho \int_{0}^{1} \left(\frac{x+3}{2} - 2x^3 \right) dx$$

$$= \rho \left\{ \frac{(x+3)^2}{4} + \frac{x^4}{4} \right\}_{-1}^{0} + \rho \left\{ \frac{(x+3)^2}{4} - \frac{x^4}{2} \right\}_{0}^{1} = \frac{9\rho}{4}.$$

The moment of inertia about the x-axis is

$$I_x = \int_{-1}^{0} \frac{\rho}{3}(y_2^3 - y_1^3)\, dx + \int_{0}^{1} \frac{\rho}{3}(y_2^3 - y_1^3)\, dx$$

$$= \frac{\rho}{3} \int_{-1}^{0} \left[\left(\frac{x+3}{2} \right)^3 - (-x^3)^3 \right] dx + \frac{\rho}{3} \int_{0}^{1} \left[\left(\frac{x+3}{2} \right)^3 - (2x^3)^3 \right] dx$$

$$= \frac{\rho}{3} \left\{ \frac{(x+3)^4}{32} + \frac{x^{10}}{10} \right\}_{-1}^{0} + \frac{\rho}{3} \left\{ \frac{(x+3)^4}{32} - \frac{4x^{10}}{5} \right\}_{0}^{1} = \frac{11\rho}{5}.$$

If r_x is the moment of gyration about the x-axis, then $\dfrac{11\rho}{5} = \dfrac{9\rho}{4}r_x^2 \implies r_x = \sqrt{44/45}$. The moment of inertia about the y-axis is

$$I_y = \int_{-1}^{0} x^2 \rho \left(\frac{x+3}{2} + x^3 \right) dx + \int_{0}^{1} x^2 \rho \left(\frac{x+3}{2} - 2x^3 \right) dx$$

$$= \frac{\rho}{2} \int_{-1}^{0} (x^3 + 3x^2 + 2x^5)\, dx + \frac{\rho}{2} \int_{0}^{1} (x^3 + 3x^2 - 4x^5)\, dx$$

$$= \frac{\rho}{2} \left\{ \frac{x^4}{4} + x^3 + \frac{x^6}{3} \right\}_{-1}^{0} + \frac{\rho}{2} \left\{ \frac{x^4}{4} + x^3 - \frac{2x^6}{3} \right\}_{0}^{1} = \frac{\rho}{2}.$$

If r_y is the radius of gyration about the y-axis, then $\dfrac{\rho}{2} = \dfrac{9\rho}{4}r_y^2 \implies r_y = \sqrt{2}/3$. The radius of gyration is the distance from a line at which a single particle of mass equal to that of the plate has the same moment of inertia as the plate itself.

20. Let us take the coplanar line to be the y-axis.
The moment of inertia about the y-axis is

$$I_y = \int_{a}^{b} x^2 \rho[f(x) - g(x)]\, dx.$$

The moment of inertia about the line $x = \bar{x}$ is

$$I = \int_{a}^{b} (x - \bar{x})^2 \rho[f(x) - g(x)]\, dx$$

$$= \int_{a}^{b} x^2 \rho[f(x) - g(x)]\, dx - 2\bar{x} \int_{a}^{b} x\rho[f(x) - g(x)]\, dx + \bar{x}^2 \int_{a}^{b} \rho[f(x) - g(x)]\, dx$$

$$= I_y - 2\bar{x}(M\bar{x}) + \bar{x}^2 M.$$

Thus, $I_y = I + M\bar{x}^2$.

22. We divide the area A into vertical rectangles of width dx, and then divide this rectangle further into horizontal rectangles of width dy. Because the polar moment of inertia of the tiny rectangle is $(x^2 + y^2)\rho\, dy\, dx$, the polar moment of inertia of the vertical rectangle is

$$\int_{g(x)}^{f(x)} \left[(x^2 + y^2)\rho\, dx\right] dy = \left\{ \rho\, dx \left(x^2 y + \frac{y^3}{3} \right) \right\}_{g(x)}^{f(x)}$$

$$= \left\{ x^2 [f(x) - g(x)] + \frac{1}{3}\{[f(x)]^3 - [g(x)]^3\} \right\} \rho\, dx.$$

The polar moment of inertia of the plate is therefore

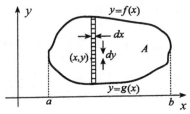

$$J_0 = \int_a^b \left\{ x^2[f(x) - g(x)] + \frac{1}{3}\{[f(x)]^3 - [g(x)]^3\} \right\} \rho\, dx$$

$$= \int_a^b x^2 \rho[f(x) - g(x)]\, dx + \int_a^b \frac{1}{3} \rho\{[f(x)]^3 - [g(x)]^3\}\, dx = I_x + I_y.$$

24. The x-coordinate of the point of intersection of
the curves is $a = 1.362\,599$. The moment of inertia is

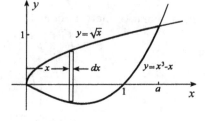

$$I = \int_0^a x^2 (2)(\sqrt{x} - x^3 + x)\, dx$$

$$= 2\int_0^a (x^{5/2} - x^5 + x^3)\, dx$$

$$= 2\left\{ \frac{2x^{7/2}}{7} - \frac{x^6}{6} + \frac{x^4}{4} \right\}_0^a = 1.278.$$

EXERCISES 7.9

2. The cone can be formed by rotating the straight
line segment shown about the x-axis. Small lengths
along the curve corresponding to lengths dx along
the x-axis are given by

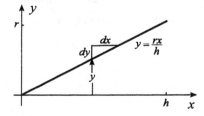

$$\sqrt{1 + \left(\frac{dy}{dx}\right)^2}\, dx = \sqrt{1 + (r/h)^2}\, dx.$$

The area of the curved portion of the cone is therefore

$$A = \int_0^h 2\pi y \sqrt{1 + (r/h)^2}\, dx = 2\pi \sqrt{1 + (r/h)^2} \int_0^h \frac{rx}{h}\, dx = \frac{2\pi r}{h^2}\sqrt{r^2 + h^2}\left\{\frac{x^2}{2}\right\}_0^h = \pi r \sqrt{r^2 + h^2}.$$

4. Cross sections of the pyramid parallel to the base are always square. At height y above the base, similar
triangles give $\|DE\|/\|EH\| = \|AC\|/\|HC\|$, or $\|DE\| = \dfrac{(h - y)(b/\sqrt{2})}{h}$.

Consequently,

$$\|FG\| = \frac{\|DF\|}{\sqrt{2}} = \frac{2\|DE\|}{\sqrt{2}}$$

$$= \sqrt{2}\left[\frac{b}{\sqrt{2}h}(h - y)\right] = \frac{b}{h}(h - y).$$

The area of the square at height y is therefore
$b^2(h - y)^2/h^2$, and the volume of the pyramid is

$$V = \int_0^h \frac{b^2}{h^2}(h - y)^2\, dy = \frac{b^2}{h^2}\left\{-\frac{1}{3}(h - y)^3\right\}_0^h = \frac{b^2 h}{3}.$$

6. (a) The number of bees in a ring of width dx at distance x from the hive is $\rho(2\pi x\, dx)$. The total number
of bees in the colony is therefore

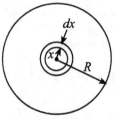

$$N = \int_0^R \rho(2\pi x\, dx)$$

$$= 2\pi \int_0^R x\left[\frac{600\,000}{31\pi R^5}(R^3 + 2R^2 x - Rx^2 - 2x^3)\right] dx$$

$$= \frac{1\,200\,000}{31 R^5}\left\{\frac{R^3 x^2}{2} + \frac{2R^2 x^3}{3} - \frac{Rx^4}{4} - \frac{2x^5}{5}\right\}_0^R = 20\,000.$$

(b) The number of bees within $R/2$ of the hive is

$$\overline{N} = \int_0^{R/2} \rho(2\pi x\, dx) = \frac{1\,200\,000}{31R^5}\left\{\frac{R^3x^2}{2} + \frac{2R^2x^3}{3} - \frac{Rx^4}{4} - \frac{2x^5}{5}\right\}_0^{R/2} = 6976,$$

or approximately 7000.

8. Cross sections of the wedge parallel to the axis of the trunk and perpendicular to the diameter are triangles. At distance x from the centre of the wedge, $y = \sqrt{625 - x^2}$, and $z = y\tan(\pi/3) = \sqrt{3}\sqrt{625 - x^2}$. The area of the triangle at position x is therefore

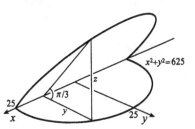

$$\frac{1}{2}yz = \frac{\sqrt{3}}{2}(625 - x^2),$$

and the volume of the wedge is

$$V = 2\int_0^{25} \frac{\sqrt{3}}{2}(625 - x^2)\, dx = \sqrt{3}\left\{625x - \frac{x^3}{3}\right\}_0^{25} = \frac{31\,250}{\sqrt{3}} \text{ cm}^3.$$

10. If we differentiate $8y^2 = x^2(1 - x^2)$ with respect to x,

$$16y\frac{dy}{dx} = 2x - 4x^3 \implies \frac{dy}{dx} = \frac{x - 2x^3}{8y}.$$

Small lengths along the curve in the first quadrant corresponding to lengths dx along the x-axis are given by

$$\sqrt{1 + \left(\frac{dy}{dx}\right)^2}\, dx = \sqrt{1 + \left(\frac{x - 2x^3}{8y}\right)^2}\, dx = \frac{\sqrt{64y^2 + x^2 - 4x^4 + 4x^6}}{8y}\, dx$$

$$= \frac{\sqrt{8x^2 - 8x^4 + x^2 - 4x^4 + 4x^6}}{8y}\, dx = \frac{3x - 2x^3}{2\sqrt{2}x\sqrt{1 - x^2}}\, dx = \frac{3 - 2x^2}{2\sqrt{2}\sqrt{1 - x^2}}\, dx.$$

The area of the surface of revolution is therefore

$$A = 2\int_0^1 2\pi y\left(\frac{3 - 2x^2}{2\sqrt{2}\sqrt{1 - x^2}}\right) dx = \sqrt{2}\pi\int_0^1 \frac{x\sqrt{1 - x^2}}{2\sqrt{2}}\left(\frac{3 - 2x^2}{\sqrt{1 - x^2}}\right) dx$$

$$= \frac{\pi}{2}\int_0^1 (3x - 2x^3)\, dx = \frac{\pi}{2}\left\{\frac{3x^2}{2} - \frac{x^4}{2}\right\}_0^1 = \frac{\pi}{2}.$$

12. For the triangle at position x, we note that by similar triangles $\|BD\|/\|CD\| = \|AO\|/\|CO\|$, or,

$$\|BD\| = (r - x)\frac{r}{r} = r - x.$$

Because the base of the triangle has length $2y = 2\sqrt{r^2 - x^2}$, the area of the triangle at x is

$$\frac{1}{2}(r - x)2\sqrt{r^2 - x^2} = (r - x)\sqrt{r^2 - x^2}.$$

The volume of the solid is therefore

$$V = 2\int_0^r (r - x)\sqrt{r^2 - x^2}\, dx$$

$$= 2r\int_0^r \sqrt{r^2 - x^2}\, dx - 2\int_0^r x\sqrt{r^2 - x^2}\, dx.$$

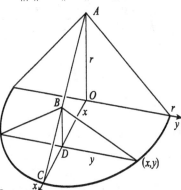

Since the first integral is one-quarter of the area of the base circle,

$$V = 2r\left(\frac{1}{4}\pi r^2\right) - 2\left\{-\frac{1}{3}(r^2 - x^2)^{3/2}\right\}_0^r = \frac{\pi r^3}{2} - \frac{2r^3}{3} = \frac{(3\pi - 4)r^3}{6}.$$

14. If x is the original length of the rod, then according to equation 7.43, the compression when M is placed on top is $Mgx/(AE)$, where $g = 9.81$. It follows that $x - \dfrac{Mgx}{AE} = L \implies x = \dfrac{AEL}{AE - Mg}$.

16. If we consider a small length dy at position y, the force on each cross section in this element is approximately the same, and equal to the weight of that part of the rod below it plus F, $\rho g(L-y)A + F$, where ρ is the density of the material in the rod. According to equation 7.43, the element dy stretches by

$$\frac{[\rho g(L-y)A + F]\,dy}{AE}.$$

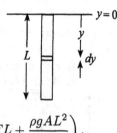

Total stretch in the rod is therefore

$$\int_0^L \frac{\rho g(L-y)A + F}{AE}\,dy = \frac{1}{AE}\left\{-\frac{\rho g A}{2}(L-y)^2 + Fy\right\}_0^L = \frac{1}{AE}\left(FL + \frac{\rho g A L^2}{2}\right).$$

The length of the rod is therefore $L + \dfrac{FL}{AE} + \dfrac{\rho g L^2}{2E}$.

18. Since the force on each cross section is Mg, the stretch of the element of width dy at position y is $[Mg/(AE)]\,dy$. Since the cross-sectional area at y is $A = \pi x^2 = \pi r^2(1 - y/L)^2$, the stretch of element dy is

$$\frac{Mg}{E\pi r^2(1 - y/L)^2}\,dy.$$

To find total stretch we should integrate this function from 0 to L. But this cannot be done because the function has an infinite discontinuity at $y = L$.

20. The force on the cross section of width dy at position y is the weight of the rod below it. To find this weight we calculate the volume by slicing,

$$\int_y^L \left[b + \frac{(a-b)}{L}y\right]^2 dy = \left\{\frac{L}{3(a-b)}\left[b + \frac{(a-b)}{L}y\right]^3\right\}_y^L$$

$$= \frac{1}{3(a-b)L^2}\left\{a^3L^3 - [bL + (a-b)y]^3\right\}.$$

The element dy therefore stretches by

$$\frac{\rho g\{a^3L^3 - [bL + (a-b)y]^3\}}{3(a-b)L^2[b + (a-b)y/L]^2 E}\,dy = \frac{\rho g\{a^3L^3 - [bL + (a-b)y]^3\}}{3E(a-b)[bL + (a-b)y]^2}\,dy.$$

Total stretch of the rod is

$$\int_0^L \frac{\rho g\{a^3L^3 - [bL + (a-b)y]^3\}}{3E(a-b)[bL + (a-b)y]^2}\,dy = \frac{\rho g}{3E(a-b)}\int_0^L \left\{\frac{a^3L^3}{[bL + (a-b)y]^2} - [bL + (a-b)y]\right\}dy$$

$$= \frac{\rho g}{3E(a-b)}\left\{\frac{-a^3L^3}{(a-b)[bL + (a-b)y]} - \frac{[bL + (a-b)y]^2}{2(a-b)}\right\}_0^L$$

$$= \frac{\rho g L^2(2a^3 - 3a^2b + b^3)}{6bE(a-b)^2}.$$

The total length of the rod is this quantity plus L.

22. Without the extra weight, the answer to the stretch of the rod in Exercise 19 is the same as that in Exercise 20 with a and b interchanged. With M added, we must add Mg to the weight of each cross section. Stretch is therefore given by the answer in Exercise 20 with a and b interchanged plus the following integral

$$\int_0^L \frac{Mg}{[a + (b-a)y/L]^2 E}\,dy = \frac{MgL^2}{E}\int_0^L \frac{1}{[aL + (b-a)y]^2}\,dy = \frac{MgL^2}{E}\left\{\frac{-1}{(b-a)[aL + (b-a)y]}\right\}_0^L = \frac{MgL}{abE}.$$

Thus, the length of the rod is $L + \dfrac{MgL}{abE} + \dfrac{\rho g L^2(2b^3 - 3ab^2 + a^3)}{6bE(a-b)^2}$.

24. (a) According to equation 7.46, deflections are given by

$$y(x) = \int_0^L \frac{1}{L\tau}[x(L-X)h(X-x) + X(L-x)h(x-X)]k\,dX$$

$$= \frac{k}{L\tau}\int_0^x X(L-x)\,dX + \frac{k}{L\tau}\int_x^L x(L-X)\,dX$$

$$= \frac{k(L-x)}{L\tau}\left\{\frac{X^2}{2}\right\}_0^x + \frac{kx}{L\tau}\left\{\frac{-(L-X)^2}{2}\right\}_x^L$$

$$= \frac{kx(L-x)}{2\tau}.$$

(b) The graph of deflections is the parabola shown above. It is symmetric about $x = L/2$ with minimum at $x = L/2$. We would expect this for constant loading.

26. (a) According to equation 7.46,

$$y(x) = \int_0^L \frac{1}{6EI}[(x-X)^3 h(x-X) - x^3 + 3Xx^2]k\,dX$$

$$= \frac{k}{6EI}\int_0^x [(x-X)^3 - x^3 + 3Xx^2]\,dX + \frac{k}{6EI}\int_x^L (3Xx^2 - x^3)\,dX$$

$$= \frac{k}{6EI}\left\{\frac{-(x-X)^4}{4} - x^3 X + \frac{3X^2 x^2}{2}\right\}_0^x + \frac{k}{6EI}\left\{\frac{3X^2 x^2}{2} - x^3 X\right\}_x^L$$

$$= \frac{kx^2(x^2 - 4Lx + 6L^2)}{24EI}.$$

(b) With uniform loading and a free end at $x = L$, maximum deflection, meaning minimum y, should occur at $x = L$. We can verify this by finding critical points of $y(x)$,

$$0 = y'(x) = \frac{k}{24EI}(4x^3 - 12Lx^2 + 12L^2 x) = \frac{4kx}{24EI}(x^2 - 3Lx + 3L^2).$$

Since $x = 0$ is the only solution (as expected), minimum $y(x)$ must occur at either $x = 0$ or $x = L$. Since $y(0) = 0$ and $y(L) < 0$, maximum deflection occurs at $x = L$.

28. (a) According to equation 7.46,

$$y(x) = \int_0^L \left[\frac{1}{6EI}(x-X)^3 h(x-X) + \frac{x^3}{6EIL^3}(-L^3 + 3LX^2 - 2X^3) + \frac{x^2}{2EIL^2}(X^3 - 2LX^2 + L^2 X)\right]k\,dX$$

$$= \frac{k}{6EIL^3}\int_0^x \left[L^3(x-X)^3 + x^3(-L^3 + 3LX^2 - 2X^3) + 3x^2 L(X^3 - 2LX^2 + L^2 X)\right]\,dX$$

$$+ \frac{k}{6EIL^3}\int_x^L \left[x^3(-L^3 + 3LX^2 - 2X^3) + 3x^2 L(X^3 - 2LX^2 + L^2 X)\right]\,dX$$

$$= \frac{k}{6EIL^3}\left\{\frac{-L^3(x-X)^4}{4} + x^3\left(-L^3 X + LX^3 - \frac{X^4}{2}\right) + 3x^2 L\left(\frac{X^4}{4} - \frac{2LX^3}{3} + \frac{L^2 X^2}{2}\right)\right\}_0^x$$

$$+ \frac{k}{6EIL^3}\left\{x^3\left(-L^3 X + LX^3 - \frac{X^4}{2}\right) + 3x^2 L\left(\frac{X^4}{4} - \frac{2LX^3}{3} + \frac{L^2 X^2}{2}\right)\right\}_x^L$$

$$= \frac{x^2(L-x)^2}{24EI}.$$

(b) Phsically we expect maximum deflection, meaning minimum y, at $x = L/2$. To confirm this, we find critical points of $y(x)$,

$$0 = y'(x) = \frac{1}{24EI}[2x(L-x)^2 - 2x^2(L-x)] = \frac{1}{12EI}x(L-x)(L-2x).$$

This gives the expected $x = 0$ and $x = L$ (the beam is horizontal at its ends), and the hoped for $x = L/2$.

30. According to the modified Torricelli law, the velocity of water through a vertical rectangle of height dy at a distance y above the bottom of the slit (see figure) is

$$v = c\sqrt{2g(H + h - y)}.$$

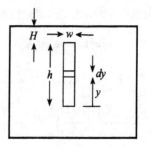

The rate at which water passes through this tiny rectangle is $v(w\,dy)$, and therefore the volume of water passing through the slit per unit time is

$$Q = \int_0^h c\sqrt{2g(H + h - y)}\, w\, dy = \sqrt{2g}cw\left\{-\frac{2}{3}(H + h - y)^{3/2}\right\}_0^h = \frac{2\sqrt{2g}cw}{3}[(H + h)^{3/2} - H^{3/2}].$$

32. Vertical cross sections of the attic parallel to the length of the roof are trapezoids. Because BGC and EGF are similar triangles, ratios of corresponding sides are equal: $\|EF\|/\|FG\| = \|BC\|/\|GC\|$, or

$$\|EF\| = \frac{x(3/2)}{5} = \frac{3x}{10}.$$

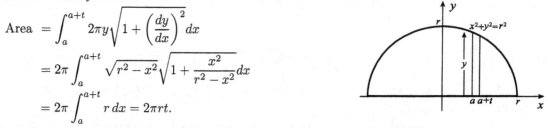

Since ABC and DEF are similar triangles, we can say that $\|DE\|/\|EF\| = \|AB\|/\|BC\|$, or

$$\|DE\| = \frac{3x}{10}\left(\frac{2}{3/2}\right) = \frac{2x}{5}.$$

The area of the trapezoid at position x is therefore

$$\|DE\|\left(\frac{1}{2}\right)[15 + (15 - 2\|EF\|)] = \frac{2x}{5}\left(\frac{1}{2}\right)\left[30 - 2\left(\frac{3x}{10}\right)\right] = \frac{3x}{25}(50 - x).$$

The volume in the attic can now be calculated as

$$V = 2\int_0^5 \frac{3x}{25}(50 - x)\, dx = \frac{6}{25}\left\{25x^2 - \frac{x^3}{3}\right\}_0^5 = 140 \text{ m}^3.$$

34. (a) The area of the peel for a slice of thickness t at any value $x = a$ is

$$\text{Area} = \int_a^{a+t} 2\pi y\sqrt{1 + \left(\frac{dy}{dx}\right)^2}\, dx$$

$$= 2\pi\int_a^{a+t} \sqrt{r^2 - x^2}\sqrt{1 + \frac{x^2}{r^2 - x^2}}\, dx$$

$$= 2\pi\int_a^{a+t} r\, dx = 2\pi rt.$$

Since this is independent of a, any slice of width t has the same peel area. If there are n slices, each of width $2r/n$, the area of peel is $2\pi r(2r/n) = 4\pi r^2/n$. This is reasonable in that this is the area of the sphere divided by the number of slices.
(b) The volume in the peel for a slice of thickness t at $x = a$ is

$$\text{Volume} = \int_a^{a+t} \{\pi(r^2 - x^2) - \pi[(r - h)^2 - x^2]\}\, dx$$

$$= \pi\int_a^{a+t} (2rh - h^2)\, dx = \pi(2rh - h^2)t.$$

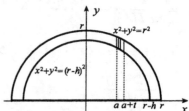

Since this is independent of a, the volume of peel is the same for each slice. This is true, however, only for slices which have holes. Clearly, slices between $r - h$ and r do not all have the same volume.

EXERCISES 7.10

2. $\displaystyle\int_3^\infty \frac{1}{(x+4)^{1/3}}\,dx = \lim_{b\to\infty}\int_3^b \frac{1}{(x+4)^{1/3}}\,dx = \lim_{b\to\infty}\left\{\frac{3}{2}(x+4)^{2/3}\right\}_3^b = \lim_{b\to\infty}\left[\frac{3}{2}(b+4)^{2/3} - \frac{3}{2}(7)^{2/3}\right] = \infty$

4. $\displaystyle\int_{-\infty}^{-4}\frac{x}{(x^2-2)^4}\,dx = \lim_{a\to-\infty}\int_a^{-4}\frac{x}{(x^2-2)^4}\,dx = \lim_{a\to-\infty}\left\{\frac{-1}{6(x^2-2)^3}\right\}_a^{-4}$

$$= \lim_{a\to-\infty}\left[-\frac{1}{6(14)^3} + \frac{1}{6(a^2-2)^3}\right] = \frac{-1}{16\,464}$$

6. $\displaystyle\int_{-\infty}^{\infty}\frac{x^3}{(x^4+5)^{1/4}}\,dx = \lim_{a\to-\infty}\int_a^0 \frac{x^3}{(x^4+5)^{1/4}}\,dx + \lim_{b\to\infty}\int_0^b \frac{x^3}{(x^4+5)^{1/4}}\,dx$

$$= \lim_{a\to-\infty}\left\{\frac{1}{3}(x^4+5)^{3/4}\right\}_a^0 + \lim_{b\to\infty}\left\{\frac{1}{3}(x^4+5)^{3/4}\right\}_0^b$$

$$= \lim_{a\to-\infty}\left[\frac{5^{3/4}-(a^4+5)^{3/4}}{3}\right] + \lim_{b\to\infty}\left[\frac{(b^4+5)^{3/4}-5^{3/4}}{3}\right]$$

Since neither of these limits exists, the integral diverges.

8. $\displaystyle\int_0^1 \frac{1}{\sqrt{1-x}}\,dx = \lim_{c\to1^-}\int_0^c \frac{1}{\sqrt{1-x}}\,dx = \lim_{c\to1^-}\left\{-2\sqrt{1-x}\right\}_0^c = \lim_{c\to1^-}(-2\sqrt{1-c}+2) = 2$

10. $\displaystyle\int_2^5 \frac{x}{\sqrt{x^2-4}}\,dx = \lim_{c\to2^+}\int_c^5 \frac{x}{\sqrt{x^2-4}}\,dx = \lim_{c\to2^+}\left\{\sqrt{x^2-4}\right\}_c^5 = \lim_{c\to2^+}(\sqrt{21}-\sqrt{c^2-4}) = \sqrt{21}$

12. $\displaystyle\int_{-\infty}^{\infty}\frac{1}{x^2}\,dx = \lim_{a\to-\infty}\int_a^{-1}\frac{1}{x^2}\,dx + \lim_{b\to0^-}\int_{-1}^b \frac{1}{x^2}\,dx + \lim_{c\to0^+}\int_c^1 \frac{1}{x^2}\,dx + \lim_{d\to\infty}\int_1^d \frac{1}{x^2}\,dx$

Since $\displaystyle\lim_{c\to0^+}\int_c^1 \frac{1}{x^2}\,dx = \lim_{c\to0^+}\left\{-\frac{1}{x}\right\}_c^1 = \lim_{c\to0^+}\left(-1+\frac{1}{c}\right) = \infty$, the integral diverges.

14. $\displaystyle\int_{-\infty}^{\pi/2}\frac{x}{(x^2-4)^2}\,dx = \lim_{a\to-\infty}\int_a^{-3}\frac{x}{(x^2-4)^2}\,dx + \lim_{b\to-2^-}\int_{-3}^b \frac{x}{(x^2-4)^2}\,dx + \lim_{c\to-2^+}\int_c^{\pi/2}\frac{x}{(x^2-4)^2}\,dx$

Since $\displaystyle\lim_{b\to-2^-}\int_{-3}^b \frac{x}{(x^2-4)^2}\,dx = \lim_{b\to-2^-}\left\{\frac{-1}{2(x^2-4)}\right\}_{-3}^b = \lim_{b\to-2^-}\left[\frac{-1}{2(b^2-4)} + \frac{1}{10}\right] = -\infty$,

the integral diverges.

16. $\displaystyle\int_{-\infty}^{\infty}\sin x\,dx = \lim_{a\to-\infty}\int_a^0 \sin x\,dx + \lim_{b\to\infty}\int_0^b \sin x\,dx$

Since $\displaystyle\lim_{b\to\infty}\int_0^b \sin x\,dx = \lim_{b\to\infty}\left\{-\cos x\right\}_0^b = \lim_{b\to\infty}(-\cos b + 1)$, and this limit does not exist, neither does the integral.

18. If we set $u = x^2 - 4$, then $du = 2x\,dx$, and

$$\int \frac{x^3}{\sqrt{x^2-4}}\,dx = \int \frac{(u+4)}{\sqrt{u}}\frac{du}{2} = \frac{1}{2}\left\{\frac{2}{3}u^{3/2} + 8\sqrt{u}\right\} + C = \frac{1}{3}(x^2-4)^{3/2} + 4\sqrt{x^2-4} + C.$$

Thus $\displaystyle\int_2^3 \frac{x^3}{\sqrt{x^2-4}}\,dx = \lim_{c\to2^+}\int_c^3 \frac{x^3}{\sqrt{x^2-4}}\,dx = \lim_{c\to2^+}\left\{\frac{1}{3}(x^2-4)^{3/2} + 4\sqrt{x^2-4}\right\}_c^3$

$$= \lim_{c\to2^+}\left[\frac{5\sqrt{5}}{3} + 4\sqrt{5} - \frac{1}{3}(c^2-4)^{3/2} - 4\sqrt{c^2-4}\right] = \frac{17\sqrt{5}}{3}.$$

20. (a) If it is possible, the area must be defined by

$$\int_1^\infty \left[1 - (1 - x^{-1/4})\right] dx = \lim_{b \to \infty} \int_1^b \frac{1}{x^{1/4}} \, dx$$

$$= \lim_{b \to \infty} \left\{ \frac{4}{3} x^{3/4} \right\}_1^b = \lim_{b \to \infty} \left(\frac{4}{3} b^{3/4} - \frac{4}{3} \right) = \infty.$$

Consequently, we cannot assign an area to the region.

(b) If it is possible, the volume must be defined by

$$\int_1^\infty \pi \left[1 - (1 - x^{-1/4})\right]^2 dx = \lim_{b \to \infty} \pi \int_1^b \frac{1}{\sqrt{x}} \, dx = \lim_{b \to \infty} \pi \{2\sqrt{x}\}_1^b = \lim_{b \to \infty} \pi(2\sqrt{b} - 2) = \infty.$$

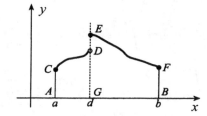

Thus, no volume can be assigned to the solid of revolution.

22. (a) If it is possible, the area must be defined by

$$\int_1^\infty \left[1 - (1 - x^{-3})\right] dx = \lim_{b \to \infty} \int_1^b \frac{1}{x^3} \, dx$$

$$= \lim_{b \to \infty} \left\{ \frac{-1}{2x^2} \right\}_1^b = \lim_{b \to \infty} \left(\frac{-1}{2b^2} + \frac{1}{2} \right) = \frac{1}{2}.$$

(b) If it is possible, the volume must be defined by

$$\int_1^\infty \pi \left[1 - (1 - x^{-3})\right]^2 dx = \lim_{b \to \infty} \pi \int_1^b \frac{1}{x^6} \, dx = \lim_{b \to \infty} \pi \left\{ \frac{-1}{5x^5} \right\}_1^b = \lim_{b \to \infty} \pi \left(\frac{-1}{5b^5} + \frac{1}{5} \right) = \frac{\pi}{5}.$$

24. $\displaystyle\int_1^\infty \frac{1}{x^p} \, dx = \lim_{b \to \infty} \int_1^b \frac{1}{x^p} \, dx = \lim_{b \to \infty} \left\{ \frac{1}{(1-p)x^{p-1}} \right\}_1^b = \lim_{b \to \infty} \left[\frac{1}{(1-p)b^{p-1}} + \frac{1}{p-1} \right] = \begin{cases} \infty, & p < 1 \\ \dfrac{1}{p-1}, & p > 1 \end{cases}$

26. $\displaystyle\int_a^b f(x) \, dx = \lim_{c \to d^-} \int_a^c f(x) \, dx + \lim_{c \to d^+} \int_c^b f(x) \, dx$

The first limit can be interpreted as area $ACDG$ were the hole to be filled in at D. The second limit is area $GEFB$. Since both areas are clearly defined, the improper integral does indeed exist.

28. No. For example, according to definition 7.49 the integral in Exercise 16 does not exist. But were we to use this definition:

$$\int_{-\infty}^\infty \sin x \, dx = \lim_{a \to \infty} \int_{-a}^a \sin x \, dx = \lim_{a \to \infty} \left\{ -\cos x \right\}_{-a}^a = \lim_{a \to \infty} \left[-\cos a + \cos(-a) \right] = \lim_{a \to \infty} (0) = 0,$$

and the improper integral would exist.

30. Since $\dfrac{x^3}{(27 - x^3)^2} \geq \dfrac{x^2}{(27 - x^3)^2} > 0$ for $1 \leq x \leq 3$, we can say that

$$\int_1^3 \frac{x^3}{(27 - x^3)^2} \, dx \geq \int_1^3 \frac{x^2}{(27 - x^3)^2} \, dx = \lim_{c \to 3^-} \int_1^c \frac{x^2}{(27 - x^3)^2} \, dx$$

$$= \lim_{c \to 3^-} \left\{ \frac{1}{3(27 - x^3)} \right\}_1^c = \lim_{c \to 3^-} \left[\frac{1}{3(27 - c^3)} - \frac{1}{78} \right] = \infty.$$

32. Since $\dfrac{\sqrt{-x}}{(x^2 + 5)^2} < \dfrac{-x}{(x^2 + 5)^2}$ for $x \leq -2$, we can say that

$$\int_{-\infty}^{-2} \frac{\sqrt{-x}}{(x^2 + 5)^2} \, dx < \int_{-\infty}^{-2} \frac{-x}{(x^2 + 5)^2} \, dx = \lim_{a \to -\infty} \int_a^{-2} \frac{-x}{(x^2 + 5)^2} \, dx$$

$$= \lim_{a \to -\infty} \left\{ \frac{1}{2(x^2 + 5)} \right\}_a^{-2} = \lim_{a \to -\infty} \left[\frac{1}{18} - \frac{1}{2(a^2 + 5)} \right] = \frac{1}{18}.$$

The given integral therefore converges.

34. $\displaystyle\int_{-\infty}^{\infty} f(x)\, \delta(x - a)\, dx = \int_{-\infty}^{\infty} f(x) \left[\lim_{\epsilon \to 0} P_\epsilon(x - a) \right] dx = \lim_{\epsilon \to 0} \int_{-\infty}^{\infty} f(x)\, P_\epsilon(x - a)\, dx$

$$= \lim_{\epsilon \to 0} \int_a^{a+\epsilon} f(x) \frac{1}{\epsilon} dx = \lim_{\epsilon \to 0} \frac{1}{\epsilon} \int_a^{a+\epsilon} f(x)\, dx.$$

If we let $F(x)$ be an antiderivative of $f(x)$, then

$$\int_{-\infty}^{\infty} f(x)\, \delta(x - a)\, dx = \lim_{\epsilon \to 0} \frac{1}{\epsilon} [F(a + \epsilon) - F(a)] = F'(a) = f(a).$$

REVIEW EXERICSES

2. (a) $\displaystyle A = \int_0^1 (2 - x - x^3)\, dx = \left\{ 2x - \frac{x^2}{2} - \frac{x^4}{4} \right\}_0^1 = \frac{5}{4}$

(b) $\displaystyle V_x = \int_0^1 \pi(y_2^2 - y_1^2)\, dx = \pi \int_0^1 [(2 - x)^2 - (x^3)^2]\, dx$

$$= \pi \left\{ -\frac{1}{3}(2 - x)^3 - \frac{x^7}{7} \right\}_0^1 = \frac{46\pi}{21}$$

$$V_y = \int_0^1 2\pi x(y_2 - y_1)\, dx = 2\pi \int_0^1 x(2 - x - x^3)\, dx = 2\pi \left\{ x^2 - \frac{x^3}{3} - \frac{x^5}{5} \right\}_0^1 = \frac{14\pi}{15}$$

(c) Since $\displaystyle A\bar{x} = \int_0^1 x(y_2 - y_1)\, dx = \frac{1}{2\pi} V_y = \frac{7}{15}$, we find $\bar{x} = \frac{7}{15}\frac{4}{5} = \frac{28}{75}$.   Since

$$A\bar{y} = \int_0^1 \frac{1}{2}(y_2 + y_1)(y_2 - y_1)\, dx = \frac{1}{2} \int_0^1 (y_2^2 - y_1^2)\, dx = \frac{1}{2\pi} V_x = \frac{23}{21}, \text{ it follows that } \bar{y} = \frac{23}{21}\frac{4}{5} = \frac{92}{105}.$$

(d) $\displaystyle I_x = \int_0^1 \frac{1}{3}(y_2^3 - y_1^3)\, dx = \frac{1}{3} \int_0^1 [(2 - x)^3 - (x^3)^3]\, dx = \frac{1}{3} \left\{ -\frac{1}{4}(2 - x)^4 - \frac{x^{10}}{10} \right\}_0^1 = \frac{73}{60}$

$$I_y = \int_0^1 x^2(2 - x - x^3)\, dx = \left\{ \frac{2x^3}{3} - \frac{x^4}{4} - \frac{x^6}{6} \right\}_0^1 = \frac{1}{4}$$

4. (a) $\displaystyle A = \int_0^1 (y + 1 - 2y)\, dy = \left\{ y - \frac{y^2}{2} \right\}_0^1 = \frac{1}{2}$

(b) $\displaystyle V_x = \int_0^1 2\pi y(x_2 - x_1)\, dy = 2\pi \int_0^1 y(1 - y)\, dy$

$$= 2\pi \left\{ \frac{y^2}{2} - \frac{y^3}{3} \right\}_0^1 = \frac{\pi}{3}$$

$$V_y = \int_0^1 \pi(x_2^2 - x_1^2)\, dy = \pi \int_0^1 [(y + 1)^2 - (2y)^2]\, dy = \pi \left\{ \frac{1}{3}(y + 1)^3 - \frac{4y^3}{3} \right\}_0^1 = \pi$$

(c) Since $\displaystyle A\bar{x} = \int_0^1 \frac{1}{2}(x_2 + x_1)(x_2 - x_1)\, dy = \frac{1}{2} \int_0^1 (x_2^2 - x_1^2)\, dy = \frac{1}{2\pi} V_y = \frac{1}{2}$, we find $\bar{x} = \frac{1}{2}(2) = 1$.

Since $\displaystyle A\bar{y} = \int_0^1 y(x_2 - x_1)\, dy = \frac{1}{2\pi} V_x = \frac{1}{6}$, it follows that $\bar{y} = \frac{1}{6}(2) = \frac{1}{3}$.

(d) $\displaystyle I_x = \int_0^1 y^2(1 - y)\, dy = \left\{ \frac{y^3}{3} - \frac{y^4}{4} \right\}_0^1 = \frac{1}{12}$

$$I_y = \int_0^1 \frac{1}{3}(x_2^3 - x_1^3)\, dy = \frac{1}{3} \int_0^1 [(y + 1)^3 - (2y)^3]\, dy = \frac{1}{3} \left\{ \frac{1}{4}(y + 1)^4 - 2y^4 \right\}_0^1 = \frac{7}{12}$$

6. $V = 2 \displaystyle\int_2^3 2\pi(y-1)(y-2)\,dy$

$\qquad = 4\pi \displaystyle\int_2^3 (y^2 - 3y + 2)\,dy$

$\qquad = 4\pi \left\{ \dfrac{y^3}{3} - \dfrac{3y^2}{2} + 2y \right\}_2^3 = \dfrac{10\pi}{3}$

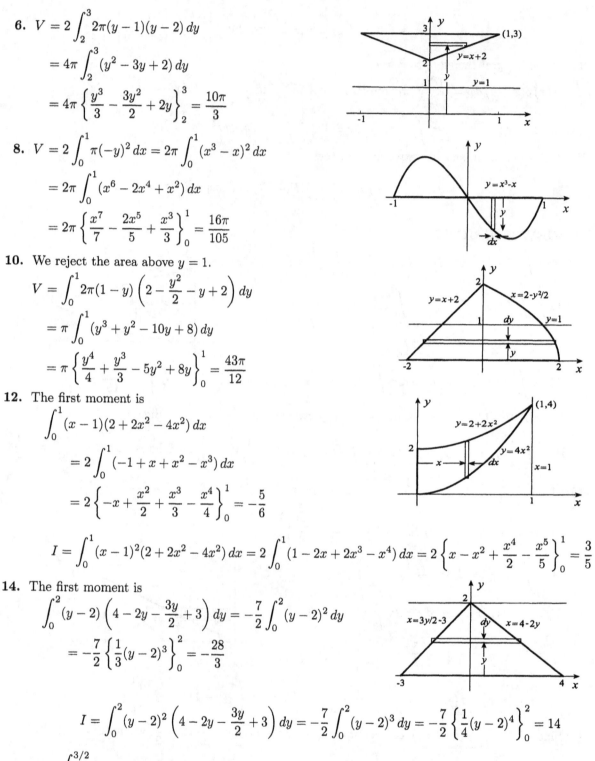

8. $V = 2 \displaystyle\int_0^1 \pi(-y)^2\,dx = 2\pi \int_0^1 (x^3 - x)^2\,dx$

$\qquad = 2\pi \displaystyle\int_0^1 (x^6 - 2x^4 + x^2)\,dx$

$\qquad = 2\pi \left\{ \dfrac{x^7}{7} - \dfrac{2x^5}{5} + \dfrac{x^3}{3} \right\}_0^1 = \dfrac{16\pi}{105}$

10. We reject the area above $y = 1$.

$\qquad V = \displaystyle\int_0^1 2\pi(1-y)\left(2 - \dfrac{y^2}{2} - y + 2\right)dy$

$\qquad = \pi \displaystyle\int_0^1 (y^3 + y^2 - 10y + 8)\,dy$

$\qquad = \pi \left\{ \dfrac{y^4}{4} + \dfrac{y^3}{3} - 5y^2 + 8y \right\}_0^1 = \dfrac{43\pi}{12}$

12. The first moment is

$\qquad \displaystyle\int_0^1 (x-1)(2 + 2x^2 - 4x^2)\,dx$

$\qquad = 2 \displaystyle\int_0^1 (-1 + x + x^2 - x^3)\,dx$

$\qquad = 2 \left\{ -x + \dfrac{x^2}{2} + \dfrac{x^3}{3} - \dfrac{x^4}{4} \right\}_0^1 = -\dfrac{5}{6}$

$\qquad I = \displaystyle\int_0^1 (x-1)^2(2 + 2x^2 - 4x^2)\,dx = 2 \int_0^1 (1 - 2x + 2x^3 - x^4)\,dx = 2 \left\{ x - x^2 + \dfrac{x^4}{2} - \dfrac{x^5}{5} \right\}_0^1 = \dfrac{3}{5}$

14. The first moment is

$\qquad \displaystyle\int_0^2 (y-2)\left(4 - 2y - \dfrac{3y}{2} + 3\right)dy = -\dfrac{7}{2}\int_0^2 (y-2)^2\,dy$

$\qquad = -\dfrac{7}{2}\left\{ \dfrac{1}{3}(y-2)^3 \right\}_0^2 = -\dfrac{28}{3}$

$\qquad I = \displaystyle\int_0^2 (y-2)^2\left(4 - 2y - \dfrac{3y}{2} + 3\right)dy = -\dfrac{7}{2}\int_0^2 (y-2)^3\,dy = -\dfrac{7}{2}\left\{ \dfrac{1}{4}(y-2)^4 \right\}_0^2 = 14$

16. $W = \displaystyle\int_0^{3/2} (3-y)1000(9.81)\pi(1)^2\,dy$

$\qquad = 9810\pi \left\{ 3y - \dfrac{y^2}{2} \right\}_0^{3/2} = 1.04 \times 10^5 \text{ J.}$

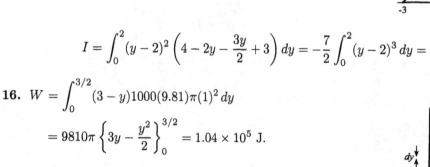

18. The surface area is twice the area of
the surface of revolution obtained by
rotating that part of $y = 2 - 2x$ in the
first quadrant about the x-axis. Small
lengths along the line are given by

$$\sqrt{1 + \left(\frac{dy}{dx}\right)^2}\, dx = \sqrt{1 + (-2)^2}\, dx = \sqrt{5}\, dx.$$

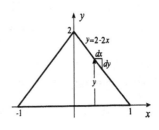

The required area is therefore

$$A = 2\int_0^1 2\pi y \sqrt{5}\, dx = 4\sqrt{5}\pi \int_0^1 (2 - 2x)\, dx = 8\sqrt{5}\pi \left\{x - \frac{x^2}{2}\right\}_0^1 = 4\sqrt{5}\pi.$$

20. Suppose x_0 is the stretch that requires W units
of work. When the spring is stretched x,
the force required to maintain this stretch
is kx. The work to stretch the spring x_0
is

$$W = \int_0^{x_0} kx\, dx = k\left\{\frac{x^2}{2}\right\}_0^{x_0} = \frac{1}{2}kx_0^2.$$

The work required to obtain a stretch of $2x_0$ is $\displaystyle \int_0^{2x_0} kx\, dx = k\left\{\frac{x^2}{2}\right\}_0^{2x_0} = 2kx_0^2 = 4W$. Thus, it
requires $3W$ more unit of work to increase the stretch from x_0 to $2x_0$.

22. $\displaystyle \int_0^3 \frac{1}{\sqrt{3 - x}}\, dx = \lim_{c \to 3^-} \int_0^c \frac{1}{\sqrt{3 - x}}\, dx = \lim_{c \to 3^-} \left\{-2\sqrt{3 - x}\right\}_0^c = \lim_{c \to 3^-} (-2\sqrt{3 - c} + 2\sqrt{3}) = 2\sqrt{3}$

24. $\displaystyle \int_{-2}^2 \frac{x}{\sqrt{4 - x^2}}\, dx = \lim_{c \to -2^+} \int_c^0 \frac{x}{\sqrt{4 - x^2}}\, dx + \lim_{d \to 2^-} \int_0^d \frac{x}{\sqrt{4 - x^2}}\, dx$

$$= \lim_{c \to -2^+} \left\{-\sqrt{4 - x^2}\right\}_c^0 + \lim_{d \to 2^-} \left\{-\sqrt{4 - x^2}\right\}_0^d$$

$$= \lim_{c \to -2^+} (\sqrt{4 - c^2} - 2) + \lim_{d \to 2^-} (2 - \sqrt{4 - d^2}) = -2 + 2 = 0$$

26. $\displaystyle \int_{-\infty}^{-3} \frac{1}{\sqrt{-x}}\, dx = \lim_{a \to -\infty} \int_a^{-3} \frac{1}{\sqrt{-x}}\, dx = \lim_{a \to -\infty} \left\{-2\sqrt{-x}\right\}_a^{-3} = \lim_{a \to -\infty} (2\sqrt{-a} - 2\sqrt{3}) = \infty$

28. $\displaystyle \int_{-\infty}^{\infty} \frac{x}{(x^2 - 1)^2}\, dx = \lim_{a \to -\infty} \int_a^{-2} \frac{x}{(x^2 - 1)^2}\, dx + \lim_{b \to -1^-} \int_{-2}^b \frac{x}{(x^2 - 1)^2}\, dx + \lim_{c \to -1^+} \int_c^0 \frac{x}{(x^2 - 1)^2}\, dx$

$$+ \lim_{d \to 1^-} \int_0^d \frac{x}{(x^2 - 1)^2}\, dx + \lim_{e \to 1^+} \int_e^2 \frac{x}{(x^2 - 1)^2}\, dx + \lim_{f \to \infty} \int_2^f \frac{x}{(x^2 - 1)^2}\, dx.$$

Since $\displaystyle \lim_{e \to 1^+} \int_e^2 \frac{x}{(x^2 - 1)^2}\, dx = \lim_{e \to 1^+} \left\{\frac{-1}{2(x^2 - 1)}\right\}_e^2 = \lim_{e \to 1^+} \left[\frac{1}{2(e^2 - 1)} - \frac{1}{6}\right] = \infty$, we conclude that the
given improper integral diverges.

30. The force of gravity on the slab shown is

$$-9.81(1000)(2x)(3)\,dy\ \text{N}.$$

Since the work to move this slab from its present position to the top of the tank is

$$(1-y)(9810)(6x)\,dy\ \text{J},$$

the total work to empty the tank is

$$W = \int_{-1}^{0} 6(9810)(1-y)\sqrt{1-y^2}\,dy$$

$$= 58\,860 \int_{-1}^{0} \sqrt{1-y^2}\,dy - 58\,860 \int_{-1}^{0} y\sqrt{1-y^2}\,dy.$$

Since the first integral represents one-quarter of the area of the circle on the end of the tank,

$$W = 58\,860 \left(\frac{1}{4}\right)\pi(1)^2 - 58\,860 \left\{-\frac{1}{3}(1-y^2)^{3/2}\right\}_{-1}^{0} = 14\,715\pi + 19\,620 = 6.58 \times 10^4\ \text{J}.$$

CHAPTER 8

EXERCISES 8.1

2. $\int xe^{-2x^2}\,dx = -\dfrac{1}{4}e^{-2x^2} + C$

4. $\int \dfrac{e^x}{1+e^x}\,dx = \ln(1+e^x) + C$

6. $\int x^2\sqrt{1-3x^3}\,dx = -\dfrac{2}{27}(1-3x^3)^{3/2} + C$

8. $\int \dfrac{x^2}{(1+x^3)^3}\,dx = \dfrac{-1}{6(1+x^3)^2} + C$

10. $\int \dfrac{1-\sqrt{x}}{\sqrt{x}}\,dx = \int (x^{-1/2}-1)\,dx = 2\sqrt{x} - x + C$

12. $\int \dfrac{x^2+2}{x^2+1}\,dx = \int \left(1 + \dfrac{1}{x^2+1}\right)dx = x + \operatorname{Tan}^{-1}x + C$

14. If we set $u = 2x+4$, then $du = 2\,dx$, and

$$\int \frac{x+3}{\sqrt{2x+4}}\,dx = \int \frac{(u-4)/2+3}{\sqrt{u}}\,\frac{du}{2} = \frac{1}{4}\int \left(\sqrt{u} + \frac{2}{\sqrt{u}}\right)du$$

$$= \frac{1}{4}\left(\frac{2}{3}u^{3/2} + 4\sqrt{u}\right) + C = \frac{1}{6}(2x+4)^{3/2} + \sqrt{2x+4} + C.$$

16. $\int \sin^3 2x \cos 2x\,dx = \dfrac{1}{8}\sin^4 2x + C$

18. If we set $u = x+5$, then $du = dx$, and

$$\int \frac{x^3}{(x+5)^2}\,dx = \int \frac{(u-5)^3}{u^2}\,du = \int \left(u - 15 + \frac{75}{u} - \frac{125}{u^2}\right)du = \frac{u^2}{2} - 15u + 75\ln|u| + \frac{125}{u} + C$$

$$= \frac{1}{2}(x+5)^2 - 15(x+5) + 75\ln|x+5| + \frac{125}{x+5} + C$$

$$= \frac{x^2}{2} - 10x + 75\ln|x+5| + \frac{125}{x+5} + D.$$

20. If we set $u = \cos 3x$, then $du = -3\sin 3x\,dx$, and

$$\int \tan 3x\,dx = \int \frac{\sin 3x}{\cos 3x}\,dx = \int \frac{1}{u}\left(-\frac{du}{3}\right) = -\frac{1}{3}\ln|u| + C = -\frac{1}{3}\ln|\cos 3x| + C = \frac{1}{3}\ln|\sec 3x| + C.$$

22. If we set $u = x^{1/4}$, or, $x = u^4$, then $dx = 4u^3\,du$, and

$$\int \frac{\sqrt{x}}{1+x^{1/4}}\,dx = \int \frac{u^2}{1+u}\,4u^3\,du = 4\int \frac{u^5}{u+1}\,du = 4\int \left(u^4 - u^3 + u^2 - u + 1 - \frac{1}{u+1}\right)du$$

$$= 4\left(\frac{u^5}{5} - \frac{u^4}{4} + \frac{u^3}{3} - \frac{u^2}{2} + u - \ln|u+1|\right) + C$$

$$= \frac{4}{5}x^{5/4} - x + \frac{4}{3}x^{3/4} - 2\sqrt{x} + 4x^{1/4} - 4\ln(x^{1/4}+1) + C.$$

24. $A = \displaystyle\int_{-1/2}^{2}\left(\dfrac{7-x}{3} - \dfrac{x^2+1}{x+1}\right)dx$

$= \displaystyle\int_{-1/2}^{2}\left(\dfrac{7}{3} - \dfrac{x}{3} - x + 1 - \dfrac{2}{x+1}\right)dx$

$= \displaystyle\int_{-1/2}^{2}\left(\dfrac{10}{3} - \dfrac{4x}{3} - \dfrac{2}{x+1}\right)dx$

$= \left\{\dfrac{10x}{3} - \dfrac{2x^2}{3} - 2\ln|x+1|\right\}_{-1/2}^{2} = \dfrac{35 - 12\ln 6}{6}$

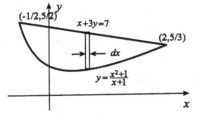

26. $\displaystyle\int_{-a}^{a} f(x)\,dx = \int_{-a}^{0} f(x)\,dx + \int_{0}^{a} f(x)\,dx$

If we set $u = -x$ and $du = -dx$ in the first integral on the right, then when $f(x)$ is an even function,

$$\int_{-a}^{a} f(x)\,dx = \int_{a}^{0} f(-u)(-du) + \int_{0}^{a} f(x)\,dx = \int_{0}^{a} f(u)\,du + \int_{0}^{a} f(x)\,dx = 2\int_{0}^{a} f(x)\,dx;$$

and when $f(x)$ is an odd function,

$$\int_{-a}^{a} f(x)\,dx = \int_{a}^{0} f(-u)(-du) + \int_{0}^{a} f(x)\,dx = \int_{0}^{a} -f(u)\,du + \int_{0}^{a} f(x)\,dx = 0.$$

28. (a) When we separate variables $\dfrac{dv}{1962 - v} = \dfrac{dt}{200}$, solutions are defined implicitly by

$$-\ln|1962 - v| = \frac{t}{200} + C \quad\Longrightarrow\quad \ln|1962 - v| = -\frac{t}{200} - C.$$

Exponentiation gives

$$|1962 - v| = e^{-C}e^{-t/200} \quad\Longrightarrow\quad 1962 - v = \pm e^{-C}e^{-t/200} = De^{-t/200},$$

where $D = \pm e^{-C}$. If we choose time $t = 0$ when descent begins, then $v(0) = 0$, and this requires $D = 1962$. Hence, $v = 1962 - 1962e^{-t/200} = 1962(1 - e^{-t/200})$ m/s.

(b) We set the velocity equal to dx/dt and integrate again,

$$x = 1962(t + 200e^{-t/200}) + E.$$

Since $x(0) = 0$, we find $E = -1962(200)$, and therefore

$$x = 1962(t + 200e^{-t/200}) - 1962(200) = 1962t + 392\,400(e^{-t/200} - 1)\ \text{m}.$$

30. If we write $\dfrac{1}{x(3 + 2x^n)} = \dfrac{A}{x} + \dfrac{Bx^{n-1}}{3 + 2x^n} = \dfrac{3A + 2Ax^n + Bx^n}{x(3 + 2x^n)}$, and equate numerators, then $3A + (2A +$ $B)x^n = 1$. This equation is satisfied for all x if we choose $A = 1/3$ and $B = -2/3$. Then

$$\int \frac{1}{x(3 + 2x^n)}\,dx = \int \left[\frac{1}{3x} - \frac{2x^{n-1}}{3(3 + 2x^n)}\right]\,dx.$$

In the second integral on the right we set $u = 3 + 2x^n$ and $du = 2nx^{n-1}\,dx$,

$$\int \frac{1}{x(3 + 2x^n)}\,dx = \frac{1}{3}\ln|x| - \frac{2}{3}\int \frac{1}{u}\left(\frac{du}{2n}\right) = \frac{1}{3}\ln|x| - \frac{1}{3n}\ln|u| + C$$

$$= \frac{1}{3}\ln|x| - \frac{1}{3n}\ln|3 + 2x^n| + C = \frac{1}{3n}\ln\left|\frac{x^n}{3 + 2x^n}\right| + C.$$

EXERCISES 8.2

2. When we set $u = x^2$, $dv = e^{2x}\,dx$, then $du = 2x\,dx$, $v = e^{2x}/2$, and

$$\int x^2 e^{2x}\,dx = \frac{x^2}{2}e^{2x} - \int 2x\frac{e^{2x}}{2}\,dx.$$

We now set $u = x$, $dv = e^{2x}\,dx$, in which case $du = dx$, $v = e^{2x}/2$, and

$$\int x^2 e^{2x}\,dx = \frac{x^2}{2}e^{2x} - \left(\frac{x}{2}e^{2x} - \int \frac{e^{2x}}{2}\,dx\right) = \frac{x^2}{2}e^{2x} - \frac{x}{2}e^{2x} + \frac{1}{4}e^{2x} + C.$$

4. When we set $u = \ln(2x)$, $dv = \sqrt{x}\,dx$, then $du = (1/x)\,dx$, $v = (2/3)x^{3/2}$, and

$$\int \sqrt{x}\ln(2x)\,dx = \frac{2}{3}x^{3/2}\ln(2x) - \int \frac{2}{3}x^{3/2}\frac{1}{x}\,dx = \frac{2}{3}x^{3/2}\ln(2x) - \frac{4}{9}x^{3/2} + C.$$

6. When we set $u = x$, $dv = \sqrt{3-x}\,dx$, then $du = dx$, $v = -\frac{2}{3}(3-x)^{3/2}$, and

$$\int x\sqrt{3-x}\,dx = -\frac{2x}{3}(3-x)^{3/2} - \int -\frac{2}{3}(3-x)^{3/2}\,dx = -\frac{2x}{3}(3-x)^{3/2} - \frac{4}{15}(3-x)^{5/2} + C.$$

8. When we set $u = x^2$, $dv = \sqrt{x+5}\,dx$, then $du = 2x\,dx$, $v = \frac{2}{3}(x+5)^{3/2}$, and

$$\int x^2\sqrt{x+5}\,dx = \frac{2}{3}x^2(x+5)^{3/2} - \int \frac{4}{3}x(x+5)^{3/2}\,dx.$$

We now set $u = x$, $dv = (x+5)^{3/2}\,dx$, in which case $du = dx$, $v = \frac{2}{5}(x+5)^{5/2}$, and

$$\int x^2\sqrt{x+5}\,dx = \frac{2}{3}x^2(x+5)^{3/2} - \frac{4}{3}\left[\frac{2}{5}x(x+5)^{5/2} - \int \frac{2}{5}(x+5)^{5/2}\,dx\right]$$

$$= \frac{2}{3}x^2(x+5)^{3/2} - \frac{8}{15}x(x+5)^{5/2} + \frac{16}{105}(x+5)^{7/2} + C.$$

10. When we set $u = x^2$, $dv = \dfrac{1}{\sqrt{2+x}}\,dx$, then $du = 2x\,dx$, $v = 2\sqrt{2+x}$, and

$$\int \frac{x^2}{\sqrt{2+x}}\,dx = 2x^2\sqrt{2+x} - \int 4x\sqrt{2+x}\,dx.$$

We now set $u = x$, $dv = \sqrt{2+x}\,dx$, in which case $du = dx$, $v = \frac{2}{3}(2+x)^{3/2}$, and

$$\int \frac{x^2}{\sqrt{2+x}}\,dx = 2x^2\sqrt{2+x} - 4\left[\frac{2x}{3}(2+x)^{3/2} - \int \frac{2}{3}(2+x)^{3/2}\,dx\right]$$

$$= 2x^2\sqrt{2+x} - \frac{8}{3}x(2+x)^{3/2} + \frac{16}{15}(2+x)^{5/2} + C.$$

12. When we set $u = \ln x$, $dv = (x-1)^2\,dx$, then $du = \dfrac{1}{x}\,dx$, $v = \frac{1}{3}(x-1)^3$, and

$$\int (x-1)^2\ln x\,dx = \frac{(x-1)^3}{3}\ln x - \int \frac{(x-1)^3}{3}\frac{1}{x}\,dx = \frac{(x-1)^3}{3}\ln x - \frac{1}{3}\int \left(x^2 - 3x + 3 - \frac{1}{x}\right)dx$$

$$= \frac{(x-1)^3}{3}\ln x - \frac{1}{3}\left(\frac{x^3}{3} - \frac{3x^2}{2} + 3x - \ln x\right) + C.$$

14. When we set $u = \operatorname{Tan}^{-1}x$, $dv = dx$, then $du = \dfrac{1}{1+x^2}\,dx$, $v = x$, and

$$\int \operatorname{Tan}^{-1}x\,dx = x\operatorname{Tan}^{-1}x - \int \frac{x}{1+x^2}\,dx = x\operatorname{Tan}^{-1}x - \frac{1}{2}\ln(1+x^2) + C.$$

16. When we set $u = e^{2x}$, $dv = \cos 3x\,dx$, then $du = 2e^{2x}\,dx$, $v = \frac{1}{3}\sin 3x$, and

$$\int e^{2x}\cos 3x\,dx = \frac{1}{3}e^{2x}\sin 3x - \int \frac{2}{3}e^{2x}\sin 3x\,dx.$$

We now set $u = e^{2x}$, $dv = \sin 3x\,dx$, in which case $du = 2e^{2x}\,dx$, $v = -\frac{1}{3}\cos 3x$, and

$$\int e^{2x}\cos 3x\,dx = \frac{1}{3}e^{2x}\sin 3x - \frac{2}{3}\left(-\frac{1}{3}e^{2x}\cos 3x - \int -\frac{2}{3}e^{2x}\cos 3x\,dx\right).$$

If we bring both integrals to the left,

$$\left(1 + \frac{4}{9}\right)\int e^{2x}\cos 3x\,dx = \frac{1}{3}e^{2x}\sin 3x + \frac{2}{9}e^{2x}\cos 3x \implies \int e^{2x}\cos 3x\,dx = \frac{1}{13}e^{2x}(3\sin 3x + 2\cos 3x) + C.$$

18. When we set $u = \ln(x^2 + 4)$, $dv = dx$, then $du = \dfrac{2x}{x^2 + 4}dx$, $v = x$, and

$$\int \ln(x^2 + 4)\,dx = x\ln(x^2 + 4) - \int \frac{2x^2}{x^2 + 4}dx = x\ln(x^2 + 4) - 2\int\left(1 - \frac{4}{x^2 + 4}\right)dx.$$

We now set $u = x/2$ and $du = dx/2$,

$$\int \ln(x^2 + 4)\,dx = x\ln(x^2 + 4) - 2x + 8\int \frac{1}{4u^2 + 4}(2\,du) = x\ln(x^2 + 4) - 2x + 4\int \frac{1}{u^2 + 1}du$$

$$= x\ln(x^2 + 4) - 2x + 4\,\mathrm{Tan}^{-1}u + C = x\ln(x^2 + 4) - 2x + 4\,\mathrm{Tan}^{-1}\!\left(\frac{x}{2}\right) + C.$$

20. $F = \displaystyle\int_0^{100} 9.81(1000)(100 - y)(2x)\,dy$

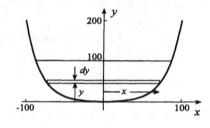

$$= 19\,620\int_0^{100}(100 - y)\frac{1}{k}\ln(y + 1)\,dy.$$

If we set $u = \ln(y + 1)$, $dv = (100 - y)\,dy$, then
$du = \dfrac{1}{y + 1}dy$, $v = -\dfrac{1}{2}(100 - y)^2$, and

$$F = \frac{19\,620}{k}\left[-\frac{1}{2}(100 - y)^2\ln(y + 1)\Big|_0^{100} - \int_0^{100} -\frac{1}{2}(100 - y)^2\frac{1}{y + 1}dy\right]$$

$$= \frac{9810}{k}\int_0^{100}\frac{10\,000 - 200y + y^2}{y + 1}dy = \frac{9810}{k}\int_0^{100}\left(y - 201 + \frac{10\,201}{y + 1}\right)dy$$

$$= \frac{9810}{k}\left\{\frac{y^2}{2} - 201y + 10\,201\ln|y + 1|\right\}_0^{100} = 5.92 \times 10^9 \text{ N}.$$

22. If we set $u = x$, $dv = \sin\dfrac{n\pi x}{L}dx$, $du = dx$, and $v = \dfrac{-L}{n\pi}\cos\dfrac{n\pi x}{L}$, then

$$\int_{-L}^{L} x\sin\frac{n\pi x}{L}\,dx = \left\{\frac{-Lx}{n\pi}\cos\frac{n\pi x}{L}\right\}_{-L}^{L} - \int_{-L}^{L}\frac{-L}{n\pi}\cos\frac{n\pi x}{L}dx$$

$$= -\frac{L^2}{n\pi}\cos n\pi - \frac{L^2}{n\pi}\cos(-n\pi) + \left\{\frac{L^2}{n^2\pi^2}\sin\frac{n\pi x}{L}\right\}_{-L}^{L} = \frac{2(-1)^{n+1}L^2}{n\pi}.$$

If we set $u = x$, $dv = \cos\dfrac{n\pi x}{L}dx$, $du = dx$, and $v = \dfrac{L}{n\pi}\sin\dfrac{n\pi x}{L}$, then

$$\int_{-L}^{L} x\cos\frac{n\pi x}{L}\,dx = \left\{\frac{Lx}{n\pi}\sin\frac{n\pi x}{L}\right\}_{-L}^{L} - \int_{-L}^{L}\frac{L}{n\pi}\sin\frac{n\pi x}{L}dx = -\left\{-\frac{L^2}{n^2\pi^2}\cos\frac{n\pi x}{L}\right\}_{-L}^{L} = 0.$$

24. If we set $u = 1 - 2x$, $dv = \sin\dfrac{n\pi x}{L}dx$, $du = -2\,dx$, and $v = \dfrac{-L}{n\pi}\cos\dfrac{n\pi x}{L}$, then

$$\int_{-L}^{L}(1 - 2x)\sin\frac{n\pi x}{L}\,dx = \left\{\frac{-L(1 - 2x)}{n\pi}\cos\frac{n\pi x}{L}\right\}_{-L}^{L} - \int_{-L}^{L}\frac{2L}{n\pi}\cos\frac{n\pi x}{L}dx$$

$$= \frac{-L}{n\pi}[(1 - 2L)\cos n\pi - (1 + 2L)\cos(-n\pi)] - \frac{2L}{n\pi}\left\{\frac{L}{n\pi}\sin\frac{n\pi x}{L}\right\}_{-L}^{L} = \frac{4(-1)^n L^2}{n\pi}.$$

If we set $u = 1 - 2x$, $dv = \cos\dfrac{n\pi x}{L}dx$, $du = -2\,dx$, and $v = \dfrac{L}{n\pi}\sin\dfrac{n\pi x}{L}$, then

$$\int_{-L}^{L}(1 - 2x)\cos\frac{n\pi x}{L}\,dx = \left\{\frac{L(1 - 2x)}{n\pi}\sin\frac{n\pi x}{L}\right\}_{-L}^{L} - \int_{-L}^{L}\frac{-2L}{n\pi}\sin\frac{n\pi x}{L}dx = \frac{2L}{n\pi}\left\{\frac{-L}{n\pi}\cos\frac{n\pi x}{L}\right\}_{-L}^{L} = 0.$$

26. When we set $u = x^{n-1}$, $dv = e^{-x}\,dx$, then $du = (n-1)x^{n-2}\,dx$, $v = -e^{-x}$, and

$$\Gamma(n) = \lim_{b\to\infty}\int_0^b x^{n-1}e^{-x}\,dx = \lim_{b\to\infty}\left[\left\{-x^{n-1}e^{-x}\right\}_0^b - \int_0^b -(n-1)x^{n-2}e^{-x}\,dx\right]$$

$$= \lim_{b\to\infty}\left[-b^{n-1}e^{-b} + (n-1)\int_0^b x^{n-2}e^{-x}\,dx\right] = (n-1)\int_0^\infty x^{n-2}e^{-x}\,dx.$$

Further integrations by parts lead to

$$\Gamma(n) = (n-1)(n-2)(n-3)\cdots(2)(1)\int_0^\infty e^{-x}\,dx = (n-1)!\lim_{b\to\infty}\int_0^b e^{-x}\,dx$$

$$= (n-1)!\lim_{b\to\infty}\left\{-e^{-x}\right\}_0^b = (n-1)!\lim_{b\to\infty}(1-e^{-b}) = (n-1)!.$$

28. $F(s) = \displaystyle\int_0^\infty t^2\,e^{-st}\,dt = \lim_{b\to\infty}\int_0^b t^2\,e^{-st}\,dt$ If we set $u = t^2$, $dv = e^{-st}\,dt$, $du = 2t\,dt$, and $v = -e^{-st}/s$,
then

$$F(s) = \lim_{b\to\infty}\left[\left\{\frac{t^2 e^{-st}}{-s}\right\}_0^b - \int_0^b \frac{2te^{-st}}{-s}\,dt\right] = \lim_{b\to\infty}\frac{b^2 e^{-bs}}{-s} + \frac{2}{s}\lim_{b\to\infty}\int_0^b te^{-st}\,dt.$$

The first limit is zero provided $s > 0$. In the integral we set $u = t$, $dv = e^{-st}\,dt$, $du = dt$, and
$v = -e^{-st}/s$, in which case

$$F(s) = \frac{2}{s}\lim_{b\to\infty}\left[\left\{\frac{te^{-st}}{-s}\right\}_0^b - \int_0^b -\frac{e^{-st}}{s}\,dt\right] = \frac{2}{s}\lim_{b\to\infty}\left(\frac{be^{-bs}}{-s}\right) + \frac{2}{s^2}\lim_{b\to\infty}\int_0^b e^{-st}\,dt$$

$$= \frac{2}{s^2}\lim_{b\to\infty}\left\{\frac{e^{-st}}{-s}\right\}_0^b = -\frac{2}{s^3}\lim_{b\to\infty}(e^{-bs}-1) = \frac{2}{s^3}.$$

30. If we set $u = t$, $dv = e^{-(s+1)t}\,dt$, $du = dt$, and $v = \dfrac{e^{-(s+1)t}}{-(s+1)}$, then

$$\int te^{-t}e^{-st}\,dt = \int te^{-(s+1)t}\,dt = -\frac{te^{-(s+1)t}}{s+1} - \int \frac{e^{-(s+1)t}}{-(s+1)}\,dt = -\frac{te^{-(s+1)t}}{s+1} - \frac{e^{-(s+1)t}}{(s+1)^2}.$$

Thus,

$$F(s) = \int_0^\infty te^{-t}e^{-st}\,dt = \lim_{b\to\infty}\int_0^b te^{-(s+1)t}\,dt = \lim_{b\to\infty}\left\{-\frac{te^{-(s+1)t}}{s+1} - \frac{e^{-(s+1)t}}{(s+1)^2}\right\}_0^b$$

$$= \lim_{b\to\infty}\left[-\frac{be^{-(s+1)b}}{s+1} - \frac{e^{-(s+1)b}}{(s+1)^2} + \frac{1}{(s+1)^2}\right] = \frac{1}{(s+1)^2}\quad\text{(provided }s > -1)$$

32. $F(\omega) = \displaystyle\int_{-\infty}^\infty f(t)e^{-i\omega t}\,dt = \int_{-L/2}^{L/2} e^{-i\omega t}\,dt = \left\{\frac{e^{-i\omega t}}{-i\omega}\right\}_{-L/2}^{L/2}$

$$= \frac{-1}{i\omega}\left(e^{-i\omega L/2} - e^{i\omega L/2}\right) = \frac{1}{i\omega}\left(e^{i\omega L/2} - e^{-i\omega L/2}\right) = \frac{2}{i\omega}\sinh\left(\frac{i\omega L}{2}\right).$$

34. $F(\omega) = \displaystyle\int_{-\infty}^\infty f(t)e^{-i\omega t}\,dt = \int_{-\infty}^\infty e^{-a|t|}e^{-i\omega t}\,dt = \int_{-\infty}^0 e^{(a-i\omega)t}\,dt + \int_0^\infty e^{-(a+i\omega)t}\,dt$

$$= \lim_{b\to-\infty}\left\{\frac{e^{(a-i\omega)t}}{a-i\omega}\right\}_b^0 + \lim_{b\to\infty}\left\{\frac{e^{-(a+i\omega)t}}{-(a+i\omega)}\right\}_0^b$$

$$= \frac{1}{a-i\omega}\lim_{b\to-\infty}[1 - e^{(a-i\omega)b}] - \frac{1}{a+i\omega}\lim_{b\to\infty}[e^{-(a+i\omega)b} - 1] = \frac{1}{a-i\omega} + \frac{1}{a+i\omega} = \frac{2a}{\omega^2 + a^2}.$$

36. If we set $u = \text{Tan}^{-1}\sqrt{x}$, $dv = dx$, $du = \dfrac{1}{2\sqrt{x}(1+x)}dx$, and $v = x$,

$$\int \text{Tan}^{-1}\sqrt{x}\,dx = x\,\text{Tan}^{-1}\sqrt{x} - \int \frac{x}{2\sqrt{x}(1+x)}dx = x\,\text{Tan}^{-1}\sqrt{x} - \frac{1}{2}\int \frac{\sqrt{x}}{1+x}dx.$$

If we now set $u = \sqrt{x}$, from which $x = u^2$, and $dx = 2u\,du$, then

$$\int \text{Tan}^{-1}\sqrt{x}\,dx = x\,\text{Tan}^{-1}\sqrt{x} - \frac{1}{2}\int \frac{u}{1+u^2}(2u\,du) = x\,\text{Tan}^{-1}\sqrt{x} - \int\left(1 - \frac{1}{1+u^2}\right)du$$

$$= x\,\text{Tan}^{-1}\sqrt{x} - u + \text{Tan}^{-1}u + C = x\,\text{Tan}^{-1}\sqrt{x} - \sqrt{x} + \text{Tan}^{-1}\sqrt{x} + C.$$

38. First we evaluate the integral of $e^x \sin x$. When we set $u = e^x$, $dv = \sin x\,dx$, then $du = e^x\,dx$, $v = -\cos x$, and

$$\int e^x \sin x\,dx = -e^x \cos x - \int -e^x \cos x\,dx.$$

We now set $u = e^x$, $dv = \cos x\,dx$, in which case $du = e^x\,dx$, $v = \sin x$, and

$$\int e^x \sin x\,dx = -e^x \cos x + \left(e^x \sin x - \int e^x \sin x\,dx\right).$$

If we bring both integrals to the left,

$$2\int e^x \sin x\,dx = e^x \sin x - e^x \cos x \quad \Longrightarrow \quad \int e^x \sin x\,dx = \frac{e^x}{2}(\sin x - \cos x) + C.$$

In the given integral we now set $u = x$, $dv = e^x \sin x\,dx$, $du = dx$, and $v = e^x(\sin x - \cos x)/2$, in which case

$$\int x\,e^x \sin x\,dx = \frac{x\,e^x(\sin x - \cos x)}{2} - \int \frac{e^x(\sin x - \cos x)}{2}dx.$$

We now need the integral of $e^x \cos x$. When we set $u = e^x$, $dv = \cos x\,dx$, $du = e^x\,dx$, $v = \sin x$, then

$$\int e^x \cos x\,dx = e^x \sin x - \int e^x \sin x\,dx.$$

We now set $u = e^x$, $dv = \sin x\,dx$, in which case $du = e^x\,dx$, $v = -\cos x$, and

$$\int e^x \cos x\,dx = e^x \sin x - \left(-e^x \cos x - \int -e^x \cos x\,dx\right).$$

If we bring both integrals to the left,

$$2\int e^x \cos x\,dx = e^x \sin x + e^x \cos x \quad \Longrightarrow \quad \int e^x \cos x\,dx = \frac{e^x}{2}(\sin x + \cos x) + C.$$

We can now say that

$$\frac{1}{2}\int e^x(\sin x - \cos x)\,dx = \frac{1}{2}\left[\frac{e^x(\sin x - \cos x)}{2} - \frac{e^x(\sin x + \cos x)}{2}\right] = -\frac{e^x \cos x}{2}.$$

Finally,

$$\int x\,e^x \sin x\,dx = \frac{x\,e^x(\sin x - \cos x)}{2} + \frac{e^x \cos x}{2} + C.$$

EXERCISES 8.3

2. $\displaystyle\int \frac{\cos x}{\sin^3 x}dx = -\frac{1}{2\sin^2 x} + C$ 　　　　　　　　**4.** $\displaystyle\int \csc^3 x \cot x\,dx = -\frac{1}{3}\csc^3 x + C$

6. $\displaystyle\int \sqrt{\tan x}\,\sec^4 x\,dx = \int \sqrt{\tan x}\,(1 + \tan^2 x)\,\sec^2 x\,dx$

$\displaystyle\qquad = \int (\tan^{1/2} x + \tan^{5/2} x)\,\sec^2 x\,dx = \frac{2}{3}\tan^{3/2} x + \frac{2}{7}\tan^{7/2} x + C$

8. $\displaystyle\int \sec^6 3x\,\tan 3x\,dx = \frac{1}{18}\sec^6 3x + C$

10. $\displaystyle\int \frac{\tan^3 x\,\sec^2 x}{\sin^2 x}\,dx = \int \frac{\sin x}{\cos^5 x}\,dx = \frac{1}{4\cos^4 x} + C = \frac{1}{4}\sec^4 x + C$

12. $\displaystyle\int \frac{\csc^2 \theta}{\cot^2 \theta}\,d\theta = \frac{1}{\cot \theta} + C = \tan \theta + C$

14. $\displaystyle\int \frac{\sec^2 x}{\sqrt{1 + \tan x}}\,dx = 2\sqrt{1 + \tan x} + C$

16. $\displaystyle\int \frac{3 + 4\csc^2 x}{\cot^2 x}\,dx = \int (3\tan^2 x + 4\sec^2 x)\,dx = \int (3\sec^2 x - 3 + 4\sec^2 x)\,dx = 7\tan x - 3x + C$

18. $\displaystyle\int \sin^4 x\,dx = \int \left(\frac{1 - \cos 2x}{2}\right)^2 dx = \frac{1}{4}\int (1 - 2\cos 2x + \cos^2 2x)\,dx$

$\displaystyle\qquad = \frac{1}{4}\int \left(1 - 2\cos 2x + \frac{1 + \cos 4x}{2}\right)dx = \frac{1}{4}\left(\frac{3x}{2} - \sin 2x + \frac{1}{8}\sin 4x\right) + C$

20. $\displaystyle\int \frac{\csc^4 x}{\cot^3 x}\,dx = \int \frac{(1 + \cot^2 x)\csc^2 x}{\cot^3 x}\,dx = \int \left(\frac{1}{\cot^3 x} + \frac{1}{\cot x}\right)\csc^2 x\,dx$

$\displaystyle\qquad = \frac{1}{2\cot^2 x} - \ln|\cot x| + C = \frac{1}{2}\tan^2 x + \ln|\tan x| + C$

22. $\displaystyle V = \int_0^{\pi/4} [\pi(1 + \tan x)^2 - \pi(1)^2]\,dx$

$\displaystyle\qquad = \pi \int_0^{\pi/4} (\tan^2 x + 2\tan x)\,dx$

$\displaystyle\qquad = \pi \int_0^{\pi/4} (\sec^2 x - 1 + 2\tan x)\,dx$

$\displaystyle\qquad = \pi\Big\{\tan x - x + 2\ln|\sec x|\Big\}_0^{\pi/4} = \pi(1 - \pi/4 + \ln 2)$

24. $\displaystyle\int \cot^4 z\,dz = \int \cot^2 z(\csc^2 z - 1)\,dz = \int (\cot^2 z\,\csc^2 z - \csc^2 z + 1)\,dz = -\frac{1}{3}\cot^3 z + \cot z + z + C$

26. $\displaystyle\int \frac{\cos^4 \theta}{1 + \sin \theta}\,d\theta = \int \frac{(1 - \sin^2 \theta)\cos^2 \theta}{1 + \sin \theta}\,d\theta = \int \frac{(1 - \sin \theta)(1 + \sin \theta)\cos^2 \theta}{1 + \sin \theta}\,d\theta = \int (1 - \sin \theta)\cos^2 \theta\,d\theta$

$\displaystyle\qquad = \int \left(\frac{1 + \cos 2\theta}{2} - \cos^2 \theta\,\sin \theta\right)d\theta = \frac{\theta}{2} + \frac{1}{4}\sin 2\theta + \frac{1}{3}\cos^3 \theta + C$

28. Using identity 1.48c, $\displaystyle\int \cos 6x\,\cos 2x\,dx = \int \frac{1}{2}(\cos 8x + \cos 4x)\,dx = \frac{1}{16}\sin 8x + \frac{1}{8}\sin 4x + C$

30. $\displaystyle\int \frac{1}{\sin x\,\cos^2 x}\,dx = \int \csc x\,\sec^2 x\,dx = \int \csc x(1 + \tan^2 x)\,dx$

$\displaystyle\qquad = \int (\csc x + \sec x\,\tan x)\,dx = \ln|\csc x - \cot x| + \sec x + C$

32. The average power is the integral of Vi over one period $2\pi/\omega$ divided by the period,

$$P_{av} = \frac{\omega}{2\pi} \int_0^{2\pi/\omega} V_m \cos(\omega t + \phi_2)\,i_m \cos(\omega t + \phi_1)\,dt$$

$$= \frac{\omega V_m i_m}{2\pi} \int_0^{2\pi/\omega} \frac{1}{2}[\cos(2\omega t + \phi_1 + \phi_2) + \cos(\phi_1 - \phi_2)]\,dt$$

$$= \frac{\omega V_m i_m}{4\pi} \left\{ \frac{1}{2\omega} \sin(2\omega t + \phi_1 + \phi_2) + t \cos(\phi_1 - \phi_2) \right\}_0^{2\pi/\omega}$$

$$= \frac{\omega V_m i_m}{4\pi} \left[\frac{1}{2\omega} \sin(4\pi + \phi_1 + \phi_2) - \frac{1}{2\omega} \sin(\phi_1 + \phi_2) + \frac{2\pi}{\omega} \cos(\phi_1 - \phi_2) \right]$$

$$= \frac{V_m i_m}{2} \cos(\phi_1 - \phi_2).$$

34. $F_{dc} = \dfrac{\omega}{2\pi} \displaystyle\int_{-\pi/\omega}^{\pi/\omega} (A \cos \omega t + B \sin \omega t) \, dt = \dfrac{\omega}{2\pi} \left\{ \dfrac{A}{\omega} \sin \omega t - \dfrac{B}{\omega} \cos \omega t \right\}_{-\pi/\omega}^{\pi/\omega} = 0$

36. $F_{dc} = \dfrac{\omega}{\pi} \displaystyle\int_{-\pi/(2\omega)}^{\pi/(2\omega)} \sin^2 \omega t \, dt = \dfrac{\omega}{\pi} \displaystyle\int_{-\pi/(2\omega)}^{\pi/(2\omega)} \left(\dfrac{1 - \cos 2\omega t}{2} \right) dt = \dfrac{\omega}{2\pi} \left\{ t - \dfrac{\sin 2\omega t}{2\omega} \right\}_{-\pi/(2\omega)}^{\pi/(2\omega)} = \dfrac{1}{2}$

EXERCISES 8.4

2. If we set $x = \dfrac{3}{\sqrt{5}} \sin \theta$, then $dx = \dfrac{3}{\sqrt{5}} \cos \theta \, d\theta$, and

$$\int \frac{1}{\sqrt{9 - 5x^2}} dx = \int \frac{1}{3 \cos \theta} \left(\frac{3}{\sqrt{5}} \right) \cos \theta \, d\theta = \frac{\theta}{\sqrt{5}} + C = \frac{1}{\sqrt{5}} \operatorname{Sin}^{-1} \left(\frac{\sqrt{5}x}{3} \right) + C.$$

4. If we set $x = 2 \sin \theta$, then $dx = 2 \cos \theta \, d\theta$, and

$$\int \frac{1}{x^2 \sqrt{4 - x^2}} dx = \int \frac{1}{4 \sin^2 \theta \, 2 \cos \theta} 2 \cos \theta \, d\theta = \frac{1}{4} \int \csc^2 \theta \, d\theta$$

$$= -\frac{1}{4} \cot \theta + C = -\frac{1}{4} \frac{\sqrt{4 - x^2}}{x} + C.$$

6. $\displaystyle\int x\sqrt{5x^2 + 3} \, dx = \frac{1}{15}(5x^2 + 3)^{3/2} + C$

8. If we set $x = \sin \theta$, then $dx = \cos \theta \, d\theta$, and

$$\int \frac{1}{1 - x^2} dx = \int \frac{1}{\cos^2 \theta} \cos \theta \, d\theta = \int \sec \theta \, d\theta = \ln |\sec \theta + \tan \theta| + C$$

$$= \ln \left| \frac{1}{\sqrt{1 - x^2}} + \frac{x}{\sqrt{1 - x^2}} \right| + C = \ln \left| \frac{1 + x}{\sqrt{(1 - x)(1 + x)}} \right| + C$$

$$= \ln \left| \frac{\sqrt{1 + x}}{\sqrt{1 - x}} \right| + C = \frac{1}{2} \ln \left| \frac{1 + x}{1 - x} \right| + C.$$

10. $\displaystyle\int \frac{x + 5}{10x^2 + 2} dx = \int \frac{x}{10x^2 + 2} dx + \frac{5}{2} \int \frac{1}{5x^2 + 1} dx$

In the second term we set $x = \dfrac{1}{\sqrt{5}} \tan \theta$ and $dx = \dfrac{1}{\sqrt{5}} \sec^2 \theta \, d\theta$,

$$\int \frac{x + 5}{10x^2 + 2} dx = \frac{1}{20} \ln(10x^2 + 2) + \frac{5}{2} \int \frac{1}{\sec^2 \theta} \left(\frac{1}{\sqrt{5}} \right) \sec^2 \theta \, d\theta$$

$$= \frac{1}{20} \ln(10x^2 + 2) + \frac{\sqrt{5}}{2} \theta + C = \frac{1}{20} \ln(5x^2 + 1) + \frac{\sqrt{5}}{2} \operatorname{Tan}^{-1}(\sqrt{5}x) + D.$$

12. If we set $x = 2 \sin \theta$, then $dx = 2 \cos \theta \, d\theta$, and

$$\int \frac{\sqrt{4 - x^2}}{x} dx = \int \frac{2 \cos \theta}{2 \sin \theta} 2 \cos \theta \, d\theta = 2 \int \frac{1 - \sin^2 \theta}{\sin \theta} d\theta$$

$$= 2 \int (\csc \theta - \sin \theta) \, d\theta = 2(\ln |\csc \theta - \cot \theta| + \cos \theta) + C$$

$$= 2 \left(\ln \left| \frac{2}{x} - \frac{\sqrt{4 - x^2}}{x} \right| + \frac{\sqrt{4 - x^2}}{2} \right) + C = 2 \ln \left| \frac{2 - \sqrt{4 - x^2}}{x} \right| + \sqrt{4 - x^2} + C.$$

14. If we set $x = 4 \sec \theta$, then $dx = 4 \sec \theta \tan \theta \, d\theta$, and

$$\int \frac{\sqrt{x^2 - 16}}{x^2} dx = \int \frac{4 \tan \theta}{16 \sec^2 \theta} 4 \sec \theta \tan \theta \, d\theta = \int \frac{\sec^2 \theta - 1}{\sec \theta} d\theta$$

$$= \int (\sec \theta - \cos \theta) \, d\theta = \ln |\sec \theta + \tan \theta| - \sin \theta + C$$

$$= \ln \left| \frac{x}{4} + \frac{\sqrt{x^2 - 16}}{4} \right| - \frac{\sqrt{x^2 - 16}}{x} + C = \ln \left| x + \sqrt{x^2 - 16} \right| - \frac{\sqrt{x^2 - 16}}{x} + D.$$

16. If we set $x = 2 \sec \theta$, then $dx = 2 \sec \theta \tan \theta \, d\theta$, and

$$\int \frac{1}{x^3 \sqrt{x^2 - 4}} dx = \int \frac{1}{8 \sec^3 \theta \, 2 \tan \theta} 2 \sec \theta \tan \theta \, d\theta = \frac{1}{8} \int \cos^2 \theta \, d\theta$$

$$= \frac{1}{8} \int \left(\frac{1 + \cos 2\theta}{2} \right) d\theta = \frac{1}{16} \left(\theta + \frac{1}{2} \sin 2\theta \right) + C$$

$$= \frac{\theta}{16} + \frac{1}{16} \sin \theta \cos \theta + C$$

$$= \frac{1}{16} \operatorname{Sec}^{-1} \left(\frac{x}{2} \right) + \frac{1}{16} \frac{\sqrt{x^2 - 4}}{x} \frac{2}{x} + C = \frac{1}{16} \operatorname{Sec}^{-1} \left(\frac{x}{2} \right) + \frac{\sqrt{x^2 - 4}}{8x^2} + C.$$

18. If we set $y = 2 \tan \theta$, then $dy = 2 \sec^2 \theta \, d\theta$, and

$$\int \frac{y^3}{\sqrt{y^2 + 4}} dy = \int \frac{8 \tan^3 \theta}{2 \sec \theta} 2 \sec^2 \theta \, d\theta = 8 \int \tan \theta \, (\sec^2 \theta - 1) \sec \theta \, d\theta$$

$$= 8 \left(\frac{\sec^3 \theta}{3} - \sec \theta \right) + C = \frac{8}{3} \left(\frac{\sqrt{y^2 + 4}}{2} \right)^3 - \frac{8 \sqrt{y^2 + 4}}{2} + C$$

$$= \frac{1}{3} (y^2 + 4)^{3/2} - 4 \sqrt{y^2 + 4} + C.$$

20. If we set $x = \tan \theta$, then $dx = \sec^2 \theta \, d\theta$, and

$$\int \frac{x^2 + 2}{x^3 + x} dx = \int \frac{x^2 + 2}{x(x^2 + 1)} dx = \int \frac{\tan^2 \theta + 2}{\tan \theta \sec^2 \theta} \sec^2 \theta \, d\theta$$

$$= \int (\tan \theta + 2 \cot \theta) \, d\theta = \ln |\sec \theta| + 2 \ln |\sin \theta| + C$$

$$= \ln |\sqrt{x^2 + 1}| + 2 \ln \left| \frac{x}{\sqrt{x^2 + 1}} \right| + C = 2 \ln |x| - \frac{1}{2} \ln (x^2 + 1) + C.$$

22. $A = 4 \displaystyle\int_0^a \frac{b}{a} \sqrt{a^2 - x^2} \, dx$

If we set $x = a \sin \theta$, then $dx = a \cos \theta \, d\theta$, and

$$A = \frac{4b}{a} \int_0^{\pi/2} a \cos \theta \, a \cos \theta \, d\theta = 4ab \int_0^{\pi/2} \left(\frac{1 + \cos 2\theta}{2} \right) d\theta$$

$$= 2ab \left\{ \theta + \frac{1}{2} \sin 2\theta \right\}_0^{\pi/2} = \pi ab.$$

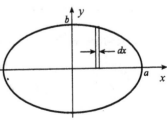

24. A plot of the ellipse is shown to the right. If we solve

$$y^2 - (2x+4)y + (2x^2 + 4x + 3) = 0$$

for y in terms of x, we obtain

$$y = \frac{2x + 4 \pm \sqrt{(2x+4)^2 - 4(2x^2 + 4x + 3)}}{2}$$

$$= \frac{2x + 4 \pm \sqrt{4 - 4x^2}}{2}$$

$$= x + 2 \pm \sqrt{1 - x^2}.$$

It follows that a rectangle of width dx at position x between the top and bottom of the ellipse has area

$$[(x + 2 + \sqrt{1 - x^2}) - (x + 2 - \sqrt{1 - x^2})]dx = 2\sqrt{1 - x^2}\, dx.$$

Since the ellipse extends from $x = -1$ to $x = 1$, its area must be $\quad A = \int_{-1}^{1} 2\sqrt{1 - x^2}\, dx.$

Setting $x = \sin\theta$ and $dx = \cos\theta\, d\theta$,

$$A = 2 \int_{-\pi/2}^{\pi/2} \cos\theta\cos\theta\, d\theta = 2 \int_{-\pi/2}^{\pi/2} \left(\frac{1 + \cos 2\theta}{2}\right) d\theta = \left\{\theta + \frac{1}{2}\sin 2\theta\right\}_{-\pi/2}^{\pi/2} = \pi.$$

26. $I = 2 \int_{0}^{\tilde{y}} y^2(\sqrt{4 - y^2} - y^2)\, dy$

In the first integral we set $y = 2\sin\theta$ and $dy = 2\cos\theta\, d\theta$. If $\tilde{\theta} = \text{Sin}^{-1}(\tilde{y}/2)$, then

$$I = 2 \int_{0}^{\tilde{\theta}} 4\sin^2\theta\, 2\cos\theta\, 2\cos\theta\, d\theta - 2\left\{\frac{y^5}{5}\right\}_{0}^{\tilde{y}}$$

$$= 32 \int_{0}^{\tilde{\theta}} \sin^2\theta\cos^2\theta\, d\theta - \frac{2\tilde{y}^5}{5}$$

$$= 32 \int_{0}^{\tilde{\theta}} \frac{1}{4}\sin^2 2\theta\, d\theta - \frac{2\tilde{y}^5}{5} = 8 \int_{0}^{\tilde{\theta}} \left(\frac{1 - \cos 4\theta}{2}\right) d\theta - \frac{2\tilde{y}^5}{5}$$

$$= 4\left\{\theta - \frac{1}{4}\sin 4\theta\right\}_{0}^{\tilde{\theta}} - \frac{2\tilde{y}^5}{5} = 4\tilde{\theta} - \sin 4\tilde{\theta} - \frac{2\tilde{y}^5}{5} = 1.053.$$

28. According to Exercise 22, the area of the ellipse is πab. If we let the required line be $y = c$, then c must satisfy the equation

$$\frac{\pi ab}{3} = 2 \int_{c}^{b} \frac{a}{b}\sqrt{b^2 - y^2}\, dy.$$

We let $y = b\sin\theta$ and $dy = b\cos\theta\, d\theta$. If $\tilde{\theta} = \text{Sin}^{-1}(c/b)$, then

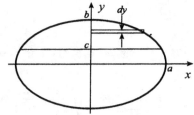

$$\frac{\pi ab}{3} = \frac{2a}{b} \int_{\tilde{\theta}}^{\pi/2} b\cos\theta\, b\cos\theta\, d\theta$$

$$= 2ab \int_{\tilde{\theta}}^{\pi/2} \left(\frac{1 + \cos 2\theta}{2}\right) d\theta = ab\left\{\theta + \frac{1}{2}\sin 2\theta\right\}_{\tilde{\theta}}^{\pi/2} = ab\left(\frac{\pi}{2} - \tilde{\theta} - \frac{1}{2}\sin 2\tilde{\theta}\right).$$

Thus, $\tilde{\theta}$ must satisfy the equation $\quad \dfrac{\pi}{3} = \dfrac{\pi}{2} - \tilde{\theta} - \dfrac{1}{2}\sin 2\tilde{\theta}$, or, $\quad 6\tilde{\theta} + 3\sin 2\tilde{\theta} - \pi = 0$. Newton's iterative procedure with $\tilde{\theta}_1 = 0.25$, $\quad \tilde{\theta}_{n+1} = \tilde{\theta}_n - \dfrac{6\tilde{\theta}_n + 3\sin 2\tilde{\theta}_n - \pi}{6 + 6\cos 2\tilde{\theta}_n}$ gives the iterations $\tilde{\theta}_2 = 0.268$, $\tilde{\theta}_3 = 0.268\,133$, $\tilde{\theta}_4 = 0.268\,133$. Hence, the required line is $y = b\sin\tilde{\theta} = 0.265b$.

30. $F = \displaystyle\int_{-r}^{r} 9.81\rho(r-y)2x\,dy = 19.62\rho \int_{-r}^{r} (r-y)\sqrt{r^2-y^2}\,dy$

If we set $y = r\sin\theta$, then $dy = r\cos\theta\,d\theta$, and

$$F = 19.62\rho \int_{-\pi/2}^{\pi/2} (r - r\sin\theta)r\cos\theta\, r\cos\theta\,d\theta$$

$$= 19.62\rho r^3 \int_{-\pi/2}^{\pi/2} (1 - \sin\theta)\cos^2\theta\,d\theta$$

$$= 19.62\rho r^3 \int_{-\pi/2}^{\pi/2} \left[\frac{1}{2}(1 + \cos 2\theta) - \cos^2\theta\,\sin\theta\right] d\theta$$

$$= 19.62\rho r^3 \left\{\frac{\theta}{2} + \frac{1}{4}\sin 2\theta + \frac{1}{3}\cos^3\theta\right\}_{-\pi/2}^{\pi/2} = 9.81\pi\rho r^3.$$

32. If we set $x = \sqrt{7}\sin\theta$, then $dx = \sqrt{7}\cos\theta\,d\theta$, and

$$\int (7 - x^2)^{3/2}\,dx = \int 7\sqrt{7}\cos^3\theta\,\sqrt{7}\cos\theta\,d\theta = 49\int \left(\frac{1 + \cos 2\theta}{2}\right)^2 d\theta$$

$$= \frac{49}{4} \int \left[1 + 2\cos 2\theta + \frac{1}{2}(1 + \cos 4\theta)\right] d\theta$$

$$= \frac{49}{4} \left(\frac{3\theta}{2} + \sin 2\theta + \frac{1}{8}\sin 4\theta\right) + C$$

$$= \frac{147}{8}\theta + \frac{49}{2}\sin\theta\,\cos\theta + \frac{49}{16}\sin 2\theta\,\cos 2\theta + C$$

$$= \frac{147}{8}\theta + \frac{49}{2}\sin\theta\,\cos\theta + \frac{49}{8}\sin\theta\,\cos\theta(1 - 2\sin^2\theta) + C$$

$$= \frac{147}{8}\mathrm{Sin}^{-1}\left(\frac{x}{\sqrt{7}}\right) + \frac{245}{8}\frac{x}{\sqrt{7}}\frac{\sqrt{7 - x^2}}{\sqrt{7}} - \frac{49}{4}\left(\frac{x}{\sqrt{7}}\right)^3\frac{\sqrt{7 - x^2}}{\sqrt{7}} + C$$

$$= \frac{147}{8}\mathrm{Sin}^{-1}\left(\frac{x}{\sqrt{7}}\right) + \frac{x}{8}(35 - 2x^2)\sqrt{7 - x^2} + C.$$

34. If we set $x = (1/2)\sec\theta$, then $dx = (1/2)\sec\theta\tan\theta\,d\theta$, and

$$\int \frac{1}{x^3(4x^2 - 1)^{3/2}}\,dx = \int \frac{1}{(1/8)\sec^3\theta\,\tan^3\theta}(1/2)\sec\theta\tan\theta\,d\theta = 4\int \frac{\cos^4\theta}{\sin^2\theta}\,d\theta$$

$$= 4\int \frac{\cos^2\theta(1 - \sin^2\theta)}{\sin^2\theta}\,d\theta = 4\int (\cot^2\theta - \cos^2\theta)\,d\theta$$

$$= 4\int \left(\csc^2\theta - 1 - \frac{1 + \cos 2\theta}{2}\right) d\theta$$

$$= 4\left(-\cot\theta - \frac{3\theta}{2} - \frac{1}{4}\sin 2\theta\right) + C$$

$$= 4\left[-\frac{1}{\sqrt{4x^2 - 1}} - \frac{3}{2}\mathrm{Sec}^{-1}(2x) - \frac{1}{2}\frac{\sqrt{4x^2 - 1}}{2x}\frac{1}{2x}\right] + C$$

$$= -6\,\mathrm{Sec}^{-1}(2x) + \frac{1 - 12x^2}{2x^2\sqrt{4x^2 - 1}} + C.$$

36. If we set $x = (1/\sqrt{3})\tan\theta$, then $dx = (1/\sqrt{3})\sec^2\theta\,d\theta$, and

$$\int \sqrt{1+3x^2}\,dx = \int \sec\theta\left(\frac{1}{\sqrt{3}}\right)\sec^2\theta\,d\theta = \frac{1}{\sqrt{3}}\int \sec^3\theta\,d\theta$$

$$= \frac{1}{2\sqrt{3}}\left(\sec\theta\tan\theta + \ln|\sec\theta + \tan\theta|\right) + C \quad \text{(using Example 8.9)}$$

$$= \frac{1}{2\sqrt{3}}\left(\sqrt{1+3x^2}\,\sqrt{3}x + \ln|\sqrt{1+3x^2} + \sqrt{3}x|\right) + C$$

$$= \frac{x}{2}\sqrt{1+3x^2} + \frac{1}{2\sqrt{3}}\ln|\sqrt{1+3x^2} + \sqrt{3}x| + C.$$

38. Differentiation of $8y^2 = x^2 - x^4$ gives

$$16y\frac{dy}{dx} = 2x - 4x^3.$$

Therefore, $\dfrac{dy}{dx} = \dfrac{x - 2x^3}{8y}$. Small lengths along that part of the curve in the first quadrant are approximated by

$$\sqrt{1 + \left(\frac{x-2x^3}{8y}\right)^2}\,dx = \sqrt{1 + \frac{(x-2x^3)^2}{64y^2}}\,dx = \sqrt{1 + \frac{x^2(1-2x^2)^2}{8x^2(1-x^2)}}\,dx$$

$$= \sqrt{\frac{9 - 12x^2 + 4x^4}{8(1-x^2)}}\,dx = \frac{3 - 2x^2}{2\sqrt{2}\sqrt{1-x^2}}\,dx.$$

The length of the curve is therefore $L = 4\displaystyle\int_0^1 \frac{3 - 2x^2}{2\sqrt{2}\sqrt{1-x^2}}\,dx$. When we set $x = \sin\theta$, and $dx = \cos\theta\,d\theta$,

$$L = \sqrt{2}\int_0^{\pi/2} \frac{3 - 2\sin^2\theta}{\cos\theta}\cos\theta\,d\theta = \sqrt{2}\int_0^{\pi/2}[3 - (1 - \cos 2\theta)]\,d\theta = \sqrt{2}\left\{2\theta + \frac{1}{2}\sin 2\theta\right\}_0^{\pi/2} = \sqrt{2}\pi.$$

40. Let the radius of the circle be r, and let the position of the line be denoted by $x = a$. Then the requirement that the second moment of area about $x = a$ be twice that about $x = 0$ can be expressed as

$$2\int_{-r}^{r}(x-a)^2\sqrt{r^2 - x^2}\,dx = 2(2)\int_{-r}^{r} x^2\sqrt{r^2 - x^2}\,dx.$$

If we set $x = r\sin\theta$ and $dx = r\cos\theta\,d\theta$ in these integrals, then

$$\int_{-\pi/2}^{\pi/2}(r\sin\theta - a)^2 r\cos\theta\, r\cos\theta\,d\theta = 2\int_{-\pi/2}^{\pi/2} r^2\sin^2\theta\, r\cos\theta\, r\cos\theta\,d\theta,$$

or,

$$0 = r^2\int_{-\pi/2}^{\pi/2}(2r^2\sin^2\theta\cos^2\theta - r^2\sin^2\theta\cos^2\theta + 2ar\cos^2\theta\sin\theta - a^2\cos^2\theta)\,d\theta$$

$$= r^2\int_{-\pi/2}^{\pi/2}\left(\frac{r^2}{4}\sin^2 2\theta + 2ar\cos^2\theta\sin\theta - a^2\cos^2\theta\right)d\theta$$

$$= r^2\int_{-\pi/2}^{\pi/2}\left[\frac{r^2}{4}\left(\frac{1 - \cos 4\theta}{2}\right) + 2ar\cos^2\theta\sin\theta - a^2\left(\frac{1 + \cos 2\theta}{2}\right)\right]d\theta$$

$$= r^2\left\{\frac{r^2}{8}\left(\theta - \frac{1}{4}\sin 4\theta\right) - \frac{2ar}{3}\cos^3\theta - \frac{a^2}{2}\left(\theta + \frac{1}{2}\sin 2\theta\right)\right\}_{-\pi/2}^{\pi/2} = r^2\left(\frac{\pi r^2}{8} - \frac{\pi a^2}{2}\right).$$

Thus, $a = r/2$.

42. (a) If we set $p = \tan\theta$ and $dp = \sec^2\theta \, d\theta$, then

$$kx + C = \int \frac{1}{\sec\theta} \sec^2\theta \, d\theta = \ln|\sec\theta + \tan\theta| = \ln|\sqrt{1+p^2} + p|.$$

Exponentiation gives $\sqrt{1+p^2} + p = De^{kx}$, where $D = e^C$. Since $p(0) = f'(0) = 0$, we obtain $D = 1$. Hence,

$$\sqrt{1+p^2} = e^{kx} - p \quad \Longrightarrow \quad 1 + p^2 = e^{2kx} - 2pe^{kx} + p^2.$$

This can be solved for $\quad p = \dfrac{dy}{dx} = \dfrac{e^{2kx} - 1}{2e^{kx}} = \dfrac{1}{2}(e^{kx} - e^{-kx}).$

(b) Integration now yields $y = \dfrac{1}{2k}(e^{kx} + e^{-kx}) + C.$

EXERCISES 8.5

2. Since $x^2 + 2x + 2 = (x+1)^2 + 1$, we set $x + 1 = \tan\theta$, in which case $dx = \sec^2\theta \, d\theta$, and

$$\int \frac{1}{\sqrt{x^2 + 2x + 2}} dx = \int \frac{1}{\sec\theta} \sec^2\theta \, d\theta$$
$$= \ln|\sec\theta + \tan\theta| + C$$
$$= \ln|\sqrt{x^2 + 2x + 2} + x + 1| + C.$$

4. Since $-x^2 + 3x - 4 = -(x - 3/2)^2 - 7/4$, we set $x - 3/2 = (\sqrt{7}/2)\tan\theta$, in which case $dx = (\sqrt{7}/2)\sec^2\theta \, d\theta$, and

$$\int \frac{1}{3x - x^2 - 4} dx = \int \frac{1}{-(7/4)\sec^2\theta}(\sqrt{7}/2)\sec^2\theta \, d\theta = -\frac{2}{\sqrt{7}}\theta + C = -\frac{2}{\sqrt{7}}\operatorname{Tan}^{-1}\left(\frac{2x-3}{\sqrt{7}}\right) + C.$$

6. Since $4x - x^2 = -(x-2)^2 + 4$, we set $x - 2 = 2\sin\theta$, in which case $dx = 2\cos\theta \, d\theta$, and

$$\int \frac{x}{(4x - x^2)^{3/2}} dx = \int \frac{2 + 2\sin\theta}{8\cos^3\theta} 2\cos\theta \, d\theta = \frac{1}{2}\int \frac{1 + \sin\theta}{\cos^2\theta} d\theta$$
$$= \frac{1}{2}\int \left(\sec^2\theta + \frac{\sin\theta}{\cos^2\theta}\right) d\theta = \frac{1}{2}\left(\tan\theta + \frac{1}{\cos\theta}\right) + C$$
$$= \frac{1}{2}\left(\frac{x-2}{\sqrt{4x - x^2}} + \frac{2}{\sqrt{4x - x^2}}\right) + C$$
$$= \frac{x}{2\sqrt{4x - x^2}} + C.$$

8. Since $x^2 + 6x + 13 = (x+3)^2 + 4$, we set $x + 3 = 2\tan\theta$, in which case $dx = 2\sec^2\theta \, d\theta$, and

$$\int \frac{2x - 3}{x^2 + 6x + 13} dx = \int \frac{2(2\tan\theta - 3) - 3}{4\sec^2\theta} 2\sec^2\theta \, d\theta = \frac{1}{2}(4\ln|\sec\theta| - 9\theta) + C$$
$$= 2\ln\left|\frac{\sqrt{x^2 + 6x + 13}}{2}\right| - \frac{9}{2}\operatorname{Tan}^{-1}\left(\frac{x+3}{2}\right) + C$$
$$= \ln(x^2 + 6x + 13) - \frac{9}{2}\operatorname{Tan}^{-1}\left(\frac{x+3}{2}\right) + D.$$

10. Since $6 + 4\ln x + (\ln x)^2 = (\ln x + 2)^2 + 2$, we set $\ln x + 2 = \sqrt{2}\tan\theta$. Then $(1/x)dx = \sqrt{2}\sec^2\theta \, d\theta$ and

$$\int \frac{1}{x\sqrt{6 + 4\ln x + (\ln x)^2}} dx = \int \frac{1}{\sqrt{2}\sec\theta}\sqrt{2}\sec^2\theta \, d\theta = \ln|\sec\theta + \tan\theta| + C$$
$$= \ln\left|\frac{\sqrt{(\ln x)^2 + 4\ln x + 6}}{\sqrt{2}} + \frac{\ln x + 2}{\sqrt{2}}\right| + C$$
$$= \ln|\sqrt{(\ln x)^2 + 4\ln x + 6} + \ln x + 2| + D.$$

12. $F = \displaystyle\int_{-5}^{-3} 9810(-y)2x\,dy = -19\,620 \int_{-5}^{-3} y\sqrt{1-(y+4)^2}\,dy$

If we set $y + 4 = \sin\theta$, then $dy = \cos\theta\,d\theta$, and

$$F = -19\,620 \int_{-\pi/2}^{\pi/2} (\sin\theta - 4)\cos\theta\,\cos\theta\,d\theta$$

$$= -19\,620 \int_{-\pi/2}^{\pi/2} [\cos^2\theta\,\sin\theta - 2(1+\cos 2\theta)]\,d\theta$$

$$= -19\,620 \left\{ -\frac{1}{3}\cos^3\theta - 2\theta - \sin 2\theta \right\}_{-\pi/2}^{\pi/2} = 39\,240\pi \text{ N.}$$

14. Since $x^2 - 2x - 3 = (x-1)^2 - 4$, we set $x - 1 = 2\sec\theta$ and $dx = 2\sec\theta\tan\theta\,d\theta$,

$$\int \sqrt{x^2 - 2x - 3}\,dx = \int 2\tan\theta\,2\sec\theta\tan\theta\,d\theta = 4\int \tan^2\theta\,\sec\theta\,d\theta$$

$$= 4\int (\sec^2\theta - 1)\sec\theta\,d\theta = 4\int (\sec^3\theta - \sec\theta)\,d\theta$$

$$= 4\left[\frac{1}{2}\ln|\sec\theta + \tan\theta| + \frac{1}{2}\sec\theta\tan\theta - \ln|\sec\theta + \tan\theta| \right] + C \text{ (see Example 8.9)}$$

$$= 2[\sec\theta\tan\theta - \ln|\sec\theta + \tan\theta|] + C$$

$$= 2\left(\frac{x-1}{2}\right)\frac{\sqrt{x^2-2x-3}}{2} - 2\ln\left| \frac{x-1}{2} + \frac{\sqrt{x^2-2x-3}}{2} \right| + C$$

$$= \frac{1}{2}(x-1)\sqrt{x^2-2x-3} - 2\ln|x - 1 + \sqrt{x^2-2x-3}| + D.$$

EXERCISES 8.6

2. $\displaystyle\int \frac{1}{y^3 + 3y^2 + 3y + 1}\,dy = \int \frac{1}{(y+1)^3}\,dy = \frac{-1}{2(y+1)^2} + C$

4. $\displaystyle\int \frac{x^2 + 2x - 4}{x^2 - 2x - 8}\,dx = \int \left(1 + \frac{4x+4}{x^2-2x-8}\right)\,dx = x + 4\int \frac{x+1}{(x-4)(x+2)}\,dx$

If we set $\dfrac{x+1}{(x-4)(x+2)} = \dfrac{A}{x-4} + \dfrac{B}{x+2}$, then $A = 5/6$ and $B = 1/6$, and

$$\int \frac{x^2 + 2x - 4}{x^2 - 2x - 8}\,dx = x + 4\int \left(\frac{5/6}{x-4} + \frac{1/6}{x+2}\right)\,dx = x + \frac{10}{3}\ln|x-4| + \frac{2}{3}\ln|x+2| + C.$$

6. If we set $\dfrac{y+1}{y(y+3)(y-2)} = \dfrac{A}{y} + \dfrac{B}{y+3} + \dfrac{C}{y-2}$, then $A = -1/6$, $B = -2/15$, $C = 3/10$, and

$$\int \frac{y+1}{y^3 + y^2 - 6y}\,dy = \int \left(\frac{-1/6}{y} - \frac{2/15}{y+3} + \frac{3/10}{y-2}\right)\,dy = -\frac{1}{6}\ln|y| - \frac{2}{15}\ln|y+3| + \frac{3}{10}\ln|y-2| + C.$$

8. If we set $\dfrac{x^3}{(x^2+2)^2} = \dfrac{Ax+B}{x^2+2} + \dfrac{Cx+D}{(x^2+2)^2}$, then $A = 1$, $B = 0$, $C = -2$, $D = 0$, and

$$\int \frac{x^3}{(x^2+2)^2}\,dx = \int \left[\frac{x}{x^2+2} - \frac{2x}{(x^2+2)^2}\right]\,dx = \frac{1}{2}\ln(x^2+2) + \frac{1}{x^2+2} + C.$$

10. $\displaystyle\int \frac{y^2}{y^2+3y+2}dy = \int\left(1 + \frac{-3y-2}{y^2+3y+2}\right)dy$ If we set $\dfrac{3y+2}{y^2+3y+2} = \dfrac{A}{y+2} + \dfrac{B}{y+1}$, then $A=4$, $B=-1$, and

$$\int \frac{y^2}{y^2+3y+2}dy = y + \int\left(\frac{-4}{y+2} + \frac{1}{y+1}\right)dy = y - 4\ln|y+2| + \ln|y+1| + C.$$

12. If we set $\dfrac{y^2+6y+4}{(y^2+4)(y^2+1)} = \dfrac{Ay+B}{y^2+4} + \dfrac{Cy+D}{y^2+1}$, then $A=-2$, $B=0$, $C=2$, $D=1$, and

$$\int \frac{y^2+6y+4}{y^4+5y^2+4}dy = \int\left(\frac{-2y}{y^2+4} + \frac{2y+1}{y^2+1}\right)dy = -\ln(y^2+4) + \ln(y^2+1) + \text{Tan}^{-1}y + C.$$

14. If we set $\dfrac{x^2+3}{(x^2+2)(x-1)(x+1)} = \dfrac{Ax+B}{x^2+2} + \dfrac{C}{x-1} + \dfrac{D}{x+1}$, then $A=0$, $B=-1/3$, $C=2/3$, $D=-2/3$, and

$$\int \frac{x^2+3}{x^4+x^2-2}dx = \int\left(\frac{-1/3}{x^2+2} + \frac{2/3}{x-1} - \frac{2/3}{x+1}\right)dx.$$

In the first term we set $x = \sqrt{2}\tan\theta$ and $dx = \sqrt{2}\sec^2\theta\,d\theta$,

$$\int \frac{x^2+3}{x^4+x^2-2}dx = -\frac{1}{3}\int \frac{1}{2\sec^2\theta}\sqrt{2}\sec^2\theta\,d\theta + \frac{2}{3}\ln|x-1| - \frac{2}{3}\ln|x+1|$$

$$= -\frac{1}{3\sqrt{2}}\theta + \frac{2}{3}\ln\left|\frac{x-1}{x+1}\right| + C = -\frac{1}{3\sqrt{2}}\text{Tan}^{-1}\left(\frac{x}{\sqrt{2}}\right) + \frac{2}{3}\ln\left|\frac{x-1}{x+1}\right| + C.$$

16. If we set $\dfrac{x^3+6}{(x-1)^2(x+2)^2} = \dfrac{A}{x-1} + \dfrac{B}{(x-1)^2} + \dfrac{C}{x+2} + \dfrac{D}{(x+2)^2}$, then $A=-5/27$, $B=7/9$, $C=32/27$, $D=-2/9$, and

$$\int \frac{x^3+6}{x^4+2x^3-3x^2-4x+4}dx = \int\left[\frac{-5/27}{x-1} + \frac{7/9}{(x-1)^2} + \frac{32/27}{x+2} - \frac{2/9}{(x+2)^2}\right]dx$$

$$= -\frac{5}{27}\ln|x-1| - \frac{7}{9(x-1)} + \frac{32}{27}\ln|x+2| + \frac{2}{9(x+2)} + C.$$

18. Separation of variables and partial fractions give

$$\int -\frac{dt}{1500} = \int \frac{dv}{v^2-2500} = \int\left(\frac{1/100}{v-50} - \frac{1/100}{v+50}\right)dv = \frac{1}{100}\int\left(\frac{1}{v-50} - \frac{1}{v+50}\right)dv.$$

Thus,

$$\frac{-t}{1500} + C = \frac{1}{100}\left(\ln|v-50| - \ln|v+50|\right) \implies -\frac{t}{15} + 100C = \ln\left|\frac{v-50}{v+50}\right|.$$

When we exponentiate,

$$\left|\frac{v-50}{v+50}\right| = e^{100C-t/15} \implies v-50 = (v+50)De^{-t/15},$$

where $D = \pm e^{100C}$. When we solve this for v, the result is $v(t) = \dfrac{50(1+De^{-t/15})}{1-De^{-t/15}}$. If we choose time $t=0$ when the car begins motion, then $v(0) = 0$, and this requires $0 = 50(1+D)/(1-D) \implies D = -1$. Hence, $v(t) = 50(1-e^{-t/15})/(1+e^{-t/15})$.

(b) If we set $u = 1 + e^{-t/15}$ and $du = -(1/15)e^{-t/15}dt$,

$$x(t) = \int \frac{50(1 - e^{-t/15})}{1 + e^{-t/15}} dt = 50 \int \frac{1 - (u - 1)}{u} \left(\frac{-15\, du}{u - 1} \right) = 750 \int \frac{u - 2}{u(u - 1)} du$$

$$= 750 \int \left(\frac{2}{u} - \frac{1}{u - 1} \right) du = 750(2 \ln |u| - \ln |u - 1|) + E$$

$$= 750[2 \ln (1 + e^{-t/15}) - \ln (e^{-t/15})] + E = 750 \left[\frac{t}{15} + 2 \ln (1 + e^{-t/15}) \right] + E.$$

If we choose $x(0) = 0$, then $0 = 750(2 \ln 2) + E \Longrightarrow E = -1500 \ln 2$, and

$$x(t) = 750 \left[\frac{t}{15} + 2 \ln (1 + e^{-t/15}) \right] - 1500 \ln 2 = 750 \left[\frac{t}{15} + 2 \ln \left(\frac{1 + e^{-t/15}}{2} \right) \right].$$

20. With $mv \dfrac{dv}{dy} = mg - kv^2$ expressed in the form $\dfrac{v\, dv}{mg - kv^2} = \dfrac{dy}{m}$, solutions are defined implicitly by

$$\frac{y}{m} + C = \int \frac{v\, dv}{mg - kv^2} = -\frac{1}{2k} \ln |mg - kv^2|.$$

When we multiply by $-2k$ and exponentiate,

$$|mg - kv^2| = e^{-2kC - 2ky/m} \quad \Longrightarrow \quad mg - kv^2 = De^{-2ky/m} \quad \Longrightarrow \quad v = \sqrt{\frac{mg}{k} - \frac{D}{k} e^{-2ky/m}},$$

where $D = \pm e^{-2kC}$. Since $v(0) = v_0$, we have $v_0 = \sqrt{\dfrac{mg}{k} - \dfrac{D}{k}} \Longrightarrow \dfrac{D}{k} = \dfrac{mg}{k} - v_0^2$. Hence,

$$v(y) = \sqrt{\frac{mg}{k} - \left(\frac{mg}{k} - v_0^2 \right) e^{-2ky/m}}.$$

The velocity of the raindrop when it strikes the earth is $\sqrt{\dfrac{mg}{k} - \left(\dfrac{mg}{k} - v_0^2 \right) e^{-2kh/m}}$.

22. The differential equation can be expressed in the form $\dfrac{dN}{N(C - N)} = \dfrac{k}{C} dt$. Partial fractions gives

$$\int \frac{k}{C} dt = \int \left(\frac{1/C}{N} + \frac{1/C}{C - N} \right) dN = \frac{1}{C} \int \left(\frac{1}{N} + \frac{1}{C - N} \right) dN.$$

When we multiply by C, solutions are defined implicitly by

$$kt + D = \ln |N| - \ln |C - N| = \ln \left| \frac{N}{C - N} \right|.$$

Exponentiation gives $\left| \dfrac{N}{C - N} \right| = e^{kt + D} \Longrightarrow N = (C - N)Ee^{kt} \Longrightarrow N = \dfrac{CEe^{kt}}{1 + Ee^{kt}}$, where $E = \pm e^{D}$.

For $N(0) = N_0$, we must have $N_0 = \dfrac{CE}{1 + E} \Longrightarrow N_0(1 + E) = CE \Longrightarrow E = \dfrac{N_0}{C - N_0}$. Hence,

$$N(t) = \frac{C \left(\dfrac{N_0}{C - N_0} \right) e^{kt}}{1 + \left(\dfrac{N_0}{C - N_0} \right) e^{kt}} = \frac{C}{1 + \left(\dfrac{C - N_0}{N_0} \right) e^{-kt}}.$$

24. We can separate the differential equation,

$$k\, dt = \frac{1}{x(N - x)} dx \quad \Longrightarrow \quad k\, dt = \left(\frac{1/N}{x} + \frac{1/N}{N - x} \right) dx.$$

Solutions are defined implicitly by

$$\frac{1}{N}\left(\ln|x| - \ln|N - x|\right) = kt + C.$$

Since x and $N - x$ are both positive, we may drop the absolute values and write

$$\ln\left(\frac{x}{N-x}\right) = N(kt + C) \quad \Longrightarrow \quad \frac{x}{N-x} = De^{Ft},$$

where we have substituted $D = e^{NC}$ and $F = kN$. Multiplication by $N - x$ gives

$$x = De^{Ft}(N - x) \quad \Longrightarrow \quad x\left(1 + De^{Ft}\right) = NDe^{Ft}.$$

Thus,

$$x(t) = \frac{NDe^{Ft}}{1 + De^{Ft}} = \frac{ND}{D + e^{-Ft}}.$$

Since $x(0) = 1$, it follows that $1 = \dfrac{ND}{1 + D} \Longrightarrow D = 1/(N - 1)$, and

$$x(t) = \frac{\dfrac{N}{N-1}}{\dfrac{1}{N-1} + e^{-Ft}} = \frac{N}{1 + (N-1)e^{-Ft}}.$$

26. We separate the differential equation and use partial fractions to write

$$\frac{1}{v_0^2 - v^2}\,dv = \frac{1}{a}\,dt \quad \Longrightarrow \quad \left[\frac{1/(2v_0)}{v_0 - v} + \frac{1/(2v_0)}{v_0 + v}\right] dv = \frac{1}{a}\,dt.$$

Solutions are defined implicitly by

$$\frac{1}{2v_0}\left[-\ln(v_0 - v) + \ln(v_0 + v)\right] = \frac{t}{a} + C \quad \Longrightarrow \quad \ln\left(\frac{v_0 + v}{v_0 - v}\right) = \frac{2v_0 t}{a} + 2v_0 C.$$

Exponentiation gives

$$\frac{v_0 + v}{v_0 - v} = e^{2v_0 t/a + 2v_0 C} = De^{2v_0 t/a},$$

where $D = e^{2v_0 C}$. We can now solve for $v(t)$,

$$v_0 + v = (v_0 - v)De^{2v_0 t/a} \quad \Longrightarrow \quad v = \frac{v_0(De^{2v_0 t/a} - 1)}{De^{2v_0 t/a} + 1}.$$

The initial condition $v(0) = 0$ requires $D = 1$, and therefore

$$v(t) = \frac{v_0(e^{2v_0 t/a} - 1)}{e^{2v_0 t/a} + 1}.$$

28. If we set $\dfrac{1}{x^5 + x^4 + 2x^3 + 2x^2 + x + 1} = \dfrac{1}{(x+1)(x^2+1)^2} = \dfrac{A}{x+1} + \dfrac{Bx + C}{x^2 + 1} + \dfrac{Dx + E}{(x^2+1)^2}$, then $A = 1/4$, $B = -1/4$, $C = 1/4$, $D = -1/2$, $E = 1/2$, and

$$\int \frac{1}{x^5 + x^4 + 2x^3 + 2x^2 + x + 1}\,dx = \int\left[\frac{1/4}{x+1} + \frac{-x/4 + 1/4}{x^2 + 1} + \frac{-x/2 + 1/2}{(x^2+1)^2}\right] dx.$$

In the very last term we set $x = \tan\theta$ and $dx = \sec^2\theta\,d\theta$, in which case

$$\int \frac{1}{(x^2+1)^2}\,dx = \int \frac{1}{\sec^4\theta}\sec^2\theta\,d\theta = \int \cos^2\theta\,d\theta = \int \left(\frac{1+\cos 2\theta}{2}\right)d\theta$$

$$= \frac{\theta}{2} + \frac{1}{4}\sin 2\theta + C = \frac{\theta}{2} + \frac{1}{2}\sin\theta\cos\theta + C$$

$$= \frac{1}{2}\mathrm{Tan}^{-1}x + \frac{1}{2}\frac{x}{x^2+1} + C.$$

Consequently,

$$\int \frac{1}{x^5+x^4+2x^3+2x^2+x+1}\,dx = \frac{1}{4}\ln|x+1| - \frac{1}{8}\ln(x^2+1) + \frac{1}{4}\mathrm{Tan}^{-1}x$$

$$+ \frac{1}{4(x^2+1)} + \frac{1}{4}\mathrm{Tan}^{-1}x + \frac{x}{4(x^2+1)} + C$$

$$= \frac{1}{4}\ln|x+1| - \frac{1}{8}\ln(x^2+1) + \frac{1}{2}\mathrm{Tan}^{-1}x + \frac{x+1}{4(x^2+1)} + C.$$

30. If we set $\dfrac{1}{x^3+1} = \dfrac{A}{x+1} + \dfrac{Bx+C}{x^2-x+1}$, then $A = 1/3$, $B = -1/3$, $C = 2/3$, and

$$\int \frac{1}{x^3+1}\,dx = \int \left(\frac{1/3}{x+1} + \frac{-x/3+2/3}{x^2-x+1}\right)dx = \frac{1}{3}\ln|x+1| + \frac{1}{3}\int \frac{-x+2}{(x-1/2)^2+3/4}\,dx.$$

In the remaining integral we set $x - 1/2 = (\sqrt{3}/2)\tan\theta$, and $dx = (\sqrt{3}/2)\sec^2\theta\,d\theta$,

$$\int \frac{1}{x^3+1}\,dx = \frac{1}{3}\ln|x+1| + \frac{1}{3}\int \frac{-1/2-(\sqrt{3}/2)\tan\theta+2}{(3/4)\sec^2\theta}\frac{\sqrt{3}}{2}\sec^2\theta\,d\theta$$

$$= \frac{1}{3}\ln|x+1| + \frac{1}{3}(\sqrt{3}\theta + \ln|\cos\theta|) + C$$

$$= \frac{1}{3}\ln|x+1| + \frac{1}{\sqrt{3}}\mathrm{Tan}^{-1}\left(\frac{2x-1}{\sqrt{3}}\right) + \frac{1}{3}\ln\left|\frac{\sqrt{3}}{2\sqrt{x^2-x+1}}\right| + C$$

$$= \frac{1}{3}\ln|x+1| + \frac{1}{\sqrt{3}}\mathrm{Tan}^{-1}\left(\frac{2x-1}{\sqrt{3}}\right) - \frac{1}{6}\ln(x^2-x+1) + D.$$

32. If we set $\dfrac{x^4+8x^3-x^2+2x+1}{x^5+x^4+x^2+x} = \dfrac{A}{x} + \dfrac{B}{x+1} + \dfrac{C}{(x+1)^2} + \dfrac{Dx+E}{x^2-x+1}$, then $A = 1$, $B = -2$, $C = 3$, $D = 2$, $E = 0$, and

$$\int \frac{x^4+8x^3-x^2+2x+1}{x^5+x^4+x^2+x}\,dx = \int \left[\frac{1}{x} - \frac{2}{x+1} + \frac{3}{(x+1)^2} + \frac{2x}{(x-1/2)^2+3/4}\right]dx.$$

In the last term we set $x - 1/2 = (\sqrt{3}/2)\tan\theta$ and $dx = (\sqrt{3}/2)\sec^2\theta\,d\theta$, in which case

$$\int \frac{2x}{(x-1/2)^2+3/4}\,dx = 2\int \frac{1/2+(\sqrt{3}/2)\tan\theta}{(3/4)\sec^2\theta}\frac{\sqrt{3}}{2}\sec^2\theta\,d\theta = \frac{2}{\sqrt{3}}(\theta + \sqrt{3}\ln|\sec\theta|) + C$$

$$= \frac{2}{\sqrt{3}}\mathrm{Tan}^{-1}\left(\frac{2x-1}{\sqrt{3}}\right) + 2\ln\left|\frac{2\sqrt{x^2-x+1}}{\sqrt{3}}\right| + C$$

$$= \frac{2}{\sqrt{3}}\mathrm{Tan}^{-1}\left(\frac{2x-1}{\sqrt{3}}\right) + \ln(x^2-x+1) + D.$$

Thus,

$$\int \frac{x^4+8x^3-x^2+2x+1}{x^5+x^4+x^2+x}\,dx = \ln|x| - 2\ln|x+1| - \frac{3}{x+1} + \frac{2}{\sqrt{3}}\mathrm{Tan}^{-1}\left(\frac{2x-1}{\sqrt{3}}\right)$$

$$+ \ln(x^2-x+1) + D.$$

34. We could use partial fractions on the integrand as it now stands, but it is easier if we first substitute $x = M^2$ and $dx = 2M\,dM$. At the same time, let us set $a = (k-1)/2$ and denote the integral by I:

$$I = \int \frac{M(1-M^2)}{M^4\left(1 + \dfrac{k-1}{2}M^2\right)}dM = \frac{1}{2}\int \frac{1-x}{x^2(1+ax)}dx = \frac{1}{2}\int\left(\frac{-a-1}{x} + \frac{1}{x^2} + \frac{a+a^2}{1+ax}\right)dx$$

$$= \frac{1}{2}\left[-(a+1)\ln x - \frac{1}{x} + (1+a)\ln(1+ax)\right] + C = -\frac{1}{2x} + \left(\frac{a+1}{2}\right)\ln\left(\frac{1+ax}{x}\right) + C$$

$$= -\frac{1}{2M^2} + \left(\frac{k+1}{4}\right)\ln\left[\frac{1 + \left(\dfrac{k-1}{2}\right)M^2}{M^2}\right] + C.$$

36. With the substitution from Exercise 35,

$$\int \sec x\,dx = \int \frac{1}{\cos x}dx = \int \frac{1+t^2}{1-t^2}\frac{2}{1+t^2}dt = 2\int\frac{1}{1-t^2}dt = 2\int\left(\frac{1/2}{1-t} + \frac{1/2}{1+t}\right)dt$$

$$= -\ln|1-t| + \ln|1+t| + C = \ln\left|\frac{1+t}{1-t}\right| + C = \ln\left|\frac{1+\tan(x/2)}{1-\tan(x/2)}\right| + C.$$

38. With the substitution from Exercise 35,

$$\int \frac{1}{1-2\cos x}dx = \int \frac{1}{1-2\left(\dfrac{1-t^2}{1+t^2}\right)}\frac{2}{1+t^2}dt = 2\int\frac{1}{3t^2-1}dt$$

$$= 2\int\left(\frac{-1/2}{\sqrt{3}t+1} + \frac{1/2}{\sqrt{3}t-1}\right)dt = -\frac{1}{\sqrt{3}}\ln|\sqrt{3}t+1| + \frac{1}{\sqrt{3}}\ln|\sqrt{3}t-1| + C$$

$$= \frac{1}{\sqrt{3}}\ln\left|\frac{\sqrt{3}t-1}{\sqrt{3}t+1}\right| + C = \frac{1}{\sqrt{3}}\ln\left|\frac{\sqrt{3}\tan(x/2)-1}{\sqrt{3}\tan(x/2)+1}\right| + C.$$

40. We must show that $\dfrac{1+\tan(x/2)}{1-\tan(x/2)} = \sec x + \tan x$.

$$\frac{1+\tan(x/2)}{1-\tan(x/2)} = \frac{1 + \dfrac{\sin(x/2)}{\cos(x/2)}}{1 - \dfrac{\sin(x/2)}{\cos(x/2)}} = \frac{\cos(x/2)+\sin(x/2)}{\cos(x/2)-\sin(x/2)} = \frac{\cos(x/2)+\sin(x/2)}{\cos(x/2)-\sin(x/2)}\frac{\cos(x/2)+\sin(x/2)}{\cos(x/2)+\sin(x/2)}$$

$$= \frac{\cos^2(x/2) + 2\sin(x/2)\cos(x/2) + \sin^2(x/2)}{\cos^2(x/2) - \sin^2(x/2)} = \frac{1+\sin x}{\cos x} = \sec x + \tan x.$$

EXERCISES 8.7

2. With the trapezoidal rule,

$$\int_2^3 \frac{1}{\sqrt{x+2}}dx \approx \frac{1/10}{2}\left(\frac{1}{\sqrt{4}} + 2\sum_{i=1}^{9}\frac{1}{\sqrt{\frac{i}{10}+4}} + \frac{1}{\sqrt{5}}\right) = \frac{1}{20}\left(\frac{1}{2} + 2\sqrt{10}\sum_{i=1}^{9}\frac{1}{\sqrt{40+i}} + \frac{1}{\sqrt{5}}\right) = 0.472\,15.$$

With Simpson's rule,

$$\int_2^3 \frac{1}{\sqrt{x+2}}dx \approx \frac{1/10}{3}\left(\frac{1}{2} + \frac{4}{\sqrt{4.1}} + \frac{2}{\sqrt{4.2}} + \cdots + \frac{2}{\sqrt{4.8}} + \frac{4}{\sqrt{4.9}} + \frac{1}{\sqrt{5}}\right) = 0.472\,14.$$

Analytically, $\displaystyle\int_2^3 \frac{1}{\sqrt{x+2}}dx = \left\{2\sqrt{x+2}\right\}_2^3 = 2\sqrt{5} - 4 \approx 0.472\,14.$

4. With the trapezoidal rule,

$$\int_0^{1/2} e^x \, dx \approx \frac{1/20}{2}\left(e^0 + 2\sum_{i=1}^9 e^{i/20} + e^{1/2}\right) = \frac{1}{40}\left(1 + 2\sum_{i=1}^9 e^{i/20} + \sqrt{e}\right) = 0.648\,86.$$

With Simpson's rule, $\int_0^{1/2} e^x \, dx \approx \dfrac{1/20}{3}\left(e^0 + 4e^{1/20} + 2e^{1/10} + \cdots + 2e^{2/5} + 4e^{9/20} + e^{1/2}\right) = 0.648\,72.$

Analytically, $\int_0^{1/2} e^x \, dx = \left\{e^x\right\}_0^{1/2} = \sqrt{e} - 1 \approx 0.648\,72.$

6. With the trapezoidal rule,

$$\int_{-3}^{-2} \frac{1}{x^3}\,dx \approx \frac{1/10}{2}\left[-\frac{1}{27} + 2\sum_{i=1}^9 \frac{1}{(-3+i/10)^3} - \frac{1}{8}\right] = \frac{1}{20}\left[-\frac{1}{27} - 2000\sum_{i=1}^9 \frac{1}{(30-i)^3} - \frac{1}{8}\right] = -0.069\,570.$$

With Simpson's rule,

$$\int_{-3}^{-2} \frac{1}{x^3}\,dx \approx \frac{1/10}{3}\left[-\frac{1}{27} + \frac{4}{(-2.9)^3} + \frac{2}{(-2.8)^3} + \cdots + \frac{2}{(-2.2)^3} + \frac{4}{(-2.1)^3} - \frac{1}{8}\right] = -0.069\,445.$$

Analytically, $\int_{-3}^{-2} \frac{1}{x^3}\,dx = \left\{-\frac{1}{2x^2}\right\}_{-3}^{-2} = -\frac{5}{72} \approx -0.069\,444.$

8. With the trapezoidal rule,

$$\int_0^1 \frac{1}{3+x^2}\,dx \approx \frac{1/10}{2}\left[\frac{1}{3} + 2\sum_{i=1}^9 \frac{1}{3+(i/10)^2} + \frac{1}{4}\right] = \frac{1}{20}\left(\frac{7}{12} + 200\sum_{i=1}^9 \frac{1}{300+i^2}\right) = 0.302\,20.$$

With Simpson's rule,

$$\int_0^1 \frac{1}{3+x^2}\,dx \approx \frac{1/10}{3}\left[\frac{1}{3} + \frac{4}{3+(1/10)^2} + \frac{2}{3+(2/10)^2} + \cdots + \frac{2}{3+(8/10)^2} + \frac{4}{3+(9/10)^2} + \frac{1}{4}\right] = 0.302\,30.$$

Analytically, we set $x = \sqrt{3}\tan\theta$ and $dx = \sqrt{3}\sec^2\theta\,d\theta$,

$$\int_0^1 \frac{1}{3+x^2}\,dx = \int_0^{\pi/6} \frac{1}{3\sec^2\theta}\sqrt{3}\sec^2\theta\,d\theta = \frac{1}{\sqrt{3}}\left\{\theta\right\}_0^{\pi/6} = \frac{\pi}{6\sqrt{3}} \approx 0.302\,30.$$

10. With the trapezoidal rule,

$$\int_0^{1/2} xe^{x^2}\,dx \approx \frac{1/20}{2}\left[0 + 2\sum_{i=1}^9 (i/20)e^{(i/20)^2} + \frac{1}{2}e^{1/4}\right] = \frac{1}{40}\left(\frac{1}{10}\sum_{i=1}^9 ie^{i^2/400} + \frac{1}{2}e^{1/4}\right) = 0.142\,21.$$

With Simpson's rule,

$$\int_0^{1/2} xe^{x^2}\,dx \approx \frac{1/20}{3}\left[0 + 4(1/20)e^{1/400} + 2(1/10)e^{1/100} + \cdots \right.$$
$$\left. + 2(2/5)e^{4/25} + 4(9/20)e^{81/400} + (1/2)e^{1/4}\right] = 0.142\,01.$$

Analytically, $\int_0^{1/2} xe^{x^2}\,dx = \left\{\dfrac{e^{x^2}}{2}\right\}_0^{1/2} = \dfrac{e^{1/4} - 1}{2} \approx 0.142\,01.$

12. With the trapezoidal rule, $\int_0^1 e^{x^2}\,dx \approx \dfrac{1/10}{2}\left[1 + 2\sum_{i=1}^9 e^{(i/10)^2} + e\right] = 1.4672.$

With Simpson's rule, $\int_0^1 e^{x^2}\,dx \approx \dfrac{1/10}{3}\left(1 + 4e^{0.01} + 2e^{0.04} + \cdots + 2e^{0.64} + 4e^{0.81} + e\right) = 1.4627.$

14. With the trapezoidal rule, and noting that $\sin(x^2)$ is an even function,

$$\int_{-1}^0 \sin(x^2)\,dx = \int_0^1 \sin(x^2)\,dx \approx \frac{1/10}{2}\left[0 + 2\sum_{i=1}^9 \sin(i/10)^2 + \sin 1\right] = 0.311\,17.$$

With Simpson's rule,

$$\int_{-1}^0 \sin(x^2)\,dx = \int_0^1 \sin(x^2)\,dx \approx \frac{1/10}{3}\left[0 + 4\sin(0.01) + 2\sin(0.04) + \cdots\right.$$
$$\left. + 2\sin(0.64) + 4\sin(0.81) + \sin 1\right] = 0.310\,26.$$

16. In equation 8.15 the error is reduced by a factor of $1/4$; in equation 8.16 it is reduced by $1/16$.

18. The length of the parabola is

$$\int_0^1 \sqrt{1 + (2x)^2}\,dx = \int_0^1 \sqrt{1 + 4x^2}\,dx \approx \frac{1/10}{3}\left[1 + 4\sqrt{1 + 4(1/10)^2} + 2\sqrt{1 + 4(1/5)^2} + \cdots\right.$$
$$\left. + 2\sqrt{1 + 4(4/5)^2} + 4\sqrt{1 + 4(9/10)^2} + \sqrt{5}\right] = 1.4789.$$

According to Exercise 8.4–39, the length of the parabola is $[2\sqrt{5} + \ln(2 + \sqrt{5})]/4$, which to four decimals is also 1.4789.

20. Using Simpson's rule, the volume in cubic metres is approximately

$$(1.8)\left(\frac{1}{3}\right)[0 + 4(6.0) + 2(7.0) + 4(6.8) + 2(5.8) + 4(4.6) + 2(3.8) + 4(3.6) + 2(3.6) + 4(3.8) + 0] = 83.76.$$

22. (a) Both rules require the value of the integrand at the lower limit of integration, but $e^x/\sqrt{x}$ is undefined at $x = 0$.
(b) If we set $u = \sqrt{x}$ and $du = 1/(2\sqrt{x})\,dx$, then

$$\int_0^4 \frac{e^x}{\sqrt{x}}\,dx = \int_0^2 e^{u^2} 2\,du = 2\int_0^2 e^{u^2}\,du,$$

and this integral is no longer improper. With Simpson's rule and 20 equal subdivisions,

$$2\int_0^2 e^{u^2}\,du \approx 2\left(\frac{1/10}{3}\right)\left[e^0 + 4e^{(0.1)^2} + 2e^{(0.2)^2} + \cdots + 2e^{(1.8)^2} + 4e^{(1.9)^2} + e^4\right] = 32.91.$$

(c) Rectangular rule 8.11 can be used since it does not require the value of $e^x/\sqrt{x}$ at $x = 0$.

24. If we set $x = 1/t$, then $dx = -(1/t^2)\,dt$, and

$$\int_1^\infty \frac{1}{1 + x^4}\,dx = \int_1^0 \frac{1}{1 + t^{-4}}\left(-\frac{1}{t^2}dt\right) = \int_0^1 \frac{t^2}{1 + t^4}\,dt.$$

The trapezoidal rule with 10 subdivisions gives

$$\int_0^1 \frac{t^2}{1 + t^4}\,dt \approx \frac{1/10}{2}\left[0 + 2\sum_{i=1}^9 \frac{(i/10)^2}{1 + (i/10)^4} + \frac{1}{2}\right] = 0.2437.$$

Simpson's rule with the same subdivision yields

$$\int_0^1 \frac{t^2}{1 + t^4}\,dt \approx \frac{1/10}{3}\left[0 + \frac{4(0.1)^2}{1 + (0.1)^4} + \frac{2(0.2)^2}{1 + (0.2)^4} + \cdots + \frac{4(0.9)^2}{1 + (0.9)^4} + \frac{1}{2}\right] = 0.2438.$$

26. If $f(x)$ is the function, then the trapezoidal rule gives

$$\int_{-1}^4 f(x)\,dx \approx \frac{1/2}{2}\{f(-1) + 2[f(-0.5) + f(0) + \cdots + f(3.5)] + f(4)\} = 2.113.$$

Simpson's rule gives

$$\int_{-1}^{4} f(x)\,dx \approx \frac{1/2}{3}[f(-1) + 4f(-0.5) + 2f(0) + \cdots + 2f(3) + 4f(3.5) + f(4)] = 1.729.$$

28. According to equation 8.16, the maximum error in approximating the definite integral of $f(x)$ from $x = a$ to $x = b$ with n equal subdivisions is given by $|S_n| = M(b-a)^5/(180n^4)$ where M is the maximum value of $|f''''(x)|$ on $a \le x \le b$. But if $f(x)$ is a cubic polynomial, $f''''(x) = 0$ for all x. Hence, $S_n = 0$. For example,

$$\int_{1}^{2} (x^3 + 1)\,dx = \left\{ \frac{x^4}{4} + x \right\}_1^2 = \frac{19}{4}.$$

Simpson's rule with 10 equal subdivisions, and $f(x) = x^3 + 1$, gives

$$\int_{1}^{2} (x^3 + 1)\,dx \approx \frac{1/10}{3}[f(1) + 4f(1.1) + 2f(1.2) + \cdots + 2f(1.8) + 4f(1.9) + f(2)] = 4.75.$$

30. (a) According to equation 8.15, the error in using the trapezoidal rule with n equal partitions is $|T_n| \le M(\pi/4)^3/(12n^2)$ where M is the maximum of the absolute value of the second derivative of $\cos x$ on $0 \le x \le \pi/4$. Since $d^2(\cos x)/dx^2 = -\cos x$, it follows that $M = 1$, and $|T_n| \le \pi^3/(768n^2)$. For $|T_n|$ to be less than 10^{-4}, we require

$$\frac{\pi^3}{768n^2} < 10^{-4} \quad \Longrightarrow \quad n > \sqrt{\frac{10^4\pi^3}{768}} = 20.09.$$

Thus, at least 21 subdivisions should be used.
(b) According to equation 8.16, the error in using Simpson's rule with n equal partitions is $|S_n| \le M(\pi/4)^5/(180n^4)$ where M is the maximum of the absolute value of the fourth derivative of $\cos x$ on $0 \le x \le \pi/4$. Since $d^4(\cos x)/dx^4 = \cos x$, it follows that $M = 1$, and $|S_n| \le (\pi/4)^5/(180n^4)$. For $|S_n|$ to be less than 10^{-4}, we require

$$\frac{\pi^5}{180(4)^5 n^4} < 10^{-4} \quad \Longrightarrow \quad n > \sqrt[4]{\frac{10^4\pi^5}{180(4)^5}} = 2.02.$$

Since n must be even, we should use at least 4 subdivisions.

32. (a) According to equation 8.15, the error in using the trapezoidal rule with n equal partitions is $|T_n| \le M(5-4)^3/(12n^2)$ where M is the maximum of the absolute value of the second derivative of $1/\sqrt{x+2}$ on $4 \le x \le 5$. Since $d^2(1/\sqrt{x+2})/dx^2 = (3/4)(x+2)^{-5/2}$, it follows that $M = (3/4)6^{-5/2}$, and $|T_n| \le (3/4)6^{-5/2}/(12n^2)$. For $|T_n|$ to be less than 10^{-4}, we require

$$\frac{1}{16(6^{5/2})n^2} < 10^{-4} \quad \Longrightarrow \quad n > \sqrt{\frac{10^4}{16(6^{5/2})}} = 2.66.$$

Thus, at least 3 subdivisions should be used.
(b) According to equation 8.16, the error in using Simpson's rule with n equal partitions is $|S_n| \le M(5-4)^5/(180n^4)$ where M is the maximum of the absolute value of the fourth derivative of $1/\sqrt{x+2}$ on $4 \le x \le 5$. Since $d^4(1/\sqrt{x+2})/dx^4 = (105/16)(x+2)^{-9/2}$, it follows that $M = (105/16)6^{-9/2}$, and $|S_n| \le (105/16)6^{-9/2}/(180n^4)$. For $|S_n|$ to be less than 10^{-4}, we require

$$\frac{105}{16(180)(6^{9/2})n^4} < 10^{-4} \quad \Longrightarrow \quad n > \sqrt[4]{\frac{105(10^4)}{16(180)(6^{9/2})}} = 0.58.$$

Since n must be even, only 2 subdivisions are needed.

REVIEW EXERCISES

2. $\displaystyle\int \frac{1}{(x+3)^2}\,dx = -\frac{1}{x+3} + C$

4. $\displaystyle\int \frac{x^2+3}{x+1}\,dx = \int\left(x - 1 + \frac{4}{x+1}\right)dx = \frac{x^2}{2} - x + 4\ln|x+1| + C$

6. If we set $u = x + 3$ and $du = dx$, then

$$\int \frac{x}{\sqrt{x+3}}\,dx = \int \frac{u-3}{\sqrt{u}}\,du = \int\left(\sqrt{u} - 3u^{-1/2}\right)du = \frac{2}{3}u^{3/2} - 6u^{1/2} + C = \frac{2}{3}(x+3)^{3/2} - 6\sqrt{x+3} + C.$$

8. If we set $u = x$, $dv = \sin x\,dx$, then $du = dx$, $v = -\cos x$, and

$$\int x\,\sin x\,dx = -x\,\cos x - \int -\cos x\,dx = -x\,\cos x + \sin x + C.$$

10. $\displaystyle\int \frac{x}{x^2+2x-3}\,dx = \int\left(\frac{3/4}{x+3} + \frac{1/4}{x-1}\right)dx = \frac{3}{4}\ln|x+3| + \frac{1}{4}\ln|x-1| + C$

12. If we set $u = \sqrt{x} + 5$, then $du = 1/(2\sqrt{x})dx$, and

$$\int \frac{2-\sqrt{x}}{\sqrt{x}+5}\,dx = \int \frac{2-(u-5)}{u}(2)(u-5)\,du = 2\int \frac{(7-u)(u-5)}{u}\,du$$

$$= 2\int\left(-\frac{35}{u} + 12 - u\right)du = 2\left(-35\ln|u| + 12u - \frac{u^2}{2}\right) + C$$

$$= -70\ln|\sqrt{x}+5| + 24(\sqrt{x}+5) - (\sqrt{x}+5)^2 + C$$

$$= -70\ln(\sqrt{x}+5) + 14\sqrt{x} - x + D.$$

14. If we set $u = e^x$, then $du = e^x\,dx$, and $\displaystyle\int \frac{e^x}{\sqrt{1-e^{2x}}}\,dx = \int \frac{1}{\sqrt{1-u^2}}\,du = \operatorname{Sin}^{-1}u + C = \operatorname{Sin}^{-1}(e^x) + C.$

16. $\displaystyle\int \frac{x}{(x^2+1)^2}\,dx = -\frac{1}{2(x^2+1)} + C$

18. If we set $u = x^2 + 1$, then $du = 2x\,dx$, and

$$\int \frac{x^3}{(x^2+1)^2}\,dx = \int \frac{u-1}{u^2}\left(\frac{du}{2}\right) = \frac{1}{2}\int\left(\frac{1}{u} - \frac{1}{u^2}\right)du = \frac{1}{2}\left(\ln|u| + \frac{1}{u}\right) + C$$

$$= \frac{1}{2}\ln(x^2+1) + \frac{1}{2(x^2+1)} + C.$$

20. $\displaystyle\int\left(\frac{x+1}{x-1}\right)^2 dx = \int\left(1 + \frac{2}{x-1}\right)^2 dx = \int\left[1 + \frac{4}{x-1} + \frac{4}{(x-1)^2}\right]dx = x + 4\ln|x-1| - \frac{4}{x-1} + C$

22. If we set $u = \operatorname{Cos}^{-1}x$, $dv = dx$, then $du = (-1/\sqrt{1-x^2})dx$, $v = x$, and

$$\int \operatorname{Cos}^{-1}x\,dx = x\,\operatorname{Cos}^{-1}x - \int \frac{-x}{\sqrt{1-x^2}}\,dx = x\,\operatorname{Cos}^{-1}x - \sqrt{1-x^2} + C.$$

24. Using identity 1.48b, $\displaystyle\int \sin x\,\cos 5x\,dx = \frac{1}{2}\int (\sin 6x - \sin 4x)\,dx = -\frac{1}{12}\cos 6x + \frac{1}{8}\cos 4x + C.$

26. Since $x^2 + 4x - 5 = (x+2)^2 - 9$, we set $x + 2 = 3\sec\theta$, in which case $dx = 3\sec\theta\,\tan\theta\,d\theta$, and

$$\int \frac{1}{\sqrt{x^2+4x-5}}\,dx = \int \frac{1}{3\tan\theta}3\sec\theta\,\tan\theta\,d\theta = \ln|\sec\theta + \tan\theta| + C$$

$$= \ln\left|\frac{x+2}{3} + \frac{\sqrt{x^2+4x-5}}{3}\right| + C$$

$$= \ln|x + 2 + \sqrt{x^2+4x-5}| + D.$$

28. If we set $u = 4 - x^2$, then $du = -2x\,dx$, and

$$\int x^3 \sqrt{4 - x^2}\,dx = \int (4 - u)\sqrt{u}\left(-\frac{du}{2}\right) = \frac{1}{2}\int (u^{3/2} - 4\sqrt{u})\,du$$

$$= \frac{1}{2}\left(\frac{2}{5}u^{5/2} - \frac{8}{3}u^{3/2}\right) + C = \frac{1}{5}(4 - x^2)^{5/2} - \frac{4}{3}(4 - x^2)^{3/2} + C.$$

30. $\displaystyle \int \frac{6x}{4 - x^2}\,dx = -3\ln|4 - x^2| + C$

32. Since $x^2 + 4x - 4 = (x + 2)^2 - 8$, we set $x + 2 = 2\sqrt{2}\sec\theta$, in which case $dx = 2\sqrt{2}\sec\theta\tan\theta\,d\theta$, and

$$\int \frac{1}{x^2 + 4x - 4}\,dx = \int \frac{1}{8\tan^2\theta}2\sqrt{2}\sec\theta\tan\theta\,d\theta = \frac{1}{2\sqrt{2}}\int \csc\theta\,d\theta = \frac{1}{2\sqrt{2}}\ln|\csc\theta - \cot\theta| + C$$

$$= \frac{1}{2\sqrt{2}}\ln\left|\frac{x + 2}{\sqrt{x^2 + 4x - 4}} - \frac{2\sqrt{2}}{\sqrt{x^2 + 4x - 4}}\right| + C$$

$$= \frac{1}{2\sqrt{2}}\ln\left|\frac{x + 2 - 2\sqrt{2}}{\sqrt{x^2 + 4x - 4}}\right| + C.$$

34. When we set $\displaystyle \frac{1}{x^4 + x^3} = \frac{1}{x^3(x + 1)} = \frac{A}{x} + \frac{B}{x^2} + \frac{C}{x^3} + \frac{D}{x + 1}$, we find that $A = 1$, $B = -1$, $C = 1$, and $D = -1$. Thus, $\displaystyle \int \frac{1}{x^4 + x^3}\,dx = \int\left(\frac{1}{x} - \frac{1}{x^2} + \frac{1}{x^3} - \frac{1}{x + 1}\right)dx = \ln|x| + \frac{1}{x} - \frac{1}{2x^2} - \ln|x + 1| + C.$

36. Since $16 - 3x + x^2 = (x - 3/2)^2 + 55/4$, we set $x - 3/2 = (\sqrt{55}/2)\tan\theta$. Then $dx = (\sqrt{55}/2)\sec^2\theta\,d\theta$, and

$$\int \frac{1}{\sqrt{16 - 3x + x^2}}\,dx = \int \frac{1}{(\sqrt{55}/2)\sec\theta}(\sqrt{55}/2)\sec^2\theta\,d\theta = \ln|\sec\theta + \tan\theta| + C$$

$$= \ln\left|\frac{2\sqrt{x^2 - 3x + 16}}{\sqrt{55}} + \frac{2x - 3}{\sqrt{55}}\right| + C$$

$$= \ln|2\sqrt{x^2 - 3x + 16} + 2x - 3| + D.$$

38. When we set $u = \text{Tan}^{-1}x$, $dv = x^2\,dx$, then $du = \dfrac{1}{1 + x^2}dx$, $v = x^3/3$, and

$$\int x^2\,\text{Tan}^{-1}x\,dx = \frac{x^3}{3}\text{Tan}^{-1}x - \frac{1}{3}\int \frac{x^3}{1 + x^2}\,dx = \frac{x^3}{3}\text{Tan}^{-1}x - \frac{1}{3}\int\left(x - \frac{x}{1 + x^2}\right)dx$$

$$= \frac{x^3}{3}\text{Tan}^{-1}x - \frac{x^2}{6} + \frac{1}{6}\ln(1 + x^2) + C.$$

40. If we set $u = \ln x$, then $du = (1/x)\,dx$, and $\displaystyle \int \frac{\ln x}{x}\,dx = \int u\,du = \frac{u^2}{2} + C = \frac{1}{2}(\ln x)^2 + C.$

42. If we set $x = 3\tan\theta$, then $dx = 3\sec^2\theta\,d\theta$, and

$$\int \frac{1}{x(9 + x^2)^2}\,dx = \int \frac{1}{3\tan\theta(81\sec^4\theta)}3\sec^2\theta\,d\theta = \frac{1}{81}\int \frac{\cos^3\theta}{\sin\theta}\,d\theta$$

$$= \frac{1}{81}\int \frac{\cos\theta(1 - \sin^2\theta)}{\sin\theta}\,d\theta = \frac{1}{81}\left(\ln|\sin\theta| + \frac{1}{2}\cos^2\theta\right) + C$$

$$= \frac{1}{81}\ln\left|\frac{x}{\sqrt{x^2 + 9}}\right| + \frac{1}{162}\left(\frac{3}{\sqrt{x^2 + 9}}\right)^2 + C$$

$$= \frac{1}{81}\ln|x| - \frac{1}{162}\ln(x^2 + 9) + \frac{1}{18(x^2 + 9)} + C.$$

44. If we set $\dfrac{x^2+2}{x^3+4x^2+4x} = \dfrac{x^2+2}{x(x+2)^2} = \dfrac{A}{x} + \dfrac{B}{x+2} + \dfrac{C}{(x+2)^2}$, we find that $A = 1/2$, $B = 1/2$, and $C = -3$. Thus,

$$\int \frac{x^2+2}{x^3+4x^2+4x}\,dx = \int \left[\frac{1/2}{x} + \frac{1/2}{x+2} - \frac{3}{(x+2)^2}\right] dx = \frac{1}{2}\ln|x| + \frac{1}{2}\ln|x+2| + \frac{3}{x+2} + C.$$

46. If we set $\dfrac{3x^2+2x+4}{x^3+x^2+4x} = \dfrac{3x^2+2x+4}{x(x^2+x+4)} = \dfrac{A}{x} + \dfrac{Bx+C}{x^2+x+4}$, we find that $A = 1$, $B = 2$, and $C = 1$. Thus,

$$\int \frac{3x^2+2x+4}{x^3+x^2+4x}\,dx = \int \left(\frac{1}{x} + \frac{2x+1}{x^2+x+4}\right) dx = \ln|x| + \ln(x^2+x+4) + C.$$

48. $\displaystyle\int \sqrt{\cot x}\,\csc^4 x\,dx = \int \sqrt{\cot x}(1+\cot^2 x)\csc^2 x\,dx = -\frac{2}{3}\cot^{3/2} x - \frac{2}{7}\cot^{7/2} x + C$

50. Since $4x - x^2 = -(x-2)^2 + 4$, we set $x - 2 = 2\sin\theta$. Then $dx = 2\cos\theta\,d\theta$, and

$$\int \frac{1}{(4x-x^2)^{3/2}}\,dx = \int \frac{1}{8\cos^3\theta}2\cos\theta\,d\theta = \frac{1}{4}\int \sec^2\theta\,d\theta = \frac{1}{4}\tan\theta + C$$

$$= \frac{1}{4}\frac{x-2}{\sqrt{4x-x^2}} + C.$$

52. With the trapezoidal rule and 10 equal partitions,

$$\int_0^1 \sqrt{\sin x}\,dx \approx \frac{1/10}{2}\left[\sqrt{\sin 0} + 2\sum_{i=1}^9 \sqrt{\sin(i/10)} + \sqrt{\sin 1}\right] = 0.636\,65.$$

With Simpson's rule, $\displaystyle\int_0^1 \sqrt{\sin x}\,dx \approx \frac{1/10}{3}[\sqrt{\sin 0} + 4\sqrt{\sin(1/10)} + 2\sqrt{\sin(1/5)} + \cdots$

$$+ 2\sqrt{\sin(4/5)} + 4\sqrt{\sin(9/10)} + \sqrt{\sin 1}] = 0.640\,41.$$

54. With the trapezoidal rule and 10 equal subdivisions,

$$\int_{-1}^3 \frac{1}{1+e^x}\,dx \approx \frac{2/5}{2}\left[\frac{1}{1+e^{-1}} + 2\sum_{i=1}^9 \frac{1}{1+e^{-1+2i/5}} + \frac{1}{1+e^3}\right] = 1.2667.$$

With Simpson's rule, $\displaystyle\int_{-1}^3 \frac{1}{1+e^x}\,dx \approx \frac{2/5}{3}\left(\frac{1}{1+e^{-1}} + \frac{4}{1+e^{-3/5}} + \frac{2}{1+e^{-1/5}} + \cdots\right.$

$$\left.+ \frac{2}{1+e^{11/5}} + \frac{4}{1+e^{13/5}} + \frac{1}{1+e^3}\right) = 1.2647.$$

56. If we set $u = x^{1/6}$, or, $x = u^6$, then $dx = 6u^5\,du$, and

$$\int \frac{1}{x^{1/3} - \sqrt{x}}\,dx = \int \frac{1}{u^2 - u^3}6u^5\,du = 6\int \frac{u^3}{1-u}\,du = 6\int \left(-u^2 - u - 1 + \frac{1}{1-u}\right) du$$

$$= 6\left(-\frac{u^3}{3} - \frac{u^2}{2} - u - \ln|1-u|\right) + C = -2\sqrt{x} - 3x^{1/3} - 6x^{1/6} - 6\ln|1 - x^{1/6}| + C.$$

58. If we set $x^2 = 4\tan\theta$, then $2x\,dx = 4\sec^2\theta\,d\theta$, and

$$\int \frac{x}{x^4+16}\,dx = \int \frac{1}{16\sec^2\theta}2\sec^2\theta\,d\theta = \frac{\theta}{8} + C = \frac{1}{8}\text{Tan}^{-1}\left(\frac{x^2}{4}\right) + C.$$

60. Since $3x - x^2 = -(x - 3/2)^2 + 9/4$, we set $x - 3/2 = (3/2)\sin\theta$, in which case $dx = (3/2)\cos\theta\,d\theta$, and

$$\int \frac{1}{(3x - x^2)^{3/2}}\,dx = \int \frac{1}{(27/8)\cos^3\theta}(3/2)\cos\theta\,d\theta = \frac{4}{9}\int \sec^2\theta\,d\theta$$

$$= \frac{4}{9}\tan\theta + C = \frac{4}{9}\left(\frac{2x - 3}{2\sqrt{3x - x^2}}\right) + C$$

$$= \frac{4x - 6}{9\sqrt{3x - x^2}} + C.$$

62. If we set $y = \sqrt{x}$ and $dy = 1/(2\sqrt{x})\,dx$, then $\int \sin\sqrt{x}\,dx = \int \sin y\,(2y\,dy)$. Now we set $u = y$, $dv = \sin y\,dy$, $du = dy$, $v = -\cos y$, and use integration by parts,

$$\int \sin\sqrt{x}\,dx = 2\left(-y\cos y - \int -\cos y\,dy\right) = -2y\cos y + 2\sin y + C = -2\sqrt{x}\cos\sqrt{x} + 2\sin\sqrt{x} + C.$$

64. Using identity 1.48b, $\int x\cos x \sin 3x\,dx = \int \frac{x}{2}(\sin 2x + \sin 4x)\,dx$. We now set $u = x$, $dv = (\sin 2x + \sin 4x)dx$, in which case $du = dx$, $v = -(1/2)\cos 2x - (1/4)\cos 4x$, and

$$\int x\cos x\sin 3x\,dx = \frac{1}{2}\left[x\left(-\frac{1}{2}\cos 2x - \frac{1}{4}\cos 4x\right) - \int\left(-\frac{1}{2}\cos 2x - \frac{1}{4}\cos 4x\right)dx\right]$$

$$= -\frac{x}{8}(2\cos 2x + \cos 4x) + \frac{1}{2}\left(\frac{1}{4}\sin 2x + \frac{1}{16}\sin 4x\right) + C.$$

66. $\int \frac{1}{1 + \cos 2x}\,dx = \int \frac{1}{1 + (2\cos^2 x - 1)}\,dx = \frac{1}{2}\int \sec^2 x\,dx = \frac{1}{2}\tan x + C$

68. $\int \sin^2 x\cos 3x\,dx = \int\left(\frac{1 - \cos 2x}{2}\right)\cos 3x\,dx = \frac{1}{2}\int\left[\cos 3x - \frac{1}{2}(\cos 5x + \cos x)\right]dx$

$$= \frac{1}{6}\sin 3x - \frac{1}{20}\sin 5x - \frac{1}{4}\sin x + C.$$

70. If we set $y = \text{Sin}^{-1}x$, then $x = \sin y$ and $dx = \cos y\,dy$. With these,

$$\int \sqrt{1 - x^2}\,\text{Sin}^{-1}x\,dx = \int \cos y\,(y)\cos y\,dy = \int y\left(\frac{1 + \cos 2y}{2}\right)dy = \frac{y^2}{4} + \frac{1}{2}\int y\cos 2y\,dy.$$

We now set $u = y$, $dv = \cos 2y\,dy$, $du = dy$, $v = (1/2)\sin 2y$, and use integration by parts,

$$\int \sqrt{1 - x^2}\,\text{Sin}^{-1}x\,dx = \frac{y^2}{4} + \frac{1}{2}\left(\frac{y}{2}\sin 2y - \int \frac{1}{2}\sin 2y\,dy\right)$$

$$= \frac{y^2}{4} + \frac{y}{4}\sin 2y + \frac{1}{8}\cos 2y + C$$

$$= \frac{y^2}{4} + \frac{y}{2}\sin y\cos y + \frac{1}{8}(1 - 2\sin^2 y) + C$$

$$= \frac{1}{4}(\text{Sin}^{-1}x)^2 + \frac{1}{2}(\text{Sin}^{-1}x)x\sqrt{1 - x^2} - \frac{1}{4}x^2 + D.$$

72. (a) If $z^2 = (1 + x)/(1 - x)$, then $z^2(1 - x) = 1 + x \implies x = (z^2 - 1)/(z^2 + 1)$, and

$$dx = \frac{(z^2 + 1)(2z) - (z^2 - 1)(2z)}{(z^2 + 1)^2}\,dz = \frac{4z}{(z^2 + 1)^2}\,dz.$$

Thus, $\int \sqrt{\frac{1 + x}{1 - x}}\,dx = \int z\frac{4z}{(z^2 + 1)^2}\,dz = 4\int \frac{z^2}{(z^2 + 1)^2}\,dz.$

We now set $z = \tan\theta$ and $dz = \sec^2\theta\,d\theta$,

$$\int \sqrt{\frac{1+x}{1-x}}\,dx = 4\int \frac{\tan^2\theta}{\sec^4\theta}\sec^2\theta\,d\theta = 4\int \sin^2\theta\,d\theta = 2\int (1-\cos 2\theta)\,d\theta$$

$$= 2\left(\theta - \frac{1}{2}\sin 2\theta\right) + C = 2\theta - 2\sin\theta\,\cos\theta + C$$

$$= 2\,\mathrm{Tan}^{-1}z - 2\frac{z}{\sqrt{z^2+1}}\frac{1}{\sqrt{z^2+1}} + C$$

$$= 2\,\mathrm{Tan}^{-1}z - \frac{2z}{z^2+1} + C$$

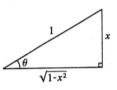

$$= 2\,\mathrm{Tan}^{-1}\sqrt{\frac{1+x}{1-x}} - \frac{2\sqrt{\frac{1+x}{1-x}}}{\frac{1+x}{1-x}+1} + C = 2\,\mathrm{Tan}^{-1}\sqrt{\frac{1+x}{1-x}} - \sqrt{1-x^2} + C.$$

(b) $\displaystyle\int \sqrt{\frac{1+x}{1-x}}\,dx = \int \sqrt{\frac{1+x}{1-x}}\frac{\sqrt{1+x}}{\sqrt{1+x}}\,dx = \int \frac{1+x}{\sqrt{1-x^2}}\,dx$

We now set $x = \sin\theta$ and $dx = \cos\theta\,d\theta$,

$$\int \sqrt{\frac{1+x}{1-x}}\,dx = \int \frac{1+\sin\theta}{\cos\theta}\cos\theta\,d\theta$$

$$= \theta - \cos\theta + C$$

$$= \mathrm{Sin}^{-1}x - \sqrt{1-x^2} + C.$$

If we set $\phi = 2\,\mathrm{Tan}^{-1}\sqrt{\dfrac{1+x}{1-x}}$, then $\dfrac{1+x}{1-x} = \tan^2(\phi/2)$. When we solve this equation for x,

$$x = \frac{\tan^2(\phi/2)-1}{\tan^2(\phi/2)+1} = \frac{\sin^2(\phi/2)-\cos^2(\phi/2)}{\sin^2(\phi/2)+\cos^2(\phi/2)} = -\cos\phi = -\sin\left(\frac{\pi}{2}-\phi\right).$$

Thus, $\dfrac{\pi}{2} - \phi = \mathrm{Sin}^{-1}(-x) = -\mathrm{Sin}^{-1}x$, or $\phi = \mathrm{Sin}^{-1}x + \pi/2$, and it follows that

$$2\,\mathrm{Tan}^{-1}\sqrt{\frac{1+x}{1-x}} - \sqrt{1-x^2} + C = \mathrm{Sin}^{-1}x + \frac{\pi}{2} - \sqrt{1-x^2} + C = \mathrm{Sin}^{-1}x - \sqrt{1-x^2} + D.$$

CHAPTER 9

EXERCISES 9.1

2. This is the parabola
$$x = (1 - y)^2 + 3(1 - y) + 4 = y^2 - 5y + 8.$$

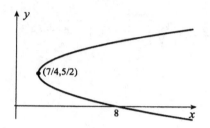

4. Since $(x + 2)^2 + (y - 3)^2 = 16 \cos^2 t + 16 \sin^2 t$ $= 16$, points lie on a circle with centre $(-2, 3)$ and radius 4. Values of t in the interval $0 \le t \le \pi$ yield only the upper semicircle.

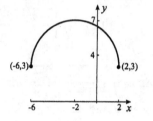

6. Since $x + y = 2t$ and $x - y = 2/t$, multiplication gives $x^2 - y^2 = 4$, a hyperbola.

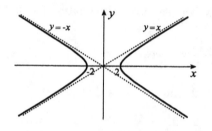

8. Since $\dfrac{(x - 1)^2}{9} + \dfrac{(y + 2)^2}{4} = \cos^2 t + \sin^2 t = 1$, points lie on an ellipse with centre $(1, -2)$. Values $0 \le t \le \pi$ yield only the top half of the ellipse.

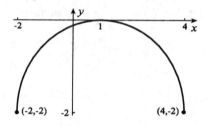

10. Sketches of x and y against t in the left and middle figures below yield the curve to the right.

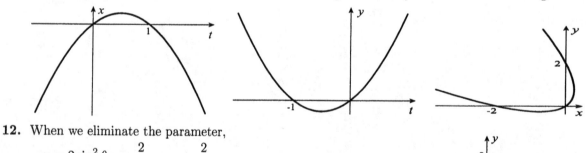

12. When we eliminate the parameter,
$$y = 2 \sin^2 \theta = \frac{2}{\csc^2 \theta} = \frac{2}{1 + \cot^2 \theta}$$
$$= \frac{2}{1 + x^2/4} = \frac{8}{4 + x^2}.$$

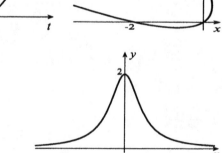

14. $\dfrac{dy}{dx} = \dfrac{dy/du}{dx/du} = \dfrac{\dfrac{(u^2 - 1)(2u) - u^2(2u)}{(u^2 - 1)^2}}{\dfrac{(u - 1)(1) - u(1)}{(u - 1)^2}} = \dfrac{-2u}{(u^2 - 1)^2} \dfrac{(u - 1)^2}{-1} = \dfrac{2u}{(u + 1)^2}$

16. Since $y = 1/x$, it follows that $dy/dx = -1/x^2$.

18. $\dfrac{dy}{dx} = \dfrac{dy/dt}{dx/dt} = \dfrac{\dfrac{(t+6)(1) - t(1)}{(t+6)^2}}{4(2t+3)^3(2)} = \dfrac{6}{(t+6)^2 8(2t+3)^3} = \dfrac{3}{4(t+6)^2(2t+3)^3}$

20. $\dfrac{dy}{dx} = \dfrac{dy/dt}{dx/dt} = \dfrac{\dfrac{-(2t+2)}{(t^2+2t-5)^2}}{\dfrac{-2t+3}{2\sqrt{-t^2+3t+5}}} = \dfrac{4(t+1)\sqrt{-t^2+3t+5}}{(2t-3)(t^2+2t-5)^2}$

22. The slope will be one when $1 = \dfrac{dy}{dx} = \dfrac{dy/dt}{dx/dt} = \dfrac{3t+1}{t^2-3}$, or, $t^2 - 3 = 3t + 1$. Thus, $0 = t^2 - 3t - 4 = (t-4)(t+1) \Longrightarrow t = -1, 4$. The required points are $(8/3, 1/2)$ and $(28/3, 28)$.

24. Since $x^2 = t - 1$ and $y^2 = t + 1$, it follows that $y^2 - x^2 = 2$. Thus, $2y\dfrac{dy}{dx} - 2x = 0 \Longrightarrow \dfrac{dy}{dx} = \dfrac{x}{y}$.

Consequently, $\dfrac{d^2y}{dx^2} = \dfrac{y(1) - x(dy/dx)}{y^2} = \dfrac{y - x(x/y)}{y^2} = \dfrac{y^2 - x^2}{y^3} = \dfrac{2}{y^3}$.

26. Since $\dfrac{dy}{dx} = \dfrac{dy/dv}{dx/dv} = \dfrac{2}{2v+2} = \dfrac{1}{v+1}$, we obtain

$$\dfrac{d^2y}{dx^2} = \dfrac{d}{dx}\left(\dfrac{1}{v+1}\right) = \dfrac{d}{dv}\left(\dfrac{1}{v+1}\right)\dfrac{dv}{dx} = \dfrac{\dfrac{d}{dv}\left(\dfrac{1}{v+1}\right)}{dx/dv} = \dfrac{\dfrac{-1}{(v+1)^2}}{2v+2} = \dfrac{-1}{2(v+1)^3}.$$

28. Since $\dfrac{(x-h)^2}{a^2} + \dfrac{(y-k)^2}{b^2} = \cos^2\theta + \sin^2\theta = 1$, the equations describe an ellipse. Values of the parameter yield all points on the ellipse exactly once.

30. By solving each equation for t and equating results $\dfrac{x - x_1}{x_2 - x_1} = \dfrac{y - y_1}{y_2 - y_1} \Longrightarrow y - y_1 = \dfrac{y_2 - y_1}{x_2 - x_1}(x - x_1)$. This is the equation of the line through P_1 and P_2. By permitting t to take on all possible values, all points on the line are obtained.

32. (a) Since $x^2 - y^2 = \sec^2\theta - \tan^2\theta = 1$, points lie on the hyperbola. Values $-\pi/2 < \theta < \pi/2$ yield only the right half of the hyperbola.
(b) Since $x^2 - y^2 = \cosh^2\phi - \sinh^2\phi = 1$, points are again on the hyperbola. Because $\cosh\phi$ is always positive, only the right half of the hyperbola is defined by the parametric equations.

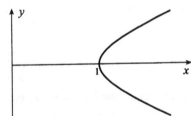

(c) Since $x^2 - y^2 = \dfrac{1}{4}\left(t + \dfrac{1}{t}\right)^2 - \dfrac{1}{4}\left(t - \dfrac{1}{t}\right)^2 = 1$, points are once again on the hyperbola. With $t > 0$, so also is x, and therefore only the right half of the hyperbola is defined by the parametric equations.

34. If we set $y = t$, then $x = \dfrac{5y^2 - y^3}{1+y} = \dfrac{5t^2 - t^3}{1+t}$.

36. Using Exercise 32, parametric equation for the hyperbola $\dfrac{x^2}{4} - \dfrac{y^2}{2} = 1$ are $x = 2\sec\theta$, $y = \sqrt{2}\tan\theta$. To get both halves of the hyperbola, we use $-\pi \le \theta \le \pi$, but do not consider $\theta = \pm\pi/2$.

38. If we differentiate $\dfrac{dy}{dx} = \dfrac{dy/dt}{dx/dt}$ with respect to x, we obtain

$$\dfrac{d^2y}{dx^2} = \dfrac{d}{dx}\left(\dfrac{dy/dt}{dx/dt}\right) = \dfrac{d}{dt}\left(\dfrac{dy/dt}{dx/dt}\right)\dfrac{dt}{dx} = \dfrac{\dfrac{\dfrac{dx}{dt}\dfrac{d^2y}{dt^2} - \dfrac{dy}{dt}\dfrac{d^2x}{dt^2}}{(dx/dt)^2}}{dx/dt} = \dfrac{\dfrac{dx}{dt}\dfrac{d^2y}{dt^2} - \dfrac{dy}{dt}\dfrac{d^2x}{dt^2}}{(dx/dt)^3}.$$

40. The area is four times that in the first
quadrant (figure to the right),

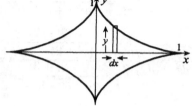

$$A = 4 \int_0^1 y \, dx = 4 \int_{\pi/2}^0 \sin^3 t (-3 \cos^2 t \sin t \, dt)$$

$$= 12 \int_0^{\pi/2} \sin^2 t (\sin t \cos t)^2 \, dt$$

$$= 12 \int_0^{\pi/2} \left(\frac{1 - \cos 2t}{2} \right) \left(\frac{\sin 2t}{2} \right)^2 \, dt$$

$$= \frac{3}{2} \int_0^{\pi/2} (\sin^2 2t - \sin^2 2t \cos 2t) \, dt = \frac{3}{2} \int_0^{\pi/2} \left(\frac{1 - \cos 4t}{2} - \sin^2 2t \cos 2t \right) \, dt$$

$$= \frac{3}{2} \left\{ \frac{t}{2} - \frac{1}{8} \sin 4t - \frac{1}{6} \sin^3 2t \right\}_0^{\pi/2} = \frac{3\pi}{8}.$$

42. The area is twice that to the right
of the y-axis (figure to the right),

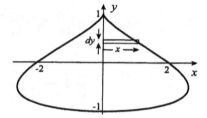

$$A = 2 \int_{-1}^1 x \, dy = 2 \int_{-\pi/2}^{\pi/2} (2 \cos t - \sin 2t)(\cos t \, dt)$$

$$= 2 \int_{-\pi/2}^{\pi/2} (1 + \cos 2t - 2 \cos^2 t \sin t) \, dt$$

$$= 2 \left\{ t + \frac{1}{2} \sin 2t + \frac{2}{3} \cos^3 t \right\}_{-\pi/2}^{\pi/2} = 2\pi.$$

44. The volume is

$$V = \int_0^{2\pi R} \pi y^2 \, dx = \pi \int_0^{2\pi} R^2 (1 - \cos \theta)^2 (R)(1 - \cos \theta) \, d\theta$$

$$= \pi R^3 \int_0^{2\pi} (1 - 3 \cos \theta + 3 \cos^2 \theta - \cos^3 \theta) \, d\theta$$

$$= \pi R^3 \int_0^{2\pi} \left[1 - 3 \cos \theta + \frac{3}{2}(1 + \cos 2\theta) - \cos \theta (1 - \sin^2 \theta) \right] \, d\theta$$

$$= \pi R^3 \left\{ \frac{5\theta}{2} - 4 \sin \theta + \frac{3}{4} \sin 2\theta + \frac{1}{3} \sin^3 \theta \right\}_0^{2\pi} = 5\pi^2 R^3.$$

46. Since small lengths along the curve are given by

$$\sqrt{\left(\frac{dx}{dt} \right)^2 + \left(\frac{dy}{dt} \right)^2} \, dt = \sqrt{(e^{-t} \cos t - e^{-t} \sin t)^2 + (-e^{-t} \sin t - e^{-t} \cos t)^2} = \sqrt{2} \, e^{-t} \, dt,$$

the length of the curve is $\int_0^1 \sqrt{2} \, e^{-t} \, dt = \sqrt{2} \left\{ -e^{-t} \right\}_0^1 = \sqrt{2}(1 - e^{-1}).$

48. Quadrupling the first quadrant length, we get

$$4 \int_0^{\pi/2} \sqrt{(-a \sin \theta)^2 + (b \cos \theta)^2} \, d\theta = 4 \int_0^{\pi/2} \sqrt{a^2 \sin^2 \theta + b^2 \cos^2 \theta} \, d\theta.$$

50. (a) If we set $u = t - 1$, then $x = (u + 1)^2 + 2(u + 1) - 1 = u^2 + 4u + 2$, and $y = (u + 1) + 5 = u + 6$,
define the same curve where $0 \le u \le 3$.
(b) If we set $v = u/3$, then $x = (3v)^2 + 4(3v) + 2 = 9v^2 + 12v + 2$, and $y = 3v + 6$ define the same curve
where $0 \le v \le 1$.

52. Parametric equations for the circle are $x = 2 \cos \theta$, $y = 2 \sin \theta$. If the particle makes two revolutions
each second, then $\theta = 4\pi t$, where $t \ge 0$. Consequently, parametric equations for the position are
$x = 2 \cos 4\pi t$, $y = 2 \sin 4\pi t$, $t \ge 0$.

54. The curve is shown to the right. Points at which the tangent line is horizontal can be found by solving

$$0 = \frac{dy}{dx} = \frac{\dfrac{dy}{dt}}{\dfrac{dx}{dt}}$$

$$= \frac{\dfrac{(1+t^2)(1-3t^2)-(t-t^3)(2t)}{(1+t^2)^2}}{\dfrac{(1+t^2)(-2t)-(1-t^2)(2t)}{(1+t^2)^2}}$$

$$= \frac{1-4t^2-t^4}{-4t}.$$

But this implies that $t^4 + 4t^2 - 1 = 0$, a quadratic equation in t^2,

$$t^2 = \frac{-4 \pm \sqrt{16+4}}{2} = -2 \pm \sqrt{5}.$$

Since t^2 must be nonnegative, it follows that $t^2 = \sqrt{5}-2$, and therefore $t = \pm\sqrt{\sqrt{5}-2}$. For these values of t, $x = 0.62$ and $y = \pm 0.30$. Points with a horizontal tangent line are therefore $(0.62, \pm 0.30)$.

56. (a) Since the length of CE is l_3,

$$(x_E - x_c)^2 + (y_E - y_c)^2 = l_3^2.$$

Substituting for x_c, and y_c gives

$$\left[x_E - d + \frac{l_2(d-l_1\cos\theta)}{\sqrt{d^2+l_1^2-2dl_1\cos\theta}} \right]^2 + \left[y_E - \frac{l_1 l_2 \sin\theta}{\sqrt{d^2+l_1^2-2dl_1\cos\theta}} \right]^2 = l_3^2.$$

Hence,

$$x_E = d - \frac{l_2(d-l_1\cos\theta)}{\sqrt{d^2+l_1^2-2dl_1\cos\theta}} \pm \sqrt{l_3^2 - \left[y_E - \frac{l_1 l_2 \sin\theta}{\sqrt{d^2+l_1^2-2dl_1\cos\theta}} \right]^2}.$$

Since x_E must be greater than x_c, we choose

$$x_E = d - \frac{l_2(d-l_1\cos\theta)}{\sqrt{d^2+l_1^2-2dl_1\cos\theta}} + \sqrt{l_3^2 - \left[y_E - \frac{l_1 l_2 \sin\theta}{\sqrt{d^2+l_1^2-2dl_1\cos\theta}} \right]^2}.$$

(b) When $l_1 = 1/2$, $l_2 = 2$, $l_3 = 4$, $d = 1$, and $y_E = 2$,

$$x_E = 1 - \frac{2[1-(1/2)\cos\theta]}{\sqrt{1+1/4-2(1/2)\cos\theta}} + \sqrt{16 - \left[2 - \frac{(1/2)(2)\sin\theta}{\sqrt{1+1/4-2(1/2)\cos\theta}} \right]^2}$$

$$= 1 - \frac{2(2-\cos\theta)}{\sqrt{5-4\cos\theta}} + \sqrt{16 - \left[2 - \frac{2\sin\theta}{\sqrt{5-4\cos\theta}} \right]^2}.$$

A plot is shown below. The estimated stroke is 1.2 m.

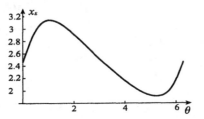

58. From the figure to the right,

$$x = \|OB\| - \|AB\| = R\theta - \|CD\|$$
$$= R\theta - b\cos(\theta - \pi/2) = R\theta - b\sin\theta,$$
$$y = \|AD\| + \|DP\| = R + b\sin(\theta - \pi/2)$$
$$= R - b\cos\theta.$$

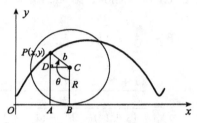

60. The graphs of x and y as functions of θ in the left and middle figures lead to the curve in the right figure.

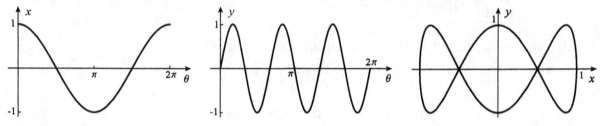

62. The graphs of x and y as functions of θ in the left and middle figures lead to the curve in the right figure.

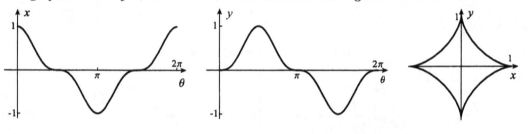

EXERCISES 9.2

2. $r = \sqrt{(-1)^2 + (\sqrt{3})^2} = 2$. Angles satisfying $\tan\theta = -\sqrt{3}$ are $-\pi/3 + n\pi$. Since the point is in the second quadrant, polar coordinates are $(2, 2\pi/3 + 2n\pi)$.

4. $r = \sqrt{(-2\sqrt{3})^2 + 2^2} = 4$. Angles satisfying $\tan\theta = -1/\sqrt{3}$ are $-\pi/6 + n\pi$. Since the point is in the second quadrant, polar coordinates are $(4, 5\pi/6 + 2n\pi)$.

6. $r = \sqrt{(-1)^2 + (-4)^2} = \sqrt{17}$. Angles satisfying $\tan\theta = 4$ are $1.33 + n\pi$. Since the point is in the third quadrant, polar coordinates are $(\sqrt{17}, -1.82 + 2n\pi)$.

8. $r = \sqrt{(-5)^2 + 2^2} = \sqrt{29}$. Angles satisfying $\tan\theta = -2/5$ are $-0.38 + n\pi$. Since the point is in the second quadrant, polar coordinates are $(\sqrt{29}, 2.76 + 2n\pi)$.

10. $(x, y) = (6 \cos(-\pi/6), 6 \sin(-\pi/6))$
$\qquad = (3\sqrt{3}, -3)$

12. $(x, y) = (3 \cos(-2.4), 3 \sin(-2.4))$
$\qquad = (-2.21, -2.03)$

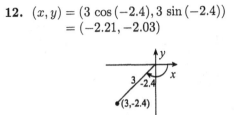

EXERCISES 9.3

2. If we set $x = r \cos\theta$ and $y = r \sin\theta$
in $y = -x$, we obtain $r \sin\theta = -r \cos\theta$.
Either $r = 0$ or $\sin\theta = -\cos\theta$. The first
describes the origin (or pole), and the second
can be expressed in the form $\tan\theta = -1$. This
is equivalent to the two half lines $\theta = 3\pi/4$
and $\theta = -\pi/4$, both of which contain $r = 0$.

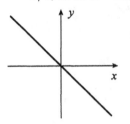

4. In polar coordinates,
$r^2 \cos^2\theta - 2r \cos\theta + r^2 \sin^2\theta - 2r \sin\theta + 1 = 0$,
or, $r^2 - 2r(\cos\theta + \sin\theta) + 1 = 0$. The curve
is more easily sketched in Cartesian coordinates
by writing its equation in the form
$(x - 1)^2 + (y - 1)^2 = 1$, a circle.

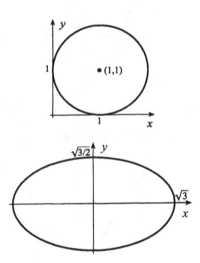

6. The equation of this ellipse in polar
coordinates is $r^2 \cos^2\theta + 2r^2 \sin^2\theta = 3$, or,
$$r^2 = \frac{3}{\cos^2\theta + 2\sin^2\theta} = \frac{3}{1 + \sin^2\theta}.$$

8. In polar coordinates, $r^2 = r - r \cos\theta$. Either $r = 0$ or $r = 1 - \cos\theta$. Since $r = 0$ also satisfies $r = 1 - \cos\theta$
(for $\theta = 0$), we need only write $r = 1 - \cos\theta$. The graph on the left leads to the curve on the right.

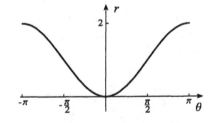

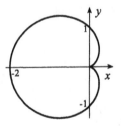

10. In polar coordinates, $r \sin\theta = \dfrac{1}{r^2 \cos^2\theta}$,
or, $r^3 = \sec^2\theta \csc\theta$. It is easily
drawn in Cartesian coordinates.

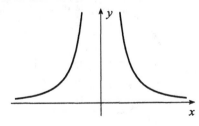

12. This equation describes a half line. Its
equation in Cartesian coordinates is
$y = (\tan 1)x, \ x \geq 0$.

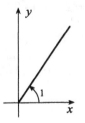

14. In Cartesian coordinates, $x^2 + y^2 = 8\sin\theta\cos\theta = 8\dfrac{y}{\sqrt{x^2+y^2}}\dfrac{x}{\sqrt{x^2+y^2}} \implies (x^2+y^2)^2 = 8xy$. The graph on the left leads to the curve on the right.

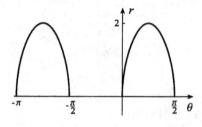
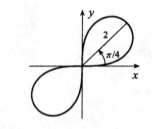

16. In Cartesian coordinates, $\sqrt{x^2+y^2} = 4\sin\theta\cos\theta = 4\dfrac{y}{\sqrt{x^2+y^2}}\dfrac{x}{\sqrt{x^2+y^2}} \implies (x^2+y^2)^{3/2} = 4xy$. The graph on the left leads to the curve on the right.

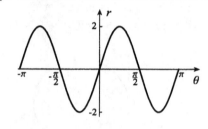
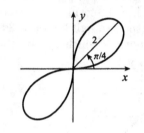

18. In Cartesian coordinates, $\sqrt{x^2+y^2} = 3 - \dfrac{4x}{\sqrt{x^2+y^2}} \implies x^2+y^2 = 3\sqrt{x^2+y^2} - 4x$. The graph on the left leads to the curve on the right.

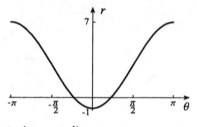
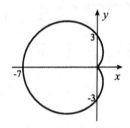

20. In Cartesian coordinates,
$$\sqrt{x^2+y^2} = \frac{\cos^2\theta}{\sin^3\theta} = \frac{x^2}{x^2+y^2}\frac{(x^2+y^2)^{3/2}}{y^3},$$
or, $y^3 = x^2$.

22. When we set $4 = 8\cos 2\theta$, we obtain
$\cos 2\theta = 1/2 \implies 2\theta = \pm\pi/3 + 2n\pi$. Thus,
$\theta = \pm\pi/6 + n\pi$. The figure indicates four points of intersection, $(2, \pm\pi/6)$ and $(2, \pm5\pi/6)$.

24. When we set $1 + \cos\theta = 2 - 2\cos\theta$, we obtain $\cos\theta = 1/3$. Thus, $\theta = \pm\mathrm{Cos}^{-1}(1/3) + 2n\pi$. The figure indicates three points of intersection, the pole $(0,\theta)$, and $(4/3, \pm\mathrm{Cos}^{-1}(1/3)) = (4/3, \pm1.23)$.

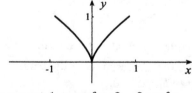
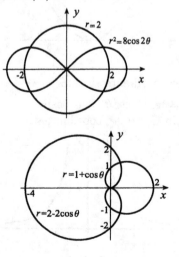

26. According to equation 9.13, the slope of the tangent line to $r = 9 \cos 2\theta$ at $\theta = \pi/6$ is

$$\frac{dy}{dx}\Big|_{\theta=\pi/6} = \frac{-18 \sin 2\theta \sin \theta + 9 \cos 2\theta \cos \theta}{-18 \sin 2\theta \cos \theta - 9 \cos 2\theta \sin \theta}\Big|_{\theta=\pi/6} = \frac{\sqrt{3}}{7}.$$

28. According to equation 9.13, the slope of the tangent line to $r = 3 - 5 \cos \theta$ at $\theta = 3\pi/4$ is

$$\frac{dy}{dx}\Big|_{\theta=3\pi/4} = \frac{5 \sin \theta \sin \theta + (3 - 5 \cos \theta) \cos \theta}{5 \sin \theta \cos \theta - (3 - 5 \cos \theta) \sin \theta}\Big|_{\theta=3\pi/4} = \frac{3}{5\sqrt{2} + 3}.$$

30. (a) According to 9.13,

$$\frac{dy}{dx}\Big|_{\theta=\pi/6} = \frac{\dfrac{3 \cos \theta \sin \theta}{(1 - \sin \theta)^2} + \dfrac{3 \cos \theta}{1 - \sin \theta}}{\dfrac{3 \cos \theta \cos \theta}{(1 - \sin \theta)^2} - \dfrac{3 \sin \theta}{1 - \sin \theta}}\Big|_{\theta=\pi/6} = \sqrt{3}.$$

(b) The equation of the curve in Cartesian coordinates is

$$\sqrt{x^2 + y^2} = \frac{3}{1 - \dfrac{y}{\sqrt{x^2 + y^2}}} = \frac{3\sqrt{x^2 + y^2}}{\sqrt{x^2 + y^2} - y} \implies \sqrt{x^2 + y^2} - y = 3.$$

When we transpose the the y, square and simplify, the result is $y = (x^2 - 9)/6$. Hence, $dy/dx = x/3$. Since the x-coordinate of the point is $x = 6(\sqrt{3}/2) = 3\sqrt{3}$, the slope of the tangent line at the point is $(3\sqrt{3})/3 = \sqrt{3}$.

32. The graph on the left leads to the curve on the right.

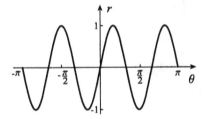 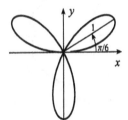

34. The graph on the left leads to the curve on the right.

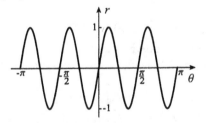 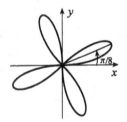

36. The graph on the left leads to the curve on the right.

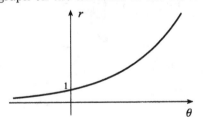 

38. The graph on the left leads to the curve on the right.

 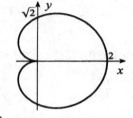

40. Maximum speed occurs when $2\pi\omega t = \pm\dfrac{\pi}{2} + n\pi \Longrightarrow t = \pm\dfrac{1}{4\omega} + \dfrac{n}{2\omega}$, where $n \geq 0$ is an integer. At these times $x = a$. Minimum speed is zero when $2\pi\omega t = n\pi \Longrightarrow t = \dfrac{n}{2\omega}$, where $n \geq 0$ is an integer. At these times, $x = a \pm b$, positions where the follower is furthest from and closest to the axis of rotation.

42. (a)   (b)

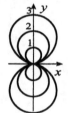 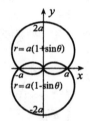

44. (a) Graphs of the cardioids are shown below.

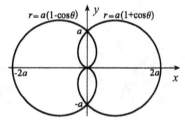

(b) Equations for $r = a(1 \pm \cos\theta)$ in Caresian coordinates are

$$\sqrt{x^2 + y^2} = a\left(1 \pm \frac{x}{\sqrt{x^2 + y^2}}\right) \quad \Longrightarrow \quad x^2 + y^2 = a(\sqrt{x^2 + y^2} \pm x).$$

Similarly, equations for the other cardioids are $x^2 + y^2 = a(\sqrt{x^2 + y^2} \pm y)$.

46. (a) We draw the curves $r = b \pm a\cos\theta$; curves $r = b \pm a\sin\theta$ are rotated $\pi/2$ radians. For the case $b = a$, see the cardioids in Exercise 44. When $a < b$, the graph on the left leads to the curve on the right.

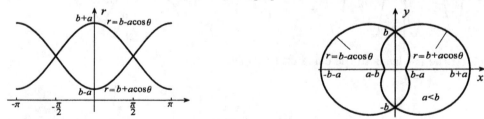

When $a > b$, the graph on the left leads to the curve on the right.

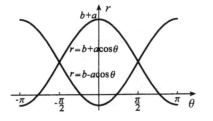

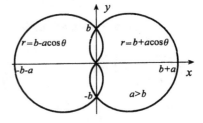

(b) Equations for $r = b \pm a \cos\theta$ in Caresian coordinates are

$$\sqrt{x^2 + y^2} = b \pm \frac{ax}{\sqrt{x^2 + y^2}} \quad \Longrightarrow \quad x^2 + y^2 = b\sqrt{x^2 + y^2} \pm ax.$$

Similarly, equations for the other curves are $x^2 + y^2 = b\sqrt{x^2 + y^2} \pm ay$.

48. The function $r = f(\theta) = a\sin n\theta$ (or $a\cos n\theta$) is $2\pi/n$ periodic. This means that the graph of $r = |a\sin n\theta|$ has $2n$ distinct parts above the θ-axis in the interval $-\pi \le \theta \le \pi$. These yield $2n$ loops (or petals) for the curve $r = |a\sin n\theta|$.

EXERCISES 9.4

2. $A = 2\displaystyle\int_{\pi/2}^{\pi} \frac{1}{2}(-6\cos\theta)^2\,d\theta$

$$= 36\int_{\pi/2}^{\pi}\left(\frac{1 + \cos 2\theta}{2}\right)d\theta$$

$$= 18\left\{\theta + \frac{1}{2}\sin 2\theta\right\}_{\pi/2}^{\pi} = 9\pi$$

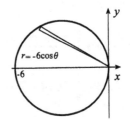

4. $A = 2\displaystyle\int_{0}^{\pi/2} \frac{1}{2}(2\sin 2\theta)\,d\theta$

$$= \left\{-\cos 2\theta\right\}_{0}^{\pi/2} = 2$$

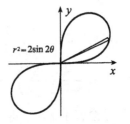

6. $A = 2\displaystyle\int_{0}^{\pi}\frac{1}{2}(2 - 2\cos\theta)^2\,d\theta$

$$= 4\int_{0}^{\pi}(1 - 2\cos\theta + \cos^2\theta)\,d\theta$$

$$= 4\int_{0}^{\pi}\left(1 - 2\cos\theta + \frac{1 + \cos 2\theta}{2}\right)d\theta$$

$$= 4\left\{\frac{3\theta}{2} - 2\sin\theta + \frac{1}{4}\sin 2\theta\right\}_{0}^{\pi} = 6\pi$$

8. $A = 2\displaystyle\int_{0}^{\pi}\frac{1}{2}(4 - 2\cos\theta)^2\,d\theta = 4\int_{0}^{\pi}(4 - 4\cos\theta + \cos^2\theta)\,d\theta$

$$= 4\int_{0}^{\pi}\left(4 - 4\cos\theta + \frac{1 + \cos 2\theta}{2}\right)d\theta$$

$$= 4\left\{\frac{9\theta}{2} - 4\sin\theta + \frac{1}{4}\sin 2\theta\right\}_{0}^{\pi} = 18\pi$$

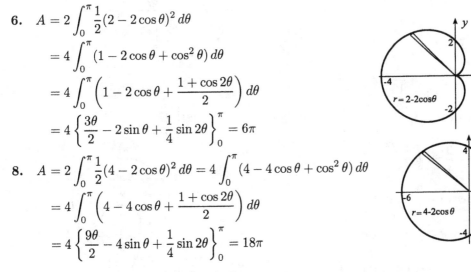

10. $A = \displaystyle\int_{-\pi/4}^{3\pi/4} \frac{1}{2} [2(\cos\theta + \sin\theta)]^2 \, d\theta$

$= 2 \displaystyle\int_{-\pi/4}^{3\pi/4} (1 + 2\cos\theta\sin\theta) \, d\theta$

$= 2 \left\{ \theta + \sin^2\theta \right\}_{-\pi/4}^{3\pi/4} = 2\pi$

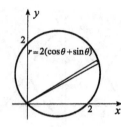

12. $A = \dfrac{\pi}{2} + 2\displaystyle\int_0^{\pi/2} \frac{1}{2}(1 - \sin\theta)^2 \, d\theta$

$= \dfrac{\pi}{2} + \displaystyle\int_0^{\pi/2} \left(1 - 2\sin\theta + \dfrac{1 - \cos 2\theta}{2} \right) d\theta$

$= \dfrac{\pi}{2} + \left\{ \dfrac{3\theta}{2} + 2\cos\theta - \dfrac{1}{4}\sin 2\theta \right\}_0^{\pi/2} = \dfrac{1}{4}(5\pi - 8)$

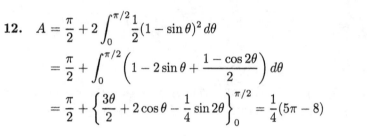

14. $A = 4\displaystyle\int_0^{\pi/2} \frac{1}{2}(2 - 2\cos\theta)^2 \, d\theta$

$= 8\displaystyle\int_0^{\pi/2} \left(1 - 2\cos\theta + \dfrac{1 + \cos 2\theta}{2} \right) d\theta$

$= 8\left\{ \dfrac{3\theta}{2} - 2\sin\theta + \dfrac{1}{4}\sin 2\theta \right\}_0^{\pi/2} = 6\pi - 16$

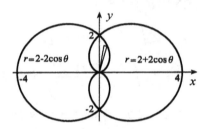

16. $A = 2\displaystyle\int_0^{\pi/3} \frac{1}{2}(1 - \cos\theta)^2 \, d\theta + 2\displaystyle\int_{\pi/3}^{\pi/2} \frac{1}{2}\cos^2\theta \, d\theta$

$= \displaystyle\int_0^{\pi/3} \left(1 - 2\cos\theta + \dfrac{1 + \cos 2\theta}{2} \right) d\theta$

$\quad + \displaystyle\int_{\pi/3}^{\pi/2} \left(\dfrac{1 + \cos 2\theta}{2} \right) d\theta$

$= \left\{ \dfrac{3\theta}{2} - 2\sin\theta + \dfrac{1}{4}\sin 2\theta \right\}_0^{\pi/3}$

$\quad + \dfrac{1}{2}\left\{ \theta + \dfrac{1}{2}\sin 2\theta \right\}_{\pi/3}^{\pi/2} = \dfrac{1}{12}(7\pi - 12\sqrt{3})$

18. $A = 2\displaystyle\int_{\tilde{\theta}}^{\pi/2} \frac{1}{2}[(4 + 3\sin\theta)^2 - 4] \, d\theta = \displaystyle\int_{\tilde{\theta}}^{\pi/2} \left[12 + 24\sin\theta + \dfrac{9}{2}(1 - \cos 2\theta) \right] d\theta$

$= \left\{ \dfrac{33\theta}{2} - 24\cos\theta - \dfrac{9}{4}\sin 2\theta \right\}_{\tilde{\theta}}^{\pi/2}$

$= \dfrac{33\pi}{4} - \dfrac{33\tilde{\theta}}{2} + 24\cos\tilde{\theta} + \dfrac{9}{4}\sin 2\tilde{\theta}$

$= \dfrac{33\pi}{4} - \dfrac{33\tilde{\theta}}{2} + 24\cos\tilde{\theta} + \dfrac{9}{2}\sin\tilde{\theta}\cos\tilde{\theta}$

$= \dfrac{33\pi}{4} + \dfrac{33}{2}\text{Sin}^{-1}\left(\dfrac{2}{3}\right) + 24\left(\dfrac{\sqrt{5}}{3}\right) + \dfrac{9}{2}\left(-\dfrac{2}{3}\right)\left(\dfrac{\sqrt{5}}{3}\right) = \dfrac{33\pi}{4} + \dfrac{33}{2}\text{Sin}^{-1}\left(\dfrac{2}{3}\right) + 7\sqrt{5}$

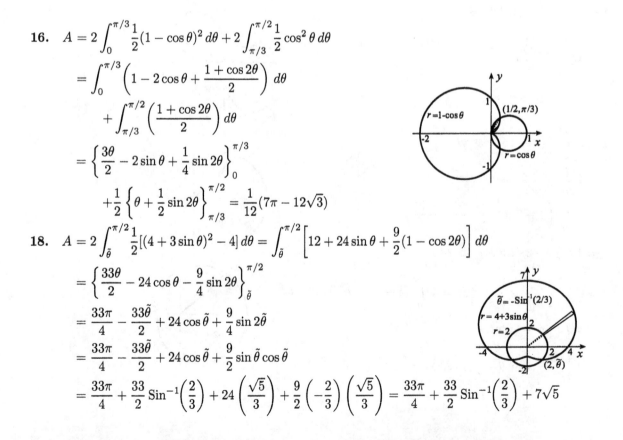

20. $A = 2\int_0^{\pi/2} \frac{1}{2}\cos^4\theta\sin^2\theta\,d\theta = \int_0^{\pi/2}\left(\frac{1+\cos 2\theta}{2}\right)\left(\frac{\sin^2 2\theta}{4}\right)d\theta$

$\qquad = \frac{1}{8}\int_0^{\pi/2}\left(\frac{1-\cos 4\theta}{2} + \sin^2 2\theta\cos 2\theta\right)d\theta$

$\qquad = \frac{1}{8}\left\{\frac{\theta}{2} - \frac{1}{8}\sin 4\theta + \frac{1}{6}\sin^3 2\theta\right\}_0^{\pi/2} = \frac{\pi}{32}$

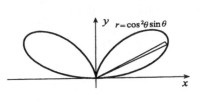

EXERCISES 9.5

2. Completion of squares leads to $(x-3/2)^2 + (y+1)^2 = 25 + 9/4 + 1 = 113/4$, a circle.

4. The presence of y^3 means that the equation describes none of the given curves.

6. The equation can be expressed in form 9.26 for a hyperbola, $\dfrac{y^2}{5/3} - \dfrac{x^2}{5/2} = 1$.

8. In the form $y = x^2/3 + 2x/3 - 4/3$, we have a parabola.

10. No x and y can make $x^2 + 2y^2 + 24$ equal to zero.

12. This is an ellipse.

14. Equation $x + 4y = 3$ represents a straight line.

16. Foci for the ellipse are on the y-axis at distances $\pm\sqrt{36-25} = \pm\sqrt{11}$ from the origin.

18. This is a parabola.

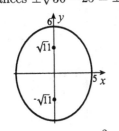

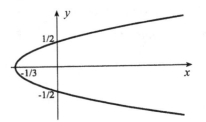

20. Foci for the ellipse $\dfrac{x^2}{16/7} + \dfrac{y^2}{16/3} = 1$ are on the y-axis at distances $\pm\sqrt{16/3 - 16/7} = \pm 8/\sqrt{21}$ from the origin.

22. This is the parabola $x = -2y^2 + 3y + 5$
$\qquad = -(2y-5)(y+1)$.

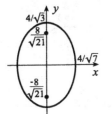

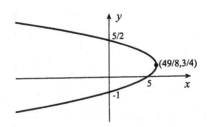

24. Foci for the ellipse $\dfrac{x^2}{2} + \dfrac{y^2}{20} = 1$ are on the y-axis at distances $\pm\sqrt{20-2} = \pm 3\sqrt{2}$ from the origin.

26. Foci for the hyperbola $\dfrac{y^2}{5} - \dfrac{x^2}{5} = 1$ are on the y-axis at distances $\pm\sqrt{5+5} = \pm\sqrt{10}$ from the origin.

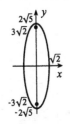

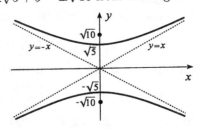

28. Foci for the hyperbola $\dfrac{x^2}{5/2} - \dfrac{y^2}{5/3} = 1$ are on the x-axis at distances $\pm\sqrt{5/2 + 5/3} = \pm 5/\sqrt{6}$ from the origin.

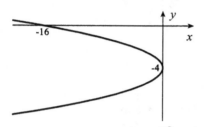

30. Foci for the ellipse $\dfrac{x^2}{2} + \dfrac{y^2}{1/8} = 1$ are on the x-axis at distances $\pm\sqrt{2 - 1/8} = \pm\sqrt{30}/4$ from the origin.

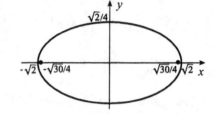

32. This is a parabola.

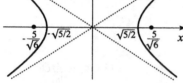

34. Foci for the hyperbola $\dfrac{(y+3)^2}{4} - \dfrac{(x-3)^2}{16} = 1$ are at the points $(3, -3 \pm 2\sqrt{5})$.

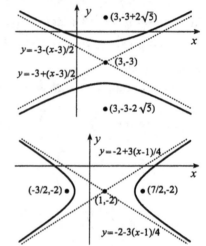

36. Foci for the the hyperbola $\dfrac{(x-1)^2}{4} - \dfrac{(y+2)^2}{9/4} = 1$ are at the points $(7/2, -2)$ and $(-3/2, -2)$.

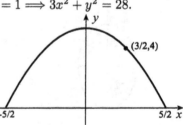

38. For an ellipse of the form $x^2/a^2 + y^2/b^2 = 1$, a and b must satisfy

$$\frac{4}{a^2} + \frac{16}{b^2} = 1 \qquad \text{and} \qquad \frac{9}{a^2} + \frac{1}{b^2} = 1.$$

These imply that $a^2 = 28/3$ and $b^2 = 28$. Hence, $3x^2/28 + y^2/28 = 1 \implies 3x^2 + y^2 = 28$.

40. With the coordinate system shown to the right, the equation of the parabola is $y = ax^2 + c$. Since $(5/2, 0)$ and $(3/2, 4)$ are points thereon,

$$0 = \frac{25a}{4} + c, \quad 4 = \frac{9a}{4} + c.$$

These imply that $a = -1$ and $c = 25/4$. Thus the height of the arch is $25/4$.

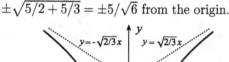

42. (a) The foci of the ellipse are $(\pm 4, 0)$, so that in equation 9.22, $a^2 - b^2 = 16$. Since the sum of the distances from a point on the ellipse to the foci is $2a$, it follows that $a = 5$, and hence $b = 3$. The required equation is therefore $x^2/25 + y^2/9 = 1 \implies 9x^2 + 25y^2 = 225$.

(b) With foci $(\pm 4, 0)$ and $2a = 10$, equation 9.20 for the ellipse is

$$\sqrt{(x+4)^2 + y^2} + \sqrt{(x-4)^2 + y^2} = 10.$$

If we transpose the second term and square both sides,

$$x^2 + 8x + 16 + y^2 = 100 - 20\sqrt{(x-4)^2 + y^2} + x^2 - 8x + 16 + y^2 \implies 100 - 16x = 20\sqrt{(x-4)^2 + y^2}.$$

Dividing by 4, and squaring, $625 - 200x + 16x^2 = 25(x^2 - 8x + 16 + y^2) \implies 9x^2 + 25y^2 = 225$.

44. For straight lines we set $A = C = 0$. For circles we set $A = C$. For parabolas of the form $y = ax^2 + bx + c$, we set $C = 0$. For parabolas of the the form $x = ay^2 + by + c$, we set $A = 0$. For ellipse 9.23 we demand that $AC > 0$. For hyperbolas 9.27 we demand that $AC < 0$.

46. If we differentiate $b^2 x^2 - a^2 y^2 = a^2 b^2$ with respect to x, $2b^2 x - 2a^2 y \dfrac{dy}{dx} = 0 \implies \dfrac{dy}{dx} = \dfrac{b^2 x}{a^2 y}$. At a point (x_0, y_0), the equation of the tangent line is

$$y - y_0 = \frac{b^2 x_0}{a^2 y_0}(x - x_0) \qquad \implies \qquad a^2 y y_0 - a^2 y_0^2 = b^2 x x_0 - b^2 x_0^2.$$

Since $b^2 x_0^2 - a^2 y_0^2 = a^2 b^2$, we may also write for the line $b^2 x x_0 - a^2 y y_0 = a^2 b^2$.

48. $A = 4 \displaystyle\int_0^a y \, dx = 4 \int_0^a \dfrac{b}{a}\sqrt{a^2 - x^2}\, dx$ If we set $x = a \sin\theta$, then $dx = a\cos\theta\, d\theta$, and

$$A = \frac{4b}{a}\int_0^{\pi/2} a\cos\theta\, a\cos\theta\, d\theta$$

$$= 4ab\int_0^{\pi/2}\left(\frac{1 + \cos 2\theta}{2}\right)d\theta$$

$$= 2ab\left\{\theta + \frac{1}{2}\sin 2\theta\right\}_0^{\pi/2} = \pi ab.$$

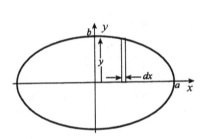

50. For the prolate spheroid,

$$V = 2\int_0^a \pi y^2\, dx = 2\pi\int_0^a \frac{b^2}{a^2}(a^2 - x^2)\, dx$$

$$= \frac{2\pi b^2}{a^2}\left\{a^2 x - \frac{x^3}{3}\right\}_0^a = \frac{4}{3}\pi ab^2.$$

For the oblate spheroid,

$$V = 2\int_0^a 2\pi x y\, dx = 4\pi\int_0^a x\frac{b}{a}\sqrt{a^2 - x^2}\, dx$$

$$= \frac{4\pi b}{a}\left\{-\frac{1}{3}(a^2 - x^2)^{3/2}\right\}_0^a = \frac{4}{3}\pi a^2 b.$$

52. If we equate coefficients in $y = ax^2 + bx + c$ and those in equation 9.18,

$$a = \frac{1}{2(q - r)}, \qquad b = \frac{-p}{q - r}, \qquad c = \frac{p^2 + q^2 - r^2}{2(q - r)}.$$

The second divided by the first gives $b/a = -2p$, or, $p = -b/(2a)$. When the third is divided by the first and p set equal to $-b/(2a)$, the result is $q^2 - r^2 = (4ac - b^2)/(4a^2)$, or,

$$\frac{4ac - b^2}{4a^2} = (q + r)(q - r) = (q + r)\frac{1}{2a}.$$

Thus,

$$q + r = \frac{4ac - b^2}{2a} \qquad \text{and} \qquad q - r = \frac{1}{2a}.$$

Addition and subtraction of these results give the expressions for q and r.

54. Exercise 15: $p = 0$; $q = (1/8)(1 - 8) = -7/8$; $r = (1/8)(-1 - 8) = -9/8$. Thus, the focus is $(0, -7/8)$, and the directrix is $y = -9/8$.

Exercise 18: $p = 1/(16/3)[1 + 4(4/3)(-1/3)] = -7/48$; $q = 0$; $r = 1/(16/3)[-1 + 4(4/3)(-1/3)] = -25/48$. Thus, the focus is $(-7/48, 0)$, and the directrix is $x = -25/48$.

Exercise 21: $p = 1/(-4)[1 + 4(-1)(1)] = 3/4$; $q = 0$; $r = 1/(-4)[-1 + 4(-1)(1)] = 5/4$. Thus, the focus is $(3/4, 0)$, and the directrix is $x = 5/4$.

Exercise 22: $p = 1/(-8)[1 + 4(-2)(5) - 9] = 6$; $q = -3/(-4)$; $r = 1/(-8)[-1 + 4(-2)(5) - 9] = 25/4$. Thus, the focus is $(6, 3/4)$, and the directrix is $x = 25/4$.

Exercise 27: $p = -6/(-2) = 3$; $q = 1/(-4)[1 + 4(-1)(-9) - 36] = -1/4$; $r = 1/(-4)[-1 + 4(-1)(-9) - 36] = 1/4$. Thus, the focus is $(3, -1/4)$, and the directrix is $y = 1/4$.

Exercise 32: With $x = -y^2 - 8y - 16$, $p = [1/(-4)][1 + 4(-1)(-16) - 64] = -1/4$; $q = 8/(-2) = -4$; $r = [1/(-4)][-1 + 4(-1)(-16) - 64] = 1/4$. Thus, the focus is $(-1/4, -4)$ and the directrix $x = 1/4$.

56. When the centre of the hyperbola is (h, k), and foci lie on $y = k$, the foci are $(h \pm c, k)$. If the difference of the distances from a point (x, y) on the hyperbola to the foci is $2a$, then the equation of the hyperbola is

$$\left| \sqrt{(x - h - c)^2 + (y - k)^2} - \sqrt{(x - h + c)^2 + (y - k)^2} \right| = 2a.$$

When we write $\sqrt{(x - h - c)^2 + (y - k)^2} = \sqrt{(x - h + c)^2 + (y - k)^2} \pm 2a$, and square both sides

$$(x - h - c)^2 + (y - k)^2 = (x - h + c)^2 + (y - k)^2 \pm 4a\sqrt{(x - h + c)^2 + (y - k)^2} + 4a^2,$$

and this equation simplifies to $\pm a\sqrt{(x - h + c)^2 + (y - k)^2} = a^2 + c(x - h)$. Squaring again leads to

$$\frac{(x - h)^2}{a^2} - \frac{(y - k)^2}{c^2 - a^2} = 1.$$

If we set $b^2 = c^2 - a^2$, then the first of equations 9.27 is obtained. The second equation in 9.27 is obtained in a similar fashion when foci are on the line $x = h$.

EXERCISES 9.6

2. Since $r = \dfrac{16/3}{1 + (5/3)\cos\theta}$, we have a hyperbola with foci on the x-axis. It crosses the x-axis at $r = 2$ when $\theta = 0$, and the y-axis at $r = 16/3$ when $\theta = \pm\pi/2$. Only the left half of the hyperbola is obtained.

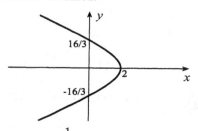

4. Since $r = \dfrac{16/5}{1 + (3/5)\cos\theta}$, we have an ellipse with foci on the x-axis. It crosses the x-axis at $r = 2$ when $\theta = 0$, and at $r = 8$ when $\theta = \pi$. The centre of the ellipse is $(-3, 0)$, and its maximum y-value is $b = \sqrt{25 - 9} = 4$.

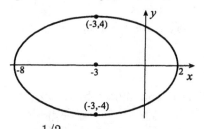

6. Since $r = \dfrac{1}{1 - (3/4)\sin\theta}$, we have an ellipse with foci on the y-axis. It crosses the y-axis at $r = 4$ when $\theta = \pi/2$, and at $r = 4/7$ when $\theta = -\pi/2$. The centre of the ellipse is $(0, 12/7)$ and its maximum x-value is $a = \sqrt{(16/7)^2 - (12/7)^2} = 4/\sqrt{7}$.

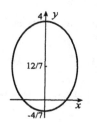

8. Since $r = \dfrac{1/2}{1 + (1/2)\sin\theta}$, we have an ellipse with foci on the y-axis. It crosses the y-axis at $r = 1/3$ when $\theta = \pi/2$, and at $r = 1$ when $\theta = -\pi/2$. The centre of the ellipse is $(0, -1/3)$ and its maximum x-value is $a = \sqrt{(2/3)^2 - (1/3)^2} = 1/\sqrt{3}$.

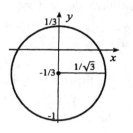

10. Since $r = \dfrac{4}{7 - 2\sin\theta} = \dfrac{4/7}{1 - (2/7)\sin\theta}$, we have
an ellipse with foci on the y-axis. It crosses the
y-axis at $r = 4/5$ when $\theta = \pi/2$, and at
$r = 4/9$ when $\theta = -\pi/2$. The centre of the
ellipse is $(0, 8/45)$ and its maximum x-value is
$a = \sqrt{(28/45)^2 - (8/45)^2} = 4\sqrt{5}/15$.

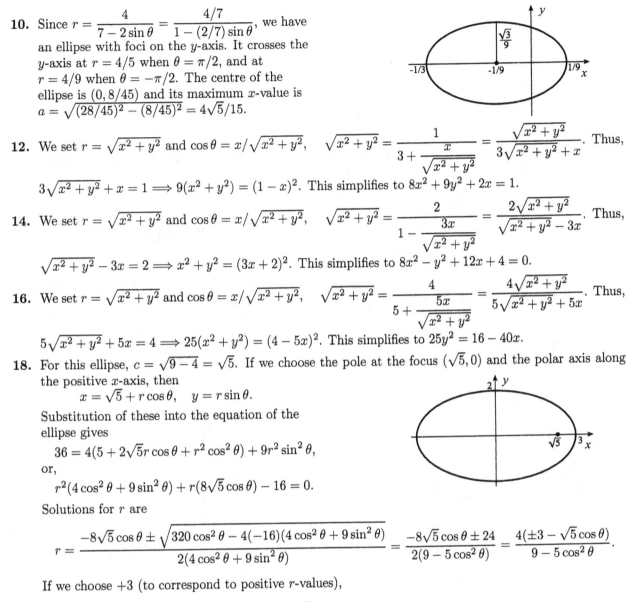

12. We set $r = \sqrt{x^2 + y^2}$ and $\cos\theta = x/\sqrt{x^2 + y^2}$, $\sqrt{x^2 + y^2} = \dfrac{1}{3 + \dfrac{x}{\sqrt{x^2+y^2}}} = \dfrac{\sqrt{x^2+y^2}}{3\sqrt{x^2+y^2}+x}$. Thus,

$3\sqrt{x^2 + y^2} + x = 1 \implies 9(x^2 + y^2) = (1 - x)^2$. This simplifies to $8x^2 + 9y^2 + 2x = 1$.

14. We set $r = \sqrt{x^2 + y^2}$ and $\cos\theta = x/\sqrt{x^2 + y^2}$, $\sqrt{x^2 + y^2} = \dfrac{2}{1 - \dfrac{3x}{\sqrt{x^2+y^2}}} = \dfrac{2\sqrt{x^2+y^2}}{\sqrt{x^2+y^2}-3x}$. Thus,

$\sqrt{x^2 + y^2} - 3x = 2 \implies x^2 + y^2 = (3x + 2)^2$. This simplifies to $8x^2 - y^2 + 12x + 4 = 0$.

16. We set $r = \sqrt{x^2 + y^2}$ and $\cos\theta = x/\sqrt{x^2 + y^2}$, $\sqrt{x^2 + y^2} = \dfrac{4}{5 + \dfrac{5x}{\sqrt{x^2+y^2}}} = \dfrac{4\sqrt{x^2+y^2}}{5\sqrt{x^2+y^2}+5x}$. Thus,

$5\sqrt{x^2 + y^2} + 5x = 4 \implies 25(x^2 + y^2) = (4 - 5x)^2$. This simplifies to $25y^2 = 16 - 40x$.

18. For this ellipse, $c = \sqrt{9 - 4} = \sqrt{5}$. If we choose the pole at the focus $(\sqrt{5}, 0)$ and the polar axis along
the positive x-axis, then
$$x = \sqrt{5} + r\cos\theta, \quad y = r\sin\theta.$$
Substitution of these into the equation of the
ellipse gives
$$36 = 4(5 + 2\sqrt{5}r\cos\theta + r^2\cos^2\theta) + 9r^2\sin^2\theta,$$
or,
$$r^2(4\cos^2\theta + 9\sin^2\theta) + r(8\sqrt{5}\cos\theta) - 16 = 0.$$
Solutions for r are

$$r = \frac{-8\sqrt{5}\cos\theta \pm \sqrt{320\cos^2\theta - 4(-16)(4\cos^2\theta + 9\sin^2\theta)}}{2(4\cos^2\theta + 9\sin^2\theta)} = \frac{-8\sqrt{5}\cos\theta \pm 24}{2(9 - 5\cos^2\theta)} = \frac{4(\pm 3 - \sqrt{5}\cos\theta)}{9 - 5\cos^2\theta}.$$

If we choose $+3$ (to correspond to positive r-values),

$$r = \frac{4(3 - \sqrt{5}\cos\theta)}{(3 - \sqrt{5}\cos\theta)(3 + \sqrt{5}\cos\theta)} = \frac{4}{3 + \sqrt{5}\cos\theta}.$$

20. For this hyperbola, $c = \sqrt{9 + 1} = \sqrt{10}$. If we choose the pole at the focus $(\sqrt{10}, 2)$ and the polar axis
parallel to the positive x-axis, then
$$x = \sqrt{10} + r\cos\theta, \quad y = 2 + r\sin\theta.$$
Substitution of these into the equation of the
hyperbola gives
$$10 + 2\sqrt{10}r\cos\theta + r^2\cos^2\theta - 9r^2\sin^2\theta = 9,$$
or,
$$r^2(\cos^2\theta - 9\sin^2\theta) + r(2\sqrt{10}\cos\theta) + 1 = 0.$$
Solutions for r are

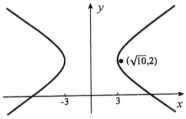

$$r = \frac{-2\sqrt{10}\cos\theta \pm \sqrt{40\cos^2\theta - 4(\cos^2\theta - 9\sin^2\theta)}}{2(\cos^2\theta - 9\sin^2\theta)} = \frac{-2\sqrt{10}\cos\theta \pm 6}{2(10\cos^2\theta - 9)} = \frac{-\sqrt{10}\cos\theta \pm 3}{10\cos^2\theta - 9}.$$

If we choose -3, then

$$r = \frac{-\sqrt{10}\cos\theta - 3}{(\sqrt{10}\cos\theta + 3)(\sqrt{10}\cos\theta - 3)} = \frac{1}{3 - \sqrt{10}\cos\theta}.$$

This gives the right half of the hyperbola. The other half is obtained when we set

$$r = \frac{-\sqrt{10}\cos\theta + 3}{10\cos^2\theta - 9} = \frac{-1}{3 + \sqrt{10}\cos\theta}.$$

22. According to equation 9.32, the distance between the foci of the ellipse determined by equation 9.29 is $2d\epsilon^2/(1 - \epsilon^2)$. For this distance to approach zero, ϵ must approach zero.

REVIEW EXERCISES

2. In Cartesian coordinates, $\sqrt{x^2 + y^2} =$

$$\frac{-y}{\sqrt{x^2 + y^2}} \implies x^2 + (y + 1/2)^2 = 1/4.$$

This is a circle.

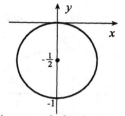

4. Since $r = \dfrac{3/2}{1 + (1/2)\sin\theta}$, the curve is an

ellipse with foci on the y-axis. It crosses

the y-axis at $r = 1$ when $\theta = \pi/2$, and at $r = 3$ when $\theta = -\pi/2$. The centre of the ellipse is $(0, -1)$, and its maximum x-value is $a = \sqrt{4 - 1} = \sqrt{3}$.

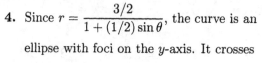

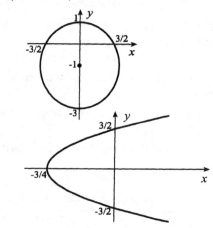

6. This is a parabola that opens to the right. It crosses the x-axis at $r = 3/4$ when $\theta = \pi$ and the y-axis at $r = 3/2$ when $\theta = \pm\pi/2$.

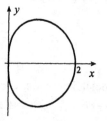

8. The graph of the function $r = 2\sqrt{\cos\theta}$ in the left figure gives the curve in the right figure.

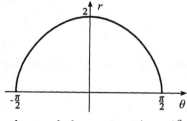

10. This is the parabola $y = 5 - 3(x - 4)^2$.

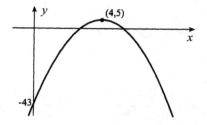

12. These equations describe the right half of of the circle $(x - 1)^2 + (y + 2)^2 = 16$.

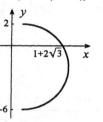

14. In polar coordinates, $r^2 = 2r$, and this describes the circle $r = 2$ and the origin $r = 0$.

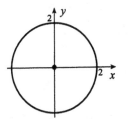

16. The graph of the function $r = 3\cos 2\theta$ in the left figure gives the curve in the right figure.

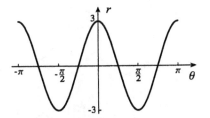

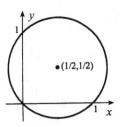

18. When squares are completed on x and y terms, $(x - 1/2)^2 + (y - 1/2)^2 = 1/2$, a circle.

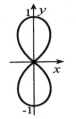

20. The graph of the function $r = \sin^2\theta - \cos^2\theta = -\cos 2\theta$ in the left figure gives the curve in the right figure.

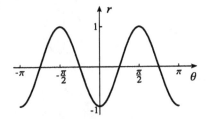

22. This is a hyperbola.

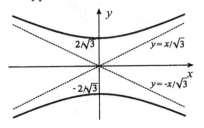

24. This is the parabola $x = (3 + y)(1 - y)$.

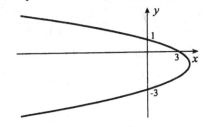

26. The graph of the function $r = \cos(\theta/2)$ in the left figure gives the curve in the right figure.

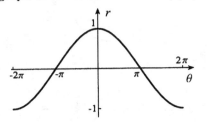

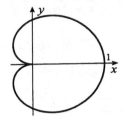

28. Completion of the square on the y term leads to
$$\frac{(y+1/2)^2}{1/4} - \frac{x^2}{1/4} = 1.$$
This is a hyperbola.

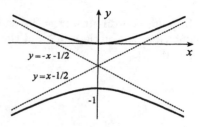

30. The graphs of x and y as functions of θ in the left figure lead to the curve in the right figure.

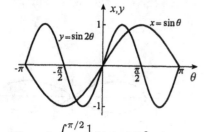

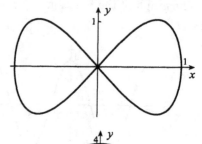

32. $A = 16\pi - 2\int_0^{\pi/2} \frac{1}{2}(4\sin 2\theta)^2 \, d\theta$

$$= 16\pi - 16 \int_0^{\pi/2} \left(\frac{1 - \cos 4\theta}{2}\right) d\theta$$

$$= 16\pi - 8\left\{\theta - \frac{1}{4}\sin 4\theta\right\}_0^{\pi/2} = 12\pi$$

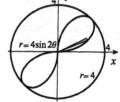

34. $A = 4\int_0^{\pi/2} \frac{1}{2}(\sin^2 \theta)^2 \, d\theta$

$$= 2\int_0^{\pi/2} \left(\frac{1 - \cos 2\theta}{2}\right)^2 d\theta$$

$$= \frac{1}{2}\int_0^{\pi/2} \left(1 - 2\cos 2\theta + \frac{1 + \cos 4\theta}{2}\right) d\theta$$

$$= \frac{1}{2}\left\{\frac{3\theta}{2} - \sin 2\theta + \frac{1}{8}\sin 4\theta\right\}_0^{\pi/2} = \frac{3\pi}{8}$$

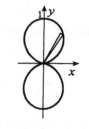

36. Since $\dfrac{dy}{dx} = \dfrac{dy/dt}{dx/dt} = \dfrac{3 - 3t^2}{3t^2 + 2}$,

$$\frac{d^2y}{dx^2} = \frac{d}{dx}\left(\frac{3 - 3t^2}{3t^2 + 2}\right) = \frac{d}{dt}\left(\frac{3 - 3t^2}{3t^2 + 2}\right)\frac{dt}{dx} = \frac{\dfrac{d}{dt}\left(\dfrac{3 - 3t^2}{3t^2 + 2}\right)}{dx/dt} = \frac{\dfrac{(3t^2 + 2)(-6t) - (3 - 3t^2)(6t)}{(3t^2 + 2)^2}}{3t^2 + 2} = \frac{-30t}{(3t^2 + 2)^3}.$$

38. By equation 9.14,

$$L = 2\int_0^\pi \sqrt{(2 + 2\cos\theta)^2 + (-2\sin\theta)^2} \, d\theta$$

$$= 4\sqrt{2} \int_0^\pi \sqrt{1 + \cos\theta} \, d\theta$$

$$= 4\sqrt{2} \int_0^\pi \sqrt{1 + [2\cos^2(\theta/2) - 1]} \, d\theta$$

$$= 8\int_0^\pi \cos(\theta/2) \, d\theta = 8\left\{2\sin(\theta/2)\right\}_0^\pi = 16.$$

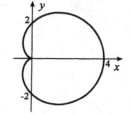

40. By equation 9.3, the length is

$$\int_0^{\pi/2} \sqrt{(e^t \cos t - e^t \sin t)^2 + (e^t \sin t + e^t \cos t)^2} \, dt = \int_0^{\pi/2} \sqrt{2} e^t \, dt = \sqrt{2}\left\{e^t\right\}_0^{\pi/2} = \sqrt{2}(e^{\pi/2} - 1).$$

CHAPTER 10

EXERCISES 10.1

2. This sequence diverges.

4. This sequence has limit 0.

6. This sequence has limit 0.

8. This sequence has limit $\lim\limits_{n\to\infty}\dfrac{n}{n^2+n+2}=\lim\limits_{n\to\infty}\dfrac{1}{n+1+2/n}=0.$

10. This sequence has limit $\pi/2$.

12. This sequence diverges.

14. This sequence has limit 0.

16. This sequence has limit $\lim\limits_{n\to\infty}\dfrac{2n+3}{n^2-5}=\lim\limits_{n\to\infty}\dfrac{2+3/n}{n-5/n}=0.$

18. This sequence has limit 0.

20. This sequence has limit $\lim\limits_{n\to\infty}\dfrac{1}{1+1/n}\mathrm{Tan}^{-1}n=\dfrac{\pi}{2}.$

22. The general term is $\dfrac{3n+1}{n^2}$.

24. The general term is $\dfrac{1+(-1)^{n+1}}{2}$.

26. The limit of the sequence $\{\ln n/\sqrt{n}\}$ as $n\to\infty$ is equal to the limit of the function $\ln x/\sqrt{x}$ as $x\to\infty$, provided the limit of the function exists. When we use L'Hôpital's rule on the limit of the function,

$$\lim_{n\to\infty}\frac{\ln n}{\sqrt{n}}=\lim_{x\to\infty}\frac{\ln x}{\sqrt{x}}=\lim_{x\to\infty}\frac{1/x}{1/(2\sqrt{x})}=\lim_{x\to\infty}\frac{2}{\sqrt{x}}=0.$$

28. The limit of the sequence $\{n\sin(4/n)\}$ as $n\to\infty$ is equal to the limit of the function $x\sin(4/x)$ as $x\to\infty$, provided the limit of the function exists. When we use L'Hôpital's rule,

$$\lim_{n\to\infty}n\sin\left(\frac{4}{n}\right)=\lim_{x\to\infty}x\sin\left(\frac{4}{x}\right)=\lim_{x\to\infty}\frac{\sin(4/x)}{1/x}=\lim_{x\to\infty}\frac{-(4/x^2)\cos(4/x)}{-1/x^2}=4.$$

30. Certainly the sequence diverges; terms get arbitrarily large for large n. On the other hand, as n increases, the difference between terms approaches $\lim\limits_{n\to\infty}[\ln n-\ln(n+1)]=\lim\limits_{n\to\infty}\ln\left(\dfrac{n}{n+1}\right)=0.$

32. The figure indicates that with initial approximation $x_1=1$, the sequence defined by Newton's iterative procedure has a limit near $-1/2$. Iteration of

$$x_1=1,\quad x_{n+1}=x_n-\frac{x_n^2+3x_n+1}{2x_n+3}$$

leads to

$x_2=0,$ $\qquad x_3=-1/3,$

$x_4=-0.381,$ $\qquad x_5=-0.381\,966,$

$x_6=-0.381\,966\,01,$ $\quad x_7=-0.381\,966\,01.$

Since $f(-0.381\,965\,95)=1.4\times10^{-7}$ and $f(-0.381\,966\,05)=-8.7\times10^{-8}$, we can

say that to seven decimals $x=-0.381\,966\,0$.

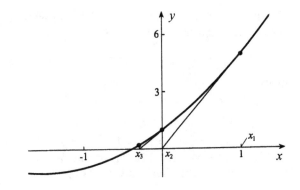

34. The figure indicates that with initial
approximation $x_1 = -1.5$, the sequence defined
by Newton's iterative procedure does not have
a limit. This is because $x_1 = -1.5$ is a
critical point of the function.

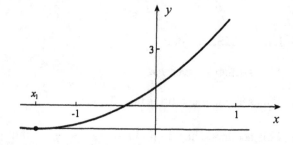

36. The figure indicates that with initial approximation $x_1 = 4$, the sequence defined by Newton's iterative
procedure has a limit near 3. Iteration of

$$x_1 = 4, \quad x_{n+1} = x_n - \frac{x_n^3 - x_n^2 + x_n - 22}{3x_n^2 - 2x_n + 1}$$

leads to

$$x_2 = 3.268, \qquad x_3 = 3.060\,9,$$
$$x_4 = 3.044\,8, \qquad x_5 = 3.044\,723\,15,$$
$$x_6 = 3.044\,723\,15.$$

Since $f(3.044\,723\,05) = -2.2 \times 10^{-6}$ and
$f(3.044\,723\,15) = 3.5 \times 10^{-8}$, we can say
that to seven decimals $x = 3.044\,723\,1$.

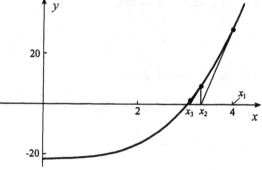

38. The figure indicates that with initial aproximation $x_1 = 2$, the sequence defined by Newton's iterative
procedure has a limit near 1. Iteration of

$$x_1 = 2, \quad x_{n+1} = x_n - \frac{x_n^5 - 3x_n + 1}{5x_n^4 - 3}$$

gives

$$x_2 = 1.649, \qquad x_3 = 1.406,$$
$$x_4 = 1.268, \qquad x_5 = 1.220,$$
$$x_6 = 1.215, \qquad x_7 = 1.214\,65,$$
$$x_8 = 1.214\,648\,04, \quad x_9 = 1.214\,648\,04.$$

Since $f(1.214\,647\,95) = -7.3 \times 10^{-7}$ and
$f(1.214\,648\,05) = 5.8 \times 10^{-8}$, we can say
that to seven decimals $x = 1.214\,648\,0$.

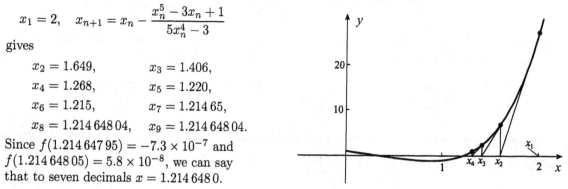

40. The figure indicates that with initial approximation $x_1 = 0$, the sequence defined by Newton's iterative
procedure has a limit near 0.3. Iteration of

$$x_1 = 0, \quad x_{n+1} = x_n - \frac{x_n^5 - 3x_n + 1}{5x_n^4 - 3}$$

gives

$$x_2 = 1/3, \qquad x_3 = 0.334\,7,$$
$$x_4 = 0.334\,734\,14, \quad x_5 = 0.334\,734\,14.$$

Since $f(0.334\,734\,05) = 2.7 \times 10^{-7}$ and
$f(0.334\,734\,15) = -2.4 \times 10^{-8}$, we can
say that to seven decimals $x = 0.334\,734\,1$.

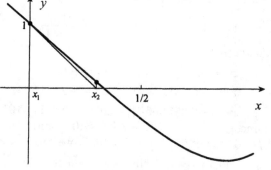

42. The figure indicates that with initial approximation $x_1 = 0.85$, the sequence defined by Newton's iterative procedure has a limit near -1.5. Iteration of

$$x_1 = 0.85, \quad x_{n+1} = x_n - \frac{x_n^5 - 3x_n + 1}{5x_n^4 - 3}$$

gives

$$x_2 = -1.987, \qquad x_3 = -1.667,$$
$$x_4 = -1.474, \qquad x_5 = -1.399,$$
$$x_6 = -1.389, \qquad x_7 = -1.388\,792\,06,$$
$$x_8 = -1.388\,791\,98, \quad x_9 = -1.388\,791\,98.$$

Since $f(-1.388\,791\,95) = 5.4 \times 10^{-7}$ and $f(-1.388\,792\,05) = -1.0 \times 10^{-6}$, we can say that to seven decimals $x = -1.388\,792\,0$.

44. Iteration of $x_1 = 2$, $\quad x_{n+1} = 2 + \dfrac{1}{x_n}$ gives

$$x_2 = 2.5, \qquad x_3 = 2.4, \qquad x_4 = 2.416\,67, \quad x_5 = 2.413\,79, \qquad x_6 = 2.414\,29,$$
$$x_7 = 2.414\,20, \quad x_8 = 2.414\,22, \quad x_9 = 2.414\,21, \quad x_{10} = 2.414\,21.$$

Since $f(2.414\,15) = -1.8 \times 10^{-4}$ and $f(2.414\,25) = 1.0 \times 10^{-4}$, it follows that to 4 decimals, $x = 2.4142$.

46. Iteration of $x_1 = 0$, $\quad x_{n+1} = \dfrac{1}{120}(x_n^4 + 20)$ gives

$$x_2 = 1/6, \quad x_3 = 0.166\,67, \quad x_4 = 0.166\,67.$$

Since $f(0.166\,65) = 2.8 \times 10^{-3}$ and $f(0.166\,75) = -9.2 \times 10^{-3}$, it follows that to 4 decimals, $x = 0.1667$.

48. Iteration of $x_1 = 0$, $\quad x_{n+1} = \dfrac{1}{2}(1 + x_n^2)^{1/3}$ gives

$$x_2 = 1/2, \qquad x_3 = 0.5386, \qquad x_4 = 0.5443, \qquad x_5 = 0.545\,17,$$
$$x_6 = 0.545\,31, \quad x_7 = 0.545\,33, \quad x_8 = 0.545\,33.$$

Since $f(0.545\,25) = -4.9 \times 10^{-4}$ and $f(0.545\,35) = 1.2 \times 10^{-4}$, it follows that to 4 decimals, $x = 0.5453$.

50. With $x_1 = 0$, and $x_{n+1} = \dfrac{x_n^4 - 3x_n^2 + 1}{3}$, iteration gives

$$x_2 = 1/3, \qquad x_3 = 0.226, \qquad x_4 = 0.283, \qquad x_5 = 0.255, \qquad x_6 = 0.270,$$
$$x_7 = 0.262, \qquad x_8 = 0.266, \qquad x_9 = 0.2642, \qquad x_{10} = 0.2652, \qquad x_{11} = 0.2647,$$
$$x_{12} = 0.2649, \quad x_{13} = 0.264\,80, \quad x_{14} = 0.264\,85, \quad x_{15} = 0.264\,83, \quad x_{16} = 0.264\,84.$$

With $f(x) = x^4 - 3x^2 - 3x + 1$, we calculate that $f(0.264\,75) = 3.9 \times 10^{-4}$ and $f(0.264\,85) = -6.6 \times 10^{-5}$. The root is therefore $x = 0.2648$ to 4 decimals.

52. With $x_1 = 0.75$, and $x_{n+1} = \sqrt{1 - \sin^2 x_n} = \sqrt{\cos^2 x_n} = \cos x_n$, iteration gives

$$x_2 = 0.732, \quad x_3 = 0.744, \quad x_4 = 0.736, \quad x_5 = 0.741, \quad x_6 = 0.738,$$
$$x_7 = 0.740, \quad x_8 = 0.7385, \quad x_9 = 0.7395.$$

With $f(x) = \sin^2 x - 1 + x^2$, we calculate that $f(0.739\,05) = -8.7 \times 10^{-5}$ and $f(0.739\,15) = 1.6 \times 10^{-4}$. The root is therefore $x = 0.7391$ to 4 decimals.

54. With $x_1 = 0.5$, and $x_{n+1} = \dfrac{e^{x_n} + e^{-x_n}}{10}$, iteration gives

$$x_2 = 0.226, \quad x_3 = 0.205, \quad x_4 = 0.2042, \quad x_5 = 0.204\,18, \quad x_6 = 0.204\,18.$$

With $f(x) = e^x + e^{-x} - 10x$, we calculate that $f(0.204\,15) = 3.2 \times 10^{-4}$ and $f(0.204\,25) = -6.4 \times 10^{-4}$. The root is therefore $x = 0.2042$ to 4 decimals.

56. (a) $d_1 = 2(0.99)(20) = 40(0.99)$ m

$d_2 = 2(0.99)[(0.99)(20)] = 40(0.99)^2$ m

$d_3 = 2(0.99)[(20)(0.99)^2] = 40(0.99)^3$ m

The pattern emerging is $d_n = 40(0.99)^n$ m.

(b) When an object falls from rest under gravity, the distance that it falls as a function of time t is given by $d = 4.905t^2$. Consequently, the time to fall from peak height between n^{th} and $(n + 1)^{\text{th}}$ bounces is given by $d_n/2 = 4.905t^2$. When this equation is solved for t, the result is $t = \sqrt{d_n/9.81}$, and therefore

$$t_n = 2\sqrt{d_n/9.81} = 2\sqrt{40(0.99)^n/9.81} = \frac{4}{\sqrt{0.981}}(0.99)^{n/2} \text{ s.}$$

58. Since each of the 12 straight line segments in the middle figure has length $P/9$,

$$P_1 = \frac{12P}{9} = \frac{4P}{3}.$$

Since each of the 48 straight line segments in the right figure has length $P/27$,

$$P_2 = \frac{48P}{27} = \frac{4^2P}{3^2}.$$

The next perimeter is $P_3 = 4(48)\dfrac{P}{81} = \dfrac{4^3P}{3^3}$. The pattern emerging is $P_n = \dfrac{4^nP}{3^n}$. The limit of P_n as $n \to \infty$ does not exist.

60. The figure shows graphs of $y = \tan x$ and $y = (e^x - e^{-x})/(e^x + e^{-x}) = \tanh x$ for $x \geq 0$. They intersect at $x = 0$ and values near 4 and 7. We use Newton's iterative procedure

$$x_{n+1} = x_n - \frac{\tan x_n - \tanh x_n}{\sec^2 x_n - \operatorname{sech}^2 x_n}$$

with $x_1 = 4$ to locate the smaller root.

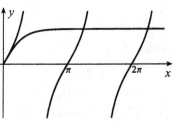

Iteration gives $x_2 = 3.932\,25$, $x_3 = 3.926\,63$, $x_4 = 3.926\,60$, $x_5 = 3.926\,60$. When we divide this by 20π, the result is 0.0625. A similar procedure gives the next natural frequency 0.1125.

62. The next two terms are 1113213211, 31131211131221. Reason as follows: The second term is 11 because there is one 1 in the first term; the third term is 21 because the second term has two 1's; the fourth term is 1211 because the third term has one 2 followed by one 1; the fifth term is 111221 because the fourth term is one 1, followed by one 2, followed by two 1's; etc.

64. The plot of the seven-point averager is shown to the right.

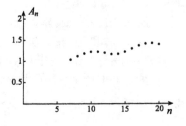

66. An explicit formula for this FIR is

$$F_n = \frac{1}{n^2}\sin\left(\frac{n}{3}\right) - \frac{2}{(n-1)^2}\sin\left(\frac{n-1}{3}\right) + \frac{3}{(n-2)^2}\sin\left(\frac{n-2}{3}\right) - \frac{4}{(n-3)^2}\sin\left(\frac{n-3}{3}\right).$$

When we substitute $n = 4, \ldots, 13$, we obtain the first 10 terms,

$-0.9712, -0.4196, -0.2461, -0.1593, -0.1059, -0.0693, -0.0430, -0.0237, -0.0096, 0.0002.$

EXERCISES 10.2

2. The limit function is $f(x) = 0$, since for each x in $0 \leq x \leq 1$,

$$\lim_{n \to \infty} \frac{n^2 x}{1 + n^3 x^2} = \lim_{n \to \infty} \frac{x}{1/n^2 + nx^2} = 0.$$

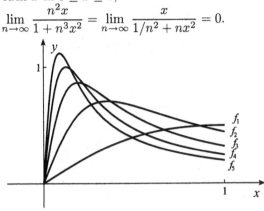

4. The limit function is $f(x) = 1/x$.

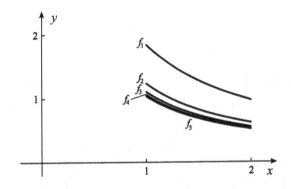

6. There is no limit function.

8. Since $f_n(0) = f_n(1) = 0$, the limit function $f(x)$ has values $f(0) = f(1) = 0$. For fixed x in $0 < x < 1$,

$$f(x) = \lim_{n \to \infty} n^2 x^n (1 - x^2) = \lim_{n \to \infty} \frac{n^2(1 - x^2)}{x^{-n}}$$
$$= \lim_{n \to \infty} \frac{2n(1 - x^2)}{-x^{-n} \ln x} = \lim_{n \to \infty} \frac{2(1 - x^2)}{x^{-n}(\ln x)^2}$$
$$= \lim_{n \to \infty} \frac{2(1 - x^2)x^n}{(\ln x)^2} = 0.$$

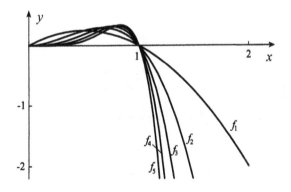

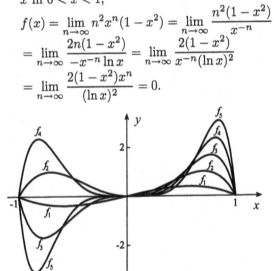

10. The limit function is $f(x) = 1$.

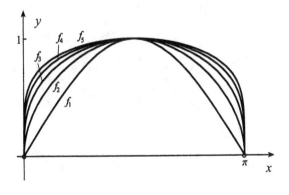

12. The limit function is $f(x) = 1$.

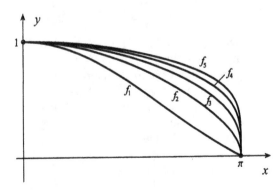

14. The limit function is $f(x) = \lim\limits_{n\to\infty} \dfrac{n^2x}{e^{nx}}$

$= \lim\limits_{n\to\infty} \dfrac{2nx}{xe^{nx}} = \lim\limits_{n\to\infty} \dfrac{2x}{x^2e^{nx}} = 0.$

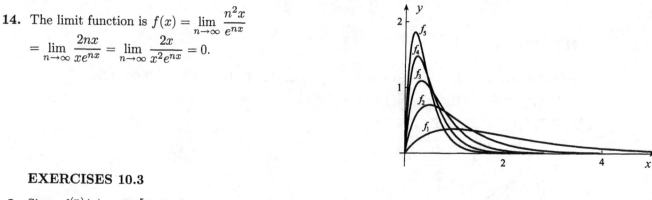

EXERCISES 10.3

2. Since $f^{(n)}(x) = 5^n e^{5x}$, Taylor's remainder formula for e^{5x} and $c = 0$ gives

$$e^{5x} = 1 + 5x + \frac{5^2}{2!}x^2 + \frac{5^3}{3!}x^3 + \cdots + \frac{5^n}{n!}x^n + R_n,$$

where $R_n = \dfrac{d^{n+1}}{dx^{n+1}}(e^{5x})_{|x=z_n} \dfrac{x^{n+1}}{(n+1)!} = \dfrac{5^{n+1}e^{5z_n}}{(n+1)!}x^{n+1}.$

If $x < 0$, then $x < z_n < 0$, and $|R_n| < 5^{n+1}|x|^{n+1}/(n+1)!$.
According to Example 10.5, $\lim_{n\to\infty}|x|^n/n! = 0$
for any x whatsoever, and therefore
$\lim_{n\to\infty} 5^{n+1}|x|^{n+1}/(n+1)! = 0$ also.
Thus, if $x < 0$, $\lim_{n\to\infty} R_n = 0$. If $x > 0$,
then $0 < z_n < x$, and

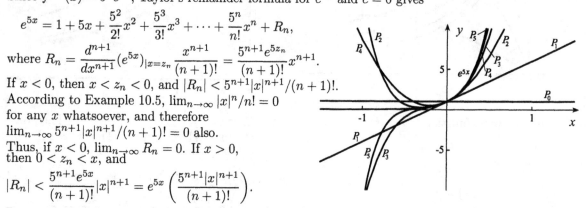

$$|R_n| < \frac{5^{n+1}e^{5x}}{(n+1)!}|x|^{n+1} = e^{5x}\left(\frac{5^{n+1}|x|^{n+1}}{(n+1)!}\right).$$

But we have just indicated that $\lim_{n\to\infty} 5^{n+1}|x|^{n+1}/(n+1)! = 0$, and therefore $\lim_{n\to\infty} R_n = 0$ for $x > 0$ also. Thus, for any x whatsoever, the sequence $\{R_n\}$ has limit 0, and the Maclaurin series for e^{5x} converges to e^{5x},

$$e^{5x} = \sum_{n=0}^{\infty} \frac{5^n}{n!}x^n, \quad -\infty < x < \infty.$$

4. Since $f(\pi/4) = \sin(\pi/4) = 1/\sqrt{2}$, $f'(\pi/4) = \cos(\pi/4) = 1/\sqrt{2}$, $f''(\pi/4) = -\sin(\pi/4) = -1/\sqrt{2}$, $f'''(\pi/4) = -\cos(\pi/4) = -1/\sqrt{2}$, $f''''(\pi/4) = \sin(\pi/4) = 1/\sqrt{2}$, etc., Taylor's remainder formula gives

$$\sin x = \frac{1}{\sqrt{2}} + \frac{1}{\sqrt{2}}(x-\pi/4) - \frac{1}{2!\sqrt{2}}(x-\pi/4)^2 - \frac{1}{3!\sqrt{2}}(x-\pi/4)^3 + \ldots + \text{term in } (x-\pi/4)^n + R_n,$$

where $R_n = \dfrac{d^{n+1}}{dx^{n+1}}(\sin x)_{|x=z_n} \dfrac{(x-\pi/4)^{n+1}}{(n+1)!}.$

The nth derivative of $\sin x$ is $\pm\sin x$ or
$\pm\cos x$, so that

$$\left|\frac{d^{n+1}}{dx^{n+1}}(\sin x)_{|x=z_n}\right| \le 1.$$

Hence, $|R_n| \le |x-\pi/4|^{n+1}/(n+1)!$. According
to Example 10.5, $\lim_{n\to\infty}|x|^n/n! = 0$
for any x whatsoever, and therefore
$\lim_{n\to\infty}|x-\pi/4|^{n+1}/(n+1)! = 0$ also. It
follows that $\lim_{n\to\infty} R_n = 0$, and the Taylor
series for $\sin x$ about $\pi/4$ therefore converges
to $\sin x$ for all x. We may write

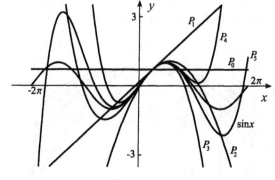

$$\sin x = \frac{1}{\sqrt{2}}\left[1 + (x-\pi/4) - \frac{1}{2!}(x-\pi/4)^2 - \frac{1}{3!}(x-\pi/4)^3 + \cdots\right], \quad -\infty < x < \infty.$$

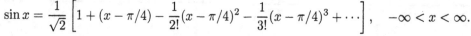

6. Since $f^{(n)}(0) = 2^n$, the Maclaurin
series for e^{2x} is

$$\sum_{n=0}^{\infty} \frac{2^n}{n!} x^n = 1 + 2x + \frac{2^2 x^2}{2!} + \cdots.$$

Plots of the polynomials suggest that
the series converges to e^{2x} for all x.

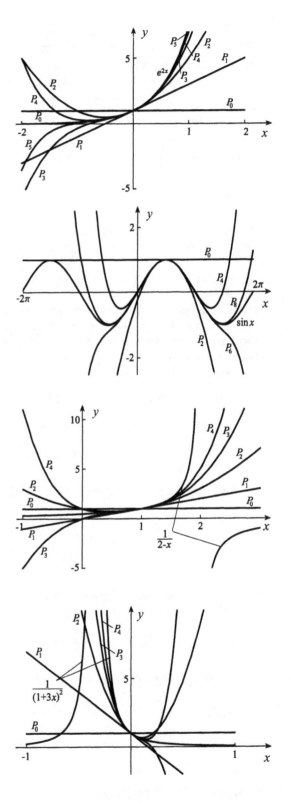

8. Since $f(\pi/2) = 1$, $f'(\pi/2) = 0$,
$f''(\pi/2) = -1$, $f'''(\pi/2) = 0$, and
$f''''(\pi/2) = 1$, the Taylor series for $\sin x$
about $x = \pi/2$ is

$$1 - \frac{(x - \pi/2)^2}{2!} + \cdots = \sum_{n=0}^{\infty} \frac{(-1)^n}{(2n)!} (x - \pi/2)^{2n}.$$

Plots of the polynomials suggest that the
series converges to $\sin x$ for all x.

10. Since $f^{(n)}(1) = n!$, the Taylor series for
$1/(2 - x)$ about $x = 1$ is

$$\sum_{n=0}^{\infty} (x - 1)^n = 1 + (x - 1) + (x - 1)^2 + \cdots.$$

Plots of the polynomials suggest that the
series converges to $1/(2 - x)$ only for
$0 < x < 2$.

12. Since $f^{(n)}(0) = (-1)^n 3^n (n + 1)!$, the Maclaurin
series for $1/(1 + 3x)^2$ is

$$\sum_{n=0}^{\infty} (-1)^n 3^n (n + 1) x^n = 1 - 6x + 3^2 (3) x^2 + \cdots.$$

Plots of the polynomials suggest that the
series converges to $1/(1 + 3x)^2$ only for
$-1/3 < x < 1/3$.

14. Calculating derivatives of the function leads to the formula $f^{(n)}(0) = \dfrac{(-1)^{n+1}3^n[1 \cdot 3 \cdot 5 \cdots (2n-3)]}{2^n}$ for $n \geq 2$, together with $f(0) = 1$ and $f'(0) = 3/2$. The Maclaurin series for $\sqrt{1+3x}$ is therefore

$$1 + \frac{3x}{2} + \sum_{n=2}^{\infty} \frac{(-1)^{n+1}3^n[1 \cdot 3 \cdot 5 \cdots (2n-3)]}{2^n \, n!}x^n.$$

Plots of the polynomials suggest that the series converges to $\sqrt{1+3x}$ only for $-1/3 \leq x \leq 1/3$.

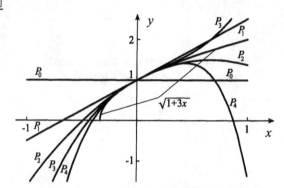

16. If I' is the open interval in which $f'(x)$ and $f''(x)$ are continuous, and we apply Taylor's remainder formula to $f(x)$ at x_0 in I', we obtain

$$f(x) = f(x_0) + f'(x_0)(x - x_0) + \frac{f''(z_1)}{2!}(x - x_0)^2 = f(x_0) + \frac{f''(z_1)}{2}(x - x_0)^2,$$

where z_1 is between x_0 and x. Suppose that $f''(x_0) > 0$. Because $f''(x)$ is continuous at x_0, there exists an open interval I containing x_0 in which $f''(x) > 0$. For any x in this interval, it follows that $f''(z_1) > 0$ also. As a result, for any x in I, $f(x) > f(x_0)$, and $f(x)$ must have a relative minimum at x_0. A similar discussion shows that when $f''(x_0) < 0$, the function has a relative maximum at x_0. If $f''(x_0) = 0$, no conclusion can be reached.

18. (a) This follows from $\displaystyle\int_c^x f'(t)\, dt = \left\{ f(t) \right\}_c^x = f(x) - f(c)$.

(b) If we set $u = f'(t)$, $du = f''(t)\, dt$, $dv = dt$, and $v = t - x$, then

$$f(x) = f(c) + \left\{ (t-x)f'(t) \right\}_c^x - \int_c^x (t-x)f''(t)\, dt = f(c) + f'(c)(x-c) + \int_c^x (x-t)f''(t)\, dt.$$

(c) If we now set $u = f''(t)$, $du = f'''(t)\, dt$, $dv = (x-t)\, dt$, and $v = -(1/2)(x-t)^2$,

$$f(x) = f(c) + f'(c)(x-c) + \left\{ -\frac{(x-t)^2 f''(t)}{2} \right\}_c^x - \int_c^x -\frac{1}{2}(x-t)^2 f'''(t)\, dt$$

$$= f(c) + f'(c)(x-c) + \frac{f''(c)}{2}(x-c)^2 + \frac{1}{2}\int_c^x (x-t)^2 f'''(t)\, dt.$$

(d) One more integration by parts should convince us that the formula is correct. If we set $u = f'''(t)$, $du = f''''(t)\, dt$, $dv = (x-t)^2\, dt$, and $v = -(1/3)(x-t)^3$,

$$f(x) = f(c) + f'(c)(x-c) + \frac{f''(c)}{2}(x-c)^2 + \frac{1}{2}\left\{ -\frac{(x-t)^3 f'''(t)}{3} \right\}_c^x - \frac{1}{2}\int_c^x -\frac{1}{3}(x-t)^3 f''''(t)\, dt$$

$$= f(c) + f'(c)(x-c) + \frac{f''(c)}{2}(x-c)^2 + \frac{f'''(c)}{3!}(x-c)^3 + \frac{1}{3!}\int_c^x (x-t)^3 f''''(t)\, dt.$$

EXERCISES 10.4

2. Since the radius of convergence is $R = \displaystyle\lim_{n \to \infty} \left| \frac{n^2}{(n+1)^2} \right| = 1$, the open interval of convergence is $-1 < x < 1$.

4. Since the radius of convergence is $R = \displaystyle\lim_{n \to \infty} \left| \frac{n^2 3^n}{(n+1)^2 3^{n+1}} \right| = \frac{1}{3}$, the open interval of convergence is $-1/3 < x < 1/3$.

6. Since the radius of convergence is $R = \displaystyle\lim_{n \to \infty} \left| \frac{(-1)^n n^3}{(-1)^{n+1}(n+1)^3} \right| = 1$, the open interval of convergence is $-4 < x < -2$.

8. Since the radius of convergence is $R = \lim\limits_{n\to\infty} \left| \dfrac{2^n \left(\dfrac{n-1}{n+2}\right)^2}{2^{n+1}\left(\dfrac{n}{n+3}\right)^2} \right| = \dfrac{1}{2}$, the open interval of convergence is

$7/2 < x < 9/2$.

10. If we set $y = x^3$, then $\sum\limits_{n=0}^{\infty}(-1)^n x^{3n} = \sum\limits_{n=0}^{\infty}(-1)^n y^n$. Since $R_y = \lim\limits_{n\to\infty}\left|\dfrac{(-1)^n}{(-1)^{n+1}}\right| = 1$, it follows that

$R_x = R_y^{1/3} = 1$. The open interval of convergence is therefore $-1 < x < 1$.

12. If we set $y = x^3$, then $\sum\limits_{n=0}^{\infty}\dfrac{1}{\sqrt{n+1}}x^{3n+1} = y^{1/3}\sum\limits_{n=0}^{\infty}\dfrac{1}{\sqrt{n+1}}y^n$. Since $R_y = \lim\limits_{n\to\infty}\left|\dfrac{1/\sqrt{n+1}}{1/\sqrt{n+2}}\right| = 1$, it

follows that $R_x = R_y^{1/3} = 1$. The open interval of convergence is therefore $-1 < x < 1$.

14. Since the radius of convergence is $R = \lim\limits_{n\to\infty}\left|\dfrac{(-e)^n/n^2}{(-e)^{n+1}/(n+1)^2}\right| = \dfrac{1}{e}$, the open interval of convergence

is $-1/e < x < 1/e$.

16. Since the radius of convergence is $R = \lim\limits_{n\to\infty}\left|\dfrac{n^n}{(n+1)^{n+1}}\right| = \lim\limits_{n\to\infty}\left(\dfrac{n}{n+1}\right)^n\left(\dfrac{1}{n+1}\right) = \dfrac{1}{e}(0) = 0$, the

series converges only for $x = 0$.

18. Since the radius of convergence is $R = \lim\limits_{n\to\infty}\left|\dfrac{n^3 3^n}{(n+1)^3 3^{n+1}}\right| = \dfrac{1}{3}$, the open interval of convergence is

$-1/3 < x < 1/3$.

20. If we set $y = x^3$, the series becomes $\sum\limits_{n=0}^{\infty} y^n/5^n$. Since $R_y = \lim\limits_{n\to\infty}\left|\dfrac{1/5^n}{1/5^{n+1}}\right| = 5$, it follows that $R_x =$

$R_y^{1/3} = 5^{1/3}$. The open interval of convergence is therefore $-5^{1/3} < x < 5^{1/3}$.

22. Using L'Hôpital's rule, $R = \lim\limits_{n\to\infty}\left|\dfrac{\dfrac{1}{n^2 \ln n}}{\dfrac{1}{(n+1)^2 \ln(n+1)}}\right| = \lim\limits_{n\to\infty}\dfrac{\ln(n+1)}{\ln n} = \lim\limits_{n\to\infty}\dfrac{1/(n+1)}{1/n} = 1$. The open

interval of convergence is therefore $-1 < x < 1$.

24. Since $R = \lim\limits_{n\to\infty}\left|\dfrac{2\cdot 4\cdot 6\cdots(2n)}{3\cdot 5\cdot 7\cdots(2n+1)}\dfrac{3\cdot 5\cdots(2n+3)}{2\cdot 4\cdots(2n+2)}\right| = \lim\limits_{n\to\infty}\dfrac{2n+3}{2n+2} = 1$, the open interval of convergence is $-1 < x < 1$.

26. $\sum\limits_{n=0}^{\infty}\dfrac{1}{4^n}x^{3n} = \sum\limits_{n=0}^{\infty}\left(\dfrac{x^3}{4}\right)^n = \dfrac{1}{1-x^3/4} = \dfrac{4}{4-x^3}$ provided $\left|\dfrac{x^3}{4}\right| < 1 \implies |x| < 4^{1/3}$

28. $\sum\limits_{n=1}^{\infty}\dfrac{1}{3^{2n}}(x-1)^n = \sum\limits_{n=1}^{\infty}\left(\dfrac{x-1}{9}\right)^n = \dfrac{\dfrac{x-1}{9}}{1-\dfrac{x-1}{9}} = \dfrac{x-1}{10-x}$ provided $\left|\dfrac{x-1}{9}\right| < 1 \implies |x-1| < 9$

30. $\sum\limits_{n=0}^{\infty}\dfrac{(-1)^n}{(2n)!}x^{4n} = \sum\limits_{n=0}^{\infty}\dfrac{(-1)^n}{(2n)!}(x^2)^{2n} = \cos(x^2)$ valid for all x

32. $\sum\limits_{n=0}^{\infty}\dfrac{(-1)^n}{3^{2n+1}(2n+1)!}x^{2n+2} = x\sum\limits_{n=0}^{\infty}\dfrac{(-1)^n}{(2n+1)!}\left(\dfrac{x}{3}\right)^{2n+1} = x\sin(x/3)$ valid for all x

34. $\sum\limits_{n=1}^{\infty}\dfrac{(-1)^n}{n!}x^n = -1 + \sum\limits_{n=0}^{\infty}\dfrac{1}{n!}(-x)^n = -1 + e^{-x}$ valid for all x

36. $\sum\limits_{n=0}^{\infty}\dfrac{2^n}{n!}(x-1/2)^n = \sum\limits_{n=0}^{\infty}\dfrac{1}{n!}(2x-1)^n = e^{2x-1}$ valid for all x

38. (a) $J_0(x) = \sum_{n=0}^{\infty} \frac{(-1)^n}{2^{2n}(n!)^2} x^{2n} = 1 - \frac{x^2}{2^2} + \frac{x^4}{2^4(2!)^2} - \frac{x^6}{2^6(3!)^2} + \frac{x^8}{2^8(4!)^2} - \cdots$

$J_1(x) = \sum_{n=0}^{\infty} \frac{(-1)^n}{2^{2n+1}(n!)(n+1)!} x^{2n+1} = \frac{x}{2} - \frac{x^3}{2^3 2!} + \frac{x^5}{2^5 2! 3!} - \frac{x^7}{2^7 3! 4!} + \frac{x^9}{2^9 4! 5!} - \cdots$

$J_m(x) = \frac{x^m}{2^m m!} - \frac{x^{m+2}}{2^{m+2}(m+1)!} + \frac{x^{m+4}}{2^{m+4}2!(m+2)!} - \frac{x^{m+6}}{2^{m+6}3!(m+3)!} + \frac{x^{m+8}}{2^{m+8}4!(m+4)!} - \cdots\cdots$

(b) $R = \lim_{n \to \infty} \left| \frac{(-1)^n}{2^{2n+m}n!(n+m)!} \frac{2^{2n+m+2}(n+1)!(n+m+1)!}{(-1)^{n+1}} \right| = \lim_{n \to \infty} 2^2(n+1)(n+m+1) = \infty$

The interval of convergence is therefore $-\infty < x < \infty$.

EXERCISES 10.5

2. $f(x) = \frac{1}{4+x^2} = \frac{1}{4}\left(\frac{1}{1+x^2/4}\right) = \frac{1}{4}\sum_{n=0}^{\infty}\left(-\frac{x^2}{4}\right)^n = \sum_{n=0}^{\infty}\frac{(-1)^n}{4^{n+1}}x^{2n}, \quad \left|-\frac{x^2}{4}\right| < 1 \Longrightarrow |x| < 2$

4. Since $e^x = \sum_{n=0}^{\infty}\frac{1}{n!}x^n, \quad -\infty < x < \infty,$ it follows that

$$e^{5x} = \sum_{n=0}^{\infty}\frac{1}{n!}(5x)^n = \sum_{n=0}^{\infty}\frac{5^n}{n!}x^n, \quad -\infty < x < \infty.$$

6. Since $f(x) = e^{1-2x} = e\, e^{-2x}$, and the Maclaurin series $e^x = \sum_{n=0}^{\infty}\frac{1}{n!}x^n$ converges for all x, it follows that

$$e^{1-2x} = e\sum_{n=0}^{\infty}\frac{1}{n!}(-2x)^n = \sum_{n=0}^{\infty}\frac{e(-1)^n 2^n}{n!}x^n, \quad -\infty < x < \infty.$$

8. $\cosh x = \frac{1}{2}(e^x + e^{-x}) = \frac{1}{2}\left[\sum_{n=0}^{\infty}\frac{1}{n!}x^n + \sum_{n=0}^{\infty}\frac{1}{n!}(-x)^n\right] = \frac{1}{2}\sum_{n=0}^{\infty}\frac{[1+(-1)^n]}{n!}x^n$

$= \frac{1}{2}\left(2 + \frac{2}{2!}x^2 + \frac{2}{4!}x^4 + \cdots\right) = 1 + \frac{1}{2!}x^2 + \frac{1}{4!}x^4 + \cdots = \sum_{n=0}^{\infty}\frac{1}{(2n)!}x^{2n}, \quad -\infty < x < \infty$

10. This function is its own Maclaurin series.

12. $\frac{1}{x+3} = \frac{1}{5+(x-2)} = \frac{1}{5\left(1+\dfrac{x-2}{5}\right)} = \frac{1}{5}\sum_{n=0}^{\infty}\left(-\frac{x-2}{5}\right)^n = \sum_{n=0}^{\infty}\frac{(-1)^n}{5^{n+1}}(x-2)^n, \quad \left|-\frac{x-2}{5}\right| < 1 \Longrightarrow$

$-3 < x < 7$

14. Long division gives

$$\frac{x^2}{3-4x} = -\frac{x}{4} - \frac{3}{16} + \frac{9/16}{3-4x} = -\frac{1}{4}(x-2) - \frac{11}{16} + \frac{9/16}{-5-4(x-2)} = -\frac{11}{16} - \frac{1}{4}(x-2) - \frac{9/80}{1+\dfrac{4(x-2)}{5}}$$

$$= -\frac{11}{16} - \frac{1}{4}(x-2) - \frac{9}{80}\sum_{n=0}^{\infty}\left[-\frac{4}{5}(x-2)\right]^n$$

$$= -\frac{11}{16} - \frac{1}{4}(x-2) - \frac{9}{80}\left[1 - \frac{4}{5}(x-2) + \sum_{n=2}^{\infty}\frac{(-1)^n 4^n}{5^n}(x-2)^n\right]$$

$$= -\frac{4}{5} - \frac{4}{25}(x-2) + \sum_{n=2}^{\infty}\frac{9(-1)^{n+1}4^{n-2}}{5^{n+1}}(x-2)^n, \quad \left|-\frac{4(x-2)}{5}\right| < 1 \Longrightarrow \frac{3}{4} < x < \frac{13}{4}$$

16. Term-by-term integration of $\dfrac{1}{1 + 2x} = \displaystyle\sum_{n=0}^{\infty} (-2x)^n = \sum_{n=0}^{\infty} (-1)^n 2^n x^n$ gives

$$\frac{1}{2} \ln |1 + 2x| = \sum_{n=0}^{\infty} \frac{(-1)^n 2^n}{n+1} x^{n+1} + C.$$

Setting $x = 0$ gives $C = 0$, and therefore $\ln |1 + 2x| = \displaystyle\sum_{n=0}^{\infty} \frac{(-1)^n 2^{n+1}}{n+1} x^{n+1}$. Since the radius of convergence of the geometric series is $1/2$, this is also the radius of convergence for the series of the logarithm function. The open interval of convergence is therefore $-1/2 < x < 1/2$, and the absolute values may be dropped,

$$\ln (1 + 2x) = \sum_{n=0}^{\infty} \frac{(-1)^n 2^{n+1}}{n+1} x^{n+1} = \sum_{n=1}^{\infty} \frac{(-1)^{n+1} 2^n}{n} x^n.$$

18. Termwise integration of

$$\frac{1}{x} = \frac{1}{2 + (x - 2)} = \frac{1}{2[1 + (x - 2)/2]} = \frac{1}{2} \sum_{n=0}^{\infty} \left(-\frac{x - 2}{2} \right)^n = \sum_{n=0}^{\infty} \frac{(-1)^n}{2^{n+1}} (x - 2)^n$$

gives $\ln |x| = \displaystyle\sum_{n=0}^{\infty} \frac{(-1)^n}{(n+1)2^{n+1}} (x - 2)^{n+1} + C$. Setting $x = 2$ gives $C = \ln 2$, and therefore

$\ln |x| = \ln 2 + \displaystyle\sum_{n=0}^{\infty} \frac{(-1)^n}{(n+1)2^{n+1}} (x - 2)^{n+1}$. Since the radius of convergence of the geometric series is 2, this is also the radius of convergence for the series of the logarithm function. The open interval of convergence is therefore $0 < x < 4$, and the absolute values may be dropped,

$$\ln x = \ln 2 + \sum_{n=0}^{\infty} \frac{(-1)^n}{(n+1)2^{n+1}} (x - 2)^{n+1} = \ln 2 + \sum_{n=1}^{\infty} \frac{(-1)^{n+1}}{n \, 2^n} (x - 2)^n.$$

20. $\dfrac{1}{x} = \dfrac{1}{4 + (x - 4)} = \dfrac{1}{4[1 + (x - 4)/4]} = \dfrac{1}{4} \displaystyle\sum_{n=0}^{\infty} \left(-\frac{x - 4}{4} \right)^n = \sum_{n=0}^{\infty} \frac{(-1)^n}{4^{n+1}} (x - 4)^n,$ provided

$\left| -\dfrac{x - 4}{4} \right| < 1 \implies 0 < x < 8$

22. With the binomial expansion 10.33b,

$$\frac{1}{(2 - x)^2} = \frac{1}{[-1 - (x - 3)]^2} = \frac{1}{[1 + (x - 3)]^2} = [1 + (x - 3)]^{-2}$$

$$= 1 - 2(x - 3) + \frac{(-2)(-3)}{2!} (x - 3)^2 + \frac{(-2)(-3)(-4)}{3!} (x - 3)^3 + \cdots$$

$$= \sum_{n=0}^{\infty} (-1)^n (n + 1)(x - 3)^n, \quad \text{provided } -1 < x - 3 < 1 \implies 2 < x < 4.$$

24. $\dfrac{1}{x^2 + 8x + 15} = \dfrac{1}{(x + 3)(x + 5)} = \dfrac{1/2}{x + 3} + \dfrac{-1/2}{x + 5} = \dfrac{1/6}{1 + x/3} - \dfrac{1/10}{1 + x/5}$

$$= \frac{1}{6} \sum_{n=0}^{\infty} \left(-\frac{x}{3} \right)^n - \frac{1}{10} \sum_{n=0}^{\infty} \left(-\frac{x}{5} \right)^n = \sum_{n=0}^{\infty} \frac{(-1)^n}{2(3^{n+1})} x^n + \sum_{n=0}^{\infty} \frac{(-1)^{n+1}}{2(5^{n+1})} x^n$$

$$= \sum_{n=0}^{\infty} \frac{(-1)^n}{2} \left(\frac{1}{3^{n+1}} - \frac{1}{5^{n+1}} \right) x^n, \quad \text{valid for } -3 < x < 3.$$

26. With the binomial expansion 10.33b,

$$\sqrt{x+3} = \sqrt{3}\sqrt{1+x/3} = \sqrt{3}\left[1 + \left(\frac{1}{2}\right)\left(\frac{x}{3}\right) + \frac{(1/2)(-1/2)}{2!}\left(\frac{x}{3}\right)^2 + \frac{(1/2)(-1/2)(-3/2)}{3!}\left(\frac{x}{3}\right)^3 + \cdots\right]$$

$$= \sqrt{3}\left[1 + \frac{x}{6} - \frac{1}{2^2 3^2 2!}x^2 + \frac{(1)(3)}{2^3 3^3 3!}x^3 - \frac{(1)(3)(5)}{2^4 3^4 4!}x^4 + \cdots\right]$$

$$= \sqrt{3}\left[1 + \frac{x}{6} - \frac{x^2}{72} + \sum_{n=3}^{\infty}\frac{(-1)^{n+1}[1\cdot 3\cdot 5\cdots(2n-3)]}{2^n 3^n n!}x^n\right]$$

$$= \sqrt{3}\left[1 + \frac{x}{6} - \frac{x^2}{72} + \sum_{n=3}^{\infty}\frac{(-1)^{n+1}[1\cdot 2\cdot 3\cdot 4\cdots(2n-3)(2n-2)]}{[2\cdot 4\cdot 6\cdots(2n-2)]6^n n!}x^n\right]$$

$$= \sqrt{3}\left[1 + \frac{x}{6} - \frac{x^2}{72} + \sum_{n=3}^{\infty}\frac{2(-1)^{n+1}(2n-2)!}{12^n n!\,(n-1)!}x^n\right]$$

$$= \sqrt{3} + \sum_{n=1}^{\infty}\frac{2\sqrt{3}(-1)^{n+1}(2n-2)!}{12^n n!\,(n-1)!}x^n, \quad \text{valid for } -1 \le \frac{x}{3} \le 1 \implies |x| \le 3.$$

28. With the binomial expansion 10.33b,

$$(1-2x)^{1/3} = [-1-2(x-1)]^{1/3} = -[1+2(x-1)]^{1/3}$$

$$= -\left\{1 + \frac{2(x-1)}{3} + \frac{(1/3)(-2/3)}{2!}[2(x-1)]^2 + \frac{(1/3)(-2/3)(-5/3)}{3!}[2(x-1)])^3 + \cdots\right\}$$

$$= -1 - \frac{2}{3}(x-1) + \frac{2^2 2}{3^2 2!}(x-1)^2 - \frac{2^3(2\cdot 5)}{3^3 3!}(x-1)^3 + \cdots$$

$$= -1 - \frac{2}{3}(x-1) + \sum_{n=2}^{\infty}\frac{(-1)^n 2^n[2\cdot 5\cdot 8\cdots(3n-4)]}{3^n n!}(x-1)^n,$$

$$= -1 + \sum_{n=1}^{\infty}\frac{(-1)^{n-1}2^n[(-1)\cdot 2\cdot 5\cdot 8\cdots(3n-4)]}{3^n n!}(x-1)^n,$$

valid for $-1 \le 2(x-1) \le 1 \implies 1/2 \le x \le 3/2$.

30. With the binomial expansion 10.33b,

$$x(1-x)^{1/3} = x\left[1 + \left(\frac{1}{3}\right)(-x) + \frac{(1/3)(-2/3)}{2!}(-x)^2 + \frac{(1/3)(-2/3)(-5/3)}{3!}(-x)^3 + \cdots\right]$$

$$= x - \frac{x^2}{3} - \frac{2}{3^2 2!}x^3 - \frac{(2)(5)}{3^3 3!}x^4 - \frac{(2)(5)(8)}{3^4 4!}x^5 + \cdots$$

$$= x - \frac{x^2}{3} - \sum_{n=3}^{\infty}\frac{2\cdot 5\cdot 8\cdots(3n-7)}{3^{n-1}(n-1)!}x^n$$

$$= x + \sum_{n=2}^{\infty}\frac{(-1)\cdot 2\cdot 5\cdot 8\cdots(3n-7)}{3^{n-1}(n-1)!}x^n, \quad \text{valid for } -1 \le x \le 1.$$

32. If we set $\sec x = \dfrac{1}{\cos x} = a_0 + a_1 x + a_2 x^2 + \cdots$, then

$$1 = \left(a_0 + a_1 x + a_2 x^2 + a_3 x^3 + \cdots\right)\left(1 - \frac{x^2}{2!} + \frac{x^4}{4!} - \frac{x^6}{6!} + \cdots\right).$$

We now multiply the power series on the right and equate coefficients:

1: $\quad 1 = a_0$

x: $\quad 0 = a_1$

x^2:　　$0 = -a_0/2! + a_2$　$\implies$　$a_2 = 1/2$

x^3:　　$0 = -a_1/2! + a_3$　$\implies$　$a_3 = 0$

x^4:　　$0 = a_0/4! - a_2/2! + a_4$　$\implies$　$a_4 = 5/24$

x^5:　　$0 = a_1/4! - a_3/2! + a_5$　$\implies$　$a_5 = 0$

x^6:　　$0 = -a_0/6! + a_2/4! - a_4/2! + a_6$　$\implies$　$a_6 = 61/720$

Thus, $\sec x = 1 + \dfrac{1}{2}x^2 + \dfrac{5}{24}x^4 + \dfrac{61}{720}x^6 + \cdots$.　Long division could also be used.

34. $\cos^2 x = \dfrac{1}{2}(1 + \cos 2x) = \dfrac{1}{2}\left[1 + \displaystyle\sum_{n=0}^{\infty} \dfrac{(-1)^n}{(2n)!}(2x)^{2n}\right] = \dfrac{1}{2}\left[1 + 1 + \displaystyle\sum_{n=1}^{\infty} \dfrac{(-1)^n 2^{2n}}{(2n)!}x^{2n}\right]$

$$= 1 + \sum_{n=1}^{\infty} \frac{(-1)^n 2^{2n-1}}{(2n)!}x^{2n}, \quad -\infty < x < \infty$$

36. The Maclaurin series for $\mathrm{Sin}^{-1}(x^2)$ can be obtained by replacing x by x^2 in the series for $\mathrm{Sin}^{-1}x$ in Example 10.26:

$$\mathrm{Sin}^{-1}(x^2) = \sum_{n=0}^{\infty} \frac{(2n)!}{(2n+1)2^{2n}(n!)^2}(x^2)^{2n+1} = \sum_{n=0}^{\infty} \frac{(2n)!}{(2n+1)2^{2n}(n!)^2}x^{4n+2}, \quad |x| < 1.$$

38. If we integrate the series $\dfrac{1}{1-x} = \displaystyle\sum_{n=0}^{\infty} x^n$, $|x| < 1$,　we obtain $-\ln|1-x| = \displaystyle\sum_{n=0}^{\infty} \dfrac{1}{n+1}x^{n+1} + C$. Substitution of $x = 0$ gives $C = 0$, and therefore　$\ln|1-x| = \displaystyle\sum_{n=0}^{\infty} \dfrac{-1}{n+1}x^{n+1} = \displaystyle\sum_{n=1}^{\infty} -\dfrac{1}{n}x^n$. The open interval of convergence is $-1 < x < 1$ so that absolute values may be dropped. If we replace x by $x/\sqrt{2}$ and $-x/\sqrt{2}$, we find

$$f(x) = \ln(1 + x/\sqrt{2}) - \ln(1 - x/\sqrt{2}) = \sum_{n=1}^{\infty} -\frac{1}{n}\left(-\frac{x}{\sqrt{2}}\right)^n - \sum_{n=1}^{\infty} -\frac{1}{n}\left(\frac{x}{\sqrt{2}}\right)^n$$

$$= \sum_{n=1}^{\infty} \left[\frac{(-1)^{n+1}}{n2^{n/2}} + \frac{1}{n2^{n/2}}\right]x^n = \sum_{n=1}^{\infty} \left[\frac{1 + (-1)^{n+1}}{n2^{n/2}}\right]x^n.$$

When n is even the coefficient of x^n is zero, and therefore

$$f(x) = \sum_{n=0}^{\infty} \frac{2}{(2n+1)2^{(2n+1)/2}}x^{2n+1} = \sum_{n=0}^{\infty} \frac{\sqrt{2}}{(2n+1)2^n}x^{2n+1}.$$

Since the added series both have open interval of convergence $-\sqrt{2} < x < \sqrt{2}$, this is the open interval of convergence for the combined series.

40. The right side of this equation is the Maclaurin series for the function identically equal to zero, and as such, its coefficients must all be zero; that is, $a_n = 0$ for all n.

42. (a) $\displaystyle\sum_{n=1}^{\infty} np(1-p)^{n-1} = p\displaystyle\sum_{n=1}^{\infty} n(1-p)^{n-1}$　If we differentiate the series $\dfrac{1}{1-x} = \displaystyle\sum_{n=0}^{\infty} x^n$, $|x| < 1$, term-by-term, we obtain $\dfrac{1}{(1-x)^2} = \displaystyle\sum_{n=0}^{\infty} nx^{n-1} = \displaystyle\sum_{n=1}^{\infty} nx^{n-1}$, $|x| < 1$. We now substitute $x = 1 - p$ into this result, $\dfrac{1}{[1-(1-p)]^2} = \displaystyle\sum_{n=1}^{\infty} n(1-p)^{n-1}$. Multiplication by p gives $\dfrac{1}{p} = \displaystyle\sum_{n=1}^{\infty} np(1-p)^{n-1}$.

(b) The probability of throwing a six is $p = 1/6$, and therefore $\displaystyle\sum_{n=1}^{\infty} np(1-p)^{n-1} = \dfrac{1}{1/6} = 6$.

44. Integrating the Maclaurin series for $\cos{(\pi t^2/2)}$ (see Example 10.21) term-by-term gives

$$C(x) = \int_0^x \left[\sum_{n=0}^{\infty} \frac{(-1)^n}{(2n)!} \left(\frac{\pi t^2}{2} \right)^{2n} \right] dt = \sum_{n=0}^{\infty} \frac{(-1)^n}{(2n)!} \frac{\pi^{2n}}{2^{2n}} \int_0^x t^{4n} \, dt$$

$$= \sum_{n=0}^{\infty} \frac{(-1)^n}{(2n)!} \frac{\pi^{2n}}{2^{2n}} \left\{ \frac{t^{4n+1}}{4n+1} \right\}_0^x = \sum_{n=0}^{\infty} \frac{(-1)^n \pi^{2n}}{(4n+1)2^{2n}(2n)!} x^{4n+1},$$

valid for $-\infty < x < \infty$. A similar procedure leads to the Maclaurin series for $S(x)$.

46. The Maclaurin series for $f(x) = xe^{-2x}$ is

$$xe^{-2x} = x \sum_{n=0}^{\infty} \frac{1}{n!} (-2x)^n = \sum_{n=0}^{\infty} \frac{(-1)^n 2^n}{n!} x^{n+1} = \sum_{n=1}^{\infty} \frac{(-1)^{n+1} 2^{n-1}}{(n-1)!} x^n.$$

But the coefficient of x^n in the Maclaurin series is $f^{(n)}(0)/n!$, and therefore

$$\frac{f^{(n)}(0)}{n!} = \frac{(-1)^{n+1} 2^{n-1}}{(n-1)!} \quad \Longrightarrow \quad f^{(n)}(0) = \frac{(-1)^{n+1} 2^{n-1} n!}{(n-1)!} = n(-1)^{n+1} 2^{n-1}.$$

48. The Taylor series for $f(x) = xe^{-x}$ about $x = 2$ is

$$xe^{-x} = [(x-2)+2]e^{-(x-2)-2} = e^{-2}[2+(x-2)] \sum_{n=0}^{\infty} \frac{(-1)^n}{n!} (x-2)^n$$

$$= e^{-2} \left[\sum_{n=0}^{\infty} \frac{2(-1)^n}{n!} (x-2)^n + \sum_{n=0}^{\infty} \frac{(-1)^n}{n!} (x-2)^{n+1} \right]$$

$$= e^{-2} \left[\sum_{n=0}^{\infty} \frac{2(-1)^n}{n!} (x-2)^n + \sum_{n=1}^{\infty} \frac{(-1)^{n+1}}{(n-1)!} (x-2)^n \right]$$

$$= e^{-2} \left[2 + \sum_{n=1}^{\infty} \frac{(-1)^n(2-n)}{n!} (x-2)^n \right].$$

But the coefficient of $(x-2)^n$ in the Taylor series is $f^{(n)}(2)/n!$, and therefore

$$\frac{f^{(n)}(2)}{n!} = \frac{(-1)^n(2-n)e^{-2}}{n!} \quad \Longrightarrow \quad f^{(n)}(2) = \frac{(-1)^n(2-n)n!}{e^2 n!} = \frac{(n-2)(-1)^{n+1}}{e^2}.$$

50. Since the Maclaurin series for e^{-x^2}, namely, $\quad e^{-x^2} = \sum_{n=0}^{\infty} \frac{1}{n!} (-x^2)^n = \sum_{n=0}^{\infty} \frac{(-1)^n}{n!} x^{2n} \quad$ contains only

even powers of x, the odd derivatives of e^{-x^2} must all be zero.

52. Using the definition of $J_m(x)$ as the Maclaurin series in Exercise 38 of Section 10.4, we may write

$$J_{m-1}(x) - J_{m+1}(x) = \sum_{n=0}^{\infty} \frac{(-1)^n}{2^{2n+m-1} n!(n+m-1)!} x^{2n+m-1} - \sum_{n=0}^{\infty} \frac{(-1)^n}{2^{2n+m+1} n!(n+m+1)!} x^{2n+m+1}.$$

We lower n by 1 in the second summation, and separate out the first term in the first summation,

$$J_{m-1}(x) - J_{m+1}(x) = \frac{1}{2^{m-1}(m-1)!} x^{m-1} + \sum_{n=1}^{\infty} \frac{(-1)^n}{2^{2n+m-1} n!(n+m-1)!} x^{2n+m-1}$$

$$+ \sum_{n=1}^{\infty} \frac{(-1)^n}{2^{2n+m-1}(n-1)!(n+m)!} x^{2n+m-1}$$

$$= \frac{1}{2^{m-1}(m-1)!} x^{m-1} + \sum_{n=1}^{\infty} \frac{(-1)^n}{2^{2n+m-1}(n-1)!(n+m-1)!} \left(\frac{1}{n} + \frac{1}{n+m} \right) x^{2n+m-1}$$

$$= \frac{1}{2^{m-1}(m-1)!}x^{m-1} + \sum_{n=1}^{\infty} \frac{(-1)^n}{2^{2n+m-1}(n-1)!(n+m-1)!}\left[\frac{2n+m}{n(n+m)}\right]x^{2n+m-1}$$

$$= \frac{m}{2^{m-1}m!}x^{m-1} + \sum_{n=1}^{\infty} \frac{(2n+m)(-1)^n}{2^{2n+m-1}n!(n+m)!}x^{2n+m-1}$$

$$= \sum_{n=0}^{\infty} \frac{(2n+m)(-1)^n}{2^{2n+m-1}n!(n+m)!}x^{2n+m-1}.$$

Term-by-term differentiation of the series for $J_m(x)$ gives $J'_m(x) = \sum_{n=0}^{\infty} \frac{(2n+m)(-1)^n}{2^{2n+m}n!(n+m)!}x^{2n+m-1}$.

Hence, $J_{m-1}(x) - J_{m+1}(x) = 2J'_m(x)$.

54. (a) If we substitute the Maclaurin series for e^x into $x = (e^x - 1)\left(1 + B_1 x + \frac{B_2}{2!}x^2 + \cdots\right)$,

$$x = \left[\left(1 + x + \frac{x^2}{2!} + \frac{x^3}{3!} + \cdots\right) - 1\right]\left(1 + B_1 x + \frac{B_2 x^2}{2!} + \cdots\right)$$

$$= \left(x + \frac{x^2}{2!} + \frac{x^3}{3!} + \frac{x^4}{4!} + \frac{x^5}{5!} + \frac{x^6}{6!} + \cdots\right)\left(1 + B_1 x + \frac{B_2 x^2}{2!} + \frac{B_3 x^3}{3!} + \frac{B_4 x^4}{4!} + \frac{B_5 x^5}{5!} + \cdots\right).$$

When we multiply the series on the right and equate coefficients of powers of x left and right:

x: $1 = 1$

x^2: $0 = \frac{1}{2!} + B_1 \implies B_1 = -\frac{1}{2}$

x^3: $0 = \frac{1}{3!} + \frac{B_1}{2!} + \frac{B_2}{2!} \implies B_2 = \frac{1}{6}$

x^4: $0 = \frac{1}{4!} + \frac{B_1}{3!} + \frac{B_2}{(2!)^2} + \frac{B_3}{3!} \implies B_3 = 0$

x^5: $0 = \frac{1}{5!} + \frac{B_1}{4!} + \frac{B_2}{2!3!} + \frac{B_3}{2!3!} + \frac{B_4}{4!} \implies B_4 = -\frac{1}{30}$

x^6: $0 = \frac{1}{6!} + \frac{B_1}{5!} + \frac{B_2}{2!4!} + \frac{B_3}{(3!)^2} + \frac{B_4}{2!4!} + \frac{B_5}{5!} \implies B_5 = 0$

(b) Suppose we set $f(x) = \frac{x}{e^x - 1} - 1 - B_1 x = \frac{x}{e^x - 1} - 1 + \frac{x}{2} = \frac{2x - 2(e^x - 1) + x(e^x - 1)}{2(e^x - 1)}$

$$= \frac{xe^x - 2e^x + x + 2}{2(e^x - 1)} = \frac{B_2}{2!}x^2 + \frac{B_3}{3!}x^3 + \cdots.$$

Since

$$f(-x) = \frac{-x}{e^{-x} - 1} - 1 - \frac{x}{2} = \frac{xe^x}{e^x - 1} - 1 - \frac{x}{2}$$

$$= \frac{2xe^x - 2(e^x - 1) - x(e^x - 1)}{2(e^x - 1)} = \frac{xe^x - 2e^x + x + 2}{2(e^x - 1)} = f(x),$$

$f(x)$ is an even function. But the Maclaurin series for $f(x)$ can represent an even function only if all odd powers are absent. In other words, $0 = B_3 = B_5 = \cdots$.

EXERCISES 10.6

2. The radius of convergence of the series is $R = \lim\limits_{n\to\infty} \left|\dfrac{n(n-1)}{(n+1)n}\right| = 1$. If we set $S(x) = \sum\limits_{n=2}^{\infty} n(n-1)x^{n-2}$,

then term-by-term integration gives $\displaystyle\int S(x)\, dx + C = \sum\limits_{n=2}^{\infty} nx^{n-1}$. A second integration leads to

$$\int \left[\int S(x)\, dx + C\right] dx + D = \sum_{n=2}^{\infty} x^n = \frac{x^2}{1-x},$$

since the series is geometric. Differentiation now gives

$$\int S(x)\, dx + C = \frac{(1-x)(2x) - x^2(-1)}{(1-x)^2} = \frac{2x - x^2}{(1-x)^2}.$$

A second differentiation provides $S(x)$,

$$S(x) = \frac{(1-x)^2(2-2x) - (2x - x^2)2(1-x)(-1)}{(1-x)^4} = \frac{2}{(1-x)^3}.$$

4. The radius of convergence of the series is $R = \lim\limits_{n\to\infty} \left|\dfrac{n^2}{(n+1)^2}\right| = 1$. If we set $S(x) = \sum\limits_{n=1}^{\infty} n^2 x^{n-1}$, then

term-by-term integration gives $\displaystyle\int S(x)\, dx + C = \sum\limits_{n=1}^{\infty} nx^n$. When $x \neq 0$, we can divide by x,

$$\frac{1}{x}\int S(x)\, dx + \frac{C}{x} = \sum_{n=1}^{\infty} nx^{n-1}.$$ Integration now gives,

$$\int \left[\frac{1}{x}\int S(x)\, dx\right] dx + C\ln|x| + D = \sum_{n=1}^{\infty} x^n = \frac{x}{1-x}.$$

If we now differentiate, $\dfrac{1}{x}\displaystyle\int S(x)\, dx + \dfrac{C}{x} = \dfrac{(1-x)(1) - x(-1)}{(1-x)^2} = \dfrac{1}{(1-x)^2}$. Multiplication by x and a further differentiation gives

$$S(x) = \frac{d}{dx}\left[\frac{x}{(1-x)^2}\right] = \frac{(1-x)^2(1) - x(2)(1-x)(-1)}{(1-x)^4} = \frac{x+1}{(1-x)^3}.$$

Since the sum of the series at $x = 0$ is 1, and this is $S(0)$, the formula $S(x) = (x+1)/(1-x)^3$ can be used for all x in $|x| < 1$.

6. The radius of convergence of the series is $R = \lim\limits_{n\to\infty} \left|\dfrac{1/(n+1)}{1/(n+2)}\right| = 1$. If we set $S(x) = \sum\limits_{n=0}^{\infty} \dfrac{1}{n+1}x^n$, then

$x\,S(x) = \sum\limits_{n=0}^{\infty} \dfrac{1}{n+1}x^{n+1}$. Term-by-term differentiation gives $\dfrac{d}{dx}[x\,S(x)] = \sum\limits_{n=0}^{\infty} x^n = \dfrac{1}{1-x}$, since the series is geometric. We now integrate,

$$x\,S(x) = \int \frac{1}{1-x}\,dx = -\ln(1-x) + C.$$

Substitution of $x = 0$ gives $C = 0$, and therefore $S(x) = -\dfrac{1}{x}\ln(1-x)$. This is valid for $-1 < x < 1$, but not at $x = 0$. It is interesting to note, however, that the limit of $S(x)$ as x approaches zero is 1 and this is the sum of the series at $x = 0$.

8. If we set $y = x^2$, the series becomes $\displaystyle\sum_{n=1}^{\infty} \frac{(-1)^n}{n} x^{2n} = \sum_{n=1}^{\infty} \frac{(-1)^n}{n} y^n$. The radius of convergence of this

series is $R_y = \displaystyle\lim_{n\to\infty} \left| \frac{(-1)^n/n}{(-1)^{n+1}/(n+1)} \right| = 1$. The radius of convergence of the original series is therefore

$R_x = 1$. If we set $S(x) = \displaystyle\sum_{n=1}^{\infty} \frac{(-1)^n}{n} x^{2n}$, then term-by-term differentiation gives

$$S'(x) = \sum_{n=1}^{\infty} 2(-1)^n x^{2n-1} = \frac{-2x}{1+x^2},$$

since the series is geometric. Integration now leads to $S(x) = -\ln(1+x^2) + C$. Since $S(0) = 0$, it follows that $C = 0$, and $S(x) = -\ln(1+x^2)$.

10. The radius of convergence of the series is $R = \displaystyle\lim_{n\to\infty} \left| \frac{(n+1)/(n+2)}{(n+2)/(n+3)} \right| = 1$. If we set

$S(x) = \displaystyle\sum_{n=0}^{\infty} \left(\frac{n+1}{n+2} \right) x^n$, and integrate, $\displaystyle\int S(x)\, dx = \sum_{n=0}^{\infty} \frac{1}{n+2} x^{n+1} + C$. Multiplication by x gives

$x \displaystyle\int S(x)\, dx = \sum_{n=0}^{\infty} \frac{1}{n+2} x^{n+2} + Cx$. Differentiation now gives

$$\frac{d}{dx} \left[x \int S(x)\, dx \right] = \sum_{n=0}^{\infty} x^{n+1} + C = \frac{x}{1-x} + C,$$

since the series is geometric. Integration now yields

$$x \int S(x)\, dx = \int \frac{x}{1-x} dx + Cx + D = -x - \ln|1-x| + Cx + D.$$

If we set $x = 0$ in this equation we find that $D = 0$. When we drop absolute values and divide by x,

$$\int S(x)\, dx = -1 - \frac{1}{x} \ln(1-x) + C, \quad x \neq 0.$$

When we differentiate this equation, we obtain $S(x) = \dfrac{1}{x^2} \ln(1-x) + \dfrac{1}{x(1-x)}$. This formula can only be used for values of x in the interval $-1 < x < 1$, but not $x = 0$. The sum at $x = 0$ is $1/2$.

12. $\displaystyle\sum_{n=0}^{\infty} \frac{(-1)^n(2n+1)}{(2n+1)!} x^{2n+1} = x \sum_{n=0}^{\infty} \frac{(-1)^n}{(2n)!} x^{2n} = x \cos x$

14. If we set $y = x^2$, the series becomes $\displaystyle\sum_{n=1}^{\infty} \frac{(2n+3)2^n}{n!} x^{2n} = \sum_{n=1}^{\infty} \frac{(2n+3)2^n}{n!} y^n$. The radius of con-

vergence of this series is $R_y = \displaystyle\lim_{n\to\infty} \left| \frac{(2n+3)2^n/n!}{(2n+5)2^{n+1}/(n+1)!} \right| = \infty$. The radius of convergence of the

original series is therefore $R_x = \infty$ also. If we set $S(x) = \displaystyle\sum_{n=1}^{\infty} \frac{(2n+3)2^n}{n!} x^{2n}$, and multiply by x^2,

$x^2 S(x) = \displaystyle\sum_{n=1}^{\infty} \frac{(2n+3)2^n}{n!} x^{2n+2}$. Integration now gives

$$\int x^2 S(x)\, dx = \sum_{n=1}^{\infty} \frac{2^n}{n!} x^{2n+3} + C = x^3 \sum_{n=1}^{\infty} \frac{1}{n!} (2x^2)^n + C = x^3 (e^{2x^2} - 1) + C.$$

We now differentiate to get

$$x^2 S(x) = 3x^2 (e^{2x^2} - 1) + x^3 (4xe^{2x^2}) \qquad \Longrightarrow \qquad S(x) = (4x^2 + 3)e^{2x^2} - 3.$$

EXERCISES 10.7

2. Taylor's remainder formula for e^x and $c = 0$ gives $e^x = 1 + x + \dfrac{x^2}{2} + \dfrac{x^3}{6} + R_3$, where

$R_3 = \dfrac{d^4}{dx^4} e^x \big|_{x=z_3} \dfrac{x^4}{4!} = e^{z_3} \dfrac{x^4}{24}$, and $0 < z_3 < x$. Since $x < 0.01$, we can say that

$$R_3 < e^x \dfrac{x^4}{24} < e^{0.01} \dfrac{(0.01)^4}{24} = 4.2 \times 10^{-10}.$$

4. According to Exercise 2, a maximum error on $0 \le x \le 0.01$ is 4.2×10^{-10}. For $-0.01 \le x < 0$, $R_3 = e^{z_3} \dfrac{x^4}{24}$ where $x < z_3 < 0$. Since $x \ge -0.01$, it follows that

$$|R_3| < e^0 \dfrac{|x|^4}{24} \le \dfrac{|-0.01|^4}{24} < 4.2 \times 10^{-10}.$$

6. Taylor's remainder formula for $\cos x$ and $c = 0$ gives $\cos x = 1 - \dfrac{x^2}{2!} + \dfrac{x^4}{4!} + R_5$, where

$R_5 = \dfrac{d^6}{dx^6} \cos x \big|_{x=z_5} \dfrac{x^6}{6!} = -(\cos z_5) \dfrac{x^6}{6!}$, and z_5 is between 0 and x. Since $|x| \le 0.1$, we can say that

$$|R_5| < (1) \dfrac{|x|^6}{6!} \le \dfrac{(0.1)^6}{6!} < 1.4 \times 10^{-9}.$$

8. The first four derivatives of $f(x) = 1/(1-x)^3$ are $f'(x) = 3/(1-x)^4$, $f''(x) = 12/(1-x)^5$, $f'''(x) = 60/(1-x)^6$, and $f''''(x) = 360/(1-x)^7$. Taylor's remainder formula for $1/(1-x)^3$ and $c = 0$ gives $\dfrac{1}{(1-x)^3} = 1 + 3x + 6x^2 + 10x^3 + R_3(x)$, where $R_3(x) = f''''(z_3) \dfrac{x^4}{4!} = \dfrac{15x^4}{(1-z_3)^7}$, and z_3 is between 0 and x. Since $|x| < 0.2$, we can say that

$$|R_3| < \dfrac{15|x|^4}{(1-0.2)^7} < \dfrac{15(0.2)^4}{(1-0.2)^7} < 0.115.$$

10. The first five derivatives of $f(x) = \ln x$ are $f'(x) = 1/x$, $f''(x) = -1/x^2$, $f'''(x) = 2/x^3$, $f''''(x) = -6/x^4$, and $f'''''(x) = 24/x^5$. Taylor's remainder formula with $c = 1$ gives

$$\ln x = (x-1) - \dfrac{1}{2}(x-1)^2 + \dfrac{1}{3}(x-1)^3 - \dfrac{1}{4}(x-1)^4 + R_4,$$

where $R_4 = f^{(5)}(z_4) \dfrac{(x-1)^5}{5!} = \dfrac{24}{z_4^5} \dfrac{(x-1)^5}{5!} = \dfrac{(x-1)^5}{5z_4^5}$ and z_4 is between 1 and x. Since $1/2 \le x \le 3/2$, we can say that

$$|R_4| < \dfrac{|x-1|^5}{5(1/2)^5} \le \dfrac{(1/2)^5}{5(1/2)^5} = 0.2.$$

12. If we set $u = x^2$ and $du = 2x\,dx$, then $\displaystyle\int_0^{1/2} \cos(x^2)\,dx = \dfrac{1}{2}\int_0^{1/4} \dfrac{\cos u}{\sqrt{u}}\,du$. Taylor's remainder formula for $\cos u$ gives

$$\cos u = 1 - \dfrac{u^2}{2!} + \dfrac{u^4}{4!} - \cdots + \dfrac{d^n(\cos u)}{du^n}\Big|_{u=0} \dfrac{u^n}{n!} + R_n(0, u),$$

where $R_n(0, u) = \dfrac{d^{n+1}(\cos u)}{du^{n+1}}\Big|_{u=z_n} \dfrac{u^{n+1}}{(n+1)!}$. Consequently,

$$\int_0^{1/2} \cos(x^2)\,dx = \dfrac{1}{2}\int_0^{1/4} \dfrac{1}{\sqrt{u}}\left[1 - \dfrac{u^2}{2!} + \dfrac{u^4}{4!} - \cdots + R_n(0, u)\right] du$$

$$= \frac{1}{2} \int_0^{1/4} \left[\frac{1}{\sqrt{u}} - \frac{u^{3/2}}{2!} + \frac{u^{7/2}}{4!} - \cdots + \frac{1}{\sqrt{u}} R_n(0, u) \right] du$$

$$= \frac{1}{2} \left\{ 2\sqrt{u} - \frac{2u^{5/2}}{5 \cdot 2!} + \frac{2u^{9/2}}{9 \cdot 4!} - \cdots \right\}_0^{1/4} + \frac{1}{2} \int_0^{1/4} \frac{1}{\sqrt{u}} R_n(0, u) \, du$$

$$= \frac{1}{2} - \frac{1}{5 \cdot 2^5 \cdot 2!} + \frac{1}{9 \cdot 2^9 \cdot 4!} - \cdots + \frac{1}{2} \int_0^{1/4} \frac{1}{\sqrt{u}} R_n(0, u) \, du.$$

Now,

$$\left| \frac{1}{2} \int_0^{1/4} \frac{1}{\sqrt{u}} R_n(0, u) \, du \right| \leq \frac{1}{2} \int_0^{1/4} \frac{1}{\sqrt{u}} |R_n(0, u)| \, du \leq \frac{1}{2} \int_0^{1/4} \frac{1}{\sqrt{u}} \frac{u^{n+1}}{(n+1)!} du$$

$$= \frac{1}{2} \int_0^{1/4} \frac{u^{n+1/2}}{(n+1)!} du = \frac{1}{2(n+1)!} \left\{ \frac{u^{n+3/2}}{n+3/2} \right\}_0^{1/4} = \frac{1}{(2n+3)(n+1)! 4^{n+3/2}}.$$

When $n = 2$, this is less than 1.9×10^{-4}. Hence, if we approximate the integral with the first two terms, namely, $\frac{1}{2} - \frac{1}{5 \cdot 2^5 \cdot 2!} = \frac{159}{320}$, then we can say that

$$\frac{159}{320} - 0.000\,19 < \int_0^{1/2} \cos(x^2) \, dx < \frac{159}{320} + 0.000\,19,$$

that is, $0.496\,685 < \int_0^{1/2} \cos(x^2) \, dx < 0.497\,065$. To three decimals, the value of the integral is 0.497.

14. If we set $w = x^2$ and $dw = 2x\, dx$, then $\int_0^{0.3} e^{-x^2} \, dx = \frac{1}{2} \int_0^{0.09} \frac{e^{-w}}{\sqrt{w}} dw$. Taylor's remainder formula applied to e^{-w} gives

$$e^{-w} = 1 - w + \frac{w^2}{2!} - \frac{w^3}{3!} + \cdots + \frac{(-1)^n w^n}{n!} + R_n(0, w)$$

where $R_n(0, w) = \frac{d^{n+1}}{dw^{n+1}}(e^{-w})|_{w=w_n} \frac{w^{n+1}}{(n+1)!} = \frac{(-1)^{n+1} e^{-w_n} w^{n+1}}{(n+1)!}$. Consequently,

$$\int_0^{0.3} e^{-x^2} \, dx = \frac{1}{2} \int_0^{0.09} \frac{1}{\sqrt{w}} \left[1 - w + \frac{w^2}{2!} - \frac{w^3}{3!} + \cdots + \frac{(-1)^n w^n}{n!} + R_n(0, w) \right] dw$$

$$= \frac{1}{2} \int_0^{0.09} \left[\frac{1}{\sqrt{w}} - \sqrt{w} + \frac{w^{3/2}}{2!} - \frac{w^{5/2}}{3!} + \cdots + \frac{(-1)^n w^{n-1/2}}{n!} + \frac{1}{\sqrt{w}} R_n(0, w) \right] dw$$

$$= \frac{1}{2} \left\{ 2\sqrt{w} - \frac{2w^{3/2}}{3} + \frac{2w^{5/2}}{5 \cdot 2!} - \frac{2w^{7/2}}{7 \cdot 3!} + \cdots + \frac{2(-1)^n w^{n+1/2}}{(2n+1)n!} \right\}_0^{0.09} + \frac{1}{2} \int_0^{0.09} \frac{R_n(0, w)}{\sqrt{w}} dw$$

$$= \sqrt{0.09} - \frac{(0.09)^{3/2}}{3} + \frac{(0.09)^{5/2}}{5 \cdot 2!} - \frac{(0.09)^{7/2}}{7 \cdot 3!} + \cdots + \frac{(-1)^n (0.09)^{n+1/2}}{(2n+1)n!} + \frac{1}{2} \int_0^{0.09} \frac{R_n(0, w)}{\sqrt{w}} dw.$$

Now,

$$\frac{1}{2} \left| \int_0^{0.09} \frac{R_n(0, w)}{\sqrt{w}} dw \right| \leq \frac{1}{2} \int_0^{0.09} \frac{1}{\sqrt{w}} \left| \frac{(-1)^{n+1} e^{-w_n} w^{n+1}}{(n+1)!} \right| dw = \frac{1}{2(n+1)!} \int_0^{0.09} e^{-w_n} w^{n+1/2} dw.$$

Since $0 < w_n < w < 0.09$, we can say $e^{-w_n} \leq 1$. Thus,

$$\frac{1}{2} \left| \int_0^{0.09} \frac{R_n(0, w)}{\sqrt{w}} dw \right| \leq \frac{1}{2(n+1)!} \left\{ \frac{2w^{n+3/2}}{2n+3} \right\}_0^{0.09} = \frac{(0.09)^{n+3/2}}{(2n+3)(n+1)!}.$$

When $n = 2$, this is less than 3.0×10^{-6}. Hence, if we approximate the integral with the first three terms, namely, $\sqrt{0.09} - \dfrac{(0.09)^{3/2}}{3} + \dfrac{(0.09)^{5/2}}{5 \cdot 2!} = 0.291\,243$, then we can say that

$$0.291\,243 - 0.000\,003 < \int_0^{0.3} e^{-x^2}\, dx < 0.291\,243 + 0.000\,003,$$

that is, $0.291\,240 < \displaystyle\int_0^{0.3} e^{-x^2}\, dx < 0.291\,246$. To three decimals, the value of the integral is 0.291.

16. $\displaystyle\lim_{x \to 0} \frac{1 - \cos x}{x^2} = \lim_{x \to 0} \frac{1}{x^2}\left[1 - \left(1 - \frac{x^2}{2!} + \frac{x^4}{4!} - \cdots\right)\right] = \lim_{x \to 0}\left(\frac{1}{2!} - \frac{x^2}{4!} + \cdots\right) = \frac{1}{2}$

18. $\displaystyle\lim_{x \to 0} \frac{\sqrt{1+x} - 1}{x} = \lim_{x \to 0} \frac{1}{x}\left[\left(1 + \frac{x}{2} - \frac{(1/2)(-1/2)}{2!}x^2 + \cdots\right) - 1\right] = \lim_{x \to 0}\left[\frac{1}{2} + \frac{x}{8} + \cdots\right] = \frac{1}{2}$

20. $\dfrac{e^x + e^{-x}}{e^x - e^{-x}} = \dfrac{1 + e^{-2x}}{1 - e^{-2x}} = \dfrac{1 + \left[1 - 2x + \dfrac{(-2x)^2}{2!} + \dfrac{(-2x)^3}{3!} + \cdots\right]}{1 - \left[1 - 2x + \dfrac{(-2x)^2}{2!} + \dfrac{(-2x)^3}{3!} + \cdots\right]} = \dfrac{2 - 2x + 2x^2 - \dfrac{4x^3}{3} + \cdots}{2x - 2x^2 + \dfrac{4x^3}{3} + \cdots}$

Long division gives $\dfrac{e^x + e^{-x}}{e^x - e^{-x}} = \dfrac{1}{x} + \dfrac{x}{3} + \cdots$.

Thus, $\displaystyle\lim_{x \to 0}\left(\frac{e^x + e^{-x}}{e^x - e^{-x}} - \frac{1}{x}\right) = \lim_{x \to 0}\left[\left(\frac{1}{x} + \frac{x}{3} + \cdots\right) - \frac{1}{x}\right] = 0$.

22. We set $u = x^3$ and consider the function $f(u) = 1/\sqrt{1+u}$ on the interval $0 < u < 1/8$. Since the nth derivative of $f(u)$ is $f^{(n)}(u) = \dfrac{(-1)^n[1 \cdot 3 \cdot 5 \cdots (2n-1)]}{2^n(1+u)^{n+1/2}}$, Taylor's remainder formula gives

$$f(u) = 1 - \frac{u}{2} + \frac{3u^2}{8} - \frac{5u^3}{16} + \cdots + \frac{(-1)^n[1 \cdot 3 \cdot 5 \cdots (2n-1)]}{2^n\, n!}u^n + R_n(0, u),$$

where $R_n(0, u) = \dfrac{f^{(n+1)}(z_n)}{(n+1)!}u^{n+1}$, and $0 < z_n < u$. Since $0 < u < 1/8$, we can say that

$$|R_n(0, u)| = \frac{1 \cdot 3 \cdot 5 \cdots (2n+1)}{2^{n+1}|1 + z_n|^{n+3/2}(n+1)!}|u|^{n+1} < \frac{1 \cdot 3 \cdot 5 \cdots (2n+1)}{2^{n+1}|1 + 0|^{n+3/2}(n+1)!}|u|^{n+1}$$

$$< \frac{1 \cdot 3 \cdot 5 \cdots (2n+1)}{2^{n+1}(n+1)!}\left(\frac{1}{8}\right)^{n+1} = \frac{1 \cdot 3 \cdot 5 \cdots (2n+1)}{2^{4n+4}(n+1)!}.$$

The smallest integer for which this is less than 10^{-4} is $n = 3$. Thus, we should approximate $1/\sqrt{1+u}$ with $1 - u/2 + 3u^2/8 - 5u^3/16$, or approximate $1/\sqrt{1+x^3}$ with

$$1 - \frac{x^3}{2} + \frac{3x^6}{8} - \frac{5x^9}{16}.$$

24. Taylor's remainder formula for $\cos^2 x = (1 + \cos 2x)/2$ gives

$$\cos^2 x = \frac{1}{2}(1 + \cos 2x) = \frac{1}{2}\left[1 + \left(1 - \frac{2^2 x^2}{2!} + \frac{2^4 x^4}{4!} - \cdots + \frac{f^{(n)}(0)}{n!}x^n + R_n(0, x)\right)\right]$$

$$= 1 - x^2 + \frac{x^4}{3} - \cdots + \frac{f^{(n)}(0)}{2n!}x^n + \frac{1}{2}R_n(0, x),$$

where $R_n(0, x) = \dfrac{f^{(n+1)}(z_n)}{(n+1)!}x^{n+1}$ and z_n is between 0 and x. Since the $(n+1)^{\text{th}}$ derivative of $f(x)$ is $\pm 2^{n+1}\sin 2x$ or $\pm 2^{n+1}\cos 2x$, and $|x| \leq 0.1$, it follows that

$$\frac{1}{2}|R_n(0, x)| \leq \frac{2^{n+1}|x|^{n+1}}{2(n+1)!} < \frac{2^n}{(n+1)!10^{n+1}}.$$

The smallest integer for which this is less than 10^{-3} is $n = 2$. Thus, the function should be approximated by $1 - x^2$.

26. If we substitute $y = f(x) = \sum_{n=0}^{\infty} a_n x^n$ into the differential equation,

$$0 = \sum_{n=0}^{\infty} n(n-1)a_n x^{n-2} + \sum_{n=0}^{\infty} na_n x^{n-1} = \sum_{n=1}^{\infty} (n+1)na_{n+1} x^{n-1} + \sum_{n=1}^{\infty} na_n x^{n-1}$$

$$= \sum_{n=1}^{\infty} [(n+1)na_{n+1} + na_n]x^{n-1}.$$

When we equate coefficients to zero:

$$(n+1)na_{n+1} + na_n = 0 \quad \Longrightarrow \quad a_{n+1} = -\frac{a_n}{n+1}, \quad n \geq 1.$$

This recursive definition implies that

$$a_2 = -\frac{a_1}{2}, \quad a_3 = -\frac{a_2}{3} = \frac{a_1}{3!}, \quad a_4 = -\frac{a_3}{4} = -\frac{a_1}{4!}, \quad \cdots .$$

Thus, $\quad y = f(x) = a_0 + a_1 \left(x - \frac{x^2}{2!} + \frac{x^3}{3!} - \frac{x^4}{4!} + \cdots \right) = a_0 + a_1 \sum_{n=1}^{\infty} \frac{(-1)^{n+1}}{n!} x^n$

$$= a_0 - a_1(e^{-x} - 1) = (a_0 + a_1) - a_1 e^{-x}.$$

28. If we substitute $y = f(x) = \sum_{n=0}^{\infty} a_n x^n$ into the differential equation,

$$0 = 4x \sum_{n=0}^{\infty} n(n-1)a_n x^{n-2} + 2 \sum_{n=0}^{\infty} na_n x^{n-1} + \sum_{n=0}^{\infty} a_n x^n$$

$$= \sum_{n=0}^{\infty} 4n(n-1)a_n x^{n-1} + \sum_{n=0}^{\infty} 2na_n x^{n-1} + \sum_{n=0}^{\infty} a_n x^n$$

$$= \sum_{n=2}^{\infty} 4n(n-1)a_n x^{n-1} + \sum_{n=1}^{\infty} 2na_n x^{n-1} + \sum_{n=1}^{\infty} a_{n-1} x^{n-1}$$

$$= (2a_1 + a_0) + \sum_{n=2}^{\infty} [(4n^2 - 4n + 2n)a_n + a_{n-1}]x^{n-1}.$$

We now equate coefficients of powers of x to zero. From the coefficient of x^0 we obtain $2a_1 + a_0 = 0$ which implies that $a_1 = -a_0/2$. From the remaining coefficients, we obtain

$$(4n^2 - 2n)a_n + a_{n-1} = 0 \quad \Longrightarrow \quad a_n = \frac{-a_{n-1}}{2n(2n-1)}, \quad n \geq 2.$$

When we iterate this recursive definition:

$$a_2 = \frac{-a_1}{4 \cdot 3} = \frac{a_0}{4!}, \quad a_3 = \frac{-a_2}{6 \cdot 5} = -\frac{a_0}{6!}, \quad \cdots$$

The solution is therefore $\quad y = f(x) = a_0 \left(1 - \frac{x}{2} + \frac{x^2}{4!} - \frac{x^3}{6!} + \cdots \right) = a_0 \sum_{n=0}^{\infty} \frac{(-1)^n}{(2n)!} x^n.$

30. If we substitute $y = f(x) = \sum_{n=0}^{\infty} a_n x^n$ into the differential equation,

$$0 = x \sum_{n=0}^{\infty} n(n-1)a_n x^{n-2} + \sum_{n=0}^{\infty} a_n x^n = \sum_{n=0}^{\infty} n(n-1)a_n x^{n-1} + \sum_{n=0}^{\infty} a_n x^n$$

$$= \sum_{n=2}^{\infty} n(n-1)a_n x^{n-1} + \sum_{n=1}^{\infty} a_{n-1} x^{n-1} = a_0 + \sum_{n=2}^{\infty} [n(n-1)a_n + a_{n-1}]x^{n-1}.$$

We now equate coefficients of powers of x to zero. From the coefficient of x^0 we obtain $a_0 = 0$. From the remaining coefficients, we obtain

$$n(n-1)a_n + a_{n-1} = 0 \implies a_n = \frac{-a_{n-1}}{n(n-1)}, \quad n \geq 2.$$

When we iterate this recursive definition:

$$a_2 = \frac{-a_1}{2 \cdot 1}, \quad a_3 = \frac{-a_2}{3 \cdot 2} = \frac{a_1}{3! \, 2!}, \quad a_4 = \frac{-a_3}{4 \cdot 3} = \frac{-a_1}{4! \, 3!}, \quad \cdots .$$

The solution is therefore

$$y = f(x) = a_1 \left(x - \frac{x^2}{2 \cdot 1} + \frac{x^3}{3! \cdot 2!} - \frac{x^4}{4! \cdot 3!} + \cdots \right) = a_1 \sum_{n=1}^{\infty} \frac{(-1)^{n+1}}{n!(n-1)!} x^n.$$

32. $K = c^2(m - m_0) = c^2 m_0 \left(\dfrac{1}{\sqrt{1 - v^2/c^2}} - 1 \right)$

$$= c^2 m_0 \left\{ \left[1 - \frac{1}{2}\left(-\frac{v^2}{c^2}\right) + \frac{(-1/2)(-3/2)}{2!}\left(-\frac{v^2}{c^2}\right)^2 + \frac{(-1/2)(-3/2)(-5/2)}{3!}\left(-\frac{v^2}{c^2}\right)^3 + \cdots \right] - 1 \right\}$$

$$= c^2 m_0 \left\{ \frac{v^2}{2c^2} + \frac{3}{8}\frac{v^4}{c^4} + \frac{5}{16}\frac{v^6}{c^6} + \cdots \right\} = \frac{1}{2}m_0 v^2 + m_0 c^2 \left(\frac{3}{8}\frac{v^4}{c^4} + \frac{5}{16}\frac{v^6}{c^6} + \cdots \right)$$

34. When we expand P_s/P_0 with the binomial expansion,

$$\frac{P_s}{P_0} = 1 + \left(\frac{k}{k-1}\right)\left(\frac{k-1}{2}\right)M_0^2 + \frac{1}{2}\left(\frac{k}{k-1}\right)\left(\frac{k}{k-1} - 1\right)\left(\frac{k-1}{2}\right)^2 M_0^4$$

$$+ \frac{1}{3!}\left(\frac{k}{k-1}\right)\left(\frac{k}{k-1} - 1\right)\left(\frac{k}{k-1} - 2\right)\left(\frac{k-1}{2}\right)^3 M_0^6 + \cdots$$

$$= 1 + \frac{k}{2}M_0^2 + \frac{k}{8}M_0^4 + \frac{k(2-k)}{48}M_0^6 + \cdots$$

$$= 1 + \frac{1}{2}M_0^2\left(\frac{\rho_0 c_0^2}{P_0}\right) + \frac{1}{8}M_0^4\left(\frac{\rho_0 c_0^2}{P_0}\right) + \frac{1}{48}M_0^6(2-k)\left(\frac{\rho_0 c_0^2}{P_0}\right) + \cdots .$$

Multiplication by P_0, and replacement of M_0^2 by V_0^2/c_0^2 in the last three terms gives

$$P_s = P_0 + \frac{1}{2}\rho_0 c_0^2\left(\frac{V_0^2}{c_0^2}\right) + \frac{1}{8}\rho_0 c_0^2\left(\frac{V_0^2}{c_0^2}\right)M_0^2 + \frac{1}{48}(2-k)\rho_0 c_0^2\left(\frac{V_0^2}{c_0^2}\right)M_0^4 + \cdots$$

$$= P_0 + \frac{1}{2}\rho_0 V_0^2 + \frac{1}{8}\rho_0 V_0^2 M_0^2 + \frac{1}{48}(2-k)\rho_0 V_0^2 M_0^4 + \cdots$$

$$= P_0 + \frac{1}{2}\rho_0 V_0^2\left[1 + \frac{M_0^2}{4} + \left(\frac{2-k}{24}\right)M_0^4 + \cdots\right].$$

36. (a) If we substitute $e^{-\beta^2/(4\alpha x)} = \displaystyle\sum_{n=0}^{\infty} \frac{1}{n!}\left(-\frac{\beta^2}{4\alpha x}\right)^n$, we obtain

$$W(\alpha, \beta) = \int_1^{\infty} \frac{1}{x} e^{-\alpha x} \sum_{n=0}^{\infty} \frac{1}{n!}\left(-\frac{\beta^2}{4\alpha x}\right)^n dx = \sum_{n=0}^{\infty} \frac{1}{n!}\left(-\frac{\beta^2}{4\alpha}\right)^n \int_1^{\infty} \frac{e^{-\alpha x}}{x^{n+1}} dx = \sum_{n=0}^{\infty} \frac{(-1)^n \beta^{2n}}{4^n \alpha^n n!} E_{n+1}(\alpha).$$

(b) We use integration by parts with $u = e^{-\alpha x}$ and $dv = \dfrac{1}{x^{n+1}} dx$,

$$E_{n+1}(\alpha) = \int_1^{\infty} \frac{e^{-\alpha x}}{x^{n+1}} dx = \left\{ -\frac{e^{-\alpha x}}{nx^n} \right\}_1^{\infty} - \int_1^{\infty} -\frac{1}{nx^n}(-\alpha)e^{-\alpha x}\, dx = \frac{e^{-\alpha}}{n} - \frac{\alpha}{n}\int_1^{\infty} \frac{e^{-\alpha x}}{x^n} dx$$

$$= \frac{1}{n}[e^{-\alpha} - \alpha E_n(\alpha)].$$

38. (a) We write $E = \dfrac{q}{4\pi\epsilon_0 x^2 \left(1 - \dfrac{d}{2x}\right)^2} - \dfrac{q}{4\pi\epsilon_0 x^2 \left(1 + \dfrac{d}{2x}\right)^2} = \dfrac{q}{4\pi\epsilon_0 x^2}\left[\left(1 - \dfrac{d}{2x}\right)^{-2} - \left(1 + \dfrac{d}{2x}\right)^{-2}\right].$

(b) If we expand each term with the binomial expansion 10.33b,

$$E = \frac{q}{4\pi\epsilon_0 x^2}\left\{\left[1 - 2\left(-\frac{d}{2x}\right) + \cdots\right] - \left[1 - 2\left(\frac{d}{2x}\right) + \cdots\right]\right\}.$$

When d is very much less than x, we omit higher order terms in d/x, and write

$$E \approx \frac{q}{4\pi\epsilon_0 x^2}\left(1 + \frac{d}{x} - 1 + \frac{d}{x}\right) = \frac{qd}{2\pi\epsilon_0 x^3}.$$

EXERCISES 10.8

2. False   The sequence $\{n\}$ is increasing but has no upper bound.

4. False   The sequence $\{-n\}$ is decreasing but has no lower bound.

6. True   An increasing sequence has a lower bound. If it also has an upper bound, then it has a limit according to Theorem 10.7.

8. True   For a sequence to be increasing and decreasing, its terms would have to satisfy $c_{n+1} > c_n$ and $c_{n+1} < c_n$ for all n. This is impossible.

10. True   This is part of the corollary to Theorem 10.7.

12. False   The sequence $\{(-1)^n/n\}$ is bounded, not monotonic, and it has limit 0.

14. False   The sequence $\{(-1)^n/n\}$ is not montonic, but it has limit 0.

16. False   The sequence $\{(-1)^n\}$ has no limit, but the sequence $\{[(-1)^n]^2\} = \{1\}$ has limit 1.

18. True   Such a sequence displays the up-down-up-down nature required of an oscillating sequence.

20. False   Each term of the sequence $\{-3^n\}$ is less than half the previous term, but the sequence does not have a limit.

22. True. The only other possibility is that the sequence $\{c_n\}$ has a limit L which is not equal to a. Suppose $L > a$ and we set $\epsilon = L - a$. Then there exists an integer N such that for $n > N$, $|c_n - L| < L - a$; that is $-(L - a) < c_n - L < L - a$. Thus, for $n > N$,

$$L - (L - a) < c_n < L + (L - a) \quad \Longrightarrow \quad a < c_n < 2L - a.$$

But this contradicts the fact that $c_n = a$ for an infinity of values of n. Hence L cannot be greater than a. A similar proof shows that L cannot be less than a.

24. False   The oscillating sequence $\{[1 + (-1)^{n+1}]/(2n)\}$ has limit zero, but terms are all nonnegative.

26. The first four terms of the sequence are $c_1 = 0$, $c_2 = 5/12$, $c_3 = 0.419\,178$, $c_4 = 0.419\,240$. The sequence appears to be increasing; that is $c_{n+1} > c_n$. This is certainly true for $n = 1$ as $c_2 > c_1$. Suppose $c_{k+1} > c_k$ for some integer k. Then, $c_{k+1}^4 > c_k^4$, and $c_{k+1}^4 + 5 > c_k^4 + 5$. Thus, $(c_{k+1}^4 + 5)/12 > (c_k^4 + 5)/12$, and this means that $c_{k+2} > c_{k+1}$. By mathematical induction then, $c_{n+1} > c_n$ for all $n \geq 1$. The first term $c_1 = 0$ must be a lower bound. We suspect that $U = 1$ is an upper bound; that is, $c_n \leq 1$. This is true for $n = 1$. Suppose $c_k \leq 1$ for some integer k. Then $c_{k+1} = (c_k^4 + 5)/12 \leq (1 + 5)/12 = 1/2 < 1$. Hence, by mathematical induction, $c_n \leq 1$ for all n.

Theorem 10.7 now implies that the sequence has a limit, call it L. By taking limits on both sides of the recursive definition of the sequence we obtain

$$\lim_{n\to\infty} c_{n+1} = \lim_{n\to\infty} \frac{1}{12}(c_n^4 + 5).$$

It follows that L must satisfy $L = (L^4 + 5)/12 \Longrightarrow 0 = L^4 - 12L + 5 = f(L)$. Since $f(0.419\,240\,5) = 6.5 \times 10^{-6}$ and $f(0.419\,241\,5) = -5.2 \times 10^{-6}$, the limit of the sequence (to six decimals) is $0.419\,241$.

28. It is clear that all terms of the sequence are positive and therefore there is no difficulty with taking square roots. The first four terms of the sequence are $c_1 = 1$, $c_2 = \sqrt{6} = 2.449$, $c_3 = 2.7294$, $c_4 = 2.7802$. The sequence appears to be increasing; that is $c_{n+1} > c_n$. This is certainly true for $n = 1$ as $c_2 > c_1$. Suppose $c_{k+1} > c_k$ for some integer k. Then, $5 + c_{k+1} > 5 + c_k$, and $\sqrt{5 + c_{k+1}} > \sqrt{5 + c_k}$. Thus, $c_{k+2} > c_{k+1}$, and by mathematical induction, $c_{n+1} > c_n$ for all $n \geq 1$. The first term $c_1 = 1$ must be a lower bound. We suspect that $U = 5$ is an upper bound; that is, $c_n \leq 5$. This is true for $n = 1$. Suppose $c_k \leq 5$ for some integer k. Then $c_{k+1} = \sqrt{5 + c_k} \leq \sqrt{5 + 5} = \sqrt{10} < 5$. Hence, by mathematical induction, $c_n \leq 5$ for all n.

Theorem 10.7 now implies that the sequence has a limit, call it L. By taking limits on both sides of the recursive definition of the sequence we obtain

$$\lim_{n \to \infty} c_{n+1} = \lim_{n \to \infty} \sqrt{5 + c_n}.$$

It follows that L must satisfy $L = \sqrt{5 + L} \implies L^2 - L - 5 = 0$. Of the two solutions $(1 \pm \sqrt{21})/2$ of this equation, only $(1 + \sqrt{21})/2$ lies between the bounds. Hence, $L = (1 + \sqrt{21})/2$.

30. It is clear that all terms of the sequence are positive and therefore there is no difficulty with taking square roots. The first four terms of the sequence are $c_1 = 3$, $c_2 = 4$, $c_3 = 4.1623$, $c_4 = 4.1878$. The sequence appears to be increasing; that is $c_{n+1} > c_n$. This is certainly true for $n = 1$ as $c_2 > c_1$. Suppose $c_{k+1} > c_k$ for some integer k. Then, $6 + c_{k+1} > 6 + c_k$, and $\sqrt{6 + c_{k+1}} > \sqrt{6 + c_k}$. Thus, $1 + \sqrt{6 + c_{k+1}} > 1 + \sqrt{6 + c_k}$, and this means that $c_{k+2} > c_{k+1}$. By mathematical induction then, $c_{n+1} > c_n$ for all $n \geq 1$. The first term $c_1 = 3$ must be a lower bound. We suspect that $U = 10$ is an upper bound; that is, $c_n \leq 10$. This is true for $n = 1$. Suppose $c_k \leq 10$ for some integer k. Then $c_{k+1} = 1 + \sqrt{6 + c_k} \leq 1 + \sqrt{6 + 10} = 5 < 10$. Hence, by mathematical induction, $c_n \leq 10$ for all n.

Theorem 10.7 now implies that the sequence has a limit, call it L. By taking limits on both sides of the recursive definition of the sequence we obtain

$$\lim_{n \to \infty} c_{n+1} = \lim_{n \to \infty} \left[1 + \sqrt{6 + c_n} \right].$$

It follows that L must satisfy $L = 1 + \sqrt{6 + L} \implies (L - 1)^2 = 6 + L$. This equation simplifies to $L^2 - 3L - 5 = 0$, and of the two solutions $(3 \pm \sqrt{29})/2$ of this equation, only $(3 + \sqrt{29})/2$ lies between the bounds. Hence, $L = (3 + \sqrt{29})/2$.

32. The first four terms of the sequence are $c_1 = 4$, $c_2 = 3$, $c_3 = 2.5858$, $c_4 = 2.4462$. Previous exercises have indicated that when it is anticipated that a sequence is monotonic and bounded, it is advantageous to first verify monotony. If a sequence is known to be increasing (or nondecreasing), then a lower bound must be the first term of the sequence. On the other hand, if a sequence is known to be decreasing (or nonincreasing), its first term is an upper bound. Unfortunately, it is not always possible to verify monotony without knowledge of bounds. This is such an example. Try to prove that this sequence is decreasing before reading the rest of this solution, and discover why the proof fails. In addition, to guarantee that all terms of the sequence are well-defined, we must know that no one of them can be greater than 5.

We begin by proving that upper and lower bounds are 4 and 0; that is, $0 \leq c_n \leq 4$. This is certainly true for $n = 1$. Suppose $0 \leq c_k \leq 4$ for some integer k. Then $0 \geq -c_k \geq -4$, from which $5 \geq 5 - c_k \geq 1$. It follows that $\sqrt{5} \geq \sqrt{5 - c_k} \geq 1$, and $-\sqrt{5} \leq -\sqrt{5 - c_k} \leq -1$. Thus, $4 - \sqrt{5} \leq 4 - \sqrt{5 - c_k} \leq 3$. This means that $0 < 4 - \sqrt{5} \leq c_{k+1} \leq 3 < 4$, and therefore by mathematical induction, $0 \leq c_n \leq 4$ for all n.

Now we verify that the sequence is decreasing, $c_{n+1} < c_n$. This is certainly true for $n = 1$ as $c_2 < c_1$. Suppose $c_{k+1} < c_k$ for some integer k. Then, $5 - c_{k+1} > 5 - c_k$. Because all terms of the sequence are between 0 and 4, both sides of this inequality are positive, and we can take square roots, $\sqrt{5 - c_{k+1}} > \sqrt{5 - c_k}$. Thus, $4 - \sqrt{5 - c_{k+1}} < 4 - \sqrt{5 - c_k}$, and this means that $c_{k+2} < c_{k+1}$. By mathematical induction, $c_{n+1} < c_n$ for all $n \geq 1$.

Theorem 10.7 now implies that the sequence has a limit, call it L. By taking limits on both sides of the recursive definition of the sequence we obtain

$$\lim_{n \to \infty} c_{n+1} = \lim_{n \to \infty} \left[4 - \sqrt{5 - c_n} \right].$$

It follows that L must satisfy $L = 4 - \sqrt{5 - L} \implies (L-4)^2 = 5 - L$. This equation simplifies to $L^2 - 7L + 11 = 0$, and of the two solutions $(7 \pm \sqrt{5})/2$, only $(7 - \sqrt{5})/2$ lies between the bounds. Hence, $L = (7 - \sqrt{5})/2$.

34. The first four terms of the sequence are $c_1 = 1$, $c_2 = 1/2$, $c_3 = 1/3$, $c_4 = 3/10$. An initial attempt at proving that this sequence is decreasing fails. We require information about its bounds. Consider proving that upper and lower bounds are 1 and 0; that is, $0 \le c_n \le 1$. This is certainly true for $n = 1$. Suppose $0 \le c_k \le 1$ for some integer k. Then, $0 \ge -2c_k \ge -2$, from which $4 \ge 4 - 2c_k \ge 2$. Inversion gives $1/4 \le 1/(4 - 2c_k) \le 1/2$, but this implies that $0 < 1/4 \le c_{k+1} \le 1/2 < 1$. By mathematical induction then, $0 \le c_n \le 1$ for all n. Now we verify that the sequence is decreasing, that its terms satisfy $c_{n+1} < c_n$. This is true for $n = 1$. Suppose $c_{k+1} < c_k$ for some integer k. Then $-2c_{k+1} > -2c_k$, and, $4 - 2c_{k+1} > 4 - 2c_k$. Because all terms of the sequence are between 0 and 1, both expressions in this inequality are positive. We may therefore invert and write $1/(4 - 2c_{k+1}) < 1/(4 - 2c_k)$; i.e., $c_{k+2} < c_{k+1}$. By mathematical induction then, $c_{n+1} < c_n$ for all n.

Because the sequence is monotonic and bounded, Theorem 10.7 guarantees that it has a limit L. By writing

$$\lim_{n \to \infty} c_{n+1} = \lim_{n \to \infty} \frac{1}{4 - 2c_n},$$

we obtain the equation $L = 1/(4 - 2L)$. Of the two solutions $1 \pm 1/\sqrt{2}$ of this equation, only $1 - 1/\sqrt{2}$ is between the bounds. Hence, $L = 1 - 1/\sqrt{2}$.

36. The first four terms of the sequence are $c_1 = 0$, $c_2 = 7/16$, $c_3 = 0.443$, $c_4 = 0.485$. An initial attempt at proving that this sequence is increasing fails. We require information about its bounds. We prove first therefore that $0 \le c_n \le 1$. This is true for $n = 1$. Suppose $0 \le c_k \le 1$ for some integer k. Then, $0 \ge -8c_k{}^2 \ge -8$, and $16 \ge 16 - 8c_k{}^2 \ge 8$. Inversion gives $1/16 \le 1/(16 - 8c_k{}^2) \le 1/8$. But then $7/16 \le 7/(16 - 8c_k{}^2) \le 7/8$, or, $0 < 7/16 \le c_{k+1} \le 7/8 < 1$. By mathematical induction then, $0 \le c_n \le 1$, and we have upper and lower bounds. Now we verify that the sequence is increasing, that its terms satisfy $c_{n+1} > c_n$. This is true for $n = 1$. Suppose $c_{k+1} > c_k$ for some integer k. Then $-8c_{k+1}{}^2 < -8c_k{}^2$, and, $16 - 8c_{k+1}{}^2 < 16 - 8c_k{}^2$. Because all terms of the sequence are between 0 and 1, both expressions in this inequality are positive. We may therefore invert and write $1/(16 - 8c_{k+1}{}^2) > 1/(16 - 8c_k{}^2)$. In other words, $7/(16 - 8c_{k+1}{}^2) > 7/(16 - 8c_k{}^2)$, or, $c_{k+2} > c_{k+1}$. By mathematical induction then, $c_{n+1} > c_n$ for all n.

Because the sequence is monotonic and bounded, Theorem 10.7 guarantees that it has a limit L. By writing

$$\lim_{n \to \infty} c_{n+1} = \lim_{n \to \infty} \frac{7}{16 - 8c_n{}^2},$$

we obtain the equation $L = 7/(16 - 8L^2) \implies 0 = 8L^3 - 16L + 7 = (2L - 1)(4L^2 + 2L - 7)$. Of the three solutions to this equation, only $1/2$ is between the bounds. Hence, $L = 1/2$.

38. It is advantageous to express the recursive definition in the form $c_{n+1} = \dfrac{3}{1 + \dfrac{2}{c_n}}$.

The first four terms of the sequence are $c_1 = 4$, $c_2 = 2$, $c_3 = 3/2$, $c_4 = 9/7$. The sequence appears to be decreasing, $c_{n+1} < c_n$. This is true for $n = 1$. Suppose $c_{k+1} < c_k$ for some integer k. Then,

$$\frac{2}{c_{k+1}} > \frac{2}{c_k} \implies 1 + \frac{2}{c_{k+1}} > 1 + \frac{2}{c_k}.$$

Thus, $\qquad \dfrac{1}{1 + \dfrac{2}{c_{k+1}}} < \dfrac{1}{1 + \dfrac{2}{c_k}} \implies \dfrac{3}{1 + \dfrac{2}{c_{k+1}}} < \dfrac{3}{1 + \dfrac{2}{c_k}};\qquad$ that is, $\qquad c_{k+2} < c_{k+1}$.

By mathematical induction, $c_{n+1} < c_n$ for all n. It follows that $U = 1$ is an upper bound, and because no term can be negative, $V = 0$. By Theorem 10.7, the sequence has a limit L that we can obtain by solving the equation $L = 3L/(2 + L)$. Of the two solutions 0 and 1 of this equation, we must choose $L = 0$.

40. It is advantageous to express the recursive definition in the form $c_{n+1} = -1 + \dfrac{6}{4 - c_n}$.

The first four terms of the sequence are $c_1 = 3/2$, $c_2 = 7/5$, $c_3 = 17/13$, $c_4 = 43/35$. It would be prudent to first verify that the sequence is decreasing in which case an upper bound would be immediate. Unfortunately, information on bounds is required to complete the proof. We therefore begin by proving that $1 \le c_n \le 2$. This is certainly true for $n = 1$. Suppose $1 \le c_k \le 2$ for some integer k. Then, $-1 \ge -c_k \ge -2$, from which $3 \ge 4 - c_k \ge 2$. Thus, $1/3 \le 1/(4 - c_k) \le 1/2$, and $2 \le 6/(4 - c_k) \le 3$. Hence, $1 \le -1 + 6/(4 - c_k) \le 2$; that is, $1 \le c_{k+1} \le 2$. By mathematical induction then, $1 \le c_n \le 2$ for all n. Now we verify that $c_{n+1} < c_n$. This is true for $n = 1$. Suppose $c_{k+1} < c_k$ for some integer k. Then $-c_{k+1} > -c_k$, and, $4 - c_{k+1} > 4 - c_k$. Since both sides are positive, we may invert,

$$\frac{1}{4 - c_{k+1}} < \frac{1}{4 - c_k} \implies -1 + \frac{6}{4 - c_{k+1}} < -1 + \frac{6}{4 - c_k}.$$

Thus, $c_{k+2} < c_{k+1}$, and by mathematical induction, $c_{n+1} < c_n$ for all n. Because the sequence is monotonic and bounded, Theorem 10.7 guarantees that it has a limit L that must satisfy the equation $L = (L + 2)/(4 - L)$. This equation reduces to $L^2 - 3L + 2 = 0$, and of the two solutions $L = 1$ and $L = 2$ only $L = 1$ could be the limit.

42. First we verify that 0 and 1 are bounds for the sequence; that is, $0 \le c_n \le 1$. This is true for $n = 1$. Suppose that $0 \le c_k \le 1$. Then, $0 \ge -c_k \ge -1$ and $0 \ge -c_k^2 \ge -1$. When we add these, $0 \ge -c_k - c_k^2 \ge -2$, and if we add 4, $4 \ge 4 - c_k - c_k^2 \ge 2$. Inverting this gives $1/4 \le 1/(4 - c_k - c_k^2) \le 1/2$. Hence, $0 < 1/4 \le c_{k+1} \le 1/2 < 1$. Consequently, by mathematical induction, $0 \le c_n \le 1$ for all $n \ge 1$.

We now verify that the sequence is decreasing by showing that $c_{n+1} < c_n$. This is true for $n = 1$ since $c_2 = 1/2 < c_1$. Suppose that $c_{k+1} < c_k$. Then $-c_{k+1} > -c_k$ and $-c_{k+1}^2 > -c_k^2$. When we add these, $-c_{k+1} - c_{k+1}^2 > -c_k - c_k^2$, and if we add 4, $4 - c_{k+1} - c_{k+1}^2 > 4 - c_k - c_k^2$. Because both sides of this inequality are positive, we may invert and reverse the sign, $1/(4 - c_{k+1} - c_{k+1}^2) < 1/(4 - c_k - c_k^2)$; that is, $c_{k+2} < c_{k+1}$. By mathematical induction, then, $c_{n+1} < c_n$ for all $n \ge 1$.

Because the sequence is monotonic and bounded, Theorem 10.7 guarantees that it has a limit that must satisfy the equation $L = 1/(4 - L - L^2)$. This equation reduces to $f(L) = L^3 + L^2 - 4L + 1 = 0$. The first nine terms of the sequence are

$$x_1 = 1 \qquad x_2 = 1/2 \qquad x_3 = 4/13 \qquad x_4 = 0.278 \qquad x_5 = 0.274$$
$$x_6 = 0.273\,903 \qquad x_7 = 0.273\,892 \qquad x_8 = 0.273\,891 \qquad x_9 = 0.273\,891.$$

Since $f(0.273\,885) = 1.8 \times 10^{-5}$ and $f(0.273\,895) = -1.4 \times 10^{-5}$, the limit is $0.273\,89$ (to 5 decimals).

44. The first four terms of the sequence are $c_1 = 2$, $c_2 = 6/5 = 1.2$, $c_3 = 1.36$, and $c_4 = 1.328$. They are oscillating. To show that the entire sequence oscillates, we calculate

$$c_{n+1} - c_n = \frac{8 - c_n}{5} - \frac{8 - c_{n-1}}{5} = -\frac{1}{5}(c_n - c_{n-1}).$$

This shows that the differences $c_{n+1} - c_n$ alternate in sign, and therefore the sequence oscillates. Because $|c_{n+1} - c_n| = |c_n - c_{n-1}|/5$, absolute values $|c_{n+1} - c_n|$ decrease and approach 0. According to Theorem 10.8 the sequence has a limit L. We find L by taking limits on both sides of the recursive definition,

$$\lim_{n \to \infty} c_{n+1} = \lim_{n \to \infty} \frac{8 - c_n}{5} \implies L = \frac{8 - L}{5} \implies L = \frac{4}{3}.$$

46. The first four terms of the sequence are $c_1 = 1$, $c_2 = 1/3$, $c_3 = 3/7$, and $c_4 = 7/17$. They are oscillating. To show that the entire sequence oscillates, we calculate

$$c_{n+1} - c_n = \frac{1}{2 + c_n} - \frac{1}{2 + c_{n-1}} = -\frac{(c_n - c_{n-1})}{(2 + c_n)(2 + c_{n-1})}.$$

Since all terms of the sequence are positive, the denominator of this expression is positive. It follows that $c_{n+1} - c_n$ has the opposite sign of $c_n - c_{n-1}$, and the sequence oscillates. To verify properties 2 and 3 of Theorem 10.8, we take absolute values in the above equation,

$$|c_{n+1} - c_n| = \frac{|c_n - c_{n-1}|}{(2 + c_n)(2 + c_{n-1})}.$$

Since all terms of the sequence are positive, it follows that

$$|c_{n+1} - c_n| < \frac{|c_n - c_{n-1}|}{(2)(2)} = \frac{1}{4}|c_n - c_{n-1}|.$$

This shows that the $|c_{n+1} - c_n|$ decrease and have limit 0. By Theorem 10.8 the sequence has a limit L that can be obtained by taking limits on both sides of the recursive definition,

$$\lim_{n \to \infty} c_{n+1} = \lim_{n \to \infty} \frac{1}{2 + c_n} \quad \Longrightarrow \quad L = \frac{1}{2 + L}.$$

Of the two solutions $-1 \pm \sqrt{2}$ of this equation, only $L = -1 + \sqrt{2}$ is positive.

48. The inequality $1 \leq c_n \leq 2$ is valid for $n = 1$. Suppose $1 \leq c_k \leq 2$. Then, $2 \leq 1 + c_k \leq 3$, and inverting, $1/2 \geq 1/(1 + c_k) \geq 1/3$. Multiplication by 3 gives $3/2 \geq 3/(1 + c_k) \geq 1$, and this states that $1 \leq c_{k+1} \leq 3/2 < 2$. Hence, by mathematical induction, $1 \leq c_n \leq 2$ for all $n \geq 1$.

The first four terms of the sequence are $c_1 = 1$, $c_2 = 3/2$, $c_3 = 6/5$, and $c_4 = 15/11$. They are oscillating. To show that the entire sequence oscillates, we calculate

$$c_{n+1} - c_n = \frac{3}{1 + c_n} - \frac{3}{1 + c_{n-1}} = -\frac{3(c_n - c_{n-1})}{(1 + c_n)(1 + c_{n-1})}.$$

Since all terms of the sequence are positive, the denominator of this expression is positive. It follows that $c_{n+1} - c_n$ has the opposite sign of $c_n - c_{n-1}$, and the sequence oscillates. To verify properties 2 and 3 of Theorem 10.8, we take absolute values in the above equation,

$$|c_{n+1} - c_n| = \frac{3|c_n - c_{n-1}|}{(1 + c_n)(1 + c_{n-1})} < \frac{3|c_n - c_{n-1}|}{(1 + 1)(1 + 1)} = \frac{3}{4}|c_n - c_{n-1}|.$$

This shows that the $|c_{n+1} - c_n|$ decrease and have limit 0. By Theorem 10.8 the sequence has a limit L that can be obtained by taking limits on both sides of the recursive definition,

$$\lim_{n \to \infty} c_{n+1} = \lim_{n \to \infty} \frac{3}{1 + c_n} \quad \Longrightarrow \quad L = \frac{3}{1 + L}.$$

Of the two solutions $(-1 \pm \sqrt{13})/2$ of this equation, only $L = (-1 + \sqrt{13})/2$ lies between 1 and 2.

50. The inequality $1/2 \leq c_n \leq 1$ is valid for $n = 1$. Suppose $1/2 \leq c_k \leq 1$. Then, $5/2 \leq 5c_k \leq 5 \Longrightarrow 9/2 \leq 2 + 5c_k \leq 7$, and inverting, $2/9 \geq 1/(2 + 5c_k) \geq 1/7$. Multiplication by 4 gives $8/9 \geq 4/(2 + 5c_k) \geq 4/7$, and this states that $1/2 < 4/7 \leq c_{k+1} \leq 8/9 < 1$. Hence, by mathematical induction, $1/2 \leq c_n \leq 1$ for all $n \geq 1$.

The first four terms of the sequence are $c_1 = 1$, $c_2 = 4/7$, $c_3 = 14/17$, and $c_4 = 17/26$. They are oscillating. To show that the entire sequence oscillates, we calculate

$$c_{n+1} - c_n = \frac{4}{2 + 5c_n} - \frac{4}{2 + 5c_{n-1}} = -\frac{20(c_n - c_{n-1})}{(2 + 5c_n)(2 + 5c_{n-1})}.$$

Since all terms of the sequence are positive, the denominator of this expression is positive. It follows that $c_{n+1} - c_n$ has the opposite sign of $c_n - c_{n-1}$, and the sequence oscillates. To verify properties 2 and 3 of Theorem 10.8, we take absolute values in the above equation,

$$|c_{n+1} - c_n| = \frac{20|c_n - c_{n-1}|}{(2 + 5c_n)(2 + 5c_{n-1})} < \frac{20|c_n - c_{n-1}|}{(2 + 5/2)(2 + 5/2)} = \frac{80}{81}|c_n - c_{n-1}|.$$

This shows that the $|c_{n+1} - c_n|$ decrease and have limit 0. By Theorem 10.8 the sequence has a limit L that can be obtained by taking limits on both sides of the recursive definition,

$$\lim_{n \to \infty} c_{n+1} = \lim_{n \to \infty} \frac{4}{2 + 5c_n} \quad \Longrightarrow \quad L = \frac{4}{2 + 5L}.$$

Of the two solutions $(-1 \pm \sqrt{21})/5$ of this equation, only $L = (-1 + \sqrt{21})/5$ is positive.

52. The first four terms of the sequence are $c_1 = 4$, $c_2 = 2.83$, $c_3 = 3.39$, and $c_4 = 3.13$. They are oscillating.

First we prove that $2 \le c_n \le 4$ which is valid for $n = 1$. Suppose $2 \le c_k \le 4$. Then, $-2 \ge -c_k \ge -4$, and multiplying by 3, $-6 \ge -3c_k \ge -12$. Adding 20 gives $14 \ge 20 - 3c_k \ge 8$. When we take square roots, $\sqrt{14} \ge \sqrt{20 - 3c_k} \ge \sqrt{8}$, and this states that $2 < \sqrt{8} \le c_{k+1} \le \sqrt{14} < 4$. Hence, by mathematical induction, $2 \le c_n \le 4$ for all $n \ge 1$. This shows that all terms of the sequence are defined. Consider now

$$(c_{n+1})^2 - (c_n)^2 = (20 - 3c_n) - (20 - 3c_{n-1}) = -3(c_n - c_{n-1}).$$

When we factor the left side into $(c_{n+1} + c_n)(c_{n+1} - c_n)$, the above equation can be rewritten in the form

$$c_{n+1} - c_n = \frac{-3(c_n - c_{n-1})}{c_{n+1} + c_n}.$$

Since all terms of the sequence are positive, the denominator of this expression is positive. It follows that $c_{n+1} - c_n$ has the opposite sign of $c_n - c_{n-1}$, and the sequence oscillates. To verify properties 2 and 3 of Theorem 10.8, we take absolute values in the above equation,

$$|c_{n+1} - c_n| = \frac{3|c_n - c_{n-1}|}{c_{n+1} + c_n} \le \frac{3|c_n - c_{n-1}|}{2 + 2} = \frac{3}{4}|c_n - c_{n-1}|.$$

This shows that the $|c_{n+1} - c_n|$ decrease and have limit 0. By Theorem 10.8 the sequence has a limit L that can be obtained by taking limits on both sides of the recursive definition,

$$\lim_{n \to \infty} c_{n+1} = \lim_{n \to \infty} \sqrt{20 - 3c_n} \quad \Longrightarrow \quad L = \sqrt{20 - 3L}.$$

Of the two solutions $(-3 \pm \sqrt{89})/2$ of this equation, only $L = (-3 + \sqrt{89})/2$ is positive.

54. The third term is $\quad c_3 = \dfrac{c_2}{4c_2 - 1} = \dfrac{\dfrac{c_1}{4c_1 - 1}}{4\left(\dfrac{c_1}{4c_1 - 1}\right) - 1} = \left(\dfrac{c_1}{4c_1 - 1}\right)\left(\dfrac{4c_1 - 1}{4c_1 - 4c_1 + 1}\right) = c_1.$ In other

words, terms of the sequence oscillate back and forth between c_1 and c_2, never converging unless $c_1 = c_2$. This occurs only when $c_1 = 0$ or $c_1 = 1/2$.

56. The first four terms of the sequence are $c_1 = 3$, $c_2 = 1.913$, $c_3 = 2.007$, and $c_4 = 1.999$. They are oscillating. First we prove that $1 \le c_n \le 3$ which is valid for $n = 1$. Suppose $1 \le c_k \le 3$. Then, $-1 \ge -c_k \ge -3$, and adding 10 gives $9 \ge 10 - c_k \ge 7$. When we take cube roots, $\sqrt[3]{9} \ge \sqrt[3]{10 - c_k} \ge \sqrt[3]{7}$, and this states that $1 < \sqrt[3]{7} \le c_{k+1} \le \sqrt[3]{9} < 3$. Hence, by mathematical induction, $1 \le c_n \le 3$ for all $n \ge 1$. Consider now

$$(c_{n+1})^3 - (c_n)^3 = (10 - c_n) - (10 - c_{n-1}) = -(c_n - c_{n-1}).$$

When we factor the left side into $(c_{n+1} - c_n)(c_{n+1}^2 + c_{n+1}c_n + c_n^2)$, the above equation can be rewritten in the form

$$c_{n+1} - c_n = \frac{-(c_n - c_{n-1})}{c_{n+1}^2 + c_{n+1}c_n + c_n^2}.$$

Since all terms of the sequence are positive, the denominator of this expression is positive. It follows that $c_{n+1} - c_n$ has the opposite sign of $c_n - c_{n-1}$, and the sequence oscillates. To verify properties 2 and 3 of Theorem 10.8, we take absolute values in the above equation,

$$|c_{n+1} - c_n| = \frac{|c_n - c_{n-1}|}{c_{n+1}^2 + c_{n+1}c_n + c_n^2} \le \frac{|c_n - c_{n-1}|}{1 + 1 + 1} = \frac{1}{3}|c_n - c_{n-1}|.$$

This shows that the $|c_{n+1} - c_n|$ decrease and have limit 0. By Theorem 10.8 the sequence has a limit L that can be obtained by taking limits on both sides of the recursive definition,

$$\lim_{n \to \infty} c_{n+1} = \lim_{n \to \infty} \sqrt[3]{10 - c_n} \quad \Longrightarrow \quad L = \sqrt[3]{10 - L} \quad \Longrightarrow \quad L = 2.$$

58. (i) Suppose $\epsilon > 0$ is any given number. Since $\lim_{n \to \infty} c_n = C$, there exists an N such that $|c_n - C| < \epsilon/|k|$ for all $n > N$. For such n,

$$|kc_n - kC| = |k||c_n - C| < |k| \left(\frac{\epsilon}{|k|} \right) = \epsilon.$$

This proves that $\lim_{n \to \infty} kc_n = kC$.

(ii) Suppose $\epsilon > 0$ is any given number. Since $\lim_{n \to \infty} c_n = C$, there exists an N_1 such that $|c_n - C| < \epsilon/2$ for all $n > N_1$. Similarly, since $\lim_{n \to \infty} d_n = D$, there exists an N_2 such that $|d_n - D| < \epsilon/2$ for all $n > N_2$. For all n greater than the larger of N_1 and N_2,

$$|(c_n + d_n) - (C + D)| = |(c_n - C) + (d_n - D)| \leq |c_n - C| + |d_n - D| < \frac{\epsilon}{2} + \frac{\epsilon}{2} = \epsilon.$$

This proves that $\lim_{n \to \infty} (c_n + d_n) = C + D$. The proof that $\lim_{n \to \infty} (c_n - d_n) = C - D$ is similar.

60. The first four terms of the sequence are $c_1 = -30$, $c_2 = -20$, $c_3 = -15$, $c_4 = -55/6$. The sequence appears to be increasing, $c_{n+1} > c_n$. This is true for $n = 1, 2$. Suppose that $c_k > c_{k-1}$ and $c_{k+1} > c_k$. Then, $c_k/3 > c_{k-1}/3$ and $c_{k+1}/2 > c_k/2$. Addition of these gives

$$\frac{c_{k+1}}{2} + \frac{c_k}{3} > \frac{c_k}{2} + \frac{c_{k-1}}{3} \quad \Longrightarrow \quad 5 + \frac{c_{k+1}}{2} + \frac{c_k}{3} > 5 + \frac{c_k}{2} + \frac{c_{k-1}}{3}.$$

This states that $c_{k+2} > c_{k+1}$, and therefore, by mathematical induction, the sequence is increasing. The first term is therefore a lower bound. We now prove that 30 is an upper bound, $c_n \leq 30$. Certainly c_1 and c_2 are both less than 30. Suppose that $c_{k-1} \leq 30$ and $c_k \leq 30$. Then

$$c_{k+1} = 5 + \frac{c_k}{2} + \frac{c_{k-1}}{3} \leq 5 + \frac{30}{2} + \frac{30}{3} = 30.$$

Hence, by mathematical induction, $c_n \leq 30$ for all $n \geq 1$. According to Theorem 10.7, the sequence has a limit that can be obtained by taking limits on both sides of the recursive definition,

$$\lim_{n \to \infty} c_{n+1} = \lim_{n \to \infty} \left(5 + \frac{c_n}{2} + \frac{c_{n-1}}{3} \right) \quad \Longrightarrow \quad L = 5 + \frac{L}{2} + \frac{L}{3} \quad \Longrightarrow \quad L = 30.$$

62. If $\lim_{n \to \infty} c_n = L$, then given $\epsilon = 1$, there exists an integer N such that for all $n > N$, $|c_n - L| < \epsilon = 1$. This is equivalent to $-1 < c_n - L < 1 \Longrightarrow L - 1 < c_n < L + 1$.

64. We concentrate only on female rabbits. After the first month, we have the original adult female and a newborn female. Hence $R_1 = 1$. After the second month, we have the original adult female, a one-month old female, and a newborn female. Hence, $R_2 = 1$. If R_n is the number of adult females after n months, then R_{n+1} is the sum of R_n and the number of one-month old females after n months. But the number of one-month old females after n months is the number of adult females R_{n-1} after $n - 1$ months. In other words, a recursive definition for the sequence is $R_1 = 1$, $R_2 = 1$, $R_{n+1} = R_n + R_{n-1}$. This is the Fibonacci sequence.

EXERCISES 10.9

2. $\displaystyle\sum_{n=1}^{\infty} \frac{2^n}{5^{n+1}} = \frac{1}{5} \sum_{n=1}^{\infty} \left(\frac{2}{5} \right)^n$, a geometric series with sum $\dfrac{1}{5} \left(\dfrac{2/5}{1 - 2/5} \right) = \dfrac{2}{15}$.

4. Since $\displaystyle\lim_{n \to \infty} \left(\frac{n}{n+1} \right)^n = \frac{1}{e}$ (see expression 1.68), the series diverges by the nth term test.

6. $\displaystyle\sum_{n=1}^{\infty} \frac{7^{n+3}}{3^{2n-2}} = \frac{7^3}{3^{-2}} \sum_{n=1}^{\infty} \left(\frac{7}{9} \right)^n$ is a geometric series with sum $7^3(3)^2 \left(\dfrac{7/9}{1 - 7/9} \right) = \dfrac{21\,609}{2}$.

8. $\displaystyle\sum_{n=1}^{\infty} \frac{\cos n\pi}{2^n} = \sum_{n=1}^{\infty} \frac{(-1)^n}{2^n} = \sum_{n=1}^{\infty} \left(-\frac{1}{2} \right)^n$ is a geometric series with sum $\dfrac{-1/2}{1 + 1/2} = -\dfrac{1}{3}$.

10. Since $\lim_{n\to\infty} \text{Tan}^{-1}n = \frac{\pi}{2}$, the series diverges by the nth term test.

12. $0.131\,313\,131\ldots = 0.13 + 0.001\,3 + 0.000\,013 + \cdots = \dfrac{13}{100} + \dfrac{13}{10\,000} + \dfrac{13}{1\,000\,000} + \cdots$

$$= \frac{13/100}{1 - 1/100} = \frac{13}{99}$$

14. $43.020\,502\,050\,205\ldots = 43 + 0.020\,5 + 0.000\,002\,05 + \cdots = 43 + \dfrac{205}{10^4} + \dfrac{205}{10^8} + \cdots$

$$= 43 + \frac{205/10^4}{1 - 1/10^4} = \frac{430\,162}{9999}$$

16. If $\sum c_n$ converges and $\sum d_n$ diverges, then $\sum (c_n + d_n)$ diverges.
Proof: Assume to the contrary that $\sum (c_n + d_n)$ converges. Let $\{C_n\}$ and $\{D_n\}$ be the sequences of partial sums for $\sum c_n$ and $\sum d_n$. It follows that $\lim_{n\to\infty} C_n$ exists, call it C, but $\lim_{n\to\infty} D_n$ does not exist. $\{C_n + D_n\}$ is the sequence of partial sums for $\sum (c_n + d_n)$, and by assumption, it has a limit, call it E. But then according to part (ii) of Theorem 10.10, the sequence $\{(C_n + D_n) - C_n\} = \{D_n\}$ must have limit $E - C$, a contradiction. Consequently, our assumption that $\sum (c_n + d_n)$ converges must be incorrect.

18. Since $\displaystyle\sum_{n=1}^{\infty} \frac{2^n}{4^n}$ and $\displaystyle\sum_{n=1}^{\infty} \frac{3^n}{4^n}$ are both geometric series with sums

$$\sum_{n=1}^{\infty} \frac{2^n}{4^n} = \frac{1/2}{1 - 1/2} = 1 \qquad \text{and} \qquad \sum_{n=1}^{\infty} \frac{3^n}{4^n} = \frac{3/4}{1 - 3/4} = 3,$$

then, by Exercise 15, $\displaystyle\sum_{n=1}^{\infty} \frac{2^n + 3^n}{4^n} = 1 + 3 = 4.$

20. Since $\displaystyle\lim_{n\to\infty} \frac{n^2 + 2^{2n}}{4^n} = \lim_{n\to\infty} \left(\frac{n^2}{4^n} + 1 \right) = 1$, the series diverges by the nth term test.

22. Since $\dfrac{1}{n(n+1)} = \dfrac{1}{n} - \dfrac{1}{n+1}$, the nth partial sum of the series is

$$S_n = \frac{1}{1 \cdot 2} + \frac{1}{2 \cdot 3} + \cdots + \frac{1}{n(n+1)} = \left(1 - \frac{1}{2}\right) + \left(\frac{1}{2} - \frac{1}{3}\right) + \left(\frac{1}{3} - \frac{1}{4}\right) + \cdots + \left(\frac{1}{n} - \frac{1}{n+1}\right)$$

$$= 1 - \frac{1}{n+1} = \frac{n}{n+1}.$$

Since $\lim_{n\to\infty} S_n = 1$, it follows that $\displaystyle\sum_{n=1}^{\infty} \frac{1}{n(n+1)} = 1.$

24. The total time taken to come to rest is

$$\sqrt{\frac{40}{9.81}} + t_1 + t_2 + t_3 + \cdots = \sqrt{\frac{40}{9.81}} + \sum_{n=1}^{\infty} \frac{4}{\sqrt{0.981}}(0.99)^{n/2} = \sqrt{\frac{40}{9.81}} + \frac{4\sqrt{0.99}/\sqrt{0.981}}{1 - \sqrt{0.99}} = 804 \text{ s.}$$

26. According to Exercise 10.1–61,

$$A_n = \frac{\sqrt{3}P^2}{36}\left(1 + \frac{1}{3} + \frac{4}{3^3} + \frac{4^2}{3^5} + \cdots + \frac{4^{n-1}}{3^{2n-1}}\right) \qquad \text{(a finite geometric series after first term)}$$

$$= \frac{\sqrt{3}P^2}{36}\left\{1 + \frac{(1/3)[1 - (4/9)^n]}{1 - 4/9}\right\} \qquad \text{(using 10.39a)}$$

$$= \frac{\sqrt{3}P^2}{180}\left[8 - 3\left(\frac{4}{9}\right)^n\right].$$

Thus, $\displaystyle\lim_{n\to\infty} A_n = \frac{\sqrt{3}P^2}{180}(8) = \frac{2\sqrt{3}P^2}{45}.$

28. (a) If we subtract $S_n = 1 + r + r^2 + \cdots + r^{n-1}$ from $T_n = 1 + 2r + 3r^2 + \cdots + nr^{n-1}$, we obtain

$$T_n - S_n = r + 2r^2 + 3r^3 + \cdots + (n-1)r^{n-1} = r[1 + 2r + 3r^2 + \cdots + (n-1)r^{n-2}] = r(T_n - nr^{n-1}).$$

When we solve this for T_n and substitute for S_n,

$$T_n = \frac{S_n - nr^n}{1 - r} = \frac{\frac{1 - r^n}{1 - r} - nr^n}{1 - r} = \frac{1 - r^n - nr^n + nr^{n+1}}{(1 - r)^2} = \frac{1 - (n+1)r^n + nr^{n+1}}{(1 - r)^2}.$$

If we now take limits as $n \to \infty$, we obtain

$$\sum_{n=1}^{\infty} nr^{n-1} = \lim_{n \to \infty} \frac{1 - (n+1)r^n + nr^{n+1}}{(1 - r)^2} = \frac{1}{(1 - r)^2}, \quad \text{provided } |r| < 1.$$

(b) If we set $S(r) = \sum_{n=1}^{\infty} nr^{n-1}$, and integrate with respect to r,

$$\int S(r)\, dr + C = \sum_{n=1}^{\infty} r^n = \frac{r}{1 - r}.$$

Differentiation now gives $S(r) = \dfrac{(1 - r)(1) - r(-1)}{(1 - r)^2} = \dfrac{1}{(1 - r)^2}.$

30. $\dfrac{2}{5} + \dfrac{4}{25} + \dfrac{6}{125} + \dfrac{8}{625} + \cdots = \dfrac{2}{5}\left(1 + \dfrac{2}{5} + \dfrac{3}{5^2} + \dfrac{4}{5^3} + \cdots\right) = \dfrac{2/5}{(1 - 1/5)^2} = \dfrac{5}{8}$

32. $\dfrac{12}{5} + \dfrac{48}{25} + \dfrac{192}{125} + \dfrac{768}{625} + \cdots = \dfrac{12}{5}\left(1 + \dfrac{4}{5} + \dfrac{16}{25} + \dfrac{64}{125} + \cdots\right) = \dfrac{12/5}{1 - 4/5} = 12$

34. The probability that the first person wins on the first toss is 1/6. The probability that the first person wins on the second toss is the product of the following three probabilities:

probability that first person does not throw a six on the first toss $= 5/6$;
probability that second person does not throw a six on first toss $= 5/6$;
probability that first person throws a six on second toss $= 1/6$.

The resultant probabilty is $(5/6)(5/6)(1/6) = 5^2/6^3$. The probability that the first person wins on the third toss is the product of the following five probabilities:

probability that first person does not throw a six on the first toss $= 5/6$;
probability that second person does not throw a six on first toss $= 5/6$;
probability that first person does not throw a six on second toss $= 5/6$.
probability that second person does not throw a six on the second toss $= 5/6$;
probability that first person throws a six on third toss $= 1/6$;

The resultant probabilty is $5^4/6^5$.
Continuation of this process leads to the following probability that the first person to toss wins

$$\frac{1}{6} + \frac{5^2}{6^3} + \frac{5^4}{6^5} + \frac{5^6}{6^7} + \cdots = \frac{1/6}{1 - 25/36} = \frac{6}{11}.$$

36. Since the radius of convergence of the series is $R = \lim\limits_{n \to \infty} \left| \dfrac{n^2 3^n}{(n+1)^2 3^{n+1}} \right| = 1/3$, the open interval of convergence is $-1/3 < x < 1/3$. At $x = 1/3$, the power series reduces to $\sum_{n=1}^{\infty} n^2$ which diverges by the nth term test. At $x = -1/3$, it reduces to $\sum_{n=1}^{\infty} (-1)^n n^2$ which also diverges by the nth term test. The interval of convergence for the series is therefore $-1/3 < x < 1/3$.

38. If we set $y = x^3$, the series becomes $\sum_{n=0}^{\infty} (-1)^n x^{3n} = \sum_{n=0}^{\infty} (-1)^n y^n$. Since the radius of convergence of this series is $R_y = \lim_{n \to \infty} \left| \dfrac{(-1)^n}{(-1)^{n+1}} \right| = 1$, the radius of convergence of the given series is $R_x = 1$. The open interval of convergence is $-1 < x < 1$. At $x = 1$, the power series reduces to $\sum_{n=0}^{\infty} (-1)^n$ which diverges by the nth term test. At $x = -1$, it reduces to $\sum_{n=0}^{\infty} 1$ which also diverges by the nth term test. The interval of convergence for the series is therefore $-1 < x < 1$.

40. (a) The minute hand moves 12 times as fast as the hour hand. While the minute hand moves through the angle $\pi/6$ radians from 12 at 1:00 to 1 at 1:05, the hour hand moves a further $(\pi/6)/12$ radians. While the minute hand moves through this angle, the hour hand moves through a further angle $[(\pi/6)/12]/12 = (\pi/6)/12^2$. Continuation of this process leads to the following angle traveled by the minute hand in catching the hour hand

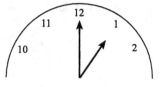

$$\frac{\pi}{6} + \frac{\pi/6}{12} + \frac{\pi/6}{12^2} + \frac{\pi/6}{12^3} + \cdots = \frac{\pi/6}{1 - 1/12} = \frac{2\pi}{11}.$$

This angle represents $\dfrac{2\pi}{11} \left(\dfrac{60}{2\pi} \right) = \dfrac{60}{11}$ minutes after 1:00.

(b) If we take time $t = 0$ at 1:00, the angle θ through which the minute hand moves in time t (in minutes) is $\theta = 2\pi t/60$. The angle ϕ that the hour hand makes with the vertical is $\phi = 2\pi t/720 + \pi/6$. These angles will be the same when $\dfrac{2\pi t}{60} = \dfrac{2\pi t}{720} + \dfrac{\pi}{6}$, the solution of which is $60/11$ minutes.

42. Suppose the length of each block is L. Taking the density of the blocks as unity, the mass of the top n blocks is nL^3. The first moment of the nth block about the y-axis is

$$L^3 \overline{x}_n = L^3 \left(\frac{L}{2} + \frac{L}{2n} \right) = \frac{L^4}{2} \left(1 + \frac{1}{n} \right).$$

The first moment of the $(n-1)^{\text{th}}$ block about the y-axis is

$$L^3 \overline{x}_{n-1} = L^3 \left[\frac{L}{2} + \frac{L}{2n} + \frac{L}{2(n-1)} \right]$$

$$= \frac{L^4}{2} \left(1 + \frac{1}{n} + \frac{1}{n-1} \right).$$

Continuing in this way, the moment of the first block about the y-axis is

$$L^3 \overline{x}_1 = L^3 \left[\frac{L}{2} + \frac{L}{2n} + \frac{L}{2(n-1)} + \cdots + \frac{L}{2} \right] = \frac{L^4}{2} \left(1 + \frac{1}{n} + \frac{1}{n-1} + \cdots + \frac{1}{2} + 1 \right).$$

The x-coordinate of the centre of mass of the top n blocks is therefore

$$\overline{x} = \frac{1}{nL^3} \left[\frac{L^4}{2} \left(1 + \frac{1}{n} \right) + \frac{L^4}{2} \left(1 + \frac{1}{n} + \frac{1}{n-1} \right) + \cdots + \frac{L^4}{2} \left(1 + \frac{1}{n} + \frac{1}{n-1} + \cdots + \frac{1}{2} + 1 \right) \right]$$

$$= \frac{L}{2n} \left[n(1) + n \left(\frac{1}{n} \right) + (n-1) \left(\frac{1}{n-1} \right) + \cdots + 2 \left(\frac{1}{2} \right) + 1(1) \right] = \frac{L}{2n}(2n) = L.$$

Thus, the centre of mass of the top n blocks is over the edge of the $(n+1)^{\text{th}}$ block. They will not tip, but they are in a state of precarious equilibrium.

The right edge of the top block sticks out the following distance over the right edge of the $(n+1)^{\text{th}}$ block

$$\frac{L}{2} + \frac{L}{4} + \frac{L}{6} + \cdots + \frac{L}{2n} = \frac{L}{2} \left(1 + \frac{1}{2} + \frac{1}{3} + \cdots + \frac{1}{n} \right).$$

This is $L/2$ times the nth partial sum of the harmonic series which we know becomes arbitrarily large as n increases. Hence, the top n blocks can be made to protrude arbitrarily far over the $(n+1)^{\text{th}}$ block.

EXERCISES 10.10

2. Since $l = \lim\limits_{n \to \infty} \dfrac{\dfrac{1}{4n-3}}{\dfrac{1}{4n}} = 1$, and $\sum\limits_{n=1}^{\infty} \dfrac{1}{4n} = \dfrac{1}{4} \sum\limits_{n=1}^{\infty} \dfrac{1}{n}$ diverges, so also does the given series (by the limit comparison test).

4. Since $l = \lim\limits_{n \to \infty} \dfrac{\dfrac{1}{5n^2-3n-1}}{\dfrac{1}{5n^2}} = 1$, and $\sum\limits_{n=1}^{\infty} \dfrac{1}{5n^2} = \dfrac{1}{5} \sum\limits_{n=1}^{\infty} \dfrac{1}{n^2}$ converges, so also does the given series (by the limit comparison test).

6. Since $l = \lim\limits_{n \to \infty} \dfrac{\dfrac{n^2}{n^4-6n^2+5}}{\dfrac{1}{n^2}} = 1$, and $\sum\limits_{n=4}^{\infty} \dfrac{1}{n^2}$ converges, so also does the given series (by the limit comparison test).

8. Since $l = \lim\limits_{n \to \infty} \dfrac{\dfrac{n-5}{n^2+3n-2}}{\dfrac{1}{n}} = 1$, and $\sum\limits_{n=6}^{\infty} \dfrac{1}{n}$ diverges, so also does the series $\sum\limits_{n=6}^{\infty} \dfrac{n-5}{n^2+3n-2}$ (by the limit comparison test). The given series therefore diverges also.

10. The function $f(x) = x^2 e^{-2x}$ is positive and continuous. Since $f'(x) = 2xe^{-2x} - 2x^2 e^{-2x} = 2xe^{-2x}(1-x)$, the function is decreasing for $x \geq 1$. Integrating by parts, and understanding that limits must be taken for the infinite limit,

$$\int_1^{\infty} x^2 e^{-2x}\, dx = \left\{\dfrac{x^2 e^{-2x}}{-2}\right\}_1^{\infty} - \int_1^{\infty} 2x\left(\dfrac{e^{-2x}}{-2}\right) dx = \dfrac{e^{-2}}{2} + \left\{\dfrac{xe^{-2x}}{-2}\right\}_1^{\infty} - \int_1^{\infty} \dfrac{e^{-2x}}{-2} dx$$

$$= \dfrac{e^{-2}}{2} + \dfrac{e^{-2}}{2} - \left\{\dfrac{e^{-2x}}{4}\right\}_1^{\infty} = \dfrac{5}{4e^2}.$$

Since the integral converges, so also does the series $\sum\limits_{n=1}^{\infty} n^2 e^{-2n}$ (by the integral test).

12. Since $l = \lim\limits_{n \to \infty} \dfrac{\dfrac{\sqrt{n+5}}{n^3+3}}{\dfrac{1}{n^{5/2}}} = 1$, and $\sum\limits_{n=1}^{\infty} \dfrac{1}{n^{5/2}}$ converges, so also does the given series (by the limit comparison test).

14. Since $\dfrac{1}{n^2 \ln n} < \dfrac{1}{n^2}$ for $n \geq 3$, and $\sum\limits_{n=3}^{\infty} \dfrac{1}{n^2}$ converges, so also does $\sum\limits_{n=3}^{\infty} \dfrac{1}{n^2 \ln n}$ (by the comparison test). The original series therefore converges also.

16. Since $l = \lim\limits_{n \to \infty} \dfrac{\dfrac{\sqrt{n^2+1}}{n^3}\mathrm{Tan}^{-1}n}{\dfrac{1}{n^2}\left(\dfrac{\pi}{2}\right)} = 1$, and $\sum\limits_{n=1}^{\infty} \dfrac{\pi}{2n^2} = \dfrac{\pi}{2} \sum\limits_{n=1}^{\infty} \dfrac{1}{n^2}$ converges, so also does the given series (by the limit comparison test).

18. Since $\dfrac{1+\ln^2 n}{n \ln^2 n} = \dfrac{1}{n \ln^2 n} + \dfrac{1}{n} > \dfrac{1}{n}$, and $\sum\limits_{n=2}^{\infty} \dfrac{1}{n}$ diverges, so also does the given series (by the comparison test).

20. Since $\dfrac{\ln(n+1)}{n+1} > \dfrac{1}{n+1}$ for $n \geq 2$, and $\displaystyle\sum_{n=2}^{\infty} \dfrac{1}{n+1}$ diverges (harmonic series with two terms missing),

so also does $\displaystyle\sum_{n=2}^{\infty} \dfrac{\ln(n+1)}{n+1}$ (by the comparison test). It follows that the given series diverges also.

22. The function $1/[x(\ln x)^{1/3}]$ is positive, continuous, and decreasing for $x \geq 2$. Since

$$\int_2^{\infty} \frac{1}{x(\ln x)^{1/3}}\,dx = \left\{\frac{3}{2}(\ln x)^{2/3}\right\}_2^{\infty} = \infty,$$

the series $\displaystyle\sum_{n=2}^{\infty} \dfrac{1}{n(\ln n)^{1/3}}$ diverges (by the integral test).

24. If we set $y = x^2$, the series becomes $\displaystyle\sum_{n=1}^{\infty} \dfrac{1}{n^2} x^{2n} = \sum_{n=1}^{\infty} \dfrac{1}{n^2} y^n$. Since the radius of convergence of this

series is $R_y = \lim\limits_{n \to \infty} \left| \dfrac{1/n^2}{1/(n+1)^2} \right| = 1$, the radius of convergence of the original series is $R_x = \sqrt{R_y} = 1$.
The open interval of convergence is $-1 < x < 1$. At the end points $x = \pm 1$, the series reduces to the
convergent series $\sum_{n=1}^{\infty} 1/n^2$. The interval of convergence is therefore $-1 \leq x \leq 1$.

26. If we set $y = (x+1)^2$, the series becomes $\displaystyle\sum_{n=0}^{\infty} \dfrac{n3^n}{(n+1)^3}(x+2)^{2n} = \sum_{n=0}^{\infty} \dfrac{n3^n}{(n+1)^3} y^n$. Since the radius of

convergence of this series is $R_y = \lim\limits_{n \to \infty} \left| \dfrac{n3^n/(n+1)^3}{(n+1)3^{n+1}/(n+2)^3} \right| = 1/3$, the radius of convergence of the
original series is $R_x = \sqrt{R_y} = 1/\sqrt{3}$. The open interval of convergence is $-2 - 1/\sqrt{3} < x < -2 + 1/\sqrt{3}$.
At the end points $x = -2 \pm 1/\sqrt{3}$, the series reduces to $\sum_{n=0}^{\infty} n/(n+1)^3$. Since $l = \lim\limits_{n \to \infty} \dfrac{n/(n+1)^3}{1/n^2} = 1$,
and $\sum_{n=1}^{\infty} 1/n^2$ converges, so also does $\sum_{n=0}^{\infty} n/(n+1)^3$ (by the limit comparison test). The interval of
convergence is therefore $-2 - 1/\sqrt{3} \leq x \leq -2 + 1/\sqrt{3}$.

28. When $p = 1$, the function $1/(x \ln x)$ is positive, continuous, and decreasing for $x \geq 2$. Since

$$\int_2^{\infty} \frac{1}{x \ln x}\,dx = \left\{\ln|\ln x|\right\}_2^{\infty} = \infty,$$

it follows by the integral test that $\displaystyle\sum_{n=2}^{\infty} \dfrac{1}{n \ln n}$ diverges.

When $p < 1$, $\quad \dfrac{1}{n(\ln n)^p} > \dfrac{1}{n \ln n}$, and therefore $\displaystyle\sum_{n=2}^{\infty} \dfrac{1}{n(\ln n)^p}$ diverges when $p < 1$ (by the comparison
test).
For $p > 1$, the function $1/[x(\ln x)^p]$ is positive, continuous, and decreasing for $x \geq 2$. Since

$$\int_2^{\infty} \frac{1}{x(\ln x)^p}\,dx = \left\{\frac{1}{(1-p)(\ln x)^{p-1}}\right\}_2^{\infty} = \frac{1}{(p-1)(\ln 2)^{p-1}},$$

the series $\displaystyle\sum_{n=2}^{\infty} \dfrac{1}{n(\ln n)^p}$ converges (by the integral test).

EXERCISES 10.11

2. Since $L = \lim\limits_{n \to \infty} \dfrac{\dfrac{1}{(n+1)!}}{\dfrac{1}{n!}} = \lim\limits_{n \to \infty} \dfrac{1}{n+1} = 0$, the series converges (by the limit ratio test).

4. Since $R = \lim\limits_{n\to\infty} \left(\dfrac{1}{n^n}\right)^{1/n} = \lim\limits_{n\to\infty} \dfrac{1}{n} = 0$, the series converges (by the limit root test).

6. Since $L = \lim\limits_{n\to\infty} \dfrac{\dfrac{(2n+2)!}{[(n+1)!]^2}}{\dfrac{(2n)!}{(n!)^2}} = \lim\limits_{n\to\infty} \dfrac{(2n+2)(2n+1)}{(n+1)^2} = 4 > 1$, the series diverges (by the limit ratio test).

8. Since $L = \lim\limits_{n\to\infty} \dfrac{\dfrac{3^{-(n+1)} + 2^{-(n+1)}}{4^{-(n+1)} + 5^{-(n+1)}}}{\dfrac{3^{-n} + 2^{-n}}{4^{-n} + 5^{-n}}} = \lim\limits_{n\to\infty} \left\{ \dfrac{2^{-(n+1)}[1 + (3/2)^{-(n+1)}]}{4^{-(n+1)}[1 + (5/4)^{-(n+1)}]} \dfrac{4^{-n}[1 + (5/4)^{-n}]}{2^{-n}[1 + (3/2)^{-n}]} \right\} = 2$, the series

diverges (by the limit ratio test).

10. Since $L = \lim\limits_{n\to\infty} \dfrac{\dfrac{2 \cdot 4 \cdots (2n+2)}{4 \cdot 7 \cdots (3n+4)}}{\dfrac{2 \cdot 4 \cdots (2n)}{4 \cdot 7 \cdots (3n+1)}} = \lim\limits_{n\to\infty} \dfrac{2n+2}{3n+4} = \dfrac{2}{3}$, the series converges (by the limit ratio test).

12. Since $L = \lim\limits_{n\to\infty} \dfrac{(n+1)(3/4)^{n+1}}{n(3/4)^n} = \dfrac{3}{4}$, the series converges (by the limit ratio test).

14. Since $\dfrac{2 \cdot 4 \cdots (2n)}{3 \cdot 5 \cdots (2n+1)} \left(\dfrac{1}{n^2}\right) = \left(\dfrac{2}{3}\right)\left(\dfrac{4}{5}\right) \cdots \left(\dfrac{2n}{2n+1}\right)\left(\dfrac{1}{n^2}\right) < \dfrac{1}{n^2}$, and $\sum\limits_{n=1}^{\infty} \dfrac{1}{n^2}$ converges, so also does

the original series (by the comparison test).

16. Since $l = \lim\limits_{n\to\infty} \dfrac{\dfrac{(n+1)^n}{n^{n+1}}}{\dfrac{e}{n}} = \lim\limits_{n\to\infty} \dfrac{1}{e}\left(\dfrac{n+1}{n}\right)^n = 1$, and $\sum\limits_{n=1}^{\infty} \dfrac{e}{n} = e \sum\limits_{n=1}^{\infty} \dfrac{1}{n}$ diverges, so also does the original

series (by the limit comparison test).

18. Since $L = \lim\limits_{n\to\infty} \dfrac{\dfrac{2^{n+1} + (n+1)^2 3^{n+1}}{4^{n+1}}}{\dfrac{2^n + n^2 3^n}{4^n}} = \lim\limits_{n\to\infty} \dfrac{1}{4}\left[\dfrac{2^{n+1} + (n+1)^2 3^{n+1}}{2^n + n^2 3^n}\right] = \lim\limits_{n\to\infty} \dfrac{1}{4}\left[\dfrac{\dfrac{2^{n+1}}{3^{n+1}} + (n+1)^2}{\dfrac{2^n}{3^{n+1}} + \dfrac{n^2}{3}}\right]$

$= 3/4$, the series converges (by the limit ratio test).

20. Since $L = \lim\limits_{n\to\infty} \dfrac{\dfrac{(2n+2)! 5^{2n+2}}{(3n+3)!}}{\dfrac{(2n)! 5^{2n}}{(3n)!}} = \lim\limits_{n\to\infty} \dfrac{25(2n+2)(2n+1)}{(3n+3)(3n+2)(3n+1)} = 0$, the series converges (by the limit

ratio test).

EXERCISES 10.12

2. Consider the series $\sum\limits_{n=1}^{\infty} \dfrac{n}{n^2+1}$. Since $l = \lim\limits_{n\to\infty} \dfrac{\dfrac{n}{n^2+1}}{\dfrac{1}{n}} = 1$, and $\sum\limits_{n=1}^{\infty} \dfrac{1}{n}$ diverges so also does $\sum\limits_{n=1}^{\infty} \dfrac{n}{n^2+1}$

(by the limit comparison test). The original series does not converge absolutely. Because the sequence $\{n/(n^2+1)\}$ is decreasing and has limit zero, the original series converges conditionally (by the alternating series test).

4. Consider the series $\sum\limits_{n=1}^{\infty} \dfrac{n^3}{3^n}$. Since $L = \lim\limits_{n\to\infty} \dfrac{\dfrac{(n+1)^3}{3^{n+1}}}{\dfrac{n^3}{3^n}} = \dfrac{1}{3}$, the series $\sum\limits_{n=1}^{\infty} \dfrac{n^3}{3^n}$ converges (by the limit ratio

test). The original series therefore converges absolutely.

6. Since $\lim_{n\to\infty} (-1)^n \dfrac{3^n}{n^3}$ does not exist, the series diverges (by the nth term test).

8. Consider the series $\sum\limits_{n=1}^{\infty} \left| \dfrac{n\sin(n\pi/4)}{2^n} \right|$. Since $\left| \dfrac{n\sin(n\pi/4)}{2^n} \right| \le \dfrac{n}{2^n}$, the series of absolute values converges

if $\sum\limits_{n=1}^{\infty} \dfrac{n}{2^n}$ converges. Because $L = \lim\limits_{n\to\infty} \dfrac{\dfrac{n+1}{2^{n+1}}}{\dfrac{n}{2^n}} = \dfrac{1}{2}$, the series $\sum\limits_{n=1}^{\infty} \dfrac{n}{2^n}$ converges (by the limit ratio test).

Consequently, the original series converges absolutely.

10. Consider the series $\sum\limits_{n=1}^{\infty} \dfrac{\sqrt{3n-2}}{n}$. Since $l = \lim\limits_{n\to\infty} \dfrac{\dfrac{\sqrt{3n-2}}{n}}{\dfrac{\sqrt{3}}{\sqrt{n}}} = 1$, and $\sum\limits_{n=1}^{\infty} \dfrac{\sqrt{3}}{\sqrt{n}}$ diverges so also does the

series $\sum\limits_{n=1}^{\infty} \dfrac{\sqrt{3n-2}}{n}$ (by the limit comparison test). The original series does not converge absolutely.

Because the sequence $\{\sqrt{3n-2}/n\}$ is nonincreasing and has limit zero, the original series converges conditionally (by the alternating series test).

12. Consider the series $\sum\limits_{n=1}^{\infty} \dfrac{\sqrt{n^2+3}}{n^2+5}$. Since $l = \lim\limits_{n\to\infty} \dfrac{\dfrac{\sqrt{n^2+3}}{n^2+5}}{\dfrac{1}{n}} = 1$, and $\sum\limits_{n=1}^{\infty} \dfrac{1}{n}$ diverges, so also does the series

$\sum\limits_{n=1}^{\infty} \dfrac{\sqrt{n^2+3}}{n^2+5}$ (by the limit comparison test). The original series does not therefore converge absolutely.

Because the sequence $\{\sqrt{n^2+3}/(n^2+5)\}$ is decreasing and has limit zero, the original series converges conditionally (by the alternating series test).

14. Consider the series $\sum\limits_{n=1}^{\infty} \left| \dfrac{\cos(n\pi/10)\mathrm{Cot}^{-1}n}{n^3+5n} \right|$. Since $\left| \dfrac{\cos(n\pi/10)\mathrm{Cot}^{-1}n}{n^3+5n} \right| < \dfrac{\pi/4}{n^3}$, and the series $\sum\limits_{n=1}^{\infty} \dfrac{\pi/4}{n^3}$

$= \dfrac{\pi}{4}\sum\limits_{n=1}^{\infty} \dfrac{1}{n^3}$ converges, so also does the series of absolute values (by the comparison test). The original

series therefore converges absolutely.

16. With radius of convergence $\lim\limits_{n\to\infty} \left| \dfrac{1/(n+1)^2}{1/(n+2)^2} \right| = 1$, the open interval of convergence is $-1 < x < 1$.

At $x = 1$, the power series reduces to a p-series $\sum\limits_{n=0}^{\infty} \dfrac{1}{(n+1)^2} = \sum\limits_{n=1}^{\infty} \dfrac{1}{n^2}$ which converges. At $x = -1$, it

reduces to $\sum\limits_{n=0}^{\infty} \dfrac{(-1)^n}{(n+1)^2}$ which converges absolutely. The interval of convergence is therefore $-1 \le x \le 1$.

18. With radius of convergence $\lim\limits_{n\to\infty} \left| \dfrac{1/\sqrt{n}}{1/\sqrt{n+1}} \right| = 1$, the open interval of convergence is $-3 < x < -1$.

At $x = -1$, the power series reduces to a p-series $\sum\limits_{n=1}^{\infty} \dfrac{1}{\sqrt{n}}$ which diverges. At $x = -3$, it reduces to

$\sum\limits_{n=1}^{\infty} \dfrac{(-1)^n}{\sqrt{n}}$ which converges conditionally. The interval of convergence is therefore $-3 \le x < -1$.

20. If we set $y = x^3$, the series becomes $\sum\limits_{n=0}^{\infty} \dfrac{1}{\sqrt{n+1}} x^{3n+1} = y^{1/3} \sum\limits_{n=0}^{\infty} \dfrac{1}{\sqrt{n+1}} y^n$. Since the radius of con-

vergence of this series is $R_y = \lim\limits_{n\to\infty} \left| \dfrac{1/\sqrt{n+1}}{1/\sqrt{n+2}} \right| = 1$, it follows that $R_x = \sqrt{R_y} = 1$ also. The open

interval of convergence is therefore $-1 < x < 1$. At $x = 1$, the power series reduces to a p-series

$\sum_{n=0}^{\infty} \frac{1}{\sqrt{n+1}} = \sum_{n=1}^{\infty} \frac{1}{\sqrt{n}}$ which diverges. At $x = -1$, it reduces to $\sum_{n=0}^{\infty} \frac{(-1)^{3n+1}}{\sqrt{n+1}} = \sum_{n=1}^{\infty} \frac{(-1)^{n+1}}{\sqrt{n}}$. This series does not converge absolutely but it does converge conditionally. Hence, the interval of convergence is $-1 \le x < 1$.

22. With radius of convergence $\lim_{n \to \infty} \left| \frac{\frac{1}{n^2 \ln n}}{\frac{1}{(n+1)^2 \ln(n+1)}} \right| = \lim_{n \to \infty} \frac{\ln(n+1)}{\ln n} = \lim_{n \to \infty} \frac{1/(n+1)}{1/n} = 1$, the open

interval of convergence of the series is $1 < x < 3$. At $x = 3$, the power series reduces to $\sum_{n=2}^{\infty} \frac{1}{n^2 \ln n}$. This

series converges by the comparison test since $1/(n^2 \ln n) < 1/n^2$ for $n \ge 3$. At $x = 1$, the power series

reduces to $\sum_{n=2}^{\infty} \frac{(-1)^n}{n^2 \ln n}$ which converges absolutely. The interval of convergence is therefore $1 \le x \le 3$.

24. If $\sum_{n=1}^{\infty} c_n$ converges absolutely, then $\sum_{n=1}^{\infty} |c_n|$ converges, and then $\lim_{n \to \infty} |c_n| = 0$. It follows that for n greater than or equal to some integer N, $0 < |c_n| < 1$. For such n, we have $|c_n|^p < |c_n|$. Consequently, $\sum_{n=N}^{\infty} |c_n^p| = \sum_{n=N}^{\infty} |c_n|^p$ converges (by the comparison test). Hence, $\sum_{n=1}^{\infty} |c_n|^p$ converges also.

EXERCISES 10.13

2. We obtain this result by setting $x = 1$ in the Maclaurin series for $\sin x$ (see Example 10.9).

4. If we set $x = -1$ in the Maclaurin series for e^x (Example 10.10),

$$e^{-1} = \sum_{n=0}^{\infty} \frac{(-1)^n}{n!} = 1 + \sum_{n=1}^{\infty} \frac{(-1)^n}{n!} \implies \frac{1}{e} - 1 = \sum_{n=1}^{\infty} \frac{(-1)^n}{n!}.$$

6. If we set $x = 2$ in the Maclaurin series for $\cos x$ (see example 10.21),

$$\cos 2 = \sum_{n=0}^{\infty} \frac{(-1)^n}{(2n)!} 2^{2n} = 1 - \frac{2^2}{2!} + \sum_{n=2}^{\infty} \frac{(-1)^n}{(2n)!} 2^{2n}.$$

Consequently,

$$\sum_{n=2}^{\infty} \frac{(-1)^{n+1}}{(2n)!} 2^{2n+3} = -8(\cos 2 - 1 + 2) = -8(1 + \cos 2).$$

8. If we set $x = 1/2$ in the Taylor series for $\ln x$ about $x = 1$ (see Example 10.22),

$$\ln(1/2) = \sum_{n=1}^{\infty} \frac{(-1)^{n+1}}{n} \left(-\frac{1}{2}\right)^n = \sum_{n=1}^{\infty} \frac{-1}{n 2^n} \implies \sum_{n=1}^{\infty} \frac{1}{n 2^n} = -\ln(1/2) = \ln 2.$$

10. The series $\sum_{n=1}^{\infty} n x^n$ converges for $-1 < x < 1$. If we set $S(x) = \sum_{n=1}^{\infty} n x^n$, then $x^{-1} S(x) = \sum_{n=1}^{\infty} n x^{n-1}$.

Term-by-term integration gives $\int \frac{S(x)}{x} dx + C = \sum_{n=1}^{\infty} x^n = \frac{x}{1-x}$, since the series is geometric. We now

differentiate with respect to x, getting $\frac{S(x)}{x} = \frac{(1-x)(1) - x(-1)}{(1-x)^2} = \frac{1}{(1-x)^2}$.

Hence, $S(x) = \sum_{n=1}^{\infty} n x^n = \frac{x}{(1-x)^2}$, and if we set $x = 1/2$, we obtain $\sum_{n=1}^{\infty} \frac{n}{2^n} = \frac{1/2}{(1-1/2)^2} = 2$.

12. Term-by-term differentiation of $\dfrac{1}{1+x^2} = 1 - x^2 + x^4 + \cdots = \displaystyle\sum_{n=0}^{\infty} (-1)^n x^{2n}$, $|x| < 1$ gives

$$\frac{-2x}{(1+x^2)^2} = \sum_{n=0}^{\infty} 2n(-1)^n x^{2n-1} \quad\Longrightarrow\quad \frac{x}{(1+x^2)^2} = \sum_{n=1}^{\infty} n(-1)^{n-1} x^{2n-1}.$$

If we set $x = 1/3$, then $\dfrac{1/3}{(1+1/9)^2} = \displaystyle\sum_{n=1}^{\infty} n(-1)^{n-1} \dfrac{1}{3^{2n-1}}$. Hence

$$\sum_{n=1}^{\infty} \frac{n(-1)^n}{3^{2n}} = -\frac{1}{3}\left[\frac{1/3}{(1+1/9)^2}\right] = -\frac{9}{100}.$$

14. The sum of the first five terms is $\displaystyle\sum_{n=1}^{5} \dfrac{n}{e^{3n}} = 0.055\,140\,9$. According to expression 10.48, the error in approximating the sum of this series with this value is less than

$$\int_5^{\infty} xe^{-3x}\,dx = \left\{-\frac{x}{3}e^{-3x} - \frac{1}{9}e^{-3x}\right\}_5^{\infty} = \frac{5}{3}e^{-15} + \frac{1}{9}e^{-15} = \frac{16}{9}e^{-15} < 6 \times 10^{-7}.$$

16. According to expression 10.48, the error in approximating the sum of this series after the 20th term is less than

$$\int_{20}^{\infty} \frac{1}{x^2}\sin\left(\frac{1}{x}\right)dx = \left\{\cos\left(\frac{1}{x}\right)\right\}_{20}^{\infty} = 1 - \cos\left(\frac{1}{20}\right) < 0.001\,25.$$

18. The sum of the first 20 terms is $-0.947\,030$. Since the series is alternating and absolute values of the terms are decreasing and have limit zero, the maximum error is the absolute value of the 21st term, $1/21^4 < 5.2 \times 10^{-6}$.

20. The sum of the first 10 terms is $0.693\,065$. The error in using this to approximate the sum of the series is

$$\sum_{n=11}^{\infty} \frac{1}{n2^n} = \frac{1}{11\cdot 2^{11}} + \frac{1}{12\cdot 2^{12}} + \frac{1}{13\cdot 2^{13}} + \cdots$$

$$< \frac{1}{11\cdot 2^{11}} + \frac{1}{11\cdot 2^{12}} + \frac{1}{11\cdot 2^{13}} + \cdots = \frac{1/(11\cdot 2^{11})}{1 - 1/2} = \frac{1}{11\cdot 2^{10}} < 9 \times 10^{-5}.$$

22. The sum of the first 20 terms is $1.067\,49$. The error in using this to approximate the sum of the series is

$$\sum_{n=22}^{\infty} \frac{2^n - 1}{3^n + n} = \frac{2^{22} - 1}{3^{22} + 22} + \frac{2^{23} - 1}{3^{23} + 23} + \cdots < \frac{2^{22}}{3^{22}} + \frac{2^{23}}{3^{23}} + \cdots = \frac{(2/3)^{22}}{1 - 2/3} < 4.01 \times 10^{-4}.$$

24. The sum of the first 100 terms is $-0.688\,172\,2$. Since the series is alternating and absolute values of the terms are decreasing with limit zero, the maximum error is the absolute value of the 101st term, $1/101$.

26. When this series is truncated after N terms, the error is

$$\sum_{n=N+1}^{\infty} \frac{1}{n^2 4^n} = \frac{1}{(N+1)^2 4^{N+1}} + \frac{1}{(N+2)^2 4^{N+2}} + \cdots < \frac{1}{(N+1)^2 4^{N+1}} + \frac{1}{(N+1)^2 4^{N+2}} + \cdots$$

$$= \frac{1}{(N+1)^2 4^{N+1}}\left(\frac{1}{1 - 1/4}\right) = \frac{1}{3(N+1)^2 4^N}.$$

This quantity is less than 10^{-4} if $3(N+1)^2 4^N > 10^4$. This occurs for $N \geq 4$.

28. (a) Using 10.48, the error is less than $\displaystyle\int_{10}^{\infty} e^{-x}\sin^2 x\,dx$. By writing the integrand in the form $e^{-x}(1-\cos 2x)/2$, and using integration by parts, we obtain the following antiderivative,

$$\int_{10}^{\infty} e^{-x}\sin^2 x\,dx = \left\{-\frac{1}{5}e^{-x}(2+\sin^2 x + 2\sin x\cos x)\right\}_{10}^{\infty}$$

$$= \frac{e^{-10}}{5}(2+\sin^2 10 + 2\sin 10\cos 10) < 2.92\times 10^{-5}.$$

(b) The error in using ten terms to approximate the sum is

$$\sum_{n=11}^{\infty} e^{-n}\sin^2 n = e^{-11}\sin^2 11 + e^{-12}\sin^2 12 + e^{-13}\sin^2 13 + \cdots$$

$$< e^{-11} + e^{-12} + e^{-13} + \cdots = \frac{e^{-11}}{1-1/e} < 2.65\times 10^{-5}.$$

The error in part (b) is better.

30. $\displaystyle\int_0^{1/2}\cos(x^2)\,dx = \int_0^{1/2}\left[1 - \frac{(x^2)^2}{2!} + \frac{(x^2)^4}{4!} - \frac{(x^2)^6}{6!} + \cdots\right]dx = \int_0^{1/2}\left[1 - \frac{x^4}{2!} + \frac{x^8}{4!} - \frac{x^{12}}{6!} + \cdots\right]dx$

$$= \left\{x - \frac{x^5}{5\cdot 2!} + \frac{x^9}{9\cdot 4!} - \frac{x^{13}}{13\cdot 6!} + \cdots\right\}_0^{1/2} = \frac{1}{2} - \frac{1}{5\cdot 2^5\cdot 2!} + \frac{1}{9\cdot 2^9\cdot 4!} - \frac{1}{13\cdot 2^{13}\cdot 6!} + \cdots.$$

This is a convergent alternating series. To find a three-decimal approximation, we calculate partial sums until two successive sums agree to three decimals:

$$S_1 = \frac{1}{2}, \qquad S_2 = S_1 - \frac{1}{5\cdot 2^5\cdot 2!} = 0.496\,78, \qquad S_3 = S_2 + \frac{1}{9\cdot 2^9\cdot 4!} = 0.496\,88.$$

Consequently, to three decimals the value of the integral is 0.497.

32. $\displaystyle\int_{-1}^{1} x^{11}\sin x\,dx = 2\int_0^1 x^{11}\left(x - \frac{x^3}{3!} + \frac{x^5}{5!} - \cdots\right)dx = 2\int_0^1\left(x^{12} - \frac{x^{14}}{3!} + \frac{x^{16}}{5!} - \cdots\right)dx$

$$= 2\left\{\frac{x^{13}}{13} - \frac{x^{15}}{15\cdot 3!} + \frac{x^{17}}{17\cdot 5!} - \cdots\right\}_0^1 = 2\left(\frac{1}{13} - \frac{1}{15\cdot 3!} + \frac{1}{17\cdot 5!} - \frac{1}{19\cdot 7!} + \cdots\right).$$

This is a convergent alternating series. To find a three-decimal approximation, we calculate partial sums until two successive sums agree to three decimals:

$$S_1 = \frac{2}{13}, \qquad\qquad\qquad\qquad S_2 = S_1 - \frac{2}{15\cdot 3!} = 0.131\,62,$$

$$S_3 = S_2 + \frac{2}{17\cdot 5!} = 0.132\,60, \qquad\qquad S_4 = S_3 - \frac{2}{19\cdot 7!} = 0.132\,58.$$

Consequently, to three decimals the value of the integral is 0.133.

34. $\displaystyle\int_0^{0.3} e^{-x^2}\,dx = \int_0^{0.3}\left[1 + (-x^2) + \frac{(-x^2)^2}{2!} + \frac{(-x^2)^3}{3!} + \cdots\right]dx = \int_0^{0.3}\left[1 - x^2 + \frac{x^4}{2!} - \frac{x^6}{3!} + \cdots\right]dx$

$$= \left\{x - \frac{x^3}{3} + \frac{x^5}{5\cdot 2!} - \frac{x^7}{7\cdot 3!} + \cdots\right\}_0^{0.3} = 0.3 - \frac{(0.3)^3}{3} + \frac{(0.3)^5}{5\cdot 2!} - \frac{(0.3)^7}{7\cdot 3!} + \cdots.$$

This is a convergent alternating series. To find a three-decimal approximation, we calculate partial sums until two successive sums agree to three decimals:

$$S_1 = 0.3, \qquad S_2 = S_1 - \frac{(0.3)^3}{3} = 0.291\,00, \qquad S_3 = S_2 + \frac{(0.3)^5}{5\cdot 2!} = 0.291\,24.$$

Consequently, to three decimals the value of the integral is 0.291.

36.
$$\int_0^{1/2} \frac{1}{x^6 - 3x^3 - 4}\, dx = \int_0^{1/2} \frac{1}{(x^3 - 4)(x^3 + 1)}\, dx = \int_0^{1/2} \left(\frac{1/5}{x^3 - 4} + \frac{-1/5}{x^3 + 1} \right) dx$$

$$= \frac{1}{5} \int_0^{1/2} \left[\frac{-1}{4(1 - x^3/4)} - \frac{1}{1 + x^3} \right] dx$$

$$= -\frac{1}{5} \int_0^{1/2} \left[\frac{1}{4} \left(1 + \frac{x^3}{4} + \frac{x^6}{4^2} + \frac{x^9}{4^3} + \cdots \right) + \left(1 - x^3 + x^6 - x^9 + \cdots \right) \right] dx$$

$$= -\frac{1}{5} \int_0^{1/2} \left[\frac{5}{4} - \left(1 - \frac{1}{4^2} \right) x^3 + \left(1 + \frac{1}{4^3} \right) x^6 - \cdots \right] dx$$

$$= -\frac{1}{5} \left\{ \frac{5x}{4} - \frac{1}{4} \left(1 - \frac{1}{4^2} \right) x^4 + \frac{1}{7} \left(1 + \frac{1}{4^3} \right) x^7 - \cdots \right\}_0^{1/2}$$

$$= \frac{1}{5} \left[-\frac{5}{4} \left(\frac{1}{2} \right) + \frac{1}{4} \left(1 - \frac{1}{4^2} \right) \left(\frac{1}{2} \right)^4 - \frac{1}{7} \left(1 + \frac{1}{4^3} \right) \left(\frac{1}{2} \right)^7 + \cdots \right].$$

This is a convergent alternating series. To find a three-decimal approximation, we calculate partial sums until two successive sums agree to three decimals:

$$S_1 = -\frac{1}{8}, \quad S_2 = S_1 + \frac{1}{20 \cdot 2^4} \left(1 - \frac{1}{4^2} \right) = -0.122\,07, \quad S_3 = S_2 - \frac{1}{35 \cdot 2^7} \left(1 + \frac{1}{4^3} \right) = -0.122\,30.$$

Consequently, to three decimals the value of the integral is -0.122.

38. If we replace the integrand by its Maclaurin series, we have

$$\frac{\sqrt{\pi}}{2} \mathrm{erf}(1) = \int_0^1 \left[1 - t^2 + \frac{(-t^2)^2}{2!} + \frac{(-t^2)^3}{3!} + \cdots \right] dt = \left\{ t - \frac{t^3}{3} + \frac{t^5}{5 \cdot 2!} - \frac{t^7}{7 \cdot 3!} + \cdots \right\}_0^1$$

$$= 1 - \frac{1}{3} + \frac{1}{5 \cdot 2!} - \frac{1}{7 \cdot 3!} + \frac{1}{9 \cdot 4!} - \cdots .$$

This is a convergent alternating series. To find a three-decimal approximation, we calculate partial sums until two successive sums agree to three decimals:

$$S_1 = 2/\sqrt{\pi} = 1.128\,38, \qquad\qquad S_2 = S_1 - (2/\sqrt{\pi})/3 = 0.755\,225,$$
$$S_3 = S_2 + (2/\sqrt{\pi})/(5 \cdot 2!) = 0.865\,09, \qquad S_4 = S_3 - (2/\sqrt{\pi})/(7 \cdot 3!) = 0.838\,23,$$
$$S_5 = S_4 + (2/\sqrt{\pi})/(9 \cdot 4!) = 0.843\,45, \qquad S_6 = S_5 - (2/\sqrt{\pi})/(11 \cdot 5!) = 0.843\,31.$$

Consequently, to three decimals the value of the integral is 0.843.

REVIEW EXERCISES

2. The first three terms are $c_1 = 1$, $c_2 = 1/\sqrt{2}$, and $c_3 = \sqrt{3/8}$. The sequence appears to be decreasing; that is, $c_{n+1} < c_n$. This is true for $n = 1$. Suppose k is some integer for which $c_{k+1} < c_k$. Then

$$c_{k+1}^2 + 1 < c_k^2 + 1 \qquad \Longrightarrow \qquad \frac{1}{2} \sqrt{c_{k+1}^2 + 1} < \frac{1}{2} \sqrt{c_k^2 + 1}.$$

In other words, $c_{k+2} < c_{k+1}$, and therefore by mathematical induction, the sequence is decreasing. It follows that $U = c_1 = 1$ is an upper bound, and clearly $V = 0$ is a lower bound. By Theorem 10.7 we conclude that the sequence has a limit L, and to find L we set

$$\lim_{n \to \infty} c_{n+1} = \lim_{n \to \infty} \frac{1}{2} \sqrt{c_n^2 + 1}.$$

This equation implies that $L = (1/2)\sqrt{L^2 + 1}$, the solution of which is $L = 1/\sqrt{3}$.

4. First we note that because the first term of the sequence is 7, and all other terms are at least 15, there is no difficulty with the square root. The first three terms are $c_1 = 7$, $c_2 = 15+\sqrt{5}$, and $c_3 = 15+\sqrt{13+\sqrt{5}}$. These terms are increasing, $c_3 > c_2 > c_1$. Suppose k is some integer for which $c_{k+1} > c_k$. Then

$$\sqrt{c_{k+1}-2} > \sqrt{c_k-2} \quad \Longrightarrow \quad 15+\sqrt{c_{k+1}-2} > 15+\sqrt{c_k-2}.$$

But this means that $c_{k+2} > c_{k+1}$, and therefore by mathematical induction, the sequence is increasing. A lower bound must be $V = c_1 = 7$. The first three terms are less than 100. Suppose k is some integer for which $c_k < 100$. Then $c_{k+1} = 15+\sqrt{c_k-2} < 15+\sqrt{100-2} < 100$, and by mathematical induction, an upper bound is $U = 100$. By Theorem 10.7, the sequence has a limit L, and to find L we set

$$\lim_{n\to\infty} c_{n+1} = \lim_{n\to\infty} (15+\sqrt{c_n-2}).$$

This equation implies that $L = 15+\sqrt{L-2}$, the solution of which is $L = (31+\sqrt{53})/2$.

6. The first four terms of the sequence are $c_1 = 6$, $c_2 = 1/29 = 0.034$, $c_3 = 0.195$, and $c_4 = 0.173$. They are oscillating. To show that the entire sequence oscillates, we calculate

$$c_{n+1} - c_n = \frac{1}{5+4c_n} - \frac{1}{5+4c_{n-1}} = -\frac{4(c_n - c_{n-1})}{(5+4c_n)(5+4c_{n-1})}.$$

Since all terms of the sequence are positive, the denominator of this expression is positive. It follows that $c_{n+1} - c_n$ has the opposite sign of $c_n - c_{n-1}$, and the sequence oscillates. To verify properties 2 and 3 of Theorem 10.8, we take absolute values in the above equation,

$$|c_{n+1} - c_n| = \frac{4|c_n - c_{n-1}|}{(5+4c_n)(5+4c_{n-1})} < \frac{4|c_n - c_{n-1}|}{(5)(5)} = \frac{4}{25}|c_n - c_{n-1}|.$$

This shows that the $|c_{n+1} - c_n|$ decrease and have limit 0. By Theorem 10.8 the sequence has a limit L that can be obtained by taking limits on both sides of the recursive definition,

$$\lim_{n\to\infty} c_{n+1} = \lim_{n\to\infty} \frac{1}{5+4c_n} \quad \Longrightarrow \quad L = \frac{1}{5+4L}.$$

Of the two solutions $(-5 \pm \sqrt{41})/8$ of this equation, only $L = (\sqrt{41} - 5)/8$ is positive.

8. When $|k| < 1$, the sequence converges to 0, and when $k = \pm 1$, it converges to 1. It diverges for all other values.

10. Since $f'(x) = \dfrac{x(1/x) - \ln x}{x^2} = \dfrac{1 - \ln x}{x^2}$, this function is decreasing ($f'(x) \le 0$) for $x \ge e$. It follows that the sequence $\{\ln(n)/n\}$ is decreasing for $n \ge 3$.

12. Since $l = \lim_{n\to\infty} \dfrac{\dfrac{n^2 + 5n + 3}{n^4 - 2n + 5}}{\dfrac{1}{n^2}} = 1$, and $\displaystyle\sum_{n=1}^{\infty} 1/n^2$ converges, so also does the original series (by the limit comparison test).

14. Since $L = \lim_{n\to\infty} \dfrac{\dfrac{(n+1)^2 + 3}{(n+1)3^{n+1}}}{\dfrac{n^2 + 3}{n3^n}} = \dfrac{1}{3}$, the series converges (by the limit ratio test).

16. Consider the series of absolute values $\displaystyle\sum_{n=1}^{\infty} \left(\dfrac{n+1}{n^2}\right)$. Since $l = \lim_{n\to\infty} \dfrac{(n+1)/n^2}{1/n} = 1$, and $\displaystyle\sum_{n=1}^{\infty} 1/n$ diverges, the original series does not converge absolutely. Because the sequence $\{(n+1)/n^2\}$ is decreasing and has limit 0, the original series converges conditionally (by the alternating series test).

18. Since $\lim_{n\to\infty} \text{Cos}^{-1}(1/n) = \pi/2$, the series diverges (by the nth term test).

20. Since $l = \lim_{n \to \infty} \dfrac{\dfrac{1}{n^2}\mathrm{Cos}^{-1}(1/n)}{\dfrac{1}{n^2}} = \dfrac{\pi}{2}$, and $\sum_{n=1}^{\infty} 1/n^2$ converges, so also does the original series (by the limit comparison test).

22. Since $L = \lim_{n \to \infty} \dfrac{\dfrac{3 \cdot 6 \cdot 9 \cdots (3n+3)}{(2n+2)!}}{\dfrac{3 \cdot 6 \cdot 9 \cdots (3n)}{(2n)!}} = \lim_{n \to \infty} \dfrac{3n+3}{(2n+2)(2n+1)} = 0$, the series converges (by the limit ratio test).

24. Because $\lim_{n \to \infty} (-1)^{n+1}\left(1 + \dfrac{1}{n}\right)^3$ does not exist, the series diverges (by the nth term test).

26. Since this is a geometric series with common ratio $10/125 = 2/25$, the series converges.

28. This is a geometric series with common ratio $1/e^\pi$, and therefore it converges.

30. Consider the series of absolute values $\sum_{n=1}^{\infty} \left| \dfrac{1}{\sqrt{n}} \cos(n\pi) \right| = \sum_{n=1}^{\infty} \dfrac{1}{\sqrt{n}}$. Since this series diverges, the original series does not converge absolutely. Because the sequence $\{1/\sqrt{n}\}$ is decreasing with limit 0, the series converges conditionally (by the alternating series test).

32. Since the radius of convergence of the series is $R = \lim_{n \to \infty} \left| \dfrac{\dfrac{1}{n^2 2^n}}{\dfrac{1}{(n+1)^2 2^{n+1}}} \right| = 2$, the open interval of convergence is $-2 < x < 2$. At $x = 2$, the power series reduces to $\sum_{n=1}^{\infty} 1/n^2$ which converges. At $x = -2$, it becomes $\sum_{n=1}^{\infty} (-1)^n/n^2$ which converges absolutely. The interval of convergence is therefore $-2 \le x \le 2$.

34. Since $R = \lim_{n \to \infty} \dfrac{1}{\sqrt[n]{1/n^n}} = \infty$, the series converges for all x.

36. Since the radius of convergence of the series is $R = \lim_{n \to \infty} \left| \dfrac{\sqrt{\dfrac{n+1}{n-1}}}{\sqrt{\dfrac{n+2}{n}}} \right| = 1$, the open interval of convergence is $-4 < x < -2$. At $x = -2$, the power series reduces to $\sum_{n=2}^{\infty} \sqrt{(n+1)/(n-1)}$ which diverges (by the nth term test). At $x = -4$, it becomes $\sum_{n=2}^{\infty} (-1)^n \sqrt{(n+1)/(n-1)}$ which also diverges. The interval of convergence is therefore $-4 < x < -2$.

38. If we set $y = x^3$, then $\sum_{n=1}^{\infty} \dfrac{2^n}{n} x^{3n} = \sum_{n=1}^{\infty} \dfrac{2^n}{n} y^n$. Since $R_y = \lim_{n \to \infty} \left| \dfrac{\dfrac{2^n}{n}}{\dfrac{2^{n+1}}{n+1}} \right| = \dfrac{1}{2}$, the radius of convergence of the original series is $R_x = R_y^{1/3} = 2^{-1/3}$. The open interval of convergence is $-2^{-1/3} < x < 2^{-1/3}$. At $x = 2^{-1/3}$, the power series reduces to the divergent harmonic series. At $x = -2^{-1/3}$, it becomes the alternating harmonic series which converges conditionally. The interval of convergence is therefore $-2^{-1/3} \le x < 2^{-1/3}$.

40. $f(x) = e^5 e^x = e^5 \sum_{n=0}^{\infty} \dfrac{1}{n!} x^n = \sum_{n=0}^{\infty} \dfrac{e^5}{n!} x^n, \quad -\infty < x < \infty$

42. When we integrate $\dfrac{1}{1+x} = \sum_{n=0}^{\infty} (-x)^n = \sum_{n=0}^{\infty} (-1)^n x^n, \ |x| < 1$, we obtain

$$\ln|1+x| = \sum_{n=0}^{\infty} \dfrac{(-1)^n}{n+1} x^{n+1} + C.$$

Substitution of $x = 0$ gives $C = 0$, and therefore $\ln|1 + x| = \sum_{n=1}^{\infty} \frac{(-1)^{n-1}}{n} x^n$. At $x = 1$, the series reduces to $\sum_{n=1}^{\infty} (-1)^{n-1}/n$ which converges conditionally. The interval of convergence is therefore $-1 < x \leq 1$. We now replace x by $2x$ and at the same time multiply by x,

$$x \ln(1 + 2x) = x \sum_{n=1}^{\infty} \frac{(-1)^{n-1}}{n} (2x)^n = \sum_{n=1}^{\infty} \frac{2^n (-1)^{n-1}}{n} x^{n+1} = \sum_{n=2}^{\infty} \frac{2^{n-1}(-1)^n}{n-1} x^n,$$

valid for $-1/2 < x \leq 1/2$.

44. Partial fractions give $f(x) = \dfrac{x}{x^2 + 4x + 3} = \dfrac{3/2}{x+3} - \dfrac{1/2}{x+1}$. Since

$$\frac{1}{x+1} = \sum_{n=0}^{\infty} (-x)^n = \sum_{n=0}^{\infty} (-1)^n x^n, \quad |x| < 1,$$

and

$$\frac{1}{x+3} = \frac{1}{3(1 + x/3)} = \frac{1}{3} \sum_{n=0}^{\infty} \left(-\frac{x}{3}\right)^n = \sum_{n=0}^{\infty} \frac{(-1)^n}{3^{n+1}} x^n, \quad |x| < 3,$$

it follows that

$$f(x) = \sum_{n=0}^{\infty} \frac{(-1)^n}{2 \cdot 3^n} x^n + \sum_{n=0}^{\infty} \frac{(-1)^{n+1}}{2} x^n = \sum_{n=0}^{\infty} \frac{(-1)^{n+1}}{2} \left(1 - \frac{1}{3^n}\right) x^n, \quad |x| < 1.$$

46. According to Exercise 42, $\quad \ln(1 + x) = \sum_{n=1}^{\infty} \frac{(-1)^{n-1}}{n} x^n$, $-1 < x \leq 1$. With this,

$$f(x) = x \ln(1 + x) + \ln(1 + x) = \sum_{n=1}^{\infty} \frac{(-1)^{n-1}}{n} x^{n+1} + \sum_{n=1}^{\infty} \frac{(-1)^{n-1}}{n} x^n$$

$$= \sum_{n=2}^{\infty} \frac{(-1)^n}{n-1} x^n + \sum_{n=1}^{\infty} \frac{(-1)^{n-1}}{n} x^n = x + \sum_{n=2}^{\infty} (-1)^{n-1} \left(\frac{-1}{n-1} + \frac{1}{n}\right) x^n$$

$$= x + \sum_{n=2}^{\infty} \frac{(-1)^n}{n(n-1)} x^n, \quad -1 < x \leq 1.$$

48. $e^{-x^2} = \sum_{n=0}^{\infty} \frac{1}{n!} (-x^2)^n = 1 - x^2 + \frac{x^4}{2!} - \frac{x^6}{3!} + \frac{x^8}{4!} - \cdots$

When this alternating series is truncated after the term in x^{2n}, the absolute value of the maximum possible error for given x is $|x|^{2n+2}/(n+1)!$. For $0 \leq x \leq 2$, the error is a maximum when $x = 2$, namely $2^{2n+2}/(n+1)!$. This error is less than 10^{-5} if $(n+1)!/2^{2n+2} > 10^5$. The smallest integer n for which this holds is $n = 18$. The terms that should be used are therefore $e^{-x^2} \approx \sum_{n=0}^{18} \frac{(-1)^n}{n!} x^{2n}$.

50. $\sqrt{1 + \sin x} = \sqrt{[\cos^2(x/2) + \sin^2(x/2)] + 2 \sin(x/2) \cos(x/2)}$

$$= \sqrt{[\cos(x/2) + \sin(x/2)]^2} = \cos(x/2) + \sin(x/2)$$

$$= \sum_{n=0}^{\infty} \frac{(-1)^n}{(2n)!} \left(\frac{x}{2}\right)^{2n} + \sum_{n=0}^{\infty} \frac{(-1)^n}{(2n+1)!} \left(\frac{x}{2}\right)^{2n+1}$$

$$= \sum_{n=0}^{\infty} \frac{(-1)^n}{2^{2n}(2n)!} x^{2n} + \sum_{n=0}^{\infty} \frac{(-1)^n}{2^{2n+1}(2n+1)!} x^{2n+1}$$

When x is not confined to $-\pi/2 \leq x \leq \pi/2$, we must write $\sqrt{1 + \sin x} = |\cos(x/2) + \sin(x/2)|$.

CHAPTER 11

EXERCISES 11.1

2. Length $= \sqrt{(1+3)^2 + (-2-2)^2 + (5-4)^2} = \sqrt{33}$

4. The diagram to the right indicates
the vertices of the cube.

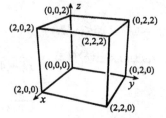

6. (a) $\sqrt{2^2 + 3^2 + (-4)^2} = \sqrt{29}$ (b) $\sqrt{3^2 + (-4)^2} = 5$ (c) $\sqrt{2^2 + (-4)^2} = 2\sqrt{5}$ (d) $\sqrt{2^2 + 3^2} = \sqrt{13}$

8. (a) $\sqrt{4^2 + 3^2} = 5$ (b) $\sqrt{3^2} = 3$ (c) $\sqrt{4^2} = 4$ (d) $\sqrt{4^2 + 3^2} = 5$

10. The lengths of the lines joining $P(1,3,5)$, $Q(-2,0,3)$, and $R(7,9,9)$ are

$$\|PQ\| = \sqrt{(-3)^2 + (-3)^2 + (-2)^2} = \sqrt{22}, \quad \|QR\| = \sqrt{(9)^2 + (9)^2 + (6)^2} = 3\sqrt{22},$$

$$\|PR\| = \sqrt{(6)^2 + (6)^2 + (4)^2} = 2\sqrt{22}.$$

Since $\|QR\| = \|PQ\| + \|PR\|$, the three points are collinear.

12. A point $P(x,y,z)$ is equidistant from $(-3,0,4)$ and $(2,1,5)$ if and only if $(x+3)^2 + (y-0)^2 + (z-4)^2 = (x-2)^2 + (y-1)^2 + (z-5)^2$ and this equation reduces to $10x + 2y + 2z = 5$. The equation should describe a plane.

14. Because $\|OJ\| = \|KA\|$, the x and y coordinates of A and J are the same, namely, $\sqrt{2}$ and $\sqrt{2}$. The coordinates of A are therefore $(\sqrt{2}, \sqrt{2}, 5)$. The length of BD is

$$\sqrt{\|BC\|^2 + \|CD\|^2} = \sqrt{(1/2)^2 + (1/2)^2} = \sqrt{2}/2.$$

The length of AL is

$$\sqrt{\|AD\|^2 - \|LD\|^2} = \sqrt{(3/4)^2 - (\sqrt{2}/4)^2} = \sqrt{7}/4.$$

Consequently, the coordinates of the remaining
corners are

$B(\sqrt{2}+1/4, \sqrt{2}-1/4, 5-\sqrt{7}/4)$,
$C(\sqrt{2}+1/4, \sqrt{2}+1/4, 5-\sqrt{7}/4)$,
$D(\sqrt{2}-1/4, \sqrt{2}+1/4, 5-\sqrt{7}/4)$,
$E(\sqrt{2}-1/4, \sqrt{2}-1/4, 5-\sqrt{7}/4)$,
$F(\sqrt{2}+1/4, \sqrt{2}-1/4, 9/2-\sqrt{7}/4)$,
$G(\sqrt{2}+1/4, \sqrt{2}+1/4, 9/2-\sqrt{7}/4)$,
$H(\sqrt{2}-1/4, \sqrt{2}+1/4, 9/2-\sqrt{7}/4)$,
$I(\sqrt{2}-1/4, \sqrt{2}-1/4, 9/2-\sqrt{7}/4)$.

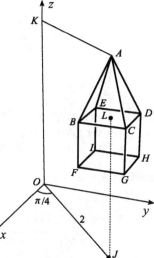

16. (a) According to Exercise 15, the required coordinates are $\left(\dfrac{1+3}{2}, \dfrac{-1+2}{2}, \dfrac{-3-4}{2}\right) = \left(2, \dfrac{1}{2}, -\dfrac{7}{2}\right)$.

(b) If the coordinates of R are (x,y,z), then, because Q is midway between P and R,

$(3,2,-4) = \left(\dfrac{x+1}{2}, \dfrac{y-1}{2}, \dfrac{z-3}{2}\right)$. When coordinates are equated, we obtain $x = 5$, $y = 5$, and $z = -5$.

18. If coordinates of the vertices are as shown in the figure, then coordinates of the midpoints of the sides are

$$P\left(\frac{x_1+x_2}{2}, \frac{y_1+y_2}{2}, \frac{z_1+z_2}{2}\right), \quad Q\left(\frac{x_2+x_3}{2}, \frac{y_2+y_3}{2}, \frac{z_2+z_3}{2}\right),$$

$$R\left(\frac{x_3+x_4}{2}, \frac{y_3+y_4}{2}, \frac{z_3+z_4}{2}\right), \quad S\left(\frac{x_4+x_1}{2}, \frac{y_4+y_1}{2}, \frac{z_4+z_1}{2}\right).$$

Midpoints of the line segments PR and QS both have coordinates

$$\left(\frac{x_1+x_2+x_3+x_4}{4}, \frac{y_1+y_2+y_3+y_4}{4}, \frac{z_1+z_2+z_3+z_4}{4}\right),$$

and the line segments therefore intersect in this point.

20. The x-coordinate of the tip of the shadow is $10 + a = 10 + 1 = 11$. The y-coordinate is 1. Let P be the point where the tip of the shadow would fall on the ground were the building not there. If z is the z-coordinate of the tip of the shadow on the wall, then from similar triangles, we may write

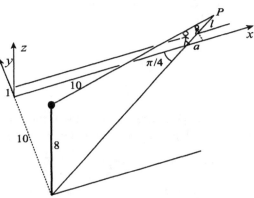

$$\frac{8}{2} = \frac{11\sqrt{2}+l}{\sqrt{2}+l} \quad \text{and} \quad \frac{8}{z} = \frac{11\sqrt{2}+l}{l}.$$

From the second of these, we obtain

$l = \dfrac{11\sqrt{2}z}{8-z}$, which substituted into the first gives

$$4\left(\sqrt{2} + \frac{11\sqrt{2}z}{8-z}\right) = 11\sqrt{2} + \frac{11\sqrt{2}z}{8-z}.$$

The solution of this equation is $z = 7/5$.

EXERCISES 11.2

See answers in text for even numbered exercises.

EXERCISES 11.3

2. $2\mathbf{w} + 3\mathbf{v} = 2(4,3,-2) + 3(-2,0,4) = (2,6,8)$

4. $\hat{\mathbf{v}} = \dfrac{\mathbf{v}}{|\mathbf{v}|} = \dfrac{(-2,0,4)}{\sqrt{20}} = \left(-\dfrac{1}{\sqrt{5}}, 0, \dfrac{2}{\sqrt{5}}\right)$

6. $|\mathbf{v}|\mathbf{v} - 2|\hat{\mathbf{v}}|\mathbf{w} = \sqrt{20}(-2,0,4) - 2(1)(4,3,-2) = (-4\sqrt{5}-8, -6, 8\sqrt{5}+4)$

8. $|3\mathbf{u}|\mathbf{v} - |-2\mathbf{v}|\mathbf{u} = 3\sqrt{46}(-2,0,4) - 2\sqrt{20}(1,3,6) = (-6\sqrt{46}-4\sqrt{5}, -12\sqrt{5}, 12\sqrt{46}-24\sqrt{5})$

10. $\dfrac{\mathbf{v}-\mathbf{w}}{|\mathbf{v}+\mathbf{w}|} = \dfrac{\mathbf{v}-\mathbf{w}}{|(-2,0,4)+(4,3,-2)|} = \dfrac{(-2,0,4)-(4,3,-2)}{|(2,3,2)|} = \dfrac{(-6,-3,6)}{\sqrt{17}} = \left(-\dfrac{6}{\sqrt{17}}, -\dfrac{3}{\sqrt{17}}, \dfrac{6}{\sqrt{17}}\right)$

12. $\mathbf{u} - \mathbf{v} = 3\hat{\mathbf{i}} - 2\hat{\mathbf{j}}$

14. $\hat{\mathbf{v}} + \hat{\mathbf{u}} = \left(\dfrac{2}{\sqrt{5}} - \dfrac{1}{\sqrt{10}}\right)\hat{\mathbf{i}} + \left(\dfrac{1}{\sqrt{5}} + \dfrac{3}{\sqrt{10}}\right)\hat{\mathbf{j}}$

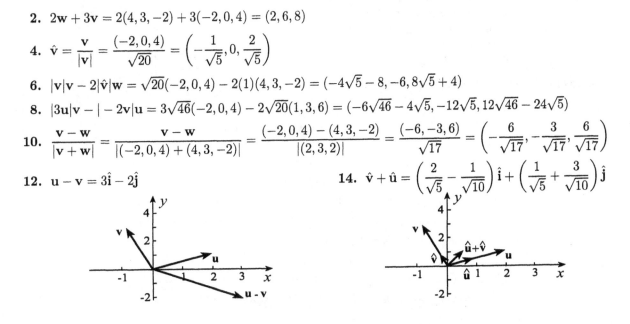

16. $5\hat{\mathbf{i}} = (5, 0, 0)$

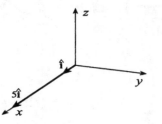

18. The vector from $(1, 1, 1)$ to $(1, 3, 5)$ is $(1, 3, 5) - (1, 1, 1) = (0, 2, 4)$. The vector of length 3 in this direction is
$$\frac{3(0, 2, 4)}{|(0, 2, 4)|} = \left(0, \frac{3}{\sqrt{5}}, \frac{6}{\sqrt{5}}\right)$$

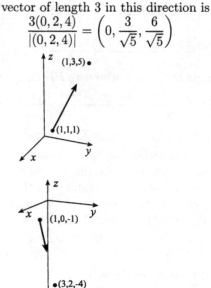

20. The vector from $(1, 0, -1)$ to $(3, 2, -4)$ is $(2, 2, -3)$. A vector half as long is $(1, 1, -3/2)$.

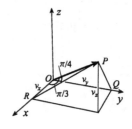

22. Let (v_x, v_y, v_z) be the components of the vector, and draw perpendiculars from the tip P of the vector to the x- and y-axes. Since triangle OPQ is right-angled at Q, $\|OQ\|/\|OP\| = \cos(\pi/4)$. Consequently,

$$v_y = \|OQ\| = \|OP\|\cos(\pi/4) = \frac{5}{2}\left(\frac{1}{\sqrt{2}}\right) = \frac{5}{2\sqrt{2}}.$$

Similarly, $v_x = \dfrac{5}{2}\cos(\pi/3) = \dfrac{5}{2}\left(\dfrac{1}{2}\right) = \dfrac{5}{4}$.

Finally, because the length of the vector is $5/2$,
$$\frac{5}{2} = \sqrt{v_x^2 + v_y^2 + v_z^2} = \sqrt{25/16 + 25/8 + v_z^2}.$$

Thus, $v_z{}^2 = \dfrac{25}{4} - \dfrac{25}{16} - \dfrac{25}{8} = \dfrac{25}{16} \implies v_z = 5/4$,

and the vector has components $(5/4, 5\sqrt{2}/4, 5/4)$.

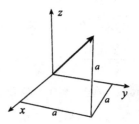

24. If the vector makes equal angles with the positive coordinate axes, its components must all be equal. If we take its components as (a, a, a), then the fact that its length is 2 requires
$$2 = \sqrt{a^2 + a^2 + a^2} = \sqrt{3}\,a \implies a = 2/\sqrt{3}.$$
The required vector is $(2/\sqrt{3}, 2/\sqrt{3}, 2/\sqrt{3})$.

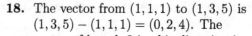

26. If we set $\mathbf{u} = (u_x, u_y, u_z)$, $\mathbf{v} = (v_x, v_y, v_z)$, and $\mathbf{w} = (w_x, w_y, w_z)$, then

$$\mathbf{u} + \mathbf{v} = (u_x + v_x, u_y + v_y, u_z + v_z) = (v_x + u_x, v_y + u_y, v_z + u_z) = \mathbf{v} + \mathbf{u};$$

$$(\mathbf{u} + \mathbf{v}) + \mathbf{w} = (u_x + v_x, u_y + v_y, u_z + v_z) + (w_x, w_y, w_z)$$
$$= (u_x + v_x + w_x, u_y + v_y + w_y, u_z + v_z + w_z)$$
$$= (u_x, u_y, u_z) + (v_x + w_x, v_y + w_y, v_z + w_z) = \mathbf{u} + (\mathbf{v} + \mathbf{w});$$

$$\lambda(\mathbf{u} + \mathbf{v}) = \lambda(u_x + v_x, u_y + v_y, u_z + v_z)$$
$$= \left(\lambda(u_x + v_x), \lambda(u_y + v_y), \lambda(u_z + v_z)\right)$$
$$= (\lambda u_x + \lambda v_x, \lambda u_y + \lambda v_y, \lambda u_z + \lambda v_z)$$
$$= (\lambda u_x, \lambda u_y, \lambda u_z) + (\lambda v_x, \lambda v_y, \lambda v_z) = \lambda\mathbf{u} + \lambda\mathbf{v};$$

$$(\lambda + \mu)\mathbf{v} = \big((\lambda + \mu)v_x, (\lambda + \mu)v_y, (\lambda + \mu)v_z\big)$$
$$= (\lambda v_x + \mu v_x, \lambda v_y + \mu v_y, \lambda v_z + \mu v_z)$$
$$= (\lambda v_x, \lambda v_y, \lambda v_z) + (\mu v_x, \mu v_y, \mu v_z) = \lambda\mathbf{v} + \mu\mathbf{v}.$$

28. All vectors lie on the cone $z = \sqrt{x^2 + y^2}$.

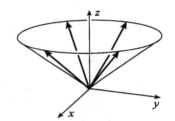

30. The slope of the curve $y = f(x) = x^2$ at $(2,4)$ is $f'(2) = 4$. A vector along the tangent line is $(1,4)$. A vector of length 3 along this line is $\mathbf{T} = \dfrac{3(1,4)}{\sqrt{1^2 + 4^2}} = \left(\dfrac{3}{\sqrt{17}}, \dfrac{12}{\sqrt{17}}\right)$.

32. The force due to the mass at $(5,1,3)$ is

$$\mathbf{F}_1 = \frac{G(5)(10)}{3^2 + (-1)^2 + 1^2}\frac{(3,-1,1)}{\sqrt{3^2 + (-1)^2 + 1^2}} = \frac{50G}{11\sqrt{11}}(3,-1,1) \text{ N.}$$

The force due to the mass at $(-1,2,1)$ is

$$\mathbf{F}_2 = \frac{G(5)(10)}{(-3)^2 + 0^2 + (-1)^2}\frac{(-3,0,-1)}{\sqrt{(-3)^2 + (-1)^2}} = \frac{5G}{\sqrt{10}}(-3,0,-1) \text{ N.}$$

The resultant force due to both masses is

$$\mathbf{F} = \mathbf{F}_1 + \mathbf{F}_2 = \frac{50G(3,-1,1)}{11\sqrt{11}} + \frac{5G(-3,0,-1)}{\sqrt{10}} = 5G\left(\frac{30}{11\sqrt{11}} - \frac{3}{\sqrt{10}}, \frac{-10}{11\sqrt{11}}, \frac{10}{11\sqrt{11}} - \frac{1}{\sqrt{10}}\right) \text{ N.}$$

34. Let T_{AC} and T_{BC} be tensions in cables AC and BC. For equilibrium when both cables are taut, x- and y-components of all forces acting at C must sum to zero:

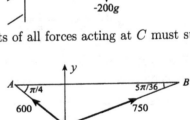

$$0 = T_{BC}\cos 5\pi/12 - T_{AC}\sin 5\pi/12,$$
$$0 = T_{BC}\sin 5\pi/12 - T_{AC}\cos 5\pi/12 - 200g,$$

where $g = 9.81$. When these are solved for T_{AC} and T_{BC}, the result, (to the nearest newton), is $T_{AC} = 586$ N and $T_{BC} = 2188$ N.

36. For equilibrium when both cables are taut, x- and y-components of all forces acting at C must sum to zero:

$$0 = 750\cos 5\pi/36 - 600\cos \pi/4 - |\mathbf{F}|\cos\theta,$$
$$0 = 750\sin 5\pi/36 + 600\sin \pi/4 - |\mathbf{F}|\sin\theta.$$

When we write

$$|\mathbf{F}|\sin\theta = 750\sin 5\pi/36 + 600/\sqrt{2},$$
$$|\mathbf{F}|\cos\theta = 750\cos 5\pi/36 - 600/\sqrt{2},$$

and divide one equation by the other,

$$\tan\theta = \frac{750\sin 5\pi/36 + 600/\sqrt{2}}{750\cos 5\pi/36 - 600/\sqrt{2}} \quad\Longrightarrow\quad \theta = 1.24 \text{ radians.}$$

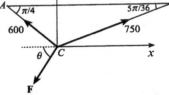

This in turn implies that $|\mathbf{F}| = 784$ N.

38. Let T_{AD}, T_{BD}, and T_{CD} denote tensions in the wires. The sum of all forces acting on the plate must be zero. There is the weight of the plate $\mathbf{W} = (0, 0, -16g)$, where $g = 9.81$, and tensions in the wires,

$$\mathbf{T}_{AD} = T_{AD}\left(\frac{\mathbf{AD}}{|\mathbf{AD}|}\right) = \frac{T_{AD}(0, -400, 600)}{\sqrt{400^2 + 600^2}} = \frac{T_{AD}(0, -2, 3)}{\sqrt{13}},$$

$$\mathbf{T}_{BD} = T_{BD}\left(\frac{\mathbf{BD}}{|\mathbf{BD}|}\right) = \frac{T_{BD}(-200, 200, 600)}{\sqrt{200^2 + 200^2 + 600^2}}$$

$$= \frac{T_{BD}(-1, 1, 3)}{\sqrt{11}},$$

$$\mathbf{T}_{CD} = T_{CD}\left(\frac{\mathbf{CD}}{|\mathbf{CD}|}\right) = \frac{T_{CD}(200, 200, 600)}{\sqrt{200^2 + 200^2 + 600^2}}$$

$$= \frac{T_{CD}(1, 1, 3)}{\sqrt{11}}.$$

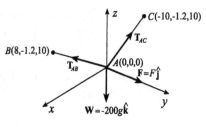

Hence, $\quad \mathbf{0} = (0, 0, -16g) + \dfrac{T_{AD}(0, -2, 3)}{\sqrt{13}} + \dfrac{T_{BD}(-1, 1, 3)}{\sqrt{11}} + \dfrac{T_{CD}(1, 1, 3)}{\sqrt{11}}.$ When we equate components to zero,

$$0 = -\frac{T_{BD}}{\sqrt{11}} + \frac{T_{CD}}{\sqrt{11}}, \quad 0 = -\frac{2T_{AD}}{\sqrt{13}} + \frac{T_{BD}}{\sqrt{11}} + \frac{T_{CD}}{\sqrt{11}}, \quad 0 = -16g + \frac{3T_{AD}}{\sqrt{13}} + \frac{3T_{BD}}{\sqrt{11}} + \frac{3T_{CD}}{\sqrt{11}}.$$

These can be solved for $T_{AD} = 16\sqrt{13}g/9$ N and $T_{BD} = T_{CD} = 16\sqrt{11}g/9$ N.

40. Let T_{AB} and T_{AC} denote tensions in the cables. The sum of all forces acting at A must be zero. There is $\mathbf{F} = F\hat{\mathbf{j}}$, $\mathbf{W} = -200g\hat{\mathbf{k}}$ $\quad (g = 9.81)$,

$$\mathbf{T}_{AB} = T_{AB}\left(\frac{\mathbf{AB}}{|\mathbf{AB}|}\right) = \frac{T_{AB}(8, -1.2, 10)}{\sqrt{8^2 + 1.2^2 + 10^2}}$$

$$= \frac{T_{AB}(20, -3, 25)}{\sqrt{1034}},$$

$$\mathbf{T}_{AC} = T_{AC}\left(\frac{\mathbf{AC}}{|\mathbf{AC}|}\right) = \frac{T_{AC}(-10, -1.2, 10)}{\sqrt{10^2 + 1.2^2 + 10^2}}$$

$$= \frac{T_{AC}(-25, -3, 25)}{\sqrt{1259}}.$$

Hence, $\quad \mathbf{0} = F\hat{\mathbf{j}} - 200g\hat{\mathbf{k}} + \dfrac{T_{AB}(20, -3, 25)}{\sqrt{1034}} + \dfrac{T_{AC}(-25, -3, 25)}{\sqrt{1259}}.$ When we equate components to zero,

$$0 = \frac{20T_{AB}}{\sqrt{1034}} - \frac{25T_{AC}}{\sqrt{1259}}, \quad 0 = F - \frac{3T_{AB}}{\sqrt{1034}} - \frac{3T_{AC}}{\sqrt{1259}}, \quad 0 = -200g + \frac{25T_{AB}}{\sqrt{1034}} + \frac{25T_{AC}}{\sqrt{1259}}.$$

These can be solved for $T_{AB} = 40\sqrt{1034}g/9$ N, $T_{AC} = 32\sqrt{1259}g/9$ N, and $F = 24g$ N.

42. Since $0\mathbf{u} - 2\mathbf{v} + \mathbf{w} = \mathbf{0}$, the vectors are linearly dependent.

44. If we set $\mathbf{0} = a\mathbf{u} + b\mathbf{v} + c\mathbf{w} = a(-1, 3, -5) + b(2, 4, -1) + c(3, 11, -7)$

$$= (-a + 2b + 3c, 3a + 4b + 11c, -5a - b - 7c),$$

then $-a + 2b + 3c = 0$, $3a + 4b + 11c = 0$, $-5a - b - 7c = 0$. This system of equations has an infinite number of solutions representable in the form $a = -c$, $b = -2c$, where c is arbitrary. The vectors are therefore linearly dependent.

46. If the coordinates of A, B, and C are (x_1, y_1, z_1), (x_2, y_2, z_2), and (x_3, y_3, z_3) respectively, then

$$D = \left(\frac{x_1 + x_2}{2}, \frac{y_1 + y_2}{2}, \frac{z_1 + z_2}{2}\right) \quad \text{and} \quad E = \left(\frac{x_1 + x_3}{2}, \frac{y_1 + y_3}{2}, \frac{z_1 + z_3}{2}\right).$$

Thus, $\mathbf{DE} = \left(\dfrac{x_3 - x_2}{2}, \dfrac{y_3 - y_2}{2}, \dfrac{z_3 - z_2}{2}\right)$ and $\mathbf{BC} = (x_3 - x_2, y_3 - y_2, z_3 - z_2)$. Clearly, $\mathbf{DE} = (1/2)\mathbf{BC}$, and the result is therefore verified.

48. If we take x- and y-components of the vector equation $M(\overline{x}, \overline{y}) = \sum_{i=1}^{n} m_i \mathbf{r}_i = \sum_{i=1}^{n} m_i(x_i, y_i)$, we obtain

$$M\overline{x} = \sum_{i=1}^{n} m_i x_1 \quad \text{and} \quad M\overline{y} = \sum_{i=1}^{n} m_i y_i. \text{ These are equations 7.31 and 7.32.}$$

50. (a) By summing vertical components of forces acting at D,

$$0 = -W + 2k(\sqrt{L^2 + y^2} - L)\sin\theta$$
$$= -W + 2k(\sqrt{L^2 + y^2} - L)\frac{y}{\sqrt{L^2 + y^2}}.$$

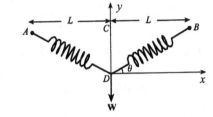

Hence, $\quad W = 2ky\left(1 - \dfrac{L}{\sqrt{L^2 + y^2}}\right).$

(b) When y is very much less than L, we use the binomial expansion to write,

$$W = 2ky\left[1 - \frac{1}{\sqrt{1 + (y/L)^2}}\right] = 2ky\left\{1 - \left[1 - \frac{1}{2}\left(\frac{y}{L}\right)^2 + \cdots\right]\right\}.$$

If we neglect higher order terms, $\quad W \approx 2ky\left(\dfrac{y^2}{2L^2}\right) = \dfrac{ky^3}{L^2}.$

EXERCISES 11.4

2. $(\mathbf{v} \cdot \mathbf{w})\mathbf{u} = [(0)(6) + (1)(-2) + (-1)(3)](2, -3, 1) = (-10, 15, -5)$

4. $2\hat{\mathbf{i}} \cdot \hat{\mathbf{u}} = (2, 0, 0) \cdot \dfrac{(2, -3, 1)}{\sqrt{4 + 9 + 1}} = \dfrac{4}{\sqrt{14}}$

6. $(3\mathbf{u} - 4\mathbf{w}) \cdot (2\hat{\mathbf{i}} + 3\mathbf{u} - 2\mathbf{v}) = (-18, -1, -9) \cdot (8, -11, 5) = -144 + 11 - 45 = -178$

8. $\dfrac{(105\mathbf{u} + 240\mathbf{v}) \cdot (105\mathbf{u} + 240\mathbf{v})}{|105\mathbf{u} + 240\mathbf{v}|^2} = \dfrac{|105\mathbf{u} + 240\mathbf{v}|^2}{|105\mathbf{u} + 240\mathbf{v}|^2} = 1$

10. $\mathbf{u} \cdot \mathbf{v} + \mathbf{v} \cdot \mathbf{w} - (\mathbf{u} + \mathbf{w}) \cdot \mathbf{v} = \mathbf{u} \cdot \mathbf{v} + \mathbf{v} \cdot \mathbf{w} - \mathbf{u} \cdot \mathbf{v} - \mathbf{w} \cdot \mathbf{v} = 0$

12. $(-3\mathbf{u}) \times (2\mathbf{v}) = -6(\mathbf{u} \times \mathbf{v}) = -6\begin{vmatrix} \hat{\mathbf{i}} & \hat{\mathbf{j}} & \hat{\mathbf{k}} \\ 3 & 1 & 4 \\ -1 & 2 & 0 \end{vmatrix} = -6[-8\hat{\mathbf{i}} - 4\hat{\mathbf{j}} + 7\hat{\mathbf{k}}] = 48\hat{\mathbf{i}} + 24\hat{\mathbf{j}} - 42\hat{\mathbf{k}}$

14. $\hat{\mathbf{u}} \times \hat{\mathbf{w}} = \dfrac{\mathbf{u}}{|\mathbf{u}|} \times \dfrac{\mathbf{w}}{|\mathbf{w}|} = \dfrac{1}{|\mathbf{u}||\mathbf{w}|}\mathbf{u} \times \mathbf{w} = \dfrac{1}{\sqrt{9 + 1 + 16}\sqrt{4 + 9 + 25}}\begin{vmatrix} \hat{\mathbf{i}} & \hat{\mathbf{j}} & \hat{\mathbf{k}} \\ 3 & 1 & 4 \\ -2 & -3 & 5 \end{vmatrix}$

$$= \dfrac{1}{\sqrt{26}\sqrt{38}}(17\hat{\mathbf{i}} - 23\hat{\mathbf{j}} - 7\hat{\mathbf{k}}) = \dfrac{1}{2\sqrt{247}}(17\hat{\mathbf{i}} - 23\hat{\mathbf{j}} - 7\hat{\mathbf{k}})$$

16. $\mathbf{u} \times (3\mathbf{v} - \mathbf{w}) = \begin{vmatrix} \hat{\mathbf{i}} & \hat{\mathbf{j}} & \hat{\mathbf{k}} \\ 3 & 1 & 4 \\ -1 & 9 & -5 \end{vmatrix} = -41\hat{\mathbf{i}} + 11\hat{\mathbf{j}} + 28\hat{\mathbf{k}}$

18. $\mathbf{u} \times \mathbf{w} - \mathbf{u} \times \mathbf{v} + \mathbf{u} \times (2\mathbf{u} + \mathbf{v}) = \mathbf{u} \times (\mathbf{w} - \mathbf{v} + 2\mathbf{u} + \mathbf{v}) = \mathbf{u} \times (\mathbf{w} + 2\mathbf{u})$

$$= \mathbf{u} \times \mathbf{w} \quad (\text{since } \mathbf{u} \times \mathbf{u} = 0)$$

$$= \begin{vmatrix} \hat{\mathbf{i}} & \hat{\mathbf{j}} & \hat{\mathbf{k}} \\ 3 & 1 & 4 \\ -2 & -3 & 5 \end{vmatrix} = 17\hat{\mathbf{i}} - 23\hat{\mathbf{j}} - 7\hat{\mathbf{k}}$$

20. $\mathbf{u} \times (\mathbf{v} \times \mathbf{w}) = \mathbf{u} \times \begin{vmatrix} \hat{\mathbf{i}} & \hat{\mathbf{j}} & \hat{\mathbf{k}} \\ -1 & 2 & 0 \\ -2 & -3 & 5 \end{vmatrix} = \mathbf{u} \times (10, 5, 7) = \begin{vmatrix} \hat{\mathbf{i}} & \hat{\mathbf{j}} & \hat{\mathbf{k}} \\ 3 & 1 & 4 \\ 10 & 5 & 7 \end{vmatrix} = (-13, 19, 5)$

22. Since $(2, 4) \cdot (-8, 4) = -16 + 16 = 0$, the vectors are perpendicular.

24. Since $(2, 3, -6) \cdot (-6, 6, 1) = -12 + 18 - 6 = 0$, the vectors are perpendicular.

26. If θ is the angle between the vectors, then

$$\theta = \text{Cos}^{-1}\left[\frac{(1, 6) \cdot (-4, 7)}{|(1, 6)| \, |(-4, 7)|}\right] = \text{Cos}^{-1}\left(\frac{38}{\sqrt{37}\sqrt{65}}\right) = 0.684 \text{ radians}.$$

28. If θ is the angle between the vectors, then

$$\theta = \text{Cos}^{-1}\left[\frac{(3, 1, -1) \cdot (-2, 1, 4)}{|(3, 1, -1)| \, |(-2, 1, 4)|}\right] = \text{Cos}^{-1}\left(\frac{-9}{\sqrt{11}\sqrt{21}}\right) = 2.20 \text{ radians}.$$

30. Since $(-2, -6, 4) = -2(1, 3, -2)$, the vectors are in opposite directions, and $\theta = \pi$.

32. One such vector is $\hat{\mathbf{j}} \times (1, -1, -9) = \begin{vmatrix} \hat{\mathbf{i}} & \hat{\mathbf{j}} & \hat{\mathbf{k}} \\ 0 & 1 & 0 \\ 1 & -1 & -9 \end{vmatrix} = (-9, 0, -1)$. All such vectors can be represented by $\lambda(9, 0, 1)$.

34. If we set $\mathbf{u} = (u_x, u_y, u_z)$, $\mathbf{v} = (v_x, v_y, v_z)$, and $\mathbf{w} = (w_x, w_y, w_z)$, then

$$\mathbf{u} \cdot \mathbf{v} = u_x v_x + u_y v_y + u_z v_z = v_x u_x + v_y u_y + v_z u_z = \mathbf{v} \cdot \mathbf{u},$$

$$\mathbf{u} \cdot (\lambda \mathbf{v} + \rho \mathbf{w}) = (u_x, u_y, u_z) \cdot (\lambda v_x + \rho w_x, \lambda v_y + \rho w_y, \lambda v_z + \rho w_z)$$
$$= u_x(\lambda v_x + \rho w_x) + u_y(\lambda v_y + \rho w_y) + u_z(\lambda v_z + \rho w_z)$$
$$= \lambda(u_x v_x + u_y v_y + u_z v_z) + \rho(u_x w_x + u_y w_y + u_z w_z) = \lambda(\mathbf{u} \cdot \mathbf{v}) + \rho(\mathbf{u} \cdot \mathbf{w}).$$

36. If α is the angle between $\mathbf{v}$ and $\hat{\mathbf{i}}$, then $\mathbf{v} \cdot \hat{\mathbf{i}} = |\mathbf{v}||\hat{\mathbf{i}}| \cos\alpha \implies \cos\alpha = \dfrac{\mathbf{v} \cdot \hat{\mathbf{i}}}{|\mathbf{v}|}$.

Since the angle between two vectors always lies between 0 and π, and such angles coincide with the principal values of the inverse cosine function, we may write that

$$\alpha = \text{Cos}^{-1}\left(\frac{\mathbf{v} \cdot \hat{\mathbf{i}}}{|\mathbf{v}|}\right) = \text{Cos}^{-1}\left(\frac{v_x}{|\mathbf{v}|}\right).$$

Similar discussions lead to the formulas for β and γ.

38. $\alpha = \text{Cos}^{-1} 0 = \pi/2$; $\beta = \text{Cos}^{-1}\left(\dfrac{1}{\sqrt{1+9}}\right) = 1.25$; $\gamma = \text{Cos}^{-1}\left(\dfrac{-3}{\sqrt{10}}\right) = 2.82$.

40. $\alpha = \text{Cos}^{-1}\left(\dfrac{-2}{\sqrt{4+9+16}}\right) = 1.95$; $\beta = \text{Cos}^{-1}\left(\dfrac{3}{\sqrt{29}}\right) = 0.980$; $\gamma = \text{Cos}^{-1}\left(\dfrac{4}{\sqrt{29}}\right) = 0.734$.

42. One example serves to show this. See Exercises 19 and 20 for an example.

44. (a) $\mathbf{u} \cdot \mathbf{v} \times \mathbf{w} = (6, -1, 0) \cdot \begin{vmatrix} \hat{\mathbf{i}} & \hat{\mathbf{j}} & \hat{\mathbf{k}} \\ 1 & 3 & 4 \\ -2 & -1 & 4 \end{vmatrix} = (6, -1, 0) \cdot (16, -12, 5) = 108$

(b) This is verified when we compare for general $\mathbf{u}$, $\mathbf{v}$, and $\mathbf{w}$,

$$\mathbf{u} \cdot \mathbf{v} \times \mathbf{w} = (u_x, u_y, u_z) \cdot \begin{vmatrix} \hat{\mathbf{i}} & \hat{\mathbf{j}} & \hat{\mathbf{k}} \\ v_x & v_y & v_z \\ w_x & w_y & w_z \end{vmatrix}$$
$$= (u_x, u_y, u_z) \cdot [(v_y w_z - v_z w_y)\hat{\mathbf{i}} + (v_z w_x - v_x w_z)\hat{\mathbf{j}} + (v_x w_y - v_y w_x)\hat{\mathbf{k}}]$$
$$= u_x(v_y w_z - v_z w_y) + u_y(v_z w_x - v_x w_z) + u_z(v_x w_y - v_y w_x)$$

and

$$\mathbf{u} \times \mathbf{v} \cdot \mathbf{w} = \begin{vmatrix} \hat{\mathbf{i}} & \hat{\mathbf{j}} & \hat{\mathbf{k}} \\ u_x & u_y & u_z \\ v_x & v_y & v_z \end{vmatrix} \cdot (w_x, w_y, w_z)$$
$$= [(u_y v_z - u_z v_y)\hat{\mathbf{i}} + (u_z v_x - u_x v_z)\hat{\mathbf{j}} + (u_x v_y - u_y v_x)\hat{\mathbf{k}}] \cdot (w_x, w_y, w_z)$$
$$= (u_y v_z - u_z v_y)w_x + (u_z v_x - u_x v_z)w_y + (u_x v_y - u_y v_x)w_z.$$

(c) The volume of the parallelepiped is the area of one of the parallelogram sides multiplied by the perpendicular distance to the opposite side. The area of the parallelogram with sides $\mathbf{v}$ and $\mathbf{w}$ in the left figure below is $|\mathbf{v}|d = |\mathbf{v}||\mathbf{w}|\sin\theta = |\mathbf{v}\times\mathbf{w}|$. The perpendicular distance between the parallel sides, one of which contains $\mathbf{v}$ and $\mathbf{w}$ (right figure), is $c = |\mathbf{u}|\sin\phi$. The volume of the parallelepiped is

$$(|\mathbf{u}|\sin\phi)|\mathbf{v}\times\mathbf{w}| = |\mathbf{u}||\mathbf{v}\times\mathbf{w}|\cos(\pi/2-\phi) = |\mathbf{u}\cdot\mathbf{v}\times\mathbf{w}|.$$

The absolute values are included to take care of the possibility that the scalar product could be negative.

(d) This result follows from the fact that the three vectors are coplanar if and only if the volume of the parallelepiped with the vectors as coterminal sides is zero.

46. If $\mathbf{u} = (u_x, u_y, u_z)$, $\mathbf{v} = (v_x, v_y, v_z)$, and $\mathbf{w} = (w_x, w_y, w_z)$, then

$$\mathbf{u}\times(\mathbf{v}\times\mathbf{w}) = \begin{vmatrix} \hat{\mathbf{i}} & \hat{\mathbf{j}} & \hat{\mathbf{k}} \\ u_x & u_y & u_z \\ v_yw_z - v_zw_y & v_zw_x - v_xw_z & v_xw_y - v_yw_x \end{vmatrix}$$

$$= [u_y(v_xw_y - v_yw_x) - u_z(v_zw_x - v_xw_z)]\hat{\mathbf{i}} + [u_z(v_yw_z - v_zw_y) - u_x(v_xw_y - v_yw_x)]\hat{\mathbf{j}}$$
$$+ [u_x(v_zw_x - v_xw_z) - u_y(v_yw_z - v_zw_y)]\hat{\mathbf{k}},$$

and

$$(\mathbf{u}\cdot\mathbf{w})\mathbf{v} - (\mathbf{u}\cdot\mathbf{v})\mathbf{w} = (u_xw_x + u_yw_y + u_zw_z)(v_x, v_y, v_z) - (u_xv_x + u_yv_y + u_zv_z)(w_x, w_y, w_z)$$

$$= [v_x(u_xw_x + u_yw_y + u_zw_z) - w_x(u_xv_x + u_yv_y + u_zv_z)]\hat{\mathbf{i}}$$
$$+ [v_y(u_xw_x + u_yw_y + u_zw_z) - w_y(u_xv_x + u_yv_y + u_zv_z)]\hat{\mathbf{j}}$$
$$+ [v_z(u_xw_x + u_yw_y + u_zw_z) - w_z(u_xv_x + u_yv_y + u_zv_z)]\hat{\mathbf{k}}.$$

Comparison of components shows that these vectors are identical.

48. If we let $\mathbf{w} = (|\mathbf{v}|\mathbf{u} + |\mathbf{u}|\mathbf{v})/||\mathbf{u}|\mathbf{v} + |\mathbf{v}|\mathbf{u}|$, then $\mathbf{w}$ is clearly a unit vector, since it is a vector divided by its own length. If we let θ be the angle between $\mathbf{u}$ and $\mathbf{v}$, then $\mathbf{u}\cdot\mathbf{v} = |\mathbf{u}||\mathbf{v}|\cos\theta$. If we let ϕ be the angle between $\mathbf{w}$ and $\mathbf{u}$, then

$$\mathbf{u}\cdot\mathbf{w} = |\mathbf{u}||\mathbf{w}|\cos\phi \quad\Longrightarrow\quad \cos\phi = \frac{|\mathbf{v}|(\mathbf{u}\cdot\mathbf{u}) + |\mathbf{u}|(\mathbf{v}\cdot\mathbf{u})}{|\mathbf{u}|(||\mathbf{u}|\mathbf{v} + |\mathbf{v}|\mathbf{u}|)} = \frac{|\mathbf{v}||\mathbf{u}|^2 + |\mathbf{u}|(\mathbf{v}\cdot\mathbf{u})}{|\mathbf{u}|(||\mathbf{u}|\mathbf{v} + |\mathbf{v}|\mathbf{u}|)} = \frac{|\mathbf{v}||\mathbf{u}| + (\mathbf{v}\cdot\mathbf{u})}{||\mathbf{u}|\mathbf{v} + |\mathbf{v}|\mathbf{u}|}.$$

Now,

$$(|\mathbf{u}||\mathbf{v}| + \mathbf{v}\cdot\mathbf{u})^2 = |\mathbf{u}|^2|\mathbf{v}|^2 + 2|\mathbf{u}||\mathbf{v}|\mathbf{v}\cdot\mathbf{u} + (\mathbf{v}\cdot\mathbf{u})^2$$
$$= |\mathbf{u}|^2|\mathbf{v}|^2 + 2|\mathbf{u}|^2|\mathbf{v}|^2\cos\theta + |\mathbf{u}|^2|\mathbf{v}|^2\cos^2\theta$$
$$= |\mathbf{u}|^2|\mathbf{v}|^2(1 + \cos\theta)^2,$$

and

$$\big||\mathbf{u}|\mathbf{v} + |\mathbf{v}|\mathbf{u}\big|^2 = (|\mathbf{u}|\mathbf{v} + |\mathbf{v}|\mathbf{u})\cdot(|\mathbf{u}|\mathbf{v} + |\mathbf{v}|\mathbf{u})$$
$$= |\mathbf{u}|^2|\mathbf{v}|^2 + |\mathbf{v}|^2|\mathbf{u}|^2 + 2|\mathbf{u}||\mathbf{v}|(\mathbf{u}\cdot\mathbf{v})$$
$$= 2|\mathbf{u}|^2|\mathbf{v}|^2(1 + \cos\theta).$$

It follows that

$$2\cos^2\phi - 1 = \frac{2[|\mathbf{u}|^2|\mathbf{v}|^2(1 + \cos\theta)^2]}{2|\mathbf{u}|^2|\mathbf{v}|^2(1 + \cos\theta)} - 1 = \cos\theta.$$

In other words $\phi = \theta/2$.

EXERCISES 11.5

2. A vector normal to the plane is $(4, 2, 3) - (2, 1, 5) = (2, 1, -2)$. The equation of the plane is therefore $0 = (2, 1, -2) \cdot (x - 2, y - 1, z - 5) = 2x + y - 2z + 5$.

4. One vector in the plane is $(3, 4, 1)$. Since $(1, -5, -2)$ is a second point in the plane, a second vector in the plane is $(2, -4, 3) - (1, -5, -2) = (1, 1, 5)$. A vector normal to the plane is

$$(3, 4, 1) \times (1, 1, 5) = \begin{vmatrix} \hat{\mathbf{i}} & \hat{\mathbf{j}} & \hat{\mathbf{k}} \\ 3 & 4 & 1 \\ 1 & 1 & 5 \end{vmatrix} = (19, -14, -1).$$

The equation of the plane is $0 = (19, -14, -1) \cdot (x - 2, y + 4, z - 3) = 19x - 14y - z - 91$.

6. Two lines determine a plane only if they are parallel or they intersect. Since these lines are parallel (a vector along each is $(3, 4, 1)$), they determine a plane. Since $(1, 0, -2)$ and $(-1, 2, -5)$ are points on the lines, a second vector in the plane is $(1, 0, -2) - (-1, 2, -5) = (2, -2, 3)$. A vector normal to the plane is

$$(3, 4, 1) \times (2, -2, 3) = \begin{vmatrix} \hat{\mathbf{i}} & \hat{\mathbf{j}} & \hat{\mathbf{k}} \\ 3 & 4 & 1 \\ 2 & -2 & 3 \end{vmatrix} = (14, -7, -14), \quad \text{as is} \quad (2, -1, -2).$$

The equation of the plane is $0 = (2, -1, -2) \cdot (x - 1, y, z + 2) = 2x - y - 2z - 6$.

8. Two lines determine a plane only if they are parallel or they intersect. The vectors $(1, 2, 4)$ and $(1, -1, 6)$ are normal to the planes $x + 2y + 4z = 21$ and $x - y + 6z = 13$, respectively. A vector along the line determined by these planes is

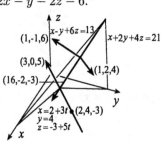

$$(1, 2, 4) \times (1, -1, 6) = \begin{vmatrix} \hat{\mathbf{i}} & \hat{\mathbf{j}} & \hat{\mathbf{k}} \\ 1 & 2 & 4 \\ 1 & -1 & 6 \end{vmatrix} = (16, -2, -3).$$

Since $(3, 0, 5)$ is a vector along the second line, the given lines are not parallel. To ensure that the lines intersect, we substitute $x = 2 + 3t$, $y = 4$, and $z = -3 + 5t$ into $x + 2y + 4z = 21$ giving $(2 + 3t) + 8 + 4(-3 + 5t) = 21 \implies t = 1$. This gives the point $(5, 4, 2)$, which also satisfies $x - y + 6z = 13$. The lines therefore intersect at this point. A vector normal to the required plane is

$$(16, -2, -3) \times (3, 0, 5) = \begin{vmatrix} \hat{\mathbf{i}} & \hat{\mathbf{j}} & \hat{\mathbf{k}} \\ 16 & -2 & -3 \\ 3 & 0 & 5 \end{vmatrix} = (-10, -89, 6).$$

The equation of the plane is therefore $0 = (10, 89, -6)(x - 2, y - 4, z + 3) = 10x + 89y - 6z - 394$.

10. (a) Since the vectors $(1, 1, -4)$ and $(2, 3, 5)$ are normal to the planes $x + y - 4z = 6$ and $2x + 3y + 5z = 10$, respectively, a vector along their line of intersection is

$$(1, 1, -4) \times (2, 3, 5) = \begin{vmatrix} \hat{\mathbf{i}} & \hat{\mathbf{j}} & \hat{\mathbf{k}} \\ 1 & 1 & -4 \\ 2 & 3 & 5 \end{vmatrix} = (17, -13, 1).$$

Since the plane is to be perpendicular to the xy-plane, $\hat{\mathbf{k}}$ must also lie in the plane, and a vector normal to the required plane is

$$(17, -13, 1) \times \hat{\mathbf{k}} = \begin{vmatrix} \hat{\mathbf{i}} & \hat{\mathbf{j}} & \hat{\mathbf{k}} \\ 17 & -13 & 1 \\ 0 & 0 & 1 \end{vmatrix} = (-13, -17, 0).$$

Since a point on the plane is $(8, -2, 0)$, the equation of the plane is $0 = (13, 17, 0) \cdot (x - 8, y + 2, z) = 13x + 17y - 70$. Similar derivations give the other two equations.

12. A vector equation is $\mathbf{r} = (x, y, z) = (1, -1, 3) + t(2, 4, -3)$. By equating components we obtain parametric equations $x = 1 + 2t$, $y = -1 + 4t$, $z = 3 - 3t$, and by solving each for t, symmetric equations are $\dfrac{x-1}{2} = \dfrac{y+1}{4} = \dfrac{z-3}{-3}$.

14. Since a vector along the line is $(3, 5, -5)$, a vector equation for the line is $\mathbf{r} = (x, y, z) = (2, -3, 4) + t(3, 5, -5)$. By equating components we obtain parametric equations $x = 2 + 3t$, $y = -3 + 5t$, $z = 4 - 5t$, and by solving each for t, symmetric equations are $\dfrac{x-2}{3} = \dfrac{y+3}{5} = \dfrac{z-4}{-5}$.

16. Since a vector along the line is $(0, 0, 1)$, a vector equation for the line is $\mathbf{r} = (x, y, z) = (1, 3, 4) + t(0, 0, 1)$. By equating components we obtain parametric equations $x = 1$, $y = 3$, $z = 4 + t$. Symmetric equations do not exist.

18. Since a vector along the line is $(1, 0, -2)$, a vector equation for the line is $\mathbf{r} = (x, y, z) = (2, 0, 3) + u(1, 0, -2)$. By equating components we obtain parametric equations $x = 2 + u$, $y = 0$, $z = 3 - 2u$, and by solving the first and last for u, partial symmetric equations are $x - 2 = \dfrac{z-3}{-2}$, $y = 0$.

20. Parametric equations for the line are $x = t$, $y = 2t - 5$, $z = 10 - 3(t) - 4(2t - 5) = 30 - 11t$. By solving each for t, symmetric equations are $x = \dfrac{y+5}{2} = \dfrac{z-30}{-11}$. A vector equation is $\mathbf{r} = (x, y, z) = (0, -5, 30) + t(1, 2, -11)$.

22. Parametric equations for the line are $x = 3 + 2t$, $y = 2 + t$, $z = -1 + 4t$. When these values are substituted into $x - y + 2z$, the result is $x - y + 2z = (3 + 2t) - (2 + t) + 2(-1 + 4t) = -1 + 9t \not\equiv -1$. Consequently, the line does not lie in the plane.

24. The equation of every plane is of the form $Ax + By + Cz + D = 0$. Because its intercept with the x-axis is a, it follows that $a = -D/A$ or $A = -D/a$. Similarly, $B = -D/b$ and $C = -D/c$. Thus, the equation of the plane is $-\dfrac{D}{a}x - \dfrac{D}{b}y - \dfrac{D}{c}z + D = 0 \implies \dfrac{x}{a} + \dfrac{y}{b} + \dfrac{z}{c} = 1$.

26. In Exercise 11.1–14, the coordinates of the corners of the birdhouse were determined as shown. The equation of face:

$FGHI$ is $z = 9/2 - \sqrt{7}/4$;

$BFIE$ is $y = \sqrt{2} - 1/4$;

$CGHD$ is $y = \sqrt{2} + 1/4$;

$BFCG$ is $x = \sqrt{2} + 1/4$;

$EIHD$ is $x = \sqrt{2} - 1/4$.

A vector normal to face ABC is

$$(4\mathbf{AB}) \times (4\mathbf{AC}) = \begin{vmatrix} \hat{\mathbf{i}} & \hat{\mathbf{j}} & \hat{\mathbf{k}} \\ 1 & -1 & -\sqrt{7} \\ 1 & 1 & -\sqrt{7} \end{vmatrix}$$

$$= (2\sqrt{7}, 0, 2).$$

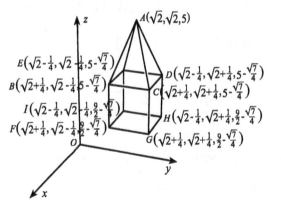

The equation of this face is therefore

$$0 = (\sqrt{7}, 0, 1) \cdot (x - \sqrt{2}, y - \sqrt{2}, z - 5)$$

$$= \sqrt{7}x + z - 5 - \sqrt{14}.$$

Similarly, equations for faces ACD, ADE, and AEB are $\sqrt{7}y + z - \sqrt{14} - 5 = 0$, $\sqrt{7}x - z = \sqrt{14} - 5$, and $\sqrt{7}y - z = \sqrt{14} - 5$.

28. (a) Since the coordinates of A and B are $(0, 1, 1)$ and $(1, 1, 1)$, it follows that $\mathbf{AB} = (1, 0, 0)$ and $\mathbf{CB} = (0, 1, 1)$. Because $\mathbf{AB} = \mathbf{OC}$ and $\mathbf{CB} = \mathbf{OA}$, $OCBA$ is a rectangle with area $|\mathbf{OA}||\mathbf{OC}| = \sqrt{2}(1)$.

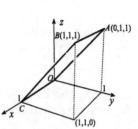

(b) The coordinates of the points A, B and C on the plane $x + y - 2z = 0$ directly above E, F, and D are $(0, 1, 1/2)$, $(1, 1, 1)$, and $(1, 0, 1/2)$, respectively (left figure below). Since $\mathbf{OA} = (0, 1, 1/2)$ is parallel to $\mathbf{CB} = (0, 1, 1/2)$, and $\mathbf{OC} = (1, 0, 1/2)$ is parallel to $\mathbf{AB} = (1, 0, 1/2)$, it follows that $OABC$ is a parallelogram with area (right figure below)

$$|\mathbf{OC}|(AD) = |\mathbf{OC}||\mathbf{OA}| \sin\phi = |\mathbf{OC} \times \mathbf{OA}| = \left\| \begin{matrix} \hat{\mathbf{i}} & \hat{\mathbf{j}} & \hat{\mathbf{k}} \\ 1 & 0 & 1/2 \\ 0 & 1 & 1/2 \end{matrix} \right\| = |(-1/2, -1/2, 1)| = \frac{\sqrt{6}}{2}.$$

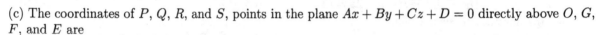

(c) The coordinates of P, Q, R, and S, points in the plane $Ax + By + Cz + D = 0$ directly above O, G, F, and E are

$$P(0, 0, -D/C), \ Q(1, 0, -(D + A)/C), \ R(1, 1, -(D + A + B)/C), \ S(0, 1, -(D + B)/C).$$

Consequently,

$$\mathbf{PQ} = (1, 0, -A/C), \ \mathbf{QR} = (0, 1, -B/C), \ \mathbf{PS} = (0, 1, -B/C), \ \mathbf{SR} = (1, 0, -A/C).$$

Because $\mathbf{PQ} = \mathbf{SR}$ and $\mathbf{PS} = \mathbf{QR}$, $PQRS$ is a parallelogram with area

$$|\mathbf{PS}|(QT) = |\mathbf{PS}||\mathbf{PQ}| \sin\phi = |\mathbf{PQ} \times \mathbf{PS}| = \left\| \begin{matrix} \hat{\mathbf{i}} & \hat{\mathbf{j}} & \hat{\mathbf{k}} \\ 1 & 0 & -A/C \\ 0 & 1 & -B/C \end{matrix} \right\|$$

$$= |(A/C, B/C, 1)|$$
$$= \sqrt{A^2/C^2 + B^2/C^2 + 1}$$
$$= \frac{\sqrt{A^2 + B^2 + C^2}}{|C|}.$$

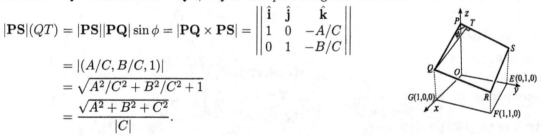

The acute angle between the xy-plane and the plane $Ax + By + Cz + D = 0$ is defined as the acute angle between their normals (see Exercise 11). Normals to these planes are $\hat{\mathbf{k}}$ and (A, B, C). If θ is the angle between these vectors, then

$$\hat{\mathbf{k}} \cdot (A, B, C) = |\hat{\mathbf{k}}||(A, B, C)| \cos\theta = (1)\sqrt{A^2 + B^2 + C^2} \cos\theta.$$

Consequently, $\cos\theta = C/\sqrt{A^2 + B^2 + C^2}$. If $C > 0$, then θ is acute and $\theta = \gamma$ where

$$\cos\gamma = \frac{C}{\sqrt{A^2 + B^2 + C^2}}.$$

If $C < 0$, then θ is obtuse and $\gamma = \pi - \theta$ where

$$\cos(\pi - \gamma) = \frac{C}{\sqrt{A^2 + B^2 + C^2}} \implies \cos\gamma = \frac{-C}{\sqrt{A^2 + B^2 + C^2}}.$$

Thus, $\cos\gamma = \dfrac{|C|}{\sqrt{A^2 + B^2 + C^2}}$ or $\sec\gamma = \dfrac{\sqrt{A^2 + B^2 + C^2}}{|C|}.$

EXERCISES 11.6

2. The area of the triangle is

$$\frac{1}{2}|\mathbf{AB} \times \mathbf{AC}| = \frac{1}{2}\begin{vmatrix} \hat{\mathbf{i}} & \hat{\mathbf{j}} & \hat{\mathbf{k}} \\ 6 & 1 & -1 \\ -5 & 2 & 1 \end{vmatrix}$$

$$= \frac{1}{2}|(3, -1, 17)| = \frac{\sqrt{299}}{2}.$$

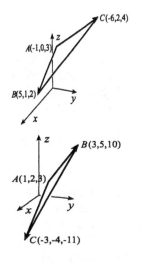

4. The area of the triangle is

$$\frac{1}{2}|\mathbf{AB} \times \mathbf{AC}| = \frac{1}{2}\begin{vmatrix} \hat{\mathbf{i}} & \hat{\mathbf{j}} & \hat{\mathbf{k}} \\ 2 & 3 & 7 \\ -4 & -6 & -14 \end{vmatrix}$$

$$= \frac{1}{2}|(0,0,0)| = 0.$$

The points must be collinear.

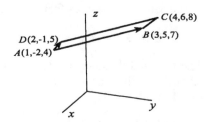

6. The area of the parallelogram is

$$|\mathbf{AB} \times \mathbf{AD}| = \begin{vmatrix} \hat{\mathbf{i}} & \hat{\mathbf{j}} & \hat{\mathbf{k}} \\ 2 & 7 & 3 \\ 1 & 1 & 1 \end{vmatrix}$$

$$= |(4, 1, -5)| = \sqrt{42}.$$

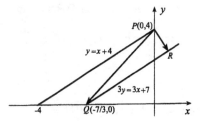

8. Since $(4, -3, 2) - (-1, 2, 3) = (5, -5, -1)$ is a vector in the required direction, the component is

$$\mathbf{v} \cdot \frac{(5, -5, -1)}{\sqrt{51}} = \frac{22}{\sqrt{51}}.$$

10. Since a vector along the line is $(-1, 2, -3)$, the component is $\mathbf{v} \cdot \dfrac{(1, -2, 3)}{\sqrt{14}} = \dfrac{18}{\sqrt{14}}$.

12. Since the lines are parallel, the required distance d is the component of $\mathbf{PQ}$ along $\mathbf{PR}$. Since the slope of the line is 1, a vector along it is $(1, 1)$, and a vector in the same direction as $\mathbf{PR}$ is $(1, -1)$. Consequently,

$$d = \left|\mathbf{PQ} \cdot \widehat{\mathbf{PR}}\right| = \left|(-7/3, -4) \cdot \frac{(1, -1)}{\sqrt{1+1}}\right| = \frac{5}{3\sqrt{2}}.$$

We could also have used formula 1.16.

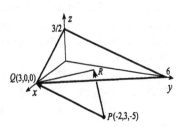

14. The required distance d is the component of $\mathbf{PQ}$ along $\mathbf{PR}$. Since a vector in the same direction as $\mathbf{PR}$ is $(2, 1, 4)$,

$$d = \left|\mathbf{PQ} \cdot \widehat{\mathbf{PR}}\right|$$

$$= \left|(5, -3, 5) \cdot \frac{(2, 1, 4)}{\sqrt{4+1+16}}\right| = \frac{27}{\sqrt{21}}.$$

16. First we confirm that the line is parallel
to the plane (else the distance is zero).
Since a vector along the line is

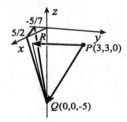

$$\begin{vmatrix} \hat{\mathbf{i}} & \hat{\mathbf{j}} & \hat{\mathbf{k}} \\ 1 & 1 & 1 \\ 2 & -1 & 1 \end{vmatrix} = (2, 1, -3),$$

and $(2, 1, -3) \cdot (2, -7, -1) = 0$, the line is
indeed parallel to the plane. Since a point
on the line is $P(3, 3, 0)$, the required distance d is the component of **PQ** along **PR**. Since a vector in
the same direction as **PR** is $(2, -7, -1)$,

$$d = \left| \mathbf{PQ} \cdot \widehat{\mathbf{PR}} \right| = \left| (-3, -3, -5) \cdot \frac{(2, -7, -1)}{\sqrt{4 + 49 + 1}} \right| = \frac{20}{3\sqrt{6}}.$$

18. First we confirm that the line is parallel to the plane
(else the distance is zero). Since a vector along the
line is $(-6, 4, -1)$ and $(-6, 4, -1) \cdot (1, 1, -2) = 0$,
the line is indeed parallel to the plane. Since a point
on the line is $P(1, 2, 0)$, the required distance is the
component of **PQ** along **PR**. Since a vector
in the same direction as **PR** is $(-1, -1, 2)$,

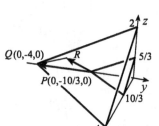

$$d = \left| \mathbf{PQ} \cdot \widehat{\mathbf{PR}} \right| = \left| (-1, -1, 0) \cdot \frac{(-1, -1, 2)}{\sqrt{6}} \right| = \frac{2}{\sqrt{6}}.$$

20. The required distance d is the component of **PQ**
along **PR**. Since a vector in the same direction
as **PR** is $(1, -1, 2)$,

$$d = \left| \mathbf{PQ} \cdot \widehat{\mathbf{PR}} \right|$$

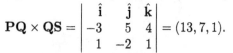

$$= \left| (0, -2/3, 0) \cdot \frac{(1, -1, 2)}{\sqrt{1 + 1 + 4}} \right| = \frac{2}{3\sqrt{6}}.$$

22. The required distance is the component of **PQ**
along **PR**. A vector perpendicular to **PQ** and **QS** is

$$\mathbf{PQ} \times \mathbf{QS} = \begin{vmatrix} \hat{\mathbf{i}} & \hat{\mathbf{j}} & \hat{\mathbf{k}} \\ -3 & 5 & 4 \\ 1 & -2 & 1 \end{vmatrix} = (13, 7, 1).$$

A vector in direction **PR** is therefore

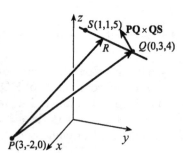

$$(\mathbf{PQ} \times \mathbf{QS}) \times \mathbf{SQ} = \begin{vmatrix} \hat{\mathbf{i}} & \hat{\mathbf{j}} & \hat{\mathbf{k}} \\ 13 & 7 & 1 \\ -1 & 2 & -1 \end{vmatrix}$$

$$= (-9, 12, 33) \quad \text{or} \quad (-3, 4, 11).$$

Thus, $\quad d = \left| \mathbf{PQ} \cdot \widehat{\mathbf{PR}} \right| = \left| (-3, 5, 4) \cdot \frac{(-3, 4, 11)}{\sqrt{(-3)^2 + (4)^2 + (11)^2}} \right| = \frac{73}{\sqrt{146}} = \frac{\sqrt{146}}{2}.$

24. Since the point $(1, 3, 3)$ is on the line, the minimum distance is 0.

26. Following Example 11.24, the distance is the
component of **RS** along **PQ**. A vector in
direction **PQ** is

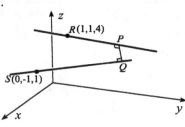

$$(1, 3, 2) \times (2, -1, 2) = \begin{vmatrix} \hat{\mathbf{i}} & \hat{\mathbf{j}} & \hat{\mathbf{k}} \\ 1 & 3 & 2 \\ 2 & -1 & 2 \end{vmatrix} = (8, 2, -7).$$

Consequently,

$$d = \left| \mathbf{RS} \cdot \widehat{\mathbf{PQ}} \right| = \left| (-1, -2, -3) \cdot \frac{(8, 2, -7)}{\sqrt{117}} \right| = \frac{9}{\sqrt{117}}.$$

28. Following Example 11.24, the distance is the component of **RS** along **PQ**. A vector in direction **PQ** is

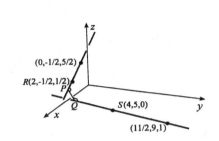

$$(-2, 0, 2) \times (3, 8, 2) = \begin{vmatrix} \hat{\mathbf{i}} & \hat{\mathbf{j}} & \hat{\mathbf{k}} \\ -2 & 0 & 2 \\ 3 & 8 & 2 \end{vmatrix} = (-16, 10, -16).$$

Consequently,

$$d = \left| \mathbf{RS} \cdot \widehat{\mathbf{PQ}} \right| = \left| (2, 11/2, -1/2) \cdot \frac{(-8, 5, -8)}{\sqrt{153}} \right| = \frac{31}{2\sqrt{153}}.$$

30. A point on the line is $Q(2, 1, 3)$ and a vector along the line is $\mathbf{v} = (-1, 2, 0)$. Using the technique of Exercise 29,

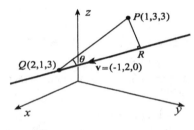

$$|\mathbf{PR}| = |\mathbf{PQ}| \sin\theta = |\mathbf{PQ}||\hat{\mathbf{v}}| \sin\theta$$

$$= |\mathbf{PQ} \times \hat{\mathbf{v}}| = \frac{1}{|\mathbf{v}|} |\mathbf{PQ} \times \mathbf{v}|$$

$$= \frac{1}{\sqrt{5}} \left\| \begin{vmatrix} \hat{\mathbf{i}} & \hat{\mathbf{j}} & \hat{\mathbf{k}} \\ 1 & -2 & 0 \\ -1 & 2 & 0 \end{vmatrix} \right\|$$

$$= \frac{1}{\sqrt{5}} |(0, 0, 0)| = 0.$$

Hence the point is on the line.

32. If coordinates of the vertices are as shown in the figure, then coordinates of the midpoints of the sides are

$$P\left(\frac{x_1 + x_2}{2}, \frac{y_1 + y_2}{2}, \frac{z_1 + z_2}{2} \right), \quad Q\left(\frac{x_2 + x_3}{2}, \frac{y_2 + y_3}{2}, \frac{z_2 + z_3}{2} \right),$$

$$R\left(\frac{x_3 + x_4}{2}, \frac{y_3 + y_4}{2}, \frac{z_3 + z_4}{2} \right), \quad S\left(\frac{x_4 + x_1}{2}, \frac{y_4 + y_1}{2}, \frac{z_4 + z_1}{2} \right).$$

Since $\mathbf{PQ} = \left(\frac{x_3 - x_1}{2}, \frac{y_3 - y_1}{2}, \frac{z_3 - z_1}{2} \right) = \mathbf{SR}$, and similarly, $\mathbf{PS} = \mathbf{QR}$, it follows that $PQRS$ is a parallelogram.

34. Since $\hat{\mathbf{v}} \cdot \hat{\mathbf{w}} = 1/2 - 1/2 = 0$, $\hat{\mathbf{v}}$ and $\hat{\mathbf{w}}$ are perpendicular. The components of **u** along $\hat{\mathbf{v}}$ and $\hat{\mathbf{w}}$ are

$$\lambda = \mathbf{u} \cdot \hat{\mathbf{v}} = \frac{2}{\sqrt{2}} + \frac{1}{\sqrt{2}} = \frac{3}{\sqrt{2}} \quad \text{and} \quad \rho = \mathbf{u} \cdot \hat{\mathbf{w}} = \frac{2}{\sqrt{2}} - \frac{1}{\sqrt{2}} = \frac{1}{\sqrt{2}}.$$

36. Since $\hat{\mathbf{u}} \cdot \hat{\mathbf{v}} = (1/\sqrt{70})(-2 + 2 + 0) = 0$, $\hat{\mathbf{u}} \cdot \hat{\mathbf{w}} = [1/(5\sqrt{14})](6 - 6 + 0) = 0$, and $\hat{\mathbf{v}} \cdot \hat{\mathbf{w}} = [1/(14\sqrt{5})](-3 - 12 + 15) = 0$, the three unit vectors are mutually perpendicular. The components of **r** along $\hat{\mathbf{u}}$, $\hat{\mathbf{v}}$, and $\hat{\mathbf{w}}$ are

$$\mathbf{r} \cdot \hat{\mathbf{u}} = \frac{1}{\sqrt{5}}(2 + 3) = \sqrt{5}; \ \mathbf{r} \cdot \hat{\mathbf{v}} = \frac{1}{\sqrt{14}}(-1 + 6 - 12) = -\frac{\sqrt{14}}{2}; \ \mathbf{r} \cdot \hat{\mathbf{w}} = \frac{1}{\sqrt{70}}(3 - 18 - 20) = -\frac{\sqrt{70}}{2}.$$

38. Since $\mathbf{v} \cdot \mathbf{w} = 6 - 6 = 0$, **v** and **w** are perpendicular. Because $\mathbf{u} \cdot \mathbf{v} = \lambda \mathbf{v} \cdot \mathbf{v} + \rho \mathbf{w} \cdot \mathbf{v} = \lambda |\mathbf{v}|^2$, it follows that $\lambda = \frac{\mathbf{u} \cdot \mathbf{v}}{|\mathbf{v}|^2} = \frac{3 - 6}{1 + 9} = -\frac{3}{10}$. Similarly, $\rho = \frac{\mathbf{u} \cdot \mathbf{w}}{|\mathbf{w}|^2} = \frac{18 + 4}{36 + 4} = \frac{11}{20}$.

40. The equation of the plane must be of the form $x - 2y + 3z + D = 0$. The distance from $P(1, 2, 3)$ to this plane is the projection of $\mathbf{PQ}$ along $\mathbf{PR}$, and hence

$$2 = |\mathbf{PQ} \cdot \widehat{\mathbf{PR}}|$$

$$= \left| (-1, -2, -D/3 - 3) \cdot \frac{(1, -2, 3)}{\sqrt{1 + 4 + 9}} \right|$$

$$= \frac{1}{\sqrt{14}} |D + 6|.$$

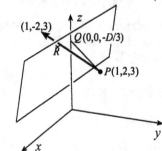

When this equation is solved, $D = -6 \pm 2\sqrt{14}$, and the two possible planes are $x - 2y + 3z = 6 \pm 2\sqrt{14}$.

EXERCISES 11.7

2. $\mathbf{M} = \mathbf{r} \times \mathbf{F} = \begin{vmatrix} \hat{\mathbf{i}} & \hat{\mathbf{j}} & \hat{\mathbf{k}} \\ -1 & 0 & 5 \\ 1 & 2 & 0 \end{vmatrix} = (-10, 5, -2)$

4. $\mathbf{M} = \mathbf{r} \times \mathbf{F} = \begin{vmatrix} \hat{\mathbf{i}} & \hat{\mathbf{j}} & \hat{\mathbf{k}} \\ -1 & -1 & -1 \\ 3 & -1 & 4 \end{vmatrix} = (-5, 1, 4)$

6. (a) Since $\mathbf{v} = (-3, 7, 1)$ is a vector along the line through $P(1, -3, 2)$ and $R(-2, 4, 3)$, the moment about the line is

Moment $= \mathbf{PQ} \times \mathbf{F} \cdot (\pm \hat{\mathbf{v}})$

$$= \pm (0, 6, 0) \times (2, 3, -4) \cdot \frac{(-3, 7, 1)}{\sqrt{59}}$$

$$= \pm \begin{vmatrix} \hat{\mathbf{i}} & \hat{\mathbf{j}} & \hat{\mathbf{k}} \\ 0 & 6 & 0 \\ 2 & 3 & -4 \end{vmatrix} \cdot \frac{(-3, 7, 1)}{\sqrt{59}}$$

$$= \pm (-24, 0, -12) \cdot \frac{(-3, 7, 1)}{\sqrt{59}} = \pm \frac{60}{\sqrt{59}}.$$

(b) Since the moment of $\mathbf{F}$ about O is

$$\mathbf{M} = \mathbf{OQ} \times \mathbf{F} = \begin{vmatrix} \hat{\mathbf{i}} & \hat{\mathbf{j}} & \hat{\mathbf{k}} \\ 1 & 3 & 2 \\ 2 & 3 & -4 \end{vmatrix} = (-18, 8, -3),$$

moments about the x-, y-, and z-axes are -18, 8, and -3, respectively.

(c) We use the point $S(2, 4, 1)$ on the line. If $\hat{\mathbf{v}}$ is a unit vector along the line, the moment of $\mathbf{F}$ about the line is

Moment $= \mathbf{SQ} \times \mathbf{F} \cdot \hat{\mathbf{v}} = (-1, -1, 1) \times (2, 3, -4) \cdot \frac{\pm(3, -2, 5)}{\sqrt{38}}$

$$= \pm \begin{vmatrix} \hat{\mathbf{i}} & \hat{\mathbf{j}} & \hat{\mathbf{k}} \\ -1 & -1 & 1 \\ 2 & 3 & -4 \end{vmatrix} \cdot \frac{(3, -2, 5)}{\sqrt{38}} = \pm(1, -2, -1) \cdot \frac{(3, -2, 5)}{\sqrt{38}} = \pm \frac{2}{\sqrt{38}}.$$

8. Since $P(0, 0, 1)$ is a point on the line and $\mathbf{v} = (1, 1, 1)$ is a vector along the line, the moment of $\mathbf{F}$ at $Q(6, -2, 1)$ is

Moment $= \pm \mathbf{PQ} \times \mathbf{F} \cdot \hat{\mathbf{v}} = \pm(6, -2, 0) \times (4, 0, -2) \cdot \frac{(1, 1, 1)}{\sqrt{3}}$

$$= \pm \begin{vmatrix} \hat{\mathbf{i}} & \hat{\mathbf{j}} & \hat{\mathbf{k}} \\ 6 & -2 & 0 \\ 4 & 0 & -2 \end{vmatrix} \cdot \frac{(1, 1, 1)}{\sqrt{3}} = \pm(4, 12, 8) \cdot \frac{(1, 1, 1)}{\sqrt{3}} = \pm 8\sqrt{3}.$$

10. Moments about the axes are M_x, M_y, and M_z.

12. Since the vector $\mathbf{PQ}$ in equation 11.46 is equal to zero, the moment is zero.

14. When the sleeve is at D, the magnitude of $\mathbf{F}$ is $|\mathbf{F}| = k[\sqrt{(1-x)^2 + 1/4} - l]$, and therefore

$\mathbf{F} = k[\sqrt{(1-x)^2 + 1/4} - l]\dfrac{(1-x), 1/2)}{\sqrt{(1-x)^2 + 1/4}}$. For a small displacement dx at position x, the amount of work done by $\mathbf{F}$ is approximately

$$\mathbf{F} \cdot (dx\,\hat{\mathbf{i}}) = k[\sqrt{(1-x)^2 + 1/4} - l]\frac{1-x}{\sqrt{(1-x)^2 + 1/4}}dx = k(1-x)\left[1 - \frac{l}{\sqrt{(1-x)^2 + 1/4}}\right]dx.$$

The total work done between B and C is therefore

$$W = \int_0^1 k(1-x)\left[1 - \frac{l}{\sqrt{(1-x)^2 + 1/4}}\right]dx = k\int_0^1\left[1 - x - \frac{l(1-x)}{\sqrt{(1-x)^2 + 1/4}}\right]dx$$

$$= k\left\{x - \frac{x^2}{2} + l\sqrt{(1-x)^2 + 1/4}\right\}_0^1 = \frac{k}{2}[1 + l(1 - \sqrt{5})] \text{ J}.$$

16. At position P, the magnitude of the force $\mathbf{F}_a$ of attraction of the asteroid on the rocket is $|\mathbf{F}_a| = \dfrac{GMm}{x^2 + R^2}$, and therefore

$\mathbf{F}_a = \dfrac{GMm}{x^2 + R^2}\dfrac{(-x, -R)}{\sqrt{x^2 + R^2}}$. For a small displacement dx at P, the work done by an equal and opposite force $\mathbf{F}$ is approximately

$$\mathbf{F} \cdot (dx\,\hat{\mathbf{i}}) = \frac{GMm}{x^2 + R^2}\frac{x}{\sqrt{x^2 + R^2}}dx$$

$$= \frac{GMmx}{(x^2 + R^2)^{3/2}}dx.$$

The total work done between A and B is therefore

$$W = \int_0^R \frac{GMmx}{(x^2 + R^2)^{3/2}}dx = GMm\left\{-\frac{1}{\sqrt{x^2 + R^2}}\right\}_0^R = \frac{GMm}{\sqrt{2}R}(\sqrt{2} - 1).$$

18. The work to stretch the two springs is three times the work to stretch the upper spring. The force necessary to hold the upper spring at position x on the x-axis against the upper spring is

$$\mathbf{F} = k(\sqrt{x^2 + l^2} - l)\frac{(x, -l)}{\sqrt{x^2 + l^2}}.$$

The work done by this force in moving the end of the spring a small amount dx along the x-axis is

$$\mathbf{F} \cdot (dx\,\hat{\mathbf{i}}) = k(\sqrt{x^2 + l^2} - l)\frac{x}{\sqrt{x^2 + l^2}}dx$$

$$= k\left(x - \frac{lx}{\sqrt{x^2 + l^2}}\right)dx.$$

The total work done in stretching both springs is therefore

$$W = 3\int_0^L k\left(x - \frac{lx}{\sqrt{x^2 + l^2}}\right)dx = 3k\left\{\frac{x^2}{2} - l\sqrt{x^2 + l^2}\right\}_0^L = 3k\left(\frac{L^2}{2} - l\sqrt{L^2 + l^2} + l^2\right).$$

20. Suppose we denote the components of $\mathbf{F}$ by
$\mathbf{F} = F_x\hat{\mathbf{i}} + F_y\hat{\mathbf{j}} + F_z\hat{\mathbf{k}}$. Then $F_x^2 + F_y^2 + F_z^2 = 200^2$.
Since the angle between $\mathbf{F}$ and $\hat{\mathbf{i}}$ is $\pi/3$ radians,

$F_x = \mathbf{F} \cdot \hat{\mathbf{i}} = |\mathbf{F}||\hat{\mathbf{i}}| \cos \pi/3 = 200(1/2) = 100.$

Similarly,

$F_z = \mathbf{F} \cdot \hat{\mathbf{k}} = |\mathbf{F}||\hat{\mathbf{k}}| \cos 5\pi/6 = 200(-\sqrt{3}/2) = -100\sqrt{3}.$

It follows that

$100^2 + F_y^2 + 3(100)^2 = 200^2 \implies F_y = 0.$

The moment of $\mathbf{F}$ about A is

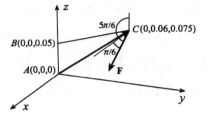

$$\mathbf{M} = (0, 0.06, 0.075) \times (100, 0, -100\sqrt{3}) = \begin{vmatrix} \hat{\mathbf{i}} & \hat{\mathbf{j}} & \hat{\mathbf{k}} \\ 0 & 0.06 & 0.075 \\ 100 & 0 & -100\sqrt{3} \end{vmatrix} = (-6\sqrt{3}, 15/2, -6) \text{ N·m.}$$

22. (a) Since $\mathbf{T} = 900\left(\dfrac{\mathbf{BD}}{|\mathbf{BD}|}\right) = \dfrac{900(2, -1, -2)}{3}$,
the moment of $\mathbf{T}$ at B about O is

$\mathbf{M} = \mathbf{OB} \times \mathbf{T} = (0, 5/2, 2) \times 300(2, -1, -2)$

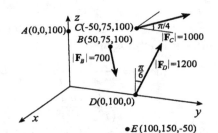

$= 300\begin{vmatrix} \hat{\mathbf{i}} & \hat{\mathbf{j}} & \hat{\mathbf{k}} \\ 0 & 5/2 & 2 \\ 2 & -1 & -2 \end{vmatrix} = 300(-3, 4, -5) \text{ N·m.}$

(b) The moment about the z-axis is -1500 N·m.
(c) Since the moment of $\mathbf{T}$ about B is zero, so also is the moment about any line through B (see Exercise 12).

24. Components of the forces acting at B, C, and D are

$\mathbf{F}_B = 700\left(\dfrac{\mathbf{BE}}{|\mathbf{BE}|}\right) = \dfrac{700(50, 75, -150)}{\sqrt{50^2 + 75^2 + 150^2}} = 100(2, 3, -6),$

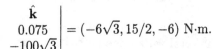
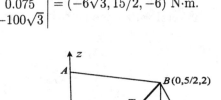

$\mathbf{F}_C = 1000\left(-\dfrac{1}{\sqrt{2}}, \dfrac{1}{\sqrt{2}}, 0\right) = 500\sqrt{2}(-1, 1, 0),$

$\mathbf{F}_D = 1200\left(0, \dfrac{1}{2}, \dfrac{\sqrt{3}}{2}\right) = 600(0, 1, \sqrt{3}).$

The sum of the moments of these forces about A is

$\mathbf{M} = \mathbf{AB} \times \mathbf{F}_B + \mathbf{AC} \times \mathbf{F}_C + \mathbf{AD} \times \mathbf{F}_D$

$= (50, 75, 0) \times 100(2, 3, -6) + (-50, 75, 0) \times 500\sqrt{2}(-1, 1, 0) + (0, 100, -100) \times 600(0, 1, \sqrt{3})$

$= 100\begin{vmatrix} \hat{\mathbf{i}} & \hat{\mathbf{j}} & \hat{\mathbf{k}} \\ 50 & 75 & 0 \\ 2 & 3 & -6 \end{vmatrix} + 500\sqrt{2}\begin{vmatrix} \hat{\mathbf{i}} & \hat{\mathbf{j}} & \hat{\mathbf{k}} \\ -50 & 75 & 0 \\ -1 & 1 & 0 \end{vmatrix} + 600\begin{vmatrix} \hat{\mathbf{i}} & \hat{\mathbf{j}} & \hat{\mathbf{k}} \\ 0 & 100 & -100 \\ 0 & 1 & \sqrt{3} \end{vmatrix}$

$= 100(-450, 300, 0) + 500\sqrt{2}(0, 0, 25) + 600(100\sqrt{3} + 100, 0, 0)$

$= 500(120\sqrt{3} + 120 - 90, 60, 25\sqrt{2}) = (60\sqrt{3} + 15, 30, 25/\sqrt{2}) \text{ N·m.}$

26. Tensions in the cable exerted on C and D are

$\mathbf{T}_C = 1349\left(\dfrac{\mathbf{CE}}{|\mathbf{CE}|}\right) = \dfrac{1349(-9/4, 9/10, 3/2)}{\sqrt{81/16 + 81/100 + 9/4}}$

$= 71(-15, 6, 10),$

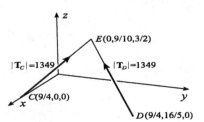

$\mathbf{T}_D = 1349\left(\dfrac{\mathbf{DE}}{|\mathbf{DE}|}\right) = \dfrac{1349(-9/4, -23/10, 3/2)}{\sqrt{81/16 + 529/100 + 9/4}}$

$= 19(-45, -46, 30).$

The moment of $\mathbf{T}_C$ about O is

$$\mathbf{M}_C = \mathbf{OC} \times \mathbf{T}_C = (9/4, 0, 0) \times 71(-15, 6, 10) = 71 \begin{vmatrix} \hat{\mathbf{i}} & \hat{\mathbf{j}} & \hat{\mathbf{k}} \\ 9/4 & 0 & 0 \\ -15 & 6 & 10 \end{vmatrix} = 71(0, -45/2, 27/2).$$

Moments of $\mathbf{T}_C$ about the coordinate axes are therefore 0, $-3195/2$ N·m, and $1917/2$ N·m. The moment of $\mathbf{T}_D$ about O is

$$\mathbf{M}_D = \mathbf{OD} \times \mathbf{T}_D = (9/4, 16/5, 0) \times 19(-45, -46, 30) = 19 \begin{vmatrix} \hat{\mathbf{i}} & \hat{\mathbf{j}} & \hat{\mathbf{k}} \\ 9/4 & 16/5 & 0 \\ -45 & -46 & 30 \end{vmatrix} = 19(96, -135/2, 81/2).$$

Moments of $\mathbf{T}_D$ about the coordinate axes are therefore 1824 N·m, $-2565/2$ N·m, and $1539/2$ N·m.

28. Since tension in the ropes is 300 N, $\mathbf{T}_{AB} = -300\hat{\mathbf{k}}$. The tension in rope AD is

$$\mathbf{T}_{AD} = 300 \left(\frac{\mathbf{AD}}{|\mathbf{AD}|} \right) = \frac{300(-1, a, -2.58)}{\sqrt{1^2 + a^2 + 2.58^2}}$$

$$= \frac{300(-1, a, -2.58)}{\sqrt{a^2 + 7.6564}}.$$

The moment of the sum of these tensions about C is

$$\mathbf{M} = \mathbf{CA} \times (2\mathbf{T}_{AB} + \mathbf{T}_{AD}) = (1, 0, 2.58) \times \left[2(-300\hat{\mathbf{k}}) + \frac{300(-1, a, -2.58)}{\sqrt{a^2 + 7.6564}} \right]$$

$$= \frac{300}{\sqrt{a^2 + 7.6564}}(1, 0, 2.58) \times (-1, a, -2\sqrt{a^2 + 7.6564} - 2.58)$$

$$= \frac{300}{\sqrt{a^2 + 7.6564}} \begin{vmatrix} \hat{\mathbf{i}} & \hat{\mathbf{j}} & \hat{\mathbf{k}} \\ 1 & 0 & 2.58 \\ -1 & a & -2\sqrt{a^2 + 7.6564} - 2.58 \end{vmatrix} = \frac{300}{\sqrt{a^2 + 7.6564}}(-2.58a, 2\sqrt{a^2 + 7.6564}, a) \text{ N·m}.$$

Since the absolute value of the x-component must be less than 375 N·m, it follows that $300(2.58a)/\sqrt{a^2 + 7.6564} \le 375$. This implies that $a \le 1.532$ m.

30. If we denote the z-coordinate of C by z, then from similar triangles AQB and CPB,

$$\frac{\|PC\|}{\|AQ\|} = \frac{\|PB\|}{\|QB\|},$$

from which

$$\frac{z}{0.75} = \frac{\sqrt{0.3^2 + 0.4^2}}{\sqrt{0.6^2 + 0.8^2}} \quad \Longrightarrow \quad z = 0.375.$$

The z-coordinate of D is then $0.375 + 0.575 = 0.95$. Components of $\mathbf{F}_1$ and $\mathbf{F}_2$ are

$$\mathbf{F}_1 = 1175 \left(\frac{\mathbf{DG}}{|\mathbf{DG}|} \right) = \frac{1175(0.45, 0.525, -0.95)}{\sqrt{0.45^2 + 0.525^2 + 0.95^2}} = 1000(0.45, 0.525, -0.95),$$

$$\mathbf{F}_2 = 870 \left(\frac{\mathbf{DH}}{|\mathbf{DH}|} \right) = \frac{870(-0.3, -0.4, -0.525)}{\sqrt{0.3^2 + 0.4^2 + 0.525^2}} = -1200(0.3, 0.4, 0.525).$$

Moments of these forces about C are

$$\mathbf{M}_1 = \mathbf{CD} \times \mathbf{F}_1 = (0, 0, 0.575) \times 1000(0.45, 0.525, -0.95) = 1000 \begin{vmatrix} \hat{\mathbf{i}} & \hat{\mathbf{j}} & \hat{\mathbf{k}} \\ 0 & 0 & 0.575 \\ 0.45 & 0.525 & -0.95 \end{vmatrix}$$

$$= (-301.875, 258.75, 0),$$

$$\mathbf{M}_2 = \mathbf{CD} \times \mathbf{F}_2 = (0, 0, 0.575) \times (-1200)(0.3, 0.4, 0.525) = -1200 \begin{vmatrix} \hat{\mathbf{i}} & \hat{\mathbf{j}} & \hat{\mathbf{k}} \\ 0 & 0 & 0.575 \\ 0.3 & 0.4 & 0.525 \end{vmatrix}$$

$$= -1200(-0.23, 0.1725, 0).$$

Moments of these forces about AB are

$$\mathbf{M}_1 \cdot \left(\frac{\mathbf{AB}}{|\mathbf{AB}|}\right) = (-301.875, 258.75, 0) \cdot \frac{(-0.6, 0.8, -0.75)}{\sqrt{0.6^2 + 0.8^2 + 0.75^2}} = 310.5 \text{ N·m},$$

$$\mathbf{M}_2 \cdot \left(\frac{\mathbf{AB}}{|\mathbf{AB}|}\right) = -1200(-0.23, 0.1725, 0) \cdot \frac{(-0.6, 0.8, -0.75)}{\sqrt{0.6^2 + 0.8^2 + 0.75^2}} = -264.96 \text{ N·m}.$$

32. The moment of $\mathbf{F}_1$ about the line of action of $\mathbf{F}_2$ is

$$\mathbf{QP} \times \mathbf{F}_1 \cdot \left(\frac{\mathbf{F}_2}{|\mathbf{F}_2|}\right),$$

and the moment of $\mathbf{F}_2$ about the line of action of $\mathbf{F}_1$ is

$$\mathbf{PQ} \times \mathbf{F}_2 \cdot \left(\frac{\mathbf{F}_1}{|\mathbf{F}_1|}\right).$$

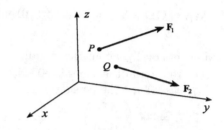

If $\mathbf{PQ} = (a, b, c)$, $\mathbf{F}_1 = (F_x, F_y, F_z)$, and $\mathbf{F}_2 = (G_x, G_y, G_z)$, then

$$\mathbf{PQ} \times \mathbf{F}_2 \cdot \mathbf{F}_1 = \begin{vmatrix} \hat{\mathbf{i}} & \hat{\mathbf{j}} & \hat{\mathbf{k}} \\ a & b & c \\ G_x & G_y & G_z \end{vmatrix} \cdot (F_x, F_y, F_z) = F_x(bG_z - cG_y) + F_y(cG_x - aG_z) + F_z(aG_y - bG_x),$$

and

$$\mathbf{QP} \times \mathbf{F}_1 \cdot \mathbf{F}_2 = \begin{vmatrix} \hat{\mathbf{i}} & \hat{\mathbf{j}} & \hat{\mathbf{k}} \\ -a & -b & -c \\ F_x & F_y & F_z \end{vmatrix} \cdot (G_x, G_y, G_z) = G_x(-bF_z + cF_y) + G_y(-cF_x + aF_z) + G_z(-aF_y + bF_x).$$

Hence, $\mathbf{PQ} \times \mathbf{F}_2 \cdot \mathbf{F}_1 = \mathbf{QP} \times \mathbf{F}_1 \cdot \mathbf{F}_2$, and because $|\mathbf{F}_1| = |\mathbf{F}_2|$, the two moments are equal.

EXERCISES 11.8

2. In the coordinate system shown, the equation of the cable is $y(x) = \dfrac{1100x^2}{2T_0}$, where T_0 is the tension in the cable at $x = 0$. Since $(a, 3)$, and $(a + 40, 9)$ are points on the cable,

$$3 = \frac{550a^2}{T_0}, \qquad 9 = \frac{550(a + 40)^2}{T_0}.$$

These imply that $a = 20(1 - \sqrt{3})$ and $T_0 = 3.93 \times 10^4$. According to the first of equations 11.47, tension in the cable at any point is $T = T_0 \sec\theta = T_0\sqrt{1 + \tan^2\theta} = T_0\sqrt{1 + (dy/dx)^2}$. This means that tension is a minimum when slope is a minimum, namely at $x = 0$. In other words, $T_0 = 3.93 \times 10^4$ N is the minimum tension in the cable. Maximum tension is at $x = a + 40 = 60 - 20\sqrt{3}$ where slope is greatest; i.e., maximum tension is

$$T_0\sqrt{1 + [y'(60 - 20\sqrt{3})]^2} = T_0\sqrt{1 + \left[\frac{1100(60 - 20\sqrt{3})}{T_0}\right]^2} = 4.82 \times 10^4 \text{ N}.$$

4. In the coordinate system shown, the equation of the cable is $y(x) = \dfrac{wx^2}{2T_0}$, where T_0 is the tension in the cable at $x = 0$. Since $(525, 94.8)$ is a point on the cable,

$$94.8 = \frac{142\,000(525)^2}{2T_0}.$$

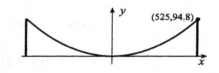

This implies that $T_0 = 2.06 \times 10^8$. According to the first of equations 11.47, tension in the cable at any point is $T = T_0 \sec\theta = T_0\sqrt{1 + \tan^2\theta} = T_0\sqrt{1 + (dy/dx)^2}$. This means that tension is a minimum when slope is a minimum, namely at $x = 0$. In other words, $T_0 = 2.06 \times 10^8$ N is the minimum tension in the cable. Maximum tension is at $x = 525$ where slope is greatest; i.e., maximum tension is

$$T_0\sqrt{1 + [y'(525)]^2} = T_0\sqrt{1 + \left[\frac{w(525)}{T_0}\right]^2} = 2.19 \times 10^8 \text{ N}.$$

6. In the coordinate system shown, the equation of the cable is $\quad y(x) = \dfrac{1000x^2}{2T_0}$,

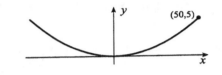

where T_0 is the tension in the cable at $x = 0$. Since $(50, 5)$ is a point on the cable,

$$5 = \frac{500(50)^2}{T_0} \implies T_0 = 250\,000.$$

Using the formula in Exercise 5(a), the length of the cable is

$$2\left\{\frac{50}{2}\sqrt{1 + \left[\frac{1000(50)}{250\,000}\right]^2} + \frac{250\,000}{2(1000)}\ln\left[\sqrt{1 + \left(\frac{1000(50)}{250\,000}\right)^2} + \frac{1000(50)}{250\,000}\right]\right\} = 100.663 \text{ m}.$$

With the two-term approximation in part (b), the length is $2(50)[1 + (2/3)(5/50)^2] = 100.667$ m.

8. When the sag is 115.8 m, the length is approximated by

$$2(639)\left[1 + \frac{2}{3}\left(\frac{115.8}{639}\right)^2\right] = 1305.98.$$

When the sag is 118.2, we find

$$2(639)\left[1 + \frac{2}{3}\left(\frac{118.2}{639}\right)^2\right] = 1307.15.$$

The difference is 1.17 m.

10. Using equation 11.51, the equation of the rope in the coordinate system shown is

$$y = \frac{T_0}{w}\left[\cosh\left(\frac{wx}{T_0}\right) - 1\right].$$

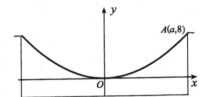

Since $A(a, 8)$ is on the curve,

$$8 = \frac{T_0}{w}\left[\cosh\left(\frac{wa}{T_0}\right) - 1\right].$$

According to Example 11.32, the length of one-half the rope is

$$20 = \frac{T_0}{w}\sinh\left(\frac{wa}{T_0}\right).$$

Finally, Example 11.33 indicates that maximum tension is at A, and therefore $350 = T_0 + 8w$. The first two equations give

$$1 = \cosh^2\left(\frac{wa}{T_0}\right) - \sinh^2\left(\frac{wa}{T_0}\right) = \left(\frac{8w}{T_0} + 1\right)^2 - \left(\frac{20w}{T_0}\right)^2 \implies \frac{16w}{T_0^2}(T_0 - 21w) = 0.$$

Thus, $T_0 = 21w$, and when this is substituted into $T_0 + 8w = 350$, we obtain $29w = 350 \implies w = 350/29$. The mass of the rope is $40(350/29)/9.81 = 49.2$ kg. With $T_0 = 350 - 8(350/29) = 7350/29$,

$$20 = \frac{7350/29}{350/29}\sinh\left(\frac{350a/29}{7350/29}\right) = 21\sinh\left(\frac{a}{21}\right) \implies a = 21\,\text{Sinh}^{-1}\left(\frac{20}{21}\right) = 17.79.$$

Hence, the horizontal distance between the buildings is $2a = 35.6$ m.

12. Using equation 11.51, the equation of the cable in the coordinate system shown is

$$y = \frac{T_0}{w}\left[\cosh\left(\frac{wx}{T_0}\right) - 1\right],$$

where $w = 4(9.81)$. Since $A(a, 37.5)$ is on the curve,

$$37.5 = \frac{T_0}{w}\left[\cosh\left(\frac{wa}{T_0}\right) - 1\right].$$

According to Example 11.32, the length of one-half the cable is $75 = \dfrac{T_0}{w}\sinh\left(\dfrac{wa}{T_0}\right)$. From these,

$$1 = \cosh^2\left(\frac{wa}{T_0}\right) - \sinh^2\left(\frac{wa}{T_0}\right) = \left(\frac{37.5w}{T_0} + 1\right)^2 - \left(\frac{75w}{T_0}\right)^2 \implies \frac{75w}{T_0^2}\left(T_0 - \frac{225w}{4}\right) = 0.$$

Thus, $T_0 = 225w/4$, and this in turn implies that

$$75 = \frac{225}{4}\sinh\left(\frac{4a}{225}\right) \implies a = \frac{225}{4}\operatorname{Sinh}^{-1}\left(\frac{4}{3}\right) = 61.797.$$

The span is therefore $2a = 123.6$ m. Maximum tension is at A where, according to Example 11.33, $T = 225(4)(9.81)/4 + 4(9.81)(37.5) = 3679$ N.

EXERCISES 11.9

2. All real t $\qquad$ **4.** $t > -4$ $\qquad$ **6.** $\dfrac{d\mathbf{u}}{dt} = \hat{\mathbf{i}} - 2t\hat{\mathbf{j}} + 2\hat{\mathbf{k}}$

8. $\dfrac{d}{dt}[g(t)\mathbf{u}(t)] = g'(t)\mathbf{u}(t) + g(t)\mathbf{u}'(t) = (6t^2 - 3)(t\hat{\mathbf{i}} - t^2\hat{\mathbf{j}} + 2t\hat{\mathbf{k}}) + (2t^3 - 3t)(\hat{\mathbf{i}} - 2t\hat{\mathbf{j}} + 2\hat{\mathbf{k}})$

$$= 2t(4t^2 - 3)\hat{\mathbf{i}} + t^2(9 - 10t^2)\hat{\mathbf{j}} + 4t(4t^2 - 3)\hat{\mathbf{k}}$$

10. $\dfrac{d}{dt}(\mathbf{u} \times t\mathbf{v}) = \dfrac{d}{dt}\begin{vmatrix} \hat{\mathbf{i}} & \hat{\mathbf{j}} & \hat{\mathbf{k}} \\ t & -t^2 & 2t \\ t & -2t^2 & 3t^3 \end{vmatrix} = \dfrac{d}{dt}[(4t^3 - 3t^5)\hat{\mathbf{i}} + (2t^2 - 3t^4)\hat{\mathbf{j}} - t^3\hat{\mathbf{k}}]$

$$= (12t^2 - 15t^4)\hat{\mathbf{i}} + (4t - 12t^3)\hat{\mathbf{j}} - 3t^2\hat{\mathbf{k}}$$

12. $\dfrac{d}{dt}(3\mathbf{u} + 4\mathbf{v}) = 3\dfrac{d\mathbf{u}}{dt} + 4\dfrac{d\mathbf{v}}{dt} = 3(\hat{\mathbf{i}} - 2t\hat{\mathbf{j}} + 2\hat{\mathbf{k}}) + 4(-2\hat{\mathbf{j}} + 6t\hat{\mathbf{k}}) = 3\hat{\mathbf{i}} - (6t + 8)\hat{\mathbf{j}} + (6 + 24t)\hat{\mathbf{k}}$

14. $\dfrac{d}{dt}[f(t)\mathbf{u} + g(t)\mathbf{v}] = f'(t)\mathbf{u} + f(t)\mathbf{u}'(t) + g'(t)\mathbf{v} + g(t)\mathbf{v}'(t)$

$$= 2t(t\hat{\mathbf{i}} - t^2\hat{\mathbf{j}} + 2t\hat{\mathbf{k}}) + (t^2 + 3)(\hat{\mathbf{i}} - 2t\hat{\mathbf{j}} + 2\hat{\mathbf{k}})$$
$$+ (6t^2 - 3)(\hat{\mathbf{i}} - 2t\hat{\mathbf{j}} + 3t^2\hat{\mathbf{k}}) + (2t^3 - 3t)(-2\hat{\mathbf{j}} + 6t\hat{\mathbf{k}})$$
$$= 9t^2\hat{\mathbf{i}} + (6t - 20t^3)\hat{\mathbf{j}} + (6 - 21t^2 + 30t^4)\hat{\mathbf{k}}$$

16. $\dfrac{d}{dt}[t(\mathbf{u} \times \mathbf{v})] = \dfrac{d}{dt}\left[t\begin{vmatrix} \hat{\mathbf{i}} & \hat{\mathbf{j}} & \hat{\mathbf{k}} \\ t & -t^2 & 2t \\ 1 & -2t & 3t^2 \end{vmatrix}\right] = \dfrac{d}{dt}[(4t^3 - 3t^5)\hat{\mathbf{i}} + (2t^2 - 3t^4)\hat{\mathbf{j}} - t^3\hat{\mathbf{k}}]$

$$= (12t^2 - 15t^4)\hat{\mathbf{i}} + (4t - 12t^3)\hat{\mathbf{j}} - 3t^2\hat{\mathbf{k}}$$

18. $\displaystyle\int [3g(t)\mathbf{v}(t) + \mathbf{u}(t)]\,dt = \int [3(2t^3 - 3t)(\hat{\mathbf{i}} - 2t\hat{\mathbf{j}} + 3t^2\hat{\mathbf{k}}) + (t\hat{\mathbf{i}} - t^2\hat{\mathbf{j}} + 2t\hat{\mathbf{k}})]\,dt$

$$= \int [(6t^3 - 8t)\hat{\mathbf{i}} + (-12t^4 + 17t^2)\hat{\mathbf{j}} + (18t^5 - 27t^3 + 2t)\hat{\mathbf{k}}]\,dt$$

$$= \left(\frac{3t^4}{2} - 4t^2\right)\hat{\mathbf{i}} + \left(-\frac{12t^5}{5} + \frac{17t^3}{3}\right)\hat{\mathbf{j}} + \left(3t^6 - \frac{27t^4}{4} + t^2\right)\hat{\mathbf{k}} + \mathbf{C}$$

20. $\mathbf{u} \times \dfrac{d\mathbf{v}}{dt} - f(t)\mathbf{u} \cdot \dfrac{d\mathbf{v}}{dt}\mathbf{v} = \mathbf{u} \times (-2\hat{\mathbf{j}} + 6t\hat{\mathbf{k}}) - f(t)\mathbf{u} \cdot (-2\hat{\mathbf{j}} + 6t\hat{\mathbf{k}})\mathbf{v}$

$$= \begin{vmatrix} \hat{\mathbf{i}} & \hat{\mathbf{j}} & \hat{\mathbf{k}} \\ t & -t^2 & 2t \\ 0 & -2 & 6t \end{vmatrix} - (t^2+3)(t\hat{\mathbf{i}} - t^2\hat{\mathbf{j}} + 2t\hat{\mathbf{k}}) \cdot (-2\hat{\mathbf{j}} + 6t\hat{\mathbf{k}})\mathbf{v}$$

$$= (-6t^3 + 4t)\hat{\mathbf{i}} - 6t^2\hat{\mathbf{j}} - 2t\hat{\mathbf{k}} - (t^2+3)(2t^2 + 12t^2)(\hat{\mathbf{i}} - 2t\hat{\mathbf{j}} + 3t^2\hat{\mathbf{k}})$$

$$= (-14t^4 - 6t^3 - 42t^2 + 4t)\hat{\mathbf{i}} + (28t^5 + 84t^3 - 6t^2)\hat{\mathbf{j}} + (-42t^6 - 126t^4 - 2t)\hat{\mathbf{k}}$$

22. If $\mathbf{u} = (u_x, u_y, u_z)$ and $\mathbf{v} = (v_x, v_y, v_z)$, then

$$\frac{d}{dt}(\mathbf{u} + \mathbf{v}) = \frac{d}{dt}[(u_x + v_x)\hat{\mathbf{i}} + (u_y + v_y)\hat{\mathbf{j}} + (u_z + v_z)\hat{\mathbf{k}}] = \left[\frac{d}{dt}(u_x + v_x)\right]\hat{\mathbf{i}} + \left[\frac{d}{dt}(u_y + v_y)\right]\hat{\mathbf{j}} + \left[\frac{d}{dt}(u_z + v_z)\right]\hat{\mathbf{k}}$$

$$= \left(\frac{du_x}{dt}\hat{\mathbf{i}} + \frac{du_y}{dt}\hat{\mathbf{j}} + \frac{du_z}{dt}\hat{\mathbf{k}}\right) + \left(\frac{dv_x}{dt}\hat{\mathbf{i}} + \frac{dv_y}{dt}\hat{\mathbf{j}} + \frac{dv_z}{dt}\hat{\mathbf{k}}\right) = \frac{d\mathbf{u}}{dt} + \frac{d\mathbf{v}}{dt}.$$

24. Using equation 11.59b, $\dfrac{d}{dt}[\mathbf{u} \cdot (\mathbf{v} \times \mathbf{w})] = \mathbf{u} \cdot \dfrac{d}{dt}(\mathbf{v} \times \mathbf{w}) + \dfrac{d\mathbf{u}}{dt} \cdot \mathbf{v} \times \mathbf{w}$, and now using equation 11.59c,

$$\frac{d}{dt}[\mathbf{u} \cdot (\mathbf{v} \times \mathbf{w})] = \mathbf{u} \cdot \left(\mathbf{v} \times \frac{d\mathbf{w}}{dt} + \frac{d\mathbf{v}}{dt} \times \mathbf{w}\right) + \frac{d\mathbf{u}}{dt} \cdot \mathbf{v} \times \mathbf{w} = \mathbf{u} \cdot \mathbf{v} \times \frac{d\mathbf{w}}{dt} + \mathbf{u} \cdot \frac{d\mathbf{v}}{dt} \times \mathbf{w} + \frac{d\mathbf{u}}{dt} \cdot \mathbf{v} \times \mathbf{w}.$$

26. If we set $\mathbf{v} = v_x(s)\hat{\mathbf{i}} + v_y(s)\hat{\mathbf{j}} + v_z(s)\hat{\mathbf{k}}$, then

$$\frac{d\mathbf{v}}{dt} = \frac{dv_x}{dt}\hat{\mathbf{i}} + \frac{dv_y}{dt}\hat{\mathbf{j}} + \frac{dv_z}{dt}\hat{\mathbf{k}} = \frac{dv_x}{ds}\frac{ds}{dt}\hat{\mathbf{i}} + \frac{dv_y}{ds}\frac{ds}{dt}\hat{\mathbf{j}} + \frac{dv_z}{ds}\frac{ds}{dt}\hat{\mathbf{k}} = \left(\frac{dv_x}{ds}\hat{\mathbf{i}} + \frac{dv_y}{ds}\hat{\mathbf{j}} + \frac{dv_z}{ds}\hat{\mathbf{k}}\right)\frac{ds}{dt} = \frac{d\mathbf{v}}{ds}\frac{ds}{dt}.$$

EXERCISES 11.10

2. If we set $x = \sqrt{2}\cos t$, then $y = \pm\sqrt{2}\sin t$. For y to increase in the first octant, we choose $y = \sqrt{2}\sin t$. Parametric and vector equations are therefore
$x = \sqrt{2}\cos t$, $y = \sqrt{2}\sin t$, $z = 4$,
$\mathbf{r} = \sqrt{2}\cos t\,\hat{\mathbf{i}} + \sqrt{2}\sin t\,\hat{\mathbf{j}} + 4\hat{\mathbf{k}}$,
$0 \le t < 2\pi$.

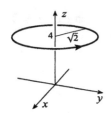

4. If we set $x = \sqrt{5}\cos t$, then $y = \pm\sqrt{5}\sin t$, and for correct direction we choose $y = \sqrt{5}\sin t$. Parametric and vector equations are therefore
$x = \sqrt{5}\cos t$, $y = \sqrt{5}\sin t$, $z = 5$,
$\mathbf{r} = \sqrt{5}\cos t\,\hat{\mathbf{i}} + \sqrt{5}\sin t\,\hat{\mathbf{j}} + 5\hat{\mathbf{k}}$,
$0 \le t < 2\pi$.

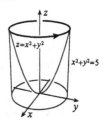

6. If we choose x as parameter by setting $x = t$, then $y = t$ and $z = \sqrt{t^2 + t^2} = \sqrt{2}|t|$. Parametric and vector equations are
$x = t$, $y = t$, $z = \sqrt{2}|t|$,
$\mathbf{r} = t(\hat{\mathbf{i}} + \hat{\mathbf{j}}) + \sqrt{2}|t|\hat{\mathbf{k}}$.

8. If we set $x = \cos t$, then $y = 1 \pm \sin t$, and for correct direction we choose $y = 1 + \sin t$. Parametric and vector equations are therefore
$x = \cos t$, $y = 1 + \sin t$,
$z = \sqrt{4 - \cos^2 t - (1 + \sin t)^2} = \sqrt{2 - 2\sin t}$,
$\mathbf{r} = \cos t\,\hat{\mathbf{i}} + (1 + \sin t)\hat{\mathbf{j}} + \sqrt{2 - 2\sin t}\,\hat{\mathbf{k}}$,
$0 \le t < 2\pi$.

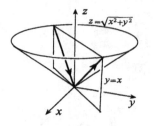

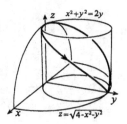

10. If we set $x = -t$ (so that x decreases as t increases), parametric and vector equations are $x = -t$, $y = t^2$, $z = \sqrt{t^2 + t^4}$, $\mathbf{r} = -t\hat{\mathbf{i}} + t^2\hat{\mathbf{j}} + \sqrt{t^2 + t^4}\,\hat{\mathbf{k}}$.

12. Since $x^2 + y^2 = 4$, the curve is two turns of a helix that rises a distance of 6π in each turn.

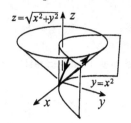

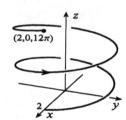

14. Since $x = t^2 - t = y^2 - y$, the curve is a parabola in the plane $z = 5$.

EXERCISES 11.11

2. Since $\mathbf{r} = t\hat{\mathbf{i}} + t^2\hat{\mathbf{j}} + t^3\hat{\mathbf{k}}$, a tangent vector is $\mathbf{T} = \dfrac{d\mathbf{r}}{dt} = \hat{\mathbf{i}} + 2t\hat{\mathbf{j}} + 3t^2\hat{\mathbf{k}}$. A unit tangent vector is

$$\hat{\mathbf{T}} = \frac{\hat{\mathbf{i}} + 2t\hat{\mathbf{j}} + 3t^2\hat{\mathbf{k}}}{\sqrt{1 + 4t^2 + 9t^4}}.$$

4. Since x decreases along the curve, we set $x = -t$ for parametric equations, in which case $y = 5 + t$, $z = t^2 - 5 - t$. A vector equation for the curve is $\mathbf{r} = -t\hat{\mathbf{i}} + (5 + t)\hat{\mathbf{j}} + (t^2 - t - 5)\hat{\mathbf{k}}$, $-5 \le t \le 0$. A tangent vector is $\mathbf{T} = \dfrac{d\mathbf{r}}{dt} = -\hat{\mathbf{i}} + \hat{\mathbf{j}} + (2t - 1)\hat{\mathbf{k}}$, and a unit tangent vector is

$$\hat{\mathbf{T}} = \frac{-\hat{\mathbf{i}} + \hat{\mathbf{j}} + (2t - 1)\hat{\mathbf{k}}}{\sqrt{1 + 1 + (2t - 1)^2}} = \frac{-\hat{\mathbf{i}} + \hat{\mathbf{j}} + (2t - 1)\hat{\mathbf{k}}}{\sqrt{4t^2 - 4t + 3}}.$$

6. Since $\mathbf{T} = \dfrac{d\mathbf{r}}{dt} = -4\sin t\,\hat{\mathbf{i}} + 6\cos t\,\hat{\mathbf{j}} + 2\cos t\,\hat{\mathbf{k}}$, the unit tangent vector at $(2\sqrt{2}, 3\sqrt{2}, \sqrt{2})$ is

$$\hat{\mathbf{T}}(\pi/4) = \frac{-2\sqrt{2}\hat{\mathbf{i}} + 3\sqrt{2}\hat{\mathbf{j}} + \sqrt{2}\hat{\mathbf{k}}}{\sqrt{8 + 18 + 2}} = \frac{-2\hat{\mathbf{i}} + 3\hat{\mathbf{j}} + \hat{\mathbf{k}}}{\sqrt{14}}.$$

8. With $x = \sqrt{2}\cos t$, $y = -\sqrt{2}\sin t$, $z = \sqrt{2}$, $0 \le t < 2\pi$, a tangent vector at $(1, 1, \sqrt{2})$ is $\mathbf{T}(7\pi/4) = (-\sqrt{2}\sin t, -\sqrt{2}\cos t, 0)_{|t=7\pi/4} = (1, -1, 0)$. Hence, $\hat{\mathbf{T}}(7\pi/4) = (1, -1, 0)/\sqrt{2}$.

10. If we set $x = 2\cos t$, then $y = 1 \pm 2\sin t$. For z to decrease when y is negative, we choose $y = 1 - 2\sin t$, and then $\mathbf{r} = 2\cos t\,\hat{\mathbf{i}} + (1 - 2\sin t)\hat{\mathbf{j}} + 2\cos t\,\hat{\mathbf{k}}$, $0 \le t < 2\pi$. Since $\mathbf{T} = d\mathbf{r}/dt = -2\sin t\,\hat{\mathbf{i}} - 2\cos t\,\hat{\mathbf{j}} - 2\sin t\,\hat{\mathbf{k}}$, $\mathbf{T}(0) = -2\hat{\mathbf{j}}$. Consequently, $\hat{\mathbf{T}}(0) = -\hat{\mathbf{j}}$.

12. Since this is a straight line segment from $(7, 0, 2)$ to $(2, 1, 6)$, its length is

$$\sqrt{5^2 + (-1)^2 + (-4)^2} = \sqrt{42}.$$

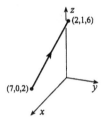

14. With equation 11.78,

$$L = \int_1^4 \sqrt{1^2 + \left(\frac{3\sqrt{t}}{2}\right)^2 + (6\sqrt{t})^2}\, dt$$

$$= \int_1^4 \sqrt{1 + 153t/4}\, dt = \left\{\frac{8}{459}\left(1 + \frac{153t}{4}\right)^{3/2}\right\}_1^4$$

$$= \frac{616\sqrt{616} - 157\sqrt{157}}{459}.$$

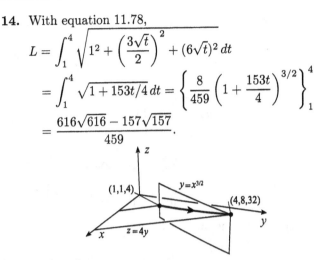

16. Since a tangent vector is

$$\mathbf{T} = \frac{dx}{dt}\hat{\mathbf{i}} + \frac{dy}{dt}\hat{\mathbf{j}} = (-\sin t + \sin t + t\cos t)\hat{\mathbf{i}} + (\cos t - \cos t + t\sin t)\hat{\mathbf{j}} = t\cos t\,\hat{\mathbf{i}} + t\sin t\,\hat{\mathbf{j}},$$

a unit tangent vector is $\hat{\mathbf{T}} = \dfrac{t\cos t\,\hat{\mathbf{i}} + t\sin t\,\hat{\mathbf{j}}}{\sqrt{t^2\cos^2 t + t^2\sin^2 t}} = \cos t\,\hat{\mathbf{i}} + \sin t\,\hat{\mathbf{j}}.$

18. (a) If we use equation 11.66, $\mathbf{T}(0) = (2t\hat{\mathbf{i}} + 3t^2\hat{\mathbf{j}} + 2t\hat{\mathbf{k}})|_{t=0} = \mathbf{0}.$

(b) A tangent vector at any point except $t = 0$ is $\mathbf{T} = 2t\hat{\mathbf{i}} + 3t^2\hat{\mathbf{j}} + 2t\hat{\mathbf{k}} = t(2\hat{\mathbf{i}} + 3t\hat{\mathbf{j}} + 2\hat{\mathbf{k}})$. It follows then that $\mathbf{S}(t) = 2\hat{\mathbf{i}} + 3t\hat{\mathbf{j}} + 2\hat{\mathbf{k}}$ must also be tangent at any point (except possibly when $t = 0$). The only way $\mathbf{S}(t)$ can assign a tangent vector to the curve continuously is for $\mathbf{S}(0) = \lim_{t\to 0}\mathbf{S}(t) = 2\hat{\mathbf{i}} + 2\hat{\mathbf{k}}.$

EXERCISES 11.12

2. From $\hat{\mathbf{T}} = \dfrac{(1, 2t, 3t^2)}{\sqrt{1 + 4t^2 + 9t^4}}$, a vector in the direction of $\hat{\mathbf{N}}$ is

$$\mathbf{N} = \frac{d\hat{\mathbf{T}}}{dt} = \frac{-(8t + 36t^3)}{2(1 + 4t^2 + 9t^4)^{3/2}}(1, 2t, 3t^2) + \frac{(0, 2, 6t)}{\sqrt{1 + 4t^2 + 9t^4}}$$

$$= \frac{1}{(1 + 4t^2 + 9t^4)^{3/2}}\left[-(4t + 18t^3)(1, 2t, 3t^2) + (1 + 4t^2 + 9t^4)(0, 2, 6t)\right]$$

$$= \frac{1}{(1 + 4t^2 + 9t^4)^{3/2}}(-4t - 18t^3, 2 - 18t^4, 6t + 12t^3).$$

Consequently, the principal normal is

$$\hat{\mathbf{N}} = \frac{(-2t - 9t^3, 1 - 9t^4, 3t + 6t^3)}{\sqrt{(2t + 9t^3)^2 + (1 - 9t^4)^2 + (3t + 6t^3)^2}} = \frac{(-2t - 9t^3, 1 - 9t^4, 3t + 6t^3)}{\sqrt{1 + 13t^2 + 54t^4 + 117t^6 + 81t^8}}.$$

The direction of the binormal is

$$\mathbf{B} = \begin{vmatrix} \hat{\mathbf{i}} & \hat{\mathbf{j}} & \hat{\mathbf{k}} \\ 1 & 2t & 3t^2 \\ -2t - 9t^3 & 1 - 9t^4 & 3t + 6t^3 \end{vmatrix}$$

$$= (6t^2 + 12t^4 - 3t^2 + 27t^6)\hat{\mathbf{i}} + (-6t^3 - 27t^5 - 3t - 6t^3)\hat{\mathbf{j}} + (1 - 9t^4 + 4t^2 + 18t^4)\hat{\mathbf{k}}$$

$$= (1 + 4t^2 + 9t^4)(3t^2\hat{\mathbf{i}} - 3t\hat{\mathbf{j}} + \hat{\mathbf{k}}).$$

Thus, $\hat{\mathbf{B}} = \dfrac{(3t^2, -3t, 1)}{\sqrt{9t^4 + 9t^2 + 1}}.$

4. With parametric equations $x = -t$, $y = 5 + t$, $z = t^2 - t - 5$, (see Exercise 11.11–4),

$$\hat{\mathbf{T}} = \frac{(-1, 1, 2t - 1)}{\sqrt{1 + 1 + (2t - 1)^2}} = \frac{(-1, 1, 2t - 1)}{\sqrt{4t^2 - 4t + 3}}.$$

A vector in the direction of $\hat{\mathbf{N}}$ is

$$\mathbf{N} = \frac{d\hat{\mathbf{T}}}{dt} = \frac{-(4t - 2)}{(4t^2 - 4t + 3)^{3/2}}(-1, 1, 2t - 1) + \frac{(0, 0, 2)}{\sqrt{4t^2 - 4t + 3}}$$

$$= \frac{2}{(4t^2 - 4t + 3)^{3/2}}\left[-(2t - 1)(-1, 1, 2t - 1) + (4t^2 - 4t + 3)(0, 0, 1)\right]$$

$$= \frac{2}{(4t^2 - 4t + 3)^{3/2}}(2t - 1, 1 - 2t, 2).$$

Consequently, the principal normal is

$$\hat{\mathbf{N}} = \frac{(2t - 1, 1 - 2t, 2)}{\sqrt{(2t - 1)^2 + (1 - 2t)^2 + 4}} = \frac{(2t - 1, 1 - 2t, 2)}{\sqrt{8t^2 - 8t + 6}}.$$

The direction of the binormal is

$$\mathbf{B} = \begin{vmatrix} \hat{\mathbf{i}} & \hat{\mathbf{j}} & \hat{\mathbf{k}} \\ -1 & 1 & 2t - 1 \\ 2t - 1 & 1 - 2t & 2 \end{vmatrix} = (2 + 4t^2 - 4t + 1)\hat{\mathbf{i}} + (4t^2 - 4t + 1 + 2)\hat{\mathbf{j}} + (-1 + 2t - 2t + 1)\hat{\mathbf{k}}$$

$$= (3 - 4t + 4t^2)(\hat{\mathbf{i}} + \hat{\mathbf{j}}).$$

Thus, $\hat{\mathbf{B}} = (\hat{\mathbf{i}} + \hat{\mathbf{j}})/\sqrt{2}$.

6. From $\hat{\mathbf{T}} = \dfrac{(-4 \sin t, 6 \cos t, 2 \cos t)}{\sqrt{16 \sin^2 t + 36 \cos^2 t + 4 \cos^2 t}} = \dfrac{(-2 \sin t, 3 \cos t, \cos t)}{\sqrt{4 + 6 \cos^2 t}}$, a vector in the direction of $\hat{\mathbf{N}}$ is

$$\mathbf{N} = \frac{d\hat{\mathbf{T}}}{dt} = \frac{6 \cos t \sin t}{(4 + 6 \cos^2 t)^{3/2}}(-2 \sin t, 3 \cos t, \cos t) + \frac{(-2 \cos t, -3 \sin t, -\sin t)}{\sqrt{4 + 6 \cos^2 t}}.$$

At $(2\sqrt{2}, 3\sqrt{2}, \sqrt{2})$, we may take $t = \pi/4$, in which case

$$\mathbf{N}(\pi/4) = \frac{3}{7\sqrt{7}}(-\sqrt{2}, 3/\sqrt{2}, 1/\sqrt{2}) + \frac{(-\sqrt{2}, -3/\sqrt{2}, -1/\sqrt{2})}{\sqrt{7}} = -\frac{4}{7\sqrt{14}}(5, 3, 1).$$

Hence, the principal normal at $(2\sqrt{2}, 3\sqrt{2}, \sqrt{2})$ is $\hat{\mathbf{N}} = -\dfrac{(5, 3, 1)}{\sqrt{35}}$. Since a tangent vector at the point is $\mathbf{T}(\pi/4) = (-\sqrt{2}, 3/\sqrt{2}, 1/\sqrt{2}) = (-2, 3, 1)/\sqrt{2}$, the direction of the binormal at the point is

$$\mathbf{B}(\pi/4) = (-2, 3, 1) \times [-(5, 3, 1)] = -\begin{vmatrix} \hat{\mathbf{i}} & \hat{\mathbf{j}} & \hat{\mathbf{k}} \\ -2 & 3 & 1 \\ 5 & 3 & 1 \end{vmatrix} = -(0, 7, -21).$$

Thus, $\hat{\mathbf{B}}(\pi/4) = (0, -1, 3)/\sqrt{10}$.

8. This curve is the circle $x^2 + y^2 = 2$ in the plane $z = \sqrt{2}$. The unit tangent and principal normal vectors at $(1, 1, \sqrt{2})$ are

$$\hat{\mathbf{T}} = \frac{(1, -1, 0)}{\sqrt{2}}, \qquad \hat{\mathbf{N}} = \frac{(-1, -1, 0)}{\sqrt{2}}.$$

The binormal is

$$\hat{\mathbf{B}} = \hat{\mathbf{T}} \times \hat{\mathbf{N}} = \frac{1}{2}\begin{vmatrix} \hat{\mathbf{i}} & \hat{\mathbf{j}} & \hat{\mathbf{k}} \\ 1 & -1 & 0 \\ -1 & -1 & 0 \end{vmatrix}$$

$$= \frac{1}{2}(0, 0, -2) = (0, 0, -1).$$

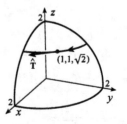

10. With the parametric equations $x = 2\cos t$, $y = 1 - 2\sin t$, $z = 2\cos t$, (see Exercise 11.11–10),

$$\hat{\mathbf{T}} = \frac{(-2\sin t, -2\cos t, -2\sin t)}{\sqrt{4\sin^2 t + 4\cos^2 t + 4\sin^2 t}} = -\frac{(\sin t, \cos t, \sin t)}{\sqrt{1 + \sin^2 t}}.$$

A vector in the direction of $\hat{\mathbf{N}}$ is

$$\mathbf{N} = \frac{d\hat{\mathbf{T}}}{dt} = \frac{\sin t \cos t}{(1 + \sin^2 t)^{3/2}}(\sin t, \cos t, \sin t) - \frac{(\cos t, -\sin t, \cos t)}{\sqrt{1 + \sin^2 t}}.$$

At $(2, 1, 2)$, $t = 0$ and $\mathbf{N}(0) = -(1, 0, 1)$. Thus, the principal normal at $(2, 1, 2)$ is $\hat{\mathbf{N}}(0) = -(1, 0, 1)/\sqrt{2}$. Since $\hat{\mathbf{T}}(0) = -(0, 1, 0)$, the binormal at $(2, 1, 2)$ is

$$\hat{\mathbf{B}}(0) = -(0, 1, 0) \times \frac{-(1, 0, 1)}{\sqrt{2}} = \frac{1}{\sqrt{2}}\begin{vmatrix} \hat{\mathbf{i}} & \hat{\mathbf{j}} & \hat{\mathbf{k}} \\ 0 & 1 & 0 \\ 1 & 0 & 1 \end{vmatrix} = \frac{1}{\sqrt{2}}(1, 0, -1).$$

12. This is a straight line in space for which $\ddot{\mathbf{r}} = \mathbf{0}$, and therefore $\kappa = 0$. Its radius of curvature is undefined.

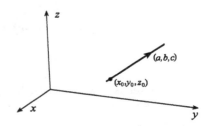

14. With $\dot{\mathbf{r}} = (e^t \cos t - e^t \sin t, e^t \sin t + e^t \cos t, 1)$ and $\ddot{\mathbf{r}} = (-2e^t \sin t, 2e^t \cos t, 0)$, we obtain

$$\dot{\mathbf{r}} \times \ddot{\mathbf{r}} = \begin{vmatrix} \hat{\mathbf{i}} & \hat{\mathbf{j}} & \hat{\mathbf{k}} \\ e^t(\cos t - \sin t) & e^t(\sin t + \cos t) & 1 \\ -2e^t \sin t & 2e^t \cos t & 0 \end{vmatrix} = (-2e^t \cos t, -2e^t \sin t, 2e^{2t}).$$

Consequently,

$$\kappa(t) = \frac{|\dot{\mathbf{r}} \times \ddot{\mathbf{r}}|}{|\dot{\mathbf{r}}|^3} = \frac{\sqrt{4e^{2t}\cos^2 t + 4e^{2t}\sin^2 t + 4e^{4t}}}{[(e^t \cos t - e^t \sin t)^2 + (e^t \sin t + e^t \cos t)^2 + 1]^{3/2}}$$

$$= \frac{2e^t\sqrt{1 + e^{2t}}}{(1 + 2e^{2t})^{3/2}},$$

and $\rho(t) = \dfrac{1}{\kappa} = \dfrac{(1 + 2e^{2t})^{3/2}}{2e^t\sqrt{1 + e^{2t}}}.$

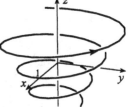

16. Since $\dot{\mathbf{r}} \times \ddot{\mathbf{r}} = \begin{vmatrix} \hat{\mathbf{i}} & \hat{\mathbf{j}} & \hat{\mathbf{k}} \\ -2\sin t & 2\cos t & 2\cos t \\ -2\cos t & -2\sin t & -2\sin t \end{vmatrix} = (0, -4, 4)$,

$$\kappa(t) = \frac{|\dot{\mathbf{r}} \times \ddot{\mathbf{r}}|}{|\dot{\mathbf{r}}|^3} = \frac{\sqrt{16 + 16}}{(4\sin^2 t + 4\cos^2 t + 4\cos^2 t)^{3/2}}$$

$$= \frac{1}{\sqrt{2}(1 + \cos^2 t)^{3/2}},$$

and $\rho(t) = \dfrac{1}{\kappa} = \sqrt{2}(1 + \cos^2 t)^{3/2}.$

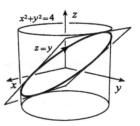

18. $\kappa(t) = \dfrac{|\dot{\mathbf{r}} \times \ddot{\mathbf{r}}|}{|\dot{\mathbf{r}}|^3} = \dfrac{|(2t, 4t^3, 2) \times (2, 12t^2, 0)|}{|(2t, 4t^3, 2)|^3} = \dfrac{1}{(4t^2 + 16t^6 + 4)^{3/2}} \begin{Vmatrix} \hat{\mathbf{i}} & \hat{\mathbf{j}} & \hat{\mathbf{k}} \\ 2t & 4t^3 & 2 \\ 2 & 12t^2 & 0 \end{Vmatrix}$

$= \dfrac{|(-24t^2, 4, 16t^3)|}{8(1 + t^2 + 4t^6)^{3/2}} = \dfrac{4\sqrt{36t^4 + 1 + 16t^6}}{8(1 + t^2 + 4t^6)^{3/2}}$

$= \dfrac{\sqrt{1 + 36t^4 + 16t^6}}{2(1 + t^2 + 4t^6)^{3/2}},$

and $\rho(t) = \dfrac{1}{\kappa} = \dfrac{2(1 + t^2 + 4t^6)^{3/2}}{\sqrt{1 + 36t^4 + 16t^6}}.$

20. For a smooth curve $\mathbf{r} = x(t)\hat{\mathbf{i}} + y(t)\hat{\mathbf{j}}$,

$\kappa(t) = \dfrac{|\dot{\mathbf{r}} \times \ddot{\mathbf{r}}|}{|\dot{\mathbf{r}}|^3} = \dfrac{|(\dot{x}, \dot{y}, 0) \times (\ddot{x}, \ddot{y}, 0)|}{|(\dot{x}, \dot{y}, 0)|^3} = \dfrac{1}{(\dot{x}^2 + \dot{y}^2)^{3/2}} \begin{Vmatrix} \hat{\mathbf{i}} & \hat{\mathbf{j}} & \hat{\mathbf{k}} \\ \dot{x} & \dot{y} & 0 \\ \ddot{x} & \ddot{y} & 0 \end{Vmatrix}$

$= \dfrac{1}{(\dot{x}^2 + \dot{y}^2)^{3/2}}|(0, 0, \dot{x}\ddot{y} - \ddot{x}\dot{y})| = \dfrac{|\dot{x}\ddot{y} - \ddot{x}\dot{y}|}{(\dot{x}^2 + \dot{y}^2)^{3/2}} = \dfrac{|\dot{y}\ddot{x} - \dot{x}\ddot{y}|}{(\dot{x}^2 + \dot{y}^2)^{3/2}}.$

22. Suppose (x, y) is a point of inflection at which $f''(x) = 0$. Because

$$f''(x) = \dfrac{\dfrac{dx}{dt}\dfrac{d^2y}{dt^2} - \dfrac{dy}{dt}\dfrac{d^2x}{dt^2}}{\left(\dfrac{dx}{dt}\right)^3} \qquad \text{(see Exercise 9.1–38),}$$

it follows from Exercise 20 that at such a point of inflection, $\kappa = 0$. If (x, y) is a point of inflection at which $f''(x)$ does not exist, then neither does the curvature.

24. (a) $\hat{\mathbf{T}} = \dfrac{(-2\sin t, 2\cos t)}{\sqrt{4\sin^2 t + 4\cos^2 t}} = (-\sin t, \cos t)$ According to Example 11.47,

$\hat{\mathbf{N}} = \text{sgn}\,[(2\cos t)(-2\cos t) - (-2\sin t)(-2\sin t)]\,\dfrac{(2\cos t, 2\sin t)}{\sqrt{4\sin^2 t + 4\cos^2 t}} = -(\cos t, \sin t).$

The binormal is $\hat{\mathbf{B}} = \hat{\mathbf{T}} \times \hat{\mathbf{N}} = \begin{vmatrix} \hat{\mathbf{i}} & \hat{\mathbf{j}} & \hat{\mathbf{k}} \\ -\sin t & \cos t & 0 \\ -\cos t & -\sin t & 0 \end{vmatrix} = (0, 0, 1).$

(b) $F_T = \mathbf{F} \cdot \hat{\mathbf{T}} = (x^2, y^2) \cdot (-\sin t, \cos t) = -x^2\sin t + y^2\cos t$

$= -(2\cos t)^2 \sin t + (2\sin t)^2 \cos t = 2\sin 2t(\sin t - \cos t)$

$F_N = \mathbf{F} \cdot \hat{\mathbf{N}} = (x^2, y^2) \cdot (-\cos t, -\sin t) = -x^2\cos t - y^2\sin t$

$= -(2\cos t)^2 \cos t - (2\sin t)^2 \sin t = -4(\cos^3 t + \sin^3 t)$

(c) $\mathbf{F} = F_T\hat{\mathbf{T}} + F_N\hat{\mathbf{N}} = 2\sin 2t(\sin t - \cos t)\hat{\mathbf{T}} - 4(\cos^3 t + \sin^3 t)\hat{\mathbf{N}}$

26. $\hat{\mathbf{T}} = \dfrac{(-\sin t, \cos t, 1)}{\sqrt{\sin^2 t + \cos^2 t + 1}} = \dfrac{(-\sin t, \cos t, 1)}{\sqrt{2}}$ A vector in the direction of $\hat{\mathbf{N}}$ is

$$\mathbf{N} = \dfrac{d\hat{\mathbf{T}}}{dt} = \dfrac{(-\cos t, -\sin t, 0)}{\sqrt{2}},$$

and therefore the principal normal is $\hat{\mathbf{N}} = -(\cos t, \sin t, 0).$

$$\hat{\mathbf{B}} = \hat{\mathbf{T}} \times \hat{\mathbf{N}} = \dfrac{1}{\sqrt{2}} \begin{vmatrix} \hat{\mathbf{i}} & \hat{\mathbf{j}} & \hat{\mathbf{k}} \\ -\sin t & \cos t & 1 \\ -\cos t & -\sin t & 0 \end{vmatrix} = \dfrac{1}{\sqrt{2}}(\sin t, -\cos t, 1).$$

$$F_T = \mathbf{F} \cdot \hat{\mathbf{T}} = (x, xy^2, 1) \cdot \frac{(-\sin t, \cos t, 1)}{\sqrt{2}} = \frac{1}{\sqrt{2}}(-x \sin t + xy^2 \cos t + 1)$$

$$= \frac{1}{\sqrt{2}}(-\cos t \, \sin t + \cos^2 t \, \sin^2 t + 1)$$

$$F_N = \mathbf{F} \cdot \hat{\mathbf{N}} = (x, xy^2, 1) \cdot (-\cos t, -\sin t, 0) = -(x \cos t + xy^2 \sin t)$$

$$= -(\cos^2 t + \cos t \, \sin^3 t)$$

$$F_B = \mathbf{F} \cdot \hat{\mathbf{B}} = (x, xy^2, 1) \cdot \frac{(\sin t, -\cos t, 1)}{\sqrt{2}} = \frac{1}{\sqrt{2}}(x \sin t - xy^2 \cos t + 1)$$

$$= \frac{1}{\sqrt{2}}(\cos t \, \sin t - \cos^2 t \, \sin^2 t + 1)$$

$$\mathbf{F} = F_T \hat{\mathbf{T}} + F_N \hat{\mathbf{N}} + F_B \hat{\mathbf{B}} = \frac{1}{\sqrt{2}}(1 - \cos t \, \sin t + \cos^2 t \, \sin^2 t)\hat{\mathbf{T}}$$

$$- \cos t(\cos t + \sin^3 t)\hat{\mathbf{N}} + \frac{1}{\sqrt{2}}(1 + \cos t \, \sin t - \cos^2 t \, \sin^2 t)\hat{\mathbf{B}}$$

28. If ϕ is the angle shown, then $\hat{\mathbf{T}} = (\cos \phi, \sin \phi)$. Consequently,

$$\kappa = \left| \frac{d\hat{\mathbf{T}}}{ds} \right| = \left| \left(-\sin \phi \frac{d\phi}{ds}, \cos \phi \frac{d\phi}{ds} \right) \right| = \left| \frac{d\phi}{ds} \right|.$$

EXERCISES 11.13

2. $\mathbf{v} = \dfrac{d\mathbf{r}}{dt} = \left(1 - \dfrac{1}{t^2}\right)\hat{\mathbf{i}} + \left(1 + \dfrac{1}{t^2}\right)\hat{\mathbf{j}} = \dfrac{1}{t^2}[(t^2 - 1)\hat{\mathbf{i}} + (t^2 + 1)\hat{\mathbf{j}}]$

$|\mathbf{v}| = \dfrac{1}{t^2}\sqrt{(t^2-1)^2 + (t^2+1)^2} = \dfrac{\sqrt{2t^4 + 2}}{t^2}$; $\quad \mathbf{a} = \dfrac{d\mathbf{v}}{dt} = \dfrac{2}{t^3}\hat{\mathbf{i}} - \dfrac{2}{t^3}\hat{\mathbf{j}} = \dfrac{2}{t^3}(\hat{\mathbf{i}} - \hat{\mathbf{j}})$

4. $\mathbf{v} = \dfrac{d\mathbf{r}}{dt} = 2t\hat{\mathbf{i}} + 2e^t(t+1)\hat{\mathbf{j}} - \dfrac{2}{t^3}\hat{\mathbf{k}}$; $\quad |\mathbf{v}| = \sqrt{4t^2 + 4e^{2t}(t+1)^2 + 4/t^6} = 2\sqrt{t^2 + e^{2t}(t+1)^2 + 1/t^6}$;

$\mathbf{a} = \dfrac{d\mathbf{v}}{dt} = 2\hat{\mathbf{i}} + 2e^t(t+2)\hat{\mathbf{j}} + \dfrac{6}{t^4}\hat{\mathbf{k}}$

6. If $\mathbf{a} = 3t^2\hat{\mathbf{i}} + (t+1)\hat{\mathbf{j}} - 4t^3\hat{\mathbf{k}}$, then $\mathbf{v} = t^3\hat{\mathbf{i}} + \left(\dfrac{t^2}{2} + t\right)\hat{\mathbf{j}} - t^4\hat{\mathbf{k}} + \mathbf{C}$. Since $\mathbf{v}(0) = \mathbf{0}$, it follows that

$\mathbf{C} = \mathbf{0}$, and $\mathbf{v} = t^3\hat{\mathbf{i}} + \left(\dfrac{t^2}{2} + t\right)\hat{\mathbf{j}} - t^4\hat{\mathbf{k}}$. Integration now gives $\mathbf{r} = \dfrac{t^4}{4}\hat{\mathbf{i}} + \left(\dfrac{t^3}{6} + \dfrac{t^2}{2}\right)\hat{\mathbf{j}} - \dfrac{t^5}{5}\hat{\mathbf{k}} + \mathbf{D}$. Since

$\mathbf{r}(0) = (1, 2, -1)$, we find $(1, 2, -1) = \mathbf{D}$, and $\mathbf{r} = \left(\dfrac{t^4}{4} + 1\right)\hat{\mathbf{i}} + \left(\dfrac{t^3}{6} + \dfrac{t^2}{2} + 2\right)\hat{\mathbf{j}} - \left(\dfrac{t^5}{5} + 1\right)\hat{\mathbf{k}}$.

8. The velocity and acceleration are $\mathbf{v} = \hat{\mathbf{i}} + 2t\hat{\mathbf{j}}$ and $\mathbf{a} = 2\hat{\mathbf{j}}$. According to equation 11.112b, the tangential component of acceleration is

$$a_T = \frac{d}{dt}|\mathbf{v}| = \frac{d}{dt}\sqrt{1 + 4t^2} = \frac{4t}{\sqrt{1 + 4t^2}}.$$

According to equation 11.113, the normal component of acceleration is

$$a_N = \sqrt{|\mathbf{a}|^2 - a_T^2} = \sqrt{4 - \frac{16t^2}{1 + 4t^2}} = \sqrt{\frac{4}{1 + 4t^2}} = \frac{2}{\sqrt{1 + 4t^2}}.$$

10. From equation 11.112b, $a_N = |\mathbf{v}| \left| \dfrac{d\hat{\mathbf{T}}}{dt} \right| = |\mathbf{v}| \left| \dfrac{d\hat{\mathbf{T}}}{ds} \right| \left| \dfrac{ds}{dt} \right| = |\mathbf{v}|\kappa|\mathbf{v}| = \dfrac{|\mathbf{v}|^2}{\rho}.$

12. (a) If $d^2y/dt^2 = 2$, then $dy/dt = 2t + C$. Since $y'(0) = 0$, $C = 0$, and $dy/dt = 2t$. Thus, $y(t) = t^2 + D$. Since $y(0) = 0$, $D = 0$, and $y(t) = t^2$. Consequently, $4t^2 = x^2$, from which $x = 2t$. Thus, $dx/dt = 2$ and $d^2x/dt^2 = 0$.

(b) If $d^2x/dt^2 = 24t^2$, then $dx/dt = 8t^3 + C$. Since $x'(0) = 0$, $C = 0$ and $dx/dt = 8t^3$. Hence, $x(t) = 2t^4 + D$. Since $x(0) = 0$, $D = 0$ and $x(t) = 2t^4$. From $4y = x^2$, we obtain $y = x^2/4 = t^8$. Differentiation now gives $dy/dt = 8t^7$ and $d^2y/dt^2 = 56t^6$.

14. If $dx/dt = 5$, then $dy/dt = 3x^2\, dx/dt - 2dx/dt = 5(3x^2 - 2)$. The acceleration of the particle is therefore

$$\mathbf{a} = \frac{d^2x}{dt^2}\hat{\mathbf{i}} + \frac{d^2y}{dt^2}\hat{\mathbf{j}} = (0)\hat{\mathbf{i}} + 5\left(6x\frac{dx}{dt}\right)\hat{\mathbf{j}} = 30x(5)\hat{\mathbf{j}} = 150x\hat{\mathbf{j}}.$$

16. Parametric equations for the particle's path are $x = h + R\cos\theta$, $y = k + R\sin\theta$, in which case $\mathbf{v} = \left(-R\sin\theta\dfrac{d\theta}{dt}\right)\hat{\mathbf{i}} + \left(R\cos\theta\dfrac{d\theta}{dt}\right)\hat{\mathbf{j}}$. Thus, $|\mathbf{v}| = \sqrt{R^2\sin^2\theta\left(\dfrac{d\theta}{dt}\right)^2 + R^2\cos^2\theta\left(\dfrac{d\theta}{dt}\right)^2} = R\dfrac{d\theta}{dt} = R\omega$.

18. (a) We choose a coordinate system so that the circle lies in the xy-plane with its centre at the origin. Then, $x = R\cos\theta$, $y = R\sin\theta$, and

$$\mathbf{v} = \left(-R\sin\theta\frac{d\theta}{dt}, R\cos\theta\frac{d\theta}{dt}\right) = \omega R(-\sin\theta, \cos\theta),$$

where $\omega = d\theta/dt$. Since $|\mathbf{v}| = \omega R$ (see Exercise 16), and $|\mathbf{v}|$ is constant, so also is ω. Hence,

$$\mathbf{a} = \omega R\left(-\cos\theta\frac{d\theta}{dt}, -\sin\theta\frac{d\theta}{dt}\right) = -\omega^2 R(\cos\theta, \sin\theta),$$

and $|\mathbf{a}| = \omega^2 R = \left(\dfrac{|\mathbf{v}|^2}{R^2}\right)R = \dfrac{|\mathbf{v}|^2}{R}$.

(b) According to Newton's universal law of gravitation, the magnitude of the force of the earth on the satellite is

$$|\mathbf{F}| = \frac{GMm}{r^2}.$$

According to Newton's second law, $\mathbf{F} = m\mathbf{a}$, and therefore

$$m|\mathbf{a}| = \frac{GMm}{r^2} \implies |\mathbf{a}| = \frac{GM}{r^2}.$$

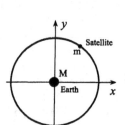

But from part (a), $|\mathbf{a}| = |\mathbf{v}|^2/r$, and therefore $\dfrac{|\mathbf{v}|^2}{r} = \dfrac{GM}{r^2}$, from which

$$|\mathbf{v}| = \sqrt{\frac{GM}{r}} = \sqrt{\frac{6.67\times 10^{-11}(4/3)\pi(6370\times 10^3)^3(5.52\times 10^3)}{6570\times 10^3}} = 7.79\times 10^3.$$

The speed of the satellite is therefore 7.79 km/s.

20. Let $\mathbf{v}_{p/a}$ be the velocity of the plane with respect to the air, $\mathbf{v}_{a/g}$ the velocity of the air with respect to the ground, and $\mathbf{v}_{p/g}$ the velocity of the plane with respect to the ground. According to Exercise 19,

$$\mathbf{v}_{p/a} + \mathbf{v}_{a/g} = \mathbf{v}_{p/g},$$

where $\mathbf{v}_{p/a} = 650[\cos(\pi/3)\hat{\mathbf{i}} + \sin(\pi/3)\hat{\mathbf{j}}] = 325(\hat{\mathbf{i}} + \sqrt{3}\hat{\mathbf{j}})$, and $\mathbf{v}_{a/g} = 40\hat{\mathbf{i}}$. Thus,

$$\mathbf{v}_{p/g} = 325(\hat{\mathbf{i}} + \sqrt{3}\hat{\mathbf{j}}) + 40\hat{\mathbf{i}} = 365\hat{\mathbf{i}} + 325\sqrt{3}\hat{\mathbf{j}} \text{ km/hr},$$

and $|\mathbf{v}_{p/g}| = \sqrt{365^2 + (325\sqrt{3})^2} = 670.9$ km/hr.

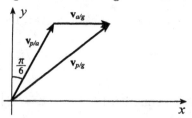

22. Let us take the positive x-direction as the direction in which the river flows. Then the velocity of the water relative to the shore is $\mathbf{v}_{w/s} = 3\hat{\mathbf{i}}$. If the canoe points in direction θ as shown, $\mathbf{v}_{c/w} = -4\sin\theta\,\hat{\mathbf{i}} + 4\cos\theta\,\hat{\mathbf{j}}$. If v is the speed of the canoe with respect to the shore, then $\mathbf{v}_{c/s} = v\hat{\mathbf{j}}$. Since $\mathbf{v}_{c/s} = \mathbf{v}_{c/w} + \mathbf{v}_{w/s}$, we obtain $v\hat{\mathbf{j}} = -4\sin\theta\,\hat{\mathbf{i}} + 4\cos\theta\,\hat{\mathbf{j}} + 3\hat{\mathbf{i}}$. When

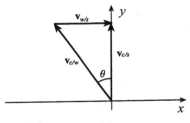

we equate components, $-4\sin\theta + 3 = 0$ and $4\cos\theta = v$. Consequently, $\theta = \mathrm{Sin}^{-1}(3/4)$ radians, and $v = 4\sqrt{1 - 9/16} = \sqrt{7}$ km/hr. To cross the river takes $(200/1000)/v = 1/(5\sqrt{7})$ hr or $12/\sqrt{7}$ min.

24. If its constant acceleration is $\mathbf{a}$, then $\mathbf{v} = \mathbf{a}t + \mathbf{C}$. If the particle starts at $t = 0$, then $\mathbf{v}(0) = \mathbf{0}$, so that $\mathbf{C} = \mathbf{0}$, and $\mathbf{v} = \mathbf{a}t$. Integration now gives $\mathbf{r} = \mathbf{a}t^2/2 + \mathbf{D}$. Since $\mathbf{r}(0) = (1, 2, 3)$, $(1, 2, 3) = \mathbf{D}$, and $\mathbf{r} = \mathbf{a}t^2/2 + (1, 2, 3)$. For the particle to be at $(4, 5, 7)$ when $t = 2$,

$$(4, 5, 7) = \frac{1}{2}\mathbf{a}(2)^2 + (1, 2, 3) \quad \Longrightarrow \quad \mathbf{a} = \frac{1}{2}[(4, 5, 7) - (1, 2, 3)] = \left(\frac{3}{2}, \frac{3}{2}, 2\right).$$

26. Since $\mathbf{v} = \dfrac{d\mathbf{r}}{dt} = \dfrac{-t}{\sqrt{1 - t^2}}\hat{\mathbf{i}} + \hat{\mathbf{j}}$ and $\mathbf{a} = \left[\dfrac{-1}{\sqrt{1 - t^2}} - \dfrac{t^2}{(1 - t^2)^{3/2}}\right]\hat{\mathbf{i}} = \dfrac{-1}{(1 - t^2)^{3/2}}\hat{\mathbf{i}}$, acceleration and velocity are perpendicular if, and when, $0 = \mathbf{v} \cdot \mathbf{a} = \left[\dfrac{-t}{\sqrt{1 - t^2}}\right]\left[\dfrac{-1}{(1 - t^2)^{3/2}}\right]$. This only occurs at $t = 0$ when the particle is at position $(3, 0)$.

28. Because the acceleration $\mathbf{a}$ is constant, $\mathbf{v} = \mathbf{a}t + \mathbf{C}$. Since $\mathbf{v}(t_0) = \mathbf{v}_0$, it follows that $\mathbf{v}_0 = \mathbf{a}t_0 + \mathbf{C}$, and $\mathbf{v} = \mathbf{a}t + \mathbf{v}_0 - \mathbf{a}t_0 = \mathbf{a}(t - t_0) + \mathbf{v}_0$. Integration now gives $\mathbf{r} = \mathbf{a}(t - t_0)^2/2 + \mathbf{v}_0 t + \mathbf{D}$. Since $\mathbf{r}(t_0) = \mathbf{r}_0$, we obtain $\mathbf{r}_0 = \mathbf{v}_0 t_0 + \mathbf{D}$, and therefore

$$\mathbf{r} = \frac{1}{2}\mathbf{a}(t - t_0)^2 + \mathbf{v}_0 t + \mathbf{r}_0 - \mathbf{v}_0 t_0 = \frac{1}{2}\mathbf{a}(t - t_0)^2 + \mathbf{v}_0(t - t_0) + \mathbf{r}_0.$$

30. The acceleration of the arrow is $\mathbf{a} = -g\hat{\mathbf{j}}$, so that $\mathbf{v} = -gt\hat{\mathbf{j}} + \mathbf{C}$. If we take $t = 0$ to be the time when the arrow leaves the bow, then when the bow is held at angle θ, and the initial speed of the arrow is v_0, $\mathbf{v}(0) = v_0(\cos\theta\,\hat{\mathbf{i}} + \sin\theta\,\hat{\mathbf{j}})$. This gives $v_0(\cos\theta\,\hat{\mathbf{i}} + \sin\theta\,\hat{\mathbf{j}}) = \mathbf{C}$. Integration of $\mathbf{v} = -gt\hat{\mathbf{j}} + \mathbf{C}$ gives $\mathbf{r} = -gt^2\hat{\mathbf{j}}/2 + \mathbf{C}t + \mathbf{D}$.

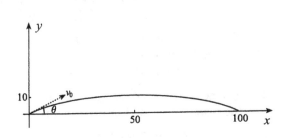

If the arrow starts from the origin, then $\mathbf{r}(0) = \mathbf{0}$, from which $\mathbf{D} = \mathbf{0}$, and therefore

$$\mathbf{r} = -\frac{1}{2}gt^2\hat{\mathbf{j}} + v_0 t(\cos\theta\,\hat{\mathbf{i}} + \sin\theta\,\hat{\mathbf{j}}) = (v_0 t\cos\theta)\hat{\mathbf{i}} + \left(-\frac{1}{2}gt^2 + v_0 t\sin\theta\right)\hat{\mathbf{j}}.$$

If T is the time for the arrow to reach maximum height, we can say that $\mathbf{r}(T) = 50\hat{\mathbf{i}} + 10\hat{\mathbf{j}}$ and $\mathbf{r}(2T) = 100\hat{\mathbf{i}}$. These imply that

$$50\hat{\mathbf{i}} + 10\hat{\mathbf{j}} = (v_0 T\cos\theta)\hat{\mathbf{i}} + \left(-\frac{1}{2}gT^2 + v_0 T\sin\theta\right)\hat{\mathbf{j}}, \quad 100\hat{\mathbf{i}} = [v_0(2T)\cos\theta]\hat{\mathbf{i}} + \left[-\frac{1}{2}g(2T)^2 + v_0(2T)\sin\theta\right]\hat{\mathbf{j}}.$$

When we equate components, we obtain four equations, three of which are independent,

$$50 = v_0 T\cos\theta, \quad 10 = -\frac{1}{2}gT^2 + v_0 T\sin\theta, \quad 0 = -2gT^2 + 2v_0 T\sin\theta.$$

We eliminate T and solve for v_0 and θ. The result is $v_0 = 37.7$ m/s and $\theta = 0.381$ radians.

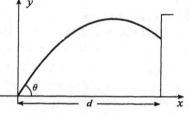

32. The acceleration of any droplet of water is
$\mathbf{a} = -g\hat{\mathbf{j}}$, so that $\mathbf{v} = -gt\hat{\mathbf{j}} + \mathbf{C}$. If we take
as $t = 0$ the time when the droplet leaves the
nozzle, then when the hose is held at angle θ,

$$\mathbf{v}(0) = S(\cos\theta\,\hat{\mathbf{i}} + \sin\theta\,\hat{\mathbf{j}}),$$

from which $S(\cos\theta\,\hat{\mathbf{i}} + \sin\theta\,\hat{\mathbf{j}}) = \mathbf{C}$. Integration of
$\mathbf{v} = -gt\hat{\mathbf{j}} + \mathbf{C}$ gives $\mathbf{r} = -gt^2\hat{\mathbf{j}}/2 + \mathbf{C}t + \mathbf{D}$. Since $\mathbf{r}(0) = 0$, it follows that $\mathbf{D} = 0$, and

$$\mathbf{r} = -\frac{1}{2}gt^2\hat{\mathbf{j}} + St(\cos\theta\,\hat{\mathbf{i}} + \sin\theta\,\hat{\mathbf{j}}) = (St\cos\theta)\hat{\mathbf{i}} + \left(-\frac{1}{2}gt^2 + St\sin\theta\right)\hat{\mathbf{j}}.$$

The droplet strikes the building where $x = d$, in which case $d = St\cos\theta$. This equation implies that
$t = d/(S\cos\theta)$, and this is the time that it strikes the building if it is fired at angle θ. The height it
reaches on the building when it is fired at angle θ is therefore

$$y(\theta) = -\frac{1}{2}g\left(\frac{d}{S\cos\theta}\right)^2 + S\sin\theta\left(\frac{d}{S\cos\theta}\right) = -\frac{gd^2}{2S^2\cos^2\theta} + d\tan\theta.$$

The problem then is to maximize $y(\theta)$ considering those values of θ which guarantee that the droplet
does strike the building. There is a smallest value, say α, below which the droplet does not reach the
wall, and a largest value, say β, beyond which the droplet also fails to reach the wall (and these values
depend on d and S). Thus, we should maximize $y(\theta)$ for $\alpha \le \theta \le \beta$ where $y(\alpha) = y(\beta) = 0$. For critical
values of $y(\theta)$, we solve

$$0 = \frac{dy}{d\theta} = \frac{gd^2}{S^2\cos^3\theta}(-\sin\theta) + d\sec^2\theta = \frac{d(-gd\sin\theta + S^2\cos\theta)}{S^2\cos^3\theta}.$$

Thus, $\tan\theta = S^2/(gd)$. We accept from this equation only the acute angle, and this must be the value
which maximizes $y(\theta)$, so that maximum $y(\theta)$ is

$$\frac{-gd^2}{2S^2}\left(\frac{S^4 + g^2d^2}{g^2d^2}\right) + \frac{dS^2}{gd} = \frac{S^4 - g^2d^2}{2gS^2}.$$

Notice that this result also implies that the water
reaches the wall if, and only if,

$$S^4 - g^2d^2 > 0 \quad \text{or} \quad S^2 > gd.$$

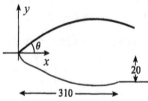

34. The acceleration of the ball after it leaves the tee is $\mathbf{a} = -g\hat{\mathbf{j}}$ so that $\mathbf{v} = -gt\hat{\mathbf{j}} + \mathbf{C}$. If we take $t = 0$ at
the instant the ball is struck, and it begins at angle θ, then

$$\mathbf{v}(0) = S(\cos\theta\,\hat{\mathbf{i}} + \sin\theta\,\hat{\mathbf{j}})$$

where $S > 0$ is its initial speed off the tee.
This implies that $\mathbf{C} = \mathbf{v}(0)$. Integration
gives

$$\mathbf{r} = -\frac{1}{2}gt^2\hat{\mathbf{j}} + \mathbf{C}t + \mathbf{D}.$$

Since $\mathbf{r}(0) = 0$, it follows that $\mathbf{D} = 0$, and

$$\mathbf{r} = -\frac{1}{2}gt^2\hat{\mathbf{j}} + St(\cos\theta\,\hat{\mathbf{i}} + \sin\theta\,\hat{\mathbf{j}}) = (St\cos\theta)\hat{\mathbf{i}} + \left(-\frac{1}{2}gt^2 + St\sin\theta\right)\hat{\mathbf{j}}.$$

According to Exercise 23, maximum range along a level fairway is attained for $\theta = \pi/4$, and in this case

$$R = \frac{2S^2(1/\sqrt{2})(1/\sqrt{2})}{g} = \frac{S^2}{g}.$$

Since maximum range is 300 m, $300 = S^2/g$, or, $S = \sqrt{300g} = 54.25$ m/s. In other words, maximum
speed of the ball off the tee is 54.25 m/s. The

ball covers a horizontal displacement of 310 m
when $310 = (54.25)t\cos\theta$, or $t = (310/54.25)\sec\theta$.
The y-displacement at this instant is

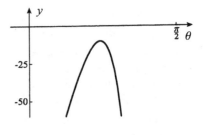

$$y = -\frac{1}{2}g\left(\frac{310}{54.25}\sec\theta\right)^2 + 54.25\left(\frac{310}{54.25}\sec\theta\right)\sin\theta$$

$$= -\frac{(310)(155)g}{(54.25)^2}\sec^2\theta + 310\tan\theta.$$

A plot of this function shows that there are angles
for which $y = -20$.

36. The acceleration of the projectile is $\mathbf{a} = -g\hat{\mathbf{j}}$, integration of which gives $\mathbf{v} = -gt\hat{\mathbf{j}} + \mathbf{C}$. If $\mathbf{v}_0 = v_0\cos(\alpha+\beta)\hat{\mathbf{i}} + v_0\sin(\alpha+\beta)\hat{\mathbf{j}}$ is the initial velocity of the projectile at time $t = 0$ when it leaves the cannon, then $\mathbf{v}_0 = \mathbf{C}$, and $\mathbf{v} = -gt\hat{\mathbf{j}} + \mathbf{v}_0$. Integration gives $\mathbf{r} = -gt^2\hat{\mathbf{j}}/2 + \mathbf{v}_0 t + \mathbf{D}$. Since $\mathbf{r}(0) = \mathbf{0}$, it follows that $\mathbf{D} = \mathbf{0}$, and $\mathbf{r}(t) = -gt^2\hat{\mathbf{j}}/2 + \mathbf{v}_0 t$. In component form,

$$x = v_0\cos(\alpha+\beta)t, \qquad y = -\frac{1}{2}gt^2 + v_0\sin(\alpha+\beta)t.$$

The projectile strikes the inclined plane at a point satisfying $y = x\tan\alpha$, and this implies that

$$-\frac{1}{2}gt^2 + v_0\sin(\alpha+\beta)t = v_0\cos(\alpha+\beta)\tan\alpha\,t \implies t = \frac{2v_0}{g}[\sin(\alpha+\beta) - \cos(\alpha+\beta)\tan\alpha].$$

Since the projectile strikes the ground horizontally, the y-component of velocity at the point of impact must be zero,

$$0 = -gt + v_0\sin(\alpha+\beta) \implies t = \frac{v_0}{g}\sin(\alpha+\beta).$$

When we equate these two expressions for t,

$$\frac{2v_0}{g}[\sin(\alpha+\beta) - \cos(\alpha+\beta)\tan\alpha] = \frac{v_0}{g}\sin(\alpha+\beta) \implies \sin(\alpha+\beta) = 2\cos(\alpha+\beta)\tan\alpha.$$

Multiplication by $\cos\alpha$ gives

$$(\sin\alpha\cos\beta + \cos\alpha\sin\beta)\cos\alpha = 2(\cos\alpha\cos\beta - \sin\alpha\sin\beta)\sin\alpha,$$

from which $\sin\alpha\cos\alpha\cos\beta = (\cos^2\alpha + 2\sin^2\alpha)\sin\beta$. This can be solved for

$$\tan\beta = \frac{\sin\alpha\cos\alpha}{\cos^2\alpha + 2\sin^2\alpha} = \frac{(1/2)\sin2\alpha}{(1 + \cos2\alpha)/2 + (1 - \cos2\alpha)} = \frac{\sin2\alpha}{3 - \cos2\alpha} \implies \beta = \text{Tan}^{-1}\left(\frac{\sin2\alpha}{3 - \cos2\alpha}\right).$$

38. (a) Since length around the circumference of the tire is given by $R\theta$ and the time rate of change of this quantity is the speed of the centre of the tire, it follows that $S = R(d\theta/dt)$. Antidifferentiation gives $\theta = St/R + C$. Since $\theta = 0$ when $t = 0$, it follows that $C = 0$ and $\theta = St/R$.
(b) Since $x = R(\theta - \sin\theta)$ and $y = R(1 - \cos\theta)$,

$$\mathbf{v} = R\left(\frac{d\theta}{dt} - \cos\theta\frac{d\theta}{dt}\right)\hat{\mathbf{i}} + \left(R\sin\theta\frac{d\theta}{dt}\right)\hat{\mathbf{j}} = S[(1 - \cos\theta)\hat{\mathbf{i}} + \sin\theta\hat{\mathbf{j}}]$$

$$|\mathbf{v}| = S\sqrt{(1 - \cos\theta)^2 + \sin^2\theta} = S\sqrt{2 - 2\cos\theta}$$

$$\mathbf{a} = S\left(\sin\theta\frac{d\theta}{dt}\hat{\mathbf{i}} + \cos\theta\frac{d\theta}{dt}\hat{\mathbf{j}}\right) = \frac{S^2}{R}(\sin\theta\hat{\mathbf{i}} + \cos\theta\hat{\mathbf{j}}).$$

(c) $a_T = \dfrac{d}{dt}|\mathbf{v}| = \dfrac{S\sin\theta}{\sqrt{2 - 2\cos\theta}}\dfrac{d\theta}{dt} = \dfrac{S^2\sin\theta}{R\sqrt{2 - 2\cos\theta}}$

$a_N = \sqrt{|\mathbf{a}|^2 - a_T^2} = \sqrt{\dfrac{S^4}{R^2} - \dfrac{S^4\sin^2\theta}{R^2(2 - 2\cos\theta)}} = \dfrac{S^2}{R}\sqrt{(1 - \cos\theta)/2}$

40. (a) $x = R + \|TV\| = R + \|UV\| - \|UT\|$

$$= R + \|PQ\| \sin\phi - (R - \|OU\|)$$

$$= \|PQ\| \sin\phi + R\cos\theta.$$

When $(\pi/2 - \theta) + \phi + \rho = \pi$, and
$\theta + 2\rho = \pi$, are solved for ρ, and
results are equated, we obtain $\phi = 3\theta/2$.
Hence, with $\|PQ\| = 2R\sin(\theta/2)$,

$$x = 2R\sin(\theta/2)\sin(3\theta/2) + R\cos\theta$$

$$= R(-\cos 2\theta + \cos\theta + \cos\theta)$$

$$= R(2\cos\theta - \cos 2\theta).$$

Furthermore,

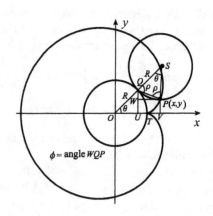

$\phi = $ angle WQP

$$y = \|UQ\| - \|QW\| = R\sin\theta - \|PQ\|\cos\phi$$

$$= R\sin\theta - 2R\sin(\theta/2)\cos(3\theta/2)$$

$$= R(\sin\theta + \sin\theta - \sin 2\theta)$$

$$= R(2\sin\theta - \sin 2\theta).$$

(b) When the point of contact has moved so that it makes angle θ with the positive x-axis, length along the stationary circle from $(R,0)$ to the point of contact is $R\theta$. Since this length changes at a rate of $R\, d\theta/dt = S$, it follows that $\theta = St/R$.

(c) $\mathbf{v} = \dfrac{d\mathbf{r}}{dt} = R(-2\sin\theta + 2\sin 2\theta)\dfrac{d\theta}{dt}\hat{\mathbf{i}} + R(2\cos\theta - 2\cos 2\theta)\dfrac{d\theta}{dt}\hat{\mathbf{j}}$

$$= 2S(\sin 2\theta - \sin\theta)\hat{\mathbf{i}} + 2S(\cos\theta - \cos 2\theta)\hat{\mathbf{j}};$$

$|\mathbf{v}| = 2S\sqrt{(\sin 2\theta - \sin\theta)^2 + (\cos\theta - \cos 2\theta)^2} = 2S\sqrt{2 - 2\sin 2\theta \sin\theta - 2\cos 2\theta \cos\theta}$

$$= 2\sqrt{2}S\sqrt{1 - \cos\theta};$$

$\mathbf{a} = 2S(2\cos 2\theta - \cos\theta)\dfrac{d\theta}{dt}\hat{\mathbf{i}} + 2S(-\sin\theta + 2\sin 2\theta)\dfrac{d\theta}{dt}\hat{\mathbf{j}}$

$$= \dfrac{2S^2}{R}[(2\cos 2\theta - \cos\theta)\hat{\mathbf{i}} + (2\sin 2\theta - \sin\theta)\hat{\mathbf{j}}].$$

(d) $a_T = \dfrac{d}{dt}|\mathbf{v}| = \dfrac{\sqrt{2}S\sin\theta}{\sqrt{1 - \cos\theta}}\dfrac{d\theta}{dt} = \dfrac{\sqrt{2}S^2\sin\theta}{R\sqrt{1 - \cos\theta}};$

$a_N = \sqrt{|\mathbf{a}|^2 - a_T{}^2} = \sqrt{\dfrac{4S^4}{R^2}[(2\cos 2\theta - \cos\theta)^2 + (2\sin 2\theta - \sin\theta)^2] - \dfrac{2S^4\sin^2\theta}{R^2(1 - \cos\theta)}}$

$$= \dfrac{S^2}{R}\sqrt{4[5 - 4\cos 2\theta\cos\theta - 4\sin 2\theta\sin\theta] - \dfrac{2\sin^2\theta}{1 - \cos\theta}}$$

$$= \dfrac{S^2}{R}\sqrt{20 - 16\cos\theta - \dfrac{2\sin^2\theta}{1 - \cos\theta}} = \dfrac{S^2}{R}\sqrt{\dfrac{20 - 36\cos\theta + 16\cos^2\theta - 2(1 - \cos^2\theta)}{1 - \cos\theta}}$$

$$= \dfrac{S^2}{R}\sqrt{\dfrac{18\cos^2\theta - 36\cos\theta + 18}{1 - \cos\theta}} = \dfrac{3\sqrt{2}S^2}{R}\sqrt{1 - \cos\theta}$$

42. Let us take the positive x-direction as the direction in which the river flows. Then the velocity of the water with respect to the shore is $\mathbf{v}_{w/s} = 3\hat{\mathbf{i}}$. If the canoe points in direction θ shown, $\mathbf{v}_{c/w} = 2\cos\theta\,\hat{\mathbf{i}} + 2\sin\theta\,\hat{\mathbf{j}}$. If v is the speed of the canoe with respect to the shore, then $\mathbf{v}_{c/s} = v\cos\phi\,\hat{\mathbf{i}} + v\sin\phi\,\hat{\mathbf{j}}$. Since $\mathbf{v}_{c/s} = \mathbf{v}_{c/w} + \mathbf{v}_{w/s}$, we obtain

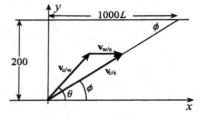

$$v\cos\phi\,\hat{\mathbf{i}} + v\sin\phi\,\hat{\mathbf{j}} = 2\cos\theta\,\hat{\mathbf{i}} + 2\sin\theta\,\hat{\mathbf{j}} + 3\hat{\mathbf{i}}.$$

When we equate components, $v \cos \phi = 2 \cos \theta + 3 = 0$ and $v \sin \phi = 2 \sin \theta$. These imply that

$$4 = 4 \cos^2 \theta + 4 \sin^2 \theta = (v \cos \phi - 3)^2 + (v \sin \phi)^2 = v^2 - 6v \cos \phi + 9.$$

Thus, $v^2 - 6v \cos \phi + 5 = 0$, the solutions of which are $v = 3 \cos \phi \pm \sqrt{9 \cos^2 \phi - 5}$. Clearly v exists only if $9 \cos^2 \phi - 5 \geq 0 \implies \cos \phi \geq \sqrt{5}/3$. But from the figure, $\cos \phi = 5L/\sqrt{1 + 25L^2}$, and therefore

$$\frac{5L}{\sqrt{1 + 25L^2}} \geq \frac{\sqrt{5}}{3} \implies L \geq \frac{\sqrt{5}}{10}.$$

When $L = \sqrt{5}/10$, there is one solution for v, namely, $v = 3 \cos \phi$, in which case

$$2 \sin \theta = 3 \cos \phi \sin \phi = 3 \left(\frac{\sqrt{5}}{3} \right) \left(\frac{2}{3} \right) \implies \theta = \mathrm{Sin}^{-1} \left(\frac{\sqrt{5}}{3} \right) = 0.841 \text{ radians.}$$

When $L > \sqrt{5}/10$, there are two solutions for v. The larger one $v = 3 \cos \phi + \sqrt{9 \cos^2 \phi - 5}$ gives the shorter travel time. For this choice,

$$2 \sin \theta = \sin \phi (3 \cos \phi + \sqrt{9 \cos^2 \phi - 5}) = \frac{1}{\sqrt{1 + 25L^2}} \left(\frac{15L}{\sqrt{1 + 25L^2}} + \sqrt{\frac{225L^2}{1 + 25L^2} - 5} \right)$$

$$= \frac{15L + \sqrt{100L^2 - 5}}{1 + 25L^2}.$$

Therefore, $\theta = \mathrm{Sin}^{-1} \left(\dfrac{15L + \sqrt{100L^2 - 5}}{2 + 50L^2} \right)$.

44. If we express the acceleration $\mathbf{a}$ of the particle in form 11.112a, write the force in the form $\mathbf{F} = \lambda(t)\mathbf{T}$, and substitute these into 11.109

$$\lambda(t)\mathbf{T} = m(a_T \hat{\mathbf{T}} + a_N \hat{\mathbf{N}}) \implies a_N = 0.$$

But then 11.112b implies that $|\mathbf{v}| \left| \dfrac{d\hat{\mathbf{T}}}{dt} \right| = 0$, from which $\dfrac{d\hat{\mathbf{T}}}{dt} = 0$, and $\hat{\mathbf{T}}$ is a constant vector. But this means that the trajectory is a straight line.

46. (a) Using the formula in part (c) of Exercise 45, the area swept out by the plane in one revolution is

$$A(t + P) - A(t) = \left[\frac{|\mathbf{C}|}{2}(t + P) + k \right] - \left[\frac{|\mathbf{C}|}{2} t + k \right] = \frac{P|\mathbf{C}|}{2}.$$

But the area of the ellipse is πab, so that $\dfrac{P|\mathbf{C}|}{2} = \pi ab \implies P = \dfrac{2\pi ab}{|\mathbf{C}|}$.

(b) If we change 11.116 to polar coordinates by setting $x = r \cos \theta$ and $y = r \sin \theta$,

$$\sqrt{x^2 + y^2} = \frac{\epsilon d}{1 + \dfrac{\epsilon x}{\sqrt{x^2 + y^2}}} \implies \sqrt{x^2 + y^2} = \epsilon(d - x).$$

Squaring gives

$$x^2 + y^2 = \epsilon^2 (d - x)^2 \implies x^2 + \frac{2d\epsilon^2 x}{1 - \epsilon^2} + \frac{y^2}{1 - \epsilon^2} = \frac{\epsilon^2 d^2}{1 - \epsilon^2}.$$

Completing the square on the x-terms results in

$$\left(x + \frac{d\epsilon^2}{1 - \epsilon^2} \right)^2 + \frac{y^2}{1 - \epsilon^2} = \frac{\epsilon^2 d^2}{1 - \epsilon^2} + \frac{d^2 \epsilon^4}{(1 - \epsilon^2)^2} = \left(\frac{\epsilon d}{1 - \epsilon^2} \right)^2.$$

This allows us to identify $a = \epsilon d/(1 - \epsilon^2)$ and $b = \epsilon d/\sqrt{1 - \epsilon^2}$. With $\epsilon = |\mathbf{b}|/(GM)$ and $d = |\mathbf{C}|^2/|\mathbf{b}|$,

$$a = \frac{|\mathbf{C}|^2/(GM)}{1 - |\mathbf{b}|^2/(G^2M^2)} = \frac{|\mathbf{C}|^2 GM}{G^2M^2 - |\mathbf{b}|^2}, \quad b = \frac{|\mathbf{C}|^2/(GM)}{\sqrt{1 - |\mathbf{b}|^2/(G^2M^2)}} = \frac{|\mathbf{C}|^2}{\sqrt{G^2M^2 - |\mathbf{b}|^2}}.$$

It follows therefore that $\dfrac{b^2}{a} = \dfrac{|\mathbf{C}|^4}{G^2M^2 - |\mathbf{b}|^2} \dfrac{G^2M^2 - |\mathbf{b}|^2}{|\mathbf{C}|^2 GM} = \dfrac{|\mathbf{C}|^2}{GM}$, and hence,

$$P^2 = \frac{4\pi^2 a^2 b^2}{|\mathbf{C}|^2} = 4\pi^2 a^2 \left(\frac{a}{GM}\right) = \frac{4\pi^2 a^3}{GM}.$$

REVIEW EXERCISES

2. $\mathbf{u} \cdot \mathbf{v} \times \mathbf{w} = \mathbf{u} \cdot \begin{vmatrix} \hat{\mathbf{i}} & \hat{\mathbf{j}} & \hat{\mathbf{k}} \\ 2 & 4 & -1 \\ 0 & 2 & 1 \end{vmatrix} = (1, 3, -2) \cdot (6, -2, 4) = 6 - 6 - 8 = -8$

4. $3\mathbf{u} \times (4\mathbf{v} - \mathbf{w}) = 3\mathbf{u} \times (8, 14, -5) = 3\begin{vmatrix} \hat{\mathbf{i}} & \hat{\mathbf{j}} & \hat{\mathbf{k}} \\ 1 & 3 & -2 \\ 8 & 14 & -5 \end{vmatrix} = 3(13, -11, -10) = (39, -33, -30)$

6. $(\mathbf{u} + \mathbf{v}) \cdot (\mathbf{r} - \mathbf{w}) = (3, 7, -3) \cdot (2, -2, -2) = 6 - 14 + 6 = -2$

8. $(\mathbf{u} \times \mathbf{v}) \times (\mathbf{r} \times \mathbf{w}) = \begin{vmatrix} \hat{\mathbf{i}} & \hat{\mathbf{j}} & \hat{\mathbf{k}} \\ 1 & 3 & -2 \\ 2 & 4 & -1 \end{vmatrix} \times \begin{vmatrix} \hat{\mathbf{i}} & \hat{\mathbf{j}} & \hat{\mathbf{k}} \\ 2 & 0 & -1 \\ 0 & 2 & 1 \end{vmatrix} = (5, -3, -2) \times (2, -2, 4)$

$$= \begin{vmatrix} \hat{\mathbf{i}} & \hat{\mathbf{j}} & \hat{\mathbf{k}} \\ 5 & -3 & -2 \\ 2 & -2 & 4 \end{vmatrix} = (-16, -24, -4)$$

10. $\dfrac{2\mathbf{r}}{\mathbf{v} \cdot \mathbf{w}} + 3(\mathbf{v} + \mathbf{u}) = \dfrac{2}{8 - 1}(2, 0, -1) + 3(3, 7, -3) = \left(\dfrac{67}{7}, 21, -\dfrac{65}{7}\right)$

For Exercises 12, 14, 16, 18, 20, 22, 24, and 26, see answers in text.

28. Since a vector along the line is $(5, -2, 1)$, parametric equations for the line are $x = 6 + 5t$, $y = 6 - 2t$, $z = 2 + t$.

30. Since $(1, 3, 2)$ is a point on the required line, parametric equations must be of the form

$$x = 1 + au, \ y = 3 + bu, \ z = 2 + cu.$$

Because the line must be perpendicular to the given line,

$$0 = (a, b, c) \cdot (1, -2, 1) = a - 2b + c.$$

Finally, since the lines must intersect, we set

$$t + 2 = 1 + au, \ 3 - 2t = 3 + bu, \ 4 + t = 2 + cu.$$

When the first two of these are solved for t and u,

$$u = \frac{2}{b + 2a} \quad \text{and} \quad t = \frac{-b}{b + 2a}.$$

Substitution into the third gives $4a + b - 2c = 0$. When this is combined with $a - 2b + c = 0$, the result is $b = 2a$ and $c = 3a$. Consequently, parametric equations for the required line are $x = 1 + au$, $y = 3 + 2au$, $z = 2 + 3au$, or, $x = 1 + v$, $y = 3 + 2v$, $z = 2 + 3v$ (where we have set $v = au$).

32. Since parametric equations for the line are $x = 4 - t$, $y = t$, $z = t$, a vector along the line (and therefore normal to the plane is $(-1, 1, 1)$. The equation of the plane is

$$0 = (-1, 1, 1) \cdot (x - 1, y - 2, z + 1) = -x + y + z.$$

34. Since the lines are not parallel, they determine a plane only if they intersect. To confirm this, we set $3t = 1 + 2t$ and $3t = 4 - t$. These both give $t = 1$ leading to the point of intersection $(3, 3, 3)$. A vector normal to the plane is

$$(3, 2, -1) \times (1, 1, 1) = \begin{vmatrix} \hat{\mathbf{i}} & \hat{\mathbf{j}} & \hat{\mathbf{k}} \\ 3 & 2 & -1 \\ 1 & 1 & 1 \end{vmatrix} = (3, -4, 1).$$

The equation of the plane is

$$0 = (3, -4, 1) \cdot (x - 3, y - 3, z - 3) = 3x - 4y + z.$$

36. The required distance d is the projection of $\mathbf{PQ}$ on the direction $\mathbf{PR}$ normal to the plane:

$$d = \left| \mathbf{PQ} \cdot \widehat{\mathbf{PR}} \right|$$
$$= \left| (-6, -2, -5) \cdot \frac{(-6, -2, 1)}{\sqrt{36 + 4 + 1}} \right|$$
$$= 35/\sqrt{41}.$$

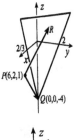

38. The distance is zero unless the line and plane are parallel. A vector along the line is $(1, -1, 1) \times (2, 1, 1) = (-2, 1, 3)$. Since $(-2, 1, 3) \cdot (3, 6, 0) = 0$, the line and plane are parallel. The required distance d is the projection of $\mathbf{PQ}$ on the direction $\mathbf{PR}$ normal to the plane:

$$d = \left| \mathbf{PQ} \cdot \widehat{\mathbf{PR}} \right| = \left| (0, -1/3, -3) \cdot \frac{(-3, -6, 0)}{\sqrt{9 + 36}} \right| = \frac{2}{3\sqrt{5}}.$$

40. According to equation 11.42,

$$\text{Area} = \frac{1}{2} \left| [(-2, 1, 0) - (1, 1, 1)] \times [(6, 3, -2) - (1, 1, 1)] \right|$$
$$= \frac{1}{2} |(-3, 0, -1) \times (5, 2, -3)| = \frac{1}{2} \left\| \begin{vmatrix} \hat{\mathbf{i}} & \hat{\mathbf{j}} & \hat{\mathbf{k}} \\ -3 & 0 & -1 \\ 5 & 2 & -3 \end{vmatrix} \right\| = \frac{1}{2} |(2, -14, -6)| = \sqrt{59}.$$

42. From $\mathbf{T} = \dfrac{d\mathbf{r}}{dt} = (2\cos t, -2\sin t, 1)$, we obtain the unit tangent vector

$$\hat{\mathbf{T}} = \frac{(2\cos t, -2\sin t, 1)}{\sqrt{4\cos^2 t + 4\sin^2 t + 1}} = \frac{(2\cos t, -2\sin t, 1)}{\sqrt{5}}.$$

A vector in the direction of $\hat{\mathbf{N}}$ is

$$\mathbf{N} = \frac{d\hat{\mathbf{T}}}{dt} = \frac{1}{\sqrt{5}}(-2\sin t, -2\cos t, 0) = -\frac{2}{\sqrt{5}}(\sin t, \cos t, 0),$$

and therefore $\hat{\mathbf{N}} = -(\sin t, \cos t, 0)$. The binormal is

$$\hat{\mathbf{B}} = \hat{\mathbf{T}} \times \hat{\mathbf{N}} = -\frac{1}{\sqrt{5}} \begin{vmatrix} \hat{\mathbf{i}} & \hat{\mathbf{j}} & \hat{\mathbf{k}} \\ 2\cos t & -2\sin t & 1 \\ \sin t & \cos t & 0 \end{vmatrix} = \frac{1}{\sqrt{5}}(\cos t, -\sin t, -2).$$

44. $\mathbf{v} = \dfrac{d\mathbf{r}}{dt} = (1, 2t, 2t)$ $|\mathbf{v}| = \sqrt{1 + 4t^2 + 4t^2} = \sqrt{1 + 8t^2}$ $\mathbf{a} = \dfrac{d\mathbf{v}}{dt} = (0, 2, 2)$

The normal component of velocity is always zero, and the tangential component of velocity is speed $|\mathbf{v}| = \sqrt{1 + 8t^2}$.

$$a_T = \frac{d}{dt}|\mathbf{v}| = \frac{8t}{\sqrt{1 + 8t^2}}, \qquad a_N = \sqrt{|\mathbf{a}|^2 - a_T^2} = \left(8 - \frac{64t^2}{1 + 8t^2}\right)^{1/2} = \frac{2\sqrt{2}}{\sqrt{1 + 8t^2}}$$

46. (a) The acceleration of the ball after it leaves the table is $\mathbf{a} = -g\hat{\mathbf{j}}$ where $g = 9.81$, so that $\mathbf{v} = -gt\hat{\mathbf{j}} + \mathbf{C}$. If we take $t = 0$ at the instant the ball leaves the table, $\mathbf{v}(0) = \hat{\mathbf{i}}/2$. This implies that $\mathbf{C} = \hat{\mathbf{i}}/2$, and

$$\mathbf{r} = -\frac{1}{2}gt^2\hat{\mathbf{j}} + \frac{t}{2}\hat{\mathbf{i}} + \mathbf{D}.$$

Since $\mathbf{r}(0) = \mathbf{0}$, it follows that $\mathbf{D} = \mathbf{0}$, and $\mathbf{r} = -\dfrac{1}{2}gt^2\hat{\mathbf{j}} + \dfrac{t}{2}\hat{\mathbf{i}}$. The ball strikes the floor when the y-component of $\mathbf{r}$ is -1,

$$-1 = -\frac{1}{2}gt^2 \quad \Longrightarrow \quad t = \sqrt{\frac{2}{g}} = \sqrt{\frac{2}{9.81}}.$$

At this instant, $\mathbf{v} = -g\sqrt{\dfrac{2}{9.81}}\hat{\mathbf{j}} + \dfrac{1}{2}\hat{\mathbf{i}} = \dfrac{1}{2}\hat{\mathbf{i}} - \sqrt{19.62}\hat{\mathbf{j}}$, and therefore its speed when it strikes the floor is

$$|\mathbf{v}| = \sqrt{1/4 + 19.62} = \sqrt{19.87} \text{ m/s.}$$

(b) Its displacement vector when it strikes the floor is

$$\mathbf{r} = -\frac{1}{2}g\left(\frac{2}{g}\right)\hat{\mathbf{j}} + \frac{1}{2}\sqrt{\frac{2}{g}}\hat{\mathbf{i}} = 0.226\hat{\mathbf{i}} - \hat{\mathbf{j}}.$$

(c) After the rebound, the acceleration of the ball is once again $\mathbf{a} = -g\hat{\mathbf{j}}$, so that $\mathbf{v} = -gt\hat{\mathbf{j}} + \mathbf{C}$. If we redefine $t = 0$ at the time of the first bounce, then

$$\mathbf{C} = \mathbf{v}(0) = \frac{4}{5}\left(\frac{1}{2}\hat{\mathbf{i}} + \sqrt{19.62}\hat{\mathbf{j}}\right).$$

Integration now gives

$$\mathbf{r} = -\frac{1}{2}gt^2\hat{\mathbf{j}} + \mathbf{C}t + \mathbf{D}.$$

If we redefine $(x, y) = (0, 0)$ at the position of the first bounce, then $\mathbf{D} = \mathbf{0}$. The ball hits the floor for the second time when the y-component of $\mathbf{r}$ is zero,

$$0 = -\frac{1}{2}gt^2 + \frac{4}{5}\sqrt{19.62}t \quad \Longrightarrow \quad t = \frac{16}{5\sqrt{19.62}}.$$

At this instant, the x-component of its displacement is $r_x = \dfrac{2}{5}\left(\dfrac{16}{5\sqrt{19.62}}\right) = 0.289$. Thus, the second bounce takes place 0.515 m from the point on the floor directly below the point it left the table.

48. In Example 11.26, it was shown that $|\mathbf{F}| = k[\sqrt{(1 - x)^2 + 1/4} - 1/2]$. A vector in the direction of $\mathbf{F}$ is $(1, 1/2) - (x, 0) = (1 - x, 1/2)$, and therefore

$$\mathbf{F} = k[\sqrt{(1 - x)^2 + 1/4} - 1/2]\frac{(1 - x, 1/2)}{\sqrt{(1 - x)^2 + 1/4}}$$

$$= k(1 - x)\left[1 - \frac{1}{\sqrt{4(1 - x)^2 + 1}}\right]\hat{\mathbf{i}} + \frac{k}{2}\left[1 - \frac{1}{\sqrt{4(1 - x)^2 + 1}}\right]\hat{\mathbf{j}} \text{ N.}$$

CHAPTER 12

EXERCISES 12.1

2. $f(a+b, a-b, ab) = (a+b)^2(a-b)^2 - (a+b)^4 + 4ab(a+b)^2 = (a+b)^2[(a-b)^2 - (a+b)^2 + 4ab] = 0$

4. For $1 - x^2 + y^2 > 0$, we must take $x^2 - y^2 < 1$. This inequality describes all points between, but not on, the branches of the hyperbola $x^2 - y^2 = 1$

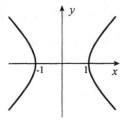

6. This function is defined for all points in space except the origin $(0,0,0)$.

For Exercises 8, 10, 12, 14, 16, 18, and 20, see answers in text.

22. Level curves are defined by $4 - \sqrt{4x^2 + y^2} = C$, or, $4x^2 + y^2 = (C-4)^2$. They are ellipses.

24. Level curves are defined by $\ln(x^2 + y^2) = C$, They are circles.

26. The volume of the box is $V = lwh$, where h is its height. Since $30 = 2hl + 2wh + 2wl$, it follows that $h = (15 - wl)/(l + w)$, and therefore $V = lw(15 - wl)/(l + w)$.

28. If P is the corner of the ellipsoid in the first octant, then

$$V = 8xyz = 8cxy\sqrt{1 - \frac{x^2}{a^2} - \frac{y^2}{b^2}}.$$

When P is not in the first octant,

$$V = 8c|xy|\sqrt{1 - \frac{x^2}{a^2} - \frac{y^2}{b^2}}.$$

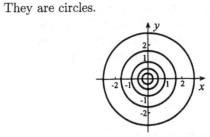

30. (a) $D(0.35, 9.0) = 0.35 + 0.9 + \dfrac{81 \cos 0.35}{9.81}\left[\sin 0.35 + \sqrt{\sin^2 0.35 + \dfrac{2(9.81)(0.5)}{81}}\right] = 7.70$ m

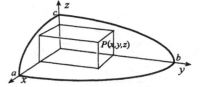

(b) Since $D(0.35, 9.9) = 0.35 + 0.9 + \dfrac{(9.9)^2 \cos 0.35}{9.81}\left[\sin 0.35 + \sqrt{\sin^2 0.35 + \dfrac{2(9.81)(0.5)}{(9.9)^2}}\right] = 8.847$, the

percentage increase is $100(8.847 - 7.70)/7.70 = 14.9$.

(c) Since $D(0.385, 9.0) = 0.35 + 0.9 + \dfrac{81 \cos 0.385}{9.81}\left[\sin 0.385 + \sqrt{\sin^2 0.385 + \dfrac{2(9.81)(0.5)}{81}}\right] = 8.042$,

the percentage increase is $100(8.042 - 7.70)/7.70 = 4.4$.

32. (a)

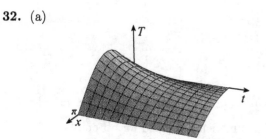

(b) The intersection curve with $x = x_0$ is a graphical history of temperature at position x_0 as a function of time. The intersection curve with $t = t_0$ is a graph of the temperature distribution throughout the rod at time t_0.

34. If x and y are the numbers of computers of models A and B, then the cost of the 100 computers is

$$C = f(x, y) = 1300x + 1200y + 1000(100 - x - y) = 100\,000 + 300x + 200y.$$

Because the computers must have at least 2000 MB of memory,

$$64x + 32y + 16(100 - x - y) \geq 2000 \quad \Longrightarrow 3x + y \geq 25.$$

Because the computers must have at least 150 GB of disk space,

$$3x + 4y + (100 - x - y) \geq 150 \quad \Longrightarrow \quad 2x + 3y \geq 50.$$

The domain of $f(x, y)$ therefore consists of all non-negative values of x and y satisfying these two inequalities. It is the points in the first quadrant above the lines in the figure to the right.

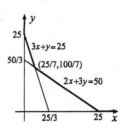

EXERCISES 12.2

2. $\displaystyle\lim_{(x,y)\to(1,1)} \frac{x^3 + 2y^3}{x^3 + 4y^3} = \frac{3}{5}$

4. $\displaystyle\lim_{(x,y,z)\to(2,3,-1)} \frac{xyz}{x^2 + y^2 + z^2} = -\frac{3}{7}$

6. $\displaystyle\lim_{(x,y,z)\to(0,\pi/2,1)} \mathrm{Tan}^{-1}\left(\frac{x}{yz}\right) = 0$

8. $\displaystyle\lim_{(x,y,z)\to(0,\pi/2,1)} \mathrm{Tan}^{-1}\left|\frac{yz}{x}\right| = \frac{\pi}{2}$

10. $\displaystyle\lim_{(x,y)\to(3,4)} \frac{|x^2 + y^2|}{x^2 + y^2} = \frac{|9 + 16|}{9 + 16} = 1$

12. $\displaystyle\lim_{(x,y)\to(2,2)} \frac{x^2 - y^2}{x - y} = \lim_{(x,y)\to(2,2)} \frac{(x+y)(x-y)}{x - y} = \lim_{(x,y)\to(2,2)} (x + y) = 4$

14. If we approach $(0, 0)$ along the line $y = mx$,

$$\lim_{(x,y)\to(0,0)} \frac{x - y}{x + y} = \lim_{x\to0} \frac{x - mx}{x + mx} = \lim_{x\to0} \frac{1 - m}{1 + m} = \frac{1 - m}{1 + m}.$$

Since this result depends on m, the original limit does not exist.

16. $\displaystyle\lim_{(x,y)\to(2,1)} \frac{(x - 2)^2(y + 1)}{x - 2} = \lim_{(x,y)\to(2,1)} (x - 2)(y + 1) = 0$

18. If we approach $(0, 0)$ along the line $y = mx$,

$$\lim_{(x,y)\to(0,0)} \frac{3x^3 - y^3}{2x^3 + 4y^3} = \lim_{x\to0} \frac{3x^3 - m^3x^3}{2x^3 + 4m^3x^3} = \lim_{x\to0} \frac{3 - m^3}{2 + 4m^3} = \frac{3 - m^3}{2 + 4m^3}.$$

Since this result depends on m, the original limit does not exist.

20. $\lim\limits_{(x,y)\to(0,0)} \operatorname{Sec}^{-1}(x^2+y^2)$ does not exist.

22. The function is discontinuous at $(0,0)$.

24. The function is discontinuous when $x=0$, or $y=0$, or $z=0$.

26. Since $f(x,y) = \dfrac{x+y}{xy(x+y)}$, the function is discontinuous when $x=0$, or $y=0$, or $x+y=0$.

28. If we approach $(0,0)$ along parabola $y=ax^2$,

$$\lim_{(x,y)\to(0,0)} \frac{x^4+y^2}{x^4-y^2} = \lim_{x\to 0} \frac{x^4+a^2x^4}{x^4-a^2x^4} = \lim_{x\to 0} \frac{1+a^2}{1-a^2} = \frac{1+a^2}{1-a^2}.$$

Since this result depends on a, the original limit does not exist.

30. If (x,y) is made to approach $(1,0)$ along the straight line $y=m(x-1)$,

$$\lim_{(x,y)\to(1,0)} \frac{(x-1)^2+y^2}{3(x-1)^2-2y^2} = \lim_{x\to 1} \frac{(x-1)^2+m^2(x-1)^2}{3(x-1)^2-2m^2(x-1)^2} = \lim_{x\to 1} \frac{1+m^2}{3-2m^2} = \frac{1+m^2}{3-2m^2}.$$

Since this result depends on m, the original limit does not exist.

32. If (x,y) is made to approach $(1,1)$ along the straight line $y-1=m(x-1)$,

$$\lim_{(x,y)\to(1,1)} \frac{x^2-2x-y^2+2y}{x^2-2x+y^2-2y+2} = \lim_{(x,y)\to(1,1)} \frac{(x-1)^2-(y-1)^2}{(x-1)^2+(y-1)^2} = \lim_{x\to 1} \frac{(x-1)^2-m^2(x-1)^2}{(x-1)^2+m^2(x-1)^2}$$

$$= \lim_{x\to 1} \frac{1-m^2}{1+m^2} = \frac{1-m^2}{1+m^2}.$$

Since this result depends on m, the original limit does not exist.

34. $\lim\limits_{(x,y)\to(1,0)} \dfrac{\sqrt{x+y}-\sqrt{x-y}}{y} = \lim\limits_{(x,y)\to(1,0)} \left(\dfrac{\sqrt{x+y}-\sqrt{x-y}}{y} \dfrac{\sqrt{x+y}+\sqrt{x-y}}{\sqrt{x+y}+\sqrt{x-y}} \right)$

$= \lim\limits_{(x,y)\to(1,0)} \dfrac{(x+y)-(x-y)}{y(\sqrt{x+y}+\sqrt{x-y})} = \lim\limits_{(x,y)\to(1,0)} \dfrac{2}{\sqrt{x+y}+\sqrt{x-y}} = 1$

36. (a) If we set $z=x-y$, then $\lim\limits_{(x,y)\to(1,1)} \dfrac{\sin(x-y)}{x-y} = \lim\limits_{z\to 0} \dfrac{\sin z}{z} = 1$.

The function is not continuous at $(1,1)$ (being undefined there).

(b) Since the value of the function along $y=x$ is always equal to 1, its limit as $(x,y)\to(1,1)$ along $y=x$ is also 1. Consequently the limit of the function as $(x,y)\to(1,1)$ is still 1. The function is now continuous at $(1,1)$.

38. False. In calculating the limit, we consider only those values of (x,y) where $f(x,y)$ is defined. The function in Exercise 36(a) is a counterexample.

EXERCISES 12.3

2. $\dfrac{\partial f}{\partial x} = 3y - 16x^3y^4$; $\quad \dfrac{\partial f}{\partial y} = 3x - 16x^4y^3$

4. $\dfrac{\partial f}{\partial x} = \dfrac{(x+y)(1)-x(1)}{(x+y)^2} - \dfrac{1}{y} = \dfrac{y^2-(x+y)^2}{y(x+y)^2} = \dfrac{-x(x+2y)}{y(x+y)^2}$;

$\dfrac{\partial f}{\partial y} = -\dfrac{x}{(x+y)^2} + \dfrac{x}{y^2} = \dfrac{-xy^2+x(x+y)^2}{y^2(x+y)^2} = \dfrac{x^2(x+2y)}{y^2(x+y)^2}$

6. $\dfrac{\partial f}{\partial x} = y\cos(xy)$; $\quad \dfrac{\partial f}{\partial y} = x\cos(xy)$

8. $\dfrac{\partial f}{\partial x} = \dfrac{1}{2}(x^2+y^2)^{-1/2}(2x) = \dfrac{x}{\sqrt{x^2+y^2}}$; $\quad \dfrac{\partial f}{\partial y} = \dfrac{1}{2}(x^2+y^2)^{-1/2}(2y) = \dfrac{y}{\sqrt{x^2+y^2}}$

10. $\dfrac{\partial f}{\partial x} = \sec^2(2x^2+y^2)(4x)$; $\quad \dfrac{\partial f}{\partial y} = \sec^2(2x^2+y^2)(2y)$

12. $\dfrac{\partial f}{\partial x} = e^{xy}(y); \quad \dfrac{\partial f}{\partial y} = e^{xy}(x)$

14. $\dfrac{\partial f}{\partial x} = \dfrac{1}{x^2 + y^2}(2x); \quad \dfrac{\partial f}{\partial y} = \dfrac{1}{x^2 + y^2}(2y)$

16. $\dfrac{\partial f}{\partial x} = \cos(ye^x)(ye^x); \quad \dfrac{\partial f}{\partial y} = \cos(ye^x)(e^x)$

18. $\dfrac{\partial f}{\partial x} = \dfrac{1}{3}[1 - \cos^3(x^2y)]^{-2/3}[-3\cos^2(x^2y)][-\sin(x^2y)](2xy) = \dfrac{2xy\cos^2(x^2y)\sin(x^2y)}{[1 - \cos^3(x^2y)]^{2/3}};$

$\dfrac{\partial f}{\partial y} = \dfrac{1}{3}[1 - \cos^3(x^2y)]^{-2/3}[-3\cos^2(x^2y)][-\sin(x^2y)](x^2) = \dfrac{x^2\cos^2(x^2y)\sin(x^2y)}{[1 - \cos^3(x^2y)]^{2/3}}$

20. $\dfrac{\partial f}{\partial x} = \dfrac{1}{\sec\sqrt{x+y}}\sec\sqrt{x+y}\tan\sqrt{x+y}\dfrac{1}{2\sqrt{x+y}} = \dfrac{\tan\sqrt{x+y}}{2\sqrt{x+y}};$

$\dfrac{\partial f}{\partial y} = \dfrac{1}{\sec\sqrt{x+y}}\sec\sqrt{x+y}\tan\sqrt{x+y}\dfrac{1}{2\sqrt{x+y}} = \dfrac{\tan\sqrt{x+y}}{2\sqrt{x+y}}$

22. $\dfrac{\partial f}{\partial z} = \dfrac{1}{1 + \dfrac{1}{(x^2+z^2)^2}}\dfrac{-1}{(x^2+z^2)^2}(2z) = \dfrac{-2z}{(x^2+z^2)^2 + 1}$

24. Since $\dfrac{\partial f}{\partial x} = \dfrac{-zt}{(x^2+y^2-t^2)^2}(2x)$, we find $\dfrac{\partial f}{\partial x}\bigg|_{(1,-1,1,-1)} = \dfrac{-(1)(-1)(2)}{(1+1-1)^2} = 2$

26. $\dfrac{\partial f}{\partial x} = \dfrac{-1}{1 + (1 + x + y + z)^2}$

28. $\dfrac{\partial f}{\partial x} = \dfrac{3x^2}{y} + \sin(yz/x) + x\cos(yz/x)\left(-\dfrac{yz}{x^2}\right) = \dfrac{3x^2}{y} + \sin(yz/x) - \dfrac{yz}{x}\cos(yz/x)$

30. $\dfrac{\partial f}{\partial z} = z\mathrm{Sin}^{-1}\left(\dfrac{x}{z}\right) + \dfrac{z^2}{2}\dfrac{1}{\sqrt{1-(x/z)^2}}\left(-\dfrac{x}{z^2}\right) + \dfrac{x}{2}\left(\dfrac{1}{2}\right)(z^2 - x^2)^{-1/2}(2z)$

$= z\mathrm{Sin}^{-1}\left(\dfrac{x}{z}\right) - \dfrac{x}{2}\dfrac{|z|}{\sqrt{z^2 - x^2}} + \dfrac{xz}{2\sqrt{z^2 - x^2}}$

$= \begin{cases} z\mathrm{Sin}^{-1}(x/z), & z > 0 \\ z\mathrm{Sin}^{-1}(x/z) + xz/\sqrt{z^2 - x^2}, & z < 0 \end{cases}$

32. Since $f(x,y,z) = \dfrac{x^3}{yz} + \dfrac{y^3}{xz} + \dfrac{z^3}{xy}$,

$x\dfrac{\partial f}{\partial x} + y\dfrac{\partial f}{\partial y} + z\dfrac{\partial f}{\partial z} = x\left(\dfrac{3x^2}{yz} - \dfrac{y^3}{x^2z} - \dfrac{z^3}{x^2y}\right) + y\left(-\dfrac{x^3}{y^2z} + \dfrac{3y^2}{xz} - \dfrac{z^3}{xy^2}\right) + z\left(-\dfrac{x^3}{yz^2} - \dfrac{y^3}{xz^2} + \dfrac{3z^2}{xy}\right)$

$= \dfrac{x^3}{yz} + \dfrac{y^3}{xz} + \dfrac{z^3}{xy} = f(x,y,z).$

34. (a) This is the normal way to calculate the derivative.

(b) This would lead to an answer of zero. We must always differentiate with respect to a variable and **then** set that variable equal to its prescribed value.

(c) This is acceptable since variables other than the one with respect to which differentiation is being performed can be specified either before or after differentiation.

(d) Same as (b)

36. The derivative vanishes if

$$0 = \dfrac{\partial F}{\partial x} = -\dfrac{2AE}{L}\left(\dfrac{L}{\sqrt{L^2 - 2hx + x^2}} - 1\right) + 2AE\left(\dfrac{h-x}{L}\right)\left[\dfrac{-(L/2)(-2h + 2x)}{(L^2 - 2hx + x^2)^{3/2}}\right]$$

$$= \dfrac{2AE}{L}\left[\dfrac{-L}{\sqrt{L^2 - 2hx + x^2}} + 1 + \dfrac{L(h - x)^2}{(L^2 - 2hx + x^2)^{3/2}}\right]$$

$$= \dfrac{2AE}{L(L^2 - 2hx + x^2)^{3/2}}\left[-L(L^2 - 2hx + x^2) + (L^2 - 2hx + x^2)^{3/2} + L(h - x)^2\right].$$

This implies that

$$(L^2 - 2hx + x^2)^{3/2} = L(L^2 - 2hx + x^2) - L(h^2 - 2hx + x^2) = L(L^2 - h^2).$$

Consequently,

$$x^2 - 2hx + L^2 = (L^3 - Lh^2)^{2/3} \implies x^2 - 2hx + L^2 - (L^3 - Lh^2)^{2/3} = 0.$$

Solutions of this quadratic are

$$x = \frac{2h \pm \sqrt{4h^2 - 4L^2 + 4(L^3 - Lh^2)^{2/3}}}{2} = h \pm (L^3 - Lh^2)^{1/3}\sqrt{1 - \frac{L^2 - h^2}{(L^3 - Lh^2)^{2/3}}}$$

$$= h \pm (L^3 - Lh^2)^{1/3}\sqrt{1 - \frac{L^2 - h^2}{L^{2/3}(L^2 - h^2)^{2/3}}} = h \pm (L^3 - Lh^2)^{1/3}\sqrt{1 - \frac{(L^2 - h^2)^{1/3}}{L^{2/3}}}$$

$$= h \pm (L^3 - Lh^2)^{1/3}\sqrt{1 - \left(1 - \frac{h^2}{L^2}\right)^{1/3}}.$$

Choosing the negative sign gives the required solution.

38. (a) From the cosine law $a = \sqrt{b^2 + c^2 - 2bc\cos A}$,

$$a_A(b, c, A) = \frac{bc\sin A}{\sqrt{b^2 + c^2 - 2bc\cos A}}.$$

(b) From the cosine law in part (a),

$$A = \text{Cos}^{-1}\left(\frac{b^2 + c^2 - a^2}{2bc}\right), \text{ and therefore,}$$

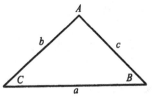

$$A_a(a, b, c) = \frac{-1}{\sqrt{1 - (b^2 + c^2 - a^2)^2/(4b^2c^2)}}\left(-\frac{a}{bc}\right)$$

$$= \frac{2a}{\sqrt{4b^2c^2 - (b^2 + c^2 - a^2)^2}}$$

(c) From the cosine law in part (a), $a_b(b, c, A) = \dfrac{b - c\cos A}{\sqrt{b^2 + c^2 - 2bc\cos A}}.$

(d) From the function in part (b),

$$A_b(a, b, c) = \frac{-1}{\sqrt{1 - (b^2 + c^2 - a^2)^2/(4b^2c^2)}}\left[\frac{2bc(2b) - (b^2 + c^2 - a^2)(2c)}{4b^2c^2}\right]$$

$$= \frac{-2bc}{\sqrt{4b^2c^2 - (b^2 + c^2 - a^2)^2}}\left[\frac{2c(b^2 + a^2 - c^2)}{4b^2c^2}\right] = \frac{c^2 - a^2 - b^2}{b\sqrt{4b^2c^2 - (b^2 + c^2 - a^2)^2}}$$

40. Let x be the distance from the end of the cylinder to any cross section in the cylinder and let L be the distance from the end of the cylinder to the piston (figure to the right). The x-component (the only component) of the velocity of gas at position x in the cylinder is

$$u = \frac{12x}{L} = \frac{12x}{0.15 + 12t}.$$

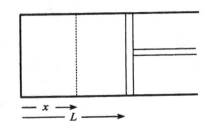

Since $\rho = \rho(t)$, the equation of continuity gives

$$\frac{\partial(\rho u)}{\partial x} + \frac{\partial\rho}{\partial t} = 0 \implies \frac{d\rho}{dt} = -\rho\frac{\partial u}{\partial x} = -\frac{12\rho}{0.15 + 12t} \implies \frac{1}{\rho}d\rho = -\frac{12}{0.15 + 12t}dt.$$

This is a separated differential equation with solutions defined implicitly by

$$\ln \rho = -\ln (0.15 + 12t) + C \qquad \Longrightarrow \qquad \rho = \frac{D}{0.15 + 12t}, \qquad (D = e^C).$$

Since $\rho(0) = 18$, it follows that $18 = D/0.15 \Longrightarrow D = 2.7$. Consequently, $\rho(t) = \dfrac{2.7}{0.15 + 12t}$ kg/m^3.

42. (a) $\dfrac{\partial u}{\partial r} = \dfrac{(1 + r^2 + 2r\cos\theta)(2r + \cos\theta) - (r^2 + r\cos\theta)(2r + 2\cos\theta)}{(1 + r^2 + 2r\cos\theta)^2} = \dfrac{2r + r^2\cos\theta + \cos\theta}{(1 + r^2 + 2r\cos\theta)^2};$

$\dfrac{1}{r}\dfrac{\partial v}{\partial \theta} = \dfrac{1}{r}\left[\dfrac{(1 + r^2 + 2r\cos\theta)(r\cos\theta) - r\sin\theta(-2r\sin\theta)}{(1 + r^2 + 2r\cos\theta)^2}\right] = \dfrac{2r + r^2\cos\theta + \cos\theta}{(1 + r^2 + 2r\cos\theta)^2};$

$\dfrac{1}{r}\dfrac{\partial u}{\partial \theta} = \dfrac{1}{r}\left[\dfrac{(1 + r^2 + 2r\cos\theta)(-r\sin\theta) - (r^2 + r\cos\theta)(-2r\sin\theta)}{(1 + r^2 + 2r\cos\theta)^2}\right] = \dfrac{(r^2 - 1)\sin\theta}{(1 + r^2 + 2r\cos\theta)^2};$

$\dfrac{\partial v}{\partial r} = \dfrac{(1 + r^2 + 2r\cos\theta)(\sin\theta) - r\sin\theta(2r + 2\cos\theta)}{(1 + r^2 + 2r\cos\theta)^2} = \dfrac{(1 - r^2)\sin\theta}{(1 + r^2 + 2r\cos\theta)^2}$

(b) $\dfrac{\partial u}{\partial r} = \dfrac{1}{2\sqrt{r}}\cos(\theta/2), \qquad \dfrac{1}{r}\dfrac{\partial v}{\partial \theta} = \dfrac{1}{\sqrt{r}}\cos(\theta/2)\left(\dfrac{1}{2}\right),$

$\dfrac{1}{r}\dfrac{\partial u}{\partial \theta} = -\dfrac{1}{\sqrt{r}}\sin(\theta/2)\left(\dfrac{1}{2}\right), \qquad \dfrac{\partial v}{\partial r} = \dfrac{1}{2\sqrt{r}}\sin(\theta/2)$

(c) $\dfrac{\partial u}{\partial r} = \dfrac{1}{r}, \qquad \dfrac{1}{r}\dfrac{\partial v}{\partial \theta} = \dfrac{1}{r}, \qquad \dfrac{1}{r}\dfrac{\partial u}{\partial \theta} = 0, \qquad \dfrac{\partial v}{\partial r} = 0$

EXERCISES 12.4

2. $\nabla f = 2xyz\hat{\mathbf{i}} + x^2 z\hat{\mathbf{j}} + x^2 y\hat{\mathbf{k}}$ **4.** $\nabla f = (2xy + y^2)\hat{\mathbf{i}} + (x^2 + 2xy)\hat{\mathbf{j}}$

6. $\nabla f = \dfrac{yz}{1 + (xyz)^2}\hat{\mathbf{i}} + \dfrac{xz}{1 + (xyz)^2}\hat{\mathbf{j}} + \dfrac{xy}{1 + (xyz)^2}\hat{\mathbf{k}} = \dfrac{1}{1 + (xyz)^2}(yz\hat{\mathbf{i}} + xz\hat{\mathbf{j}} + xy\hat{\mathbf{k}})$

8. $\nabla f = e^{x+y+z}\hat{\mathbf{i}} + e^{x+y+z}\hat{\mathbf{j}} + e^{x+y+z}\hat{\mathbf{k}} = e^{x+y+z}(\hat{\mathbf{i}} + \hat{\mathbf{j}} + \hat{\mathbf{k}})$

10. $\nabla f = \dfrac{-x}{(x^2 + y^2 + z^2)^{3/2}}\hat{\mathbf{i}} + \dfrac{-y}{(x^2 + y^2 + z^2)^{3/2}}\hat{\mathbf{j}} + \dfrac{-z}{(x^2 + y^2 + z^2)^{3/2}}\hat{\mathbf{k}} = -\dfrac{x\hat{\mathbf{i}} + y\hat{\mathbf{j}} + z\hat{\mathbf{k}}}{(x^2 + y^2 + z^2)^{3/2}}$

12. Since $\nabla f = -\sin(x + y + z)(\hat{\mathbf{i}} + \hat{\mathbf{j}} + \hat{\mathbf{k}})$, the gradient at the point $(-1, 1, 1)$ is $= -(\sin 1)(\hat{\mathbf{i}} + \hat{\mathbf{j}} + \hat{\mathbf{k}})$.

14. $\nabla f_{|(2,2)} = e^{-x^2 - y^2}(-2x\hat{\mathbf{i}} - 2y\hat{\mathbf{j}})_{|(2,2)} = -2e^{-8}(2\hat{\mathbf{i}} + 2\hat{\mathbf{j}}) = -4e^{-8}(\hat{\mathbf{i}} + \hat{\mathbf{j}})$

16. $\nabla F = A\hat{\mathbf{i}} + B\hat{\mathbf{j}} + C\hat{\mathbf{k}}$ But according to Theorem 11.4, this vector is perpendicular to the plane.

18. $\nabla(fg) = \dfrac{\partial}{\partial x}(fg)\hat{\mathbf{i}} + \dfrac{\partial}{\partial y}(fg)\hat{\mathbf{j}} + \dfrac{\partial}{\partial z}(fg)\hat{\mathbf{k}} = \left(f\dfrac{\partial g}{\partial x} + g\dfrac{\partial f}{\partial x}\right)\hat{\mathbf{i}} + \left(f\dfrac{\partial g}{\partial y} + g\dfrac{\partial f}{\partial y}\right)\hat{\mathbf{j}} + \left(f\dfrac{\partial g}{\partial z} + g\dfrac{\partial f}{\partial z}\right)\hat{\mathbf{k}}$

$= f\left(\dfrac{\partial g}{\partial x}\hat{\mathbf{i}} + \dfrac{\partial g}{\partial y}\hat{\mathbf{j}} + \dfrac{\partial g}{\partial z}\hat{\mathbf{k}}\right) + g\left(\dfrac{\partial f}{\partial x}\hat{\mathbf{i}} + \dfrac{\partial f}{\partial y}\hat{\mathbf{j}} + \dfrac{\partial f}{\partial z}\hat{\mathbf{k}}\right) = f\nabla g + g\nabla f$

It looks like the product rule for differentiation.

20. The slope dy/dx of the tangent line at any point (x, y) on this curve is defined by

$$3x^2 + y + x\dfrac{dy}{dx} + 4y^3\dfrac{dy}{dx} = 0,$$

or,

$$\dfrac{dy}{dx} = -\dfrac{3x^2 + y}{x + 4y^3}.$$

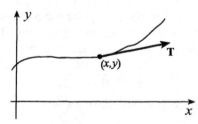

It follows that a vector along the tangent line at (x, y) is $\mathbf{T} = (x + 4y^3, -3x^2 - y)$. A vector perpendicular to $\mathbf{T}$ is $\mathbf{N} = (3x^2 + y, x + 4y^3)$. But $\nabla F = (3x^2 + y)\hat{\mathbf{i}} + (x + 4y^3)\hat{\mathbf{j}}$, and therefore $\nabla F = \mathbf{N}$.

22. Since $|x - y|$ fails to have derivatives with respect to x and y when $y = x$, it follows that ∇F is not defined at the points $(x, x, 0)$.

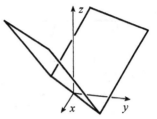

24. If $\nabla f = (2xy - y)\hat{\mathbf{i}} + (x^2 - x)\hat{\mathbf{j}}$, then $\dfrac{\partial f}{\partial x} = 2xy - y$ and $\dfrac{\partial f}{\partial y} = x^2 - x$. From the first equation, we can say that $f(x, y) = x^2 y - xy + \phi(y)$, where $\phi(y)$ is any differentiable function of y. To determine $\phi(y)$ we substitute this expression for $f(x, y)$ into the second equation, $x^2 - x + d\phi/dy = x^2 - x$. Consequently, $d\phi/dy = 0$, and therefore $\phi(y) = C$, a constant. Thus, $f(x, y) = x^2 y - xy + C$.

26. If $\nabla f = yz\hat{\mathbf{i}} + (xz + 2yz)\hat{\mathbf{j}} + (xy + y^2)\hat{\mathbf{k}}$, then $\dfrac{\partial f}{\partial x} = yz$, $\dfrac{\partial f}{\partial y} = xz + 2yz$, $\dfrac{\partial f}{\partial z} = xy + y^2$. From the first, $f(x, y, z) = xyz + \phi(y, z)$, where $\phi(y, z)$ is any function with first partial derivatives. Substitution of this into the second equation gives

$$xz + \frac{\partial \phi}{\partial y} = xz + 2yz \quad \Longrightarrow \quad \frac{\partial \phi}{\partial y} = 2yz.$$

Consequently, $\phi(y, z) = y^2 z + \psi(z)$, and $f(x, y, z) = xyz + y^2 z + \psi(z)$. Substitution of this into the third equation requires $\psi(z)$ to satisfy

$$xy + y^2 + \frac{d\psi}{dz} = xy + y^2 \quad \Longrightarrow \quad \frac{d\psi}{dz} = 0.$$

Thus, $\psi(z) = C$, and $f(x, y, z) = xyz + y^2 z + C$.

28. If $\nabla f = \nabla g$, then $\dfrac{\partial f}{\partial x} = \dfrac{\partial g}{\partial x}$ and $\dfrac{\partial f}{\partial y} = \dfrac{\partial g}{\partial y}$. The first requires $f(x, y) = g(x, y) + \phi(y)$, which substituted into the second gives

$$\frac{\partial g}{\partial y} + \frac{d\phi}{dy} = \frac{\partial g}{\partial y} \quad \text{or} \quad \frac{d\phi}{dy} = 0.$$

Thus, $\phi(y) = C$, a constant, and $f(x, y) = g(x, y) + C$.

EXERCISES 12.5

2. From $\dfrac{\partial f}{\partial y} = -\dfrac{2x}{y^2} + 12x^3 y^3$ and $\dfrac{\partial^2 f}{\partial y^2} = \dfrac{4x}{y^3} + 36x^3 y^2$, we obtain $\dfrac{\partial^3 f}{\partial y^3} = -\dfrac{12x}{y^4} + 72x^3 y$.

4. Since $\dfrac{\partial f}{\partial z} = xye^{x+y+z} + xyze^{x+y+z} = xy(1 + z)e^{x+y+z}$,

$$\frac{\partial^2 f}{\partial y \partial z} = x(1 + z)e^{x+y+z} + xy(1 + z)e^{x+y+z} = x(1 + z)(1 + y)e^{x+y+z}.$$

6. Since $\dfrac{\partial f}{\partial y} = e^{x+y} + \dfrac{2x^2}{y^3}$, we find $\dfrac{\partial^2 f}{\partial x \partial y} = e^{x+y} + \dfrac{4x}{y^3}$, and $\dfrac{\partial^3 f}{\partial x^2 \partial y} = e^{x+y} + \dfrac{4}{y^3}$.

8. Since $\dfrac{\partial f}{\partial z} = 2x^2 z + 2y^2 z$, we have $\dfrac{\partial^2 f}{\partial y \partial z} = 4yz$, and $\dfrac{\partial^3 f}{\partial x \partial y \partial z} = 0$. This must also be its value at $(1, 0, -1)$.

10. Since $\dfrac{\partial f}{\partial z} = \dfrac{1}{\sqrt{x^2 + y^2 + z^2}} \dfrac{z}{\sqrt{x^2 + y^2 + z^2}} = \dfrac{z}{x^2 + y^2 + z^2}$,

$$\frac{\partial^2 f}{\partial z^2} = \frac{1}{x^2 + y^2 + z^2} - \frac{2z^2}{(x^2 + y^2 + z^2)^2} = \frac{x^2 + y^2 - z^2}{(x^2 + y^2 + z^2)^2}.$$

12. From $\dfrac{\partial f}{\partial x} = \dfrac{1}{1+(y/x)^2}\left(\dfrac{-y}{x^2}\right) = \dfrac{x^2}{x^2+y^2}\left(\dfrac{-y}{x^2}\right) = \dfrac{-y}{x^2+y^2}$, we obtain $\dfrac{\partial^2 f}{\partial x^2} = \dfrac{2xy}{(x^2+y^2)^2}$.

14. From $\dfrac{\partial f}{\partial y} = \dfrac{1}{\sqrt{1-(x^2+y^2)^{-2}}}\dfrac{-2y}{(x^2+y^2)^2} = \dfrac{x^2+y^2}{\sqrt{(x^2+y^2)^2-1}}\dfrac{-2y}{(x^2+y^2)^2} = \dfrac{-2y}{(x^2+y^2)\sqrt{(x^2+y^2)^2-1}}$,

we obtain $\dfrac{\partial^2 f}{\partial x\partial y} = \dfrac{4xy}{(x^2+y^2)^2\sqrt{(x^2+y^2)^2-1}} + \dfrac{y(2)(x^2+y^2)(2x)}{(x^2+y^2)[(x^2+y^2)^2-1]^{3/2}}$.

Thus, $\dfrac{\partial^2 f}{\partial x\partial y}\bigg|_{(-2,-2)} = \dfrac{16}{8^2\sqrt{8^2-1}} + \dfrac{(-2)(2)(8)(-4)}{8[8^2-1]^{3/2}} = \dfrac{127}{756\sqrt{7}}$.

16. $\dfrac{\partial f}{\partial x} = 8x^7 y^9 z^{10}$; $\dfrac{\partial^2 f}{\partial x^2} = 56x^6 y^9 z^{10}$; $\cdots$; $\dfrac{\partial^8 f}{\partial x^8} = 8!\,y^9 z^{10}$

18. $\dfrac{\partial f}{\partial x} = -\sin(x+y^3)$; $\dfrac{\partial^2 f}{\partial x^2} = -\cos(x+y^3)$; $\dfrac{\partial^3 f}{\partial x^3} = \sin(x+y^3)$; $\dfrac{\partial^4 f}{\partial x^3\partial y} = 3y^2\cos(x+y^3)$

20. Since $\dfrac{\partial f}{\partial y} = \dfrac{x}{xy\sqrt{x^2 y^2-1}} = \dfrac{1}{y\sqrt{x^2 y^2-1}}$, we find $\dfrac{\partial^2 f}{\partial x\partial y} = \dfrac{-xy^2}{y(x^2 y^2-1)^{3/2}} = \dfrac{-xy}{(x^2 y^2-1)^{3/2}}$.

22. $\dfrac{\partial u}{\partial x} = 1 + ze^{y/x}(-y/x^2) = 1 - (yz/x^2)e^{y/x}$; $\dfrac{\partial u}{\partial y} = 1 + ze^{y/x}(1/x) = 1 + (z/x)e^{y/x}$;

$\dfrac{\partial u}{\partial z} = e^{y/x}$; $\dfrac{\partial^2 u}{\partial x^2} = 2(yz/x^3)e^{y/x} - (yz/x^2)e^{y/x}(-y/x^2) = [yz(2x+y)/x^4]e^{y/x}$;

$\dfrac{\partial^2 u}{\partial x\partial y} = -(z/x^2)e^{y/x} - (yz/x^2)e^{y/x}(1/x) = -[z(x+y)/x^3]e^{y/x}$;

$\dfrac{\partial^2 u}{\partial y^2} = (z/x)e^{y/x}(1/x) = (z/x^2)e^{y/x}$; $\dfrac{\partial^2 u}{\partial y\partial z} = (1/x)e^{y/x}$; $\dfrac{\partial^2 u}{\partial z^2} = 0$;

$\dfrac{\partial^2 u}{\partial x\partial z} = e^{y/x}(-y/x^2) = -(y/x^2)e^{y/x}$.

Thus,

$$x^2\dfrac{\partial^2 u}{\partial x^2} + y^2\dfrac{\partial^2 u}{\partial y^2} + z^2\dfrac{\partial^2 u}{\partial z^2} + 2xy\dfrac{\partial^2 u}{\partial x\partial y} + 2yz\dfrac{\partial^2 u}{\partial y\partial z} + 2xz\dfrac{\partial^2 u}{\partial x\partial z}$$

$$= \dfrac{yz(2x+y)}{x^2}e^{y/x} + \dfrac{y^2 z}{x^2}e^{y/x} + 0 - \dfrac{2yz(x+y)}{x^2}e^{y/x} + \dfrac{2yz}{x}e^{y/x} - \dfrac{2yz}{x}e^{y/x} = 0.$$

24. From $\dfrac{\partial f}{\partial x} = \dfrac{2x}{x^2+y^2}$, we obtain $\dfrac{\partial^2 f}{\partial x^2} = \dfrac{(x^2+y^2)(2) - 2x(2x)}{(x^2+y^2)^2} = \dfrac{2(y^2-x^2)}{(x^2+y^2)^2}$, and similarly,

$\dfrac{\partial^2 f}{\partial y^2} = \dfrac{2(x^2-y^2)}{(x^2+y^2)^2}$. Since $\dfrac{\partial^2 f}{\partial x^2} + \dfrac{\partial^2 f}{\partial y^2} = 0$, and second partial derivatives are continuous except at $(0,0)$, the function $f(x,y)$ is harmonic in any region not containing $(0,0)$.

26. $\dfrac{\partial f}{\partial x} = 6xyz + y$; $\dfrac{\partial^2 f}{\partial x^2} = 6yz$; $\dfrac{\partial f}{\partial y} = 3x^2 z - 3y^2 z + x$; $\dfrac{\partial^2 f}{\partial y^2} = -6yz$; $\dfrac{\partial f}{\partial z} = 3x^2 y - y^3$; $\dfrac{\partial^2 f}{\partial z^2} = 0$

Since $\dfrac{\partial^2 f}{\partial x^2} + \dfrac{\partial^2 f}{\partial y^2} + \dfrac{\partial^2 f}{\partial z^2} = 6yz - 6yz + 0 = 0$, and all second partial derivatives are continuous, $f(x,y,z)$ is harmonic in all space.

28. From $\dfrac{\partial f}{\partial x} = 3x^2 y^3 z^3$, we find $\dfrac{\partial^2 f}{\partial x^2} = 6xy^3 z^3$. Similarly, $\dfrac{\partial^2 f}{\partial y^2} = 6x^3 yz^2$ and $\dfrac{\partial^2 f}{\partial z^2} = 6x^3 y^3 z$. Since

$\dfrac{\partial^2 f}{\partial x^2} + \dfrac{\partial^2 f}{\partial y^2} + \dfrac{\partial^2 f}{\partial z^2} = 6xyz(y^2 z^2 + x^2 z^2 + x^2 y^2)$, and this is not zero in any region of space, the function is not harmonic.

30. From $\dfrac{\partial V}{\partial x} = GM\left[\dfrac{-x}{(x^2+y^2+z^2)^{3/2}}\right]$, we obtain

$$\dfrac{\partial^2 V}{\partial x^2} = GM\left[\dfrac{-1}{(x^2+y^2+z^2)^{3/2}} + \dfrac{3x^2}{(x^2+y^2+z^2)^{5/2}}\right] = GM\left[\dfrac{2x^2-y^2-z^2}{(x^2+y^2+z^2)^{5/2}}\right].$$

Similarly, $\dfrac{\partial^2 V}{\partial y^2} = GM \left[\dfrac{2y^2 - x^2 - z^2}{(x^2 + y^2 + z^2)^{5/2}} \right]$ and $\dfrac{\partial^2 V}{\partial z^2} = GM \left[\dfrac{2z^2 - x^2 - y^2}{(x^2 + y^2 + z^2)^{5/2}} \right]$, and therefore

$$\frac{\partial^2 V}{\partial x^2} + \frac{\partial^2 V}{\partial y^2} + \frac{\partial^2 V}{\partial z^2} = 0 \qquad \text{(except at } (0,0,0)\text{)}.$$

32. Suppose the cross-sectional area of the rod is A and its density is ρ. Then $F(x) = \rho g A(L - x)$, and $y(x)$ must satisfy

$$0 = E \frac{d^2 y}{dx^2} + \rho g A(L - x) \quad \Longrightarrow \quad \frac{d^2 y}{dx^2} = -\frac{\rho g A}{E}(L - x) \quad \Longrightarrow \quad \frac{dy}{dx} = \frac{\rho g A}{2E}(L - x)^2 + C.$$

Since $y'(L) = 0$, it follows that $C = 0$. Thus, $\dfrac{dy}{dx} = \dfrac{\rho g A}{2E}(L - x)^2$, and a second integration gives $y = -\dfrac{\rho g A}{6E}(L - x)^3 + D$. The condition $y(0) = 0$ implies that $D = \dfrac{\rho g A L^3}{6E}$, and therefore displacements of cross sections are given by $y(x) = -\dfrac{\rho g A}{6E}(L - x)^3 + \dfrac{\rho g A L^3}{6E}$. Since $y(L) = \dfrac{\rho g A L^3}{6E}$, it follows that the length of the bar is $L + \dfrac{\rho g A L^3}{6E}$.

34. (a) Since second partial derivatives are $\dfrac{\partial^2 y}{\partial x^2} = -\lambda^2 y$ and $\dfrac{\partial^2 y}{\partial t^2} = -c^2 \lambda^2 y$, it follows that $y(x,t)$ does indeed satisfy equation 12.13a.

(b) The condition $y(0, t) = 0$ implies that $0 = B(C \sin c\lambda t + D \cos c\lambda t)$ for all t. This requires $B = 0$. With $B = 0$, condition $y_x(L, t) = 0$ implies that $0 = A\lambda \cos \lambda L(C \sin c\lambda t + D \cos c\lambda t)$. Since A cannot be zero, nor can λ, and the term in brackets cannot be equal to 0 for all t, we must set $\cos \lambda L = 0$. But this implies that $\lambda L = (2n - 1)\pi/2$, where n is an integer; that is, $\lambda = (2n - 1)\pi/(2L)$.

(c) With $B = 0$ and $g(x) = 0$, initial condition 12.13c requires $0 = y_t(x, 0) = A \sin \lambda x(Cc\lambda) \implies C = 0$.

(d) With $B = 0$ and $f(x) = 0$, initial condition 12.13b requires $0 = y(x, 0) = A \sin \lambda x(D) \implies D = 0$.

36. With $F(x) = -9.81\rho$, a constant, two integrations of $d^2 y/dx^2 = 9.81\rho/T$ give $y(x) = 4.905\rho x^2/T + Ax + B$. The conditions $y(0) = 0 = y(L)$ require $B = 0$ and $A = -4.905\rho L/T$. Thus, $y(x) = 4.905\rho x(x - L)/T$, a parabola. In Exercise 35 it was assumed that that the string experiences only small displacements. This results in a constant force for gravity. No such assumption is made in Example 3.39.

38. (a) Since $\dfrac{\partial^2 y}{\partial t^2} = -(3\pi^2 c^2/L^2) \sin(\pi x/L) \cos(\pi ct/L)$ and $\dfrac{\partial^2 y}{\partial x^2} = -(3\pi^2/L^2) \sin(\pi x/L) \cos(\pi ct/L)$, it follows that $y(x, t)$ satisfies the partial differential equation. It is straightforward to check that it satisfies the remaining conditions.

(b) A plot of the surface is shown to the left below. A cross section of the surface with a plane $x = x_0$ gives a graphical history of the displacement of the point x_0 in the string. A cross section $t = t_0$ gives the position of the string at time t_0.

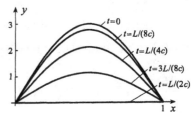

(c) Plots are shown to the right above.

40. (a) If differentiation and summation operations can be interchanged, it is a matter of showing that the function $z(x,t) = \sin(2n-1)\pi x \cos(2n-1)\pi ct$ satisfies the condition for any positive integer n. Since $\partial^2 z/\partial t^2 = -(2n-1)^2\pi^2 c^2 z(x,t)$ and $\partial^2 z/\partial x^2 = -(2n-1)^2\pi^2 z(x,t)$, it follows that $z(x,t)$ satisfies the partial differential equation. It is obvious that $z(x,t)$ satisfies the boundary conditions $z(0,t) = z(L,t) = 0$ and the initial condition $z_t(x,0) = 0$.
(b) Plots are shown to the right.

42. (a) The boundary condition $T(0,t) = 0$ requires $0 = B$. With $B = 0$, the condition $T(L,t) = 0$ necessitates $0 = A\sin\lambda L e^{-k\lambda^2 t}$. Since A cannot be equal to zero, we must set $\sin\lambda L = 0$ and this requires $\lambda L = n\pi \implies \lambda = n\pi/L$, where n is an integer.
(b) Plots are shown to the right. The $t = 0$ plot does appear to be the parabola $x(1-x)$.

44. When $F(x)$ has constant value F, two integrations of the differential equation give $T(x) = -Fx^2/(2k) + Ax + B$. The boundary conditions require $T_0 = T(0) = B$ and $T_L = T(L) = -FL^2/(2k) + AL + B$. These give $T(x) = T_0 + (T_L - T_0)x/L + Fx(L-x)/(2k)$.

46. Integration of $\dfrac{d^2 y}{dx^2} = -\dfrac{\tau(x)}{E}$ gives $\dfrac{dy}{dx} = -\dfrac{1}{E}\displaystyle\int \tau(x)\,dx + C$. To incorporate the boundary condtion at $x = L$, we write this in the form

$$y'(x) = -\frac{1}{E}\int_0^x \tau(u)\,du + C.$$

The condition $y'(L) = 0$ now implies that $0 = -\dfrac{1}{E}\displaystyle\int_0^L \tau(u)\,du + C$. This gives the indicated value for C. A second integration gives

$$y(x) = Cx - \frac{1}{E}\int\left[\int_0^x \tau(u)\,du\right]dx + D.$$

To incorporate the boundary condition at $x = 0$, we write this in the form

$$y(x) = Cx - \frac{1}{E}\int_0^x\left[\int_0^v \tau(u)\,du\right]dv + D.$$

The condition $y(0) = 0$ now implies that $D = 0$, and therefore

$$y(x) = Cx - \frac{1}{E}\int_0^x\int_0^v \tau(u)\,du\,dv.$$

48. (a) Since $\dfrac{\partial^2 u}{\partial x^2} + \dfrac{\partial^2 u}{\partial y^2} = 2 - 2 = 0$, the function is harmonic.

(b) According to Exercise 41 of Section 12.3, $\dfrac{\partial v}{\partial y} = \dfrac{\partial u}{\partial x} = 2x$, $\dfrac{\partial v}{\partial x} = -\dfrac{\partial u}{\partial y} = 2y$. From the first $v(x,y) = 2xy + \phi(x)$, which substituted into the second gives $2y + d\phi/dx = 2y$. Thus, $\phi(x) = C$, a constant, and $v(x,y) = 2xy + C$.

50. Since $\dfrac{\partial f}{\partial x} = n(x^2 + y^2 + z^2)^{n-1}(2x)$, we find

$$\frac{\partial^2 f}{\partial x^2} = 2n(x^2 + y^2 + z^2)^{n-1} + 2nx(n-1)(x^2 + y^2 + z^2)^{n-2}(2x)$$

$$= 2n(x^2 + y^2 + z^2)^{n-2}[x^2 + y^2 + z^2 + 2(n-1)x^2]$$

$$= 2n(x^2 + y^2 + z^2)^{n-2}[(2n-1)x^2 + y^2 + z^2].$$

Similarly, $\dfrac{\partial^2 f}{\partial y^2} = 2n(x^2 + y^2 + z^2)^{n-2}[x^2 + (2n-1)y^2 + z^2]$, and

$\dfrac{\partial^2 f}{\partial z^2} = 2n(x^2 + y^2 + z^2)^{n-2}[x^2 + y^2 + (2n-1)z^2]$. The function satisfies Laplace's equation if

$$0 = \frac{\partial^2 f}{\partial x^2} + \frac{\partial^2 f}{\partial y^2} + \frac{\partial^2 f}{\partial z^2} = 2n(x^2 + y^2 + z^2)^{n-2}[(2n+1)x^2 + (2n+1)y^2 + (2n+1)z^2]$$

$$= 2n(2n+1)(x^2 + y^2 + z^2)^{n-1}.$$

Thus, $n = 0$ or $n = -1/2$. When $n = 0$, the function is equal to 1, and it is harmonic in all space. When $n = -1/2$, the function is $1/\sqrt{x^2 + y^2 + z^2}$, and it is harmonic in any region not containing the origin.

EXERCISES 12.6

2. In general, $\dfrac{\partial z}{\partial t} = \dfrac{\partial z}{\partial x}\dfrac{\partial x}{\partial t} + \dfrac{\partial z}{\partial y}\dfrac{\partial y}{\partial t}$, and specifically,

$$\frac{\partial z}{\partial t} = \left(2xe^y + \frac{y}{x}\right)(-s^2 \sin t) + (x^2 e^y + \ln x)\left[\frac{4(2t)}{(t^2 + 2s)\sqrt{(t^2 + 2s)^2 - 1}}\right].$$

4. In general, $\dfrac{dz}{du} = \dfrac{\partial z}{\partial x}\dfrac{dx}{du} + \dfrac{\partial z}{\partial y}\dfrac{dy}{du} + \dfrac{\partial z}{\partial v}\dfrac{dv}{du}$, and specifically,

$$\frac{dz}{du} = (2xyv^3)(3u^2 + 2) + (x^2 v^3)\left(\frac{2u}{u^2 + 1}\right) + (3x^2 yv^2)(ue^u + e^u)$$

$$= xv^2\left[2yv(3u^2 + 2) + \frac{2xuv}{u^2 + 1} + 3xye^u(u+1)\right].$$

6. In general, $\dfrac{\partial z}{\partial t} = \dfrac{dz}{dx}\dfrac{dx}{dy}\dfrac{\partial y}{\partial t}$, and specifically,

$$\frac{\partial z}{\partial t} = (3^{x+2} \ln 3)(2y)[-\csc(r^2 + t)\cot(r^2 + t)].$$

8. In general, $\dfrac{\partial x}{\partial y} = \dfrac{\partial x}{\partial r}\dfrac{dr}{dy} + \dfrac{\partial x}{\partial s}\dfrac{\partial s}{\partial y} + \dfrac{\partial x}{\partial t}\dfrac{\partial t}{\partial y}$, and specifically,

$$\frac{\partial x}{\partial y} = (2rs^2 t^2)(-5y^{-6}) + (2r^2 st^2)\left[\frac{-2y}{(y^2 + z^2)^2}\right] + (2r^2 s^2 t)\left(\frac{-2}{y^3}\right)$$

$$= -2rst\left[\frac{5st}{y^6} + \frac{2rty}{(y^2 + z^2)^2} + \frac{2rs}{y^3}\right].$$

10. In general,

$$\frac{dz}{dt} = \frac{\partial z}{\partial x}\frac{dx}{dv}\frac{dv}{dt} + \frac{\partial z}{\partial y}\frac{dy}{dt} + \frac{\partial z}{\partial u}\frac{\partial u}{\partial x}\frac{dx}{dv}\frac{dv}{dt} + \frac{\partial z}{\partial u}\frac{\partial u}{\partial y}\frac{dy}{dt}$$

$$= \left(\frac{\partial z}{\partial x} + \frac{\partial z}{\partial u}\frac{\partial u}{\partial x}\right)\frac{dx}{dv}\frac{dv}{dt} + \left(\frac{\partial z}{\partial y} + \frac{\partial z}{\partial u}\frac{\partial u}{\partial y}\right)\frac{dy}{dt},$$

and specifically,

$$\frac{dz}{dt} = \left\{2x + 2u\left[\frac{-2x}{(x^2 - y^2)^2}\right]\right\}(3v^2 - 6v)e^t + \left\{2y + 2u\left[\frac{2y}{(x^2 - y^2)^2}\right]\right\}4e^{4t}$$

$$= 6xve^t(v-2)\left[1 - \frac{2u}{(x^2 - y^2)^2}\right] + 8ye^{4t}\left[1 + \frac{2u}{(x^2 - y^2)^2}\right].$$

12. From $\dfrac{dx}{dt} = \dfrac{\partial x}{\partial y}\dfrac{dy}{dt} + \dfrac{\partial x}{\partial t} = (2y+t)(t^2 e^t + 2te^t) + (y-2t)$

$$= te^t(t^2 + 2yt + 4y + 2t) + y - 2t,$$

$$\frac{d^2 x}{dt^2} = \frac{\partial}{\partial y}\left(\frac{dx}{dt}\right)\frac{dy}{dt} + \frac{\partial}{\partial t}\left(\frac{dx}{dt}\right)$$

$$= [te^t(2t+4)+1](t^2 e^t + 2te^t) + [te^t(2t+2y+2)+(te^t+e^t)(t^2+2yt+4y+2t)-2]$$

$$= 2t^2(t+2)^2 e^{2t} + (6t^2 + 6t + 8ty + t^3 + 2yt^2 + 4y)e^t - 2.$$

14. From $\dfrac{dz}{dv} = \dfrac{\partial z}{\partial x}\dfrac{dx}{dv} + \dfrac{\partial z}{\partial y}\dfrac{dy}{dv}$

$$= y\cos(xy)(-3\sin v) + x\cos(xy)(4\cos v)$$

$$= \cos(xy)(-3y\sin v + 4x\cos v),$$

$$\frac{d^2 z}{dv^2} = \frac{\partial}{\partial x}\left(\frac{dz}{dv}\right)\frac{dx}{dv} + \frac{\partial}{\partial y}\left(\frac{dz}{dv}\right)\frac{dy}{dv} + \frac{\partial}{\partial v}\left(\frac{dz}{dv}\right)$$

$$= [-y\sin(xy)(-3y\sin v + 4x\cos v) + 4\cos v\,\cos(xy)](-3\sin v)$$

$$+ [-x\sin(xy)(-3y\sin v + 4x\cos v) - 3\sin v\,\cos(xy)](4\cos v)$$

$$+ \cos(xy)(-3y\cos v - 4x\sin v)$$

$$= (24xy\sin v\cos v - 16x^2\cos^2 v - 9y^2\sin^2 v)\sin(xy)$$

$$-(24\sin v\cos v + 3y\cos v + 4x\sin v)\cos(xy)$$

$$= -(3y\sin v - 4x\cos v)^2\sin(xy) - (24\sin v\cos v + 3y\cos v + 4x\sin v)\cos(xy).$$

16. From the schematic, $\dfrac{\partial u}{\partial x} = \dfrac{du}{dr}\dfrac{\partial r}{\partial x} = \dfrac{du}{dr}\dfrac{x}{\sqrt{x^2+y^2+z^2}},$

and similarly,

$$\frac{\partial u}{\partial y} = \frac{du}{dr}\frac{y}{\sqrt{x^2+y^2+z^2}}, \quad \frac{\partial u}{\partial z} = \frac{du}{dr}\frac{z}{\sqrt{x^2+y^2+z^2}}.$$

Consequently,

$$\left(\frac{\partial u}{\partial x}\right)^2 + \left(\frac{\partial u}{\partial y}\right)^2 + \left(\frac{\partial u}{\partial z}\right)^2 = \left(\frac{du}{dr}\right)^2\left(\frac{x^2}{x^2+y^2+z^2} + \frac{y^2}{x^2+y^2+z^2} + \frac{z^2}{x^2+y^2+z^2}\right) = \left(\frac{du}{dr}\right)^2.$$

18. The volume V of a right circular cone in terms of its radius r and height h is $V = \pi r^2 h/3$. From the schematic,

$$\frac{dV}{dt} = \frac{\partial V}{\partial r}\frac{dr}{dt} + \frac{\partial V}{\partial h}\frac{dh}{dt} = \frac{2}{3}\pi rh\frac{dr}{dt} + \frac{1}{3}\pi r^2\frac{dh}{dt}.$$

When $r = 10$ and $h = 20$,

$$\frac{dV}{dt} = \frac{2}{3}\pi(10)(20)(1) + \frac{1}{3}\pi(10)^2(-2) = \frac{200\pi}{3}\ \text{cm}^3/\text{min}.$$

Multivariable calculus is not needed since $V = \pi r^2 h/3$ can be differentiated with respect to t using the product and power rules, $\dfrac{dV}{dt} = \dfrac{1}{3}\pi\left(2r\dfrac{dr}{dt}h + r^2\dfrac{dh}{dt}\right)$, and this is the same result as above.

20. Newton's universal law of gravitation states that $F = GMm/r^2$, where $G = 6.67 \times 10^{-11}$, $M =$ mass of earth, $m =$ mass of rocket, and $r =$ distance from centre of earth to rocket.

$$\frac{dF}{dt} = \frac{\partial F}{\partial m}\frac{dm}{dt} + \frac{\partial F}{\partial r}\frac{dr}{dt} = \frac{GM}{r^2}\frac{dm}{dt} - \frac{2GMm}{r^3}\frac{dr}{dt}$$

$$= \frac{GM}{r^3}\left(r\frac{dm}{dt} - 2m\frac{dr}{dt}\right).$$

With $M = (4/3)\pi(\text{radius of earth})^3(\text{density of earth}) = (4/3)\pi(6.37 \times 10^6)^3(5.52 \times 10^3)$, we find that when $r = 6.47 \times 10^6$,

$$\frac{dF}{dt} = 6.67 \times 10^{-11} \left(\frac{4}{3}\right) \pi \frac{(6.37 \times 10^6)^3(5.52 \times 10^3)}{(6.47 \times 10^6)^3} \left[6.47 \times 10^6(-50) - 2(12 \times 10^6)(2 \times 10^3)\right]$$

$$= -7.11 \times 10^4 \text{ N/s.}$$

22. (a) Since $f(tx, ty) = (tx)^2 + (tx)(ty) + 3(ty)^2 = t^2(x^2 + xy + 3y^2) = t^2 f(x, y)$, the function is positively homogeneous of degree 2.

(b) Since $f(tx, ty) = (tx)^2(ty) + (tx)(ty) - 2(tx)(ty)^2 = t^2(tx^2y + xy - 2txy^2)$, the function is not homogeneous.

(c) Since $f(tx, ty, tz) = (tx)^2 \sin[ty/(tz)] + (ty)^2 + (ty)^3/(tz) = t^2[x^2 \sin(y/z) + y^2 + y^3/z]$, the function is positively homogeneous of degree 2.

(d) Since $f(tx, ty, tz) = (tx)e^{ty/(tz)} - (tx)(ty)(tz) = t(xe^{y/z} - t^2xyz)$, the function is not homogeneous.

(e) Since $f(ux, uy, uz, ut) = (ux)^4 + (uy)^4 + (uz)^4 + (ut)^4 - (ux)(uy)(uz)(ut) = u^4(x^4 + y^4 + z^4 + t^4 - xyzt)$, the function is positively homogeneous of degree 4.

(f) Since $f(ux, uy, uz, ut) = e^{u^2x^2 + u^2y^2}[(uz)^2 + (ut)^2] = u^2(z^2 + t^2)e^{u^2(x^2+y^2)}$, the function is not homogeneous.

(g) Since $f(tx, ty, tz) = \cos(t^2xy) \sin(t^2yz)$, the function is not homogeneous.

(h) Since $f(tx, ty) = \sqrt{t^2x^2 + t^2xy + t^2y^2} e^{ty/(tx)}(2t^2x^2 - 3t^2y^2) = t^3\sqrt{x^2 + xy + y^2}e^{y/x}(2x^2 - 3y^2)$, the function is positively homogeneous of degree 3.

24.

$$\frac{\partial V}{\partial r} = \frac{V_1 - V_2}{\pi\left[1 + \left(\frac{2Rr\sin\theta}{R^2 - r^2}\right)^2\right]} \left[\frac{(R^2 - r^2)(2R\sin\theta) - 2Rr\sin\theta(-2r)}{(R^2 - r^2)^2}\right] = \frac{2(V_1 - V_2)R(R^2 + r^2)\sin\theta}{\pi[(R^2 - r^2)^2 + 4R^2r^2\sin^2\theta]}$$

$$\frac{\partial^2 V}{\partial r^2} = \frac{4(V_1 - V_2)Rr\sin\theta}{\pi[(R^2 - r^2)^2 + 4R^2r^2\sin^2\theta]} - \frac{2(V_1 - V_2)R(R^2 + r^2)\sin\theta[-4r(R^2 - r^2) + 8R^2r\sin^2\theta]}{\pi[(R^2 - r^2)^2 + 4R^2r^2\sin^2\theta]^2}$$

$$\frac{\partial V}{\partial\theta} = \frac{V_1 - V_2}{\pi\left[1 + \left(\frac{2Rr\sin\theta}{R^2 - r^2}\right)^2\right]} \left(\frac{2Rr\cos\theta}{R^2 - r^2}\right) = \frac{2(V_1 - V_2)Rr(R^2 - r^2)\cos\theta}{\pi[(R^2 - r^2)^2 + 4R^2r^2\sin^2\theta]}$$

$$\frac{\partial^2 V}{\partial\theta^2} = \frac{-2(V_1 - V_2)Rr(R^2 - r^2)\sin\theta}{\pi[(R^2 - r^2)^2 + 4R^2r^2\sin^2\theta]} - \frac{2(V_1 - V_2)Rr(R^2 - r^2)\cos\theta(8R^2r^2\sin\theta\cos\theta)}{\pi[(R^2 - r^2)^2 + 4R^2r^2\sin^2\theta]^2}$$

When these are substituted into the left side of equation 12.24, and considerable algebra is performed, the result is zero.

26. If we set $s = x - y$, then

$$\frac{\partial f}{\partial y} = \frac{df}{ds}\frac{\partial s}{\partial y} = \frac{df}{ds}(-1) \quad \text{and} \quad \frac{\partial f}{\partial x} = \frac{df}{ds}\frac{\partial s}{\partial x} = \frac{df}{ds}(1).$$

Hence, $\dfrac{\partial f}{\partial y} = -\dfrac{\partial f}{\partial x}$.

28. If we set $s = x + y$, then $u = xf(s) + yg(s)$. From the schematic,

$$\frac{\partial u}{\partial x} = \frac{\partial u}{\partial s}\frac{\partial s}{\partial x} + \frac{\partial u}{\partial x}\bigg)_{y,s}$$

$$= \left[x\frac{df}{ds} + y\frac{dg}{ds}\right](1) + f(s) = xf'(s) + yg'(s) + f(s),$$

and similarly, $\dfrac{\partial u}{\partial y} = xf'(s) + yg'(s) + g(s)$. The schematic for $\partial u/\partial x$ and $\partial u/\partial y$ gives

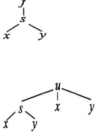

$$\frac{\partial^2 u}{\partial x^2} = \frac{\partial}{\partial s}\left(\frac{\partial u}{\partial x}\right)\frac{\partial s}{\partial x} + \frac{\partial}{\partial x}\left(\frac{\partial u}{\partial x}\right)_{y,s}$$

$$= [xf''(s) + yg''(s) + f'(s)](1) + f'(s);$$

$$\frac{\partial^2 u}{\partial x\partial y} = \frac{\partial}{\partial s}\left(\frac{\partial u}{\partial y}\right)\frac{\partial s}{\partial x} + \frac{\partial}{\partial x}\left(\frac{\partial u}{\partial y}\right)_{y,s}$$

$$= [xf''(s) + yg''(s) + g'(s)](1) + f'(s);$$

$$\frac{\partial^2 u}{\partial y^2} = \frac{\partial}{\partial s}\left(\frac{\partial u}{\partial y}\right)\frac{\partial s}{\partial y} + \frac{\partial}{\partial y}\left(\frac{\partial u}{\partial y}\right)_{x,s} = [xf''(s) + yg''(s) + g'(s)](1) + g'(s);$$

Thus, $\dfrac{\partial^2 u}{\partial x^2} - 2\dfrac{\partial^2 u}{\partial x\partial y} + \dfrac{\partial^2 u}{\partial y^2} = 0.$

30. From the schematic for $u = x^2 f(v)$, $v = y/x$,

$$\frac{\partial u}{\partial x} = \frac{\partial u}{\partial x}\Big)_v + \frac{\partial u}{\partial v}\frac{\partial v}{\partial x} = 2xf(v) + x^2 f'(v)\left(\frac{-y}{x^2}\right),$$

$$\frac{\partial u}{\partial y} = \frac{\partial u}{\partial v}\frac{\partial v}{\partial y} = x^2 f'(v)\left(\frac{1}{x}\right).$$

Hence,

$$x\frac{\partial u}{\partial x} + y\frac{\partial u}{\partial y} = x[2xf(v) - yf'(v)] + y[xf'(v)] = 2x^2 f(v) = 2u.$$

This is also an immediate consequence of Euler's theorem since u is positively homogeneous of degree 2.

32. The schematic to the right describes the functional situation $f(x,y) = F[u(x,y), v(x,y)]$ where $u = u(x,y) = (x+y)/2$ and $v = v(x,y) = (x-y)/2$. It gives

$$\frac{\partial f}{\partial x} = \frac{\partial F}{\partial u}\frac{\partial u}{\partial x} + \frac{\partial F}{\partial v}\frac{\partial v}{\partial x} = \frac{1}{2}\frac{\partial F}{\partial u} + \frac{1}{2}\frac{\partial F}{\partial v}.$$

The schematic for $\partial f/\partial x$ leads to

$$\frac{\partial^2 f}{\partial x^2} = \frac{\partial}{\partial u}\left(\frac{\partial f}{\partial x}\right)\frac{\partial u}{\partial x} + \frac{\partial}{\partial v}\left(\frac{\partial f}{\partial v}\right)\frac{\partial v}{\partial x}$$

$$= \left(\frac{1}{2}\frac{\partial^2 F}{\partial u^2} + \frac{1}{2}\frac{\partial^2 F}{\partial u\partial v}\right)\left(\frac{1}{2}\right) + \left(\frac{1}{2}\frac{\partial^2 F}{\partial v\partial u} + \frac{1}{2}\frac{\partial^2 F}{\partial v^2}\right)\left(\frac{1}{2}\right)$$

$$= \frac{1}{4}\left(\frac{\partial^2 F}{\partial u^2} + 2\frac{\partial^2 F}{\partial u\partial v} + \frac{\partial^2 F}{\partial v^2}\right).$$

Similarly, $\dfrac{\partial^2 f}{\partial y^2} = \dfrac{1}{4}\left(\dfrac{\partial^2 F}{\partial u^2} - 2\dfrac{\partial^2 F}{\partial u\partial v} + \dfrac{\partial^2 F}{\partial v^2}\right)$. Hence, $\quad 0 = \dfrac{\partial^2 f}{\partial x^2} - \dfrac{\partial^2 f}{\partial y^2} = \dfrac{\partial^2 F}{\partial u\partial v}.$

34. The biharmonic equation can be expressed in the form

$$0 = \left(\frac{\partial^4}{\partial x^4} + 2\frac{\partial^4}{\partial x^2\partial y^2} + \frac{\partial^4}{\partial y^4}\right)V = \left(\frac{\partial^2}{\partial x^2} + \frac{\partial^2}{\partial y^2}\right)^2 V.$$

But according to Example 12.19, the operator equivalent to $\nabla^2 = \partial^2/\partial x^2 + \partial^2/\partial y^2$ in polar coordinates is

$$\nabla^2 = \frac{\partial^2}{\partial r^2} + \frac{1}{r}\frac{\partial}{\partial r} + \frac{1}{r^2}\frac{\partial^2}{\partial \theta^2}.$$

Hence, the biharmonic equation in polar coordinates is

$$\left(\frac{\partial^2}{\partial r^2} + \frac{1}{r}\frac{\partial}{\partial r} + \frac{1}{r^2}\frac{\partial^2}{\partial \theta^2}\right)\left(\frac{\partial^2\Phi}{\partial r^2} + \frac{1}{r}\frac{\partial\Phi}{\partial r} + \frac{1}{r^2}\frac{\partial^2\Phi}{\partial \theta^2}\right) = 0.$$

36. From the schematic,

$$\frac{\partial f}{\partial x} = \frac{df}{dr}\frac{\partial r}{\partial x} = \frac{xf'(r)}{\sqrt{x^2+y^2+z^2}} = \frac{xf'(r)}{r}.$$

Similarly, $\dfrac{\partial f}{\partial y} = \dfrac{yf'(r)}{r}$ and $\dfrac{\partial f}{\partial z} = \dfrac{zf'(r)}{r}$. Thus,

$$\nabla f = \frac{\partial f}{\partial x}\hat{\mathbf{i}} + \frac{\partial f}{\partial y}\hat{\mathbf{j}} + \frac{\partial f}{\partial z}\hat{\mathbf{k}} = \frac{xf'(r)}{r}\hat{\mathbf{i}} + \frac{yf'(r)}{r}\hat{\mathbf{j}} + \frac{zf'(r)}{r}\hat{\mathbf{k}} = \frac{f'(r)}{r}(x\hat{\mathbf{i}} + y\hat{\mathbf{j}} + z\hat{\mathbf{k}}).$$

38. If we set $u = x^2 - y^2$ and $v = 2xy$, the schematic to the right gives

$$\frac{\partial F}{\partial x} = \frac{\partial f}{\partial u}\frac{\partial u}{\partial x} + \frac{\partial f}{\partial v}\frac{\partial v}{\partial x} = 2x\frac{\partial f}{\partial u} + 2y\frac{\partial f}{\partial v}.$$

The schematic for $\partial F/\partial x$ now gives

$$\frac{\partial^2 F}{\partial x^2} = \frac{\partial}{\partial u}\left(\frac{\partial F}{\partial x}\right)\frac{\partial u}{\partial x} + \frac{\partial}{\partial v}\left(\frac{\partial F}{\partial x}\right)\frac{\partial v}{\partial x} + \frac{\partial}{\partial x}\left(\frac{\partial F}{\partial x}\right)_{u,v,y}$$

$$= \left(2x\frac{\partial^2 f}{\partial u^2} + 2y\frac{\partial^2 f}{\partial u \partial v}\right)(2x) + \left(2x\frac{\partial^2 f}{\partial v \partial u} + 2y\frac{\partial^2 f}{\partial v^2}\right)(2y) + 2\frac{\partial f}{\partial u}$$

$$= 4x^2\frac{\partial^2 f}{\partial u^2} + 8xy\frac{\partial^2 f}{\partial u \partial v} + 4y^2\frac{\partial^2 f}{\partial v^2} + 2\frac{\partial f}{\partial u}.$$

A similar derivation gives $\dfrac{\partial^2 F}{\partial y^2} = 4y^2\dfrac{\partial^2 f}{\partial u^2} - 8xy\dfrac{\partial^2 f}{\partial u \partial v} + 4x^2\dfrac{\partial^2 f}{\partial v^2} - 2\dfrac{\partial f}{\partial u}$. Hence,

$$\frac{\partial^2 F}{\partial x^2} + \frac{\partial^2 F}{\partial y^2} = 4(x^2+y^2)\frac{\partial^2 f}{\partial u^2} + 4(x^2+y^2)\frac{\partial^2 f}{\partial v^2} = 4(x^2+y^2)\left(\frac{\partial^2 f}{\partial u^2} + \frac{\partial^2 f}{\partial v^2}\right) = 0,$$

since $f(u,v)$ is harmonic.

40. In the proof of Theorem 12.3, differentiation of equation 12.25 with respect to t gave

$$nt^{n-1}f(x,y,z) = x\frac{\partial f}{\partial u} + y\frac{\partial f}{\partial v} + z\frac{\partial f}{\partial w}.$$

Since the same schematic in the theorem applies to $\partial f/\partial u$, $\partial f/\partial v$, and $\partial f/\partial w$, differentiation of this equation with respect to t gives

$$n(n-1)t^{n-2}f(x,y,z) = x\frac{\partial}{\partial t}\left(\frac{\partial f}{\partial u}\right) + y\frac{\partial}{\partial t}\left(\frac{\partial f}{\partial v}\right) + z\frac{\partial}{\partial t}\left(\frac{\partial f}{\partial w}\right)$$

$$= x\left[\frac{\partial}{\partial u}\left(\frac{\partial f}{\partial u}\right)\frac{\partial u}{\partial t} + \frac{\partial}{\partial v}\left(\frac{\partial f}{\partial u}\right)\frac{\partial v}{\partial t} + \frac{\partial}{\partial w}\left(\frac{\partial f}{\partial u}\right)\frac{\partial w}{\partial t}\right]$$

$$+ y\left[\frac{\partial}{\partial u}\left(\frac{\partial f}{\partial v}\right)\frac{\partial u}{\partial t} + \frac{\partial}{\partial v}\left(\frac{\partial f}{\partial v}\right)\frac{\partial v}{\partial t} + \frac{\partial}{\partial w}\left(\frac{\partial f}{\partial v}\right)\frac{\partial w}{\partial t}\right]$$

$$+ z\left[\frac{\partial}{\partial u}\left(\frac{\partial f}{\partial w}\right)\frac{\partial u}{\partial t} + \frac{\partial}{\partial v}\left(\frac{\partial f}{\partial w}\right)\frac{\partial v}{\partial t} + \frac{\partial}{\partial w}\left(\frac{\partial f}{\partial w}\right)\frac{\partial w}{\partial t}\right]$$

$$= x\left[x\frac{\partial^2 f}{\partial u^2} + y\frac{\partial^2 f}{\partial v \partial u} + z\frac{\partial^2 f}{\partial w \partial u}\right] + y\left[x\frac{\partial^2 f}{\partial u \partial v} + y\frac{\partial^2 f}{\partial v^2} + z\frac{\partial^2 f}{\partial w \partial v}\right]$$

$$+ z\left[x\frac{\partial^2 f}{\partial u \partial w} + y\frac{\partial^2 f}{\partial v \partial w} + z\frac{\partial^2 f}{\partial w^2 u}\right].$$

If we set $t = 1$, we obtain the identity

$$n(n-1)f(x,y,z) = x^2\frac{\partial^2 f}{\partial x^2} + y^2\frac{\partial^2 f}{\partial y^2} + z^2\frac{\partial^2 f}{\partial z^2} + 2xy\frac{\partial^2 f}{\partial x \partial y} + 2yz\frac{\partial^2 f}{\partial y \partial z} + 2zx\frac{\partial^2 f}{\partial x \partial z}.$$

EXERCISES 12.7

2. If we set $F(x, y) = (x + y)^2 - 2x$, then $\dfrac{dy}{dx} = -\dfrac{\dfrac{\partial(F)}{\partial(x)}}{\dfrac{\partial(F)}{\partial(y)}} = -\dfrac{F_x}{F_y} = -\dfrac{2(x + y) - 2}{2(x + y)} = \dfrac{1 - x - y}{x + y}.$

4. If we set $F(x, y) = \sin(x + y) + y^2 - 12x^2 - y$, then

$$\frac{dy}{dx} = -\frac{\dfrac{\partial(F)}{\partial(x)}}{\dfrac{\partial(F)}{\partial(y)}} = -\frac{F_x}{F_y} = -\frac{\cos(x + y) - 24x}{\cos(x + y) + 2y - 1} = \frac{24x - \cos(x + y)}{\cos(x + y) + 2y - 1}.$$

6. If we set $F(x, y, z) = x^2 z^2 + yz + 3x - 4$, then

$$\frac{\partial z}{\partial x} = -\frac{\dfrac{\partial(F)}{\partial(x)}}{\dfrac{\partial(F)}{\partial(z)}} = -\frac{F_x}{F_z} = -\frac{2xz^2 + 3}{2x^2 z + y}; \qquad \frac{\partial z}{\partial y} = -\frac{\dfrac{\partial(F)}{\partial(y)}}{\dfrac{\partial(F)}{\partial(z)}} = -\frac{F_y}{F_z} = -\frac{z}{2x^2 z + y}.$$

8. If we set $F(x, y, z) = \text{Tan}^{-1}(yz) - xz$, then $\dfrac{\partial z}{\partial x} = -\dfrac{\dfrac{\partial(F)}{\partial(x)}}{\dfrac{\partial(F)}{\partial(z)}} = -\dfrac{F_x}{F_z} = -\dfrac{-z}{\dfrac{y}{1 + y^2 z^2} - x} = \dfrac{z(1 + y^2 z^2)}{y - x(1 + y^2 z^2)},$

and $\dfrac{\partial z}{\partial y} = -\dfrac{\dfrac{\partial(F)}{\partial(y)}}{\dfrac{\partial(F)}{\partial(z)}} = -\dfrac{F_y}{F_z} = -\dfrac{\dfrac{z}{1 + y^2 z^2}}{\dfrac{y}{1 + y^2 z^2} - x} = \dfrac{z}{x(1 + y^2 z^2) - y}.$

10. If we set $F(x, t, z) = \sin(x + t) - \sin(x - t) - z$, then

$$\frac{\partial x}{\partial t} = -\frac{\dfrac{\partial(F)}{\partial(t)}}{\dfrac{\partial(F)}{\partial(x)}} = -\frac{F_t}{F_x} = -\frac{\cos(x + t) + \cos(x - t)}{\cos(x + t) - \cos(x - t)} = \frac{\cos(x + t) + \cos(x - t)}{\cos(x - t) - \cos(x + t)}.$$

12. If we set $F(x, y, z) = x^2 + y^2 - z^2 + 2xy - 1$ and $G(x, y) = x^3 + y^3 - 5y - 4$, then

$$\frac{dz}{dx} = -\frac{\dfrac{\partial(F, G)}{\partial(y, x)}}{\dfrac{\partial(F, G)}{\partial(y, z)}} = -\frac{\begin{vmatrix} F_y & F_x \\ G_y & G_x \end{vmatrix}}{\begin{vmatrix} F_y & F_z \\ G_y & G_z \end{vmatrix}} = -\frac{\begin{vmatrix} 2y + 2x & 2x + 2y \\ 3y^2 - 5 & 3x^2 \end{vmatrix}}{\begin{vmatrix} 2y + 2x & -2z \\ 3y^2 - 5 & 0 \end{vmatrix}}$$

$$= -\frac{3x^2(2y + 2x) - (2x + 2y)(3y^2 - 5)}{2z(3y^2 - 5)} = \frac{(x + y)(3x^2 - 3y^2 + 5)}{z(5 - 3y^2)}.$$

14. If we set $F(x, y, z, u, v) = x^2 - y \cos(uv) + z^2$, $G(x, y, z, u, v) = x^2 + y^2 - \sin(uv) + 2z^2 - 2$, and $H(x, y, z, u, v) = xy - \sin u \cos v + z$, then

$$\frac{\partial x}{\partial u} = -\frac{\dfrac{\partial(F, G, H)}{\partial(u, y, z)}}{\dfrac{\partial(F, G, H)}{\partial(x, y, z)}} = -\frac{\begin{vmatrix} F_u & F_y & F_z \\ G_u & G_y & G_z \\ H_u & H_y & H_z \end{vmatrix}}{\begin{vmatrix} F_x & F_y & F_z \\ G_x & G_y & G_z \\ H_x & H_y & H_z \end{vmatrix}} = -\frac{\begin{vmatrix} yv \sin(uv) & -\cos(uv) & 2z \\ -v \cos(uv) & 2y & 4z \\ -\cos u \cos v & x & 1 \end{vmatrix}}{\begin{vmatrix} 2x & -\cos(uv) & 2z \\ 2x & 2y & 4z \\ y & x & 1 \end{vmatrix}}.$$

When $x = y = 1$, $u = \pi/2$, $v = z = 0$, $\dfrac{\partial x}{\partial u} = - \begin{vmatrix} 0 & \text{-}1 & 0 \\ 0 & 2 & 0 \\ 0 & 1 & 1 \\ 2 & \text{-}1 & 0 \\ 2 & 2 & 0 \\ 1 & 1 & 1 \end{vmatrix} = 0.$

16. $\dfrac{dz}{dt} = \dfrac{\partial z}{\partial x}\dfrac{dx}{dt} + \dfrac{\partial z}{\partial y}\dfrac{dy}{dt} = e^x \cos y \dfrac{dx}{dt} - e^x \sin y \dfrac{dy}{dt}$ If we set $F(x,t) = x^3 + e^x - t^2 - t - 1$ and

$G(y,t) = yt^2 + y^2 t - t + y$, then $\dfrac{dx}{dt} = -\dfrac{\dfrac{\partial(F)}{\partial(t)}}{\dfrac{\partial(F)}{\partial(x)}} = -\dfrac{F_t}{F_x} = -\dfrac{-2t-1}{3x^2 + e^x} = \dfrac{2t+1}{3x^2 + e^x}$, and

$\dfrac{dy}{dt} = -\dfrac{\dfrac{\partial(G)}{\partial(t)}}{\dfrac{\partial(G)}{\partial(y)}} = -\dfrac{G_t}{G_y} = -\dfrac{2yt + y^2 - 1}{t^2 + 2yt + 1} = \dfrac{1 - 2yt - y^2}{1 + 2yt + t^2}.$

Consequently, $\dfrac{dz}{dt} = e^x \cos y \left(\dfrac{2t+1}{3x^2 + e^x}\right) - e^x \sin y \left(\dfrac{1-2yt-y^2}{1+2yt+t^2}\right).$

18. $\dfrac{\partial z}{\partial y} = \dfrac{\partial z}{\partial u}\dfrac{\partial u}{\partial y} + \dfrac{\partial z}{\partial v}\dfrac{\partial v}{\partial y} = [3u^2 v + v \cos(uv)]\dfrac{\partial u}{\partial y} + [u^3 + u\cos(uv)]\dfrac{\partial v}{\partial y}$

If we set $F(x,u,v) = e^u \cos v - x$, $G(y,u,v) = e^u \sin v - y$, then

$\dfrac{\partial u}{\partial y} = -\dfrac{\dfrac{\partial(F,G)}{\partial(y,v)}}{\dfrac{\partial(F,G)}{\partial(u,v)}} = -\dfrac{\begin{vmatrix} F_y & F_v \\ G_y & G_v \end{vmatrix}}{\begin{vmatrix} F_u & F_v \\ G_u & G_v \end{vmatrix}} = -\dfrac{\begin{vmatrix} 0 & -e^u \sin v \\ -1 & e^u \cos v \end{vmatrix}}{\begin{vmatrix} e^u \cos v & -e^u \sin v \\ e^u \sin v & e^u \cos v \end{vmatrix}} = \dfrac{e^u \sin v}{e^{2u}} = e^{-u} \sin v,$

and $\dfrac{\partial v}{\partial y} = -\dfrac{\dfrac{\partial(F,G)}{\partial(u,y)}}{\dfrac{\partial(F,G)}{\partial(u,v)}} = -\dfrac{\begin{vmatrix} F_u & F_y \\ G_u & G_y \end{vmatrix}}{e^{2u}} = -\dfrac{\begin{vmatrix} e^u \cos v & 0 \\ e^u \sin v & -1 \end{vmatrix}}{e^{2u}} = e^{-u} \cos v.$

Thus, $\dfrac{\partial z}{\partial y} = [3u^2 v + v \cos(uv)]e^{-u} \sin v + [u^3 + u\cos(uv)]e^{-u} \cos v.$

20. If we set $F(x,u,v) = x - u^2 + v^2$, $G(y,u,v) = y - 2uv$, then

$\dfrac{\partial u}{\partial x} = -\dfrac{\dfrac{\partial(F,G)}{\partial(x,v)}}{\dfrac{\partial(F,G)}{\partial(u,v)}} = -\dfrac{\begin{vmatrix} F_x & F_v \\ G_x & G_v \end{vmatrix}}{\begin{vmatrix} F_u & F_v \\ G_u & G_v \end{vmatrix}} = -\dfrac{\begin{vmatrix} 1 & 2v \\ 0 & -2u \end{vmatrix}}{\begin{vmatrix} -2u & 2v \\ -2v & -2u \end{vmatrix}} = \dfrac{2u}{4u^2 + 4v^2} = \dfrac{u}{2(u^2 + v^2)}.$

The chain rule now gives

$\dfrac{\partial^2 u}{\partial x^2} = \dfrac{\partial}{\partial u}\left(\dfrac{\partial u}{\partial x}\right)\dfrac{\partial u}{\partial x} + \dfrac{\partial}{\partial v}\left(\dfrac{\partial u}{\partial x}\right)\dfrac{\partial v}{\partial x}$

$= \left[\dfrac{2(u^2 + v^2) - u(4u)}{4(u^2 + v^2)^2}\right]\dfrac{\partial u}{\partial x} + \left[\dfrac{-2uv}{2(u^2 + v^2)^2}\right]\dfrac{\partial v}{\partial x}.$

Since $\dfrac{\partial v}{\partial x} = -\dfrac{\dfrac{\partial(F,G)}{\partial(u,x)}}{\dfrac{\partial(F,G)}{\partial(u,v)}} = -\dfrac{\begin{vmatrix} F_u & F_x \\ G_u & G_x \end{vmatrix}}{4(u^2 + v^2)} = -\dfrac{\begin{vmatrix} -2u & 1 \\ -2v & 0 \end{vmatrix}}{4(u^2 + v^2)} = \dfrac{-2v}{4(u^2 + v^2)} = \dfrac{-v}{2(u^2 + v^2)},$

we obtain $\dfrac{\partial^2 u}{\partial x^2} = \dfrac{v^2 - u^2}{2(u^2 + v^2)^2}\dfrac{u}{2(u^2 + v^2)} - \dfrac{uv}{(u^2 + v^2)^2}\dfrac{-v}{2(u^2 + v^2)} = \dfrac{3uv^2 - u^3}{4(u^2 + v^2)^3}.$

22. If we set $F(x, y, s, t) = x^2 - 2y^2s^2t - 2st^2 - 1$ and $G(x, y, s, t) = x^2 + 2y^2s^2t + 5st^2 - 1$, then

$$\frac{\partial t}{\partial y} = -\frac{\dfrac{\partial(F, G)}{\partial(s, y)}}{\dfrac{\partial(F, G)}{\partial(s, t)}} = -\frac{\begin{vmatrix} F_s & F_y \\ G_s & G_y \end{vmatrix}}{\begin{vmatrix} F_s & F_t \\ G_s & G_t \end{vmatrix}} = -\frac{\begin{vmatrix} -4y^2st - 2t^2 & -4ys^2t \\ 4y^2st + 5t^2 & 4ys^2t \end{vmatrix}}{\begin{vmatrix} -4y^2st - 2t^2 & -2y^2s^2 - 4st \\ 4y^2st + 5t^2 & 2y^2s^2 + 10st \end{vmatrix}}$$

$$= -\frac{4ys^2t(-4y^2st - 2t^2 + 4y^2st + 5t^2)}{4y^2st(-2y^2s^2 - 10st + 2y^2s^2 + 4st) + t^2(-4y^2s^2 - 20st + 10y^2s^2 + 20st)}$$

$$= \frac{-12ys^2t^3}{-24y^2s^2t^2 + 6y^2s^2t^2} = \frac{2t}{3y}.$$

Thus, $\dfrac{\partial^2 t}{\partial y^2} = \dfrac{2}{3y}\dfrac{\partial t}{\partial y} - \dfrac{2t}{3y^2} = \dfrac{2}{3y}\left(\dfrac{2t}{3y}\right) - \dfrac{2t}{3y^2} = -\dfrac{2t}{9y^2}.$

EXERCISES 12.8

2. The vector that joins $(3, 2, 1)$ to $(3, 1, -1)$ is $\mathbf{v} = (0, -1, -2)$. At the point $(-1, 1, -1)$,

$$D_{\mathbf{v}}f = \nabla f_{|(-1,1,-1)} \cdot \hat{\mathbf{v}} = (2xy + z, x^2, x)_{|(-1,1,-1)} \cdot \frac{(0, -1, -2)}{\sqrt{1+4}} = (-3, 1, -1) \cdot \frac{(0, -1, -2)}{\sqrt{5}} = \frac{1}{\sqrt{5}}.$$

4. The vector from $(1, 1, 1)$ to $(-1, -2, 3)$ is $\mathbf{v} = (-2, -3, 2)$. At the point $(1, 1, 1)$,

$$D_{\mathbf{v}}f = \nabla f_{|(1,1,1)} \cdot \hat{\mathbf{v}} = \left[\frac{1}{xy + yz + xz}(y + z, x + z, x + y)\right]_{|(1,1,1)} \cdot \frac{(-2, -3, 2)}{\sqrt{4 + 9 + 4}}$$

$$= \frac{1}{3}(2, 2, 2) \cdot \frac{(-2, -3, 2)}{\sqrt{17}} = \frac{-2}{\sqrt{17}}.$$

6. A vector along $3x + 4y = -2$ in the direction of decreasing y is $\mathbf{v} = (4, -3)$. At the point $(2, -2)$,

$$D_{\mathbf{v}}f = \nabla f_{|(2,-2)} \cdot \hat{\mathbf{v}} = [\cos(x + y)(1, 1)]_{|(2,-2)} \cdot \frac{(4, -3)}{\sqrt{16 + 9}} = (1, 1) \cdot \frac{(4, -3)}{5} = \frac{1}{5}.$$

8. Since parametric equations for the line are $x = -2t - 1$, $y = t$, $z = \dfrac{1}{2}(2 + 2t + 1 + t) = \dfrac{3}{2} + \dfrac{3}{2}t$, a vector along the line in the direction of decreasing z is $\mathbf{v} = (4, -2, -3)$. At the point $(1, -1, 0)$,

$$D_{\mathbf{v}}f = \nabla f_{|(1,-1,0)} \cdot \hat{\mathbf{v}} = (2xy + z^2, x^2 + 2yz, y^2 + 2xz)_{|(1,-1,0)} \cdot \frac{(4, -2, -3)}{\sqrt{16 + 4 + 9}}$$

$$= (-2, 1, 1) \cdot \frac{(4, -2, -3)}{\sqrt{29}} = \frac{-13}{\sqrt{29}}.$$

10. Since the slope of the curve at $(-1, 3)$ is -9, a tangent vector at the point is $\mathbf{T} = (-1, 9)$. The required rate of change is therefore

$$D_{\mathbf{T}}f = \nabla f_{|(-1,3)} \cdot \hat{\mathbf{T}} = (2x, 1)_{|(-1,3)} \cdot \hat{\mathbf{T}} = (-2, 1) \cdot \frac{(-1, 9)}{\sqrt{82}} = \frac{11}{\sqrt{82}}.$$

12. Since parametric equations for the curve are $x = t$, $y = -\sqrt{t^2 - 3}$, $z = t$, a tangent vector at $(2, -1, 2)$ is $\mathbf{T}(2) = \left(1, \dfrac{-t}{\sqrt{t^2 - 3}}, 1\right)_{|t=2} = (1, -2, 1)$. At the point $(2, -1, 2)$ then,

$$D_{\mathbf{T}}f = \nabla f_{|(2,-1,2)} \cdot \hat{\mathbf{T}} = (2xy + y^3z, x^2 + 3xy^2z, xy^3)_{|(2,-1,2)} \cdot \frac{(1, -2, 1)}{\sqrt{1 + 4 + 1}}$$

$$= (-6, 16, -2) \cdot \frac{(1, -2, 1)}{\sqrt{6}} = \frac{-40}{\sqrt{6}}.$$

14. The function increases most rapidly in the direction

$$\nabla f_{|(2,1/2)} = (2y + 1/x, 2x + 1/y)_{|(2,1/2)} = (3/2, 6),$$

or, $(2/3)(3/2, 6) = (1, 4)$. The rate of change in this direction is $\sqrt{9/4 + 36} = \sqrt{153}/2$.

16. The function increases most rapidly in the direction

$$\nabla f_{|(1,-3,2)} = \left[\frac{1}{(x^2 + y^2 + z^2)^{3/2}}(x, y, z) \right]_{|(1,-3,2)} = \frac{1}{14\sqrt{14}}(1, -3, 2),$$

or, $(1, -3, 2)$. The rate of change in this direction is $\dfrac{1}{14\sqrt{14}}\sqrt{1 + 9 + 4} = \dfrac{1}{14}$.

18. The function increases most rapidly in the direction

$$\nabla f_{|(1,1)} = (ye^{xy} + xy^2 e^{xy}, xe^{xy} + x^2 ye^{xy})_{|(1,1)} = (2e, 2e),$$

or, $(1, 1)$. The rate of change in this direction is $\sqrt{4e^2 + 4e^2} = 2\sqrt{2}e$.

20. At the point $(1, -1)$ and in the direction $\hat{\mathbf{v}} = a\hat{\mathbf{i}} + b\hat{\mathbf{j}}$,

$$D_{\hat{\mathbf{v}}}f = \nabla f_{|(1,-1)} \cdot \hat{\mathbf{v}} = (2xy, x^2 + 3y^2)_{|(1,-1)} \cdot \hat{\mathbf{v}} = (-2, 4) \cdot (a, b) = -2a + 4b.$$

(a) The directional derivative vanishes if $0 = -2a + 4b$. Because $\hat{\mathbf{v}}$ is a unit vector, we also know that $a^2 + b^2 = 1$. Thus, $1 = (2b)^2 + b^2 = 5b^2$. This implies that $b = \pm 1/\sqrt{5}$ and $a = \pm 2/\sqrt{5}$. The required directions are therefore $\pm(2, 1)$. This is to be expected since in a direction perpendicular to ∇f, the rate of change should be zero, and $\pm(2, 1)$ are both perpendicular to $(-2, 4)$.

(b) The rate of change is 1 if, and when, $1 = -2a + 4b$. Substitution from this equation into $a^2 + b^2 = 1$ gives $1 = \left(\dfrac{4b - 1}{2}\right)^2 + b^2 \implies 20b^2 - 8b - 3 = 0$. Solutions of this equation are $b = (2 \pm \sqrt{19})/10$, and these give $a = (-1 \pm 2\sqrt{19})/10$. The required directions are therefore $(-1 \pm 2\sqrt{19}, 2 \pm \sqrt{19})$.

(c) The rate of change is 20, if, and when $20 = -2a + 4b$. Substitution from this equation into $a^2 + b^2 = 1$ gives $1 = (2b - 10)^2 + b^2 \longrightarrow 5b^2 - 40b + 99 = 0$. Since this quadratic equation has no real solutions, there are no directions in which the rate of change of the function is equal to 20. This is also clear from the fact that the maximum rate of change is $|\nabla f| = 2\sqrt{5}$.

22. (a) Yes. In any direction perpendicular to the gradient of the function, its rate of change is zero.

(b) Not necessarily. If the gradient of the function at the point has length 2 say, then the maximum rate of change for all directions is 2. Hence, for no direction could it be equal to 3. On the other hand, if the length of the gradient is 4, then values of the rate of change would vary between -4 and 4 and hence there would exist directions in which it is equal to 3.

24. The distance to the origin is given by $d = \sqrt{x^2 + y^2 + z^2}$. The required derivative is

$$\begin{aligned}
D_{\mathbf{T}}d = \nabla d \cdot \hat{\mathbf{T}} &= \frac{1}{\sqrt{x^2 + y^2 + z^2}}(x\hat{\mathbf{i}} + y\hat{\mathbf{j}} + z\hat{\mathbf{k}}) \cdot \frac{(-2\sin t, 2\cos t, 3)}{\sqrt{4\sin^2 t + 4\cos^2 t + 9}} \\
&= \frac{1}{\sqrt{x^2 + y^2 + z^2}}\frac{-2x\sin t + 2y\cos t + 3z}{\sqrt{13}} \\
&= \frac{1}{\sqrt{4\cos^2 t + 4\sin^2 t + 9t^2}}\frac{-4\cos t \sin t + 4\sin t \cos t + 9t}{\sqrt{13}} \\
&= \frac{9t}{\sqrt{13}\sqrt{4 + 9t^2}}.
\end{aligned}$$

When $t = 0$, the rate of change is 0. This is expected since the curve is a helix, and when $t = 0$, the point $(2, 0, 0)$ is the closest point to the origin. Hence, the distance should have a minimum and its derivative should vanish.

26. A tangent vector along C : $x = t^2$, $y = t$, $z = t^2$ is $\mathbf{T} = (2t, 1, 2t)$. At any point on C,

$$D_{\mathbf{T}}f = \nabla f \cdot \hat{\mathbf{T}} = (2x, -2y, 2z) \cdot \frac{(2t, 1, 2t)}{\sqrt{1 + 8t^2}} = \frac{2(2xt - y + 2zt)}{\sqrt{1 + 8t^2}}.$$

For this derivative to vanish, $0 = 2xt - y + 2zt = 2t^3 - t + 2t^3 = t(4t^2 - 1)$. Thus, $t = 0, \pm 1/2$, and these values give the points $(0, 0, 0)$ and $(1/4, \pm 1/2, 1/4)$.

28. It is the negative of the rate of change in the other direction.

30. Let (a, b) be the gradient of $f(x, y)$ at the point (x_0, y_0). Then,

$$3 = D_{\hat{\mathbf{i}} + 2\hat{\mathbf{j}}}f_{|(x_0, y_0)} = \nabla f_{|(x_0, y_0)} \cdot \frac{(1, 2)}{\sqrt{5}} = (a, b) \cdot \frac{(1, 2)}{\sqrt{5}} = \frac{a + 2b}{\sqrt{5}}, \quad \text{and}$$

$$-1 = D_{-2\hat{\mathbf{i}} - \hat{\mathbf{j}}}f_{|(x_0, y_0)} = \nabla f_{|(x_0, y_0)} \cdot \frac{(-2, -1)}{\sqrt{5}} = (a, b) \cdot \frac{(-2, -1)}{\sqrt{5}} = \frac{-2a - b}{\sqrt{5}}.$$

These imply that $a = -\sqrt{5}/3$ and $b = 5\sqrt{5}/3$. The rate of change in direction $2\hat{\mathbf{i}} + 3\hat{\mathbf{j}}$ is therefore

$$\left(-\frac{\sqrt{5}}{3}, \frac{5\sqrt{5}}{3} \right) \cdot \frac{(2, 3)}{\sqrt{13}} = \frac{\sqrt{65}}{3}.$$

32. Since $\nabla f = (3x^2y^2, 2x^3y)$, the first directional derivative at any point (x, y) in direction $\mathbf{v} = (1, -2)$ is

$$D_{\mathbf{v}}f = (3x^2y^2, 2x^3y) \cdot \frac{(1, -2)}{\sqrt{5}} = \frac{1}{\sqrt{5}}(3x^2y^2 - 4x^3y).$$

The second directional derivative is

$$D_{\mathbf{v}}(D_{\mathbf{v}}f)_{|(1,1)} = \nabla \left[\frac{1}{\sqrt{5}}(3x^2y^2 - 4x^3y) \right]_{|(1,1)} \cdot \frac{(1, -2)}{\sqrt{5}} = \frac{1}{5}(6xy^2 - 12x^2y, 6x^2y - 4x^3)_{|(1,1)} \cdot (1, -2)$$

$$= \frac{1}{5}(-6, 2) \cdot (1, -2) = -2.$$

34. Since a tangent vector to the curve at any point is $\mathbf{T} = R(1 - \cos\theta, \sin\theta)$, a unit tangent vector is

$$\hat{\mathbf{T}} = \frac{(1 - \cos\theta, \sin\theta)}{\sqrt{(1 - \cos\theta)^2 + \sin^2\theta}} = \frac{(1 - \cos\theta, \sin\theta)}{\sqrt{2 - 2\cos\theta}}.$$

The rate of change of the distance $d = \sqrt{x^2 + y^2}$ from the origin to the stone is

$$D_{\hat{\mathbf{T}}}d = \nabla d \cdot \hat{\mathbf{T}} = \frac{x\hat{\mathbf{i}} + y\hat{\mathbf{j}}}{\sqrt{x^2 + y^2}} \cdot \hat{\mathbf{T}} = \frac{x(1 - \cos\theta) + y\sin\theta}{\sqrt{x^2 + y^2}\sqrt{2 - 2\cos\theta}}.$$

When $\theta = \pi/2$, we have $x = R(\pi/2 - 1)$, $y = R$, and

$$D_{\hat{\mathbf{T}}}d_{|\theta = \pi/2} = \frac{R(\pi/2 - 1)(1) + R(1)}{\sqrt{R^2(\pi/2 - 1)^2 + R^2}\sqrt{2}} = \frac{\pi}{\sqrt{2}\sqrt{8 - 4\pi + \pi^2}}.$$

When $\theta = \pi$, we have $x = \pi R$, $y = 2R$, and

$$D_{\hat{\mathbf{T}}}d_{|\theta = \pi} = \frac{\pi R(2) + 2R(0)}{\sqrt{\pi^2 R^2 + 4R^2}\sqrt{4}} = \frac{\pi}{\sqrt{4 + \pi^2}}.$$

(b) Since $\nabla(y) = \hat{\mathbf{j}}$, the rate of change of the y-coordinate is

$$D_{\hat{\mathbf{T}}}y = \hat{\mathbf{j}} \cdot \frac{(1 - \cos\theta, \sin\theta)}{\sqrt{2 - 2\cos\theta}} = \frac{\sin\theta}{\sqrt{2 - 2\cos\theta}}.$$

When $\theta = \pi/2$, $D_{\hat{\mathbf{T}}}y_{|\theta = \pi/2} = 1/\sqrt{2}$, and when $\theta = \pi$, $D_{\hat{\mathbf{T}}}y_{|\theta = \pi} = 0$.

(c) Since $\nabla(x) = \hat{\mathbf{i}}$, the rate of change of the x-coordinate is

$$D_{\hat{\mathbf{T}}}x = \hat{\mathbf{i}} \cdot \frac{(1 - \cos\theta, \sin\theta)}{\sqrt{2 - 2\cos\theta}} = \frac{1 - \cos\theta}{\sqrt{2 - 2\cos\theta}}.$$

When $\theta = \pi/2$, $D_{\hat{\mathbf{T}}}x|_{\theta = \pi/2} = 1/\sqrt{2}$, and when $\theta = \pi$, $D_{\hat{\mathbf{T}}}x|_{\theta = \pi} = 2/\sqrt{4} = 1$.

EXERCISES 12.9

2. Since a tangent vector at $(1,1,1)$ is $\dfrac{d\mathbf{r}}{dt}_{|t=1} = (1, 2t, 3t^2)_{|t=1} = (1, 2, 3)$, parametric equations for the tangent line are $x = 1 + u$, $y = 1 + 2u$, $z = 1 + 3u$.

4. With parametric equations $x = t$, $y = t^2$, $z = t$, a tangent vector at the point $(-2, 4, -2)$ is $\dfrac{d\mathbf{r}}{dt}_{|t=-2}$ $= (1, 2t, 1)_{|t=-2} = (1, -4, 1)$. Parametric equations for the tangent line are $x = -2 + u$, $y = 4 - 4u$, $z = -2 + u$.

6. Since a tangent vector at $(1,5,1)$ is $\dfrac{d\mathbf{r}}{dt}_{|t=1} = (-2t, 2, 1)_{|t=1} = (-2, 2, 1)$, parametric equations for the tangent line are $x = 1 - 2u$, $y = 5 + 2u$, $z = 1 + u$.

8. Because a vector normal to the curve at $(1, 4)$ is

$$\nabla(x^2y^3 + xy - 68)_{|(1,4)} = (2xy^3 + y, 3x^2y^2 + x)_{|(1,4)} = (132, 49),$$

a vector tangent to the curve is $(49, -132)$. The slope of the tangent line is is therefore $-132/49$, and its equations are $y - 4 = -\dfrac{132}{49}(x - 1)$, $z = 0 \implies 132x + 49y = 328$, $z = 0$.

10. Since a tangent vector at $(1, 0, 0)$ is

$$\frac{d\mathbf{r}}{dt}_{|t=0} = (-e^{-t}\cos t - e^{-t}\sin t, -e^{-t}\sin t + e^{-t}\cos t, 1)_{|t=0} = (-1, 1, 1),$$

parametric equations for the tangent line are $x = 1 - u$, $y = u$, $z = u$.

12. Since normals to the surfaces at $(2, -\sqrt{5}, -1)$ are $\nabla(y^2 + z^2 - 6)_{|(2, -\sqrt{5}, -1)} = (0, 2y, 2z)_{|(2, -\sqrt{5}, -1)} = (0, -2\sqrt{5}, -2)$ and $\nabla(x + z - 1)_{|(2, -\sqrt{5}, -1)} = (1, 0, 1)$, a vector along the tangent line is

$$(0, \sqrt{5}, 1) \times (1, 0, 1) = \begin{vmatrix} \hat{\mathbf{i}} & \hat{\mathbf{j}} & \hat{\mathbf{k}} \\ 0 & \sqrt{5} & 1 \\ 1 & 0 & 1 \end{vmatrix} = (\sqrt{5}, 1, -\sqrt{5}).$$

Because $(1, 1/\sqrt{5}, -1)$ is also a tangent vector, parametric equations for the tangent line are $x = 2 + t$, $y = -\sqrt{5} + t/\sqrt{5}$, $z = -1 - t$.

14. Because a tangent vector at $(4, 1, \sqrt{17})$ is $\dfrac{d\mathbf{r}}{dt}_{|t=4} = (1, 0, t/\sqrt{1+t^2})_{|t=4} = (1, 0, 4/\sqrt{17})$, as is the vector $(\sqrt{17}, 0, 4)$, parametric equations for the tangent line are $x = 4 + \sqrt{17}u$, $y = 1$, $z = \sqrt{17} + 4u$.

16. With parametric equations $x = t^2 + t^3$, $y = t - t^4$, $z = t$, a tangent vector at the point $(12, -14, 2)$ is $\dfrac{d\mathbf{r}}{dt}_{|t=2} = (2t + 3t^2, 1 - 4t^3, 1)_{|t=2} = (16, -31, 1)$. Parametric equations for the tangent line are $x = 12 + 16u$, $y = -14 - 31u$, $z = 2 + u$.

18. Since normals to the surfaces at $(0, 1, 1)$ are $\nabla(2x^2 + y^2 + 2y - 3)_{|(0,1,1)} = (4x, 2y + 2, 0)_{|(0,1,1)} = (0, 4, 0)$ and $\nabla(x - z + 1)_{|(0,1,1)} = (1, 0, -1)$, a vector along the tangent line is

$$(0, 1, 0) \times (1, 0, -1) = \begin{vmatrix} \hat{\mathbf{i}} & \hat{\mathbf{j}} & \hat{\mathbf{k}} \\ 0 & 1 & 0 \\ 1 & 0 & -1 \end{vmatrix} = (-1, 0, -1),$$

as is $(1, 0, 1)$. Parametric equations for the tangent line are $x = t$, $y = 1$, $z = 1 + t$.

20. Since a tangent vector at $(0, 2\pi, 4\pi)$ is $\dfrac{d\mathbf{r}}{dt}\Big|_{t=2\pi} = (\sin t + t\cos t, \cos t - t\sin t, 2)_{|t=2\pi} = (2\pi, 1, 2)$, parametric equations for the tangent line are $x = 2\pi u$, $y = 2\pi + u$, $z = 4\pi + 2u$.

22. Since a normal to the tangent plane is

$$\nabla(x - x^2 + y^3 z)_{|(2,-1,-2)} = (1 - 2x, 3y^2 z, y^3)_{|(2,-1,-2)} = (-3, -6, -1),$$

as is $(3, 6, 1)$, the equation of the tangent plane is

$$0 = (3, 6, 1) \cdot (x - 2, y + 1, z + 2) = 3x + 6y + z + 2.$$

24. Because the surface is a plane, the tangent plane is the surface itself, $x + y + z = 4$.

26. Since a normal to the tangent plane is $\nabla(x^2 + y^2 + 2y - 1)_{|(1,0,3)} = (2x, 2y + 2, 0)_{|(1,0,3)} = (2, 2, 0)$, as is $(1, 1, 0)$, the equation of the tangent plane is $0 = (1, 1, 0) \cdot (x - 1, y, z - 3) = x + y - 1$.

28. A vector tangent to the curve at $(1, 1, 1)$ is

$$\mathbf{T} = \nabla(x^2 - y^2 + z^2 - 1)_{|(1,1,1)} \times \nabla(xy + xz - 2)_{|(1,1,1)} = (2x, -2y, 2z)_{|(1,1,1)} \times (y + z, x, x)_{|(1,1,1)}$$

$$= (2, -2, 2) \times (2, 1, 1) = \begin{vmatrix} \hat{\mathbf{i}} & \hat{\mathbf{j}} & \hat{\mathbf{k}} \\ 2 & -2 & 2 \\ 2 & 1 & 1 \end{vmatrix} = (-4, 2, 6).$$

A vector normal to the surface at $(1, 1, 1)$ is

$$\mathbf{n} = \nabla(xyz - x^2 - 6y + 6)_{|(1,1,1)} = (yz - 2x, xz - 6, xy)_{|(1,1,1)} = (-1, -5, 1).$$

Since $\mathbf{T} \cdot \mathbf{n} = 4 - 10 + 6 = 0$, the vectors are perpendicular, and the curve is tangent to the surface at $(1, 1, 1)$.

30. Since a vector along the curve is $\mathbf{T} = \nabla(x + y + z - 4) \times \nabla(x - y + z - 2) = \begin{vmatrix} \hat{\mathbf{i}} & \hat{\mathbf{j}} & \hat{\mathbf{k}} \\ 1 & 1 & 1 \\ 1 & -1 & 1 \end{vmatrix} = (2, 0, -2)$,

we find that at $(3, 1, 0)$,

$$D_{\hat{\mathbf{T}}}f = \nabla f_{|(3,1,0)} \cdot \hat{\mathbf{T}} = (4x, 2yz^2, 2y^2 z)_{|(3,1,0)} \cdot \frac{(1, 0, -1)}{\sqrt{2}} = (12, 0, 0) \cdot \frac{(1, 0, -1)}{\sqrt{2}} = \frac{12}{\sqrt{2}} = 6\sqrt{2}.$$

32. Since $f(x, y, z) = 0$ everywhere on the curve, its directional derivative must also be zero.

34. A vector normal to the surface at (x_0, y_0, z_0) is $\nabla\left(\dfrac{x^2}{a^2} + \dfrac{y^2}{b^2} + \dfrac{z^2}{c^2} - 1\right)_{|(x_0, y_0, z_0)} = \left(\dfrac{2x_0}{a^2}, \dfrac{2y_0}{b^2}, \dfrac{2z_0}{c^2}\right)$. Hence, the equation of the tangent plane is

$$0 = \left(\frac{x_0}{a^2}, \frac{y_0}{b^2}, \frac{z_0}{c^2}\right) \cdot (x - x_0, y - y_0, z - z_0) = \frac{x_0 x}{a^2} + \frac{y_0 y}{b^2} + \frac{z_0 z}{c^2} - \left(\frac{x_0^2}{a^2} + \frac{y_0^2}{b^2} + \frac{z_0^2}{c^2}\right).$$

Since (x_0, y_0, z_0) is on the ellipsoid, $x_0^2/a^2 + y_0^2/b^2 + z_0^2/c^2 = 1$, and the equation of the plane reduces to $x_0 x/a^2 + y_0 y/b^2 + z_0 z/c^2 = 1$.

36. A normal vector to the surface is $\nabla(4x^2 + 4y^2 - z^2) = (8x, 8y, -2z)$, as is $(4x, 4y, -z)$. The tangent plane is parallel to $x - y + 2z = 3$, which has normal $(1, -1, 2)$, if, and where, $(4x, 4y, -z) = \lambda(1, -1, 2)$ for some λ. This requires $4x = \lambda$, $4y = -\lambda$, and $-z = 2\lambda$. Substitution of $x = \lambda/4$, $y = -\lambda/4$, and $z = -2\lambda$ into the equation of the surface gives $4\lambda^2 = 4\left(\dfrac{\lambda^2}{16}\right) + 4\left(\dfrac{\lambda^2}{16}\right)$. The only solution of this equation is $\lambda = 0$. But this implies that $x = y = z = 0$, and this is unacceptable since there is no tangent plane to the surface at $(0, 0, 0)$.

38. A normal vector to the paraboloid at any point $P(x, y, z)$ is

$$\mathbf{n} = \nabla(x^2 + y^2 - 1 - z) = (2x, 2y, -1).$$

This vector coincides with **OP** if $(x, y, z) = \lambda(2x, 2y, -1)$, or, $x = 2\lambda x$, $y = 2\lambda y$, $z = -\lambda$. When these are combined with $z = x^2 + y^2 - 1$, the points obtained are $(0, 0, -1)$ and $x^2 + y^2 = 1/2$, $z = -1/2$.

EXERCISES 12.10

2. For critical points we solve $0 = \dfrac{\partial f}{\partial x} = 3y - 3x^2$, $0 = \dfrac{\partial f}{\partial y} = 3x - 3y^2$. Solutions are $(0, 0)$ and $(1, 1)$.

$$\frac{\partial^2 f}{\partial x^2} = -6x, \qquad \frac{\partial^2 f}{\partial x \partial y} = 3, \qquad \frac{\partial^2 f}{\partial y^2} = -6y$$

At $(0, 0)$, $B^2 - AC = 9 - 0$, and therefore $(0, 0)$ yields a saddle point. At $(1, 1)$, $B^2 - AC = 9 - (-6)(-6) = -27$, and $A = -6$, and therefore $(1, 1)$ gives a relative maximum.

4. For critical points we solve $0 = \dfrac{\partial f}{\partial x} = 2xy^2 + 3$, $0 = \dfrac{\partial f}{\partial y} = 2x^2 y$. Because there are no solutions of these equations, there are no critical points.

6. For critical points we solve $0 = \dfrac{\partial f}{\partial x} = \sin y$, $0 = \dfrac{\partial f}{\partial y} = x \cos y$. Solutions are $(0, n\pi)$, where n is an integer.

$$\frac{\partial^2 f}{\partial x^2} = 0, \qquad \frac{\partial^2 f}{\partial x \partial y} = \cos y, \qquad \frac{\partial^2 f}{\partial y^2} = -x \sin y$$

At $(0, n\pi)$, $B^2 - AC = [\cos(n\pi)]^2 - 0 > 0$, and therefore all critical points yield saddle points.

8. For critical points we solve $0 = \dfrac{\partial f}{\partial x} = 2x - 2y$, $0 = \dfrac{\partial f}{\partial y} = -2x + 2y$. All points on the line $y = x$ are critical. Because $f(x, y) = (x - y)^2$, it is clear that each of these points gives a relative minimum.

10. For critical points we solve

$$0 = \frac{\partial f}{\partial x} = 4x^3 y^3, \qquad 0 = \frac{\partial f}{\partial y} = 3x^4 y^2.$$

Every point on the x- and y-axes is critical, and at each of these points $f(x, y) = 0$. The diagram to the right showing the sign of $f(x, y)$ in the four quadrants indicates that the points $(0, y)$ for $y > 0$ yield relative minima; $(0, y)$ for $y < 0$ yield relative maxima; and $(x, 0)$ yield saddle points.

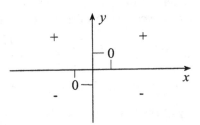

12. If $x > 0$, then $\partial f / \partial x = 1$, and if $x < 0$, then $\partial f / \partial x = -1$. Consequently, there are no critical points at which $0 = \partial f / \partial x = \partial f / \partial y$. However $\partial f / \partial x$ does not exist when $x = 0$, and therefore every point on the y-axis is critical. They cannot give saddle points because $\partial f / \partial x$ does not exist at these points. Since the cross-section of the surface $z = |x| + y^2$ with the plane $x = 0$ is the parabola $z = y^2$, $x = 0$ (see figure to the right), no critical point except possible $(0, 0)$ can yield a relative maximum or minimum. Finally, because $f(0, 0) = 0$, and $f(x, y) > 0$ for $(x, y) \neq (0, 0)$, it follows that $(0, 0)$ gives a relative minimum.

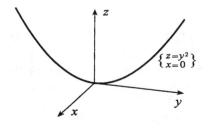

14. For critical points we solve $0 = \dfrac{\partial f}{\partial x} = 4x^3 - 2x = 2x(2x^2 - 1)$, $0 = \dfrac{\partial f}{\partial y} = 4y^3 - 2y = 2y(2y^2 - 1)$. These give $x = 0, \pm 1/\sqrt{2}$ and $y = 0, \pm 1/\sqrt{2}$, and therefore critical points are $(0,0)$, $(0, \pm 1/\sqrt{2})$, $(\pm 1/\sqrt{2}, 0)$, $(1/\sqrt{2}, \pm 1/\sqrt{2})$, $(-1/\sqrt{2}, \pm 1/\sqrt{2})$. With $f_{xx} = 12x^2 - 2$, $f_{xy} = 0$, and $f_{yy} = 12y^2 - 2$, we construct the following table.

	A	B	C	$B^2 - AC$	Classification
$(0,0)$	-2	0	-2	-4	relative maximum
$(0, \pm 1/\sqrt{2})$	-2	0	4	8	saddle points
$(\pm 1/\sqrt{2}, 0)$	4	0	-2	8	saddle points
$(1/\sqrt{2}, \pm 1/\sqrt{2})$	4	0	4	-16	relative minima
$(-1/\sqrt{2}, \pm 1/\sqrt{2})$	4	0	4	-16	relative minima

16. For critical points we solve

$$0 = \frac{\partial f}{\partial x} = 2xy^2z^2 + 2xt^2 + 3, \quad 0 = \frac{\partial f}{\partial y} = 2x^2yz^2, \quad 0 = \frac{\partial f}{\partial z} = 2x^2y^2z, \quad 0 = \frac{\partial f}{\partial t} = 2x^2t.$$

There are no solutions to these equations.

18. For critical points we solve

$$0 = \frac{\partial f}{\partial x} = yze^{x^2+y^2+z^2} + 2x^2yze^{x^2+y^2+z^2} = yz(1 + 2x^2)e^{x^2+y^2+z^2},$$

$$0 = \frac{\partial f}{\partial y} = xz(1 + 2y^2)e^{x^2+y^2+z^2}, \qquad 0 = \frac{\partial f}{\partial z} = xy(1 + 2z^2)e^{x^2+y^2+z^2}.$$

All points on the coordinate axes satisfy these equations.

20. If $f(x,y)$ is harmonic in D, it has continuous second partial derivatives in D and $\partial^2 f/\partial x^2 + \partial^2 f/\partial y^2 = 0$ therein. For $f(x,y)$ to have a relative maximum or minimum at a point (x,y) in D, its first partial derivatives must vanish there. In addition,

$$0 > B^2 - AC = \left(\frac{\partial^2 f}{\partial x \partial y}\right)^2 - \left(\frac{\partial^2 f}{\partial x^2}\right)\left(\frac{\partial^2 f}{\partial y^2}\right) = \left(\frac{\partial^2 f}{\partial x \partial y}\right)^2 - \left(\frac{\partial^2 f}{\partial x^2}\right)\left(-\frac{\partial^2 f}{\partial x^2}\right) = \left(\frac{\partial^2 f}{\partial x \partial y}\right)^2 + \left(\frac{\partial^2 f}{\partial x^2}\right)^2,$$

an impossibility if f_{xx} or f_{xy} does not vanish.

22. For critical points we solve $0 = \dfrac{\partial f}{\partial x} = 4x^3 + 3y^2$, $0 = \dfrac{\partial f}{\partial y} = 6xy + 2y$. The only solutions are $(0,0)$ and $(-1/3, \pm 2/9)$.

$$\frac{\partial^2 f}{\partial x^2} = 12x^2, \qquad \frac{\partial^2 f}{\partial x \partial y} = 6y, \qquad \frac{\partial^2 f}{\partial y^2} = 6x + 2$$

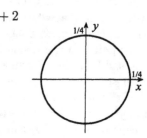

At $(-1/3, \pm 2/9)$, $B^2 - AC = 36(4/81) - (12/9)(0) > 0$, and therefore $(-1/3, \pm 2/9)$ yield saddle points. At $(0,0)$, $B^2 - AC = 0$ and the test fails. Now $f(0,0) = 0$, and in the circle shown to the right, $f(x,y) = x^4 + y^2(1 + 3x) > 0$ (except at $(0,0)$). This implies that $(0,0)$ gives a relative minimum.

EXERCISES 12.11

2. For critical points of $f(x,y)$, we solve
$$0 = \frac{\partial f}{\partial x} = 2x + 1, \quad 0 = \frac{\partial f}{\partial y} = 6y + 1.$$

The only solution is $(-1/2, -1/6)$ at which
$f(-1/2, -1/6) = \boxed{-1/3}$.

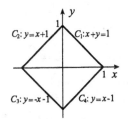

On C_1, $z = f(x,y)$ becomes
$$z = F(x) = x^2 + x + 3(1-x)^2 + (1-x)$$
$$= 4x^2 - 6x + 4, \quad 0 \le x \le 1.$$

For critical points of this function, we solve $0 = F'(x) = 8x - 6$. The only critical point is $x = 3/4$ at which $F(3/4) = \boxed{7/4}$.

On C_2, $z = f(x,y)$ is $z = F(x) = x^2 + x + 3(x+1)^2 + (x+1) = 4x^2 + 8x + 4$, $-1 \le x \le 0$. For critical points of this function, we solve $0 = F'(x) = 8x + 8$. At the critical point $x = -1$, $F(-1) = \boxed{0}$.

On C_3, $z = f(x,y)$ is $z = F(x) = x^2 + x + 3(-x-1)^2 + (-x-1) = 4x^2 + 6x + 2$, $-1 \le x \le 0$. For critical points of this function, we solve $0 = F'(x) = 8x + 6$. At the critical point $x = -3/4$, $F(-3/4) = \boxed{-1/4}$.

On C_4, $z = f(x,y)$ is $z = F(x) = x^2 + x + 3(x-1)^2 + (x-1) = 4x^2 - 4x + 2$, $0 \le x \le 1$. For critical points we solve $0 = F'(x) = 8x - 4$. At the critical point $x = 1/2$, $F(1/2) = \boxed{1}$.

Finally, at the remaining three corners of R, $f(1,0) = \boxed{2}$, $f(0,1) = \boxed{4}$, $f(0,-1) = \boxed{2}$. Maximum and minimum values of $f(x,y)$ on R are therefore 4 and $-1/3$.

4. For critical points of $f(x,y)$, we solve
$$0 = \frac{\partial f}{\partial x} = 2xy + y^2, \quad 0 = \frac{\partial f}{\partial y} = x^2 + 2xy + 1.$$

Solutions are $(\pm 1/\sqrt{3}, \mp 2/\sqrt{3})$ at which
$f(1/\sqrt{3}, -2/\sqrt{3}) = \boxed{-4\sqrt{3}/9}$, $f(-1/\sqrt{3}, 2/\sqrt{3}) = \boxed{4\sqrt{3}/9}$.

On C_1, $z = f(x,y)$ becomes

$$z = F(y) = y + y^2 + y = 2y + y^2, \quad -1 \le y \le 1.$$

For critical points of this function, we solve $0 = F'(y) = 2 + 2y$. The only critical point is $y = -1$ at which $F(-1) = \boxed{-1}$.

On C_2, $z = f(x,y)$ is $z = F(x) = x^2 + x + 1$, $-1 \le x \le 1$. For critical points of this function, we solve $0 = F'(x) = 2x + 1$. At the critical point $x = -1/2$, $F(-1/2) = \boxed{3/4}$.

On C_3, $z = f(x,y)$ becomes $z = F(y) = y - y^2 + y = 2y - y^2$, $-1 \le y \le 1$. For critical points, we solve $0 = F'(y) = 2 - 2y$. At the critical point $y = 1$, $F(1) = \boxed{1}$.

On C_4, $z = f(x,y)$ is $z = F(x) = -x^2 + x - 1$, $-1 \le x \le 1$. For critical points we solve $0 = F'(x) = -2x + 1$. At the critical point $x = 1/2$, $F(1/2) = \boxed{-3/4}$.

Finally, we evaluate $f(x,y)$ at the remaining two corners, $f(1,1) = \boxed{3}$ and $f(-1,-1) = \boxed{-3}$. Maximum and minimum values of $f(x,y)$ on R are therefore 3 and -3.

6. For critical points of $f(x,y)$, we solve
$$0 = \frac{\partial f}{\partial x} = 3x^2 - 3, \quad 0 = \frac{\partial f}{\partial y} = 2y + 2.$$

Solutions are $(\pm 1, -1)$ both of which are outside the region. On the edge $x = 0$,

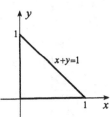

$$f(0,y) = F(y) = y^2 + 2y, \quad 0 \le y \le 1.$$

It has a critical point when $0 = 2y + 2 \implies y = -1$, which we reject.

On $y = 0$, $f(x,0) = G(x) = x^3 - 3x$, $0 \le x \le 1$, which has critical points when $0 = 3x^2 - 3 \implies x = \pm 1$. The value of $G(x)$ at $x = 1$ is $G(1) = \boxed{-2}$.

On $x + y = 1$, $f(x,y) = H(x) = x^3 - 3x + (1-x)^2 + 2(1-x) = x^3 + x^2 - 7x + 3$, $0 \le x \le 1$. It has critical points when $0 = H'(x) = 3x^2 + 2x - 7$. Neither of the solutions $x = (-1 \pm \sqrt{22})/3$ lie in the interval $0 \le x \le 1$.

We have evaluated $f(x, y)$ at vertex $(1, 0)$ of the triangle. Its values at the remaining two vertices are $f(0, 1) = \boxed{3}$ and $f(0, 0) = \boxed{0}$. Maximum and minimum values are therefore 3 and -2.

8. For critical points of $f(x, y)$, we solve

$$0 = \frac{\partial f}{\partial x} = 3x^2 - 3, \quad 0 = \frac{\partial f}{\partial y} = 3y^2 - 3.$$

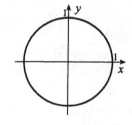

The solutions $(\pm 1, 1)$ and $(\pm 1, -1)$ are exterior to the circle. On the edge of the circle we set

$$x = \cos t, \ y = \sin t, \ 0 \le t \le 2\pi,$$

in which case

$$f(x, y) = F(t) = \cos^3 t + \sin^3 t - 3\cos t - 3\sin t + 2, \quad 0 \le t \le 2\pi.$$

For critical points we solve

$$\begin{aligned}
0 = F'(t) &= -3\cos^2 t \, \sin t + 3\sin^2 t \, \cos t + 3\sin t - 3\cos t \\
&= 3\sin t \, \cos t(\sin t - \cos t) + 3(\sin t - \cos t) \\
&= 3(\sin t \, \cos t + 1)(\sin t - \cos t).
\end{aligned}$$

Setting the factor $\sin t \, \cos t + 1 = 0$ leads to $\sin 2t = -2$, an impossibility. The other possibility is to set $\sin t - \cos t = 0$, which leads to $t = \pi/4$, and $t = 5\pi/4$. Since

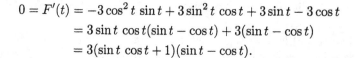

$$F(0) = F(2\pi) = \boxed{0}, \quad F(\pi/4) = \boxed{\frac{2\sqrt{2} - 5}{\sqrt{2}}}, \quad F(5\pi/4) = \boxed{\frac{2\sqrt{2} + 5}{\sqrt{2}}},$$

maximum and minimum values are $(2\sqrt{2} + 5)/\sqrt{2}$ and $(2\sqrt{2} - 5)/\sqrt{2}$.

10. The distance D from O to any point $P(x, y, z)$ on the plane is given by

$$\begin{aligned}
D^2 &= x^2 + y^2 + z^2 \\
&= (6 - y + 2z)^2 + y^2 + z^2.
\end{aligned}$$

For critical points of D^2 we solve

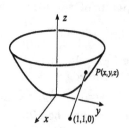

$$0 = \frac{\partial D^2}{\partial y} = -2(6 - y + 2z) + 2y, \quad 0 = \frac{\partial D^2}{\partial z} = 4(6 - y + 2z) + 2z.$$

The only solution is $y = 1$, $z = -2$. Since y and z can take on all possible values, and D^2 becomes infinite for large values of y or z, the critical point must minimize D^2. The closest point is therefore $(1, 1, -2)$.

12. The distance D from $(1, 1, 0)$ to any point $P(x, y, z)$ on the surface is given by

$$\begin{aligned}
D^2 &= (x - 1)^2 + (y - 1)^2 + z^2 \\
&= (x - 1)^2 + (y - 1)^2 + (x^2 + y^2)^2.
\end{aligned}$$

For critical points of D^2, we solve

$$0 = \frac{\partial D^2}{\partial x} = 2(x - 1) + 4x(x^2 + y^2),$$

$$0 = \frac{\partial D^2}{\partial y} = 2(y - 1) + 4y(x^2 + y^2).$$

The only solution is $x = 1/2$, $y = 1/2$. Since x and y can take on all possible values, and D^2 becomes infinite for large values of x and y, the critical point must minimize D^2. The closest point is therefore $(1/2, 1/2, 1/2)$.

14. For critical points of $V(x, y)$, we solve

$$0 = \frac{\partial V}{\partial x} = 48y - 96x^2, \quad 0 = \frac{\partial V}{\partial y} = 48x - 48y.$$

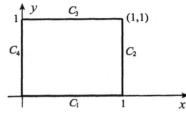

At the critical points $(0, 0)$ and $(1/2, 1/2)$,

$$V(0, 0) = \boxed{0}, \quad V(1/2, 1/2) = \boxed{2}.$$

On C_1, $V = -32x^3$, $0 \le x \le 1$. This function has a critical point at $x = 0$ corresponding to $(0, 0)$.

On C_2, $V = 48y - 32 - 24y^2$, $0 \le y \le 1$. For critical points, $0 = dV/dy = 48 - 48y$. At the critical point $y = 1$, $V = \boxed{-8}$.

On C_3, $V = 48x - 32x^3 - 24$, $0 \le x \le 1$. For critical points, $0 = dV/dx = 48 - 96x^2$. At the critical point $x = 1/\sqrt{2}$, $V = \boxed{8(2\sqrt{2} - 3)}$.

On C_4, $V = -24y^2$, $0 \le y \le 1$. The only critical point of this function is $y = 0$, corresponding to $(0, 0)$.

Finally, at the remaining two corners, $V(1, 0) = \boxed{-32}$ and $V(0, 1) = \boxed{-24}$. Maximum and minimum values of $V(x, y)$ are therefore 2 and -32.

16. If k is the cost per square centimetre for lining the side of the tank, the total cost is

$$C = k(2hw + 2hl) + 3k(wl).$$

Because the tank must hold 1000 L,

$$10^6 = lwh,$$

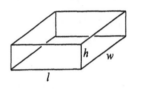

and therefore

$$C = C(w, h) = k\left[2hw + 2h\left(\frac{10^6}{wh}\right) + 3w\left(\frac{10^6}{wh}\right)\right] = k\left[2hw + \frac{2 \times 10^6}{w} + \frac{3 \times 10^6}{h}\right].$$

This function must be minimized for all points in the first quadrant of the wh-plane. For critical points of $C(w, h)$,

$$0 = \frac{\partial C}{\partial w} = k\left[2h - \frac{2 \times 10^6}{w^2}\right], \qquad 0 = \frac{\partial C}{\partial h} = k\left[2w - \frac{3 \times 10^6}{h^2}\right].$$

The only critical point is $w = 100(2/3)^{1/3}$, $h = 100(3/2)^{2/3}$. Since C becomes infinite as $h \to 0$ or $w \to 0$, or h or w become infinite, it follows that the critical point must minimize C. The required dimensions are therefore $w = 100(2/3)^{1/3}$ cm, $h = 100(3/2)^{2/3}$ cm, $l = 100(2/3)^{1/3}$ cm.

18. If $A(x_1, y_1, z_1)$, $B(x_2, y_2, z_2)$, and $C(x_3, y_3, z_3)$ are vertices of a triangle, and $P(x, y, z)$ is any other point, then the sum of the squares of the distances from P to A, B, and C is

$$\begin{aligned} D &= \|PA\|^2 + \|PB\|^2 + \|PC\|^2 \\ &= (x - x_1)^2 + (y - y_1)^2 + (z - z_1)^2 + (x - x_2)^2 + (y - y_2)^2 + (z - z_2)^2 \\ &\quad + (x - x_3)^2 + (y - y_3)^2 + (z - z_3)^2. \end{aligned}$$

For critical points of this function, we solve

$$0 = \frac{\partial D}{\partial x} = 2(x - x_1) + 2(x - x_2) + 2(x - x_3),$$

$$0 = \frac{\partial D}{\partial y} = 2(y - y_1) + 2(y - y_2) + 2(y - y_3),$$

$$0 = \frac{\partial D}{\partial z} = 2(z - z_1) + 2(z - z_2) + 2(z - z_3).$$

The solution is $x = \frac{1}{3}(x_1 + x_2 + x_3)$, $y = \frac{1}{3}(y_1 + y_2 + y_3)$, $z = \frac{1}{3}(z_1 + z_2 + z_3)$, the centroid of the triangle (see Exercise 43 in Section 7.7). Since D becomes infinite as the point (x, y) moves farther and farther away from the triangle, it follows that this one, and only one, critical point must minimize D.

20. The volume obtained from a point $P(x, y, z)$ on that part of the ellipsoid in the first octant is

$$V = 8xyz = 8cxy\sqrt{1 - x^2/a^2 - y^2/b^2}.$$

This function must be maximized in the first quadrant portion R of the ellipse $x^2/a^2 + y^2/b^2 = 1$. For critical points of V, we solve

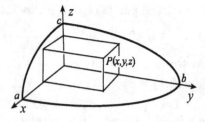

$$0 = \frac{\partial V}{\partial x} = 8cy\sqrt{1 - x^2/a^2 - y^2/b^2}$$
$$- \frac{8cx^2y/a^2}{\sqrt{1 - x^2/a^2 - y^2/b^2}},$$

$$0 = \frac{\partial V}{\partial y} = 8cx\sqrt{1 - x^2/a^2 - y^2/b^2} - \frac{8cxy^2/b^2}{\sqrt{1 - x^2/a^2 - y^2/b^2}}.$$

The only solution of these equations inside R is $x = a/\sqrt{3}$, $y = b/\sqrt{3}$. Since $V = 0$ on the three parts of the boundary of R, it follows that V must be maximized at this critical point, and therefore the dimensions of the largest box are $2a/\sqrt{3} \times 2b/\sqrt{3} \times 2c/\sqrt{3}$.

22. We write $f(x, y) = F(x) = x^2 - (1 - x^2) = 2x^2 - 1$, $-1 \le x \le 1$. For critical points of $F(x)$, we solve $0 = F'(x) = 4x \implies x = 0$. Since $F(-1) = 1$, $F(0) = -1$, and $F(1) = 1$, maximum and minimum values are ± 1.

24. Since $y^2 = (1 - |x|)^2$ on the edges of the square, we can express $f(x, y)$ in terms of x alone,

$$f(x, y) = F(x) = x^2 - (1 - |x|)^2$$
$$= 2|x| - 1, \quad -1 \le x \le 1.$$

There are no points at which the derivative of this function vanishes, but it does not exist at the critical point $x = 0$. Since $F(-1) = 1$, $F(0) = -1$, and $F(1) = 1$, maximum and minimum values are ± 1.

26. We may write

$$A^2 = \frac{P}{2}\left(\frac{P}{2} - x\right)\left(\frac{P}{2} - y\right)\left(\frac{P}{2} + x + y - P\right) = \frac{P}{2}\left(\frac{P}{2} - x\right)\left(\frac{P}{2} - y\right)\left(x + y - \frac{P}{2}\right).$$

This function must be maximized for those points (x, y) in the triangle R shown to the right. For critical points of A^2,

$$0 = \frac{\partial(A^2)}{\partial x} = -\frac{P}{2}\left(\frac{P}{2} - y\right)\left(x + y - \frac{P}{2}\right) + \frac{P}{2}\left(\frac{P}{2} - x\right)\left(\frac{P}{2} - y\right),$$

$$0 = \frac{\partial(A^2)}{\partial y} = -\frac{P}{2}\left(\frac{P}{2} - x\right)\left(x + y - \frac{P}{2}\right) + \frac{P}{2}\left(\frac{P}{2} - x\right)\left(\frac{P}{2} - y\right).$$

The only solution of these equations inside R is $x = y = P/3$. Since $A^2 = 0$ on the boundary of R, it follows that A is maximized when $x = y = P/3$, in which case $z = P/3$ also.

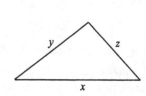

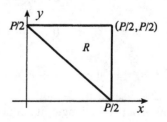

28. Because the function is 2π-periodic in x and y, we need only show that the inequality is valid in the square $R : 0 \leq x \leq 2\pi,\ 0 \leq y \leq 2\pi$. We shall show that maximum and minimum values of the function $f(x, y) = \cos x + \cos y + \sin x \sin y$ in R are ± 2. Critical points of $f(x, y)$ are given by

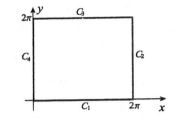

$$0 = \frac{\partial f}{\partial x} = -\sin x + \cos x \sin y, \qquad 0 = \frac{\partial f}{\partial y} = -\sin y + \sin x \cos y.$$

If we solve the first for $\sin x$ and substitute into the second,

$$0 = -\sin y + (\cos x \sin y)\cos y = \sin y(\cos x \cos y - 1).$$

For $\sin y$ to vanish, y must be 0, π, or 2π. The first and third of these are boundaries of R which will be treated later. If $y = \pi$, then $\sin x = 0$, from which x must be 0, π, or 2π. Again, the first and third of these are boundaries of R. In other words, we obtain a critical point of $f(x, y)$ in R to be (π, π) at which $f(\pi, \pi) = -2$. The other possibility for critical points interior to R is to set $\cos x \cos y = 1$. This can be true only if both x and y are equal to 1 or -1. This implies that both x and y must be equal to 0, π, or 2π, and we are led to the same critical point (π, π) inside R. On the boundary C_1, $f(x, y) = 1 + \cos x$, $0 \leq x \leq 2\pi$. The maximum and minimum values are 2 and 0. The same results are obtained on the remaining three boundaries C_2, C_3, and C_4. Thus, maximum and minimum values of $f(x, y)$ on the square are ± 2, and our proof is complete.

30. If we set $P(x, y) = F(y) = kx^\alpha y^{1-\alpha} = k\left(\dfrac{C - By}{A}\right)^\alpha y^{1-\alpha}$, $0 \leq y \leq C/B$, then the minimum value of the function occurs at the end points. The maximum must occur at a critical point. For critical points we solve

$$0 = F'(y) = k\left[\left(\frac{C - By}{A}\right)^\alpha (1 - \alpha)y^{-\alpha} - \frac{\alpha B}{A}\left(\frac{C - By}{A}\right)^{\alpha - 1} y^{1-\alpha}\right]$$

$$= k\left(\frac{C - By}{A}\right)^{\alpha - 1} y^{-\alpha}\left[\left(\frac{C - By}{A}\right)(1 - \alpha) - \frac{\alpha B}{A}y\right].$$

Ths solution is $y = C(1 - \alpha)/B = C\beta/B$. The corresponding x-value is $x = C\alpha/A$.

32. For critical points of $f(x, y, z)$ inside the sphere we solve

$$0 = \frac{\partial f}{\partial x} = y + z, \quad 0 = \frac{\partial f}{\partial y} = x, \quad 0 = \frac{\partial f}{\partial z} = x.$$

For the line of critical points $(0, y, -y)$, $f(0, y, -y) = \boxed{0}$. On the boundary $S : x^2 + y^2 + z^2 = 1$ of the region,

$$f(x, y, z) = F(y, z) = \pm(y + z)\sqrt{1 - y^2 - z^2}, \quad y^2 + z^2 \leq 1.$$

For critical points of $F(y, z)$ inside $y^2 + z^2 = 1$, we solve

$$0 = \frac{\partial F}{\partial y} = \pm\sqrt{1 - y^2 - z^2} \mp \frac{y(y + z)}{\sqrt{1 - y^2 - z^2}},$$

$$0 = \frac{\partial F}{\partial z} = \pm\sqrt{1 - y^2 - z^2} \mp \frac{z(y + z)}{\sqrt{1 - y^2 - z^2}}.$$

Solutions are $(y, z) = (\pm 1/2, \pm 1/2)$ at which $F(\pm 1/2, \pm 1/2) = \boxed{\pm 1/\sqrt{2}}$. On the boundary $C : y^2 + z^2 = 1$ of S, $f(x, y, z) = \boxed{0}$. Consequently, maximum and minimum values of $f(x, y, z)$ are $\pm 1/\sqrt{2}$.

EXERCISES 12.12

2. The constraint $x^2 + 2y^2 + 4z^2 = 9$ defines a **closed** surface (an ellipsoid). We define the Lagrangian

$$L(x, y, z, \lambda) = 5x - 2y + 3z + 4 + \lambda(x^2 + 2y^2 + 4z^2 - 9).$$

For critical points of L, we solve

$$0 = \frac{\partial L}{\partial x} = 5 + 2\lambda x, \quad 0 = \frac{\partial L}{\partial y} = -2 + 4\lambda y, \quad 0 = \frac{\partial L}{\partial z} = 3 + 8\lambda z, \quad 0 = \frac{\partial L}{\partial \lambda} = x^2 + 2y^2 + 4z^2 - 9.$$

Critical points (x, y, z) are $(\pm 10/\sqrt{13}, \mp 2/\sqrt{13}, \pm 3/(2\sqrt{13}))$. Since $f(\pm 10/\sqrt{13}, \mp 2/\sqrt{13}, \pm 3/(2\sqrt{13})) = (8 \pm 9\sqrt{13})/2$, these are the maximum and minimum values of $f(x, y, z)$.

4. The constraint $x^2 + y^2 + z^2 = 9$ is a **closed** surface (a sphere). We define the Lagrangian

$$L(x, y, z, \lambda) = x^3 + y^3 + z^3 + \lambda(x^2 + y^2 + z^2 - 9).$$

For critical points of L, we solve

$$0 = \frac{\partial L}{\partial x} = 3x^2 + 2\lambda x, \quad 0 = \frac{\partial L}{\partial y} = 3y^2 + 2\lambda y, \quad 0 = \frac{\partial L}{\partial z} = 3z^2 + 2\lambda z, \quad 0 = \frac{\partial L}{\partial \lambda} = x^2 + y^2 + z^2 - 9.$$

Critical points (x, y, z) are $(\pm 3, 0, 0)$, $(0, \pm 3, 0)$, $(0, 0, \pm 3)$, $(0, \pm 3/\sqrt{2}, \pm 3/\sqrt{2})$, $(\pm 3/\sqrt{2}, 0, \pm 3/\sqrt{2})$, $(\pm 3/\sqrt{2}, \pm 3/\sqrt{2}, 0)$, $(\pm\sqrt{3}, \pm\sqrt{3}, \pm\sqrt{3})$. Since $f(x, y, z) = \pm 27$ at the first six critical points, $f(x, y, z) = \pm 27/\sqrt{2}$ at the second set of six critical points, and $f(\pm\sqrt{3}, \pm\sqrt{3}, \pm\sqrt{3}) = \pm 9\sqrt{3}$, maximum and minimum values of $f(x, y, z)$ are ± 27.

6. The constraints $x^2 + y^2 = 1$, $z = y$ define a **closed** curve. We define the Lagrangian

$$L(x, y, z, \lambda, \mu) = x^2 y + z + \lambda(x^2 + y^2 - 1) + \mu(z - y).$$

For critical points of L, we solve

$$0 = \frac{\partial L}{\partial x} = 2xy + 2\lambda x, \quad 0 = \frac{\partial L}{\partial y} = x^2 + 2\lambda y - \mu, \quad 0 = \frac{\partial L}{\partial z} = 1 + \mu,$$

$$0 = \frac{\partial L}{\partial \lambda} = x^2 + y^2 - 1, \quad 0 = \frac{\partial L}{\partial \mu} = z - y.$$

Solutions (x, y, z) of these equations are $(0, \pm 1, \pm 1)$, $(1/\sqrt{3}, \pm\sqrt{2/3}, \pm\sqrt{2/3})$, and $(-1/\sqrt{3}, \pm\sqrt{2/3}, \pm\sqrt{2/3})$. Since $f(0, \pm 1, \pm 1) = \pm 1$, $f(1/\sqrt{3}, \pm\sqrt{2/3}, \pm\sqrt{2/3}) = \pm\sqrt{32/27}$, and $f(-1/\sqrt{3}, \pm\sqrt{2/3}, \pm\sqrt{2/3}) = \pm\sqrt{32/27}$, maximum and minimum values are $\pm\sqrt{32/27}$.

8. The constraints $x^2 + y^2 = 1$, $z = \sqrt{x^2 + y^2}$ define the **closed** curve $x^2 + y^2 = 1$, $z = 1$, so that we write alternatively, $f(x, y, z) = F(x, y) = xy - x^2$ subject to $x^2 + y^2 = 1$. We define the Lagrangian $L(x, y, \lambda) = xy - x^2 + \lambda(x^2 + y^2 - 1)$. For critical points of L, we solve

$$0 = \frac{\partial L}{\partial x} = y - 2x + 2\lambda x, \quad 0 = \frac{\partial L}{\partial y} = x + 2\lambda y, \quad 0 = \frac{\partial L}{\partial \lambda} = x^2 + y^2 - 1.$$

Solutions (x, y) of these equations are $\left(\dfrac{\pm\sqrt{2 - \sqrt{2}}}{2}, \dfrac{\pm\sqrt{2 + \sqrt{2}}}{2}\right)$, $\left(\dfrac{\pm\sqrt{2 + \sqrt{2}}}{2}, \dfrac{\mp\sqrt{2 - \sqrt{2}}}{2}\right)$.
Since values of $F(x, y)$ at these points are $(-1 \pm \sqrt{2})/2$, maximum and minimum values are $(\sqrt{2} - 1)/2$ and $-(\sqrt{2} + 1)/2$.

10. The distance D from $(1, 1, 0)$ to any point $P(x, y, z)$ is given by $D^2 = (x - 1)^2 + (y - 1)^2 + z^2$. This function must be minimized subject to the constraint $z = x^2 + y^2$. If we define the Lagrangian $L(x, y, z, \lambda) = (x - 1)^2 + (y - 1)^2 + z^2 + \lambda(x^2 + y^2 - z)$, its critical points are given by

$$0 = \frac{\partial L}{\partial x} = 2(x - 1) + 2\lambda x, \quad 0 = \frac{\partial L}{\partial y} = 2(y - 1) + 2\lambda y, \quad 0 = \frac{\partial L}{\partial z} = 2z - \lambda, \quad 0 = \frac{\partial L}{\partial \lambda} = x^2 + y^2 - z.$$

The solution for (x, y, z) is $(1/2, 1/2, 1/2)$. Since D^2 becomes infinite as x, y, and z become infinite, it follows that this point must minimize D^2 (or D).

12. Since A is maximized when A^2 is maximized, we define the Lagrangian

$$L(x, y, z, \lambda) = \frac{P}{2}\left(\frac{P}{2} - x\right)\left(\frac{P}{2} - y\right)\left(\frac{P}{2} - z\right) + \lambda(x + y + z - P).$$

Critical points of L are given by

$$0 = \frac{\partial L}{\partial x} = -\frac{P}{2}\left(\frac{P}{2} - y\right)\left(\frac{P}{2} - z\right) + \lambda, \quad 0 = \frac{\partial L}{\partial y} = -\frac{P}{2}\left(\frac{P}{2} - x\right)\left(\frac{P}{2} - z\right) + \lambda,$$

$$0 = \frac{\partial L}{\partial z} = -\frac{P}{2}\left(\frac{P}{2} - x\right)\left(\frac{P}{2} - y\right) + \lambda, \quad 0 = \frac{\partial L}{\partial \lambda} = x + y + z - P.$$

The only solution of these equations is $x = y = z = P/3$; or, any two of x, y, and z equal to $P/2$ and the third equal to zero. In the latter case, the triangle has degenerated to a straight line with $A = 0$. Since this case represents the bounds for possible values of x, y, and z, it follows that $x = y = z = P/3$ must maximize A.

14. The volume of the silo is

$$V = \pi(6)^2 H + \frac{1}{3}\pi(6)^2 h = 12\pi(h + 3H).$$

Since the area of the silo must be 200 m^2,

$$200 = 2\pi(6)H + \pi(6)\sqrt{36 + h^2}.$$

We define the Lagrangian

$$L(h, H, \lambda) = 12\pi(h + 3H) + \lambda(12\pi H + 6\pi\sqrt{36 + h^2} - 200).$$

Its critical points are given by

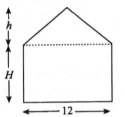

$$0 = \frac{\partial L}{\partial h} = 12\pi + \frac{6\pi\lambda h}{\sqrt{36 + h^2}}, \quad 0 = \frac{\partial L}{\partial H} = 36\pi + 12\pi\lambda, \quad 0 = \frac{\partial L}{\partial \lambda} = 12\pi H + 6\pi\sqrt{36 + h^2} - 200.$$

The solution of these equations for h and H is $h = 12/\sqrt{5}$ and $H = (50\sqrt{5} - 27\pi)/(3\sqrt{5}\pi)$. Clearly $h \geq 0$ and the constraint requires $h \leq \sqrt{4 \times 10^4 - 36^2\pi^2}/(6\pi)$. Since

$$V_{|h=0} = 260.7, \quad V_{|h=12/\sqrt{5}} = 347.1, \quad V_{|h=\sqrt{4\times10^4-36^2\pi^2}/(6\pi)} = 329.9,$$

it follows that V is maximized for $h = 12/\sqrt{5}$ m and $H = (50\sqrt{5} - 27\pi)/(3\sqrt{5}\pi)$ m.

16. We define the Lagrangian $L(x, y, \lambda) = |x - y| + \lambda(x^2 + y^2 - 1)$. For critical points of L, we solve

$$0 = \frac{\partial L}{\partial x} = \frac{|x - y|}{x - y} + 2\lambda x, \quad 0 = \frac{\partial L}{\partial y} = -\frac{|x - y|}{x - y} + 2\lambda y, \quad 0 = \frac{\partial L}{\partial \lambda} = x^2 + y^2 - 1.$$

Critical points (x, y) are $(\pm 1/\sqrt{2}, \mp 1/\sqrt{2})$. The derivatives do not exist when $y = x$ and this leads to the additional critical points $(\pm 1/\sqrt{2}, \pm 1/\sqrt{2})$. Since $f(\pm 1/\sqrt{2}, \pm 1/\sqrt{2}) = 0$ and $f(\pm 1/\sqrt{2}, \mp 1/\sqrt{2}) = \sqrt{2}$, maximum and minimum values of $f(x, y)$ are $\sqrt{2}$ and 0.

18. We define the Lagrangian $L(x, y, \lambda) = |x - 2y| + \lambda(|x| + |y| - 1)$. For critical points of L, we solve

$$0 = \frac{\partial L}{\partial x} = \frac{|x - 2y|}{x - 2y} + \frac{\lambda|x|}{x}, \quad 0 = \frac{\partial L}{\partial y} = -2\frac{|x - 2y|}{x - 2y} + \frac{\lambda|y|}{y}, \quad 0 = \frac{\partial L}{\partial \lambda} = |x| + |y| - 1.$$

There are no solutions of these equations. Since the partial derivative with respect to x fails to exist at $x = 0$ and when $x = 2y$, and the derivative with respect to y does not exist at $y = 0$ or when $x = 2y$, critical points are $(0, \pm 1)$, $(\pm 1, 0)$, and $(\pm 2/3, \pm 1/3)$. Since $f(\pm 1, 0) = 1$, $f(0, \pm 1) = 2$, and $f(\pm 2/3, \pm 1/3) = 0$, maximum and minimum values of $f(x, y)$ are 2 and 0.

20. The distance D between points $P(x, y)$ on $F(x, y) = 0$ and $Q(X, Y)$ on $G(x, y) = 0$ is given by $D^2 = (x - X)^2 + (y - Y)^2$, where $F(x, y) = 0$ and $G(X, Y) = 0$. We define the Lagrangian

$$L(x, y, X, Y, \lambda, \mu) = (x - X)^2 + (y - Y)^2 + \lambda F(x, y) + \mu G(X, Y).$$

For critical points of this function,

$$0 = \frac{\partial L}{\partial x} = 2(x - X) + \lambda F_x, \quad 0 = \frac{\partial L}{\partial y} = 2(y - Y) + \lambda F_y, \quad 0 = \frac{\partial L}{\partial X} = -2(x - X) + \mu G_X,$$

$$0 = \frac{\partial L}{\partial Y} = -2(y - Y) + \mu G_Y, \quad 0 = \frac{\partial L}{\partial \lambda} = F(x, y), \quad 0 = \frac{\partial L}{\partial \mu} = G(X, Y).$$

If $P(x_0, y_0)$ and $Q(X_0, Y_0)$ are the points that minimize D^2, then they must satisfy these equations. In particular, the first four give

$$0 = 2(x_0 - X_0) + \lambda F_x(x_0, y_0), \quad 0 = 2(y_0 - Y_0) + \lambda F_y(x_0, y_0),$$

$$0 = -2(x_0 - X_0) + \mu G_X(X_0, Y_0), \quad 0 = -2(y_0 - Y_0) + \mu G_Y(X_0, Y_0).$$

From the first two equations, we obtain $\dfrac{y_0 - Y_0}{x_0 - X_0} = \dfrac{F_y(x_0, y_0)}{F_x(x_0, y_0)}$. But $(y_0 - Y_0)/(x_0 - X_0)$ is the slope of the line joining $P(x_0, y_0)$ and $Q(X_0, Y_0)$, and the slope of the tangent line to $F(x, y) = 0$ at (x_0, y_0) is $-F_x(x_0, y_0)/F_y(x_0, y_0)$. Hence, these lines are perpendicular. The last two equations indicate that PQ is perpendicular to the tangent line to $G(x, y) = 0$ at Q.

22. The volume of a right circular cylinder is $V = \pi r^2 h$, and were there no constraints on r and h, this function would be considered for all points in the first quadrant of the rh-plane. However, r and h must satisfy a constraint that geometrically can be interpreted as a curve in the rh-plane. What we must do then is minimize $V = \pi r^2 h$, considering only those points (r, h) on the curve defined by the constraint. Clearly there is only one independent variable in the problem—either r or h, but not both. If we choose r as the independent variable, then we note from the constraint that as h becomes very large, r approaches $2.4048/\sqrt{k}$. Since there is no upper bound on r, we can state that the values of r to be considered in the minimization of V are $r > 2.4048/\sqrt{k}$.

To find critical points of V we introduce the Lagrangian

$$L(r, h, \lambda) = \pi r^2 h + \lambda \left[\left(\frac{2.4048}{r} \right)^2 + \left(\frac{\pi}{h} \right)^2 - k \right],$$

and first solve the equations

$$0 = \frac{\partial L}{\partial r} = 2\pi r h + \lambda \left[\frac{-2(2.4048)^2}{r^3} \right],$$

$$0 = \frac{\partial L}{\partial h} = \pi r^2 + \lambda \left(\frac{-2\pi^2}{h^3} \right),$$

$$0 = \frac{\partial L}{\partial \lambda} = \left(\frac{2.4048}{r} \right)^2 + \left(\frac{\pi}{h} \right)^2 - k.$$

If we solve each of the first two equations for λ and equate the resulting expressions, we have

$$\frac{\pi r^4 h}{2.4048^2} = \frac{r^2 h^3}{2\pi}.$$

Since neither r nor h can be zero, we divide by $r^2 h$:

$$\frac{\pi r^2}{2.4048^2} = \frac{h^2}{2\pi} \quad \Longrightarrow \quad r = \frac{2.4048 h}{\sqrt{2\pi}}.$$

Substitution of this result into the constraint equation gives

$$\left(\frac{\sqrt{2}\pi}{h}\right)^2 + \left(\frac{\pi}{h}\right)^2 = k,$$

and this equation can be solved for $h = \pi\sqrt{3/k}$. This gives

$$r = \frac{2.4048}{\sqrt{2}\pi} \frac{\pi\sqrt{3}}{\sqrt{k}} = 2.4048\sqrt{3/(2k)}.$$

We have obtained therefore only one critical point (r, h) at which the derivatives of L vanish. The only values of r and h at which the derivatives of L do not exist are $r = 0$ and $h = 0$, but these must be rejected since the constraint requires both r and h to be positive.

To finish the problem we note that

$$\lim_{r\to\infty} V = \infty, \qquad \lim_{r\to2.4048/\sqrt{k}^+} V = \lim_{h\to\infty} V = \infty.$$

It follows, therefore, that the single critical point at which $r = 2.4048\sqrt{3/(2k)}$ and $h = \pi\sqrt{3/k}$ must give the absolute minimum value of $V(r, h)$.

24. For critical points of $f(x, y)$, we solve $\quad 0 = \dfrac{\partial f}{\partial x} = 6x + 2y$, $0 = \dfrac{\partial f}{\partial y} = 2x - 2y$. At the only solution $(0, 0)$, $f(0, 0) = 5$. On the **closed** boundary $4x^2 + 9y^2 = 36$ of the region, we define the Lagrangian $L(x, y, \lambda) = 3x^2 + 2xy - y^2 + 5 + \lambda(4x^2 + 9y^2 - 36)$. For its critical points,

$$0 = \frac{\partial L}{\partial x} = 6x + 2y + 8\lambda x, \quad 0 = \frac{\partial L}{\partial y} = 2x - 2y + 18\lambda y, \quad 0 = \frac{\partial L}{\partial \lambda} = 4x^2 + 9y^2 - 36.$$

The solutions (x, y) of these equations are $(\pm 0.55086, \mp 1.96600)$ and $(\pm 2.94899, \pm 0.36724)$. Since $f(\pm 0.55086, \mp 1.96600) = -0.12$ and $f(\pm 2.94899, \pm 0.36724) = 33.12$, maximum and minimum values of $f(x, y)$ are 33.12 and -0.12.

26. For critical points of $f(x, y, z)$ we solve $\quad 0 = \dfrac{\partial f}{\partial x} = y + z$, $0 = \dfrac{\partial f}{\partial y} = x$, $0 = \dfrac{\partial f}{\partial z} = x$. For the line of critical points $(0, y, -y)$, we evaluate $f(0, y, -y) = \boxed{0}$. On the **closed** boundary $x^2 + y^2 + z^2 = 1$, we define the Lagrangian $L(x, y, z, \lambda) = xy + xz + \lambda(x^2 + y^2 + z^2 - 1)$. For critical points of L,

$$0 = \frac{\partial L}{\partial x} = y + z + 2\lambda x, \quad 0 = \frac{\partial L}{\partial y} = x + 2\lambda y, \quad 0 = \frac{\partial L}{\partial z} = x + 2\lambda z, \quad 0 = \frac{\partial L}{\partial \lambda} = x^2 + y^2 + z^2 - 1.$$

The solutions of these equations for (x, y, z) are $(0, \pm 1/\sqrt{2}, \mp 1/\sqrt{2})$, $(1/\sqrt{2}, \pm 1/2, \pm 1/2)$, and $(-1/\sqrt{2}, \pm 1/2, \pm 1/2)$. Since $f(0, \pm 1/\sqrt{2}, \mp 1/\sqrt{2}) = \boxed{0}$, $f(1/\sqrt{2}, \pm 1/2, \pm 1/2) = \boxed{\pm 1/\sqrt{2}}$, and $f(-1/\sqrt{2}, \pm 1/2, \pm 1/2) = \boxed{\mp 1/\sqrt{2}}$, maximum and minimum values of $f(x, y, z)$ are $\pm 1/\sqrt{2}$.

28. The distance D from the origin to any point (x, y) on the ellipse is given by $D^2 = x^2 + y^2$, subject to $3x^2 + 4xy + 6y^2 = 140$. Ends of the major and minor axes maximize and minimize this function. To find these points we define the Lagrangian $L(x, y, \lambda) = x^2 + y^2 + \lambda(3x^2 + 4xy + 6y^2 - 140)$. Critical points are given by

$$0 = \frac{\partial L}{\partial x} = 2x + \lambda(6x + 4y), \quad 0 = \frac{\partial L}{\partial y} = 2y + \lambda(4x + 12y), \quad 0 = \frac{\partial L}{\partial \lambda} = 3x^2 + 4xy + 6y^2 - 140.$$

When the first two are solved for λ and the expressions equated, $\dfrac{-x}{3x + 2y} = \dfrac{-y}{2x + 6y}$, which simplifies to $0 = 2x^2 + 3xy - 2y^2 = (2x - y)(x + 2y)$. Thus, $y = 2x$ or $x = -2y$. These lead to the four points $(\pm 2, \pm 4)$ and $(\pm 2\sqrt{14}, \mp \sqrt{14})$. Since $D^2(\pm 2, \pm 4) = 20$ and $D^2(\pm 2\sqrt{14} \mp \sqrt{14}) = 70$, the ends of the major axis are $(\pm 2\sqrt{14}, \mp \sqrt{14})$, and the ends of the minor axis are $(\pm 2, \pm 4)$.

30. (a) The distance D from the origin to any point (x, y) on the folium is given by

$$D^2(t) = x^2 + y^2 = \frac{9a^2t^2}{(1+t^3)^2} + \frac{9a^2t^4}{(1+t^3)^2} = \frac{9a^2(t^2+t^4)}{(1+t^3)^2}, \quad 0 \le t < \infty.$$

For critical points we solve

$$\begin{aligned}
0 = \frac{dD^2}{dt} &= 9a^2 \left[\frac{(1+t^3)^2(2t+4t^3) - (t^2+t^4)2(1+t^3)(3t^2)}{(1+t^3)^4} \right] \\
&= \frac{9a^2}{(1+t^3)^3}(2t + 4t^3 + 2t^4 + 4t^6 - 6t^4 - 6t^6) \\
&= \frac{-18a^2t(t-1)(t^4+t^3+3t^2+t+1)}{(1+t^3)^3}.
\end{aligned}$$

Since the quartic polynomial has no positive solutions, the only critical points are $t = 0$ and $t = 1$. Since

$$D^2(0) = 0, \quad D^2(1) = \frac{9a^2}{2}, \quad \lim_{t \to \infty} D^2 = 0,$$

it follows that distance is maximized at the point $(3a/2, 3a/2)$.

(b) An implicit definition of the curve is

$$\begin{aligned}
x^3 + y^3 &= \frac{27a^3t^3}{(1+t^3)^3} + \frac{27a^3t^6}{(1+t^3)^3} = \frac{27a^3t^3(1+t^3)}{(1+t^3)^3} \\
&= \frac{27a^3t^3}{(1+t^3)^2} = 3a \left(\frac{3at}{1+t^3} \right) \left(\frac{3at^2}{1+t^3} \right) = 3axy.
\end{aligned}$$

To maximize $D^2 = x^2 + y^2$ subject to $x^3 + y^3 = 3axy$, we define the Lagrangian

$$L(x, y, \lambda) = x^2 + y^2 + \lambda(x^3 + y^3 - 3axy).$$

For critical points of L, we solve

$$0 = \frac{\partial L}{\partial x} = 2x + \lambda(3x^2 - 3ay), \quad 0 = \frac{\partial L}{\partial y} = 2y + \lambda(3y^2 - 3ax), \quad 0 = \frac{\partial L}{\partial \lambda} = x^3 + y^3 - 3axy.$$

When x times the second equation is subtracted from y times the first,

$$\begin{aligned}
0 &= \lambda(3x^2y - 3xy^2 - 3ay^2 + 3ax^2) = 3\lambda[xy(x-y) + a(x-y)(x+y)] \\
&= 3\lambda(x-y)[xy + a(x+y)].
\end{aligned}$$

Thus, $y = x$ or $xy + a(x+y) = 0$. The second equation cannot be satisfied for positive x and y. The only critical point is obtained from $x^3 + x^3 - 3ax^2 = 0$. The solution is $x = 3a/2$. Since the curve is **closed**, the maximum value of D^2 must be at $(3a/2, 3a/2)$.

EXERCISES 12.13

2. Least squares estimates for parameters a and b in a linear function $P = aA + b$ must satisfy equations similar to 12.71. For the tabular values these equations are

$$1482a + 130b = 13\,663, \qquad 130a + 13b = 1315.$$

The solution is $a = 2.8187$ and $b = 72.967$.

4. (a) The plot is to the right.

(b) For critical points of $S(a, b, c)$ we solve

$$0 = \frac{\partial S}{\partial a} = \sum_{i=1}^{16} 2(ax_i{}^2 + bx_i + c - \overline{y_i})(x_i{}^2),$$

$$0 = \frac{\partial S}{\partial b} = \sum_{i=1}^{16} 2(ax_i{}^2 + bx_i + c - \overline{y_i})(x_i),$$

$$0 = \frac{\partial S}{\partial c} = \sum_{i=1}^{16} 2(ax_i{}^2 + bx_i + c - \overline{y_i}).$$

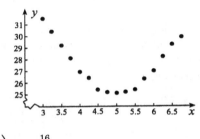

These can be rewritten in the form

$$\left(\sum_{i=1}^{16} x_i{}^4\right) a + \left(\sum_{i=1}^{16} x_i{}^3\right) b + \left(\sum_{i=1}^{16} x_i{}^2\right) c = \sum_{i=1}^{16} x_i{}^2 \overline{y_i},$$

$$\left(\sum_{i=1}^{16} x_i{}^3\right) a + \left(\sum_{i=1}^{16} x_i{}^2\right) b + \left(\sum_{i=1}^{16} x_i\right) c = \sum_{i=1}^{16} x_i \overline{y_i},$$

$$\left(\sum_{i=1}^{16} x_i{}^2\right) a + \left(\sum_{i=1}^{16} x_i\right) b + 16c = \sum_{i=1}^{16} \overline{y_i}.$$

(c) From the tabular values,

$$12\,117.5a + 2164.5b + 401.5c = 10\,979.4, \quad 2164.5a + 401.5b + 78c = 2133.47, \quad 401.5a + 78b + 16c = 439.4.$$

The solution is $a = 1.6653$, $b = -16.642$, and $c = 66.802$.

6. The sum of the squares of the differences between observed and predicted values is

$$S = S(a, b) = \sum_{i=1}^{8} (a + bQ_i^2 - \overline{H}_i)^2,$$

where $(Q_i, \overline{H}_i)$ are the points in the table. For critical points of S, we solve

$$0 = \frac{\partial S}{\partial b} = \sum_{i=1}^{8} 2(a + bQ_i^2 - \overline{H}_i)(Q_i^2), \qquad 0 = \frac{\partial S}{\partial a} = \sum_{i=1}^{8} 2(a + bQ_i^2 - \overline{H}_i).$$

These can rewritten in the form

$$\left(\sum_{i=1}^{8} Q_i^4\right) b + \left(\sum_{i=1}^{8} Q_i^2\right) a = \sum_{i=1}^{n} Q_i^2 \overline{H}_i, \qquad \left(\sum_{i=1}^{8} Q_i^2\right) b + 8a = \sum_{i=1}^{8} \overline{H}_i.$$

From the table, these become

$$2.19185 \times 10^8 b + 34728.8a = 632647, \qquad 34728.8b + 8a = 197.2.$$

The solution is $a = 38.82$ and $b = 0.003265$; that is, the least squares quadratic is $H = 38.82 - 0.003265Q^2$.

8. (a) When we take logarithms of y-values in the given table, we obtain

x	0.5	1.0	1.5	2.0	2.5	3.0	3.5	4.0	4.5	5.0	5.5
y	4.94	5.19	5.44	5.67	5.90	6.12	6.34	6.51	6.67	6.91	7.11

The plot of these to the right indicates
that a straight line fit is indeed reasonable.
(b) Equations for a and B corresponding to
12.71 are

$$\left(\sum_{i=1}^{11} x_i^2\right) a + \left(\sum_{i=1}^{11} x_i\right) B = \sum_{i=1}^{11} x_i \overline{Y}_i,$$

$$\left(\sum_{i=1}^{11} x_i\right) a + (11)b = \sum_{i=1}^{11} \overline{Y}_i,$$

from which

$$126.5a + 33B = 212.149, \quad 33a + 11B = 66.795.$$

The solution of these is $a = 0.43$ and $B = 4.79$. Consequently, the least-squares estimates give

$$\ln y = Y = 4.79 + 0.43x.$$

When we take exponentials,

$$y = e^{4.79 + 0.43x} = 120.3e^{0.43x}.$$

10. To find $y = f(v) = a - bv$, the number of kilometres per litre for a truck travelling at speed v, we use least squares for the 18 data points in the table. Equations for a and b are

$$-\left(\sum_{i=1}^{18} v_i^2\right) b + \left(\sum_{i=1}^{18} v_i\right) a = \sum_{i=1}^{18} v_i \overline{y}_i, \qquad -\left(\sum_{i=1}^{18} v_i\right) b + 18a = \sum_{i=1}^{18} \overline{y}_i.$$

These become

$$-147\,000b + 1620a = 3379, \quad -1620b + 18a = 37.71.$$

The solution is $a = 3.2125$ and $b = 0.0124167$. We can now substitute these into the formula $v = a/(b + \sqrt{bp/w})$ in Exercise 59 of Section 4.7,

$$v = \frac{3.2125}{0.0124167 + \sqrt{\dfrac{0.0124167(0.6)}{20}}} = 101.3 \text{ kilometres per hour.}$$

12. (a) The data points of Y against X are reasonably collinear.
(b) If a line $Y = aX + B$ is to fit the
points in the plot, then a and B must
satisfy equations similar to 12.71 where
B replaces b. They are

$$53.721\,796a + 16.077\,273B = 81.047\,068,$$
$$16.077\,273a + 5B = 24.904\,540.$$

The solution is $a = 0.477\,603\,27$ and $B = 3.445\,196\,4$.

Thus, $Y = \ln y = 0.477\,603\,27X + 3.445\,196\,4 = 0.477\,603\,27 \ln x + 3.445\,196\,4$, or, $y = 31.35x^{0.4776}$.

14. (a) A plot is shown to the right.

(b) If we take logarithms of $N = be^{at}$, $\ln N = at + \ln b$, and define $n = \ln N$ and $B = \ln b$, then, $n = at + B$. Least-squares estimates for a and B are given by

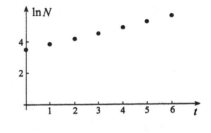

$$\left(\sum_{i=1}^{7} t_i^2\right) a + \left(\sum_{i=1}^{7} t_i\right) B = \sum_{i=1}^{7} t_i \overline{n}_i,$$

$$\left(\sum_{i=1}^{7} t_i\right) a + 7B = \sum_{i=1}^{7} \overline{n}_i.$$

These become

$$91a + 21B = 105.2312, \qquad 21a + 7B = 31.7587,$$

the solution of which is $a = 0.35554$ and $B = 3.47034$. Therefore,

$$n = 0.35554t + 3.47034 \implies \ln N = 0.35554t + 3.47034 \implies N = e^{0.355554t + 3.47034} = 32.1476e^{0.35554t}.$$

16. (a) To find a least-squares estimate for b, we form the sum of squares $S(b) = \sum_{i=1}^{n} \left(\overline{y}_i - \dfrac{b}{x_i^2}\right)^2$. For critical points of $S(b)$, we solve

$$0 = \frac{dS}{db} = \sum_{i=1}^{n} 2\left(\overline{y}_i - \frac{b}{x_i^2}\right)\left(-\frac{1}{x_i^2}\right) = \sum_{i=1}^{n} \frac{b}{x_i^4} - \sum_{i=1}^{n} \frac{\overline{y}_i}{x_i^2}.$$

Thus, $b = \left(\sum_{i=1}^{n} \dfrac{\overline{y}_i}{x_i^2}\right) \Big/ \left(\sum_{i=1}^{n} \dfrac{1}{x_i^4}\right)$.

(b) For the data in the table,

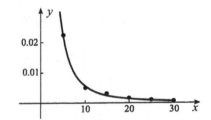

$$b = \frac{9.669\,874 \times 10^{-4}}{1.729\,797\,7 \times 10^{-3}} = 0.559\,017\,65.$$

The least squares approximation for the curve is therefore $y = \dfrac{0.5590}{x^2}$.

(c) The curve and data are plotted to the right.

18. If we set $Y = 1/y = ax + b$, then least-squares estimates for a and b are given by equations similar to 12.71,

$$255a + 35b = 28.285\,127, \quad 35a + 5b = 3.979\,676\,5.$$

The solution is $a = 0.042\,739\,2$ and $b = 0.496\,761\,3$, and therefore

$$Y = \frac{1}{y} = 0.042\,739\,2x + 0.496\,761\,3 \implies y = \frac{1}{0.04274x + 0.4968}.$$

EXERCISES 12.14

2. If we set $z = f(x, y)$, then $dz = \dfrac{\partial f}{\partial x}dx + \dfrac{\partial f}{\partial y}dy = \dfrac{y}{1 + x^2 y^2}dx + \dfrac{x}{1 + x^2 y^2}dy = \dfrac{1}{1 + x^2 y^2}(y\,dx + x\,dy).$

4. If we set $u = f(x, y, z)$, then $du = \dfrac{\partial f}{\partial x}dx + \dfrac{\partial f}{\partial y}dy + \dfrac{\partial f}{\partial z}dz = [yz\cos(xyz) - 2xy^2 z^2]\,dx$

$$+ [xz\cos(xyz) - 2x^2 yz^2]\,dy + [xy\cos(xyz) - 2x^2 y^2 z]\,dz.$$

6. If we set $z = f(x, y)$, then

$$dz = \frac{\partial f}{\partial x}dx + \frac{\partial f}{\partial y}dy = \frac{y}{\sqrt{1 - x^2 y^2}}dx + \frac{x}{\sqrt{1 - x^2 y^2}}dy = \frac{1}{\sqrt{1 - x^2 y^2}}(y\,dx + x\,dy).$$

8. If we set $u = f(x, y, z, t)$, then

$$du = \frac{\partial f}{\partial x}dx + \frac{\partial f}{\partial y}dy + \frac{\partial f}{\partial z}dz + \frac{\partial f}{\partial t}dt = (y + t)\,dx + (x + z)\,dy + (y + t)\,dz + (z + x)\,dt.$$

10. If we set $u = f(x, y, z, t)$, then

$$du = \frac{\partial f}{\partial x}dx + \frac{\partial f}{\partial y}dy + \frac{\partial f}{\partial z}dz + \frac{\partial f}{\partial t}dt$$

$$= 2xe^{x^2+y^2+z^2-t^2}\,dx + 2ye^{x^2+y^2+z^2-t^2}\,dy + 2ze^{x^2+y^2+z^2-t^2}\,dz - 2te^{x^2+y^2+z^2-t^2}\,dt$$

$$= 2e^{x^2+y^2+z^2-t^2}(x\,dx + y\,dy + z\,dz - t\,dt)$$

12. $dV = \dfrac{\partial V}{\partial a}da + \dfrac{\partial V}{\partial b}db = \dfrac{4}{3}\pi b^2\,da + \dfrac{8}{3}\pi ab\,db = \dfrac{4}{3}\pi b(b\,da + 2a\,db)$

Since the percentage changes in a and b are 1%, it follows that $100\dfrac{da}{a} = 100\dfrac{db}{b} = 1$, and therefore the approximate percentage change in V is

$$100\frac{dV}{V} = 100\frac{(4/3)\pi b(b\,da + 2a\,db)}{4\pi ab^2/3} = 100\left(\frac{da}{a} + 2\frac{db}{b}\right) = 1 + 2 = 3.$$

EXERCISES 12.15

2. By setting $v_x = x - c$ and $v_y = y - d$ in the expression for $F'''(0)$,

$$F'''(0) = f_{xxx}(c, d)(x - c)^3 + 3f_{xxy}(c, d)(x - c)^2(y - d) + 3f_{xyy}(c, d)(x - c)(y - d)^2 + f_{yyy}(c, d)(y - d)^3.$$

When this is substituted into $F'''(0)t^3/3!$ and t is set equal to 1, the cubic terms are

$$\frac{1}{3!}[f_{xxx}(c, d)(x - c)^3 + 3f_{xxy}(c, d)(x - c)^2(y - d) + 3f_{xyy}(c, d)(x - c)(y - d)^2 + f_{yyy}(c, d)(y - d)^3].$$

4. $e^{2x-3y} = e^5 e^{2(x-1)-3(y+1)} = e^5 \sum_{n=0}^{\infty} \frac{1}{n!}[2(x - 1) - 3(y + 1)]^n$

$$= \sum_{n=0}^{\infty}\sum_{r=0}^{n} \frac{e^5}{n!}\binom{n}{r}2^r(x - 1)^r(-3)^{n-r}(y + 1)^{n-r} = \sum_{n=0}^{\infty}\sum_{r=0}^{n} \frac{e^5(-1)^{n-r}2^r 3^{n-r}}{(n - r)!r!}(x - 1)^r(y + 1)^{n-r}$$

6. Since $\ln(1 + x) = x - x^2/2 + x^3/3 - x^4/4 + \cdots$,

$$\ln(1 + x^2 + y^2) = x^2 + y^2 - \frac{(x^2 + y^2)^2}{2} + \frac{(x^2 + y^2)^3}{3} - \cdots$$

$$= \sum_{n=1}^{\infty} \frac{(-1)^{n+1}}{n}(x^2 + y^2)^n = \sum_{n=1}^{\infty} \frac{(-1)^{n+1}}{n}\sum_{r=0}^{n}\binom{n}{r}x^{2r}y^{2(n-r)}$$

$$= \sum_{n=1}^{\infty}\sum_{r=0}^{n} \frac{(-1)^{n+1}(n - 1)!}{(n - r)!\,r!}x^{2r}y^{2(n-r)}$$

8. $\dfrac{xy^2}{1 + y^2} = [(x + 1) - 1]\dfrac{y^2}{1 + y^2} = [(x + 1) - 1]\sum_{n=0}^{\infty}(-1)^n y^{2n+2}$

10. With $f(2, 1) = \sqrt{1 + 2} = \sqrt{3}$, $f_x(2, 1) = \dfrac{y}{2\sqrt{1 + xy}}\Big|_{(2,1)} = \dfrac{1}{2\sqrt{3}}$,

$$f_y(2, 1) = \frac{x}{2\sqrt{1 + xy}}\Big|_{(2,1)} = \frac{1}{\sqrt{3}}, \qquad f_{xx}(2, 1) = \frac{-y^2}{4(1 + xy)^{3/2}}\Big|_{(2,1)} = -\frac{1}{12\sqrt{3}},$$

$$f_{xy}(2, 1) = \left[\frac{1}{2\sqrt{1 + xy}} - \frac{xy}{4(1 + xy)^{3/2}}\right]\Big|_{(2,1)} = \frac{1}{3\sqrt{3}}, \qquad f_{yy}(2, 1) = \frac{-x^2}{4(1 + xy)^{3/2}}\Big|_{(2,1)} = -\frac{1}{3\sqrt{3}},$$

$$\sqrt{1+xy} = \sqrt{3} + \frac{1}{2\sqrt{3}}(x-2) + \frac{1}{\sqrt{3}}(y-1)$$

$$+ \frac{1}{2!}\left[-\frac{1}{12\sqrt{3}}(x-2)^2 + \frac{2}{3\sqrt{3}}(x-2)(y-1) - \frac{1}{3\sqrt{3}}(y-1)^2\right] + \cdots$$

$$= \frac{1}{24\sqrt{3}}\left[72 + 12(x-2) + 24(y-1) - (x-2)^2 + 8(x-2)(y-1) - 4(y-1)^2\right] + \cdots .$$

12. With $f(0,1) = 0$,

$$f_x(0,1) = [2(x+y)\ln(x+y) + x + y]_{|(0,1)} = 1,$$
$$f_y(0,1) = [2(x+y)\ln(x+y) + x + y]_{|(0,1)} = 1,$$
$$f_{xx}(0,1) = [2\ln(x+y) + 2 + 1]_{|(0,1)} = 3,$$
$$f_{xy}(0,1) = [2\ln(x+y) + 2 + 1]_{|(0,1)} = 3,$$
$$f_{yy}(0,1) = [2\ln(x+y) + 2 + 1]_{|(0,1)} = 3,$$

$$(x+y)^2\ln(x+y) = x + (y-1) + \frac{1}{2!}[3x^2 + 6x(y-1) + 3(y-1)^2] + \cdots$$

$$= \frac{1}{2}[2x + 2(y-1) + 3x^2 + 6x(y-1) + 3(y-1)^2] + \cdots .$$

14. The first six terms vanish since the function is its own Taylor series about $(0,0)$.

16. If the operator is defined as

$$\left[(x-c)\frac{\partial}{\partial x} + (y-d)\frac{\partial}{\partial y}\right]^n f(c,d) = \sum_{r=0}^{n}\binom{n}{r}(x-c)^{n-r}(y-d)^r\frac{\partial^n f(x,y)}{\partial x^{n-r}\partial y^r}_{|(c,d)},$$

then equation 12.76 becomes

$$f(x,y) = \sum_{n=0}^{\infty}\frac{1}{n!}\left[(x-c)\frac{\partial}{\partial x} + (y-d)\frac{\partial}{\partial y}\right]^n f(c,d).$$

REVIEW EXERCISES

2. From $\dfrac{\partial f}{\partial y} = \dfrac{2y}{x^2+y^2+z^2}$, we obtain $\dfrac{\partial^2 f}{\partial y^2} = \dfrac{(x^2+y^2+z^2)(2) - 2y(2y)}{(x^2+y^2+z^2)^2} = \dfrac{2(x^2-y^2+z^2)}{(x^2+y^2+z^2)^2}.$

4. If we set $F(x,y,z) = z^2 x + \operatorname{Tan}^{-1}z + y - 3x$, then

$$\frac{\partial z}{\partial x} = -\frac{\dfrac{\partial(F)}{\partial(x)}}{\dfrac{\partial(F)}{\partial(z)}} = -\frac{F_x}{F_z} = -\frac{z^2 - 3}{2xz + \dfrac{1}{1+z^2}} = \frac{(3-z^2)(1+z^2)}{1 + 2xz + 2xz^3}.$$

6. $\dfrac{df}{dt} = \dfrac{\partial f}{\partial x}\dfrac{dx}{dt} + \dfrac{\partial f}{\partial y}\dfrac{dy}{dt}$

$$= (2x - ye^{xy})(3t^2 + 3) + (2y - xe^{xy})(\ln t + 1)$$

8. If we set $F(x,y,u,v) = u^2 + v^2 - xy - 5$ and $G(x,u,v) = 3u - 2v + x^2 u - 2v^3$, then

$$\frac{\partial u}{\partial x} = -\frac{\dfrac{\partial(F,G)}{\partial(x,v)}}{\dfrac{\partial(F,G)}{\partial(u,v)}} = -\frac{\begin{vmatrix} F_x & F_v \\ G_x & G_v \end{vmatrix}}{\begin{vmatrix} F_u & F_v \\ G_u & G_v \end{vmatrix}} = -\frac{\begin{vmatrix} -y & 2v \\ 2xu & -2-6v^2 \end{vmatrix}}{\begin{vmatrix} 2u & 2v \\ 3+x^2 & -2-6v^2 \end{vmatrix}}$$

$$= -\frac{y(2+6v^2) - 4xuv}{-2u(2+6v^2) - 2v(3+x^2)} = \frac{y(1+3v^2) - 2xuv}{u(2+6v^2) + v(3+x^2)}.$$

10. $\dfrac{df}{dt} = \dfrac{\partial f}{\partial x}\dfrac{dx}{dt} + \dfrac{\partial f}{\partial y}\dfrac{dy}{dt}$

$\quad\quad = (y - 2x)(te^t + e^t) + (x - 2y)(-te^{-t} + e^{-t})$

$\quad\quad = e^t(t+1)(y - 2x) + e^{-t}(1 - t)(x - 2y)$

12. If we set $F(x, r, \theta) = r\cos\theta - x$ and $G(y, r, \theta) = r\sin\theta - y$, then

$$\frac{\partial r}{\partial x} = -\frac{\dfrac{\partial(F,G)}{\partial(x,\theta)}}{\dfrac{\partial(F,G)}{\partial(r,\theta)}} = -\frac{\begin{vmatrix} F_x & F_\theta \\ G_x & G_\theta \end{vmatrix}}{\begin{vmatrix} F_r & F_\theta \\ G_r & G_\theta \end{vmatrix}} = -\frac{\begin{vmatrix} -1 & -r\sin\theta \\ 0 & r\cos\theta \end{vmatrix}}{\begin{vmatrix} \cos\theta & -r\sin\theta \\ \sin\theta & r\cos\theta \end{vmatrix}} = \frac{r\cos\theta}{r} = \cos\theta.$$

14. $\dfrac{\partial u}{\partial r} = \dfrac{\partial u}{\partial x}\dfrac{\partial x}{\partial r} + \dfrac{\partial u}{\partial y}\dfrac{\partial y}{\partial r}$

$\quad\quad = (2x - 3x^2y^2)\cos\theta + (-2yx^3)\sin\theta$

16. $\dfrac{du}{dt} = \dfrac{\partial u}{\partial x}\dfrac{dx}{dt} + \dfrac{\partial u}{\partial z}\dfrac{dz}{dt} = \left(\dfrac{1}{z^2} + \dfrac{2z}{x^3}\right)(3t^2) + \left(-\dfrac{2x}{z^3} - \dfrac{1}{x^2}\right)\left(-\dfrac{3}{t^4}\right)$

$\dfrac{d^2u}{dt^2} = \dfrac{\partial}{\partial x}\left(\dfrac{du}{dt}\right)\dfrac{dx}{dt} + \dfrac{\partial}{\partial z}\left(\dfrac{du}{dt}\right)\dfrac{dz}{dt} + \dfrac{\partial}{\partial t}\left(\dfrac{du}{dt}\right)$

$\quad = \left(-\dfrac{18zt^2}{x^4} + \dfrac{6}{z^3t^4} - \dfrac{6}{x^3t^4}\right)(3t^2) + \left(-\dfrac{6t^2}{z^3} + \dfrac{6t^2}{x^3} - \dfrac{18x}{z^4t^4}\right)\left(-\dfrac{3}{t^4}\right)$

$\quad\quad + \left(\dfrac{1}{z^2} + \dfrac{2z}{x^3}\right)(6t) + \left(\dfrac{2x}{z^3} + \dfrac{1}{x^2}\right)\left(-\dfrac{12}{t^5}\right)$

$\quad = \dfrac{6}{x^4t^8z^4}(-9z^5t^{12} + 6zt^6x^4 - 6xt^6z^4 + 9x^5 + t^9z^2x^4 + 2z^5xt^9 - 4x^5zt^3 - 2x^2t^3z^4)$

18. If we set $F(x, y, z) = xz - x^2z^3 + y^2 - 3$, then $\dfrac{\partial z}{\partial x} = -\dfrac{\dfrac{\partial(F)}{\partial(x)}}{\dfrac{\partial(F)}{\partial(z)}} = -\dfrac{F_x}{F_z} = -\dfrac{z - 2xz^3}{x - 3x^2z^2} = \dfrac{2xz^3 - z}{x - 3x^2z^2}.$

$\dfrac{\partial^2z}{\partial x^2} = \dfrac{\partial}{\partial x}\left(\dfrac{\partial z}{\partial x}\right) + \dfrac{\partial}{\partial z}\left(\dfrac{\partial z}{\partial x}\right)\dfrac{\partial z}{\partial x}$

$\quad = \left[\dfrac{(x - 3x^2z^2)(2z^3) - (2xz^3 - z)(1 - 6xz^2)}{(x - 3x^2z^2)^2}\right]$

$\quad\quad + \left[\dfrac{(x - 3x^2z^2)(6xz^2 - 1) - (2xz^3 - z)(-6x^2z)}{(x - 3x^2z^2)^2}\right]\left(\dfrac{2xz^3 - z}{x - 3x^2z^2}\right)$

$\quad = \dfrac{1}{(x - 3x^2z^2)^3}[x(1 - 3xz^2)(2xz^3 - 6x^2z^5 - 2xz^3 + z + 12x^2z^5 - 6xz^3)$

$\quad\quad + z(2xz^2 - 1)(6x^2z^2 - x - 18x^3z^4 + 3x^2z^2 + 12x^3z^4 - 6x^2z^2)]$

$\quad = \dfrac{xz[(1 - 3xz^2)(1 + 6x^2z^4 - 6xz^2) + (1 - 2xz^2)(1 + 6x^2z^4 - 3xz^2)]}{(x - 3x^2z^2)^3}$

20. $\dfrac{\partial u}{\partial t} = \dfrac{\partial u}{\partial x}\dfrac{\partial x}{\partial t} + \dfrac{\partial u}{\partial y}\dfrac{\partial y}{\partial t} + \dfrac{\partial u}{\partial t}\Big)_{x,y}$

$= \left(yt^2 - \dfrac{3y}{\sqrt{1 - x^2 y^2}} \right)(2v^2 t - 2)$

$\quad + \left(xt^2 - \dfrac{3x}{\sqrt{1 - x^2 y^2}} \right)(v \sec^2 t) + 2xyt$

$= 2y(v^2 t - 1)\left(t^2 - \dfrac{3}{\sqrt{1 - x^2 y^2}} \right) + xv \sec^2 t\left(t^2 - \dfrac{3}{\sqrt{1 - x^2 y^2}} \right) + 2xyt$

22. With $\dfrac{\partial u}{\partial x} = 4x + y$, $\dfrac{\partial^2 u}{\partial x^2} = 4$, $\dfrac{\partial^2 u}{\partial x \partial y} = 1$, $\dfrac{\partial u}{\partial y} = -6y + x$, $\dfrac{\partial^2 u}{\partial y^2} = -6$,

$$x^2 \frac{\partial^2 u}{\partial x^2} + 2xy\frac{\partial^2 u}{\partial x \partial y} + y^2\frac{\partial^2 u}{\partial y^2} = 4x^2 + 2xy - 6y^2 = 2u.$$

24. If we set $s = x^2 - y^2$ and $t = y^2 - x^2$, then

$\dfrac{\partial f}{\partial x} = \dfrac{\partial f}{\partial s}\dfrac{\partial s}{\partial x} + \dfrac{\partial f}{\partial t}\dfrac{\partial t}{\partial x} = \dfrac{\partial f}{\partial s}(2x) + \dfrac{\partial f}{\partial t}(-2x)$

$\dfrac{\partial f}{\partial y} = \dfrac{\partial f}{\partial s}\dfrac{\partial s}{\partial y} + \dfrac{\partial f}{\partial t}\dfrac{\partial t}{\partial y} = \dfrac{\partial f}{\partial s}(-2y) + \dfrac{\partial f}{\partial t}(2y).$

Consequently,

$$y\frac{\partial f}{\partial x} + x\frac{\partial f}{\partial y} = y\left(2x\frac{\partial f}{\partial s} - 2x\frac{\partial f}{\partial t} \right) + x\left(-2y\frac{\partial f}{\partial s} + 2y\frac{\partial f}{\partial t} \right) = 0.$$

26. With $\nabla f_{|(1,0,1)} = (2x, 2y, 2z)_{|(1,0,1)} = (2,0,2)$, and a unit vector $\hat{\mathbf{v}} = (1,-1,2)/\sqrt{1+1+4}$ in the direction from $(1,0,1)$ to $(2,-1,3)$, $D_{\hat{\mathbf{v}}}f = (2,0,2)\cdot\dfrac{(1,-1,2)}{\sqrt{6}} = \sqrt{6}$.

28. With parametric equations, $x = 2t - 1$, $y = -t$, $z = 5 - 3t$ for the line, a vector along the line is $\mathbf{T} = (2,-1,-3)$. Since $\nabla f_{|(1,-1,2)} = (2x, 1, -2)_{|(1,-1,2)} = (2,1,-2)$,

$$D_{\mathbf{T}}f = (2,1,-2)\cdot\frac{(2,-1,-3)}{\sqrt{4+1+9}} = \frac{9}{\sqrt{14}}.$$

30. A vector tangent to the curve at $(0,1,1)$ is

$$\mathbf{T} = \nabla(x^2 + y^2 + z^2 - 2)_{|(0,1,1)} \times \nabla(z - y)_{|(0,1,1)} = \begin{vmatrix} \hat{\mathbf{i}} & \hat{\mathbf{j}} & \hat{\mathbf{k}} \\ 0 & 2 & 2 \\ 0 & -1 & 1 \end{vmatrix} = (4,0,0).$$

Since $\nabla f_{|(0,1,1)} = (2yz - 2x, 2xz, 2xy - 2z)_{|(0,1,1)} = (2,0,-2)$, $D_{\mathbf{T}}f = (2,0,-2)\cdot(1,0,0) = 2$.

32. Since a vector normal to the plane is $\nabla(x^2 - y^2 + z^3)_{|(-1,3,2)} = (2x, -2y, 3z^2)_{|(-1,3,2)} = (-2,-6,12)$, its equation is $0 = (1,3,-6)\cdot(x+1, y-3, z-2) = x + 3y - 6z + 4$.

34. Since a tangent vector to the curve at $(2,0,6)$ is $\mathbf{T} = \dfrac{d\mathbf{r}}{dt}\Big|_{t=1} = (2t, 2t, 3t^2 + 5)_{|t=1} = (2,2,8)$, parametric equations for the tangent line are $x = 2 + u$, $y = u$, $z = 6 + 4u$.

36. A tangent vector to the curve at $(1,1,1)$ is

$\nabla(xy - z)_{|(1,1,1)} \times \nabla(x^2 + y^2 - 2)_{|(1,1,1)}$

$= (y, x, -1)_{|(1,1,1)} \times (2x, 2y, 0)_{|(1,1,1)} = \begin{vmatrix} \hat{\mathbf{i}} & \hat{\mathbf{j}} & \hat{\mathbf{k}} \\ 1 & 1 & -1 \\ 2 & 2 & 0 \end{vmatrix} = (2,-2,0).$

Parametric equations for the tangent line are $x = 1 + t$, $y = 1 - t$, $z = 1$.

38. For critical points we solve $0 = \dfrac{\partial f}{\partial x} = ye^x$, $0 = \dfrac{\partial f}{\partial y} = e^x$. There are no solutions to these equations.

40. For critical points we solve $0 = \dfrac{\partial f}{\partial x} = 4x(x^2 + y^2 - 1)$, $0 = \dfrac{\partial f}{\partial y} = 4y(x^2 + y^2 - 1)$. The solutions are $(0,0)$ and every point on the circle $x^2 + y^2 = 1$. Since $f(x,y) = 0$ for every point on $x^2 + y^2 = 1$, but is otherwise positive, each of these critical points yields a relative minimum.

$$\frac{\partial^2 f}{\partial x^2} = 4(x^2 + y^2 - 1) + 8x^2, \qquad \frac{\partial^2 f}{\partial x \partial y} = 8xy, \qquad \frac{\partial^2 f}{\partial y^2} = 4(x^2 + y^2 - 1) + 8y^2.$$

At $(0,0)$, $B^2 - AC = 0 - (-4)(-4) < 0$, and $A = -4$, and therefore $(0,0)$ gives a relative maximum.

42. For critical points of $f(x,y)$ we solve $0 = \dfrac{\partial f}{\partial x} = y$, $0 = \dfrac{\partial f}{\partial y} = x$. At the critical point $(0,0)$, $f(0,0) = \boxed{0}$. On the boundary of the region we set $x = \cos t$, $y = \sin t$, in which case $f(x,y)$ becomes

$$z(t) = \cos t \sin t = \frac{1}{2}\sin 2t, \quad 0 \le t \le 2\pi.$$

For critical points of $z(t)$, we set $0 = dz/dt = \cos 2t$. At the critical points $t = \pi/4, 3\pi/4, 5\pi/4$, and $7\pi/4$,

$$z(\pi/4) = z(5\pi/4) = \boxed{1/2}, \qquad z(3\pi/4) = z(7\pi/4) = \boxed{-1/2}.$$

Finally, $z(0) = z(2\pi) = \boxed{0}$. Maximum and minimum values are therefore $\pm 1/2$.

44. The distance D from the origin to any point $P(x,y)$ on the curve is given by

$$D^2 = x^2 + y^2 = x^2 + (1 - x^2 - x^4) = 1 - x^4.$$

This function must be maximized and minimized on the interval $|x| \le \sqrt{(\sqrt{5} - 1)/2}$. Since the only critical point of D^2 is $x = 0$, we evaluate

$$D^2(0) = 1, \qquad D^2\left(\pm\sqrt{(\sqrt{5} - 1)/2}\right) = (\sqrt{5} - 1)/2.$$

The closest and farthest points are therefore $(\pm\sqrt{(\sqrt{5} - 1)/2}, 0)$ and $(0, \pm 1)$ respectively.

46. If the farmer plants x hectares of corn, y hectares of potatoes, and z hectares of sunflowers, his losses are

$$L = pax^2 + qby^2 + rcz^2.$$

Since $x + y + z = 100$,

$$L = pax^2 + qby^2 + rc(100 - x - y)^2.$$

This function must be minimized for (x,y) in the triangle shown. For critical points of L, we solve

$$0 = \frac{\partial L}{\partial x} = 2apx - 2rc(100 - x - y),$$

$$0 = \frac{\partial L}{\partial y} = 2bqy - 2rc(100 - x - y).$$

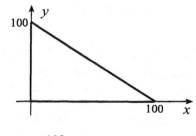

The solution of these equations is $x_0 = \dfrac{100bcqr}{acpr + abpq + bcqr}$, $y_0 = \dfrac{100acpr}{acpr + abpq + bcqr}$. When $x = 0$,

$$L = qby^2 + rc(100 - y)^2, \quad 0 \le y \le 100.$$

For critical points of this function, $0 = \dfrac{dL}{dy} = 2qby - 2rc(100 - y)$. The solution is $y_1 = 100rc/(cr + bq)$.

Similarly, when $y = 0$, we obtain the critical point $x_1 = 100rc/(cr + ap)$ of

$$L = pax^2 + rc(100 - x)^2, \quad 0 \le x \le 100.$$

When $x + y = 100$,

$$L = pax^2 + qb(100 - x)^2, \quad 0 \le x \le 100.$$

The critical point of this function is $x_2 = 100bq/(bq + ap)$.

We now evaluate L at each of these critical points and the corners of the triangle:

$$L(x_0, y_0) = \frac{10^4 abcpqr}{acpr + abpq + bcqr}, \quad L(y_1) = \frac{10^4 bcqr}{cr + bq}, \quad L(x_1) = \frac{10^4 acpr}{cr + ap}, \quad L(x_2) = \frac{10^4 abpq}{ap + bq},$$

$$L(0, 100) = 10^4 bq, \quad L(100, 0) = 10^4 ap, \quad L(0, 0) = 10^4 cr.$$

Notice now that:

$$\frac{1}{L(x_0, y_0)} = 10^{-4} \left(\frac{1}{bq} + \frac{1}{cr} + \frac{1}{ap} \right),$$

$$\frac{1}{L(x_1)} = 10^{-4} \left(\frac{1}{ap} + \frac{1}{cr} \right), \quad \frac{1}{L(y_1)} = 10^{-4} \left(\frac{1}{bq} + \frac{1}{cr} \right), \quad \frac{1}{L(x_2)} = 10^{-4} \left(\frac{1}{bq} + \frac{1}{ap} \right),$$

$$\frac{1}{L(0, 100)} = \frac{10^{-4}}{bq}, \quad \frac{1}{L(100, 0)} = \frac{10^{-4}}{ap}, \quad \frac{1}{L(0, 0)} = \frac{10^{-4}}{cr}.$$

It follows that $1/L(x_0, y_0)$ is the largest of these numbers, and therefore $L(x_0, y_0)$ is the smallest value for L. Thus L is minimized when $x = x_0$, $y = y_0$, and

$$z = 100 - x_0 - y_0 = \frac{100abpq}{acpr + abpq + bcqr}.$$

48. With $f(1, \pi/4) = 1/\sqrt{2}$,

$$f_x(1, \pi/4) = [3x^2 \sin(x^2 y) + 2x^4 y \cos(x^2 y)]_{|(1, \pi/4)} = \frac{3}{\sqrt{2}} + \frac{2\pi}{4} \frac{1}{\sqrt{2}} = \frac{3\sqrt{2}}{2} + \frac{\sqrt{2}\pi}{4},$$

$$f_y(1, \pi/4) = x^5 \cos(x^2 y)_{|(1, \pi/4)} = \frac{1}{\sqrt{2}},$$

$$f_{xx}(1, \pi/4) = [6x \sin(x^2 y) + 14x^3 y \cos(x^2 y) - 4x^5 y^2 \sin(x^2 y)]_{|(1, \pi/4)}$$

$$= \frac{6}{\sqrt{2}} + 14 \left(\frac{\pi}{4} \right) \frac{1}{\sqrt{2}} - 4 \left(\frac{\pi^2}{16} \right) \frac{1}{\sqrt{2}} = \frac{6\sqrt{2}}{2} + \frac{7\sqrt{2}\pi}{4} - \frac{\sqrt{2}\pi^2}{8},$$

$$f_{xy}(1, \pi/4) = [5x^4 \cos(x^2 y) - 2x^6 y \sin(x^2 y)]_{|(1, \pi/4)}$$

$$= \frac{5}{\sqrt{2}} - 2 \left(\frac{\pi}{4} \right) \frac{1}{\sqrt{2}} = \frac{5}{\sqrt{2}} - \frac{\sqrt{2}\pi}{4},$$

$$f_{yy}(1, \pi/4) = -x^7 \sin(x^2 y)_{|(1, \pi/4)} = -\frac{1}{\sqrt{2}},$$

$$x^3 \sin(x^2 y) = \frac{1}{\sqrt{2}} + \left(\frac{3\sqrt{2}}{2} + \frac{\sqrt{2}\pi}{4} \right) (x - 1) + \frac{1}{\sqrt{2}} (y - \pi/4)$$

$$+ \frac{1}{2} \left[\left(\frac{6\sqrt{2}}{2} + \frac{7\sqrt{2}\pi}{4} - \frac{\sqrt{2}\pi^2}{8} \right) (x - 1)^2 + 2 \left(\frac{5\sqrt{2}}{2} - \frac{\sqrt{2}\pi}{4} \right) (x - 1)(y - \pi/4) \right.$$

$$\left. - \frac{1}{\sqrt{2}} (y - \pi/4)^2 \right] + \cdots.$$

CHAPTER 13

EXERCISES 13.1

2. $\displaystyle\int_{-3}^{3}\int_{-\sqrt{18-2y^2}}^{\sqrt{18-2y^2}} x\,dx\,dy = \int_{-3}^{3}\left\{\frac{x^2}{2}\right\}_{-\sqrt{18-2y^2}}^{\sqrt{18-2y^2}} dy = \int_{-3}^{3} 0\,dy = 0$

4. $\displaystyle\int_{-1}^{0}\int_{y}^{2}(1+y)^2\,dx\,dy = \int_{-1}^{0}\left\{x(1+y)^2\right\}_{y}^{2}dy = \int_{-1}^{0}(2+3y-y^3)\,dy = \left\{2y+\frac{3y^2}{2}-\frac{y^4}{4}\right\}_{-1}^{0} = \frac{3}{4}$

6. $\displaystyle\int_{1}^{2}\int_{1}^{y}e^{x+y}\,dx\,dy = \int_{1}^{2}\left\{e^{x+y}\right\}_{1}^{y}dy = \int_{1}^{2}(e^{2y}-e^{y+1})\,dy = \left\{\frac{e^{2y}}{2}-e^{y+1}\right\}_{1}^{2} = \frac{e^2(1-e)^2}{2}$

8. $\displaystyle\int_{-1}^{1}\int_{x}^{2x}(xy+x^3y^3)\,dy\,dx = \int_{-1}^{1}\left\{\frac{xy^2}{2}+\frac{x^3y^4}{4}\right\}_{x}^{2x}dx = \frac{1}{4}\int_{-1}^{1}(6x^3+15x^7)\,dx = \frac{1}{4}\left\{\frac{3x^4}{2}+\frac{15x^8}{8}\right\}_{-1}^{1} = 0$

10. $\displaystyle\int_{1}^{2}\int_{x}^{2x}\frac{1}{(x+y)^3}\,dy\,dx = \int_{1}^{2}\left\{\frac{-1}{2(x+y)^2}\right\}_{x}^{2x}dx = \frac{5}{72}\int_{1}^{2}\frac{1}{x^2}\,dx = \frac{5}{72}\left\{-\frac{1}{x}\right\}_{1}^{2} = \frac{5}{144}$

12. $\displaystyle\int_{-1}^{1}\int_{1}^{e}\frac{y}{x}\,dx\,dy = \int_{-1}^{1}\left\{y\ln|x|\right\}_{1}^{e}dy = \int_{-1}^{1}y\,dy = \left\{\frac{y^2}{2}\right\}_{-1}^{1} = 0$

14. $\displaystyle\int_{0}^{2}\int_{x^2}^{2x^2} x\cos y\,dy\,dx = \int_{0}^{2}\left\{x\sin y\right\}_{x^2}^{2x^2}dx = \int_{0}^{2}x[\sin 2x^2 - \sin x^2]\,dx$

$$= \left\{-\frac{1}{4}\cos 2x^2 + \frac{1}{2}\cos x^2\right\}_{0}^{2} = -0.54$$

16. $\displaystyle\int_{0}^{1}\int_{0}^{y^3}\frac{1}{1+y^2}\,dx\,dy = \int_{0}^{1}\left\{\frac{x}{1+y^2}\right\}_{0}^{y^3}dy = \int_{0}^{1}\frac{y^3}{1+y^2}\,dy = \int_{0}^{1}\left(y-\frac{y}{1+y^2}\right)dy$

$$= \left\{\frac{y^2}{2}-\frac{1}{2}\ln(y^2+1)\right\}_{0}^{1} = \frac{1-\ln 2}{2}$$

18. $\displaystyle\int_{0}^{2}\int_{-x}^{x}(8-2x^2)^{3/2}\,dy\,dx = \int_{0}^{2}\left\{y(8-2x^2)^{3/2}\right\}_{-x}^{x}dx = 2\int_{0}^{2}x(8-2x^2)^{3/2}\,dx$

$$= 2\left\{-\frac{(8-2x^2)^{5/2}}{10}\right\}_{0}^{2} = \frac{128\sqrt{2}}{5}$$

20. $\displaystyle\int_{-9}^{0}\int_{0}^{x^2\sqrt{9+x}}dy\,dx = \int_{-9}^{0}\left\{y\right\}_{0}^{x^2\sqrt{9+x}}dx = \int_{-9}^{0}x^2\sqrt{9+x}\,dx$ If we set $u = 9+x$, then $du = dx$, and

$$\int_{-9}^{0}\int_{0}^{x^2\sqrt{9+x}}dy\,dx = \int_{0}^{9}(u-9)^2\sqrt{u}\,du = \int_{0}^{9}(81\sqrt{u}-18u^{3/2}+u^{5/2})\,du$$

$$= \left\{\frac{162u^{3/2}}{3}-\frac{36u^{5/2}}{5}+\frac{2u^{7/2}}{7}\right\}_{0}^{9} = \frac{11\,664}{35}.$$

22. $\displaystyle\int_{-1}^{0}\int_{y}^{0}x\sqrt{x^2+y^2}\,dx\,dy = \int_{-1}^{0}\left\{\frac{1}{3}(x^2+y^2)^{3/2}\right\}_{y}^{0}dy$

$$= \frac{1}{3}\int_{-1}^{0}(2\sqrt{2}-1)y^3\,dy = \frac{2\sqrt{2}-1}{3}\left\{\frac{y^4}{4}\right\}_{-1}^{0} = \frac{1-2\sqrt{2}}{12}$$

24. $\displaystyle\int_{0}^{1}\int_{0}^{\mathrm{Cos}^{-1}x}x\cos y\,dy\,dx = \int_{0}^{1}\left\{x\sin y\right\}_{0}^{\mathrm{Cos}^{-1}x}dx = \int_{0}^{1}x\sqrt{1-x^2}\,dx = \left\{-\frac{1}{3}(1-x^2)^{3/2}\right\}_{0}^{1} = \frac{1}{3}$

26. If we set $u = x^2 - y^2$, then $du = 2x\,dx$, and

$$\int_0^1 \int_{\sqrt{2}y}^{\sqrt{y^2+y}} x^3 \sqrt{x^2 - y^2}\,dx\,dy = \int_0^1 \int_{y^2}^{y} (u + y^2)\sqrt{u}\frac{du}{2}\,dy = \frac{1}{2}\int_0^1 \int_{y^2}^{y} (u^{3/2} + y^2\sqrt{u})\,du\,dy$$

$$= \frac{1}{2}\int_0^1 \left\{ \frac{2u^{5/2}}{5} + \frac{2y^2 u^{3/2}}{3} \right\}_{y^2}^{y} dy = \frac{1}{15}\int_0^1 (3y^{5/2} + 5y^{7/2} - 8y^5)\,dy$$

$$= \frac{1}{15}\left\{ \frac{6y^{7/2}}{7} + \frac{10y^{9/2}}{9} - \frac{4y^6}{3} \right\}_0^1 = \frac{8}{189}.$$

28. $\int_{-2}^0 \int_y^0 \frac{x}{\sqrt{x^2 + y^2}}\,dx\,dy = \int_{-2}^0 \left\{ \sqrt{x^2 + y^2} \right\}_y^0 dy = \int_{-2}^0 (\sqrt{2} - 1)y\,dy = (\sqrt{2} - 1)\left\{ \frac{y^2}{2} \right\}_{-2}^0 = 2(1 - \sqrt{2})$

30. If we set $y = x\tan\theta$, then $dy = x\sec^2\theta\,d\theta$, and

$$\int_0^1 \int_0^x \sqrt{x^2 + y^2}\,dy\,dx = \int_0^1 \int_0^{\pi/4} x\sec\theta\,x\sec^2\theta\,d\theta\,dx = \int_0^1 \int_0^{\pi/4} x^2\sec^3\theta\,d\theta\,dx$$

$$= \int_0^1 \frac{x^2}{2}\left\{ \sec\theta\,\tan\theta + \ln|\sec\theta + \tan\theta| \right\}_0^{\pi/4} dx \quad \text{(see Example 8.9)}$$

$$= \frac{\sqrt{2} + \ln(\sqrt{2} + 1)}{2}\int_0^1 x^2\,dx = \frac{\sqrt{2} + \ln(\sqrt{2} + 1)}{6}.$$

32. From the continuity equation,

$$\frac{\partial v}{\partial y} = -\frac{\partial u}{\partial x} = -2x.$$

Integration gives $v(x, y) = -2xy + f(x)$, where $f(x)$ is any differentiable function of x.

34. From the continuity equation,

$$\frac{\partial u}{\partial x} = -\frac{\partial v}{\partial y} = \frac{-xy}{\sqrt{x^2 + y^2}}.$$

Integration gives $u(x, y) = -y\sqrt{x^2 + y^2} + f(y)$, where $f(y)$ is any differentiable function of y.

36. Stream functions must satisfy

$$\frac{\partial \psi}{\partial x} = y, \qquad \frac{\partial \psi}{\partial y} = x.$$

Integration of the first gives $\psi(x, y) = xy + f(y)$, where $f(y)$ is any differentiable function of y. Substitution of this into the second equation requires $x + f'(y) = x \implies f(y) = C$, where C is a constant. Thus, $\psi(x, y) = xy + C$.

38. Stream functions must satisfy

$$\frac{\partial \psi}{\partial x} = -x\sqrt{x^2 + y^2}, \qquad \frac{\partial \psi}{\partial y} = -y\sqrt{x^2 + y^2}.$$

Integration of the first gives $\psi(x, y) = -\frac{1}{3}(x^2 + y^2)^{3/2} + f(y)$, where $f(y)$ is any differentiable function of y. Substitution of this into the second equation requires

$$-y\sqrt{x^2 + y^2} + f'(y) = -y\sqrt{x^2 + y^2} \implies f(y) = C,$$

where C is a constant. Thus, $\psi(x, y) = -(1/3)(x^2 + y^2)^{3/2} + C$.

EXERCISES 13.2

2. $\displaystyle\iint_R (4 - x^2 - y)\, dA = \int_0^2 \int_0^{4-x^2} (4 - x^2 - y)\, dy\, dx$

$$= \int_0^2 \left\{ 4y - x^2 y - \frac{y^2}{2} \right\}_0^{4-x^2} dx$$

$$= \frac{1}{2} \int_0^2 (16 - 8x^2 + x^4)\, dx$$

$$= \frac{1}{2} \left\{ 16x - \frac{8x^3}{3} + \frac{x^5}{5} \right\}_0^2 = \frac{128}{15}$$

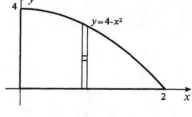

4. $\displaystyle\iint_R xy^2\, dA = \int_{-1}^2 \int_{-1-y}^{1-y^2} xy^2\, dx\, dy = \int_{-1}^2 \left\{ \frac{x^2 y^2}{2} \right\}_{-1-y}^{1-y^2} dy$

$$= \frac{1}{2} \int_{-1}^2 (y^6 - 3y^4 - 2y^3)\, dy$$

$$= \frac{1}{2} \left\{ \frac{y^7}{7} - \frac{3y^5}{5} - \frac{y^4}{2} \right\}_{-1}^2 = -\frac{621}{140}$$

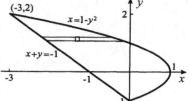

6. $\displaystyle\iint_R (x + y)\, dA = \int_{-3}^3 \int_{-\sqrt{9-x^2}}^{\sqrt{9-x^2}} (x + y)\, dy\, dx$

$$= \int_{-3}^3 \left\{ xy + \frac{y^2}{2} \right\}_{-\sqrt{9-x^2}}^{\sqrt{9-x^2}} dx$$

$$= 2 \int_{-3}^3 x\sqrt{9 - x^2}\, dx = 2 \left\{ -\frac{1}{3} (9 - x^2)^{3/2} \right\}_{-3}^3 = 0$$

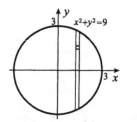

8. $\displaystyle\iint_R (xy + y^2 - 3x^2)\, dA = \int_1^2 \int_{-y}^y (xy + y^2 - 3x^2)\, dx\, dy$

$$= \int_1^2 \left\{ \frac{x^2 y}{2} + xy^2 - x^3 \right\}_{-y}^y dy$$

$$= \int_1^2 (0)\, dy = 0$$

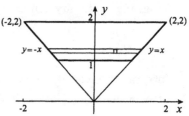

10. $\displaystyle\iint_R (x + y)\, dA = \int_{-2}^2 \int_{y^2}^{\sqrt{12+y^2}} (x + y)\, dx\, dy = \int_{-2}^2 \left\{ \frac{x^2}{2} + xy \right\}_{y^2}^{\sqrt{12+y^2}} dy$

$$= \frac{1}{2} \int_{-2}^2 (12 + y^2 - 2y^3 - y^4 + 2y\sqrt{12 + y^2})\, dy$$

$$= \frac{1}{2} \left\{ 12y + \frac{y^3}{3} - \frac{y^4}{2} - \frac{y^5}{5} + \frac{2}{3} (12 + y^2)^{3/2} \right\}_{-2}^2$$

$$= 304/15$$

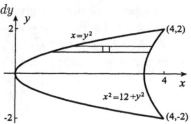

12. $\displaystyle\iint_R y^2\, dA = \int_{1/2}^1 \int_0^{1/(y\sqrt{y^2+12})} y^2\, dx\, dy$

$$= \int_{1/2}^1 \left\{ xy^2 \right\}_0^{1/(y\sqrt{y^2+12})} dy$$

$$= \int_{1/2}^1 \frac{y}{\sqrt{y^2 + 12}}\, dy = \left\{ \sqrt{y^2 + 12} \right\}_{1/2}^1 = \sqrt{13} - \frac{7}{2}$$

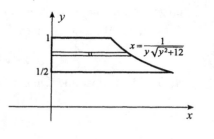

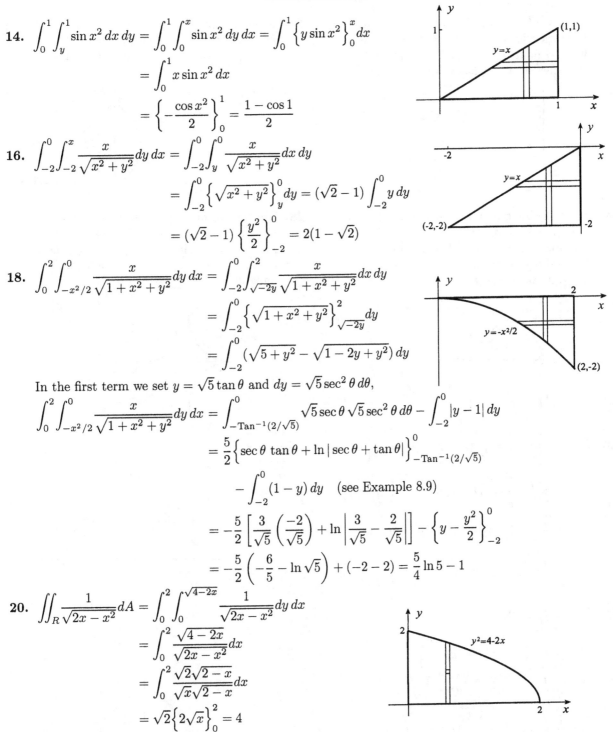

14. $\displaystyle\int_0^1\int_y^1 \sin x^2\,dx\,dy = \int_0^1\int_0^x \sin x^2\,dy\,dx = \int_0^1 \Big\{y\sin x^2\Big\}_0^x dx$

$$= \int_0^1 x\sin x^2\,dx$$

$$= \Big\{-\frac{\cos x^2}{2}\Big\}_0^1 = \frac{1-\cos 1}{2}$$

16. $\displaystyle\int_{-2}^0\int_{-2}^x \frac{x}{\sqrt{x^2+y^2}}\,dy\,dx = \int_{-2}^0\int_y^0 \frac{x}{\sqrt{x^2+y^2}}\,dx\,dy$

$$= \int_{-2}^0 \Big\{\sqrt{x^2+y^2}\Big\}_y^0 dy = (\sqrt{2}-1)\int_{-2}^0 y\,dy$$

$$= (\sqrt{2}-1)\Big\{\frac{y^2}{2}\Big\}_{-2}^0 = 2(1-\sqrt{2})$$

18. $\displaystyle\int_0^2\int_{-x^2/2}^0 \frac{x}{\sqrt{1+x^2+y^2}}\,dy\,dx = \int_{-2}^0\int_{\sqrt{-2y}}^2 \frac{x}{\sqrt{1+x^2+y^2}}\,dx\,dy$

$$= \int_{-2}^0 \Big\{\sqrt{1+x^2+y^2}\Big\}_{\sqrt{-2y}}^2 dy$$

$$= \int_{-2}^0 (\sqrt{5+y^2} - \sqrt{1-2y+y^2})\,dy$$

In the first term we set $y = \sqrt{5}\tan\theta$ and $dy = \sqrt{5}\sec^2\theta\,d\theta$,

$$\int_0^2\int_{-x^2/2}^0 \frac{x}{\sqrt{1+x^2+y^2}}\,dy\,dx = \int_{-\mathrm{Tan}^{-1}(2/\sqrt{5})}^0 \sqrt{5}\sec\theta\,\sqrt{5}\sec^2\theta\,d\theta - \int_{-2}^0 |y-1|\,dy$$

$$= \frac{5}{2}\Big\{\sec\theta\,\tan\theta + \ln|\sec\theta+\tan\theta|\Big\}_{-\mathrm{Tan}^{-1}(2/\sqrt{5})}^0$$

$$- \int_{-2}^0 (1-y)\,dy \quad \text{(see Example 8.9)}$$

$$= -\frac{5}{2}\Big[\frac{3}{\sqrt{5}}\Big(\frac{-2}{\sqrt{5}}\Big) + \ln\Big|\frac{3}{\sqrt{5}} - \frac{2}{\sqrt{5}}\Big|\Big] - \Big\{y-\frac{y^2}{2}\Big\}_{-2}^0$$

$$= -\frac{5}{2}\Big(-\frac{6}{5} - \ln\sqrt{5}\Big) + (-2-2) = \frac{5}{4}\ln 5 - 1$$

20. $\displaystyle\iint_R \frac{1}{\sqrt{2x-x^2}}\,dA = \int_0^2\int_0^{\sqrt{4-2x}} \frac{1}{\sqrt{2x-x^2}}\,dy\,dx$

$$= \int_0^2 \frac{\sqrt{4-2x}}{\sqrt{2x-x^2}}\,dx$$

$$= \int_0^2 \frac{\sqrt{2}\sqrt{2-x}}{\sqrt{x}\sqrt{2-x}}\,dx$$

$$= \sqrt{2}\Big\{2\sqrt{x}\Big\}_0^2 = 4$$

22. Since x^2y^2 is an even function of y, and the region is symmetric about the x-axis, we may double the value of the integral over that part of the region above the x-axis.

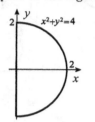

24. Since the integrand is an odd function of x, and the region is symmetric about the y-axis, the value of the integral is zero.

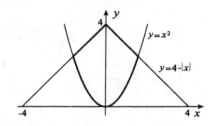

26. Since the integrand is an even function of x and y, and the region is symmetric about the origin, we may double the value of the integral over that part of the region to the right of the y-axis.

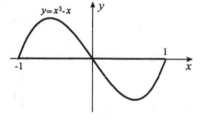

28. The first term of the integrand is an odd function of y and the second term is an odd function of x. Since the region is symmetric about the x- and y-axes, the value of the integral is zero.

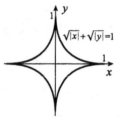

30. Since $\text{Area}(R) = \int_0^1 (2 - y^2 - y)\, dy = \left\{ 2y - \dfrac{y^3}{3} - \dfrac{y^2}{2} \right\}_0^1 = \dfrac{7}{6},$

$$\overline{f} = \frac{6}{7} \int_0^1 \int_y^{2-y^2} (x+y)\, dx\, dy = \frac{6}{7} \int_0^1 \left\{ \frac{x^2}{2} + xy \right\}_y^{2-y^2} dy$$

$$= \frac{3}{7} \int_0^1 (4 + 4y - 7y^2 - 2y^3 + y^4)\, dy$$

$$= \frac{3}{7} \left\{ 4y + 2y^2 - \frac{7y^3}{3} - \frac{y^4}{2} + \frac{y^5}{5} \right\}_0^1 = \frac{101}{70}.$$

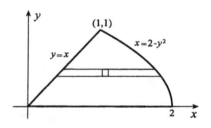

32. $\overline{f} = \dfrac{1}{2} \displaystyle\iint_R e^{x+y}\, dA$

$$= \frac{1}{2} \int_{-1}^0 \int_{-1-x}^{x+1} e^{x+y}\, dy\, dx + \frac{1}{2} \int_0^1 \int_{x-1}^{1-x} e^{x+y}\, dy\, dx$$

$$= \frac{1}{2} \int_{-1}^0 \left\{ e^{x+y} \right\}_{-1-x}^{x+1} dx + \frac{1}{2} \int_0^1 \left\{ e^{x+y} \right\}_{x-1}^{1-x} dx$$

$$= \frac{1}{2} \int_{-1}^0 (e^{2x+1} - e^{-1})\, dx + \frac{1}{2} \int_0^1 (e - e^{2x-1})\, dx$$

$$= \frac{1}{2} \left\{ \frac{1}{2} e^{2x+1} - \frac{x}{e} \right\}_{-1}^0 + \frac{1}{2} \left\{ ex - \frac{1}{2} e^{2x-1} \right\}_0^1 = \frac{e^2 - 1}{2e}$$

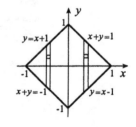

34. $\text{Average} = \dfrac{1}{(4)(10)} \displaystyle\int_{45}^{55} \int_8^{12} 10\,000 x^{0.7} y^{0.3}\, dy\, dx = 250 \int_{45}^{55} \left\{ \dfrac{x^{0.7} y^{1.3}}{1.3} \right\}_8^{12} dx$

$$= \frac{250(12^{1.3} - 8^{1.3})}{1.3} \left\{ \frac{x^{1.7}}{1.7} \right\}_{45}^{55} = 307\,973$$

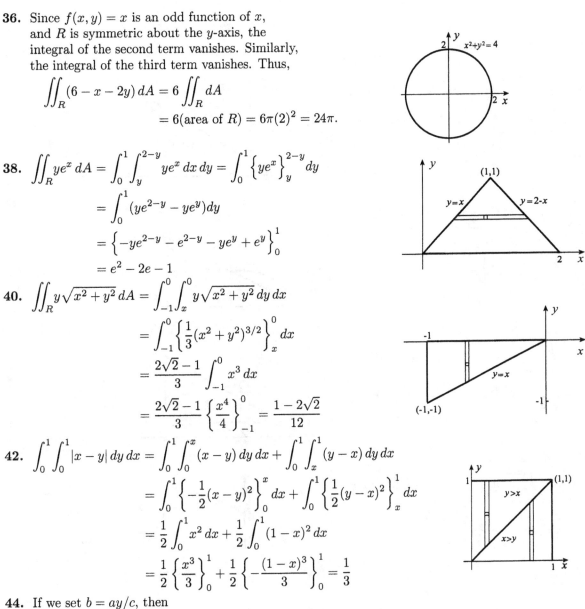

36. Since $f(x, y) = x$ is an odd function of x,
and R is symmetric about the y-axis, the
integral of the second term vanishes. Similarly,
the integral of the third term vanishes. Thus,

$$\iint_R (6 - x - 2y)\, dA = 6 \iint_R dA$$
$$= 6(\text{area of } R) = 6\pi(2)^2 = 24\pi.$$

38. $\displaystyle \iint_R ye^x\, dA = \int_0^1 \int_y^{2-y} ye^x\, dx\, dy = \int_0^1 \left\{ ye^x \right\}_y^{2-y} dy$

$$= \int_0^1 (ye^{2-y} - ye^y)\, dy$$
$$= \left\{ -ye^{2-y} - e^{2-y} - ye^y + e^y \right\}_0^1$$
$$= e^2 - 2e - 1$$

40. $\displaystyle \iint_R y\sqrt{x^2 + y^2}\, dA = \int_{-1}^0 \int_x^0 y\sqrt{x^2 + y^2}\, dy\, dx$

$$= \int_{-1}^0 \left\{ \frac{1}{3}(x^2 + y^2)^{3/2} \right\}_x^0 dx$$
$$= \frac{2\sqrt{2} - 1}{3} \int_{-1}^0 x^3\, dx$$
$$= \frac{2\sqrt{2} - 1}{3} \left\{ \frac{x^4}{4} \right\}_{-1}^0 = \frac{1 - 2\sqrt{2}}{12}$$

42. $\displaystyle \int_0^1 \int_0^1 |x - y|\, dy\, dx = \int_0^1 \int_0^x (x - y)\, dy\, dx + \int_0^1 \int_x^1 (y - x)\, dy\, dx$

$$= \int_0^1 \left\{ -\frac{1}{2}(x - y)^2 \right\}_0^x dx + \int_0^1 \left\{ \frac{1}{2}(y - x)^2 \right\}_x^1 dx$$
$$= \frac{1}{2} \int_0^1 x^2\, dx + \frac{1}{2} \int_0^1 (1 - x)^2\, dx$$
$$= \frac{1}{2} \left\{ \frac{x^3}{3} \right\}_0^1 + \frac{1}{2} \left\{ -\frac{(1 - x)^3}{3} \right\}_0^1 = \frac{1}{3}$$

44. If we set $b = ay/c$, then

$$n = \frac{2n_c L}{\pi} \int_0^{2d} (1 - b^2) \int_0^\infty \frac{x^2}{(1 + x^2)(x^2 + b^2)}\, dx\, dy,$$

and partial fractions gives

$$n = \frac{2n_c L}{\pi} \int_0^{2d} (1 - b^2) \int_0^\infty \left[\frac{1/(1 - b^2)}{1 + x^2} - \frac{b^2/(1 - b^2)}{b^2 + x^2} \right] dx\, dy$$
$$= \frac{2n_c L}{\pi} \int_0^{2d} \left\{ \text{Tan}^{-1} x - b\, \text{Tan}^{-1}\!\left(\frac{x}{b}\right) \right\}_0^\infty dy$$
$$= \frac{2n_c L}{\pi} \int_0^{2d} \left(\frac{\pi}{2} - \frac{b\pi}{2} \right) dy$$
$$= n_c L \int_0^{2d} \left(1 - \frac{ay}{c} \right) dy = n_c L \left\{ y - \frac{ay^2}{2c} \right\}_0^{2d}$$
$$= n_c L \left(2d - \frac{2ad^2}{c} \right) = 2n_c dL \left(1 - \frac{ad}{c} \right).$$

EXERCISES 13.3

2. $A = \int_{-1}^{6} \int_{x^2}^{5x+6} dy\,dx = \int_{-1}^{6} (5x + 6 - x^2)\,dx$

$\qquad = \left\{ \dfrac{5x^2}{2} + 6x - \dfrac{x^3}{3} \right\}_{-1}^{6} = \dfrac{343}{6}$

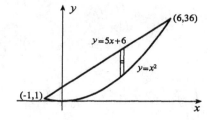

4. $A = 2 \int_{0}^{2} \int_{x^3+8}^{4x+8} dy\,dx = 2 \int_{0}^{2} (4x - x^3)\,dx$

$\qquad = 2 \left\{ 2x^2 - \dfrac{x^4}{4} \right\}_{0}^{2} = 8$

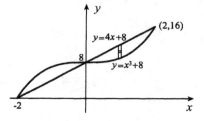

6. $A = \int_{0}^{2} \int_{xe^{-x}}^{x} dy\,dx = \int_{0}^{2} (x - xe^{-x})\,dx$

$\qquad = \left\{ \dfrac{x^2}{2} + xe^{-x} + e^{-x} \right\}_{0}^{2} = 1 + 3e^{-2}$

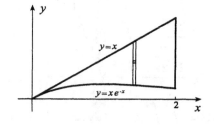

8. $A = \int_{-3}^{4} \int_{y(y-2)}^{12-y} dx\,dy = \int_{-3}^{4} (12 - y^2 + y)\,dy$

$\qquad = \left\{ 12y - \dfrac{y^3}{3} + \dfrac{y^2}{2} \right\}_{-3}^{4} = \dfrac{343}{6}$

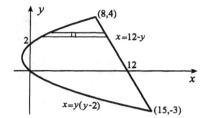

10. $A = \int_{-5/3}^{-1/3} \int_{1-x}^{2x+6} dy\,dx + \int_{-1/3}^{0} \int_{1-x}^{5-x} dy\,dx + \int_{0}^{4/3} \int_{2x+1}^{5-x} dy\,dx$

$\qquad = \int_{-5/3}^{-1/3} (3x + 5)\,dx + \int_{-1/3}^{0} 4\,dx + \int_{0}^{4/3} (4 - 3x)\,dx$

$\qquad = \left\{ \dfrac{3x^2}{2} + 5x \right\}_{-5/3}^{-1/3} + \left\{ 4x \right\}_{-1/3}^{0} + \left\{ 4x - \dfrac{3x^2}{2} \right\}_{0}^{4/3} = \dfrac{20}{3}$

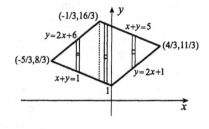

12. $V = 2 \int_{0}^{3} \int_{0}^{2\sqrt{9-x^2}/3} 2\pi y\,dy\,dx = 2\pi \int_{0}^{3} \left\{ y^2 \right\}_{0}^{2\sqrt{9-x^2}/3} dx$

$\qquad = \dfrac{8\pi}{9} \int_{0}^{3} (9 - x^2)\,dx = \dfrac{8\pi}{9} \left\{ 9x - \dfrac{x^3}{3} \right\}_{0}^{3} = 16\pi$

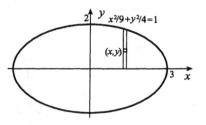

14. $V = 2\int_0^2 \int_{2x^2}^{x^2+4} 2\pi y\, dy\, dx = 2\pi \int_0^2 \left\{ y^2 \right\}_{2x^2}^{x^2+4} dx$

$= 2\pi \int_0^2 (16 + 8x^2 - 3x^4)\, dx$

$= 2\pi \left\{ 16x + \frac{8x^3}{3} - \frac{3x^5}{5} \right\}_0^2 = \frac{1024\pi}{15}$

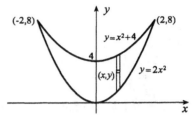

16. $V = \int_0^1 \int_{y^3}^{2-y^2} 2\pi(1-y)\, dx\, dy = 2\pi \int_0^1 \left\{ x(1-y) \right\}_{y^3}^{2-y^2} dy$

$= 2\pi \int_0^1 (2 - 2y - y^2 + y^4)\, dy$

$= 2\pi \left\{ 2y - y^2 - \frac{y^3}{3} + \frac{y^5}{5} \right\}_0^1 = \frac{26\pi}{15}$

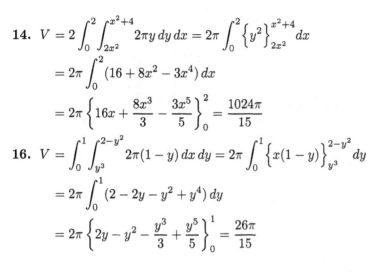

18. $V = 2\int_0^3 \int_0^{3y-y^2} 2\pi(4-y)\, dx\, dy = 4\pi \int_0^3 \left\{ x(4-y) \right\}_0^{3y-y^2} dy$

$= 4\pi \int_0^3 (12y - 7y^2 + y^3)\, dy$

$= 4\pi \left\{ 6y^2 - \frac{7y^3}{3} + \frac{y^4}{4} \right\}_0^3 = 45\pi$

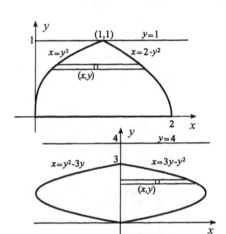

20. $V = \int_0^2 \int_{y^2/4+1}^{4-y} 2\pi(y+1)\, dx\, dy = 2\pi \int_0^2 \left\{ x(y+1) \right\}_{y^2/4+1}^{4-y} dy$

$= \frac{\pi}{2} \int_0^2 (12 + 8y - 5y^2 - y^3)\, dy$

$= \frac{\pi}{2} \left\{ 12y + 4y^2 - \frac{5y^3}{3} - \frac{y^4}{4} \right\}_0^2 = \frac{34\pi}{3}$

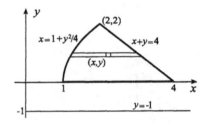

22. $A = \int_1^6 \int_{-x^2}^{x/\sqrt{x+3}} dy\, dx = \int_1^6 \left(\frac{x}{\sqrt{x+3}} + x^2 \right) dx$

If we set $u = x+3$ and $du = dx$,

$A = \int_4^9 \left(\frac{u-3}{\sqrt{u}} \right) du + \left\{ \frac{x^3}{3} \right\}_1^6 = \int_4^9 \left(\sqrt{u} - \frac{3}{\sqrt{u}} \right) du + \frac{215}{3}$

$= \left\{ \frac{2}{3}u^{3/2} - 6\sqrt{u} \right\}_4^9 + \frac{215}{3} = \frac{235}{3}$

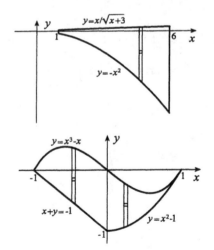

24. $A = \int_{-1}^0 \int_{-1-x}^{x^3-x} dy\, dx + \int_0^1 \int_{x^2-1}^{x^3-x} dy\, dx$

$= \int_{-1}^0 (x^3+1)\, dx + \int_0^1 (x^3 - x - x^2 + 1)\, dx$

$= \left\{ \frac{x^4}{4} + x \right\}_{-1}^0 + \left\{ \frac{x^4}{4} - \frac{x^2}{2} - \frac{x^3}{3} + x \right\}_0^1 = \frac{7}{6}$

26. $A = 4 \displaystyle\int_1^2 \int_0^{\sqrt{4-x^2}} dy\, dx = 4 \int_1^2 \sqrt{4-x^2}\, dx$

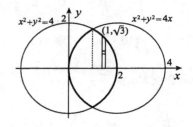

If we set $x = 2\sin\theta$ and $dx = 2\cos\theta\, d\theta$,

$A = 4 \displaystyle\int_{\pi/6}^{\pi/2} 2\cos\theta\, 2\cos\theta\, d\theta = 16 \int_{\pi/6}^{\pi/2} \left(\frac{1+\cos 2\theta}{2}\right) d\theta$

$= 8 \left\{ \theta + \dfrac{1}{2}\sin 2\theta \right\}_{\pi/6}^{\pi/2} = \dfrac{8\pi}{3} - 2\sqrt{3}$

28. $A = 2 \displaystyle\int_{-9}^0 \int_0^{x^2\sqrt{9+x}} dy\, dx = 2 \int_{-9}^0 x^2\sqrt{9+x}\, dx$

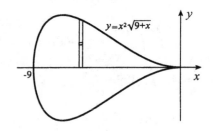

If we set $u = 9 + x$ and $du = dx$,

$A = 2 \displaystyle\int_0^9 (u-9)^2 \sqrt{u}\, du$

$= 2 \displaystyle\int_0^9 (u^{5/2} - 18u^{3/2} + 81u^{1/2})\, du$

$= 2 \left\{ \dfrac{2u^{7/2}}{7} - \dfrac{36u^{5/2}}{5} + \dfrac{162u^{3/2}}{3} \right\}_0^9 = \dfrac{23\,328}{35}$

30. $A = \displaystyle\int_0^4 \int_0^{(4-x)/(x+2)^2} dy\, dx = \int_0^4 \frac{4-x}{(x+2)^2}\, dx$

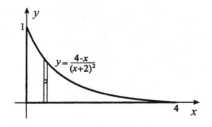

$= \displaystyle\int_0^4 \left[\frac{6}{(x+2)^2} - \frac{1}{x+2} \right] dx$

$= \left\{ -\dfrac{6}{x+2} - \ln|x+2| \right\}_0^4 = 2 - \ln 3$

32. We reject the area below $y = -1$ to obtain

$V = 2\pi(1)^2(1) + 2 \displaystyle\int_1^{\sqrt{2}} \int_{x^2-2}^0 2\pi(y+1)\, dy\, dx$

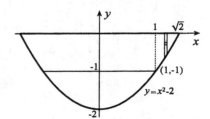

$= 2\pi + 4\pi \displaystyle\int_1^{\sqrt{2}} \left\{ \frac{1}{2}(y+1)^2 \right\}_{x^2-2}^0 dx$

$= 2\pi + 2\pi \displaystyle\int_1^{\sqrt{2}} (-x^4 + 2x^2)\, dx$

$= 2\pi + 2\pi \left\{ -\dfrac{x^5}{5} + \dfrac{2x^3}{3} \right\}_1^{\sqrt{2}} = \dfrac{16\pi(\sqrt{2}+1)}{15}$

34. $V = \displaystyle\int_{-1}^0 \int_{\sqrt{4+12y^2}}^{20y+24} 2\pi(-y)\, dx\, dy = 2\pi \int_{-1}^0 \left\{ -xy \right\}_{\sqrt{4+12y^2}}^{20y+24} dy$

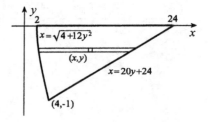

$= 2\pi \displaystyle\int_{-1}^0 (y\sqrt{4+12y^2} - 20y^2 - 24y)\, dy$

$= 2\pi \left\{ \dfrac{1}{36}(4+12y^2)^{3/2} - \dfrac{20y^3}{3} - 12y^2 \right\}_{-1}^0 = \dfrac{68\pi}{9}$

36. $\displaystyle A = 2\int_0^{2/3}\int_0^{\sqrt{6x-x^2}} dy\,dx + 2\int_{2/3}^{2}\int_0^{\sqrt{4-x^2}} dy\,dx$

$\displaystyle \quad = 2\int_0^{2/3}\sqrt{6x-x^2}\,dx + 2\int_{2/3}^{2}\sqrt{4-x^2}\,dx$

$\displaystyle \quad = 2\int_0^{2/3}\sqrt{9-(x-3)^2}\,dx + 2\int_{2/3}^{2}\sqrt{4-x^2}\,dx$

If we set $x - 3 = 3\sin\theta$ in the first integral,
and $x = 2\sin\phi$ in the second,

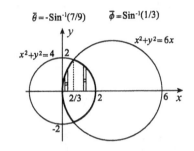

$\bar\theta = \text{-}\mathrm{Sin}^{\text{-}1}(7/9) \qquad \bar\phi = \mathrm{Sin}^{\text{-}1}(1/3)$

$\displaystyle A = 2\int_{-\pi/2}^{\bar\theta} 3\cos\theta\, 3\cos\theta\, d\theta + 2\int_{\bar\phi}^{\pi/2} 2\cos\phi\, 2\cos\phi\, d\phi$

$\displaystyle \quad = 18\int_{-\pi/2}^{\bar\theta}\left(\frac{1+\cos 2\theta}{2}\right)d\theta + 8\int_{\bar\phi}^{\pi/2}\left(\frac{1+\cos 2\phi}{2}\right)d\phi$

$\displaystyle \quad = 9\left\{\theta + \frac{1}{2}\sin 2\theta\right\}_{-\pi/2}^{\bar\theta} + 4\left\{\phi + \frac{1}{2}\sin 2\phi\right\}_{\bar\phi}^{\pi/2}$

$\displaystyle \quad = 9\left(\bar\theta + \sin\bar\theta\,\cos\bar\theta + \frac{\pi}{2}\right) + 4\left(\frac{\pi}{2} - \bar\phi - \sin\bar\phi\,\cos\bar\phi\right) = 5.38.$

38. $\displaystyle V = \iint_R 2\pi\frac{|2x - y + 3|}{\sqrt{5}}\,dA = \frac{2\pi}{\sqrt{5}}\int_{-1}^{3}\int_{x^2}^{2x+3}(2x - y + 3)\,dy\,dx$

$\displaystyle \quad = \frac{2\pi}{\sqrt{5}}\int_{-1}^{3}\left\{(2x+3)y - \frac{y^2}{2}\right\}_{x^2}^{2x+3} dx$

$\displaystyle \quad = \frac{2\pi}{\sqrt{5}}\int_{-1}^{3}\left[\frac{1}{2}(2x+3)^2 + \frac{x^4}{2} - 2x^3 - 3x^2\right]dx$

$\displaystyle \quad = \frac{2\pi}{\sqrt{5}}\left\{\frac{1}{12}(2x+3)^3 + \frac{x^5}{10} - \frac{x^4}{2} - x^3\right\}_{-1}^{3} = \frac{512\pi}{15\sqrt{5}}$

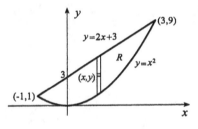

40. $\displaystyle V = \iint_R 2\pi\frac{|x+y+1|}{\sqrt{2}}\,dA = \sqrt{2}\pi\int_0^1\int_{2y}^{y+1}(x+y+1)\,dx\,dy$

$\displaystyle \quad = \sqrt{2}\pi\int_0^1\left\{\frac{1}{2}(x+y+1)^2\right\}_{2y}^{y+1} dy$

$\displaystyle \quad = \frac{\pi}{\sqrt{2}}\int_0^1\left[(2y+2)^2 - (3y+1)^2\right]dy$

$\displaystyle \quad = \frac{\pi}{\sqrt{2}}\left\{\frac{1}{6}(2y+2)^3 - \frac{1}{9}(3y+1)^3\right\}_0^1 = \frac{7\sqrt{2}\pi}{6}$

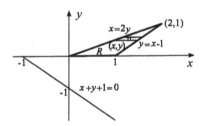

EXERCISES 13.4

2. $\displaystyle F = 2\int_0^6\int_{x^2/9-4}^{0}\rho g(-y)\,dy\,dx \quad (g = 9.81)$

$\displaystyle \quad = -2\rho g\int_0^6\left\{\frac{y^2}{2}\right\}_{x^2/9-4}^{0} dx = \rho g\int_0^6\left(\frac{x^2}{9} - 4\right)^2 dx$

$\displaystyle \quad = \frac{\rho g}{81}\int_0^6(x^4 - 72x^2 + 1296)\,dx$

$\displaystyle \quad = \frac{\rho g}{81}\left\{\frac{x^5}{5} - 24x^3 + 1296x\right\}_0^6 = \frac{256\rho g}{5}$

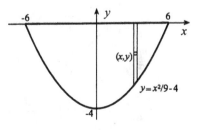

4. $F = 2 \displaystyle\int_0^4 \int_{3x/4-6}^{-3} \rho g(-y)\, dy\, dx \quad (g = 9.81)$

$= -2\rho g \displaystyle\int_0^4 \left\{ \frac{y^2}{2} \right\}_{3x/4-6}^{-3} dx = \rho g \int_0^4 \left[\left(\frac{3x}{4} - 6 \right)^2 - 9 \right] dx$

$= \rho g \left\{ \dfrac{4}{9} \left(\dfrac{3x}{4} - 6 \right)^3 - 9x \right\}_0^4 = 48\rho g$

6. $F = \displaystyle\int_0^5 \int_{2x/5-8}^0 \rho g(-y)\, dy\, dx \quad (g = 9.81)$

$= -\rho g \displaystyle\int_0^5 \left\{ \frac{y^2}{2} \right\}_{2x/5-8}^0 dx = \frac{\rho g}{2} \int_0^5 \left(\frac{2x}{5} - 8 \right)^2 dx$

$= \dfrac{\rho g}{2} \left\{ \dfrac{5}{6} \left(\dfrac{2x}{5} - 8 \right)^3 \right\}_0^5 = \dfrac{370\rho g}{3}$

8. $F = 2 \displaystyle\int_0^5 \int_{-\sqrt{25-x^2}}^0 \rho g(-y)\, dy\, dx \quad (g = 9.81)$

$= -2\rho g \displaystyle\int_0^5 \left\{ \frac{y^2}{2} \right\}_{-\sqrt{25-x^2}}^0 dx = \rho g \int_0^5 (25 - x^2)\, dx$

$= \rho g \left\{ 25x - \dfrac{x^3}{3} \right\}_0^5 = \dfrac{250\rho g}{3}$

10. $F = 2 \displaystyle\int_0^{18} \int_{x^2/36-9}^0 \rho g(-y)\, dy\, dx \quad (g = 9.81)$

$= -2\rho g \displaystyle\int_0^{18} \left\{ \frac{y^2}{2} \right\}_{x^2/36-9}^0 dx = \rho g \int_0^{18} \left(\frac{x^4}{1296} - \frac{x^2}{2} + 81 \right) dx$

$= 9810 \left\{ \dfrac{x^5}{1296 \cdot 5} - \dfrac{x^3}{6} + 81x \right\}_0^{18} = 7.63 \times 10^6 \text{ N}$

12. $F = \displaystyle\int_{-10/\sqrt{29}}^0 \int_{-2x/5-\sqrt{29}}^{5x/2} \rho g(-y)\, dy\, dx \quad (g = 9.81)$

$\qquad + \displaystyle\int_0^{10/\sqrt{29}} \int_{5x/2-\sqrt{29}}^{-2x/5} \rho g(-y)\, dy\, dx$

$= -\rho g \displaystyle\int_{-10/\sqrt{29}}^0 \left\{ \frac{y^2}{2} \right\}_{-2x/5-\sqrt{29}}^{5x/2} dx$

$\qquad -\rho g \displaystyle\int_0^{10/\sqrt{29}} \left\{ \frac{y^2}{2} \right\}_{5x/2-\sqrt{29}}^{-2x/5} dx$

$= \dfrac{\rho g}{2} \displaystyle\int_{-10/\sqrt{29}}^0 \left[\left(-\frac{2x}{5} - \sqrt{29} \right)^2 - \frac{25x^2}{4} \right] dx$

$\qquad + \dfrac{\rho g}{2} \displaystyle\int_0^{10/\sqrt{29}} \left[\left(\frac{5x}{2} - \sqrt{29} \right)^2 - \frac{4x^2}{25} \right] dx$

$= \dfrac{\rho g}{2} \left\{ \dfrac{5}{6} \left(\dfrac{2x}{5} + \sqrt{29} \right)^3 - \dfrac{25x^3}{12} \right\}_{-10/\sqrt{29}}^0 + \dfrac{\rho g}{2} \left\{ \dfrac{2}{15} \left(\dfrac{5x}{2} - \sqrt{29} \right)^3 - \dfrac{4x^3}{75} \right\}_0^{10/\sqrt{29}} = 5\sqrt{29}\rho g$

14. $F = \displaystyle\int_{-2}^{0} \rho g(-y)5\,dy \quad (g = 9.81)$

$= -5\rho g\left\{\dfrac{y^2}{2}\right\}_{-2}^{0} = 10\rho g$

16. $F = 2\displaystyle\int_{-r}^{r}\int_{0}^{\sqrt{r^2-y^2}} \rho g(r-y)\,dx\,dy \quad (g = 9.81)$

$= 2\rho g\displaystyle\int_{-r}^{r}\left\{x(r-y)\right\}_{0}^{\sqrt{r^2-y^2}}dy = 2\rho g\displaystyle\int_{-r}^{r}(r-y)\sqrt{r^2-y^2}\,dy$

If we set $y = r\sin\theta$ and $dy = r\cos\theta\,d\theta$ in the first term,

$F = 2\rho g\displaystyle\int_{-\pi/2}^{\pi/2} r(r\cos\theta)r\cos\theta\,d\theta + 2\rho g\left\{\dfrac{1}{3}(r^2-y^2)^{3/2}\right\}_{-r}^{r}$

$= 2\rho g r^3\displaystyle\int_{-\pi/2}^{\pi/2}\left(\dfrac{1+\cos 2\theta}{2}\right)d\theta = \rho g r^3\left\{\theta + \dfrac{\sin 2\theta}{2}\right\}_{-\pi/2}^{\pi/2} = \pi\rho g r^3$

18. According to Exercise 40 in Section 7.7, the force on the plate is $F = \rho g h(\pi r^2) = \pi\rho g h r^2$. By symmetry, $x_c = 0$. If we integrate over the right-half of the circle and double the result,

$Fy_c = 2\displaystyle\int_{-r}^{r}\int_{0}^{\sqrt{r^2-y^2}} y\rho g(h-y)\,dx\,dy$

$= 2\rho g\displaystyle\int_{-r}^{r}\left\{xy(h-y)\right\}_{0}^{\sqrt{r^2-y^2}}dy$

$= 2\rho g\displaystyle\int_{-r}^{r}(hy\sqrt{r^2-y^2} - y^2\sqrt{r^2-y^2})\,dy.$

If we set $y = r\sin\theta$ and $dy = r\cos\theta\,d\theta$ in the last term,

$y_c = \dfrac{2\rho g h}{F}\left\{-\dfrac{1}{3}(r^2-y^2)^{3/2}\right\}_{-r}^{r} - \dfrac{2\rho g}{F}\displaystyle\int_{-\pi/2}^{\pi/2} r^2\sin^2\theta\,r\cos\theta\,r\cos\theta\,d\theta$

$= -\dfrac{2\rho g r^4}{F}\displaystyle\int_{-\pi/2}^{\pi/2}\left(\dfrac{1-\cos 4\theta}{8}\right)d\theta = -\dfrac{\rho g r^4}{4F}\left\{\theta - \dfrac{\sin 4\theta}{4}\right\}_{-\pi/2}^{\pi/2} = -\dfrac{\rho g\pi r^4}{4(\pi\rho g h r^2)} = -\dfrac{r^2}{4h}.$

20. The fluid force on the triangle is

$F = 2\displaystyle\int_{0}^{\sqrt{3}L/2}\int_{0}^{y/\sqrt{3}} \rho g\left(\dfrac{\sqrt{3}L}{2} - y\right)dx\,dy$

$= \rho g\displaystyle\int_{0}^{\sqrt{3}L/2}\left\{x(\sqrt{3}L - 2y)\right\}_{0}^{y/\sqrt{3}}dy$

$= \dfrac{\rho g}{\sqrt{3}}\displaystyle\int_{0}^{\sqrt{3}L/2}(\sqrt{3}Ly - 2y^2)\,dy$

$= \dfrac{\rho g}{\sqrt{3}}\left\{\dfrac{\sqrt{3}Ly^2}{2} - \dfrac{2y^3}{3}\right\}_{0}^{\sqrt{3}L/2} = \dfrac{\rho g L^3}{8}.$

By symmetry, $x_c = 0$, and

$y_c = \dfrac{2}{F}\displaystyle\int_{0}^{\sqrt{3}L/2}\int_{0}^{y/\sqrt{3}} \rho g y\left(\dfrac{\sqrt{3}L}{2} - y\right)dx\,dy = \dfrac{\rho g}{F}\displaystyle\int_{0}^{\sqrt{3}L/2}\left\{xy(\sqrt{3}L - 2y)\right\}_{0}^{y/\sqrt{3}}dy$

$= \dfrac{\rho g}{\sqrt{3}F}\displaystyle\int_{0}^{\sqrt{3}L/2}(\sqrt{3}Ly^2 - 2y^3)\,dy = \dfrac{\rho g}{\sqrt{3}F}\left\{\dfrac{Ly^3}{\sqrt{3}} - \dfrac{y^4}{2}\right\}_{0}^{\sqrt{3}L/2}$

$= \dfrac{\sqrt{3}\rho g L^4}{32F} = \dfrac{\sqrt{3}\rho g L^4}{32}\dfrac{8}{\rho g L^3} = \dfrac{\sqrt{3}L}{4}.$

22. The force on the triangle is

$$F = \int_{-L}^{0} \int_{0}^{ly/L+l} \rho g(-y)\, dx\, dy = -\rho g \int_{-L}^{0} \left\{ xy \right\}_{0}^{ly/L+l} dy$$

$$= -\rho g \int_{-L}^{0} y\left(\frac{ly}{L} + l\right) dy = -\rho g \left\{ \frac{ly^3}{3L} + \frac{ly^2}{2} \right\}_{-L}^{0} = \frac{\rho g l L^2}{6}.$$

According to equations 13.30,

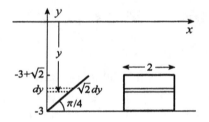

$$x_c = \frac{1}{F} \int_{-L}^{0} \int_{0}^{ly/L+l} \rho g(-y)x\, dx\, dy = -\frac{\rho g}{F} \int_{-L}^{0} \left\{ \frac{x^2 y}{2} \right\}_{0}^{ly/L+l} dy$$

$$= -\frac{\rho g}{2F} \int_{-L}^{0} y\left(\frac{l^2 y^2}{L^2} + \frac{2l^2 y}{L} + l^2\right) dy = -\frac{\rho g l^2}{2F} \left\{ \frac{y^4}{4L^2} + \frac{2y^3}{3L} + \frac{y^2}{2} \right\}_{-L}^{0}$$

$$= \frac{\rho g l^2 L^2}{24 F} = \frac{\rho g l^2 L^2}{24} \frac{6}{\rho g l L^2} = \frac{l}{4},$$

$$y_c = \frac{1}{F} \int_{-L}^{0} \int_{0}^{ly/L+l} \rho g(-y)y\, dx\, dy = -\frac{\rho g}{F} \int_{-L}^{0} \left\{ xy^2 \right\}_{0}^{ly/L+l} dy$$

$$= -\frac{\rho g}{F} \int_{-L}^{0} y^2 \left(\frac{ly}{L} + l\right) dy = -\frac{\rho g l}{F} \left\{ \frac{y^4}{4L} + \frac{y^3}{3} \right\}_{-L}^{0} = -\frac{\rho g l L^3}{12F} = -\frac{\rho g l L^3}{12} \frac{6}{\rho g l L^2} = -\frac{L}{2}.$$

24. $F = \int_{-3}^{-3+\sqrt{2}} \rho g(-y)(2)(\sqrt{2}\, dy)$ $(g = 9.81)$

$$= -\sqrt{2}\rho g \left\{ y^2 \right\}_{-3}^{-3+\sqrt{2}}$$

$$= 9.00 \times 10^4 \text{ N}$$

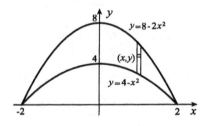

EXERCISES 13.5

2. $A = 2 \int_{0}^{2} \int_{4-x^2}^{8-2x^2} dy\, dx = 2 \int_{0}^{2} (4 - x^2)\, dx$

$$= 2 \left\{ 4x - \frac{x^3}{3} \right\}_{0}^{2} = \frac{32}{3}$$

By symmetry, $\bar{x} = 0$. Since

$$A\bar{y} = 2 \int_{0}^{2} \int_{4-x^2}^{8-2x^2} y\, dy\, dx = 2 \int_{0}^{2} \left\{ \frac{y^2}{2} \right\}_{4-x^2}^{8-2x^2} dx$$

$$= 3 \int_{0}^{2} (16 - 8x^2 + x^4)\, dx = 3 \left\{ 16x - \frac{8x^3}{3} + \frac{x^5}{5} \right\}_{0}^{2} = \frac{256}{5},$$

we obtain $\bar{y} = \frac{256}{5} \frac{3}{32} = \frac{24}{5}.$

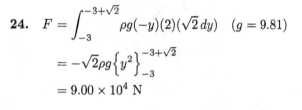

4. $A = \displaystyle\int_1^4 \int_{4/x}^{5-x} dy\, dx = \int_1^4 (5 - x - 4/x)\, dx = \left\{ 5x - \dfrac{x^2}{2} - 4\ln|x| \right\}_1^4 = \dfrac{15}{2} - 4\ln 4$

From

$A\bar{x} = \displaystyle\int_1^4 \int_{4/x}^{5-x} x\, dy\, dx = \int_1^4 \left\{ xy \right\}_{4/x}^{5-x} dx$

$= \displaystyle\int_1^4 (5x - x^2 - 4)\, dx = \left\{ \dfrac{5x^2}{2} - \dfrac{x^3}{3} - 4x \right\}_1^4 = \dfrac{9}{2},$

we obtain $\bar{x} = \dfrac{9}{2} \dfrac{2}{15 - 8\ln 4} = \dfrac{9}{15 - 16\ln 2}$. Since

$A\bar{y} = \displaystyle\int_1^4 \int_{4/x}^{5-x} y\, dy\, dx = \int_1^4 \left\{ \dfrac{y^2}{2} \right\}_{4/x}^{5-x} dx = \dfrac{1}{2}\int_1^4 \left[(5-x)^2 - \dfrac{16}{x^2} \right] dx = \dfrac{1}{2}\left\{ -\dfrac{1}{3}(5-x)^3 + \dfrac{16}{x} \right\}_1^4 = \dfrac{9}{2},$

$\bar{y}$ is also equal to $9/(15 - 16\ln 2)$.

6. $A = \dfrac{1}{8}\pi(2)^2 = \dfrac{\pi}{2}$ Since

$A\bar{x} = \displaystyle\int_0^{\sqrt{2}} \int_x^{\sqrt{4-x^2}} x\, dy\, dx$

$= \displaystyle\int_0^{\sqrt{2}} \left\{ xy \right\}_x^{\sqrt{4-x^2}} dx = \int_0^{\sqrt{2}} (x\sqrt{4-x^2} - x^2)\, dx$

$= \left\{ -\dfrac{1}{3}(4-x^2)^{3/2} - \dfrac{x^3}{3} \right\}_0^{\sqrt{2}} = \dfrac{4}{3}(2 - \sqrt{2}),$

it follows that $\bar{x} = \dfrac{4}{3}(2 - \sqrt{2})\dfrac{2}{\pi} = \dfrac{8(2 - \sqrt{2})}{3\pi}$. Since

$A\bar{y} = \displaystyle\int_0^{\sqrt{2}} \int_x^{\sqrt{4-x^2}} y\, dy\, dx = \int_0^{\sqrt{2}} \left\{ \dfrac{y^2}{2} \right\}_x^{\sqrt{4-x^2}} dx = \int_0^{\sqrt{2}} (2 - x^2)\, dx = \left\{ 2x - \dfrac{x^3}{3} \right\}_0^{\sqrt{2}} = \dfrac{4\sqrt{2}}{3},$

we obtain $\bar{y} = \dfrac{4\sqrt{2}}{3}\dfrac{2}{\pi} = \dfrac{8\sqrt{2}}{3\pi}$.

8. $A = \displaystyle\int_0^1 \int_{4y-4y^2}^{y+3} dx\, dy = \int_0^1 (3 - 3y + 4y^2)\, dy$

$= \left\{ 3y - \dfrac{3y^2}{2} + \dfrac{4y^3}{3} \right\}_0^1 = \dfrac{17}{6}$

Since

$A\bar{x} = \displaystyle\int_0^1 \int_{4y-4y^2}^{y+3} x\, dx\, dy = \int_0^1 \left\{ \dfrac{x^2}{2} \right\}_{4y-4y^2}^{y+3} dy$

$= \dfrac{1}{2}\displaystyle\int_0^1 [(3+y)^2 - 16y^2 + 32y^3 - 16y^4]\, dy = \dfrac{1}{2}\left\{ \dfrac{(3+y)^3}{3} - \dfrac{16y^3}{3} + 8y^4 - \dfrac{16y^5}{5} \right\}_0^1 = \dfrac{59}{10},$

we find $\bar{x} = \dfrac{59}{10}\dfrac{6}{17} = \dfrac{177}{85}$. Since

$A\bar{y} = \displaystyle\int_0^1 \int_{4y-4y^2}^{y+3} y\, dx\, dy = \int_0^1 \left\{ xy \right\}_{4y-4y^2}^{y+3} dy = \int_0^1 (3y - 3y^2 + 4y^3)\, dy = \left\{ \dfrac{3y^2}{2} - y^3 + y^4 \right\}_0^1 = \dfrac{3}{2},$

we obtain $\bar{y} = \dfrac{3}{2}\dfrac{6}{17} = \dfrac{9}{17}$.

10. $A = \int_0^1 \int_x^{2x} dy\, dx + \int_1^3 \int_x^{(x+3)/2} dy\, dx$

$= \int_0^1 x\, dx + \frac{1}{2}\int_1^3 (3-x)\, dx$

$= \left\{\frac{x^2}{2}\right\}_0^1 + \frac{1}{2}\left\{-\frac{1}{2}(3-x)^2\right\}_1^3 = \frac{3}{2}$

Since

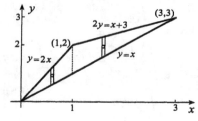

$A\bar{x} = \int_0^1 \int_x^{2x} x\, dy\, dx + \int_1^3 \int_x^{(x+3)/2} x\, dy\, dx$

$= \int_0^1 x^2\, dx + \frac{1}{2}\int_1^3 (3x - x^2)\, dx = \left\{\frac{x^3}{3}\right\}_0^1 + \frac{1}{2}\left\{\frac{3x^2}{2} - \frac{x^3}{3}\right\}_1^3 = 2,$

it follows that $\bar{x} = 2 \cdot \frac{2}{3} = \frac{4}{3}$. With

$A\bar{y} = \int_0^1 \int_x^{2x} y\, dy\, dx + \int_1^3 \int_x^{(x+3)/2} y\, dy\, dx = \int_0^1 \left\{\frac{y^2}{2}\right\}_x^{2x} dx + \int_1^3 \left\{\frac{y^2}{2}\right\}_x^{(x+3)/2} dx$

$= \frac{3}{2}\int_0^1 x^2\, dx + \frac{1}{2}\int_1^3 \left[\frac{1}{4}(x+3)^2 - x^2\right] dx = \frac{3}{2}\left\{\frac{x^3}{3}\right\}_0^1 + \frac{1}{2}\left\{\frac{1}{12}(x+3)^3 - \frac{x^3}{3}\right\}_1^3 = \frac{5}{2},$

we obtain $\bar{y} = (5/2)(2/3) = 5/3$.

12. $I = \int_{-4}^0 \int_{(y-4)/2}^y y^2\, dx\, dy = \int_{-4}^0 \left\{xy^2\right\}_{(y-4)/2}^y dy$

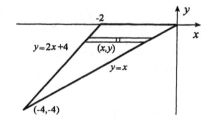

$= \frac{1}{2}\int_{-4}^0 (4y^2 + y^3)\, dy = \frac{1}{2}\left\{\frac{4y^3}{3} + \frac{y^4}{4}\right\}_{-4}^0 = \frac{32}{3}$

14. $I = \int_{-1}^2 \int_{x^2-4}^{2x-x^2} (x+2)^2\, dy\, dx$

$= \int_{-1}^2 \left\{y(x+2)^2\right\}_{x^2-4}^{2x-x^2} dx$

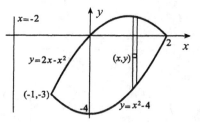

$= 2\int_{-1}^2 (8 + 12x + 2x^2 - 3x^3 - x^4)\, dx$

$= 2\left\{8x + 6x^2 + \frac{2x^3}{3} - \frac{3x^4}{4} - \frac{x^5}{5}\right\}_{-1}^2 = \frac{603}{10}$

16. Moment $= 2\int_0^1 \int_{x^2-1}^{2-2x^2} \rho(y+2)\, dy\, dx$

$= 2\rho\int_0^1 \left\{\frac{1}{2}(y+2)^2\right\}_{x^2-1}^{2-2x^2} dx$

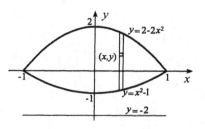

$= 3\rho\int_0^1 (5 - 6x^2 + x^4)\, dx$

$= 3\rho\left\{5x - 2x^3 + \frac{x^5}{5}\right\}_0^1 = \frac{48\rho}{5}$

18. $I_{xy} = \int_0^1 \int_{x^3}^{x^2} xy\rho\, dy\, dx$

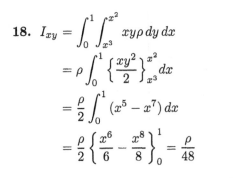

$= \rho \int_0^1 \left\{ \frac{xy^2}{2} \right\}_{x^3}^{x^2} dx$

$= \frac{\rho}{2} \int_0^1 (x^5 - x^7)\, dx$

$= \frac{\rho}{2} \left\{ \frac{x^6}{6} - \frac{x^8}{8} \right\}_0^1 = \frac{\rho}{48}$

20. $I_{xy} = \int_{-2}^{-1} \int_{-3y-2}^{-2y} xy\rho\, dx\, dy + \int_{-1}^0 \int_{-y}^{-2y} xy\rho\, dx\, dy$

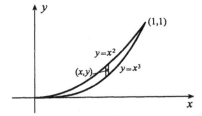

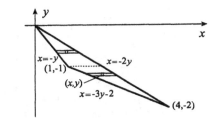

$= \rho \int_{-2}^{-1} \left\{ \frac{x^2 y}{2} \right\}_{-3y-2}^{-2y} dy + \rho \int_{-1}^0 \left\{ \frac{x^2 y}{2} \right\}_{-y}^{-2y} dy$

$= \frac{\rho}{2} \int_{-2}^{-1} (-5y^3 - 12y^2 - 4y)\, dy + \frac{3\rho}{2} \int_{-1}^0 y^3\, dy$

$= \frac{\rho}{2} \left\{ -\frac{5y^4}{4} - 4y^3 - 2y^2 \right\}_{-2}^{-1} + \frac{3\rho}{2} \left\{ \frac{y^4}{4} \right\}_{-1}^0$

$= -2\rho$

22. $A = 2 \int_0^1 \int_{-x^2}^2 dy\, dx + 2 \left(\frac{1}{2} \right)(3)(3)$

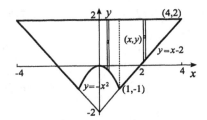

$= 2 \int_0^1 (2 + x^2)\, dx + 9$

$= 2 \left\{ 2x + \frac{x^3}{3} \right\}_0^1 + 9 = \frac{41}{3}$

By symmetry, $\bar{x} = 0$. Since

$A\bar{y} = 2 \int_0^1 \int_{-x^2}^2 y\, dy\, dx + 2 \int_1^4 \int_{x-2}^2 y\, dy\, dx$

$= 2 \int_0^1 \left\{ \frac{y^2}{2} \right\}_{-x^2}^2 dx + 2 \int_1^4 \left\{ \frac{y^2}{2} \right\}_{x-2}^2 dx$

$= \int_0^1 (4 - x^4)\, dx + \int_1^4 [4 - (x-2)^2]\, dx = \left\{ 4x - \frac{x^5}{5} \right\}_0^1 + \left\{ 4x - \frac{1}{3}(x-2)^3 \right\}_1^4 = \frac{64}{5},$

we obtain $\bar{y} = \frac{64}{5} \frac{3}{41} = \frac{192}{205}$.

24. Since the area inside an ellipse is π multiplied by the product of half the major and minor axes,

$A = \pi(4)(2\sqrt{3}) - \pi(1)^2 = \pi(8\sqrt{3} - 1).$

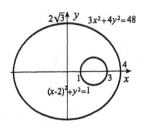

By symmetry, $\bar{y} = 0$. Since the first moment
of the area about the y-axis is that of the
ellipse less that of the circle,

$A\bar{x} = 0 - 2\pi(1)^2 = -2\pi.$

Thus, $\bar{x} = \frac{-2\pi}{\pi(8\sqrt{3} - 1)} = \frac{-2}{8\sqrt{3} - 1}.$

26. $A = \int_{-7}^{2} \int_{(x^2-4)/15}^{\sqrt{2-x}} dy\, dx$

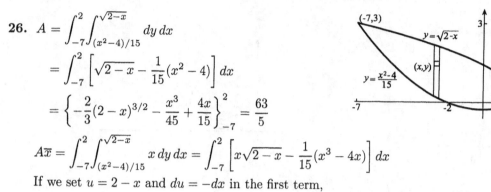

$$= \int_{-7}^{2} \left[\sqrt{2-x} - \frac{1}{15}(x^2-4) \right] dx$$

$$= \left\{ -\frac{2}{3}(2-x)^{3/2} - \frac{x^3}{45} + \frac{4x}{15} \right\}_{-7}^{2} = \frac{63}{5}$$

$A\bar{x} = \int_{-7}^{2} \int_{(x^2-4)/15}^{\sqrt{2-x}} x\, dy\, dx = \int_{-7}^{2} \left[x\sqrt{2-x} - \frac{1}{15}(x^3 - 4x) \right] dx$

If we set $u = 2 - x$ and $du = -dx$ in the first term,

$$A\bar{x} = \int_{9}^{0} (2-u)\sqrt{u}(-du) - \frac{1}{15}\left\{ \frac{x^4}{4} - 2x^2 \right\}_{-7}^{2} = \left\{ \frac{4}{3}u^{3/2} - \frac{2}{5}u^{5/2} \right\}_{0}^{9} - \frac{1}{15}\left\{ \frac{x^4}{4} - 2x^2 \right\}_{-7}^{2} = -\frac{549}{20}.$$

Thus, $\bar{x} = -(549/20)(5/63) = -61/28$. Since

$$A\bar{y} = \int_{-7}^{2} \int_{(x^2-4)/15}^{\sqrt{2-x}} y\, dy\, dx = \int_{-7}^{2} \left\{ \frac{y^2}{2} \right\}_{(x^2-4)/15}^{\sqrt{2-x}} dx = \frac{1}{2}\int_{-7}^{2} \left[2 - x - \frac{1}{225}(x^4 - 8x^2 + 16) \right] dx$$

$$= \frac{1}{450}\left\{ 450x - \frac{225x^2}{2} - \frac{x^5}{5} + \frac{8x^3}{3} - 16x \right\}_{-7}^{2} = \frac{7263}{500},$$

we find $\bar{y} = (7263/500)(5/63) = 807/700$.

28. $I = \iint_R (x^2 + y^2)\rho\, dA = 4\rho \int_{0}^{a/2} \int_{0}^{b/2} (x^2 + y^2)\, dy\, dx$

$$= 4\rho \int_{0}^{a/2} \left\{ x^2 y + \frac{y^3}{3} \right\}_{0}^{b/2} dx = 4\rho \int_{0}^{a/2} \left(\frac{bx^2}{2} + \frac{b^3}{24} \right) dx$$

$$= 4\rho \left\{ \frac{bx^3}{6} + \frac{b^3 x}{24} \right\}_{0}^{a/2} = \frac{\rho ab}{12}(a^2 + b^2)$$

30. $I = \int_{-2}^{1} \int_{y^2}^{2-y} (x+1)^2\, dx\, dy = \int_{-2}^{1} \left\{ \frac{1}{3}(x+1)^3 \right\}_{y^2}^{2-y} dy$

$$= \frac{1}{3}\int_{-2}^{1} [(3-y)^3 - y^6 - 3y^4 - 3y^2 - 1]\, dy$$

$$= \frac{1}{3}\left\{ -\frac{1}{4}(3-y)^4 - \frac{y^7}{7} - \frac{3y^5}{5} - y^3 - y \right\}_{-2}^{1} = \frac{4761}{140}$$

32. If we take directed distances to the right of the line $x + y = 1$ as positive, then the directed distance from the line to the area $dy\, dx$ at point (x, y) is $(x+y-1)/\sqrt{2}$. The first moment of the plate about the line is

$$M = \int_{-1}^{2} \int_{y^2-2}^{y} \frac{x+y-1}{\sqrt{2}}\, dx\, dy = \frac{1}{\sqrt{2}}\int_{-1}^{2} \left\{ \frac{(x+y-1)^2}{2} \right\}_{y^2-2}^{y} dy$$

$$= \frac{1}{2\sqrt{2}}\int_{-1}^{2} [(2y-1)^2 - y^4 - 2y^3 + 5y^2 + 6y - 9]\, dy$$

$$= \frac{1}{2\sqrt{2}}\left\{ \frac{(2y-1)^3}{6} - \frac{y^5}{5} - \frac{y^4}{2} + \frac{5y^3}{3} + 3y^2 - 9y \right\}_{-1}^{2} = -\frac{81\sqrt{2}}{40}.$$

The second moment about the line is

$$I = \int_{-1}^{2} \int_{y^2-2}^{y} \frac{(x+y-1)^2}{2}\, dx\, dy = \frac{1}{2}\int_{-1}^{2} \left\{ \frac{(x+y-1)^3}{3} \right\}_{y^2-2}^{y} dy$$

$$= \frac{1}{6} \int_{-1}^{2} [(2y-1)^3 - y^6 - 3y^5 + 6y^4 + 17y^3 - 18y^2 - 27y + 27] \, dy$$

$$= \frac{1}{6} \left\{ \frac{(2y-1)^4}{8} - \frac{y^7}{7} - \frac{y^6}{2} + \frac{6y^5}{5} + \frac{17y^4}{4} - 6y^3 - \frac{27y^2}{2} + 27y \right\}_{-1}^{2} = \frac{1863}{280}.$$

34. If we take directed distances to the right of the line $2x + y = 3$ as positive, then the directed distance from the line to the area $dy \, dx$ at point (x, y) is $(2x + y - 3)/\sqrt{5}$. The first moment of the plate about the line is

$$M = \int_0^1 \int_{x^3}^{\sqrt{x}} \frac{2x + y - 3}{\sqrt{5}} \, dy \, dx$$

$$= \frac{1}{\sqrt{5}} \int_0^1 \left\{ \frac{(2x + y - 3)^2}{2} \right\}_{x^3}^{\sqrt{x}} \, dx$$

$$= \frac{1}{2\sqrt{5}} \int_0^1 (-x^6 - 4x^4 + 6x^3 + 4x^{3/2} + x - 6\sqrt{x}) \, dx$$

$$= \frac{1}{2\sqrt{5}} \left\{ -\frac{x^7}{7} - \frac{4x^5}{5} + \frac{3x^4}{2} + \frac{8x^{5/2}}{5} + \frac{x^2}{2} - 4x^{3/2} \right\}_0^1 = -\frac{47\sqrt{5}}{350}.$$

The second moment about the line is

$$I = \int_0^1 \int_{x^3}^{\sqrt{x}} \frac{(2x + y - 3)^2}{5} \, dy \, dx = \frac{1}{5} \int_0^1 \left\{ \frac{(2x + y - 3)^3}{3} \right\}_{x^3}^{\sqrt{x}} \, dx$$

$$= \frac{1}{15} \int_0^1 (-x^9 - 6x^7 + 9x^6 - 12x^5 + 36x^4 - 27x^3 + 6x^2 - 9x + 12x^{5/2} - 35x^{3/2} + 27\sqrt{x}) \, dx$$

$$= \frac{1}{15} \left\{ -\frac{x^{10}}{10} - \frac{3x^8}{4} + \frac{9x^7}{7} - 2x^6 + \frac{36x^5}{5} - \frac{27x^4}{4} + 2x^3 - \frac{9x^2}{2} + \frac{24x^{7/2}}{7} - 14x^{5/2} + 18x^{3/2} \right\}_0^1 = \frac{89}{350}.$$

36. (a) $I_{xy} = \int_0^b \int_0^{h(b-x)/b} xy\rho \, dy \, dx$

$$= \rho \int_0^b \left\{ \frac{xy^2}{2} \right\}_0^{h(b-x)/b} \, dx$$

$$= \frac{\rho h^2}{2b^2} \int_0^b (b^2 x - 2bx^2 + x^3) \, dx$$

$$= \frac{\rho h^2}{2b^2} \left\{ \frac{b^2 x^2}{2} - \frac{2bx^3}{3} + \frac{x^4}{4} \right\}_0^b = \frac{\rho b^2 h^2}{24}$$

(b) The centre of mass is $(b/3, h/3)$. The product moment of inertia about horizontal and vertical lines through this point is

$$I = \int_0^b \int_0^{h(b-x)/b} \left(x - \frac{b}{3} \right) \left(y - \frac{h}{3} \right) \rho \, dy \, dx = \rho \int_0^b \left\{ \frac{1}{2} \left(x - \frac{b}{3} \right) \left(y - \frac{h}{3} \right)^2 \right\}_0^{h(b-x)/b} \, dx$$

$$= \frac{\rho h^2}{18b^2} \int_0^b (9x^3 - 15bx^2 + 7b^2 x - b^3) \, dx = \frac{\rho h^2}{18b^2} \left\{ \frac{9x^4}{4} - 5bx^3 + \frac{7b^2 x^2}{2} - b^3 x \right\}_0^b = -\frac{\rho b^2 h^2}{72}.$$

38. Suppose we choose the point as the origin and axes along the principal axes of the plate. Then one principal axis has slope zero while the slope of the other is undefined. According to equation 13.39, this occurs when $I_{xy} = 0$. This can also be seen by taking $\theta = 0$ or $\theta = \pi/2$ in Exercise 45.

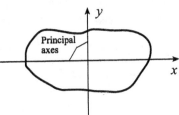

40. Since

$$I_x = I_y = \int_0^a \int_0^a y^2 \rho \, dy \, dx = \rho \int_0^a \left\{ \frac{y^3}{3} \right\}_0^a dx$$

$$= \frac{\rho a^3}{3} \left\{ x \right\}_0^a = \frac{\rho a^4}{3},$$

and

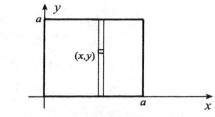

$$I_{xy} = \int_0^a \int_0^a xy\rho \, dy \, dx = \rho \int_0^a \left\{ \frac{xy^2}{2} \right\}_0^a dx = \frac{\rho a^2}{2} \left\{ \frac{x^2}{2} \right\}_0^a = \frac{\rho a^4}{4},$$

slopes of the principal axes are defined by equation 13.39,

$$m = \frac{I_x - I_y}{2I_{xy}} \pm \sqrt{1 + \left(\frac{I_x - I_y}{2I_{xy}} \right)^2} = \pm 1.$$

Principal axes are therefore the lines $y = \pm x$. According to equation 13.40, principal moments of inertia are

$$\frac{I_x + I_y}{2} \pm \sqrt{\left(\frac{I_x - I_y}{2} \right)^2 + (I_{xy})^2} = I_x \pm I_{xy} = \frac{7\rho a^4}{12}, \frac{\rho a^4}{12}.$$

42. Since $\lim_{m \to \pm\infty} I(m) = I_y$, we must show that

$$\frac{I_x + I_y}{2} - \sqrt{\left(\frac{I_x - I_y}{2} \right)^2 + (I_{xy})^2} \leq I_y \leq \frac{I_x + I_y}{2} + \sqrt{\left(\frac{I_x - I_y}{2} \right)^2 + (I_{xy})^2}.$$

But this is equivalent to

$$-\sqrt{\left(\frac{I_x - I_y}{2} \right)^2 + (I_{xy})^2} \leq \frac{I_y - I_x}{2} \leq \sqrt{\left(\frac{I_x - I_y}{2} \right)^2 + (I_{xy})^2},$$

which is valid for any I_x and I_y.

44. Suppose the x-axis is chosen as the axis of symmetry. Choose the y-axis through any point on the line. According to Exercise 43, the product moment of inertia about the origin for the coordinate axes is zero. Consequently the moment of inertia about any line through the origin with slope m is

$$I(m) = \frac{1}{m^2 + 1}(I_x + m^2 I_y) = I_y + \frac{I_x - I_y}{m^2 + 1},$$

(see equation 13.38). If $I_x > I_y$, then this is an even function of m, decreasing from $I(0) = I_x$ to $\lim_{m \to \infty} I(m) = I_y$; that is, principal axes are $x = 0$ and $y = 0$. If $I_x < I_y$, then this even function increases from $I(0) = I_x$ to $\lim_{m \to \infty} I(m) = I_y$, and once again I_x and I_y are principal moments of inertia. If $I_x = I_y$, then $I(m) = I_y$ for all m, in which case all pairs of perpendicular lines through the origin are principal axes.

46. If we orient the area so that the axis of rotation is the y-axis, then

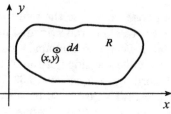

$$V = \iint_R 2\pi x \, dA = 2\pi \iint_R x \, dA$$

$$= 2\pi(A\bar{x}) = (2\pi\bar{x})A.$$

48. If we orient the area so that the line is the
y-axis, then

$$I = \iint_R x^2 \rho \, dA = \rho \iint_R [(x - \overline{x}) + \overline{x}]^2 \, dA$$

$$= \rho \iint_R [(x - \overline{x})^2 + 2\overline{x}(x - \overline{x}) + \overline{x}^2] \, dA$$

$$= \iint_R \rho(x - \overline{x})^2 \, dA + 2\overline{x} \iint_R x\rho \, dA - \overline{x}^2 \iint_R \rho \, dA$$

$$= \iint_R \rho(x - \overline{x})^2 \, dA + 2\overline{x}(M\overline{x}) - \overline{x}^2 M = \iint_R \rho(x - \overline{x})^2 \, dA + M\overline{x}^2.$$

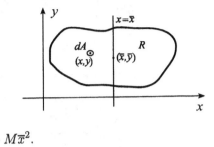

50. $I_{\overline{x}\,\overline{y}} = \iint_R (x - \overline{x})(y - \overline{y})\rho \, dA$

$$= \iint_R xy\rho \, dA - \overline{x} \iint_R y\rho \, dA$$

$$- \overline{y} \iint_R x\rho \, dA + \iint_R \overline{x}\,\overline{y}\rho \, dA$$

$$= I_{xy} - \overline{x}(M\overline{y}) - \overline{y}(M\overline{x}) + \overline{x}\,\overline{y}(M) = I_{xy} - M\overline{x}\,\overline{y}$$

EXERCISES 13.6

2. Area $= \iint_{S_{xy}} \sqrt{1 + \left(\dfrac{\partial z}{\partial x}\right)^2 + \left(\dfrac{\partial z}{\partial y}\right)^2}\, dA = \iint_{S_{xy}} \sqrt{1 + \left(\dfrac{1}{3}\right)^2 + \left(\dfrac{2}{3}\right)^2}\, dA$

$$= \frac{\sqrt{14}}{3} \iint_{S_{xy}} dA = \frac{\sqrt{14}}{3}(\text{Area of } S_{xy})$$

$$= \frac{\sqrt{14}}{3}\frac{1}{2}(2)(4) = \frac{4\sqrt{14}}{3}$$

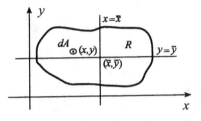

4. Area $= \iint_{S_{xy}} \sqrt{1 + \left(\dfrac{\partial z}{\partial x}\right)^2 + \left(\dfrac{\partial z}{\partial y}\right)^2}\, dA$

$$= \iint_{S_{xy}} \sqrt{1 + \left(\frac{y}{\sqrt{2xy}}\right)^2 + \left(\frac{x}{\sqrt{2xy}}\right)^2}\, dA$$

$$= \iint_{S_{xy}} \left(\frac{x + y}{\sqrt{2xy}}\right) dA = \frac{1}{\sqrt{2}} \int_1^2 \int_1^3 \left(\frac{\sqrt{x}}{\sqrt{y}} + \frac{\sqrt{y}}{\sqrt{x}}\right) dy\, dx$$

$$= \frac{1}{\sqrt{2}} \int_1^2 \left\{2\sqrt{xy} + \frac{2y^{3/2}}{3\sqrt{x}}\right\}_1^3 dx = \frac{1}{\sqrt{2}} \int_1^2 \left[2(\sqrt{3} - 1)\sqrt{x} + \frac{6\sqrt{3} - 2}{3\sqrt{x}}\right] dx$$

$$= \frac{1}{\sqrt{2}} \left\{\frac{4}{3}(\sqrt{3} - 1)x^{3/2} + \frac{4(3\sqrt{3} - 1)\sqrt{x}}{3}\right\}_1^2 = \frac{4}{3}(5\sqrt{3} - 2\sqrt{6} - 3 + \sqrt{2})$$

6. Area $= \iint_{S_{xy}} \sqrt{1 + \left(\dfrac{\partial z}{\partial x}\right)^2 + \left(\dfrac{\partial z}{\partial y}\right)^2}\, dA$

$$= \iint_{S_{xy}} \sqrt{1 + \left(\frac{3\sqrt{x}}{2}\right)^2 + \left(\frac{3\sqrt{y}}{2}\right)^2}\, dA$$

$$= \frac{1}{2} \iint_{S_{xy}} \sqrt{4 + 9x + 9y}\, dA$$

$$= \frac{1}{2} \int_0^1 \int_0^{1-x} \sqrt{4 + 9x + 9y} \, dy \, dx = \frac{1}{2} \int_0^1 \left\{ \frac{2}{27} (4 + 9x + 9y)^{3/2} \right\}_0^{1-x} dx$$

$$= \frac{1}{27} \int_0^1 [13\sqrt{13} - (4 + 9x)^{3/2}] \, dx = \frac{1}{27} \left\{ 13\sqrt{13}x - \frac{2(4 + 9x)^{5/2}}{45} \right\}_0^1 = \frac{247\sqrt{13} + 64}{1215}$$

8. We quadruple the area in the first octant.

$$\text{Area} = 4 \iint_{S_{yz}} \sqrt{1 + \left(\frac{\partial x}{\partial y}\right)^2 + \left(\frac{\partial x}{\partial z}\right)^2} \, dA$$

$$= 4 \iint_{S_{yz}} \sqrt{1 + \left(\frac{y}{2}\right)^2 + \left(\frac{z}{2}\right)^2} \, dA$$

$$= 2 \iint_{S_{yz}} \sqrt{4 + y^2 + z^2} \, dA$$

$$= 2 \int_0^4 \int_0^{\sqrt{16-y^2}} \sqrt{4 + y^2 + z^2} \, dz \, dy$$

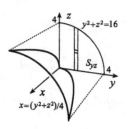

10. We quadruple the area in the first octant.

$$\text{Area} = 4 \iint_{S_{xy}} \sqrt{1 + \left(\frac{\partial z}{\partial x}\right)^2 + \left(\frac{\partial z}{\partial y}\right)^2} \, dA$$

$$= 4 \iint_{S_{xy}} \sqrt{1 + [4x(x^2 + y^2)]^2 + [4y(x^2 + y^2)]^2} \, dA$$

$$= 4 \int_0^{\sqrt{2}} \int_0^{\sqrt{2-x^2}} \sqrt{1 + 16(x^2 + y^2)^3} \, dy \, dx$$

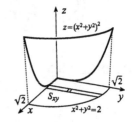

12. $$\text{Area} = \iint_{S_{xy}} \sqrt{1 + \left(\frac{\partial z}{\partial x}\right)^2 + \left(\frac{\partial z}{\partial y}\right)^2} \, dA$$

$$= \iint_{S_{xy}} \sqrt{1 + \left(\frac{1}{1 + x + y}\right)^2 + \left(\frac{1}{1 + x + y}\right)^2} \, dA$$

$$= \int_0^1 \int_0^{1-x^2} \sqrt{1 + \frac{2}{(1 + x + y)^2}} \, dy \, dx$$

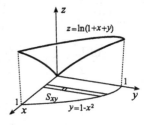

14. We quadruple the area in the first octant.

$$\text{Area} = 4 \iint_{S_{xy}} \sqrt{1 + \left(\frac{\partial z}{\partial x}\right)^2 + \left(\frac{\partial z}{\partial y}\right)^2} \, dA$$

$$= 4 \iint_{S_{xy}} \sqrt{1 + \left(\frac{hx}{r\sqrt{x^2 + y^2}}\right)^2 + \left(\frac{hy}{r\sqrt{x^2 + y^2}}\right)^2} \, dA$$

$$= \frac{4\sqrt{r^2 + h^2}}{r} \iint_{S_{xy}} dA = \frac{4\sqrt{r^2 + h^2}}{r} (\text{Area of } S_{xy})$$

$$= \frac{4\sqrt{r^2 + h^2}}{r} \left(\frac{\pi r^2}{4}\right) = \pi r \sqrt{r^2 + h^2}$$

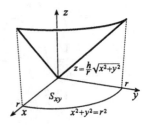

16. We double the area of the upper half.

$$\text{Area} = 2 \iint_{S_{xy}} \sqrt{1 + \left(\frac{\partial z}{\partial x}\right)^2 + \left(\frac{\partial z}{\partial y}\right)^2} \, dA = 2 \iint_{S_{xy}} \sqrt{1 + \left(\frac{-1}{2\sqrt{y-x}}\right)^2 + \left(\frac{1}{2\sqrt{y-x}}\right)^2} \, dA$$

$$= \sqrt{2} \iint_{S_{xy}} \sqrt{2 + \frac{1}{y-x}} \, dA$$

$$= \sqrt{2} \int_{-1/\sqrt{2}}^{1/\sqrt{2}} \int_{x}^{\sqrt{1-x^2}} \sqrt{2 + \frac{1}{y-x}} \, dy \, dx$$

$$+ \sqrt{2} \int_{-1}^{-1/\sqrt{2}} \int_{-\sqrt{1-x^2}}^{\sqrt{1-x^2}} \sqrt{2 + \frac{1}{y-x}} \, dy \, dx$$

18. $\text{Area} = \iint_{S_{xy}} \sqrt{1 + \left(\frac{\partial z}{\partial x}\right)^2 + \left(\frac{\partial z}{\partial y}\right)^2} \, dA$

$$= \iint_{S_{xy}} \sqrt{1 + (3x^2)^2 + (3y^2)^2} \, dA$$

$$= \int_0^1 \int_{1-x}^{2-x} \sqrt{1 + 9x^4 + 9y^4} \, dy \, dx$$

$$+ \int_1^2 \int_0^{2-x} \sqrt{1 + 9x^4 + 9y^4} \, dy \, dx$$

20. If S_{xy} is the region of the xy-plane bounded by the lines $x = 2$, $y = 0$, and $y = x$, then

$$\text{Area} = \iint_{S_{xy}} \sqrt{1 + \left(\frac{\partial z}{\partial x}\right)^2 + \left(\frac{\partial z}{\partial y}\right)^2} \, dA = \iint_{S_{xy}} \sqrt{1 + (4x)^2 + (3)^2} \, dA$$

$$= \int_0^2 \int_0^x \sqrt{10 + 16x^2} \, dy \, dx = \int_0^2 x\sqrt{10 + 16x^2} \, dx$$

$$= \left\{\frac{1}{48}(10 + 16x^2)^{3/2}\right\}_0^2 = \frac{1}{24}(37\sqrt{74} - 5\sqrt{10}).$$

EXERCISES 13.7

2. $\iint_R x \, dA = \int_0^{\pi/2} \int_0^{2\sin\theta} r\cos\theta \, r \, dr \, d\theta = \int_0^{\pi/2} \left\{\frac{r^3}{3}\cos\theta\right\}_0^{2\sin\theta} d\theta$

$$= \frac{8}{3} \int_0^{\pi/2} \sin^3\theta \cos\theta \, d\theta$$

$$= \frac{8}{3} \left\{\frac{\sin^4\theta}{4}\right\}_0^{\pi/2} = \frac{2}{3}$$

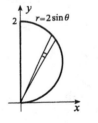

4. $\iint_R \frac{1}{\sqrt{x^2 + y^2}} \, dA = 2 \int_0^{\pi/3} \int_2^{4\cos\theta} \frac{1}{r} r \, dr \, d\theta$

$$= 2 \int_0^{\pi/3} (4\cos\theta - 2) \, d\theta$$

$$= 4\left\{2\sin\theta - \theta\right\}_0^{\pi/3} = \frac{4}{3}(3\sqrt{3} - \pi)$$

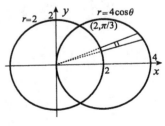

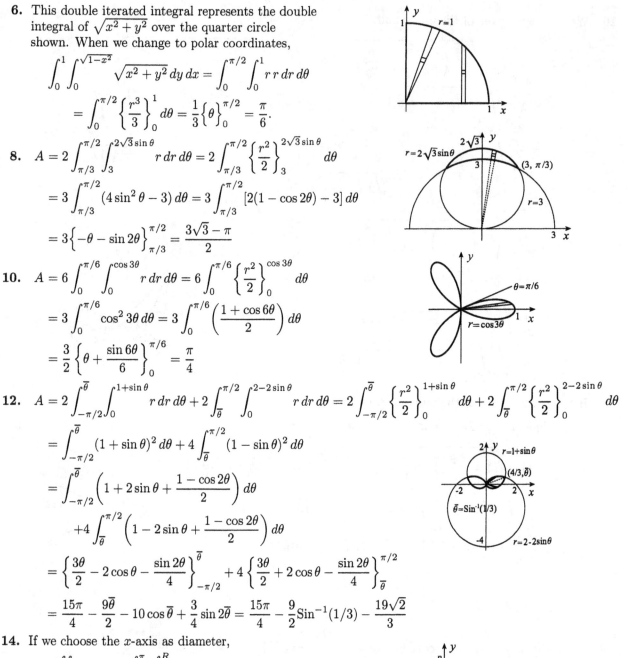

6. This double iterated integral represents the double integral of $\sqrt{x^2 + y^2}$ over the quarter circle shown. When we change to polar coordinates,

$$\int_0^1 \int_0^{\sqrt{1-x^2}} \sqrt{x^2 + y^2}\, dy\, dx = \int_0^{\pi/2} \int_0^1 r\, r\, dr\, d\theta$$

$$= \int_0^{\pi/2} \left\{ \frac{r^3}{3} \right\}_0^1 d\theta = \frac{1}{3} \{\theta\}_0^{\pi/2} = \frac{\pi}{6}.$$

8. $A = 2\int_{\pi/3}^{\pi/2} \int_3^{2\sqrt{3}\sin\theta} r\, dr\, d\theta = 2\int_{\pi/3}^{\pi/2} \left\{ \frac{r^2}{2} \right\}_3^{2\sqrt{3}\sin\theta} d\theta$

$$= 3\int_{\pi/3}^{\pi/2} (4\sin^2\theta - 3)\, d\theta = 3\int_{\pi/3}^{\pi/2} [2(1 - \cos 2\theta) - 3]\, d\theta$$

$$= 3\left\{ -\theta - \sin 2\theta \right\}_{\pi/3}^{\pi/2} = \frac{3\sqrt{3} - \pi}{2}$$

10. $A = 6\int_0^{\pi/6} \int_0^{\cos 3\theta} r\, dr\, d\theta = 6\int_0^{\pi/6} \left\{ \frac{r^2}{2} \right\}_0^{\cos 3\theta} d\theta$

$$= 3\int_0^{\pi/6} \cos^2 3\theta\, d\theta = 3\int_0^{\pi/6} \left(\frac{1 + \cos 6\theta}{2} \right) d\theta$$

$$= \frac{3}{2}\left\{ \theta + \frac{\sin 6\theta}{6} \right\}_0^{\pi/6} = \frac{\pi}{4}$$

12. $A = 2\int_{-\pi/2}^{\bar{\theta}} \int_0^{1+\sin\theta} r\, dr\, d\theta + 2\int_{\bar{\theta}}^{\pi/2} \int_0^{2-2\sin\theta} r\, dr\, d\theta = 2\int_{-\pi/2}^{\bar{\theta}} \left\{ \frac{r^2}{2} \right\}_0^{1+\sin\theta} d\theta + 2\int_{\bar{\theta}}^{\pi/2} \left\{ \frac{r^2}{2} \right\}_0^{2-2\sin\theta} d\theta$

$$= \int_{-\pi/2}^{\bar{\theta}} (1 + \sin\theta)^2\, d\theta + 4\int_{\bar{\theta}}^{\pi/2} (1 - \sin\theta)^2\, d\theta$$

$$= \int_{-\pi/2}^{\bar{\theta}} \left(1 + 2\sin\theta + \frac{1 - \cos 2\theta}{2} \right) d\theta$$

$$\quad + 4\int_{\bar{\theta}}^{\pi/2} \left(1 - 2\sin\theta + \frac{1 - \cos 2\theta}{2} \right) d\theta$$

$$= \left\{ \frac{3\theta}{2} - 2\cos\theta - \frac{\sin 2\theta}{4} \right\}_{-\pi/2}^{\bar{\theta}} + 4\left\{ \frac{3\theta}{2} + 2\cos\theta - \frac{\sin 2\theta}{4} \right\}_{\bar{\theta}}^{\pi/2}$$

$$= \frac{15\pi}{4} - \frac{9\bar{\theta}}{2} - 10\cos\bar{\theta} + \frac{3}{4}\sin 2\bar{\theta} = \frac{15\pi}{4} - \frac{9}{2}\text{Sin}^{-1}(1/3) - \frac{19\sqrt{2}}{3}$$

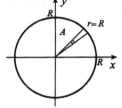

14. If we choose the x-axis as diameter,

$$I = \iint_A y^2\, dA = \int_{-\pi}^{\pi} \int_0^R r^2 \sin^2\theta\, r\, dr\, d\theta$$

$$= \int_{-\pi}^{\pi} \left\{ \frac{r^4}{4} \sin^2\theta \right\}_0^R d\theta = \frac{R^4}{4} \int_{-\pi}^{\pi} \left(\frac{1 - \cos 2\theta}{2} \right) d\theta$$

$$= \frac{R^4}{8} \left\{ \theta - \frac{\sin 2\theta}{2} \right\}_{-\pi}^{\pi} = \frac{\pi R^4}{4}$$

16. We quadruple the area in the first octant.

$$A = 4 \iint_{S_{xy}} \sqrt{1 + \left(\frac{\partial z}{\partial x}\right)^2 + \left(\frac{\partial z}{\partial y}\right)^2}\, dA = 4 \iint_{S_{xy}} \sqrt{1 + (2x)^2 + (2y)^2}\, dA$$

$$= 4 \int_0^{\pi/2} \int_0^2 \sqrt{1 + 4r^2}\, r\, dr\, d\theta = 4 \int_0^{\pi/2} \left\{ \frac{(1 + 4r^2)^{3/2}}{12} \right\}_0^2 d\theta$$

$$= \frac{17\sqrt{17} - 1}{3} \{\theta\}_0^{\pi/2} = \frac{(17\sqrt{17} - 1)\pi}{6}$$

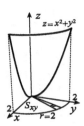

18. We quadruple the area in the first octant.

$$A = 4 \iint_{S_{xy}} \sqrt{1 + \left(\frac{\partial z}{\partial x}\right)^2 + \left(\frac{\partial z}{\partial y}\right)^2}\, dA = 4 \iint_{S_{xy}} \sqrt{1 + (y)^2 + (x)^2}\, dA$$

$$= 4 \int_0^{\pi/2} \int_0^3 \sqrt{1 + r^2}\, r\, dr\, d\theta = 4 \int_0^{\pi/2} \left\{ \frac{(1 + r^2)^{3/2}}{3} \right\}_0^3 d\theta$$

$$= 4 \frac{10\sqrt{10} - 1}{3} \{\theta\}_0^{\pi/2} = \frac{2\pi}{3}(10\sqrt{10} - 1)$$

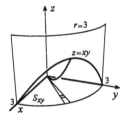

20. We quadruple the area in the first octant.

$$A = 4 \iint_{S_{xy}} \sqrt{1 + \left(\frac{\partial z}{\partial x}\right)^2 + \left(\frac{\partial z}{\partial y}\right)^2}\, dA$$

$$= 4 \iint_{S_{xy}} \sqrt{1 + (2x)^2 + (-2y)^2}\, dA$$

$$= 4 \int_0^{\pi/2} \int_1^2 \sqrt{1 + 4r^2}\, r\, dr\, d\theta$$

$$= 4 \int_0^{\pi/2} \left\{ \frac{(1 + 4r^2)^{3/2}}{12} \right\}_1^2 d\theta$$

$$= \frac{17\sqrt{17} - 5\sqrt{5}}{3} \{\theta\}_0^{\pi/2} = \frac{(17\sqrt{17} - 5\sqrt{5})\pi}{6}.$$

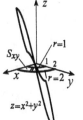

22. If R is that part of the cardioid to the right of the y-axis, then

$$V = \iint_R 2\pi x\, dA = 2\pi \int_{-\pi/2}^{\pi/2} \int_0^{1+\sin\theta} r\cos\theta\, r\, dr\, d\theta$$

$$= 2\pi \int_{-\pi/2}^{\pi/2} \left\{ \frac{r^3}{3}\cos\theta \right\}_0^{1+\sin\theta} d\theta$$

$$= \frac{2\pi}{3} \int_{-\pi/2}^{\pi/2} (1 + \sin\theta)^3 \cos\theta\, d\theta = \frac{2\pi}{3} \left\{ \frac{1}{4}(1 + \sin\theta)^4 \right\}_{-\pi/2}^{\pi/2} = \frac{8\pi}{3}$$

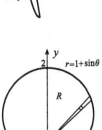

24. The equation of the inner surface of the shell is $r = 700$ in polar coordinates. The equation of the right-half of the outer surface of the shell is $r = 710 - 10\theta/\pi$. The volume of the shell is the product of its length 5000 cm and the cross-sectional area shown,

$$V = 5000(2) \int_0^{\pi/2} \int_{700}^{710 - 10\theta/\pi} r\, dr\, d\theta = 10\,000 \int_0^{\pi/2} \left\{ \frac{r^2}{2} \right\}_{700}^{710 - 10\theta/\pi} d\theta$$

$$= 5000 \int_0^{\pi/2} \left[\left(710 - \frac{10\theta}{\pi}\right)^2 - 490\,000 \right] d\theta = 5000 \left\{ -\frac{\pi}{30}\left(710 - \frac{10\theta}{\pi}\right)^3 - 490\,000\theta \right\}_0^{\pi/2}$$

$$= 8.29 \times 10^7 \text{ cc.}$$

26. If we rotate $x^2 + y^2 \le R^2$ about $x = R$,

$$V = \iint_A 2\pi(R-x)\,dA = 2\pi \int_{-\pi}^{\pi} \int_0^R (R - r\cos\theta)\,r\,dr\,d\theta$$

$$= 2\pi \int_{-\pi}^{\pi} \left\{ \frac{Rr^2}{2} - \frac{r^3}{3}\cos\theta \right\}_0^R d\theta = 2\pi R^3 \int_{-\pi}^{\pi} \left(\frac{1}{2} - \frac{1}{3}\cos\theta \right) d\theta$$

$$= 2\pi R^3 \left\{ \frac{\theta}{2} - \frac{1}{3}\sin\theta \right\}_{-\pi}^{\pi} = 2\pi^2 R^3.$$

28. The force on q due to the charge $\rho\,dA$ in dA has magnitude $\dfrac{q\rho\,dA}{4\pi\epsilon_0 s^2}$. Since x- and y-components of contributions from all parts of the plate cancel, only the z-components survive, and for the contribution from dA, the z-component is $\dfrac{q\rho\,dA}{4\pi\epsilon_0 s^2}\cos\psi$, where ψ is the angle between the z-axis and the line joining P and dA. The total force therefore has z-component

$$F_z = \iint_A \frac{q\rho\cos\psi}{4\pi\epsilon_0 s^2}\,dA = \frac{q\rho}{4\pi\epsilon_0} \iint_A \frac{d}{s^3}\,dA = \frac{q\rho d}{4\pi\epsilon_0} \int_{-\pi}^{\pi} \int_0^R \frac{1}{(r^2+d^2)^{3/2}}\,r\,dr\,d\theta$$

$$= \frac{q\rho d}{4\pi\epsilon_0} \int_{-\pi}^{\pi} \left\{ \frac{-1}{\sqrt{r^2+d^2}} \right\}_0^R d\theta = \frac{q\rho d}{4\pi\epsilon_0}\left(\frac{1}{d} - \frac{1}{\sqrt{R^2+d^2}} \right)(2\pi) = \frac{q\rho}{2\epsilon_0}\left(1 - \frac{d}{\sqrt{R^2+d^2}} \right).$$

As the radius of the plate becomes very large, $\displaystyle\lim_{R\to\infty} F_z = \frac{q\rho}{2\epsilon_0}$.

30. $A = 2 \displaystyle\int_0^{\pi/3} \int_0^{4-2\cos\theta} r\,dr\,d\theta + 2\int_{\pi/3}^{\pi/2} \int_0^{6\cos\theta} r\,dr\,d\theta$

$$= 2 \int_0^{\pi/3} \left\{ \frac{r^2}{2} \right\}_0^{4-2\cos\theta} d\theta + 2\int_{\pi/3}^{\pi/2} \left\{ \frac{r^2}{2} \right\}_0^{6\cos\theta} d\theta = 4\int_0^{\pi/3} (2-\cos\theta)^2\,d\theta + 36\int_{\pi/3}^{\pi/2}\cos^2\theta\,d\theta$$

r=4-2cosθ

(3,π/3) r=6cosθ

$$= 4\int_0^{\pi/3} \left(4 - 4\cos\theta + \frac{1+\cos 2\theta}{2} \right) d\theta$$

$$\quad + 36\int_{\pi/3}^{\pi/2} \left(\frac{1+\cos 2\theta}{2} \right) d\theta$$

$$= 4\left\{ \frac{9\theta}{2} - 4\sin\theta + \frac{\sin 2\theta}{4} \right\}_0^{\pi/3} + 18\left\{ \theta + \frac{\sin 2\theta}{2} \right\}_{\pi/3}^{\pi/2}$$

$$= 9\pi - 12\sqrt{3}$$

r=1+cosθ

32. $A = 2\displaystyle\int_0^{\pi} \int_0^{1+\cos\theta} r\,dr\,d\theta = \int_0^{\pi} (1+\cos\theta)^2\,d\theta$

$$= \int_0^{\pi} \left(1 + 2\cos\theta + \frac{1+\cos 2\theta}{2} \right) d\theta$$

$$= \left\{ \frac{3\theta}{2} + 2\sin\theta + \frac{\sin 2\theta}{4} \right\}_0^{\pi} = \frac{3\pi}{2}$$

By symmetry, $\overline{y} = 0$. Since

$$A\overline{x} = 2\int_0^{\pi} \int_0^{1+\cos\theta} r\cos\theta\, r\,dr\,d\theta = 2\int_0^{\pi} \left\{ \frac{r^3}{3}\cos\theta \right\}_0^{1+\cos\theta} d\theta = \frac{2}{3}\int_0^{\pi} (1+\cos\theta)^3\cos\theta\,d\theta$$

$$= \frac{2}{3}\int_0^{\pi} \left[\cos\theta + 3\left(\frac{1+\cos 2\theta}{2} \right) + 3\cos\theta(1-\sin^2\theta) + \frac{1}{4}\left(1 + 2\cos 2\theta + \frac{1+\cos 4\theta}{2} \right) \right] d\theta$$

$$= \frac{2}{3}\left\{ 4\sin\theta + \frac{15\theta}{8} + \sin 2\theta - \sin^3\theta + \frac{\sin 4\theta}{32} \right\}_0^{\pi} = \frac{5\pi}{4},$$

we find $\overline{x} = \dfrac{5\pi}{4}\dfrac{2}{3\pi} = \dfrac{5}{6}$.

34. $I = \iint_R \sqrt{\dfrac{1-x^2-y^2}{1+x^2+y^2}}\, dA = \int_0^1 \int_{-\pi}^{\pi} \sqrt{\dfrac{1-r^2}{1+r^2}}\, r\, d\theta\, dr = 2\pi \int_0^1 \sqrt{\dfrac{1-r^2}{1+r^2}}\, r\, dr$

If we set $u = \sqrt{1+r^2}$, then $2u\, du = 2r\, dr$, and

$$I = 2\pi \int_1^{\sqrt{2}} \sqrt{\dfrac{1-(u^2-1)}{u^2}}\, u\, du = 2\pi \int_1^{\sqrt{2}} \sqrt{2-u^2}\, du.$$

If we now set $u = \sqrt{2}\sin\phi$ and $du = \sqrt{2}\cos\phi\, d\phi$, then

$$I = 2\pi \int_{\pi/4}^{\pi/2} \sqrt{2}\cos\phi \sqrt{2}\cos\phi\, d\phi = 4\pi \int_{\pi/4}^{\pi/2}\left(\dfrac{1+\cos 2\phi}{2}\right) d\phi$$

$$= 2\pi \left\{\phi + \dfrac{\sin 2\phi}{2}\right\}_{\pi/4}^{\pi/2} = \dfrac{\pi(\pi-2)}{2}.$$

36. The volume of blood flowing through any cross-section of the larger blood vessel per unit time is

$$\int_{-\pi}^{\pi}\int_0^R V_{\max}\left[1 - \left(\dfrac{r}{R}\right)^2\right] r\, dr\, d\theta = V_{\max}\int_{-\pi}^{\pi}\left\{\dfrac{r^2}{2} - \dfrac{r^4}{4R^2}\right\}_0^R d\theta = \dfrac{R^2 V_{\max}}{4}\left\{\theta\right\}_{-\pi}^{\pi} = \dfrac{\pi R^2 V_{\max}}{2}.$$

Similarly, the volume of blood flowing through any cross-section of the smaller blood vessel is

$$\int_{-\pi}^{\pi}\int_0^{R_1} U_{\max}\left[1 - \left(\dfrac{r}{R_1}\right)^2\right] r\, dr\, d\theta = \dfrac{\pi R_1^2 U_{\max}}{2}.$$

When we equate these and set $R_1 = \alpha R$,

$$\dfrac{\pi R^2 V_{\max}}{2} = \dfrac{\pi \alpha^2 R^2 U_{\max}}{2} \quad \Longrightarrow \quad U_{\max} = \dfrac{V_{\max}}{\alpha^2}.$$

EXERCISES 13.8

2. $\displaystyle\iiint_V x\, dV = \int_0^4 \int_0^{4-x} \int_0^{4-x-y} x\, dz\, dy\, dx$

$\displaystyle = \int_0^4 \int_0^{4-x} x(4-x-y)\, dy\, dx$

$\displaystyle = \int_0^4 \left\{x(4-x)y - \dfrac{xy^2}{2}\right\}_0^{4-x} dx$

$\displaystyle = \dfrac{1}{2}\int_0^4 (16x - 8x^2 + x^3)\, dx$

$\displaystyle = \dfrac{1}{2}\left\{8x^2 - \dfrac{8x^3}{3} + \dfrac{x^4}{4}\right\}_0^4 = \dfrac{32}{3}$

4. $\displaystyle\iiint_V xy\, dV = \int_{-1}^{1}\int_{-\sqrt{1-x^2}}^{\sqrt{1-x^2}}\int_0^{\sqrt{1-x^2-y^2}} xy\, dz\, dy\, dx$

$\displaystyle = \int_{-1}^{1}\int_{-\sqrt{1-x^2}}^{\sqrt{1-x^2}} xy\sqrt{1-x^2-y^2}\, dy\, dx$

$\displaystyle = \int_{-1}^{1}\left\{-\dfrac{x}{3}(1-x^2-y^2)^{3/2}\right\}_{-\sqrt{1-x^2}}^{\sqrt{1-x^2}} dx = 0$

6. $\displaystyle\iiint_V (x^2 + 2z)\, dV = \int_{-2}^{2}\int_{x^2}^{4}\int_{0}^{4-y} (x^2 + 2z)\, dz\, dy\, dx = \int_{-2}^{2}\int_{x^2}^{4}\left\{x^2 z + z^2\right\}_0^{4-y} dy\, dx$

$\displaystyle = \int_{-2}^{2}\int_{x^2}^{4} [x^2(4-y) + (4-y)^2]\, dy\, dx$

$\displaystyle = \int_{-2}^{2}\left\{-\frac{x^2}{2}(4-y)^2 - \frac{(4-y)^3}{3}\right\}_{x^2}^{4} dx$

$\displaystyle = \frac{1}{6}\int_{-2}^{2}(128 - 48x^2 + x^6)\, dx$

$\displaystyle = \frac{1}{6}\left\{128x - 16x^3 + \frac{x^7}{7}\right\}_{-2}^{2} = \frac{1024}{21}$

8. $\displaystyle\iiint_V xyz\, dV = \int_{0}^{1}\int_{0}^{\sqrt{1-x^2}}\int_{x^2+y^2}^{\sqrt{x^2+y^2}} xyz\, dz\, dy\, dx = \int_{0}^{1}\int_{0}^{\sqrt{1-x^2}}\left\{\frac{xyz^2}{2}\right\}_{x^2+y^2}^{\sqrt{x^2+y^2}} dy\, dx$

$\displaystyle = \frac{1}{2}\int_{0}^{1}\int_{0}^{\sqrt{1-x^2}} (x^3 y + xy^3 - x^5 y - 2x^3 y^3 - xy^5)\, dy\, dx$

$\displaystyle = \frac{1}{2}\int_{0}^{1}\left\{\frac{x^3 y^2}{2} + \frac{xy^4}{4} - \frac{x^5 y^2}{2} - \frac{x^3 y^4}{2} - \frac{xy^6}{6}\right\}_0^{\sqrt{1-x^2}} dx$

$\displaystyle = \frac{1}{24}\int_{0}^{1} [3x(1-x^2)^2 - 2x(1-x^2)^3]\, dx$

$\displaystyle = \frac{1}{24}\left\{-\frac{1}{2}(1-x^2)^3 + \frac{1}{4}(1-x^2)^4\right\}_0^{1} = \frac{1}{96}$

10. $\displaystyle\iiint_V (x + y + z)\, dV = \int_{0}^{1}\int_{z}^{2-z}\int_{0}^{1} (x + y + z)\, dx\, dy\, dz = \int_{0}^{1}\int_{z}^{2-z}\left\{\frac{1}{2}(x + y + z)^2\right\}_0^{1} dy\, dz$

$\displaystyle = \frac{1}{2}\int_{0}^{1}\int_{z}^{2-z} [(1 + y + z)^2 - (y + z)^2]\, dy\, dz$

$\displaystyle = \frac{1}{2}\int_{0}^{1}\left\{\frac{1}{3}(1 + y + z)^3 - \frac{1}{3}(y + z)^3\right\}_z^{2-z} dz$

$\displaystyle = \frac{1}{6}\int_{0}^{1} [19 + 8z^3 - (1 + 2z)^3]\, dz$

$\displaystyle = \frac{1}{6}\left\{19z + 2z^4 - \frac{(1+2z)^4}{8}\right\}_0^{1} = \frac{11}{6}$

12. $\displaystyle\iiint_V x^2 y\, dV = \int_{0}^{2}\int_{0}^{3\sqrt{4-x^2}/2}\int_{x^2/4+y^2/9}^{1} x^2 y\, dz\, dy\, dx$

$\displaystyle = \int_{0}^{2}\int_{0}^{3\sqrt{4-x^2}/2} x^2 y\left(1 - \frac{x^2}{4} - \frac{y^2}{9}\right) dy\, dx$

$\displaystyle = \int_{0}^{2}\left\{x^2\left(1 - \frac{x^2}{4}\right)\frac{y^2}{2} - \frac{x^2 y^4}{36}\right\}_0^{3\sqrt{4-x^2}/2} dx$

$\displaystyle = \frac{9}{64}\int_{0}^{2} (16x^2 - 8x^4 + x^6)\, dx$

$\displaystyle = \frac{9}{64}\left\{\frac{16x^3}{3} - \frac{8x^5}{5} + \frac{x^7}{7}\right\}_0^{2} = \frac{48}{35}$

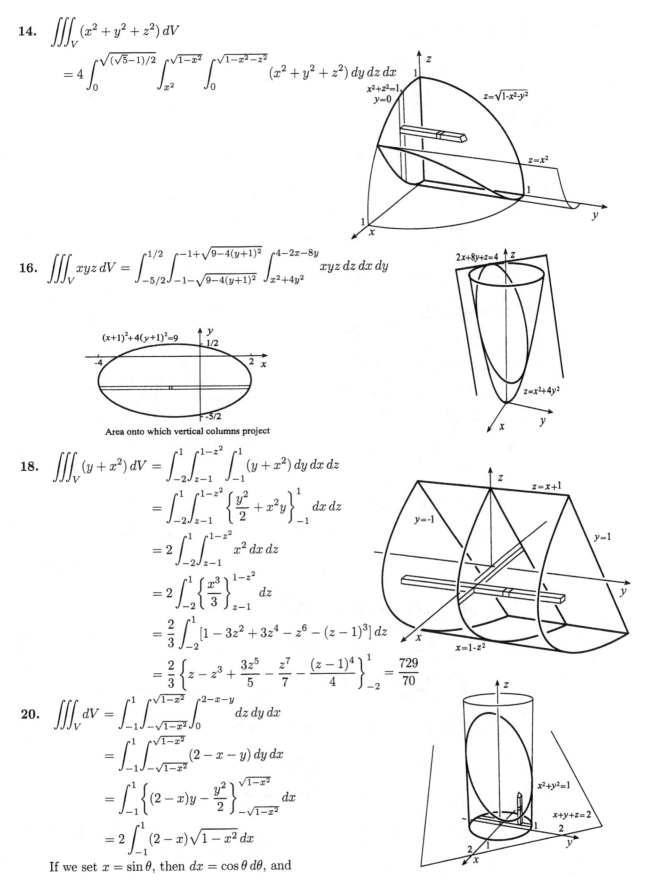

14. $\displaystyle\iiint_V (x^2 + y^2 + z^2)\, dV$

$$= 4 \int_0^{\sqrt{(\sqrt{5}-1)/2}} \int_{x^2}^{\sqrt{1-x^2}} \int_0^{\sqrt{1-x^2-z^2}} (x^2 + y^2 + z^2)\, dy\, dz\, dx$$

16. $\displaystyle\iiint_V xyz\, dV = \int_{-5/2}^{1/2} \int_{-1-\sqrt{9-4(y+1)^2}}^{-1+\sqrt{9-4(y+1)^2}} \int_{x^2+4y^2}^{4-2x-8y} xyz\, dz\, dx\, dy$

Area onto which vertical columns project

18. $\displaystyle\iiint_V (y + x^2)\, dV = \int_{-2}^{1} \int_{z-1}^{1-z^2} \int_{-1}^{1} (y + x^2)\, dy\, dx\, dz$

$$= \int_{-2}^{1} \int_{z-1}^{1-z^2} \left\{ \frac{y^2}{2} + x^2 y \right\}_{-1}^{1} dx\, dz$$

$$= 2 \int_{-2}^{1} \int_{z-1}^{1-z^2} x^2\, dx\, dz$$

$$= 2 \int_{-2}^{1} \left\{ \frac{x^3}{3} \right\}_{z-1}^{1-z^2} dz$$

$$= \frac{2}{3} \int_{-2}^{1} [1 - 3z^2 + 3z^4 - z^6 - (z-1)^3]\, dz$$

$$= \frac{2}{3} \left\{ z - z^3 + \frac{3z^5}{5} - \frac{z^7}{7} - \frac{(z-1)^4}{4} \right\}_{-2}^{1} = \frac{729}{70}$$

20. $\displaystyle\iiint_V dV = \int_{-1}^{1} \int_{-\sqrt{1-x^2}}^{\sqrt{1-x^2}} \int_0^{2-x-y} dz\, dy\, dx$

$$= \int_{-1}^{1} \int_{-\sqrt{1-x^2}}^{\sqrt{1-x^2}} (2 - x - y)\, dy\, dx$$

$$= \int_{-1}^{1} \left\{ (2 - x)y - \frac{y^2}{2} \right\}_{-\sqrt{1-x^2}}^{\sqrt{1-x^2}} dx$$

$$= 2 \int_{-1}^{1} (2 - x)\sqrt{1 - x^2}\, dx$$

If we set $x = \sin\theta$, then $dx = \cos\theta\, d\theta$, and

$$\iiint_V dV = 2 \int_{-\pi/2}^{\pi/2} (2 - \sin\theta) \cos\theta \, \cos\theta \, d\theta$$

$$= 2 \int_{-\pi/2}^{\pi/2} (1 + \cos 2\theta - \cos^2\theta \sin\theta) \, d\theta = 2 \left\{ \theta + \frac{\sin 2\theta}{2} + \frac{\cos^3\theta}{3} \right\}_{-\pi/2}^{\pi/2} = 2\pi.$$

22. Because of the symmetry, integrals of x and y vanish. We multiply the integral of z over the first octant volume by 4,

$$\iiint_V (x + y + z) \, dV = \iiint_V z \, dV = 4 \int_0^{\sqrt{2}} \int_0^{\sqrt{2-x^2}} \int_{y^2/2 - x^2/2}^{1-x^2} z \, dz \, dy \, dx$$

$$= 4 \int_0^{\sqrt{2}} \int_0^{\sqrt{2-x^2}} \left\{ \frac{z^2}{2} \right\}_{y^2/2 - x^2/2}^{1-x^2} dy \, dx = \frac{1}{2} \int_0^{\sqrt{2}} \int_0^{\sqrt{2-x^2}} (4 - 8x^2 + 3x^4 - y^4 + 2x^2 y^2) \, dy \, dx$$

$$= \frac{1}{2} \int_0^{\sqrt{2}} \left\{ 4y - 8x^2 y + 3x^4 y - \frac{y^5}{5} + \frac{2x^2 y^3}{3} \right\}_0^{\sqrt{2-x^2}} dx$$

$$= \frac{1}{2} \int_0^{\sqrt{2}} \left[4\sqrt{2-x^2} - 8x^2\sqrt{2-x^2} + 3x^4\sqrt{2-x^2} - \frac{1}{5}(2-x^2)^{5/2} + \frac{2x^2}{3}(2-x^2)^{3/2} \right] dx.$$

Columns project onto interior of circle

If we set $x = \sqrt{2}\sin\theta$, then $dx = \sqrt{2}\cos\theta \, d\theta$, and

$$\iiint_V (x + y + z) \, dV = \frac{1}{2} \int_0^{\pi/2} \left(4\sqrt{2}\cos\theta - 16\sqrt{2}\sin^2\theta \cos\theta + 12\sqrt{2}\sin^4\theta \cos\theta - \frac{4\sqrt{2}}{5}\cos^5\theta \right.$$

$$\left. + \frac{8\sqrt{2}}{3}\sin^2\theta \cos^3\theta \right) \sqrt{2}\cos\theta \, d\theta$$

$$= 4 \int_0^{\pi/2} \left[\cos^2\theta - \sin^2 2\theta + 3\left(\frac{\sin^2 2\theta}{4}\right)\left(\frac{1 - \cos 2\theta}{2}\right) \right.$$

$$\left. - \frac{1}{5}\left(\frac{1 + \cos 2\theta}{2}\right)^3 + \frac{2}{3}\left(\frac{\sin^2 2\theta}{4}\right)\left(\frac{1 + \cos 2\theta}{2}\right) \right] d\theta$$

$$= 4 \int_0^{\pi/2} \left\{ \frac{1 + \cos 2\theta}{2} - \left(\frac{1 - \cos 4\theta}{2}\right) + \frac{3}{8}\left[\frac{1 - \cos 4\theta}{2} - \sin^2 2\theta \cos 2\theta \right] \right.$$

$$- \frac{1}{40}\left[1 + 3\cos 2\theta + \frac{3}{2}(1 + \cos 4\theta) + \cos 2\theta(1 - \sin^2 2\theta) \right]$$

$$\left. + \frac{1}{12}\left(\frac{1 - \cos 4\theta}{2} + \sin^2 2\theta \cos 2\theta\right) \right\} d\theta$$

$$= 4 \left\{ \frac{\theta}{2} + \frac{\sin 2\theta}{4} - \frac{\theta}{2} + \frac{\sin 4\theta}{8} + \frac{3\theta}{16} - \frac{3\sin 4\theta}{64} - \frac{\sin^3 2\theta}{16} \right.$$

$$\left. - \frac{\theta}{40} - \frac{3\sin 2\theta}{80} - \frac{3\theta}{80} - \frac{3\sin 4\theta}{320} - \frac{\sin 2\theta}{80} + \frac{\sin^3 2\theta}{240} + \frac{\theta}{24} - \frac{\sin 4\theta}{96} + \frac{\sin^3 2\theta}{72} \right\}_0^{\pi/2}$$

$$= \pi/3.$$

24. We quadruple the integral over the first octant volume.

$$\iiint_V (x^2 + y^2 + z^2)\, dV = 4\int_0^1 \int_0^{\sqrt{1-x^2}} \int_0^{\sqrt{x^2+y^2}/2} (x^2+y^2+z^2)\, dz\, dy\, dx$$

$$+ 4\int_0^1 \int_{\sqrt{1-x^2}}^{\sqrt{4/3-x^2}} \int_{\sqrt{x^2+y^2-1}}^{\sqrt{x^2+y^2}/2} (x^2+y^2+z^2)\, dz\, dy\, dx$$

$$+ 4\int_1^{2/\sqrt{3}} \int_0^{\sqrt{4/3-x^2}} \int_{\sqrt{x^2+y^2-1}}^{\sqrt{x^2+y^2}/2} (x^2+y^2+z^2)\, dz\, dy\, dx$$

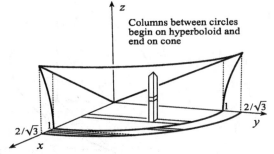

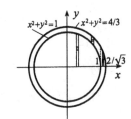

EXERCISES 13.9

2. $V = \int_0^1 \int_{x^2}^{\sqrt{x}} \int_0^2 dy\, dz\, dx = \int_0^1 \int_{x^2}^{\sqrt{x}} 2\, dz\, dx$

$= 2\int_0^1 (\sqrt{x} - x^2)\, dx$

$= 2\left\{ \dfrac{2x^{3/2}}{3} - \dfrac{x^3}{3} \right\}_0^1 = \dfrac{2}{3}$

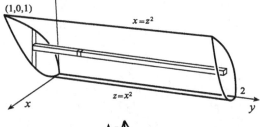

4. $V = \int_{-2}^2 \int_0^{4-x^2} \int_0^{6-x-y} dz\, dy\, dx$

$= \int_{-2}^2 \int_0^{4-x^2} (6 - x - y)\, dy\, dx$

$= \int_{-2}^2 \left\{ (6-x)y - \dfrac{y^2}{2} \right\}_0^{4-x^2} dx$

$= \dfrac{1}{2}\int_{-2}^2 (32 - 8x - 4x^2 + 2x^3 - x^4)\, dx$

$= \dfrac{1}{2}\left\{ 32x - 4x^2 - \dfrac{4x^3}{3} + \dfrac{x^4}{2} - \dfrac{x^5}{5} \right\}_{-2}^2 = \dfrac{704}{15}$

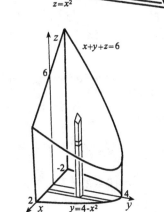

6. $V = \displaystyle\int_0^3 \int_{y/3}^{4-y} \int_0^{4-y-z} dx\, dz\, dy = \int_0^3 \int_{y/3}^{4-y} (4 - y - z)\, dz\, dy$

$= \displaystyle\int_0^3 \left\{ (4-y)z - \frac{z^2}{2} \right\}_{y/3}^{4-y} dy$

$= \dfrac{1}{18} \displaystyle\int_0^3 [7y^2 - 24y + 9(4-y)^2]\, dy$

$= \dfrac{1}{18} \left\{ \dfrac{7y^3}{3} - 12y^2 - 3(4-y)^3 \right\}_0^3 = 8$

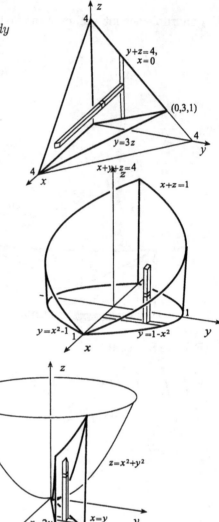

8. We double the volume to the right of the xz-plane.

$V = 2 \displaystyle\int_{-1}^1 \int_0^{1-x^2} \int_0^{1-x} dz\, dy\, dx$

$= 2 \displaystyle\int_{-1}^1 \int_0^{1-x^2} (1-x)\, dy\, dx$

$= 2 \displaystyle\int_{-1}^1 (1-x)(1-x^2)\, dx$

$= 2 \left\{ x - \dfrac{x^2}{2} - \dfrac{x^3}{3} + \dfrac{x^4}{4} \right\}_{-1}^1 = \dfrac{8}{3}$

10. $V = \displaystyle\int_0^1 \int_{x/2}^x \int_0^{x^2+y^2} dz\, dy\, dx$

$= \displaystyle\int_0^1 \int_{x/2}^x (x^2 + y^2)\, dy\, dx$

$= \displaystyle\int_0^1 \left\{ x^2 y + \dfrac{y^3}{3} \right\}_{x/2}^x dx$

$= \dfrac{19}{24} \displaystyle\int_0^1 x^3\, dx = \dfrac{19}{24} \left\{ \dfrac{x^4}{4} \right\}_0^1 = \dfrac{19}{96}$

12. We double the volume to the right of the xz-plane.

$V = 2 \displaystyle\int_0^1 \int_0^{1-y} \int_z^{3-z} dx\, dz\, dy$

$= 2 \displaystyle\int_0^1 \int_0^{1-y} (3 - 2z)\, dz\, dy$

$= 2 \displaystyle\int_0^1 \left\{ -\dfrac{1}{4}(3-2z)^2 \right\}_0^{1-y} dy$

$= \dfrac{1}{2} \displaystyle\int_0^1 [9 - (2y+1)^2]\, dy$

$= \dfrac{1}{2} \left\{ 9y - \dfrac{1}{6}(2y+1)^3 \right\}_0^1 = \dfrac{7}{3}$

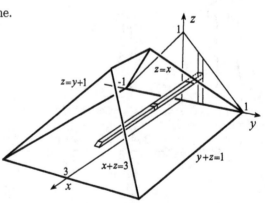

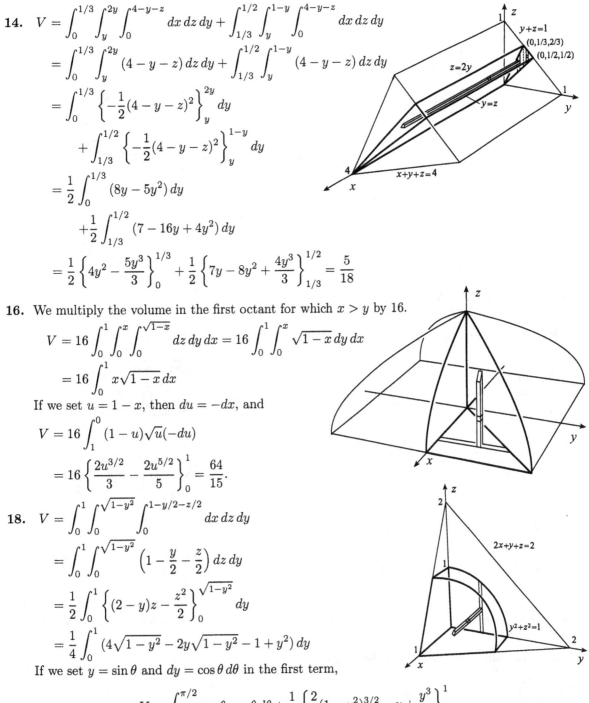

14.
$$V = \int_0^{1/3} \int_y^{2y} \int_0^{4-y-z} dx\, dz\, dy + \int_{1/3}^{1/2} \int_y^{1-y} \int_0^{4-y-z} dx\, dz\, dy$$

$$= \int_0^{1/3} \int_y^{2y} (4-y-z)\, dz\, dy + \int_{1/3}^{1/2} \int_y^{1-y} (4-y-z)\, dz\, dy$$

$$= \int_0^{1/3} \left\{ -\frac{1}{2}(4-y-z)^2 \right\}_y^{2y} dy$$

$$+ \int_{1/3}^{1/2} \left\{ -\frac{1}{2}(4-y-z)^2 \right\}_y^{1-y} dy$$

$$= \frac{1}{2} \int_0^{1/3} (8y - 5y^2)\, dy$$

$$+ \frac{1}{2} \int_{1/3}^{1/2} (7 - 16y + 4y^2)\, dy$$

$$= \frac{1}{2} \left\{ 4y^2 - \frac{5y^3}{3} \right\}_0^{1/3} + \frac{1}{2} \left\{ 7y - 8y^2 + \frac{4y^3}{3} \right\}_{1/3}^{1/2} = \frac{5}{18}$$

16. We multiply the volume in the first octant for which $x > y$ by 16.

$$V = 16 \int_0^1 \int_0^x \int_0^{\sqrt{1-x}} dz\, dy\, dx = 16 \int_0^1 \int_0^x \sqrt{1-x}\, dy\, dx$$

$$= 16 \int_0^1 x\sqrt{1-x}\, dx$$

If we set $u = 1 - x$, then $du = -dx$, and

$$V = 16 \int_1^0 (1-u)\sqrt{u}(-du)$$

$$= 16 \left\{ \frac{2u^{3/2}}{3} - \frac{2u^{5/2}}{5} \right\}_0^1 = \frac{64}{15}.$$

18.
$$V = \int_0^1 \int_0^{\sqrt{1-y^2}} \int_0^{1-y/2-z/2} dx\, dz\, dy$$

$$= \int_0^1 \int_0^{\sqrt{1-y^2}} \left(1 - \frac{y}{2} - \frac{z}{2} \right) dz\, dy$$

$$= \frac{1}{2} \int_0^1 \left\{ (2-y)z - \frac{z^2}{2} \right\}_0^{\sqrt{1-y^2}} dy$$

$$= \frac{1}{4} \int_0^1 \left(4\sqrt{1-y^2} - 2y\sqrt{1-y^2} - 1 + y^2 \right) dy$$

If we set $y = \sin\theta$ and $dy = \cos\theta\, d\theta$ in the first term,

$$V = \int_0^{\pi/2} \cos\theta \cos\theta\, d\theta + \frac{1}{4} \left\{ \frac{2}{3}(1-y^2)^{3/2} - y + \frac{y^3}{3} \right\}_0^1$$

$$= \int_0^{\pi/2} \left(\frac{1 + \cos 2\theta}{2} \right) d\theta - \frac{1}{3} = \frac{1}{2} \left\{ \theta + \frac{\sin 2\theta}{2} \right\}_0^{\pi/2} - \frac{1}{3} = \frac{\pi}{4} - \frac{1}{3}$$

20.   Since Volume $= \displaystyle\iiint_V dV = \int_0^1 \int_0^{1-x} \int_0^{1-x-y} dz\,dy\,dx$

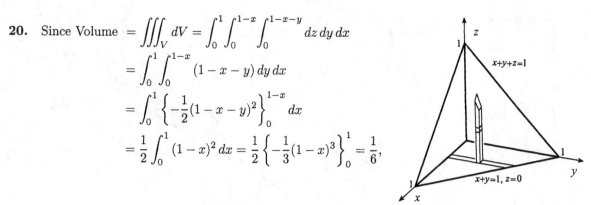

$$= \int_0^1 \int_0^{1-x} (1-x-y)\,dy\,dx$$

$$= \int_0^1 \left\{ -\frac{1}{2}(1-x-y)^2 \right\}_0^{1-x} dx$$

$$= \frac{1}{2}\int_0^1 (1-x)^2\,dx = \frac{1}{2}\left\{ -\frac{1}{3}(1-x)^3 \right\}_0^1 = \frac{1}{6},$$

$$\overline{f} = 6\iiint_V xy\,dV = 6\int_0^1 \int_0^{1-x} \int_0^{1-x-y} xy\,dz\,dy\,dx = 6\int_0^1 \int_0^{1-x} xy(1-x-y)\,dy\,dx$$

$$= 6\int_0^1 \left\{ x(1-x)\frac{y^2}{2} - \frac{xy^3}{3} \right\}_0^{1-x} dx = \int_0^1 (x - 3x^2 + 3x^3 - x^4)\,dx$$

$$= \left\{ \frac{x^2}{2} - x^3 + \frac{3x^4}{4} - \frac{x^5}{5} \right\}_0^1 = \frac{1}{20}$$

22.   Since Volume $= \dfrac{1}{2}(2)(2)(1) = 2,$

$$\overline{f} = \frac{1}{2}\iiint_V (x^2+y^2+z^2)\,dV$$

$$= \frac{1}{2}\int_0^2 \int_{2-y}^2 \int_0^1 (x^2+y^2+z^2)\,dx\,dz\,dy$$

$$= \frac{1}{2}\int_0^2 \int_{2-y}^2 \left\{ \frac{x^3}{3} + x(y^2+z^2) \right\}_0^1 dz\,dy$$

$$= \frac{1}{6}\int_0^2 \int_{2-y}^2 [1 + 3(y^2+z^2)]\,dz\,dy$$

$$= \frac{1}{6}\int_0^2 \left\{ z(1+3y^2) + z^3 \right\}_{2-y}^2 dy$$

$$= \frac{1}{6}\int_0^2 [8 + y + 3y^3 - (2-y)^3]\,dy = \frac{1}{6}\left\{ 8y + \frac{y^2}{2} + \frac{3y^4}{4} + \frac{(2-y)^4}{4} \right\}_0^2 = \frac{13}{3}$$

24.   If we set $x^2 - y^2 = 4 - x^2 - y^2$, then $2x^2 = 4$ or $x = \pm\sqrt{2}$. This implies that the curve of intersection of the surfaces divides into two parts, two parabolas $z = 2 - y^2$, $x = \pm\sqrt{2}$ in parallel planes. There is no bounded volume.

26.   We double the volume to the right of the xz-plane.

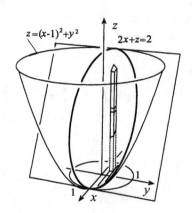

$$V = 2\int_0^1 \int_{-\sqrt{1-y^2}}^{\sqrt{1-y^2}} \int_{(x-1)^2+y^2}^{2-2x} dz\,dx\,dy$$

$$= 2\int_0^1 \int_{-\sqrt{1-y^2}}^{\sqrt{1-y^2}} (1-x^2-y^2)\,dx\,dy$$

$$= 2\int_0^1 \left\{ x(1-y^2) - \frac{x^3}{3} \right\}_{-\sqrt{1-y^2}}^{\sqrt{1-y^2}} dy$$

$$= \frac{8}{3}\int_0^1 (1-y^2)^{3/2}\,dy$$

If we set $y = \sin\theta$, then $dy = \cos\theta\,d\theta$, and

$$V = \frac{8}{3} \int_0^{\pi/2} \cos^4\theta\, d\theta = \frac{8}{3} \int_0^{\pi/2} \left(\frac{1 + \cos 2\theta}{2}\right)^2 d\theta$$

$$= \frac{2}{3} \int_0^{\pi/2} \left(1 + 2\cos 2\theta + \frac{1 + \cos 4\theta}{2}\right) d\theta = \frac{2}{3} \left\{\frac{3\theta}{2} + \sin 2\theta + \frac{1}{8}\sin 4\theta\right\}_0^{\pi/2} = \frac{\pi}{2}.$$

EXERCISES 13.10

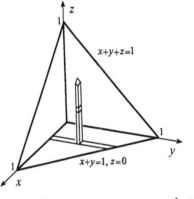

2. $\quad M = \int_0^1 \int_0^{1-x} \int_0^{1-x-y} \rho\, dz\, dy\, dx = \rho \int_0^1 \int_0^{1-x} (1 - x - y)\, dy\, dx$

$$= \rho \int_0^1 \left\{-\frac{1}{2}(1 - x - y)^2\right\}_0^{1-x} dx$$

$$= \frac{\rho}{2} \int_0^1 (1 - x)^2\, dx = \frac{\rho}{2} \left\{-\frac{1}{3}(1 - x)^3\right\}_0^1 = \frac{\rho}{6}$$

Since $\quad M\bar{x} = \int_0^1 \int_0^{1-x} \int_0^{1-x-y} x\rho\, dz\, dy\, dx = \rho \int_0^1 \int_0^{1-x} x(1 - x - y)\, dy\, dx = \rho \int_0^1 \left\{x(1 - x)y - \frac{xy^2}{2}\right\}_0^{1-x} dx$

$$= \frac{\rho}{2} \int_0^1 (x - 2x^2 + x^3)\, dx = \frac{\rho}{2} \left\{\frac{x^2}{2} - \frac{2x^3}{3} + \frac{x^4}{4}\right\}_0^1 = \frac{\rho}{24},$$

it follows by symmetry that $\bar{x} = \bar{y} = \bar{z} = \dfrac{\rho}{24} \dfrac{6}{\rho} = \dfrac{1}{4}.$

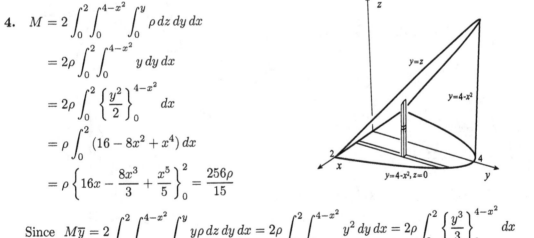

4. $\quad M = 2 \int_0^2 \int_0^{4-x^2} \int_0^y \rho\, dz\, dy\, dx$

$$= 2\rho \int_0^2 \int_0^{4-x^2} y\, dy\, dx$$

$$= 2\rho \int_0^2 \left\{\frac{y^2}{2}\right\}_0^{4-x^2} dx$$

$$= \rho \int_0^2 (16 - 8x^2 + x^4)\, dx$$

$$= \rho \left\{16x - \frac{8x^3}{3} + \frac{x^5}{5}\right\}_0^2 = \frac{256\rho}{15}$$

Since $\quad M\bar{y} = 2 \int_0^2 \int_0^{4-x^2} \int_0^y y\rho\, dz\, dy\, dx = 2\rho \int_0^2 \int_0^{4-x^2} y^2\, dy\, dx = 2\rho \int_0^2 \left\{\frac{y^3}{3}\right\}_0^{4-x^2} dx$

$$= \frac{2\rho}{3} \int_0^2 (64 - 48x^2 + 12x^4 - x^6)\, dx = \frac{2\rho}{3} \left\{64x - 16x^3 + \frac{12x^5}{5} - \frac{x^7}{7}\right\}_0^2 = \frac{4096\rho}{105},$$

it follows that $\bar{y} = \dfrac{4096\rho}{105} \dfrac{15}{256\rho} = \dfrac{16}{7}.$ By symmetry, $\bar{x} = 0.$ We find that $\bar{z} = 8/7$ since

$$M\bar{z} = 2 \int_0^2 \int_0^{4-x^2} \int_0^y z\rho\, dz\, dy\, dx = 2\rho \int_0^2 \int_0^{4-x^2} \left\{\frac{z^2}{2}\right\}_0^y dy\, dx = \rho \int_0^2 \int_0^{4-x^2} y^2\, dy\, dx = \frac{1}{2}M\bar{y}.$$

6. $\displaystyle I = \int_0^1 \int_0^1 \int_0^1 (y^2 + z^2)\rho\, dz\, dy\, dx = \rho \int_0^1 \int_0^1 \left\{ y^2 z + \frac{z^3}{3} \right\}_0^1 dy\, dx$

$\displaystyle = \frac{\rho}{3} \int_0^1 \int_0^1 (3y^2 + 1)\, dy\, dx$

$\displaystyle = \frac{\rho}{3} \int_0^1 \left\{ y^3 + y \right\}_0^1 dx = \frac{2\rho}{3} \left\{ x \right\}_0^1 = \frac{2\rho}{3}$

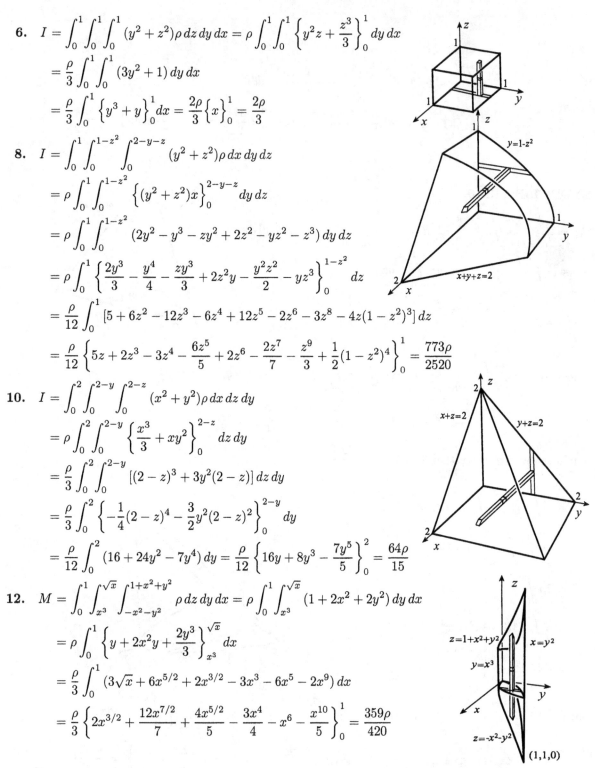

8. $\displaystyle I = \int_0^1 \int_0^{1-z^2} \int_0^{2-y-z} (y^2 + z^2)\rho\, dx\, dy\, dz$

$\displaystyle = \rho \int_0^1 \int_0^{1-z^2} \left\{ (y^2 + z^2)x \right\}_0^{2-y-z} dy\, dz$

$\displaystyle = \rho \int_0^1 \int_0^{1-z^2} (2y^2 - y^3 - zy^2 + 2z^2 - yz^2 - z^3)\, dy\, dz$

$\displaystyle = \rho \int_0^1 \left\{ \frac{2y^3}{3} - \frac{y^4}{4} - \frac{zy^3}{3} + 2z^2 y - \frac{y^2 z^2}{2} - yz^3 \right\}_0^{1-z^2} dz$

$\displaystyle = \frac{\rho}{12} \int_0^1 [5 + 6z^2 - 12z^3 - 6z^4 + 12z^5 - 2z^6 - 3z^8 - 4z(1-z^2)^3]\, dz$

$\displaystyle = \frac{\rho}{12} \left\{ 5z + 2z^3 - 3z^4 - \frac{6z^5}{5} + 2z^6 - \frac{2z^7}{7} - \frac{z^9}{3} + \frac{1}{2}(1-z^2)^4 \right\}_0^1 = \frac{773\rho}{2520}$

10. $\displaystyle I = \int_0^2 \int_0^{2-y} \int_0^{2-z} (x^2 + y^2)\rho\, dx\, dz\, dy$

$\displaystyle = \rho \int_0^2 \int_0^{2-y} \left\{ \frac{x^3}{3} + xy^2 \right\}_0^{2-z} dz\, dy$

$\displaystyle = \frac{\rho}{3} \int_0^2 \int_0^{2-y} [(2-z)^3 + 3y^2(2-z)]\, dz\, dy$

$\displaystyle = \frac{\rho}{3} \int_0^2 \left\{ -\frac{1}{4}(2-z)^4 - \frac{3}{2}y^2(2-z)^2 \right\}_0^{2-y} dy$

$\displaystyle = \frac{\rho}{12} \int_0^2 (16 + 24y^2 - 7y^4)\, dy = \frac{\rho}{12} \left\{ 16y + 8y^3 - \frac{7y^5}{5} \right\}_0^2 = \frac{64\rho}{15}$

12. $\displaystyle M = \int_0^1 \int_{x^3}^{\sqrt{x}} \int_{-x^2 - y^2}^{1 + x^2 + y^2} \rho\, dz\, dy\, dx = \rho \int_0^1 \int_{x^3}^{\sqrt{x}} (1 + 2x^2 + 2y^2)\, dy\, dx$

$\displaystyle = \rho \int_0^1 \left\{ y + 2x^2 y + \frac{2y^3}{3} \right\}_{x^3}^{\sqrt{x}} dx$

$\displaystyle = \frac{\rho}{3} \int_0^1 (3\sqrt{x} + 6x^{5/2} + 2x^{3/2} - 3x^3 - 6x^5 - 2x^9)\, dx$

$\displaystyle = \frac{\rho}{3} \left\{ 2x^{3/2} + \frac{12x^{7/2}}{7} + \frac{4x^{5/2}}{5} - \frac{3x^4}{4} - x^6 - \frac{x^{10}}{5} \right\}_0^1 = \frac{359\rho}{420}$

Since

$\displaystyle M\bar{x} = \int_0^1 \int_{x^3}^{\sqrt{x}} \int_{-x^2 - y^2}^{1 + x^2 + y^2} x\rho\, dz\, dy\, dx = \rho \int_0^1 \int_{x^3}^{\sqrt{x}} x(1 + 2x^2 + 2y^2)\, dy\, dx$

$\displaystyle = \rho \int_0^1 \left\{ xy + 2x^3 y + \frac{2xy^3}{3} \right\}_{x^3}^{\sqrt{x}} dx = \frac{\rho}{3} \int_0^1 (3x^{3/2} + 6x^{7/2} + 2x^{5/2} - 3x^4 - 6x^6 - 2x^{10})\, dx$

$$= \frac{\rho}{3}\left\{\frac{6x^{5/2}}{5} + \frac{4x^{9/2}}{3} + \frac{4x^{7/2}}{7} - \frac{3x^5}{5} - \frac{6x^7}{7} - \frac{2x^{11}}{11}\right\}_0^1 = \frac{1693\rho}{3465},$$

we obtain $\overline{x} = \dfrac{1693\rho}{3465}\dfrac{420}{359\rho} = \dfrac{6772}{11\,847}$. Since

$$M\overline{y} = \int_0^1 \int_{x^3}^{\sqrt{x}} \int_{-x^2-y^2}^{1+x^2+y^2} y\rho\,dz\,dy\,dx = \rho\int_0^1 \int_{x^3}^{\sqrt{x}} y(1+2x^2+2y^2)\,dy\,dx$$

$$= \rho\int_0^1 \left\{\frac{y^2}{2} + x^2 y^2 + \frac{y^4}{2}\right\}_{x^3}^{\sqrt{x}} dx = \frac{\rho}{2}\int_0^1 (x + x^2 + 2x^3 - x^6 - 2x^8 - x^{12})\,dx$$

$$= \frac{\rho}{2}\left\{\frac{x^2}{2} + \frac{x^3}{3} + \frac{x^4}{2} - \frac{x^7}{7} - \frac{2x^9}{9} - \frac{x^{13}}{13}\right\}_0^1 = \frac{365\rho}{819},$$

we find $\overline{y} = \dfrac{365\rho}{819}\dfrac{420}{359\rho} = \dfrac{7300}{14\,001}$. Since the top and bottom surfaces have exactly the same shape, it follows that $\overline{z} = 1/2$.

14. $M = 8\displaystyle\int_0^2 \int_0^x \int_x^2 \rho\,dz\,dy\,dx = 8\rho\int_0^2 \int_0^x (2-x)\,dy\,dx$

$$= 8\rho\int_0^2 (2x - x^2)\,dx = 8\rho\left\{x^2 - \frac{x^3}{3}\right\}_0^2 = \frac{32\rho}{3}$$

By symmetry, $\overline{x} = \overline{y} = 0$. Since

$$M\overline{z} = 8\int_0^2 \int_0^x \int_x^2 z\rho\,dz\,dy\,dx$$

$$= 8\rho\int_0^2 \int_0^x \left\{\frac{z^2}{2}\right\}_x^2 dy\,dx = 4\rho\int_0^2 \int_0^x (4 - x^2)\,dy\,dx$$

$$= 4\rho\int_0^2 (4x - x^3)\,dx = 4\rho\left\{2x^2 - \frac{x^4}{4}\right\}_0^2 = 16\rho,$$

we find $\overline{z} = 16\rho\dfrac{3}{32\rho} = \dfrac{3}{2}$.

16. $I = 8\displaystyle\int_0^a \int_0^{\sqrt{a^2-x^2}} \int_0^{\sqrt{a^2-x^2}} (y^2 + z^2)\rho\,dy\,dz\,dx$

$$= 8\rho\int_0^a \int_0^{\sqrt{a^2-x^2}} \left\{\frac{y^3}{3} + yz^2\right\}_0^{\sqrt{a^2-x^2}} dz\,dx$$

$$= \frac{8\rho}{3}\int_0^a \int_0^{\sqrt{a^2-x^2}} [(a^2-x^2)^{3/2} + 3z^2\sqrt{a^2-x^2}]\,dz\,dx$$

$$= \frac{8\rho}{3}\int_0^a \left\{(a^2-x^2)^{3/2} z + z^3\sqrt{a^2-x^2}\right\}_0^{\sqrt{a^2-x^2}} dx$$

$$= \frac{16\rho}{3}\int_0^a (a^4 - 2a^2 x^2 + x^4)\,dx = \frac{16\rho}{3}\left\{a^4 x - \frac{2a^2 x^3}{3} + \frac{x^5}{5}\right\}_0^a = \frac{128\rho a^5}{45}$$

18. The distance from the volume $dz\,dy\,dx$ at point (x, y, z) to the plane $x + y + z = 1$ is $|x + y + z - 1|/\sqrt{3}$. If we take distances from those points on the origin side of the plane as negative, then the required first moment is

$$\int_0^3 \int_0^{6-2z} \int_0^{12-2y-4z} \frac{x+y+z-1}{\sqrt{3}} \rho\,dx\,dy\,dz$$

$$= \frac{\rho}{\sqrt{3}} \int_0^3 \int_0^{6-2z} \left\{ \frac{(x+y+z-1)^2}{2} \right\}_0^{12-2y-4z} dy\,dz$$

$$= \frac{\rho}{2\sqrt{3}} \int_0^3 \int_0^{6-2z} [(11-y-3z)^2 - (y+z-1)^2]\,dy\,dz$$

$$= \frac{\rho}{2\sqrt{3}} \int_0^3 \left\{ \frac{(11-y-3z)^3}{-3} - \frac{(y+z-1)^3}{3} \right\}_0^{6-2z} dz$$

$$= \frac{\rho}{6\sqrt{3}} \int_0^3 [-2(5-z)^3 + (11-3z)^3 + (z-1)^3]\,dz$$

$$= \frac{\rho}{6\sqrt{3}} \left\{ \frac{(5-z)^4}{2} - \frac{(11-3z)^4}{12} + \frac{(z-1)^4}{4} \right\}_0^3 = 51\sqrt{3}\rho.$$

20. The product moment of inertia I_{xy} is

$$I_{xy} = \int_0^2 \int_x^{2x} \int_0^{2-x} xy\rho\,dz\,dy\,dx = \rho \int_0^2 \int_x^{2x} \left\{ xyz \right\}_0^{2-x} dy\,dx$$

$$= \rho \int_0^2 \int_x^{2x} xy(2-x)\,dy\,dx$$

$$= \rho \int_0^2 \left\{ \frac{x(2-x)y^2}{2} \right\}_x^{2x} dx$$

$$= \frac{3\rho}{2} \int_0^2 (2x^3 - x^4)\,dx = \frac{3\rho}{2} \left\{ \frac{x^4}{2} - \frac{x^5}{5} \right\}_0^2 = \frac{12\rho}{5}.$$

The other two are

$$I_{yz} = \int_0^2 \int_x^{2x} \int_0^{2-x} yz\rho\,dz\,dy\,dx = \rho \int_0^2 \int_x^{2x} \left\{ \frac{yz^2}{2} \right\}_0^{2-x} dy\,dx = \frac{\rho}{2} \int_0^2 \int_x^{2x} y(2-x)^2\,dy\,dx$$

$$= \frac{\rho}{2} \int_0^2 \left\{ \frac{(2-x)^2 y^2}{2} \right\}_x^{2x} dx = \frac{3\rho}{4} \int_0^2 (4x^2 - 4x^3 + x^4)\,dx = \frac{3\rho}{4} \left\{ \frac{4x^3}{3} - x^4 + \frac{x^5}{5} \right\}_0^2 = \frac{4\rho}{5},$$

$$I_{xz} = \int_0^2 \int_x^{2x} \int_0^{2-x} xz\rho\,dz\,dy\,dx = \rho \int_0^2 \int_x^{2x} \left\{ \frac{xz^2}{2} \right\}_0^{2-x} dy\,dx = \frac{\rho}{2} \int_0^2 \int_x^{2x} x(2-x)^2\,dy\,dx$$

$$= \frac{\rho}{2} \int_0^2 \left\{ x(2-x)^2 y \right\}_x^{2x} dx = \frac{\rho}{2} \int_0^2 (4x^2 - 4x^3 + x^4)\,dx = \frac{\rho}{2} \left\{ \frac{4x^3}{3} - x^4 + \frac{x^5}{5} \right\}_0^2 = \frac{8\rho}{15}.$$

22. If we orient the volume so that the line is the x-axis, then

$$I_x = \iiint_V (y^2 + z^2)\rho\,dV = \iiint_V \{[(y - \bar{y}) + \bar{y}]^2 + [(z - \bar{z}) + \bar{z}]^2\}\,dV$$

$$= \iiint_V [(y - \bar{y})^2 + 2\bar{y}(y - \bar{y}) + \bar{y}^2$$

$$+ (z - \bar{z})^2 + 2\bar{z}(z - \bar{z}) + \bar{z}^2]\,dV$$

$$= \iiint_V [(y - \bar{y})^2 + (z - \bar{z})^2]\rho\,dV + 2\bar{y} \iiint_V y\rho\,dV$$

$$- \bar{y}^2 \iiint_V \rho\,dV + 2\bar{z} \iiint_V z\rho\,dV - \bar{z}^2 \iiint_V \rho\,dV$$

$$= I_{\bar{x}} + 2\bar{y}(M\bar{y}) - \bar{y}^2(M) + 2\bar{z}(M\bar{z}) - \bar{z}^2(M) = I_{\bar{x}} + M(\bar{y}^2 + \bar{z}^2).$$

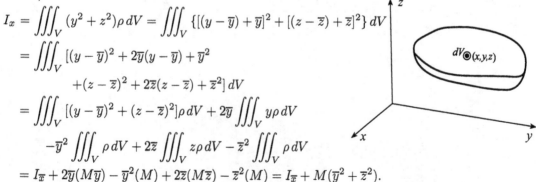

24. Let H be the height of the can and h be
the depth of pop. Let m and M be the
mass of the pop and can, respectively. Let
A be the cross-sectional area of the can and
pop and ρ be the density of the pop. If z
is the centre of mass of can plus pop, then
$(m+M)z = m(h/2) + M(H/2)$. Hence,

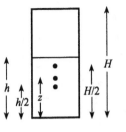

$$z = \frac{mh + MH}{2(m+M)} = \frac{(\rho Ah)h + MH}{2(\rho Ah + M)},$$

where $0 < h < H$. For critical points of z as a function of h, we solve

$$0 = \frac{dz}{dh} = \frac{(\rho Ah + M)(2\rho Ah) - (\rho Ah^2 + MH)(\rho A)}{2(\rho Ah + M)^2} = \frac{\rho A(\rho Ah^2 + 2Mh - MH)}{2(\rho Ah + M)^2}.$$

Solutions are $h = \dfrac{-2M \pm \sqrt{4M^2 + 4\rho AMH}}{2\rho A}$, only the positive root being acceptable. Since $z(0) = z(H) = H/2$, and there is only one critical point, it follows that this critical point must yield a minimum for z. We could substitute the critical value of h into the function $z(h)$ to find the minimum. Instead, notice that if we substitute $\rho Ah = M(H - 2h)/h$, then the minimum value is

$$z = \frac{M(H - 2h) + MH}{(2M/h)(H - 2h) + 2M} = \frac{2M(H - h)}{(2M/h)(H - 2h + h)} = h;$$

that is, the centre of mass is in the surface of the pop.

26. Suppose we let r and h be the radius and
height of that part of the cone under water.
For buoyancy, the weight of the water displaced
must be equal to the weight of the cone,

$$\frac{1}{3}\pi r^2 h(1000)g = \frac{1}{3}\pi R^2 H(800)g \implies r^2 h = \frac{4}{5}R^2 H.$$

By similar triangles, $r/h = R/H \implies r = Rh/H$, and therefore

$$\left(\frac{Rh}{H}\right)^2 h = \frac{4}{5}R^2 H \implies h = \left(\frac{4}{5}\right)^{1/3} H \quad \text{and} \quad r = \left(\frac{4}{5}\right)^{1/3} R.$$

To find the centre of buoyancy we require the centre of mass of a right-circular cone of water with radius $r = (4/5)^{1/3}R$ and height $h = (4/5)^{1/3}H$. Such a cone with apex at the origin has equation $z = (H/R)\sqrt{x^2 + y^2}$. The mass of the displaced water is $M = (1000/3)\pi R^2 H$ kg. If $\bar{z}$ is the z-coordinate of its centre of mass, then

$$M\bar{z} = 4\int_0^{(4/5)^{1/3}R} \int_0^{\sqrt{(4/5)^{2/3}R^2 - x^2}} \int_{(H/R)\sqrt{x^2+y^2}}^{(4/5)^{1/3}H} z(1000)\, dz\, dy\, dx$$

$$= 4000 \int_0^{(4/5)^{1/3}R} \int_0^{\sqrt{(4/5)^{2/3}R^2 - x^2}} \left\{\frac{z^2}{2}\right\}_{(H/R)\sqrt{x^2+y^2}}^{(4/5)^{1/3}H} dy\, dx$$

$$= 2000 \int_0^{(4/5)^{1/3}R} \int_0^{\sqrt{(4/5)^{2/3}R^2 - x^2}} \left[\left(\frac{4}{5}\right)^{2/3} H^2 - \frac{H^2}{R^2}(x^2 + y^2)\right] dy\, dx.$$

If we transform this double iterated integral to polar coordinates,

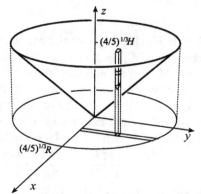

$$M\overline{z} = 2000 \int_0^{\pi/2} \int_0^{(4/5)^{1/3}R} \left[\left(\frac{4}{5}\right)^{2/3} H^2 - \frac{H^2 r^2}{R^2} \right] r\, dr\, d\theta$$

$$= 2000 \int_0^{\pi/2} \left\{ \left(\frac{4}{5}\right)^{2/3} \frac{H^2 r^2}{2} - \frac{H^2 r^4}{4R^2} \right\}_0^{(4/5)^{1/3}R} d\theta$$

$$= 500(4/5)^{4/3} H^2 R^2 \left\{ \theta \right\}_0^{\pi/2} = 250(4/5)^{4/3}\pi H^2 R^2.$$

Thus, $\overline{z} = 250(4/5)^{4/3}\pi H^2 R^2 \dfrac{3}{1000\pi R^2 H} = \dfrac{3(4/5)^{4/3}H}{4}$. The centre of buoyancy of the floating cone is

therefore $\left(\dfrac{4}{5}\right)^{1/3} H - \dfrac{3}{4}\left(\dfrac{4}{5}\right)^{4/3} H = \dfrac{2}{5}\left(\dfrac{4}{5}\right)^{1/3} H$ below the surface.

EXERCISES 13.11

2. The equation is $r = 1$.
It is symmetric about the z-axis.

4. The equation is $r\cos\theta + r\sin\theta = 5$.
It is not symmetric about the z-axis.

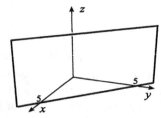

6. The equation is $z = r^2 \cos^2\theta$.
It is not symmetric about the z-axis.

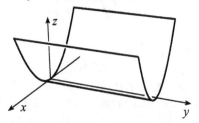

8. The equation is $4z = r^2$.
It is symmetric about the z-axis.

10. The equation is $r^2 = 1 + z^2$.
It is symmetric about the z-axis.

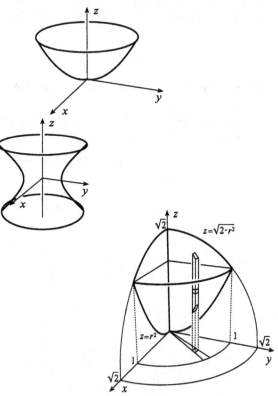

12. We quadruple the volume in the first octant.

$$V = 4 \int_0^{\pi/2} \int_0^1 \int_{r^2}^{\sqrt{2-r^2}} r\, dz\, dr\, d\theta$$

$$= 4 \int_0^{\pi/2} \int_0^1 \left(r\sqrt{2-r^2} - r^3 \right) dr\, d\theta$$

$$= 4 \int_0^{\pi/2} \left\{ -\frac{1}{3}(2-r^2)^{3/2} - \frac{r^4}{4} \right\}_0^1 d\theta$$

$$= \frac{8\sqrt{2}-7}{3}\left\{ \theta \right\}_0^{\pi/2} = \frac{(8\sqrt{2}-7)\pi}{6}$$

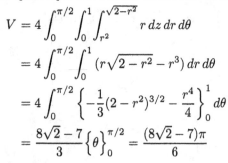

14. We quadruple the volume in the first octant.

$$V = 4 \int_0^{\pi/2} \int_0^{\sqrt{2}} \int_{r^2}^{4-r^2} r \, dz \, dr \, d\theta$$

$$= 4 \int_0^{\pi/2} \int_0^{\sqrt{2}} r(4 - 2r^2) \, dr \, d\theta$$

$$= 4 \int_0^{\pi/2} \left\{ 2r^2 - \frac{r^4}{2} \right\}_0^{\sqrt{2}} d\theta = 8 \{\theta\}_0^{\pi/2} = 4\pi$$

16. We multiply the first octant volume by eight.

$$V = 8 \int_0^{\pi/2} \int_1^2 \int_0^{\sqrt{4-r^2}} r \, dz \, dr \, d\theta$$

$$= 8 \int_0^{\pi/2} \int_1^2 r \sqrt{4 - r^2} \, dr \, d\theta$$

$$= 8 \int_0^{\pi/2} \left\{ -\frac{1}{3}(4 - r^2)^{3/2} \right\}_1^2 d\theta$$

$$= 8\sqrt{3} \{\theta\}_0^{\pi/2} = 4\sqrt{3}\pi$$

18. The six triple iterated integrals are

$$\int_{-\pi}^{\pi} \int_0^3 \int_0^{1+r^2} f(r\cos\theta, r\sin\theta, z) \, r \, dz \, dr \, d\theta,$$

$$\int_0^3 \int_{-\pi}^{\pi} \int_0^{1+r^2} f(r\cos\theta, r\sin\theta, z) \, r \, dz \, d\theta \, dr,$$

$$\int_0^3 \int_0^{1+r^2} \int_{-\pi}^{\pi} f(r\cos\theta, r\sin\theta, z) \, r \, d\theta \, dz \, dr,$$

$$\int_0^1 \int_0^3 \int_{-\pi}^{\pi} f(r\cos\theta, r\sin\theta, z) \, r \, d\theta \, dr \, dz + \int_1^{10} \int_{\sqrt{z-1}}^3 \int_{-\pi}^{\pi} f(r\cos\theta, r\sin\theta, z) \, r \, d\theta \, dr \, dz,$$

$$\int_{-\pi}^{\pi} \int_0^1 \int_0^3 f(r\cos\theta, r\sin\theta, z) \, r \, dr \, dz \, d\theta + \int_{-\pi}^{\pi} \int_1^{10} \int_{\sqrt{z-1}}^3 f(r\cos\theta, r\sin\theta, z) \, r \, dr \, dz \, d\theta,$$

$$\int_0^1 \int_{-\pi}^{\pi} \int_0^3 f(r\cos\theta, r\sin\theta, z) \, r \, dr \, d\theta \, dz + \int_1^{10} \int_{-\pi}^{\pi} \int_{\sqrt{z-1}}^3 f(r\cos\theta, r\sin\theta, z) \, r \, dr \, d\theta \, dz$$

20. We multiply the moment of inertia about the z-axis of that part in the first octant by eight.

$$I_z = 8 \int_0^R \int_0^{\pi/2} \int_0^{\sqrt{R^2-r^2}} r^2 \rho \, r \, dz \, d\theta \, dr$$

$$= 8\rho \int_0^R \int_0^{\pi/2} r^3 \sqrt{R^2 - r^2} \, d\theta \, dr$$

$$= 4\pi\rho \int_0^R r^3 \sqrt{R^2 - r^2} \, dr$$

If we set $u = R^2 - r^2$, then $du = -2r \, dr$, and

$$I_z = 4\pi\rho \int_{R^2}^0 (R^2 - u)\sqrt{u} \left(-\frac{du}{2} \right)$$

$$= 2\pi\rho \left\{ \frac{2}{3} R^2 u^{3/2} - \frac{2}{5} u^{5/2} \right\}_0^{R^2} = \frac{8\pi\rho R^5}{15}.$$

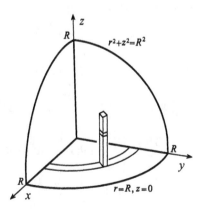

22. The limits define the first octant volume inside the sphere $x^2 + y^2 + z^2 = 81$. The value of the triple iterated integral is therefore given by

$$\int_0^9 \int_0^{\pi/2} \int_0^{\sqrt{81-r^2}} \frac{1}{r} r \, dz \, d\theta \, dr$$

$$= \int_0^9 \int_0^{\pi/2} \sqrt{81 - r^2} \, d\theta \, dr = \frac{\pi}{2} \int_0^9 \sqrt{81 - r^2} \, dr.$$

If we set $r = 9 \sin\phi$, then $dr = 9 \cos\phi \, d\phi$, and

$$\int_0^9 \int_0^{\pi/2} \int_0^{\sqrt{81-r^2}} \frac{1}{r} r \, dz \, d\theta \, dr = \frac{\pi}{2} \int_0^{\pi/2} 9 \cos\phi \, 9 \cos\phi \, d\phi$$

$$= \frac{81\pi}{2} \int_0^{\pi/2} \left(\frac{1 + \cos 2\phi}{2} \right) d\phi = \frac{81\pi}{4} \left\{ \phi + \frac{\sin 2\phi}{2} \right\}_0^{\pi/2} = \frac{81\pi^2}{8}.$$

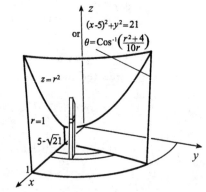

24. The limits define the volume under $z = x^2 + y^2$, above $z = 0$, and bounded on the sides by the cylinders $x^2 + y^2 = 1$ and $(x-5)^2 + y^2 = 21$, and the xz-plane. The value of the triple iterated integral is therefore given by

$$\int_{5-\sqrt{21}}^1 \int_0^{\theta(r)} \int_0^{r^2} r \sin\theta \, r \, dz \, d\theta \, dr$$

$$= \int_{5-\sqrt{21}}^1 \int_0^{\theta(r)} r^4 \sin\theta \, d\theta \, dr$$

$$= \int_{5-\sqrt{21}}^1 \left\{ -r^4 \cos\theta \right\}_0^{\theta(r)} dr$$

$$= \int_{5-\sqrt{21}}^1 \left(r^4 - \frac{r^5}{10} - \frac{2r^3}{5} \right) dr$$

$$= \left\{ \frac{r^5}{5} - \frac{r^6}{60} - \frac{r^4}{10} \right\}_{5-\sqrt{21}}^1 = 0.084.$$

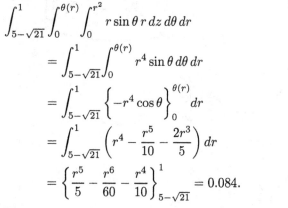

26. The moment of inertia of the upper leg about a line through its centre of mass G_U is

$$I_{G_U} = (0.137)(73) \left(\frac{0.07^2}{4} + \frac{0.45^2}{12} \right) = 0.181 \text{ kg·m}^2.$$

Similarly,

$$I_{G_L} = (0.06)(73) \left(\frac{0.05^2}{4} + \frac{0.5^2}{12} \right) = 0.094 \text{ kg·m}^2.$$

Since $HG_U = 0.225$ and $HG_L = \sqrt{0.45^2 + 0.25^2 - 2(0.45)(0.25) \cos(\pi/3)} = 0.391$, the moment of inertia of the leg about the hip is

$$[0.181 + (0.137)(73)(0.225)^2] + [0.094 + (0.06)(73)(0.391)^2] = 1.45 \text{ kg·m}^2.$$

28. $M = 2 \int_0^{\pi/2} \int_0^{2\cos\theta} \int_0^r \rho r \, dz \, dr \, d\theta = 2\rho \int_0^{\pi/2} \int_0^{2\cos\theta} r^2 \, dr \, d\theta$

$$= 2\rho \int_0^{\pi/2} \left\{ \frac{r^3}{3} \right\}_0^{2\cos\theta} d\theta = \frac{16\rho}{3} \int_0^{\pi/2} \cos^3\theta \, d\theta$$

$$= \frac{16\rho}{3} \int_0^{\pi/2} \cos\theta (1 - \sin^2\theta) \, d\theta$$

$$= \frac{16\rho}{3} \left\{ \sin\theta - \frac{1}{3} \sin^3\theta \right\}_0^{\pi/2} = \frac{32\rho}{9}$$

By symmetry, $\bar{y} = 0$. Since

$$M\bar{x} = 2\int_0^{\pi/2}\int_0^{2\cos\theta}\int_0^r r\cos\theta\,\rho r\,dz\,dr\,d\theta = 2\rho\int_0^{\pi/2}\int_0^{2\cos\theta} r^3\cos\theta\,dr\,d\theta$$

$$= 2\rho\int_0^{\pi/2}\left\{\frac{r^4}{4}\cos\theta\right\}_0^{2\cos\theta}\,d\theta = 8\rho\int_0^{\pi/2}\cos^5\theta\,d\theta = 8\rho\int_0^{\pi/2}\cos\theta(1-\sin^2\theta)^2\,d\theta$$

$$= 8\rho\int_0^{\pi/2}\cos\theta(1-2\sin^2\theta+\sin^4\theta)\,d\theta = 8\rho\left\{\sin\theta - \frac{2}{3}\sin^3\theta + \frac{1}{5}\sin^5\theta\right\}_0^{\pi/2} = \frac{64\rho}{15},$$

it follows that $\bar{x} = \dfrac{64\rho}{15}\dfrac{9}{32\rho} = \dfrac{6}{5}$. Since

$$M\bar{z} = 2\int_0^{\pi/2}\int_0^{2\cos\theta}\int_0^r z\rho r\,dz\,dr\,d\theta = 2\rho\int_0^{\pi/2}\int_0^{2\cos\theta}\left\{\frac{rz^2}{2}\right\}_0^r\,dr\,d\theta$$

$$= \rho\int_0^{\pi/2}\int_0^{2\cos\theta} r^3\,dr\,d\theta = \rho\int_0^{\pi/2}\left\{\frac{r^4}{4}\right\}_0^{2\cos\theta}\,d\theta = 4\rho\int_0^{\pi/2}\cos^4\theta\,d\theta$$

$$= 4\rho\int_0^{\pi/2}\left(\frac{1+\cos 2\theta}{2}\right)^2\,d\theta = \rho\int_0^{\pi/2}\left(1 + 2\cos 2\theta + \frac{1+\cos 4\theta}{2}\right)\,d\theta$$

$$= \rho\left\{\frac{3\theta}{2} + \sin 2\theta + \frac{1}{8}\sin 4\theta\right\}_0^{\pi/2} = \frac{3\pi\rho}{4},$$

we obtain $\bar{z} = \dfrac{3\pi\rho}{4}\dfrac{9}{32\rho} = \dfrac{27\pi}{128}$.

30. The volume bounded by the planes and cylinder is

$$V = \int_0^{2\pi}\int_0^R\int_{my}^{my+h} r\,dz\,dr\,d\theta = \int_0^{2\pi}\int_0^R r(h)\,dr\,d\theta = h(\pi R^2).$$

32. We quadruple the first octant volume.

$$V = 4\int_0^{\pi/4}\int_0^{\sqrt{\cos 2\theta}}\int_0^{r^2} r\,dz\,dr\,d\theta$$

$$= 4\int_0^{\pi/4}\int_0^{\sqrt{\cos 2\theta}} r^3\,dr\,d\theta = 4\int_0^{\pi/4}\left\{\frac{r^4}{4}\right\}_0^{\sqrt{\cos 2\theta}}\,d\theta$$

$$= \int_0^{\pi/4}\cos^2 2\theta\,d\theta = \int_0^{\pi/4}\left(\frac{1+\cos 4\theta}{2}\right)\,d\theta$$

$$= \frac{1}{2}\left\{\theta + \frac{1}{4}\sin 4\theta\right\}_0^{\pi/4} = \frac{\pi}{8}$$

34. $V = \displaystyle\int_{\operatorname{Tan}^{-1}(1/2)}^{\pi/4}\int_0^1\int_0^{\sqrt{1-r^2}} r\,dz\,dr\,d\theta$

$$= \int_{\operatorname{Tan}^{-1}(1/2)}^{\pi/4}\int_0^1 r\sqrt{1-r^2}\,dr\,d\theta$$

$$= \int_{\operatorname{Tan}^{-1}(1/2)}^{\pi/4}\left\{-\frac{1}{3}(1-r^2)^{3/2}\right\}_0^1\,d\theta$$

$$= \frac{1}{3}\{\theta\}_{\operatorname{Tan}^{-1}(1/2)}^{\pi/4} = \frac{1}{3}\left[\frac{\pi}{4} - \operatorname{Tan}^{-1}(1/2)\right]$$

36. We quadruple the first octant volume.

$$V = 4 \int_0^{\pi/2} \int_0^{2\sin\theta} \int_0^{r^2} r \, dz \, dr \, d\theta$$

$$= 4 \int_0^{\pi/2} \int_0^{2\sin\theta} r^3 \, dr \, d\theta = 4 \int_0^{\pi/2} \left\{ \frac{r^4}{4} \right\}_0^{2\sin\theta} d\theta$$

$$= \int_0^{\pi/2} 16 \sin^4\theta \, d\theta = 16 \int_0^{\pi/2} \left(\frac{1 - \cos 2\theta}{2} \right)^2 d\theta$$

$$= 4 \int_0^{\pi/2} \left(1 - 2\cos 2\theta + \frac{1 + \cos 4\theta}{2} \right) d\theta$$

$$= 4 \left\{ \frac{3\theta}{2} - \sin 2\theta + \frac{1}{8} \sin 4\theta \right\}_0^{\pi/2} = 3\pi$$

$z = r^2$

$r = 2\sin\theta$

38. We quadruple the first octant volume.

$$V = 4 \int_0^{\pi/2} \int_0^1 \int_0^{e^{-r^2}} r \, dz \, dr \, d\theta = 4 \int_0^{\pi/2} \int_0^1 re^{-r^2} \, dr \, d\theta$$

$$= 4 \int_0^{\pi/2} \left\{ -\frac{1}{2} e^{-r^2} \right\}_0^1 d\theta$$

$$= -2(e^{-1} - 1) \left\{ \theta \right\}_0^{\pi/2} = \pi(1 - 1/e)$$

$r = 1$

$z = e^{-r^2}$

40. $$V = \int_{-\pi/4}^{3\pi/4} \int_0^{\cos\theta + \sin\theta} \int_{r^2}^{r\cos\theta + r\sin\theta} r \, dz \, dr \, d\theta$$

$$= \int_{-\pi/4}^{3\pi/4} \int_0^{\cos\theta + \sin\theta} (r^2 \cos\theta + r^2 \sin\theta - r^3) \, dr \, d\theta$$

$$= \int_{-\pi/4}^{3\pi/4} \left\{ \frac{r^3}{3} \cos\theta + \frac{r^3}{3} \sin\theta - \frac{r^4}{4} \right\}_0^{\cos\theta + \sin\theta} d\theta$$

$$= \frac{1}{12} \int_{-\pi/4}^{3\pi/4} [4(\cos\theta + \sin\theta)^4 - 3(\cos\theta + \sin\theta)^4] \, d\theta$$

$$= \frac{1}{12} \int_{-\pi/4}^{3\pi/4} (\cos^4\theta + 4\cos^3\theta\sin\theta + 6\cos^2\theta\sin^2\theta + 4\cos\theta\sin^3\theta + \sin^4\theta) \, d\theta$$

$$= \frac{1}{12} \int_{-\pi/4}^{3\pi/4} [(\cos^2\theta + \sin^2\theta)^2 + 4(\cos^3\theta\sin\theta + \cos^2\theta\sin^2\theta + \cos\theta\sin^3\theta)] \, d\theta$$

$$= \frac{1}{12} \int_{-\pi/4}^{3\pi/4} [1 + 4(\cos^3\theta\sin\theta + \cos\theta\sin^3\theta) + (\sin 2\theta)^2] \, d\theta$$

$$= \frac{1}{12} \int_{-\pi/4}^{3\pi/4} \left[1 + 4(\cos^3\theta\sin\theta + \cos\theta\sin^3\theta) + \frac{1 - \cos 4\theta}{2} \right] d\theta$$

$$= \frac{1}{12} \left\{ \frac{3\theta}{2} - \cos^4\theta + \sin^4\theta - \frac{1}{8}\sin 4\theta \right\}_{-\pi/4}^{3\pi/4} = \frac{\pi}{8}$$

$z = r(\cos\theta + \sin\theta)$

$z = r^2$

$r = \cos\theta + \sin\theta$

42.
$$\iiint_V \sqrt{x^2+y^2+z^2}\, dV = 4\int_0^{\pi/2}\int_0^3\int_0^z \sqrt{r^2+z^2}\, r\, dr\, dz\, d\theta$$

$$= 4\int_0^{\pi/2}\int_0^3 \left\{\frac{1}{3}(r^2+z^2)^{3/2}\right\}_0^z dz\, d\theta$$

$$= \frac{4}{3}\int_0^{\pi/2}\int_0^3 (2\sqrt{2}z^3 - z^3)\, dz\, d\theta$$

$$= \frac{4(2\sqrt{2}-1)}{3}\int_0^{\pi/2}\left\{\frac{z^4}{4}\right\}_0^3 d\theta$$

$$= 27(2\sqrt{2}-1)\left\{\theta\right\}_0^{\pi/2} = \frac{27\pi(2\sqrt{2}-1)}{2}$$

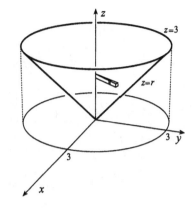

EXERCISES 13.12

2. The equation is $\Re \sin\phi = 1$. See figure for Exercise 13.11-2.

4. The equation is $\Re = 4\csc\phi\cot\phi$. See figure for Exercise 13.11-8.

6. The equation is $\Re^2 = -\sec 2\phi$. See figure for Exercise 13.11-10.

8. We quadruple the volume in the first octant.

$$V = 4\int_0^{\pi/2}\int_0^{\pi/4}\int_0^1 \Re^2 \sin\phi\, d\Re\, d\phi\, d\theta$$

$$= \frac{4}{3}\int_0^{\pi/2}\int_0^{\pi/4} \sin\phi\, d\phi\, d\theta$$

$$= \frac{4}{3}\int_0^{\pi/2}\left\{-\cos\phi\right\}_0^{\pi/4} d\theta$$

$$= \frac{2(2-\sqrt{2})}{3}\left\{\theta\right\}_0^{\pi/2} = \frac{(2-\sqrt{2})\pi}{3}$$

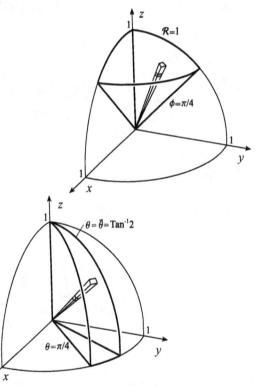

10.
$$V = \int_{\pi/4}^{\bar\theta}\int_0^{\pi/2}\int_0^1 \Re^2 \sin\phi\, d\Re\, d\phi\, d\theta$$

$$= \frac{1}{3}\int_{\pi/4}^{\bar\theta}\int_0^{\pi/2} \sin\phi\, d\phi\, d\theta$$

$$= \frac{1}{3}\int_{\pi/4}^{\bar\theta}\left\{-\cos\phi\right\}_0^{\pi/2} d\theta$$

$$= \frac{1}{3}\left\{\theta\right\}_{\pi/4}^{\bar\theta} = \frac{1}{3}(\mathrm{Tan}^{-1}2 - \pi/4)$$

12. We quadruple the volume in the first octant.

$$V = 4 \int_0^{\pi/2} \int_{\bar\phi}^{\pi/2} \int_0^{2\csc\phi} \Re^2 \sin\phi \, d\Re \, d\phi \, d\theta$$

$$= 4 \int_0^{\pi/2} \int_{\bar\phi}^{\pi/2} \left\{ \frac{\Re^3}{3} \sin\phi \right\}_0^{2\csc\phi} d\phi \, d\theta$$

$$= \frac{32}{3} \int_0^{\pi/2} \int_{\bar\phi}^{\pi/2} \csc^2\phi \, d\phi \, d\theta$$

$$= \frac{32}{3} \int_0^{\pi/2} \left\{ -\cot\phi \right\}_{\bar\phi}^{\pi/2} d\theta$$

$$= \frac{64}{3} \left\{ \theta \right\}_0^{\pi/2} = \frac{32\pi}{3}$$

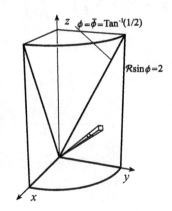

14. We multiply the moment of inertia of the first octant portion of the sphere about the z-axis by eight.

$$I_z = 8 \int_0^{\pi/2} \int_0^{\pi/2} \int_0^R (\Re^2 \sin^2\phi) \rho \Re^2 \sin\phi \, d\Re \, d\phi \, d\theta$$

$$= 8\rho \int_0^{\pi/2} \int_0^{\pi/2} \left\{ \frac{\Re^5}{5} \sin^3\phi \right\}_0^R d\phi \, d\theta$$

$$= \frac{8\rho R^5}{5} \int_0^{\pi/2} \int_0^{\pi/2} (1 - \cos^2\phi) \sin\phi \, d\phi \, d\theta$$

$$= \frac{8\rho R^5}{5} \int_0^{\pi/2} \left\{ -\cos\phi + \frac{\cos^3\phi}{3} \right\}_0^{\pi/2} d\theta$$

$$= \frac{16\rho R^5}{15} \left\{ \theta \right\}_0^{\pi/2} = \frac{8\pi\rho R^5}{15}.$$

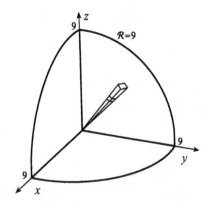

16. Using the figure in Exercise 14,

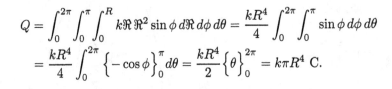

$$Q = \int_0^{2\pi} \int_0^\pi \int_0^R k\Re \, \Re^2 \sin\phi \, d\Re \, d\phi \, d\theta = \frac{kR^4}{4} \int_0^{2\pi} \int_0^\pi \sin\phi \, d\phi \, d\theta$$

$$= \frac{kR^4}{4} \int_0^{2\pi} \left\{ -\cos\phi \right\}_0^\pi d\theta = \frac{kR^4}{2} \left\{ \theta \right\}_0^{2\pi} = k\pi R^4 \text{ C}.$$

18. The limits define the first octant volume inside the sphere $x^2 + y^2 + z^2 = 81$. The value of the triple iterated integral is therefore given by

$$\int_0^{\pi/2} \int_0^{\pi/2} \int_0^9 \frac{1}{\Re^2} \Re^2 \sin\phi \, d\Re \, d\phi \, d\theta$$

$$= 9 \int_0^{\pi/2} \int_0^{\pi/2} \sin\phi \, d\phi \, d\theta$$

$$= 9 \int_0^{\pi/2} \left\{ -\cos\phi \right\}_0^{\pi/2} d\theta$$

$$= 9 \left\{ \theta \right\}_0^{\pi/2} = \frac{9\pi}{2}$$

20. $V = 4 \displaystyle\int_0^{\pi/2} \int_0^{\overline\phi} \int_0^R \Re^2 \sin\phi \, d\Re \, d\phi \, d\theta$

$= \dfrac{4R^3}{3} \displaystyle\int_0^{\pi/2} \int_0^{\overline\phi} \sin\phi \, d\phi \, d\theta$

$= \dfrac{4R^3}{3} \displaystyle\int_0^{\pi/2} \Big\{ -\cos\phi \Big\}_0^{\overline\phi} d\theta$

$= \dfrac{4R^3}{3} (1 - \cos\overline\phi) \Big\{ \theta \Big\}_0^{\pi/2}$

$= \dfrac{2\pi R^3}{3} \left(1 - \dfrac{k}{\sqrt{1+k^2}} \right)$

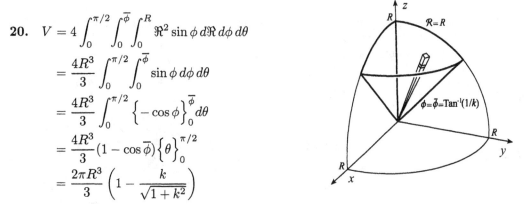

22. (a) Let ρ_b and ρ_w represent the densities of the ball and water. The magnitude of the force of gravity on the ball is $(4/3)\pi R^3 \rho_b g$ where R is its radius, and $g > 0$ is the acceleration due to gravity. Since this must be equal to the weight of water displaced by the half-submerged ball, $\dfrac{4}{3}\pi R^3 \rho_b g = \dfrac{2}{3}\pi R^3 \rho_w g$. This equation implies that $\rho_b = \rho_w/2$.

(b) In the diagram, we let the plane $z = R/2$ represent the surface of the water. The volume of ball above water is given by

$4 \displaystyle\int_0^{\pi/2} \int_0^{\pi/3} \int_{(R/2)\sec\phi}^R \Re^2 \sin\phi \, d\Re \, d\phi \, d\theta$

$= 4 \displaystyle\int_0^{\pi/2} \int_0^{\pi/3} \Big\{ \dfrac{\Re^3}{3} \sin\phi \Big\}_{(R/2)\sec\phi}^R d\phi \, d\theta$

$= \dfrac{R^3}{6} \displaystyle\int_0^{\pi/2} \int_0^{\pi/3} (8\sin\phi - \tan\phi \sec^2\phi) \, d\phi \, d\theta$

$= \dfrac{R^3}{6} \displaystyle\int_0^{\pi/2} \Big\{ -8\cos\phi - \dfrac{\tan^2\phi}{2} \Big\}_0^{\pi/3} d\theta$

$= \dfrac{5R^3}{12} \Big\{ \theta \Big\}_0^{\pi/2} = \dfrac{5\pi R^3}{24}.$

The force required to keep the ball at this position is equal to the extra weight of water (above that in (a)) displaced; i.e., $\left(\dfrac{2}{3}\pi R^3 - \dfrac{5}{24}\pi R^3 \right) \rho_w g = \dfrac{11}{24}\pi \rho_w g R^3.$

24. (a) Since $s^2 = \Re^2 + d^2 - 2\Re d \cos\phi$,

$$V = \int_{-\pi}^{\pi} \int_0^{\pi} \int_0^R \dfrac{\rho}{4\pi\epsilon_0 s} \Re^2 \sin\phi \, d\Re \, d\phi \, d\theta = \dfrac{\rho}{4\pi\epsilon_0} \int_{-\pi}^{\pi} \int_0^{\pi} \int_0^R \dfrac{\Re^2 \sin\phi}{\sqrt{\Re^2 + d^2 - 2\Re d \cos\phi}} d\Re \, d\phi \, d\theta.$$

(b) In order to change ϕ to s we first write $V = \dfrac{\rho}{4\pi\epsilon_0} \displaystyle\int_{-\pi}^{\pi} \int_0^R \int_0^{\pi} \dfrac{\Re^2 \sin\phi}{\sqrt{\Re^2 + d^2 - 2\Re d \cos\phi}} d\phi \, d\Re \, d\theta$. If $s^2 = \Re^2 + d^2 - 2\Re d \cos\phi$, then $2s \, ds = 2\Re d \sin\phi \, d\phi$, and

$$V = \dfrac{\rho}{4\pi\epsilon_0} \int_{-\pi}^{\pi} \int_0^R \int_{d-\Re}^{d+\Re} \dfrac{\Re^2}{s} \left(\dfrac{s \, ds}{\Re d} \right) d\Re \, d\theta = \dfrac{\rho}{4\pi\epsilon_0 d} \int_{-\pi}^{\pi} \int_0^R \int_{d-\Re}^{d+\Re} \Re \, ds \, d\Re \, d\theta.$$

(c) $V = \dfrac{\rho}{4\pi\epsilon_0 d} \displaystyle\int_{-\pi}^{\pi} \int_0^R \Big\{ \Re s \Big\}_{d-\Re}^{d+\Re} d\Re \, d\theta = \dfrac{\rho}{2\pi\epsilon_0 d} \int_{-\pi}^{\pi} \int_0^R \Re^2 \, d\Re \, d\theta$

$= \dfrac{\rho}{2\pi\epsilon_0 d} \displaystyle\int_{-\pi}^{\pi} \Big\{ \dfrac{\Re^3}{3} \Big\}_0^R d\theta = \dfrac{\rho R^3}{6\pi\epsilon_0 d} \Big\{ \theta \Big\}_{-\pi}^{\pi} = \dfrac{\rho R^3}{3\epsilon_0 d}$

Since $Q = (4/3)\pi R^3 \rho$, $\dfrac{1}{4\pi\epsilon_0} \dfrac{Q}{d} = \dfrac{1}{4\pi\epsilon_0 d} \left(\dfrac{4}{3}\pi R^3 \rho \right) = \dfrac{\rho R^3}{3\epsilon_0 d}$, and therefore $V = \dfrac{1}{4\pi\epsilon_0} \dfrac{Q}{d}.$

26. We can always choose a coordinate system so that the point is on the z-axis. Symmetry makes it clear that x- and y-components of the force vanish.
The contribution to the z-component of the force due to the mass in dV is

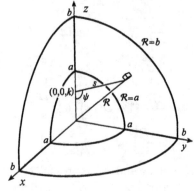

$$-\frac{Gm\rho dV}{s^2}\cos\psi = -\frac{Gm\rho}{s^2}\left(\frac{k^2+s^2-\Re^2}{2ks}\right)dV.$$

Therefore

$$F_z = \int_0^\pi \int_a^b \int_{-\pi}^{\pi} -\frac{Gm\rho}{2ks^3}(k^2+s^2-\Re^2)\Re^2\sin\phi\,d\theta\,d\Re\,d\phi$$

$$= -\frac{Gm\rho\pi}{k}\int_0^\pi \int_a^b \left(\frac{s^2+k^2-\Re^2}{s^3}\right)\Re^2\sin\phi\,d\Re\,d\phi$$

$$= -\frac{Gm\rho\pi}{k}\int_a^b \int_0^\pi \left(\frac{s^2+k^2-\Re^2}{s^3}\right)\Re^2\sin\phi\,d\phi\,d\Re.$$

If we set $s = \sqrt{\Re^2+k^2-2k\Re\cos\phi}$ in the inner integral, then $2s\,ds = 2k\Re\sin\phi\,d\phi$, and

$$F_z = -\frac{Gm\rho\pi}{k}\int_a^b \int_{\Re-k}^{\Re+k} \left(\frac{s^2+k^2-\Re^2}{s^3}\right)\Re^2\sin\phi\left(\frac{s\,ds}{k\Re\sin\phi}\right)d\Re$$

$$= -\frac{Gm\rho\pi}{k^2}\int_a^b \int_{\Re-k}^{\Re+k} \Re\left(\frac{s^2+k^2-\Re^2}{s^2}\right)ds\,d\Re = -\frac{Gm\rho\pi}{k^2}\int_a^b \Re\left\{s-\frac{k^2-\Re^2}{s}\right\}\Big|_{\Re-k}^{\Re+k}d\Re$$

$$= -\frac{Gm\rho\pi}{k^2}\int_a^b \Re\left[\Re+k-\frac{k^2-\Re^2}{\Re+k}-(\Re-k)+\frac{k^2-\Re^2}{\Re-k}\right]d\Re = 0.$$

EXERCISES 13.13

2. (a) This is a change to polar coordinates,

$$\iint_R xy\,dA = \int_{\pi/6}^{\pi/3}\int_0^{1-\cos\theta} r\cos\theta\, r\sin\theta\, r\,dr\,d\theta = \int_{\pi/6}^{\pi/3}\int_0^{1-\cos\theta} r^3\cos\theta\sin\theta\,dr\,d\theta.$$

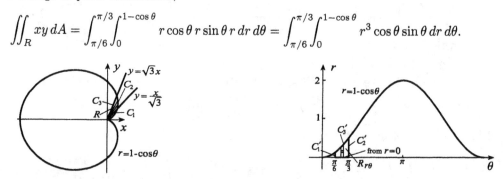

(b) Alternatively, region R in the xy-plane is mapped to the region $R_{r\theta}$ in the $r\theta$-plane shown above.
With $\dfrac{\partial(x,y)}{\partial(r,\theta)} = \begin{vmatrix} \cos\theta & -r\sin\theta \\ \sin\theta & r\cos\theta \end{vmatrix} = r$, equation 13.70 gives

$$\iint_R xy\,dA = \iint_{R_{r\theta}} r\cos\theta\, r\sin\theta\left|\frac{\partial(x,y)}{\partial(r,\theta)}\right|dr\,d\theta = \int_{\pi/6}^{\pi/3}\int_0^{1-\cos\theta} r^3\cos\theta\sin\theta\,dr\,d\theta.$$

4. Region R in the xy-plane is mapped to the rectangle R_{uv} in the uv-plane shown below.

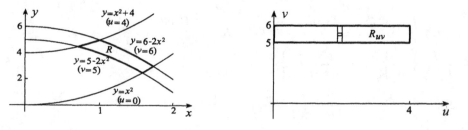

With $\dfrac{\partial(x,y)}{\partial(u,v)} = \dfrac{1}{\dfrac{\partial(u,v)}{\partial(x,y)}} = \dfrac{1}{\begin{vmatrix} -2x & 1 \\ 4x & 1 \end{vmatrix}} = -\dfrac{1}{6x}$, equation 13.70 gives

$$\iint_R (x^2 + y)\, dA = \iint_{R_{uv}} (x^2 + y)\left| \frac{\partial(x,y)}{\partial(u,v)} \right| dv\, du = \frac{1}{6}\iint_{R_{uv}} \left(\frac{x^2 + y}{x} \right) dv\, du$$

$$= \frac{1}{6}\int_0^4 \int_5^6 \left(\sqrt{\frac{v-u}{3}} + \frac{(v+2u)/3}{\sqrt{(v-u)/3}} \right) dv\, du = \frac{1}{6\sqrt{3}}\int_0^4 \int_5^6 \frac{2v+u}{\sqrt{v-u}}\, dv\, du.$$

6. (a) This is a change to cylindrical coordinates,

$$\iiint_V (x^2 + y^2)\, dV = \int_0^{\pi/2}\int_0^3\int_0^r r^2\, r\, dz\, dr\, d\theta = \int_0^{\pi/2}\int_0^3\int_0^r r^3\, dz\, dr\, d\theta.$$

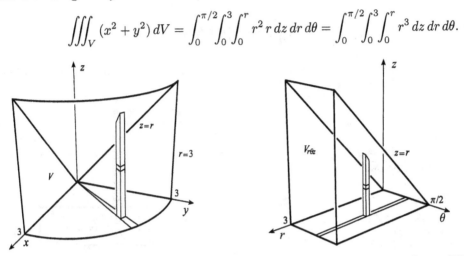

(b) Region V in xyz-space is mapped to the region $V_{r\theta z}$ in $r\theta z$-space shown above. With

$\dfrac{\partial(x,y,z)}{\partial(r,\theta,z)} = \begin{vmatrix} \cos\theta & -r\sin\theta & 0 \\ \sin\theta & r\cos\theta & 0 \\ 0 & 0 & 1 \end{vmatrix} = r$, equation 13.73 gives

$$\iiint_V (x^2 + y^2)\, dV = \iiint_{V_{r\theta z}} r^2 \left| \frac{\partial(x,y,z)}{\partial(r,\theta,z)} \right| dz\, dr\, d\theta = \int_0^{\pi/2}\int_0^3\int_0^r r^3\, dz\, dr\, d\theta.$$

8. (a) This is a change to spherical coordinates,

$$\iiint_V x^2 y^2 z\, dV = \int_{-\pi}^{\pi}\int_0^{\pi/4}\int_0^2 (\Re^2 \sin^2\phi \cos^2\theta)(\Re^2 \sin^2\phi \sin^2\theta)(\Re\cos\phi)\,\Re^2\sin\phi\, d\Re\, d\phi\, d\theta$$

$$= \int_{-\pi}^{\pi}\int_0^{\pi/4}\int_0^2 \Re^7 \sin^5\phi \cos\phi \sin^2\theta \cos^2\theta\, d\Re\, d\phi\, d\theta.$$

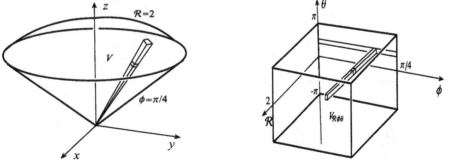

(b) Region V in xyz-space is mapped to the region $V_{\Re\phi\theta}$ in $\Re\phi\theta$-space shown above. With

$$\frac{\partial(x,y,z)}{\partial(\Re,\phi,\theta)} = \begin{vmatrix} \sin\phi\cos\theta & \Re\cos\phi\cos\theta & -\Re\sin\phi\sin\theta \\ \sin\phi\sin\theta & \Re\cos\phi\sin\theta & \Re\sin\phi\cos\theta \\ \cos\phi & -\Re\sin\phi & 0 \end{vmatrix}$$

$$= \cos\phi(\Re^2\sin\phi\cos\phi\cos^2\theta + \Re^2\sin\phi\cos\phi\sin^2\theta)$$
$$+ \Re\sin\phi(\Re\sin^2\phi\cos^2\theta + \Re\sin^2\phi\sin^2\theta)$$
$$= \Re^2\cos^2\phi\sin\phi + \Re^2\sin^3\phi = \Re^2\sin\phi,$$

equation 13.73 gives

$$\iiint_V x^2 y^2 z\, dV = \iiint_{V_{\Re\phi\theta}} x^2 y^2 z \left| \frac{\partial(x,y,z)}{\partial(\Re,\phi,\theta)} \right| d\Re\, d\phi\, d\theta$$

$$= \iiint_{V_{\Re\phi\theta}} (\Re^2\sin^2\phi\cos^2\theta)(\Re^2\sin^2\phi\sin^2\theta)(\Re\cos\phi)\Re^2\sin\phi\, d\Re\, d\phi\, d\theta$$

$$= \int_{-\pi}^{\pi}\int_0^{\pi/4}\int_0^2 \Re^7\sin^5\phi\cos\phi\sin^2\theta\cos^2\theta\, d\Re\, d\phi\, d\theta.$$

10. If we let $u = x^2 - y^2$ and $v = x^2 + y^2$, then the region R in the xy-plane is mapped to the rectangle R_{uv} in the uv-plane shown below.

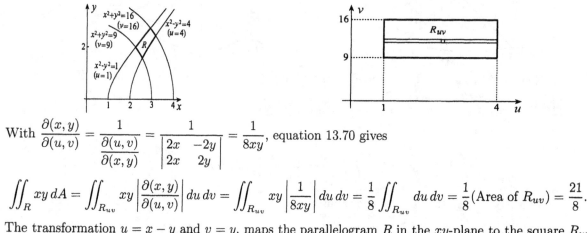

With $\dfrac{\partial(x,y)}{\partial(u,v)} = \dfrac{1}{\dfrac{\partial(u,v)}{\partial(x,y)}} = \dfrac{1}{\begin{vmatrix} 2x & -2y \\ 2x & 2y \end{vmatrix}} = \dfrac{1}{8xy}$, equation 13.70 gives

$$\iint_R xy\, dA = \iint_{R_{uv}} xy \left| \frac{\partial(x,y)}{\partial(u,v)} \right| du\, dv = \iint_{R_{uv}} xy \left| \frac{1}{8xy} \right| du\, dv = \frac{1}{8} \iint_{R_{uv}} du\, dv = \frac{1}{8}(\text{Area of } R_{uv}) = \frac{21}{8}.$$

12. The transformation $u = x - y$ and $v = y$, maps the parallelogram R in the xy-plane to the square R_{uv} in the uv-plane shown below.

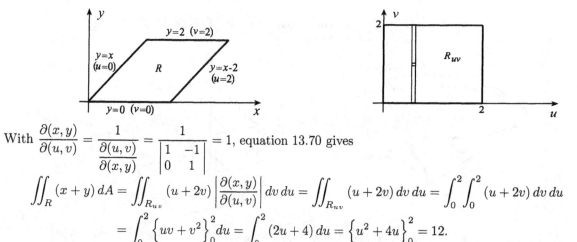

With $\dfrac{\partial(x,y)}{\partial(u,v)} = \dfrac{1}{\dfrac{\partial(u,v)}{\partial(x,y)}} = \dfrac{1}{\begin{vmatrix} 1 & -1 \\ 0 & 1 \end{vmatrix}} = 1$, equation 13.70 gives

$$\iint_R (x+y)\, dA = \iint_{R_{uv}} (u+2v) \left| \frac{\partial(x,y)}{\partial(u,v)} \right| dv\, du = \iint_{R_{uv}} (u+2v)\, dv\, du = \int_0^2\int_0^2 (u+2v)\, dv\, du$$

$$= \int_0^2 \left\{ uv + v^2 \right\}_0^2 du = \int_0^2 (2u+4)\, du = \left\{ u^2 + 4u \right\}_0^2 = 12.$$

14. The transformation maps the first octant volume V bounded by the surfaces to the box V_{uvw} in uvw-space shown below.

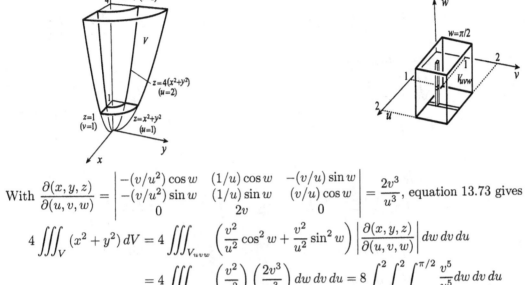

With $\dfrac{\partial(x,y,z)}{\partial(u,v,w)} = \begin{vmatrix} -(v/u^2)\cos w & (1/u)\cos w & -(v/u)\sin w \\ -(v/u^2)\sin w & (1/u)\sin w & (v/u)\cos w \\ 0 & 2v & 0 \end{vmatrix} = \dfrac{2v^3}{u^3}$, equation 13.73 gives

$$4\iiint_V (x^2+y^2)\,dV = 4\iiint_{V_{uvw}} \left(\frac{v^2}{u^2}\cos^2 w + \frac{v^2}{u^2}\sin^2 w\right)\left|\frac{\partial(x,y,z)}{\partial(u,v,w)}\right|\,dw\,dv\,du$$

$$= 4\iiint_{V_{uvw}} \left(\frac{v^2}{u^2}\right)\left(\frac{2v^3}{u^3}\right)\,dw\,dv\,du = 8\int_1^2 \int_1^2 \int_0^{\pi/2} \frac{v^5}{u^5}\,dw\,dv\,du$$

$$= 8\int_1^2 \int_1^2 \left\{\frac{v^5 w}{u^5}\right\}_0^{\pi/2}\,dv\,du = 4\pi\int_1^2 \left\{\frac{v^6}{6u^5}\right\}_1^2\,du = 42\pi\left\{-\frac{1}{4u^4}\right\}_1^2 = \frac{315\pi}{32}.$$

16. The transformation maps the triangle R in the xy-plane to the triangle R_{uv} in the uv-plane shown below.

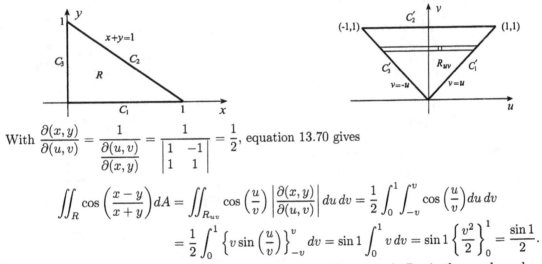

With $\dfrac{\partial(x,y)}{\partial(u,v)} = \dfrac{1}{\dfrac{\partial(u,v)}{\partial(x,y)}} = \dfrac{1}{\begin{vmatrix} 1 & -1 \\ 1 & 1 \end{vmatrix}} = \dfrac{1}{2}$, equation 13.70 gives

$$\iint_R \cos\left(\frac{x-y}{x+y}\right)dA = \iint_{R_{uv}} \cos\left(\frac{u}{v}\right)\left|\frac{\partial(x,y)}{\partial(u,v)}\right|\,du\,dv = \frac{1}{2}\int_0^1 \int_{-v}^v \cos\left(\frac{u}{v}\right)du\,dv$$

$$= \frac{1}{2}\int_0^1 \left\{v\sin\left(\frac{u}{v}\right)\right\}_{-v}^v\,dv = \sin 1\int_0^1 v\,dv = \sin 1\left\{\frac{v^2}{2}\right\}_0^1 = \frac{\sin 1}{2}.$$

18. The transformation maps the region R in the xy-plane to the rectangle R_{uv} in the uv-plane shown below.

With $\dfrac{\partial(x,y)}{\partial(u,v)} = \dfrac{1}{\dfrac{\partial(u,v)}{\partial(x,y)}} = \dfrac{1}{\begin{vmatrix} \dfrac{2(y^2-x^2)}{(x^2+y^2)^2} & \dfrac{-4xy}{(x^2+y^2)^2} \\[2mm] \dfrac{-4xy}{(x^2+y^2)^2} & \dfrac{2(x^2-y^2)}{(x^2+y^2)^2} \end{vmatrix}} = -\dfrac{(x^2+y^2)^2}{4}$, equation 13.70 gives

$$\iint_R \frac{1}{(x^2+y^2)^2}\, dA = \iint_{R_{uv}} \frac{1}{(x^2+y^2)^2}\left|\frac{\partial(x,y)}{\partial(u,v)}\right| dv\, du = \frac{1}{4}\int_{1/3}^{1/2}\int_{1/4}^{1} dv\, du = \frac{1}{4}\left(\frac{3}{4}\right)\left(\frac{1}{6}\right) = \frac{1}{32}.$$

20. (a) We require two iterated integrals for direct evaluation,

$$\iiint_V (x+y+z)\, dV = \int_0^1\int_{1-x}^{1+x}\int_{2x+y}^{2x+y+1}(x+y+z)\, dz\, dy\, dx + \int_1^2\int_{x-1}^{3-x}\int_{2x+y}^{2x+y+1}(x+y+z)\, dz\, dy\, dx$$

$$= \int_0^1\int_{1-x}^{1+x}\left\{\frac{(x+y+z)^2}{2}\right\}_{2x+y}^{2x+y+1} dy\, dx + \int_1^2\int_{x-1}^{3-x}\left\{\frac{(x+y+z)^2}{2}\right\}_{2x+y}^{2x+y+1} dy\, dx$$

$$= \frac{1}{2}\int_0^1\int_{1-x}^{1+x}(6x+4y+1)\, dy\, dx + \frac{1}{2}\int_1^2\int_{x-1}^{3-x}(6x+4y+1)\, dy\, dx$$

$$= \frac{1}{2}\int_0^1\left\{\frac{(6x+4y+1)^2}{8}\right\}_{1-x}^{1+x} dx + \frac{1}{2}\int_1^2\left\{\frac{(6x+4y+1)^2}{8}\right\}_{x-1}^{3-x} dx$$

$$= \frac{1}{16}\int_0^1\left[(10x+5)^2 - (2x+5)^2\right] dx + \frac{1}{16}\int_1^2\left[(2x+13)^2 - (10x-3)^2\right] dx$$

$$= \frac{1}{16}\left\{\frac{(10x+5)^3}{30} - \frac{(2x+5)^3}{6}\right\}_0^1 + \frac{1}{16}\left\{\frac{(2x+13)^3}{6} - \frac{(10x-3)^3}{30}\right\}_1^2 = 11.$$

(b) The transformation maps the region V in xyz-space to the box V_{uvw} in uvw-space shown above. The Jacobian of the transformation is $\dfrac{\partial(x,y,z)}{\partial(u,v,w)} = \dfrac{1}{\dfrac{\partial(u,v,w)}{\partial(x,y,z)}} = \dfrac{1}{\begin{vmatrix} 1 & -1 & 0 \\ 1 & 1 & 0 \\ -2 & -1 & 1 \end{vmatrix}} = \dfrac{1}{2}$. Equation 13.73 gives

$$\iiint_V (x+y+z)\, dV = \iiint_{V_{uvw}}(x+y+z)\left|\frac{\partial(x,y,z)}{\partial(u,v,w)}\right| dw\, dv\, du$$

$$= \frac{1}{2}\iiint_{V_{uvw}}\left(\frac{u+v}{2} + \frac{v-u}{2} + w + \frac{u}{2} + \frac{3v}{2}\right) dw\, dv\, du$$

$$= \frac{1}{4}\int_{-1}^{1}\int_1^3\int_0^1 (2w+u+5v)\, dw\, dv\, du = \frac{1}{4}\int_{-1}^1\int_1^3\left\{w^2+uw+5vw\right\}_0^1 dv\, du$$

$$= \frac{1}{4}\int_{-1}^1\int_1^3(1+u+5v)\, dv\, du = \frac{1}{4}\int_{-1}^1\left\{\frac{(1+u+5v)^2}{10}\right\}_1^3 du$$

$$= \frac{1}{40}\int_{-1}^1\left[(u+16)^2 - (u+6)^2\right] du = \frac{1}{40}\left\{\frac{(u+16)^3}{3} - \frac{(u+6)^3}{3}\right\}_{-1}^1 = 11.$$

EXERCISES 13.14

2. With Leibnitz's rule, $F'(x) = \int_1^x \left(\dfrac{2x}{y^2}\right) dy + \left(\dfrac{x^2}{x^2} + e^x\right)(1) = \left\{-\dfrac{2x}{y}\right\}_1^x + 1 + e^x = e^x + 2x - 1$. If we

evaluate the integral, $F(x) = \left\{-\dfrac{x^2}{y} + e^y\right\}_1^x = -x + e^x + x^2 - e$, in which case $F'(x) = -1 + e^x + 2x$.

4. With Leibnitz's rule, $F'(x) = \int_{x^2}^{x^3-1} dy + [x + (x^3 - 1)\ln(x^3 - 1)](3x^2) - [x + x^2\ln(x^2)](2x)$

$$= x^3 - 1 - x^2 + 3x^2[x + (x^3 - 1)\ln(x^3 - 1)] - 2x[x + x^2\ln(x^2)]$$

$$= 4x^3 - 3x^2 - 1 + 3x^2(x^3 - 1)\ln(x^3 - 1) - 2x^3\ln(x^2).$$

If we evaluate the integral,

$$F(x) = \left\{xy + \dfrac{y^2}{2}\ln y - \dfrac{y^2}{4}\right\}_{x^2}^{x^3-1} = x(x^3 - 1) + \dfrac{1}{2}(x^3 - 1)^2\ln(x^3 - 1) - \dfrac{1}{4}(x^3 - 1)^2 - x^3$$

$$- \dfrac{x^4}{2}\ln(x^2) + \dfrac{x^4}{4},$$

in which case $F'(x) = 4x^3 - 1 + 3x^2(x^3 - 1)\ln(x^3 - 1) + \dfrac{1}{2}(x^3 - 1)(3x^2) - \dfrac{1}{2}(x^3 - 1)(3x^2)$

$$-3x^2 - 2x^3\ln(x^2) - x^3 + x^3$$

$$= 4x^3 - 3x^2 - 1 + 3x^2(x^3 - 1)\ln(x^3 - 1) - 2x^3\ln(x^2).$$

6. $F(x) = \left\{\dfrac{y^4}{4}\ln y - \dfrac{y^4}{16} + x^3 e^y\right\}_x^{2x} = (4\ln 2)x^4 + \dfrac{15x^4}{4}\ln x - \dfrac{15x^4}{16} + x^3(e^{2x} - e^x)$, and therefore

$$F'(x) = (16\ln 2)x^3 + 15x^3\ln x + \dfrac{15x^3}{4} - \dfrac{15x^3}{4} + 3x^2(e^{2x} - e^x) + x^3(2e^{2x} - e^x)$$

$$= x^3(16\ln 2 + 15\ln x + 2e^{2x} - e^x) + 3x^2(e^{2x} - e^x).$$

8. With Leibnitz's rule,

$$F'(x) = \int_{\sin x}^{e^x} 0\, dy + \sqrt{1 + e^{3x}}(e^x) - \sqrt{1 + \sin^3 x}(\cos x) = e^x\sqrt{1 + e^{3x}} - \cos x\sqrt{1 + \sin^3 x}.$$

10. Since $\dfrac{dy}{dx} = \dfrac{1}{2}\int_0^x f(t)(e^{x-t} + e^{t-x})\, dt$, it follows that $\dfrac{d^2y}{dx^2} = \dfrac{1}{2}\int_0^x f(t)(e^{x-t} - e^{t-x})\, dt + \dfrac{1}{2}f(x)(2)(1)$,

and therefore

$$\dfrac{d^2y}{dx^2} - y = \dfrac{1}{2}\int_0^x f(t)(e^{x-t} - e^{t-x})\, dt + f(x) - \dfrac{1}{2}\int_0^x f(t)(e^{x-t} - e^{t-x})\, dt = f(x).$$

12. Differentiation of $\displaystyle\int_0^b \dfrac{1}{1 + ax}\, dx = \dfrac{1}{a}\ln(1 + ab)$ with respect to a gives

$$\int_0^b \dfrac{-x}{(1 + ax)^2}\, dx = -\dfrac{1}{a^2}\ln(1 + ab) + \dfrac{1}{a}\dfrac{b}{1 + ab} \implies \int_0^b \dfrac{x}{(1 + ax)^2}\, dx = \dfrac{1}{a^2}\ln(1 + ab) - \dfrac{b}{a(1 + ab)}.$$

14. If we write $\displaystyle\int_0^b \dfrac{1}{a^2 + x^2}\, dx = \dfrac{1}{a}\mathrm{Tan}^{-1}\left(\dfrac{b}{a}\right)$, then differentiation with respect to a gives

$$\int_0^b \dfrac{-2a}{(a^2 + x^2)^2}\, dx = -\dfrac{1}{a^2}\mathrm{Tan}^{-1}\left(\dfrac{b}{a}\right) + \dfrac{1}{a}\dfrac{1}{1 + b^2/a^2}\left(\dfrac{-b}{a^2}\right),$$

or,

$$\int_0^b \dfrac{1}{(a^2 + x^2)^2}\, dx = \dfrac{1}{2a^3}\mathrm{Tan}^{-1}\left(\dfrac{b}{a}\right) + \dfrac{b}{2a^2(a^2 + b^2)}.$$

Another differentiation gives

$$\int_0^b \frac{-4a}{(a^2+x^2)^3}\,dx = -\frac{3}{2a^4}\,\mathrm{Tan}^{-1}\!\left(\frac{b}{a}\right) + \frac{1}{2a^3}\frac{1}{1+b^2/a^2}\left(\frac{-b}{a^2}\right) - \frac{b(8a^3+4ab^2)}{4a^4(a^2+b^2)^2},$$

or,

$$\int_0^b \frac{1}{(a^2+x^2)^3}\,dx = -\frac{1}{4a}\left[-\frac{3}{2a^4}\,\mathrm{Tan}^{-1}\!\left(\frac{b}{a}\right) - \frac{b}{2a^3(a^2+b^2)} - \frac{b(2a^2+b^2)}{a^3(a^2+b^2)^2}\right]$$

$$= \frac{3}{8a^5}\,\mathrm{Tan}^{-1}\!\left(\frac{b}{a}\right) + \frac{b(3b^2+5a^2)}{8a^4(a^2+b^2)^2}.$$

Thus, $\displaystyle\int \frac{1}{(a^2+x^2)^3}\,dx = \frac{3}{8a^5}\,\mathrm{Tan}^{-1}\!\left(\frac{x}{a}\right) + \frac{x(3x^2+5a^2)}{8a^4(a^2+x^2)^2} + C.$

16. If we set $F(a) = \displaystyle\int_0^\pi \frac{\ln(1+a\cos x)}{\cos x}\,dx$, then $F'(a) = \displaystyle\int_0^\pi \frac{1}{1+a\cos x}\,dx.$

To evaluate this integral we let $t = \tan\dfrac{x}{2}$. Then $\cos x = \dfrac{1-t^2}{1+t^2}$, $dx = \dfrac{2}{1+t^2}\,dt$ (see Exercise 35 in Section 8.6), and

$$F'(a) = \int_0^\infty \frac{1}{1+a\left(\dfrac{1-t^2}{1+t^2}\right)}\frac{2}{1+t^2}\,dt = 2\int_0^\infty \frac{1}{1+t^2+a(1-t^2)}\,dt = 2\int_0^\infty \frac{1}{(a+1)+(1-a)t^2}\,dt.$$

We now set $t = \sqrt{(1+a)/(1-a)}\,\tan\theta$ and $dt = \sqrt{(1+a)/(1-a)}\,\sec^2\theta\,d\theta,$

$$F'(a) = 2\int_0^{\pi/2} \frac{1}{(a+1)+(a+1)\tan^2\theta}\sqrt{\frac{1+a}{1-a}}\,\sec^2\theta\,d\theta = 2\left\{\frac{\theta}{\sqrt{1-a^2}}\right\}_0^{\pi/2} = \frac{\pi}{\sqrt{1-a^2}}.$$

Hence, $F(a) = \pi\,\mathrm{Sin}^{-1}a + C$. Since $F(0)=0$, it follows that $0=C$, and $F(a)=\pi\,\mathrm{Sin}^{-1}a.$

18. We calculate: $\dfrac{\partial T}{\partial t} = e^{-(1-x)^2/(4t)}\left(\dfrac{x-1}{4t^{3/2}}\right) + e^{-(1+x)^2/(4t)}\left(\dfrac{-x-1}{4t^{3/2}}\right),$

$$\frac{\partial T}{\partial x} = e^{-(1-x)^2/(4t)}\left(\frac{-1}{2\sqrt{t}}\right) + e^{-(1+x)^2/(4t)}\left(\frac{1}{2\sqrt{t}}\right),$$

$$\frac{\partial^2 T}{\partial x^2} = e^{-(1-x)^2/(4t)}\left(\frac{x-1}{4t^{3/2}}\right) + e^{-(1+x)^2/(4t)}\left(\frac{-1-x}{4t^{3/2}}\right).$$

Thus, $\partial T/\partial t = \partial^2 T/\partial x^2.$

REVIEW EXERCISES

2. $\displaystyle\iiint_V xyz\,dV = \int_0^1 \int_0^3 \int_0^y xyz\,dz\,dy\,dx$

$$= \int_0^1 \int_0^3 \left\{\frac{xyz^2}{2}\right\}_0^y dy\,dx$$

$$= \frac{1}{2}\int_0^1 \int_0^3 xy^3\,dy\,dx$$

$$= \frac{1}{2}\int_0^1 \left\{\frac{xy^4}{4}\right\}_0^3 dx$$

$$= \frac{81}{8}\int_0^1 x\,dx = \frac{81}{8}\left\{\frac{x^2}{2}\right\}_0^1 = \frac{81}{16}$$

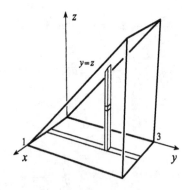

4. $\displaystyle\iiint_V (x^2 - y^3)\, dV = \int_0^1 \int_0^1 \int_0^{xy} (x^2 - y^3)\, dz\, dy\, dx = \int_0^1 \int_0^1 xy(x^2 - y^3)\, dy\, dx$

$$= \int_0^1 \left\{ \frac{x^3 y^2}{2} - \frac{xy^5}{5} \right\}_0^1 dx$$

$$= \frac{1}{10} \int_0^1 (5x^3 - 2x)\, dx$$

$$= \frac{1}{10} \left\{ \frac{5x^4}{4} - x^2 \right\}_0^1 = \frac{1}{40}$$

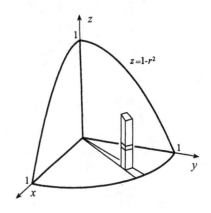

6. $\displaystyle\iint_R y\, dA = \int_0^3 \int_{(x-1)^2}^{x+1} y\, dy\, dx$

$$= \int_0^3 \left\{ \frac{y^2}{2} \right\}_{(x-1)^2}^{x+1} dx$$

$$= \frac{1}{2} \int_0^3 [(x+1)^2 - (x-1)^4]\, dx$$

$$= \frac{1}{2} \left\{ \frac{1}{3}(x+1)^3 - \frac{1}{5}(x-1)^5 \right\}_0^3 = \frac{36}{5}$$

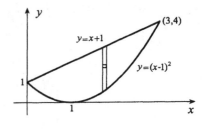

8. Since R is symmetric about the x-axis and $x^2 y$ is an odd function of y, $\displaystyle\iint_R x^2 y\, dA = 0$.

10. $\displaystyle\iint_R (xy - x^2 y^2)\, dA = 2 \int_0^1 \int_{2x^2}^{4-2x^2} -x^2 y^2\, dy\, dx = -2 \int_0^1 \left\{ \frac{x^2 y^3}{3} \right\}_{2x^2}^{4-2x^2} dx$

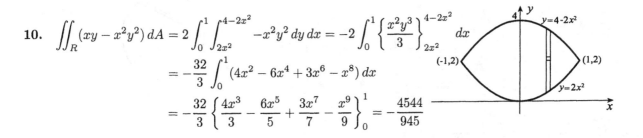

$$= -\frac{32}{3} \int_0^1 (4x^2 - 6x^4 + 3x^6 - x^8)\, dx$$

$$= -\frac{32}{3} \left\{ \frac{4x^3}{3} - \frac{6x^5}{5} + \frac{3x^7}{7} - \frac{x^9}{9} \right\}_0^1 = -\frac{4544}{945}$$

12. Integrals of x and y yield zero. We quadruple the integral of z over the first octant volume.

$$\iiint_V (x + y + z)\, dV = \iiint_V z\, dV$$

$$= 4 \int_0^{\pi/2} \int_0^1 \int_0^{1-r^2} z\, r\, dz\, dr\, d\theta$$

$$= 4 \int_0^{\pi/2} \int_0^1 \left\{ \frac{rz^2}{2} \right\}_0^{1-r^2} dr\, d\theta$$

$$= 2 \int_0^{\pi/2} \int_0^1 r(1 - r^2)^2\, dr\, d\theta$$

$$= 2 \int_0^{\pi/2} \left\{ -\frac{1}{6}(1 - r^2)^3 \right\}_0^1 d\theta$$

$$= \frac{1}{3} \{\theta\}_0^{\pi/2} = \frac{\pi}{6}$$

14. We multiply the first octant volume by 8.

$$\iiint_V dV = 8 \int_0^{\pi/2} \int_0^2 \int_0^r r\, dz\, dr\, d\theta$$

$$= 8 \int_0^{\pi/2} \int_0^2 r^2\, dr\, d\theta$$

$$= 8 \int_0^{\pi/2} \left\{ \frac{r^3}{3} \right\}_0^2 d\theta$$

$$= \frac{64}{3} \left\{ \theta \right\}_0^{\pi/2} = \frac{32\pi}{3}$$

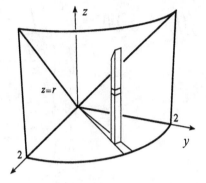

16. We quadruple the integral over the volume in the first octant.

$$\iiint_V (x^2 + y^2 + z^2)\, dV = 4 \int_0^{\pi/2} \int_0^{\pi/2} \int_0^1 \mathcal{R}^2\, \mathcal{R}^2 \sin\phi\, d\mathcal{R}\, d\phi\, d\theta$$

$$= 4 \int_0^{\pi/2} \int_0^{\pi/2} \left\{ \frac{\mathcal{R}^5}{5} \right\}_0^1 \sin\phi\, d\phi\, d\theta$$

$$= \frac{4}{5} \int_0^{\pi/2} \left\{ -\cos\phi \right\}_0^{\pi/2} d\theta$$

$$= \frac{4}{5} \left\{ \theta \right\}_0^{\pi/2} = \frac{2\pi}{5}$$

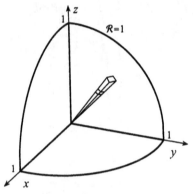

18. $\displaystyle \iint_R (x^2 + y^2)\, dA = 2 \int_0^{\pi/2} \int_0^{2\cos\theta} r^2\, r\, dr\, d\theta = 2 \int_0^{\pi/2} \left\{ \frac{r^4}{4} \right\}_0^{2\cos\theta} d\theta$

$$= 8 \int_0^{\pi/2} \cos^4\theta\, d\theta$$

$$= 2 \int_0^{\pi/2} \left(1 + 2\cos 2\theta + \frac{1 + \cos 4\theta}{2} \right) d\theta$$

$$= 2 \left\{ \frac{3\theta}{2} + \sin 2\theta + \frac{1}{8} \sin 4\theta \right\}_0^{\pi/2} = \frac{3\pi}{2}$$

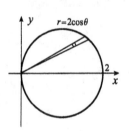

20. $\displaystyle \iint_R \frac{1}{x^2 + y^2}\, dA = \int_1^2 \int_0^x \frac{1}{x^2 + y^2}\, dy\, dx$

If we set $y = x\tan\theta$, then $dy = x\sec^2\theta\, d\theta$, and

$$\iint_R \frac{1}{x^2 + y^2}\, dA = \int_1^2 \int_0^{\pi/4} \frac{1}{x^2 \sec^2\theta} x\sec^2\theta\, d\theta\, dx$$

$$= \int_1^2 \left\{ \frac{\theta}{x} \right\}_0^{\pi/4} dx = \frac{\pi}{4} \int_1^2 \frac{1}{x}\, dx = \frac{\pi}{4} \left\{ \ln|x| \right\}_1^2 = \frac{\pi}{4} \ln 2.$$

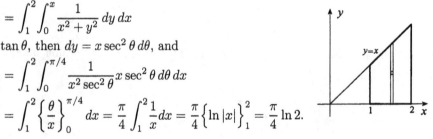

22. See answer in text.

24. $\displaystyle A = 4 \int_0^2 \int_0^{x\sqrt{4-x^2}} dy\, dx$

$$= 4 \int_0^2 x\sqrt{4 - x^2}\, dx$$

$$= 4 \left\{ -\frac{1}{3}(4 - x^2)^{3/2} \right\}_0^2 = \frac{32}{3}$$

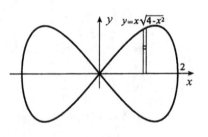

26. We quadruple the volume in the first octant.

$$V = 4 \int_0^1 \int_{1+x^2}^2 \int_0^{2-y} dz\, dy\, dx$$

$$= 4 \int_0^1 \int_{1+x^2}^2 (2-y)\, dy\, dx$$

$$= 4 \int_0^1 \left\{ -\frac{1}{2}(2-y)^2 \right\}_{1+x^2}^2 dx$$

$$= 2 \int_0^1 (1 - 2x^2 + x^4)\, dx$$

$$= 2 \left\{ x - \frac{2x^3}{3} + \frac{x^5}{5} \right\}_0^1 = \frac{16}{15}$$

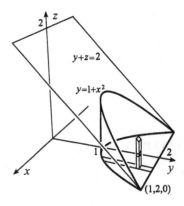

28. We quadruple the area in the first octant.

$$A = 4 \iint_{S_{xy}} \sqrt{1 + \left(\frac{\partial z}{\partial x}\right)^2 + \left(\frac{\partial z}{\partial y}\right)^2}\, dA$$

$$= 4 \iint_{S_{xy}} \sqrt{1 + (-8x)^2 + (-8y)^2}\, dA$$

$$= 4 \int_0^{\pi/2} \int_0^{1/2} \sqrt{1 + 64r^2}\, r\, dr\, d\theta$$

$$= 4 \int_0^{\pi/2} \left\{ \frac{1}{192}(1 + 64r^2)^{3/2} \right\}_0^{1/2} d\theta$$

$$= \frac{17\sqrt{17} - 1}{48} \{\theta\}_0^{\pi/2} = \frac{(17\sqrt{17} - 1)\pi}{96}$$

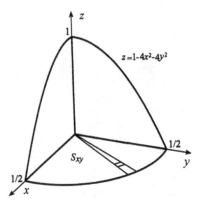

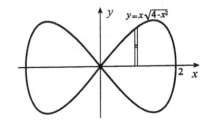

30. $$V_x = 2 \int_0^2 \int_0^{x\sqrt{4-x^2}} 2\pi y\, dy\, dx = 2\pi \int_0^2 \left\{ y^2 \right\}_0^{x\sqrt{4-x^2}} dx$$

$$= 2\pi \int_0^2 x^2(4 - x^2)\, dx = 2\pi \left\{ \frac{4x^3}{3} - \frac{x^5}{5} \right\}_0^2 = \frac{128\pi}{15}$$

$$V_y = 2 \int_0^2 \int_0^{x\sqrt{4-x^2}} 2\pi x\, dy\, dx = 4\pi \int_0^2 x^2 \sqrt{4 - x^2}\, dx$$

If we set $x = 2\sin\theta$ and $dx = 2\cos\theta\, d\theta$, then

$$V_y = 4\pi \int_0^{\pi/2} 4\sin^2\theta\, 2\cos\theta\, 2\cos\theta\, d\theta = 64\pi \int_0^{\pi/2} \left(\frac{\sin 2\theta}{2}\right)^2 d\theta$$

$$= 16\pi \int_0^{\pi/2} \left(\frac{1 - \cos 4\theta}{2}\right) d\theta = 8\pi \left\{ \theta - \frac{1}{4}\sin 4\theta \right\}_0^{\pi/2} = 4\pi^2$$

32. $$I = \int_0^1 \int_0^{(1-x^3)^{1/3}} x^2\, dy\, dx$$

$$= \int_0^1 x^2(1 - x^3)^{1/3}\, dx$$

$$= \left\{ -\frac{1}{4}(1 - x^3)^{4/3} \right\}_0^1 = \frac{1}{4}$$

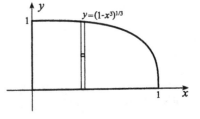

34. $\displaystyle \bar{f} = \frac{1}{\pi (2)^2} \int_{-\pi}^{\pi} \int_0^2 r^2\, r\, dr\, d\theta$

$\displaystyle = \frac{1}{4\pi} \int_{-\pi}^{\pi} \left\{ \frac{r^4}{4} \right\}_0^2 d\theta$

$\displaystyle = \frac{1}{\pi} \left\{ \theta \right\}_{-\pi}^{\pi} = 2$

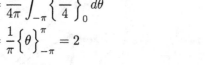

36. We quadruple the moment of inertia of the first octant volume.

$\displaystyle I = 4 \int_0^{\pi/2} \int_0^1 \int_0^{\sqrt{1+z^2}} r^2\, \rho\, r\, dr\, dz\, d\theta$

$\displaystyle = 4\rho \int_0^{\pi/2} \int_0^1 \left\{ \frac{r^4}{4} \right\}_0^{\sqrt{1+z^2}} dz\, d\theta$

$\displaystyle = \rho \int_0^{\pi/2} \int_0^1 (1 + 2z^2 + z^4)\, dz\, d\theta$

$\displaystyle = \rho \int_0^{\pi/2} \left\{ z + \frac{2z^3}{3} + \frac{z^5}{5} \right\}_0^1 d\theta$

$\displaystyle = \frac{28\rho}{15} \left\{ \theta \right\}_0^{\pi/2} = \frac{14\pi\rho}{15}$

38. $\displaystyle A = 2 \int_0^{\pi/2} \int_0^{\sin\theta \cos^2\theta} r\, dr\, d\theta$

$\displaystyle = 2 \int_0^{\pi/2} \left\{ \frac{r^2}{2} \right\}_0^{\sin\theta \cos^2\theta} d\theta$

$\displaystyle = \int_0^{\pi/2} \sin^2\theta \cos^4\theta\, d\theta$

$\displaystyle = \int_0^{\pi/2} \left(\frac{\sin 2\theta}{2} \right)^2 \left(\frac{1 + \cos 2\theta}{2} \right) d\theta$

$\displaystyle = \frac{1}{8} \int_0^{\pi/2} \left(\frac{1 - \cos 4\theta}{2} + \sin^2 2\theta \cos 2\theta \right) d\theta = \frac{1}{8} \left\{ \frac{\theta}{2} - \frac{1}{8}\sin 4\theta + \frac{1}{6}\sin^3 2\theta \right\}_0^{\pi/2} = \frac{\pi}{32}$

By symmetry, $\bar{x} = 0$. Since $\displaystyle A\bar{y} = 2 \int_0^{\pi/2} \int_0^{\sin\theta \cos^2\theta} r\sin\theta\, r\, dr\, d\theta = 2 \int_0^{\pi/2} \left\{ \frac{r^3}{3} \sin\theta \right\}_0^{\sin\theta \cos^2\theta} d\theta$

$\displaystyle = \frac{2}{3} \int_0^{\pi/2} \sin^4\theta \cos^6\theta\, d\theta = \frac{2}{3} \int_0^{\pi/2} \left(\frac{\sin 2\theta}{2} \right)^4 \left(\frac{1 + \cos 2\theta}{2} \right) d\theta$

$\displaystyle = \frac{1}{48} \int_0^{\pi/2} \left[\left(\frac{1 - \cos 2\theta}{2} \right)^2 + \sin^4 2\theta \cos 2\theta \right] d\theta$

$\displaystyle = \frac{1}{48} \int_0^{\pi/2} \left[\frac{1}{4}\left(1 - 2\cos 2\theta + \frac{1 + \cos 4\theta}{2} \right) + \sin^4 2\theta \cos 2\theta \right] d\theta$

$\displaystyle = \frac{1}{48} \left\{ \frac{3\theta}{8} - \frac{1}{4}\sin 2\theta + \frac{1}{32}\sin 4\theta + \frac{1}{10}\sin^5 2\theta \right\}_0^{\pi/2} = \frac{\pi}{256},$

it follows that $\displaystyle \bar{y} = \frac{\pi}{256} \frac{32}{\pi} = \frac{1}{8}.$

40. $M = 2 \displaystyle\int_0^1 \int_0^x \int_0^{\sqrt{1-x^2}} \rho \, dz \, dy \, dx$

$= 2\rho \displaystyle\int_0^1 \int_0^x \sqrt{1-x^2} \, dy \, dx$

$= 2\rho \displaystyle\int_0^1 x\sqrt{1-x^2} \, dx$

$= 2\rho \left\{ -\dfrac{1}{3}(1-x^2)^{3/2} \right\}_0^1 = \dfrac{2\rho}{3}$

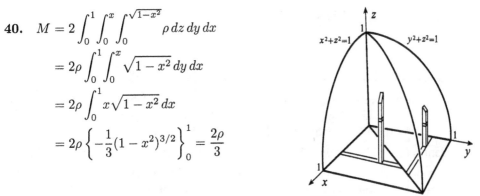

$M\overline{x} = \displaystyle\int_0^1 \int_0^x \int_0^{\sqrt{1-x^2}} x\rho \, dz \, dy \, dx + \int_0^1 \int_0^y \int_0^{\sqrt{1-y^2}} x\rho \, dz \, dx \, dy$

$= \rho \displaystyle\int_0^1 \int_0^x x\sqrt{1-x^2} \, dy \, dx + \rho \int_0^1 \int_0^y x\sqrt{1-y^2} \, dx \, dy$

$= \rho \displaystyle\int_0^1 x^2 \sqrt{1-x^2} \, dx + \dfrac{\rho}{2} \int_0^1 y^2 \sqrt{1-y^2} \, dy = \dfrac{3\rho}{2} \int_0^1 x^2 \sqrt{1-x^2} \, dx$

If we now set $x = \sin\theta$ and $dx = \cos\theta \, d\theta$, then

$$M\overline{x} = \dfrac{3\rho}{2} \int_0^{\pi/2} \sin^2\theta \cos\theta \cos\theta \, d\theta = \dfrac{3\rho}{2} \int_0^{\pi/2} \left(\dfrac{\sin 2\theta}{2} \right)^2 d\theta$$

$$= \dfrac{3\rho}{8} \int_0^{\pi/2} \left(\dfrac{1-\cos 4\theta}{2} \right) d\theta = \dfrac{3\rho}{16} \left\{ \theta - \dfrac{1}{4}\sin 4\theta \right\}_0^{\pi/2} = \dfrac{3\pi\rho}{32},$$

we obtain $\overline{x} = \dfrac{3\pi\rho}{32} \dfrac{3}{2\rho} = \dfrac{9\pi}{64}$. By symmetry, $\overline{y} = \overline{x} = 9\pi/64$. Since

$$M\overline{z} = 2 \int_0^1 \int_0^x \int_0^{\sqrt{1-x^2}} z\rho \, dz \, dy \, dx = 2\rho \int_0^1 \int_0^x \left\{ \dfrac{z^2}{2} \right\}_0^{\sqrt{1-x^2}} dy \, dx$$

$$= \rho \int_0^1 \int_0^x (1-x^2) \, dy \, dx = \rho \int_0^1 (x-x^3) \, dx = \rho \left\{ \dfrac{x^2}{2} - \dfrac{x^4}{4} \right\}_0^1 = \dfrac{\rho}{4},$$

we find $\overline{z} = \dfrac{\rho}{4} \dfrac{3}{2\rho} = \dfrac{3}{8}$.

42. We quadruple the first octant volume.

$V = 4 \displaystyle\int_0^2 \int_0^{y(2-y)} \int_0^{\sqrt{y^2(2-y)^2-x^2}} dz \, dx \, dy$

$= 4 \displaystyle\int_0^2 \int_0^{y(2-y)} \sqrt{y^2(2-y)^2 - x^2} \, dx \, dy$

If we set $x = y(2-y)\sin\theta$ and $dx = y(2-y)\cos\theta \, d\theta$,

$V = 4 \displaystyle\int_0^2 \int_0^{\pi/2} y(2-y)\cos\theta \, y(2-y)\cos\theta \, d\theta \, dy$

$= 4 \displaystyle\int_0^2 \int_0^{\pi/2} y^2(2-y)^2 \left(\dfrac{1+\cos 2\theta}{2} \right) d\theta \, dy = 2 \int_0^2 \left\{ y^2(2-y)^2 \left(\theta + \dfrac{\sin 2\theta}{2} \right) \right\}_0^{\pi/2} dy$

$= \pi \displaystyle\int_0^2 (4y^2 - 4y^3 + y^4) \, dy = \pi \left\{ \dfrac{4y^3}{3} - y^4 + \dfrac{y^5}{5} \right\}_0^2 = \dfrac{16\pi}{15}.$

CHAPTER 14

EXERCISES 14.1

2. Closed, connected   **4.** Connected   **6.** Closed, connected   **8.** Open

10. For interior, exterior, and boundary points replace circle with sphere in planar definitions. Open, closed, connected, and domain definitions are identical. A domain is simply-connected if every closed curve in the domain is the boundary of a surface that contains only points of the domain.

12. Closed, connected   **14.** Connected   **16.** Open   **18.** Open, connected, simply-connected domain

20. To be open a set must not contain any of its boundary points. To be closed it must contain all of its boundary points. The only way to satisfy both conditions is for the set to have no boundary points. The whole plane is the only nonempty set that has no boundary points.

22. $\nabla f = -\dfrac{1}{2}(x^2 + y^2 + z^2)^{-3/2}(2x\hat{\mathbf{i}} + 2y\hat{\mathbf{j}} + 2z\hat{\mathbf{k}}) = -(x^2 + y^2 + z^2)^{-3/2}(x\hat{\mathbf{i}} + y\hat{\mathbf{j}} + z\hat{\mathbf{k}})$

24. $\nabla f_{|(1,2)} = (3x^2 y - 2\cos y)\hat{\mathbf{i}} + (x^3 + 2x\sin y)\hat{\mathbf{j}}_{|(1,2)} = (6 - 2\cos 2)\hat{\mathbf{i}} + (1 + 2\sin 2)\hat{\mathbf{j}}$

26. $\nabla \cdot \mathbf{F} = 2e^y + 0 - 2x^2 y = 2(e^y - x^2 y)$

28. $\nabla \cdot \mathbf{F} = 2x\cos(x^2 + y^2 + z^2) - \sin(y + z)$

30. $\nabla \cdot \mathbf{F}_{(1,1,1)} = (2xy^3 - 3x + 2z)_{(1,1,1)} = 2 - 3 + 2 = 1.$

32. $\nabla \cdot \mathbf{F} = \dfrac{\partial}{\partial x}\left(\dfrac{x}{\sqrt{x^2 + y^2 + z^2}}\right) + \dfrac{\partial}{\partial y}\left(\dfrac{y}{\sqrt{x^2 + y^2 + z^2}}\right) + \dfrac{\partial}{\partial z}\left(\dfrac{z}{\sqrt{x^2 + y^2 + z^2}}\right)$

$$= \left[\dfrac{1}{\sqrt{x^2 + y^2 + z^2}} - \dfrac{x^2}{(x^2 + y^2 + z^2)^{3/2}}\right] + \left[\dfrac{1}{\sqrt{x^2 + y^2 + z^2}} - \dfrac{y^2}{(x^2 + y^2 + z^2)^{3/2}}\right]$$

$$+ \left[\dfrac{1}{\sqrt{x^2 + y^2 + z^2}} - \dfrac{z^2}{(x^2 + y^2 + z^2)^{3/2}}\right]$$

$$= \dfrac{3(x^2 + y^2 + z^2) - (x^2 + y^2 + z^2)}{(x^2 + y^2 + z^2)^{3/2}} = \dfrac{2}{\sqrt{x^2 + y^2 + z^2}}$$

34. $\nabla \times \mathbf{F} = \begin{vmatrix} \hat{\mathbf{i}} & \hat{\mathbf{j}} & \hat{\mathbf{k}} \\ \partial/\partial x & \partial/\partial y & \partial/\partial z \\ x^2 z & 12xyz & 32y^2 z^4 \end{vmatrix} = (64yz^4 - 12xy)\hat{\mathbf{i}} + x^2\hat{\mathbf{j}} + 12yz\hat{\mathbf{k}}$

36. $\nabla \times \mathbf{F} = \begin{vmatrix} \hat{\mathbf{i}} & \hat{\mathbf{j}} & \hat{\mathbf{k}} \\ \partial/\partial x & \partial/\partial y & \partial/\partial z \\ x^2 & y^2 & z^2 \end{vmatrix} = \mathbf{0}$

38. $\nabla \times \mathbf{F}_{|(2,0)} = \begin{vmatrix} \hat{\mathbf{i}} & \hat{\mathbf{j}} & \hat{\mathbf{k}} \\ \partial/\partial x & \partial/\partial y & \partial/\partial z \\ y & -x & 0 \end{vmatrix}_{|(2,0)} = (-2)\hat{\mathbf{k}}$

40. $\nabla \times \mathbf{F} = \begin{vmatrix} \hat{\mathbf{i}} & \hat{\mathbf{j}} & \hat{\mathbf{k}} \\ \partial/\partial x & \partial/\partial y & \partial/\partial z \\ \mathrm{Sec}^{-1}(x+y) & \mathrm{Csc}^{-1}(y+x) & 0 \end{vmatrix}$

$$= \left(\dfrac{-1}{(y+x)\sqrt{(y+x)^2 - 1}} - \dfrac{1}{(x+y)\sqrt{(x+y)^2 - 1}}\right)\hat{\mathbf{k}} = \dfrac{-2}{(x+y)\sqrt{(x+y)^2 - 1}}\hat{\mathbf{k}}$$

42. (a) For $f(x, y, z)$, the equation $\nabla \cdot \nabla f = 0$ can be written in the form

$$0 = \nabla \cdot \left(\dfrac{\partial f}{\partial x}\hat{\mathbf{i}} + \dfrac{\partial f}{\partial y}\hat{\mathbf{j}} + \dfrac{\partial f}{\partial z}\hat{\mathbf{k}}\right) = \dfrac{\partial^2 f}{\partial x^2} + \dfrac{\partial^2 f}{\partial y^2} + \dfrac{\partial^2 f}{\partial z^2},$$

which is equation 12.12.

(b) Since $\dfrac{\partial f}{\partial x} = \dfrac{-x}{(x^2 + y^2 + z^2)^{3/2}}$, it follows that

$$\frac{\partial^2 f}{\partial x^2} = \frac{-1}{(x^2 + y^2 + z^2)^{3/2}} + \frac{3x^2}{(x^2 + y^2 + z^2)^{5/2}} = \frac{2x^2 - y^2 - z^2}{(x^2 + y^2 + z^2)^{5/2}}.$$

Similarly, $\dfrac{\partial^2 f}{\partial y^2} = \dfrac{2y^2 - x^2 - z^2}{(x^2 + y^2 + z^2)^{5/2}}$ and $\dfrac{\partial^2 f}{\partial z^2} = \dfrac{2z^2 - y^2 - x^2}{(x^2 + y^2 + z^2)^{5/2}}$. When these second partial derivatives are added together, the result is zero.

44. From the schematic,

$$\frac{d\rho}{dt} = \frac{\partial \rho}{\partial t} + \frac{\partial \rho}{\partial x}\frac{dx}{dt} + \frac{\partial \rho}{\partial y}\frac{dy}{dt} + \frac{\partial \rho}{\partial z}\frac{dz}{dt}$$

$$= \frac{\partial \rho}{\partial t} + \left(\frac{\partial \rho}{\partial x}\hat{\mathbf{i}} + \frac{\partial \rho}{\partial y}\hat{\mathbf{j}} + \frac{\partial \rho}{\partial z}\hat{\mathbf{k}} \right) \cdot \left(\frac{dx}{dt}\hat{\mathbf{i}} + \frac{dy}{dt}\hat{\mathbf{j}} + \frac{dz}{dt}\hat{\mathbf{k}} \right)$$

$$= \frac{\partial \rho}{\partial t} + \nabla \rho \cdot \frac{d\mathbf{r}}{dt}.$$

46. If $\nabla f = \mathbf{F}$, then $\dfrac{\partial f}{\partial x} = 3x^2y^2 + 3$, $\dfrac{\partial f}{\partial y} = 2x^3y + 2$. From the first of these, $f(x, y) = x^3y^2 + 3x + v(y)$, which substituted into the second requires $2x^3y + v'(y) = 2x^3y + 2$. Thus, $v(y) = 2y + C$, where C is a constant, and $f(x, y) = x^3y^2 + 3x + 2y + C$.

48. If $\nabla f = \mathbf{F}$, then $\dfrac{\partial f}{\partial x} = \dfrac{1}{x + y}$, $\dfrac{\partial f}{\partial y} = \dfrac{1}{x + y}$. From the first, $f(x, y) = \ln|x + y| + v(y)$, which substituted into the second requires $1/(x + y) + v'(y) = 1/(x + y)$. Thus, $v(y) = C$, where C is a constant, and $f(x, y) = \ln|x + y| + C$.

50. If $\nabla f = x\hat{\mathbf{i}} + y\hat{\mathbf{j}} + z\hat{\mathbf{k}}$, then $\dfrac{\partial f}{\partial x} = x$, $\dfrac{\partial f}{\partial y} = y$, $\dfrac{\partial f}{\partial z} = z$. From the first, $f(x, y, z) = x^2/2 + v(y, z)$, which substituted into the second requires $\partial v/\partial y = y$. Thus, $v(y, z) = y^2/2 + w(z)$, and $f(x, y, z) = x^2/2 + y^2/2 + w(z)$. Substitution into the third equation gives $w'(z) = z$. Thus, $w(z) = z^2/2 + C$, where C is a constant, and $f(x, y, z) = x^2/2 + y^2/2 + z^2/2 + C$.

52. If $\nabla f = (1 + x + y + z)^{-1}(\hat{\mathbf{i}} + \hat{\mathbf{j}} + \hat{\mathbf{k}})$, then

$$\frac{\partial f}{\partial x} = \frac{1}{1 + x + y + z}, \quad \frac{\partial f}{\partial y} = \frac{1}{1 + x + y + z}, \quad \frac{\partial f}{\partial z} = \frac{1}{1 + x + y + z}.$$

From the first, $f(x, y, z) = \ln|1 + x + y + z| + v(y, z)$, which substituted into the second requires

$$\frac{1}{1 + x + y + z} + \frac{\partial v}{\partial y} = \frac{1}{1 + x + y + z}.$$

Thus, $v(y, z) = w(z)$, and $f(x, y, z) = \ln|1 + x + y + z| + w(z)$. Substitution into the third equation gives $\dfrac{1}{1 + x + y + z} + w'(z) = \dfrac{1}{1 + x + y + z}$. Thus, $w(z) = C$, and $f(x, y, z) = \ln|1 + x + y + z| + C$.

54. If $\nabla f = (1 + x^2y^2)^{-1}(y\hat{\mathbf{i}} + x\hat{\mathbf{j}}) + z\hat{\mathbf{k}}$, then $\dfrac{\partial f}{\partial x} = \dfrac{y}{1 + x^2y^2}$, $\dfrac{\partial f}{\partial y} = \dfrac{x}{1 + x^2y^2}$, $\dfrac{\partial f}{\partial z} = z$. From the first, $f(x, y, z) = \text{Tan}^{-1}(xy) + v(y, z)$, which substituted into the second requires

$$\frac{x}{1 + x^2y^2} + \frac{\partial v}{\partial y} = \frac{x}{1 + x^2y^2}.$$

Thus, $v(y, z) = w(z)$, and $f(x, y, z) = \text{Tan}^{-1}(xy) + w(z)$. Substitution into the third equation gives $w'(z) = z$. Thus, $w(z) = z^2/2 + C$, where C is a constant, and $f(x, y, z) = \text{Tan}^{-1}(xy) + z^2/2 + C$.

56. (a) $\mathbf{F}$ is irrotational if

$$\mathbf{0} = \nabla \times \mathbf{F} = \begin{vmatrix} \hat{\mathbf{i}} & \hat{\mathbf{j}} & \hat{\mathbf{k}} \\ \partial/\partial x & \partial/\partial y & \partial/\partial z \\ x^2 + 2y + az & bx - 3y - z & 4x + cy + 2z \end{vmatrix} = (c+1)\hat{\mathbf{i}} + (a-4)\hat{\mathbf{j}} + (b-2)\hat{\mathbf{k}}.$$

Thus, $a = 4$, $b = 2$, $c = -1$.

(b) If $\nabla f = \mathbf{F}$, then $\dfrac{\partial f}{\partial x} = x^2 + 2y + 4z$, $\dfrac{\partial f}{\partial y} = 2x - 3y - z$, $\dfrac{\partial f}{\partial z} = 4x - y + 2z$. From the first,

$f(x, y, z) = x^3/3 + 2xy + 4xz + v(y, z)$, which substituted into the second requires $2x + \dfrac{\partial v}{\partial y} = 2x - 3y - z$.

Thus, $v(y, z) = -3y^2/2 - yz + w(z)$, and $f(x, y, z) = x^3/3 + 2xy + 4xz - 3y^2/2 - yz + w(z)$. Substitution into the third equation gives $4x - y + w'(z) = 4x - y + 2z$. Thus, $w(z) = z^2 + C$, where C is a constant, and $f(x, y, z) = x^3/3 + 2xy + 4xz - 3y^2/2 - yz + z^2 + C$.

58. (a) $\nabla V = -\mathbf{E} = -\dfrac{\mathbf{F}}{Q} = -\dfrac{q}{4\pi\epsilon_0 |\mathbf{r}|^3}\mathbf{r} = \dfrac{-q}{4\pi\epsilon_0} \dfrac{x\hat{\mathbf{i}} + y\hat{\mathbf{j}} + z\hat{\mathbf{k}}}{(x^2 + y^2 + z^2)^{3/2}}$. This implies that

$$\frac{\partial V}{\partial x} = \frac{-q}{4\pi\epsilon_0}\frac{x}{(x^2 + y^2 + z^2)^{3/2}}, \quad \frac{\partial V}{\partial y} = \frac{-q}{4\pi\epsilon_0}\frac{y}{(x^2 + y^2 + z^2)^{3/2}}, \quad \frac{\partial V}{\partial z} = \frac{-q}{4\pi\epsilon_0}\frac{z}{(x^2 + y^2 + z^2)^{3/2}}.$$

From the first, $V(x, y, z) = \dfrac{q}{4\pi\epsilon_0\sqrt{x^2 + y^2 + z^2}} + w(y, z)$, which substituted into the second requires

$$\frac{-q}{4\pi\epsilon_0}\frac{y}{(x^2 + y^2 + z^2)^{3/2}} + \frac{\partial w}{\partial y} = \frac{-q}{4\pi\epsilon_0}\frac{y}{(x^2 + y^2 + z^2)^{3/2}}.$$

Thus, $w(y, z) = k(z)$, and $V(x, y, z) = \dfrac{q}{4\pi\epsilon_0\sqrt{x^2 + y^2 + z^2}} + k(z)$. Substitution into the third equation

gives $\dfrac{-q}{4\pi\epsilon_0}\dfrac{z}{(x^2 + y^2 + z^2)^{3/2}} + k'(z) = \dfrac{-q}{4\pi\epsilon_0}\dfrac{z}{(x^2 + y^2 + z^2)^{3/2}}$. Hence, $k(z) = C$, where C is a constant,

and $V(x, y, z) = \dfrac{q}{4\pi\epsilon_0\sqrt{x^2 + y^2 + z^2}} + C$.

(b) $\nabla V = -\mathbf{E} = -\dfrac{\mathbf{F}}{Q} = -\dfrac{\sigma}{2\epsilon_0}\hat{\mathbf{k}}$. This implies that $\dfrac{\partial V}{\partial x} = 0$, $\dfrac{\partial V}{\partial y} = 0$, $\dfrac{\partial V}{\partial z} = -\dfrac{\sigma}{2\epsilon_0}$.

These require $V(x, y, z) = -\dfrac{\sigma z}{2\epsilon_0} + C$.

60. Exercise 42(a) indicates that if $f(x, y, z)$ satisfies Laplace's equation, then $\nabla \cdot (\nabla f) = 0$. In other words, ∇f is solenoidal. On the other hand, for any function whatsoever, equation 14.14 indicates that $\nabla \times (\nabla f) = \mathbf{0}$; i.e., ∇f is irrotational.

62. Since $\nabla \cdot \mathbf{F} = 1 + 1 - 2 = 0$, $\mathbf{F}$ is solenoidal. Because $\mathbf{F}(tx, ty, tz) = t\mathbf{F}(x, y, z)$, we use the formula in Exercise 61(c) to obtain

$$\mathbf{v} = \frac{1}{3}\mathbf{F} \times \mathbf{r} = \frac{1}{3}\begin{vmatrix} \hat{\mathbf{i}} & \hat{\mathbf{j}} & \hat{\mathbf{k}} \\ x & y & -2z \\ x & y & z \end{vmatrix} = \frac{1}{3}(3yz\hat{\mathbf{i}} - 3xz\hat{\mathbf{j}}) = yz\hat{\mathbf{i}} - xz\hat{\mathbf{j}}.$$

64. Since $\nabla \cdot \mathbf{F} = 4x - 2y + 2y - 4x = 0$, $\mathbf{F}$ is solenoidal. Because $\mathbf{F}(tx, ty, tz) = t^2\mathbf{F}(x, y, z)$, we use the formula in Exercise 61(c) to obtain

$$\mathbf{v} = \frac{1}{4}\mathbf{F} \times \mathbf{r} = \frac{1}{4}\begin{vmatrix} \hat{\mathbf{i}} & \hat{\mathbf{j}} & \hat{\mathbf{k}} \\ 2x^2 & -y^2 & 2yz - 4xz \\ x & y & z \end{vmatrix}$$

$$= \frac{1}{4}\left[(-y^2z - 2y^2z + 4xyz)\hat{\mathbf{i}} + (2xyz - 4x^2z - 2x^2z)\hat{\mathbf{j}} + (2x^2y + xy^2)\hat{\mathbf{k}}\right]$$

$$= \frac{1}{4}\left[(4xyz - 3y^2z)\hat{\mathbf{i}} + (2xyz - 6x^2z)\hat{\mathbf{j}} + (2x^2y + xy^2)\hat{\mathbf{k}}\right].$$

EXERCISES 14.2

2. $\displaystyle\oint_C (x^2 + y^2)\, ds = \int_{C_1} (x^2 + y^2)\, ds + \int_{C_2} (x^2 + y^2)\, ds$

$$+ \int_{C_3} (x^2 + y^2)\, ds + \int_{C_4} (x^2 + y^2)\, ds$$

$$= \int_{-1}^{1} (1 + t^2)\, dt + \int_{-1}^{1} (t^2 + 1)\, dt$$

$$+ \int_{-1}^{1} (1 + t^2)\, dt + \int_{-1}^{1} (t^2 + 1)\, dt$$

$$= 4 \int_{-1}^{1} (1 + t^2)\, dt = 4 \left\{ t + \frac{t^3}{3} \right\}_{-1}^{1} = \frac{32}{3}$$

$C_2: x = -t, y = 1, -1 \leq t \leq 1$

$C_3: x = -1,\ y = -t,\ -1 \leq t \leq 1$

$C_1: x = 1,\ y = t,\ -1 \leq t \leq 1$

$C_4: x = t, y = -1, -1 \leq t \leq 1$

4. With parametric equations $C:\ x = 1 + 2t,\ y = 2,\ z = -1 + 6t,\ 0 \leq t \leq 1$,

$$\int_C (x^2 + yz)\, ds = \int_0^1 [(1 + 2t)^2 + 2(-1 + 6t)] \sqrt{2^2 + 0 + 6^2}\, dt$$

$$= 2\sqrt{10} \left\{ \frac{1}{6}(1 + 2t)^3 + \frac{1}{6}(-1 + 6t)^2 \right\}_0^1 = \frac{50\sqrt{10}}{3}$$

6. With parametric equations $C:\ x = 1 - 4t,\ y = -1/2 + 4t,\ z = 1/2,\ 0 \leq t \leq 1$,

$$\int_C x^2 yz\, ds = \int_0^1 (1 - 4t)^2 \left(-\frac{1}{2} + 4t\right)\left(\frac{1}{2}\right) \sqrt{(-4)^2 + (4)^2}\, dt = \sqrt{2} \int_0^1 (1 - 4t)^2 (8t - 1)\, dt$$

$$= \sqrt{2} \int_0^1 (128t^3 - 80t^2 + 16t - 1)\, dt = \sqrt{2} \left\{ 32t^4 - \frac{80t^3}{3} + 8t^2 - t \right\}_0^1 = \frac{37\sqrt{2}}{3}$$

8. $\displaystyle L = \int_C ds = \int_0^{12\pi} \sqrt{(-3\sin t)^2 + (3\cos t)^2 + [3/(4\pi)]^2}\, dt$

$$= \int_0^{12\pi} \sqrt{9 + \frac{9}{16\pi^2}}\, dt = \frac{3}{4\pi} \sqrt{16\pi^2 + 1} \left\{ t \right\}_0^{12\pi} = 9\sqrt{1 + 16\pi^2}\ \text{cm}$$

10. With parametric equations $C:\ y = x^2,\ z = 1 - x^2,\ 0 \leq x \leq 1$,

$$\int_C xz\, ds = \int_0^1 x(1 - x^2)\sqrt{1^2 + (2x)^2 + (-2x)^2}\, dx = \int_0^1 x(1 - x^2)\sqrt{1 + 8x^2}\, dx.$$

If we set $u = 1 + 8x^2$ and $du = 16x\, dx$, then

$$\int_C xz\, ds = \int_1^9 \left[1 - \left(\frac{u - 1}{8}\right) \right] \sqrt{u} \left(\frac{du}{16}\right) = \frac{1}{128} \int_1^9 (9\sqrt{u} - u^{3/2})\, du = \frac{1}{128} \left\{ 6u^{3/2} - \frac{2}{5}u^{5/2} \right\}_1^9 = \frac{37}{80}.$$

12. With parametric equations $C:\ x = 1 - t,\ y = 9t/14,\ z = 1 + 4t/7,\ 0 \leq t \leq 1$,

$$\int_C x\sqrt{y + z}\, ds = \int_0^1 (1 - t)\sqrt{\frac{9t}{14} + 1 + \frac{4t}{7}} \sqrt{(-1)^2 + \left(\frac{9}{14}\right)^2 + \left(\frac{4}{7}\right)^2}\, dt = \frac{\sqrt{341}}{14\sqrt{14}} \int_0^1 (1 - t)\sqrt{14 + 17t}\, dt.$$

If we set $u = 14 + 17t$ and $du = 17\, dt$, then

$$\int_C x\sqrt{y + z}\, ds = \frac{\sqrt{341}}{14\sqrt{14}} \int_{14}^{31} \left[1 - \left(\frac{u - 14}{17}\right) \right] \sqrt{u} \left(\frac{du}{17}\right) = \frac{\sqrt{341}}{4046\sqrt{14}} \int_{14}^{31} (31\sqrt{u} - u^{3/2})\, du$$

$$= \frac{\sqrt{341}}{4046\sqrt{14}} \left\{ \frac{62u^{3/2}}{3} - \frac{2u^{5/2}}{5} \right\}_{14}^{31} = 0.78.$$

14. With parametric equations $C: y = x$, $z = 1 + x^4$, $-1 \le x \le 1$,

$$\int_C (x+y)z\, ds = \int_{-1}^{1} (x+x)(1+x^4)\sqrt{1+1+(4x^3)^2}\, dx = 2\int_{-1}^{1} (x+x^5)\sqrt{2+16x^6}\, dx = 0,$$

since the integrand is an odd function of x.

16. With parametric equations $C: x = t^2$, $y = -t$, $z = -t^3$, $0 \le t \le 2$,

$$\int_C (2y+9z)\, ds = \int_0^2 (-2t-9t^3)\sqrt{(2t)^2+1+(-3t^2)^2}\, dt = -\int_0^2 (2t+9t^3)\sqrt{1+4t^2+9t^4}\, dt.$$

If we set $u = 1 + 4t^2 + 9t^4$ and $du = (8t+36t^3)\, dt = 4(2t+9t^3)\, dt$, then

$$\int_C (2y+9z)\, ds = -\int_1^{161} \sqrt{u}\left(\frac{du}{4}\right) = -\left\{\frac{1}{6}u^{3/2}\right\}_1^{161} = \frac{1-161\sqrt{161}}{6}.$$

18. When we use x as parameter along the curve,

$$A = \int_C 2\pi y\, ds = 2\pi \int_1^2 x^3\sqrt{1+(3x^2)^2}\, dx = 2\pi\int_1^2 x^3\sqrt{1+9x^4}\, dx = 2\pi\left\{\frac{(1+9x^4)^{3/2}}{54}\right\}_1^2$$
$$= \frac{(145\sqrt{145}-10\sqrt{10})\pi}{27}.$$

20. We double the area obtained by rotating the first quadrant part of the curve.

$$A = 2\int_C 2\pi y\, ds \quad \text{where}$$

$$ds = \sqrt{1+\left[\frac{1}{2\sqrt{2}}\left(\sqrt{1-x^2}-\frac{x^2}{\sqrt{1-x^2}}\right)\right]^2}\, dx$$

$$= \sqrt{1+\frac{1}{8}\left(\frac{1-2x^2}{\sqrt{1-x^2}}\right)^2}\, dx$$

$$= \sqrt{\frac{8(1-x^2)+(1-2x^2)^2}{8(1-x^2)}}\, dx = \frac{3-2x^2}{2\sqrt{2}\sqrt{1-x^2}}\, dx$$

$$y = \frac{x\sqrt{1-x^2}}{2\sqrt{2}}$$

Thus, $A = 4\pi\displaystyle\int_0^1 \frac{x\sqrt{1-x^2}}{2\sqrt{2}}\frac{3-2x^2}{2\sqrt{2}\sqrt{1-x^2}}\, dx = \frac{\pi}{2}\int_0^1 (3x-2x^3)\, dx = \frac{\pi}{2}\left\{\frac{3x^2}{2}-\frac{x^4}{2}\right\}_0^1 = \frac{\pi}{2}.$

22. Using x as the parameter along the curve,

$$\int_C xy\, ds = \int_0^{1/2} x^4\sqrt{1+(3x^2)^2}\, dx = \int_0^{1/2} x^4\sqrt{1+9x^4}\, dx$$

$$= \int_0^{1/2} x^4\left[1+\frac{1}{2}(9x^4)+\frac{(1/2)(-1/2)}{2}(9x^4)^2+\cdots\right]dx$$

$$= \int_0^{1/2}\left(x^4+\frac{9}{2}x^8-\frac{9^2}{2^2 2!}x^{12}+\frac{9^3 3}{2^3 3!}x^{16}-\frac{9^4 3\cdot 5}{2^4 4!}x^{20}+\cdots\right)dx$$

$$= \left\{\frac{x^5}{5}+\frac{x^9}{2}-\frac{9^2 x^{13}}{2^2 2!13}+\frac{9^3 3x^{17}}{2^3 3!17}-\frac{9^4 3\cdot 5x^{21}}{2^4 4!21}+\cdots\right\}_0^{1/2}$$

$$= \frac{1}{5\cdot 2^5}+\frac{1}{2^{10}}-\frac{9^2}{2^{15}2!13}+\frac{9^3 3}{2^{20}3!17}-\frac{9^4 3\cdot 5}{2^{25}4!21}+\frac{9^5 3\cdot 5\cdot 7}{2^{30}5!25}+\cdots.$$

Because this series is alternating (after the first term), and absolute values of terms decrease and approach zero, the sum is between any two consecutive partial sums. Since the sum of the first two terms is $0.007\,22$ and the sum of the first three terms is $0.007\,13$, we can say that to three decimals, the value of the integral is 0.007.

24. With parametric equations $C: \ x = 2\cos t, \ y = 2\sin t, \ -\pi < t \le \pi$,

$$\overline{f} = \frac{1}{2\pi(2)} \int_C x^2 y^2 \, ds = \frac{1}{4\pi} \int_{-\pi}^{\pi} 4\cos^2 t \, 4\sin^2 t \, \sqrt{4\sin^2 t + 4\cos^2 t} \, dt = \frac{8}{\pi} \int_{-\pi}^{\pi} \cos^2 t \, \sin^2 t \, dt$$

$$= \frac{8}{\pi} \int_{-\pi}^{\pi} \left(\frac{\sin 2t}{2}\right)^2 dt = \frac{2}{\pi} \int_{-\pi}^{\pi} \left(\frac{1 - \cos 4t}{2}\right) dt = \frac{1}{\pi} \left\{ t - \frac{\sin 4t}{4} \right\}_{-\pi}^{\pi} = 2.$$

26. With parametric equations $C: \ y = x^2, \ z = x^2, \ 0 \le x \le 1$, the length of the curve is

$$L = \int_0^1 \sqrt{1 + (2x)^2 + (2x)^2} \, dx = \int_0^1 \sqrt{1 + 8x^2} \, dx. \text{ If we set } x = [1/(2\sqrt{2})]\tan\theta, \text{ then}$$

$$L = \int_0^{\overline{\theta}} \sec\theta \frac{1}{2\sqrt{2}} \sec^2\theta \, d\theta \quad (\overline{\theta} = \text{Tan}^{-1}(2\sqrt{2}))$$

$$= \frac{1}{4\sqrt{2}} \left\{ \sec\theta \tan\theta + \ln|\sec\theta + \tan\theta| \right\}_0^{\overline{\theta}} \quad \text{(see Example 8.9)}$$

$$= \frac{1}{4\sqrt{2}} [6\sqrt{2} + \ln(3 + 2\sqrt{2})].$$

The average value of the function is $\quad \overline{f} = \frac{1}{L} \int_0^1 xyz \, ds = \frac{1}{L} \int_0^1 x^5 \sqrt{1 + 8x^2} \, dx.$ If we set $u = 1 + 8x^2$,

then $du = 16x \, dx$, and

$$\overline{f} = \frac{1}{L} \int_1^9 \left(\frac{u-1}{8}\right)^2 \sqrt{u} \left(\frac{du}{16}\right) = \frac{1}{1024L} \int_1^9 (u^{5/2} - 2u^{3/2} + \sqrt{u}) \, du$$

$$= \frac{1}{1024L} \left\{ \frac{2u^{7/2}}{7} - \frac{4u^{5/2}}{5} + \frac{2u^{3/2}}{3} \right\}_1^9 = 0.242.$$

28. $A = 4 \displaystyle\int_C \sqrt{x^2 + y^2} \, ds$ where C is that part of the lemniscate in the first quadrant. According to formula 9.14,

$$ds = \sqrt{r^2 + \left(\frac{dr}{d\theta}\right)^2} \, d\theta.$$

Hence,

$$A = 4\int_0^{\pi/4} r\sqrt{r^2 + \left(\frac{dr}{d\theta}\right)^2} \, d\theta$$

$$= 4\int_0^{\pi/4} \sqrt{\cos 2\theta} \sqrt{\cos 2\theta + \frac{\sin^2 2\theta}{\cos 2\theta}} \, d\theta = 4\int_0^{\pi/4} d\theta = \pi.$$

$r^2 = \cos 2\theta$

30. With $ds = \sqrt{r^2 + \left(\dfrac{dr}{d\theta}\right)^2} \, d\theta = \sqrt{(2 - \sin\theta)^2 + (-\cos\theta)^2} \, d\theta = \sqrt{5 - 4\sin\theta} \, d\theta$, we obtain

$$\int_C \frac{x}{\sqrt{x^2 + y^2}} \, ds = \int_0^{\pi/2} \frac{r\cos\theta}{r} \sqrt{5 - 4\sin\theta} \, d\theta = \left\{ -\frac{1}{6}(5 - 4\sin\theta)^{3/2} \right\}_0^{\pi/2} = \frac{5\sqrt{5} - 1}{6}.$$

32. With $ds = \sqrt{r^2 + \left(\dfrac{dr}{d\theta}\right)^2} \, d\theta = \sqrt{e^{2\theta} + e^{2\theta}} \, d\theta = \sqrt{2}e^\theta \, d\theta$,

$$\int_C xy \, ds = \int_0^{2\pi} r\cos\theta \, r\sin\theta \, \sqrt{2}e^\theta \, d\theta = \sqrt{2}\int_0^{2\pi} \sin\theta \cos\theta \, e^{3\theta} \, d\theta = \frac{1}{\sqrt{2}} \int_0^{2\pi} e^{3\theta} \sin 2\theta \, d\theta$$

$$= \frac{1}{\sqrt{2}} \left\{ \frac{e^{3\theta}}{13}(3\sin 2\theta - 2\cos 2\theta) \right\}_0^{2\pi} = \frac{\sqrt{2}(1 - e^{6\pi})}{13}.$$

34. With parametric equations $C: y = 1 - x, z = 2x^2 - 2x + 1, -1 \le x \le 1,$

$$I = \int_{-1}^{1} [x^2(1-x) + 2x^2 - 2x + 1]\sqrt{1 + (-1)^2 + (4x - 2)^2}\, dx$$

$$= \int_{-1}^{1} (1 - 2x + 3x^2 - x^3)\sqrt{6 - 16x + 16x^2}\, dx = \sqrt{2}\int_{-1}^{1} (1 - 2x + 3x^2 - x^3)\sqrt{3 - 8x + 8x^2}\, dx.$$

If we denote the integrand by $f(x)$, then Simpson's rule with 10 equal subdivisions gives

$$I \approx \sqrt{2}\left(\frac{1/5}{3}\right)[f(-1) + 4f(-0.8) + 2f(-0.6) + \cdots + 2f(0.6) + 4f(0.8) + f(1)] = 17.08.$$

EXERCISES 14.3

2. Using x as parameter along the curve,

$$\int_C x\, dx + yz\, dy + x^2\, dz = \int_{-1}^{2} [x\, dx + xx^2\, dx + x^2(2x\, dx)] = \int_{-1}^{2} (x + 3x^3)\, dx = \left\{\frac{x^2}{2} + \frac{3x^4}{4}\right\}_{-1}^{2} = \frac{51}{4}.$$

4. With parametric equations $C: x = -2 + 3t, y = 3 - 3t, z = 3 - 3t, 0 \le t \le 1,$ for the straight line,

$$\int_C x^2\, dx + y^2\, dy + z^2\, dz = \int_0^1 [(-2 + 3t)^2(3\, dt) + (3 - 3t)^2(-3\, dt) + (3 - 3t)^2(-3\, dt)]$$

$$= 3\int_0^1 [(-2 + 3t)^2 - 2(3 - 3t)^2]\, dt = 3\left\{\frac{1}{9}(-2 + 3t)^3 + \frac{2}{9}(3 - 3t)^3\right\}_0^1 = -15.$$

6. $\oint_C x^2 y\, dx + (x - y)\, dy = \int_{C_1} x^2 y\, dx + (x - y)\, dy + \int_{C_2} x^2 y\, dx + (x - y)\, dy$

$$= \int_{-2}^{1} [(1 - t^2)^2 t(-2t\, dt) + (1 - t^2 - t)\, dt]$$

$$+ \int_0^3 [t^2(1 - t)(-dt) + (-t - 1 + t)(-dt)]$$

$$= \int_{-2}^{1} (-2t^6 + 4t^4 - 3t^2 - t + 1)\, dt + \int_0^3 (t^3 - t^2 + 1)\, dt$$

$$= \left\{-\frac{2t^7}{7} + \frac{4t^5}{5} - t^3 - \frac{t^2}{2} + t\right\}_{-2}^{1} + \left\{\frac{t^4}{4} - \frac{t^3}{3} + t\right\}_0^3 = -\frac{99}{140}$$

8. With parametric equations $C: x = 1 - y, z = 2y^2 - 2y + 1, 0 \le y \le 2,$

$$\int_C y\, dx + x\, dy + z\, dz = \int_0^2 [y(-dy) + (1 - y)\, dy + (2y^2 - 2y + 1)(4y - 2)\, dy]$$

$$= \int_0^2 [-2y + 1 + (2y^2 - 2y + 1)(4y - 2)]\, dy = \left\{-y^2 + y + \frac{(2y^2 - 2y + 1)^2}{2}\right\}_0^2 = 10.$$

10. $\oint_C y^2\, dx + x^2\, dy = \int_{C_1} y^2\, dx + x^2\, dy + \int_{C_2} y^2\, dx + x^2\, dy + \int_{C_3} y^2\, dx + x^2\, dy + \int_{C_4} y^2\, dx + x^2\, dy$

$$= \int_0^1 [(1 - t)^2\, dt + t^2(-dt)] + \int_{-1}^{0} [(t + 1)^2(-dt) + t^2(-dt)]$$

$$+ \int_{-1}^{0} [t^2(-dt) + (t + 1)^2\, dt] + \int_0^1 [t^2\, dt + (t - 1)^2\, dt]$$

$$= \int_0^1 2(t - 1)^2\, dt + \int_{-1}^{0} (-2t^2)\, dt$$

$$= \left\{\frac{2(t - 1)^3}{3}\right\}_0^1 + \left\{-\frac{2t^3}{3}\right\}_{-1}^{0} = 0$$

12. (a) $\displaystyle\int_C xy\,dx + x^2\,dy = \int_0^{\pi/2} [3\cos t(3\sin t)(-3\sin t\,dt) + 9\cos^2 t(3\cos t\,dt)]$

$$= 27\int_0^{\pi/2}[-\sin^2 t\,\cos t + \cos t(1-\sin^2 t)]\,dt = 27\left\{-\frac{2}{3}\sin^3 t + \sin t\right\}_0^{\pi/2} = 9$$

(b) $\displaystyle\int_C xy\,dx + x^2\,dy = \int_0^3\left[\sqrt{9-y^2}(y)\left(\frac{-y}{\sqrt{9-y^2}}dy\right) + (9-y^2)\,dy\right] = \int_0^3 (9-2y^2)\,dy$

$$= \left\{9y - \frac{2y^3}{3}\right\}_0^3 = 9$$

14. With parametric equations $C:\ x = a\cos t,\ y = b\sin t,\ -\pi \le t < \pi$,

$$W = \oint_C \mathbf{F}\cdot d\mathbf{r} = \oint_C x\,dx + y\,dy = \int_{-\pi}^{\pi}[a\cos t(-a\sin t\,dt) + b\sin t(b\cos t\,dt)]$$

$$= (b^2 - a^2)\int_{-\pi}^{\pi}\sin t\,\cos t\,dt = (b^2 - a^2)\left\{\frac{\sin^2 t}{2}\right\}_{-\pi}^{\pi} = 0.$$

16. With parametric equations $C:\ x = 2 + \cos t,\ y = -\sin t,\ 0 \le t < 4\pi$,

$$\oint_C (x^2 + 2y^2)\,dy = \int_0^{4\pi}[(2+\cos t)^2 + 2(-\sin t)^2](-\cos t\,dt)$$

$$= -\int_0^{4\pi}(4 + 4\cos t + \cos^2 t + 2\sin^2 t)\cos t\,dt = -\int_0^{4\pi}(5 + 4\cos t + \sin^2 t)\cos t\,dt$$

$$= -\int_0^{4\pi}[5\cos t + 2(1 + \cos 2t) + \sin^2 t\,\cos t]\,dt = -\left\{5\sin t + 2t + \sin 2t + \frac{1}{3}\sin^3 t\right\}_0^{4\pi}$$

$$= -8\pi.$$

18. With parametric equations $C:\ x = t - 1,\ y = 1 + 2t^2,\ z = t,\ 1 \le t \le 2$,

$$\int_C x^2 y\,dx + y\,dy + \sqrt{1-x^2}\,dz = \int_1^2[(t-1)^2(1+2t^2)\,dt + (1+2t^2)(4t\,dt) + \sqrt{1-(t-1)^2}\,dt]$$

$$= \int_1^2[2t^4 + 4t^3 + 3t^2 + 2t + 1 + \sqrt{1-(t-1)^2}]\,dt.$$

In the last term we set $t - 1 = \sin\theta$ and $dt = \cos\theta\,d\theta$,

$$\int_C x^2 y\,dx + y\,dy + \sqrt{1-x^2}\,dz = \left\{\frac{2t^5}{5} + t^4 + t^3 + t^2 + t\right\}_1^2 + \int_0^{\pi/2}\cos^2\theta\,d\theta$$

$$= \frac{192}{5} + \int_0^{\pi/2}\left(\frac{1+\cos 2\theta}{2}\right)d\theta = \frac{192}{5} + \left\{\frac{\theta}{2} + \frac{\sin 2\theta}{4}\right\}_0^{\pi/2} = \frac{192}{5} + \frac{\pi}{4}.$$

20. The line integral along C is equal to the sum of the line integrals along C_1, C_2, and C_3; that is,

$$\int_C \frac{x^3}{(1+x^4)^3}dx + y^2 e^y\,dy + \frac{z}{\sqrt{1+z^2}}dz$$

$$= \int_0^1 \frac{x^3}{(1+x^4)^3}dx + \int_{-1}^0 y^2 e^y\,dy$$

$$+ \int_1^2 \frac{z}{\sqrt{1+z^2}}dz$$

$$= \left\{\frac{-1}{8(1+x^4)^2}\right\}_0^1 + \left\{y^2 e^y - 2y e^y + 2e^y\right\}_{-1}^0$$

$$+ \left\{\sqrt{1+z^2}\right\}_1^2$$

$$= \frac{67}{32} + \sqrt{5} - \sqrt{2} - \frac{5}{e}.$$

22. At position x, the magnitude of the force $\mathbf{F}_1$ of q_1 on q_3 is $|\mathbf{F}_1| = \dfrac{q_1 q_3}{4\pi\epsilon_0[(x-5)^2 + 25]}$. Since a unit vector in the direction of $\mathbf{F}_1$ is

$$\hat{\mathbf{F}}_1 = \frac{(x-5, -5)}{\sqrt{(x-5)^2 + 25}},$$

it follows that

$$\mathbf{F}_1 = \frac{q_1 q_3}{4\pi\epsilon_0[(x-5)^2 + 25]^{3/2}}(x-5, -5).$$

Similarly, the force $\mathbf{F}_2$ of q_2 on q_3 is

$$\mathbf{F}_2 = \frac{q_2 q_3}{4\pi\epsilon_0[(x+2)^2 + 9]^{3/2}}(x+2, -3).$$

The work done by these forces is

$$W = \int_C (\mathbf{F}_1 + \mathbf{F}_2) \cdot d\mathbf{r} = \int_C (\mathbf{F}_1 + \mathbf{F}_2) \cdot (dx\,\hat{\mathbf{i}})$$

$$= \int_C \left\{ \frac{q_1 q_3}{4\pi\epsilon_0[(x-5)^2 + 25]^{3/2}}(x-5) + \frac{q_2 q_3}{4\pi\epsilon_0[(x+2)^2 + 9]^{3/2}}(x+2) \right\} dx$$

$$= \int_{-1}^{1} \left\{ \frac{q_1 q_3}{4\pi\epsilon_0[(-t-5)^2 + 25]^{3/2}}(-t-5) + \frac{q_2 q_3}{4\pi\epsilon_0[(2-t)^2 + 9]^{3/2}}(2-t) \right\} (-dt)$$

$$= \left\{ \frac{-q_1 q_3}{4\pi\epsilon_0 \sqrt{(t+5)^2 + 25}} + \frac{-q_2 q_3}{4\pi\epsilon_0 \sqrt{(2-t)^2 + 9}} \right\}_{-1}^{1}$$

$$= \frac{1}{4\pi\epsilon_0}\left[q_1 q_3 \left(\frac{1}{\sqrt{41}} - \frac{1}{\sqrt{61}} \right) + q_2 q_3 \left(\frac{1}{3\sqrt{2}} - \frac{1}{\sqrt{10}} \right) \right].$$

24. Using points A and D we obtain $k_2 = 1000$ and $k_1 = 20\,000$. Designating the four parts of the cycle starting at A by C_1, C_2, C_3, and C_4,

$$W = \int_{C_1} P\,dV + \int_{C_2} P\,dV + \int_{C_3} P\,dV + \int_{C_4} P\,dV = \int_{1/10}^{1/100} \frac{k_2}{V}\,dV + \int_{1/100}^{1/5} 10^5\,dV + \int_{1/5}^{2} \frac{k_1}{V}\,dV + \int_{2}^{1/10} 10^4\,dV$$

$$= k_2 \Big\{ \ln V \Big\}_{1/10}^{1/100} + 10^5 \left(\frac{1}{5} - \frac{1}{100} \right) + k_1 \Big\{ \ln V \Big\}_{1/5}^{2} + 10^4 \left(\frac{1}{10} - 2 \right) = 4.4 \times 10^4 \text{ J.}$$

26. Using points A and D we obtain $k_2 = 15.0$ and $k_1 = 66.6$. Designating the four parts of the cycle starting at A by C_1, C_2, C_3, and C_4,

$$W = \int_{C_1} P\,dV + \int_{C_2} P\,dV + \int_{C_3} P\,dV + \int_{C_4} P\,dV$$

$$= \int_{20\times10^{-4}}^{2\times10^{-4}} \frac{k_2}{V^{1.4}}\,dV + \int_{2\times10^{-4}}^{5.75\times10^{-4}} (23 \times 10^5)\,dV + \int_{5.75\times10^{-4}}^{20\times10^{-4}} \frac{k_1}{V^{1.4}}\,dV + 0$$

$$= k_2 \left\{ \frac{-1}{0.4V^{0.4}} \right\}_{20\times10^{-4}}^{2\times10^{-4}} + 23 \times 10^5 (5.75 \times 10^{-4} - 2 \times 10^{-4}) + k_1 \left\{ \frac{-1}{04.V^{0.4}} \right\}_{5.75\times10^{-4}}^{20\times10^{-4}} = 1.5 \times 10^3 \text{ J.}$$

28. Designating the four parts of the cycle by C_1, C_2, C_3, and C_4,

$$W = \int_{C_1} P\,dV + \int_{C_2} P\,dV + \int_{C_3} P\,dV + \int_{C_4} P\,dV$$

$$= 0 + \int_{V_1}^{V_2} P_2\,dV + \int_{V_2}^{V_3} CV^{-\gamma}\,dV + \int_{V_3}^{V_2} P_3\,dV$$

$$= P_2(V_2 - V_1) + C\left\{ \frac{V^{1-\gamma}}{1-\gamma} \right\}_{V_2}^{V_3} + P_3(V_2 - V_3)$$

$$= P_2(V_2 - V_1) + \frac{C}{1-\gamma}(V_3^{1-\gamma} - V_2^{1-\gamma}) + P_3(V_2 - V_3).$$

Since $P_2 V_2^\gamma = C = P_3 V_3^\gamma$, it follows that $P_3 = P_2 (V_2/V_3)^\gamma$, and

$$W = P_2(V_2 - V_1) + \frac{P_2 V_2^\gamma}{1 - \gamma}(V_3^{1-\gamma} - V_2^{1-\gamma}) + P_2 \left(\frac{V_2}{V_3}\right)^\gamma (V_2 - V_3).$$

30. When we use y as the parameter along the curve, the value of the line integral is

$$I = \int_{-1}^{1} y^2 y^3 (2y\, dy) + \tan y^2\, dy + e^{y^3}(3y^2\, dy) = \int_{-1}^{1} (2y^6 + 3y^2 e^{y^3} + \tan y^2)\, dy$$

$$= \left\{ \frac{2y^7}{7} + e^{y^3} \right\}_{-1}^{1} + 2 \int_{0}^{1} \tan y^2\, dy.$$

If we use Simpson's rule with 10 equal subdivisions on the remaining integral

$$I \approx \left(\frac{2}{7} + e\right) - \left(-\frac{2}{7} + e^{-1}\right) + \frac{1/5}{3}\big[\tan(0) + 4\tan(0.01) + 2\tan(0.04)$$

$$+ \cdots + 2\tan(0.64) + 4\tan(0.81) + \tan(1)\big] = 3.719.$$

32. $\displaystyle \int_C xyz\, dy = \int_{-1}^{1} \left(\frac{1-t^2}{1+t^2}\right) \left[\frac{t(1-t^2)}{1+t^2}\right] t \left[\frac{(1+t^2)(1-3t^2) - (t-t^3)(2t)}{(1+t^2)^2}\right] dt$

$$= \int_{-1}^{1} \frac{t^2(1-t^2)^2(1 - 4t^2 - t^4)}{(1+t^2)^4}\, dt$$

If we denote the integrand by $f(t)$, then Simpson's rule with 10 equal subdivisions gives

$$\int_C xyz\, dy \approx \frac{2/10}{3}[f(-1) + 4f(-0.8) + 2f(-0.6) + \cdots + 2f(0.6) + 4f(0.8) + f(1)]$$

$$= -4.26 \times 10^{-4}.$$

34. Since $x = r\cos\theta$, $y = r\sin\theta$, and $r = \theta$, it follows that $x = \theta\cos\theta$ and $y = \theta\sin\theta$. Hence,

$$\int_C y\, dx + x\, dy = \int_0^\pi \theta\sin\theta(\cos\theta - \theta\sin\theta)\, d\theta + \theta\cos\theta(\sin\theta + \theta\cos\theta)\, d\theta$$

$$= \int_0^\pi [2\theta\sin\theta\cos\theta + \theta^2(\cos^2\theta - \sin^2\theta)]\, d\theta$$

$$= \int_0^\pi [\theta\sin 2\theta + \theta^2\cos 2\theta]\, d\theta = \left\{\frac{\theta^2}{2}\sin 2\theta\right\}_0^\pi = 0.$$

36. When $C: x = x(t)$, $y = y(t)$, $z = z(t)$, $\alpha \le t \le \beta$ are parametric equations for C, then parametric equations for $-C$ are $x = x(t)$, $y = y(t)$, $z = z(t)$, $\beta \ge t \ge \alpha$. To obtain an increasing parameter along $-C$, we set $u = -t$, in which case

$$-C: \quad x = x(-u), \quad y = y(-u), \quad z = z(-u), \quad -\beta \le u \le -\alpha.$$

If $\mathbf{F} = P\hat{\mathbf{i}} + Q\hat{\mathbf{j}} + R\hat{\mathbf{k}}$, then the value of the line integral along $-C$ can be expressed as a definite integral with respect to u:

$$\int_{-C} \mathbf{F} \cdot d\mathbf{r} = \int_{-C} P\, dx + Q\, dy + R\, dz$$

$$= \int_{-\beta}^{-\alpha} \left\{ P[x(-u), y(-u), z(-u)]\frac{dx}{du} + Q[x(-u), y(-u), z(-u)]\frac{dy}{du} \right.$$

$$\left. + R[x(-u), y(-u), z(-u)]\frac{dz}{du} \right\} du.$$

If we now change variables of integration by setting $t = -u$,

$$\frac{dx}{du} = \frac{dx}{dt}\frac{dt}{du} = -\frac{dx}{dt},$$

and similarly for dy/du and dz/du. Consequently,

$$\int_{-C} \mathbf{F} \cdot d\mathbf{r} = \int_{\beta}^{\alpha} \left\{ P[x(t), y(t), z(t)] \left(-\frac{dx}{dt} \right) + Q[x(t), y(t), z(t)] \left(-\frac{dy}{dt} \right) \right.$$

$$\left. + R[x(t), y(t), z(t)] \left(-\frac{dz}{dt} \right) \right\} (-dt)$$

$$= -\int_{\alpha}^{\beta} \left\{ P[x(t), y(t), z(t)] \frac{dx}{dt} + Q[x(t), y(t), z(t)] \frac{dy}{dt} \right.$$

$$\left. + R[x(t), y(t), z(t)] \frac{dz}{dt} \right\} dt$$

$$= -\int_{C} \mathbf{F} \cdot d\mathbf{r}.$$

38. (a) Since the centre of the circle has coordinates $(R\theta, R)$, the direction of the unit force is

$$(R\theta - R\theta + R\sin\theta, R - R + R\cos\theta) = R(\sin\theta, \cos\theta).$$

Hence, the unit force has components $\mathbf{F} = (\sin\theta, \cos\theta)$. The work done during a half revolution is

$$W = \int_{C} \mathbf{F} \cdot d\mathbf{r} = \int_{C} \sin\theta \, dx + \cos\theta \, dy = \int_{0}^{\pi} [\sin\theta(R - R\cos\theta) \, d\theta + \cos\theta(R\sin\theta) \, d\theta$$

$$= R\int_{0}^{\pi} \sin\theta \, d\theta = R\left\{ -\cos\theta \right\}_{0}^{\pi} = 2R.$$

(b) The work done by the vertical component is

$$W = \int_{C} \cos\theta \hat{\mathbf{j}} \cdot d\mathbf{r} = \int_{C} \cos\theta \, dy = \int_{0}^{\pi} \cos\theta(R\sin\theta) \, d\theta = \left\{ \frac{R}{2}\sin^2\theta \right\}_{0}^{\pi} = 0.$$

EXERCISES 14.4

2. Since $\nabla(x^3 + xy) = (3x^2 + y)\hat{\mathbf{i}} + x\hat{\mathbf{j}}$, the line integral is independent of path in space, and

$$\int_{C} (3x^2 + y) \, dx + x \, dy = \left\{ x^3 + xy \right\}_{(2,1,5)}^{(-3,2,4)} = -43.$$

4. Since $\nabla(x^3yz - 2z^2) = 3x^2yz\hat{\mathbf{i}} + x^3z\hat{\mathbf{j}} + (x^3y - 4z)\hat{\mathbf{k}}$, the line integral is independent of path in space, and

$$\int_{C} 3x^2yz \, dx + x^3z \, dy + (x^3y - 4z) \, dz = \left\{ x^3yz - 2z^2 \right\}_{(-1,-1,1)}^{(1,1,-1)} = -2.$$

6. Since $\nabla(y\sin x) = y\cos x\,\hat{\mathbf{i}} + \sin x\,\hat{\mathbf{j}}$, the line integral is independent of path in the xy-plane, and

$$\oint_{C} y\cos x \, dx + \sin x \, dy = 0.$$

8. Since $\nabla(xy + z^2/2) = y\hat{\mathbf{i}} + x\hat{\mathbf{j}} + z\hat{\mathbf{k}}$, the line integral is independent of path in space, and

$$\int_{C} y \, dx + x \, dy + z \, dz = \left\{ xy + \frac{z^2}{2} \right\}_{(1,0,1)}^{(-1,2,5)} = 10.$$

10. Since $\nabla(x^3y^3) = 3x^2y^3\hat{\mathbf{i}} + 3x^3y^2\hat{\mathbf{j}}$, the line integral is independent of path in the xy-plane, and

$$\int_{C} 3x^2y^3 \, dx + 3x^3y^2 \, dy = \left\{ x^3y^3 \right\}_{(0,1)}^{(1,e)} = e^3.$$

12. No. It may be independent of path. All we know is that Theorem 14.4 fails to imply independence of path.

14. Since $\nabla(xy\tan x + z) = y(\tan x + x\sec^2 x)\hat{\mathbf{i}} + x\tan x\,\hat{\mathbf{j}} + \hat{\mathbf{k}}$, the line integral is certainly independent of path in the domain $-1.1 < x < 1.1$ containing the curve $x^2 + y^2 = 1$. Hence

$$\oint_C y(\tan x + x\sec^2 x)\,dx + x\tan x\,dy + dz = 0.$$

16. Since $\nabla(xz/y) = (z/y)\hat{\mathbf{i}} - (xz/y^2)\hat{\mathbf{j}} + (x/y)\hat{\mathbf{k}}$, the line integral is independent of path in any domain which not containing points in the xz-plane. Hence

$$\oint_C \frac{zy\,dx - xz\,dy + xy\,dz}{y^2} = 0.$$

18. Since $\nabla\left(\dfrac{-1}{(x-3)(y+5)} + \ln|z+4|\right) = \dfrac{1}{(x-3)^2(y+5)}\hat{\mathbf{i}} + \dfrac{1}{(x-3)(y+5)^2}\hat{\mathbf{j}} + \dfrac{1}{z+4}\hat{\mathbf{k}}$, the line integral is independent of path in any domain not containing points in the planes $x = 3$, $y = -5$, and $z = -4$. Thus,

$$\int_C \frac{1}{(x-3)^2(y+5)}\,dx + \frac{1}{(x-3)(y+5)^2}\,dy + \frac{1}{z+4}\,dz = \left\{\frac{-1}{(x-3)(y+5)} + \ln|z+4|\right\}_{0,0,0}^{(2,2,2)}$$

$$= \frac{8}{105} + \ln(3/2).$$

20. Since

$$\frac{\partial}{\partial x}\left(\frac{-x}{x^2+y^2}\right) = \frac{\partial}{\partial y}\left(\frac{y}{x^2+y^2}\right)$$

in the simply-connected domain shown, the line integral is independent of path therein. We may therefore replace C with the semicircle
C' : $x = \cos t$, $y = \sin t$, $0 \le t \le \pi$.
Hence,

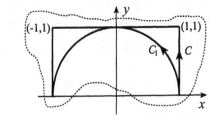

$$\int_C \frac{y}{x^2+y^2}\,dx - \frac{x}{x^2+y^2}\,dy = \int_{C'} \frac{y}{x^2+y^2}\,dx - \frac{x}{x^2+y^2}\,dy$$

$$= \int_0^\pi \frac{\sin t(-\sin t\,dt) - \cos t(\cos t\,dt)}{1} = -\int_0^\pi dt = -\pi.$$

22. A quick calculation shows that each of the partial derivatives $\dfrac{\partial}{\partial y}\left(\dfrac{y}{x^2+y^2}\right)$ and $\dfrac{\partial}{\partial x}\left(\dfrac{-x}{x^2+y^2}\right)$ is equal

to $\dfrac{x^2 - y^2}{(x^2+y^2)^2}$. These derivatives are equal in each of the domains specified. Since domains $x > 0$, $x < 0$, $y > 0$, and $y < 0$ are simply-connected, the line integral is independent of path therein. The domain $x^2 + y^2 > 0$ is not simply-connected. The line integral around C : $x = \cos t$, $y = \sin t$, $0 \le t \le 2\pi$ is

$$\oint_C \frac{y\,dx - x\,dy}{x^2+y^2} = \int_0^{2\pi} \sin t(-\sin t\,dt) - \cos t(\cos t\,dt) = -2\pi.$$

By Corollary 2 to Theorem 14.3, the line integral is not independent of path in $x^2 + y^2 > 0$.

24. Since $\nabla(e^{x^2y}) = (2xye^{x^2y})\hat{\mathbf{i}} + x^2e^{x^2y}\hat{\mathbf{j}}$, the line integral $\int 2xye^{x^2y}\,dx + x^2e^{x^2y}dy$ is independent of path in space, and its value around the given curve is zero. The given line integral therefore reduces to

$$\oint_C x^2y\,dx = \int_0^{2\pi} 4\cos^2 t(-\sin t)(-2\sin t\,dt)$$

$$= 8\int_0^{2\pi} \left(\frac{\sin 2t}{2}\right)^2 dt$$

$$= 2\int_0^{2\pi}\left(\frac{1-\cos 4t}{2}\right)dt = \left\{t - \frac{\sin 4t}{4}\right\}_0^{2\pi} = 2\pi.$$

EXERCISES 14.5

2. Since $\nabla \times \mathbf{F} = \begin{vmatrix} \hat{\mathbf{i}} & \hat{\mathbf{j}} & \hat{\mathbf{k}} \\ \partial/\partial x & \partial/\partial y & \partial/\partial z \\ mx & xy & 0 \end{vmatrix} = y\hat{\mathbf{k}}$, $\mathbf{F}$ is not conservative.

4. Since $\nabla(-mgz) = \mathbf{F}$, $\mathbf{F}$ is conservative with potential energy function $U(z) = mgz$. $\mathbf{F}$ is the force of gravity on a mass m.

6. A normal vector to the equipotential surface $U(x, y, z) = C$ at any point $P(x, y, z)$ on the surface is $\nabla(U - C) = \nabla U$. But $\mathbf{F} = -\nabla U$, and therefore $\mathbf{F}$ is normal to the surface at P.

8. The magnitude of the force is

$$|\mathbf{F}| = k\left(\sqrt{x^2 + y^2 + z^2} - L\right),$$

where k is the spring constant. Since $\mathbf{F}$ is directed toward the origin, it follows that

$$\mathbf{F} = -k\left(\sqrt{x^2 + y^2 + z^2} - L\right)\frac{(x\hat{\mathbf{i}} + y\hat{\mathbf{j}} + z\hat{\mathbf{k}})}{\sqrt{x^2 + y^2 + z^2}}.$$

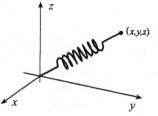

Since $\nabla\left[-\dfrac{k}{2}\left(\sqrt{x^2 + y^2 + z^2} - L\right)^2\right] = \mathbf{F}$ in the domain $x^2 + y^2 + z^2 > 0$, it follows that the force is conservative.

10. (a) Suppose we take the school at the origin. When a student is at position (x, y, z), the magnitude of the force is $|\mathbf{F}| = \dfrac{d}{x^2 + y^2 + z^2}$, where d a constant. Since the force is toward the origin,

$$\mathbf{F} = \frac{-d}{x^2 + y^2 + z^2}\frac{(x\hat{\mathbf{i}} + y\hat{\mathbf{j}} + z\hat{\mathbf{k}})}{\sqrt{x^2 + y^2 + z^2}} = \frac{-d(x\hat{\mathbf{i}} + y\hat{\mathbf{j}} + z\hat{\mathbf{k}})}{(x^2 + y^2 + z^2)^{3/2}}.$$

In the domain $x^2 + y^2 + z^2 > 10\,000$, $\nabla\left(\dfrac{d}{\sqrt{x^2 + y^2 + z^2}}\right) = \mathbf{F}$, and therefore $\mathbf{F}$ is conservative.

(b) Suppose the donut shop is at position (a, b, c). Then the force is

$$\mathbf{F} = \frac{-d}{[(x - a)^2 + (y - b)^2 + (z - c)^2]^{3/2}}[(x - a)\hat{\mathbf{i}} + (y - b)\hat{\mathbf{j}} + (z - c)\hat{\mathbf{k}}].$$

Since $\nabla\left(\dfrac{d}{\sqrt{(x - a)^2 + (y - b)^2 + (z - c)^2}}\right) = \mathbf{F}$, the force is still conservative.

12. (a) The curl of $\mathbf{F}$ is

$$\nabla \times \mathbf{F} = \begin{vmatrix} \hat{\mathbf{i}} & \hat{\mathbf{j}} & \hat{\mathbf{k}} \\ \partial/\partial x & \partial/\partial y & \partial/\partial z \\ f(r)x & f(r)y & f(r)z \end{vmatrix} = \left(z\frac{\partial f}{\partial y} - y\frac{\partial f}{\partial z}\right)\hat{\mathbf{i}} + \left(x\frac{\partial f}{\partial z} - z\frac{\partial f}{\partial x}\right)\hat{\mathbf{j}} + \left(y\frac{\partial f}{\partial x} - x\frac{\partial f}{\partial y}\right)\hat{\mathbf{k}}.$$

The x-component is $z\left(\dfrac{df}{dr}\dfrac{\partial r}{\partial y}\right) - y\left(\dfrac{df}{dr}\dfrac{\partial r}{\partial z}\right) = \dfrac{df}{dr}\left(\dfrac{zy}{\sqrt{x^2+y^2+z^2}} - \dfrac{yz}{\sqrt{x^2+y^2+z^2}}\right) = 0$. Similarly, the y- and z-components of $\nabla \times \mathbf{F}$ vanish, and $\nabla \times \mathbf{F} = \mathbf{0}$. According to Theorem 14.4, $\mathbf{F}$ is conservative in any simply-connected domain that does not contain the origin.

(b) $\displaystyle\int_C \mathbf{F} \cdot d\mathbf{r} = \int_C f(r)\mathbf{r} \cdot d\mathbf{r} = \int_C f(r)(x\,dx + y\,dy + z\,dz)$

Since $r = \sqrt{x^2+y^2+z^2}$, it follows that

$$dr = \frac{x\,dx}{\sqrt{x^2+y^2+z^2}} + \frac{y\,dy}{\sqrt{x^2+y^2+z^2}} + \frac{z\,dz}{\sqrt{x^2+y^2+z^2}} = \frac{x\,dx + y\,dy + z\,dz}{r},$$

and therefore $\displaystyle\int_C \mathbf{F} \cdot d\mathbf{r} = \int_a^b f(r)r\,dr.$

(c) The electrostatic force in Example 14.14 and the gravitational force in Exercise 5 are radially symmetric.

EXERCISES 14.6

2. With Green's theorem,

$$\oint_C (x^2 + 2y^2)\,dy = \iint_R 2x\,dA$$
$$= 2(\text{First moment of } R \text{ about } y\text{-axis})$$
$$= 2(\text{Area of R})(\overline{x}) = 2(\pi)(2) = 4\pi.$$

4. By Green's theorem,

$$\oint_C xy^3\,dx + x^2\,dy = \iint_R (2x - 3xy^2)\,dA = \int_{-\sqrt{3}}^{\sqrt{3}}\int_{\sqrt{1+y^2}}^{2} (2x - 3xy^2)\,dx\,dy$$

$$= \int_{-\sqrt{3}}^{\sqrt{3}}\left\{x^2 - \frac{3x^2y^2}{2}\right\}_{\sqrt{1+y^2}}^{2}\,dy$$

$$= \int_{-\sqrt{3}}^{\sqrt{3}}\left(3 - \frac{11y^2}{2} + \frac{3y^4}{2}\right)\,dy$$

$$= \left\{3y - \frac{11y^3}{6} + \frac{3y^5}{10}\right\}_{-\sqrt{3}}^{\sqrt{3}} = \frac{2\sqrt{3}}{5}.$$

6. By Green's theorem,

$$\oint_C 2\mathrm{Tan}^{-1}\left(\frac{y}{x}\right)dx + \ln(x^2+y^2)\,dy$$

$$= \iint_R \left[\frac{2x}{x^2+y^2} - \frac{2}{1+y^2/x^2}\left(\frac{1}{x}\right)\right]dA$$

$$= 0.$$

8. By Green's theorem,

$$\oint_C (x^3 + y^3)\,dx + (x^3 - y^3)\,dy = \iint_R (3x^2 - 3y^2)\,dA$$

$$= 12\iint_{R_1} (x^2 - y^2)\,dA \quad (R_1 = \text{first quadrant part of } R)$$

$$= 12\int_0^{1/2}\int_0^{1-2x} (x^2 - y^2)\,dy\,dx = 12\int_0^{1/2}\left\{x^2y - \frac{y^3}{3}\right\}_0^{1-2x}\,dx$$

$$= 4\int_0^{1/2}[3x^2 - 6x^3 - (1-2x)^3]\,dx = 4\left\{x^3 - \frac{3x^4}{2} + \frac{(1-2x)^4}{8}\right\}_0^{1/2} = -\frac{3}{8}.$$

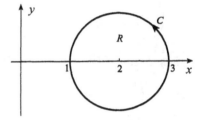

10. By Green's theorem,

$$\oint_C (xy^2 + 2x)\, dx + (x^2y + y + x^2)\, dy$$

$$= \iint_R (2xy + 2x - 2xy)\, dA$$

$$= 2(2) \int_0^3 \int_0^{\sqrt{4+x^2}} x\, dy\, dx = 4 \int_0^3 \Big\{ xy \Big\}_0^{\sqrt{4+x^2}}\, dx$$

$$= 4 \int_0^3 x\sqrt{4+x^2}\, dx = 4 \left\{ \frac{(4+x^2)^{3/2}}{3} \right\}_0^3 = \frac{4}{3}(13\sqrt{13} - 8).$$

12. Since $\nabla \times \mathbf{F} = \begin{vmatrix} \hat{\mathbf{i}} & \hat{\mathbf{j}} & \hat{\mathbf{k}} \\ \partial/\partial x & \partial/\partial y & \partial/\partial z \\ P & Q & 0 \end{vmatrix} = \left(\frac{\partial Q}{\partial x} - \frac{\partial P}{\partial y} \right) \hat{\mathbf{k}},$

$$\oint_C \mathbf{F} \cdot d\mathbf{r} = \iint_R \left(\frac{\partial Q}{\partial x} - \frac{\partial P}{\partial y} \right) dA = \iint_R (\nabla \times \mathbf{F}) \cdot \hat{\mathbf{k}}\, dA.$$

14. $A = \dfrac{1}{2} \oint_C x\, dy - y\, dx$

$$= \frac{1}{2} \int_{-\pi}^{\pi} [a\cos t(b\cos t\, dt) - b\sin t(-a\sin t\, dt)]$$

$$= \frac{ab}{2} \int_{-\pi}^{\pi} dt = \pi ab$$

16. $A = \dfrac{1}{2} \oint_C x\, dy - y\, dx = \dfrac{1}{2} \int_0^{2\pi} \cos^3\theta(3\sin^2\theta\,\cos\theta\, d\theta) - \sin^3\theta(-3\cos^2\theta\,\sin\theta\, d\theta)$

$$= \frac{3}{2} \int_0^{2\pi} (\cos^4\theta\,\sin^2\theta + \sin^4\theta\,\cos^2\theta)\, d\theta$$

$$= \frac{3}{2} \int_0^{2\pi} \cos^2\theta\,\sin^2\theta\, d\theta = \frac{3}{2} \int_0^{2\pi} \frac{1}{4} \sin^2 2\theta\, d\theta$$

$$= \frac{3}{8} \int_0^{2\pi} \left(\frac{1 - \cos 4\theta}{2} \right) d\theta = \frac{3}{16} \left\{ \theta - \frac{\sin 4\theta}{4} \right\}_0^{2\pi} = \frac{3\pi}{8}$$

18. $A = \dfrac{1}{2} \oint_C x\, dy - y\, dx$

$$= \frac{1}{2} \int_0^{2\pi} (2\cos t + \cos 2t)(2\cos t - 2\cos 2t)\, dt$$

$$\qquad -(2\sin t - \sin 2t)(-2\sin t - 2\sin 2t)\, dt$$

$$= \int_0^{2\pi} (2\cos^2 t - \cos t\,\cos 2t - \cos^2 2t + 2\sin^2 t$$

$$\qquad + \sin t\,\sin 2t - \sin^2 2t)\, dt$$

$$= \int_0^{2\pi} (1 - \cos 3t)\, dt = \left\{ t - \frac{\sin 3t}{3} \right\}_0^{2\pi} = 2\pi$$

20. $A = \dfrac{1}{2} \begin{bmatrix} 1 & 0 \\ 0 & 1 \\ -1 & 0 \\ 0 & -1 \\ 1 & 0 \end{bmatrix}$

$$= \frac{1}{2}[(1+1) - (-1-1)] = 2$$

22. $A = \dfrac{1}{2} \begin{bmatrix} 2 & -2 \\ 5 & 6 \\ -2 & 1 \\ 1 & -3 \\ 2 & -2 \end{bmatrix}$

$= \dfrac{1}{2}[(12 + 5 + 6 - 2) - (-10 - 12 + 1 - 6)] = 24$

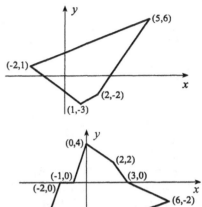

24. $A = \dfrac{1}{2} \begin{bmatrix} 0 & 4 \\ -1 & 0 \\ -2 & 0 \\ -3 & -4 \\ 0 & -5 \\ 6 & -2 \\ 3 & 0 \\ 2 & 2 \\ 0 & 4 \end{bmatrix} = \dfrac{1}{2}[(8 + 15 + 6 + 8) - (-4 - 30 - 6)] = 77/2$

26. By Green's theorem,

$$\oint_C (3x^2y^3 - x^2y)\,dx + (xy^2 + 3x^3y^2)\,dy$$

$$= \iint_R (y^2 + 9x^2y^2 - 9x^2y^2 + x^2)\,dA = \iint_R (x^2 + y^2)\,dA$$

$$= 4\int_0^{\pi/2}\int_0^3 r^2\,r\,dr\,d\theta = 4\int_0^{\pi/2}\left\{\frac{r^4}{4}\right\}_0^3 d\theta = 81\left\{\theta\right\}_0^{\pi/2} = \frac{81\pi}{2}.$$

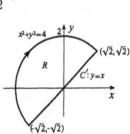

28. Since $\displaystyle\int_{C'} (x - y)(dx + dy) = 0$, we may write

$$\int_C (x - y)(dx + dy) = \oint_{C+C'} (x - y)(dx + dy)$$

$$= -\iint_R (1 + 1)\,dA$$

$$= -2(\text{Area of } R) = -2(2\pi) = -4\pi.$$

30. (a) If we draw curves Γ and Γ' as shown, then P and Q have continuous first partial derivatives in a domain that contains R_1 and its boundary, and also in a domain that contains R_2 and its boundary. If we apply Green's theorem to these regions, denoting their boundaries by $\beta(R_1)$ and $\beta(R_2)$,

$$\oint_{\beta(R_1)} P\,dx + Q\,dy = \iint_{R_1} \left(\frac{\partial Q}{\partial x} - \frac{\partial P}{\partial y}\right) dA,$$

$$\oint_{\beta(R_2)} P\,dx + Q\,dy = \iint_{R_2} \left(\frac{\partial Q}{\partial x} - \frac{\partial P}{\partial y}\right) dA.$$

When these results are added,

$$\oint_{\beta(R_1)} P\,dx + Q\,dy + \oint_{\beta(R_2)} P\,dx + Q\,dy = \iint_{R_1} \left(\frac{\partial Q}{\partial x} - \frac{\partial P}{\partial y}\right) dA + \iint_{R_2} \left(\frac{\partial Q}{\partial x} - \frac{\partial P}{\partial y}\right) dA.$$

Now $R_1 + R_2 = R$. Furthermore, tracing $\beta(R_1)$ and $\beta(R_2)$ in the directions indicated is equivalent to tracing C and C' in the directions indicated, plus Γ and Γ' each traversed once in one direction and then in the reverse direction. Consequently,

$$\oint_C P\,dx + Q\,dy + \oint_{C'} P\,dx + Q\,dy = \iint_R \left(\frac{\partial Q}{\partial x} - \frac{\partial P}{\partial y}\right) dA.$$

(b) In this case we draw curves Γ_i and Γ_i' from C_i to C as shown. This divides R into regions R_i and R' to which we apply Green's theorem,

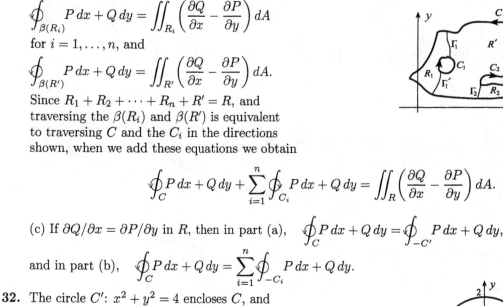

$$\oint_{\beta(R_i)} P\,dx + Q\,dy = \iint_{R_i}\left(\frac{\partial Q}{\partial x} - \frac{\partial P}{\partial y}\right)dA$$

for $i = 1,\dots,n$, and

$$\oint_{\beta(R')} P\,dx + Q\,dy = \iint_{R'}\left(\frac{\partial Q}{\partial x} - \frac{\partial P}{\partial y}\right)dA.$$

Since $R_1 + R_2 + \cdots + R_n + R' = R$, and traversing the $\beta(R_i)$ and $\beta(R')$ is equivalent to traversing C and the C_i in the directions shown, when we add these equations we obtain

$$\oint_C P\,dx + Q\,dy + \sum_{i=1}^{n}\oint_{C_i} P\,dx + Q\,dy = \iint_R\left(\frac{\partial Q}{\partial x} - \frac{\partial P}{\partial y}\right)dA.$$

(c) If $\partial Q/\partial x = \partial P/\partial y$ in R, then in part (a), $\displaystyle\oint_C P\,dx + Q\,dy = \oint_{-C'} P\,dx + Q\,dy,$

and in part (b), $\displaystyle\oint_C P\,dx + Q\,dy = \sum_{i=1}^{n}\oint_{-C_i} P\,dx + Q\,dy.$

32. The circle C': $x^2 + y^2 = 4$ encloses C, and everywhere except at $(0,0)$,

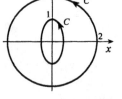

$$\frac{\partial}{\partial x}\left[\frac{x^3}{(x^2+y^2)^2}\right] = \frac{\partial}{\partial y}\left[\frac{-x^2 y}{(x^2+y^2)^2}\right].$$

Hence, by Exercise 30(c), (and using the parametric equations C' : $x = 2\cos t$, $y = 2\sin t$, $-\pi < t \le \pi$),

$$\oint_C \frac{-x^2 y\,dx + x^3\,dy}{(x^2+y^2)^2} = \oint_{C'} \frac{-x^2 y\,dx + x^3\,dy}{(x^2+y^2)^2}$$

$$= \int_{-\pi}^{\pi} \frac{-4\cos^2 t\, 2\sin t(-2\sin t\,dt) + (2\cos t)^3(2\cos t\,dt)}{16}$$

$$= \int_{-\pi}^{\pi} (\cos^2 t\,\sin^2 t + \cos^4 t)\,dt = \int_{-\pi}^{\pi}\cos^2 t\,dt$$

$$= \int_{-\pi}^{\pi}\left(\frac{1+\cos 2t}{2}\right)dt = \frac{1}{2}\left\{t + \frac{\sin 2t}{2}\right\}_{-\pi}^{\pi} = \pi.$$

34. Suppose C is a curve enclosing the origin in the counterclockwise sense. It is always possible to find a circle C' of radius $r > 0$ centred at the origin which is interior to C. Since $\partial Q/\partial x = \partial P/\partial y$ in a domain containing C and C' and the area between them, it follows by Exercise 30(c) that

$$\oint_C \frac{-y\,dx + x\,dy}{x^2+y^2} = \oint_{C'} \frac{-y\,dx + x\,dy}{x^2+y^2}$$

$$= \int_{-\pi}^{\pi} \frac{(-r\sin t)(-r\sin t\,dt) + r\cos t(r\cos t\,dt)}{r^2}$$

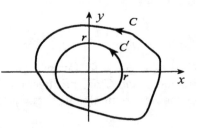

$$= \int_{-\pi}^{\pi} dt = 2\pi,$$

where we have used the parametric equations C' : $x = r\cos t$, $y = r\sin t$, $-\pi < t \le \pi$. If C encloses the origin in the opposite direction, then the value of the line integral is -2π.

36. According to Exercise 13,

$$\oint_C \frac{\partial P}{\partial n} ds = \oint_C \nabla P \cdot \hat{\mathbf{n}} \, ds = \iint_R \nabla \cdot \nabla P \, dA = \iint_R \left[\frac{\partial}{\partial x}\left(\frac{\partial P}{\partial x}\right) + \frac{\partial}{\partial y}\left(\frac{\partial P}{\partial y}\right) \right] dA = \iint_R \nabla^2 P \, dA.$$

If P satisfies Laplace's equation in R, then $\nabla^2 P = 0$, and $\oint_C \dfrac{\partial P}{\partial n} ds = 0$.

38. If we reverse the roles of P and Q in Exercise 37,

$$\oint_C \frac{Q \partial P}{\partial n} ds = \iint_R Q\nabla^2 P \, dA + \iint_R \nabla Q \cdot \nabla P \, dA.$$

When we subtract this result from that in Exercise 37, we obtain

$$\oint_C \left(P\frac{\partial Q}{\partial n} - Q\frac{\partial P}{\partial n} \right) ds = \iint_R (P\nabla^2 Q - Q\nabla^2 P) \, dA.$$

EXERCISES 14.7

2. $\displaystyle \iint_S (x^2 + y^2)z \, dS$

$$= \iint_{S_{xy}} (x^2 + y^2)(x + y)\sqrt{1 + \left(\frac{\partial z}{\partial x}\right)^2 + \left(\frac{\partial z}{\partial y}\right)^2} \, dA$$

$$= \iint_{S_{xy}} (x^2 + y^2)(x + y)\sqrt{1 + (1)^2 + (1)^2} \, dA$$

$$= \sqrt{3} \int_0^1 \int_0^{1-x} (x^3 + xy^2 + x^2 y + y^3) \, dy \, dx$$

$$= \sqrt{3} \int_0^1 \left\{ x^3 y + \frac{xy^3}{3} + \frac{x^2 y^2}{2} + \frac{y^4}{4} \right\}_0^{1-x} dx$$

$$= \frac{\sqrt{3}}{12} \int_0^1 [4x - 6x^2 + 12x^3 - 10x^4 + 3(1-x)^4] \, dx$$

$$= \frac{\sqrt{3}}{12} \left\{ 2x^2 - 2x^3 + 3x^4 - 2x^5 - \frac{3}{5}(1-x)^5 \right\}_0^1 = \frac{2\sqrt{3}}{15}$$

4. $\displaystyle \iint_S xy \, dS = \iint_{S_{xy}} xy\sqrt{1 + \left(\frac{\partial z}{\partial x}\right)^2 + \left(\frac{\partial z}{\partial y}\right)^2} \, dA$

$$= \iint_{S_{xy}} xy\sqrt{1 + \left(\frac{x}{\sqrt{x^2 + y^2}}\right)^2 + \left(\frac{y}{\sqrt{x^2 + y^2}}\right)^2} \, dA$$

$$= \sqrt{2} \iint_{S_{xy}} xy \, dA = \sqrt{2} \int_0^1 \int_0^{\sqrt{1-x^2}} xy \, dy \, dx$$

$$= \sqrt{2} \int_0^1 \left\{ \frac{xy^2}{2} \right\}_0^{\sqrt{1-x^2}} dx = \frac{1}{\sqrt{2}} \int_0^1 (x - x^3) \, dx$$

$$= \frac{1}{\sqrt{2}} \left\{ \frac{x^2}{2} - \frac{x^4}{4} \right\}_0^1 = \frac{1}{4\sqrt{2}}$$

6.
$$\iint_S \sqrt{4y+1}\, dS = \iint_{S_{xz}} \sqrt{4x^2+1}\sqrt{1+\left(\frac{\partial y}{\partial x}\right)^2+\left(\frac{\partial y}{\partial z}\right)^2}\, dA = \iint_{S_{xz}} \sqrt{4x^2+1}\sqrt{1+(2x)^2}\, dA$$

$$= \iint_{S_{xz}} (1+4x^2)\, dA = \int_0^{\sqrt{2}-1} \int_0^{1-2x-x^2} (1+4x^2)\, dz\, dx$$

$$= \int_0^{\sqrt{2}-1} (1+4x^2)(1-2x-x^2)\, dx$$

$$= \int_0^{\sqrt{2}-1} (1-2x+3x^2-8x^3-4x^4)\, dx$$

$$= \left\{ x - x^2 + x^3 - 2x^4 - \frac{4x^5}{5} \right\}_0^{\sqrt{2}-1} = \frac{44\sqrt{2}-61}{5}$$

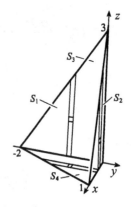

8.
$$\iint_S (x+y)\, dS = \iint_{S_1} (x+y)\, dS + \iint_{S_2} (x+y)\, dS$$
$$+ \iint_{S_3} (x+y)\, dS + \iint_{S_4} (x+y)\, dS$$
$$= \iint_{S_{1xy}} (x+y)\sqrt{1+(-3)^2+(3/2)^2}\, dA$$
$$+ \iint_{S_{2xz}} x\, dA + \iint_{S_{3yz}} y\, dA + \iint_{S_{4xy}} (x+y)\, dA.$$

Since $S_{1xy} = S_{4xy}$,

$$\iint_S (x+y)\, dS = \frac{9}{2}\iint_{S_{1xy}} (x+y)\, dA + \iint_{S_{2xz}} x\, dA + \iint_{S_{3yz}} y\, dA$$

$$= \frac{9}{2}\int_0^1 \int_{2x-2}^0 (x+y)\, dy\, dx + \int_0^1 \int_0^{3-3x} x\, dz\, dx + \int_{-2}^0 \int_0^{3+3y/2} y\, dz\, dy$$

$$= \frac{9}{2}\int_0^1 \left\{ xy + \frac{y^2}{2} \right\}_{2x-2}^0 dx + \int_0^1 \left\{ xz \right\}_0^{3-3x} dx + \int_{-2}^0 \left\{ yz \right\}_0^{3+3y/2} dy$$

$$= \frac{9}{4}\int_0^1 [-4x^2+4x-4(x-1)^2]\, dx + 3\int_0^1 (x-x^2)\, dx + \frac{3}{2}\int_{-2}^0 (2y+y^2)\, dy$$

$$= \frac{9}{4}\left\{ \frac{-4x^3}{3} + 2x^2 - \frac{4(x-1)^3}{3} \right\}_0^1 + 3\left\{ \frac{x^2}{2} - \frac{x^3}{3} \right\}_0^1 + \frac{3}{2}\left\{ y^2 + \frac{y^3}{3} \right\}_{-2}^0 = -3.$$

10. If S is that portion of the cone in the first octant,

$$\text{Area} = 4\iint_S dS = 4\iint_{S_{xy}} \sqrt{1+\left(\frac{\partial z}{\partial x}\right)^2+\left(\frac{\partial z}{\partial y}\right)^2}\, dA$$

$$= 4\iint_{S_{xy}} \sqrt{1+\left(\frac{hx}{R\sqrt{x^2+y^2}}\right)^2+\left(\frac{hy}{R\sqrt{x^2+y^2}}\right)^2}\, dA$$

$$= \frac{4\sqrt{R^2+h^2}}{R}\iint_{S_{xy}} dA = \frac{4\sqrt{R^2+h^2}}{R}(\text{Area of } S_{xy})$$

$$= \frac{4\sqrt{R^2+h^2}}{R}\left(\frac{1}{4}\pi R^2\right) = \pi R\sqrt{R^2+h^2}.$$

12. $\displaystyle\iint_S xyz\,dS = \iint_{S_{xz}} xyz\sqrt{1 + \left(\frac{\partial y}{\partial x}\right)^2 + \left(\frac{\partial y}{\partial z}\right)^2}\,dA$

$\displaystyle = \iint_{S_{xz}} xyz\sqrt{1 + \left(\frac{-1}{4\sqrt{9-x}}\right)^2}\,dA$

$\displaystyle = \iint_{S_{xz}} xyz\sqrt{\frac{145 - 16x}{16(9-x)}}\,dA$

$\displaystyle = \frac{1}{4}\int_0^9\int_0^3 xz\frac{\sqrt{9-x}}{2}\sqrt{\frac{145-16x}{9-x}}\,dz\,dx$

$\displaystyle = \frac{1}{8}\int_0^9\left\{x\sqrt{145-16x}\,\frac{z^2}{2}\right\}_0^3 dx = \frac{9}{16}\int_0^9 x\sqrt{145-16x}\,dx$

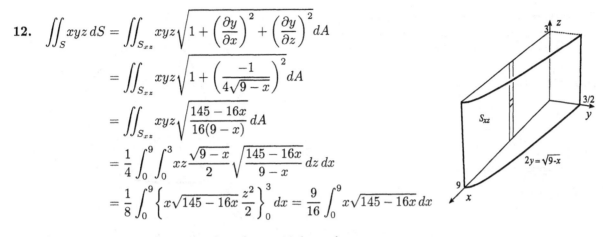

If we now set $u = 145 - 16x$, then $du = -16\,dx$, and

$\displaystyle \iint_S xyz\,dS = \frac{9}{16}\int_{145}^1\left(\frac{145-u}{16}\right)\sqrt{u}\left(\frac{du}{-16}\right) = \frac{-9}{4096}\left\{\frac{290u^{3/2}}{3} - \frac{2u^{5/2}}{5}\right\}_{145}^1 = \frac{3(145^{5/2} - 361)}{5120}.$

14. We quadruple the integral over that part of the surface in the first octant.

$\displaystyle \iint_S z\,dS = 4\iint_{S_{xy}} z\sqrt{1 + \left(\frac{\partial z}{\partial x}\right)^2 + \left(\frac{\partial z}{\partial y}\right)^2}\,dA$

$\displaystyle = 4\iint_{S_{xy}} z\sqrt{1 + \left(\frac{x}{z}\right)^2 + \left(\frac{y}{z}\right)^2}\,dA$

$\displaystyle = 4\iint_{S_{xy}} z\sqrt{\frac{x^2+y^2+z^2}{z^2}}\,dA = 4\iint_{S_{xy}}\sqrt{2x^2+2y^2-1}\,dA$

$\displaystyle = 4\int_0^{\pi/2}\int_1^{\sqrt{2}}\sqrt{2r^2-1}\,r\,dr\,d\theta = 4\int_0^{\pi/2}\left\{\frac{(2r^2-1)^{3/2}}{6}\right\}_1^{\sqrt{2}}d\theta$

$\displaystyle = \frac{2(3\sqrt{3}-1)}{3}\left\{\theta\right\}_0^{\pi/2} = \frac{\pi(3\sqrt{3}-1)}{3}$

16. We quadruple the integral over that part of the surface in the first octant.

$\displaystyle \iint_S x^2\,dS = 4\iint_{S_{xy}} x^2\sqrt{1 + \left(\frac{\partial z}{\partial x}\right)^2 + \left(\frac{\partial z}{\partial y}\right)^2}\,dA$

$\displaystyle = 4\iint_{S_{xy}} x^2\sqrt{1 + y^2 + x^2}\,dA$

$\displaystyle = 4\int_0^2\int_0^{\pi/2} r^2\cos^2\theta\sqrt{1+r^2}\,r\,d\theta\,dr$

$\displaystyle = 4\int_0^2\int_0^{\pi/2} r^3\sqrt{1+r^2}\left(\frac{1+\cos 2\theta}{2}\right)d\theta\,dr$

$\displaystyle = 2\int_0^2\left\{r^3\sqrt{1+r^2}\left(\theta + \frac{\sin 2\theta}{2}\right)\right\}_0^{\pi/2}dr = \pi\int_0^2 r^3\sqrt{1+r^2}\,dr$

If we set $u = 1 + r^2$, then $du = 2r\,dr$, and

$\displaystyle \iint_S x^2\,dS = \pi\int_1^5 (u-1)\sqrt{u}\left(\frac{du}{2}\right) = \frac{\pi}{2}\int_1^5 (u^{3/2} - \sqrt{u})\,du = \frac{\pi}{2}\left\{\frac{2u^{5/2}}{5} - \frac{2u^{3/2}}{3}\right\}_1^5 = \frac{(50\sqrt{5}+2)\pi}{15}.$

18. The surface integral over S is eight times that over that part of the the upper hemisphere $z = \sqrt{R^2 - x^2 - y^2}$ in the first octant. Since

$$dS = \sqrt{1 + \left(\frac{\partial z}{\partial x}\right)^2 + \left(\frac{\partial z}{\partial y}\right)^2} \, dA$$

$$= \sqrt{1 + \frac{x^2}{R^2 - x^2 - y^2} + \frac{y^2}{R^2 - x^2 - y^2}} \, dA$$

$$= \frac{R}{\sqrt{R^2 - x^2 - y^2}} \, dA,$$

it follows that

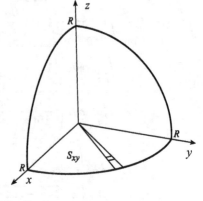

$$\iint_S dS = 8 \iint_{S_{xy}} \frac{R}{\sqrt{R^2 - x^2 - y^2}} \, dA = 8R \int_0^{\pi/2} \int_0^R \frac{1}{\sqrt{R^2 - r^2}} \, r \, dr \, d\theta$$

$$= 8R \int_0^{\pi/2} \left\{ -\sqrt{R^2 - r^2} \right\}_0^R d\theta = 8R^2 \left\{ \theta \right\}_0^{\pi/2} = 4\pi R^2.$$

Alternatively, using area element 14.56,

$$\iint_S dS = 8 \int_0^{\pi/2} \int_0^{\pi/2} R^2 \sin\phi \, d\phi \, d\theta = 8R^2 \int_0^{\pi/2} \left\{ -\cos\phi \right\}_0^{\pi/2} d\theta = 8R^2 \left\{ \theta \right\}_0^{\pi/2} = 4\pi R^2.$$

20. The hemisphere projects one-to-one onto the circle $S_{xy} : x^2 + y^2 \leq 9$, $z = 0$. Since

$$dS = \sqrt{1 + \left(\frac{\partial z}{\partial x}\right)^2 + \left(\frac{\partial z}{\partial y}\right)^2} \, dA$$

$$= \sqrt{1 + \frac{x^2}{9 - x^2 - y^2} + \frac{y^2}{9 - x^2 - y^2}} \, dA$$

$$= \frac{3}{\sqrt{9 - x^2 - y^2}} \, dA,$$

it follows that

$$\iint_S (x^2 - y^2) \, dS = \iint_{S_{xy}} \frac{3(x^2 - y^2)}{\sqrt{9 - x^2 - y^2}} \, dA = 3 \int_0^3 \int_{-\pi}^\pi \frac{(r^2 \cos^2\theta - r^2 \sin^2\theta)}{\sqrt{9 - r^2}} \, r \, d\theta \, dr$$

$$= 3 \int_0^3 \int_{-\pi}^\pi \frac{r^3}{\sqrt{9 - r^2}} \cos 2\theta \, d\theta \, dr = 3 \int_0^3 \left\{ \frac{r^3}{\sqrt{9 - r^2}} \frac{\sin 2\theta}{2} \right\}_{-\pi}^\pi dr = 0.$$

Alternatively, using $dS = R^2 \sin\phi \, d\phi \, d\theta$, with $R = 3$,

$$\iint_S (x^2 - y^2) \, dS = \int_{-\pi}^\pi \int_0^{\pi/2} (9 \sin^2\phi \cos^2\theta - 9 \sin^2\phi \sin^2\theta) \, 9 \sin\phi \, d\phi \, d\theta$$

$$= 81 \int_{-\pi}^\pi \int_0^{\pi/2} \sin^3\phi \cos 2\theta \, d\phi \, d\theta = 81 \int_{-\pi}^\pi \int_0^{\pi/2} \sin\phi (1 - \cos^2\phi) \cos 2\theta \, d\phi \, d\theta$$

$$= 81 \int_{-\pi}^\pi \left\{ \left(-\cos\phi + \frac{\cos^3\phi}{3} \right) \cos 2\theta \right\}_0^{\pi/2} d\theta = 54 \left\{ \frac{\sin 2\theta}{2} \right\}_{-\pi}^\pi = 0.$$

22. The first octant part of the surface projects one-to-one onto the area $S_{xy} : 3R^2 \leq x^2 + y^2 \leq 4R^2$ in the first quadrant of the xy-plane. Since

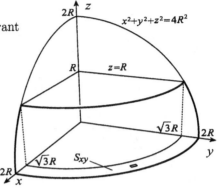

$$dS = \sqrt{1 + \left(\frac{\partial z}{\partial x}\right)^2 + \left(\frac{\partial z}{\partial y}\right)^2}\, dA$$

$$= \sqrt{1 + \frac{x^2}{4R^2 - x^2 - y^2} + \frac{y^2}{4R^2 - x^2 - y^2}}\, dA$$

$$= \frac{2R}{\sqrt{4R^2 - x^2 - y^2}}\, dA,$$

it follows that

$$\iint_S \frac{1}{x^2 + y^2}\, dS = 4 \iint_{S_{xy}} \frac{1}{x^2 + y^2}\, \frac{2R}{\sqrt{4R^2 - x^2 - y^2}}\, dA$$

$$= 8R \int_{\sqrt{3}R}^{2R} \int_0^{\pi/2} \frac{1}{r^2\sqrt{4R^2 - r^2}}\, r\, d\theta\, dr = 4\pi R \int_{\sqrt{3}R}^{2R} \frac{1}{r\sqrt{4R^2 - r^2}}\, dr.$$

If we set $r = 2R\sin\phi$ and $dr = 2R\cos\phi\, d\phi$, then

$$\iint_S \frac{1}{x^2 + y^2}\, dS = 4\pi R \int_{\pi/3}^{\pi/2} \frac{1}{2R\sin\phi\, 2R\cos\phi}\, 2R\cos\phi\, d\phi = 2\pi \int_{\pi/3}^{\pi/2} \csc\phi\, d\phi$$

$$= 2\pi\left\{ \ln|\csc\phi - \cot\phi| \right\}_{\pi/3}^{\pi/2} = \pi\ln 3.$$

Alternatively, using area element 14.56,

$$\iint_S \frac{1}{x^2 + y^2}\, dS = 4 \int_0^{\pi/2} \int_{\pi/3}^{\pi/2} \frac{1}{4R^2\sin^2\phi\cos^2\theta + 4R^2\sin^2\phi\sin^2\theta}\, 4R^2\sin\phi\, d\phi\, d\theta$$

$$= 4 \int_0^{\pi/2} \int_{\pi/3}^{\pi/2} \csc\phi\, d\phi\, d\theta = 4 \int_0^{\pi/2} \left\{ \ln|\csc\phi - \cot\phi| \right\}_{\pi/3}^{\pi/2} d\theta = \pi\ln 3.$$

24. If S projects one-to-one onto S_{xy}, then an element of area dS on S is related to its projection dA in the xy-plane by $dS = \sqrt{1 + \left(\dfrac{\partial z}{\partial x}\right)^2 + \left(\dfrac{\partial z}{\partial y}\right)^2}\, dA$, where

$$\frac{\partial z}{\partial x} = -\frac{\dfrac{\partial(F)}{\partial(x)}}{\dfrac{\partial(F)}{\partial(z)}} = -\frac{F_x}{F_z} \quad \text{and} \quad \frac{\partial z}{\partial y} = -\frac{\dfrac{\partial(F)}{\partial(y)}}{\dfrac{\partial(F)}{\partial(z)}} = -\frac{F_y}{F_z}.$$

Thus, $dS = \sqrt{1 + \left(-\dfrac{F_x}{F_z}\right)^2 + \left(-\dfrac{F_y}{F_z}\right)^2}\, dA = \dfrac{\sqrt{(F_x)^2 + (F_y)^2 + (F_z)^2}}{|F_z|}\, dA = \dfrac{|\nabla F|}{|F_z|}\, dA$, and

$$\iint_S f(x, y, z)\, dS = \iint_{S_{xy}} f[x, y, g(x, y)]\frac{|\nabla F|}{|F_z|}\, dA,$$

where $z = g(x, y)$ on S.

EXERCISES 14.8

2. Since $\hat{\mathbf{n}} = \dfrac{\nabla(y - x^2)}{|\nabla(y - x^2)|} = \dfrac{(-2x, 1, 0)}{\sqrt{4x^2 + 1}}$,

$$\iint_S (yz^2\hat{\mathbf{i}} + ye^x\hat{\mathbf{j}} + x\hat{\mathbf{k}}) \cdot \hat{\mathbf{n}}\, dS = \iint_{S_{xz}} \frac{-2xyz^2 + ye^x}{\sqrt{4x^2 + 1}} \sqrt{1 + \left(\frac{\partial y}{\partial x}\right)^2 + \left(\frac{\partial y}{\partial z}\right)^2}\, dA$$

$$= \iint_{S_{xz}} \frac{-2x^3z^2 + x^2e^x}{\sqrt{4x^2 + 1}} \sqrt{1 + (2x)^2}\, dA$$

$$= \int_{-2}^{2} \int_0^1 (x^2e^x - 2x^3z^2)\, dz\, dx = \int_{-2}^{2} \left\{ x^2ze^x - \frac{2x^3z^3}{3} \right\}_0^1 dx$$

$$= \int_{-2}^{2} \left(x^2e^x - \frac{2x^3}{3} \right) dx = \left\{ x^2e^x - 2xe^x + 2e^x - \frac{x^4}{6} \right\}_{-2}^{2}$$

$$= 2e^2 - 10e^{-2}.$$

4. Since $\hat{\mathbf{n}} = \dfrac{\nabla(x^2 + y^2 - z)}{|\nabla(x^2 + y^2 - z)|} = \dfrac{(2x, 2y, -1)}{\sqrt{1 + 4x^2 + 4y^2}}$,

$$\iint_S (yz\hat{\mathbf{i}} + zx\hat{\mathbf{j}} + xy\hat{\mathbf{k}}) \cdot \hat{\mathbf{n}}\, dS$$

$$= \iint_{S_{xy}} \frac{2xyz + 2xyz - xy}{\sqrt{1 + 4x^2 + 4y^2}} \sqrt{1 + \left(\frac{\partial z}{\partial x}\right)^2 + \left(\frac{\partial z}{\partial y}\right)^2}\, dA$$

$$= \iint_{S_{xy}} \frac{4xy(x^2 + y^2) - xy}{\sqrt{1 + 4x^2 + 4y^2}} \sqrt{1 + (2x)^2 + (2y)^2}\, dA$$

$$= \int_{-1}^{1} \int_{-1}^{1} (4x^3y + 4xy^3 - xy)\, dy\, dx$$

$$= \int_{-1}^{1} \left\{ 2x^3y^2 + xy^4 - \frac{xy^2}{2} \right\}_{-1}^{1} dx = 0.$$

6. Since $\hat{\mathbf{n}} = \dfrac{\nabla(x^2 + y^2 - z^2)}{|\nabla(x^2 + y^2 - z^2)|}$

$$= \frac{(2x, 2y, -2z)}{\sqrt{4x^2 + 4y^2 + 4z^2}} = \frac{(x, y, -z)}{\sqrt{2}\, z},$$

$$\iint_S (x\hat{\mathbf{i}} + y\hat{\mathbf{j}}) \cdot \hat{\mathbf{n}}\, dS = \iint_{S_{xy}} \frac{x^2 + y^2}{\sqrt{2}\, z} \sqrt{1 + \left(\frac{\partial z}{\partial x}\right)^2 + \left(\frac{\partial z}{\partial y}\right)^2}\, dA$$

$$= \frac{1}{\sqrt{2}} \iint_{S_{xy}} \frac{x^2 + y^2}{z} \sqrt{1 + \left(\frac{x}{\sqrt{x^2 + y^2}}\right)^2 + \left(\frac{y}{\sqrt{x^2 + y^2}}\right)^2}\, dA$$

$$= \frac{1}{\sqrt{2}} \iint_{S_{xy}} \frac{x^2 + y^2}{\sqrt{x^2 + y^2}} \sqrt{2}\, dA = 4 \int_0^{\pi/2} \int_0^1 r\, r\, dr\, d\theta = 4 \int_0^{\pi/2} \left\{ \frac{r^3}{3} \right\}_0^1 d\theta = \frac{4}{3}\left\{ \theta \right\}_0^{\pi/2} = \frac{2\pi}{3}.$$

8. Since $\hat{n} = \dfrac{\nabla(2 - x^2 - y^2 - z)}{|\nabla(2 - x^2 - y^2 - z)|} = \dfrac{(-2x, -2y, -1)}{\sqrt{1 + 4x^2 + 4y^2}}$,

$$\iint_S (x^2 y\hat{i} + xy\hat{j} + z\hat{k}) \cdot \hat{n} \, dS$$

$$= \iint_{S_{xy}} \frac{-2x^3 y - 2xy^2 - z}{\sqrt{1 + 4x^2 + 4y^2}} \sqrt{1 + \left(\frac{\partial z}{\partial x}\right)^2 + \left(\frac{\partial z}{\partial y}\right)^2} \, dA$$

$$= \iint_{S_{xy}} \frac{-(2x^3 y + 2xy^2 + 2 - x^2 - y^2)}{\sqrt{1 + 4x^2 + 4y^2}} \sqrt{1 + (-2x)^2 + (-2y)^2} \, dA$$

$$= -\iint_{S_{xy}} [2x^3 y + 2xy^2 + 2 - (x^2 + y^2)] \, dA.$$

Because the first two terms are odd functions of x and S_{xy} is symmetric about the y-axis, their double integral vanishes, and

$$\iint_S (x^2 y\hat{i} + xy\hat{j} + z\hat{k}) \cdot \hat{n} \, dS = -\int_{-\pi}^{\pi} \int_0^{\sqrt{2}} (2 - r^2) r \, dr \, d\theta = -\int_{-\pi}^{\pi} \left\{ r^2 - \frac{r^4}{4} \right\}_0^{\sqrt{2}} d\theta = -\left\{ \theta \right\}_{-\pi}^{\pi} = -2\pi.$$

10. Since $\hat{n} = \dfrac{\nabla(x^2 + y^2 + z^2 - 4)}{|\nabla(x^2 + y^2 + z^2 - 4)|} = \dfrac{(2x, 2y, 2z)}{\sqrt{4x^2 + 4y^2 + 4z^2}} = \dfrac{(x, y, z)}{2}$,

$$\iint_S (x\hat{i} + y\hat{j}) \cdot \hat{n} \, dS$$

$$= \iint_{S_{xy}} \frac{x^2 + y^2}{2} \sqrt{1 + \left(\frac{\partial z}{\partial x}\right)^2 + \left(\frac{\partial z}{\partial y}\right)^2} \, dA$$

$$= \frac{1}{2} \iint_{S_{xy}} (x^2 + y^2) \sqrt{1 + \left(\frac{-x}{z}\right)^2 + \left(\frac{-y}{z}\right)^2} \, dA$$

$$= \frac{1}{2} \iint_{S_{xy}} (x^2 + y^2) \left(\frac{2}{z}\right) dA$$

$$= \iint_{S_{xy}} \frac{x^2 + y^2}{\sqrt{4 - x^2 - y^2}} \, dA = \int_0^{\sqrt{3}} \int_{-\pi}^{\pi} \frac{r^2}{\sqrt{4 - r^2}} r \, d\theta \, dr = 2\pi \int_0^{\sqrt{3}} \frac{r^3}{\sqrt{4 - r^2}} \, dr.$$

If we set $u = 4 - r^2$ and $du = -2r \, dr$, then

$$\iint_S (x\hat{i} + y\hat{j}) \cdot \hat{n} \, dS = 2\pi \int_4^1 \frac{(4 - u)}{\sqrt{u}} \left(-\frac{du}{2}\right) = \pi \int_1^4 \left(\frac{4}{\sqrt{u}} - \sqrt{u}\right) du = \pi \left\{ 8\sqrt{u} - \frac{2u^{3/2}}{3} \right\}_1^4 = \frac{10\pi}{3}.$$

12. Since $\hat{n} = \dfrac{\nabla(x^2 + y^2 + z^2 - 1)}{|\nabla(x^2 + y^2 + z^2 - 1)|} = \dfrac{(2x, 2y, 2z)}{\sqrt{4x^2 + 4y^2 + 4z^2}} = (x, y, z)$,

$$\iint_S (y\hat{i} - x\hat{j} + \hat{k}) \cdot \hat{n} \, dS$$

$$= \iint_{S_{xy}} (yx - xy + z) \sqrt{1 + \left(\frac{\partial z}{\partial x}\right)^2 + \left(\frac{\partial z}{\partial y}\right)^2} \, dA$$

$$= \iint_{S_{xy}} z \sqrt{1 + \left(\frac{-x}{z}\right)^2 + \left(\frac{-y}{z}\right)^2} \, dA$$

$$= \iint_{S_{xy}} dA = \text{Area of } S_{xy}.$$

Since S_{xy} is the region inside the ellipse $2x^2 + 4(y - 1/2)^2 = 1$,

$$\iint_S (y\hat{\mathbf{i}} - x\hat{\mathbf{j}} + \hat{\mathbf{k}}) \cdot \hat{\mathbf{n}}\, dS = \pi \left(\frac{1}{\sqrt{2}}\right)\left(\frac{1}{2}\right) = \frac{\pi}{2\sqrt{2}}.$$

14. The surface projects one-to-one onto the area S_{xy} in the xy-plane shown. Since

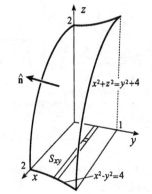

$$dS = \sqrt{1 + \left(\frac{\partial z}{\partial x}\right)^2 + \left(\frac{\partial z}{\partial y}\right)^2}\, dA$$

$$= \sqrt{1 + \frac{x^2}{4 + y^2 - x^2} + \frac{y^2}{4 + y^2 - x^2}}\, dA$$

$$= \sqrt{\frac{4 + 2y^2}{4 + y^2 - x^2}}\, dA,$$

and $\hat{\mathbf{n}} = \dfrac{\nabla(x^2 + z^2 - y^2 - 4)}{|\nabla(x^2 + z^2 - y^2 - 4)|} = \dfrac{(2x, -2y, 2z)}{\sqrt{4x^2 + 4y^2 + 4z^2}} = \dfrac{(x, -y, z)}{\sqrt{x^2 + y^2 + 4 + y^2 - x^2}} = \dfrac{(x, -y, z)}{\sqrt{4 + 2y^2}},$

it follows that

$$\iint_S (x^2\hat{\mathbf{i}} + xy\hat{\mathbf{j}} + xz\hat{\mathbf{k}}) \cdot \hat{\mathbf{n}}\, dS = \iint_{S_{xy}} \frac{x^3 - xy^2 + xz^2}{\sqrt{4 + 2y^2}} \sqrt{\frac{4 + 2y^2}{4 + y^2 - x^2}}\, dA$$

$$= \iint_{S_{xy}} \frac{4x}{\sqrt{4 + y^2 - x^2}}\, dA = 4\int_0^1 \int_0^{\sqrt{4+y^2}} \frac{x}{\sqrt{4 + y^2 - x^2}}\, dx\, dy$$

$$= 4\int_0^1 \left\{-\sqrt{4 + y^2 - x^2}\right\}_0^{\sqrt{4+y^2}}\, dy = 4\int_0^1 \sqrt{4 + y^2}\, dy.$$

If we set $y = 2\tan\theta$ and $dy = 2\sec^2\theta\, d\theta$, then

$$\iint_S (x^2\hat{\mathbf{i}} + xy\hat{\mathbf{j}} + xz\hat{\mathbf{k}}) \cdot \hat{\mathbf{n}}\, dS = 4\int_0^{\overline{\theta}} 2\sec\theta\, 2\sec^2\theta\, d\theta \quad (\overline{\theta} = \text{Tan}^{-1}(1/2))$$

$$= 8\left\{\sec\theta \tan\theta + \ln|\sec\theta + \tan\theta|\right\}_0^{\overline{\theta}} \quad \text{(see Example 8.9)}$$

$$= 2\sqrt{5} + 8\ln[(\sqrt{5} + 1)/2].$$

16. We divide S into two parts

$S_1 : x^2 + y^2/4 + z^2 = 1, \ z \geq 0,$
$S_2 : x^2 + y^2/4 + z^2 = 1, \ z \leq 0,$

both of which project onto the ellipse

$$S_{xy} : x^2 + \frac{y^2}{4} \leq 1, \ z = 0.$$

We have shown one-quarter of S_1 and S_{xy} in the figure. On S_1, $\partial z/\partial x = -x/z$ and $\partial z/\partial y = -y/(4z)$, so that

$$dS = \sqrt{1 + \left(\frac{x^2}{z^2}\right) + \left(\frac{y^2}{16z^2}\right)}\, dA = \frac{\sqrt{16z^2 + 16x^2 + y^2}}{4z}\, dA.$$

Since $\hat{\mathbf{n}} = \dfrac{\nabla(x^2 + y^2/4 + z^2 - 1)}{|\nabla(x^2 + y^2/4 + z^2 - 1)|} = \dfrac{(2x, y/2, 2z)}{\sqrt{4x^2 + y^2/4 + 4z^2}} = \dfrac{(4x, y, 4z)}{\sqrt{16x^2 + y^2 + 16z^2}},$

$$\iint_{S_1} (yx\hat{\mathbf{i}} + y^2\hat{\mathbf{j}} + yz\hat{\mathbf{k}}) \cdot \hat{\mathbf{n}}\, dS = \iint_{S_{xy}} \frac{(4x^2y + y^3 + 4yz^2)}{\sqrt{16x^2 + y^2 + 16z^2}} \frac{\sqrt{16z^2 + 16x^2 + y^2}}{4z}\, dA$$

$$= \iint_{S_{xy}} \frac{y}{z}\, dA = \iint_{S_{xy}} \frac{y}{\sqrt{1 - x^2 - y^2/4}}\, dA.$$

On S_2, $dS = \dfrac{\sqrt{16z^2 + 16x^2 + y^2}}{-4z}\, dA$ and $\hat{\mathbf{n}} = \dfrac{(4x, y, 4z)}{\sqrt{16x^2 + y^2 + 16z^2}}$, and therefore

$$\iint_{S_2} (yx\hat{\mathbf{i}} + y^2\hat{\mathbf{j}} + yz\hat{\mathbf{k}}) \cdot \hat{\mathbf{n}}\, dS = \iint_{S_{xy}} \frac{y}{-z}\, dA = \iint_{S_{xy}} \frac{y}{\sqrt{1 - x^2 - y^2/4}}\, dA.$$

Hence, $\displaystyle\iint_{S} (yx\hat{\mathbf{i}} + y^2\hat{\mathbf{j}} + yz\hat{\mathbf{k}}) \cdot \hat{\mathbf{n}}\, dS = 2 \iint_{S_{xy}} \frac{y}{\sqrt{1 - x^2 - y^2/4}}\, dA = 0$, because the integrand is an odd function of y and S_{xy} is symmetric about the x-axis.

18. (a) Since $\hat{\mathbf{n}} = \dfrac{\nabla(x^2 + y^2 + z - 9)}{|\nabla(x^2 + y^2 + z - 9)|}$

$$= \frac{(2x, 2y, 1)}{\sqrt{1 + 4x^2 + 4y^2}},$$

$\displaystyle\iint_{S} (y\hat{\mathbf{i}} - x\hat{\mathbf{j}} + z\hat{\mathbf{k}}) \cdot \hat{\mathbf{n}}\, dS$

$$= \iint_{S_{xy}} \frac{(2yx - 2xy + z)}{\sqrt{1 + 4x^2 + 4y^2}} \sqrt{1 + \left(\frac{\partial z}{\partial x}\right)^2 + \left(\frac{\partial z}{\partial y}\right)^2}\, dA$$

$$= \iint_{S_{xy}} \frac{z}{\sqrt{1 + 4x^2 + 4y^2}} \sqrt{1 + (-2x)^2 + (-2y)^2}\, dA$$

$$= \iint_{S_{xy}} (9 - x^2 - y^2)\, dA.$$

If we set up polar coordinates with the pole at $(0, -1)$ and polar axis parallel to the positive x-axis, then $x = r\cos\theta$ and $y = -1 + r\sin\theta$, and

$$\iint_{S} (y\hat{\mathbf{i}} - x\hat{\mathbf{j}} + z\hat{\mathbf{k}}) \cdot \hat{\mathbf{n}}\, dS = \int_{-\pi}^{\pi} \int_{0}^{\sqrt{10}} [9 - r^2\cos^2\theta - (-1 + r\sin\theta)^2]\, r\, dr\, d\theta$$

$$= \int_{-\pi}^{\pi} \int_{0}^{\sqrt{10}} (8 + 2r\sin\theta - r^2)\, r\, dr\, d\theta = \int_{-\pi}^{\pi} \left\{ 4r^2 + \frac{2r^3\sin\theta}{3} - \frac{r^4}{4} \right\}_{0}^{\sqrt{10}} d\theta$$

$$= \int_{-\pi}^{\pi} \left(15 + \frac{20\sqrt{10}}{3}\sin\theta \right) d\theta = \left\{ 15\theta - \frac{20\sqrt{10}}{3}\cos\theta \right\}_{-\pi}^{\pi} = 30\pi.$$

(b) With $\hat{\mathbf{n}} = \dfrac{\nabla(z - 2y)}{|\nabla(z - 2y)|} = \dfrac{(0, -2, 1)}{\sqrt{5}}$,

$$\iint_{S} (y\hat{\mathbf{i}} - x\hat{\mathbf{j}} + z\hat{\mathbf{k}}) \cdot \hat{\mathbf{n}}\, dS = \iint_{S_{xy}} \frac{(2x + z)}{\sqrt{5}} \sqrt{1 + \left(\frac{\partial z}{\partial x}\right)^2 + \left(\frac{\partial z}{\partial y}\right)^2}\, dA$$

$$= \frac{1}{\sqrt{5}} \iint_{S_{xy}} (2x + 2y)\sqrt{1 + (2)^2}\, dA = 2 \int_{-\pi}^{\pi} \int_{0}^{\sqrt{10}} (r\cos\theta - 1 + r\sin\theta)\, r\, dr\, d\theta$$

$$= 2 \int_{-\pi}^{\pi} \left\{ \frac{r^3}{3}(\cos\theta + \sin\theta) - \frac{r^2}{2} \right\}_{0}^{\sqrt{10}} d\theta = 2 \int_{-\pi}^{\pi} \left[\frac{10\sqrt{10}}{3}(\cos\theta + \sin\theta) - 5 \right] d\theta$$

$$= 2 \left\{ \frac{10\sqrt{10}}{3}(\sin\theta - \cos\theta) - 5\theta \right\}_{-\pi}^{\pi} = -20\pi.$$

20. (a) The force on the end $S_1 : x = 0$ is in the
negative x-direction and has magnitude

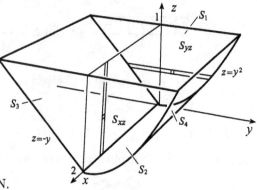

$$F_1 = \int_0^1 \int_{-z}^{\sqrt{z}} 9810(1-z)\, dy\, dz$$

$$= 9810 \int_0^1 \left\{ y(1-z) \right\}_{-z}^{\sqrt{z}} dz$$

$$= 9810 \int_0^1 (\sqrt{z} - z^{3/2} + z - z^2)\, dz$$

$$= 9810 \left\{ \frac{2z^{3/2}}{3} - \frac{2z^{5/2}}{5} + \frac{z^2}{2} - \frac{z^3}{3} \right\}_0^1 = 4251 \text{ N}.$$

Thus, $\mathbf{F}_1 = -4251\hat{\mathbf{i}}$ N. The force on $S_2 : x = 2$ is $\mathbf{F}_2 = 4251\hat{\mathbf{i}}$ N. Forces on all parts of $S_3 : y = -z$ are
in the same direction, namely $-\hat{\mathbf{j}} - \hat{\mathbf{k}}$. The magnitude of the force is

$$F_3 = \iint_{S_3} P\, dS = \iint_{S_3} 9810(1-z)\, dS = 9810 \iint_{S_{xz}} (1-z)\sqrt{1 + \left(\frac{\partial y}{\partial x}\right)^2 + \left(\frac{\partial y}{\partial z}\right)^2}\, dA$$

$$= 9810 \iint_{S_{xz}} (1-z)\sqrt{1+1}\, dA = 9810\sqrt{2} \int_0^2 \int_0^1 (1-z)\, dz\, dx$$

$$= 9810\sqrt{2} \int_0^2 \left\{ z - \frac{z^2}{2} \right\}_0^1 dx = 4905\sqrt{2}\left\{ x \right\}_0^2 = 9810\sqrt{2} \text{ N}.$$

Thus, $\mathbf{F}_3 = 9810\sqrt{2}\left(\dfrac{-\hat{\mathbf{j}} - \hat{\mathbf{k}}}{\sqrt{2}} \right) = -9810(\hat{\mathbf{j}} + \hat{\mathbf{k}})$ N. The force on an element dS on $S_4 : z = y^2$ points in
the direction normal to S_4. At a point (x, y, z), the unit downward normal is $\hat{\mathbf{n}} = (0, 2y, -1)/\sqrt{1 + 4y^2}$.
The magnitude of the force on dS is $P\, dS = 9810(1-z)\, dS$. The force on dS at (x, y, z) is therefore
$\dfrac{9810(1-z)\, dS(0, 2y, -1)}{\sqrt{1+4y^2}}$. The y-component of the total force on S_4 is

$$F_{4_y} = \iint_{S_4} \frac{9810(1-z)(2y)}{\sqrt{1+4y^2}}\, dS = 19\,620 \iint_{S_{xz}} \frac{y(1-z)}{\sqrt{1+4y^2}} \sqrt{1 + \left(\frac{\partial y}{\partial x}\right)^2 + \left(\frac{\partial y}{\partial z}\right)^2}\, dA$$

$$= 19\,620 \iint_{S_{xz}} \frac{y(1-z)}{\sqrt{1+4y^2}} \sqrt{1 + \left(\frac{1}{2\sqrt{z}}\right)^2}\, dA = 19\,620 \int_0^2 \int_0^1 \left(\frac{1-z}{2}\right)\, dz\, dx$$

$$= 9810 \int_0^2 \left\{ z - \frac{z^2}{2} \right\}_0^1 dx = 4905\left\{ x \right\}_0^2 = 9810 \text{ N}.$$

The z-component of the total force on S_4 is

$$F_{4_z} = \iint_{S_4} \frac{9810(1-z)(-1)}{\sqrt{1+4y^2}}\, dS = 9810 \iint_{S_{xz}} \frac{(z-1)}{\sqrt{1+4y^2}} \sqrt{1 + \left(\frac{1}{2\sqrt{z}}\right)^2}\, dA$$

$$= 9810 \int_0^2 \int_0^1 \left(\frac{z-1}{2\sqrt{z}}\right)\, dz\, dx = 4905 \int_0^2 \left\{ \frac{2z^{3/2}}{3} - 2\sqrt{z} \right\}_0^1 dx = -6540\left\{ x \right\}_0^2 = -13\,080 \text{ N}.$$

Thus, $\mathbf{F}_4 = 9810\hat{\mathbf{j}} - 13\,080\hat{\mathbf{k}}$ N.

(b) The sum of the four forces is $\mathbf{F}_1 + \mathbf{F}_2 + \mathbf{F}_3 + \mathbf{F}_4 = -22\,890\,\hat{\mathbf{k}}$ N. The magnitude of the weight of the
water in the trough is

$$\int_0^1 \int_{-z}^{\sqrt{z}} \int_0^2 9810\, dx\, dy\, dz = 9810 \int_0^1 \int_{-z}^{\sqrt{z}} \left\{ x \right\}_0^2 dy\, dx = 19\,620 \int_0^1 \left\{ y \right\}_{-z}^{\sqrt{z}} dz$$

$$= 19\,620 \int_0^1 (\sqrt{z} + z)\, dz = 19\,620 \left\{ \frac{2z^{3/2}}{3} + \frac{z^2}{2} \right\}_0^1 = 22\,890 \text{ N}.$$

22. The magnitude of the fluid force on S is

$$\iint_S P\,dS = \iint_{R_{xy}} -\rho g z \sqrt{1 + \left(\frac{\partial z}{\partial x}\right)^2 + \left(\frac{\partial z}{\partial y}\right)^2}\,dA$$

$$= -\rho g \iint_{R_{xy}} z\sqrt{1 + \left(\frac{-A}{C}\right)^2 + \left(\frac{-B}{C}\right)^2}\,dA$$

$$= -\frac{\rho g\sqrt{A^2 + B^2 + C^2}}{|C|} \iint_{R_{xy}} \left(\frac{-D - Ax - By}{C}\right)dA$$

$$= \frac{\rho g\sqrt{A^2 + B^2 + C^2}}{C|C|} \iint_{R_{xy}} (D + Ax + By)\,dA.$$

A unit normal vector to S is $\pm(A, B, C)/\sqrt{A^2 + B^2 + C^2}$, where the plus or minus is chosen depending on the sign of C and whether we consider the top or bottom of S. The force on S is

$$\left[\frac{\rho g\sqrt{A^2 + B^2 + C^2}}{C|C|} \iint_{R_{xy}} (D + Ax + By)\,dA\right] \frac{\pm(A, B, C)}{\sqrt{A^2 + B^2 + C^2}}.$$

The magnitude of the z-component of this force is $F_z = \dfrac{\rho g}{C} \displaystyle\iint_{R_{xy}} (D + Ax + By)\,dA$. On the other hand, the weight of the column of fluid above S is

$$W = \iint_{R_{xy}} \int_{(-D-Ax-By)/C}^{0} \rho g\,dz\,dA = \iint_{R_{xy}} \rho g\left(\frac{D + Ax + By}{C}\right)dA = \frac{\rho g}{C}\iint_{R_{xy}} (D + Ax + By)\,dA.$$

Consequently, the magnitude of F_z is equal to W.

24. The total amount of blood per second is

$$\iint_S \mathbf{F}\cdot\hat{\mathbf{n}}\,dS = \iint_S e^{-y}\,dS.$$

The same amount of diffusion occurs in each of the four octants. If S_{xy} is the projection of the first octant part of S in the xy-plane, then

$$\iint_S \mathbf{F}\cdot\hat{\mathbf{n}}\,dS = 4\iint_{S_{xy}} e^{-y}\sqrt{1 + \left(\frac{\partial z}{\partial x}\right)^2 + \left(\frac{\partial z}{\partial y}\right)^2}\,dA$$

$$= 4\iint_{S_{xy}} e^{-y}\sqrt{1 + \left(-\frac{x}{z}\right)^2}\,dA = 4\iint_{S_{xy}} \frac{e^{-y}}{\sqrt{1 - x^2}}\,dA$$

$$= 4\int_0^1\int_0^2 \frac{e^{-y}}{\sqrt{1 - x^2}}\,dy\,dx = 4\int_0^1 \left\{\frac{-e^{-y}}{\sqrt{1 - x^2}}\right\}_0^2 dx = 4(1 - e^{-2})\int_0^1 \frac{1}{\sqrt{1 - x^2}}\,dx$$

$$= 4(1 - e^{-2})\left\{\mathrm{Sin}^{-1}x\right\}_0^1 = 2\pi(1 - e^{-2}).$$

EXERCISES 14.9

2. By the divergence theorem, $\displaystyle\oiint_S (x^2\hat{\mathbf{i}} + y^2\hat{\mathbf{j}} + z^2\hat{\mathbf{k}})\cdot\hat{\mathbf{n}}\,dS = \iiint_V (2x + 2y + 2z)\,dV$. Since the triple integral is twice the sum of the first moments of the sphere about the coordinate planes, it must have value zero.

4. By the divergence theorem,

$$\oiint_S [(z^2 - x)\hat{\mathbf{i}} - xy\hat{\mathbf{j}} + 3z\hat{\mathbf{k}}] \cdot \hat{\mathbf{n}}\, dS = \iiint_V (-1 - x + 3)\, dV$$

$$= \int_0^3 \int_{-2}^2 \int_0^{4-y^2} (2 - x)\, dz\, dy\, dx$$

$$= \int_0^3 \int_{-2}^2 (2 - x)(4 - y^2)\, dy\, dx$$

$$= \int_0^3 \left\{ (2 - x)\left(4y - \frac{y^3}{3}\right) \right\}_{-2}^2 dx = \frac{32}{3}\left\{ 2x - \frac{x^2}{2} \right\}_0^3 = 16.$$

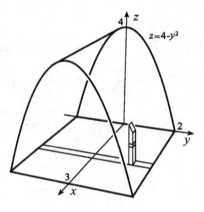

6. By the divergence theorem,

$$\oiint_S (x\hat{\mathbf{i}} + y\hat{\mathbf{j}} + 2z\hat{\mathbf{k}}) \cdot \hat{\mathbf{n}}\, dS = \iiint_V (1 + 1 + 2)\, dV$$

$$= 4\int_{-\pi}^{\pi} \int_0^{\sqrt{3}} \int_0^{2r^2\cos^2\theta + r^2\sin^2\theta} r\, dz\, dr\, d\theta$$

$$= 4\int_{-\pi}^{\pi} \int_0^{\sqrt{3}} r(2r^2\cos^2\theta + r^2\sin^2\theta)\, dr\, d\theta$$

$$= 4\int_{-\pi}^{\pi} \left\{ \frac{r^4}{2}\cos^2\theta + \frac{r^4}{4}\sin^2\theta \right\}_0^{\sqrt{3}} d\theta = 9\int_{-\pi}^{\pi} (2\cos^2\theta + \sin^2\theta)\, d\theta$$

$$= 9\int_{-\pi}^{\pi} \left(1 + \frac{1 + \cos 2\theta}{2}\right) d\theta = 9\left\{ \frac{3\theta}{2} + \frac{\sin 2\theta}{4} \right\}_{-\pi}^{\pi} = 27\pi.$$

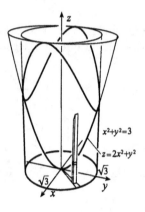

8. By the divergence theorem,

$$\oiint_S (2x^2y\hat{\mathbf{i}} - y^2\hat{\mathbf{j}} + 4xz^2\hat{\mathbf{k}}) \cdot \hat{\mathbf{n}}\, dS = \iiint_V (4xy - 2y + 8xz)\, dV$$

$$= 2\int_0^3 \int_0^{\sqrt{9-y^2}} \int_0^2 (2xy - y + 4xz)\, dx\, dz\, dy$$

$$= 2\int_0^3 \int_0^{\sqrt{9-y^2}} \left\{ x^2y - xy + 2x^2z \right\}_0^2 dz\, dy$$

$$= 2\int_0^3 \int_0^{\sqrt{9-y^2}} (2y + 8z)\, dz\, dy$$

$$= 4\int_0^3 \left\{ yz + 2z^2 \right\}_0^{\sqrt{9-y^2}} dy = 4\int_0^3 (y\sqrt{9 - y^2} + 18 - 2y^2)\, dy$$

$$= 4\left\{ -\frac{1}{3}(9 - y^2)^{3/2} + 18y - \frac{2y^3}{3} \right\}_0^3 = 180.$$

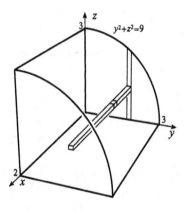

10. By the divergence theorem,

$$\oiint_S (x^3\hat{\mathbf{i}} + y^3\hat{\mathbf{j}} - z^3\hat{\mathbf{k}}) \cdot \hat{\mathbf{n}}\, dS = \iiint_V (3x^2 + 3y^2 - 3z^2)\, dV$$

$$= 3\int_{-\pi}^{\pi} \int_0^2 \int_r^{6-r^2} (r^2 - z^2)\, r\, dz\, dr\, d\theta$$

$$= 3\int_{-\pi}^{\pi} \int_0^2 \left\{ r^3z - \frac{rz^3}{3} \right\}_r^{6-r^2} dr\, d\theta$$

$$= \int_{-\pi}^{\pi} \int_0^2 (r^7 - 21r^5 - 2r^4 + 126r^3 - 216r)\, dr\, d\theta$$

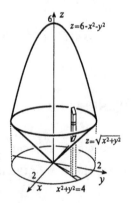

$$= \int_{-\pi}^{\pi} \left\{ \frac{r^8}{8} - \frac{7r^6}{2} - \frac{2r^5}{5} + \frac{63r^4}{2} - 108r^2 \right\}_0^2 d\theta$$

$$= \frac{-664}{5} \left\{ \theta \right\}_{-\pi}^{\pi} = -\frac{1328\pi}{5}.$$

12. By the divergence theorem,

$$\oiint_S (xy\hat{\mathbf{i}} + z^2\hat{\mathbf{k}}) \cdot \hat{\mathbf{n}} \, dS = \iiint_V (y + 2z) \, dV$$

$$= \int_0^2 \int_x^{2x} \int_0^{6-x-y} (y + 2z) \, dz \, dy \, dx$$

$$+ \int_2^3 \int_x^{6-x} \int_0^{6-x-y} (y + 2z) \, dz \, dy \, dx$$

$$= \int_0^2 \int_x^{2x} \left\{ yz + z^2 \right\}_0^{6-x-y} dy \, dx$$

$$+ \int_2^3 \int_x^{6-x} \left\{ yz + z^2 \right\}_0^{6-x-y} dy \, dx$$

$$= \int_0^2 \int_x^{2x} [y(6 - x - y) + (6 - x - y)^2] \, dy \, dx + \int_2^3 \int_x^{6-x} [y(6 - x - y) + (6 - x - y)^2] \, dy \, dx$$

$$= \int_0^2 \left\{ 3y^2 - \frac{xy^2}{2} - \frac{y^3}{3} - \frac{(6 - x - y)^3}{3} \right\}_x^{2x} dx + \int_2^3 \left\{ 3y^2 - \frac{xy^2}{2} - \frac{y^3}{3} - \frac{(6 - x - y)^3}{3} \right\}_x^{6-x} dx$$

$$= \int_0^2 \left[9x^2 - \frac{23x^3}{6} - \frac{(6 - 3x)^3}{3} + \frac{(6 - 2x)^3}{3} \right] dx$$

$$+ \int_2^3 \left[3(6 - x)^2 - \frac{(6 - x)^3}{3} + \frac{(6 - 2x)^3}{3} - 18x + 3x^2 + \frac{x^3}{3} \right] dx$$

$$= \left\{ 3x^3 - \frac{23x^4}{24} + \frac{(6 - 3x)^4}{36} - \frac{(6 - 2x)^4}{24} \right\}_0^2$$

$$+ \left\{ -(6 - x)^3 + \frac{(6 - x)^4}{12} - \frac{(6 - 2x)^4}{24} - 9x^2 + x^3 + \frac{x^4}{12} \right\}_2^3 = \frac{57}{2}.$$

14. If we create a closed surface by including with S the surface $S' : z = 2$, $x^2 + y^2 \leq 4$, then

$$\iint_S (xy\hat{\mathbf{i}} - yz\hat{\mathbf{j}} + x^2 z\hat{\mathbf{k}}) \cdot \hat{\mathbf{n}} \, dS + \iint_{S'} (xy\hat{\mathbf{i}} - yz\hat{\mathbf{j}} + x^2 z\hat{\mathbf{k}}) \cdot \hat{\mathbf{n}} \, dS = -\iiint_V (y - z + x^2) \, dV$$

provided $\hat{\mathbf{n}} = -\hat{\mathbf{k}}$ on S'. Now,

$$\iiint_V (y - z + x^2) \, dV = \int_{-\pi}^{\pi} \int_0^2 \int_r^2 (-z + r^2 \cos^2 \theta) \, r \, dz \, dr \, d\theta$$

$$= \int_{-\pi}^{\pi} \int_0^2 r \left\{ -\frac{z^2}{2} + zr^2 \cos^2 \theta \right\}_r^2 dr \, d\theta$$

$$= \int_{-\pi}^{\pi} \int_0^2 \left(-2r + 2r^3 \cos^2 \theta + \frac{r^3}{2} - r^4 \cos^2 \theta \right) dr \, d\theta$$

$$= \int_{-\pi}^{\pi} \left\{ -r^2 + \frac{r^4 \cos^2 \theta}{2} + \frac{r^4}{8} - \frac{r^5 \cos^2 \theta}{5} \right\}_0^2 d\theta$$

$$= \int_{-\pi}^{\pi} \left(-4 + 8 \cos^2 \theta + 2 - \frac{32}{5} \cos^2 \theta \right) d\theta$$

$$= \int_{-\pi}^{\pi} \left[-2 + \frac{4(1 + \cos 2\theta)}{5} \right] d\theta = \left\{ -\frac{6\theta}{5} + \frac{2 \sin 2\theta}{5} \right\}_{-\pi}^{\pi} = -\frac{12\pi}{5}.$$

We now calculate that

$$\iint_{S'} (xy\hat{\mathbf{i}} - yz\hat{\mathbf{j}} + x^2 z\hat{\mathbf{k}}) \cdot \hat{\mathbf{n}}\, dS = \iint_{S'} -x^2 z\, dS = -2 \iint_{S'} x^2\, dS = -2 \iint_{S'_{xy}} x^2\, dA$$

$$= -2 \int_{-\pi}^{\pi} \int_0^2 r^2 \cos^2\theta\, r\, dr\, d\theta = -2 \int_{-\pi}^{\pi} \left\{ \frac{r^4}{4} \cos^2\theta \right\}_0^2 d\theta = -8 \int_{-\pi}^{\pi} \cos^2\theta\, d\theta$$

$$= -8 \int_{-\pi}^{\pi} \left(\frac{1 + \cos 2\theta}{2} \right) d\theta = -4 \left\{ \theta + \frac{\sin 2\theta}{2} \right\}_{-\pi}^{\pi} = -8\pi.$$

Thus, $\iint_S (xy\hat{\mathbf{i}} - yz\hat{\mathbf{j}} + x^2 z\hat{\mathbf{k}}) \cdot \hat{\mathbf{n}}\, dS = \dfrac{12\pi}{5} + 8\pi = \dfrac{52\pi}{5}.$

16. By the divergence theorem, $\dfrac{1}{3} \oiint_S \mathbf{r} \cdot \hat{\mathbf{n}}\, dS = \dfrac{1}{3} \iiint_V \nabla \cdot \mathbf{r}\, dV = \dfrac{1}{3} \iiint_V (1+1+1)\, dV = \iiint_V dV = V.$

18. Using the discussion in Example 14.26, we can state that the total buoyant force must be

$$\iint_S (9.81\rho z\hat{\mathbf{k}}) \cdot (-\hat{\mathbf{n}})\, dS$$

where S is the submerged portion of the surface. Suppose we remove that part of the object above the fluid surface and denote by S' the remaining part of the solid in the fluid surface. Since

$$\iint_{S'} (9.81\rho z\hat{\mathbf{k}}) \cdot (-\hat{\mathbf{n}})\, dS = 0,$$

we may write that

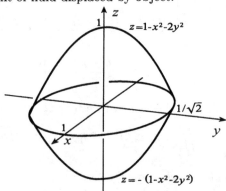

$$\iint_S (9.81\rho z\hat{\mathbf{k}}) \cdot (-\hat{\mathbf{n}})\, dS + \iint_{S'} (9.81\rho z\hat{\mathbf{k}}) \cdot (-\hat{\mathbf{n}})\, dS = \iiint_V \nabla \cdot (9.81\rho z\hat{\mathbf{k}})\, dV,$$

or, $\iint_S (9.81\rho z\hat{\mathbf{k}}) \cdot (-\hat{\mathbf{n}})\, dS = \iiint_V 9.81\rho\, dV = 9.81\rho(\text{Volume of object below fluid surface})$

$= \text{Weight of fluid displaced by object.}$

20. By the divergence theorem,

$$\oiint_S [(x+y)^2\hat{\mathbf{i}} + x^2 y\hat{\mathbf{j}} - x^2 z\hat{\mathbf{k}}] \cdot \hat{\mathbf{n}}\, dS$$

$$= -\iiint_V [2(x+y) + x^2 - x^2]\, dV$$

$$= -2 \iiint_V (x+y)\, dV.$$

Since the triple integral is the sum of the first moments of V about the yz- and xz-coordinate planes, its value must be zero.

22. By the divergence theorem, $\oiint_S \mathbf{B} \cdot \hat{\mathbf{n}}\, dS = \iiint_V \nabla \cdot \mathbf{B}\, dV = \iiint_V \nabla \cdot (\nabla \times \mathbf{A})\, dV.$ But according to equation 14.15, $\nabla \cdot (\nabla \times \mathbf{A}) = 0$, and hence the required result follows immediately.

24. By the divergence theorem, $\oiint_S \nabla P \cdot \hat{\mathbf{n}}\, dS = \iiint_V \nabla \cdot \nabla P\, dV = \iiint_V \nabla^2 P\, dV.$ If $\nabla^2 P = 0$ in V, then $\oiint_S \nabla P \cdot \hat{\mathbf{n}}\, dS = 0.$

26. If we reverse the roles of P and Q in Exercise 25, $\oiint_S Q\nabla P \cdot \hat{\mathbf{n}}\, dS = \iiint_V (Q\nabla^2 P + \nabla Q \cdot \nabla P)\, dV.$ When we subtract this result from that in Exercise 25, we obtain

$$\oiint_S (P\nabla Q - Q\nabla P) \cdot \hat{\mathbf{n}}\, dS = \iiint_V (P\nabla^2 Q - Q\nabla^2 P)\, dV.$$

EXERCISES 14.10

2. According to Stokes's theorem,

$$\oint_C y^2\,dx + xy\,dy + xz\,dz$$

$$= \iint_S \nabla \times (y^2, xy, xz) \cdot \hat{\mathbf{n}}\,dS$$

where S is any surface with C as boundary. Now,

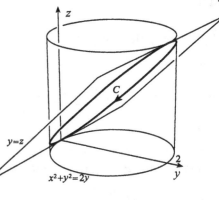

$$\nabla \times (y^2, xy, xz) = \begin{vmatrix} \hat{\mathbf{i}} & \hat{\mathbf{j}} & \hat{\mathbf{k}} \\ \partial/\partial x & \partial/\partial y & \partial/\partial z \\ y^2 & xy & xz \end{vmatrix} = (0, -z, -y).$$

If we choose S as that part of $z = y$ bounded by C, then $\hat{\mathbf{n}} = (0, -1, 1)/\sqrt{2}$, and

$$\oint_C y^2\,dx + xy\,dy + xz\,dz = \iint_S (0, -z, -y) \cdot \frac{(0, -1, 1)}{\sqrt{2}}\,dS = \frac{1}{\sqrt{2}} \iint_S (z - y)\,dS = 0.$$

4. By Stokes's theorem, $\quad \oint_C \mathbf{F} \cdot d\mathbf{r} = \iint_S \nabla \times \mathbf{F} \cdot \hat{\mathbf{n}}\,dS$

where S is any surface with C as boundary and $\mathbf{F} = (2xy + y)\hat{\mathbf{i}} + (x^2 + xy - 3y)\hat{\mathbf{j}} + 2xz\hat{\mathbf{k}}$. Now

$$\nabla \times \mathbf{F} = \begin{vmatrix} \hat{\mathbf{i}} & \hat{\mathbf{j}} & \hat{\mathbf{k}} \\ \partial/\partial x & \partial/\partial y & \partial/\partial z \\ 2xy + y & x^2 + xy - 3y & 2xz \end{vmatrix}$$

$$= -2z\hat{\mathbf{j}} + (y - 1)\hat{\mathbf{k}}.$$

If we choose S as that part of the plane $z = 4$ inside C, then $\hat{\mathbf{n}} = \pm\hat{\mathbf{k}}$, depending on the direction along C, and $\nabla \times \mathbf{F} \cdot \hat{\mathbf{n}} = \pm(y - 1)$. Hence

$$\oint_C \mathbf{F} \cdot d\mathbf{r} = \pm \iint_S (y - 1)\,dS = \pm \left[\iint_{S_{xy}} y\,dA - \iint_{S_{xy}} dA \right] = \pm(0 - 16\pi) = \pm 16\pi.$$

6. By Stokes's theorem,

$$\oint_C y\,dx + x\,dy + (x^2 + y^2 + z^2)\,dz$$

$$= \iint_S \nabla \times (y, x, x^2 + y^2 + z^2) \cdot \hat{\mathbf{n}}\,dS$$

where S is any surface with C as boundary. Now

$$\nabla \times (y, x, x^2 + y^2 + z^2) = \begin{vmatrix} \hat{\mathbf{i}} & \hat{\mathbf{j}} & \hat{\mathbf{k}} \\ \partial/\partial x & \partial/\partial y & \partial/\partial z \\ y & x & x^2 + y^2 + z^2 \end{vmatrix}$$

$$= (2y, -2x, 0).$$

If we choose S as that part of $z = xy$ inside C, then $\hat{\mathbf{n}} = (y, x, -1)/\sqrt{1 + x^2 + y^2}$. Thus,

$$\oint_C y\,dx + x\,dy + (x^2 + y^2 + z^2)\,dz = \iint_S \frac{2y^2 - 2x^2}{\sqrt{1 + x^2 + y^2}}\,dS$$

$$= 2 \iint_{S_{xy}} \frac{y^2 - x^2}{\sqrt{1 + x^2 + y^2}} \sqrt{1 + \left(\frac{\partial z}{\partial x}\right)^2 + \left(\frac{\partial z}{\partial y}\right)^2}\,dA$$

$$= 2 \iint_{S_{xy}} \frac{y^2 - x^2}{\sqrt{1 + y^2 + x^2}} \sqrt{1 + y^2 + x^2}\,dA = 2 \iint_{S_{xy}} y^2\,dA - 2 \iint_{S_{xy}} x^2\,dA = 0$$

since these integrals are equal.

8. By Stokes's theorem, $\oint_C y\,dx + z\,dy + x\,dz = \iint_S \nabla \times (y,z,x) \cdot \hat{\mathbf{n}}\,dS$ where S is any surface with C as boundary. Now

$$\nabla \times (y,z,x) = \begin{vmatrix} \hat{\mathbf{i}} & \hat{\mathbf{j}} & \hat{\mathbf{k}} \\ \partial/\partial x & \partial/\partial y & \partial/\partial z \\ y & z & x \end{vmatrix}$$

$$= (-1,-1,-1).$$

If we choose S as that part of $x + y = 2b$ inside C, then $\hat{\mathbf{n}} = (1,1,0)/\sqrt{2}$, and,

$$\oint_C y\,dx + z\,dy + x\,dz$$

$$= \iint_S \frac{-2}{\sqrt{2}}\,dS = -\sqrt{2}\iint_S dS$$

$$= -\sqrt{2}(\text{Area of } S) = -2\sqrt{2}\pi b^2.$$

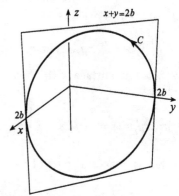

10. By Stokes's theorem, $\oint_C (x+y)^2\,dx + (x+y)^2\,dy + yz^3\,dz = \iint_S \nabla \times ((x+y)^2, (x+y)^2, yz^3) \cdot \hat{\mathbf{n}}\,dS$ where S is any surface with C as boundary. Now

$$\nabla \times ((x+y)^2, (x+y)^2, yz^3) = \begin{vmatrix} \hat{\mathbf{i}} & \hat{\mathbf{j}} & \hat{\mathbf{k}} \\ \partial/\partial x & \partial/\partial y & \partial/\partial z \\ (x+y)^2 & (x+y)^2 & yz^3 \end{vmatrix}$$

$$= (z^3, 0, 0).$$

If we choose S as that part of $z = \sqrt{x^2 + y^2}$ inside C, then

$$\hat{\mathbf{n}} = \frac{\pm \nabla(x^2 + y^2 - z^2)}{|\nabla(x^2 + y^2 - z^2)|} = \frac{\pm(2x, 2y, -2z)}{\sqrt{4x^2 + 4y^2 + 4z^2}} = \frac{\pm(x, y, -z)}{\sqrt{2}z},$$

the sign depending on the direction along C. Hence,

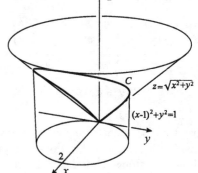

$$\oint_C (x+y)^2\,dx + (x+y)^2\,dy + yz^3\,dz = \pm \iint_S \frac{xz^3}{\sqrt{2}z}\,dS = \frac{\pm 1}{\sqrt{2}}\iint_S xz^2\,dS$$

$$= \frac{\pm 1}{\sqrt{2}}\iint_{S_{xy}} x(x^2 + y^2)\sqrt{1 + \left(\frac{\partial z}{\partial x}\right)^2 + \left(\frac{\partial z}{\partial y}\right)^2}\,dA$$

$$= \frac{\pm 1}{\sqrt{2}}\iint_{S_{xy}} x(x^2 + y^2)\sqrt{1 + \left(\frac{x}{\sqrt{x^2 + y^2}}\right)^2 + \left(\frac{y}{\sqrt{x^2 + y^2}}\right)^2}\,dA$$

$$= \pm \iint_{S_{xy}} x(x^2 + y^2)\,dA = \pm 2\int_0^{\pi/2}\int_0^{2\cos\theta} r^4\cos\theta\,dr\,d\theta = \pm 2\int_0^{\pi/2}\left\{\frac{r^5}{5}\cos\theta\right\}_0^{2\cos\theta}\,d\theta$$

$$= \frac{\pm 64}{5}\int_0^{\pi/2}\cos^6\theta\,d\theta = \frac{\pm 64}{5}\int_0^{\pi/2}\left(\frac{1 + \cos 2\theta}{2}\right)^3\,d\theta$$

$$= \frac{\pm 8}{5}\int_0^{\pi/2}\left[1 + 3\cos 2\theta + \frac{3}{2}(1 + \cos 4\theta) + \cos 2\theta(1 - \sin^2 2\theta)\right]\,d\theta$$

$$= \frac{\pm 8}{5}\left\{\frac{5\theta}{2} + 2\sin 2\theta + \frac{3\sin 4\theta}{8} - \frac{\sin^3 2\theta}{6}\right\}_0^{\pi/2} = \pm 2\pi.$$

12. The curve of intersection lies in the plane $z = 1$. If we choose S as that part of the plane interior to C, then

$$\oint_C y^3\, dx - x^3\, dy + xyz\, dz$$

$$= \iint_S \nabla \times (y^3, -x^3, xyz) \cdot \hat{n}\, dS.$$

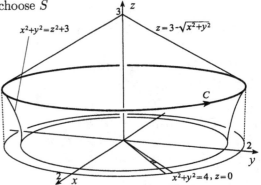

Now, $\nabla \times (y^3, -x^3, xyz) = \begin{vmatrix} \hat{i} & \hat{j} & \hat{k} \\ \partial/\partial x & \partial/\partial y & \partial/\partial z \\ y^3 & -x^3 & xyz \end{vmatrix}$

$$= (xz, -yz, -3x^2 - 3y^2).$$

Since $\hat{n} = \hat{k}$ on S,

$$\oint_C y^3\, dx - x^3\, dy + xyz\, dz = \iint_S (-3x^2 - 3y^2)\, dS = -3 \iint_{S_{xy}} (x^2 + y^2)\, dA$$

$$= -3 \int_{-\pi}^{\pi} \int_0^2 r^2\, r\, dr\, d\theta = -3 \int_{-\pi}^{\pi} \left\{ \frac{r^4}{4} \right\}_0^2 d\theta = -12 \{\theta\}_{-\pi}^{\pi} = -24\pi.$$

14. If we choose S as the upper part of the sphere bounded by C, then

$$\oint_C -2y^3 x^2\, dx + x^3 y^2\, dy + z\, dz$$

$$= \iint_S \nabla \times (-2y^3 x^2, x^3 y^2, z) \cdot \hat{n}\, dS.$$

Now,

$$\nabla \times (-2y^3 x^2, x^3 y^2, z) = \begin{vmatrix} \hat{i} & \hat{j} & \hat{k} \\ \partial/\partial x & \partial/\partial y & \partial/\partial z \\ -2y^3 x^2 & x^3 y^2 & z \end{vmatrix}$$

$$= (0, 0, 9x^2 y^2),$$

and $\hat{n} = \dfrac{(x, y, z)}{\sqrt{x^2 + y^2 + z^2}} = \dfrac{(x, y, z)}{2}$. Thus,

$$\oint_C -2y^3 x^2\, dx + x^3 y^2\, dy + z\, dz = \iint_S \frac{z}{2}(9x^2 y^2)\, dS = \frac{9}{2} \iint_S x^2 y^2 \sqrt{4 - x^2 - y^2}\, dS$$

$$= \frac{9}{2} \iint_{S_{xy}} x^2 y^2 \sqrt{4 - x^2 - y^2} \sqrt{1 + \frac{x^2}{4 - x^2 - y^2} + \frac{y^2}{4 - x^2 - y^2}}\, dA$$

$$= 9 \iint_{S_{xy}} x^2 y^2\, dA = 36 \int_0^1 \int_0^{\sqrt{4 - 4y^2}} x^2 y^2\, dx\, dy$$

$$= 36 \int_0^1 \left\{ \frac{x^3 y^2}{3} \right\}_0^{\sqrt{4 - 4y^2}} dy = 96 \int_0^1 y^2 (1 - y^2)^{3/2}\, dy.$$

If we set $y = \sin\theta$ and $dy = \cos\theta\, d\theta$, then

$$\oint_C -2y^3 x^2\, dx + x^3 y^2\, dy + z\, dz = 96 \int_0^{\pi/2} \sin^2\theta \cos^3\theta \cos\theta\, d\theta = 96 \int_0^{\pi/2} \sin^2\theta \cos^4\theta\, d\theta$$

$$= 96 \int_0^{\pi/2} \frac{\sin^2 2\theta}{4} \left(\frac{1 + \cos 2\theta}{2} \right) d\theta$$

$$= 12 \int_0^{\pi/2} \left(\frac{1 - \cos 4\theta}{2} + \sin^2 2\theta \cos 2\theta \right) d\theta$$

$$= 12 \left\{ \frac{\theta}{2} - \frac{\sin 4\theta}{8} + \frac{\sin^3 2\theta}{6} \right\}_0^{\pi/2} = 3\pi.$$

16. Both surfaces have the curve $C : x^2 + y^2 = 1$, $z = 0$ as boundary. Consequently, by Stokes's theorem, both surface integrals are equal to $\oint_C \mathbf{F} \cdot d\mathbf{r}$ and are therefore equal to each other.

REVIEW EXERICSES

2. $\nabla \cdot \mathbf{F} = 3x^2y + x^2/y^2$

4. $\nabla \times \mathbf{F} = \begin{vmatrix} \hat{\mathbf{i}} & \hat{\mathbf{j}} & \hat{\mathbf{k}} \\ \partial/\partial x & \partial/\partial y & \partial/\partial z \\ x+y+z & x+y+z & x+y+z \end{vmatrix} = (1-1)\hat{\mathbf{i}} + (1-1)\hat{\mathbf{j}} + (1-1)\hat{\mathbf{k}} = \mathbf{0}$

6. $\nabla \cdot \mathbf{F} = ye^x + ze^y + xe^z$

8. $\nabla f = \dfrac{1}{\sqrt{1 - (x+y)^2}}(\hat{\mathbf{i}} + \hat{\mathbf{j}})$

10. $\nabla \times \mathbf{F} = \begin{vmatrix} \hat{\mathbf{i}} & \hat{\mathbf{j}} & \hat{\mathbf{k}} \\ \partial/\partial x & \partial/\partial y & \partial/\partial z \\ \mathrm{Cot}^{-1}(xyz) & 0 & 0 \end{vmatrix} = \dfrac{-xy}{1+x^2y^2z^2}\hat{\mathbf{j}} + \dfrac{xz}{1+x^2y^2z^2}\hat{\mathbf{k}}$

12. $\displaystyle\iint_S (x^2 + yz)\,dS = \iint_{S_{xy}} [x^2 + y(2 - x - y)]\sqrt{1 + \left(\dfrac{\partial z}{\partial x}\right)^2 + \left(\dfrac{\partial z}{\partial y}\right)^2}\,dA$

$\displaystyle = \iint_{S_{xy}} (x^2 + 2y - xy - y^2)\sqrt{1 + (-1)^2 + (-1)^2}\,dA$

$\displaystyle = \sqrt{3} \int_0^2 \int_0^{2-x} (x^2 + 2y - xy - y^2)\,dy\,dx$

$\displaystyle = \sqrt{3} \int_0^2 \left\{ x^2y + y^2 - \dfrac{xy^2}{2} - \dfrac{y^3}{3} \right\}_0^{2-x} dx$

$\displaystyle = \dfrac{\sqrt{3}}{6} \int_0^2 [24x^2 - 9x^3 - 12x + 6(2-x)^2 - 2(2-x)^3]\,dx$

$\displaystyle = \dfrac{\sqrt{3}}{6} \left\{ 8x^3 - \dfrac{9x^4}{4} - 6x^2 - 2(2-x)^3 + \dfrac{(2-x)^4}{2} \right\}_0^2 = 2\sqrt{3}$

14. Since $\hat{\mathbf{n}} = \dfrac{\nabla(x^2 + y^2 - z)}{|\nabla(x^2 + y^2 - z)|} = \dfrac{(2x, 2y, -1)}{\sqrt{1 + 4x^2 + 4y^2}}$,

$\displaystyle\iint_S (x\hat{\mathbf{i}} + y\hat{\mathbf{j}}) \cdot \hat{\mathbf{n}}\,dS$

$\displaystyle = \iint_{S_{xy}} \dfrac{2x^2 + 2y^2}{\sqrt{1 + 4x^2 + 4y^2}}\sqrt{1 + \left(\dfrac{\partial z}{\partial x}\right)^2 + \left(\dfrac{\partial z}{\partial y}\right)^2}\,dA$

$\displaystyle = \iint_{S_{xy}} \dfrac{2(x^2 + y^2)}{\sqrt{1 + 4x^2 + 4y^2}}\sqrt{1 + (2x)^2 + (2y)^2}\,dA$

$\displaystyle = 2\iint_{S_{xy}} (x^2 + y^2)\,dA = 2\int_{-\pi}^{\pi}\int_0^1 r^2 \, r\,dr\,d\theta$

$\displaystyle = 2\int_{-\pi}^{\pi} \left\{ \dfrac{r^4}{4} \right\}_0^1 d\theta = \dfrac{1}{2}\{\theta\}_{-\pi}^{\pi} = \pi.$

16. With parametric equations $C : x = -t$, $y = \sqrt{1 + t^2}$, $z = \sqrt{1 - 2t^2}$, $-1/\sqrt{2} \le t \le 1/\sqrt{2}$,

$\displaystyle\int_C xy\,dx + xz\,dz = \int_{-1/\sqrt{2}}^{1/\sqrt{2}} \left[-t\sqrt{1 + t^2}(-dt) - t\sqrt{1 - 2t^2}\left(\dfrac{-2t}{\sqrt{1 - 2t^2}}\right)dt \right]$

$\displaystyle = \int_{-1/\sqrt{2}}^{1/\sqrt{2}} (t\sqrt{1 + t^2} + 2t^2)\,dt = \left\{ \dfrac{1}{3}(1 + t^2)^{3/2} + \dfrac{2t^3}{3} \right\}_{-1/\sqrt{2}}^{1/\sqrt{2}} = \dfrac{\sqrt{2}}{3}.$

18. By Green's theorem,

$$\oint_C 2xy^3\,dx + (3x^2y^2 + x^2)\,dy$$

$$= \iint_R (6xy^2 + 2x - 6xy^2)\,dA$$

$$= 2\iint_R x\,dA = 2\bar{x}(\text{Area of } R) = 2(1)\pi(1)^2 = 2\pi.$$

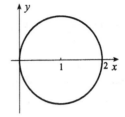

20. We quadruple the integral over that part of the surface in the first octant.

$$\iint_S (x^2 + y^2)\,dS = 4\iint_{S_{xy}} (x^2 + y^2)\sqrt{1 + \left(\frac{\partial z}{\partial x}\right)^2 + \left(\frac{\partial z}{\partial y}\right)^2}\,dA$$

$$= 4\iint_{S_{xy}} (x^2 + y^2)\sqrt{1 + \left(-\frac{x}{z}\right)^2 + \left(-\frac{y}{z}\right)^2}\,dA$$

$$= 4\iint_{S_{xy}} (x^2 + y^2)\frac{\sqrt{6}}{z}\,dA$$

$$= 4\sqrt{6}\iint_{S_{xy}} \frac{x^2 + y^2}{\sqrt{6 - x^2 - y^2}}\,dA$$

$$= 4\sqrt{6}\int_0^{\sqrt{2}}\int_0^{\pi/2} \frac{r^2}{\sqrt{6 - r^2}}\,r\,d\theta\,dr$$

$$= 2\sqrt{6}\pi\int_0^{\sqrt{2}} \frac{r^3}{\sqrt{6 - r^2}}\,dr$$

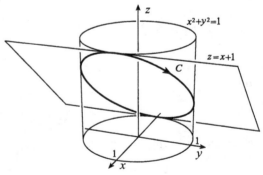

If we set $u = 6 - r^2$ and $du = -2r\,dr$, then

$$\iint_S (x^2 + y^2)\,dS = 2\sqrt{6}\pi\int_6^4 \frac{6 - u}{\sqrt{u}}\left(-\frac{du}{2}\right) = \sqrt{6}\pi\int_6^4\left(-\frac{6}{\sqrt{u}} + \sqrt{u}\right)du$$

$$= \sqrt{6}\pi\left\{-12\sqrt{u} + \frac{2u^{3/2}}{3}\right\}_6^4 = \frac{8\pi(18 - 7\sqrt{6})}{3}.$$

22. By Stokes's theorem,

$$\oint_C (x^2\hat{\mathbf{i}} + y\hat{\mathbf{j}} - xz\hat{\mathbf{k}})\cdot d\mathbf{r} = \iint_S \nabla\times(x^2, y, -xz)\cdot\hat{\mathbf{n}}\,dS$$

where S is any surface with C as boundary. Now

$$\nabla\times(x^2, y, -xz) = \begin{vmatrix} \hat{\mathbf{i}} & \hat{\mathbf{j}} & \hat{\mathbf{k}} \\ \partial/\partial x & \partial/\partial y & \partial/\partial z \\ x^2 & y & -xz \end{vmatrix}$$

$$= (0, z, 0).$$

If we choose S as that part of $z = x + 1$ inside C, then $\hat{\mathbf{n}} = (-1, 0, 1)/\sqrt{2}$, and

$$\oint_C (x^2\hat{\mathbf{i}} + y\hat{\mathbf{j}} - xz\hat{\mathbf{k}})\cdot d\mathbf{r} = \iint_S 0\,dS = 0.$$

24. By Stokes's theorem, $\oint_C y\,dx + 2x\,dy - 3z^2\,dz = \iint_S \nabla \times (y, 2x, -3z^2) \cdot \hat{\mathbf{n}}\,dS$ where S is any surface with C as boundary. Now,

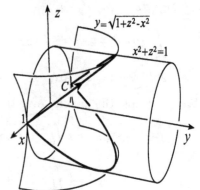

$$\nabla \times (y, 2x, -3z^2) = \begin{vmatrix} \hat{\mathbf{i}} & \hat{\mathbf{j}} & \hat{\mathbf{k}} \\ \partial/\partial x & \partial/\partial y & \partial/\partial z \\ y & 2x & -3z^2 \end{vmatrix} = (0, 0, 1).$$

If we choose S as that part of $y = \sqrt{1 + z^2 - x^2}$ inside C, then

$$\hat{\mathbf{n}} = \frac{\nabla(z^2 - x^2 - y^2 + 1)}{|\nabla(z^2 - x^2 - y^2 + 1)|} = \frac{(-2x, -2y, 2z)}{\sqrt{4x^2 + 4y^2 + 4z^2}}$$

$$= \frac{(-x, -y, z)}{\sqrt{x^2 + y^2 + z^2}}.$$

Hence, $\oint_C y\,dx + 2x\,dy - 3z^2\,dz = \iint_S \frac{z}{\sqrt{x^2 + y^2 + z^2}}\,dS$

$$= \iint_{S_{xz}} \frac{z}{\sqrt{x^2 + y^2 + z^2}} \sqrt{1 + \left(\frac{\partial y}{\partial x}\right)^2 + \left(\frac{\partial y}{\partial z}\right)^2}\,dA$$

$$= \iint_{S_{xz}} \frac{z}{\sqrt{x^2 + y^2 + z^2}} \sqrt{1 + \left(-\frac{x}{y}\right)^2 + \left(\frac{z}{y}\right)^2}\,dA$$

$$= \iint_{S_{xz}} \frac{z}{y}\,dA = \iint_{S_{xz}} \frac{z}{\sqrt{1 + z^2 - x^2}}\,dA = 0,$$

since the integrand is an odd function of z and S_{xz} is symmetric about the x-axis.

26. If S' is that part of the xy-plane bounded by $x^2 + y^2 = 1$, $z = 0$, then

$$\iint_{S'} (x^2 yz\hat{\mathbf{i}} - x^2 yz\hat{\mathbf{j}} - xyz^2\hat{\mathbf{k}}) \cdot \hat{\mathbf{n}}\,dS = 0.$$

By the divergence theorem,

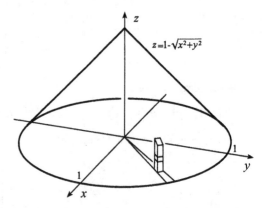

$$\oiint_{S+S'} (x^2 yz\hat{\mathbf{i}} - x^2 yz\hat{\mathbf{j}} - xyz^2\hat{\mathbf{k}}) \cdot \hat{\mathbf{n}}\,dS$$

$$= \iiint_V (2xyz - x^2 z - 2xyz)\,dV$$

Thus,

$$\iint_S (x^2 yz\hat{\mathbf{i}} - x^2 yz\hat{\mathbf{j}} - xyz^2\hat{\mathbf{k}}) \cdot \hat{\mathbf{n}}\,dS = -\iiint_V x^2 z\,dV - \iint_{S'} (x^2 yz\hat{\mathbf{i}} - x^2 yz\hat{\mathbf{j}} - xyz^2\hat{\mathbf{k}}) \cdot \hat{\mathbf{n}}\,dS$$

$$= -\int_{-\pi}^{\pi} \int_0^1 \int_0^{1-r} zr^2 \cos^2\theta\, r\,dz\,dr\,d\theta = -\int_{-\pi}^{\pi} \int_0^1 \left\{\frac{z^2 r^3 \cos^2\theta}{2}\right\}_0^{1-r}\,dr\,d\theta$$

$$= -\frac{1}{2}\int_{-\pi}^{\pi} \int_0^1 (r^3 - 2r^4 + r^5) \cos^2\theta\,dr\,d\theta$$

$$= -\frac{1}{2}\int_{-\pi}^{\pi} \left\{\frac{r^4}{4} - \frac{2r^5}{5} + \frac{r^6}{6}\right\}_0^1 \cos^2\theta\,d\theta$$

$$= \frac{-1}{120}\int_{-\pi}^{\pi} \left(\frac{1 + \cos 2\theta}{2}\right)\,d\theta = \frac{-1}{240}\left\{\theta + \frac{\sin 2\theta}{2}\right\}_{-\pi}^{\pi} = \frac{-\pi}{120}.$$

28. $\displaystyle\iint_S y\,dS = \iint_{S_{xz}} y\sqrt{1+\left(\frac{\partial y}{\partial x}\right)^2 + \left(\frac{\partial y}{\partial z}\right)^2}\,dA$

$\displaystyle\qquad = \iint_{S_{xz}} y\sqrt{1+[1/(2y)]^2}\,dA = \frac{1}{2}\iint_{S_{xz}}\sqrt{4y^2+1}\,dA$

$\displaystyle\qquad = \frac{1}{2}\iint_{S_{xz}}\sqrt{4(x-1)+1}\,dA = \frac{1}{2}\iint_{S_{xz}}\sqrt{4x-3}\,dA$

$\displaystyle\qquad = \frac{1}{2}\int_1^2\int_0^{2-x}\sqrt{4x-3}\,dz\,dx = \frac{1}{2}\int_1^2(2-x)\sqrt{4x-3}\,dx$

If we set $u = 4x-3$ and $du = 4\,dx$ in the second term,

$\displaystyle\iint_S y\,dS = \left\{\frac{1}{6}(4x-3)^{3/2}\right\}_1^2 - \frac{1}{2}\int_1^5\left(\frac{u+3}{4}\right)\sqrt{u}\left(\frac{du}{4}\right)$

$\displaystyle\qquad = \frac{1}{6}(5\sqrt{5}-1) - \frac{1}{32}\left\{\frac{2u^{5/2}}{5} + 2u^{3/2}\right\}_1^5 = \frac{25\sqrt{5}-11}{120}.$

30. Since $\displaystyle\hat{\mathbf{n}} = \frac{\nabla(x^2+y^2-z^2+1)}{|\nabla(x^2+y^2-z^2+1)|} = \frac{(2x,2y,-2z)}{\sqrt{4x^2+4y^2+4z^2}} = \frac{(x,y,-z)}{\sqrt{x^2+y^2+z^2}},$

$\displaystyle\iint_S(x\hat{\mathbf{i}}+y\hat{\mathbf{j}})\cdot\hat{\mathbf{n}}\,dS = \iint_{S_{xy}}\frac{x^2+y^2}{\sqrt{x^2+y^2+z^2}}\sqrt{1+\left(\frac{\partial z}{\partial x}\right)^2 + \left(\frac{\partial z}{\partial y}\right)^2}\,dA$

$\displaystyle\qquad = \iint_{S_{xy}}\frac{x^2+y^2}{\sqrt{x^2+y^2+z^2}}\sqrt{1+\left(\frac{x}{z}\right)^2 + \left(\frac{y}{z}\right)^2}\,dA$

$\displaystyle\qquad = \iint_{S_{xy}}\frac{x^2+y^2}{z}\,dA$

$\displaystyle\qquad = \iint_{S_{xy}}\frac{x^2+y^2}{\sqrt{1+x^2+y^2}}\,dA$

$\displaystyle\qquad = \int_0^{\sqrt{3}}\int_{-\pi}^{\pi}\frac{r^2}{\sqrt{1+r^2}}r\,d\theta\,dr = 2\pi\int_0^{\sqrt{3}}\frac{r^3}{\sqrt{1+r^2}}\,dr.$

If we set $u = 1+r^2$ and $du = 2r\,dr$, then

$\displaystyle\iint_S(x\hat{\mathbf{i}}+y\hat{\mathbf{j}})\cdot\hat{\mathbf{n}}\,dS = 2\pi\int_1^4\frac{u-1}{\sqrt{u}}\left(\frac{du}{2}\right) = \pi\int_1^4\left(\sqrt{u}-\frac{1}{\sqrt{u}}\right)du = \pi\left\{\frac{2u^{3/2}}{3}-2\sqrt{u}\right\}_1^4 = \frac{8\pi}{3}.$

32. $\displaystyle\nabla(|\mathbf{r}|^n) = \nabla[(x^2+y^2+z^2)^{n/2}] = \frac{n}{2}(x^2+y^2+z^2)^{n/2-1}(2x\hat{\mathbf{i}}+2y\hat{\mathbf{j}}+2z\hat{\mathbf{k}})$

$\displaystyle\qquad = n(x^2+y^2+z^2)^{(n-2)/2}(x\hat{\mathbf{i}}+y\hat{\mathbf{j}}+z\hat{\mathbf{k}}) = n|\mathbf{r}|^{n-2}\mathbf{r}$

CHAPTER 15

EXERCISES 15.1

2. For the given function, $\dfrac{dy}{dx} - \dfrac{y^2}{x^2} = \dfrac{(1+Cx)(1) - x(C)}{(1+Cx)^2} - \dfrac{x^2}{x^2(1+Cx)^2} = 0.$

4. For the given function, $\dfrac{d^2y}{dx^2} + 9y = (-9C_1 \sin 3x - 9C_2 \cos 3x) + 9(C_1 \sin 3x + C_2 \cos 3x) = 0.$

6. For the given function,

$$2\frac{d^2y}{dx^2} - 8\frac{dy}{dx} + 9y = 2[4C_1 e^{2x} \cos(x/\sqrt{2}) - (4/\sqrt{2})C_1 e^{2x} \sin(x/\sqrt{2}) - (C_1/2)e^{2x} \cos(x/\sqrt{2})$$
$$+ 4C_2 e^{2x} \sin(x/\sqrt{2}) + (4/\sqrt{2})C_2 e^{2x} \cos(x/\sqrt{2}) - (C_2/2)e^{2x} \sin(x/\sqrt{2})]$$
$$- 8[2C_1 e^{2x} \cos(x/\sqrt{2}) - (C_1/\sqrt{2})e^{2x} \sin(x/\sqrt{2}) + 2C_2 e^{2x} \sin(x/\sqrt{2})$$
$$+ (C_2/\sqrt{2})e^{2x} \cos(x/\sqrt{2})] + 9[C_1 e^{2x} \cos(x/\sqrt{2}) + C_2 e^{2x} \sin(x/\sqrt{2})] = 0.$$

8. For the given function,

$$2\frac{d^2y}{dx^2} - 16\frac{dy}{dx} + 32y = 2[-(1/2)e^{4x} + 2(C_2 - x/2)4e^{4x} + 16(C_1 + C_2 x - x^2/4)e^{4x}]$$
$$- 16[(C_2 - x/2)e^{4x} + 4(C_1 + C_2 x - x^2/4)e^{4x}] + 32(C_1 + C_2 x - x^2/4)e^{4x} = -e^{4x}.$$

10. For the given function,

$$x^2\frac{d^2y}{dx^2} + x\frac{dy}{dx} + (x^2 - 1/4)y = x^2\left(-\frac{C_1 \sin x}{\sqrt{x}} - \frac{C_1 \cos x}{x^{3/2}} + \frac{3C_1 \sin x}{4x^{5/2}} - \frac{C_2 \cos x}{\sqrt{x}}\right.$$
$$\left. + \frac{C_2 \sin x}{x^{3/2}} + \frac{3C_2 \cos x}{4x^{5/2}}\right) + x\left(\frac{C_1 \cos x}{\sqrt{x}} - \frac{C_1 \sin x}{2x^{3/2}} - \frac{C_2 \sin x}{\sqrt{x}} - \frac{C_2 \cos x}{2x^{3/2}}\right)$$
$$+ x^2\left(\frac{C_1 \sin x}{\sqrt{x}} + \frac{C_2 \cos x}{\sqrt{x}}\right) - \frac{1}{4}\left(\frac{C_1 \sin x}{\sqrt{x}} + \frac{C_2 \cos x}{\sqrt{x}}\right) = 0.$$

12. For $y(0) = 2$ and $y(\pi/2) = 3$, constants C_1 and C_2 must satisfy $2 = C_2$ and $3 = -C_1$. Thus, $y(x) = -3\sin 3x + 2\cos 3x$.

14. For $y(1) = 1$ and $y(2) = 2$, constants C_1 and C_2 must satisfy

$$1 = C_1 \sin 3 + C_2 \cos 3 \qquad \text{and} \qquad 2 = C_1 \sin 6 + C_2 \cos 6.$$

The solution of these equations is $C_1 = (2\cos 3 - \cos 6)/\sin 3$ and $C_2 = (\sin 6 - 2\sin 3)/\sin 3$, and therefore $y(x) = [(2\cos 3 - \cos 6)\sin 3x + (\sin 6 - 2\sin 3)\cos 3x]/\sin 3$.

16. Integration with respect to x gives $y(x) = \displaystyle\int \frac{1}{9 + x^2}dx$, and if we set $x = 3\tan\theta$ and $dx = 3\sec^2\theta\, d\theta$,

then, $y(x) = \displaystyle\int \frac{1}{9\sec^2\theta}3\sec^2\theta\, d\theta = \frac{1}{3}\theta + C = \frac{1}{3}\text{Tan}^{-1}(x/3) + C.$

18. Integration with respect to x gives $\dfrac{dy}{dx} = \displaystyle\int x\ln x\, dx = \frac{x^2}{2}\ln x - \frac{x^2}{4} + C.$ A further integration now

yields $y(x) = \displaystyle\int \left(\frac{x^2}{2}\ln x - \frac{x^2}{4} + C\right)dx = \frac{x^3}{6}\ln x - \frac{5x^3}{36} + Cx + D.$

20. (a) Since the slope of the curve is also the slope of the string,

$$\frac{dy}{dx} = -\frac{\sqrt{L^2 - x^2}}{x}.$$

(b) If we integrate with respect to x,

$$y = -\int \frac{\sqrt{L^2 - x^2}}{x}dx.$$

We now set $x = L\sin\theta$ and $dx = L\cos\theta\, d\theta$,

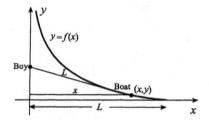

$$y = -\int \frac{L\cos\theta}{L\sin\theta} L\cos\theta\,d\theta = -L\int \frac{1-\sin^2\theta}{\sin\theta}\,d\theta = L\int (\sin\theta - \csc\theta)\,d\theta$$

$$= L[-\cos\theta - \ln|\csc\theta - \cot\theta|] + C = -L\left(\frac{\sqrt{L^2-x^2}}{L} + \ln\left|\frac{L}{x} - \frac{\sqrt{L^2-x^2}}{x}\right|\right) + C$$

$$= -\sqrt{L^2-x^2} - L\ln\left|\frac{L-\sqrt{L^2-x^2}}{x}\right| + C = -\sqrt{L^2-x^2} - L\ln\left|\frac{L-\sqrt{L^2-x^2}}{x}\frac{L+\sqrt{L^2-x^2}}{L+\sqrt{L^2-x^2}}\right| + C$$

$$= -\sqrt{L^2-x^2} - L\ln\left|\frac{x}{L+\sqrt{L^2-x^2}}\right| + C = L\ln\left|\frac{L+\sqrt{L^2-x^2}}{x}\right| - \sqrt{L^2-x^2} + C.$$

Since $y = 0$ when $x = L$, it follows that $C = 0$, and

$$y = L\ln\left(\frac{L+\sqrt{L^2-x^2}}{x}\right) - \sqrt{L^2-x^2}.$$

22. For this function, $\dfrac{dy}{dx} - 3x^2(y-1)^2 = \dfrac{3x^2}{(x^3+C)^2} - 3x^2(x^3+C)^{-2} = 0$. A singular solution is $y(x) \equiv 1$.

24. (a) Differentiation of $y^2 = Cx$ (implicitly) with respect to x gives $2y\dfrac{dy}{dx} = C$. If we substitute $C = y^2/x$, we obtain

$$2y\frac{dy}{dx} = \frac{y^2}{x} \implies 2x\frac{dy}{dx} = y.$$

(b) The curves are the parabolas shown to the right.
(c) For that solution passing through (x_0, y_0), C must satisfy $y_0^2 = Cx_0$. Thus, $C = y_0^2/x_0$, except when $x_0 = 0$.

26. If we integrate both sides of the differential equation with respect to x, we obtain the general solution

$$y(x) = \begin{cases} -1/x + C, & x < 0 \\ -1/x + D, & x > 0 \end{cases}.$$

(a) For the solution to satisfy $y(1) = 1$, it is necessary for $D = 2$. Constant C is undetermined.
(b) For the solution to satisfy $y(-1) = 2$, it is necessary for $C = 1$. Constant D is undetermined.
(c) For the solution to satisfy $y(1) = 1$ and $y(-1) = 2$, we choose $C = 1$ and $D = 2$.

EXERCISES 15.2

2. This equation can be separated, $\dfrac{1}{2-y}\,dy = 2x\,dx$, (provided $y \neq 2$), and therefore a one-parameter family of solutions is defined implicitly by $-\ln|2-y| = x^2 + C$. This equation can be solved for $y(x) = 2 + De^{-x^2}$, where $D = \pm e^{-C}$. The function $y = 2$ is a solution of the differential equation. If we allow D to be equal to zero, this solution is contained in the family, and it is not, therefore, a singular solution.

4. When this equation is separated, $\dfrac{1}{3y+2}\,dy = dx$, (provided $y \neq -2/3$). A one-parameter family of solutions is defined implicitly by $(1/3)\ln|3y+2| = x + C$. This equation can be solved for $y(x) = De^{3x} - 2/3$, where $D = \pm(1/3)e^{3C}$. The function $y = -2/3$ is a solution of the differential equation. If we allow D to be equal to zero, this solution is contained in the family, and it is not, therefore, a singular solution.

6. This equation can be separated, $\dfrac{x^2\,dx}{1-x} = \dfrac{y^2\,dy}{1+y}$, (provided $y \neq -1$), and a one-parameter family of solutions is therefore defined implicitly by $\displaystyle\int \dfrac{x^2}{1-x}\,dx = \int \dfrac{y^2}{1+y}\,dy$. To integrate, we write

$$\int\left(-x - 1 + \frac{1}{1-x}\right)dx = \int\left(y - 1 + \frac{1}{y+1}\right)dy,$$

and hence, $-\dfrac{x^2}{2} - x - \ln|x-1| + C = \dfrac{y^2}{2} - y + \ln|y+1|$. The function $y = -1$ is a solution of the differential equation, and because it is not contained in the one-parameter family, it is a singular solution.

8. When we separate this equation, $y^2\,dy = \left(\dfrac{1-x^2 e^x}{x}\right)dx$, and a one-parameter family of solutions is defined implicitly by $\dfrac{y^3}{3} = \ln|x| - \displaystyle\int xe^x\,dx = \ln|x| - xe^x + e^x + C$. Explicitly we obtain $y(x) = [D + 3\ln|x| - 3xe^x + 3e^x]^{1/3}$, where $D = 3C$.

10. When we separate this equation, $\dfrac{dy}{1+y^2} = \dfrac{dx}{1+x^2}$, and a one-parameter family of solutions is defined implicitly by $\operatorname{Tan}^{-1}y = \operatorname{Tan}^{-1}x + C$. If we apply the tangent function to both sides of $C = \operatorname{Tan}^{-1}y - \operatorname{Tan}^{-1}x$, we find

$$\tan C = D = \tan\left(\operatorname{Tan}^{-1}y - \operatorname{Tan}^{-1}x\right) = \frac{\tan\left(\operatorname{Tan}^{-1}y\right) - \tan\left(\operatorname{Tan}^{-1}x\right)}{1 + \tan\left(\operatorname{Tan}^{-1}y\right)\tan\left(\operatorname{Tan}^{-1}x\right)} = \frac{y - x}{1 + xy}.$$

When this equation is solved for y the result is $y(x) = \dfrac{x + D}{1 - Dx}$.

12. We separate this equation, $\left(\dfrac{y-1}{y}\right)dy = \left(\dfrac{x+1}{x}\right)dx$, (provided $y \neq 0$). A one-parameter family of solutions is defined implicitly by $y - \ln|y| = x + \ln|x| + C$. When we take exponentials, $xy = De^{y-x}$, where $D = \pm e^{-C}$. To satisfy $y(1) = 2$, we must have $(1)(2) = De^{2-1}$. Thus, $D = 2/e$, and $xy = 2e^{y-x-1}$.

14. When we separate variables, $\dfrac{1}{1+y^2}\,dy = 2x\,dx$. A one-parameter family of solutions is defined implicitly by $\operatorname{Tan}^{-1}y = x^2 + C$. Thus, $y(x) = \tan\left(x^2 + C\right)$. To satisfy $y(2) = 4$, we must have $4 = \tan\left(4 + C\right)$. This implies that $C = \operatorname{Tan}^{-1}4 - 4 + n\pi$, where n is an integer, and hence

$$y(x) = \tan\left(x^2 + \operatorname{Tan}^{-1}4 - 4 + n\pi\right) = \frac{\tan\left(x^2 - 4\right) + \tan\left(\operatorname{Tan}^{-1}4 + n\pi\right)}{1 - \tan\left(x^2 - 4\right)\tan\left(\operatorname{Tan}^{-1}4 + n\pi\right)} = \frac{4 + \tan\left(x^2 - 4\right)}{1 - 4\tan\left(x^2 - 4\right)}.$$

16. (a) Since the girl's speed is proportional to x^2, $dx/dt = kx^2$. Separation of this equation gives

$$\frac{1}{x^2}\,dx = k\,dt,$$

a one-parameter family of solutions of which is defined implicitly by $-1/x = kt + C$. Thus, $x = -1/(kt + C)$. If we choose time $t = 0$ when $x = 6$, then $6 = -1/C$ and $x = 6/(1 - 6kt)$ km.

(b) The girl reaches school when $x = 0$, but this only happens after an infinitely long time.

18. If $T(t)$ represents the temperature of the water as a function of time t, then according to Newton's law of cooling, $\dfrac{dT}{dt} = k(T - 20)$, where $k < 0$ is a constant. Separation of variables leads to $\dfrac{1}{T - 20}\,dT = k\,dt$, and therefore $\ln|T - 20| = kt + C$. When we solve for T, the result is $T(t) = 20 + De^{kt}$. If we choose time $t = 0$ when $T = 80$, then $80 = 20 + D$. Thus, $D = 60$, and $T(t) = 20 + 60e^{kt}$. Because $T(2) = 60$, it follows that $60 = 20 + 60e^{2k}$, from which $k = (1/2)\ln(2/3) = -0.203$. Finally then, $T(t) = 20 + 60e^{-0.203t}$, or, $T(t) = 20 + 60e^{(t/2)\ln(2/3)} = 20 + 60(2/3)^{t/2}$.

20. If $A(t)$ represents the amount of drug in the body at time t (in hours), then $\dfrac{dA}{dt} = kA$, where $k < 0$ is a constant. Separation of variables gives $\dfrac{1}{A}\, dA = k\, dt$, and therefore $\ln |A| = kt + C$, or, $A = De^{kt}$. If A_0 is the size of the original dose injected at time $t = 0$, then $A_0 = D$, and $A = A_0 e^{kt}$. Since $A(1) = 0.95 A_0$, it follows that $0.95 A_0 = A_0 e^{k}$. Thus, $k = \ln(0.95)$, and $A = A_0 e^{t \ln(0.95)}$. The dose decreases to $A_0/2$ when $A_0/2 = A_0 e^{t \ln(0.95)}$, the solution of which is $t = -\ln 2 / \ln(0.95) = 13.51$ h.

22. Because the total rate of change dA/dt of the amount of glucose in the bloodstream is the rate at which it is added less the rate at which it is used up, $\dfrac{dA}{dt} = R - kA$, where $k > 0$ is a constant. This equation can be separated, $\dfrac{1}{R - kA}\, dA = dt$, and therefore a one-parameter family of solutions is defined implicitly by $-(1/k) \ln |R - kA| = t + C$. When this equation is solved for A, the result is $A(t) = R/k + De^{-kt}$. If we choose time $t = 0$ when $A = A_0$, then $A_0 = R/k + D$, and

$$A(t) = \frac{R}{k} + \left(A_0 - \frac{R}{k} \right) e^{-kt} = \frac{R}{k} \left(1 - e^{-kt} \right) + A_0 e^{-kt}.$$

24. If x is the number of grams of C at time t, then $x/2$ came from each of A and B. This means that there is $10 - x/2$ grams of A and $15 - x/2$ grams of B remaining. Hence, the rate dx/dt at which C is formed is related to x by

$$\frac{dx}{dt} = K \left(10 - \frac{x}{2} \right) \left(15 - \frac{x}{2} \right) = \frac{K}{4}(20 - x)(30 - x) = k(20 - x)(30 - x),$$

where we have set $k = K/4$. If we choose time $t = 0$ when A and B are brought together, then $x(t)$ must also satisfy $x(0) = 0$. We separate the equation and use partial fractions to write it in the form

$$k\, dt = \frac{1}{(20 - x)(30 - x)}\, dx = \left(\frac{1/10}{20 - x} - \frac{1/10}{30 - x} \right) dx.$$

A one-parameter family of solutions is defined implicitly by

$$kt + C = \frac{1}{10} \left[-\ln(20 - x) + \ln(30 - x) \right].$$

Absolute values are unnecessary because x cannot exceed 20. We now solve this equation for x by writing

$$10(kt + C) = \ln \left(\frac{30 - x}{20 - x} \right), \quad \text{and exponentiating,} \quad \frac{30 - x}{20 - x} = De^{10kt},$$

where $D = e^{10C}$. Cross multiplying gives $(20 - x)De^{10kt} = 30 - x$, and therefore $x = \dfrac{30 - 20De^{10kt}}{1 - De^{10kt}}$. The initial condition $x(0) = 0$ requires $D = 3/2$, in which case

$$x = \frac{30 - 30e^{10kt}}{1 - (3/2)e^{10kt}} = \frac{60 \left(1 - e^{10kt} \right)}{2 - 3e^{10kt}} \text{ g} \,.$$

26. In Section 5.5 we showed that if V is the volume of water in the left sphere, then

$$\frac{dV}{dt} = A(y)\frac{dy}{dt}$$

where $A(y)$ is the surface area of the water. Since water leaves this container at rate

$$\frac{a}{3}\sqrt{2gh} = \frac{a}{3}\sqrt{2g(2y)} = \frac{2a}{3}\sqrt{gy},$$

it follows that

$$A(y)\frac{dy}{dt} = -\frac{2a}{3}\sqrt{gy}.$$

Because $A(y) = \pi x^2 = \pi(R^2 - y^2)$,

$$\pi(R^2 - y^2)\frac{dy}{dt} = -\frac{2a}{3}\sqrt{gy} \quad\Longrightarrow\quad \frac{R^2 - y^2}{\sqrt{y}}\,dy = -\frac{2a\sqrt{g}}{3\pi}\,dt,$$

a separated equation. A one-parameter family of solutions is defined implicitly by

$$2R^2\sqrt{y} - \frac{2}{5}y^{5/2} = -\frac{2a\sqrt{g}t}{3\pi} + C.$$

If we choose $t = 0$ when $y = R$, then $2R^2\sqrt{R} - \frac{2}{5}R^{5/2} = C$, and

$$2R^2\sqrt{y} - \frac{2}{5}y^{5/2} = -\frac{2a\sqrt{g}t}{3\pi} + \frac{8}{5}R^{5/2}.$$

The water levels are the same in both spheres when $y = 0$, in which case

$$0 = -\frac{2a\sqrt{g}t}{3\pi} + \frac{8}{5}R^{5/2} \quad\Longrightarrow\quad t = \frac{12\pi R^{5/2}}{5a\sqrt{g}}.$$

28. Volume of water in the lock above the downstream level is $V = (8)(16)(2 - h)$. Consequently,

$$\frac{dV}{dt} = -128\frac{dh}{dt}.$$

Since water enters the lock at $0.04\sqrt{2gh}$ m^3/s, it follows that

$$-128\frac{dh}{dt} = 0.04\sqrt{2gh} \quad\Longrightarrow\quad \frac{1}{\sqrt{h}}\,dh = -\frac{\sqrt{2g}}{3200}\,dt,$$

a separated equation. A one-parameter family of solutions is defined implicitly by

$$2\sqrt{h} = -\frac{\sqrt{2g}t}{3200} + C.$$

If valve A is opened at $t = 0$ (when $h = 2$), then $2\sqrt{2} = C$, and

$$2\sqrt{h} = -\frac{\sqrt{2g}t}{3200} + 2\sqrt{2}.$$

The upstream gate is opened when $h = 0.02$ in which case

$$2\sqrt{0.02} = -\frac{\sqrt{2g}t}{3200} + 2\sqrt{2} \quad\Longrightarrow\quad t = -\frac{3200}{\sqrt{2g}}(2\sqrt{0.02} - 2\sqrt{2}) = 1839.$$

Operation of the lock takes a little over 30 minutes.

30. The differential equation is separable, $\dfrac{1}{2gh - v^2}dv = \dfrac{1}{2L}dt$, so that a one-parameter of solutions is defined implicitly by

$$\frac{t}{2L} + C = \int \frac{1}{2gh - v^2}dv = \int \left(\frac{\frac{1}{2\sqrt{2gh}}}{\sqrt{2gh} + v} + \frac{\frac{1}{2\sqrt{2gh}}}{\sqrt{2gh} - v} \right) dv$$

$$= \frac{1}{2\sqrt{2gh}} \left(\ln|\sqrt{2gh} + v| - \ln|\sqrt{2gh} - v| \right) = \frac{1}{2\sqrt{2gh}} \ln\left| \frac{\sqrt{2gh} + v}{\sqrt{2gh} - v} \right|.$$

Exponentiation gives

$$\frac{\sqrt{2gh} + v}{\sqrt{2gh} - v} = De^{\sqrt{2gh}t/L} \;\Longrightarrow\; \sqrt{2gh} + v = De^{\sqrt{2gh}t/L}(\sqrt{2gh} - v) \;\Longrightarrow\; v = \frac{D\sqrt{2gh}e^{\sqrt{2gh}t/L} - \sqrt{2gh}}{1 + De^{\sqrt{2gh}t/L}}.$$

Since $v(0) = 0$,

$$0 = \frac{D\sqrt{2gh} - \sqrt{2gh}}{1 + D} \qquad \Longrightarrow \qquad D = 1.$$

Thus, $v(t) = \dfrac{\sqrt{2gh}(e^{\sqrt{2gh}\,t/L} - 1)}{e^{\sqrt{2gh}\,t/L} + 1}$.

32. The differential equation can be separated, and partial fractions leads to

$$\frac{1}{A}\,dA = \frac{M^2 - 1}{M\left[\left(\dfrac{k-1}{2}\right)M^2 + 1\right]}\,dM = \left[\frac{(k+1)M/2}{\left(\dfrac{k-1}{2}\right)M^2 + 1} - \frac{1}{M}\right]dM.$$

A one-parameter family of solutions is defined implicitly by

$$\ln A = \left(\frac{k+1}{2}\right)\left(\frac{1}{k-1}\right)\ln\left[\left(\frac{k-1}{2}\right)M^2 + 1\right] - \ln M + C,$$

from which

$$A = \frac{D\left[\left(\dfrac{k-1}{2}\right)M^2 + 1\right]^{(k+1)/(2k-2)}}{M}.$$

The condition $A(1) = A_0$ gives

$$A_0 = D\left[\left(\frac{k-1}{2}\right) + 1\right]^{(k+1)/(2k-2)} = D\left(\frac{k+1}{2}\right)^{(k+1)/(2k-2)},$$

from which

$$A = \frac{A_0}{M}\frac{\left[\left(\dfrac{k-1}{2}\right)M^2 + 1\right]^{(k+1)/(2k-2)}}{\left(\dfrac{k+1}{2}\right)^{(k+1)/(2k-2)}} = \frac{A_0}{M}\left[\frac{(k-1)M^2 + 2}{k+1}\right]^{(k+1)/(2k-2)}.$$

34. If we set $v = y/x$ or $y = vx$, then $\dfrac{dy}{dx} = v + x\dfrac{dv}{dx}$. Substitution into the differential equation gives

$$v + x\frac{dv}{dx} = f(v) \qquad \Longrightarrow \qquad \frac{1}{f(v) - v}\,dv = \frac{1}{x}\,dx,$$

which is separated.

36. We write $\dfrac{dy}{dx} = \dfrac{x^2 - y^2}{xy} = \dfrac{1 - (y/x)^2}{y/x}$, a homogeneous differential equation. If we set $v = y/x$, or $y = vx$,

then $dy/dx = v + x\,dv/dx$, and $v + x\dfrac{dv}{dx} = \dfrac{1 - v^2}{v}$. This can be separated in the form

$$\frac{1}{x}\,dx = \frac{1}{\dfrac{1 - v^2}{v} - v}\,dv = \frac{v}{1 - 2v^2}\,dv.$$

A one-parameter family of solutions is defined implicitly by $-(1/4)\ln|1 - 2v^2| = \ln|x| + C$. Exponentiation of both sides leads to $2v^2 = (x^4 - D)/x^4$, and when we substitute $v = y/x$, the solution reduces to $x^2(x^2 - 2y^2) = D$.

38. When we write $\dfrac{dy}{dx} = \dfrac{y/x+1}{y/x-1}$, the differential equation is clearly homogeneous. We therefore set $v = y/x$,

or, $y = vx$, in which case $dy/dx = v + x\,dv/dx$, and $v + x\dfrac{dv}{dx} = \dfrac{v+1}{v-1}$. This can be separated in the form

$\dfrac{v-1}{-v^2+2v+1}\,dv = \dfrac{1}{x}\,dx$. A one-parameter family of solutions of this equation is defined implicitly by $-(1/2)\ln|-v^2+2v+1| = \ln|x| + C$. When this equation is exponentiated, $-v^2 + 2v + 1 = D/x^2$, and substitution of $v = y/x$ gives $x^2 + 2xy - y^2 = D$.

40. We write $\dfrac{dy}{dx} = \dfrac{e^{-y/x} + (y/x)^2}{y/x}$, a homogeneous differential equation. When we set $v = y/x$, or, $y = vx$,

then $dy/dx = v + x\,dv/dx$, and $v + x\dfrac{dv}{dx} = \dfrac{e^{-v}+v^2}{v} \implies ve^v\,dv = \dfrac{1}{x}\,dx$, a separated differential equation with one-parameter family of solutions defined implicitly by $ve^v - e^v = \ln|x| + C$. Substitution of $v = y/x$ leads to $e^{y/x}(y-x) = x\ln|x| + Cx$.

42. Let $P(x_0, y_0)$ be any point on the required curve $y = f(x)$. The equation of the tangent line at (x_0, y_0) is

$$y - y_0 = f'(x_0)(x - x_0).$$

The y-intercept of this line is $y_0 - f'(x_0)x_0$.
Since $\|OQ\|^2 = \|PQ\|^2$,

$$[y_0 - f'(x_0)x_0]^2 = x_0^2 + [f'(x_0)x_0]^2, \text{ or,}$$
$$y_0^2 - 2f'(x_0)x_0y_0 + [f'(x_0)]^2x_0^2 = x_0^2 + [f'(x_0)x_0]^2.$$

Thus, $y_0^2 - x_0^2 = 2x_0y_0 f'(x_0)$. Since this must be valid at every point on the curve, we drop the subscripts and set $f'(x) = dy/dx$,

$$y^2 - x^2 = 2xy\frac{dy}{dx} \implies \frac{dy}{dx} = \frac{y^2 - x^2}{2xy}.$$

In this homogeneous differential equation, we set $y = vx$ and $dy/dx = v + x\,dv/dx$,

$$v + x\frac{dv}{dx} = \frac{v^2x^2 - x^2}{2vx^2} = \frac{v^2 - 1}{2v}.$$

Thus, $x\dfrac{dv}{dx} = \dfrac{v^2-1}{2v} - v = -\dfrac{v^2+1}{2v} \implies \dfrac{2v}{v^2+1}\,dv = -\dfrac{1}{x}\,dx$. A one-parameter family of solutions is defined implicitly by $\ln(v^2+1) = -\ln|x| + C$, or, $v^2 + 1 = D/x$. Substitution of $v = y/x$ now gives $x^2 + y^2 = Dx$. Since $(1,2)$ is on the curve, it follows that $1 + 4 = D$, and the required curve is $y = \sqrt{5x - x^2}$.

44. If $y(t)$ represents the amount of trypsin at any given time, then

$$\frac{dy}{dt} = ky[A - (y - y_0)] = ky(A + y_0 - y),$$

where k is a constant. This equation can be separated,

$$k\,dt = \frac{1}{y(A + y_0 - y)}\,dy = \left(\frac{\dfrac{1}{A+y_0}}{y} + \frac{\dfrac{1}{A+y_0}}{A + y_0 - y}\right)dy.$$

Thus, $\left(\dfrac{1}{y} + \dfrac{1}{A + y_0 - y}\right)dy = k(A + y_0)\,dt$, a one-parameter family of solutions of which is defined implicitly by $\ln y - \ln(A + y_0 - y) = k(A + y_0)t + C$. When this equation is exponentiated and solved for y, the result is

$$y(t) = \frac{D(A + y_0)e^{k(A+y_0)t}}{1 + De^{k(A+y_0)t}}.$$

Since $y(0) = y_0$, it follows that $y_0 = \dfrac{D(A+y_0)}{1+D}$, from which $D = y_0/A$. Hence,

$$y(t) = \frac{\dfrac{y_0}{A}(A+y_0)e^{k(A+y_0)t}}{1 + \dfrac{y_0}{A}e^{k(A+y_0)t}} = \frac{y_0(A+y_0)}{y_0 + Ae^{-k(A+y_0)t}}.$$

46. If we set $v = x + y$, then $dv/dx = 1 + dy/dx$, and $\dfrac{dv}{dx} - 1 = v$. This equation can be separated, $\dfrac{1}{v+1}\,dv = dx$, a one-parameter family of solutions of which is defined implicitly by $\ln|v+1| = x + C$. Exponentiation and substitution of $v = x + y$ leads to $y = De^x - x - 1$.

48. If we set $v = 2x + 3y$, then $dv/dx = 2 + 3\,dy/dx$, and $\dfrac{1}{3}\dfrac{dv}{dx} - \dfrac{2}{3} = \dfrac{1}{v}$. This equation can be separated, $\dfrac{v}{3+2v}\,dv = dx$, a one-parameter family of solutions of which is defined implicitly by

$$x + C = \int\left(\frac{1}{2} - \frac{3/2}{2v+3}\right)dv = \frac{v}{2} - \frac{3}{4}\ln|2v+3|.$$

Substitution of $v = 2x + 3y$ gives $6y - 3\ln|4x + 6y + 3| = D$.

50. When we substitute for k and C, and separate variables,

$$\frac{dN}{N(1 - N/10^6)} = dt \implies \frac{10^6\,dN}{N(10^6 - N)} = dt \implies \left(\frac{1}{N} + \frac{1}{10^6 - N}\right)dN = dt.$$

A one-parameter family of solutions is defined implicitly by

$$\ln|N| - \ln|10^6 - N| = t + D \implies \ln\left|\frac{N}{10^6 - N}\right| = t + D \implies \frac{N}{10^6 - N} = \pm e^{t+D} = Ee^t,$$

where $E = \pm e^D$. Consequently, $N = (10^6 - N)Ee^t \implies N = \dfrac{10^6 Ee^t}{1 + Ee^t} = \dfrac{10^6}{1 + Fe^{-t}}$, where $F = 1/E$. From $N(0) = 100$, we obtain $100 = 10^6/(1 + F) \implies F = 9999$. The number of bacteria is therefore $N(t) = 10^6/(1 + 9999e^{-t})$.

52. The differential equation is separable, $\dfrac{dw}{aw^{2/3} - bw} = dt$, in which case a one-parameter family of solutions is defined implicitly by

$$\int \frac{1}{aw^{2/3} - bw}\,dw = t + C.$$

If we set $u = w^{1/3} \implies w = u^3$, and $dw = 3u^2\,du$, then

$$t + C = \int \frac{3u^2}{au^2 - bu^3}\,du = 3\int \frac{1}{a - bu}\,du = -\frac{3}{b}\ln|a - bu|.$$

Consequently,

$$\ln|a - bu| = -\frac{b}{3}(t + C) \implies |a - bu| = e^{-b(t+C)/3} \implies u = \frac{a}{b} + De^{-bt/3},$$

where $D = \pm(1/b)e^{-bC/3}$. Since $u = w^{1/3}$, we find $w = (a/b + De^{-bt/3})^3$. From $w(0) = w_0$, we obtain $w_0 = (a/b + D)^3 \implies D = w_0^{1/3} - a/b$. Finally, then

$$w(t) = \left[\frac{a}{b} + \left(w_0^{1/3} - \frac{a}{b}\right)e^{-bt/3}\right]^3 = \left[\frac{a}{b}\left(1 - e^{-bt/3}\right) + w_0^{1/3}e^{-bt/3}\right]^3.$$

54. When H is a constant, the differential becomes

$$\frac{dv}{dt} = \frac{K}{n}\left[\frac{v^2 - (D - \gamma - H)v - DH}{v(v - D)}\right] \quad \Longrightarrow \quad \frac{v(v - D)}{v^2 - (D - \gamma - H)v - DH}\,dv = \frac{K}{n}\,dt.$$

a separated differential equation. A one-parameter family of solutions is defined implicitly by

$$\int\left[1 + \frac{-(H + \gamma)v + DH}{v^2 - (D - \gamma - H)v - DH}\right]\,dv = \frac{K}{n}t + C.$$

The roots of the quadratic $v^2 - (D - \gamma - H)v - DH = 0$ are

$$v = \frac{(D - \gamma - H) \pm \sqrt{(D - \gamma - H)^2 + 4DH}}{2}.$$

Let us denote them by r_1 and r_2 where r_1 uses the positive radical and r_2 the negative one. Then

$$\frac{Kt}{n} + C = \int\left[1 + \frac{DH - (H + \gamma)v}{(v - r_1)(v - r_2)}\right]\,dv = v + \int\left[\frac{\dfrac{DH - (H + \gamma)r_1}{r_1 - r_2}}{v - r_1} + \frac{\dfrac{DH - (H + \gamma)r_2}{r_2 - r_1}}{v - r_2}\right]\,dv$$

$$= v + \frac{1}{r_1 - r_2}\left\{[DH - (H + \gamma)r_1]\ln|v - r_1| - [DH - (H + \gamma)r_2]\ln|v - r_2|\right\}.$$

For $v(0) = 0$,

$$\frac{1}{r_1 - r_2}\left\{[DH - (H + \gamma)r_1]\ln r_1 - [DH - (H + \gamma)r_2]\ln r_2\right\} = C.$$

Thus, $v(t)$ is defined implicitly by

$$\frac{Kt}{n} = v + \frac{1}{r_1 - r_2}\left\{[DH - (H + \gamma)r_1]\ln|v - r_1| - [DH - (H + \gamma)r_2]\ln|v - r_2|\right\}$$

$$- \frac{1}{r_1 - r_2}\left\{[DH - (H + \gamma)r_1]\ln r_1 - [DH - (H + \gamma)r_2]\ln r_2\right\}$$

$$= v + \frac{1}{r_1 - r_2}\left\{[DH - (H + \gamma)r_1]\ln\left|\frac{v - r_1}{r_1}\right| - [DH - (H + \gamma)r_2]\ln\left|\frac{v - r_2}{r_2}\right|\right\}.$$

56. If $S(t)$ represents the amount of drug in the dog as a function of time t, then $\dfrac{dS}{dt} = kS$, where $k < 0$ is a constant. Separation of this equation gives $\dfrac{1}{S}\,dS = k\,dt$, a one-parameter family of solutions of which is defined by $\ln S = kt + C$. Exponentiation gives $S = De^{kt}$. If S_0 represents the amount injected at time $t = 0$, then $S_0 = D$, and $S = S_0 e^{kt}$. Since $S = S_0/2$ when $t = 5$, it follows that $S_0/2 = S_0 e^{5k}$, and this implies that $k = -(1/5)\ln 2$. At the end of the one hour operation, the amount of drug in the body must be 400 mg, and therefore $400 = S_0 e^k$. Thus, $S_0 = 400e^{-k} = 459.5$ mg.

58. We can separate this equation, $\left(\dfrac{y^6 - 1}{y^4}\right)\,dy = (x^2 + 2)\,dx$, and therefore a one-parameter family of solutions is defined implicitly by $\dfrac{y^3}{3} + \dfrac{1}{3y^3} = \dfrac{x^3}{3} + 2x + C$. For $y(1) = 1$, we must have $1/3 + 1/3 = 1/3 + 2 + C$. Thus, $C = -5/3$, and $\dfrac{y^3}{3} + \dfrac{1}{3y^3} = \dfrac{x^3}{3} + 2x - \dfrac{5}{3}$. Multiplication by $3y^3$ gives $y^3(x^3 + 6x - 5) = 1 + y^6$.

EXERCISES 15.3

2. An integrating factor is $e^{\int 2/x\,dx} = e^{2\ln|x|} = x^2$. When the differential equation is multiplied by x^2,

$$x^2\frac{dy}{dx} + 2xy = 6x^5 \qquad \Longrightarrow \qquad \frac{d}{dx}(yx^2) = 6x^5.$$

Integration now gives $yx^2 = x^6 + C \implies y = x^4 + C/x^2$.

4. An integrating factor is $e^{\int \cot x\,dx} = e^{\ln|\sin x|} = |\sin x|$. For either $\sin x < 0$ or $\sin x > 0$, multiplication of the differential equation by $|\sin x|$ gives

$$\sin x\frac{dy}{dx} + y\cos x = 5\sin x e^{\cos x} \qquad \Longrightarrow \qquad \frac{d}{dx}(y\sin x) = 5\sin x\, e^{\cos x}.$$

Integration gives $y\sin x = -5e^{\cos x} + C \implies y = \csc x(C - 5e^{\cos x})$.

6. If we write $\dfrac{dy}{dx} - \dfrac{2}{x+1}y = 2$, an integrating factor is $e^{\int -2/(x+1)\,dx} = e^{-2\ln|x+1|} = 1/(x+1)^2$. When we multiply the differential equation by $1/(x+1)^2$,

$$\frac{1}{(x+1)^2}\frac{dy}{dx} - \frac{2}{(x+1)^3}y = \frac{2}{(x+1)^2} \qquad \Longrightarrow \qquad \frac{d}{dx}\left[\frac{y}{(x+1)^2}\right] = \frac{2}{(x+1)^2}.$$

Integration now gives $\dfrac{y}{(x+1)^2} = \dfrac{-2}{x+1} + C \implies y = -2(x+1) + C(x+1)^2$.

8. Since $dy/dx - y = e^{2x}$, an integrating factor is $e^{\int -dx} = e^{-x}$. When we multiply the differential equation by this integrating factor,

$$e^{-x}\frac{dy}{dx} - ye^{-x} = e^x \qquad \Longrightarrow \qquad \frac{d}{dx}(ye^{-x}) = e^x.$$

Integration gives $ye^{-x} = e^x + C \implies y = e^{2x} + Ce^x$.

10. Since $\dfrac{dy}{dx} + \left(\dfrac{2-3x^2}{x^3}\right)y = 1$, an integrating factor is $e^{\int \frac{2-3x^2}{x^3}\,dx} = e^{-1/x^2 - 3\ln|x|} = \dfrac{e^{-1/x^2}}{|x|^3}$. For either $x < 0$ or $x > 0$, multiplication of the differential equation by this factor gives

$$\frac{1}{x^3}e^{-1/x^2}\frac{dy}{dx} + \left(\frac{2-3x^2}{x^6}\right)e^{-1/x^2}y = \frac{1}{x^3}e^{-1/x^2} \qquad \Longrightarrow \qquad \frac{d}{dx}\left(\frac{y}{x^3}e^{-1/x^2}\right) = \frac{1}{x^3}e^{-1/x^2}.$$

Integration gives $\dfrac{y}{x^3}e^{-1/x^2} = \dfrac{1}{2}e^{-1/x^2} + C \implies y = x^3/2 + Cx^3e^{1/x^2}$.

12. If we write $\dfrac{dy}{dx} - (2\cot 2x)y = 1 - 2x\cot 2x - 2\csc 2x$, an integrating factor is $e^{\int -2\cot 2x\,dx} = e^{-\ln|\sin 2x|} = |\csc 2x|$. For either $\csc 2x < 0$ or $\csc 2x > 0$, multiplication of the differential equation by $|\csc 2x|$ leads to

$$\csc 2x\frac{dy}{dx} - 2y\cot 2x\csc 2x = \csc 2x - 2x\cot 2x\csc 2x - 2\csc^2 2x,$$

or,

$$\frac{d}{dx}(y\csc 2x) = \csc 2x - 2x\cot 2x\csc 2x - 2\csc^2 2x.$$

Integration gives $y\csc 2x = x\csc 2x + \cot 2x + C \implies y = x + \cos 2x + C\sin 2x$.

14. Since $dy/dx + y = e^x \sin x$, an integrating factor is $e^{\int dx} = e^x$. Multiplication of the differential equation by this factor gives

$$e^x \frac{dy}{dx} + ye^x = e^{2x} \sin x \qquad \Longrightarrow \qquad \frac{d}{dx}(ye^x) = e^{2x} \sin x.$$

Integration now yields $ye^x = \frac{1}{5}(2e^{2x} \sin x - e^{2x} \cos x) + C \implies y = e^x(2 \sin x - \cos x)/5 + Ce^{-x}$. For $y(0) = -1$, we must have $-1 = -1/5 + C$. Thus, $y = e^x(2 \sin x - \cos x)/5 - (4/5)e^{-x}$.

16. If we write $dx/dy + (1/y)x = y^2$, the differential equation is linear in $x = x(y)$. An integrating factor is $e^{\int (1/y) \, dy} = e^{\ln |y|} = |y|$. For either $y < 0$ or $y > 0$, multiplication by $|y|$ yields

$$y \frac{dx}{dy} + x = y^3 \qquad \Longrightarrow \qquad \frac{d}{dy}(yx) = y^3.$$

Integration gives $yx = \dfrac{y^4}{4} + C$, an implicit definition for solutions.

18. If we set $z = 1/y$, then $dz/dx = (-1/y^2)dy/dx$, and $-y^2 \dfrac{dz}{dx} + y = y^2 e^x$. Division by $-y^2$ gives $\dfrac{dz}{dx} - \dfrac{1}{y} = -e^x \implies \dfrac{dz}{dx} - z = -e^x$. An integrating factor for this equation is $e^{\int -dx} = e^{-x}$. When we multiply the differential equation by this factor,

$$e^{-x} \frac{dz}{dx} - ze^{-x} = -1 \qquad \Longrightarrow \qquad \frac{d}{dx}(ze^{-x}) = -1.$$

Integration now yields

$$ze^{-x} = -x + C \implies z = (C - x)e^x \implies \frac{1}{y} = (C - x)e^x \implies y = \frac{e^{-x}}{C - x}.$$

20. If we set $z = 1/y$, then $dz/dx = -(1/y^2)dy/dx$, and $-y^2 \dfrac{dz}{dx} - y = -(x^2 + 2x)y^2$. Division by $-y^2$ gives $\dfrac{dz}{dx} + \dfrac{1}{y} = x^2 + 2x \implies \dfrac{dz}{dx} + z = x^2 + 2x$. An integrating factor for this equation is $e^{\int dx} = e^x$. Multiplication by this factor now yields

$$e^x \frac{dz}{dx} + ze^x = (x^2 + 2x)e^x \qquad \Longrightarrow \qquad \frac{d}{dx}(ze^x) = (x^2 + 2x)e^x.$$

This can be integrated to give

$$ze^x = \int (x^2 + 2x)e^x \, dx = x^2 e^x + C \implies z = \frac{1}{y} = x^2 + Ce^{-x} \implies y = \frac{1}{x^2 + Ce^{-x}}.$$

22. If we set $z = y^{-3}$, then $dz/dx = -(3/y^4)dy/dx$, and $-\dfrac{y^4}{3} \dfrac{dz}{dx} + y \tan x = y^4 \sin x$. Multiplication by $-3y^{-4}$ gives

$$\frac{dz}{dx} - \frac{3 \tan x}{y^3} = -3 \sin x \qquad \Longrightarrow \qquad \frac{dz}{dx} - (3 \tan x)z = -3 \sin x.$$

An integrating factor for this equation is $e^{\int -3 \tan x \, dx} = e^{3 \ln |\cos x|} = |\cos x|^3$. For either $\cos x < 0$ or $\cos x > 0$, multiplication by $|\cos x|^3$ gives

$$\cos^3 x \frac{dz}{dx} - 3z \cos^2 x \sin x = -3 \cos^3 x \sin x \qquad \Longrightarrow \qquad \frac{d}{dx}(z \cos^3 x) = -3 \sin x \cos^3 x.$$

Integration now yields

$$z \cos^3 x = \frac{3}{4} \cos^4 x + C \implies z = \frac{3}{4} \cos x + C \sec^3 x.$$

When we replace z with $1/y^3$,

$$\frac{1}{y^3} = \frac{3}{4}\cos x + C\sec^3 x \quad \Longrightarrow \quad y = \left(\frac{3}{4}\cos x + C\sec^3 x\right)^{-1/3}.$$

24. If $S(t)$ represents the number of grams of sugar in the tank as a function of time t, then dS/dt, the rate of change of S with respect to t, must be the rate at which sugar is added less the rate at which it is removed. The rate at which sugar is added is 2 g/min. Since the concentration of sugar at time t is $S/(10^5 + 100t)$ g/mL, the rate at which sugar is removed at time t is $100S/(10^5 + 100t)$ g/min. Thus,

$$\frac{dS}{dt} = 2 - \frac{100S}{10^5 + 100t}, \quad \Longrightarrow \quad \frac{dS}{dt} + \frac{S}{1000 + t} = 2.$$

An integrating factor for this equation is $e^{\int 1/(1000+t)\,dt} = e^{\ln|1000+t|} = 1000 + t$. Multiplication of the differential equation by $1000 + t$ gives

$$(1000 + t)\frac{dS}{dt} + S = 2(1000 + t) \quad \Longrightarrow \quad \frac{d}{dt}[(1000 + t)S] = 2(1000 + t) \quad \Longrightarrow \quad (1000 + t)S = (1000 + t)^2 + C.$$

Since $S(0) = 4000$, it follows that $(1000)(4000) = (1000)^2 + C$. Thus, $C = 3 \times 10^6$, and

$$S = 1000 + t + \frac{3 \times 10^6}{1000 + t} \text{ g.}$$

26. If $S(t)$ represents the number of grams of salt in the tank as a function of time t, then dS/dt, the rate of change of S with respect to t, must be the rate at which salt is added less the rate at which it is removed. The rate at which salt is added is 0.2 g/s. Since the concentration of salt at time t is $S/(10^6 + 5t)$ g/mL, the rate at which salt is removed at time t is $5S/(10^6 + 5t)$ g/s. Thus,

$$\frac{dS}{dt} = \frac{1}{5} - \frac{5S}{10^6 + 5t} \quad \Longrightarrow \quad \frac{dS}{dt} + \frac{5S}{10^6 + 5t} = \frac{1}{5}.$$

An integrating factor for this equation is $e^{\int 5/(10^6 + 5t)\,dt} = e^{\ln|10^6 + 5t|} = 10^6 + 5t$. Multiplication by $10^6 + 5t$ gives

$$(10^6 + 5t)\frac{dS}{dt} + 5S = \frac{1}{5}(10^6 + 5t) \quad \Longrightarrow \quad \frac{d}{dt}[(10^6 + 5t)S] = \frac{1}{5}(10^6 + 5t).$$

Integration now yields

$$(10^6 + 5t)S = \frac{1}{50}(10^6 + 5t)^2 + C \quad \Longrightarrow \quad S(t) = \frac{1}{50}(10^6 + 5t) + \frac{C}{10^6 + 5t}.$$

Since $S(0) = 5000$, it follows that
$5000 = \dfrac{10^6}{50} + \dfrac{C}{10^6}$. This gives
$C = -15 \times 10^9$, and therefore
$$S(t) = \frac{1}{50}(10^6 + 5t) - \frac{15 \times 10^9}{10^6 + 5t} \text{ g.}$$
A graph is shown to the right; it is asymptotic to the line $S = (10^6 + 5t)/50$.

28. If $S(t)$ represents the number of grams of salt in the tank as a function of time t, then dS/dt, the rate of change of S with respect to t, must be the rate at which salt is added less the rate at which it is removed. The rate at which salt is added is 0.2 g/s. Since the concentration of salt at time t is $S/(10^6 - 10t)$ g/mL, the rate at which salt is removed at time t is $20S/(10^6 - 10t)$ g/s. Thus,

$$\frac{dS}{dt} = \frac{1}{5} - \frac{20S}{10^6 - 10t} \quad \Longrightarrow \quad \frac{dS}{dt} + \frac{2S}{10^5 - t} = \frac{1}{5},$$

valid for $0 < t < 10^5$. An integrating factor is $e^{\int 2/(10^5-t)\,dt} = e^{-2\ln(10^5-t)} = \dfrac{1}{(10^5-t)^2}$. When we multiply by this factor, the differential equation becomes

$$\frac{1}{(10^5-t)^2}\frac{dS}{dt} + \frac{2S}{(10^5-t)^3} = \frac{1}{5(10^5-t)^2} \quad\Longrightarrow\quad \frac{d}{dt}\left[\frac{S}{(10^5-t)^2}\right] = \frac{1}{5(10^5-t)^2}.$$

Integration now gives

$$\frac{S}{(10^5-t)^2} = \frac{1}{5(10^5-t)} + C \quad\Longrightarrow\quad S(t) = \frac{10^5-t}{5} + C(10^5-t)^2.$$

Since $S(0) = 5000$, it follows that $5000 = \dfrac{10^5}{5} + C(10^{10})$, and therefore $C = -15/10^7$. Thus,

$$S(t) = 20\,000 - \frac{t}{5} - \frac{15}{10^7}(10^5-t)^2.$$

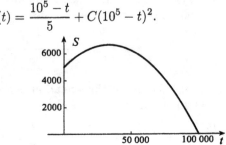

30. If $t = 0$ when the oven is turned on, its temperature is $T_o(t) = 20 + 36t$ for $0 \le t \le 5$, and 200 for $t > 5$. If $T(t)$ is the temperature of the potato, then

$$\frac{dT}{dt} = k(T_o - T), \quad T(0) = 20.$$

For $0 \le t \le 5$, this becomes $dT/dt = -kT + k(20+36t)$. An integrating factor for this linear first-order differential equation is e^{kt} so that

$$e^{kt}\frac{dT}{dt} + ke^{kt}T = k(20+36t)e^{kt} \quad\Longrightarrow\quad \frac{d}{dt}(Te^{kt}) = k(20+36t)e^{kt} \quad\Longrightarrow\quad Te^{kt} = k\int(20+36t)e^{kt}\,dt + C.$$

Integration by parts leads to

$$T = ke^{-kt}\left(\frac{20e^{kt}}{k} - \frac{36e^{kt}}{k^2} + \frac{36te^{kt}}{k}\right) + Ce^{-kt} = 20 - \frac{36}{k} + 36t + Ce^{-kt}.$$

For $T(0) = 20$, we have $20 = 20 - 36/k + C \Longrightarrow C = 36/k$, and

$$T(t) = 20 - \frac{36}{k} + 36t + \frac{36}{k}e^{-kt} = 20 + 36t + \frac{36}{k}(e^{-kt}-1).$$

For $t > 5$,

$$\frac{dT}{dt} = k(200 - T), \quad T(5) = 200 + \frac{36}{k}(e^{-5k}-1).$$

This equation is separable,

$$\frac{dT}{200-T} = k\,dt \quad\Longrightarrow\quad -\ln|200-T| = kt + C \quad\Longrightarrow\quad T = 200 + De^{-kt},$$

where $D = \pm e^{-C}$. The temperature at $t = 5$ requires

$$200 + De^{-5k} = 200 + \frac{36}{k}(e^{-5k}-1) \quad\Longrightarrow\quad D = \frac{36}{k}(1 - e^{5k}).$$

Hence, for $t > 5$, temperature is $T(t) = 200 + (36/k)(1 - e^{5k})e^{-kt}$.

32. The energy balance equation is

$$\left(\frac{100}{t+1}\right)(4190)(10) + 2000 = \left(\frac{100}{t+1}\right)(4190)T + 100(4190)\frac{dT}{dt} \implies \frac{dT}{dt} + \frac{T}{t+1} = \frac{2}{419} + \frac{10}{t+1}.$$

An integrating factor is $e^{\int [1/(t+1)]\,dt} = e^{\ln(t+1)} = t+1$, so that the differential equation can be expressed in the form

$$(t+1)\frac{dT}{dt} + T = \frac{2}{419}(t+1) + 10 \implies \frac{d}{dt}[(t+1)T] = \frac{2}{419}(t+1) + 10.$$

Integration now gives

$$(t+1)T = 10t + \frac{1}{419}(t+1)^2 + C \implies T(t) = \frac{10t}{t+1} + \frac{1}{419}(t+1) + \frac{C}{t+1}.$$

Since $T(0) = 10$, we find $10 = \frac{1}{419} + C \implies C = \frac{4189}{419}$.

Temperature of the water is therefore

$$T(t) = \frac{10t}{t+1} + \frac{t+1}{419} + \frac{4189}{419(t+1)}$$
$$= \frac{4190t + 4189}{419(t+1)} + \frac{t+1}{419}.$$

34. The energy balance equation is

$$(0.03)(4190)(10e^{-t}) + 2000 = (0.03)(4190)T + 100(4190)\frac{dT}{dt} \implies \frac{dT}{dt} + \frac{3T}{10\,000} = \frac{2}{419} + \frac{3e^{-t}}{1000}.$$

An integrating factor is $e^{\int (3/10\,000)\,dt} = e^{3t/10\,000}$, so that the differential equation can be expressed in the form

$$e^{3t/10\,000}\frac{dT}{dt} + \frac{3Te^{3t/10\,000}}{10\,000} = \left(\frac{2}{419} + \frac{3e^{-t}}{1000}\right)e^{3t/10\,000} \implies \frac{d}{dt}(Te^{3t/10\,000}) = \left(\frac{2}{419} + \frac{3e^{-t}}{1000}\right)e^{3t/10\,000}.$$

Integration gives

$$Te^{3t/10\,000} = \frac{20\,000}{1257}e^{3t/10\,000} - \frac{30}{9997}e^{-9997t/10\,000} + C \implies T(t) = \frac{20\,000}{1257} - \frac{30}{9997}e^{-t} + Ce^{-3t/10\,000}.$$

Since $T(0) = 10$, we find $10 = \frac{20\,000}{1257} - \frac{30}{9997} + C,$

and this implies that $C = -\frac{74\,240\,000}{12\,566\,229}$.

Temperature of the water is therefore

$$T(t) = \frac{20\,000}{1257} - \frac{30}{9997}e^{-t} - \frac{74\,240\,000}{12\,566\,229}e^{-3t/10\,000}.$$

36. The differential equation is linear first-order,

$$\frac{di}{dt} + \frac{Ri}{L} = \frac{R}{L}[h(t) - h(t-1)],$$

with integrating factor $e^{\int (R/L)\,dt} = e^{Rt/L}.$ Multiplying the differential equation by this factor leads to

$$\frac{d}{dt}\left[ie^{Rt/L}\right] = \frac{R}{L}[h(t) - h(t-1)]e^{Rt/L}.$$

Integration for $0 < t < 1$ yields

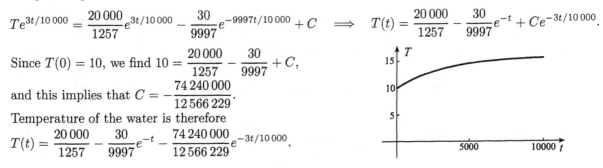

$$ie^{Rt/L} = \frac{R}{L}\int e^{Rt/L}dt = e^{Rt/L} + D.$$

Division by $e^{Rt/L}$ gives $i(t) = 1 + De^{-Rt/L}$. The initial condition $i(0) = 0$ gives $0 = 1 + D$, so that for $0 < t < 1$,

$$i(t) = 1 - e^{-Rt/L}.$$

When $t > 1$, integration gives

$$ie^{Rt/L} = \frac{R}{L}\int (1-1)e^{Rt/L}dt = E \qquad \Longrightarrow \qquad i = Ee^{-Rt/L}.$$

To evaluate E we demaind that the current
be continuous at $t = 1$. This requires
$\lim\limits_{t\to 1^-} i(t) = \lim\limits_{t\to 1^+} i(t)$, or

$1 - e^{-R/L} = Ee^{-R/L} \quad \Longrightarrow \quad E = e^{R/L} - 1.$

The pulse response function is

$$i(t) = \begin{cases} 1 - e^{-Rt/L}, & 0 \le t \le 1 \\ (e^{R/L} - 1)e^{-Rt/L}, & t > 1 \end{cases}.$$

Its graph is shown to the right.

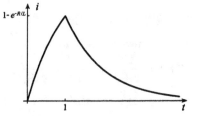

38. (a) Since $\dfrac{dI}{dt} + \dfrac{1}{RC}I = \dfrac{1}{R}\dfrac{dE}{dt} = \dfrac{\omega}{R}E_0\cos(\omega t)$, an integrating factor is $e^{\int [1/(RC)]\,dt} = e^{t/(RC)}$. Multiplication by this factor gives

$$e^{t/(RC)}\frac{dI}{dt} + \frac{1}{RC}Ie^{t/(RC)} = \frac{\omega}{R}E_0 e^{t/(RC)}\cos(\omega t) \quad \Longrightarrow \quad \frac{d}{dt}\left[Ie^{t/(RC)}\right] = \frac{\omega}{R}E_0 e^{t/(RC)}\cos(\omega t).$$

Integration now yields

$$Ie^{t/(RC)} = \frac{\omega E_0}{R}\left\{ \frac{e^{t/(RC)}}{\omega^2 + 1/(RC)^2}[(1/(RC))\cos(\omega t) + \omega\sin(\omega t)] \right\} + A.$$

Thus, $I(t) = Ae^{-t/(RC)} + \dfrac{\omega E_0}{R[\omega^2 + 1/(RC)^2]}\{[1/(RC)]\cos(\omega t) + \omega\sin(\omega t)\}$. If we set

$$\frac{1}{RC}\cos(\omega t) + \omega\sin(\omega t) = B\sin(\omega t - \phi) = B\sin(\omega t)\cos\phi - B\cos(\omega t)\sin\phi,$$

then $B\cos\phi = \omega$ and $-B\sin\phi = 1/(RC)$. These equations imply that

$$B = \sqrt{\omega^2 + \frac{1}{(RC)^2}} \qquad \text{and} \qquad \tan\phi = \frac{-1}{\omega CR} \quad \Longrightarrow \quad \phi = \text{Tan}^{-1}\left(\frac{-1}{\omega CR}\right).$$

Hence, $I(t) = Ae^{-t/(RC)} + \dfrac{\omega E_0}{R[\omega^2 + 1/(RC)^2]}\sqrt{\omega^2 + 1/(RC)^2}\sin(\omega t - \phi)$

$$= Ae^{-t/(RC)} + \frac{E_0}{Z}\sin(\omega t - \phi),$$

where $Z = \dfrac{R\sqrt{\omega^2 + 1/(RC)^2}}{\omega} = \sqrt{R^2 + \dfrac{1}{\omega^2 C^2}}.$

(b) If $I(0) = I_0$, then

$$I_0 = A + \frac{E_0}{Z}\sin(-\phi) = A + \frac{E_0}{Z}\frac{1}{\sqrt{1 + \omega^2 C^2 R^2}}$$

$$= A + \frac{E_0}{\omega CZ\sqrt{R^2 + 1/(\omega C)^2}} = A + \frac{E_0}{\omega CZ^2},$$

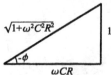

and therefore $A = I_0 - E_0/(\omega CZ^2)$.

40. If we substitute $Q = \dfrac{S}{K(1-x)} - \dfrac{xI}{1-x}$ into the differential equation,

$$\frac{dS}{dt} = I - \frac{S}{K(1-x)} + \frac{xI}{1-x} \quad \Longrightarrow \quad \frac{dS}{dt} + \frac{S}{K(1-x)} = \frac{I}{1-x} = \frac{I_0 e^{-\lambda t}}{1-x}.$$

An integrating factor is $e^{t/b}$, where $b = K(1-x)$. Multiplication of the differential equation by $e^{t/b}$ gives

$$e^{t/b}\frac{dS}{dt} + \frac{Se^{t/b}}{b} = \frac{I_0 e^{t/b}e^{-\lambda t}}{1-x} \quad \Longrightarrow \quad \frac{d}{dt}[Se^{t/b}] = \frac{I_0}{1-x}e^{(-\lambda + 1/b)t}.$$

Integration gives

$$Se^{t/b} = \frac{I_0}{(1-x)(-\lambda + 1/b)}e^{(-\lambda + 1/b)t} + C \quad \Longrightarrow \quad S(t) = Ce^{-t/b} + \frac{I_0}{(1-x)(-\lambda + 1/b)}e^{-\lambda t}.$$

Thus,

$$Q(t) = \frac{C}{b}e^{-t/b} + \frac{I_0}{b(1-x)(-\lambda + 1/b)}e^{-\lambda t} - \frac{xI_0 e^{-\lambda t}}{1-x}.$$

Since $Q(0) = 0$,

$$0 = \frac{C}{b} + \frac{I_0}{b(1-x)(-\lambda + 1/b)} - \frac{xI_0}{1-x}.$$

This gives C, and

$$Q(t) = \left[\frac{xI_0}{1-x} - \frac{I_0}{b(1-x)(-\lambda + 1/b)}\right]e^{-t/b} + \left[\frac{-xI_0}{1-x} + \frac{I_0}{b(1-x)(-\lambda + 1/b)}\right]e^{-\lambda t}$$

$$= I_0\left[\frac{1}{(1-x)(1-\lambda b)} - \frac{x}{1-x}\right](e^{-\lambda t} - e^{-t/b}).$$

EXERCISES 15.4

2. Since x is explicitly missing, we set $\dfrac{dy}{dx} = v$ and $\dfrac{d^2y}{dx^2} = \dfrac{dv}{dx} = \dfrac{dv}{dy}\dfrac{dy}{dx} = v\dfrac{dv}{dy}$,

$$2yv\frac{dv}{dy} = 1 + v^2.$$

This equation can be separated, $\dfrac{2v}{1+v^2}\,dv = \dfrac{1}{y}\,dy$, and a one-parameter family of solutions is defined implicitly by $\ln(1+v^2) = \ln|y| + C$. When this equation is solved for v, $v = \dfrac{dy}{dx} = \pm\sqrt{Dy - 1}$. This equation can also be separated, $\dfrac{1}{\sqrt{Dy-1}}\,dy = \pm dx$, and a two-parameter family of solutions is defined implicitly by $2\sqrt{Dy-1} = \pm Dx + F$.

4. Since y is explicitly missing, we set $\dfrac{dy}{dx} = v$ and $\dfrac{d^2y}{dx^2} = \dfrac{dv}{dx}$, giving $x^2\dfrac{dv}{dx} = v^2$. This equation can be separated, $\dfrac{1}{v^2}\,dv = \dfrac{1}{x^2}\,dx$, and a one-parameter family of solutions is defined implicitly by $-1/v = -1/x - C$. We now solve for v, $v = \dfrac{dy}{dx} = \dfrac{x}{Cx+1}$, and integrate to obtain

$$y = \int \frac{x}{Cx+1}\,dx = \int\left(\frac{1}{C} + \frac{-1/C}{Cx+1}\right)dx = \frac{x}{C} - \frac{1}{C^2}\ln|Cx+1| + D$$

$$= Ex - E^2\ln\left|\frac{x}{E} + 1\right| + D = Ex - E^2\ln|x + E| + F.$$

6. When we substitute $\dfrac{dy}{dx} = v$ and $\dfrac{d^2y}{dx^2} = \dfrac{dv}{dx} = \dfrac{dv}{dy}\dfrac{dy}{dx} = v\dfrac{dv}{dy}$, we obtain $v\dfrac{dv}{dy} = (1+v^2)^{3/2}$. This equation can be separated, $\dfrac{v}{(1+v^2)^{3/2}}\,dv = dy$, a one-parameter family of solutions of which is defined implicitly by $-1/\sqrt{1+v^2} = y + C$. When this is solved for v,

$$v = \frac{dy}{dx} = \pm\frac{\sqrt{1-(y+C)^2}}{y+C} \implies \frac{y+C}{\sqrt{1-(y+C)^2}}\,dy = \pm dx.$$

Integration gives $-\sqrt{1-(y+C)^2} = \pm x + D$. By squaring this equation, the solution can be rewritten in the form $(x+E)^2 + (y+C)^2 = 1$.

8. Since x is explicitly missing, we set $\dfrac{dy}{dx} = v$ and $\dfrac{d^2y}{dx^2} = \dfrac{dv}{dx} = \dfrac{dv}{dy}\dfrac{dy}{dx} = v\dfrac{dv}{dy}$, giving $v\dfrac{dv}{dy} = vy$. Division by v and integration gives

$$v = \frac{y^2}{2} + C \implies \frac{dy}{dx} = \frac{y^2}{2} + \frac{D}{2} \quad (D = 2C).$$

Separation now yields $\dfrac{1}{D+y^2}\,dy = \dfrac{1}{2}\,dx \implies \dfrac{x}{2} + E = \displaystyle\int \dfrac{1}{D+y^2}\,dy$. If $D = 0$, then $x/2 + E = -1/y$. If $D > 0$, we set $y = \sqrt{D}\tan\theta$ and $dy = \sqrt{D}\sec^2\theta\,d\theta$,

$$\frac{x}{2} + E = \int \frac{\sqrt{D}\sec^2\theta}{D\sec^2\theta}\,d\theta = \frac{1}{\sqrt{D}}\theta = \frac{1}{\sqrt{D}}\mathrm{Tan}^{-1}\left(\frac{y}{\sqrt{D}}\right) = F\,\mathrm{Tan}^{-1}(Fy).$$

If $D < 0$, we set $D = -F^2$ and use partial fractions to write

$$\frac{x}{2} + E = \int \frac{1}{y^2 - F^2}\,dy = \int\left(\frac{\frac{-1}{2F}}{y+F} + \frac{\frac{1}{2F}}{y-F}\right)dy$$

$$= -\frac{1}{2F}\ln|y+F| + \frac{1}{2F}\ln|y-F| = \frac{1}{2F}\ln\left|\frac{y-F}{y+F}\right|.$$

10. If we set $\dfrac{dy}{dx} = v$ and $\dfrac{d^2y}{dx^2} = \dfrac{dv}{dx}$, then $\left(\dfrac{dv}{dx}\right)^2 = 1 + v^2$. This equation can be separated, $\dfrac{1}{\sqrt{1+v^2}}\,dv = \pm dx$, and a one-parameter family of solutions is defined implicitly by

$$\pm x + C = \int \frac{1}{\sqrt{1+v^2}}\,dv \quad \text{(and if we set } v = \tan\theta,\ dv = \sec^2\theta\,d\theta\text{)},$$

$$= \int \frac{\sec^2\theta}{\sec\theta}\,d\theta = \int \sec\theta\,d\theta = \ln|\sec\theta + \tan\theta| = \ln|\sqrt{1+v^2} + v|.$$

We now exponentiate, $\sqrt{1+v^2} + v = De^{\pm x}$, transpose the v and square, $1 + v^2 = D^2e^{\pm 2x} - 2Dve^{\pm x} + v^2$. We can now solve for v,

$$v = \frac{dy}{dx} = \frac{D^2e^{\pm 2x} - 1}{2De^{\pm x}} = \frac{1}{2}\left(De^{\pm x} - \frac{1}{D}e^{\mp x}\right).$$

In either case dy/dx is of the form $\dfrac{dy}{dx} = \dfrac{1}{2}\left(De^x - \dfrac{1}{D}e^{-x}\right)$, and hence, $y = \dfrac{1}{2}\left(De^x + \dfrac{1}{D}e^{-x}\right) + E$.

12. We could use substitution 15.28 but it simpler to write

$$\frac{d}{dr}\left(r\frac{dT}{dr}\right) = k \implies r\frac{dT}{dr} = kr + C \implies \frac{dT}{dr} = k + \frac{C}{r} \implies T = kr + C\ln r + D.$$

For $T(a) = T_a$ and $T(b) = T_b$, C and D must satisfy $T_a = ka + C\ln a + D$ and $T_b = kb + C\ln b + D$. These can be solved for $C = [T_b - T_a - k(b-a)]/\ln(b/a)$ and $D = [T_a\ln b - T_b\ln a + k(b\ln a - a\ln b)]/\ln(b/a)$.

EXERCISES 15.5

2. The condition $v(0) = v_0$ implies that $C = (1/k)\ln|kv_0 - mg|$. Hence

$$\frac{1}{k}\ln|kv - mg| = -\frac{t}{m} + \frac{1}{k}\ln|kv_0 - mg| \implies \ln\left|\frac{kv - mg}{kv_0 - mg}\right| = -\frac{kt}{m} \implies \left|\frac{kv - mg}{kv_0 - mg}\right| = e^{-kt/m}.$$

Since terminal velocity occurs when $dv/dt = 0$, it follows from the differential equation that $mg - kv = 0$. Terminal velocity is therefore $v = mg/k$. If the initial velocity v_0 is less than terminal velocity, then v will always be less than terminal velocity. In this case, both $kv - mg < 0$ and $kv_0 - mg < 0$, the quotient being positive. If v_0 is greater than terminal velocity, v will always be greater than terminal velocity. In this case, both $kv - mg > 0$ and $kv_0 - mg > 0$, as is their quotient. In both cases then, we may write

$$\frac{kv - mg}{kv_0 - mg} = e^{-kt/m} \implies kv - mg = (kv_0 - mg)e^{-kt/m} \implies v = \frac{mg}{k} + \left(v_0 - \frac{mg}{k}\right)e^{-kt/m}.$$

4. Let us measure x as positive in the direction of motion taking $x = 0$ and $t = 0$ at the instant the brakes are applied. Because the coefficient of kinetic friction is less than one, we can say that the x-component of the force of friction has magnitude less than $9.81m$, where m is the mass of the car. If we use this as the magnitude of the frictional force, then because this is the maximum possible, we will be finding the maximum possible speed before the brakes were applied. In other words, we are testifying for the defence. Newton's second law for the x-component of the acceleration dv/dt of the car gives

$$m\frac{dv}{dt} = -9.81m,$$

and this can be integrated for $v(t) = -9.81t + C$. If we set $v = v_0$ at time $t = 0$, then $v_0 = C$, and

$$v = \frac{dx}{dt} = -9.81t + v_0.$$

Integration gives $x(t) = -4.905t^2 + v_0 t + D$. Because $x(0) = 0$, it follows that $D = 0$, and

$$x(t) = -4.905t^2 + v_0 t.$$

The car comes to a stop when $0 = v = -9.81t + v_0$, and this implies that $t = v_0/9.81$. Since $x = 9$ at this instant,

$$9 = -4.905\left(\frac{v_0}{9.81}\right)^2 + v_0\left(\frac{v_0}{9.81}\right).$$

The solution of this equation is $v_0 = 13.29$ m/s or $v_0 = 47.8$ km/hr. Thus, the maximum possible speed of the car was 47.8 km/hr.

6. (a) Let us take y as positive downward with $y = 0$ and time $t = 0$ when motion begins. Newton's second gives

$$m\frac{dv}{dt} = mg - kv \implies \frac{1}{mg - kv}dv = \frac{1}{m}dt,$$

a separated differential equation. A one-parameter family of solutions is defined implicitly by

$$-\frac{1}{k}\ln|mg - kv| = \frac{t}{m} + C \implies \ln|mg - kv| = -\frac{kt}{m} - kC \implies mg - kv = De^{-kt/m}.$$

For $v(0) = 0$, we find $mg = D$, and therefore

$$mg - kv = mge^{-kt/m} \implies v(t) = \frac{mg}{k}(1 - e^{-kt/m}).$$

Terminal velocity is mg/k. Velocity is 95% of this when

$$0.95\frac{mg}{k} = \frac{mg}{k}(1 - e^{-kt/m}) \implies e^{-kt/m} = 1 - 0.95 \implies t = \frac{m}{k}\ln 20.$$

(b) Integration of velocity gives

$$y(t) = \frac{mg}{k}\left(t + \frac{m}{k}e^{-kt/m}\right) + D.$$

Since $y(0) = 0$, it follows that $0 = (mg/k)(m/k) + D$. Hence, $D = -m^2g/k^2$ and

$$y(t) = \frac{mg}{k}\left(t + \frac{m}{k}e^{-kt/m}\right) - \frac{m^2g}{k^2} = \frac{mgt}{k} - \frac{m^2g}{k^2}(1 - e^{-kt/m}).$$

If we substitute $t = (m/k)\ln 20$, we obtain distance fallen as

$$y = \frac{m^2g}{k^2}\ln 20 - \frac{m^2g}{k^2}(1 - e^{-\ln 20}) = \frac{m^2g}{k^2}\left(\ln 20 - \frac{19}{20}\right).$$

8. (a) When M is at position x, and is moving to the left, the spring force is $-kx$ and the frictional force is μMg. Newton's second law therefore gives

$$M\frac{d^2x}{dt^2} = -kx + \mu Mg, \quad x(0) = x_0, \quad v(0) = -v_0,$$

and this equation is valid until M comes to an instantaneous stop for the first time.

(b) If we set $\dfrac{dx}{dt} = v$ and $\dfrac{d^2x}{dt^2} = \dfrac{dv}{dt} = \dfrac{dv}{dx}\dfrac{dx}{dt} = v\dfrac{dv}{dx}$, then $Mv\dfrac{dv}{dx} = -kx + \mu Mg$, subject to $v(x_0) = -v_0$. This equation can be separated, $Mv\,dv = (-kx + \mu Mg)\,dx$, and a one-parameter family of solutions is defined implicitly by $\dfrac{Mv^2}{2} = -\dfrac{kx^2}{2} + \mu Mgx + C$. The initial condition requires $\dfrac{Mv_0^2}{2} = -\dfrac{kx_0^2}{2} + \mu Mgx_0 + C$. Thus,

$$\frac{Mv^2}{2} = -\frac{kx^2}{2} + \mu Mgx + \frac{Mv_0^2}{2} + \frac{kx_0^2}{2} - \mu Mgx_0 \implies \frac{k}{2}(x_0^2 - x^2) = \frac{M}{2}(v^2 - v_0^2) + \mu Mg(x_0 - x).$$

The left side is the loss of stored energy in the spring at x relative to that initially; $M(v^2 - v_0^2)/2$ is the gain in kinetic energy at x; and $\mu Mg(x_0 - x)$ is the work done against friction when m moves from x_0 to x.

(c) If we set $x = x^*$ when $v = 0$, then x^* is defined implicitly by

$$\frac{k}{2}(x_0^2 - x^{*2}) = -\frac{Mv_0^2}{2} + \mu Mg(x_0 - x^*).$$

This is a quadratic equation in x^*, $kx^{*2} - 2\mu Mgx^* + (2\mu Mgx_0 - kx_0^2 - Mv_0^2) = 0$, with solutions

$$x^* = \frac{2\mu Mg \pm \sqrt{4\mu^2M^2g^2 - 4k(2\mu Mgx_0 - kx_0^2 - Mv_0^2)}}{2k}$$
$$= \frac{\mu Mg \pm \sqrt{\mu^2M^2g^2 - k(2\mu Mgx_0 - kx_0^2 - Mv_0^2)}}{k}.$$

Whether the mass stops to the left of, to the right of, or on the origin depends on the quantity $2\mu Mgx_0 - kx_0^2 - Mv_0^2$. When it is negative, the sum $(1/2)kx_0^2 + (1/2)Mv_0^2$ of the initial energy stored in the spring $(1/2)kx_0^2$ and the initial kinetic energy of the mass $(1/2)Mv_0^2$ is greater than the work done against friction μMgx_0 as the mass travels from $x = x_0$ to $x = 0$. The mass therefore stops at position

$$x^* = \frac{\mu Mg - \sqrt{\mu^2M^2g^2 + k(kx_0^2 + Mv_0^2 - 2\mu Mgx_0)}}{k}$$

to the left of the origin. When $2\mu Mgx_0 - kx_0^2 - Mv_0^2 = 0$, initial spring energy and kinetic energy are just sufficient to bring the mass back to the origin ($x^* = 0$). Finally, when $2\mu Mgx_0 - kx_0^2 - Mv_0^2 > 0$, there is not sufficient initial energy to return the mass to the origin; it stops at position

$$x^* = \frac{\mu Mg - \sqrt{\mu^2M^2g^2 - k(2\mu Mgx_0 - kx_0^2 - Mv_0^2)}}{k} > 0.$$

10. Let us choose y as positive downward taking $y = 0$ and $t = 0$ when the mass is released. Newton's second law for the acceleration of the mass is

$$(1)\frac{dv}{dt} = g - \frac{v^2}{500},$$

where $g = 9.81$. This equation is separable, $\frac{1}{v^2 - 500g}\, dv = -\frac{1}{500}\, dt$, and a one-parameter family of solutions is defined implicitly by

$$-\frac{t}{500} + C = \int \frac{1}{v^2 - 500g}\, dv = \int \left(\frac{\frac{1}{2\sqrt{500g}}}{v - \sqrt{500g}} + \frac{\frac{-1}{2\sqrt{500g}}}{v + \sqrt{500g}} \right) dv$$

$$= \frac{1}{2\sqrt{500g}} \left(\ln |v - \sqrt{500g}| - \ln |v + \sqrt{500g}| \right).$$

When this equation is solved for v, the result is $v = \dfrac{\sqrt{500g}[1 + De^{-2\sqrt{g/500}t}]}{1 - De^{-2\sqrt{g/500}t}}$. Since $v(0) = 100$, it

follows that $100 = \dfrac{\sqrt{500g}[1 + D]}{1 - D}$, and therefore $D = (100 - \sqrt{500g})/(100 + \sqrt{500g}) = 0.176$. Thus,

$v(t) = 70.0 \left(\dfrac{1 + 0.176e^{-0.280t}}{1 - 0.176e^{-0.280t}} \right).$

12. If we choose y as positive downward, then integration of the differential equation $m\, dv/dt = mg - kv^2$ as in Example 15.9 leads to

$$v(t) = \frac{V(1 + De^{-2kVt/m})}{1 - De^{-2kVt/m}},$$

where $V = \sqrt{mg/k}$ is the terminal velocity of the body. The initial velocity $v(0) = v_0$ requires

$$v_0 = \frac{V(1 + D)}{1 - D} \quad \Longrightarrow \quad v_0(1 - D) = V(1 + D) \quad \Longrightarrow \quad D = \frac{v_0 - V}{v_0 + V}.$$

Hence, $v(t) = \dfrac{V \left[1 + \left(\dfrac{v_0 - V}{v_0 + V} \right) e^{-2kVt/m} \right]}{1 - \left(\dfrac{v_0 - V}{v_0 + V} \right) e^{-2kVt/m}} = \dfrac{V \left[1 - \left(\dfrac{V - v_0}{V + v_0} \right) e^{-2kVt/m} \right]}{1 + \left(\dfrac{V - v_0}{V + v_0} \right) e^{-2kVt/m}}.$

14. If we choose y as positive upward, the differential equation describing the velocity of the body is

$$m\frac{dv}{dt} = -mg - kv^2 \quad \Longrightarrow \quad \frac{dv}{v^2 + mg/k} = -\frac{k}{m}\, dt \quad \Longrightarrow \quad \sqrt{\frac{k}{mg}} \mathrm{Tan}^{-1}\left(\frac{v}{\sqrt{mg/k}} \right) = -\frac{kt}{m} + C.$$

Hence, $v(t) = \sqrt{\dfrac{mg}{k}} \tan \left(D - \sqrt{\dfrac{kg}{m}}t \right)$, where $D = \sqrt{mg/k}\, C$. The initial condition $v(0) = v_0$ requires

$v_0 = \sqrt{mg/k} \tan D \Longrightarrow D = \mathrm{Tan}^{-1}[\sqrt{k/(mg)}\, v_0]$. Thus,

$$v(t) = \sqrt{\frac{mg}{k}} \tan \left[\mathrm{Tan}^{-1}\left(\sqrt{\frac{k}{mg}} v_0 \right) - \sqrt{\frac{kg}{m}}t \right].$$

Maximum height is attained when

$$0 = v(t) = \sqrt{\frac{mg}{k}} \tan \left[\mathrm{Tan}^{-1}\left(\sqrt{\frac{k}{mg}} v_0 \right) - \sqrt{\frac{kg}{m}}t \right] \quad \Longrightarrow \quad \mathrm{Tan}^{-1}\left(\sqrt{\frac{k}{mg}} v_0 \right) - \sqrt{\frac{kg}{m}}t = n\pi,$$

where n is an integer. Solving for t gives $t = \sqrt{\dfrac{m}{kg}}\left[\text{Tan}^{-1}\left(\sqrt{\dfrac{k}{mg}}\,v_0\right) - n\pi\right]$. When we choose $n = 0$

to obtain the smallest positive solution, $t = \sqrt{\dfrac{m}{kg}}\,\text{Tan}^{-1}\left(\sqrt{\dfrac{k}{mg}}\,v_0\right)$.

16. (a) Let us choose y as positive upward taking $y = 0$ and $t = 0$ when the mass is released. Newton's

second law for acceleration during ascent is $m\dfrac{dv}{dt} = -mg - kv^2$. This is valid only during ascent since

air resistance during descent is kv^2. If we set $V = \sqrt{mg/k}$, then

$$-\frac{m}{k}\frac{dv}{dt} = V^2 + v^2 \quad\Longrightarrow\quad \frac{1}{v^2 + V^2}\,dv = -\frac{k}{m}\,dt \quad\Longrightarrow\quad \frac{1}{V}\text{Tan}^{-1}\left(\frac{v}{V}\right) = -\frac{kt}{m} + C.$$

Since $v(0) = v_0$, we obtain $\dfrac{1}{V}\text{Tan}^{-1}\left(\dfrac{v_0}{V}\right) = C$. Thus,

$$\frac{1}{V}\text{Tan}^{-1}\left(\frac{v}{V}\right) = -\frac{kt}{m} + \frac{1}{V}\text{Tan}^{-1}\left(\frac{v_0}{V}\right).$$

When we solve this equation for v, the result is $v(t) = V\tan\left[\text{Tan}^{-1}\left(\dfrac{v_0}{V}\right) - \dfrac{kVt}{m}\right]$. Once again this is

only valid during ascent since air resistance is kv^2 (rather than $-kv^2$) on descent.

(b) Integration of the velocity gives $y = \dfrac{m}{k}\ln\left|\cos\left[\text{Tan}^{-1}\left(\dfrac{v_0}{V}\right) - \dfrac{kVt}{m}\right]\right| + D$. Since $y(0) = 0$, we find

that $0 = \dfrac{m}{k}\ln\left|\cos\left[\text{Tan}^{-1}\left(\dfrac{v_0}{V}\right)\right]\right| + D$, and therefore $D = (m/k)\ln\left(\sqrt{v_0^2 + V^2}/V\right)$. The height of the
mass is

$$y = \frac{m}{k}\ln\left|\cos\left[\text{Tan}^{-1}\left(\frac{v_0}{V}\right) - \frac{kVt}{m}\right]\right| + \frac{m}{k}\ln\left(\frac{\sqrt{v_0^2 + V^2}}{V}\right).$$

Maximum height occurs when $0 = v = V\tan\left[\text{Tan}^{-1}\left(\dfrac{v_0}{V}\right) - \dfrac{kVt}{m}\right]$, and the solution of this equation is

$t = \dfrac{m}{kV}\text{Tan}^{-1}\left(\dfrac{v_0}{V}\right)$. Maximum height is therefore

$$\frac{m}{k}\ln\left|\cos\left[\text{Tan}^{-1}\left(\frac{v_0}{V}\right) - \text{Tan}^{-1}\left(\frac{v_0}{V}\right)\right]\right| + \frac{m}{k}\ln\left(\frac{\sqrt{v_0^2 + V^2}}{V}\right) = \frac{m}{k}\ln\left(\frac{\sqrt{v_0^2 + V^2}}{V}\right).$$

18. The initial-value problem for the motion
of the projectile (figure to the right) is

$$m\frac{d^2\mathbf{r}}{dt^2} = -mg\hat{\mathbf{j}} - \beta\mathbf{v},$$

subject to initial displacement $\mathbf{r}(0) = \mathbf{0}$ and initial
velocity $\mathbf{r}'(0) = \mathbf{v}_0 = v_0\cos\theta\,\hat{\mathbf{i}} + v_0\sin\theta\,\hat{\mathbf{j}}$.
We can solve this differential equation in vector
form, or in component form. The component scalar
differential equations are identical, so let us save space by integrating vectorially. When we write

$$\frac{d\mathbf{v}}{dt} + \frac{\beta}{m}\mathbf{v} = -g\hat{\mathbf{j}},$$

we have a linear first-order differential equation with integrating factor $e^{\int (\beta/m)dt} = e^{\beta t/m}$. When we
multiply the differential equation by $e^{\beta t/m}$, it can be expressed in the form

$$\frac{d}{dt}\left(\mathbf{v}e^{\beta t/m}\right) = -ge^{\beta t/m}\hat{\mathbf{j}} \quad\Longrightarrow\quad \mathbf{v}e^{\beta t/m} = -\frac{mg}{\beta}e^{\beta t/m}\hat{\mathbf{j}} + \mathbf{C} \quad\Longrightarrow\quad \mathbf{v} = -\frac{mg}{\beta}\hat{\mathbf{j}} + \mathbf{C}e^{-\beta t/m}.$$

The initial velocity condition requires $\mathbf{v_0} = -\dfrac{mg}{\beta}\hat{\mathbf{j}} + \mathbf{C}$, and therefore

$$\mathbf{v}(t) = -\frac{mg}{\beta}\hat{\mathbf{j}} + \left(\mathbf{v_0} + \frac{mg}{\beta}\hat{\mathbf{j}}\right)e^{-\beta t/m}.$$

Integrating once again gives

$$\mathbf{r}(t) = -\frac{mgt}{\beta}\hat{\mathbf{j}} - \frac{m}{\beta}\left(\mathbf{v_0} + \frac{mg}{\beta}\hat{\mathbf{j}}\right)e^{-\beta t/m} + \mathbf{D}.$$

For $\mathbf{r}(0) = \mathbf{0}$, we must have $\mathbf{0} = -\dfrac{m}{\beta}\left(\mathbf{v_0} + \dfrac{mg}{\beta}\hat{\mathbf{j}}\right) + \mathbf{D}$. Thus,

$$\mathbf{r}(t) = -\frac{mgt}{\beta}\hat{\mathbf{j}} - \frac{m}{\beta}\left(\mathbf{v_0} + \frac{mg}{\beta}\hat{\mathbf{j}}\right)e^{-\beta t/m} + \frac{m}{\beta}\left(\mathbf{v_0} + \frac{mg}{\beta}\hat{\mathbf{j}}\right)\hat{\mathbf{j}} = -\frac{mgt}{\beta}\hat{\mathbf{j}} + \frac{m}{\beta}\left(\mathbf{v_0} + \frac{mg}{\beta}\hat{\mathbf{j}}\right)(1 - e^{-\beta t/m}).$$

20. Newton's second law for acceleration gives

$$m\frac{dv}{dt} = mg\sin\alpha.$$

Integration gives $v = (g\sin\alpha)t + C$. Since $v(0) = 0$, it follows that $C = 0$, and

$$v = \frac{dx}{dt} = (g\sin\alpha)t.$$

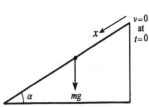

Integration now gives $x = (g\sin\alpha)t^2/2 + E$. If we choose $x = 0$ when motion begins, then $E = 0$, and $x = (g\sin\alpha)t^2/2$. The time for the mass to travel distance D is given by $D = (g\sin\alpha)t^2/2 \implies t = \sqrt{2D/(g\sin\alpha)}$. The speed of the mass at this time is $g\sin\alpha\sqrt{\dfrac{2D}{g\sin\alpha}} = \sqrt{2gD\sin\alpha}$.

22. Since $dy/dt = 2x + 4$ and $dx/dt = (3 - y)/2$, it follows that

$$\frac{dy}{dx} = \frac{2x + 4}{(3 - y)/2} \implies (3 - y)\,dy = 4(x + 2)\,dx,$$

a separated equation. A one-parameter family of solutions is defined implicitly by

$$3y - \frac{y^2}{2} = 4\left(\frac{x^2}{2} + 2x\right) + C.$$

Since the electron passes through $(0, 3)$, we must have $9 - 9/2 = C$, and therefore

$$3y - \frac{y^2}{2} = 2x^2 + 8x + \frac{9}{2} \implies 4x^2 + y^2 + 16x - 6y + 9 = 0.$$

This is an ellipse.

EXERCISES 15.6

2. Since $L(y_1 + y_2) = 15x(y_1 + y_2) = 15xy_1 + 15xy_2 = L(y_1) + L(y_2)$, and
$$L(cy_1) = 15x(cy_1) = c(15xy_1) = cL(y_1),$$
the operator L is linear.

4. Since $L(y_1 + y_2) = \lim\limits_{x\to 3}(y_1 + y_2) = \lim\limits_{x\to 3} y_1 + \lim\limits_{x\to 3} y_2 = L(y_1) + L(y_2)$, and
$$L(cy_1) = \lim\limits_{x\to 3} cy_1 = c\lim\limits_{x\to 3} y_1 = cL(y_1),$$
the operator L is linear.

6. Since $L(y_1 + y_2) = \dfrac{d}{dx}(y_1 + y_2) = \dfrac{dy_1}{dx} + \dfrac{dy_2}{dx} = L(y_1) + L(y_2)$, and

$$L(cy_1) = \dfrac{d}{dx}(cy_1) = c\,\dfrac{dy_1}{dx} = cL(y_1),$$

the operator L is linear.

8. Since $L(y_1 + y_2) = \displaystyle\int (y_1 + y_2)\,dx = \int y_1\,dx + \int y_2\,dx = L(y_1) + L(y_2)$, and

$$L(cy_1) = \int cy_1\,dx = c\int y_1\,dx = cL(y_1),$$

the operator L is linear.

10. Since $L(y_1 + y_2) = (y_1 + y_2)^{1/3}$, but $L(y_1) + L(y_2) = y_1^{1/3} + y_2^{1/3}$, the operator L is not linear.

12. This equation is linear, and in operator notation

$$\phi(x, D)y = x^2 \qquad \text{where } \phi(x, D) = 2xD^2 + (x^3 - 5).$$

14. This equation is linear, and in operator notation

$$\phi(x, D)y = 10\sin x \qquad \text{where } \phi(x, D) = xD^3 + 3xD^2 - 2D + 1.$$

16. Because of the term $y\,d^3y/dx^3$, the equation is not linear.

18. Because of the yy'' term, this equation is not linear.

20. This equation is linear, and in operator notation

$$\phi(D)y = \ln x \qquad \text{where } \phi(D) = D^4 + D^2 - 1.$$

22. If $y_1(x)$ and $y_2(x)$ are any two functions in S, and c is a constant,

$$L(y_1 + y_2) = \int_0^{2\pi} [y_1(x) + y_2(x)]\cos nx\,dx = \int_0^{2\pi} y_1(x)\,\cos nx\,dx + \int_0^{2\pi} y_2(x)\,\cos nx\,dx$$
$$= L(y_1) + L(y_2),$$
$$L(cy_1) = \int_0^{2\pi} cy_1(x)\,\cos nx\,dx = c\int_0^{2\pi} y_1(x)\,\cos nx\,dx = cL(y_1).$$

Thus, L is a linear operator.

EXERCISES 15.7

2. Since $y_1' + y_1\tan x = -\sin x + \cos x\tan x = 0$, $y_1(x)$ is a solution. Because the equation is linear and homogeneous, a general solution is $y(x) = C\cos x$.

4. Since $2y_1'' - 16y_1' + 32y_1 = 96e^{4x} - 192e^{4x} + 96e^{4x} = 0$, and

$$2y_2'' - 16y_2' + 32y_2 = -4(16xe^{4x} + 8e^{4x}) + 32(4xe^{4x} + e^{4x}) - 64xe^{4x} = 0,$$

$y_1(x)$ and $y_2(x)$ are solutions of the equation. Because the equation is linear and homogeneous, a general solution is $y(x) = D_1(3e^{4x}) + D_2(-2xe^{4x}) = (C_1 + C_2x)e^{4x}$.

6. Since $2y_1'' - 8y_1' + 9y_1 = 2[4e^{2x}\cos(x/\sqrt{2}) - (2\sqrt{2})e^{2x}\sin(x/\sqrt{2}) - (1/2)e^{2x}\cos(x/\sqrt{2})]$
$$-8[2e^{2x}\cos(x/\sqrt{2}) - (1/\sqrt{2})e^{2x}\sin(x/\sqrt{2})] + 9e^{2x}\cos(x/\sqrt{2})$$
$$= 0,$$

and similarly, $2y_2'' - 8y_2' + 9y_2 = 0$, $y_1(x)$ and $y_2(x)$ are solutions of the equation. Because the equation is linear and homogeneous, a general solution is $y(x) = C_1e^{2x}\cos(x/\sqrt{2}) + C_2e^{2x}\sin(x/\sqrt{2})$.

8. Since $4y_1 + xy_1' + x^2y_1'' = 4\cos(2\ln x) + x[(-2/x)\sin(2\ln x)]$
$$+x^2[(2/x^2)\sin(2\ln x) - (4/x^2)\cos(2\ln x)]$$
$$= 0,$$

and similarly, $4y_2 + xy_2' + x^2y_2'' = 0$, $y_1(x)$ and $y_2(x)$ are solution of the equation. Because the equation is linear and homogeneous, a general solution is $y(x) = C_1\cos(2\ln x) + C_2\sin(2\ln x)$.

10. If the $y_i(x)$ are linearly dependent, there exists constants C_i, not all zero, such that

$$C_1y_1(x) + C_2y_2(x) + \cdots + C_ny_n(x) = 0.$$

When we differentiate this equation $n-1$ times, we obtain $n-1$ more equations:

$$\begin{array}{ccccccc} C_1y_1'(x) & + & C_2y_2'(x) & + & \cdots & + & C_ny_n'(x) & = & 0, \\ C_1y_1''(x) & + & C_2y_2''(x) & + & \cdots & + & C_ny_n''(x) & = & 0, \\ \vdots & & \vdots & & & & \vdots & & \\ C_1y_1^{(n-1)}(x) & + & C_2y_2^{(n-1)}(x) & + & \cdots & + & C_ny_n^{(n-1)}(x) & = & 0. \end{array}$$

Because this system of n linear equations in $C_1, \ldots, C_n$ has a nontrivial solution, it follows that the determinant of its coefficients must vanish; that is,

$$W(y_1, \ldots, y_n) = \begin{vmatrix} y_1 & y_2 & \cdots & y_n \\ y_1' & y_2' & \cdots & y_n' \\ \vdots & \vdots & & \vdots \\ y_1^{(n-1)} & y_2^{(n-1)} & \cdots & y_n^{(n-1)} \end{vmatrix} = 0 \quad \text{on } I.$$

12. Since $W(x, 2x - 3x^2, x^2) = \begin{vmatrix} x & 2x - 3x^2 & x^2 \\ 1 & 2 - 6x & 2x \\ 0 & -6 & 2 \end{vmatrix} = x[2(2-6x) + 12x] - [2(2x - 3x^2) + 6x^2] = 0,$

the functions are linearly dependent.

14. Since $W(x, xe^x, x^2e^x) = \begin{vmatrix} x & xe^x & x^2e^x \\ 1 & (x+1)e^x & (x^2+2x)e^x \\ 0 & (x+2)e^x & (x^2+4x+2)e^x \end{vmatrix}$, which at $x = 1$ reduces to $\begin{vmatrix} 1 & e & e \\ 1 & 2e & 3e \\ 0 & 3e & 7e \end{vmatrix} = e^2,$

the functions are therefore linearly independent.

EXERCISES 15.8

2. The auxiliary equation is $0 = 2m^2 - 16m + 32 = 2(m-4)^2$ with solutions $m = 4, 4$. A general solution of the differential equation is therefore $y(x) = (C_1 + C_2x)e^{4x}$.

4. The auxiliary equation is $0 = m^2 + 2m - 2$ with solutions $m = -1 \pm \sqrt{3}$. A general solution of the differential equation is therefore $y(x) = C_1e^{-(1+\sqrt{3})x} + C_2e^{(-1+\sqrt{3})x}$.

6. The auxiliary equation is $0 = m^3 - 3m^2 + m - 3 = (m-3)(m^2+1)$ with solutions $m = 3, \pm i$. A general solution of the differential equation is therefore $y(x) = C_1e^{3x} + C_2\cos x + C_3\sin x$.

8. The auxiliary equation is $0 = m^3 - 6m^2 + 12m - 8 = (m-2)^3$ with solutions $m = 2, 2, 2$. A general solution of the differential equation is therefore $y(x) = (C_1 + C_2x + C_3x^2)e^{2x}$.

10. The auxiliary equation is $0 = m^4 + 5m^2 + 4 = (m^2+1)(m^2+4)$ with solutions $m = \pm i, \pm 2i$. A general solution of the differential equation is therefore $y(x) = C_1\cos x + C_2\sin x + C_3\cos 2x + C_4\sin 2x$.

12. The auxiliary equation is $0 = m^4 + 16 = (m^2 + 4i)(m^2 - 4i)$. To solve $m^2 = 4i$, we set $m = a + bi$, so that $4i = (a+bi)^2 = (a^2 - b^2) + 2abi$. When we equate real and imaginary parts, $a^2 - b^2 = 0$ and $2ab = 4$, These imply that $a = b = \pm\sqrt{2}$. Thus, $m = \pm\sqrt{2}(1+i)$. From $m^2 = -4i$, we obtain $m = \pm\sqrt{2}(1-i)$. A general solution of the differential equation is therefore

$$y(x) = e^{\sqrt{2}x}[C_1\cos(\sqrt{2}x) + C_2\sin(\sqrt{2}x)] + e^{-\sqrt{2}x}[C_3\cos(\sqrt{2}x) + C_4\sin(\sqrt{2}x)].$$

14. For this general solution, roots of the auxiliary equation had to be $m = -2 \pm 4i$, and therefore

$$\phi(m) = (m + 2 + 4i)(m + 2 - 4i) = m^2 + 4m + 20.$$

A possible differential equation is therefore $y'' + 4y' + 20y = 0$.

16. For this general solution, the roots of the auxiliary equation had to be $m = 1 \pm \sqrt{2}i$, $1 \pm \sqrt{2}i$, and therefore

$$\phi(m) = (m - 1 + \sqrt{2}i)^2(m - 1 - \sqrt{2}i)^2 = (m^2 - 2m + 3)^2 = m^4 - 4m^3 + 10m^2 - 12m + 9.$$

A possible differential equation is therefore $y'''' - 4y''' + 10y'' - 12y' + 9y = 0$.

18. For this $y(x)$ to be a solution, the roots of the auxiliary equation must be $m = -1$, $-2 \pm 4i$, and therefore $\phi(m) = (m + 1)(m + 2 + 4i)(m + 2 - 4i) = m^3 + 5m^2 + 24m + 20$. It follows that

$$m^3 + 5m^2 + 24m + 20 = m^3 + am^2 + bm + c,$$

and we conclude that $a = 5$, $b = 24$, and $c = 20$.

20. The auxiliary equation is $m^2 + \lambda = 0$. Solutions depend on whether λ is positive, negative, or zero. We consider all three cases. When $\lambda < 0$, solutions are $m = \pm\sqrt{-\lambda}$, in which case $y(x) = C_1 e^{\sqrt{-\lambda}x} + C_2 e^{-\sqrt{-\lambda}x}$. The boundary conditions require C_1 and C_2 to satisfy

$$0 = y'(0) = \sqrt{-\lambda}\, C_1 - \sqrt{-\lambda}\, C_2, \qquad 0 = y'(4) = \sqrt{-\lambda}\, C_1 e^{4\sqrt{-\lambda}} - \sqrt{-\lambda}\, C_2 e^{-4\sqrt{-\lambda}}.$$

The only solution of these is $C_1 = C_2 = 0$, and therefore $y(x) = 0$.

When $\lambda = 0$, the auxiliary equation has a double root $m = 0$, in which case $y(x) = C_1 + C_2 x$. The boundary conditions require C_1 and C_2 to satisfy

$$0 = y'(0) = C_2, \qquad 0 = y'(4) = C_2.$$

Thus, $\lambda_0 = 0$ is an eigenvalue with eigenfunction $y_0(x) = C_1$.

When $\lambda > 0$, roots of the auxiliary equation are $m = \pm\sqrt{\lambda}i$, in which case $y(x) = C_1 \cos\sqrt{\lambda}x + C_2 \sin\sqrt{\lambda}x$. The boundary conditions require C_1 and C_2 to satisfy

$$0 = y'(0) = \sqrt{\lambda}\, C_2, \qquad 0 = y'(4) = -\sqrt{\lambda}\, C_1 \sin 4\sqrt{\lambda} + \sqrt{\lambda}\, C_2 \cos 4\sqrt{\lambda}.$$

With $C_2 = 0$, the second of these implies that $C_1 \sin 4\sqrt{\lambda} = 0$. Since we cannot set $C_1 = 0$, else $y(x)$ would again be zero, we must set $\sin 4\sqrt{\lambda} = 0 \implies 4\sqrt{\lambda} = n\pi$, where $n \neq 0$ is an integer. Thus, additional eigenvalues of the Sturm-Liouville system are $\lambda_n = n^2\pi^2/16$, with corresponding eigenfunctions $y_n(x) = C_1 \cos(n\pi x/4)$.

22. The auxiliary equation is $m^2 + \lambda = 0$. Solutions depend on whether λ is positive, negative, or zero. We consider all three cases. When $\lambda < 0$, solutions are $m = \pm\sqrt{-\lambda}$, in which case $y(x) = C_1 e^{\sqrt{-\lambda}x} + C_2 e^{-\sqrt{-\lambda}x}$. The boundary conditions require C_1 and C_2 to satisfy

$$0 = y'(0) = \sqrt{-\lambda}\, C_1 - \sqrt{-\lambda}\, C_2, \qquad 0 = y(5) = C_1 e^{5\sqrt{-\lambda}} + C_2 e^{-5\sqrt{-\lambda}}.$$

The only solution of these is $C_1 = C_2 = 0$, and therefore $y(x) = 0$.

When $\lambda = 0$, the auxiliary equation has a double root $m = 0$, in which case $y(x) = C_1 + C_2 x$. The boundary conditions require C_1 and C_2 to satisfy

$$0 = y'(0) = C_2, \qquad 0 = y(5) = C_1 + 5C_2.$$

Once again, the only solution is $C_1 = C_2 = 0$, from which $y(x) = 0$.

When $\lambda > 0$, roots of the auxiliary equation are $m = \pm\sqrt{\lambda}i$, in which case $y(x) = C_1 \cos\sqrt{\lambda}x + C_2 \sin\sqrt{\lambda}x$. The boundary conditions require C_1 and C_2 to satisfy

$$0 = y'(0) = \sqrt{\lambda}\, C_2, \qquad 0 = y(5) = C_1 \cos 5\sqrt{\lambda} + C_2 \sin 5\sqrt{\lambda}.$$

With $C_2 = 0$, the second of these implies that $C_1 \cos 5\sqrt{\lambda} = 0$. Since we cannot set $C_1 = 0$, else $y(x)$ would again be zero, we must set $\cos 5\sqrt{\lambda} = 0 \implies 5\sqrt{\lambda} = (2n - 1)\pi/2$, where n is an integer. Thus, eigenvalues of the Sturm-Liouville system are $\lambda_n = (2n-1)^2\pi^2/100$, with corresponding eigenfunctions $y_n(x) = C_1 \cos\left[(2n-1)\pi x/10\right]$.

24. (a) When the mass is at position (x, y), the force acting on it is

$$\mathbf{F} = k\sqrt{x^2+y^2}\left(\frac{-x\hat{\mathbf{i}} - y\hat{\mathbf{j}}}{\sqrt{x^2+y^2}}\right) = -k(x\hat{\mathbf{i}} + y\hat{\mathbf{j}}).$$

According to Newton's second law, the acceleration
is given by

$$M\frac{d^2\mathbf{r}}{dt^2} = -k(x\hat{\mathbf{i}} + y\hat{\mathbf{j}}).$$

When we equate components,

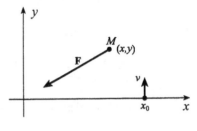

$$M\frac{d^2x}{dt^2} = -kx, \quad M\frac{d^2y}{dt^2} = -ky.$$

The auxiliary equation for each of these differential equations is $Mm^2 + k = 0$ with solutions $m = \pm\sqrt{k/M}\,i$. If we set $\omega = \sqrt{k/M}$, then

$$x(t) = A\cos\omega t + B\sin\omega t, \qquad y(t) = C\cos\omega t + D\sin\omega t.$$

With the initial conditions $x(0) = x_0$, $x'(0) = 0$, $y(0) = 0$, and $y'(0) = v$,

$$x_0 = A, \quad 0 = \omega B, \quad 0 = C, \quad v = \omega D.$$

Thus, $x = x_0\cos\omega t$ and $y = (v/\omega)\sin\omega t$ define the path of the mass parametrically. Eliminating t gives

$$\left(\frac{x}{x_0}\right)^2 + \left(\frac{\omega y}{v}\right)^2 = 1 \quad\Longrightarrow\quad \frac{x^2}{x_0^2} + \frac{ky^2}{Mv^2} = 1,$$

an ellipse.

(b) In this case, the differential equations are

$$M\frac{d^2x}{dt^2} = kx, \quad M\frac{d^2y}{dt^2} = ky.$$

The auxiliary equation for each of these differential equations is $Mm^2 - k = 0$ with solutions $m = \pm\sqrt{k/M}$. If we set $\omega = \sqrt{k/M}$, then

$$x(t) = Ae^{\omega t} + Be^{-\omega t}, \qquad y(t) = Ce^{\omega t} + De^{-\omega t}.$$

The initial conditions require

$$x_0 = A + B, \quad 0 = \omega A - \omega B, \quad 0 = C + D, \quad v = \omega C - \omega D.$$

These give $A = B = x_0/2$ and $C = -D = v/(2\omega)$. Thus, parametric equations for the path of the mass
are

$$x = \frac{x_0}{2}e^{\omega t} + \frac{x_0}{2}e^{-\omega t} = \frac{x_0}{2}(e^{\omega t} + e^{-\omega t}), \quad y = \frac{v}{2\omega}e^{\omega t} - \frac{v}{2\omega}e^{-\omega t} = \frac{v}{2\omega}(e^{\omega t} - e^{-\omega t}).$$

When t is eliminated, we obtain

$$\left(\frac{2x}{x_0}\right)^2 - \left(-\frac{2\omega y}{v}\right)^2 = (e^{\omega t} + e^{-\omega t})^2 - (e^{\omega t} - e^{-\omega t})^2 = 4 \quad\Longrightarrow\quad \frac{x^2}{x_0^2} - \frac{ky^2}{Mv^2} = 1,$$

a hyperbola. The mass moves along the right half of this hyperbola.

EXERCISES 15.9

2. The auxiliary equation is $0 = m^2 + 2m - 2$ with solutions $m = -1 \pm \sqrt{3}$. A general solution of the associated homogeneous equation is $y_h(x) = C_1 e^{-(1+\sqrt{3})x} + C_2 e^{(-1+\sqrt{3})x}$. By operators,

$$y_p = \frac{1}{D^2 + 2D - 2}(x^2 e^{-x}) = e^{-x}\frac{1}{(D-1)^2 + 2(D-1) - 2}(x^2) = e^{-x}\frac{1}{D^2 - 3}(x^2)$$

$$= \frac{e^{-x}}{-3}\frac{1}{1 - D^2/3}(x^2) = \frac{e^{-x}}{-3}\left(1 + \frac{D^2}{3} + \cdots\right)x^2 = \frac{e^{-x}}{-3}\left(x^2 + \frac{2}{3}\right).$$

By undetermined coefficients, $y_p = Ax^2 e^{-x} + Bxe^{-x} + Ce^{-x}$. Subsitution into the differential equation gives

$$(2Ae^{-x} - 4Axe^{-x} + Ax^2 e^{-x} - 2Be^{-x} + Bxe^{-x} + Ce^{-x}) + 2(2Axe^{-x} - Ax^2 e^{-x} + Be^{-x}$$
$$- Bxe^{-x} - Ce^{-x}) - 2(Ax^2 e^{-x} + Bxe^{-x} + Ce^{-x}) = x^2 e^{-x}.$$

When we equate coefficients of $x^2 e^{-x}$, xe^{-x}, and e^{-x}:

$$-3A = 1, \qquad -3B = 0, \qquad 2A - 3C = 0.$$

Thus, $A = -1/3$, $B = 0$, and $C = -2/9$ and once again $y_p = (-1/3)x^2 e^{-x} - (2/9)e^{-x}$. A general solution of the differential equation is therefore $y(x) = C_1 e^{-(1+\sqrt{3})x} + C_2 e^{(-1+\sqrt{3})x} - e^{-x}(3x^2 + 2)/9$.

4. The auxiliary equation is $0 = m^4 + 2m^2 + 1 = (m^2 + 1)^2$ with solutions $m = \pm i, \pm i$. A general solution of the associated homogeneous equation is $y_h(x) = (C_1 + C_2 x)\cos x + (C_3 + C_4 x)\sin x$. By operators,

$$y_p = \frac{1}{D^4 + 2D^2 + 1}\cos 2x = \text{Re}\left(\frac{1}{D^4 + 2D^2 + 1}e^{2ix}\right)$$

$$= \text{Re}\left[e^{2ix}\frac{1}{(D+2i)^4 + 2(D+2i)^2 + 1}(1)\right] = \text{Re}\left[e^{2ix}\frac{1}{9 + \cdots}(1)\right] = \text{Re}\left[\frac{e^{2ix}}{9}\right] = \frac{1}{9}\cos 2x.$$

By undetermined coefficients, $y_p = A\cos 2x + B\sin 2x$. Substitution into the differential equation gives

$$(16A\cos 2x + 16B\sin 2x) + 2(-4A\cos 2x - 4B\sin 2x) + (A\cos 2x + B\sin 2x) = \cos 2x.$$

When we equate coefficients of $\cos 2x$ and $\sin 2x$, we obtain $9A = 1$, and $9B = 0$. Thus, $A = 1/9$ and $B = 0$, and $y_p = (1/9)\cos 2x$. A general solution of the differential equation is

$$y(x) = (C_1 + C_2 x)\cos x + (C_3 + C_4 x)\sin x + (1/9)\cos 2x.$$

6. The auxiliary equation is $0 = m^4 + 5m^2 + 4 = (m^2 + 4)(m^2 + 1)$ with solutions $m = \pm i, \pm 2i$. A general solution of the associated homogneous equation is $y_h(x) = C_1\cos x + C_2\sin x + C_3\cos 2x + C_4\sin 2x$. By operators,

$$y_p = \frac{1}{D^4 + 5D^2 + 4}e^{-2x} = e^{-2x}\frac{1}{(D-2)^4 + 5(D-2)^2 + 4}(1) = e^{-2x}\frac{1}{40 + \cdots}(1) = \frac{1}{40}e^{-2x}.$$

By undetermined coefficients, $y_p = Ae^{-2x}$. Substitution into the differential equation gives

$$(16Ae^{-2x}) + 5(4Ae^{-2x}) + 4(Ae^{-2x}) = e^{-2x}.$$

This equation implies that $A = 1/40$, and a general solution of the differential equation is

$$y(x) = C_1\cos x + C_2\sin x + C_3\cos 2x + C_4\sin 2x + (1/40)e^{-2x}.$$

8. The auxiliary equation is $0 = 2m^2 + 16m + 82$ with solutions $m = -4 \pm 5i$. A general solution of the associated homogeneous equation is $y_h(x) = e^{-4x}(C_1 \cos 5x + C_2 \sin 5x)$. By operators,

$$y_p = \frac{1}{2D^2 + 16D + 82}(-2e^{2x}\sin x) = -\frac{1}{D^2 + 8D + 41}\mathrm{Im}[e^{(2+i)x}] = -\mathrm{Im}\left[\frac{1}{D^2 + 8D + 41}e^{(2+i)x}\right]$$

$$= -\mathrm{Im}\left[e^{(2+i)x}\frac{1}{(D+2+i)^2 + 8(D+2+i) + 41}(1)\right] = -\mathrm{Im}\left[e^{(2+i)x}\frac{1}{60 + 12i + \cdots}(1)\right]$$

$$= -\frac{1}{12}\mathrm{Im}\left[e^{(2+i)x}\frac{1}{5+i}\frac{5-i}{5-i}\right] = \frac{-1}{12}\mathrm{Im}\left[e^{(2+i)x}\frac{5-i}{26}\right]$$

$$= \frac{-1}{312}\mathrm{Im}\left[e^{2x}(\cos x + i\sin x)(5-i)\right] = -\frac{e^{2x}}{312}(-\cos x + 5\sin x).$$

By undetermined coefficients, $y_p = Ae^{2x}\sin x + Be^{2x}\cos x$. Substitution into the differential equation gives

$$2(4Ae^{2x}\sin x + 4Ae^{2x}\cos x - Ae^{2x}\sin x + 4Be^{2x}\cos x - 4Be^{2x}\sin x - Be^{2x}\cos x)$$
$$+ 16(2Ae^{2x}\sin x + Ae^{2x}\cos x + 2Be^{2x}\cos x - Be^{2x}\sin x)$$
$$+ 82(Ae^{2x}\sin x + Be^{2x}\cos x) = -2e^{2x}\sin x.$$

When we equate coefficients of $e^{2x}\sin x$ and $e^{2x}\cos x$:

$$120A - 24B = -2, \qquad 120B + 24A = 0.$$

These imply that $A = -5/312$ and $B = 1/312$, and once again $y_p = e^{2x}(\cos x - 5\sin x)/312$. A general solution of the differential equation is therefore

$$y(x) = e^{-4x}(C_1 \cos 5x + C_2 \sin 5x) + e^{2x}(\cos x - 5\sin x)/312.$$

10. The auxiliary equation is $0 = m^2 - 4m + 5$ with solutions $m = 2 \pm i$. A general solution of the associated homogeneous equation is $y_h(x) = e^{2x}(C_1 \cos x + C_2 \sin x)$. By operators,

$$y_p = \frac{1}{D^2 - 4D + 5}(x\cos x) = \mathrm{Re}\left[\frac{1}{D^2 - 4D + 5}xe^{ix}\right]$$

$$= \mathrm{Re}\left[e^{ix}\frac{1}{(D+i)^2 - 4(D+i) + 5}(x)\right] = \mathrm{Re}\left[e^{ix}\frac{1}{D^2 + (-4+2i)D + (4-4i)}(x)\right]$$

$$= \mathrm{Re}\left[\frac{e^{ix}}{4-4i}\frac{1}{1 + \frac{(-4+2i)D + D^2}{4-4i}}(x)\right] = \mathrm{Re}\left\{\frac{e^{ix}}{4-4i}\left[1 - \left(\frac{(-4+2i)D + D^2}{4-4i}\right) + \cdots\right]x\right\}$$

$$= \mathrm{Re}\left[\frac{e^{ix}}{4(1-i)}\left(x + \frac{4-2i}{4-4i}\right)\right] = \mathrm{Re}\left[\frac{e^{ix}}{4}\left(\frac{x}{1-i}\frac{1+i}{1+i} + \frac{4-2i}{-8i}\right)\right]$$

$$= \mathrm{Re}\left[\frac{\cos x + i\sin x}{4}\left(\frac{x(1+i)}{2} + \frac{1+2i}{4}\right)\right] = \frac{x}{8}(\cos x - \sin x) + \frac{1}{16}(\cos x - 2\sin x).$$

By undetermined coefficients, $y_p = Ax\cos x + Bx\sin x + C\cos x + D\sin x$. Substitution into the differential equation gives

$$(-2A\sin x - Ax\cos x + 2B\cos x - Bx\sin x - C\cos x - D\sin x)$$
$$- 4(A\cos x - Ax\sin x + B\sin x + Bx\cos x - C\sin x + D\cos x)$$
$$+ 5(Ax\cos x + Bx\sin x + C\cos x + D\sin x) = x\cos x.$$

When we equate coefficients of $x\cos x$, $x\sin x$, $\cos x$, and $\sin x$:

$$4A - 4B = 1, \quad 4A + 4B = 0, \quad -4A + 2B + 4C - 4D = 0, \quad -2A - 4B + 4C + 4D = 0.$$

These imply that $A = 1/8$, $B = -1/8$, $C = 1/16$, and $D = -1/8$, giving the same y_p as above. A general solution of the differential equation is therefore

$$y(x) = e^{2x}(C_1 \cos x + C_2 \sin x) + x(\cos x - \sin x)/8 + (\cos x - 2\sin x)/16.$$

12. The auxiliary equation is $0 = m^3 + 9m^2 + 27m + 27 = (m+3)^2$ with solutions $m = -3, -3, -3$. A general solution of the associated homogeneous equation is $y_h(x) = (C_1 + C_2 x + C_3 x^2)e^{-3x}$. Undetermined coefficients suggests $y_p(x) = Axe^{3x} + Be^{3x} + Cx\cos x + Dx\sin x + E\cos x + F\sin x$.

14. The auxiliary equation is $0 = 2m^3 - 6m^2 - 12m + 16 = 2(m-1)(m-4)(m+2)$ with solutions $m = 1, -2, 4$. A general solution of the associated homogeneous equation is $y_h(x) = C_1 e^x + C_2 e^{-2x} + C_3 e^{4x}$. Undetermined coefficients suggests $y_p(x) = Ax^2 e^x + Bxe^x + Cx^3 + Dx^2 + Ex + F + G\cos x + H\sin x$.

16. According to the operator shift theorem, $\phi(D)\{e^{px}g(x)\} = e^{px}[\phi(D+p)g(x)]$, and therefore $e^{px}g(x) = \frac{1}{\phi(D)}\{e^{px}[\phi(D+p)g(x)]\}$. If we set $f(x) = \phi(D+p)g(x)$, in which case $g(x) = \frac{1}{\phi(D+p)}f(x)$, then

$$e^{px}\frac{1}{\phi(D+p)}f(x) = \frac{1}{\phi(D)}\{e^{px}f(x)\}.$$

18. The auxiliary equation is $0 = 2m^2 - 4m + 3$ with solutions $m = 1 \pm (1/\sqrt{2})i$. A general solution of the associated homogeneous equation is $y_h(x) = e^x[C_1 \cos(x/\sqrt{2}) + C_2 \sin(x/\sqrt{2})]$. By operators,

$$
\begin{aligned}
y_p &= \frac{1}{2D^2 - 4D + 3}(\cos x \sin 2x) = \frac{1}{2D^2 - 4D + 3}\left[\frac{1}{2}(\sin 3x + \sin x)\right] \\
&= \frac{1}{2}\mathrm{Im}\left[\frac{1}{2D^2 - 4D + 3}e^{3ix}\right] + \frac{1}{2}\mathrm{Im}\left[\frac{1}{2D^2 - 4D + 3}e^{ix}\right] \\
&= \frac{1}{2}\mathrm{Im}\left[e^{3ix}\frac{1}{2(D+3i)^2 - 4(D+3i) + 3}(1)\right] + \frac{1}{2}\mathrm{Im}\left[e^{ix}\frac{1}{2(D+i)^2 - 4(D+i) + 3}(1)\right] \\
&= \frac{1}{2}\mathrm{Im}\left[\frac{e^{3ix}}{-15 - 12i}\right] + \frac{1}{2}\mathrm{Im}\left[\frac{e^{ix}}{1 - 4i}\right] = -\frac{1}{6}\mathrm{Im}\left[\frac{e^{3ix}}{5+4i}\frac{5-4i}{5-4i}\right] + \frac{1}{2}\mathrm{Im}\left[\frac{e^{ix}}{1-4i}\frac{1+4i}{1+4i}\right] \\
&= \frac{-1}{6}\mathrm{Im}\left[\frac{(\cos 3x + i\sin 3x)(5-4i)}{41}\right] + \frac{1}{2}\mathrm{Im}\left[\frac{(\cos x + i\sin x)(1+4i)}{17}\right] \\
&= -\frac{1}{246}(-4\cos 3x + 5\sin 3x) + \frac{1}{34}(4\cos x + \sin x).
\end{aligned}
$$

A general solution of the differential equation is therefore

$$y(x) = e^x[C_1 \cos(x/\sqrt{2}) + C_2 \sin(x/\sqrt{2})] + (4\cos 3x - 5\sin 3x)/246 + (4\cos x + \sin x)/34.$$

20. The auxiliary equation $m^2 + 9 = 0$ has solutions $m = \pm 3i$. A general solution of the associated homogeneous equation is $y_h(x) = C_1 \cos 3x + C_2 \sin 3x$. By operators,

$$y_p = \frac{1}{D^2 + 9}\{x[\mathrm{Im}(e^{3ix}) + \mathrm{Re}(e^{3ix})]\} = \mathrm{Im}\left[\frac{1}{D^2 + 9}(xe^{3ix})\right] + \mathrm{Re}\left[\frac{1}{D^2 + 9}(xe^{3ix})\right].$$

Consider then $\quad \frac{1}{D^2 + 9}(xe^{3ix}) = e^{3ix}\frac{1}{(D+3i)^2 + 9}(x) = e^{3ix}\frac{1}{D^2 + 6iD}(x)$

$$
\begin{aligned}
&= e^{3ix}\frac{1}{6iD[1 + D/(6i)]}(x) = e^{3ix}\frac{1}{6iD}\left(1 - \frac{D}{6i} + \cdots\right)x \\
&= e^{3ix}\frac{1}{6iD}\left(x - \frac{1}{6i}\right) = -\frac{i}{6}e^{3ix}\left(\frac{x^2}{2} + \frac{ix}{6}\right) = \frac{1}{36}e^{3ix}(x - 3ix^2).
\end{aligned}
$$

Thus, $y_p(x) = \dfrac{1}{36}(x\sin 3x - 3x^2 \cos 3x) + \dfrac{1}{36}(x\cos 3x + 3x^2 \sin 3x)$, and

$$y(x) = C_1 \cos 3x + C_2 \sin 3x + \frac{x}{36}(\cos 3x + \sin 3x) + \frac{x^2}{12}(\sin 3x - \cos 3x).$$

For $y(0) = 0$ and $y'(0) = 0$, we must have $0 = C_1$ and $0 = 3C_2 + 1/36$. Hence,

$$y(x) = -\frac{1}{108} \sin 3x + \frac{x}{36}(\cos 3x + \sin 3x) + \frac{x^2}{12}(\sin 3x - \cos 3x).$$

22. Since $\dfrac{dy}{dx} = \dfrac{dy}{dz}\dfrac{dz}{dx} = \dfrac{dy}{dz} \Big/ \dfrac{dx}{dz} = \dfrac{1}{x}\dfrac{dy}{dz}$, we obtain

$$\frac{d^2y}{dx^2} = \frac{d}{dx}\left(\frac{1}{x}\frac{dy}{dz}\right) = -\frac{1}{x^2}\frac{dy}{dz} + \frac{1}{x}\frac{d}{dz}\left(\frac{dy}{dz}\right)\frac{dz}{dx} = -\frac{1}{x^2}\frac{dy}{dz} + \frac{1}{x^2}\frac{d^2y}{dz^2}.$$

Thus, $x\dfrac{dy}{dx} = \dfrac{dy}{dz}$ and $x^2\dfrac{d^2y}{dx^2} = \dfrac{d^2y}{dz^2} - \dfrac{dy}{dz}$. When we substitute these into the differential equation,

$$\frac{d^2y}{dz^2} - \frac{dy}{dz} + a\frac{dy}{dz} + by = F(e^z) \quad\Longrightarrow\quad \frac{d^2y}{dz^2} + (a-1)\frac{dy}{dz} + by = F(e^z),$$

a linear differential equation with constant coefficients.

24. If we set $x = e^z$ and use the results of Exercise 22, $\quad 1 = \dfrac{d^2y}{dz^2} - \dfrac{dy}{dz} + \dfrac{dy}{dz} + 4y = \dfrac{d^2y}{dz^2} + 4y$. The auxiliary equation is $m^2 + 4 = 0$ with solutions $m = \pm 2i$. A general solution of the associated homogeneous equation is $y_h(z) = C_1 \cos 2z + C_2 \sin 2z$. Since $y_p(z) = 1/4$,

$$y(z) = C_1 \cos 2z + C_2 \sin 2z + 1/4 \quad\Longrightarrow\quad y(x) = C_1 \cos(2\ln x) + C_2 \sin(2\ln x) + 1/4.$$

EXERCISES 15.10

2. With the coordinate system of Figure 15.11, the differential equation describing the position $x(t)$ of the mass is

$$\frac{1}{5}\frac{d^2x}{dt^2} + \frac{3}{2}\frac{dx}{dt} + 10x = 4\sin 10t \quad\Longrightarrow\quad 2x'' + 15x' + 100x = 40\sin 10t,$$

subject to $x(0) = 0$, $x'(0) = 0$. The auxiliary equation is $0 = 2m^2 + 15m + 100$ with solutions $m = (-15 \pm 5\sqrt{23}i)/4$. A general solution of the associated homogeneous equation is

$$x_h(t) = e^{-15t/4}[C_1 \cos(5\sqrt{23}t/4) + C_2 \sin(5\sqrt{23}t/4)].$$

A particular solution of the differential equation is

$$x_p(t) = \frac{1}{2D^2 + 15D + 100}\left[40\,\text{Im}(e^{10it})\right] = 40\,\text{Im}\left[\frac{1}{2D^2 + 15D + 100}(e^{10it})\right]$$

$$= 40\,\text{Im}\left[e^{10it}\frac{1}{2(D+10i)^2 + 15(D+10i) + 100}(1)\right] = 40\,\text{Im}\left(\frac{e^{10it}}{-100 + 150i}\right)$$

$$= \frac{4}{5}\text{Im}\left(\frac{e^{10it}}{-2+3i}\frac{-2-3i}{-2-3i}\right) = -\frac{4}{5}\text{Im}\left[\frac{(2+3i)(\cos 10t + i\sin 10t)}{13}\right]$$

$$= -\frac{4}{65}(3\cos 10t + 2\sin 10t).$$

Thus, $x(t) = e^{-15t/4}[C_1 \cos(5\sqrt{23}t/4) + C_2 \sin(5\sqrt{23}t/4)] - (4/65)(3\cos 10t + 2\sin 10t)$. To satisfy the initial conditions, we must have $0 = C_1 - 12/65$ and $0 = -15C_1/4 + 5\sqrt{23}C_2/4 - 16/13$. These imply that $C_1 = 12/65$ and $C_2 = 20/(13\sqrt{23})$, and therefore

$$x(t) = e^{-15t/4}\{(12/65)\cos(5\sqrt{23}t/4) + [20/(13\sqrt{23})]\sin(5\sqrt{23}t/4)\}$$
$$- (4/65)(3\cos 10t + 2\sin 10t) \text{ m}.$$

4. The differential equation describing charge $Q(t)$ on the capacitor is

$$(1)\frac{d^2Q}{dt^2} + 100\frac{dQ}{dt} + \frac{1}{0.02}Q = 0 \quad \Longrightarrow \quad Q'' + 100Q' + 50Q = 0,$$

subject to $Q(0) = 5$ and $Q'(0) = 0$. The auxiliary equation is $m^2 + 100m + 50 = 0$ with solutions $m = -0.50, -99.50$. A general solution of the differential equation is therefore $Q(t) = C_1 e^{-0.50t} + C_2 e^{-99.50t}$. To satisfy the initial conditions, we must have $5 = C_1 + C_2$ and $0 = -0.50C_1 - 99.50C_2$. These imply that $C_1 = 5.03$ and $C_2 = -0.0253$, and therefore $Q(t) = 5.03e^{-0.50t} - 0.0253e^{-99.50t}$ C.

6. With the coordinate system of Figure 15.11, the differential equation describing the position of M is $M\frac{d^2x}{dt^2} + kx = 0$. According to equation 15.63, the stretch in the spring at equilibrium is Mg/k, and therefore the initial conditions are $x(0) = Mg/k$ and $x'(0) = 0$. The auxiliary equation is $0 = Mm^2 + k$ with solutions $m = \pm\sqrt{k/M}i$. Thus, $x(t) = C_1 \cos\left(\sqrt{k/M}t\right) + C_2 \sin\left(\sqrt{k/M}t\right)$. To satisfy the initial conditions, $Mg/k = C_1$ and $0 = \sqrt{k/M}C_2$. Thus, $x(t) = (Mg/k)\cos\left(\sqrt{k/M}t\right)$.

8. (a) Since the x-component of the force of friction when the mass is moving to the left is $1/2$ N, the differential equation describing the position $x(t)$ of the mass from the time it starts until it comes to a stop for the first time is

$$\frac{1}{2}\frac{d^2x}{dt^2} + 18x = \frac{1}{2} \quad \Longrightarrow \quad x'' + 36x = 1,$$

subject to $x(0) = 1/4$ and $x'(0) = 0$.
(b) The auxiliary equation is $m^2 + 36 = 0$ with solutions $m = \pm 6i$, and therefore $x(t) = C_1 \cos 6t + C_2 \sin 6t + 1/36$. To satisfy the initial conditions, we must have $1/4 = C_1 + 1/36$ and $0 = 6C_2$. Thus, $x(t) = (2/9)\cos 6t + 1/36$. Since $v(t) = (-4/3)\sin 6t$, the mass comes to rest for the first time when $6t = \pi$, and at this time, its position is $x = (2/9)\cos\pi + 1/36 = -7/36$ m. At this position, the spring force is $18(7/36) = 7/2$ N. Because this is greater than the $1/2$ N friction force, further motion will occur.

10. (a) If $\beta = 0$, then $M\frac{d^2x}{dt^2} + kx = 0$. The auxiliary equation is $Mm^2 + k = 0$ with solutions $m = \pm\sqrt{k/M}i$. Thus, $x(t) = C_1 \cos\left(\sqrt{k/M}t\right) + C_2 \sin\left(\sqrt{k/M}t\right)$.
(b) If $\beta \neq 0$ and $\beta^2 - 4kM < 0$, the auxiliary equation $Mm^2 + \beta m + k = 0$ has solutions

$$m = \frac{-\beta \pm \sqrt{\beta^2 - 4kM}}{2M} = -\frac{\beta}{2M} \pm \frac{\sqrt{4kM - \beta^2}}{2M}i.$$

Thus, $x(t) = e^{-\beta t/(2M)}(C_1 \cos\omega t + C_2 \sin\omega t)$, where $\omega = \sqrt{4kM - \beta^2}/(2M)$.

(c) If $\beta \neq 0$ and $\beta^2 - 4kM > 0$, the auxiliary equation has solutions $m = \dfrac{-\beta \pm \sqrt{\beta^2 - 4kM}}{2M}$, and therefore

$$x(t) = C_1 e^{(-\beta + \sqrt{\beta^2 - 4kM})t/(2M)} + C_2 e^{(-\beta - \sqrt{\beta^2 - 4kM})t/(2M)} = e^{-\beta t/(2M)}(C_1 e^{\omega t} + C_2 e^{-\omega t}),$$

where $\omega = \sqrt{\beta^2 - 4kM}/(2M)$.
(d) If $\beta \neq 0$ and $\beta^2 - 4kM = 0$, the auxiliary equation has solutions $m = -\beta/(2M), -\beta/(2M)$, and therefore $x(t) = (C_1 + C_2 t)e^{-\beta t/(2M)}$.

12. With the coordinate system of Figure 15.11, the differential equation describing the position $x(t)$ of the mass is

$$\frac{1}{10}\frac{d^2x}{dt^2} + 4000x = 3\cos 200t \quad \Longrightarrow \quad x'' + 40\,000x = 30\cos 200t,$$

subject to $x(0) = 0$, $x'(0) = 10$. The auxiliary equation is $m^2 + 40\,000 = 0$ with solutions $m = \pm 200i$. A general solution of the associated homogeneous equation is $x_h(t) = C_1 \cos 200t + C_2 \sin 200t$. Substituting a particular solution of the form $x_p = At\cos 200t + Bt\sin 200t$ into the differential equation gives

$$(-400A\sin 200t - 40\,000At\cos 200t + 400B\cos 200t - 40\,000Bt\sin 200t)$$
$$+ 40\,000(At\cos 200t + Bt\sin 200t) = 30\cos 200t.$$

This implies that $A = 0$ and $B = 3/40$, so that $x(t) = C_1\cos 200t + C_2\sin 200t + (3t/40)\sin 200t$. The initial conditions require $0 = C_1$ and $10 = 200C_2$. Thus, $x(t) = (1/20 + 3t/40)\sin 200t$ m. Displacements become unbounded as t gets large.

14. The differential equation describing the position of the mass is $M\dfrac{d^2x}{dt^2} + kx = A\cos\omega t$. Solutions of the auxiliary equation $Mm^2 + k = 0$ are $m = \pm\sqrt{k/M}\,i$. Hence the general solution of the associated homogeneous equation is $x(t) = C_1\cos\sqrt{k/M}\,t + C_2\sin\sqrt{k/M}\,t$. Resonance occurs when $\sqrt{k/M} = \omega$.

16. (a) Substituting a particular solution of the form $x_p(t) = B\cos\omega t + C\sin\omega t$ into the differential equation gives

$$M(-\omega^2 B\cos\omega t - \omega^2 C\sin\omega t) + \beta(-\omega B\sin\omega t + \omega C\cos\omega t) + k(B\cos\omega t + C\sin\omega t) = A\cos\omega t.$$

When we equate coefficients of $\cos\omega t$ and $\sin\omega t$, we obtain

$$(k - M\omega^2)B + \beta\omega C = A, \qquad -\beta\omega B + (k - M\omega^2)C = 0.$$

Solutions of these are $B = \dfrac{A(k - M\omega^2)}{(k - M\omega^2)^2 + \beta^2\omega^2}$, $C = \dfrac{A\beta\omega}{(k - M\omega^2)^2 + \beta^2\omega^2}$. The particular solution is therefore

$$x_p(t) = \frac{A}{(k - M\omega^2)^2 + \beta^2\omega^2}[(k - M\omega^2)\cos\omega t + \beta\omega\sin\omega t].$$

(b) If we set $(k - M\omega^2)\cos\omega t + \beta\omega\sin\omega t = R\sin(\omega t + \phi) = R(\sin\omega t\cos\phi + \cos\omega t\sin\phi)$, and equate coefficients of $\sin\omega t$ and $\cos\omega t$,

$$k - M\omega^2 = R\sin\phi, \qquad \beta\omega = R\cos\phi.$$

These imply that $R^2 = (k - M\omega^2)^2 + \beta^2\omega^2$, and therefore

$$\sin\phi = \frac{k - M\omega^2}{\sqrt{(k - M\omega^2)^2 + \beta^2\omega^2}}, \qquad \cos\phi = \frac{\beta\omega}{\sqrt{(k - M\omega^2)^2 + \beta^2\omega^2}}.$$

(c) The amplitude $x(t)$ is a maximum when $(k - M\omega^2)^2 + \beta^2\omega^2$ is smallest. To determine the value of ω that yields the minimum, we solve

$$0 = 2(k - M\omega^2)(-2M\omega) + 2\beta^2\omega = 2\omega[-2M(k - M\omega^2) + \beta^2].$$

The nonzero solution is $\omega = \sqrt{k/M - \beta^2/(2M^2)}$. The amplitude at this value of ω is

$$\frac{A}{\sqrt{\left[k - M\left(\dfrac{k}{M} - \dfrac{\beta^2}{2M^2}\right)\right]^2 + \beta^2\left(\dfrac{k}{M} - \dfrac{\beta^2}{2M^2}\right)}} = \frac{2AM}{\beta\sqrt{4kM - \beta^2}}.$$

18. If x measures displacement of the platform from its equilibrium position, then the differential equation for the combined motion is

$$\left(\frac{W + w}{g}\right)\frac{d^2x}{dt^2} + \beta\frac{dx}{dt} + kx = 0.$$

The auxiliary equation is $\left(\dfrac{W + w}{g}\right)m^2 + \beta m + k = 0$ with solutions

$$m = \frac{-\beta \pm \sqrt{\beta^2 - 4k(W + w)/g}}{2(W + w)/g}.$$

Oscillations occur for large w, and for small values of w no oscillations occur. The largest value of w for no oscillations occurs when

$$\beta^2 - \frac{4k(W+w)}{g} = 0 \implies w = \frac{\beta^2 g}{4k} - W.$$

20. (a) The differential equation is

$$200\frac{d^2y}{dt^2} + 3000\frac{dy}{dt} + 50\,000y = 5000\sin\frac{\pi vt}{40} \implies \frac{d^2y}{dt^2} + 15\frac{dy}{dt} + 250y = 25\sin\frac{\pi vt}{40}.$$

The auxiliary equation is $m^2 + 15m + 250 = 0$ with solutions

$$m = \frac{-15 \pm \sqrt{15^2 - 4(250)}}{2} = \frac{-15 \pm 5\sqrt{31}i}{2}.$$

Hence, $y_h(t) = e^{-15t/2}\left(C_1\cos\frac{5\sqrt{31}t}{2} + C_2\sin\frac{5\sqrt{31}t}{2}\right).$ A particular solution is of the form

$y_p(t) = A\cos\dfrac{\pi vt}{40} + B\sin\dfrac{\pi vt}{40}.$ Substituting into the differential equation,

$$\left(-\frac{\pi^2 v^2 A}{1600}\cos\frac{\pi vt}{40} - \frac{\pi^2 v^2 B}{1600}\sin\frac{\pi vt}{40}\right) + 15\left(\frac{-\pi vA}{40}\sin\frac{\pi vt}{40} + \frac{\pi vB}{40}\cos\frac{\pi vt}{40}\right)$$

$$+ 250\left(A\cos\frac{\pi vt}{40} + B\sin\frac{\pi vt}{40}\right) = 25\sin\frac{\pi vt}{40}.$$

Equating coefficients gives

$$-\frac{\pi^2 v^2 A}{1600} + \frac{3\pi vB}{8} + 250A = 0, \qquad -\frac{\pi^2 v^2 B}{1600} - \frac{3\pi vA}{8} + 250B = 25.$$

The solution is

$$A = \frac{-24\,000\,000\pi v}{(400\,000 - \pi^2 v^2)^2 + 360\,000\pi^2 v^2}, \qquad B = \frac{40\,000(400\,000 - \pi^2 v^2)}{(400\,000 - \pi^2 v^2)^2 + 360\,000\pi^2 v^2}.$$

Thus, $y(t) = e^{-15t/2}\left(C_1\cos\dfrac{5\sqrt{31}t}{2} + C_2\sin\dfrac{5\sqrt{31}t}{2}\right) + A\cos\dfrac{\pi vt}{40} + B\sin\dfrac{\pi vt}{40}.$ The initial conditions $y(0) = 0$ and $y'(0) = 0$ require

$$0 = C_1 + A, \qquad 0 = -\frac{15}{2}C_1 + \frac{5\sqrt{31}}{2}C_2 + \frac{\pi vB}{40}.$$

These give $C_1 = -A$ and $C_2 = -(300A + \pi vB)/(100\sqrt{31})$. When $v = 10$, the solution is

$$y(t) = e^{-15t/2}\left(0.00473\cos\frac{5\sqrt{31}t}{2} - 0.00310\sin\frac{5\sqrt{31}t}{2}\right) - 0.00473\cos\frac{\pi t}{4} + 0.100\sin\frac{\pi t}{4}.$$

The amplitude of the steady-state part is

$$\sqrt{(0.00473)^2 + (0.100)^2} = 0.100.$$

(b) When $v = 20$, the solution is

$$y(t) = e^{-15t/2}\left(0.00953\cos\frac{5\sqrt{31}t}{2} - 0.00616\sin\frac{5\sqrt{31}t}{2}\right) - 0.00953\cos\frac{\pi t}{2} + 0.100\sin\frac{\pi t}{2}.$$

The amplitude of the steady-state part is

$$\sqrt{(0.00953)^2 + (0.100)^2} = 0.100.$$

22. (a) The force field associated with the given potential is $\mathbf{F} = -\nabla V = (-36y + 96x)\hat{\mathbf{i}} - 36x\hat{\mathbf{j}}$. By Newton's second law,

$$3\left(\frac{d^2x}{dt^2}\hat{\mathbf{i}} + \frac{d^2y}{dt^2}\hat{\mathbf{j}}\right) = (-36y + 96x)\hat{\mathbf{i}} - 36x\hat{\mathbf{j}}.$$

When we equate components,

$$\frac{d^2x}{dt^2} = -12y + 32x, \qquad \frac{d^2y}{dt^2} = -12x.$$

(b) If we substitute from the second of the differential equations in part (a) into the first, we obtain

$$-\frac{1}{12}\frac{d^4y}{dt^4} = -12y + 32\left(-\frac{1}{12}\frac{d^2y}{dt^2}\right) \qquad \Longrightarrow \qquad \frac{d^4y}{dt^4} - 32\frac{d^2y}{dt^2} - 144y = 0.$$

The auxiliary equation is $0 = m^4 - 32m^2 - 144 = (m^2 - 36)(m^2 + 4)$ with solutions $m = \pm 6, \pm 2i$. Hence,

$$y(t) = C_1 e^{6t} + C_2 e^{-6t} + C_3 \cos 2t + C_4 \sin 2t.$$

The second equation gives

$$x(t) = -\frac{1}{12}\frac{d^2y}{dt^2} = -3C_1 e^{6t} - 3C_2 e^{-6t} + \frac{1}{3}C_3 \cos 2t + \frac{1}{3}C_4 \sin 2t.$$

The initial conditions $x(0) = 10$, $x'(0) = 0$, $y(0) = -10$, and $y'(0) = 0$ require

$$10 = -3C_1 - 3C_2 + \frac{C_3}{3}, \qquad 0 = -18C_1 + 18C_2 + \frac{2C_4}{3},$$

$$-10 = C_1 + C_2 + C_3, \qquad 0 = 6C_1 - 6C_2 + 2C_4.$$

These give $C_1 = -2$, $C_2 = -2$, $C_3 = -6$, and $C_4 = 0$. Thus,

$$x(t) = 6e^{6t} + 6e^{-6t} - 2\cos 2t, \qquad y(t) = -2e^{6t} - 2e^{-6t} - 6\cos 2t.$$

(c) A plot of the curve is shown to the right. It appears to be a straight line with slope -3. This is not the case, however. It looks this way only because the exponential function e^{6t} is so dominant, the oscillations of the cosine terms are obliterated.

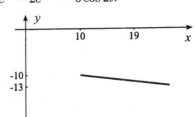

24. (a) We write $\dfrac{d^2h}{dx^2} - \dfrac{K_v}{KbB}h = -\dfrac{K_v H_0}{KbB}$. The auxiliary equation $m^2 - K_v/(KbB) = 0$ has real roots $\pm\sqrt{K_v/(KbB)}$. If we let $m = \sqrt{K_v/(KbB)}$, then a general solution of the differential equation is

$$h(x) = C_1 e^{mx} + C_2 e^{-mx} + H_0.$$

The boundary conditions require

$$H_0 = C_1 + C_2 + H_0, \qquad b = C_1 e^{mL} + C_2 e^{-mL} + H_0,$$

from which $C_1 = -C_2 = \dfrac{b - H_0}{e^{mL} - e^{-mL}}$. Thus,

$$h(x) = \frac{b - H_0}{e^{mL} - e^{-mL}}(e^{mx} - e^{-mx}) + H_0.$$

(b) With $K_v = 10^{-8}$, $K = 10^{-6}$, $b = 100$, $B = 1$, $H_0 = 125$, and $L = 1000$,

$$h(x) = 125 - \frac{25}{e^{10} - e^{-10}}(e^{x/100} - e^{-x/100}).$$

The plot is indeed relatively flat for $0 \le x \le 600$.

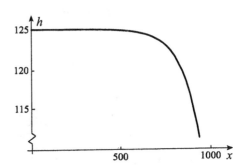

26. Suppose the mass of the chain is M so that its mass per unit length is M/a. When the length of chain hanging from the edge of the table is y, then

$$M\frac{d^2y}{dt^2} = \frac{Mgy}{a}.$$

This differential equation is subject to the initial conditions $y(0) = b$ and $y'(0) = 0$, provided $t = 0$ is taken at the instant motion begins. The differential equation is linear with auxiliary equation $m^2 - g/a = 0 \Longrightarrow m = \pm\sqrt{g/a}$. A general solution is therefore $y(t) = C_1 e^{\sqrt{g/a}\,t} + C_2 e^{-\sqrt{g/a}\,t}$. The initial conditions require

$$b = C_1 + C_2, \quad 0 = \sqrt{\frac{g}{a}}C_1 - \sqrt{\frac{g}{a}}C_2 \quad\Longrightarrow\quad C_1 = C_2 = b/2.$$

Thus, $y(t) = \dfrac{b}{2}(e^{\sqrt{g/a}\,t} + e^{-\sqrt{g/a}\,t})$. The chain slides off the table when $y = a$ in which case

$$a = \frac{b}{2}(e^{\sqrt{g/a}\,t} + e^{-\sqrt{g/a}\,t}) \quad\Longrightarrow\quad e^{2\sqrt{g/a}\,t} - \frac{2a}{b}e^{\sqrt{g/a}\,t} + 1 = 0.$$

This is a quadratic in $e^{\sqrt{g/a}\,t}$ with solutions

$$e^{\sqrt{g/a}\,t} = \frac{2a/b \pm \sqrt{4a^2/b^2 - 4}}{2} = \frac{1}{b}(a \pm \sqrt{a^2 - b^2}) \quad\Longrightarrow t = \sqrt{\frac{a}{g}}\ln\left(\frac{a \pm \sqrt{a^2 - b^2}}{b}\right).$$

It is straightforward to verify that $(a - \sqrt{a^2 - b^2})/b < 1$ in which case t would be negative, an unacceptable value. Hence, $t = \sqrt{\dfrac{a}{g}}\ln\left(\dfrac{a + \sqrt{a^2 - b^2}}{b}\right)$.

28. (a) The auxiliary equation $m^4 + P/(EI)m^2 = 0$ has solutions $m = 0, 0 \pm \sqrt{P/(EI)}i$. A general solution of the differential equation is

$$y(x) = C_1 \cos\sqrt{\frac{P}{EI}}x + C_2 \sin\sqrt{\frac{P}{EI}}x + C_3 x + C_4.$$

(b) If we set $\mu = \sqrt{P/(EI)}$ for simplicity in notation, the boundary conditions require

$$0 = y(0) = C_1 + C_4, \quad 0 = y'(0) = \mu C_2 + C_3, \quad 0 = y(L) = C_1 \cos\mu L + C_2 \sin\mu L + C_3 L + C_4,$$

$$0 = y'(L) = -\mu C_1 \sin\mu L + \mu C_2 \cos\mu L + C_3.$$

If we solve the first two for C_4 and C_3 and substitute into the last two, the result is

$$0 = C_1 \cos\mu L + C_2 \sin\mu L - \mu L C_2 - C_1, \quad 0 = -\mu C_1 \sin\mu L + \mu C_2 \cos\mu L - \mu C_2,$$

or,

$$C_1(\cos\mu L - 1) + C_2(\sin\mu L - \mu L) = 0, \quad C_1(\sin\mu L) + C_2(1 - \cos\mu L) = 0.$$

If we solve each for C_1 and equate results,

$$-\frac{C_2(\sin\mu L - \mu L)}{\cos\mu L - 1} = -\frac{C_2(1 - \cos\mu L)}{\sin\mu L},$$

or,

$$0 = C_2(\sin^2\mu L - \mu L \sin\mu L + \cos^2\mu L - 2\cos\mu L + 1) = C_2(2 - 2\cos\mu L - \mu L \sin\mu L)$$

$$= C_2\left[2 - 2\left(1 - 2\sin^2\frac{\mu L}{2}\right) - 2\mu L \sin\frac{\mu L}{2}\cos\frac{\mu L}{2}\right] = 2C_2 \sin\frac{\mu L}{2}\left(2\sin\frac{\mu L}{2} - \mu L \cos\frac{\mu L}{2}\right).$$

There are three possibilities:

Case 1: $C_2 = 0$ — In this case, $C_3 = 0$, $C_4 = -C_1$, and $C_1(\cos \mu L - 1) = 0 = C_1 \sin \mu L$. We cannot set $C_1 = 0$, else $C_4 = 0$ also, and $y(x) \equiv 0$. Hence we must set $\cos \mu L = 1$ and $\sin \mu L = 0$. These imply that $\mu L = 2n\pi$, where n is an integer, and therefore $\sqrt{P/(EI)}L = 2n\pi \implies P = 4n^2\pi^2 EI/L^2$. The smallest positive P is $P = 4\pi^2 EI/L^2$.

Case 2: $\sin \dfrac{\mu L}{2} = 0$ — In this case, $\mu L/2 = n\pi$, where n is an integer, and $\sqrt{P/(EI)}L/2 = n\pi \implies P = 4n^2\pi^2 EI/L^2$. The smallest positive P is once again $P = 4\pi^2 EI/L^2$.

Case 3: $2\sin \dfrac{\mu L}{2} - \mu L \cos \dfrac{\mu L}{2} = 0$ — In this case, $\mu L/2$ must satisfy the equation

$$\tan \frac{\mu L}{2} = \frac{\mu L}{2}.$$

This equation must be solved numerically. The smallest positive solution is $\mu L/2 = 4.49$. Hence, $\sqrt{P/(EI)}L/2 = 4.49 \implies P = 4(4.49)^2 EI/L^2 = 80.6EI/L^2$. This value is larger than that in Cases 1 and 2. Thus, the Euler buckling load is $P = 4\pi^2 EI/L^2$.

REVIEW EXERCISES

2. This equation can be separated $y\, dy = \left(\dfrac{x+1}{x}\right)\, dx$, and a one-parameter family of solutions is therefore defined implicitly by $\dfrac{y^2}{2} = x + \ln|x| + C \implies y = \pm\sqrt{2(x + \ln|x|) + D}$.

4. An integrating factor for this linear first-order equation is $e^{\int 4\, dx} = e^{4x}$. When we multiply the differential equation by this factor,

$$e^{4x}\frac{dy}{dx} + 4ye^{4x} = x^2 e^{4x} \implies \frac{d}{dx}(ye^{4x}) = x^2 e^{4x}.$$

Integration now gives

$$ye^{4x} = \int x^2 e^{4x}\, dx = \frac{1}{4}x^2 e^{4x} - \frac{x}{8}e^{4x} + \frac{1}{32}e^{4x} + C \implies y(x) = Ce^{-4x} + x^2/4 - x/8 + 1/32.$$

6. The auxiliary equation is $0 = m^2 + 3m + 4$ with solutions $m = (-3 \pm \sqrt{7}i)/2$, and therefore $y(x) = e^{-3x/2}[C_1 \cos(\sqrt{7}x/2) + C_2 \sin(\sqrt{7}x/2)] + 1/2$.

8. If we set $dy/dx = v$ and $d^2y/dx^2 = dv/dx$, then $\dfrac{dv}{dx} + \dfrac{1}{x}v = x$, and multiplication by x results in $x^2 = x\dfrac{dv}{dx} + v = \dfrac{d}{dx}(xv)$. Integration now gives $xv = x^3/3 + C$, and hence $\dfrac{dy}{dx} = v = \dfrac{x^2}{3} + \dfrac{C}{x}$. Integration with respect to x now gives $y(x) = \dfrac{x^3}{9} + C\ln|x| + D$.

10. An integrating factor for this linear first-order equation is $e^{\int 2x\, dx} = e^{x^2}$. When we multiply the differential equation by this factor,

$$e^{x^2/2}\frac{dy}{dx} + 2xe^{x^2/2}y = 2x^3 e^{x^2/2} \implies \frac{d}{dx}(ye^{x^2}) = 2x^3 e^{x^2}.$$

Integration now gives

$$ye^{x^2} = \int 2x^3 e^{x^2}\, dx = x^2 e^{x^2} - e^{x^2} + C \implies y(x) = x^2 - 1 + Ce^{-x^2}.$$

12. The auxiliary equation $0 = m^2 - 4m + 4 = (m-2)^2$ has solutions $m = 2, 2$. A general solution of the associated homogeneous equation is $y_h(x) = (C_1 + C_2 x)e^{2x}$. A particular solution is

$$y_p(x) = \frac{1}{(D-2)^2}\text{Im}(e^{ix}) = \text{Im}\left[\frac{1}{(D-2)^2}e^{ix}\right] = \text{Im}\left[e^{ix}\frac{1}{(D-2+i)^2}(1)\right]$$

$$= \text{Im}\left(e^{ix}\frac{1}{3-4i}\right) = \text{Im}\left(\frac{e^{ix}}{3-4i}\frac{3+4i}{3+4i}\right) = \frac{1}{25}(4\cos x + 3\sin x).$$

Thus, $y(x) = (C_1 + C_2 x)e^{2x} + (4\cos x + 3\sin x)/25$.

14. The auxiliary equation $0 = m^2 + 4$ has solutions $m = \pm 2i$. A general solution of the associated homogeneous equation is $y_h(x) = C_1 \cos 2x + C_2 \sin 2x$. A particular solution is

$$y_p(x) = \frac{1}{D^2+4}\text{Im}(e^{2ix}) = \text{Im}\left[\frac{1}{D^2+4}e^{2ix}\right] = \text{Im}\left[e^{2ix}\frac{1}{(D+2i)^2+4}(1)\right]$$

$$= \text{Im}\left[e^{2ix}\frac{1}{D(D+4i)}(1)\right] = \text{Im}\left(\frac{e^{2ix}}{4i}x\right) = -\frac{x}{4}\text{Im}(ie^{2ix}) = -\frac{x}{4}\cos 2x.$$

Thus, $y(x) = (C_1 - x/4)\cos 2x + C_2 \sin 2x$.

16. This equation can be separated $\frac{2}{y}\,dy = -\frac{(x+1)^2}{x}\,dx$, (provided $y \neq 0$), and a one-parameter family of solutions is defined implicitly by

$$2\ln|y| = -\int\left(x + 2 + \frac{1}{x}\right)dx = -\left(\frac{x^2}{2} + 2x + \ln|x|\right) + C.$$

Exponentiation leads to $xy^2 = De^{-2x-x^2/2}$. The function $y = 0$ is a solution of the differential equation, but because it can be obtained if we permit $D = 0$, it is not a singular solution.

18. The auxiliary equation $0 = m^2 + 2m + 4$ has solutions $m = -1 \pm \sqrt{3}i$. A general solution of the associated homogeneous equation is $y_h(x) = e^{-x}[C_1 \cos(\sqrt{3}x) + C_2 \sin(\sqrt{3}x)]$. A particular solution is

$$y_p(x) = \frac{1}{D^2+2D+4}\text{Re}(e^{-x}e^{\sqrt{3}ix}) = \text{Re}\left[\frac{1}{D^2+2D+4}e^{(-1+\sqrt{3}i)x}\right]$$

$$= \text{Re}\left[e^{(-1+\sqrt{3}i)x}\frac{1}{(D-1+\sqrt{3}i)^2+2(D-1+\sqrt{3}i)+4}(1)\right]$$

$$= \text{Re}\left[e^{(-1+\sqrt{3}i)x}\frac{1}{D(D+2\sqrt{3}i)}(1)\right] = \text{Re}\left[\frac{e^{(-1+\sqrt{3}i)x}}{2\sqrt{3}i}x\right]$$

$$= -\frac{xe^{-x}}{2\sqrt{3}}\text{Re}(ie^{\sqrt{3}ix}) = \frac{x}{2\sqrt{3}}e^{-x}\sin(\sqrt{3}x).$$

Thus, $y(x) = e^{-x}[C_1 \cos(\sqrt{3}x) + C_2 \sin(\sqrt{3}x)] + (\sqrt{3}/6)xe^{-x}\sin(\sqrt{3}x)$.

20. An integrating factor for this linear first-order equation in $x(y)$, $\frac{dx}{dy} + 3x = -2y^2$, is $e^{\int 3\,dy} = e^{3y}$. Multiplication by e^{3y} gives

$$e^{3y}\frac{dx}{dy} + 3xe^{3y} = -2y^2 e^{3y} \quad \Longrightarrow \quad \frac{d}{dy}(xe^{3y}) = -2y^2 e^{3y}.$$

Integrate with respect to y now yields

$$xe^{3y} = \int -2y^2 e^{3y}\,dy = -\frac{2y^2}{3}e^{3y} + \frac{4y}{9}e^{3y} - \frac{4}{27}e^{3y} + C \quad \Longrightarrow \quad x = -2y^2/3 + 4y/9 - 4/27 + Ce^{-3y}.$$

22. The auxiliary equation $0 = m^2 - 8m - 9 = (m-9)(m+1)$ has solutions $m = -1, 9$. A general solution of the associated homogeneous equation is $y_h(x) = C_1 e^{-x} + C_2 e^{9x}$. A particular solution is

$$y_p(x) = \frac{1}{D^2 - 8D - 9}(2x+4) = \frac{1}{-9\left(1 + \frac{8D - D^2}{9}\right)}(2x+4)$$

$$= -\frac{1}{9}\left[1 - \left(\frac{8D - D^2}{9}\right) + \cdots\right](2x+4) = -\frac{1}{9}\left[(2x+4) - \frac{16}{9}\right].$$

Thus, $y(x) = C_1 e^{-x} + C_2 e^{9x} - 2x/9 - 20/81$. For $y(0) = 3$ and $y'(0) = 7$, we must have $3 = C_1 + C_2 - 20/81$ and $7 = -C_1 + 9C_2 - 2/9$. These imply that $C_1 = 11/5$ and $C_2 = 424/405$, and therefore $y(x) = \frac{11}{5}e^{-x} + \frac{424}{405}e^{9x} - \frac{2x}{9} - \frac{20}{81}$.

24. An integrating factor for this linear first-order equation is $e^{\int 2/x\, dx} = e^{2\ln|x|} = x^2$. Multiplication of the differential equation by this factor gives

$$x^2 \frac{dy}{dx} + 2xy = x^2 \sin x \quad\Longrightarrow\quad \frac{d}{dx}(x^2 y) = x^2 \sin x.$$

We now integrate to obtain

$$x^2 y = \int x^2 \sin x\, dx = -x^2 \cos x + 2x \sin x + 2\cos x + C.$$

For $y(1) = 1$, we must have $1 = -\cos 1 + 2\sin 1 + 2\cos 1 + C$. Thus,

$$y(x) = -\cos x + \frac{2}{x}\sin x + \frac{2}{x^2}\cos x + \frac{1}{x^2}(1 - \cos 1 - 2\sin 1).$$

26. (a) According to Archimedes' principle, the buoyant force due to fluid pressure is the weight of fluid displaced by the wood, $(1.0 \times 10^{-6})(900)(9.81) = 8.829 \times 10^{-3}$ N. The total force due to fluid and gravity has magnitude $8.829 \times 10^{-3} - (1.0 \times 10^{-6})(500)(9.81) = 3.924 \times 10^{-3}$ N.

(b) If y measures distance from the bottom, then $\frac{1}{2000}\frac{d^2 y}{dt^2} = 3.924 \times 10^{-3} - 2\frac{dy}{dt}$. We may integrate this equation, $\frac{1}{2000}\frac{dy}{dt} + 2y = 3.924 \times 10^{-3}t + C$. Since $y'(0) = 0 = y(0)$, it follows that C must also be zero, and $\frac{dy}{dt} + 4000y = 7.848t$. An integrating factor for this equation is $e^{\int 4000\, dt} = e^{4000t}$. Multiplication by this factor gives

$$e^{4000t}\frac{dy}{dt} + 4000e^{4000t}y = 7.848te^{4000t} \quad\Longrightarrow\quad \frac{d}{dt}(ye^{4000t}) = 7.848te^{4000t}.$$

Integration yields

$$ye^{4000t} = 7.848\left(\frac{t}{4000}e^{4000t} - \frac{1}{16 \times 10^6}e^{4000t}\right) + D.$$

Consequently, $y(t) = 1.962 \times 10^{-3}t - 4.905 \times 10^{-7} + De^{-4000t}$. Because $y(0) = 0$, it follows that $0 = -4.905 \times 10^{-7} + D$, and therefore $y(t) = 1.962 \times 10^{-3}t + 4.905 \times 10^{-7}(e^{-4000t} - 1)$ m.

28. (a),(b) Let us take $y = 0$ on the bridge and y positive downward. During free fall, $\frac{1}{100}\frac{dv}{dt} = \frac{1}{100}(9.81)$. Thus, $v(t) = 9.81t + C$, and the condition $v(0) = 0$ requires $C = 0$. Integration now gives $y(t) = 4.905t^2 + D$. Because $y(0) = 0$, we obtain $D = 0$, and $y(t) = 4.905t^2$. The stone strikes the water when $50 = 4.905t^2$, and this equation implies that $t = \sqrt{50/4.905}$ s. At this instant, its velocity is $v = 9.81\sqrt{50/4.905} = \sqrt{50(19.62)}$. When the stone is in the water,

$$\frac{1}{100}\frac{dv}{dt} = \frac{1}{100}(9.81) - \frac{v}{5} \quad\Longrightarrow\quad \frac{dv}{dt} + 20v = 9.81.$$

An integrating factor for this equation is $e^{\int 20\,dt} = e^{20t}$, so that

$$e^{20t}\frac{dv}{dt} + 20ve^{20t} = 9.81e^{20t} \quad\Longrightarrow\quad \frac{d}{dt}(ve^{20t}) = 9.81e^{20t}.$$

Integration gives

$$ve^{20t} = 0.4905e^{20t} + C \quad\Longrightarrow\quad v(t) = 0.4905 + Ce^{-20t}.$$

Since $v(\sqrt{50/4.905}) = 0.9\sqrt{50(19.62)}$,

$$0.9\sqrt{50(19.62)} = 0.4905 + Ce^{-20\sqrt{50/4.905}},$$

and this equation implies that $C = 1.494 \times 10^{29}$. Integration of $v(t)$ yields $y(t) = 0.4905t - \dfrac{C}{20}e^{-20t} + D$.

Because $y(\sqrt{50/4.905}) = 50$, we have $50 = 0.4905\sqrt{50/4.905} - \dfrac{1.494 \times 10^{29}}{20}e^{-20\sqrt{50/4.905}} + D$, and this requires $D = 49.82$. Thus, $y(t) = 0.4905t - 7.47 \times 10^{27}e^{-20t} + 49.82$.

(c) The stone reaches the bottom when $60 = 0.4905t - 7.47 \times 10^{27}e^{-20t} + 49.82$. The solution of this equation is 20.75 seconds.